The Lung
Development, Aging and the Environment

The Lung
Development, Aging and the Environment

Second Edition

Edited by

Richard Harding
Department of Anatomy and Developmental Biology,
Monash University,
Clayton, VIC,
Australia

Kent E. Pinkerton
Department of Pediatrics, School of Medicine,
Department of Anatomy, Physiology and Cell Biology,
School of Veterinary Medicine,
Center for Health and the Environment,
California National Primate Research Center,
John Muir Institute of the Environment,
University of California – Davis,
Davis, CA, USA

AMSTERDAM • BOSTON • HEIDELBERG • LONDON • NEW YORK • OXFORD • PARIS
SAN DIEGO • SAN FRANCISCO • SINGAPORE • SYDNEY • TOKYO

Academic Press is an imprint of Elsevier

Academic Press is an imprint of Elsevier
32 Jamestown Road, London NW1 7BY, UK
525 B Street, Suite 1800, San Diego, CA 92101-4495, USA
225 Wyman Street, Waltham, MA 02451, USA
The Boulevard, Langford Lane, Kidlington, Oxford OX5 1GB, UK

First edition 2004

Notices
Knowledge and best practice in this field are constantly changing. As new research and experience broaden our understanding, changes in research methods, professional practices, or medical treatment may become necessary.

Practitioners and researchers must always rely on their own experience and knowledge in evaluating and using any information, methods, compounds, or experiments described herein. In using such information or methods they should be mindful of their own safety and the safety of others, including parties for whom they have a professional responsibility.

To the fullest extent of the law, neither the Publisher nor the authors, contributors, or editors, assume any liability for any injury and/or damage to persons or property as a matter of products liability, negligence or otherwise, or from any use or operation of any methods, products, instructions, or ideas contained in the material herein.

ISBN: 978-0-12-799941-8

British Library Cataloguing in Publication Data
A catalogue record for this book is available from the British Library

Library of Congress Cataloging in Publication Data
A catalog record for this book is available from the Library of Congress

For information on all Academic Press publications visit our website at http://store.elsevier.com/

Working together
to grow libraries in
developing countries

www.elsevier.com • www.bookaid.org

Contents

Part I
Critical Events in Normal Development and Aging

1. Lung Progenitor Cell Specification and Morphogenesis

Munemasa Mori and Wellington V. Cardoso

2. Development of Airway Epithelium

Charles Plopper and Michelle Fanucchi

3. Development of the Innervation of the Lower Airways: Structure and Function

Nicolle J. Domnik, Ernest Cutz and John T. Fisher

4. The Formation of Pulmonary Alveoli

Stephen E. McGowan

5. Pulmonary Vascular Development

Rosemary Jones, Diane E. Capen and Lynne Reid

26. Effect of Environment and Aging on the Pulmonary Surfactant System

Sandra Orgeig, Janna L. Morrison and Christopher B. Daniels

27. Environmental Determinants of Lung Aging

Francis H.Y. Green and Kent E. Pinkerton

Contributors

Steven H. Abman The Pediatric Heart Lung Center, Departments of Pediatrics, University of Colorado Denver, Anschutz Medical Campus and Children's Hospital Colorado, Aurora, CO, USA

Kurt H. Albertine Departments of Pediatrics, Medicine, and Neurobiology & Anatomy, University of Utah School of Medicine, Salt Lake City, UT, USA

Diane E. Capen Department of Anesthesia, Critical Care and Pain Medicine, Massachusetts General Hospital, Charlestown, MA, USA

Wellington V. Cardoso Columbia Center for Human Development, Pulmonary Allergy & Critical Care Medicine, Department of Medicine, Columbia University Medical Center, New York, NY, USA

Jocelyn Claude Center for Health and the Environment, University of California Davis, Davis, CA, USA

Candace M. Crowley Department of Anatomy, Physiology, & Cell Biology, School of Veterinary Medicine, University of California – Davis, Davis, CA, USA

Ernest Cutz Division of Pathology, Department of Paediatric Laboratory Medicine, The Hospital for Sick Children, Toronto, ON, Canada; Department of Laboratory Medicine and Pathobiology, University of Toronto, Toronto, ON, Canada

Christopher B. Daniels Barbara Hardy Institute, University of South Australia, Adelaide, SA, Australia

Reuben B. Dodson The Pediatric Heart Lung Center, Departments of Surgery, University of Colorado Denver, Anschutz Medical Campus and Children's Hospital Colorado, Aurora, CO, USA

Nicolle J. Domnik Department of Biomedical and Molecular Sciences, Physiology Program, Queen's University, Kingston, ON, Canada

Michelle Fanucchi School of Veterinary Medicine, Department of Anatomy, Physiology and Cell Biology, University of California – Davis, Davis, CA, USA

Michelle V. Fanucchi Department of Environmental Health Sciences, School of Public Health, University of Alabama at Birmingham, Birmingham, AL, USA

John T. Fisher Department of Biomedical and Molecular Sciences, Physiology Program, Queen's University, Kingston, ON, Canada; Department of Medicine, Queen's University, Kingston, ON, Canada

Csaba Galambos The Pediatric Heart Lung Center, Departments of Pathology, University of Colorado Denver, Anschutz Medical Campus and Children's Hospital Colorado, Aurora, CO, USA

Laurel J. Gershwin University of California – Davis, Davis, Veterinary Medicine (PMI), Davis, CA, USA

Rakesh Ghosh Division of Environmental Health, Keck School of Medicine, University of Southern California, Los Angeles, CA, USA

Francis H.Y. Green Department of Pathology & Laboratory Medicine, University of Calgary, Calgary, Alberta, Canada

Richard Harding Department of Anatomy and Developmental Biology, Monash University, Clayton, VIC, Australia

Matt J. Herring Department of Anatomy, Physiology and Cell Biology, School of Veterinary Medicine, Center for Health and the Environment, California National Primate Research Center, University of California – Davis, Davis, CA, USA

Irva Hertz-Picciotto Department of Public Health Sciences, University of California – Davis, Davis, CA, USA

Stuart B. Hooper The Ritchie Centre, MIMR-PHI Institute of Medical Research, and The Department of Obstetrics and Gynaecology, Monash University, Clayton, VIC, Australia

Connie C.W. Hsia Department of Internal Medicine, University of Texas Southwestern Medical Center, Dallas, TX, USA

Dallas M. Hyde   Department of Anatomy, Physiology and Cell Biology, School of Veterinary Medicine, Center for Health and the Environment, California National Primate Research Center, University of California – Davis, Davis, CA, USA

Rosemary Jones   Harvard Medical School and Department of Anesthesia, Critical Care and Pain Medicine, Massachusetts General Hospital, Boston, MA, USA

Lisa A. Joss-Moore   Department of Pediatrics, University of Utah, Salt Lake City, UT, USA

Marcus J. Kitchen   School of Physics, Monash University, Clayton, VIC, Australia

Steven R. Kleeberger   Laboratory of Respiratory Biology, National Institute of Environmental Health Sciences, National Institutes of Health Research Triangle Park, NC, USA

Robert H. Lane   Department of Pediatrics, Medical College of Wisconsin, WI, USA

Gert S. Maritz   Department of Physiological Sciences, University of the Western Cape, Bellville, South Africa

Robert De Matteo   Department of Anatomy and Developmental Biology, Monash University, Clayton, VIC, Australia

Zachary McCaw   Laboratory of Respiratory Biology, National Institute of Environmental Health Sciences, National Institutes of Health Research Triangle Park, NC, USA

Annie R.A. McDougall   The Ritchie Centre, MIMR-PHI Institute of Medical Research, and The Department of Obstetrics and Gynaecology, Monash University, Clayton, VIC, Australia

Stephen E. McGowan   Department of Veterans Affairs Research Service, Department of Internal Medicine, University of Iowa Carver College of Medicine, Iowa City, IA, USA

Lisa A. Miller   Department of Anatomy, Physiology, & Cell Biology, School of Veterinary Medicine, University of California – Davis, Davis, CA, USA

Munemasa Mori   Columbia Center for Human Development, Pulmonary Allergy & Critical Care Medicine, Department of Medicine, Columbia University Medical Center, New York, NY, USA

Janna L. Morrison   School of Pharmacy & Medical Sciences, Sansom Institute for Health Research, University of South Australia, Adelaide, SA, Australia

Jennifer L. Nichols   Oak Ridge Institute for Science and Education, Office of Research and Development, U.S. Environmental Protection Agency, Research Triangle Park, NC, USA

Sandra Orgeig   School of Pharmacy & Medical Sciences, Sansom Institute for Health Research, University of South Australia, Adelaide, SA, Australia

Kent E. Pinkerton   Department of Pediatrics, School of Medicine, Department of Anatomy, Physiology and Cell Biology, School of Veterinary Medicine, Center for Health and the Environment, California National Primate Research Center, John Muir Institute of the Environment, University of California – Davis, Davis, CA, USA

Charles Plopper   School of Veterinary Medicine, Department of Anatomy, Physiology and Cell Biology, University of California – Davis, Davis, CA, USA

Lynne Reid   Department of Pathology, Harvard Medical School Children's Hospital, Boston, MA, USA

Megan O'Reilly   Department of Pediatrics and Women and Children's Health Research Institute, University of Alberta, Edmonton, AB, Canada

Melissa L. Siew   The Ritchie Centre, MIMR-PHI Institute of Medical Research, Monash University, VIC, Australia

Suzette Smiley-Jewell   Center for Health and the Environment, University of California – Davis, Davis, CA, USA

Foula Sozo   Department of Anatomy and Developmental Biology, Monash University, Melbourne, VIC, Australia

Lucy C. Sullivan   Department of Microbiology and Immunology, The University of Melbourne, Melbourne, VIC, Australia

Arjan B. te Pas   Division of Neonatology, Department of Pediatrics, Leiden University Medical Centre, Leiden, The Netherlands

Bernard Thébaud   Sprott Centre for Stem Cell Research, Ottawa Hospital Research Institute, Ottawa, ON, Canada; Division of Neonatology, Department of Pediatrics, Children's Hospital of Eastern Ontario (CHEO) and CHEO Research Institute, Ottawa, ON, Canada

Kirsten C. Verhein   Laboratory of Respiratory Biology, National Institute of Environmental Health Sciences, National Institutes of Health Research Triangle Park, NC, USA

Megan J. Wallace   The Ritchie Centre, MIMR-PHI Institute of Medical Research, and The Department of Obstetrics and Gynaecology, Monash University, Clayton, VIC, Australia

Ewald R. Weibel Institute of Anatomy, University of Bern, Bern, Switzerland

Jonathan H. Widdicombe Department of Physiology & Membrane Biology, University of California – Davis, Davis, CA, USA

Jingyi Xu Center for Health and the Environment, University of California – Davis, Davis, CA, USA; Affiliated Zhongshan Hospital of Dalian University, Dalian, China

Cuneyt Yilmaz Department of Internal Medicine, University of Texas Southwestern Medical Center, Dallas, TX, USA

Bradley A. Yoder Departments of Pediatrics, University of Utah School of Medicine, Salt Lake City, UT, USA

Introduction

The lung is essential to our health and well-being throughout life. From the moment of birth, the lung is totally responsible for providing our tissues with oxygen from the atmosphere and eliminating carbon dioxide. In addition to gas exchange, the lung plays an important role in immunity and other protective functions. Indeed our ability to achieve our physical and mental potential throughout our entire life span is strongly influenced by the efficient functioning of our lungs. Since the first edition of this book was published, research into the normal functions of the lung and disease states of the lung throughout the life cycle has expanded enormously, resulting in a much greater understanding of these processes. This burgeoning field of research has clearly shown that both environmental and genetic factors operating during the early stages of life can alter the risk of impaired lung function and respiratory health later in life. The principal objectives of the book are, firstly, to concisely present current concepts of normal processes involved in the development, maturation and aging of the lung, and secondly, to integrate the growing body of evidence regarding the influence of the environment and genetic factors on lung structure and function and on respiratory health in later life. A third objective is to review novel treatments for the diseased lung. These are important topics for current review as respiratory illness is a major contributor to morbidity and mortality at all stages of life, and new technologies are offering improved treatments.

With the increasing use of molecular and cellular technologies, our understanding of the biological processes involved in the development of the respiratory organs has expanded tremendously. As a result, new concepts regarding the control of lung development and the early-life origins of lung disease have evolved rapidly; this is especially true of obstructive lung diseases. In parallel with our greater understanding of normal development is the realization that a wide range of environmental factors can impact upon the genetic program of lung development, both before and after birth. Many such factors can result in persistent alterations in lung structure and function that can, in turn, lead to an increased susceptibility to respiratory illness through all stages of postnatal life. A large body of epidemiological data indicates that early life events such as premature birth, restricted growth, respiratory infections or exposure to allergens can predispose the individual to airway dysfunction and common respiratory disorders such as asthma and chronic obstructive airway disease (COPD), increasing the risk of death from respiratory causes. It is also evident that genetic polymorphisms can affect an individual's susceptibility to a range of environmental factors such as allergens, cigarette smoke, nutrient restriction and infection, and these variations are only now becoming better understood.

With the increasing interest in early-life origins of ill health and the role of the environment in human biology, we believe it is timely to review the recent scientific literature covering these important health issues in relation to lung biology. Our purpose is to integrate current knowledge of the impact of environmental factors that can influence lung development, susceptibility to respiratory illness, and the rate of aging of the lung. Each of these aspects is directly relevant to an understanding of respiratory health, a matter that is likely to become increasingly important in an aging population. This book addresses two general questions: during the development of the respiratory system, what are the critical events that lead to the complex mature organ system? And, what is the impact of environmental and genetic factors on these developmental events?

In both the original and second editions of this book, we aimed at making it accessible not only to those actively researching lung biology, but also to those with a broad interest in human health. Our hope is that this book will be of value to everyone concerned with respiratory health, including thoracic physicians, respiratory scientists, members of the pharmaceutical industry, toxicological and environmental regulators, pediatricians, perinatologists, and gerontologists.

Richard Harding
Kent E. Pinkerton

Critical Events in Normal Development and Aging

Chapter 1

Lung Progenitor Cell Specification and Morphogenesis

Munemasa Mori and Wellington V. Cardoso

Columbia Center for Human Development, Pulmonary Allergy & Critical Care Medicine, Department of Medicine, Columbia University Medical Center, New York, NY, USA

INTRODUCTION

The respiratory system represents a major interface of the body with the external environment, playing a crucial role in efficient clearing and conduction of air, and in promoting efficient gas-exchange for metabolic needs. To achieve these goals the respiratory system in mammalians has evolved into a highly complex system of branching epithelial and vascular structures that connects to a vast network of alveolar gas-exchanging units. The generation of this complex organ involves multiple steps and encompasses events that span prenatal and postnatal life (Figure 1).[1] Overall the process includes specification of respiratory progenitors, expansion and patterning of the epithelial progenitors as they interact with neighbor cells to form the distinct lung regions and generate the specific cell types that populate the airways and alveoli.

This chapter focuses on the mechanisms that regulate the initial events leading to cell fate specification and formation of the embryonic lung, and how signaling molecules present at early developmental stages influence this process. The data reviewed here have been generated largely in mouse models; thus, this species will be used as a reference throughout the text.

ONSET OF LUNG DEVELOPMENT

During early stages of embryonic development, after gastrulation, endodermal cells undergo extensive morphogenetic movements to form the primitive gut tube. A number of signaling molecules and transcription factors start to be expressed in the endoderm in overlapping but distinct domains along the anterior-posterior (A-P) axis of the gut tube. This roughly subdivides the gut endoderm into three regions: the foregut, midgut, and hindgut. The foregut is the most anterior (cranial) region of this tube, while the midgut and hindgut are at more posterior regions, towards the caudal end of the embryo. From these regions organ-specific domains arise and undergo morphogenesis to form organ primordia.[2,3]

Specification of Respiratory Progenitors

The progenitor cells of the lung and trachea originate from the foregut endoderm. Other foregut derivatives include the thymus, thyroid, esophagus, stomach liver, and pancreas.[2] By midgestation, in the mouse at embryonic days E8.0–9.0, the progenitor cells for some of these organs can be recognized by regional expression of representative transcription factors in the foregut endoderm. For instance, the homeodomain protein gene Nkx2-1 (also known as thyroid transcription factor 1 [Ttf1] or T/EBP) is expressed in the thyroid and respiratory primordia.[4] Hex (hematopoietically expressed homeobox) is expressed in the thyroid and liver primordia,[5] while Pax8 and Pdx1 (pancreas-duodenal-associated homeobox gene) are found in the thyroid[6] and pancreatic primordia, respectively.[7] Endodermal development is influenced not only by locally expressed transcription factors, but also by soluble factors that diffuse from adjacent cell layers to the endoderm.

The earliest sign of endoderm specification into the respiratory lineage is the local expression of Nkx2-1 in the mid-region of the foregut endoderm, in mice at around E9. Analysis of Nkx2-1 null mice reveals the crucial role of this gene in lung progenitor cell fate.[8] In these mutants there is no evidence of lung epithelial cell differentiation, as assessed by marker genes typically found in the lung, and only a few ciliated cells are present. Expression of Wnt2 and Wnt2b in the foregut mesoderm is essential for specification of the respiratory epithelial progenitors. There is no Nkx2-1-expressing cells in the prospective lung region of the foregut of Wnt2/2b double null mice.[9]

There is also evidence that Fgf1 and Fgf2 secreted from the adjacent cardiac mesoderm influence the fate of foregut endoderm and can induce Nkx2-1 expression in endodermal progenitors.[10] Indeed, FGF2 is critical for the efficient derivation of Nkx2-1-expressing progenitor cells in mouse ES cell cultures.[11,12,13]

The Lung. http://dx.doi.org/10.1016/B978-0-12-799941-8.00001-8

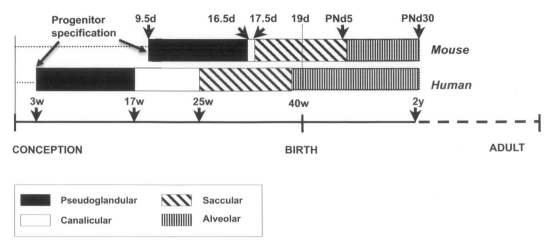

FIGURE 1   Stages of lung development in humans and mice. Specification of respiratory progenitors and overall lung development initiates earlier in humans compared to mice. During the pseudoglandular stage most branching morphogenesis occurs and the lung has a gland-like appearance with epithelial tubules separated by thick mesenchyme; during the canalicular stage, airway branching is completed, the mesenchyme becomes thinner leading to an approximation between the epithelial tubules and blood vessels. During the saccular stage the distal lung expands to form primitive saccules, and type I and type II cells differentiate. During the alveolar stage, septation of saccules gives rise to mature alveoli. d: day; PN: postnatal; w: week; y: years.

Formation of the Lung Primordium

Primary lung buds and tracheal primordium are identified in humans around the fourth week of embryonic life. However in species such as the mouse or rat, primordial lungs emerge much later, at midgestation (embryonic days E9.5 and E11.5, respectively).[1] Endodermal buds form from each ventro-lateral side of the foregut and invade the adjacent mesoderm; these buds then grow caudally and ventrally connecting at the midline to form the primordial lung. At the site where the primary buds connect (future carina), the trachea develops.[14]

Formation of the primary lung buds requires expansion of the Nkx2-1 expressing lung progenitor cells by activation of Fgfr2b signaling. At E9.5, Fgf10 is locally induced in the foregut mesoderm at sites of prospective lung bud formation.[15,16] Fgf10 induces budding by binding to and activating Fgfr2 signaling in the endoderm.[17] *Fgfr2* is expressed throughout the foregut endoderm. Fgf10 is a chemoattractant and a proliferation factor for epithelial cells.[18,19] The mechanism elicited by Fgf10-Fgfr2 appears to be a rather general strategy to form buds. For example, mice lacking Fgf10 or its receptor Fgfr2b do not have lungs or thyroid.[20,21]

Lung agenesis has been also reported when retinoic acid (RA) signaling is disrupted during organogenesis. RA is the active form of vitamin A, a key regulator of cellular functions in multiple systems. The RA effects are mediated by two families of nuclear receptors, RARs and RXRs, which are expressed throughout lung development.[22–27]

Studies using genetic and pharmacological models to modulate RA signaling at the onset of lung development show that RA is not required to specify respiratory progenitors. However, it is crucial to control Fgf10 expression required to expand the initial population of lung progenitors and form the lung primordium. For this, RA signaling is strongly activated in the foregut mesoderm where it suppresses expression of the Wnt inhibitor Dkk1. This allows activation of the canonical Wnt pathway. Moreover, RA inhibits Tgfβ signaling. The RA effect on Wnt and Tgfβ signaling leads to proper mesodermal Fgf10 expression, which is required for formation of the lung primordium (Figure 2D).[28] These studies suggest that the failure of this mechanism is likely to be the molecular basis of the lung agenesis classically reported in vitamin A deficiency.

Formation of the Trachea

There is morphological and genetic evidence suggesting that the trachea and lungs originate by independent processes. In mice, formation of lung buds precedes tracheal formation. A striking observation from *Fgf10* knockout mice is the absence of lungs in the presence of a well formed and apparently normal trachea.[20] This suggests that, once specified, tracheal and lung progenitors undergo overlapping but also distinct mechanisms.

Tracheal formation is tightly connected to Dorsal-Ventral (D-V) patterning of the endoderm, a mechanism that ultimately leads to separation of the trachea from the neighbor esophagus. Bmp signaling is critical for expansion of the ventral foregut endoderm and tracheal development. Bmp4 and its receptors (Bmpr1a, Bmpr1b) are prominently expressed in ventral foregut mesoderm and endoderm, respectively. Disruption of Bmp4 or Bmp receptors results in the tracheal agenesis/atresia and ectopic primary lung buds.[29]

Bmp signaling appears to control D-V patterning by balancing expression of Nkx2-1 (ventral: tracheal progenitors) and Sox2/p63-expressing endodermal progenitors (dorsal: prospective esophagus). Loss of Bmpr1a/1b reduces

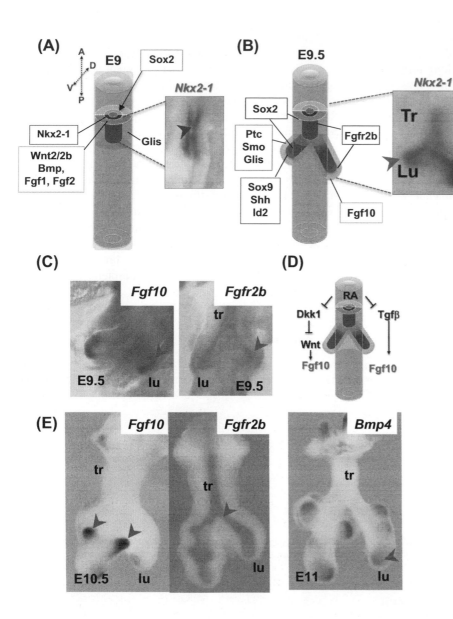

FIGURE 2 Progenitor cell specification and formation of the lung (Lu) and tracheal (Tr) primordia in mice: (A) Lung progenitor specification. At E9.0 lung and tracheal progenitors arise from the ventral foregut endoderm (purple) and are identified collectively by Nkx2-1 expression (panels: in situ hybridization). Wnt2/2b, Bmp4, Fgf2 and Fgf1 in the adjacent mesoderm (light blue) regulate this process. Sox2 endodermal expression predominantly in the dorsal region of the foregut (gray). (B) Primary lung bud formation: at E9.5 local Fgf10 (green) expression in the foregut mesoderm activates Fgfr2b in the endoderm to expand the lung progenitors and form primary lung buds. Proximal (Sox2) and distal (Sox9, Id2, Shh) domains in the lung epithelium. Increased Nkx2-1 and Shh in distal epithelium. Ptc, Smu and Glis in lung mesenchyme. (C) ISH of *Fgf10* and *Fgfr2b* during primary lung bud formation at E9.5. (D) Retinoic acid (RA)-dependent network at the onset of lung development. RA signaling in the foregut mesoderm suppresses Dkk1 to allow Wnt signaling and inhibits Tgfβ signaling. The balanced activity of Wnt and Tgfβ leads to proper Fgf10 expression required for formation of the lung primordium. (E) Secondary bud formation: Expression of *Fgf10* in the lung mesenchyme activates Fgfr2b signaling and budding in the lung epithelium; *Bmp4* is induced in the distal epithelium by Fgf10-Fgr2b. A, B, C, E: whole mount in situ hybridization. Arrowheads point to signal in each panel. A, anterior; P, posterior; V, ventral; D, dorsal.

expression of Nkx2-1 and expands the Sox2/p63 dorsal domain. Thus, during early development Bmp signaling is critical for expansion of the tracheal progenitors.

Genetic studies also implicate the Gli family of zinc finger transcription factors in early lung development. Glis transduce signaling by Sonic hedgehog (Shh, discussed below) and are expressed in the foregut mesoderm and later in the developing lung mesenchyme.[30,31] When *Gli2* and *Gli3* are simultaneously inactivated in knockout mice, no lungs or trachea are formed and other foregut derivatives, such as stomach and pancreas, are hypoplastic.[32] This phenotype is intriguing because it is more severe than that found in *Shh* null mice,[33] suggesting that Gli 2 and 3 may be shared with other pathways.

Several mechanisms have been proposed to explain how the tracheal tube forms and separates from the developing foregut. It is currently accepted that once lung buds form and fuse in the midline, a septum growing from caudal to cranial regions separates tracheal and esophageal compartments. Alternatively it is thought that separation occurs by fusion of endodermal ridges growing from each side of the foregut; as they meet in the midline, two tubes form.[34] In addition, a mechanism involving local activation of programmed cell death in the endoderm has been proposed.[35]

Tracheo–esophageal fistula, a relatively common abnormality of human tracheal development, results from partial to complete lack of separation of the respiratory tract from the esophagus.[36] This abnormality has been reported in a number of knockout mice, including *Shh-/-*,[33] *Nkx2-1-/-*,[8] and *Gli2-/-;Gli3+/-*.[32] Retinoids are also essential for normal tracheal development because in Vitamin A deficient rat embryos and *RARalpha* and *beta* double null mice, tracheoesophageal fistula is observed.[26,37]

BRANCHING MORPHOGENESIS

Once the secondary buds arise from the lung primordium, the epithelial tubules undergo branching morphogenesis to generate the bronchial tree.[38] The process involves bud outgrowth, bud elongation, and subdivision of the terminal units by reiterated budding and by formation of clefts between buds. In mice branching initiates at E10.5 and extends to around E17, when saccule formation initiates (Figure 1).[1]

During branching *Fgf10* is expressed in a dynamic fashion in the lung mesenchyme where distal epithelial buds form (Figures 2 & 3). The unique pattern of expression of *Fgf10* in the early lung suggests that Fgf10 is involved in the spatial control of lung bud formation.[16,18] Epithelial-mesenchymal interactions play a key role in branching morphogenesis.[39] The exchange of signals between epithelial and mesenchymal

cell layers of nascent buds establishes feedback loops that control airway size, branching modes, and cell fate.

Sonic Hedgehog (Shh) is an important signaling molecule expressed in the epithelium in a proximal-distal (P-D) gradient with the highest levels at the distal tips (Figure 3). Shh signals through its receptor, Patched (Ptch) and Smoothened (Smoth), and the transcription factors Gli1-3 in the mesenchyme.[31,40] Shh signaling controls mesenchymal gene expression and cell survival.[33] Lungs from *Shh* null mice show disrupted airway branching and resemble rudimentary sacs. Interestingly in these mice *Fgf10* expression is de-repressed and becomes diffuse.[33] Thus, Shh in the distal bud may function to locally inhibit *Fgf10* expression in the mesenchyme and prevent widespread distribution of *Fgf10* signals. Fgf signaling and airway branching are also controlled by a family of cysteine-rich proteins collectively

FIGURE 3 Branching morphogenesis and differentiation of the developing lung epithelium. (A) Diagram showing representative molecular regulators of branching morphogenesis in the lung epithelium and mesenchyme (inset: E11.5 lung). (B) Diagram on left: bud formation during branching resulting from activation of Fgfr2b signaling in the epithelium (blue) by Fgf10 in mesenchymal cells (yellow). Panels on right: cleft formation (arrows) during branching resulting from mesenchymal accumulation of Tgfb1 at branch points (marked by expression of Tgfbi) and Tgfb-mediated local inhibition of proliferation in the epithelium. Establishment of proximal and distal domains in the lung epithelium marked by Sox2 and Bmp4 expression, respectively (d1, d2: day 1 and 2). (C) Diagram representing airway and alveolar epithelial cell types and their molecular markers.

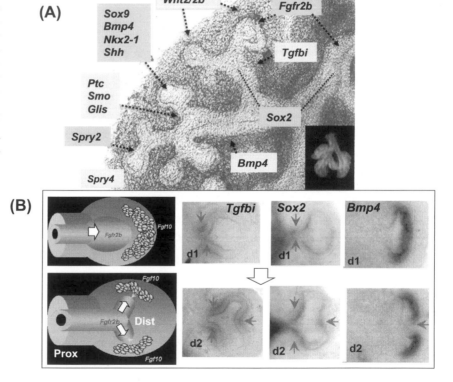

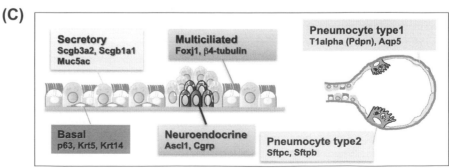

called Sprouty (Spry). *Spry* gene mutation in flies results in increased number of tracheal branches.[41,42] *Spry2* and *Spry4* are expressed in the epithelium and mesenchyme of the developing distal lung, respectively.[43–45] Disruption of *Spry2* in lung cultures stimulates branching while *Spry2* overexpression in the distal lung epithelium of transgenic mice inhibits branching and epithelial cell proliferation.[45]

Branching is also accomplished by formation of clefts in distal buds. The process is associated with local activation of Tgfb signaling in the epithelium at branch points, inhibiting proliferation locally.[46] Tgfb signaling is also activated in the mesenchyme where it suppresses Fgf10 expression and induces synthesis of extracellular matrix (ECM) components (Figure 3).[47] The dynamic pattern of Tgfb activation during branching is well illustrated by the distribution of Tgfbi (Tgf beta-induced or BigH3) in the mesenchyme associated with the stalk of distal buds (Figures 2 & 3).[48,49]

The correct patterning of this highly complex three-dimensional structure depends on a combination of branching modes[50] and input from multiple other signals including microRNAs.[51,52]

Left-Right Asymmetry

One of the least understood and most intriguing patterning events in organogenesis is the establishment of left-right (L-R) asymmetry. Left and right lungs have highly stereotypical but different branching patterns and number of lobes, which vary according to species. L-R patterning of the lung is linked to the general body plan and is actually initiated well before lungs are formed. *Lefty 1* and *2*, *nodal*, and *Pitx-2* have been identified as major regulators of L-R asymmetry in viscera.[53] When expression of these transcription factors is disrupted in mice, laterality defects known as pulmonary isomerisms are found.[54] These defects are characterized by abnormally symmetric lungs. In wild type mice the left and right lungs consist of one and four lobes, respectively. However, in *Lefty-1 -/-* null mice single-lobed lungs are found on each side.[54] Paradoxically, while branching is influenced by Lefty 1, this regulator is not expressed by the developing lung. Like *Lefty 2* and *nodal*, *Lefty 1* is expressed only during a short window of time around E8–8.5, on the left side of the prospective floor plate and lateral plate mesoderm. This suggests that some patterning decisions have already occurred when organ primordia arise. Other signaling molecules such as Shh, RA, Gli, and activin receptor IIb have been implicated in L-R asymmetry in the lung.[32,55,56]

ESTABLISHMENT OF PROXIMAL-DISTAL CELL FATE AND DIFFERENTIATION

Although defined cellular phenotypes are recognized largely at late gestation, molecular features of differentiation can be detected much earlier, while the airways are still branching.

One of the first signs of the establishment of differences in cell fate along the proximal-distal (P-D) axis of the developing lung epithelium is the expression of Sox2 in the proximal region (airway) and Sox9 and Sftpc (Surfactant-associated-protein C) in the distal regions (distal buds).[57,58]

The balance of P-D cell fate in the epithelium is tightly regulated by signals, such as Bmp and Wnt. During branching, Bmp4-Bmp receptors are expressed and activated at the tip bud epithelium (Figure 2).[59,60] Bmp signaling appears to restrict cell proliferation and promote distal cell fate in the bud epithelium. Epithelial disruption of Bmp signaling in the lung of transgenic mice results in proximalization, an expansion of the proximal domain at the costs of the distal.[60] Fgf10 controls *Bmp4* levels. Expression of *Bmp4*-Bmpr is induced in the distal epithelium by Fgf10.[19,61] Bmp signaling is also controlled by antagonists, such as *Noggin*,[62] *Chordin*, and the Cerberus-related factor *Cer1*, [60,46] all expressed in the developing lung.

Wnt signaling is critically required in the developing lung epithelium to maintain distal cell fate in branching airways. Analysis of canonical Wnt reporter (TOPGAL) mice shows activity in the distal lung buds undergoing branching.[63] Forced activation of Wnt signaling in the lung epithelium results in ectopic expansion of the distal domain at proximal sites.[64,65,66] By contrast, preventing activation of Wnt signaling by overexpressing the Wnt antagonist Dickkopf 1 (Dkk1) or disrupting beta catenin expression leads to proximalization of the lung.[67]

From E14.5 onwards a number of molecular markers of differentiation start to be identified in the airway epithelium, indicating commitment to a program of differentiation to specific airway cell types. These markers include Foxj1 (multiciliated), Scgb3a2 (secretory, Clara), Ascl1 (neuroendocrine). By E16.5–18.5 morphologic features of differentiation become apparent in these cells as they start to express Beta-tubulin 4 (multiciliated), Scgb1a1/Clara cell secretory protein (secretory, Clara), Cgrp (neuroendocrine). Differentiation of the distal epithelium into type I and type II cells occurs when the lungs undergo sacculation (in mice ~E17).[1] Type I cells become characteristically flat and express markers, such as Aquaporin 5 and T1alpha.[66,68,69] Type II cells become cuboidal, expressing various surfactant proteins and form lamellar bodies, cytoplasmic inclusions that store surfactant material.[70] Development of the gas-exchange region of the lung is completed postnatally with septation of the primitive saccules through the process of alveolization.[71,72]

CONCLUSIONS

Over the past decade there have been major advances in the understanding of the mechanisms that generate lung progenitors and regulate growth and differentiation of the lung. This was facilitated by the wide use of increasingly sophisticated approaches for genetic manipulation, lineage tracing, genome-wide screening, and imaging. The knowledge gained from these studies has provided the basis for the understanding of the role of developmental pathways in the pathogenesis of lung disease and how these pathways influence injury-repair-regeneration.

The accumulated information has been used as the road map for the in vitro generation of lung epithelial cell types through directed differentiation of embryonic stem (ES) and inducible pluripotent stem (iPS) cells. This has opened an entire new field of investigation and new perspectives for the potential use of cell-based therapies in tissue engineering and regenerative medicine in lung disease.

REFERENCES

1. Ten Have-Opbroek AA. The development of the lung in mammals: an analysis of concepts and findings. *Am J Anat* 1981;**162**:201–19.
2. Grapin-Botton A. Endoderm specification. *StemBook* 2008. http://dx.doi.org/10.3824/stembook.1.30.1.
3. Wells JM, Melton DA. Vertebrate endoderm development. *Annu Rev Cell Dev Biol* 1999;**15**:393–410.
4. Lazzaro D, Price M, de Felice M, Di Lauro R. The transcription factor TTF-1 is expressed at the onset of thyroid and lung morphogenesis and in restricted regions of the foetal brain. *Development* 1991;**113**:1093–104.
5. Jung J, Zheng M, Goldfarb M, Zaret KS. Initiation of mammalian liver development from endoderm by fibroblast growth factors. *Science* 1999;**284**:1998–2003.
6. Fagman H, Nilsson M. Morphogenetics of early thyroid development. *J Mol Endocrinol* 2010;**46**:R33–42.
7. Offield MF, Jetton TL, Labosky PA, Ray M, Stein RW, et al. *PDX-1 is required for pancreatic outgrowth and differentiation of the rostral duodenum,*1996;**995**:983–995.
8. Minoo P, Su G, Drum H, Bringas P, Kimura S. Defects in tracheoesophageal and lung morphogenesis in Nkx2.1(-/-) mouse embryos. *Dev Biol* 1999;**209**:60–71.
9. Goss AM, Tian Y, Tsukiyama T, Cohen ED, Zhou D, et al. Wnt2/2b and beta-catenin signaling are necessary and sufficient to specify lung progenitors in the foregut. *Dev Cell* 2009;**17**:290–8.
10. Serls AE, Doherty S, Parvatiyar P, Wells JM, Deutsch GH. Different thresholds of fibroblast growth factors pattern the ventral foregut into liver and lung. *Development* 2005;**132**:35–47.
11. Longmire TA, Ikonomou L, Hawkins F, Christodoulou C, Cao Y, Jean JC, et al. Efficient derivation of purified lung and thyroid progenitors from embryonic stem cells. *Cell Stem Cell* 2012;**10**:398–411.
12. Kadzik RS, Morrisey EE. Directing lung endoderm differentiation in pluripotent stem cells. *Cell Stem Cell* 2012;**10**:355–61.
13. Huang SXL, et al. Efficient generation of lung and airway epithelial cells from human pluripotent stem cells. *Nat Biotechnol* 2014;**32**:84–91.
14. Spooner BS, Wessells NK. Mammalian lung development: interactions in primordium formation and bronchial morphogenesis. *J Exp Zool* 1970;**175**:445–54.
15. Orr-Urtreger A, Bedford MT, Burakova T, Arman E, Zimmer Y, Yayon A, et al. Developmental localization of the splicing alternatives of fibroblast growth factor receptor-2 (FGFR2). *Dev Biol* 1993;**158**:475–86.
16. Bellusci S, Grindley J, Emoto H, Itoh N, Hogan BL. Fibroblast growth factor 10 (FGF10) and branching morphogenesis in the embryonic mouse lung. *Development* 1997;**124**:4867–78.
17. Ohuchi H, Ori Y, Yamasaki M, Harada H, Sekine K, Kato S, et al. FGF10 acts as a major ligand for FGF receptor 2 IIIb in mouse multiorgan development. *Biochem Biophys Res Commun* 2000;**277**:643–9.
18. Park WY, Miranda B, Lebeche D, Hashimoto G, Cardoso WV. FGF-10 is a chemotactic factor for distal epithelial buds during lung development. *Dev Biol* 1998;**201**:125–34.
19. Weaver M, Dunn NR, Hogan BL. Bmp4 and Fgf10 play opposing roles during lung bud morphogenesis. *Development* 2000;**127**:2695–704.
20. Sekine K, Ohuchi H, Fujiwara M, Yamasaki M, Yoshizawa T, Sato T, et al. Fgf10 is essential for limb and lung formation. *Nat Genet* 1999;**21**:138–41.
21. De Moerlooze L, Spencer-Dene B, Revest J. M., Hajihosseini, M., Rosewell, I., Dickson C. An important role for the IIIb isoform of fibroblast growth factor receptor 2 (FGFR2) in mesenchymal-epithelial signalling during mouse organogenesis. *Development* 2000;**127**:483–92.
22. Kastner P, Mark M, Ghyselinck N, Krezel W, Dupé V, Grondona JM. Genetic evidence that the retinoid signal is transduced by heterodimeric RXR/RAR functional units during mouse development. *Development* 1997;**124**:313–26.
23. Niederreither K, Subbarayan V, Dollé P, Chambon P. Embryonic retinoic acid synthesis is essential for early mouse post-implantation development. *Nat Genet* 1999;**21**:444–8.
24. Dollé P, Ruberte E, Leroy P, Morriss-Kay G, Chambon P. Retinoic acid receptors and cellular retinoid binding proteins. I. A systematic study of their differential pattern of transcription during mouse organogenesis. *Development* 1990;**110**:1133–51.
25. Malpel S, Mendelsohn C, Cardoso WV. Regulation of retinoic acid signaling during lung morphogenesis. *Development* 2000;**127**:3057–67.
26. Mendelsohn C, et al. Function of the retinoic acid receptors (RARs) during development (II). Multiple abnormalities at various stages of organogenesis in RAR double mutants. *Development* 1994;**120**:2749–71.
27. Wilson JG, Roth CB, Warkany J. An analysis of the syndrome of malformations induced by maternal vitamin A deficiency. Effects of restoration of vitamin A at various times during gestation. *Am J Anat* 1953;**92**:189–217.
28. Chen F, Desai TJ, Qian J, Niederreither K, Lü J, Cardoso WV. Inhibition of Tgf beta signaling by endogenous retinoic acid is essential for primary lung bud induction. *Development* 2007;**134**:2969–79.
29. Domyan ET, Ferretti E, Throckmorton K, Mishina Y, Nicolis SK, Sun X. Signaling through BMP receptors promotes respiratory identity in the foregut via repression of Sox2. *Development* 2011;**138**:971–81.
30. Hui C-C, Slusarski D, Platt KA, Holmgren R, Joyner AL. Expression of three mouse homologs of the Drosophila segment polarity gene cubitus interruptus, Gli, Gli-2, and Gli-3, in ectoderm - and mesoderm - derived tissues suggests multiple roles during postimplantation development. *Dev Biol* 1994;**162**:402–13.
31. Grindley JC, Bellusci S, Perkins D, Hogan BL. Evidence for the involvement of the Gli gene family in embryonic mouse lung development. *Dev Biol* 1997;**188**:337–48.
32. Motoyama J, Liu J, Mo R, Ding Q, Post M, Hui CC. Essential function of Gli2 and Gli3 in the formation of lung, trachea and oesophagus. *Nat Genet* 1998;**20**:54–7.
33. Pepicelli CV, Lewis PM. & McMahon, a P. Sonic hedgehog regulates branching morphogenesis in the mammalian lung. *Curr Biol* 1998;**8**:1083–6.
34. Sutliff KS, Hutchins GM. Septation of the respiratory and digestive tracts in human embryos: crucial role of the tracheoesophageal sulcus. *Anat Rec* 1994;**238**:237–47.

35. Zhou B, Hutson JM, Farmer PJ, Hasthorpe S, Myers NA, Liu M. Apoptosis in tracheoesophageal embryogenesis in rat embryos with or without adriamycin treatment. *J Pediatr Surg* 1999;**34**:872–5. discussion 876.

36. Shaw-Smith C. Oesophageal atresia, tracheo-oesophageal fistula, and the VACTERL association: review of genetics and epidemiology. *J Med Genet* 2006;**43**:545–54.

37. Dickman ED, Thaller C, Smith SM. Temporally-regulated retinoic acid depletion produces specific neural crest, ocular and nervous system defects. *Development* 1997;**124**:3111–21.

38. Hogan BLM. *Morphogenesis* 1999;**96**:225–33.

39. Shannon JM. Induction of alveolar type II cell differentiation in fetal tracheal epithelium by grafted distal lung mesenchyme. *Dev Biol* 1994;**166**:600–14.

40. Bellusci S, Furuta Y, Rush MG, Henderson R, Winnier G, Hogan BL. Involvement of Sonic hedgehog (Shh) in mouse embryonic lung growth and morphogenesis. *Development* 1997;**124**:53–63.

41. Hacohen N, Kramer S, Sutherland D, Hiromi Y, Krasnow MA. sprouty encodes a novel antagonist of FGF signaling that patterns apical branching of the Drosophila airways. *Cellule* 1998;**92**:253–63.

42. Minowada G, Jarvis LA, Chi CL, Neubüser A, Sun X, Hacohen N, et al. Vertebrate Sprouty genes are induced by FGF signaling and can cause chondrodysplasia when overexpressed. *Development* 1999;**126**:4465–75.

43. Tefft JD, Lee M, Smith S, Leinwand M, Zhao J, Bringas P, et al. Conserved function of mSpry-2, a murine homolog of Drosophila sprouty, which negatively modulates respiratory organogenesis. *Curr Biol* 1999;**9**:219–22.

44. De Maximy AA, Nakatake Y, Moncada S, Itoh N, Thiery JP, Bellusci S. Cloning and expression pattern of a mouse homologue of drosophila sprouty in the mouse embryo. *Mech Dev* 1999;**81**:213–6.

45. Mailleux AA, Tefft D, Ndiaye D, Itoh N, Thiery JP, Warburton D, et al. Evidence that SPROUTY2 functions as an inhibitor of mouse embryonic lung growth and morphogenesis. *Mech Dev* 2001;**102**:81–94.

46. Massagué J, Chen Y. Controlling TGF- β signaling. *Genes Dev* 2000;**14**:627–44.

47. Heine UI, Munoz EF, Flanders KC, Roberts AB, Sporn MB. Colocalization of TGF-beta 1 and collagen I and III, fibronectin and glycosaminoglycans during lung branching morphogenesis. *Development* 1990;**109**:29–36.

48. Lü J, Qian J, Izvolsky KI, Cardoso WV. Global analysis of genes differentially expressed in branching and non-branching regions of the mouse embryonic lung. *Dev Biol* 2004;**273**:418–35.

49. Serra R, Moses HL. pRb is necessary for inhibition of N-myc expression by TGF-beta 1 in embryonic lung organ cultures. *Development* 1995;**121**:3057–66.

50. Metzger RJ, Klein OD, Martin GR, Krasnow MA. The branching programme of mouse lung development. *Nature* 2008;**453**:745–50.

51. Lu Y, Thomson JM, Wong HYF, Hammond SM, Hogan BLM. Transgenic over-expression of the microRNA miR-17-92 cluster promotes proliferation and inhibits differentiation of lung epithelial progenitor cells. *Dev Biol* 2007;**310**:442–53.

52. Tian Y, Zhang Y, Hurd L, Hannenhalli S, Liu F, Lu MM, et al. Regulation of lung endoderm progenitor cell behavior by miR302/367. *Development* 2011;**138**:1235–45.

53. Mercola M, Levin M. Left-right asymmetry determination in vertebrates. *Annu Rev Cell Dev Biol* 2001;**17**:779–805.

54. Meno C, Shimono A, Saijoh Y, Yashiro K, Mochida K, Ohishi S, et al. Lefty-1 is required for left-right determination as a regulator of lefty-2 and nodal. *Cell* 1998;**94**:287–97.

55. Tsukui T, Capdevila J, Tamura K, Ruiz-Lozano P, Rodriguez-Esteban C, Yonei-Tamura S, et al. Multiple left-right asymmetry defects in Shh(-/-) mutant mice unveil a convergence of the shh and retinoic acid pathways in the control of Lefty-1. *Proc Natl Acad Sci USA* 1999;**96**:11376–81.

56. Oh SP, Li E. The signaling pathway mediated by the type IIB activin receptor controls axial patterning and lateral asymmetry in the mouse. *Genes Dev* 1997;**11**:1812–26.

57. Cornett B, Snowball J, Varisco BM, Lang R, Whitsett J, Sinner D. Wntless is required for peripheral lung differentiation and pulmonary vascular development. *Dev Biol* 2013;**379**:38–52.

58. Chang DR, Martinez Alanis D, Miller RK, Ji H, Akiyama H, McCrea PD, et al. Lung epithelial branching program antagonizes alveolar differentiation. *Proc Natl Acad Sci USA* 2013;**110**:18042–51.

59. Bellusci S, Henderson R, Winnier G, Oikawa T, Hogan BL. Evidence from normal expression and targeted misexpression that bone morphogenetic protein (Bmp-4) plays a role in mouse embryonic lung morphogenesis. *Development* 1996;**122**:1693–702.

60. Weaver M, Yingling JM, Dunn NR, Bellusci S, Hogan BL. Bmp signaling regulates proximal-distal differentiation of endoderm in mouse lung development. *Development* 1999;**126**:4005–15.

61. Lebeche D, Malpel S, Cardoso WV. Fibroblast growth factor interactions in the developing lung. *Mech Dev* 1999;**86**:125–36.

62. Brunet LJ, McMahon JA, McMahon AP, Harland RM. Noggin, cartilage morphogenesis, and joint formation in the mammalian skeleton. *Science* 1998;**280**:1455–7.

63. Al Alam D, Green M, Tabatabai Irani R, Parsa S, Danopoulos S, Sala FG, et al. Contrasting expression of canonical Wnt signaling reporters TOPGAL, BATGAL and Axin2(LacZ) during murine lung development and repair. *PLoS One* 2011;**6**: e23139.

64. Mucenski ML, Nation JM, Thitoff AR, Besnard V, Xu Y, Wert SE, et al. *beta -Catenin regulates differentiation of respiratory epithelial cells in vivo*, 2005;**3039**:971–979.

65. Hashimoto S, Chen H, Que J, Brockway BL, Drake JA, Snyder JC, et al. β-Catenin-SOX2 signaling regulates the fate of developing airway epithelium. *J Cell Sci* 2012;**125**:932–42.

66. Borok Z, Danto SI, Lubman RL, Cao Y, Williams MC, Crandall E. D. Modulation of t1alpha expression with alveolar epithelial cell phenotype in vitro. *Am J Physiol* 1998;**275**:L155–64.

67. Volckaert T, Campbell A, Dill E, Li C, Minoo P, De Langhe S. Localized Fgf10 expression is not required for lung branching morphogenesis but prevents differentiation of epithelial progenitors. *Development* 2013;**140**:3731–42.

68. Williams MC, Cao Y, Hinds A, Rishi AK, Wetterwald A. T1 alpha protein is developmentally regulated and expressed by alveolar type I cells, choroid plexus, and ciliary epithelia of adult rats. *Am J Respir Cell Mol Biol* 1996;**14**:577–85.

69. Borok Z, Lubman RL, Danto SI, Zhang XL, Zabski SM, King LS, et al. Keratinocyte growth factor modulates alveolar epithelial cell phenotype in vitro: expression of aquaporin 5. *Am J Respir Cell Mol Biol* 1998;**18**:554–61.

70. Zhou L, Dey CR, Wert SE, Whitsett JA. Arrested lung morphogenesis in transgenic mice bearing an SP-C-TGF-beta 1 chimeric gene. *Dev Biol* 1996;**175**:227–38.

71. Schittny JC, Mund SI, Stampanoni M. Evidence and structural mechanism for late lung alveolarization. *Am J Physiol Lung Cell Mol Physiol* 2008;**294**:246–54.

72. Maeda Y, Davé V, Whitsett JA. Transcriptional control of lung morphogenesis. *Physiol Rev* 2007;**87**:219–44.

Chapter 2

Development of Airway Epithelium

Charles Plopper and Michelle Fanucchi

School of Veterinary Medicine, Department of Anatomy, Physiology and Cell Biology, University of California – Davis, Davis, CA, USA

INTRODUCTION

A number of developmental processes are involved in the establishment of the tracheobronchial airway tree. The pattern of branching of the airways including the angle of branching and the proportions of daughter branches in relation to parent airway appears to be established relatively early by the process of branching morphogenesis. As summarized in detail in Chapter 1, this process is initiated with the earliest formation of respiratory tract structures in the thorax in the embryonic period and continues for a substantial period of time during early gestation. It is heavily dependent on epithelial-mesenchymal contact and continual interaction to regulate the rate and pattern of formation. The composition of the wall of the airways in adults varies substantially between different segments, with most of the differences being highly polarized from more proximal airways to more distal airways. The major components of the wall include: (1) the surface lining epithelium with its associated derivative, the submucosal gland; (2) the basement membrane zone, including basal lamina and an extended population of fibroblasts; and (3) bundles of smooth muscle and cartilage. The distribution of all of these components varies substantially within the airway tree in adults. The entire wall is invested with a large number of nerves that appear to be in two separate distributional patterns, one associated with the epithelial surface and another associated with the glands and smooth muscle in the submucosa and adventitia. As detailed in Chapter 5, the formation of the nerves occurs early in development once the pattern of the airway tree has been laid down. The presence of nerves in the wall, however, does not establish that they have processes that extend into the epithelial compartment or directly to the smooth muscle. It is not clear when this occurs, but it apparently occurs during the differentiation process. Chapter 5 also addresses airway smooth muscle and establishes that it is differentiated early in development once the basic pattern of the wall airway has been laid down. Once the basic

geometric pattern of the airways has been established, they undergo substantial enlargement through longitudinal and circumferential growth. The tremendous increase in cell and tissue mass necessary to accomplish growth relies on active proliferation of resident cell populations and the ability of the same cell populations to synthesize and secrete matrix components. How these processes are established and regulated and how they are balanced with forces promoting differentiation of the same cell populations is not understood. Further, these complex processes continue for a substantial period of time after birth. The temporal pattern for the differentiation processes varies significantly by species, but always moves in a proximal to distal direction with time. What this means is that during pre- and postnatal development of the airways, different airway generations will be in different stages of development. At any given time point, more proximal airway generations will be more differentiated than more distal generations. Because the other aspects of airway development have been defined in Chapters 1 and 5, this chapter will emphasize the epithelium and its pattern of differentiation and what is known about regulation of the differentiation process.

DIFFERENCES IN PHENOTYPIC EXPRESSION IN ADULTS

This chapter is organized on the premise that understanding the development of cellularly and architecturally complex organ systems such as the respiratory system, especially in the case of tracheobronchial airways, requires definition of changes based on specific airway sites and clear distinction of the timing of events in these sites. One of the major considerations in evaluating the potential toxicity of environmental contaminants for the developmental process is understanding which of the compartments is in which stage of development and differentiation at the time of exposure. Further understanding of airway development and the mechanisms that regulate it needs to be based on (1)

The Lung. http://dx.doi.org/10.1016/B978-0-12-799941-8.00002-X

a clear understanding of the architectural organization and microenvironment related characteristics that are expressed in differentiated systems in adults and (2) on how these microenvironments respond to toxic stressors in adults. Previous studies have clearly established that two of the major classes of respiratory toxicants, oxidant air pollutants and bioactivated polyaromatic hydrocarbons, produce patterns of acute cytotoxicity that are highly site- and cell-selective. To define the complexity of the respiratory toxic response, we have compared multiple sites with profoundly different responses to oxidant air pollutants and bioactivated cytotoxicants: respiratory mucosa and olfactory mucosa in the nasal cavity, the trachea, and proximal, mid-level bronchi and distal bronchioles in the lungs.[1-14] It is now well-recognized that the respiratory system of adult mammals contains over 40 different cell phenotypes distributed within a large number of distinct microenvironments. Using microdissection approaches our group has defined the complexity and microenvironment-dependent nature of phenotypic expression for potential target cell populations within the different airway sites.[2,15-32] Virtually every aspect of the composition of the wall of the airways, including epithelium and glands, smooth muscle, cartilage, varies by species. This is especially true for the airway epithelium, which is highly varied in any one species depending on precisely where in the airway tree the cell populations are examined for these characteristics: the composition of the cell populations lining the luminal surface (Table 1); the composition of the secretory product found within these epithelial populations (Tables 1, 2, 3). Where the same phenotypes are present in many different airways, their relative abundance and proportion of the luminal surface occupied by the specific phenotypes may vary tremendously.

What this means is that the organization of tracheal epithelium is very different from that of terminal and respiratory bronchioles in the same animal. When different species are compared on an airway-by-airway basis it is clear that the same types of variability exist. In fact in many species, individuals that are free of respiratory disease have very different cell populations in the same microenvironment than do other equally healthy species. This is also true for the potential of the metabolism of xenobiotics either by an activation system (cytochrome P450 monooxygenases) or a variety of detoxification and antioxidant systems (see Chapter 12). This also applies to the distribution of submucosal glands, with many species having glands extensively down the airway tree as far as small bronchioles, whereas in other species they are restricted to the most proximal portions of the trachea. Cartilage is not a prominent feature of the conducting airways distal to the trachea in most species the

TABLE 1 Carbohydrate content of tracheal epithelium

Species	Cell Type	Abundance	Carbohydrate Content		
			PAS	AB	HID
Hamster	Clara	+++	+	−	−
	Mucous	+	+	+	−
Rat	Serous	+++	+	−	−
	Mucous	+	+	+	−
Mouse	Mucous	+	+	+	−
	Clara	+++	+/−	−	−
Rabbit	Mucous	+	+	+	+
	Clara	+++	+/−	−	−
Dog	Mucous	++	+	+	+
Cat	Mucous	++	+	+	+ and −
	Serous	+	ND	ND	ND
Pig	Mucous	++	+	+	+ and −
Sheep	Mucous	++	+	+	+ and −
Rhesus	Mucous	++	+	+	+
Human	Mucous	+++	+	+	+ and −

Source: Reproduced from references 18, 33–37.

TABLE 2 Carbohydrate content of tracheal submucosal glands

Species	Gland Abundance	Cell Type	Carbohydrate Content		
			PAS	AB	HID
Hamster	+/−	Mucous	+	+	−
Rat	+	Serous	+	−	−
		Mucous	+	+	+ and −
Mouse	+/−	Serous	+	−	−
		Mucous	+	+	+ and −
Rabbit	+/−	Mucous	+	+	+
Dog	++	Serous	+	−	−
		Mucous	+	+	+ and −
Cat	++++	Serous	+	−	−
		Mucous	+	+	+
Pig	++	Serous	+	−	−
		Mucous	+	+	+ and −
Sheep	++	Serous	+	−	−
		Mucous	+	+	+
Rhesus	++	Serous	+	−	−
		Mucous	+	+	+
Human	+++	Serous	+	−	−
		Mucous	+	+	+ and −

size of rabbits or smaller, but is found extensively throughout the intrapulmonary airways in larger species, including humans. The distribution and organization of smooth muscle appears to be relatively site specific. The complexity of the cellular organization within even a restricted portion, i.e., the bronchial airways, of a complex organ such as the lungs emphasizes the need for highly precise sampling methodology.

This need is further emphasized by the wide variability in local exposure dose created by the architectural complexity of the tracheobronchial airway tree itself.[5,9,44] As would be expected from a highly complex cellular organization, the metabolic potential of cell populations in different microenvironments within the respiratory system varies widely. The principal enzyme system for xenobiotic bioactivation, the cytochrome P450 monooxygenases, has broad variability in isozyme expression, substrate specificity, and level of activity.[2,45–52] This is also true for the enzyme systems involved in detoxification, especially the glutathione S-transferases and epoxide hydrolases.[46,51–54]

The cells in each of these different microenvironments also manage their glutathione pools very differently.[9,52,55,56] The pattern of heterogeneity of metabolic function appears to be relatively unique for each species of mammal. Inflammatory responses generated by acute exposure to oxidant air pollutants also vary greatly by site within the tracheobronchial airway tree.[3,57] The biological uniqueness of the cell populations in local airway microenvironments is further emphasized by the fact that when epithelial populations are cultured with the surrounding matrix intact, they maintain the same phenotypic expression and response to toxicants that would be expected if they were still resident within the intact animal.[10,12,58,59] This complexity emphasizes the need for precise sampling to establish meaningful cellular and metabolic profiles and to validate them for patterns of cytotoxicity. They have been used for definition of local cytotoxicity,[12,17,30] metabolism,[60] maintenance of biological function in vitro,[10,12] and definition of local exposure dose.[5,9,44] They have even been validated for obtaining nucleic acids for definition of gene expression at the level

TABLE 3 Lectin reactivity in airway luminal epithelium and submucosal glands

Species/Airway	Cell Type		Lectin							
		Lectin Sugar Specificity:	LCA Man Glc GlcNAc	WGA NANA GlcNAc	BSA1 Gal	DBA GalNAc	SBA Gal GalNAc	PNA Gal GalNAc	RCA Gal GalNAc	UEA1 Fuc
Luminal epithelium										
Human bronchi	Mucous		−	++++	+	+	+++	++++	++++	++++
Sheep trachea-bronchioles	Mucous M1				−	+++	++	+++	++++	++++
	Mucous M2				−	+++	++	+++		++++
	Mucous M3				+++	+++	−	−		−
Rat trachea	Mucous		−		−	+	+	−(+++)*	+	+++
Rhesus trachea	Mucous		++++	++++	++++	++++	−	−(+++)	−(+++)	+
Submucosal glands										
Human bronchi	Mucous		−	+	+	+	+++	++++	+++	++++
	Serous		+++	+++	−	−	++	+	++	−
Sheep trachea-bronchi	Mucous M4				−	++++	++	−		++
	Serous				−	+++	++	−		++
Mouse trachea	Mucous		−		−	−	+	−(+++)	−	+++
	Serous				−	++++	++++	++++	++++	−
Rat trachea	Mucous		++		−	++	++	−(+++)	−	+++
	Serous				−	−	−	−(++++)	−	−
Rhesus trachea	Mucous		++++	++++	+++	−	−(++++)	−(++++)	−	++++
	Serous		++	++	+	−	−(++)	−(++)		+

Abbreviations: BSA1, *Bandeirea simplicifolia*; DBA, *Dolichos biflorus*; Fuc, fucose; Gal, galactose; GalNAc, N-acetylgalactosaminee; Glc, galactose; **LCA**, **Lotus tetragonolobus**; Man, mannose; NANA, N-acetylneuraminic acid (sialic acid); PNA, *Arachis hypogea*; RCA, *Ricinus communis*; SBA, *Glycine max*; UEA1, *Ulex europeus*; WGA, wheat germ agglutinin
¹Reaction in parenthesis is after neuraminidase treatment.
Source: Reproduced from references 18, 36, 38–43.

of the local microenvironment using the two species we propose to evaluate through center support.[61]

OVERALL DEVELOPMENT OF AIRWAYS

Early Branching Morphogenesis

As outlined in Chapter 1, the early formation of the airways and the subsequent development of submucosal glands are produced by the process of branching morphogenesis. In essence, this involves the differential growth of an epithelial tube into an associated mesenchymal derivative containing both cells and matrix. The composition of the matrix appears to dictate where the growing tube will divide. The bifurcation process itself is produced by focal differences in proliferation and programmed cell death to produce rapid growth in areas adjacent to sites of no growth. The no growth sites appear to be associated with bands of newly formed collagen and elastin. Each branching of this growing tube is regulated by a variety of cytokines and growth factors, as outlined in Chapter 1. Subsequent development of the other components that form the wall in adults occurs at later times in specific airways. It appears to move in a proximal to distal pattern following the branching of the epithelial tube. For the formation of the airway tree this process is thought to be complete prior to birth and varies from species to species as to the percentage of gestation during which the process is complete. As outlined in Chapter 1, subsequent branching produces alveolar septation in alveolar spaces. Once the general pattern of the tree has been established, subsequent developmental processes are essentially growth in two directions: either longitudinally to extend the length of the tube or circumferentially to increase its diameter. What regulates these processes and how they are associated with differentiation and growth of the constituents of the wall is not clear and has not been carefully evaluated. This would be of particular significance given the substantial impact that the size and angles of the airways have on the flow of air during the respiratory cycle. In most mammalian species the majority of the growth of the airways is a postnatal event. This suggests that for an extended period after birth, these growth events are susceptible to perturbations by environmental contaminants.

Respiratory Bronchioles

The most distal airways, located at the junction between the gas exchange area and tracheobronchial airway tree, form an extensive transitional zone in the human lung. This zone, exceeding three generations of branching in humans, is characterized by intermixing of alveolar epithelium, simple cuboidal epithelium mixed with the pseudostratified cuboidal epithelium (with basal, mucus, and ciliated cells) found in more proximal airways (see[62] for a review). The respiratory bronchioles are extensive (exceeding three generations) in humans, macaques, dogs, cats, and ferrets. In rhesus

monkeys, and possibly in other primates, including humans, the two epithelial populations, bronchiolar and alveolar, are distributed on opposite sides of the airway in relation to the position of the pulmonary arteriole.[63] A pseudostratified population with ciliated cells lines numerous generations of respiratory bronchioles on the side adjacent to the pulmonary arteriole. The alveolarized areas are surrounded by a simple cuboidal bronchiolar epithelial population on the side opposite the arteriole. In the majority of mammalian species, the bronchiolar epithelium occupies the proximal portion of the transitional bronchiole, and alveolar gas exchange epithelium lines the distal portion. This is the case for mice, hamsters, rats, guinea pigs, rabbits, pigs, sheep, cattle, and horses.[64]

The composition of the peribronchiolar region associated with Clara cells includes the presence of smooth muscle adjacent to the basal lamina, extensive collagen interspersed with elastin, and few capillaries. Those capillaries that are present are not closely associated with the epithelial basal lamina. The principal vessel in the area is the pulmonary arteriole. In contrast, the alveolar portions of this transitional zone generally include a substantial capillary bed closely applied to the basal lamina of the alveolar epithelial populations. Although the matrix composition of the alveolar gas exchange portions of the lung have been studied in some detail, the same is not true for the matrix associated with the bronchioles.[65]

In fetal animals, where the majority of epithelial cells are poorly differentiated or undifferentiated, the boundary between the epithelium lining presumptive distal conducting airway and that lining future gas exchange regions is relatively easily defined in some species.[65–68] The features include differences in epithelial configuration and modifications in the surrounding mesenchymally derived components. Most of these components, including smooth muscle and fibroblast-like cells, appear to mature somewhat more quickly then do the associated epithelium.[63,69] The morphogenesis of the respiratory bronchiole during fetal lung development has been studied in detail in only one species: rhesus monkeys.[69] The respiratory bronchiole begins as a tube lined by glycogen-filled cuboidal cells intermixed with an occasional ciliated cell. Alveolarization begins in the most proximal aspect of the respiratory bronchiole, at approximately 60% gestation in rhesus monkeys and in humans. The alveolarization appears as a formation of outpocketings into surrounding extracellular matrix. The outpocketings, which are lined by cuboidal epithelium, occur only on the side of the potential respiratory bronchiole opposite the pulmonary arteriole. They begin at the same time that secondary septa are forming in the distal acinus. Outpocketing or alveolarization occurs over a very short period of time (5 days) in rhesus monkeys. As alveolarization progresses from proximal to distal in the potential respiratory bronchiole, the epithelial cells also differentiate. By 67% gestation, ciliated cells are confined to the epithelium adjacent to the pulmonary arteriole, and the cytodifferentiation of the epithelial cells characteristic of alveoli is beginning in the outpocketings. Contacts between epithelium and underlying fibroblastic cells are observed for a very brief period in regions of respiratory bronchiole development. Epithelium of proximal generations of respiratory bronchiole differentiates earlier than more distal generations, but much later than in the trachea.

SUBMUCOSAL GLANDS

The developmental events involved in formation of submucosal glands have been well described for a number of species, including rats,[70] opossums,[71] ferrets,[72] rhesus monkeys,[73] and humans.[74,75] The sequence of events in

humans has been characterized sub-grossly,[76,77] and histologically.[74,75,78,79] The ultrastructure and histochemistry of gland development have been characterized in the most detail in rhesus monkeys.[73] In rhesus monkeys, most of the process occurs in the fetus between the end of the pseudoglandular stage and the beginning of the terminal sac stage of development. Gland development implies four phases summarized in Figure 1: (1) the formation of buds by projections of undifferentiated cells from the maturing surface epithelium; (2) the outgrowth and branching of these buds into cylinders of undifferentiated cells; (3) the differentiation of mucus cells in proximal tubules associated with proliferation of tubules and acini and with undifferentiated cells

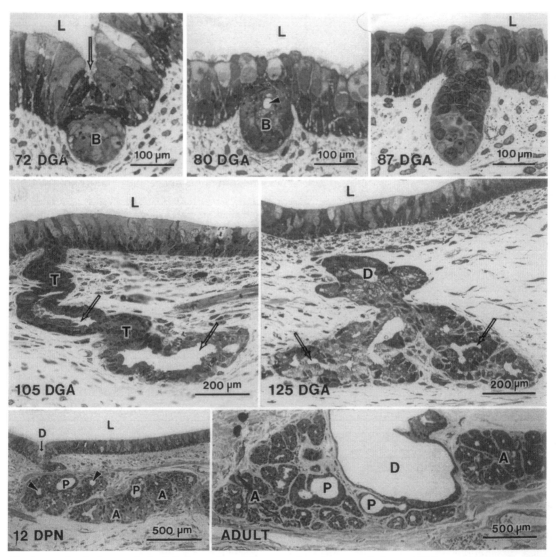

FIGURE 1   Morphogenesis of submucosal glands in the trachea of rhesus monkeys. In very early fetuses (72 days gestational age (DGA)), the initial phase is projection of buds (B) from the luminal (L) surface epithelium into the surrounding matrix accompanied by an invagination (arrow) of the surface epithelium. As fetuses age (80 and 87 DGA) the buds extend further into submucosal connective tissue, with an apparent lumen (arrowhead) and continue until the formation of a cylindrical projection. In midgestation (105 DGA), the tube (T) branches extensively into the matrix with a patent lumen (arrow) apparent throughout. In late gestation (125 DGA) and early postnatal age (12 DPN), differentiation between the duct (D) and more peripheral secretory structures including proximal tubes (P) and large numbers of secretory acini (A) are evident. Continuing growth includes expansion of a center area's proximal tubular structures and marked enlargement of the ducts in adults.

distally; and (4) differentiation of serous cells in peripheral tubules and acini, with continued proliferation in most distal areas. The cells forming gland buds are not basal cells, as first thought, but rather an undifferentiated cell similar to the surface epithelium (Figure 1).[73] Connective tissue appears to play a role in this process, as evidenced primarily through the presence of cartilage plates in the areas of initial bud formation. Glands appear first at the junction of cartilage plate and smooth muscle, followed by areas over cartilage plates and then in the area over smooth muscle. The secretory cell population differentiates in a centrifugal pattern, with nearly mature cells lining proximal tubules and immature cells in more distal portions. Mucus cells in the proximal portion of the gland develop before serous cells. Glandular mucus cells and serous cells differentiate at different times during development and through a different sequence of events.[73]

EPITHELIAL DIFFERENTIATION

Overview

Of the over 40 different cellular phenotypes that have been identified in the lungs of adult mammals, the differentiation of the epithelial cells lining the air passages appears to be the most critical in the successful function of the lung in adults. At least eight of these cell phenotypes line the tracheobronchial conducting airways, including ciliated cells, basal cells, mucus goblet cells, serous cells, Clara cells, small mucus granule cells, brush cells, neuroendocrine cells, and a number of undifferentiated or partially differentiated phenotypes that have not been well characterized. The abundance and distribution of these cell types within the conducting airway tree vary by position within the tree and by species. The pattern of differentiation of the tracheal epithelial lining has been characterized for a large number of species.[80] The general pattern appears to be the same for most species in terms of which phenotypes are identified earliest during development and which differentiate later (summarized in Figure 2). The critical difference between differentiation of these epithelial cells during development is the percentage of intrauterine life in which the differentiation occurs. The epithelium of the trachea is the earliest of all the epithelial populations to differentiate. In some species it is relatively differentiated prior to birth, and in other species the majority of the differentiation occurs postnatally. In most species, with the possible exception of ferrets, ciliated cells differentiate first. Nonciliated cells with secretory granules appear next. Basal and small mucus granule cells appear last. There is a polarity in the differentiation of ciliated cells in the trachea, with the epithelium over the smooth muscle undergoing ciliogenesis earlier than that on the cartilaginous side. The reverse appears to be true for the nonciliated secretory cells, with secretory granules appearing on the cartilaginous side of the trachea first.

Pattern in Trachea and Bronchi

The ultrastructural features of overall tracheal epithelial differentiation in developing fetuses have been described in rabbits,[81] mice,[82] hamsters,[83,84] and rats.[79] In view of the diversity in the airways in different species, small laboratory mammals may not be adequate models for the study of human tracheobronchial epithelium.[79] The most extensive study of the development of the mucus cell was performed on the trachea of rhesus monkeys,[85] and is reviewed here. Gestation for rhesus monkeys averages 168 days, with the stages of lung development as follows: embryonic period, 21 to 55 days gestational age (DGA); pseudoglandular, 56 to 80 DGA; canalicular, 80 to 130 DGA; and terminal sac, 131 to term.[86]

In the youngest fetuses, all cells appear as illustrated in Figure 3A. The cells are columnar and the apices of most of the cells reach the luminal surface. Nuclei have little heterochromatin, and the cytoplasm is filled from base to apex with glycogen. The few organelles present are located in the apex of the cell and included short narrow strands of granular endoplasmic reticulum (GER), small spherical mitochondria, and a small Golgi apparatus located adjacent to the lateral surface of the cell (Figure 3A). These cells were present in the epithelial lining in the youngest animal in the embryonic stage to the middle of the canalicular phase.

In fetuses near the end of the embryonic period, many cells similar to those at younger ages, but containing larger numbers of apical organelles, are observed. The organelles include spherical mitochondria and increased amounts of GER with dilated cisternae. The cisternae of the Golgi apparatus are dilated and surrounded by enlarged membrane-bound vacuoles, and glycogen is concentrated near the nucleus and intermixed with the organelles. These cells are observed in fetuses up to early in the canalicular stage.

Through most of the pseudoglandular stage, most nonciliated cells had increased numbers of apical membrane-bound secretory granules containing a flocculent matrix, with a small electron-dense spherical core (Figure 3B). Most of the remaining cytoplasm is still filled with glycogen. The cytoplasm surrounding the glycogen is more electron dense than in younger ages and occupied more of the apical portion of the cell. The nuclei exhibit prominent nucleoli and small patches of heterochromatin. The mitochondria exhibit noncircular profiles and appeared to be tubular. The amount of GER appears to be the same as in younger ages, but the cisternae were no longer dilated. The Golgi apparatus is surrounded by vacuoles of various sizes. The luminal surface of

FIGURE 2 Diagrammatic comparison of the pattern of differentiation and maturation of the principle cell phenotypes observed in the tracheobronchial airways, with an emphasis on epithelium in trachea. This compares the timing of events in relation to parturition (open arrow) for each species. The dotted line indicates the duration of time during gestation from the initial observation of the formation of cilia to parturition. It is represented both in terms of the percentage of gestation over which differentiation occurs in utero and the actual number of days it takes. For most species, a significant portion of differentiation is postnatal. The proportion of time during gestation when these events occur is very species specific.

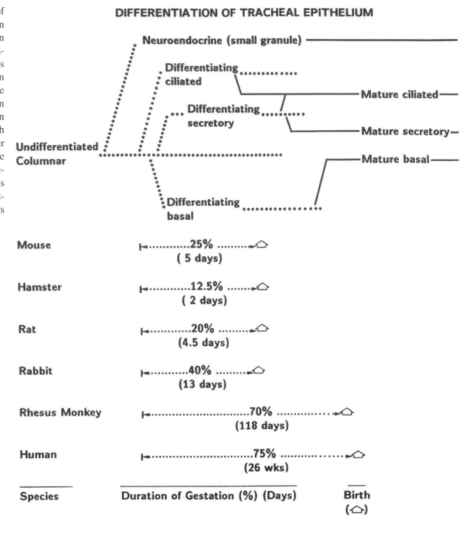

these cells is covered by long, regular microvilli. In somewhat older animals (late pseudoglandular), the apices of a large proportion of the secretory cells are filled with spherical granules (Figure 3C). Most of these cells have abundant cytoplasmic glycogen, most of which is basal to the nucleus. Apical to the nucleus, glycogen is interspersed among organelles and granules. The cells appear more fusiform than at younger ages, being wide at the luminal side and narrow at the base (Figure 3C). In fetuses from midcanalicular stage and older, secretory cells containing cytoplasmic glycogen are rare and, when observed, the glycogen content was minimal. From this time to parturition, only two forms of secretory cells are observed. Both cells have little cytoplasmic glycogen, and the cytoplasm was condensed. There is a distinct variation in the abundance of apical secretory granules in these cells, ranging from very few, in cells with a narrow cytoplasm and few organelles, to cells with an abundance of these granules (Figure 3D). The cytoplasm of these cells contains small mitochondria and varying amounts of GER. The Golgi apparatus is located on the apical side of the

nucleus and showed variable degrees of activity. In cells with more granules (Figure 3D), the Golgi apparatus is larger, has more cisternal stacks, and has larger and more numerous adjacent vesicles. Long, regular microvilli are a characteristic feature of the surface of the secretory cells. There is considerable variability in the abundance of these cellular forms between 105 days and parturition. In the earlier ages, they are of approximately equal abundance. Near parturition, most of the secretory cells resemble that in Figure 3D. Some of the cells have an even larger percentage of their cytoplasm occupied by granules than illustrated in Figure 3D.

In the postnatal period, most of the secretory cells have an abundance of electron-lucent granules filling their apical cytoplasm. The majority of these granules have small electron-dense cores. A few have large electron-dense biphasic cores, as is observed in the adult. In general, the nucleus and its surrounding cytoplasm are restricted to the basal portion of the cell and the Golgi apparatus, and other organelles occupy a small percentage of the cytoplasm. Through 134 days postparturition, there are, however, a few secretory cells

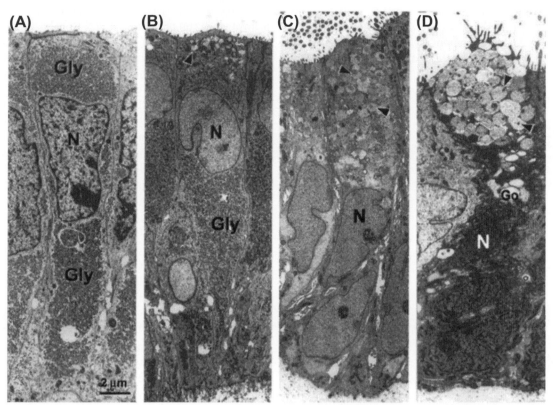

FIGURE 3 Differentiation of mucus goblet cells in trachea of fetal rhesus monkeys. A: early in gestation (46 DGA) columnar cells contain large pools of glycogen (Gly) and a central nucleus (N). B: with continued age, columnar cells taper at the base and begin to accumulate membrane bound secretory granules (arrowheads) in apical cytoplasm near the lumenal surface (62 DGA). C: by approximately 50 % of gestation (90 DGA) columnar cells have markedly tapered bases, numerous apical secretory granules (arrowhead) and little cytoplasmic glycogen. D: in the perinatal period (141 DGA) many of the cells have a prominent Golgi apparatus (Go) and an abundance of secretory granules in their apical cytoplasm (arrowhead).

the cytoplasm of which contained abundant organelles and a variable number of secretory granules, as is observed in the late fetal period (Figure 3D). By 134 days of postnatal age, nearly all of the secretory cells have a configuration similar to that observed in adults. The cytoplasm is filled with electron-lucent secretory granules that appeared to distend the cell's cytoplasm. The nucleus is compressed at the basal portion of the cell, and organelles are minimal. In most cases, the cytoplasmic granules contain a biphasic core. The central part of the core is the most electron-dense portion of the granules.

Differentiation of mucus glycoprotein biosynthesis and secretion develops slightly behind the other aspects of cellular differentiation. Prior to the presence of secretory granules the principal material reacting with periodic acid-Schiff (PAS) is the large store of glycogen surrounding the nucleus (Figure 4). Once granules appear, based on ultrastructure, the contents of the granules resemble relatively closely that identified with adults (Figure 4). These granules are not only PAS positive, but also positive for alcian blue (AB), indicating acidic groups and for high iron diamine (HID) indicating that they are sulfated. The range of distribution of these patterns of stain reactivity varies by airway depending on the mix in the adult. In core granules the sulfated material is generally

identified in the center, in uncored granules it tends to be on the periphery. As the cells fill with secretory product, the distribution of staining reaction tends to follow closely that of the granules. The pattern by which sugars are expressed during differentiation is quite variable, depending on the specific end group (Table 4). The majority of sugars are expressed during the early phases of biosynthesis and granule formation. Others are expressed much later and all are generally, if they will be expressed in the adult, present in cells with a differentiated secretory apparatus by birth. With the exception of N-acetylgalactosamine (PNA lectin) all the others are more prominent in postnatal animals than in neonatal. This particular terminal sugar appears to be expressed in the early phases of mucus cell differentiation and its expression is suppressed, at least in rhesus monkeys, in late gestation and early postnatal life. Use of monoclonal antibodies established against mucus antigens indicates that very early in differentiation the composition of the core proteins for the secretory product may not be incorporated into the granules that initially form (Figure 5). Their incorporation appears to be somewhat later during the differentiation process.

Tracheobronchial epithelium continually renews itself. To identify the progenitor cell types that are involved in the self-renewal in vivo, the traditional approach is to carry out mitotic index and nuclear labeling studies. For the nuclear labeling study, the incorporation of {³H} thymidine or bromdeoxyuridine is used. Using these approaches, most of the

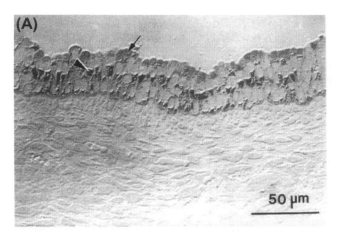

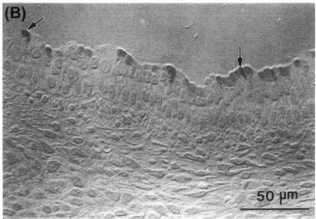

FIGURE 4   Comparison of the distribution of reaction products for alcian blue/periodic acid-Schiff (AB/PAS) and high iron diamine (HID) in the tracheal epithelium of fetal rhesus monkeys during the period when granule formation is first observed (62 DGA). The granules found in the apex of the cell are primarily AB/PAS positive (A) (arrow). Highly PAS-positive material is found on the apical and basal portions of the nucleus (arrowhead). The majority (but not all) of the AB positive material is also sulfated (B) (arrow).

data suggest that less than 1% of the epithelial cell population is involved in cell proliferation.[87–91] Both basal and secretory cell types are capable of incorporating these nucleotide precursors and mitosis, whereas ciliated cells are considered to be terminally differentiated and incapable of division.[92] In fact, only under exceptional circumstances are the ciliated cells of isolated hamster trachea capable of synthesizing DNA, as evidenced by the incorporation of {³H} thymidine.[93] Differentiation and proliferation normally are inversely related. Based on this view, a number of investigators[94,95] suggest that it is the basal cell type that serves as the stem cells, or the progenitor cell type that is involved in normal maintenance as well as in the regeneration and redifferentiation of bronchial epithelium after injury. However, this view is inconsistent with data obtained from the developmental studies and studies of injury/repair results. In the developing tracheas of a number of animal species, including humans and nonhuman primates, basal cells are derived from an undifferentiated columnar epithelium.[96] Furthermore, the appearance of the basal cell type in the tracheal surface lining layer occurs after the appearance of ciliated and nonciliated secretory cell types.[97] Furthermore, in the growing intrapulmonary airways,[98,99] the basal cell type is not found in the smallest airway.[96] In the injury models, such as the mechanical and toxic gases exposure model, hyperproliferation is seen in the secretory cell type, but not in the basal cell type.[91,100–102] These results point out that it is less likely for the basal cell type to serve as a progenitor cell type that initiates the growth of airway epithelium and the repair of epithelial damage.[96]

Studies of the repopulation of epithelial cells on denuded tracheal grafts have been used to assess the "progenitor" nature of various bronchial epithelial cell types. Denuded tracheal grafts generally are produced by removing the lining epithelial layer by repeated freezing and thawing of tracheal grafts, a technique developed several years ago by Nettesheim and his colleagues.[103] Using this technique, combined with the cell separation technique, Hook and his colleagues[104,105] have demonstrated the repopulation of a mucociliary epithelium in the denuded tracheal graft by

TABLE 4 Lectin reactivity in developing rhesus monkey trachea

Lectin	Days Gestational Age							
	50	60	70	80	90	135	155	18 d PN
LCA	−	−	−	+	+	+	+	+
UEA 1	−	−	−	+	+	+	+	+
SWGA	−	+	+	+	+	+	+	+
BSA 1	−	−	−	−	−	−	+	+
PNA	−	+	+	+	+	−	−	−
DBA	−	−	−	−	−	−	−	−

For abbreviations, see Table 3.

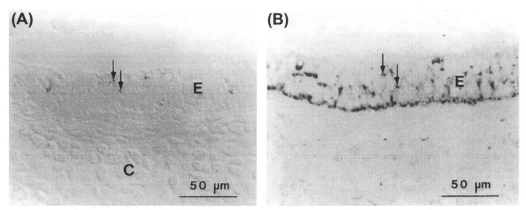

FIGURE 5 Comparison of the distribution of immunoreactive mucin with an antibody that reacts with all mucus cells in adult rhesus monkeys and the distribution of AB/PAS – positive material from serial sections of trachea of fetal rhesus monkey when secretory granules are just beginning to form (50 DGA). A: the immunoreactive secretory product is in highly focal areas of a small number of positive cells (arrow). B: on section serial to A, it is clear that these sites are also positive for PAS (arrow). Cartilage (C) is negative.

enriched basal cell population from rabbits and rats. These experiments clearly demonstrate the polypotent nature of the basal cell type. However, there are several deficiencies in these experiments. First of all, the definition of basal cell type is based on the ultrastructural picture and the immunohistochemical stain. It is well known, however, that secretory cells lose their differentiated features upon cell isolation and culturing in vitro. The degranulated secretory cells may resemble the basal cell type, and the morphologic tools used in these studies cannot distinguish satisfactorily the basal one from the degranulated secretory cell type in dissociated and isolated cell preparations. Furthermore, for the preparations in these studies, the purity of basal cell type population is only 90%. When Johnson et al.[106] used flow cytometry to isolate basal cells, they found that basal cells from rat trachea had a colony-forming efficiency of 0.6%, whereas secretory cells and unsorted cells had efficiencies of 3.4% and 2.6%, respectively. From these results, the authors concluded that basal cells had less proliferative activity than did secretory cells. It is therefore difficult to conclude from these tracheal graft repopulation studies that basal cell type is the progenitor cell type responsible for the initiation of airway epithelial cell growth and the repair to response to injury.

Pattern in Bronchioles

The process of cytodifferentiation of the nonciliated cells of distal bronchioles entails substantial rearrangement, loss, and biogenesis of cellular organelles. Up to late fetal age, terminal bronchioles are lined by simple cuboidal to columnar epithelium composed of glycogen-filled nonciliated cells with few organelles. The shifts in cellular components with time for species in which the predominant cellular constituent in adults is agranular endoplasmic reticulum (AER), such as in mice, hamsters, rats, and rabbits, are summarized in Figure 6. The pattern is essentially similar for these species. What varies from species to species is the timing of these

events. The first event is a dramatic loss in cytoplasmic glycogen. In rabbits, this drop is from approximately 70% of cytoplasmic volume to less than 10% cytoplasmic volume in adults. A similar substantial loss occurs in rats, hamsters, and mice. In rabbits, this loss begins immediately prior to birth and continues for up to 4 weeks of postnatal age.[107] A similar change occurs in mice.[108] In rats, the loss of cytoplasmic glycogen begins at birth and drops to adult levels within the first week of postnatal life.[109] In hamsters, cytoplasmic glycogen is not detectable immediately after birth.[110] Associated with the drop in cellular glycogen is a substantial biogenesis of membranous organelles, especially AER. Smooth endoplasmic reticulum is not detected in nonciliated cells until immediately prior to birth in rabbits (Figure 6).[111] At birth, fewer than 20% of the cells contain greater than 10% AER. By 2 weeks, in almost 70% of the nonciliated cells, AER occupies greater than 10% (up to 50%) of the cell volume. The adult configuration is reached at approximately 28 days postnatally in rabbits. In mice, the adult configuration of AER is reached at approximately 3 weeks postnatally.[108] Granular endoplasmic reticulum in prenatal animals is approximately twice as abundant in rabbit Clara cells as it is in rats (Figure 6).[112] The decrease in cellular abundance of GER occurs gradually in rabbits and is still double the adult configuration (2% of cell volume) at 4 weeks postnatally, but in rats the level decreases by 50% immediately postpartum and is at or near the adult configuration (less than 1%) by 10 days postnatally. The situation for rats and mice appears similar to that for rabbits, but for hamsters GER is near the adult configuration immediately postpartum.[110,113] Secretory granule appearance also varies by species. The earliest at which secretory granules are detected in the Clara cells of rabbits and mice is within the first week of postnatal life, whereas in rats and hamsters granules are abundant prenatally. In rabbits as well as mice, granule abundance resembling adult levels occurs by 21 days postnatally. In rats, granule abundance reaches adult abundance by 7 days postnatally and is at adult configuration

FIGURE 6 Diagrammatic comparison of Clara cellular organization during pre- and postnatal differentiation with correlation to morphometric analysis of cytoplasmic content of smooth endoplasmic reticulum from morphometric measurements in lungs of rabbit. The morphometric data represents the percentage of the cell population with less than 10% of the cytoplasm occupied by smooth endoplasmic reticulum and the percentage of the population with each 10% increase in abundance up to >50% of cytoplasmic volume in adult animals.

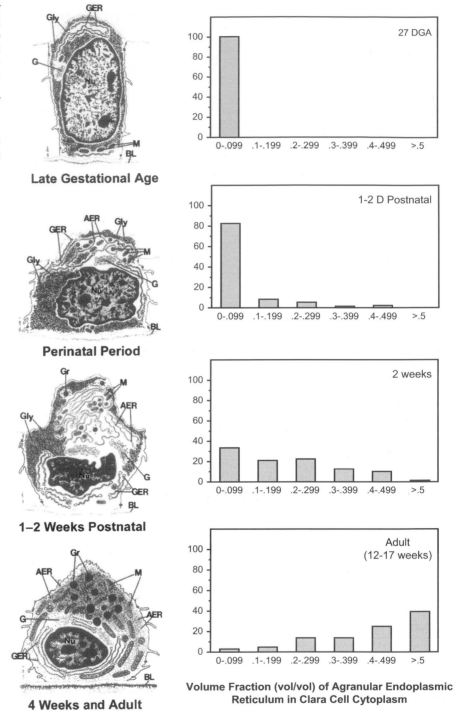

immediately postpartum in the hamster. The only species in which Clara cell differentiation has been characterized fully where the adult Clara cell population does not have an abundance of AER is rhesus monkeys.[53] In that species, the loss of cytoplasmic glycogen and an increase, rather than a decrease, in GER occurs over a substantial period both prenatally and postnatally. Studies in humans suggest that developmental events for Clara cells are similar to those in rhesus monkeys,

but may extend longer than the 6 months to a year (postnatally) required for differentiation of all of the nonciliated cells in terminal respiratory bronchioles of monkeys.

As summarized in Chapter 11, the expression of cytochrome P450 monooxygenases (CYP) in Clara cells during their differentiation has been evaluated in a number of species.[112–114] Protein for the NADPH P450 reductase and CYP2B is detected earliest, with the reductase somewhat later than CYP2B in rabbits. CYP4B is detected 2–3 days of age later (Chapter 11). The initial distribution is in the most apical border of a small percentage of the nonciliated cell population. During the period in which the amount of detectable protein increases, the distribution changes in two ways. First, an immunologically detectable protein is found in an increasing proportion of the nonciliated cells as animals become older. Second, the distribution of detectable protein within an individual cell increases from the apex to the base with increasing age. The youngest age at which intracellular protein can be detected immunohistochemically varies substantially within these three species. Protein becomes detectable in hamsters approximately 3–4 days prior to birth and reaches the distribution and intensity observed in the adult by 3 days postnatally. CYB4B is not detectable before 1 day postnatally, but is at adult levels shortly thereafter. In rabbits and rats, the timing is somewhat different. NADPH reductase is found initially just prior to birth in rabbits, and CYP2B and 4B are not observed until after birth. All of these proteins have an adult distribution and intensity

by 28 days postnatally. In rats, CYP2B, CYP4B, and reductase are detected in the first 2–3 days of postnatal life and are apparently at adult densities and distributions by 21 days postnatally. CYPIAI is not detectable prenatally in rats but can be detected in increasing, but small amounts until it reaches adult levels at approximately 21 days postnatally. Intracellular expression of protein precedes the appearance and increase in the abundance of AER by 2–4 days in each of these species. Activity for these proteins is first detected approximately 2–3 days after the protein is immunologically detectable within Clara cells. The activity studies have been done with whole lung homogenates and reflect potential activity from other cell populations as well as from Clara cells. While both the AER abundance and antigenic protein intensity reach the adult configuration in approximately 3 to 4 weeks in rats and rabbits, the activity for these isozymes is still considerably below that for adults. This suggests that the functionality of these proteins continues to increase after the protein density and organelle composition have reached adult levels of expression. Table 5 summarizes the relationship

TABLE 5 Development of AER, P450 reductase, and monoxygenase enzymes in rabbit lung

Assay	27-28 DGA	1-2 DPN	7 DPN	14 DPN	28 DPN	Adult
	Amount of Activity					
	% of Adult Value					
SER[a,b]	0.2	8.2	8.2	30.1	64.5	100
P450 reductase						
Immunohistochemistry[b,c]	±	+	++	++	++++	++++
Western blot	+	+	++	++	++++	++++
P450 isozyme 2B						
Immunohistochemistry[b]	0	±	+	++	++++	++++
Western blot	0	0	±	+	++++	++++
P450 isozyme 4B						
Immunohistochemistry[b]	0	+	++	++	++++	++++
Western blot	0	±	+	++	+++	++++
Microsomal P450[d]	0	0	20.9	44.5	56.0	100
P450 activity						
Ethoxyresorufin[e]	0	10.8	11.7	14.9	59.3	100
Pentoxyresorufin[f]	0	6.5	8.1	29.9	51.8	100

[a]Average adult cell volume for AER is 43.9 ±3.5%.
[b]Bronchiolar epithelium.
[c]Symbols indicate staining intensities in relation to adult (++++).
[d]Average adult level of microsomal P450 is 0.575 ± 0.238 nmol/mg of protein.
[e]Average adult ethoxyresorufin O-dealkylase activity is 47.99 ± 14.69 pmol/mg of protein/min.
[f]Average adult pentoxyresorufin O-dealkylase activity is 64.07 ± 64.51 pmol/mg of protein/min.

between changes in AER abundance, expression of immunoreactive protein, and microsomal P450 activity for rabbits. The timing for rats is somewhat shifted to the left for postnatal time points and to the right for perinatal ones compared with rabbits.

The pattern of expression of Clara cell secretory protein is similar to that of the cytochrome P450 monooxygenase system in relation to the appearance of cellular organelles. While there is substantial interspecies variability in the timing of expression, the general pattern is similar, at least for the four species studied in most detail: rats, rabbits, hamsters, and mice.[108,111,115-119] The protein appears earliest in the central or apical portion of a few cells per bronchiole, and the number of cells in which antigen can be detected increases with increasing age. In hamsters, the secretory protein antigen can be detected in a number of cells by the beginning of the last trimester of pregnancy and reaches the adult configuration in terms of density and number of cells labeled at about 3–4 days postnatally. In rats, a small proportion of the cells are labeled prenatally, and the distribution observed in the adult is present at about 7 days postnatally. This adult configuration occurs between 3 and 4 weeks in rabbits, and the earliest detectable signal in the Clara cells is immediately prior to birth. The timing in mice and rats is similar to that in rabbits. Immunoreactive protein has also been detected in late fetal humans, but when the distribution resembles adults has not been determined. Intracellular expression of the protein follows the changes in GER and is closely related to the first appearance and increase in the abundance of secretory granules. Western blotting of this protein indicates that it is present earlier in lung homogenate than its appearance in bronchiolar Clara cells suggests. This is because secretory cells in proximal airways express the protein much earlier and, in general, are more differentiated in the perinatal period than are secretory cells of bronchioles. In hamsters, the situation is the inverse, with the bronchioles differentiating in this respect prior to the bronchi.

REGULATION OF DIFFERENTIATION

Recent reviews have highlighted the complexity of our current state of knowledge regarding the regulation of epithelial differentiation in development and disease, with emphasis on proximal-to-distal differences in the roles of different phenotypes, in addition to basal cells, as progenitors.[120-122] At least eight lipid mediators, six cytokines, eight chemokines, and more than eight growth factors (including TGF-β, a number of EGFs, PDGF, FGF, and IGF) are expressed by bronchial epithelium. They appear to play roles in influencing neighboring nonepithelial populations, as well as epithelial function and differentiated expression.[121]

Trachea and Bronchi

Defining the regulatory mechanisms for maturation of fetal and neonatal airway epithelium is an area of active research with studies focused primarily in two areas: in vitro studies using isolated cell populations and in vitro genetically manipulated mice. At this point, the factors that control differentiation and the mechanisms for these controls are poorly understood, especially as they operate in vivo. Three areas of regulatory control have been explored in some detail: the interleukins, vitamin A and growth factors, and transcription factors.

In vivo studies have been minimal for direct treatment. However, the effects of epidermal growth factor (EGF) on lung development have been examined in rhesus monkeys. EGF treatment in utero markedly stimulates the maturation of the tracheal secretory apparatus, including both the tracheal surface and submucosal glands.[123] The secretory apparatus is more differentiated in that there are more mucus cells, increased secretory product stored in the epithelium and glands, and increased quantities of secretory product in lavage and amniotic fluid. By contrast, treatment with triamicinalone, a glucocorticoid, induces maturation of the gas exchange area,[124] but does not affect the maturation of the secretory apparatus (Table 6).

Based on studies primarily in mice, it appears that most of the differentiation of proximal airway epithelium is under the regulation of two transcriptional factors, the homeodomain transcriptional factor NKX 2.1 (otherwise known as TTF-1) and winged-helix family transcription factor HNF-3/forkhead homolog-4 (HFH-4).[111] The former appears to promote the initial branching of the tracheal bud but also promotes other aspects of tracheal differentiation, possibly

TABLE 6 Effect of EGF and triamcinolone acetonide (TAC) on total glycoconjugate detectable in the trachea of fetal rhesus monkeys

Days Gestational Age	Treatment	Total Secretory Product (mm³ × 10³/mm²) (Mean ± 1 SD)
128	None	0.48 ± 0.37
150	None	1.36* ± 0.33
128	EGF	1.77* ± 0.28
150	TAC (1 mg/day)	1.27 ± 1.35
150	TAC (10 mg/day)	0.75 ± 0.34

*$p < 0.05$ compared with 128 DGA control.

via the FGF pathway. The key regulatory mechanisms by which these transcription factors modulate differentiation of epithelium in proximal airways is not clearly understood. There appears to be a few key differentiation processes that may be regulated by specific transcription factors. Some studies have suggested that NKX 2.1 promotes differentiation of Clara Cells, but other studies suggest that tracheal mucous cell formation is independent of this transcription factor.[112] These developmental studies were based on mice after gene manipulation; however, healthy adult mice normally do not have differentiated mucous goblet cell. In contrast, it appears that HNF-3/forkhead homolog-4 transcription factor is critical for the differentiation of ciliated cells in respiratory epithelial populations,[113,114] A number of epithelial mesenchymal interactions appear to be critical to the differentiation to the epithelium. Two primary ones are through HNF-3 and GATA6 where they may serve as regulators of cell-cell communication activities.

The two interleukins that appear to have the most impact on transdifferentiation of airway epithelium in proximal airways of mice from a Clara cell phenotype to a mucus cell phenotype are IL9 and IL13,[125,126] also Gomperts, 2007 and Zehn, 2007. In vitro studies have established that IL13 depresses ciliated cell differentiation and promotes mucus – secreting cells by inhibition of foxj1 expression, as does STAT6.[125] Inhibition of TGF-α and EGFR expression have similar effects both by the depression of transcription factor FOX A2, but apparently by different mechanisms.[126] When the IL13 gene is attached to a promoter for the Clara cell 10KDA protein, it actively promotes the expression of mucus cells of the proximal airways of mice.[116–118] In all these cases the airways of transgenic mice have three characteristic features: mucus metaplasia, eosinophilic inflammation, and airway hyper responsiveness. In vivo studies have also shown that inhibition of IL13 blocks allergen induced effects on the airways of mice. On the other hand, direct administration of IL13 to the airways can produce the same effects as allergen.[119,120] For both IL9 and IL13, a variety of cellular responses occur, including elevation of eosinophiles, lymphocytes, mast cells and subepithelial collagen in airways walls.[115,119,120] While most of these events appear to be regulated through the IL13 R alpha 1 subunit in combination with the IL4 receptor, they do not appear to be modulated by factors that regulate the matrix changes nor influence inflammatory cell populations.[118] Defining exactly how IL13 produces these changes has been difficult. It is clear that for poorly differentiated airway epithelial cells in vitro, IL13 induces a dramatically different pattern of gene expression than is observed in airway smooth muscle cells or lung fibroblasts.[121] The four major transcription factors increased were OTF2, HSP factor 4, Id-3, and NRF-1. There was elevated phosphorylation of STAT 6 and some increases in factors related to extracellular matrix production.

Three factors, Notch, SPDEF, and beta-catenin, appear to play roles in the programming of early progenitor cells for ciliated or secrectory cell differentiation.[127–132] The mice whose alterations inactivate Notch signaling conditionally either by inhibiting Notch-ligand binding or the transcriptional factor for Notch signaling have airways absent of Clara cells and populated by ciliated cells and neuroendocrine cells.[127] Inhibition of in a central Notch pathway component Pofut1 in TGFb3-Cre expressing mice produced animals whose airways had marked mucus goblet cell metaplasia, few Clara cells, and abundant ciliated cells.[128] Notch signaling also appears to be critical for the selection of the Clara cell, as opposed to the ciliated cell, lineage during development.[130] In addition, expression of SAM-pointed domain-containing Ets-factor (SPDEF) regulates transcriptional factors promoting mucus goblet cell differentiation from Clara cells. This appears to be by enhancing expression of genes regulating goblet cell differentiation and protein glycosylation, including Foxa3, Agr2, and Gent3.[129] Depletion of SPDEF also inhibits goblet cell hyperplasia associated with allergen exposure.[129] Proliferation of basal cells and their differentiation into ciliated cells appears to also be regulated by beta-catenin expression, both in vivo and in vitro.[131,132] Elevated beta-catenin expression promotes proliferation and differentiation into ciliated cells while stabilization or inhibition of expression reduces both.

Tracheobronchial epithelial cells, like many other epithelial cells, lose their differentiated functions upon culturing in vitro. However, the loss of differentiated functions, at least in primary tracheobronchial epithelial culture, is transient. In repopulation studies using cultured cells,[103] undifferentiated rabbit tracheal epithelial cells maintained long-term in culture are able to repopulate the grafts and form a new mucociliary epithelium.[133,134] Epithelial cells, despite dedifferentiation in culture, apparently maintain their intrinsic differentiated potential, which is expressed if an appropriate environment is provided: hormonal requirements, vitamin A supplement, and collagen gel substratum.[135,136] Based on the amino acid and carbohydrate composition analyses of the in vitro secretory products as compared with the in vivo mucin products purified from sputum and epithelial cell layer, it appears that cultured tracheal epithelial cells from a number of species are able to secrete authentic mucin.[137] Critical to use of in vitro models for epithelial differentiation has been the development of the Whitcutt chamber to grow airway epithelial cells between air and a liquid medium interface,[132] based on the premise that airway epithelial cells in vivo are usually located between air and a liquid interface.[138–142] Using this chamber, columnarized formation of cultured epithelial cells was observed, and with further development of mucociliary differentiation in culture.[139,143] including both human and monkey tracheobronchial epithelial cells.[124,125] Scanning electron microscopy demonstrates extensive ciliary features on the culture surface, and transmission electron microscopy has demonstrated the formation of abundant mucus-secreting granules and the columnarized features with a two- to four-cell layer. The basal cell layer is compressed and resembles basal cells in vivo.

Tracheobronchial epithelium is a vitamin A-targeted tissue.[144–147] The epithelium requires vitamin A for the preservation and induction of the expression of differentiated functions. Keratinizing squamous metaplasia of mucociliary epithelium occurs with vitamin A deficiency along with a reduction in the synthesis of mucus glycoproteins. The administration of vitamin A or its synthetic derivatives (retinoids) reverses this phenomenon. Excess vitamin A can convert stratified skin epithelium in chick embryos to an epithelium containing mucus-secreting granules.[148] Vitamin A treatment enhances the proliferation of small mucus-granule cell type in primary hamster tracheal epithelial cultures.[149–151] However, vitamin A does enhance DNA synthesis of basal cells of keratinocyte cultures.[152] There is no evidence that vitamin A inhibits squamous cell proliferation. Vitamin A and its derivatives clearly play a role in the differentiation and expression of mucin genes in human tracheal bronchial epithelial cells,[146] at least four mucin genes (MUC2, MUC5 AC, MUC5 B, MUC7) are retinoic (RA) or retinol-dependent while MUCi, MUC4, and MUC 8 are not. Regulation of mucin genes by retinoic acids appears to be mediated by retinoic acid receptors RAR α and γ. Two other regulators of cellular function interact closely with RA to modulate mucin genes: thyroid hormone (T_3) and epidermal growth factor (EGF). T_3 inhibits mucin gene expression, particularly MUC5 AC, apparently through competitive inhibition of receptor responses through the thyroid receptors by inhibiting gene transcription. While EGF is thought to stimulate mucin expression and secretion in cultured airways of some species, especially the rat, it has an inhibitory effect in human bronchial cells in culture, but does suggest that EGF's impact may in fact be retinoic acid dependent.

Regulation in Bronchioles

Factors regulating Clara cell differentiation are not well understood. The postnatal nature of the majority of the cytodifferentiation process in most species suggests that it is independent of the hormones associated with pregnancy and parturition. The fact that the timing varies by as much as 2–3 weeks in different species would further suggest that the process may be under regulation of a variety of factors that act in different temporal sequences and with different levels of influence in different species. A number of mediators have been shown to stimulate cytodifferentiation of type II alveolar epithelial cell and produce architectural rearrangements of lung connective tissue elements to promote gas exchange, including corticosteroids, thyroid hormone, epidermal growth factor, and cyclic AMP.[153]

Whether all of these mediators influence Clara cell differentiation is not known. The best studied are the glucocorticoids, especially dexamethasone. Treatment in the perinatal period retards Clara cell differentiation as evidenced by an increase in cytoplasmic glycogen and minimal alterations in organelles in both rats and mice.[154,155] Dexamethasone administered either prenatally or immediately postnatally elevates the surfactant protein messenger ribonucleic acid (mRNA) levels in lungs of rats of all ages, producing this elevation in both alveolar type II cells and Clara cells.[156] Glucocorticoid administered to pregnant rabbits appears to have a stimulatory effect on the differentiation of secretory potential in fetal Clara cells by elevating the amount of the uteroglobinlike Clara cell secretory protein.[157,158] Dexamethasone administered to pregnant rabbits also has a stimulatory effect on the pulmonary cytochrome P450 system in fetuses, based on measurements of whole lung microsomes.[159–161] While glycogenolysis is retarded by dexamethasone treatment, glycogen, epinephrine, and 8-bromo-cAMP produce a rapid drop in Clara cell glycogen content.[109]

One of the factors that appears to have the most impact on Clara cell differentiation is injury during the developmental period, in which normal differentiation occurs. Normal differentiation is characterized by loss of glycogen and appearance of secretory granules, and by differentiation of Clara cells into ciliated cells, even in the absence of frank injury to either ciliated or Clara cells. Postnatal exposure to compounds that injure the respiratory system retard Clara cell differentiation. Hyperoxia during the early postnatal period inhibits differentiation (Massaro et al., unpublished data).[162,163] Injury by treatment with 4-ipomeanol impedes Clara cell differentiation even for a short term after treatment is discontinued.[164] Not only are Clara cells in postnatal animals more susceptible to injury than are Clara cells in adults, but the expression of the P450 system in the post-treatment period is markedly reduced. In rats, exposure to cigarette smoke of either the pregnant mother or the newborn accelerates the appearance of one cytochrome P450 monooxygenase isozyme, CYP 1A 1, but not CYP2B.[114] The increased P450 expression is primarily in the Clara cell population and is not found in either alveolar type II cells or in the vascular endothelium, both targets for inducers in adult animals. Other factors besides postnatal hyperoxia, including maternal undernutrition during the last 5 days of pregnancy, retard Clara cell differentiation, but these effects appear to be reversible with time.[109,162–164]

There is considerable indirect evidence to suggest that a number of growth factors, including TGF-α, EGE basic fibroblast growth factor (FGF), insulin-like growth factors, and platelet-derived growth factor, may play roles in regulating bronchiolar epithelial differentiation.[165,166] The EGF receptor (EGFr) has been detected in bronchiolar epithelium throughout pre- and postnatal lung development in rats and humans.[167,168] EGFr has also been detected in human lung at midgestation,[169,170] and has been detected in human and rat fetal lung extracts.[167,171] Both ligands of EGFr, as well as TGF-α and EGF, have been detected immunohistochemically in bronchiolar epithelium in a number of species. EGF is barely detectable in bronchiolar epithelium of fetal humans (first and second trimesters), but is present in postnatal human lung.[172] EGF has been reported in homogenates of lung from late fetal (21-day gestational age) and adult rats,[173] and immunoreactive protein has been detected in bronchiolar epithelium throughout fetal development in lambs and mice.[174,175] TGF-α has been detected in bronchiolar epithelium of midgestational humans.[176] It can be extracted and mRNA can be detected in fetal rat lung homogenates.[177] Late fetal (21-day gestational age) and adult rat lung contains EGF.[19] Platelet-derived growth factor receptor has also been detected in bronchiolar epithelium during most of the prenatal stages of lung development.[169,170] Basic FGF and its receptor are found in bronchiolar epithelium during most of fetal rat lung development.[178] Both the FGF receptor and the protein appear to colocalize in the epithelium and adjacent interstitial compartments. There is some suggestion that insulin-like growth factors are involved in aspects of epithelial development in bronchioles.[179] These growth factors may play a role in autocrine regulation because both receptors and the proteins themselves appear within the bronchiolar epithelium. They also may play a paracrine role because growth factor protein appears to be distributed to interstitial cell components, fibroblasts, and smooth muscle surrounding bronchiolar epithelium, during various stages of lung development. At present, there is no direct evidence that any of these factors influence bronchiolar epithelial maturation. There is, however, evidence that pharmacological doses of EGF alters branching morphogenesis in mice,[180] enhances differentiation of alveolar type II cells in fetal rabbits, monkeys, and sheep,[80,181,182] and alters the differentiation of tracheal epithelium in rhesus monkeys.[123]

REFERENCES

1. Paige RC, Wong V, Plopper CG. Long-term exposure to ozone increases acute pulmonary centriacinar injury by 1-nitronaphthalene: Ii. Quantitative histopathology. *J Pharmacol Exp Ther* 2000;**295**: 942–50.
2. Fanucchi MV, Murphy ME, Buckpitt AR, Philpot RM, Plopper CG. Pulmonary cytochrome p450 monooxygenase and clara cell differentiation in mice. *Am J Respir Cell Mol Biol* 1997;**17**:302–14.
3. Hyde DM, Hubbard WC, Wong V, Wu R, Pinkerton K, Plopper CG. Ozone-induced acute tracheobronchial epithelial injury: Relationship to granulocyte emigration in the lung. *Am J Respir Cell Mol Biol* 1992;**6**:481–97.
4. Paige R, Wong V, Plopper C. Dose-related airway-selective epithelial toxicity of 1-nitronaphthalene in rats. *Toxicol Appl Pharmacol* 1997;**147**:224–33.
5. Pinkerton KE, Plopper CG, Mercer RR, Roggli VL, Patra AL, Brody AR, Crapo JD. Airway branching patterns influence asbestos fiber location and the extent of tissue injury in the pulmonary parenchyma. *Lab Invest* 1986;**55**(6):688–95.
6. Plopper C, Suverkropp C, Morin D, Nishio S, Buckpitt A. Relationship of cytochrome p-450 activity to clara cell cytotoxicity: I. Histopathologic comparison of the respiratory tract of mice, rats and hamsters after parenteral administration of naphthalene. *J Pharmacol Exp Ther* 1992;**261**:353–63.
7. Plopper C, Macklin J, Nishio S, Hyde D, Buckpitt A. Relationship of cytochrome p-450 activity to clara cell cytotoxicity iii. Morphometric comparison of changes in the epithelial populations of terminal bronchioles and lobar bronchi in mice, hamsters, and rats after parenteral administration of napthalene. *Lab Invest* 1992;**67**:553–65.
8. Plopper CG, Chu FP, Haselton CJ, Peake J, Wu J, Pinkerton KE. Dose-dependent tolerance to ozone i. Tracheobronchial epithelial reorganization in rats after 20 months exposure. *Am J Pathol* 1994;**144**(2):404–20.
9. Plopper CG, Hatch GE, Wong V, Duan X, Weir AJ, Tarkington BK, Devlin RB, Becker S, Buckpitt AR. Relationship of inhaled ozone concentration to acute tracheobronchial epithelial injury, site-specific ozone dose, and glutathione depletion in rhesus monkeys. *Am J Respir Cell Mol Biol* 1998;**19**:387–99.
10. Postlethwait EM, Joad JP, Hyde DM, Schelegle ES, Bric JM, Weir AJ, Putney LF, Wong VJ, Velsor LW, Plopper CG. Three-dimensional mapping of ozone-induced acute cytoxicity in tracheobronchial airways of isolated perfused rat lung. *Am J Respir Cell Mol Biol* 2000;**22**:191–9.
11. Van Winkle LS, Isaac JM, Plopper CG. Distribution of epidermal growth factor receptor and ligands during bronchiolar epithelial repair from naphthalene-induced clara cell injury in the mouse. *Am J Pathol* 1997;**151**:443–59.
12. Van Winkle LS, Johnson ZA, Nishio SJ, Brown CD, Plopper CG. Early events in naphthalene-induced acute clara cell toxicity. Comparison of membrane permeability and ultrastructure [in process citation]. *Am J Respir Cell Mol Biol* 1999;**21**:44–53.
13. West JA, Chichester CH, Buckpitt AR, Tyler NK, Brennan P, Helton C, Plooper CG. Hetergeneity of clara cell glutathione. A poosible basis for differences in cellular responses to pulmonary cytotoxicants. *Am J Respir Cell Mol Biol* 2000;**23**:27–36.
14. Wilson DW, Plopper CG, Dungworth DL. The response of the macaque tracheobronchial epithelium to acute ozone injury. *Am J Pathol* 1984;**116**:193–206.
15. Plopper CG, Heidsiek JG, Weir AJ. St. George JA, Hyde DM. Tracheobronchial epithelium in the adult rhesus monkey: A quantitative histochemical and ultrastructural study. *Am J Anat* 1989;**184**:31–40.
16. Mariassy A, Plopper C. Tracheobronchial epithelium of the sheep: I. Quantitative light-microscopic study of epithelial cell abundance, and distribution. *Anat Rec* 1983;**205**:263–75.
17. Plopper C, Halsebo J, Berger W, Sonstegard K, Nettesheim P. Distribution of nonciliated bronchiolar epithelial (clara) cells in intra- and extrapulmonary airways of the rabbit. *Exp Lung Res* 1983;**4**:79–98.
18. Plopper C. St. George J, Nishio S, Etchison J, Nettesheim P. Carbohydrate cytochemistry of tracheobronchial airway epithelium of the rabbit. *J Histochem Cytochem* 1984;**32**:209–18.
19. St. George J, Plopper C, Etchison J, Dungworth D. An immunocytochemical/histochemical approach to tracheobronchial mucus characterization in the rabbit. *Am Rev Respir Dis* 1984;**130**:124–7.

20. Mariassy AT, Plopper CG. Tracheobronchial epithelium of the sheep: Ii. Ultrastructural and morphometric analysis of the epithelial secretory cell types. *Anat Rec* 1984;**209**:523–34.

21. Wilson D, Plopper CG, Hyde DM. The tracheobronchial epithelium of the bonnet monkey (macaca radiata): A quantitative ultrastructural study. *Am J Anat* 1984;**171**:25–40.

22. St.George JA, Cranz DL, Zicker S, Etchison JR, Dungworth DL, Plopper CG. An immunohistochemical characterization of rhesus monkey respiratory secretions using monoclonal antibodies. *Am Rev Respir Dis* 1985;**132**:556–63.

23. Heidsiek JG, Hyde DM, Plopper CG, St. George JA. Quantitative histochemistry of mucosubstance in tracheal epithelium of the macaque monkey. *J Histochem Cytochem* 1987;**35**:435–42.

24. Plopper C, Cranz D, Kemp L, Serabjit-Singh C, Philpot R. Immunohistochemical demonstration of cytochrome p-450 monooxygenase in clara cells throughout the tracheobronchial airways of the rabbit. *Exp Lung Res* 1987;**13**:59–68.

25. Mariassy AT, St. George JA, Nishio SJ, Plopper CG. Tracheobronchial epithelium of the sheep: Iii. Carbohydrate histochemical and cytochemical characterization of secretory epithelial cells. *Anat Record* 1988;**221**:540–9.

26. Mariassy AT, Plopper CG, St. George JA, Wilson DW. Tracheobronchial epithelium of the sheep: Iv. Lectin histochemical characterization of secretory epithelial cells. *Anat Rec* 1988;**222**:49–59.

27. Evans MJ, Plopper CG. The role of basal cells in adhesion of columnar epithelium to airway basement membrane. *Am Rev Respir Dis* 1988;**138**:481–3.

28. Evans M, Cox RA, Shami SG, Wilson B, Plopper CG. The role of basal cells in attachment of columnar cells to the basal lamina of the trachea. *Am J Respir Cell Mol Biol* 1989;**1**:463–9.

29. Dodge DE, Rucker RB, Singh G, Plopper CG. Quantitative comparison of intracellular concentration and volume of clara cell 10 kd protein in rat bronchi and bronchioles based on laser scanning confocal microscopy. *J Histochem Cytochem* 1993;**41**(8):1171–83.

30. Avadhanam KP, Plopper CG, Pinkerton KE. Mapping the distribution of neuroepithelial bodies of the rat lung. A whole-mount immunohistochemical approach. *Am J Pathol* 1997;**150**:851–9.

31. Fanuchi MF, Buckpitt AR, Murphy ME, D.H. S, Hammock BD, Plopper CG. Development of phase ii xenobiotic metabolizing enzymes in differentiating murie clara cells. *Toxicol Appl Pharmacol* 2000;**168**:253–67.

32. Fanucchi M, Buckpitt A, Murphy ME, Plopper CG. Naphthalene cytotoxicity of differentiating clara cells in neonatal mice. *Toxicol Appl Pharm* 1997;**144**:96–104.

33. Jeffery P. Structure and function of mucus-secreting cells of cat and goose airway epithelium. In: *Respiratory tract mucus*. New York: Elsevier-North Holland; 1977. pp. 5–19.

34. Emura M, Mohr U. Morphological studies on the development of tracheal epithelium in the syrian golden hamster. I. Light microscopy. *Versuchstierk* 1975;**17**:14. 10.

35. McCarthy C, Reid L. Acid mucopolysaccharide in the bronchial tree in the mouse and rat (sialomucin and sulphate). *Q J Exp Physiol* 1964;**49**:81–4.

36. Jones R, Baskerville A, Reid L. Histochemical identification of glycoproteins in pig bronchial epithelium:(a) normal and (b) hypertrophied from enzootic pneumonia. *J Pathol* 1975;**116**:1–11.

37. St George JA, Nishio SJ, Plopper CG. Carbohydrate cytochemistry of rhesus monkey tracheal epithelium. *Anat Rec* 1984;**210**:293–302.

38. Mariassy AT, George JA, Nishio SJ, Plopper CG. Tracheobronchial epithelium of the sheep: Iii. Carbohydrate histochemical and cytochemical characterization of secretory epithelial cells. *Anat Rec* 1988;**221**:540–9.

39. McCarthy C, Reid L. Intracellular mucopolysaccharides in the normal human bronchial tree. *Q J Exp Physiol Cogn Med Sci* 1964;**49**:85–94.

40. McCarthy C, Reid LM. Acid mucopolysaccharides in the bronchial tree in the mouse and rat (sialomucins and sulphate). *Q J Exp Physiol* 1964;**49**:81–4.

41. StGeorge JA, Nishio SJ, Plopper CG. Carbohydrate cytochemistry of the rhesus monkey tracheal epithelium. *Anat Rec* 1984;**210**:293–302.

42. Spicer S, Chakrin L, Wardell J, Kendrick W. Histochemistry of mucosubstances in the canine and human respiratory tract. *Lab Invest* 1971;**25**:483–90.

43. Lamb D, Reid LM. Acidic glycoproteins produced by the mucous cells of the bronchial submucosal glands in the fetus and child: A histochemical autoradiographic study. *Br J Dis Chest* 1972;**66**:248–53.

44. Pinkerton KE, Gallen JT, Mercer RR, Wong VC, Plopper CG, Tarkington BK. Aerosolized flourescent microspheres detected in the lung using confocal scanning laser microscopy. *Microsc Res Tech* 1993;**26**:437–43.

45. Buckpitt A, Buonarati M, Avey L, Chang A, Morin D, Plopper C. Relationship of cytochrome p-450 activity to clara cell cytotoxicity: Ii. Comparison of stereoselectivity of naphthalene epoxidation in lung and nasal mucosa of mouse, hamster, rat and rhesus monkey. *J Pharmacol Exp Ther* 1992;**261**:364–72.

46. Buckpitt A, Chang A, Weir A, Van Winkle L, Duan X, Philpot R, et al. Relationship of cytochrome p450 activity to clara cell cytotoxicity. Iv. Metabolism of naphthalene and naphthalene oxide in microdissected airways from the mouse, rat, and hamster. *Mol Pharmacol* 1995;**47**:74–81.

47. Lee C, Watt KC, Chang AM, Plopper CG, Buckpitt AR, Pinkerton KE. Site-selective differences in cytochrome p450 isoform activities. Comparison of expression in rat and rhesus monkey lung and induction in rats. *Drug Metab Dispos* 1998;**26**:396–400.

48. Paige RC, Royce FH, Plopper CG, Buckpitt AR. Long-term exposure to ozone increases acute pulmonary centriacinar injury by 1-nitronaphthalene: I. Region-specific enzyme activity. *J Pharmacol Exp Ther* 2000;**295**:934–41.

49. Plopper CG. Pulmonary bronchiolar epithelial cytotoxicity: Microanatomical considerations. In: Gram TE, editor. *Metabolic Activation and Toxicity of Chemical Agents to Lung Tissue and Cells*. New York: Pergamon Press; 1993. pp. 1–24.

50. Seaton M, Plopper C, Bond J. 1,3-butadiene metabolism by lung airways isolated from mice and rats. *Toxicology* 1996;**113**:314–7.

51. Watt KC, Plopper CG, Weir AJ, Tarkington B, Buckpitt AR. Cytochrome p450 2e1 in rat tracheobronchial airways: Response to ozone exposure. *Toxicol Appl Pharmacol* 1998;**149**:195–202.

52. Watt KC, Morin DM, Kurth MJ, Mercer RS, Plopper CG, Buckpitt AR. Glutathione conjugation of electrophilic metabolites of 1-nitronaphthalene in rat tracheobronchial airways and liver: Identification by mass spectrometry and proton nuclear magnetic resonance spectroscopy [in process citation]. *Chem Res Toxicol* 1999;**12**:831–9.

53. Duan X, Buckpitt AR, Plopper CG. Variation in antioxidant enzyme activites in anatomic subcompartments within rat and rhesus monkey lung. *Toxicol Appl Pharmacol* 1993;**123**:73–82.

54. Plopper CG, Duan X, Buckpitt AR, Pinkerton KE. Dose-dependent tolerance to ozone iv. Site-specific elevation in antioxidant enzymes in the lungs of rats exposed for 90 days or 20 months. *Toxicol Appl Pharmacol* 1994;**127**:124–31.

55. West JA, Buckpitt AR, Plopper CG. Elevated airway gsh resynthesis confers protection to clara cells from naphthalane injury in mice made tolerant by repeated exposures. *J Pharmacol Exp Ther* 2000;**297**:516–23.

56. Duan X, Plopper C, Brennan P, Buckpitt A. Rates of glutathione synthesis in lung subcompartments of mice and monkeys: Possible role in species and site selective injury. *J Pharmacol Exp Ther* 1996;**277**:1402–9.

57. Hyde DM, Miller LA, McDonald RJ, Stovall MY, Wong V, Pinkerton KE, et al. Neutrophils enhance clearance of necrotic epithial cells in ozone-induced lung injury in rehsus monkeys. *Am J Physiol Lung Cell Mol Physiol* 1999;**277**:L1190–1198.

58. Van Winkle LS, Buckpitt AR, Plopper CG. Maintenance of differentiated murine clara cells in microdissected airway cultures. *Am J Respir Cell Mol Biol* 1996;**14**:586–98.

59. Van Winkle L, Isaac J, Plopper C. Repair of naphthalene-injured microdissected airways in vitro. *Am J Respir Cell Mol Biol* 1996; **15**:1–8.

60. Plopper CG, Chang AM, Pang A, Buckpitt AR. Use of microdissected airways to define metabolism and cytotoxicity in murine bronchiolar epithelium. *Exp Lung Res* 1991;**17**:197–212.

61. Royce FR, Van Winkle LS, Yin J, Plopper CG. Comparison of regional variability in lung-specific gene expression using a novel method for rna isolation from lung subcompartments of rats and mice. *Am J Pathol* 1996;**148**:1779–86.

62. Plopper CG, ten Have-Opbroek AAW. Anatomical and histological classification of the bronchioles. In: Epler GR, editor. *Diseases of the bronchioles*. New York: Raven Press, Ltd.; 1994. pp. 15–25.

63. Tyler NK, Hyde DM, Hendrickx AG, Plopper CG. Cytodifferentiation of two epithelial populations of the respiratory bronchiole during fetal lung development in the rhesus monkey. *Anat Rec* 1989;**225**:297–309.

64. Plopper CG, Hyde DM. Epithelial cells of bronchioles. In: Parent RA, editor. *Treatise on pulmonary toxicology comparative biology of the normal lung*. Boca Raton, FL: CRC Press, Inc; 1992. pp. 85–92.

65. Sannes PL. Basement membrane and extracellular matrix. In: Parent RA, editor. *Comparative biology of the normal lung*. Boca Raton, FL: CRC Press; 1 ed. 1992. pp. 129–44.

66. ten Have-Opbroek AAW. The structural composition of the pulmonary acinus in the mouse. *Anat Embryol* 1986;**174**:49–57.

67. ten Have-Opbroek AAW. Lung development in the mouse embryo. *Exp Lung Res* 1991;**17**:111–30.

68. ten Have-Opbroek AAW, Otto-Verberne CJM, Dubbeldam JA, Dykman JH. The proximal border of the human respiratory unit, as shown by scanning and transmission electron microscopy and light microscopical cytochemistry. *Anat Rec* 1991;**229**:339–54.

69. Tyler NK, Hyde DM, Hendrickx AG, Plopper CG. Morphogenesis of the respiratory bronchiole in rhesus monkey lungs. *Am J Anat* 1988;**182**(3):215–23.

70. Smolich JJ, Stratford BF, Maloney JE, Ritchie BC. New features in the development of the submucosal gland of the respiratory tract. *J Anat* 1978;**127**(2):223–38.

71. Krause W, Leeson C. The postnatal development of the respiratory system of the opossum i. Light and scanning electron microscopy. *Am J Anat* 1973;**137**(3):337–56.

72. Leigh M, Gambling T, Carson J, Collier A, Wood R, Boat T. Postnatal development of tracheal surface epithelium and submucosal glands in the ferret. *Exp Lung Res* 1986;**10**:153–69.

73. Plopper CG, Weir AJ, Nishio SJ, Cranz DL. St. George JA. Tracheal submucosal gland development in the rhesus monkey, macaca mulatta: Ultrasturcture and histochemistry. *Anat Embryol* 1986;**174**:167–78.

74. Thurlbeck W, Benjamin B, Reid L. Development and distribution of mucous glands in the fetal human trachea. *Br J Dis Chest* 1961;**55**:54–64.

75. Bucher U, Reid LM. Development of the mucus-secreting elements in human lung. *Thorax* 1961;**16**:219–25.

76. Tos M. Development of the tracheal glands in man. Number, density, structure, shape, and distribution of mucous glands elucidated by quantitative studies of whole mounts. *Acta Pathol Microbiol Scand* 1966;**68**. Suppl 185:183+.

77. Tos M. Distribution and situation of the mucous glands in the main bronchus of human foetuses. *Anat Anz Bd* 1968;**123**:481–95.

78. Lamb D, Reid LM. Acidic glycoproteins produced by the mucous cells of the bronchial submucosal glands in the fetus and child: A histochemical autoradiographic study. *Br J Dis Chest* 1972;**66**: 248–53.

79. Jeffery P, Reid L. Ultrastructure of airway epithelium and submucosal glands during development. In: Hodson, editor. *Development of the lung*. New York: Marcel Dekker; 1977. pp. 87–134.

80. Plopper CG, St. George JA, Cardoso W, Wu R, Pinkerton K, Buckpitt AR. Development of airway epithelium: Patterns of expression for markers of differentiation. *Chest* 1992;**101**:2S–5S.

81. Leeson TS. The development of the trachea in the rabbit, with particular reference to its fine structure. *Anat Anz Bd* 1961;**110**:214–23.

82. Kawamata S, Fujita H. Fine structural aspects of the development and aging of the tracheal epithelium of mice. *Arch Histol Jap* 1983;**46**:355–72.

83. McDowell EM, Newkirk C, Coleman B. Development of hamster tracheal epithelium: I. A quantitative morphologic study in the fetus. *Anat Rec* 1985;**213**:429–47.

84. Emura M, Mohr U. Morphological studies on the development of tracheal epithelium in the syrian golden hamster. I. Light microscopy. *Zeitschrift Versuchstierk Bd* 1975;**17**:14–26.

85. Plopper C, Alley J, Weir A. Differentiation of tracheal epithelium during fetal lung maturation in the rhesus monkey, macaca mulatta. *Am J Anat* 1986;**175**:59–72.

86. Boyden EA. The development of the lung in the pig-tail monkey (macaca nemestrina, l.). *Anat Rec* 1976;**186**:15–38.

87. Lane BP, Gordon RE, Upton AC. Regeneration of rat tracheal epithelium after mechanical injury. I. The relationship between mitotic activity and cellular differentiation. *Proc Society Exp Biol Med* 1974;**145**:1139–44.

88. Boren HG, Paradise LJ. *Pathogenesis and Therapy of Lung Cancer*. New York: Marcel Dekker; 1978.

89. Donnelly GM, Haack DG, Heird CS. Tracheal epithelium: Cell kinetics and differentiation in normal rat tissue. *Cell Tissue Kinet* 1982;**15**:119–30.

90. Keenan KP, Combs JW, McDowell EM. Regeneration of hamster tracheal epithelium after mechanical injury i. Focal lesions: Quantitative morphologic study of cell proliferation. *Virchows Arch (Cell Pathol)* 1982;**41**:193–214.

91. Evans MJ, Shami SG. Lung cell kinetics. In: Massaro D, editor. *Lung cell biology*. New York: Marcel Dekker; 1989. pp. 1–36.

92. McDowell EM, Trump BF. Histogenesis of preneoplastic and neoplastic lesions in tracheobronchial epithelium. *Surv Synth Pathol Res* 1983;**2**:235–79.

93. Rutten AJL, Beems RB, Wilmer JWGM, Feron VJ. Ciliated cells in vitamin a-deprived culture hamster tracheal epithelium to divide. *In Vitro Cell Dev Biol Anim* 1996;**24**:931–5.

94. Chopra DP. Squamous metaplasia in organ culture of vitamin a-deficient hamster trachea: Cytokinetic and ultrastructural alterations. *J Natl Cancer Inst* 1982;**69**:895–905.

95. Jeffery P, Ayers M, Rogers D. The mechanisms and control of bronchial mucous cell hyperplasia. *Mucus health disease ii* 1982: 399–409. New York.

96. Evans MJ, Moller PC. Biology of airway basal cells. *Exp Lung Res* 1991;**17**:513–31.

97. Plopper CG, St. George J, Pinkerton KE, Tyler N, Mariassy A, Wilson D, et al. Tracheobronchial epithelium in vivo: Composition, differentiation and response to hormones. In: Thomassen DG, Nettesheim P, editors. *Biology, toxicology, and carcinogenesis of respiratory epithelium.* 1st ed. Washington DC: Hemisphere; 1990. pp. 6–22.

98. Hislop A, Muri DCF, Jacobsen M, Simon G, Reid L. Postnatal growth and function of the pre-acinar airways. *Thorax* 1972;**27**:265–74.

99. Burrington JD. Tracheal growth and healing. *J Thorac Cardiovasc Surg* 1978;**76**:453–8.

100. Basbaum C, Jany B. Plasticity in the airway epithelium. *Am J Phys* 1990;**259**:L38–46.

101. Keenan KP, Wilson TS, McDowell EM. Regeneration of hamster tracheal epithelium after mechanical injury iv. Histochemical, immunocytochemical and ultrastructural studies. *Virchows Arch (Cell Pathol)* 1983;**43**:213–40.

102. Johnson NF, Hubbs AF. Epithelial progenitor cells in the rat trachea. *Am J Respir Cell Mol Biol* 1990;**3**:579–85.

103. Terzaghi M, Nettesheim P, Williams ML. Repopulation of denuded tracheal grafts with normal, preneoplastic, and neoplastic epithlial cell population. *Cancer Res* 1978;**38**:4546–53.

104. Inayama Y, Hook GER, Brody AR, Cameron GS, Jetten AM, Gilmore LB, Gray T, Nettesheim P. The differentiation potential of tracheal basal cells. *Lab Invest* 1988;**58**:706–17.

105. Inayama Y, Hook GE, Brody AR, Jetten AM, Gray T, Mahler J, Nettesheim P. In vitro and in vivo growth and differentiation of clones of tracheal basal cells. *Am J Pathol* 1989;**134**(3):539–49.

106. Johnson NF, Hubbs AF, Thomassen DG. Epithelial progenitor cells in the rat respiratory tract. In: Thomassen DG, Nettesheim P, editors. *Biology, toxicology and carcinogenesis of respiratory epithelium.* Washington, DC: Hemisphere; 1990. pp. 88–98.

107. Plopper C, Alley J, Serabjit-Singh C, Philpot R. Cytodifferentiation of the nonciliated bronchiolar epithelial (clara) cell during rabbit lung maturation: An ultrastructural and morphometric study. *Am J Anat* 1983;**167**:329–57.

108. ten Have-Opbroek AAW, De Vries ECP. Clara cell differentiation in the mouse: Ultrastructural morphology and cytochemistry for surfactant protein a and clara cell 10 kd protein. *Microsc Res Tech* 1993;**26**:400–11.

109. Massaro GD. Nonciliated bronchiolar epithelial (clara) cells. In: Massaro D, editor. *Lung cell biology.* New York: Marcel Dekker; 1989. pp. 81–114.

110. Ito T, Newkirk C, Strum JM, McDowell EM. Modulation of glycogen stores in epithelial cells during airway development in syrian golden hamsters: A histochemical study comparing concanavalin a binding with the periodic acid-schiff reaction. *J Histochem Cytochem* 1990;**38**(5):691–7.

111. Cardoso W, Stewart LG, Pinkerton KE, Ji C, Hook GER, Singh G, et al. Secretory product expression during clara cell differentiation in the rabbit and rat. *Am J Phys* 1993;**8**(6):L543–52.

112. Plopper CG, Weir AJ, Morin D, Chang A, Philpot RM, Buckpitt AR. Postnatal changes in the expression and distribution of pulmonary cytochrome p450 monooxygenases during clara cell differentiation in the rabbit. *Mol Pharmacol* 1993;**44**:51–61.

113. Strum JM, Singh G, Katyal SL, McDowell EM. Immunochemical localization of clara cell protein by light and electron microscopy in conducting airways of fetal and neonatal hamster lung. *Anat Rec* 1990;**227**:77–86.

114. Ji CM, Plopper CG, Witschi HP, Pinkerton KE. Exposure to sidestream cigarette smoke alters bronchiolar epithelial cell differentiation in the postnatal rat lung. *Am J Respir Cell Mol Biol* 1994;**11**:312–20.

115. Strum J, Ito T, Philpot R, DeSanti A, McDowell E. The immunocytochemical detection of cytochrome p-450 monooxygenase in the lungs of fetal,neonatal, and adult hamsters. *Am J Respir Cell Mol Biol* 1990;**2**:493–501.

116. Katyal SL, Singh G, Brown WE, Kennedy AL, Squeglia N, Wong-Chong ML. Clara cell secretory (10 kdaltons) protein: Amino acid and cdna nucleotide sequences, and developmental expression. *Prog Respir Res* 1990;**25**:29–35.

117. Strum JM, Compton RS, Katyal SL, Singh G. The regulated expression of mrna for clara cell protein in the developing airways of the rat, as revealed by tissue *in situ* hybridization. *Tissue Cell* 1992;**24**:461–71.

118. Singh G, Katyal SK. Secretory proteins of clara cells and type ii cells. In: Parent RA, editor. *Comparative biology of the normal lung.* Boca Raton, FL: CRC Press; 1992. pp. 93–108.

119. Singh G, Katyal SL, Wong-Chong ML. A quantitative assay for a clara cell-specific protein and its application in the study of development of pulmonary airways in the rat. *Ped Res* 1986;**20**:802–5.

120. Rock JR, Hogan BL. Epithelial progenitor cells in lung development, maintenance, repair, and disease. *Annu Rev Cell Dev Biol* 2011;**27**:493–512.

121. Knight DA, Holgate ST. The airway epithelium: Structural and functional properties in health and disease. *Respirology* 2003;**8**:432–46.

122. Rackley CR, Stripp BR. Building and maintaining the epithelium of the lung. *J Clin Invest* 2012;**122**:2724–30.

123. St.George JA, Read LC, Cranz DL, Tarantal AF, George-Nascimento C, Plopper CG. Effect of epidermal growth factor on the fetal development of the tracheobronchial secretory apparatus in rhesus monkey. *Am J Respir Cell Mol Biol* 1991;**4**:95–101.

124. Bunton TE, Plopper CG. Triamcinolone-induced structural alterations in the development of the lung of the fetal rhesus macaque. *Am J Obstet Gynecol* 1984;**148**(2):203–15.

125. Gomperts BN, Kim LJ, Flaherty SA, Hackett BP. Il-13 regulates cilia loss and foxj1 expression in human airway epithelium. *Am J Respir Cell Mol Biol* 2007;**37**:339–46.

126. Zhen G, Park SW, Nguyenvu LT, Rodriguez MW, Barbeau R, Paquet AC, et al. Il-13 and epidermal growth factor receptor have critical but distinct roles in epithelial cell mucin production. *Am J Respir Cell Mol Biol* 2007;**36**:244–53.

127. Tsao PN, Vasconcelos M, Izvolsky KI, Qian J, Lu J, Cardoso WV. Notch signaling controls the balance of ciliated and secretory cell fates in developing airways. *Development* 2009;**136**:2297–307.

128. Tsao PN, Wei SC, Wu MF, Huang MT, Lin HY, Lee MC, et al. Notch signaling prevents mucous metaplasia in mouse conducting airways during postnatal development. *Development* 2011;**138**:3533–43.

129. Chen G, Korfhagen TR, Xu Y, Kitzmiller J, Wert SE, Maeda Y, et al. Spdef is required for mouse pulmonary goblet cell differentiation and regulates a network of genes associated with mucus production. *J Clin Invest* 2009;**119**:2914–24.

130. Morimoto M, Liu Z, Cheng HT, Winters N, Bader D, Kopan R. Canonical notch signaling in the developing lung is required for determination of arterial smooth muscle cells and selection of clara versus ciliated cell fate. *J Cell Sci* 2010;**123**:213–24.

131. Giangreco A, Lu L, Vickers C, Teixeira VH, Groot KR, Butler CR, et al. Beta-catenin determines upper airway progenitor cell fate and preinvasive squamous lung cancer progression by modulating epithelial-mesenchymal transition. *J Pathol* 2012;**226**:575–87.

132. Brechbuhl HM, Ghosh M, Smith MK, Smith RW, Li B, Hicks DA, et al. Beta-catenin dosage is a critical determinant of tracheal basal cell fate determination. *Am J Pathol* 2011;**179**:367–79.

133. Wu R, Groelke JW, Chang LY, Porter ME, Smith D, Nettesheim P. Effects of hormones on the multiplication and differentiation of tracheal epithelial cells in culture. In: Sirbasku D, Sato GH, Pardee A, editors. *Growth of cells in hormonally defined media.* Cold Spring Harbor: Cold Spring Harbor Laboratory; 1982. pp. 641–56.

134. Wu R, Smith D. Continuous multiplication of rabbit tracheal epithelial cells in a defined hormone-supplemented medium. *In Vitro* 1982;**18**:800–12.

135. Wu R. In vitro differentiation of airway epithelial cells. In: Schiff LJ, editor. *In vitro models of respiratory epithelium.* Boca Raton, FL: CRC Press; 1986. pp. 1–26.

136. Robinson CB, Wu R. Culture of conducting airway epithelial cells in serum-free medium. *J Tiss Cult Meth* 1991;**13**:95–102.

137. Wu R, Plopper CG, Cheng PW. Mucin-like glycoprotein secreted by cultured hamster tracheal epithelial cells. Biochemical and immunological characterization. *Biochem J* 1991;**277**:713–8.

138. Wu R, Sato GH, Whitcutt MJ. Developing differentiated epithelial cell cultures: Airway epithelial cells. *Fundam Appl Toxicol* 1986;**6**:580 90.

139. Adler K, Schwarz J, Whitcutt M, Wu R. A new chamber system for maintaining differentiated guinea pig respiratory epithelial cells between air and liquid phases. *Bio Techniques* 1987;**5**:462–5.

140. Whitcutt J, Adler K, Wu R. A biphasic culture system for maintaining polarity of differentiation of cultured respiratory tract epithelial cells. *In Vitro Cell Dev Biol* 1988;**24**:420–8.

141. deJong PM, Van Strekenburg MAJA, Hesseling SC, Kempenaar JA, Mulder AA, Mommaas AM, et al. Ciliogenesis in human bronchial epithelial cells cultured at the air-liquid interface. *Am J Respir Cell Mol Biol* 1994;**10**:271–7.

142. Gray T, Guzman K, Davis C, Abdullah L, Nettesheim P. Mucociliary differentiation of serially passaged normal human tracheobronchial epithelial cells. *Am J Respir Cell Mol Biol* 1996;**14**:104–12.

143. Whitcutt J, Adler K, Wu R. A biphasic culture system for maintaining polarity of differentiation of cultured respiratory tract epithelial cells. *In Vitro Cell Dev Biol* 1988;**24**:420–8.

144. Wolbach SB, Howe PR. Tissue changes following deprivaton of fat-soluble a vitamin. *J Exp Med* 1926;**42**:753–81.

145. Wong YC, Buck RC. An electronic microscopic study of metaplasia of the rat tracheal epithelium in vitamin a deficiency. *Lab Invest* 1971;**24**:55–66.

146. Harris CC, Silverman T, Jackson R, Boren HG. Proliferation of tracheal epithelial cells in normal and vitamin a-deficient syrian golden hamsters. *J Natl Cancer Inst* 1973;**51**:1059–62.

147. Sporn MI, Clamon GH, Dunlop NJ, Newton DL, Smith JM, Saffiotti U. Activity of vitamin a analogues in cell cultures of mouse epidermis and organ cultures of hamster trachea. *Nature* 1975;**253**:47–9.

148. Fell HB, Mellanby E. Metaplasia produced in cultures of chick ectoderm by high vitamin a. *J Physiol* 1953;**119**:470–88.

149. McDowell EM, Ben T, Coleman B, Chang S, Newkirk C, DeLuca LM. Effects of retinoic acid on the growth and morphology of hamster tracheal epithelial cells in primary culture. *Virchows Arch B Cell Pathol* 1987;**54**:38–51.

150. De Luca LM, McDowell EM. Effects of vitamin a status on hamster tracheal epithelium in vivo and in vitro. *Food Nutr Bull* 1989;**11**:20–4.

151. McDowell EM, DeSanti AM, Newkirk C, Strum JM. Effects of vitamin a-deficiency and inflammation on the conducting airway epithelium of syrian golden hamsters. *Virchows Arch B Cell Pathol* 1990;**59**:231–42.

152. Kopan R, Fuchs E. The use of retinoic acid to probe the relation between hyperproliferation-associated keratins and cell proliferation in normal and malignant epidermal cells. *J Cell Biol* 1989;**109**:209–307.

153. Smith BT. Lung maturation in the fetal rat: Acceleration by injection of fibroblast-pneumonocyte factor. *Science* 1979;**204**:1094–5.

154. Sepulveda J, Velasquez BJ. Study of the influence of na-872 (ambroxol) and dexamethasone on the differentiation of clara cells in albino mice. *Respiration* 1982;**43**:363–8.

155. Massaro D, Massaro G. Dexamethasone accelerates postnatal alveolar wall thinning and alters wall composition. *Am J Phys* 1986;**251**:R218–24.

156. Phelps DS, Floros J. Dexamethasone in vivo raises surfactant protein b mrna in alveolar and bronchiolar epithelium. *Am J Phys* 1991;**260**:L146–52.

157. Fernandez-Renau D, Lombardero M, Nieto A. Glucocorticoid-dependent uteroglobin synthesis and uteroglobulin mrna levels in rabbit lung explants cultured in vitro. *Eur J Biochem* 1984;**144**:523 7.

158. Lombardero M, Nieto A. Glucocorticoid and developmental regulation of uteroglobin synthesis in rabbit lung. *Biochem J* 1981;**200**:487–94.

159. Devereux TR, Fouts JR. Effect of pregnancy or treatment with certain steroids on n, n-dimethylaniline demethylation and n-oxidation by rabbit liver or lung microsomes. *Drug Metab Dispos* 1975;**3**:254–8.

160. Devereux TR, Fouts JR. Effect of dexamethasone treatment on n, n-dimethylaniline demethylation and n-oxidation in pulmonary microsomes from pregnant and fetal rabbits. *Biochem Pharmacol* 1977;**27**:1007–8.

161. Fouts JR, Devereux TR. Developmental aspects of hepatic and extrahepatic drug-metabolizing enzyme systems: Microsomal enzymes and components in rabbit liver and lung during the first month of life. *J Pharmacol Exp Ther* 1972;**183**:458–68.

162. Massaro GD, McCoy L, Massaro D. Development of bronchiolar epithelium: Time course of response to oxygen and recovery. *Am J Phys* 1988;**254**:R755–60.

163. Massaro G, Olivier J, Massaro D. Brief perinatal hypoxia impairs postnatal development of the bronchiolar epithelium. *Am J Phys* 1989;**257**:L80–5.

164. Massaro GD, McCoy L, Massaro D. Hyperoxia reversibly suppresses development of bronchiolar epithelium. *Am J Phys* 1986;**251**:r1045–50.

165. Jetten AM. Growth and differentiation factors in tracheobronchial epithelium. *Am J Physiol Lung Cell Mol Physiol* 1991;**260**:L361–73.

166. Kelley J. Cytokines of the lung. *Am Rev Respir Dis* 1990;**141**: 765–88.

167. Strandjord TP, Clark JG, Madtes DK. Expression of tgf-a, egf, and egf receptor in fetal rat lung. *Lung Cell Mol Physiol* 1994;**267**: L384–9.

168. Johnson MD, Gray ME, Carpenter G, Pepinsky RB, Stahlman MT. Ontogeny of epidermal growth factor receptor and lipocortin- 1 in fetal and neonatal human lungs. *Hum Pathol* 1990;**21**:182–91.

169. Han RN, Liu J, Tanswell AK, Post M. Ontogeny of platelet-derived growth factor receptor in fetal rat lung. *Microsc Res Tech* 1993;**26**:381–8.

170. Caniggia I, Liu J, Han R, Buch S, Funa K, Tanswell K, Post M. Fetal lung epithelial cells express receptors for platelet-derived growth factor. *Am J Respir Cell Mol Biol* 1993;**9**:54–63.

171. Nexo E, Kryger-Baggesen N. The receptor for epidermal growth factor is present in human fetal kidney, liver and lung. *Regul Pept* 1989;**26**:1–8.

172. Stahlman MT, Orth DN, Gray ME. Immunocytochemical localization of epidermal growth factor in the developing human respiratory system and in acute and chronic lung disease in the neonate. *Lab Invest* 1989;**60**(4):539–47.

173. Raaberg L, Seier Poulsen S, Nexo E. Epidermal growth factor in the rat lung. *Histochemistry* 1991;**95**:471–5.

174. Johnson MD, Gray ME, Carpenter G, Pepinsky RB, Sundell H, Stahlman MT. Ontogeny of epidermal growth factor receptor/kinase and of lipocortin-1 in the ovine lung. *Pediatr Res* 1989;**25**(5):535–41.

175. Snead M, Luo W, Oliver P, Nakamura M, Don-Wheeler G, Bessem C, et al. Localization of epidermal growth factor precursor in tooth and lung during embryonic mouse development. *Dev Biol* 1989;**134**: 420–9.

176. Strandjord TP, Clark JG, Hodson WA, Schmidt RA, Madtes DK. Expression of transforming growth factor apha in mid-gestation human fetal lung. *Am J Respir Cell Mol Biol* 1993;**8**:266–72.

177. Kida K, Utsuyama M, Takizawa T, Thurlbeck WM. Changes in lung morphologic features and elasticity caused by streptozotocin-induced diabetes mellitus in growing rats. *Am Rev Respir Dis* 1983;**128**: 125–31.

178. Han R, Liu J, Tanswell A, Post M. Expression of basic fibroblast growth factor and receptor: Immunolocalization studies in developing rat fetal lung. *Pediatr Res* 1992;**31**:435–40.

179. Stiles AD, D'Ercole AJ. The insulin-like growth factors and the lung. *Am J Respir Cell Mol Biol* 1990;**3**:93–100.

180. Warburton D, Seth R, Shum L, Horcher P, Hall F, Werb Z, et al. Epigenetic role of epidermal growth factor expression and signalling in embryonic mouse lung morphogenesis. *Dev Biol* 1992;**149**:123–33.

181. Catterton WZ, Escobedo MB, Sexson WR, Gray ME, Sundell HW, Stahlman MT. Effect of epidermal growth factor on lung maturation in fetal rabbits. *Ped Res* 1979;**13**:104–8.

182. Sundell HW, Gray ME, Serenius FS, Escobedo MB, Stahlman MT. Effects of epidermal growth factor on lung maturation in fetal lambs. *Am J Pathol* 1980;**100**:707–26.

Chapter 3

Development of the Innervation of the Lower Airways: Structure and Function

Nicolle J. Domnik*, Ernest Cutz[†,‡] and John T. Fisher*[,§]

*Department of Biomedical and Molecular Sciences, Physiology Program, Queen's University, Kingston, ON; [†]Division of Pathology, Department of Paediatric Laboratory Medicine, The Hospital for Sick Children, Toronto, ON; [‡]Department of Laboratory Medicine and Pathobiology, University of Toronto, Toronto, ON; [§]Department of Medicine, Queen's University, Kingston, ON, Canada

INTRODUCTION

The innervation of the lung and the airway smooth muscle (ASM) are important components of the developing lung, and both are present in the epithelial tubules of the embryonic lung bud shortly after evagination from the foregut. Whereas the ASM is functionally competent shortly after it is laid down, the growth and maturation of the innervation largely follows the morphological stages of lung development. In the human lung, the primary pattern of branching of the bronchial tree is established during the pseudoglandular stage, followed by elongation of the conducting airways and increased vascularization of the lung periphery during the canalicular stage. Development within the acini (thinning of the epithelial cells, expansion of the air spaces) occurs during the saccular stage.

Knowledge of the overall innervation of the fetal airways was restricted historically, because nerves were mostly detected only in thin sections by light microscopy, revealing cross-sections or short lengths of a nerve pathway that may be many centimetres in length, or from electron microscopy. These methods led to the description of a limited number of ganglia and nerve bundles in the peribronchial region of large airways of human fetal lungs from 9 to 40 weeks gestation.[1] An additional consequence of the limitations of past imaging techniques was that studies of the functional neurophysiologic behavior of afferent and efferent nerves preceded detailed knowledge of the morphology of airway innervation. Recent confocal microscopy has revealed many hundreds of ganglia, together with their connecting nerve pathways and fine branches, supplying the ASM. In the present chapter we provide an update of the advances in the knowledge of the development of airway innervation since this chapter was first published in 2004.[2] Our goal is to explore the development, as well as the structural and functional aspects,

of the conducting airways. The revised first section of the chapter remains a tribute to the outstanding work of late Dr. Malcolm Sparrow, who, along with his talented trainees,[2] contributed so much to the modern appreciation of the functional elements of lower airway development.

ANATOMY, MORPHOLOGY, AND DISTRIBUTION IN THE PRENATAL LUNG

The ontogeny of the pulmonary innervation in relation to the developing airways remains a primary focus of attention. Neural tissue is a dominant feature of the fetal lung and undergoes dramatic development during gestation; the stages of neural development can be graphically documented by confocal microscopy, using the immunofluorescence method on whole-mount preparations of lungs, lobes, and airway segments scanned by optical sectioning through the entire thickness of the sample. The use of specific markers of neural tissue, in conjunction with markers for ASM and epithelial tubules, provides three-dimensional information with detailed images of the network of nerves and ganglia that envelop the lung primordia. As lung development proceeds through gestation to postnatal life, comprehensive maps of the pathways of the nerves to their target tissues emerge with respect to airway innervation. Neural tissue and ASM persist in a dynamic state throughout gestation and postnatal life. Here we review evidence from the embryonic lung of the mouse, followed by human and pig.

Origin of the Innervation: The Fetal Mouse Lung

In the developing murine lung, the trachea and lung buds appear ~ embryonic day 9 (E9),[3,4] with the development of the innervation from E9.5 to E16.5 (mouse; weeks 5–17

The Lung. http://dx.doi.org/10.1016/B978-0-12-799941-8.00003-1

in human) representing the early pseudoglandular stage.[5] During this period, branching morphogenesis is at its peak, and every 24 hours sees a striking change in lung structure and in the maturation of the innervation that accompanies branching. In mice, two lung buds begin to evaginate from the foregut at E10,[6] whereas in humans and most mammals the lung develops from a single bud.

Transient neural progenitor cells, neural crest-derived cells (NCC), are present in the foregut subsequent to the closure of the neural tube and prior to the formation of lung buds (E10.5). NCC migrate into the lung, via the foregut and developing vagus nerve, where they differentiate into either intrinsic pulmonary neurons, occurring more frequently in early generations of airways, with fewer in the distal airways[7,8], or into glia[8–11]; note that NCCs are present in lung buds prior to vagal innervation, suggesting that primary migration occurs via the foregut. NCC migration is presumably aided by some combination of chemoattractive growth factors such as the neurotrophins NT3, NT4, nerve growth factor (NGF), and brain-derived neurotrophic factor (BDNF) signaling through high-affinity receptors (i.e., tyrosine-kinases TrkA, TrkB, and TrkC) and/or low-affinity receptors (i.e., p75^{NTR}).[12] This migration has been demonstrated in mouse lung by immunostaining whole mounts of foregut, including lung buds, from E10 onward (pseudoglandular stage).[13] At approximately E10.5, most NCCs are in apposition with the esophagus, with increasing numbers found in the trachea and lung buds, becoming increasingly associated with the developing bronchi and epithelial tubules, as gestation continues through E11.5 and thereafter; the NCC in lung buds can appear both as individual cells as well as interconnected chains.[8] NCCs can be identified using antibodies against protein gene product 9.5 (PGP 9.5; a general neural marker) and NCC-specific markers, including phox2b (a transcription factor expressed in NCC nuclei[14]) and p75^{NTR} (localized to the cell membrane of NCCs and their processes[13–15]). Neurons are also identified with an antibody to PGP 9.5,[16] which stains mature neurons and nerve fibres but not the precursors (i.e., NCC).

At E10, PGP 9.5- and p75^{NTR}-positive nerve fibres run along the dorsal aspect of the foregut. Among these fibres are many migrating NCC with phox2b-positive nuclei and p75^{NTR}-positive membranes with vagal processes on either side of the foregut. At this early stage, the emerging lung buds are largely free of NCC, albeit with a few solitary NCC at the base of the lung buds with occasional processes directed into the bud. Some NCC in the foregut have matured sufficiently to show PGP 9.5 staining, whereas the cells in the lung buds remain negative for this neuronal marker.

Recently, genetic models have been employed in order to understand neural development.[8,17,18] For example, experiments to examine the placodal- vs. neural crest-derived origins of vagal ganglia relied on a Wnt-Cre/R26R model, in which β-galactosidase is exclusively expressed by neurons originating in the neural crest, coupled with retrograde labeling (1,1'-dioctadecyl-3,3,3',3'-tetramethylindocarbocyanine perchlorate; DiI) of pulmonary vagal nerve fibers. This approach revealed that the murine vagus, which displays a single anatomic ganglion, does indeed contain both elements of the otherwise distinct nodose and jugular ganglia, i.e., placodal-derived and neural crest-derived neurons, respectively.[18]

This led to the characterization of both neural crest-derived and placodal crest-derived nociceptive vagal populations (C-fibres), each of which expresses a unique complement of neural markers. The complement belonging to the neural crest (jugular)-derived vagal population is similar to that of pulmonary fibres originating in the dorsal root ganglia (which also originate in the neural crest[18]).

These findings were confirmed and extended by studies using the Wnt1-Cre/R26R-EYFP mutant mouse[8] model, designed to trace neural crest lineage. Earlier studies showed that the intrinsic neurons of the lungs originate from the neural crest in the chicken (via visualization of expressed rearranged during transfection (RET) receptor)[10] and human (via visualization of expressed RET co-receptor, GFRα1).[9] They can also be traced throughout development via their expression of yellow fluorescent protein (YFP; see Figure 1). Studies employing these double-knockout mice confirmed the migration of NCCs from the region of the esophagus (starting ~E10.5) into the developing lung. The critical role for the proto-oncogenic glial cell-line-derived neurotrophic factor (GDNF; a ligand of RET) as a chemoattractant was assessed by the targeted growth and migration of YFP-positive cells towards GDNF-impregnated beads in organotypic lung culture.[8] The organotypic lung culture technique was described by Tollet et al. in the laboratory of Dr. Malcolm Sparrow[13,19] and was used to study spontaneous contractile activity in developing airways, potentially an important stimulus to lung growth.[13] RET-based signaling does not appear to alter the net innervation density (as assessed via imaging in homozygous RET knockout and GFRα1 knockout vs. wildtype mice), suggesting potential redundancy in receptor expression; however, conflicting evidence exists as to the impact of knockout on the number of neurons, with data supporting both a decrease and no change in neural marker (TuJ1)-positive neurons in the lungs of RET knockout mice.[7,8]

A novel optical projection tomography (OPT) technology has been used to examine 3-dimensional development of NCCs at E14.5 (Figure 2), illustrating YFP-positive NCCs along extra- and intrapulmonary airways, vagi, esophagus, and a decreasing density of NCCs from the proximal to distal airways.[8] This imaging technique provides more detailed and comprehensive structural information about tissue architecture, including airways and their innervation. For example, OPT has been recently used in the investigation

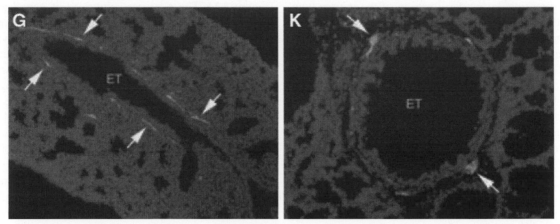

FIGURE 1   Sections through a Wnt1-Cre, YFP embryo, both immunostained with anti-green fluorescence protein. Left panel (G): longitudinal section at E18. Right panel (K): transverse cross-section at postnatal day 0. The images show the juxtaposition of neural elements/ganglia derived from NCCs around the airway wall. (This figure is reproduced in color in the color plate section.) *(Modified, with permission, from[8].)*

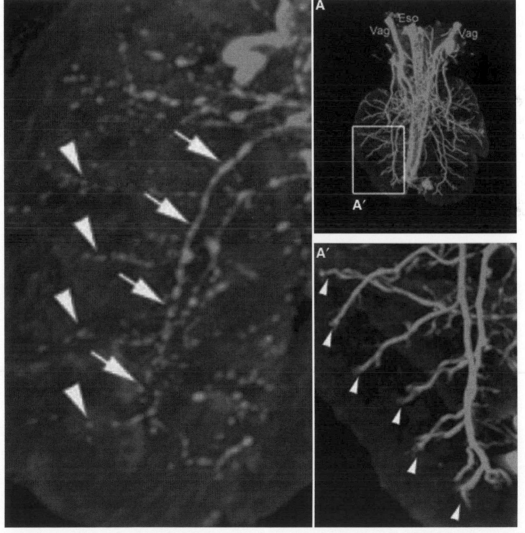

FIGURE 2   Sections through lungs from Wnt1-Cre, YFP embryos at E14.5. Sections are immunostained with anti-green fluorescence protein and imaged with 3-D optical projection tomography (OPT; adapted, with permission, from[8]). Left panel: combined OPT and YFP staining showing nerve trunk projection of the large (arrows) as well as fine, peripheral (arrowheads) branching airways. Right panels: Combined OPT and neural marker TuJ1 staining showing the overall vagal innervation of the developing lungs (upper panel) as well as a magnified view (lower panel) of the bronchi and their consistently associated nerve fibres (This figure is reproduced in color in the color plate section.) (Vag, vagus nerve; Eso, esophagus).

of gene expression during development of the murine lung bud in both normal and pathological conditions.[20–23] Recent application of a computerized quantification and visualization method that combines existing software solutions with a custom program provides an accurate, comprehensive mapping of airway epithelium and its innervation throughout the entire murine bronchial tree.[24]

By E11, neural tissue along the foregut condenses into two large nerve trunks, the vagus nerves, which stain strongly for PGP 9.5.[13] Labeling of NCC has uncovered the developmental origin of afferent, extrathoracic neurons of the vagal jugular ganglion as being derived from the neural crest, versus an epibranchial placodal origin for nodose ganglia, which appear to project to intrapulmonary locations.[18,25–28] Although the murine vagus nerve possesses a single, or fused, vagal ganglion (Figure 3), the evidence from mutant mice, which express β-galactosidase only in neurons derived from NCCs, suggests that the single ganglion comprises NCC-derived neurons in the rostral aspect of the ganglion, and non-NCC-derived cells (presumably placodal) in the mid and caudal aspects.[18]

As with the nerves of the jugular ganglion, the nerves of the dorsal root ganglion (DRG) are derived from the neural crest.[28] In mature mice, the jugular- and nodose-originating nociceptive C-fibres of the vagus can be distinguished by their respective complement of cell markers (e.g., receptors) and their resulting response profiles in the presence of purinergic agonists (e.g., ATP); jugular C-fibres express exclusively P2X3 purinergic receptors, whereas nodose C-fibres express both P2X3 and P2X3 purinergic receptors, resulting in a persistent, rather than transient, inward current when exposed to ATP.[18]

Neural processes positive for PGP 9.5 and p75[NTR] reach from the vagi to the trachea and primary bronchi

(Figure 4A). The vagi are comprised of neural processes and migrating NCCs (Figure 4B). These processes are likely to comprise both afferent fibres originating from vagal and spinal sensory ganglia, and preganglionic efferent fibres that will ultimately synapse on NCCs once they have completed migration. Neural processes from the vagi to the primary bronchi (Figure 4C) and the dorsal trachea contain migrating NCCs. Many NCCs are present on the dorsal trachea, located over the trachealis muscle, and some are on the ventral surface of the proximal primary bronchi, in the process of aggregating into large ganglia.

By E12 the lobular organization of the murine lung is complete, with one large left lobe and four smaller right lobes. A large nerve plexus is present on the ventral side of the lung near the hilum (Figure 5A). It originates from the vagus[13] and is comprised of nerve fibres and large ganglia-like clusters of NCCs, with numerous cells in each cluster. From these ganglia, nerves positive for PGP 9.5 (Figure 5B) and p75[NTR] (Figure 5C) extend along the bronchi, following the smooth muscle covered tubules. NCCs also migrate along these nerve tracts but lag behind the growth of the nerve axons (Figure 5C). Superimposed confocal projections of both the neural tissue and the ASM reveal their close relationship.

By E13 the neuronal precursors lying over the dorsal trachea have matured to form a PGP 9.5-positive network of thin nerve trunks interconnected by small ganglia, giving fine fibres that penetrate the smooth muscle layer. By E14 this plexus is more extensive, comprising larger ganglia and more numerous thick nerve trunks with multiple connections to the vagi (Figure 6A). Small nerves from the ganglia branch into many fine varicose fibres that run along the smooth muscle bundles. The ganglia vary greatly in size, and many large ganglia contain over 100 cell bodies positive

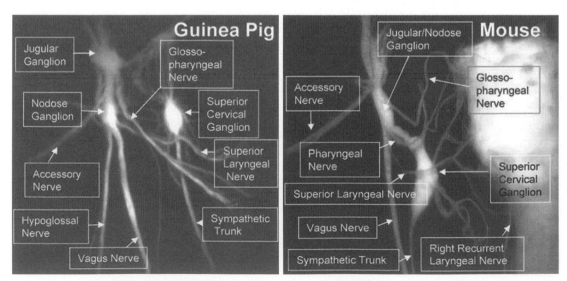

FIGURE 3   Dissected view of the vagal ganglia in the guinea pig (left) and mouse (right), illustrating distinct jugular and nodose ganglia in the guinea pig and a superficially fused, or single, jugular-nodose ganglion complex in the mouse. *(Reproduced with permission from[18].)*

for PGP9.5 (Figure 6A) and their nuclei positive for phox2b (Figure 6B). The axons in the nerve bundles connecting the ganglia stain strongly for GFRα1, the receptor for GDNF (see following text). The innervation from the vagus to the main ganglia lying on the dorsal trachea, ventral hilum, and left lobe are shown schematically in Figure 6C. During this early pseudoglandular phase, most nerves follow the smooth muscle-covered tubules, but some nerves course through the mesenchyme towards the lung cap, where, by E13, they form varicose terminal arborisations.[13]

Among the first neurotransmitters to appear in the foregut is calcitonin gene-related peptide (CGRP) at E12.[29] By E13, neuronal nitric oxide synthase (nNOS) can be demonstrated in the lung by NADPH-diaphorase activity in nerves associated with the airways and blood vessels. At E15, immunostaining reveals the presence of nNOS in neurons

and fibres along the trachea, extending from the hilum to the bronchioles,[30] as well as broad expression of the neurotrophin receptor TrkB, including marked expression in the neuroepithelial bodies (NEBs; see following text) and pulmonary nerves, which declines with age (TrkB measured at E15, 3 mo., and 6 mo.).[12] Interestingly, NEBs (discussed in greater detail in following text) are absent, and neural density is decreased, in mice homozygous for the knockout of the TrkB receptor, as compared with heterozygous mice or wildtypes.[12] The impact of neurotrophic factors on NEBs and their innervation is further shown by a decrease in NEB-related P2X3- and P2X3-purinergic receptor-positive innervation in NT4 knockout mice.[31] TrkB (activated by BDNF and NT4) expression has been specifically shown on C-fibres of placodal (nodose) origin in pre-, peri-, and postnatal life, while TrkA (activated by NGF) is expressed

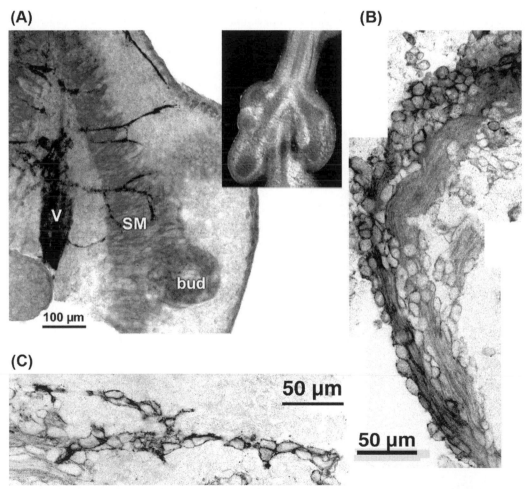

FIGURE 4   Embryonic mouse lung at E11. Panel A: a confocal projection showing a ventral view of the right upper half of the lung stained for nerves (black) with the protein gene product 9.5 (PGP 9.5). This also stained the undifferentiated epithelium of the tubules and growing end buds (grey). The ASM that covers the tubules is stained with α-actin (dark grey). The carina lies at the top of the figure. The left vagus (V) sends out nerve processes to the ASM covering the left lobar bronchus. Some extend towards the mesenchymal cap. Inset at right shows videomicrograph of ventral side. Panel B: a single optical section through the vagus nerve showing that it contains neural crest cells and many axons running between them (stained with an antibody to p75[NTR] that is positive for cell membranes and axons). Panel C: nerve fibres running from the vagus into the lung (see A) comprise processes and NCC cells (stained for p75[NTR]).

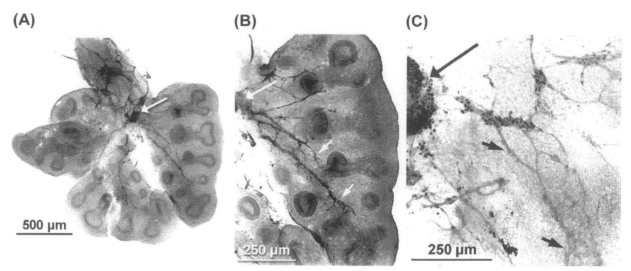

(A) 500 µm

(B) 250 µm

(C) 250 µm

FIGURE 5 Embryonic mouse lung at E12. Panel A: a confocal projection (ventral view) showing the lobular organization (the accessory lobe and the vagi have been removed) with the epithelial tubules in longitudinal section. The first two laterals of the left lobe reveal the end buds in the process of dividing. The undifferentiated epithelium of the tubules, particularly their end buds, are immunoreactive to PGP 9.5 (grey-black). PGP 9.5 diffusely stains ganglia (black) connected by nerve trunks and fibres in the ventral hilum (long arrow). Panel B: PGP 9.5-positive nerve fibres issue from the large ganglion (long arrow) at the base of the left pulmonary bronchi and reach along the left lobar bronchus (short arrows) and along some of the laterals but no PGP 9.5-positive cells and ganglia are present along the tubules. Panel C: the large ganglion at the base of the left lobar bronchus (long arrow) contains many neural crest cells (NCC) with phox2b-positive nuclei (black) and p75^{NTR}-positive membranes (grey). The NCC migrate along the p75^{NTR} nerve fibres (short arrows) that grow along the lobar bronchus and laterals. The cells lag behind the growth of fibres; the majority have only reached as far as the first lateral and a few as far as the second lateral. (A and B can be viewed in color, see,[13] *courtesy: American Association of Anatomists.*)

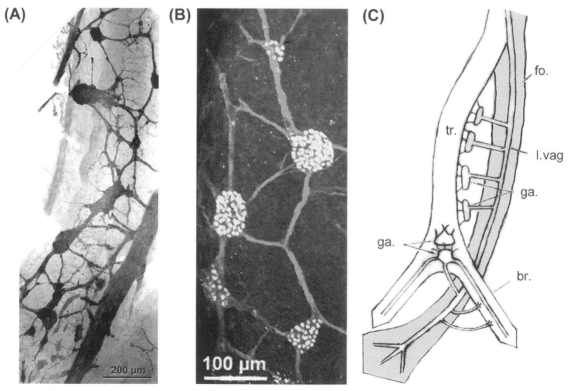

(A) 200 µm

(B) 100 µm

(C) fo.
tr.
l.vag
ga.
ga.
br.

FIGURE 6 Embryonic mouse trachea at E14. Panel A: PGP 9.5-positive (black) network of ganglia connected by thick bundles to the vagus (running diagonally at lower right). Nerves from ganglia spread over smooth muscle on the surface of trachea (upper part of panel). Panel B: ganglia with phox2b-positive nuclei (white) and nerve trunks staining for GFRα1 (grey) lying over the dorsal trachea. Panel C: scheme showing the innervation from the vagus to ganglia lying on the dorsal trachea and ventral hilum. Main nerve trunks to the lobes arise from the latter. Oblique ventral view. tr., trachea; fo., foregut; ga., ganglia; l. vag., left vagus; br., bronchus.

by approximately 50% of neural crest-derived, or jugular, C-fibres.[18] Neurons derived from each ganglion are thought to innervate distinct structures in the lung, with studies in dogs,[32] guinea pigs, and mice[18] demonstrating neural crest-derived neurons exhibiting extrapulmonary, or bronchial, targets, and placodal neurons exhibiting (intra)pulmonary targets.

GDNF has been identified as the most important neurotrophic factor in the development of the enteric nervous system[33] and may play a similar role during lung development.[19] In the gut of mice lacking GDNF or RET (receptor for GDNF), all neurons below the oesophagus and proximal stomach are absent[34] but it is not known whether lung neurons are affected. In cultured lung explants at E12, neurons survive and display proliferation, differentiation and continued migration along the developing smooth muscle-covered tubules.[19] A characteristic of these explants is the formation of a layer of α-actin-positive cells (possibly ASM precursors) that grows out from the lung periphery and attracts nerves that grow onto this layer. When cultured in GDNF-supplemented medium, the amount of neural tissue on this layer increases 14-fold. The neural tissue consists of a high-density network of nerve trunks and large ganglia and contains many PGP 9.5-positive cells, indicating that migration, proliferation, and differentiation of neuronal precursors, as well as neurite extension, have taken place as a direct result of stimulation by GDNF. This suggests that GDNF is a chemoattractant to both nerves and NCC. GDNF-impregnated beads attract nerve outgrowth from cultured lung explants and in some instances NCC surround the treated beads. The membranes and nerve processes of the NCC are positive for the GDNF-receptor, GFRα1 (Figure 6B), suggesting that nerves and NCC are guided by GDNF. The presence of GDNF-mRNA has been demonstrated in the mesenchyme adjacent to the fetal mouse epithelial tubules,[35] possibly in the smooth muscle, which thus may play an important role in attracting nerve fibres and migrating NCC.

Mapping the Innervation: Fetal Pig and Human Lung

The rapid development of lung innervation in mice during the pseudoglandular stage (E10–E14) contrasts with that of large mammals where the equivalent time period is from week 3 to week 8 in the pig and week 5 to week 17 in the human.[36,37] In mice at E14 and thereafter, the use of confocal microscopy has been limited due to technical problems: reduced signal emission at increasing depth of scanning, and a decrease in antibody penetration (see aforementioned for recent advances in imaging). These problems can be overcome by removal of the lung cap, mesenchyme, and pulmonary vascular tissue leaving the bronchial tree fully exposed, which is only feasible in larger mammals.[38–40] Thus, the entire bronchial tree can be progressively scanned

with the confocal microscope at high resolution. Using this approach, montages of near complete bronchial trees in the pseudoglandular stage (about 6 mm long) from fetal pigs[36,38] (Figure 7), and smaller lengths of subsegmental airways from fetal humans[39] can be assembled. These clearly display the organization of nerves and ganglia and their relationship to the ASM, the glands, and blood vessels; fine detail is also shown at selected sites.[38] Thus, the development of the innervation from the embryonic lung bud through to postnatal life is revealed.

The structural characteristics and distribution of the nerves are similar in the three species (mouse, pig, and human) at comparable developmental stages and likewise the ASM. The muscle bundles are orientated around the airways perpendicular to their long axis (i.e., between the C-shaped cartilage rings) from the trachea through to the base of the epithelial buds, and this arrangement persists into postnatal life. The innervation of the porcine and human bronchial tree from the adventitia to the epithelium has been reported from early gestation through postnatal life.[38–41]

Pseudoglandular Stage

The main characteristics of this stage are:

(i) Chains of forming ganglia interconnected by thick nerve trunks to each other and to the vagus lying over the ASM of the dorsal trachea and the ventral surface of the hilum.

(ii) In general, two thick main nerve trunks extend from the hilum along each airway to the growing tips. These lie above the ASM supported by the mesenchyme. In the fetal pig at 5.5 weeks gestation, proximal trunks (~50 μm in diameter) run about 40–60 μm above the ASM, progressively thinning distally over a length of 4 mm to ~20 μm in diameter and 15–20 μm from the ASM. The nerve trunks terminate as thin bundles in the collar of ASM that surrounds the epithelial buds.[38,42] Along the length of the trunks, branches descend towards the smooth muscle and break up into small bundles. From these, fine varicose fibres issue that spread over the muscle ending in arborizations ~1 μm from muscle cells, suggesting functional innervation. At this stage the varicose fibres are randomly distributed on and in the smooth muscle (Figure 7B) but later become oriented along the smooth muscle bundles.[38,40]

(iii) Immature ganglia are present along the main trunks, from which nerve branches radiate out to connect with many other smaller ganglia that form a network covering the airway wall. Figure 7C shows this innervation in the distal airways of a fetal human lung at 7.5 weeks gestation. The mean distance between ganglia[40] is 64 ± 18 μm (n = 87), very similar to the pig (70 μm at comparable gestation). Ganglia also lie at most airway branch points and give rise to smaller trunks that follow

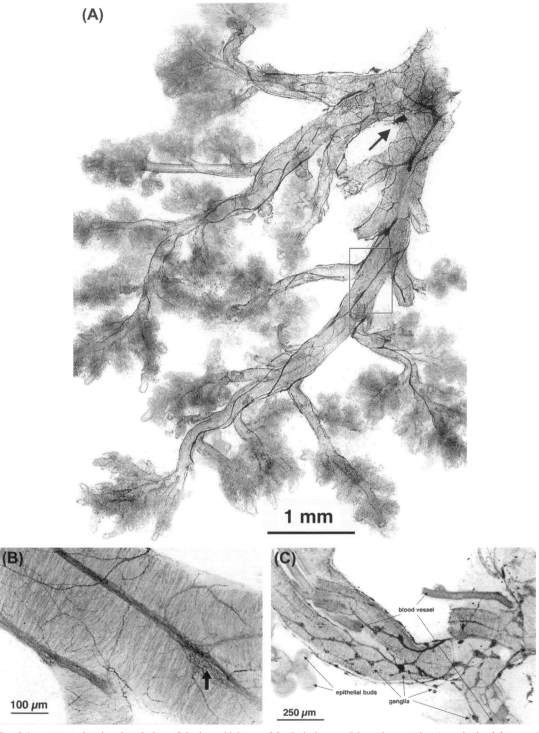

(A)

1 mm

(B)

100 μm

(C)

blood vessel

epithelial buds

ganglia

250 μm

FIGURE 7   Panel A: montage showing dorsal view of the bronchial tree of fetal pig lung at 5.8 weeks gestation (pseudoglandular stage) with nerves and ganglia stained for synaptic vesicle protein (SV2, black) and for airway smooth muscle with smooth muscle myosin (grey). It has been optically sectioned to show the outline of the airway wall. Nerve trunks run down the length of the airways to the epithelial buds. Arrow shows a large ganglion. Ganglia were seen at airway bifurcations and at the branching points of the nerve trunks. Box shows an immature ganglion, also seen in panel B. Panel B: two major nerve trunks stained with SV2 (black) giving rise to a fine network of varicose processes overlying the airway smooth muscle (grey). At this stage the varicose fibres are randomly distributed on and in the smooth muscle, some located within 1μm from the muscle cells. The accumulation of cell bodies (arrow) is a precursor ganglion present at the bifurcation point of the airway. The cell profiles in the ganglion can be distinguished by the SV2 positive nerve fibres lying around them. (Panels A and B are from,[38] *courtesy: American Thoracic Society.*) Panel C: the innervation and airway smooth muscle in the developing airways of a fetal human lung at 58 days of gestation (pseudoglandular stage). The field shows branching epithelial tubules in the periphery of a lobe. Nerves and ganglia are stained for PGP 9.5 (black) and form a network overlying the airway smooth muscle stained for α-actin (grey). The circumferential arrangement of the muscle bundles around the epithelial tubule is faintly seen in each of the above, where it is perpendicular to the long axis of the tubule.

the airways as they proceed distally. Proximal ganglia are large (e.g., > 300 cell bodies at 5.5 weeks gestation), whereas distal ganglia are small and ultimately comprise a few neurons. Individual neurons within ganglia show different intensities of PGP 9.5 staining, indicating variance in their type or maturity.

(iv) PGP 9.5 diffusely stains nerve trunks, but many cell profiles of Schwann cells remain unstained (revealed using an antibody to the Schwann cell marker S–100). Staining for synaptic vesicle protein 2 (SV2), a component of the membranes of vesicles in varicosities, reveals individual varicose fibres in nerve trunks indicating that vesicle traffic is prolific at this stage of development.[38,42] This abundance of SV2 positive fibres decreases with ongoing maturation; by postnatal life, varicose fibres are restricted to the distal nerve bundles and the fine fibres that lie on and in the ASM. Staining for neurofilament sharply defines a small proportion of individual fibres in a trunk; these can be traced along the tubules, where several terminate in the collar of smooth muscle that surrounds the base of the epithelial bud.[42] The low proportion of neurofilament-positive fibres in the nerve trunks may reflect the level of maturity of these nerves, as the proportion of neurofilament-positive neural tissue increases as gestation progresses.[36]

In addition to the aforementioned main characteristics, expression of Ca^{2+}-sensing receptor (CaSR) has been shown in mouse airway epithelium during hypercalcaemic fetal development, specifically in the pseudoglandular stage (CaSR detected from E10.5 to E18.5, with peak expression at E12.5); during this time CaSR has been implicated in negatively regulating the cellular proliferation of airway branching and morphogenesis, while regulating lung fluid secretion in a $[Ca^{2+}]$-dependant manner.[43,44] The pseudoglandular stage has also recently been shown to have luminal airway epithelial expression of L-type ($Ca_V1.2$, $Ca_V1.3$) and R-type ($Ca_V2.3$) voltage-gated Ca^{2+} channels in both the human and the mouse.[45] A functional role for these receptors was shown via their blockade (nifedipine and SNX-482), which resulted in a removal of the usual inhibitory effect of Ca^{2+} on airway morphogenesis.[45]

Canalicular Stage

With airway growth there is increasing spatial separation of the ganglia. The large ganglia lying on the central airways that form nodes at nerve junctions undergo a 4-fold increase in separation to ~254 μm. The ganglia vary greatly in size—large ones are 120 μm at their greatest width and contain as many as 200 neurons of ~11 μm diameter. Many of those lying on the trunks gradually become displaced laterally to become attached by a stem, with nerves radiating out from them over the airways.[36]

The bronchial vasculature becomes more prominent, with arterioles running adjacent to the trunks and around the ganglia. Nerve fibres penetrate the submucosal glands. By midterm, ganglia have condensed and become compact and spherical. Figure 8A shows a montage of nerve tracts in the subsegmental airways of an 18-week fetal human lung.[39] Large nerve trunks run the entire length of airways, reducing in diameter from 45 μm to <20 μm distally, with many ganglia attached to them from which nerves issue to connect with a network of smaller ganglia lying closer to the airway surface. A high power view (Figure 8B) shows a fine plexus of nerves containing many small ganglia lying close to the ASM. Mucosal nerves are now abundant; they arise from branches of the adventitial nerves that penetrate the ASM layer at intervals where they run in parallel bundles in the lamina propria along the length of the airway.[39] The development of the mucosal vascular circulation is now well advanced.[41] At this point the lung is well endowed with the beginnings of a neural network that can serve the afferent and efferent functions of the vagus nerve.

Most neurotransmitters make their appearance during the canalicular stage (in human fetal lung 16–26 weeks,[37] pigs 7–13 weeks,[36] rats 18–19 days, and mice 16.6–17.4 days[4]). In rats at day E17, CGRP is present in NEBs in the epithelium. At 18 days, CGRP nerve fibres are present in the trachea, stem bronchi, and proximal intrapulmonary airways, mainly lying below the epithelium, and by days E19–20 fine fibres are seen on the bronchial smooth muscle and around blood vessels in the adventitia.[46] In mouse lung, nitrergic neurons can be detected in airways as early as E13 using an assay for NADPH diaphorase; by E15, nNOS expression is present in airway neurons and fibres.[30]

Functional cholinergic transmission has been demonstrated (as airway narrowing) in the pseudoglandular stage, indicating that some fibres are already cholinergic.[42] In humans, evidence for the presence of cholinergic neurons by 10–12 weeks was reported using acetylcholinesterase,[11] which may not be a reliable marker for the presence of acetylcholine.[47] Choline acetyltransferase (ChAT) is a specific marker for acetylcholine (ACh),[48] and in fetal pig lungs both ChAT–positive neurons and fibres are present in the trachea and on the ASM of the peripheral airways at the early canalicular stage. ChAT does not stain the very fine terminal varicose fibres that SV2 reveals, indicating that it may not be sensitive enough to detect very low levels of ACh.[36]

In human lung, vasoactive intestinal peptide (VIP) and substance P (SP) positive fibres are present at 16 weeks in ASM, and thin fibres containing CGRP start to ascend from the basement membrane of the epithelium.[11] The latter are likely to be the sensory C-fibres seen in postnatal life.[41,49,50] SP and CGRP are present in the axons of the nerve trunks

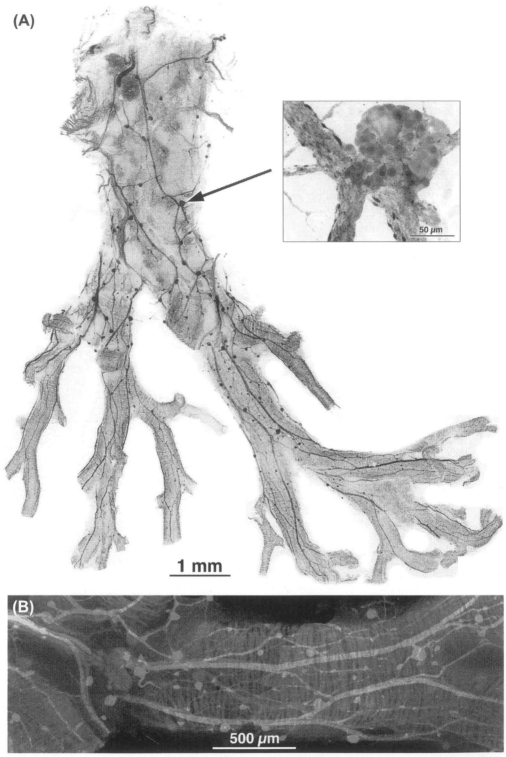

FIGURE 8   Panel A: a montage showing the peribronchial innervation of a segmental bronchus and its branches from human fetal lung (18 weeks gestation, canalicular stage). Nerves are green (PGP 9.5 stain) and smooth muscle red (α-actin). Nerve trunks extend to the most distal airways. Ganglia are present along the trunks and at the divisions of nerve bundles. The box inset at right shows a higher power projection of a ganglion at the junction of several nerve trunks (arrow). Panel B: a higher power view of the straight region on the lower right hand side of the montage show-ing the disposition of the nerves, ganglia (green), and airway smooth muscle (red). PGP 9.5 stained a plexus of fine nerves containing many small ganglia. Arterioles (red) of the bronchial circulation can be seen accompanying the larger nerve trunks. (This figure is reproduced in color in the color plate section.) *(Reproduced from,*[39] *courtesy: American Thoracic Society.)*

running in the airway wall adventitia at midterm[40]; by the beginning of the saccular stage VIP, tyrosine hydroxylase (TH), and NPY are also present.

Saccular Stage

In humans, the saccular stage runs through most of the third trimester, when further maturation of the ganglia and nerves occurs. The processes of glial cells increasingly surround neurons in the ganglia, and the axons in nerve trunks and bundles.[36] This glial ensheathment may contribute to restricting intraganglionic communication between adjacent neurons.[51] Separation of the ganglia greatly increases as airways lengthen and widen. Neurotransmitters are fully expressed now, with strong immunostaining of neurons and their axons[40]; neurofilament is also expressed in many neurons and axons. The perikarya are located mainly in the periphery of the ganglion with many neurite structures in the centre. Neurons appear to contain one major axon and, therefore, correspond to Dogiel type 1 neurons. Some neurons show strong PGP9.5 staining of the nucleus only while others exhibit a faint homogeneous staining throughout the perikaryon.[36]

The bronchial mucosal circulation, which is rudimentary at the end of the pseudoglandular stage, increases in complexity during the canalicular stage, becoming a well-developed network of microvessels. The mucosa is now richly innervated with nerve bundles and varicose fibres running the length of the airways which, in fetal pig, stain for NOS, SP, CGRP, VIP, TH, and NPY.[40] The presence of the neuropeptides SP and CGRP is indicative of afferent nerves. Although it appears that many nerve bundles follow arterioles, immunohistochemical evidence suggests the opposite, as the neural tissue can be stained earlier (i.e., in the pseudoglandular stage) than the bronchial vessels that are first demonstrable in the canalicular stage, where they run contiguously with the nerves.

Function of the Airway Innervation During Fetal Life

Whether the innervation plays a functional role during specific events in fetal lung development is unknown. With recent insight into the organization of the bronchial innervation, it should be feasible to study neurotransmission in the ganglia and nerves in the lung either excised from the fetus, or in situ. At birth, a range of sensory reflexes are present (e.g., Hering–Breuer reflex). Afferent mechanoreceptors, such as slowly adapting stretch receptors located in or adjacent to the ASM, fire at a low rate in liquid-filled fetal lung[52]; however, despite the reasonable assumption that afferent and efferent nerves are capable of function well before birth, little is known about the pathways of afferent nerves and their receptors in fetal airways.

Afferent fibres must already be present when NCC and nerve processes migrate from the vagi at the formation of the lung bud. They doubtless make up a major part of the nerve trunks in the fetal lung as more than 70% of the nerves in the vagus that innervate the lung are afferent.[53] These are the fibres that pass through the ganglia as they extend distally to their sensory receptor endings. Markers of C-fibre sensory nerves (i.e., SP and CGRP) are seen in fibres in lamina propria of rats at mid-term[46] and in the epithelium in humans,[11] but the apical plexus of C-fibre nerve endings in the epithelium that constitute the receptive fields in postnatal pigs (see Figure 9B) and humans[41] were not observed. Surprisingly, mechanoreceptors (large tree-like arborizations 100–200 µm long) have not been recognized in fetal airways, but more focussed searching in ASM and lamina propria of airways may reveal them. It is likely that the application of advanced genetic labeling and computational/imaging techniques will provide novel insight into the subtypes of lung afferents present during fetal and postnatal life.[54]

Summary: Ontogeny of Lung Innervation Before Birth

Neural tissue and ASM are integral components of the primordial lung in which epithelial tubules (the future bronchial tree) are enveloped in a network of precursor ganglia and loose bundles of nerve fibres. These ganglia comprise flat patches of neural crest cells that have migrated along nerve processes that issue from the vagi. They lie over the wall of the epithelial tubules supported by the mesenchyme, and are interconnected by nerve bundles. ASM is laid down at the base of the epithelial buds that are sites of new tubule growth, which occurs through an epithelial–mesenchymal interaction.[55] Thus, as the epithelial tubules elongate, the ASM forms a continuous layer extending from the trachea to the growing tips. Small nerve bundles branch from the nerve network and descend to the ASM; the ASM is functionally mature shortly after it is formed as the terminal tubules show rhythmic contractions in situ.[56–58] GDNF is a likely neurotrophic factor that acts as a chemo-attractant for nerves in lung explants,[19] and GDNF receptors (GFRα1) are present on the nerve processes in vivo, but whether GDNF is expressed by ASM is not yet known. By the end of the pseudoglandular stage, when branching is virtually complete, most precursor neural tissue has completed proliferation and is differentiating into mature neurons. In the canalicular stage, ganglia develop a more compact, spherical shape, and come to lie away from the nerve bundles. Arterioles of the bronchial circulation appear adjacent to the nerve trunks and nerve bundles. The mucosal innervation becomes established followed by the mucosal vasculature. The chemical coding of neurons and their fibres occurs during this stage. In the saccular stage (most of the third

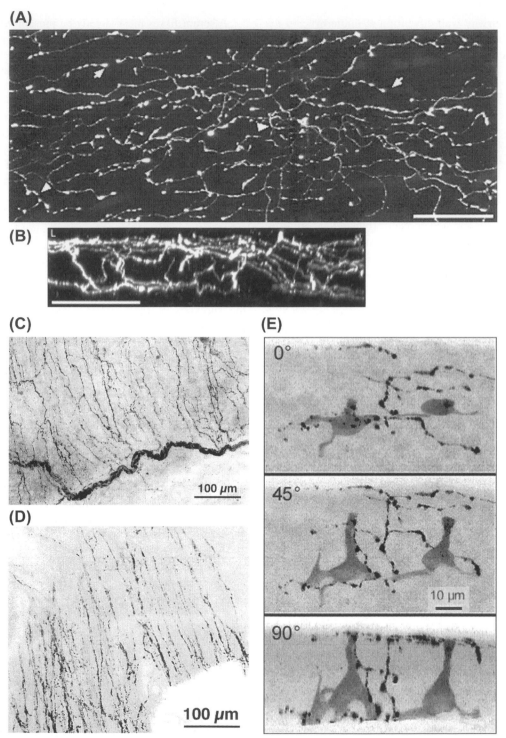

FIGURE 9   Panel A: confocal projection showing SP immunoreactive nerves in the bronchial epithelium of a young pig. Nerve endings have one or more enlarged terminal varicosities (arrows). Some nerves have curved profiles (arrow heads) where they encircle goblet cells (not stained). Image depth from lumen 21μm. Bar, 50 μm. Panel B: projected cross-sections show SP nerves in the epithelium arranged in two plexi. An apical plexus lies immediately below the luminal surface (lumen, L) with fibres descending to the base of the epithelium to form a second lateral layer. Cross-sections were reconstructed from a confocal z-series imaged from the lumen surface with optical sections 0.5 μm. Loss of resolution occurs in reconstructing side views, making the nerves appear thicker than they are. Panels C and D: confocal projections of the nerves overlying the ASM viewed from the adventitial surface of the bronchus of an adult mouse (C) and of a 54 year human (D). Varicose nerves run along the ASM bundles that lie around the circumference arranged perpendicular to the long axis of the airway (muscle bundles not shown to avoid obscuring nerves). To view double staining in color see references[13] and.[40] Panel E: varicose nerves in close proximity to pulmonary neuroendocrine cells (PNEC) in the epithelium of a human bronchus. Both nerves and PNEC are revealed with PGP9.5. "0°" lumen view, 90° side view. The latter shows that the cell body of the PNEC stands at the base of the epithelium and the apex at the lumen. Processes extend from along the basement membrane and others upwards towards the lumen.

trimester), lung growth is rapid with greater spatial separation of ganglia, and their neurons becoming progressively ensheathed by glial cell processes, as do the axons in nerve trunks and bundles.

ANATOMY, MORPHOLOGY, AND DISTRIBUTION IN THE POSTNATAL LUNG

Early Postnatal Period to Late Adulthood

Overviews and higher power views obtained by 3-D imaging of the adventitial (peribronchial) and mucosal innervation have been obtained using airways of young pigs,[38,41] humans,[39] mice,[13,19] and rats.[59] The peribronchial nerve plexus has been designated as "extrachondrial" and "subchondrial" in the bronchi, where the main trunks lie outside the cartilage plates with a lesser layer of nerves between the cartilage and ASM.[60] In the bronchioles the distinction disappears. Because only a single nerve plexus originates over the ASM in the fetal lung, the peribronchial (adventitial) nerves can be regarded as a single entity. With postnatal growth this plexus continues to comprise two major nerve trunks running the length of the airways. Large bundles frequently branch off and rejoin the main trunk or lesser branches.[38,60] The density of the nerve fibres in this plexus is sufficiently great as to give the appearance of a continuous network.[38] However in terminal airways the nerves thin out revealing arborizations of varicose fibres. In the fetus some nerve bundles from this adventitial plexus penetrate between the bundles of the ASM to initiate the development of the mucosal plexus.[39]

After birth in pigs, large nerve bundles run the length of the lamina propria lying close to the ASM border. Smaller bundles comprising both sensory and motor nerves[41,61] branch off to run parallel with the bronchial arterioles, while others continue upward to enter the epithelium where they form a basal plexus of fibres lying just above the basement membrane. In pigs and humans, fibres ascend from this basal plexus between the epithelial cells to within a few microns of the lumen where they arborize forming thin (~1 μm diameter) varicose fibres that spread out (up to 120 μm in length) over the apical epithelium (Figure 9A). Each varicose fibre terminates in one or more swollen varicosities (~3 μm diameter) that contain SP and CGRP.[41,50] These are the sensory endings of C-fibres[41,61] that respond to chemical and mechanical stimuli.[26] They have been demonstrated immunochemically in thin sections in most species; they contain SP, CGRP, and neurokinin A.[62,63] When confocal projections of the SP nerves in the epithelium of pigs are reconstructed, the basal and apical plexuses are readily seen (Figure 9B). A dense plexus of varicose fibres containing SP in rat tracheal epithelium has been demonstrated, but an apical plexus was not seen.[59]

Advances made using 3-D imaging techniques have given new insight into airway innervation, adding to the classic images of lower airway nerves, neurons, and afferent mechanoreceptors that form the foundation of current understanding of lower airway innervation.[60,64] In the last 30 years, the use of specific antibodies to autonomic nerve neurotransmitters has enabled identification of the chemical coding of nerves in the airway wall. The absence of a suitable, robust antibody to stain parasympathetic cholinergic nerves has presented a major problem due to (a) the lack of specificity of staining for acetylcholinesterase[47] and (b) the antibody to choline acetyltransferase[48] lacking the sensitivity to reveal fine varicose nerve fibres.[36] However, an antibody to vesicular acetylcholine transporter protein has been successfully used in rat intestine,[65] and has been used to stain both sensory and motor nerve fibres innervating the airways of the rat[66,67] and mouse.[68] Thus, gradually, a large body of information has been obtained on the neurotransmitters found in the neurons in the parasympathetic ganglia, and in nerves of the lower airways. Of note, recent studies have begun to shed new light on the airway intramucosal nerve ending morphology of presumptive slowly adapting mechanoreceptors; specifically, in identification of the smooth muscle-associated receptors, SMARs.[69] SMARs are myelinated, sensory fibres of vagal origin that express vesicular glutamate transporters (VGLUTs) 1 and 2 and Na$^+$/K$^+$-ATPase α3.[69-71] The observation that SMAR innervation holds parallels, in terms of its neurochemical characteristics, to select nerve fibre populations innervating NEBs has prompted an interesting discussion as to the identity of these sensory fibres, and a potential relationship between epithelial NEBs and subepithelial, or intra-ASM, SMARs (Figure 10).[70,72-75] More work is needed in order to definitively unravel the morphological identity of the various pulmonary mechanosensors. There are many specialised reviews available on neurotransmitters in the nerves in ASM,[76] glands, and goblet cells,[77] immune tissue in the airway wall,[78] bronchial vasculature,[79] as well as their suggested roles in co-transmission and neuromodulation.[80,81]

Density of Innervation

With the realization that nerves are abundant in the fetal and postnatal airways in at least five species of mammals (humans,[39] pigs,[38] mice,[13] rats,[59] and dogs[40]), the question can be posed—what is their functional significance? Firstly, the innervation needs to be quantified so that comparisons can be made between tissues (e.g., ASM vs. epithelium), and species (e.g., rat vs. pig). Nerve densities in the trachea of postnatal rats[59] and bronchi/bronchioles of pigs[38,41] have been estimated by point counting. The studies show that the total innervation of the ASM is ~2-fold denser than in the epithelium, and the densities in pigs are about twice that of rats. Substance P nerves comprise 90% (rat[59]) and 94%

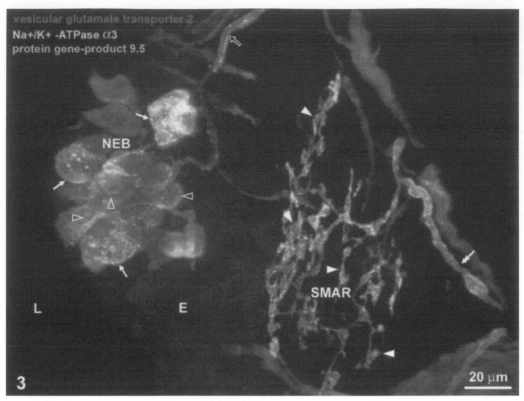

FIGURE 10 Sectioned rat lung immunocytochemically stained for VGLUT2 (red), PGP 9.5 (blue), and Na+/K+-ATPaseα3 (green) showing VGLUT2-positive nerve terminals innervating a PGP 9.5-positive NEB (open arrowheads) and a subepithelial SMAR (filled arrowheads; E: epithelium). Select NEB innervation is also shown expressing Na+/K+-ATPaseα3; it (single-headed arrow) is distinct from the Na+/K+-ATPase α3-positive innervation of the SMAR (double-headed arrow). (This figure is reproduced in color in the color plate section.) *(Reproduced with permission from[70].)*

(pig[41]) of the total epithelial nerves. While nerves have not been quantified in the fetal lung, montages of the fetal airway innervation[13,38,39] show that their abundance approaches that of postnatal life. By birth, the varicose fibres covering the ASM are oriented in the direction of the muscle bundles, with single fibres running along most muscle bundles in the young pig,[38] adult mouse (Figure 9C), and human infant[39] and adult (Figure 9D). Thus nerve density is maintained during the considerable growth that occurs during development, indicating that nerves continue to extend over the expanding surface of the airway wall as growth proceeds. This arrangement of nerve fibres running along the muscle bundle is probably the main determinant of the ultimate nerve density attained in ASM.

Orientation of Airway Smooth Muscle Bundles

In the fetal lungs of humans, pigs, and mice, muscle bundles encircle the airways and lie perpendicular to the long axis of the airways. This orientation is maintained into postnatal life in these species as well as in postnatal rat and young dog.[40] An ultrastructural study reporting that ASM has a pitch of approximately 30° in adult humans is at variance with the confocal microscopic studies of the central

and distal airways.[82] At branching points the orientation of some ASM bundles varies widely, presumably to suit the local airway wall architecture, and occasionally bundles lie almost parallel with the length of the airway.[83]

Efferent Nerves: Long Preganglionic and Short Postganglionic Fibres

In the adult lungs of large mammals the distribution and morphology of the innervation becomes increasingly difficult to characterize because of the sheer size of the lung and airway tree. The thickness of the tissue layers, particularly connective tissue and cartilage, makes the adventitial nerves and ganglia difficult to expose. Furthermore the density of ganglia becomes reduced with growth so that they become difficult to locate. Indeed, ganglia are considered to be absent beyond the third generation airways, a view largely based on early reports.[60,84,85] From this it is assumed that long postganglionic nerves run from the ganglia in the central airways along the length of the bronchial tree to the terminal airways.[51,60] The central ganglia may, therefore, play a key role in regulating airway function[51]; however, this view is not compatible with studies on the development of airway innervation.[36,38,39] Ganglia are shown to extend to the 9–10th generation in the canalicular stage in 18 week fetal human lungs (see Figure 8A).

In the lung of mid-gestation fetal pigs, ganglia extend to distal generations of 50 μm diameter, which is the limit of dissection.[36,40] These ganglia are mature (proximal) or maturing (distal) and there is no a priori reason to believe that they might disappear from the lungs (e.g., via apoptosis) at this late stage of development. It is reasonable to assume that ganglia lie chiefly in a thin layer surrounding the airway wall; the extent of their separation with lung growth is then a function of the increase in the surface area of the airway wall and the increase in length of the airways. It should be possible to obtain estimates for these parameters to determine ganglia separation in the adult, and thereby the probability of finding them in thin sections. In summary, studying the ontogeny of neural development reveals long preganglionic fibres and short postganglionic fibres, which is consistent with other tissues innervated by autonomic nerves.

The Pulmonary Neuroendocrine Cell System and Its Innervation

Definition, General Morphologic Features, and Distribution

The pulmonary neuroendocrine cell (PNEC) system is comprised of solitary PNECs and innervated clusters of 5–15 PNECs (i.e., NEBs); in conjunction with Clara-like cells, NEBs form the so-called "NEB microenvironment".[86] The PNEC system derives its name from its constituent neural and endocrine cell phenotype. The general morphologic, immunohistochemical, and ultrastructural features of PNECs/NEBs have been the subject of several recent reviews,[68,87–90] and therefore selected pertinent highlights are discussed herein. PNECs and NEBs express amines (serotonin; 5-HT), a variety of peptides (i.e., bombesin in humans, CGRP in rodents), and a host of neuroendocrine markers (i.e., chromogranin A, SV2).[87,88] PNECs/NEBs are highly conserved and found in the respiratory systems of amphibians, birds, reptiles, and fish.[91,92] In mammals, PNECs are distributed from the nasal respiratory mucosa[93] to the terminal airways and alveoli.[94] By contrast, NEBs are confined to the intrapulmonary airways, frequently at bifurcation points.[95] NECs are abundant in adult human lungs and are homogeneously distributed in the epithelium[96] at a density of ~250/mm², which is higher than previously reported.[97] In humans, relative NEB density decreases with age, becoming increasingly difficult to image or isolate in the adult lung.[97]

Development of PNECs/NEBs

PNECs are the first cell type to differentiate within the primitive airway epithelium; they originate from the endoderm that lines the tubules in the fetal lung early in gestation.[91,95,98,99] In mice, this period of differentiation is represented by the pseudoglandular stage[91,100]; in rabbits, by the late glandular and early canalicular (i.e., circa E21) stages, prior to extensive airway branching.[98] Labeling studies in mice have shown that, while PNECs never co-label (CGRP^{CreER} line; CreER (estrogen receptor) induced by tamoxifen at varying stages of embryonic development) with Clara or ciliated cells, co-labeling with alveolar type I and type II cells does occur between E12.5 and E14.5, suggesting that PNECs and alveolar cells may share a common origin or may be derived from separate but similar progenitors; PNECs seem to be well differentiated compared to other cell types by approximately E15.5.[91] The initiation of PNEC differentiation is critically dependent on the proneural basic helix-loop-helix (bHLH) gene, mammalian homolog of the achete-scute complex (Mash-1), which is uniformly expressed by PNECs by E15 and without which (as observed in Mash-1 knockout mice) PNECs fail to develop.[101,102] Differentiation is also influenced by oxygenation status, with hypoxia (5% in fetal mouse lung organ culture) enhancing branching morphogenesis and resulting in hyperplasia, but reducing both overall PNEC numbers as well as Mash-1 expression. Interestingly, the effects of hypoxia on the airway epithelium appears to be both PNEC/NEB-specific and temporal in nature, as hypoxia did not change the expression or frequency of differentiating Clara cells, and did not impact on PNECs/NEBs in samples taken subsequent to E15.[102] Downstream of Mash-1, the transcription factor Prox-1 correlates with, but is not required for, PNEC differentiation and maturation.[110] Deficiency of another bHLH transcription factor, NeuroD, results in an increase in NEB number concomitant with a decrease in solitary PNEC number.[103]

Ontogenetic studies of aortic arch development across vertebrates (mammals, amphibians, and fish) have pointed to a common phylogenetic origin of O_2 sensing cells and their locations (Figure 11; for review, see[92]). It has been suggested that the change from a diffuse system of isolated chemosensory cells (e.g., NECs of fish) to a more focal chemosensory system (e.g., the carotid body of mammals) may have mirrored the transition from water-breathing to air-breathing.[104] Chemosensation in fish, for example, is performed by innervated single neuroendocrine cells (NECs), which may monitor both environmental and physiologic changes in gas composition due to their placement in gill arch filaments and lamellae.[92] It is interesting to speculate whether NEBs, which display both diffuse and focal characteristics, and which are proposed to be of particular importance in the perinatal period prior to carotid body maturation, might represent an intermediary phase.

PNEC/NEB Innervation and Its Development

While earlier studies investigating NEB innervation focussed on the rabbit[105–107] and rat,[69,108,109] more recent efforts have increasingly concentrated on the mouse,[68] due

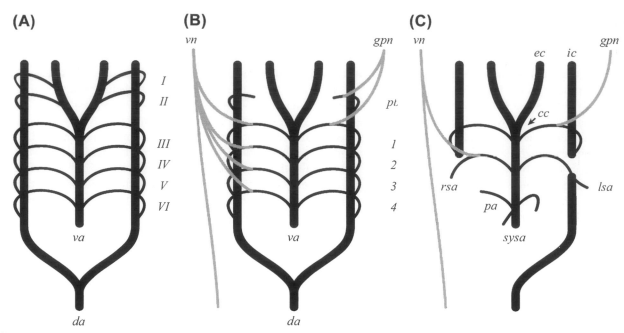

FIGURE 11   Phylogenetic development of vertebrate (teleost and mammal) aortic arches showing innervation (in blue; bilateral, but shown unilaterally in figure) by the glossopharyngeal (*gpn*) and vagus (*vn*) nerves (*taken with permission from*[92] *as adapted from*[104] *and*[292]). Panel A: common embryonic structure consisting of six aortic arches (I-IV, corresponding to named structures and arches 1–4 in Panels B and C) connecting the ventral (*va*) and paired dorsal (*pa*) aortae. Panel B: in teleosts there is loss of the first, and modification of the second into an innervated (*gpn*) non-respiratory pseudobranch (*pb*), aortic arch. The aortic arches III–VI correspond to mature gill arches 1–4, innervated by the *gpn* (1) and *vn* (1–4). Panel C: in mammals the ventral aorta becomes the common carotid artery which gives rise to the internal (ic) and external (ec) carotid arteries.

to the usefulness of mutant murine models in both structural and functional studies. These studies have shown three dominant NEB innervation subtypes: a sensory, vagal component; a DRG-derived component[110]; and an intrinsic, intraepithelial, nitrergic (nNOS-positive) component (Figure 12, Table 1).[68,111] Further heterogeneity exists within each of these populations, with respect to the expression of markers. For example, the vagal afferent component is divided into two subpopulations: nerve fibres staining positive for VGLUT 1 (predominant in mice) and VGLUT 2 (predominant in rats), Na+/K+-ATPase α3, and calbindin D-28 (CB), which ramify between and form basket-like structures surrounding NEBs, and nerve fibres staining positive for P2X$_2$ and P2X$_3$ purinergic ATP receptors.[68,112] These markers have been associated with mechanoreception,[68,69] with Na+/K+-ATPase α3 being potentially associated with electrophysiologically characterized SARs.[74,113] Both the VGLUT/CB positive (innervating ~ 20% of NEBs) and the P2X-positive (innervating ~ 24% of NEBs) populations, which are myelinated in the adult but lose myelination immediately approaching the NEB, display variable expression of vesicular acetylcholine transporter (VAChT)[68]; the developmental profile is unknown. Further, both the VGLUT- and P2X-positive populations express the mechanosensitive, inhibitory K$_{2P}$ channel, TRAAK,[112] which suggests a role in termination or adaptation of mechanoreceptor action potential

firing during lung inflation. The expression of markers (VGLUT) associated with excitatory glutamatergic signaling may also indicate dual afferent/efferent or afferent/regulatory function in these vagal nerves.[68] While rat vagal fibres have been traced to the nodose ganglion, the jugular-nodose complex of the mouse has prevented their origins from being specifically defined to date.[68] Any individual NEB may be contacted by any combination of the nerve populations described earlier, supporting the hypothesis that NEBs are a heterogeneous population of multimodal sensory receptors.

The development of the innervation of PNECs/NEBs has been investigated in fetal and neonatal lungs of rabbits using multi-label immunohistochemistry employing the pan-neural marker SV2, 5-HT (a marker for PNEC/NEB cells), and smooth muscle actin (SMA) or cytokeratin for anatomical (ASM and airway epithelium, respectively) landmarks.[98] Single and small clusters of PNECs first appear with variable innervation at E18, with a significant increase in the number of PNECs and innervated, NEB-like corpuscles at E21 (Figure 13).[98] The overall number of PNECs/NEBs, as well as the density of mucosal, submucosal, and intracorpuscular pulmonary innervation, increased throughout gestation and development, peaking at postnatal day 2 (P2), at which time the majority of rabbit PNECs/NEBs possessed extensive neural contacts, in keeping with their multifunctional role in the developing lung and

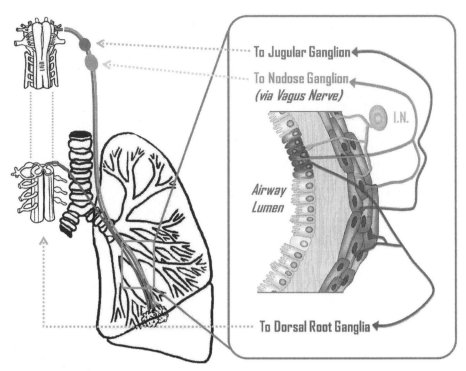

FIGURE 12 Schematic diagram illustrating the innervation of the pulmonary airways by heterogeneous populations of extrinsic neurons of the vagi (blue, to jugular ganglion; green, to nodose ganglion) and dorsal root ganglia (purple). Intrinsic neurons (I.N.; orange), in turn innervated by extrinsic neurons (not shown), are also present. Both the airway smooth muscle and epithelial cells, such as the neuroepithelial bodies (shown as non-ciliated cells in green (Clara-like cells) and purple (neuroendocrine cells)), receive innervation from both intrinsic and extrinsic sources. (This figure is reproduced in color in the color plate section.)

during neonatal adaptation (Figure 13).[98] Though conclusive evidence as to whether PNEC/NEB differentiation is dependent on their innervation does not yet exist, the tendency for intramucosal nerve endings to selectively contact PNECs/NEBs suggests that PNECs/NEBs may guide their innervation through the release of neurotrophic factors.[98] In week-old hamsters, the total volume of the NEBs is approximately equivalent to that of the carotid body and parathyroid glands, suggesting an important regulatory role,[114] and PNEC/NEB density is at a peak during the perinatal period in most species.[98,115,116]

Ultrastructural studies have shown that PNECs and NEBs already become innervated in fetal life,[99,117,118] and nerve terminals with synaptic contacts have been described at the base of the NEC in infant bronchial epithelium.[119] In fetal human lungs at mid-gestation, cholinergic axon terminals were shown deep within NEBs.[118] Some terminals exhibited vesicle profiles indicative of adrenergic fibres and formed gap junctions with adjacent cells within the NEB. However, the majority of axons are sensory, as demonstrated by a loss of NEB innervation after unilateral vagotomy.[120]

As in mammals, the gills and NECs of fish receive extrinsic (e.g., vagal) and intrinsic innervation[121]; this innervation can be serotonergic, nitrergic, and/or catecholaminergic,[122] with different NECs receiving innervation from different compliments of nerves.[92] In rats, labeling neurons in the nodose ganglia with DiI enabled tracing of sensory afferents to NEBs.[123] These nerves do not contain the sensory neuropeptide CGRP, in contrast to the afferents that supply the C-fibre endings in the epithelium.

Proposed Function(s) and Pathobiology of PNECs/NEBs

Solitary PNECs can be flask-shaped, reaching from the basement membrane to the airway lumen ("open" type), or elongated, with lateral cytoplasmic processes extending along the basement membrane that may not contact the airway lumen ("closed" type); select PNECs exhibit both types of processes.[95] Similarly, both "open"- and "closed"-type NECs have been reported in the gills of fish.[124] It is speculated that PNEC morphology may reflect specific functions e.g., gas (O_2, CO_2) or volatile substance detection vs. mechanoreception of the flask-shaped and lateral PNEC, respectively.[95,125] PNEC/NEB cells show wide species variation in their profile of bioactive amines and peptides, namely 5-HT, CGRP, chromogranin A, and bombesin.[126] The NECs of fish gills display a similar, albeit heterogeneous (e.g., variably expressing 5-HT and the purinergic $P2X_3$ receptor) profile,[121,127] with 5-HT the dominant neurotransmitter in these cells.[128]

Of interest are recent observations of volatile-sensing properties of select PNECs in the epithelium of human trachea and bronchi, but apparently not in other species.[129] Different olfactory receptors have been identified at the gene and protein level in PNECs in primary cell culture of airway epithelium. The expression of olfactory receptors in some, but not all, PNECs, and their absence from NEB cells, supports the concept of heterogeneity in the functional PNEC/NEB cell subtypes. Hence, the capacity of PNECs to respond to volatile agents (e.g., perfumes,

TABLE 1 Summary of the extrinsic (vagal and dorsal root ganglionic) and intrinsic innervation of murine NEBs, including known neural markers, neurotransmitters, and receptors, and putative functions. Notable interspecies differences are included in italics below, subsequent to a legend of used abbreviations. (*Reproduced with permission from*[88]).

Pulmonary Location of Nerve Terminals	Neural Markers (NM), Neurotransmitters (NT), and Neurotransmitter Receptors (NTR)	Proposed Function
Vagal[68,71,293]		
Arborizations within and surrounding NEBs[68,71]	**NM** • [1]TRAAK[112] • [2]VAChT[68] • VGLUT1[68,71] • VGLUT2[68,71,294] **NTR** • [3]Either P_xX_3[108,109] (NTR) or Na/K-ATPase (α3 subunit)[71,109,112,293] **Other** • CB[68,71,108,293]	Sensory[68,71] • Chemoreceptive • Mechanoreceptive
[4]Arborizations within NEBs[68]	**NM** • [1]TRAAK[112] • [1]VAChT[68] **NTR** • $P2X_2$[68] • $P2X_3$[68]	Sensory[68,71] • Chemoreceptive • Mechanoreceptive
Subepithelial plexus at base of NEB; synapses on intrinsic nitrergic fibres (V)[68]	**NT** • CGRP[68] • ± SP[68] These fibres can be delicate (smooth) or varicose	Sensory[68,71] • Chemoreceptive • Mechanoreceptive
DRG (T1-T6)[68,71]		
Nerve plexus at base of NEB (subepithelial)[68,71,293]	**NT** • CGRP[68,71,293] • ± SP[68,71,293] (In the mouse, DRG-originating fibres may/may not express SP; all other characteristics remain unchanged) In the rat, this population expresses: **NM** • VGLUT2[71] **NTR** • TRPV1[71]	Sensory • Mechanoreceptive
Intrinsic to alveolar interstitium and airway ganglia[68,71,293]		
Mouse: Subepithelial (base of NEB); Rat: within NEB[68,71]	• nNOS[68,71] and/or • VIP[68,71] **(NT)**	

Legend: CB = Calbindin D-28k, CGRP = Calcitonin Gene-Related Peptide, DRG = Dorsal Root Ganglion, ir = Immunoreactive, nNOS = Neuronal Nitric Oxide Synthase, P2X = Purinergic ATP receptors, SP = Substance P, TRAAK = TWIK-related, arachidonic-acid-stimulated K+ ion channel (K2Psubfamily), TRPV1 = Transient Receptor Potential Channel (Vanilloid Family, Type 1), VAChT = Vesicular Acetylcholine Transporter, VGLUT = Vesicular Glutamate Transporters, VIP = Vasoactive Intestinal Peptide.
[1]*Only mice were tested.*
[2]*VAChT present in select VGLUT/CB-ir fibres in the mouse*[68]
[3]*In rat, P_xX_3*[108,109] *or Na/K-ATPase (α3 subunit) can be expressed on VGLUT/CB-ir fibres.*[71,109,112,293] *In mouse, Na/K-ATPase (α3 subunit) seems to be expressed by TRAAK-ir fibres.*
[4]*This category pertains to mice only.*

industrial solvents, pollutants) expands the repertoire of external stimuli acting on these polymodal airway chemosensors[129] (Figure 14).

Intraepithelial NEBs are thought to represent hypoxia-sensitive chemoreceptors,[105,130–132] as an O_2-sensitive K^+ channel coupled to an O_2-sensing protein has been demonstrated in their cell membranes at the luminal surface.[131] Similarly, K^+ current through K_B channels has been shown to dominate NEC responsiveness to hypoxia in zebrafish.[133] This critical O_2 sensor has been identified to be an NADPH oxidase complex

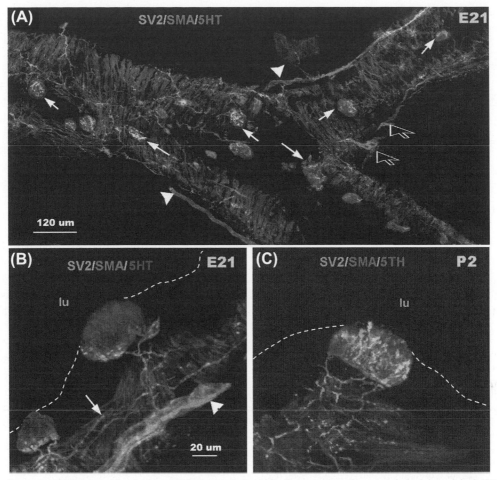

FIGURE 13   Development of NEB innervation in the rabbit before and after birth. Pan-neural marker synaptic vesicle protein 2 (SV2), smooth muscle marker smooth muscle actin (SMA), and neuroendocrine marker serotonin (5-HT) indicate ASM, pulmonary innervation, and NEB cells, respectively. The airway lumen is indicated by "lu." Panel A shows a relatively low-magnification cross-section of a large AW, with both neural and neuroendocrine (e.g., arrows indicating NEBs) populations shown. Panel B shows two closely apposed, innervated NEBs at high magnification at E21, the innervating fibres of which seem to be continuous (arrow) and derived from a single, large nerve trunk (arrowhead) in areas, suggesting a parallel connection between these NEBs. In contrast with Panel B, Panel C shows the increased density of innervation (especially intra-NEB arborization) of a NEB at P2. (This figure is reproduced in color in the color plate section.) *(Adapted with permission from[98].)*

in mice, without which, as shown in oxidase-deficient mice, responses to hypoxia fail to occur in vitro[134] and in vivo.[135] Further, exposure to hypoxia evokes action potentials and reversibly reduces outward K+ current in monkey NEBs.[136] In an ex vivo rabbit lung slice preparation (neonatal), exposure to hypoxia (15–20 mmHg PO_2) elicited an almost 50% (~47%) decrease in the whole cell, slowly inactivating current in Kv 3.4 and Kv 4.3-expressing NEB cells.[132] Of note, nicotine also supressed this A-type K+ current in neonatal rabbit NEBs, suggesting a potential mechanism or explanation for the correlation between early nicotine exposure and subsequent pulmonary pathology.[132] The evidence for a potential role for NEBs as airway O_2 sensors has been recently reviewed.[88,95,137]

Early studies suggested that NEBs in neonatal rabbit lungs are sensitive to both hypoxia and hypercapnia[105]; however, these findings could not be replicated.[138] Subsequent investigations have found that hypercapnia/acidosis has significant effects on the secretory response of NEB cells and on the proliferation of NEB cell-derived tumour lines.[139] Recent studies provide direct evidence that CO_2/[H+] stimulates 5-HT release from intact NEB cells in neonatal hamster lung and in a NEB cell-derived tumour cell line, used as a representative cell model (E. Cutz et al., unpublished observations). Exposure to combined hypoxia and hypercapnia yields an augmentation of the response, confirming polymodal sensing properties of airway NEB sensors. Further, CO_2/[H+]-evoked 5-HT secretion requires hydrolysis by carbonic anhydrase (CA), and voltage-gated entry of extracellular Ca^{2+}. Expression of mRNA for several CA isozymes was also found, particularly CA II (at both the mRNA and protein level), and CA II siRNA-mediated knockdown suppressed CO_2/[H+]-induced release of 5-HT, confirming the critical role of CA in mediating CO_2/[H+] effects in these cells (E. Cutz et al., unpublished observations). Additional evidence for a polymodal chemosensitive role for NECs comes from experiments demonstrating

Airway Lumen

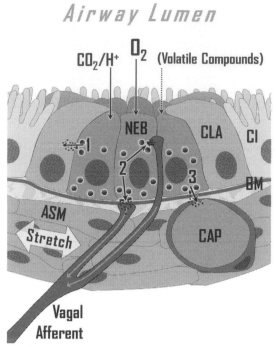

FIGURE 14   Schematic of a NEB showing its location within the airway epithelium (CLA: Clara-like cell, CI: Ciliated cell, BM: basement membrane, ASM: airway smooth muscle, CAP: capillary), innervation by a vagal afferent nerve fibre, and the complement of stimuli to which PNECs and NEBs are known to respond: hypoxia, hypercarbia/acidosis, mechanical stretch, and (in the case of solitary PNECs only, not NEBs) volatile substances. Activation by a stimulus leads to degranulation of PNEC/NEB cell dense core vesicles (containing, for example, 5-HT or ATP; shown as black circle with white edge), and one of three types of signaling: (1) paracrine (i.e., direct interaction with neighbouring cells, such as the illustrated Clara-like cell); (2) neural (e.g., with vagal afferents contacting the base of the NEB or ramifying between NEB cells; or (3) endocrine (i.e., via release of bioactive substances into the circulation). (This figure is reproduced in color in the color plate section.)

inhibition of fish NEC K_B current,[140,141] as well increased intracellular Ca^{2+} ($[Ca^{2+}]_i$),[142] in response to hypercapnia. Taken together, these studies show that NEB cells belong to a specialized group of excitable cells (analogous to carotid body glomus cells) representing polymodal airway sensors monitoring the gas composition (PO_2/PCO_2/pH/volatile substances) of the airway lumen.

Further support for a functional role for (P)NEC O_2 sensing comes from studies in zebrafish, which exhibit mature hypoxic responses only after extrinsic innervation of their NECs.[143] Application of the 5-HT_2 receptor antagonist, ketanserin, abolishes the hypoxic ventilatory response early in zebrafish development (by 10 days postfertilization), implicating 5-HT in hypoxia signaling.[144] Chronic exposure to hypoxia induces hypertrophy and hyperplasia of serotonergic and non-serotonergic NECs of the zebrafish, respectively.[133]

As alluded to earlier in the discussion of SMARs, evidence suggesting a mechano-sensitive role for NEBs is based primarily on recent morphological studies.[69–71,145] However, earlier in vitro studies[125] demonstrated release of 5-HT via

mechano-sensitive channels in cultures of PNECs/NEBs isolated from rabbit fetal lung subjected to cyclic stretch simulating fetal breathing movements. This is consistent with the important role of mechanical forces in lung development, which may be partially mediated by PNECs/NEBs and their amine and peptide products.[146] Recent studies have demonstrated the release of ATP from NEB cells via a $[Ca^{2+}]_i$-dependant mechanism in the mouse lung slice preparation using a hypo-osmotic stimulus.[145,147] It is believed that the ATP signal generated by NEBs in the presence of stretch is received and mediated via $P2X_{2/3}$ ATP receptors on NEB innervation (discussed in more detail in following text).[68] In mice, myelinated NEB innervation of vagal origin (fibres and terminals) also expresses the intrinsically mechanosensitive, two-pore K^+ channel (K_{2P}) TWIK-related arachidonic-acid-stimulated K^+ channel, TRAAK,[112,148] further suggesting a role in mechanosensation.

An additional role proposed for PNECs and NEBs involves the regulation of epithelial cell growth (including development, where cytoplasmic 5-HT release in response to simulated fetal breathing movements may impact on lung fluid reabsorption[149] as well as epithelial and mesenchymal cell growth[150]) and regeneration through a paracrine mechanism, whereby bioactive NEB peptides are released into the surrounding epithelium and the lamina propria.[95,151,152] It has been further suggested that PNECs and NEBs form a related stem cell niche[151,153]; however, conflicting evidence now exists as to the role for PNECs in Clara cell proliferation subsequent to naphthalene-induced lung injury.[91,151,153] PNECs, as well as seemingly PNEC-derived Clara and ciliated cells (in contrast with no shared lineage in normal development), proliferate after naphthalene-induced lung injury. However, PNECs may be redundant in Clara cell regeneration in this particular model of lung injury, as Clara cell regeneration was maintained even after prior PNEC ablation.[91] Interestingly, PNECs seem to be unable to regenerate from any source subsequent to ablation.[91]

Though the nature of the PNEC and Clara-like cell relationship is poorly understood, select interactions have been demonstrated. For example, in an ex vivo mouse lung slice model, application of extracellular $[K^+]_o$ caused a rapid, concentration-dependant increase in intracellular calcium ($[Ca^{2+}]_i$) in NEB cells, followed by a similar increase in the adjacent Clara-like cells.[86] This interaction was shown to be mediated via ATP released from NEB cells, acting on P2Y2 receptors in Clara-like cells.[154]

Further evidence for a unique regulatory role for NEBs comes from recent work using RT-PCR as well as immunocytochemical staining to demonstrate exclusive NEB expression of the extracellular G-protein coupled CaSR (GPRC2A) in the pulmonary epithelium beginning at postnatal day 14 (mice[155]). The CaSR has a role in the regulation of proliferation and specification of adult stem cells, as well as in developmental cell growth and differentiation.[156]

Previously CaSR expression was not found during the terminal saccular or postnatal periods in the mouse lung.[155] These earlier negative findings can be attributed to indiscriminate, rather than NEB-targeted, sampling of lung tissue; given the relative representation and density of NEBs in the whole lung, such sampling could have masked the expression of CaSR by means of dilution.[155] In contrast to these studies in mice, suggesting a postnatal expression profile of CaSR localized to NEBs, immunohistochemical studies in rats employing prior retrograde labeling of vagal afferents, including sensory pulmonary afferents, showed CaSR expression both in the cell bodies of bronchopulmonary vagal afferent fibres, as well as the trachea and lung parenchyma.[157] Potential reasons for this discrepancy and its implications remain to be investigated.

A functional role for CaSR in NEBs, as well as vagal C-fibres, has also been suggested. Increases in $[Ca^{2+}]_o$ caused a selective increase in NEB $[Ca^{2+}]_i$; interestingly, extracellular ATP-mediated activation (measured by increased $[Ca^{2+}]_i$) of Clara and Clara-like cells exclusively results in similar activation ($[Ca^{2+}]_i$) of NEBs in the presence of a calcimimetic agent (NPS-R568); this suggests a CaSR-dependant interaction between these cell types.[155] Further, blockade of the TRPC family of channels inhibited this interaction, suggesting a modulatory role for TRPC channels, including the TRPC5 channel implicated in the osmo- and mechano-sensation of NEBs.[147,155] In regard to C-fibres, both strong activation and blockade of CaSR modulate the capcaisin-, but not ATP- or 5-HT-, evoked whole-cell inward current.[157] Although CaSR is expressed by peripheral neurons (e.g., superior cervical ganglion sympathetic fibres,[158] C-fibres[157]) during the peri- and postnatal periods, to the best of our knowledge, the potential expression of CaSR by NEB-specific neurons has not been investigated.

The physiological role of NECs and NEBs has begun to emerge, although the specific role of the innervation is not yet well understood. It has been proposed that the nerve endings at the base of the NEC subserve an axon reflex, presumably in the NEB itself and possibly to deeper tissues such as ASM.[117] There may also be local reflex connections through peripheral ganglia. Hypoxia detected by the O_2 sensor in the NEC is presumed to release mediators that stimulate vagal afferents, but no central nervous reflexes have been identified. Whether they exert only intraganglionic effects remains to be shown. Recordings from single afferent fibres from NEB have not been made, and previous studies on the effect of hypoxia on vagal afferents from C-fibres, RARs, and SARs have been negative.[159] While advances in microscopic techniques will surely continue to shed light on the morphological basis for many of the suggested functions of NEC innervation, coupling of such visualizations with functional data is crucial to fully elucidate the role of NECs and NEBs in vivo.

Studies of a variety of neonatal/pediatric and adult pulmonary disease conditions have revealed abnormalities in the number and distribution of PNECs/NEBs, including hypo- and hyper-plasia. These changes have been described in several perinatal conditions, including pulmonary hypoplasia,[95] bronchopulmonary dysplasia (BPD),[95,160] congenital central hypoventilation syndrome,[95] sudden infant death syndrome,[95,161] and neuroendocrine cell hyperplasia of infancy.[95,162,163] In the adult, conditions include the preneoplastic diffuse idiopathic PNEC hyperplasia,[164,165] the presumed origin of small cell lung cancer,[91] for which a causal link has been established via the PNEC inactivation of multiple tumor suppressors (p53, Rb, Pten).

Transgenic mouse models may be useful tools to further define the pathobiology of PNECs/NEBs, and their role in various pulmonary disorders. Loss of function models include Mash-1 knockout mice that fail to develop PNECs/NEBs, but their usefulness is limited as homozygous mice die shortly after birth.[101] Interestingly, heterozygous mice (Mash-1[+/-]) are viable, show a significantly reduced number of PNECs/NEBs, and exhibit an abnormal pattern of breathing.[166,167] The previously noted NADPH oxidase deficiency model (gp91 phox[-/-] mice), representative of a loss of function of the NEB cell-based O_2 sensor protein, shows abrogated O_2 sensing properties both in vitro and in vivo, while NEB cell structure and numbers remain intact.[135] A possible gain of function model is the prolyl-hydroxylase (PHD-1) knockout mouse model, which shows striking hyperplasia of NEBs that persists into adulthood.[168] PHDs regulate the stability of hypoxia-inducible factors (HIFs) in an O_2-dependent manner, and function as an intrinsic O_2 sensor. The PHD-1 knockout mouse model appears to be a unique model of congenital NEB hyperplasia. The effect of PHD-1 deficiency on NEB O_2 sensing function has not yet been investigated. Abnormalities in PNEC/NEB number and distribution have been described, with significantly reduced ASM mass and airway innervation, in a cystic fibrosis transmembrane regulator knockout (Cftr[-/-]) mouse model.[169] Although the role of Cftr in chloride transport and the pathogenesis of cystic fibrosis (CF) is well known, its involvement in PNEC/NEB development or function is presently unknown.

ONTOGENY AND REFLEX CONTROL OF AIRWAY SMOOTH MUSCLE: FUNCTIONAL CONSEQUENCES

The functional consequences of ASM activation are typically considered in relation to the effects of airway narrowing on airway resistance.[170] Increased resistance leading to increased work of breathing and eventual respiratory failure are two features associated with inflammatory neonatal lung diseases, such as BPD.[171,172] BPD consists of a range of pathophysiologic changes including non-specific airway hyper-responsiveness.[171–173] That ASM innervation is involved in these events can be inferred from the action of muscarinic antagonists or β2 agonists both of which cause bronchodilation in BPD.[174-176] Indeed, the action of muscarinic antagonists suggests that the activity of vagal innervation to ASM, which relies on excitatory cholinergic pathways, is significantly increased.[175] Although the pathophysiologic approach to understanding ASM function illustrated above has shed much light on the ontogeny of airway function, considerable data suggest that the ability to regulate the contractile function of ASM is a normal characteristic of both neonatal and prenatal life (see following text). Furthermore, elucidation of mechanisms involved in the signal transduction and the control of intracellular contractile machinery of smooth muscle has begun to improve understanding of the ontogeny of ASM function. The following section highlights the functional consequences of ASM activation on the pre- and postnatal lung, and the insight provided by recent advances in muscarinic receptor biology and signal transduction mechanisms controlling the molecular machinery responsible for smooth muscle shortening.

Prenatal Airway Smooth Muscle Function

Some evidence was present in the 1930s for contractile function in human fetal airways.[177] This observation, however, was largely forgotten until a series of elegant studies of fetal airway narrowing.[42,56,57,178-183] These began with studies showing contractile function in airways from first trimester human airways studied in vitro[57,184] and were followed by a series of studies on fetal airways from pigs and humans.[42,56,179-183] These reports highlighted both the ability of airways to narrow in response to cholinergic agonists, and the loss of airway contractile function in the presence of the non-specific muscarinic receptor antagonist, atropine. In the fetal pig during the pseudoglandular phase, the airways exhibit contractile activity[42]; furthermore, airway narrowing is observed during exposure to cholinergic agonists and to field stimulation, both of which are blocked by atropine. The response to field stimulation was also eliminated by tetrodotoxin, suggesting that the functional cholinergic innervation appeared very early in the developmental sequence of events related to the lung.[42]

Perhaps the greatest challenge of studies describing fetal bronchial motility is the interpretation of the physiologic role of the "precocial" development of vagal innervation to ASM. The peristalsis-like waves of activity observed in these preparations were reminiscent of those reported for the gut, but without a functional correlate. It was speculated that contractile activity may indeed be important for normal lung development by providing local stretch of the airway wall.[42] Stretch is both a powerful and important stimulus for lung growth[185-188] (see Chapter 8) and contractions of the fetal airways may be part of a coordinated series of events that lead to appropriate alveolar development.[42,185-188] As outlined above with respect to the innervation of the pre- and postnatal lung, there is abundant evidence showing

airway innervation as well as considerable ASM.[36,38,42,50,181] Vagal innervation to the lung is also thought to be critically important for the successful transition from fetal life to air breathing at birth, as it has been shown that the successful transition to air breathing at birth is dependent on intact vagal innervation to the lung[189] and that loss of innervation in utero leads to altered surfactant function.[190] These studies illustrate other potentially critical physiologic roles for prenatal vagal innervation.

Postnatal Airway Smooth Muscle Function

Studies of the postnatal function of ASM predate those described above for fetal airways and, as such, they were somewhat preoccupied with demonstrating the presence and impact of innervation to ASM in vivo[191-196] or in vitro.[179-181,197-209] Direct activation of vagal efferent nerves to ASM in the newborn narrows and stiffens the airways.[183,193,210-212] Considering the increased compliance of neonatal airways[182,206,213,214] in vitro, it is surprising that the change in lung resistance in vivo is not greater than that seen in the adult (for review, see[215]). The physiologic wisdom of this dichotomy is reflected by both in vitro and in vivo studies. In the former, ASM contraction is sufficient to retard the normal collapse or narrowing of airways subjected to collapsing pressure.[183,210,212] The responses to maximal activation of vagal efferent fibres or to injected cholinergic agonists in vivo are modest in the newborn.[193,206,216] This may reflect enhanced parenchymal interdependence in the newborn,[217-219] which could provide some level of protection against airway closure compared to the adult.

Reflex Control of Airway Smooth Muscle in the Newborn Lung

The ASM of the newborn is activated in response to various stimuli,[170] including chemoreceptor[220] or irritant stimuli.[221-223] Reflex responses rely upon the function of peripheral receptors, both within and outside the lung, integration of such feedback by medullary centres and neurotransmission from vagal nuclei and ganglia to the airways.[215] There is currently little insight into the impact of such integrated behavior in the fetal lung.

In the newborn, studies performed in vivo show that chemoreceptor stimulation by inhalation of CO_2 results in a modest reflex bronchoconstriction,[192] presumably reflecting the increased drive to breathe associated with chemoreflex stimulation.[224] Indeed, chemoreflex stimulation appears to cause a parallel output to both respiratory skeletal and smooth muscles.[224] Unlike the adult,[225,226] hypoxia does not generally cause an increase in bronchomotor tone in the newborn.[192,227] This probably reflects the fact that in the newborn hypoxia can cause a decrease in respiratory drive due to a concomitant decrease in metabolic rate.[217,228-229]

Therefore, the lack of bronchomotor effects associated with hypoxia in the newborn[192,227] is consistent with the parallel output hypothesis for respiratory skeletal muscle and ASM.[224]

The reflex responses to irritant C-fibre stimuli (e.g., capsaicin, lactic acid) are suggestive of protective airway responses aimed at limiting airway injury. The capsaicin response relies on the use of a vanilloid compound to activate the capsaicin receptor transient receptor potential vanilloid 1 (TRPV1)[49,230]; lactic acid as a stimulus is more physiological,[49,222,223,231] especially in the newborn.[222] The reflex narrowing and stiffening of tracheobronchial airways[221–223] causes inhaled irritants to be impacted at large airway branch points and protects the airways from dynamic collapse during rapid shallow breathing. In the newborn lamb, C-fibre afferents also alter the pattern of breathing and laryngeal muscles during irritant stimuli such as capsaicin or pulmonary edema.[232,233] Thus, the neonatal response to lung irritation relies on the integrated efferent control of both upper and lower airway muscles, which is initiated by C-fibre afferents. Whether such reflex actions are present in the fetus remains to be seen. One might speculate that such control would be present, since robust reflex responses occur in neonatal sheep exposed to C-fibre/TRPV1 stimulants such as capsaicin or acid.[234–236] The control of laryngeal aperture is an active component of fetal breathing that influences the egress of pulmonary fetal liquid (see Chapter 8 [Wallace et al.]).

MUSCARINIC RECEPTORS IN THE LUNG

Muscarinic receptors are members of the heterotrimeric G protein-coupled receptor (GPCR) family for which five muscarinic receptor subtypes have been cloned.[237–239] Even-numbered muscarinic GPCRs (M2 and M4) are coupled to Gi/Go and exert their action through inhibitory effects on adenylate cyclase, whereas odd-numbered muscarinic GPCRs (M1, M3, and M5) are coupled to Gq/G11.[240,241] In ASM cells, the latter act through stimulatory effects on phospholipase C and inositol phosphate generation[215,239] to cause airway narrowing. Desensitization of GPCR-mediated effects is due to G protein receptor kinases (GRK) and protein kinase C.[242]

It is widely recognized that muscarinic receptors play a critical role in cholinergic responses of the heart, lung, and CNS.[215,239,243–246] Muscarinic receptor antagonists provide some insight into receptor subtype distribution, but they lack a high degree of selectivity and therefore possess unwanted side effects.[215,239,247,248] This continues to hamper the ability to assign specific actions to specific muscarinic receptors or to identify specific interactions between muscarinic receptor subtypes. Mutant mice have been engineered to lack specific muscarinic receptor subtypes, so the question of the physiologic role of different receptor subtypes continues to receive significant attention,[239,243–246] mostly in the adult.

Muscarinic receptors are important in asthma, where an exaggerated, vagally mediated, reflex bronchoconstriction and airway hyper-responsiveness to muscarinic acetylcholine receptor (mAChR) agonists are present.[249–251] Furthermore, the magnitude of neural muscarinic activation to the lung plays a critical role in determining whether airway closure occurs (J. Fisher et al., unpublished data). These findings and others[251] highlight the potential for exaggerated vagal efferent activity in asthma and the importance of understanding the control of signal transduction in ASM. Based on the effects of "selective" pharmacological antagonists in vivo the cardiopulmonary actions of ACh in the newborn display some differences from the adult.[247] This, plus in vitro studies, has suggested differences in receptor subtype expression or action compared to the adult.[247,252]

Three of the five cloned muscarinic receptors have been classified as having cardiopulmonary actions.[253–256] The M1, M2, and M3 receptors have all been linked to the vagal efferent pathway to ASM,[215] and M4 has been found in lung tissue.[257–259] M1 receptors have been suggested to be located in the neural tissue[253,254] and airway sympathetic ganglia of some species.[253,254] M1 receptors have also been suggested to be present in parasympathetic ganglia of some species,[260,261] including humans,[254,255,262] but not of others.[253,263,264]

Prejunctional M2 receptors, presumed to be located on postganglionic vagal axons, are autoinhibitory and reduce the magnitude of vagally mediated bronchoconstriction in vivo.[253,261,265–272] Viral infection[273] and experimentally induced airway inflammation cause a downregulation of prejunctional M2 receptors that leads to exaggerated vagally-mediated bronchoconstriction.[274,275] Myocardial M2 receptors are thought to mediate the bradycardia associated with vagal stimulation.[239,246]

The M3 receptor is thought to mediate contraction of ASM and mucus secretion,[255,264,270,276] and it is the receptor subtype target for anti-cholinergic bronchodilators.[175,253–255,277] Use of these non-specific antagonists causes bronchodilation in normal subjects, asthmatics, COPD patients, and in infants with BPD.[175,253,254,277,278] The M3 receptor is also linked to vasodilation mediated by NO.[239] M4 receptors have been detected in rabbit lung, but their functional significance remains unknown.[255,257–259]

Muscarinic Receptors: Their Development and Pharmacological Antagonists

The developmental physiology of muscarinic receptors has relied on selective antagonists and suggests that M receptor differentiation is present but is developmentally regulated.[216,252] Typically such approaches compare the ability of several "selective" muscarinic antagonists to differentially inhibit airway versus cardiac events associated with vagal stimulation in vivo.[216,248] Indeed, it is only by comparing multiple antagonists that one can begin to assign function based on selective antagonists in vivo, especially with M3 and M1 antagonists.[261,263,265,266]

Physiologic studies of neonatal ASM suggest that muscarinic receptor subtypes are at least partially differentiated at birth.[216,252] Studies using semiselective muscarinic receptor antagonists suggest that M3 receptors in ASM are most likely responsible for airway narrowing.[216] However, M2 receptor responses are either absent or sufficiently weak that the autoinhibitory function of prejunctional M2 receptors on ACh release that is normally seen in the adult is not apparent.[216] This is supported by indirect studies of M2 receptor pharmacology.[216,252] M1 receptors do not appear to be functionally significant.[216,252] Much remains to be learned about the pre- and postnatal maturation of muscarinic receptor subtypes. For example, studies of cardiac tissue suggest that mRNA or protein for various muscarinic receptors are present in areas that were traditionally thought to be exclusively M2 receptor in function. Similar molecular studies have yet to be carried out on pre- or postnatal airways. M2 receptors influence the proliferative effects of transforming growth factor β1, TGF-β1, in the adult lung,[279] and this potential mechanism has yet to be explored in the developing lung, where TGF-β influences branching morphogenesis.[280,281] Interestingly, the use of mutant mice lacking functional M1, M2, or M3 receptors[240,282] appears to have been applied globally only to cilia beat frequency and epithelial development, where M1–M3 receptor subtypes appear to play no critical roles in epithelial development.[283] Emerging "biased signaling" pathways (see following text) for muscarinic receptors[284] may prove to be useful tools to examine the role of muscarinic receptor function in the developing lung.

Recent appreciation of the presence of a wider spectrum of GPCR signaling has led to a re-examination of both adrenergic[285] and muscarinic signaling pathways[284] for the lung. In addition to agonist-activated G protein-coupled signaling, it is now clear that intracellular signaling pathways are recruited via the arrestin family of proteins, initially characterized for their role in receptor internalization.[285–287] Thus, agonist activation of lung GPCRs has the potential to mobilize traditional signal transduction mechanisms, such as ASM contraction via M3 receptor coupling to mobilize intracellular calcium release,[241,271] as well as arrestin signaling, which potentially leads to proliferation of ASM.[285] Muscarinic receptor subtypes have been implicated in airway remodelling during inflammatory disease in the adult lung, where lack of M3 receptor function prevents the normal increased ASM and airway wall changes associated with chronic inflammation.[288] This paradigm of airway plasticity has yet to be applied to the preterm and term lung.

A robust, non-neural source of ACh signaling is now fully acknowledged in the lung.[289–291] The machinery for the production and release of ACh from airway epithelial cells represents a paracrine signaling mechanism, which may be linked to cell proliferation, inflammation, and lung pathologies such as cystic fibrosis.[289,290] This control system significantly extends the traditional efferent sources and actions of ACh signaling via vagal innervation to the lung, and it is highly likely that muscarinic receptors will be found to play a role in lung development.

ACKNOWLEDGMENTS

Supported by the Canadian Institutes of Health Research (CIHR: JT Fisher - MOP81211; E Cutz - MOP 15270), the Queen's University Spear Endowment Fund for Respiratory Research (JT Fisher), the Ontario Thoracic Society (JT Fisher), and NSERC PGS/MSFSS (NJ Domnik).

REFERENCES

1. Loosli CG, Hung K-S. Development of pulmonary innervation. In: Hodson WA, editor. *Development of the Lung*. New York: Marcel Dekker; 1977. pp. 269–309.
2. Sparrow MP, Weichselbaum M, Tollet J, McFawn PK, Fisher JT. Development of the Airway Innervation. In: Harding R, Pinkerton KE, Plopper CG, editors. *The Lung: Development, Aging and the Environment*. 1st ed. London: Elsevier Academic Press; 2004. pp. 33–53.
3. Cardoso WV, Lu J. Regulation of early lung morphogenesis: questions, facts and controversies. *Development* 2006;**133**:1611–24.
4. Ten Have-Opbroek AA. The development of the lung in mammals: an analysis of concepts and findings. *Am J Anat* 1981;**162**:201–19.
5. Riccardi D, Brennan SC, Chang W. The extracellular calcium-sensing receptor, CaSR, in fetal development. *Best Pract Res Clin Endocrinol Metab* 2013;**27**:443–53.
6. Spooner BS, Wessells NK. Mammalian lung development: interactions in primordium formation and bronchial morphogenesis. *J Exp Zool* 1970;**175**:445–54.
7. Langsdorf A, Radzikinas K, Kroten A, Jain S, Ai X. Neural crest cell origin and signals for intrinsic neurogenesis in the mammalian respiratory tract. *Am J Respir Cell Mol Biol* 2011;**44**:293–301.
8. Freem LJ, Escot S, Tannahill D, Druckenbrod NR, Thapar N, Burns AJ. The intrinsic innervation of the lung is derived from neural crest cells as shown by optical projection tomography in Wnt1-Cre;YFP reporter mice. *J Anat* 2010;**217**:651–64.
9. Burns AJ, Thapar N, Barlow AJ. Development of the neural crest-derived intrinsic innervation of the human lung. *Am J Respir Cell Mol Biol* 2008;**38**:269–75.
10. Burns AJ, Delalande JM. Neural crest cell origin for intrinsic ganglia of the developing chicken lung. *Dev Biol* 2005;**277**:63–79.
11. Dey RD, Hung K-S. Development of innervation in the lung. In: McDonald JA, editor. *Lung Growth and Development*. New York: Marcel Dekker; 1997. pp. 244–65.
12. Garcia-Suarez O, Perez-Pinera P, Laura R, Germana A, Esteban I, Cabo R, et al. TrkB is necessary for the normal development of the lung. *Respir Physiol Neurobiol* 2009;**167**:281–91.

13. Tollet J, Everett AW, Sparrow MP. Spatial and temporal distribution of nerves, ganglia, and smooth muscle during the early pseudoglandular stage of fetal mouse lung development. *Dev Dyn* 2001;**221**:48–60.

14. Young HM, Ciampoli D, Hsuan J, Canty AJ. Expression of Ret-, p75(NTR)-, Phox2a-, Phox2b-, and tyrosine hydroxylase-immunoreactivity by undifferentiated neural crest-derived cells and different classes of enteric neurons in the embryonic mouse gut. *Dev Dyn* 1999;**216**:137–52.

15. Chalazonitis A, Rothman TP, Chen J, Gershon MD. Age-dependent differences in the effects of GDNF and NT-3 on the development of neurons and glia from neural crest-derived precursors immunoselected from the fetal rat gut: expression of GFRalpha-1 in vitro and in vivo. *Dev Biol* 1998;**204**:385–406.

16. Thompson RJ, Doran JF, Jackson P, Dhillon AP, Rode J. PGP 9.5 – a new marker for vertebrate neurons and neuroendocrine cells. *Brain Res* 1983;**278**:224–8.

17. Jiang X, Rowitch DH, Soriano P, McMahon AP, Sucov HM. Fate of the mammalian cardiac neural crest. *Development* 2000;**127**:1607–16.

18. Nassenstein C, Taylor-Clark TE, Myers AC, Ru F, Nandigama R, Bettner W, Undem BJ. Phenotypic distinctions between neural crest and placodal derived vagal C-fibres in mouse lungs. *J Physiol* 2010;**588**:4769–83.

19. Tollet J, Everett AW, Sparrow MP. Development of neural tissue and airway smooth muscle in fetal mouse lung explants: a role for glial-derived neurotrophic factor in lung innervation. *Am J Respir Cell Mol Biol* 2002;**26**:420–9.

20. Sato H, Murphy P, Giles S, Bannigan J, Takayasu H, Puri P. Visualizing expression patterns of Shh and Foxf1 genes in the foregut and lung buds by optical projection tomography. *Pediatr Surg Int* 2008;**24**:3–11.

21. Sato H, Murphy P, Hajduk P, Takayasu H, Kitagawa H, Puri P. Sonic hedgehog gene expression in nitrofen induced hypoplastic lungs in mice. *Pediatr Surg Int* 2009;**25**:967–71.

22. Hajduk P, Murphy P, Puri P. Mesenchymal expression of Tbx4 gene is not altered in Adriamycin mouse model. *Pediatr Surg Int* 2010;**26**:407–11.

23. Hajduk P, Murphy P, Puri P. Fgf10 gene expression is delayed in the embryonic lung mesenchyme in the adriamycin mouse model. *Pediatr Surg Int* 2010;**26**:23–7.

24. Scott GD, Fryer AD, Jacoby DB. Quantifying nerve architecture in murine and human airways using three-dimensional computational mapping. *Am J Respir Cell Mol Biol* 2013;**48**:10–6.

25. Undem BJ, Chuaychoo M-G, Lee D, Weinreich A, Myers C, Kollarik M. Subtypes of vagal afferent C-fibres in guinea-pig lungs. *J Physiol* 2004;**556**:905–17.

26. Riccio MM, Kummer W, Biglari B, Myers AC, Undem BJ. Interganglionic segregation of distinct vagal afferent fibre phenotypes in guinea-pig airways. *J Physiol* 1996;**496**(Pt 2):521–30.

27. Fisher JT. The TRPV1 ion channel: implications for respiratory sensation and dyspnea. *Respir Physiol Neurobiol* 2009;**167**:45–52.

28. Baker CV, Schlosser G. The evolutionary origin of neural crest and placodes. *J Exp Zool B Mol Dev Evol* 2005;**304**:269–73.

29. Tharakan T, Kirchgessner AL, Baxi LV, Gershon MD. Appearance of neuropeptides and NADPH-diaphorase during development of the enteropancreatic innervation. *Brain Res Dev Brain Res* 1995;**84**:26–38.

30. Guembe L, Villaro AC. Histochemical demonstration of neuronal nitric oxide synthase during development of mouse respiratory tract. *Am J Respir Cell Mol Biol* 1999;**20**:342–51.

31. Oztay F, Brouns I, Pintelon I, Raab M, Neuhuber W, Timmermans JP, Adriaensen D. Neurotrophin-4 dependency of intraepithelial vagal sensory nerve terminals that selectively contact pulmonary NEBs in mice. *Histol Histopathol* 2010;**25**:975–84.

32. Coleridge HM, Coleridge JC. Impulse activity in afferent vagal C-fibres with endings in the intrapulmonary airways of dogs. *Respir Physiol* 1977;**29**:125–42.

33. Young HM, Hearn CJ, Farlie PG, Canty AJ, Thomas PQ, Newgreen DF. GDNF is a chemoattractant for enteric neural cells. *Dev Biol* 2001;**229**:503–16.

34. Durbec P, Marcos-Gutierrez CV, Kilkenny C, Grigoriou M, Wartiowaara K, Suvanto P, et al. GDNF signalling through the Ret receptor tyrosine kinase. *Nature* 1996;**381**:789–93.

35. Towers PR, Woolf AS, Hardman P. Glial cell line-derived neurotrophic factor stimulates ureteric bud outgrowth and enhances survival of ureteric bud cells in vitro. *Exp Nephrol* 1998;**6**:337–51.

36. Weichselbaum M, Sparrow MP. A confocal microscopic study of the formation of ganglia in the airways of fetal pig lung. *Am J Respir Cell Mol Biol* 1999;**21**:607–20.

37. Burri PH. Structural aspects of prenatal and postnatal development and growth of the lung. In: McDonald JA, editor. *Lung Growth and Development*. New York: Marcel Dekker; 1997. pp. 1–35.

38. Weichselbaum M, Everett AW, Sparrow MP. Mapping the innervation of the bronchial tree in fetal and postnatal pig lung using antibodies to PGP 9.5 and SV2. *Am J Respir Cell Mol Biol* 1996;**15**:703–10.

39. Sparrow MP, Weichselbaum M, McCray PB. Development of the innervation and airway smooth muscle in human fetal lung. *Am J Respir Cell Mol Biol* 1999;**20**:550–60.

40. Weichselbaum M. *The structure and distribution of the innervation of the developing lung: confocal miscroscope study*; 2001. PhD Thesis, University of Western Australia.

41. Lamb JP, Sparrow MP. Sensory innervation in the bronchial mucosa of the pig. *Int Tachykinin* 2000. 17-10-2000.

42. Sparrow MP, Warwick SP, Everett AW. Innervation and function of the distal airways in the developing bronchial tree of fetal pig lung. *Am J Respir Cell Mol Biol* 1995;**13**:518–25.

43. Finney BA, del Moral PM, Wilkinson WJ, Cayzac S, Cole M, Warburton D, et al. Regulation of mouse lung development by the extracellular calcium-sensing receptor, CaR. *J Physiol* 2008;**586**:6007–19.

44. Riccardi D, Finney BA, Wilkinson WJ, Kemp PJ. Novel regulatory aspects of the extracellular Ca2+-sensing receptor, CaR. *Pflugers Arch* 2009;**458**:1007–22.

45. Brennan SC, Finney BA, Lazarou M, Rosser AE, Scherf C, Adriaensen D, et al. Fetal calcium regulates branching morphogenesis in the developing human and mouse lung: involvement of voltage-gated calcium channels. *PLoS One* 2013;**8**:e80294.

46. Cadieux A, Springall DR, Mulderry PK, Rodrigo J, Ghatei MA, Terenghi G, et al. Occurrence, distribution and ontogeny of CGRP immunoreactivity in the rat lower respiratory tract: effect of capsaicin treatment and surgical denervations. *Neuroscience* 1986;**19**:605–27.

47. Butcher L. Acetylcholine histochemistry. In: ABaT. Hoekfeld, editor. *Handbook of Chemical Neuroanatomy*. Amsterdam: Elsevier; 1983. pp. 1–49.

48. Schemann M, Sann H, Schaaf C, Mader M. Identification of cholinergic neurons in enteric nervous system by antibodies against choline acetyltransferase. *Am J Physiol* 1993;**265**:G1005–9.

49. Lee LY, Pisarri TE. Afferent properties and reflex functions of bronchopulmonary C-fibres. *Respir Physiol* 2001;**125**:47–65.

50. Sparrow MP, Weichselbaum M. Structure and function of the adventitial and mucosal nerve plexuses of the bronchial tree in the developing lung. *Clin Exp Pharmacol Physiol* 1997;**24**:261–8.

51. Undem BJ, Myers AC. Autonomic ganglia. In: Barnes PJ, editor. *Autonomic Control of the Respiratory System*. United Kingdom: Harwood Academic Publications; 1997. pp. 87–118.

52. Ponte J, Purves MJ. Types of afferent nervous activity which may be measured in the vagus nerve of the sheep foetus. *J Physiol* 1973;**229**:51–76.

53. Agostoni E, Chinnock JE. De Burgh Daly M, Murray JG. Functional and histological studies of the vagus nerve and its branches to the heart, lungs and abdominal viscera in the cat. *J Physiol* 1957;**135**:182–205.

54. Wu H, Williams J, Nathans J. Morphologic diversity of cutaneous sensory afferents revealed by genetically directed sparse labeling. *Elife* 2012;**1**:e00181.

55. Warburton D, Zhao J, Berberich MA, Bernfield M. Molecular embryology of the lung: then, now, and in the future. *Am J Physiol* 1999;**276**:L697–704.

56. Sparrow MP, Warwick SP, Mitchell HW. Foetal airway motor tone in prenatal lung development of the pig. *Eur Respir J* 1994;**7**:1416–24.

57. McCray Jr PB. Spontaneous contractility of human fetal airway smooth muscle. *Am J Respir Cell Mol Biol* 1993;**8**:573–80.

58. Schittny JC, Miserocchi G, Sparrow MP. Spontaneous peristaltic airway contractions propel lung liquid through the bronchial tree of intact and fetal lung explants. *Am J Respir Cell Mol Biol* 2000;**23**:11–8.

59. Baluk P, Nadel JA, McDonald DM. Substance P-immunoreactive sensory axons in the rat respiratory tract: a quantitative study of their distribution and role in neurogenic inflammation. *J Comp Neurol* 1992;**319**:586–98.

60. Larsell O. The ganglia, plexus and nerve-terminations of the mammalian lung and pleura pulmonis. *J Comp Neurol* 1922;**35**:97–132.

61. Sparrow AK, Sparrow MP. The spatial relationship of CGRP nerves in the mucosal nerve plexus to the bronchial circulation in the pig lung. *Eur Resp J* 1996;**9**:290s.

62. Lundberg JM, Hokfelt T, Martling CR, Saria A, Cuello C. Substance P-immunoreactive sensory nerves in the lower respiratory tract of various mammals including man. *Cell Tissue Res* 1984;**235**:251–61.

63. Barnes PJ. Neurogenic inflammation in the airways. *Respir Physiol* 2001;**125**:145–54.

64. Larsell O, Dow RS. The innervation of the human lung. *Am J Anat* 1933;**52**:125–46.

65. Li ZS, Fox-Threlkeld JE, Furness JB. Innervation of intestinal arteries by axons with immunoreactivity for the vesicular acetylcholine transporter (VAChT). *J Anat* 1998;**192**:107–17.

66. Van GJ, Brouns I, Burnstock G, Timmermans JP, Adriaensen D. Quantification of neuroepithelial bodies and their innervation in fawn-hooded and Wistar rat lungs. *Am J Respir Cell Mol Biol* 2004;**30**:20–30.

67. Pintelon I, Brouns I, Van Genechten J, Scheuermann DW, Timmermans JP, Adriaensen D. Pulmonary expression of the vesicular acetylcholine transporter with special reference to neuroepithelial bodies [abstract]. *Auton Neurosci* 2003;**106**:47.

68. Brouns I, Oztay F, Pintelon I, De Proost I, Lembrechts R, Timmermans JP, et al. Neurochemical pattern of the complex innervation of neuroepithelial bodies in mouse lungs. *Histochem Cell Biol* 2009;**131**:55–74.

69. Brouns I, De Proost I, Pintelon I, Timmermans JP, Adriaensen D. Sensory receptors in the airways: neurochemical coding of smooth muscle-associated airway receptors and pulmonary neuroepithelial body innervation. *Auton Neurosci* 2006;**126–127**:307–19.

70. Adriaensen D, Brouns I, Pintelon I, De Proost I, Timmermans JP. Evidence for a role of neuroepithelial bodies as complex airway sensors: comparison with smooth muscle-associated airway receptors. *J Appl Physiol* 2006;**101**:960–70.

71. Brouns I, Pintelon I, De Proost I, Alewaters R, Timmermans JP, Adriaensen D. Neurochemical characterisation of sensory receptors in airway smooth muscle: comparison with pulmonary neuroepithelial bodies. *Histochem Cell Biol* 2006;**125**:351–67.

72. Adriaensen D, Brouns I, Pintelon I, De Proost I, Timmermans JP. Reply to Yu. *J Appl Physiol* 2007;**102**:1728.

73. Yu J, Lin SX, Zhang JW, Walker JF. Pulmonary nociceptors are potentially connected with neuroepithelial bodies. *Adv Exp Med Biol* 2006;**580**:301–6.

74. Yu J, Zhang J, Wang Y, Fan F, Yu A. Neuroepithelial bodies not connected to pulmonary slowly adapting stretch receptors. *Respir Physiol Neurobiol* 2004;**144**:1–14.

75. Yu J. Are neuroepithelial bodies a part of pulmonary slowly adapting receptors? *J Appl Physiol* 2007;**102**:1727.

76. Black JL. Innervation of airway smooth muscle. In: Barnes PJ, editor. *Autonomic Control of the Respiratory System*. United Kingdom: Harwood Academic Publications; 1997. pp. 185–200.

77. Rogers DF. Motor control of airway goblet cells and glands. *Respir Physiol* 2001;**125**:129–44.

78. Undem BJ, Kajekar R, Hunter DD, Myers AC. Neural integration and allergic disease. *J Allergy Clin Immunol* 2000;**106**:S213–20.

79. Rogers DF, Barnes PJ. Neural control of the airway vasculature. In: Barnes PJ, editor. *Autonomic Control of the Respiratory System*. United Kingdom: Harwood Academic Publications; 1997. pp. 185–200.

80. Barnes PJ. Neuromodulation in airways. In: Barnes PJ, editor. *Autonomic Control of the Respiratory System*. United Kingdom: Harwood Academic Press; 1997. pp. 139–84.

81. Barnes PJ. Airway neuropeptides. In: Busse WM, Holgate ST, editors. *Asthma and Rhinitis*. Oxford: Blackwell Scientific; 2000. pp. 891–908.

82. Ebina M, Yaegashi H, Takahashi T, Motomiya M, Tanemura M. Distribution of smooth muscles along the bronchial tree. A morphometric study of ordinary autopsy lungs. *Am Rev Respir Dis* 1990;**141**:1322–6.

83. Schelegle ES, Green JF. An overview of the anatomy and physiology of slowly adapting pulmonary stretch receptors. *Respir Physiol* 2001;**125**:17–31.

84. Honjin R. On the ganglia and nerves of the lower respiratory tract of the mouse. *J Morph* 1954;**95**:263–87.

85. Honjin R. On the nerve supply of the mouse with special reference to the structure of the peripheral vegetative system. *J Comp Neurol* 1956;**105**:587–625.

86. De Proost I, Pintelon I, Brouns I, Kroese AB, Riccardi D, Kemp PJ, et al. Functional live cell imaging of the pulmonary neuroepithelial body microenvironment. *Am J Respir Cell Mol Biol* 2008;**39**:180–9.

87. Domnik NJ, Cutz E. Pulmonary neuroepithelial bodies as airway sensors: putative role in the generation of dyspnea. *Curr Opin Pharmacol* 2011;**11**:211–7.

88. Cutz E, Pan J, Yeger H, Domnik NJ, Fisher JT. Recent advances and contraversies on the role of pulmonary neuroepithelial bodies as airway sensors. *Semin Cell Dev Biol* 2013;**24**:40–50.

89. Burnstock G, Brouns I, Adriaensen D, Timmermans JP. Purinergic signaling in the airways. *Pharmacol Rev* 2012;**64**:834–68.

90. Brouns I, Pintelon I, Timmermans JP, Adriaensen D. Novel insights in the neurochemistry and function of pulmonary sensory receptors. *Adv Anat Embryol Cell Biol* 2012;**211**:1–115.

91. Song H, Yao E, Lin C, Gacayan R, Chen MH, Chuang PT. Functional characterization of pulmonary neuroendocrine cells in lung development, injury, and tumorigenesis. *Proc Natl Acad Sci USA* 2012;**109**:17531–6.

92. Zachar PC, Jonz MG. Neuroepithelial cells of the gill and their role in oxygen sensing. *Respir Physiol Neurobiol* 2012;**184**:301–8.

93. Johnson EW, Eller PM, Jafek BW. Protein gene product 9.5-like and calbindin-like immunoreactivity in the nasal respiratory mucosa of perinatal humans. *Anat Rec* 1997;**247**:38–45.

94. Adriaensen D, Scheuermann DW. Neuroendocrine cells and nerves of the lung. *Anat Rec* 1993;**236**:70–85.

95. Cutz E, Yeger H, Pan J. Pulmonary neuroendocrine cell system in pediatric lung disease-recent advances. *Pediatr Dev Pathol* 2007;**10**:419–35.

96. Weichselbaum M, Sparrow MP, Thompson PJ, Knight DA. A confocal microscopy study of pulmonary neuroendocrine cells in human adult airway epithelium. *Cairns: Thorac Soc Aust N Z* 2002.

97. Gosney JR, Sissons MC, O'Malley JA. Quantitative study of endocrine cells immunoreactive for calcitonin in the normal adult human lung. *Thorax* 1985;**40**:866–9.

98. Pan J, Yeger H, Cutz E. Innervation of pulmonary neuroendocrine cells and neuroepithelial bodies in developing rabbit lung. *J Histochem Cytochem* 2004;**52**:379–89.

99. Sorokin SP, Hoyt RF. Neuroepithelial bodies and solitary granule-cells. In: Massaro D, editor. *Lung Cell Biology*. New York: Marcel Dekker; 1989. pp. 91–344.

100. Rawlins EL, Clark CP, Xue Y, Hogan BL. The Id2+ distal tip lung epithelium contains individual multipotent embryonic progenitor cells. *Development* 2009;**136**:3741–5.

101. Borges M, Linnoila RI, van de Velde HJ, Chen H, Nelkin BD, Mabry M, et al. An achaete-scute homologue essential for neuroendocrine differentiation in the lung. *Nature* 1997;**386**:852–5.

102. McGovern S, Pan J, Oliver G, Cutz E, Yeger H. The role of hypoxia and neurogenic genes (Mash-1 and Prox-1) in the developmental programming and maturation of pulmonary neuroendocrine cells in fetal mouse lung. *Lab Invest* 2010;**90**:180–95.

103. Neptune ER, Podowski M, Calvi C, Cho JH, Garcia JG, Tuder R, et al. Targeted disruption of NeuroD, a proneural basic helix-loop-helix factor, impairs distal lung formation and neuroendocrine morphology in the neonatal lung. *J Biol Chem* 2008;**283**:21160–9.

104. Milsom WK, Burleson ML. Peripheral arterial chemoreceptors and the evolution of the carotid body. *Respir Physiol Neurobiol* 2007;**157**:4–11.

105. Lauweryns JM, Cokelaere M, Deleersynder M, Liebens M. Intrapulmonary neuro-epithelial bodies in newborn rabbits. Influence of hypoxia, hyperoxia, hypercapnia, nicotine, reserpine, L-DOPA and 5-HTP. *Cell Tissue Res* 1977;**182**:425–40.

106. Lauweryns JM, Van LA. Morphometric analysis of hypoxia-induced synaptic activity in intrapulmonary neuroepithelial bodies. *Cell Tissue Res* 1982;**226**:201–14.

107. Lauweryns JM, Van Lommel AT, Dom RJ. Innervation of rabbit intrapulmonary neuroepithelial bodies. Quantitative and qualitative ultrastructural study after vagotomy. *J Neurol Sci* 1985;**67**:81–92.

108. Brouns I, Van GJ, Hayashi H, Gajda M, Gomi T, Burnstock G, et al. Dual sensory innervation of pulmonary neuroepithelial bodies. *Am J Respir Cell Mol Biol* 2003;**28**:275–85.

109. Brouns I, Adriaensen D, Burnstock G, Timmermans JP. Intraepithelial vagal sensory nerve terminals in rat pulmonary neuroepithelial bodies express P2X(3) receptors. *Am J Respir Cell Mol Biol* 2000;**23**:52–61.

110. Springall DR, Cadieux A, Oliveira H, Su H, Royston D, Polak JM. Retrograde tracing shows that CGRP-immunoreactive nerves of rat trachea and lung originate from vagal and dorsal root ganglia. *J Auton Nerv Syst* 1987;**20**:155–66.

111. Adriaensen D, Brouns I, Van Genechten J, Timmermans JP. Functional morphology of pulmonary neuroepithelial bodies: extremely complex airway receptors. *Anat Rec* 2003;**270A**:25–40.

112. Lembrechts R, Pintelon I, Schnorbusch K, Timmermans JP, Adriaensen D, Brouns I. Expression of mechanogated two-pore domain potassium channels in mouse lungs: special reference to mechanosensory airway receptors. *Histochem Cell Biol* 2011;**136**:371–85.

113. Yu J, Wang YF, Zhang JW. Structure of slowly adapting pulmonary stretch receptors in the lung periphery. *J Appl Physiol* 2003;**95**:385–93.

114. Bolle T, Van LA, Lauweryns JM. Stereological estimation of number and volume of pulmonary neuroepithelial bodies (NEBs) in neonatal hamster lungs. *Microsc Res Tech* 1999;**44**:190–4.

115. Van LA, Lauweryns JM. Postnatal development of the pulmonary neuroepithelial bodies in various animal species. *J Auton Nerv Syst* 1997;**65**:17–24.

116. Cho T, Chan W, Cutz E. Distribution and frequency of neuro-epithelial bodies in post-natal rabbit lung: quantitative study with monoclonal antibody against serotonin. *Cell Tissue Res* 1989;**255**:353–62.

117. Stahlman MT, Gray ME. Ontogeny of neuroendocrine cells in human fetal lung. I. An electron microscopic study. *Lab Invest* 1984;**51**:449–63.

118. Stahlman MT, Gray ME. Immunogold EM localization of neurochemicals in human pulmonary neuroendocrine cells. *Microsc Res Tech* 1997;**37**:77–91.

119. Lauweryns JM, Peuskens JC, Cokelaere M. Argyrophil, fluorescent and granulated (peptide and amine producing?) AFG cells in human infant bronchial epithelium. Light and electron microscopic studies. *Life Sci I* 1970;**9**:1417–29.

120. Van LA, Lauweryns JM. Neuroepithelial bodies in the Fawn Hooded rat lung: morphological and neuroanatomical evidence for a sensory innervation. *J Anat* 1993;**183**:553–66.

121. Jonz MG, Nurse CA. Peripheral chemoreceptors in air- versus water-breathers. *Adv Exp Med Biol* 2012;**758**:19–27.

122. Evans DH, Piermarini PM, Choe KP. The multifunctional fish gill: dominant site of gas exchange, osmoregulation, acid-base regulation, and excretion of nitrogenous waste. *Physiol Rev* 2005;**85**:97–177.

123. Adriaensen D, Timmermans JP, Brouns I, Berthoud HR, Neuhuber WL, Scheuermann DW. Pulmonary intraepithelial vagal nodose afferent nerve terminals are confined to neuroepithelial bodies: an anterograde tracing and confocal microscopy study in adult rats. *Cell Tissue Res* 1998;**293**:395–405.

124. Zaccone G, Fasulo S, Ainis L, Licata A. Paraneurons in the gills and airways of fishes. *Microsc Res Tech* 1997;**37**:4–12.

125. Pan J, Copland I, Post M, Yeger H, Cutz E. Mechanical stretch-induced serotonin release from pulmonary neuroendocrine cells: implications for lung development. *Am J Physiol Lung Cell Mol Physiol* 2006;**290**:L185–93.

126. Polak JM, Becker KL, Cutz E, Gail DB, Goniakowska-Witalinska L, Gosney JR, et al. Lung endocrine cell markers, peptides, and amines. *Anat Rec* 1993;**236**:169–71.

127. Jonz MG, Nurse CA. Neuroepithelial cells and associated innervation of the zebrafish gill: a confocal immunofluorescence study. *J Comp Neurol* 2003;**461**:1–17.

128. Perry S, Jonz MG, Gilmour KM. Oxygen sensing and the hypoxic ventillatory response. In: Richards JG, Farrell AP, Brauner CJ, editors. *Fish Physiology*, vol. 27. New York: Academic Press; 2009. pp. 193–253.

129. Gu X, Karp PH, Brody SL, Pierce RA, Welsh MJ, Holtzman MJ, et al. Chemosensory functions for pulmonary neuroendocrine cells. *Am J Respir Cell Mol Biol* 2014;**50**:637–46.

130. Fu XW, Nurse CA, Wong V, Cutz E. Hypoxia-induced secretion of serotonin from intact pulmonary neuroepithelial bodies in neonatal rabbit. *J Physiol* 2002;**539**:503–10.

131. Youngson C, Nurse C, Yeger H, Cutz E. Oxygen sensing in airway chemoreceptors. *Nature* 1993;**365**:153–5.

132. Fu XW, Nurse C, Cutz E. Characterization of slowly inactivating KV{alpha} current in rabbit pulmonary neuroepithelial bodies: effects of hypoxia and nicotine. *Am J Physiol Lung Cell Mol Physiol* 2007;**293**:L892–902.

133. Jonz MG, Fearon IM, Nurse CA. Neuroepithelial oxygen chemoreceptors of the zebrafish gill. *J Physiol* 2004;**560**:737–52.

134. Fu XW, Wang D, Nurse CA, Dinauer MC, Cutz E. NADPH oxidase is an O2 sensor in airway chemoreceptors: evidence from K+ current modulation in wild-type and oxidase-deficient mice. *Proc Natl Acad Sci USA* 2000;**97**:4374–9.

135. Kazemian P, Stephenson R, Yeger H, Cutz E. Respiratory control in neonatal mice with NADPH oxidase deficiency. *Respir Physiol* 2001;**126**:89–101.

136. Fu XW, Spindel ER. Recruitment of GABA(A) receptors in chemoreceptor pulmonary neuroepithelial bodies by prenatal nicotine exposure in monkey lung. *Adv Exp Med Biol* 2009;**648**:439–45.

137. Cutz E, Fu XW, Yeger H, Pan J, Nurse CA. Oxygen sensing in mammalian pulmonary neuroepithelial bodies. In: Zaccone G, Cutz E, Adriaensen D, Nurse CA, Mauceri A, editors. *Airway Chemoreceptors in Vertebrates: Structure, Evolution and Function*. Enfield, NH: Science Publishers; 2009. pp. 269–90.

138. Lauweryns JM, Tierens A, Decramer M. Influence of hypercapnia on rabbit intrapulmonary neuroepithelial bodies: microfluorimetric and morphometric study. *Eur Respir J* 1990;**3**:182–6.

139. Ebina M, Hoyt Jr RF, McNelly NA, Sorokin SP, Linnoila RI. Effects of hydrogen and bicarbonate ions on endocrine cells in fetal rat lung organ cultures. *Am J Physiol* 1997;**272**:L178–86.

140. Qin Z, Lewis JE, Perry SF. Zebrafish (Danio rerio) gill neuroepithelial cells are sensitive chemoreceptors for environmental CO2. *J Physiol* 2010;**588**:861–72.

141. Perry SF, Abdallah S. Mechanisms and consequences of carbon dioxide sensing in fish. *Respir Physiol Neurobiol* 2012;**184**:309–15.

142. Abdallah SJ, Perry SF, Jonz MG. CO(2) signaling in chemosensory neuroepithelial cells of the zebrafish gill filaments: role of intracellular Ca(2+) and pH. *Adv Exp Med Biol* 2012;**758**:143–8.

143. Jonz MG, Nurse CA. Development of oxygen sensing in the gills of zebrafish. *J Exp Biol* 2005;**208**:1537–49.

144. Shakarchi K, Zachar PC, Jonz MG. Serotonergic and cholinergic elements of the hypoxic ventilatory response in developing zebrafish. *J Exp Biol* 2013;**216**:869–80.

145. De Proost I, Pintelon I, Wilkinson WJ, Goethals S, Brouns I, Van Nassauw L, et al. Purinergic signaling in the pulmonary neuroepithelial body microenvironment unraveled by live cell imaging. *FASEB J* 2009;**23**:1153–60.

146. Kitterman JA. The effects of mechanical forces on fetal lung growth. *Clin Perinatol* 1996;**23**:727–40.

147. Lembrechts R, Brouns I, Schnorbusch K, Pintelon I, Timmermans JP, Adriaensen D. Neuroepithelial bodies as mechanotransducers in the intrapulmonary airway epithelium: involvement of TRPC5. *Am J Respir Cell Mol Biol* 2012;**47**:315–23.

148. Maingret F, Fosset M, Lesage F, Lazdunski M, Honore E. TRAAK is a mammalian neuronal mechano-gated K+ channel. *J Biol Chem* 1999;**274**:1381–7.

149. Chua BA, Perks AM. The pulmonary neuroendocrine system and drainage of the fetal lung: effects of serotonin. *Gen Comp Endocrinol* 1999;**113**:374–87.

150. Seuwen K, Pouyssegur J. Serotonin as a growth factor. *Biochem Pharmacol* 1990;**39**:985–90.

151. Reynolds SD, Giangreco A, Power JH, Stripp BR. Neuroepithelial bodies of pulmonary airways serve as a reservoir of progenitor cells capable of epithelial regeneration. *Am J Pathol* 2000;**156**:269–78.

152. Linnoila RI. Functional facets of the pulmonary neuroendocrine system. *Lab Invest* 2006;**86**:425–44.

153. Reynolds SD, Hong KU, Giangreco A, Mango GW, Guron C, Morimoto Y, et al. Conditional clara cell ablation reveals a self-renewing progenitor function of pulmonary neuroendocrine cells. *Am J Physiol Lung Cell Mol Physiol* 2000;**278**:L1256–63.

154. De Proost I, Pintelon I, Wilkinson WJ, Goethals S, Brouns I, Van Nassauw L, et al. ATP released from pulmonary neuroepithelial bodies activates Clara-like cells in the NEB microenvironment via P2Y$_2$ receptors. *FASEB J* 2008;**22**:929.4.

155. Lembrechts R, Brouns I, Schnorbusch K, Pintelon I, Kemp PJ, Timmermans JP, et al. Functional expression of the multimodal extracellular calcium-sensing receptor in pulmonary neuroendocrine cells. *J Cell Sci* 2013;**126**:4490–501.

156. Riccardi D, Kemp PJ. The calcium-sensing receptor beyond extracellular calcium homeostasis: conception, development, adult physiology, and disease. *Annu Rev Physiol* 2012;**74**:271–97.

157. Gu Q, Vysotskaya ZV, Moss CR, Kagira MK, Gilbert CA. Calcium-sensing receptor in rat vagal bronchopulmonary sensory neurons regulates the function of the capsaicin receptor TRPV1. *Exp Physiol* 2013;**98**:1631–42.

158. Vizard TN, O'Keeffe GW, Gutierrez H, Kos CH, Riccardi D, Davies AM. Regulation of axonal and dendritic growth by the extracellular calcium-sensing receptor. *Nat Neurosci* 2008;**11**:285–91.

159. Coleridge HM, Coleridge JC. Reflexes evoked from the tracheobronchial tree and lungs. In: Cherniack NS, Widdicombe JG, editors. *Handbook of Physiology, Section 3: The Respiratory System, Vol. II: Control of Breathing Part I*. Washington, DC: American Physiological Society; 1986. pp. 395–429.

160. Ambalavanan N, Carlo WA. Bronchopulmonary dysplasia: new insights. *Clin Perinatol* 2004;**31**:613–28.

161. Cutz E, Perrin DG, Pan J, Haas EA, Krous HF. Pulmonary neuroendocrine cells and neuroepithelial bodies in sudden infant death syndrome: potential markers of airway chemoreceptor dysfunction. *Pediatr Dev Pathol* 2007;**10**:106–16.

162. Doan ML, Elidemir O, Dishop MK, Zhang H, Smith EO, Black PG, et al. Serum KL-6 differentiates neuroendocrine cell hyperplasia of infancy from the inborn errors of surfactant metabolism. *Thorax* 2009;**64**:677–81.

163. Young LR, Brody AS, Inge TH, Acton JD, Bokulic RE, Langston C, et al. Neuroendocrine cell distribution and frequency distinguish neuroendocrine cell hyperplasia of infancy from other pulmonary disorders. *Chest* 2011;**139**:1060–71.

164. Coletta EN, Voss LR, Lima MS, Arakaki JS, Camara J, D'Andretta NC, et al. Diffuse idiopathic pulmonary neuroendocrine cell hyperplasia accompanied by airflow obstruction. *J Bras Pneumol* 2009;**35**:489–94.

165. Gosney JR, Williams IJ, Dodson AR, Foster CS. Morphology and antigen expression profile of pulmonary neuroendocrine cells in reactive proliferations and diffuse idiopathic pulmonary neuroendocrine cell hyperplasia (DIPNECH). *Histopathology* 2011;**59**:751–62.

166. Ito T, Udaka N, Yazawa T, Okudela K, Hayashi H, Sudo T, et al. Basic helix-loop-helix transcription factors regulate the neuroendocrine differentiation of fetal mouse pulmonary epithelium. *Development* 2000;**127**:3913–21.

167. Dauger S, Renolleau S, Vardon G, Nepote V, Mas C, Simonneau M, et al. Ventilatory responses to hypercapnia and hypoxia in Mash-1 heterozygous newborn and adult mice. *Pediatr Res* 1999;**46**:535–42.

168. Pan J, Yeger H, Ratcliffe P, Bishop T, Cutz E. Hyperplasia of pulmonary neuroepithelial bodies (NEB) in lungs of prolyl hydroxylase -1(PHD-1) deficient mice. *Adv Exp Med Biol* 2012;**758**:149–55.

169. Pan J, Luk C, Kent G, Cutz E, Yeger H. Pulmonary neuroendocrine cells, airway innervation, and smooth muscle are altered in Cftr null mice. *Am J Respir Cell Mol Biol* 2006;**35**:320–6.

170. Waldron MA, Fisher JT. Neural control of airway smooth muscle in the newborn. In: Haddad G, Farber JP, editors. *Developmental Neurobiology of Breathing*. New York: Marcel Dekker; 1991. pp. 483–518.

171. Merritt TA, Northway WH, Boynton BR. *Bronchopulmonary Dysplasia*. Boston: Blackwell Scientific Publications; 1985.

172. Northway WH. Bronchopulmonary dysplasia. *Biomed Pharmacother* 1991;**45**:323.

173. Motoyama EK, Fort MD, Klesh KW, Mutich RL, Guthrie RD. Early onset of airway reactivity in premature infants with bronchopulmonary dysplasia. *Am Rev Respir Dis* 1987;**136**:50–7.

174. Kao LC, Warburton D, Platzker AC, Keens TG. Effect of isoproterenol inhalation on airway resistance in chronic bronchopulmonary dysplasia. *Pediatrics* 1984;**73**:509–14.

175. Brundage KL, Mohsini KG, Froese AB, Fisher JT. Bronchodilator response to ipratropium bromide in infants with bronchopulmonary dysplasia. *Am Rev Respir Dis* 1990;**142**:1137–42.

176. Denjean A, Guimaraes H, Migdal M, Miramand JL, Dehan M, Gaultier C. Dose-related bronchodilator response to aerosolized salbutamol (albuterol) in ventilator-dependent premature infants. *J Pediatr* 1992;**120**:974–9.

177. Sollmann T, Gilbert AJ. Microscopic observations of bronchiolar reactions. *J Pharmacol Exp Ther* 1937;**61**:272–85.

178. Spina D, Shah S, Harrison S. Modulation of sensory nerve function in the airways. *Trends Pharmacol Sci* 1998;**19**:460–6.

179. Sparrow MP, Mitchell HW. Contraction of smooth muscle of pig airway tissues from before birth to maturity. *J Appl Physiol* 1990;**68**:468–77.

180. Mitchell HW, Sparrow MP, Tagliaferri RP. Inhibitory and excitatory responses to field stimulation in fetal and adult pig airway. *Pediatr Res* 1990;**28**:69–74.

181. Booth RJ, Sparrow MP, Mitchell HW. Early maturation of force production in pig tracheal smooth muscle during fetal development. *Am J Respir Cell Mol Biol* 1992;**7**:590–7.

182. McFawn PK, Mitchell HW. Bronchial compliance and wall structure during development of the immature human and pig lung. *Eur Respir J* 1997;**10**:27–34.

183. McFawn PK, Mitchell HW. Effect of transmural pressure on preloads and collapse of immature bronchi. *Eur Respir J* 1997;**10**:322–9.

184. Richards IS, Kulkarni A, Brooks SM. Human fetal tracheal smooth muscle produces spontaneous electromechanical oscillations that are Ca2+ dependent and cholinergically potentiated. *Dev Pharmacol Ther* 1991;**16**:22–8.

185. Liu M, Skinner SJ, Xu J, Han RN, Tanswell AK, Post M. Stimulation of fetal rat lung cell proliferation in vitro by mechanical stretch. *Am J Physiol* 1992;**263**:L376–83.

186. Liu M, Qin Y, Liu J, Tanswell AK, Post M. Mechanical strain induces pp60src activation and translocation to cytoskeleton in fetal rat lung cells. *J Biol Chem* 1996;**271**:7066–71.

187. Liu M, Tanswell AK, Post M. Mechanical force-induced signal transduction in lung cells. *Am J Physiol* 1999;**277**:L667–83.

188. Liu M, Post M. Invited review: mechanochemical signal transduction in the fetal lung. *J Appl Physiol* 2000;**89**:2078–84.

189. Wong KA, Bano A, Rigaux A, Wang B, Bharadwaj B, Schurch S, et al. Pulmonary vagal innervation is required to establish adequate alveolar ventilation in the newborn lamb. *J Appl Physiol* 1998;**85**:849–59.

190. Hasan SU, Lalani S, Remmers JE. Significance of vagal innervation in perinatal breathing and gas exchange. *Respir Physiol* 2000;**119**:133–41.

191. Schwieler GH, Douglas JS, Bouhuys A. Postnatal development of autonomic efferent innervation in the rabbit. *Am J Physiol* 1970;**219**:391–7.

192. Waldron MA, Fisher JT. Differential effects of CO2 and hypoxia on bronchomotor tone in the newborn dog. *Respir Physiol* 1988;**72**:271–82.

193. Fisher JT, Brundage KL, Waldron MA, Connelly BJ. Vagal cholinergic innervation of the airways in newborn cat and dog. *J Appl Physiol* 1990;**69**:1525–31.

194. Tepper RS. Maturation affects the maximal pulmonary response to methacholine in rabbits. *Pediatr Pulmonol* 1993;**16**:48–53.

195. Tepper RS, Gunst SJ, Doerschuk CM, Shen X, Bray W. Effect of transpulmonary pressure on airway closure in immature and mature rabbits. *J Appl Physiol* 1995;**78**:505–12.

196. Tepper RS, Du T, Styhler A, Ludwig M, Martin JG. Increased maximal pulmonary response to methacholine and airway smooth muscle in immature compared with mature rabbits. *Am J Respir Crit Care Med* 1995;**151**:836–40.

197. Hayashi S, Toda N. Age-related alterations in the response of rabbit tracheal smooth muscle to agents. *J Pharmacol Exp Ther* 1980;**214**:675–81.

198. Duncan PG, Douglas JS. Influences of gender and maturation on responses of guinea-pig airway tissues to LTD4. *Eur J Pharmacol* 1985;**112**:423–7.

199. Panitch HB, Allen JL, Ryan JP, Wolfson MR, Shaffer TH. A comparison of preterm and adult airway smooth muscle mechanics. *J Appl Physiol* 1989;**66**:1760–5.

200. Murphy TM, Mitchell RW, Blake JS, Mack MM, Kelly EA, Munoz NM, et al. Expression of airway contractile properties and acetylcholinesterase activity in swine. *J Appl Physiol* 1989;**67**:174–80.

201. Mitchell RW, Murphy TM, Kelly E, Leff AR. Maturation of acetyl-cholinesterase expression in tracheal smooth muscle contraction. *Am J Physiol* 1990;**259**:L130–5.

202. Murphy TM, Mitchell RW, Phillips IJ, Leff AR. Ontogenic expression of acetylcholinesterase activity in trachealis of young swine. *Am J Physiol* 1991;**261**:L322–6.

203. Sauder RA, McNicol KJ, Stecenko AA. Effect of age on lung mechanics and airway reactivity in lambs. *J Appl Physiol* 1986;**61**:2074–80.

204. Ikeda K, Mitchell RW, Guest KA, Seow CY, Kirchhoff CF, Murphy TM, et al. Ontogeny of shortening velocity in porcine trachealis. *Am J Physiol* 1992;**262**:L280–5.

205. Murphy TM, Mitchell RW, Halayko A, Roach J, Roy L, Kelly EA, et al. Effect of maturational changes in myosin content and morphometry on airway smooth muscle contraction. *Am J Physiol* 1991;**260**:L471–80.

206. Fisher JT. Airway smooth muscle contraction at birth: in vivo versus in vitro comparisons to the adult. *Can J Physiol Pharmacol* 1992;**70**:590–6.

207. Stevens EL, Uyehara CF, Southgate WM, Nakamura KT. Furosemide differentially relaxes airway and vascular smooth muscle in fetal, newborn, and adult guinea pigs. *Am Rev Respir Dis* 1992;**146**:1192–7.

208. Southgate WM, Pichoff BE, Stevens EL, Balaraman V, Uyehara CF, Nakamura KT. Ontogeny of epithelial modulation of airway smooth muscle function in the guinea pig. *Pediatr Pulmonol* 1993;**15**:105–10.

209. Fayon M, Ben-Jebria A, Elleau C, Carles D, Demarquez JL, Savineau JP, Marthan R. Human airway smooth muscle responsiveness in neonatal lung specimens. *Am J Physiol* 1994;**267**:L180–6.

210. Penn RB, Wolfson MR, Shaffer TH. Effect of tracheal smooth muscle tone on collapsibility of immature airways. *J Appl Physiol* 1988;**65**:863–9.

211. Mitchell HW, McFawn PK, Sparrow MP. Increased narrowing of bronchial segments from immature pigs. *Eur Respir J* 1992;**5**:207–12.

212. Bhutani VK, Koslo RJ, Shaffer TH. The effect of tracheal smooth muscle tone on neonatal airway collapsibility. *Pediatr Res* 1986;**20**:492–5.

213. Bhutani VK, Rubenstein SD, Shaffer TH. Pressure–volume relationships of tracheae in fetal newborn and adult rabbits. *Respir Physiol* 1981;**43**:221–31.

214. Bhutani VK, Rubenstein D, Shaffer TH. Pressure-induced deformation in immature airways. *Pediatr Res* 1981;**15**:829–32.

215. Fisher JT, Haxhiu MA, Martin RJ. Regulation of lower airway function. In: Polin RA, Fox WW, editors. *Fetal and Neonatal Physiology*. Philadelphia, PA: W.B. Saunders; 1998. pp. 1060–70.

216. Fisher JT, Brundage KL, Anderson JW. Cardiopulmonary actions of muscarinic receptor subtypes in the newborn dog. *Can J Physiol Pharmacol* 1996;**74**:603–13.

217. Mortola JP. *Respiratory Physiology of Newborn Mammals: A Comparative Perspective*. Baltimore: The Johns Hopkins University Press; 2001.

218. Fisher JT, Mortola JP. Statics of the respiratory system and growth: an experimental and allometric approach. *Am J Physiol* 1981;**241**:R336–41.

219. Mortola JP. Dynamics of breathing in newborn mammals. *Physiol Rev* 1987;**67**:187–243.

220. Waldron MA, Fisher JT. Differential effects of CO2 and hypoxia on bronchomotor tone in the newborn dog. *Respir Physiol* 1988;**72**:271–82.

221. Anderson JW, Fisher JT. Capsaicin-induced reflex bronchoconstriction in the newborn. *Respir Physiol* 1993;**93**:13–27.

222. Nault MA, Vincent SG, Fisher JT. Mechanisms of capsaicin- and lactic acid-induced bronchoconstriction in the newborn dog. *J Physiol* 1999;**515**:567–78.

223. Marantz MJ, Vincent SG, Fisher JT. Role of vagal C-fibre afferents in the bronchomotor response to lactic acid in the newborn dog. *J Appl Physiol* 2001;**90**:2311–8.

224. Richardson CA, Herbert DA, Mitchell RA. Modulation of pulmonary stretch receptors and airway resistance by parasympathetic efferents. *J Appl Physiol* 1984;**57**:1842–9.

225. Nadel JA, Widdicombe JG. Effect of changes in blood gas tensions and carotid sinus pressure on tracheal volume and total lung resistance to airflow. *J Physiol* 1962;**163**:13–33.

226. Green M, Widdicombe JG. The effects of ventilation of dogs with different gas mixtures on airway calibre and lung mechanics. *J Physiol* 1966;**186**:363–81.

227. Fisher JT, Waldron MA, Armstrong CJ. Effects of hypoxia on lung mechanics in the newborn cat. *Can J Physiol Pharmacol* 1987;**65**:1234–8.

228. Mortola JP, Tenney SM. Effects of hyperoxia on ventilatory and metabolic rates of newborn mice. *Respir Physiol* 1986;**63**:267–74.

229. Mortola JP, Rezzonico R. Metabolic and ventilatory rates in newborn kittens during acute hypoxia. *Respir Physiol* 1988;**73**:55–67.

230. Caterina MJ, Schumacher MA, Tominaga M, Rosen TA, Levine JD, Julius D. The capsaicin receptor: a heat-activated ion channel in the pain pathway. *Nature* 1997;**389**:816–24.

231. Lee LY, Morton RF, Lundberg JM. Pulmonary chemoreflexes elicited by intravenous injection of lactic acid in anesthetized rats. *J Appl Physiol* 1996;**81**:2349–57.

232. Diaz V, Dorion D, Kianicka I, Letourneau P, Praud JP. Vagal afferents and active upper airway closure during pulmonary edema in lambs. *J Appl Physiol* 1999;**86**:1561–9.

233. Diaz V, Dorion D, Renolleau S, Letourneau P, Kianicka I, Praud JP. Effects of capsaicin pretreatment on expiratory laryngeal closure during pulmonary edema in lambs. *J Appl Physiol* 1999;**86**:1570–7.

234. Reix P, Duvareille C, Letourneau P, Pouliot M, Samson N, Niyonsenga T, et al. C-fibre blockade influence on non-nutritive swallowing in full-term lambs. *Respir Physiol Neurobiol* 2006;**152**:27–35.

235. Roulier S, Arsenault J, Reix P, Dorion D, Praud JP. Effects of C fibre blockade on cardiorespiratory responses to laryngeal stimulation in concious lambs. *Respir Physiol Neurobiol* 2003;**136**:13–23.

236. St-Hilaire M, Samson N, Duvareille C, Praud JP. Laryngeal stimulation by an acid solution in the pre-term lamb. *Adv Exp Med Biol* 2008;**605**:154–8.

237. Hosey MM. Diversity of structure, signaling and regulation within the family of muscarinic cholinergic receptors. *FASEB J* 1992;**6**:845–52.

238. Hulme EC, Birdsall NJ, Buckley NJ. Muscarinic receptor subtypes. *Annu Rev Pharmacol Toxicol* 1990;**30**:633–73.

239. Wess J. Molecular biology of muscarinic acetylcholine receptors. *Crit Rev Neurobiol* 1996;**10**:69–99.

240. Wess J, Eglen RM, Gautam D. Muscarinic acetylcholine receptors: mutant mice provide new insights for drug development. *Nat Rev Drug Discov* 2007;**6**:721–33.

241. Eglen RM. Overview of muscarinic receptor subtypes. *Handb Exp Pharmacol* 2012:3–28.

242. Pitcher JA, Freedman NJ, Lefkowitz RJ. G protein-coupled receptor kinases. *Annu Rev Biochem* 1998;**67**:653–92.

243. Gomeza J, Zhang L, Kostenis E, Felder C, Bymaster F, Brodkin J, et al. Enhancement of D1 dopamine receptor-mediated locomotor stimulation in M(4) muscarinic acetylcholine receptor knockout mice. *Proc Natl Acad Sci USA* 1999;**96**:10483–8.

244. Gomeza J, Shannon H, Kostenis E, Felder C, Zhang L, Brodkin J, et al. Pronounced pharmacologic deficits in M2 muscarinic acetylcholine receptor knockout mice. *Proc Natl Acad Sci USA* 1999;**96**:1692–7.

245. Shapiro MS, Loose MD, Hamilton SE, Nathanson NM, Gomeza J, Wess J, et al. Assignment of muscarinic receptor subtypes mediating G-protein modulation of Ca(2+) channels by using knockout mice. *Proc Natl Acad Sci USA* 1999;**96**:10899–904.

246. Stengel PW, Gomeza J, Wess J, Cohen ML. M(2) and M(4) receptor knockout mice: muscarinic receptor function in cardiac and smooth muscle in vitro. *J Pharmacol Exp Ther* 2000;**292**:877–85.

247. Fisher JT, Brundage KL, Anderson JW. Cardiopulmonary actions of muscarinic receptor subtypes in the newborn dog. *Can J Physiol Pharmacol* 1996;**74**:603–13.

248. Fisher JT, Froese AB, Brundage KL. Physiological basis for the use of muscarine antagonists in bronchopulmonary dysplasia. *Arch Pediatr* 1995;**2**(Suppl 2.):163S–71S.

249. Julia-Serda G, Molfino NA, Chapman KR, McClean PA, Zamel N, Slutsky AS, et al. Heterogeneous airway tone in asthmatic subjects. *J Appl Physiol* 1992;**73**:2328–32.

250. Molfino NA, Slutsky AS, Hoffstein V, McClean PA, Rebuck AS, Drazen JM, et al. Changes in cross-sectional airway areas induced by methacholine, histamine, and LTC4 in asthmatic subjects. *Am Rev Respir Dis* 1992;**146**:577–80.

251. Molfino NA, Slutsky AS, Julia-Serda G, Hoffstein V, Szalai JP, Chapman KR, et al. Assessment of airway tone in asthma. Comparison between double lung transplant patients and healthy subjects. *Am Rev Respir Dis* 1993;**148**:1238–43.

252. Haxhiu-Poskurica B, Ernsberger P, Haxhiu MA, Miller MJ, Cattarossi L, Martin RJ. Development of cholinergic innervation and muscarinic receptor subtypes in piglet trachea. *Am J Physiol* 1993;**264**:L606–14.

253. Maclagan J, Barnes PJ. Muscarinic pharmacology of the airways. *Trends Pharmacol Sci* 1989; (Suppl):88–92.

254. Barnes PJ. Muscarinic receptor subtypes: implications for therapy. *Agents Actions Suppl* 1993;**43**:243–52.

255. Barnes PJ. Muscarinic receptor subtypes in airways. *Life Sci* 1993;**52**:521–7.

256. Harvey RD. Muscarinic receptor agonists and antagonists: effects on cardiovascular function. *Handb Exp Pharmacol* 2012:299–316.

257. Lazareno S, Buckley NJ, Roberts FF. Characterization of muscarinic M4 binding sites in rabbit lung, chicken heart, and NG108-15 cells. *Mol Pharmacol* 1990;**38**:805–15.

258. Dorje F, Levey AI, Brann MR. Immunological detection of muscarinic receptor subtype proteins (m1-m5) in rabbit peripheral tissues. *Mol Pharmacol* 1991;**40**:459–62.

259. Mak JC, Haddad EB, Buckley NJ, Barnes PJ. Visualization of muscarinic m4 mRNA and M4 receptor subtype in rabbit lung. *Life Sci* 1993;**53**:1501–8.

260. Maclagan J, Fryer AD, Faulkner D. Identification of M1 muscarinic receptors in pulmonary sympathetic nerves in the guinea-pig by use of pirenzepine. *Br J Pharmacol* 1989;**97**:499–505.

261. Beck KC, Vettermann J, Flavahan NA, Rehder K. Muscarinic M1 receptors mediate the increase in pulmonary resistance during vagus nerve stimulation in dogs. *Am Rev Respir Dis* 1987;**136**:1135–9.

262. Lammers JW, Minette P, McCusker M, Barnes PJ. The role of pirenzepine-sensitive (M1) muscarinic receptors in vagally mediated bronchoconstriction in humans. *Am Rev Respir Dis* 1989;**139**:446–9.

263. Maclagan J, Faulkner D. Effect of pirenzepine and gallamine on cardiac and pulmonary muscarinic receptors in the rabbit. *Br J Pharmacol* 1989;**97**:506–12.

264. Eltze M, Galvan M. Involvement of muscarinic M2 and M3, but not of M1 and M4 receptors in vagally stimulated contractions of rabbit bronchus/trachea. *Pulm Pharmacol* 1994;**7**:109–20.

265. Fryer AD, Maclagan J. Muscarinic inhibitory receptors in pulmonary parasympathetic nerves in the guinea-pig. *Br J Pharmacol* 1984;**83**:973–8.

266. Blaber LC, Fryer AD, Maclagan J. Neuronal muscarinic receptors attenuate vagally-induced contraction of feline bronchial smooth muscle. *Br J Pharmacol* 1985;**86**:723–8.

267. Faulkner D, Fryer AD, Maclagan J. Postganglionic muscarinic inhibitory receptors in pulmonary parasympathetic nerves in the guinea-pig. *Br J Pharmacol* 1986;**88**:181–7.

268. Ito Y, Yoshitomi T. Autoregulation of acetylcholine release from vagus nerve terminals through activation of muscarinic receptors in the dog trachea. *Br J Pharmacol* 1988;**93**:636–46.

269. Minette PA, Barnes PJ. Prejunctional inhibitory muscarinic receptors on cholinergic nerves in human and guinea pig airways. *J Appl Physiol* 1988;**64**:2532–7.

270. Watson N, Barnes PJ, Maclagan J. Actions of methoctramine, a muscarinic M2 receptor antagonist, on muscarinic and nicotinic cholinoceptors in guinea-pig airways in vivo and in vitro. *Br J Pharmacol* 1992;**105**:107–12.

271. Fisher JT, Vincent SG, Gomeza J, Yamada M, Wess J. Loss of vagally mediated bradycardia and bronchoconstriction in mice lacking M2 or M3 muscarinic acetylcholine receptors. *FASEB J* 2004;**18**:711–3.

272. Scott GD, Fryer AD. Role of parasympathetic nerves and muscarinic receptors in allergy and asthma. *Chem Immunol Allergy* 2012;**98**:48–69.

273. Fryer AD, Jacoby DB. Parainfluenza virus infection damages inhibitory M2 muscarinic receptors on pulmonary parasympathetic nerves in the guinea-pig. *Br J Pharmacol* 1991;**102**:267–71.

274. Schultheis AH, Bassett DJ, Fryer AD. Ozone-induced airway hyperresponsiveness and loss of neuronal M2 muscarinic receptor function. *J Appl Physiol* 1994;**76**:1088–97.

275. Gambone LM, Elbon CL, Fryer AD. Ozone-induced loss of neuronal M2 muscarinic receptor function is prevented by cyclophosphamide. *J Appl Physiol* 1994;**77**:1492–9.

276. Janssen LJ, Daniel EE. Pre- and postjunctional muscarinic receptors in canine bronchi. *Am J Physiol* 1990;**259**:L304–14.

277. Gross NJ. Ipratropium bromide. *N Engl J Med* 1988;**319**:486–94.

278. De TA, Yernault JC, Rodenstein D. Effects of vagal blockade on lung mechanics in normal man. *J Appl Physiol* 1979;**46**:217–26.

279. Oenema TA, Mensink G, Smedinga L, Halayko AJ, Zaagsma J, Meurs H, et al. Cross-talk between transforming growth factor-beta(1) and muscarinic M(2) receptors augments airway smooth muscle proliferation. *Am J Respir Cell Mol Biol* 2013;**49**:18–27.

280. Xing Y, Li C, Hu L, Tiozzo C, Li M, Chai Y, et al. Mechanisms of TGFbeta inhibition of LUNG endodermal morphogenesis: the role of TbetaRII, Smads, Nkx2.1 and Pten. *Dev Biol* 2008;**320**:340–50.

281. Wu S, Kasisomayajula K, Peng J, Bancalari E. Inhibition of JNK enhances TGF-beta1-activated Smad2 signaling in mouse embryonic lung. *Pediatr Res* 2009;**65**:381–6.

282. Wess J. Muscarinic acetylcholine receptor knockout mice: novel phenotypes and clinical implications. *Annu Rev Pharmacol Toxicol* 2004;**44**:423–50.

283. Klein MK, Haberberger RV, Hartmann P, Faulhammer P, Lips KS, Krain B, et al. Muscarinic receptor subtypes in cilia-driven transport and airway epithelial development. *Eur Respir J* 2009;**33**:1113–21.

284. Wess J, Nakajima K, Jain S. Novel designer receptors to probe GPCR signaling and physiology. *Trends Pharmacol Sci* 2013;**34**:385–92.

285. Penn RB, Bond RA, Walker JK. GPCRs and arrestins in airways: implications for asthma. *Handb Exp Pharmacol* 2014;**219**:387–403.

286. Walker JK, Gainetdinov RR, Feldman DS, McFawn PK, Caron MG, Lefkowitz RJ, et al. G protein-coupled receptor kinase 5 regulates airway responses induced by muscarinic receptor activation. *Am J Physiol Lung Cell Mol Physiol* 2004;**286**:L312–9.

287. Reiter E, Ahn S, Shukla AK, Lefkowitz RJ. Molecular mechanism of beta-arrestin-biased agonism at seven-transmembrane receptors. *Annu Rev Pharmacol Toxicol* 2012;**52**:179–97.

288. Kistemaker LE, Bos ST, Mudde WM, Hylkema MN, Hiemstra PS, Wess J, et al. Muscarinic M3 receptors contribute to allergen-induced airway remodeling in mice. *Am J Respir Cell Mol Biol* 2013. in press.

289. Wessler I, Bittinger F, Kamin W, Zepp F, Meyer E, Schad A, et al. Dysfunction of the non-neuronal cholinergic system in the airways and blood cells of patients with cystic fibrosis. *Life Sci* 2007;**80**:2253–8.

290. Kummer W, Lips KS, Pfeil U. The epithelial cholinergic system of the airways. *Histochem Cell Biol* 2008;**130**:219–34.

291. Hollenhorst MI, Lips KS, Wolff M, Wess J, Gerbig S, Takats Z, et al. Luminal cholinergic signalling in airway lining fluid: a novel mechanism for activating chloride secretion via Ca(2)(+)-dependent Cl(-) and K(+) channels. *Br J Pharmacol* 2012;**166**:1388–402.

292. Jonz MG, Nurse CA. Oxygen-sensitive neuroepithelial cells in the gills of aquatic vertebrates. In: Zaccone G, Cutz E, Adriaensen D, Nurse CA, Mauceri A, editors. *Airway Chemoreceptors in the Vertebrates: Structure, Evolution and Function*. Enfield, NH: Science Publishers; 2009. pp. 1–30.

293. Adriaensen D, Timmermans JP. Purinergic signalling in the lung: important in asthma and COPD? *Curr Opin Pharmacol* 2004;**4**:207–14.

294. Brouns I, Pintelon I, Van GJ, De P I, Timmermans JP, Adriaensen D. Vesicular glutamate transporter 2 is expressed in different nerve fibre populations that selectively contact pulmonary neuroepithelial bodies. *Histochem Cell Biol* 2004;**121**:1–12.

Chapter 4

The Formation of Pulmonary Alveoli

Stephen E. McGowan

Department of Veterans Affairs Research Service, Department of Internal Medicine, University of Iowa Carver College of Medicine, Iowa City, IA, USA

INTRODUCTION

Respiration involves phasic movement of air through the conducting airways in and out of the gas-exchange portion of the lung, i.e., the respiratory bronchioles, alveolar ducts, alveolar sacs, and alveoli.[1] Alveoli are microscopic, thin-walled sacs that facilitate the exchange of gases between the inspired air and capillaries in the alveolar wall. The alveolar surface area is enormous, approximately 100 m[2] in an adult human and represents the contribution of approximately 300×10^6 alveoli.[1–3]

The alveolar wall in the adult lung consists of a narrow connective tissue core that contains fibroblasts, pericytes and capillary endothelial cells plus extracellular matrix components, most importantly, those comprising the elastic fiber. The alveolar epithelium is made up of two cell types, i.e., alveolar type I (AT1) cells and alveolar type II (AT2) cells. Alveolar type I cells are thin, flattened cells that, together with the capillary endothelial cell and the fused basal laminae of the capillary and epithelium, form the gas-blood exchange surface. Alveolar type I cells cover about 90% of the alveolar surface area.[4] The AT2 cell is more compact, occupies less than 10% of the alveolar surface, and is frequently located in the corner of an alveolus. The AT2 cell is a progenitor for renewal of the damaged alveolar epithelium, as AT1 cells do not divide.[5] Alveolar type II cells secrete pulmonary surfactant, a lipoprotein that spreads on the alveolar aqueous lining layer and reduces its surface tension.[6]

In the human, the formation of pulmonary alveoli begins at the end of gestation and continues after birth.[1] In other species, alveolarization can be predominantly postnatal (in rats and mice) or predominantly prenatal (in guinea pigs and rabbits).[7] Regardless of the timing, alveolarization involves a similar process in all mammals. An impairment of alveolarization has been implicated in the pathogenesis of bronchopulmonary dysplasia (BPD), a disease that affects prematurely born human infants.[8] Damage to alveoli, with a resulting decrease in gas-exchange surface area, is also involved in the pathogenesis of emphysema.[9]

Since writing the prior version of this chapter for the first edition of this book in 2002, extensive progress has been made in understanding the origins of alveolar epithelial and mesenchymal cells, paracrine signaling between them, the cell biology of the AT2 cells, and regulation of elastic-fiber assembly. This revised chapter will focus on these aspects of alveolar development and the reader is referred to the prior version for details about mechanisms that were already firmly established in 2002. Additional excellent reviews on topics that were covered in more detail in the prior version have been published and are resources for understanding the biochemistry and physiology of surfactant,[6,10,11] the transcriptional control of lung development,[12] glucocorticoids,[13,14] and how these concepts are translated into the treatment of infants with bronchopulmonary dysplasia.[15,16]

Alveolarization is a relatively late event in lung development, occurring after branching morphogenesis has laid down the conducting airway system.[17] However, some of the regulatory factors that control alveolarization are also involved in lung bud formation and branching morphogenesis, which will be briefly reviewed.

STAGES OF LUNG DEVELOPMENT

Lung Bud Formation and Branching

The lung begins as a diverticulum from the embryonic foregut, at about four weeks post-fertilization in the human.[1] The foregut and its diverticula (such as the lung) are lined with epithelial cells that are derived from the endodermal germ layer. The lung diverticulum is covered with splanchnic mesoderm, which gives rise to the connective tissue components of the lung (i.e., the cartilages, smooth muscle, blood vessels, etc.). The lung diverticulum almost immediately displays aspects of the right/left asymmetry characteristic of the adult lung. Specifically, lung buds in the human initially form three diverticula in the right lung bud and two in the left that correspond to the lobes of the adult lungs.

The Lung. http://dx.doi.org/10.1016/B978-0-12-799941-8.00004-3

The developmental process most characteristic of very early lung development is branching morphogenesis. The endoderm-lined ducts of the early lung buds undergo a stereotypic dichotomous branching process that gives rise initially to the primary bronchi, then the secondary (lobular) bronchi, the tertiary (segmental) bronchi, and so forth.[12] Branching morphogenesis in early lung development involves the stabilization of the linear portion of the distal ducts, the creation of a cleft region at the rounded tips of the ducts and growth on either side of the cleft, a process that results in branching of the terminal portion of the duct and in this way creation of a new order of airways. The regulation of branching morphogenesis involves epithelial-mesenchymal interactions via extracellular matrix components and growth factors.[12]

Pseudoglandular Phase

Repeated dichotomous branching is characteristic of the first phase of lung development, called the pseudoglandular phase, which occurs in the human from about 6 to 16 weeks of gestation and days 11.5 through 15.5 in the mouse.[1,18] Almost all of the pulmonary conducting airways are created during this process, about 20–22 orders of airways in the human.[17] Differentiation of the epithelium of the conducting airways also commences during this early phase of lung development.[1] The most distal aspects of the branching duct system are eventually remodeled giving rise to the alveolar region of the lung. Epithelial cells in the distal region remain undifferentiated during this phase of lung development; they are tall columnar cells with no specialized features apparent other than large pools of intracellular glycogen.[19]

Canalicular Phase

Remodeling the distal portions of the branching duct system to gas-exchange units begins during the canalicular stage of lung development. This stage of lung development is characterized by a tremendous increase in the number of capillaries in the connective tissue between the terminal ducts and the beginning of the differentiation of the presumptive alveolar epithelium.[19] In the human, this stage of lung development lasts from 16 to 24 weeks of gestation and from days 15.5 through 17.5 in the mouse.[1,18] Some of the distal epithelial cells become more cuboidal and begin to synthesize and secrete surfactant into the amniotic fluid. The newly differentiated type II cells are characterized by decreased intracellular glycogen pools, microvilli on their apical surface, and the appearance of lamellar bodies, the intracellular organelle that stores pulmonary surfactant.[19] At the same time, capillaries in the interstitium induce the overlying epithelial cells to flatten and differentiate into AT1 cells. Factors that regulate vasculogenesis and angiogenesis in the developing lung include vascular endothelial growth factor (VEGF), fibroblast growth factors (FGFs), and their receptors.[20] Type I and type II alveolar epithelial cells differentiate in response to a number of regulatory factors.[21] Thus, some aspects of alveolar development, in particular alveolar epithelial cell differentiation, begin during the canalicular stage of lung development. However, in the human, true alveoli do not begin to form until much closer to birth.[1]

Saccular Phase

The saccular stage is the last phase of fetal lung development and occurs from about 24 weeks of gestation to term in the human and from E17.5 through P4 in mice. During the saccular phase, the terminal portions of the branching duct system (the terminal sacs) enlarge and may undergo further branching to give rise to alveolar ducts and alveolar sacs.[1] In the human, some true alveoli form late in gestation, however, it is thought that the bulk of alveolar formation (~85%) occurs after birth.[1] Premature infants born with lungs in the late canalicular/early saccular stages can survive with surfactant therapy and ventilation, although they are at risk to develop BPD.[8]

At birth, gas exchange occurs in the air sacs residing at the termini of the conducting airways that arose through branching-morphogenesis. Alveolarization involves subdivision and remodeling of the sacs, which then become the alveolar ducts. Because alveoli arise from the terminal air sacs, their genesis is determined by characteristics of the epithelial and mesenchymal cells that previously resided at the distal tips of the airways *in utero*. The expansion and differentiation of these distal epithelial and mesenchymal cells has been studied using lineage-tracing in mice. This approach requires identifying a spatially and temporally regulated gene and using its promoter-regulatory region to drive expression of a marker, which persists in the progeny of the parent cells.[22] The marker enables one to follow the expansion of the parent population, trace the movement of the progeny away from the parental location, and to localize proteins that signify differentiation of the progeny. The marker is typically a fluorescent protein that is expressed after activation by Cre-recombinase, which is driven by the spatially and temporally regulated gene. In some cases, the Cre-recombinase is also controlled by an inducible tetracycline or estrogen responsive element. Because the unique marker gene has been permanently activated in a subset of cells, these tools enable one to pinpoint the time when and the location where the parent cell population is expanding, and then trace the fate of their progeny through subsequent growth and differentiation.

The Formation of Alveoli (Alveolarization)

The process of alveolarization has been described in the greatest detail in mice, rats, and humans.[1] During the late canalicular and terminal saccular stages of prenatal lung development, when the terminal conducting airways are enlarging rather than branching, there is a progressive loss of the mesenchymal cells that separate capillaries from the epithelium lining the future air spaces. This yields a rudimentary gas exchange surface that can support the earliest respirations. Shortly after birth, in both mice and humans, the surface area of the air-blood interface begins to increase markedly as the terminal saccules become the alveolar ducts and these in turn give rise to alveolar sacs and the alveoli. Major paracrine interactions, which regulate the formation of saccules and alveoli, are summarized in Figure 1.

Secondary Septation

Alveolar formation in the developing lung has been divided into several phases, which probably reflect different processes. The first phase, termed secondary septation or alveolarization, is characterized by the formation of new alveolar septa by septation of the terminal sacs. Septation is initiated by the protrusion of secondary crests from primary septa. The primary septa, which are the walls of the terminal sacs, are comprised of a central core that contains fibroblasts and other connective tissue components, surrounded on either side by capillaries and the epithelium of the terminal sac.[23] These same components are also found in the secondary crest. Capillaries are interconnected in the primary septa, but the two capillary layers formed in the secondary septa

during septation initially have few interconnections. On postnatal days 2 to 4 in the rat, fibroblasts proliferate at the site of origin of the secondary crest, and this causes the secondary septum to lengthen and project perpendicularly into the alveolar sac.[24] The secondary crests are usually adjacent to elastic fibers and arise where the capillaries in the primary septum can be folded up.[1,25] The tips of the secondary crests are occupied by myofibroblastic cells that are located next to elastic fibers.[26] After postnatal day 4 in the rat, fibroblast proliferation decreases in the proximal aspect of the secondary septum, but persists at a higher level in the distal septal tips.[24] Lengthening of the secondary septae is accompanied by an increase in the mass of lamellar bodies in type II alveolar epithelial cells and a corresponding increase in surfactant secretion.[1] The levels of surfactant produced in the lung increase in parallel with increasing surface area in the distal portion of the developing lung.[27]

Remodeling of the Alveolar Wall

The second phase of alveolarization is marked by the further lengthening and thinning of the secondary septae, primarily via the loss of interstitial mesenchymal cells and extensive capillary remodeling.[1] During the remodeling of the vascular elements in the secondary septae, the original dual capillary system becomes a single capillary system

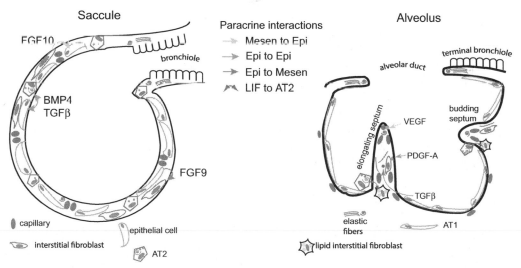

FIGURE 1   Paracrine signaling regulates cellular proliferation and differentiation. In the saccular phase, epithelial (Epi) cells are fewer and less attenuated, although some possess lamellar bodies (small dots) characteristic of alveolar type II (AT2) cells. Epithelial cells continue to proliferate, through Nmyc and/or Kras-mediated signaling initiated by fibroblast growth factor-10 (FGF10) and through bone morphogenetic protein-4 (BMP4) acting through anaplastic lymphoma kinase-3 (Alk3). Transforming growth factor-β attenuates proliferation as epithelial cells begin to flatten. Epithelial and other mesenchymal (Mesen) cells signal to mesenchymal cells through FGF9, which maintains proliferation. Capillaries form but are fewer, compared to alveoli, and maintain a paired configuration. During the alveolar phase, epithelial cells continue to differentiate with an increase in lamellar bodies, and formation of tubular myelin (small spirals) distinctive of AT2 cells. Some epithelial cells differentiate and spread, becoming alveolar type I (AT1) cells. Platelet-derived growth factor-A (PDGF-A) promotes interstitial fibroblasts proliferation. Some fibroblasts migrate within the elongating alveolar septum, whereas others (lipid interstitial fibroblasts) remain at the septal base. Elastic fibers become more numerous during the alveolar phase and are most abundant at the entry ring, where the alveolar duct empties into the alveoli. Vascular endothelial growth factor (VEGF) fosters new capillaries, which are more diffusely distributed along the narrowing alveolar epithelial–endothelial diffusion barrier. Lipid interstitial fibroblasts provide lipids and signal to differentiating AT2 cells. Directed by TGFβs and other factors, some interstitial fibroblasts acquire myofibroblast characteristics, containing alpha-smooth muscle actin (αSMA) and synthesizing tropoelastin. (This figure is reproduced in color in the color plate section.)

in which the capillaries are interconnected through a process that has been termed intussusceptive microvascular growth.[28] This involves the fusion of adjacent capillaries by interconnecting endothelium-lined tissue pillars. During the period of exuberant new septal formation, the surface area of the lung increases to the 1.6th power of lung volume.[23]

Growth

Traditional morphometric approaches revealed that alveolarization continues through the third postnatal week in rats and through year 2 in humans.[1] The lung then enters the third phase of alveolar formation, a growth phase when the gas-exchange surface area increases to the 0.7th power of lung volume, a factor consistent with an isotropic expansion of the alveoli.[23] This process is accompanied by an increase in the volume of air space in the lung, at the expense of alveolar septal tissue mass. During the growth phase of alveolarization, the volume density of AT1 and AT2 epithelial cells increases roughly in proportion to the increase in lung volume.[1] In mammals, this phase of lung development usually ends prior to the termination of the growth of long bones and of the increase in lean body mass.

This view has been revised with the advent of new technology, including synchrotron radiation X-ray tomographic microscopy and hyperpolarized helium magnetic resonance image. After correcting for the change in volume due to lung growth Mund and associates estimated the length density of septal tips from birth to 125 days in mice by quantifying the "lifting-off" of new septa, which are observed three dimensionally as low ridges, from pre-existing ones.[29] The investigators divided septation into two phases: postnatal day-2 (P2) through P14, and P14 through adulthood. During the first phase, secondary septa lift off the pre-existing primary septa (saccules at birth) and during the second phase they ascend from already formed secondary septa, while there is continued lengthening of those formed in the first phase. The increase in new alveolar septa (the change in the number of free septal tips between two time points) was compared to the increase in lung volume over time. The comparison showed that the growth in septal number was greater than the growth in volume from P4 through P7, equal from P7 to P36, when the formation of new septa ceased. After normalizing to the number of free septal tips in adults, they calculated that 10% of alveolar septa were present at birth, 50% were formed between P4 and P14, and an additional 40% formed from P14 through P36.[29]

Following right pneumonectomy, it was reported that as the human left lung expanded to fill most of the residual right pleural cavity, the radial dimensions of the terminal airways approximated the expected dimensions in a normal adult human lung.[30] This argued against expansion from the stretching of pre-existing acini. Furthermore, the alveolar depth was significantly below normal, suggesting that new (and smaller) alveoli had formed to populate the acinar space. Although the conditions driving alveolar growth after pneumonectomy are at least quantitatively different from those present during development, qualitatively similar mechanical and biochemical factors probably apply to both.[30]

Interspecies Comparison of Alveolarization

Rats and Mice

As previously stated, alveolarization largely occurs postnatally in rats, mice, and humans. At birth, the gas-exchange portion of the lung consists primarily of immature terminal saccules with some secondary septa. Elastin is a critical structural protein in the primary and secondary septa. Tropoelastin (TE, the soluble precursor of elastin) mRNA is present in the developing lung during the pseudoglandular stage of development but only in the walls of airways and blood vessels.[31] Elastin synthesis in the primary alveolar septa begins during the canalicular stage of lung development.[32] Alveolar septation is accompanied by an approximately 4 to 6-fold increase in parenchymal desmosine (a cross-linked amino acid that is unique to elastin and is a marker for elastin fiber deposition) and an approximately 6-fold increase in hydroxyproline residues, which are primarily found in collagen.[33,34] Postnatal elastin accumulation has also been studied in mice, which have a developmental pattern similar to that observed in rats, with the most abundant deposition of cross-linked elastin occurring between postnatal days 9 and 20.[35]

Humans

Human newborn lungs contain approximately 50×10^6 alveoli (around 18% of the alveolar number found in adults), data suggestive that alveolar septation starts in utero and finishes after birth in the human.[2] Postnatal alveolar septation occurs at a more constant rate in humans than in rats; children do not undergo the rapid burst in septal outgrowth that is observed during early postnatal life in rats. In children, septation ends sometime between 2 and 5 years of age.[1] In humans, peripheral lung desmosine can be detected during the final 10 weeks of gestation, but the major ~6-fold increase in its concentration occurs during the first 2 years of life, during alveolar secondary septal formation.[36] During this same period, lung hydroxyproline, a marker for collagen, increases approximately 3-fold when adjusted to dry lung weight.[34] Despite the differences in the kinetics and duration of the alveolar phases in the two species, the fold enlargement of alveolar surface area from birth to adulthood is very similar in rats (21.4-fold) and humans (20.5-fold).[2]

Other Mammals

Unlike rats, mice, and humans, certain other mammals develop majority of their alveoli prior to birth. In rodents such as the rabbit and guinea pig, the majority of the internal surface area of the lung is acquired prenatally.[37] Unlike humans, monkeys develop nearly all of their alveoli prior to birth ($26.2 \times 10^6/cm^2$ in newborns compared to $26.6 \times 10^6/cm^2$ in the adult).[38] Similarly, in sheep, alveolar septation primarily occurs prenatally.[39,40] Secondary septa in fetal sheep double in number during the final quarter of gestation and alveolarization is accompanied by both thinning of the alveolar septa and maturation of capillaries, events which occur primarily postnatally in humans and rats. Alveolarization in the sheep fetus is accompanied by a 20 to 30% increase in parenchymal elastin and an approximately 10% increase in parenchymal collagen over the same interval.[39, 40]

Development of the Alveolar Epithelium

The differentiated alveolar epithelium consists of two unique cell types, the alveolar type I (AT1) and alveolar type II (AT2) cells.[19] Both of these cell types arise from the endoderm-derived epithelial cells that line the distal portion of the branching duct system in the developing lung. The differentiation of alveolar epithelial cells commences in the human fetus prior to the formation of true alveoli, which is primarily a postnatal event.

Alveolar Epithelial Progenitors

Alveolar type II cells can divide and give rise to new AT2 cells or differentiate into AT1 cells.[41,42] There is little evidence that type I cells can divide; therefore, most investigators consider the type I cell to be terminally differentiated. When the lung epithelium is injured, AT2 cells divide and cover the injured area and this is followed by the gradual reappearance of type I cells.[5,41,42]

Following publication of the prior version in the first edition of this book, new mouse models and tools for lineage tracing have expanded our understanding of the origins of AT2 and AT1 cells. The traditional view is that AT2 cells arise from the distal tubules during the canalicular and saccular stages and are the major source for AT2 and AT1 cells during secondary septation. Fate mapping of inhibitor of differentiation-2-expressing (Id2)+ progenitors supports this view.[43] Earlier work identified a distinct population of epithelial cells at the termini of the distal tubes during the pseudoglandular phase.[44] These cells exhibited high levels of bone-morphogenetic protein-4 Bmp4, sonic hedgehog (Shh), Nmyc-proto-oncogene, sex-determining region-Y, box9 (Sox9), and Id2. Distally, the cells exhibited increased Nmyc-driven proliferation compared to epithelial cells located in the more proximal regions of the airways. Wnt7a and Wnt2 both contribute to proliferation, although the proportions of mesenchymal and epithelial cells remain normal, resulting in small but structurally intact terminal airway units at birth. Wnt5a may provide a brake on proliferation mediated by the other two Wnt ligands.[45]

Lineage-tracing using genes activated during the pseudoglandular stage has defined their progeny during the saccular and alveolar stages. Using Id2 driven by an inducible Cre-recombinase to mark progeny with the fluorescent label dTomato (dTom), Rawlins and associates showed that epithelial cells that were labeled at embryonic day (E)11.5 gave rise to all epithelial lineages (Clara, ciliated, AT2, and AT1) except neuroendrocrine bodies.[43] When tamoxifen was used to induce Cre at E16.5 (canalicular stage) only AT2 and AT1 Id2 progeny (dTom- labeled) were observed. These data demonstrate that Id2-progenitors are multipotent and divide asymmetrically, with later divisions restricted to alveolar epithelial cells. With further asymmetric division, the proximal epithelial cells become more lineage restricted. Notch signaling is critical for determining whether Id2+ progeny differentiate into secretory (Clara) cells, which require active notch signaling, or into ciliated cells that differentiate in the absence of Notch signaling.[46,47]

Although this is the dominant paradigm during development, two additional sources of AT1 cells have been identified in injured adult lungs, and could influence secondary septation in diseased lungs (bronchopulmonary dysplasia, BPD): (a) the bronchoalveolar duct junction (BADJ) and (b) distal epithelial cells bearing α6β4 integrin. Epithelial progenitors at the BADJ, sometimes called bronchoalveolar stem cells (BASC), contribute to the alveolar epithelial surface during response to alveolar injury.[48,49] Their phenotype has been characterized using flow cytometry as CD31-, CD45- (neither endothelial nor derived from bone marrow stromal cells), EpCam+ (epithelial cell adhesion molecule, a transmembrane glycoprotein involved in homotypic cell-cell adhesion), Sca1+ (stem cell antigen-1), α6-integrin (CD49b)+, and integrin-β4 (CD104)+. Bronchoalveolar stem cells expressing both α6 and β4-integrins also express Clara cell 10 protein (CCSP, uteroglobin, SCGB1A1, Clara cell secretory protein, CCSP), typical of Clara cells. A subset of BASCs, which lacks integrin β4, gives rise to AT2 cells.[48]

Although this pathway may be activated in acutely damaged lungs, it probably does not contribute to normal alveolar development. Rawlins and associates developed a CC10CreER knock-in which enabled them to lineage-trace cells with characteristics of Clara cells.[50] Reasoning that if the BADJ contributed to AT2 cells, they pulse labeled with tamoxifen at E18.5 and analyzed Cre-mediated lineage-labeling at P2, P21, and in adults. They observed a progressive increase in the number of labeled Clara and ciliated cells in the terminal airways, but none of these stained for pro-SP-C (a marker of AT2 cells). In contrast only 2–5% of pro-SPC staining AT2 cells labeled and did not expand. This suggests that unlabeled AT2 cells proliferated at the same rate as the few lineage-labeled AT2 cells, arguing that Clara cells do not give rise to alveolar AT2 cells during normal saccular development.

Barkauskas expanded on the observations of Rock and Rawlins that used a Clara cell specific, estrogen-sensitive (CC10CreER) lineage tracing of progeny from the BADJ in adult mouse lungs after injury.[50–52] In this study Barkauskas and associates used a SP-C and estrogen driven Cre construct (SPC-CreER) to study clonal expansion of lineage-labeled AT2 cells. In mice that bore both Cre-driven constructs, they observed that new AT2s predominantly repopulated injured regions from other AT2 cells rather than progenitors at the BADJ (although both sources contributed).

At P21 Rawlins and coworkers also examined the progeny from dams which received the estrogen-congener tamoxifen on day E18.5, and their litters were maintained in a hyperoxic environment from P0 through P7, and then transferred to normoxic conditions.[50] They did not observe a progressive increase in labeled AT2 cells proximate to the BADJ, nor did hyperoxia influence the number of labeled cells. There was an increase in AT2-proliferation after hyperoxia, suggesting that AT2 cells proliferated and then differentiated to AT1 cells. The lineage-labeled cells increased in both the trachea and bronchi during postnatal lung growth, suggesting Clara cells served as progenitors in these more proximal regions.

The second source in adults, integrin α6β4+ epithelial progenitors, has been observed in mouse and human lungs.[53,54] These progenitors are not limited to the BADJ and are scattered more diffusely along the distal alveolar walls. They can form colonies ex-vivo and divide asymmetrically with some cells retaining progenitor characteristics whereas others acquire either CC10 or SP-C.

Regulation of Alveolar Epithelial Cell Differentiation

Many factors regulate the differentiation of the alveolar epithelium in the fetal lung. They include transcription factors, hormones, growth factors, regulatory agents such as cyclic AMP, neurotransmitters, and physical forces such as stretch. Bioinformatic and genomic approaches have broadened the list and identified regulatory networks, with specific transcription factors at their hubs.[55,56]

Epithelial–mesenchymal interactions are well-established mediators of lung development and are also likely to be specifically involved in alveolarization.[12,57] The composition of the extracellular matrix beneath the alveolar epithelium has dramatic effects on the differentiation status of alveolar epithelial cells.[58] In addition, the growth of the pulmonary vasculature, in particular the capillary network

forms in the primary and secondary septal walls, influences the differentiation state of the overlying epithelium, probably promoting an AT1 cell phenotype in the cells that directly overlie capillaries.[1,59] Alveolar type II cells are localized preferentially over cables of elastin in the alveolar wall.[60]

Differentiation of the Alveolar Type I Cell

The thin flattened AT1 cell is an important component of the air-blood barrier. It overlies capillaries in the alveolar wall and comprises the vast majority of the surface area of the alveolar wall (~90%). Commonly used markers of AT1 cells are T1α (podoplanin) and aquaporin5 (Aqup5). T1α is the best characterized marker and is a plasma membrane protein.[61] In the lung, it is expressed only in AT1 cells, but T1α is also present in epithelial cells of the ciliary body in the eye and of the choroid plexus of the brain.[62] T1α is induced in type II cells undergoing a phenotypic transition to a type I-like appearance in vitro.[63] Aquaporins are a family of water channels that are expressed ubiquitously. However in the distal portion of the human lung, aquaporins 4 and 5 seem to be relatively restricted to the AT1 cell while aquaporin 3 is expressed in the AT2 cell.[64]

Based on labeling studies following inhalational injury, it has long been held that resident AT2 cells can proliferate and then differentiate into AT1 cells.[5] This process is regulated by an interplay between bone morphogenetic protein-4 (BMP4) and transforming growth factor-β (TGFβ). BMP4 favors retention of the AT2 phenotype (production of surfactant proteins), whereas TGFβs favor differentiation into AT1 cells (expression of aquaporin 5 and podoplanin, T1α).[65] BMP4 belongs to the TGFβ-superfamily and signals through distinct type 2 receptors (BMPR2 versus TGFβR2) and a more limited set of type1 receptors (anaplastic lymphoma kinase, Alk5 for TGFβs and Alk3 for BMP4) and different regulatory Smads (2,3 for TGFβs and 1, 3, 5 for BMP4).[66] Adult rodents have been used to characterize this pathway and its contribution to secondary septal formation remains unclear.[67] However, this pathway may be relevant to BPD where the epithelial barrier is compromised and injured from hyperoxia and positive pressure ventilation.

Gap junctions are interposed between adjacent alveolar type I and II cells in the alveolar wall.[68] The expression of relatively unique complexes of connexins have been described in alveolar type I and type II cells.[69] Alveolar type I cells are also characterized by abundant caveolae and the expression of caveolin 1 protein while AT2 cells possess few caveolae and express little caveolin 1 protein.[70] Caveolae are plasma membrane structures that may be important in mediating the transport of materials across a cell and also may concentrate signaling mediators.[71]

Claudins are a family of tetraspan transmembrane proteins that contribute to the extracellular portion of the tight junctions of alveolar epithelium, most prominently in AT1 cells.[72] They interact with intracellular zona-occludins-1 (Z0-1), -2 and other scaffolding proteins that tether the tight junctions to the actin cytoskeleton and with other transmembrane proteins such as occludins. Their structure is maintained by critical extracellular intramolecular disulfide bonds that are required for heterotypic or homotypic claudin-claudin juxta-cellular connections. Claudins physically connect the extracellular tight junctions on adjacent cells whereas occludins stabilize claudins and regulate their turnover. Claudins-3, -4, and -18 are the predominant claudins found in the alveolar epithelium, with claudin-3 being more abundant on AT2 than AT1 cells and increasing the alveolar permeability between AT1 and AT2 cells. In contrast, claudin-18 strengthens the tight junctions and reduces barrier permeability. The abundance of claudin-4 is more dynamically regulated and its gene expression is more sensitive to agents or diseases that cause epithelial injury. Paracellular permeability is influenced by the types and proportions of the various claudins that comprise the tight junctions; the balance is disrupted in diseases such as cystic fibrosis. The pattern of claudin gene-expression changes during fetal lung development where claudin-3 is gradually superseded by claudin-4.[73] The grainyhead like-2 gene is is an important developmental regulator of claudin-4 and epithelial (E-) cadherin gene-expression, and is required for the formation and maintenance of both adherens and tight junctions.[74]

Differentiation of the Alveolar Type II Cell

Alveolar type II cells are usually cuboidal in shape and occupy only a small portion of the alveolar surface area, ~7%.[3] However, there are about twice as many AT2 cells as AT1 cells in the lung.[75] Alveolar type II cell differentiation begins as early as at about 24 weeks gestation in the human (and about day 18 of gestation in the mouse fetus).[1,12] Differentiated AT2 cells produce pulmonary surfactant, a lipoprotein substance that is required for proper lung function P.[10] Premature human newborns frequently do not have adequate numbers of differentiated AT2 cells and thus do not produce sufficient amounts of pulmonary surfactant. Inadequate amounts of surfactant can lead to neonatal respiratory distress syndrome (RDS), which until recently was a leading cause of death in premature newborns.[76,77] Intratracheal instillation of exogenous surfactant has reduced early mortality in prematurely born infants, but those that survive may develop chronic lung disease of the newborn with persistent alveolar structural abnormalities.[78]

Surfactant Phospholipids

Surfactant is comprised of about 80% phospholipids, 10% cholesterol, and 10% protein.[10] The most abundant class of phospholipids in pulmonary surfactant is phosphatidylcholine, in particular dipalmitoylphosphatidylcholine (DPPC), which is the primary surface-tension lowering component in surfactant.[10] Prenatally phosphatidyglycerol (PG) and phosphatidylinositol (PI) comprise a larger fraction of surfactant phospholipids but they diminish greatly by or after birth.[10,79] The fatty acids which contribute to the surfactant lipid pool are synthesized *de novo* in AT2 cells (particularly during late gestation) from glycogen or lactate or are acquired from the circulation or other cells within their niche (lipofibroblasts).[10,80] The enzymatic activity of fatty acid synthase, an important regulatory enzyme, is increased by keratinocyte growth factor (FGF7), glucocorticoids, and decreased by thyroid hormone and TGFβ.[81] Approximately 25% of PC is derived from *de novo* synthesis, whereas the remainder comes from remodeling of lysoPC.[10]

The synthesis of surfactant phospholipids in differentiating AT2 cells is accompanied by several morphologic changes in the type II cell.[19] First, lamellar bodies (LBs), which are the intracellular storage form of pulmonary surfactant, appear in the cytoplasm.[19] LBs serve as reservoirs for lipids that are delivered from the ER via the Golgi and where they are mixed with surfactant proteins, which are arriving from the multivesicular body (MVB).[6] As yet unspecified, intracellular phospholipid transfer proteins shuttle lipids to the outer membrane of the LB where the ATP-binding cassette (ABC) transporter, ABCA3 is required for entry of lipids into the LB. ABCA-null mice lack mature LB and alveolar surfactant is markedly reduced.[82] Lipids may travel through the cytosol bound to phospholipid transfer proteins (PLTPs) or alternatively there may be fusion between the outer LB and the ER.[6] Exocytosis of surfactant from the LB is regulated through a variety of signaling pathways including purinergic, protein kinase A, and through calcium shifts. Following exocytosis, the LB contents are mixed with surfactant protein-A (SP-A) which is instrumental in tubular myelin formation and in determining the size of surfactant aggregates in the alveolar layer, which resides in the extracellular air-space.[6,11] Secreted lamellar bodies undergo a structural transformation to form tubular myelin, a surfactant intermediate that is thought to give rise to the monolayer of surfactant that lines the alveolar aqueous lining layer and reduces its surface tension.[19] Tubular myelin contains regions with a more uniform mixture of lipids and proteins, as well as less ordered regions where the concentrations of the two components vary as new constituents are added and others are removed for recycling. Surfactant surface tension lowering activity is reduced by unfavorable extracellular mixing, which alters the distribution of lipids and proteins, oxidation, or by aqueous dilution. The surface tension lowering capacity is restored through surfactant recycling and re-packaging. Studies using atomic force or epifluorescence microscopy suggest that the extracellular surfactant layer folds upon itself as the alveolar volume decreases during expiration and unfolds again during inspiration.[6,83] The surfactant proteins mediate interactions between the folded lipid layers and between the lipids and the cell surface.

Another characteristic morphologic change in the differentiating AT2 cell is the disappearance of the large glycogen pools, which are characteristic of the tall columnar undifferentiated precursor cell that lines the terminal portion of the branching duct system in the developing lung.[19]

Surfactant Proteins

SP-A and SP-D

The surfactant-associated proteins are an important component of pulmonary surfactant.[84] Four surfactant associated proteins have been identified to date: SP-A, -B, -C, and -D.[84] All are expressed in AT2 cells and all are developmentally regulated.[84] SP-A is a ~35 kDa glycoprotein and is the most abundant surfactant protein, which along with calcium is instrumental in converting surfactant within the lamellar bodies to tubular myelin on the alveolar epithelial surface.[85] SPA is also critical for surfactant recycling. SP-A along with SP-D, a ~43 kDa glycoprotein, are members of the Collectin (collagen containing C-type lectin) family of proteins, which enhance innate immunity by binding to a variety of microorganisms.[85] SP-A gene-deleted mice have normal lung structure and function but are more susceptible to infection with several pulmonary pathogens than intact wild-type mice.[86] SP-D gene-deleted mice have abnormalities in surfactant metabolism and accumulate large amounts of surfactant in their alveoli.[87] The surface tension lowering properties of their surfactant appear to be normal.[87]

SP-B and SP-C

The more hydrophobic surfactant proteins, SP-B and SP-C, are the most instrumental in surface tension lowering and interact with the lipid components of surfactant. SP-B is a small (~6.5 kDa), extremely hydrophobic protein that facilitates the spreading of surfactant proteins on the alveolar surface.[6] SP-B is synthesized as a high molecular weight precursor and cleaved to the active molecule in the multivesicular body, where it facilitates the organization of surfactant membranes and the transport of the surfactant proteins into the lamellar body.[6,88] In the absence of SP-B, lamellar bodies develop vesiculations. SP-B gene-deleted mice die immediately after birth due to respiratory distress.[89] In humans, genetic mutations in the SP-B gene cause congenital alveolar proteinosis, a lethal condition in human newborns that can only be treated by lung transplantation.[84] The deficiency in SP-B production in these patients is also associated with a defect in the intracellular processing of the SP-C precursor protein.[84] In the human, SP-B is expressed in AT2 cells and in Clara cells of the conducting airways.[6]

SP-C is another low molecular weight (~5 kDa), extremely hydrophobic surfactant protein.[6] SPC is an integral membrane protein of the ER and the portion that resides in the ER stabilizes the hydrophobic protein prior to transfer to the Golgi where it is palmitoylated. SP-C gene-deleted mice have almost no phenotypic differences from wild-type mice other than a tendency of their surfactant to have atypical biophysical properties.[90] In humans, SP-C is expressed only in AT2 cells of the lung.[6]

PARACRINE SIGNALING TO EPITHELIAL CELLS

Fibroblast Growth Factor-10 (FGF10) Promotes Epithelial Proliferation

Signals from the mesenchyme regulate epithelial proliferation and differentiation during development of the airways, saccules and alveoli. FGF10, which is produced by the surrounding sub-mesothelial mesenchyme, is an important paracrine signal that regulates the expansion and differentiation of the distal tubular epithelium.[45] FGF10 favors proliferation and migration of the epithelium and the acquisition of branch points, whereas a loss of FGF10 function leads to fewer progenitor cells and branch points. Overexpression of FGF10 leads to differentiation-arrest and an excess of progenitors. In the embryo, FGF10 expression progressively increases from E11.5 to E18.5.[91]

Although the effects of FGF10 on epithelia have been most extensively studied during the pseudoglandular stage (its absence leads to diminished branching and absent lobes), FGF10-signaling also impacts the saccular and alveolar stages.[92] Diminished FGF10 reduces epithelial proliferation, Wnt canonical signaling, and the number of thyroid transcription factor-1 (TTF1) and SP-B positive cells at birth.[45] At postnatal day 2, mice that are haploinsufficient for FGF10 exhibit very dilated saccules, mesenchymal thinning, and diminished pulmonary capillary density. There is also a marked decrease in alpha-smooth muscle actin (αSMA, ACTA2) in the saccules at P1.[45] FGF10 is also produced by the mesenchyme surrounding the airways and blood vessels. In the saccular walls, FGF10-producing cells also express the adipose-differentiation related protein (ADRP, perilipin-2), which is found in adipocytes.[45] During the alveolar phase, FGF10 expression is limited to the lipid-laden interstitial fibroblasts.[93]

FGF10 binds to FGFR2b and when a dominant negative decoy-protein was used to disrupt this receptor in mice during E14.5-E18.5, postnatal alveolar septation was arrested.[94,95] There was a concomitant decrease in the progression of PDGFRα+ alveolar fibroblasts to myofibroblasts.[95] Conversely, allelic depletion of FGFR2c (the receptor on mesenchymal cells for FGF9) expression caused mesenchymal cells to undergo autocrine stimulation by FGF10.[92] This results in increased mesenchymal proliferation, but diminished differentiation (diminished αSMA, elastin and fibronectin). As was observed in the dominant negative FGFR2b model, the neonatal gas exchange region was dilated and septa were diminished.

Sox9 Restrains Epithelial Differentiation until Airway Branching Has Been Optimized

During branching, cells in the distal tips proliferate, are repositioned, and extend along the proximal-distal axis. During the canalicular phase, the epithelial cells begin to flatten with shrinkage of the surrounding mesenchyme, leading to cystic areas at the termini of the ducts. With remodeling, the enlarged and thinner distal tubules form the terminal air sacs that are characteristic of the saccular phase. Sox9 is expressed in the distal tubule epithelium that surrounds the terminal sacs.[58] Targeted (Shh-Cre) deletion of epithelial Sox9 (loss of function, LOF) produces fewer but dilated sacs whereas Sox9 gain of function (GOF) increases the abundance of these cells, which eventually occupy saccular lumens at E16.5.[96] Epithelial cells expressing Sox9 were more proliferative and Sox9 was most abundant where the bifurcated branches were growing away from the branch point. Sox9 GOF inhibited differentiation and as the number of Sox9+ cells was increased at E18.5 fewer had differentiated into AT2 (SP-C) and AT1 (Aqup5) cells, whereas Sox9 LOF led to premature differentiation. In addition to increasing epithelial proliferation, Sox9 also played critical roles in formation of the basal lamina, the microtubular cytoskeleton, and in supporting epithelial migration.

Using Shh-Cre or Sox2-Cre to specifically lineage-trace epithelial cells in the distal and proximal airways, respectively, Chang and associates observed that Sox9+ epithelial cells maintained a less differentiated state (fewer expressed Sp-B or Aqup5).[96] Sox9 expression declined as differentiation increased and the distal tubules expanded into saccules. Active Sox9 expression was limited to the distal tubules where the epithelia were actively proliferating and expression was absent more proximally, where differentiation had occurred. Targeting the Sox9+ cells to over-express *Kras* dampened differentiation and increased proliferation, producing the same cystic structures that were observed by Rockich and associates.[58] In Sox9 LOF lungs, Chang and associates observed an increase in mesenchyme surrounding the distal tubules and a higher level of FGF10.[96] Concordantly, they observed that Sox9 gene expression required functional FGFR2-signaling through Kras, showing that Sox9 is a downstream target of FGFs. Inactivation of Sox9 enabled differentiation to commence as early as E12.5 within the most proximal portion of the developing lung. They proposed that Sox9 is an important brake on differentiation and suppresses the formation of saccules

until branching has sufficiently progressed to enable a full complement of terminal airways. They hypothesized that by so doing, Sox9 serves as an important integrator between the branching and saccular phases of lung development and ensures that the saccular complement is adequate to support respiration at birth.

DEVELOPMENT OF THE ALVEOLAR INTERSTITIUM

During fetal lung development, the interstitial mesenchyme in between the epithelium-lined branching ducts plays an important inductive role in conducting airway and terminal sac formation. During late gestation the mesenchyme gradually becomes more attenuated as there is a progressively closer apposition of the epithelium of the terminal sacs and the vasculature. During alveolarization, the interstitium must assume the function of providing structural support for the gas-exchange unit, which postnatally is under the phasic mechanical stress of respiratory movement.

Interstitial Fibroblasts

The interstitial fibroblast (IF) is the major synthetic cell in the interstitium. It is thought to produce much of the extracellular matrix (ECM) in the alveolar interstitium and to provide metabolic substrates to the epithelium. During alveolarization, four functions characterize IF development, i.e., proliferation, migration, synthesis of ECM components, and apoptosis. The IF are not synchronized in these functions, and since the IF population is heterogenous, some IF may be proliferating while others are migrating or involved in synthesis of ECM. Two populations of alveolar interstitial cells were observed during electron micrographic studies of the rodent lung parenchyma and were termed lipid-interstitial cells and non-lipid interstitial cells.[97,98] In addition to the presence or absence of characterizing lipid droplets, these two populations were notable for their rich endoplasmic reticulum and lacunae surrounding microfibrils and immature elastic fibers. When isolated from rats or mice during the first two postnatal weeks, the interstitial cells have characteristics of fibroblasts (lack cadherins specific to endothelial or epithelial cells), contain vimentin and/or alpha-smooth muscle actin (αSMA) intermediate filaments, and abundantly transcribe tropoelastin and fibrillar collagen mRNA.

Function and Signaling of Interstitial Fibroblast Progenitors

Some of the IF have characteristics of bi-potent progenitors.[99,100] McQualter and associates studied stem cell antigen-1 (Sca1) positive or negative cells that lacked cluster of differentiation-45 (CD45, marker of the hematopoietic lineage) and CD31 (marker of endothelial cells).[99] The CD45-, CD31-, Sca1+ cells had a gene-expression profile characteristic of fibroblasts and when cultured under permissive conditions, they accumulated lipid droplets. Whereas the CD45-, CD31-, Sca1- cells were comprised of both PDGFRα+ (CD140a+) and PDGFRα- cells, all of the CD45-, CD31-, Sca1+ cells were PDGFRα+ and CD34+. As mice approached adulthood, the CD45-, CD31-, Sca1+ mesenchymal population increased as a proportion of total CD45-, CD31- cells, indicating that the progenitors persist despite aging.

This same population of mesenchymal cells was subsequently shown to support the differentiation of CD45-, CD31, Sca1+ epithelial cells.[48] The epithelial cells were marked by epithelial cell-adhesion molecule (EpCam, CD326), and a lower abundance of Sca1 compared to the mesenchymal cells. When the EpCam+, Sca1^{low} epithelial cells were co-cultured in Matrigel with the EpCam-, Sca1^{high} mesenchymal cells, they differentiated into cells, that produce either mucin (MUC5AC) or surfactant protein (pro SP-C), typical of airway or alveolar epithelial cells, respectively, as well as Clara cells expressing Clara-cell specific protein (CCSP, CC10). Differentiated epithelial cells were also obtained if the Sca1+ mesenchymal cells were replaced with fibroblast growth factor 10 (FGF10) or hepatocyte growth factor (HGF). This has been observed by others using lineage-tracing and co-culturing lineage-specific (SPC-Cre) adult alveolar epithelial cells with PDGFRα-expressing IF.[51] The PDGFRα-expressing LF accumulated neutral lipids and promoted AT2 differentiation in vitro.

PDGFRα-Expressing Mesenchymal Progenitors Are Required for Alveolarization

By germ cell deletion of PDGF-A, Boström and associates were the first to observe that disruption of PDGF-A/PDGFRα signaling virtually eliminated the formation of secondary alveolar septa.[101] By the second postnatal week, the mice that did not express PDGF-A exhibited dilated airspaces (the residual sacs that only formed alveolar ducts), and a striking absence of αSMA and elastic fibers in the septal walls. They proposed that PDGFRα-expressing precursor cells failed to move out of existing primary septa and spread into the airspaces to form the secondary septa.[102]

During secondary septation, the Sca-1+, CD45-, PDGFRα-expressing mesenchymal cell-population expands rapidly and localizes to the distal portions of the elongating septa.[103] At postnatal days 4 and 8, two subpopulations PDGFRα-expressing LF are observed, based on the intensity of GFP-fluorescence. Firstly, cells with a lower level of GFP-fluorescence (less PDGFRα-gene expression), which accumulate more neutral lipid droplets, are less proliferative at P4, are less likely to co-localize with αSMA, and are more often found at the septal base rather than the tip.[104,105] The abundance of the GFP-low population diminishes by P12. And secondly, a subpopulation of mesenchymal cells with a higher level of GFP-expression and which more abundantly express αSMA. Compared to IF, which do not express PDGFRα, the PDGFRα-GFP+ IF are more likely to express sonic hedgehog (Shh) and exhibit primary cilia, which are critical for Shh-mediated signaling. Sca-1 is expressed in both PDGFRα- and PDGFRα+ fibroblasts, but the Sca1+, PDGFRα+ fibroblasts remain more proliferative at P12 and concurrently express αSMA, whereas the PDGFRα- LF exhibit reduced αSMA.[103] The PDGFRα-expressing IF persist in adult mice, participate in lung growth following unilateral pneumonectomy and can support epithelial proliferation and differentiation after alveolar damage.[95,51]

The Lipid-Laden IF (LIF)

Lipid Accumulation and Expression of Thy-1

Lipid-laden IF (LIF) are first evident in rat lungs at gestational day 16, and the triglyceride content of whole rat lung tissue increases 3-fold between gestational day 17 and 19.[106] Lung triglyceride content increases another 2.5-fold between gestational day 21 and postnatal day 1, and then peaks during the second postnatal week.[107] The abundance of LIF in the lung follows the same time-course.[107] Like adipocytes, LIF express lipoprotein lipase, fatty acid transporter, and intracellular lipid binding proteins and LDL and VLDL receptors are able to accumulate neutral lipids when purified triglycerides are added to the culture medium.[108–110] In addition to triglycerides, the LIF accumulate retinyl esters, which may serve as an endogenous source of more active retinoids (retinol or retinoic acid).[111,112] The LIF possess retinoic acid and retinoid-X receptors and the receptor gene expression changes during secondary septation. Endogenous retinoids increase elastin gene expression in explant cultures of lung obtained from rats on gestational day 19 while exogenous RA increases elastin gene expression in cultured LIF.[113] Mice that have a null deletion of RARγ, and are lacking one allele of RXRα, have diminished levels of TE mRNA in their LIF at postnatal day 10 and contain less elastin in their lungs at postnatal day 28.[114]

During lung development, mesenchymal cells lie in close apposition to epithelium and play a central role in the growth and differentiation of epithelial cells into AT2 cells, the site of pulmonary surfactant synthesis.[115,116] The appearance of LIF during the canalicular stage coincides with the appearance of surfactant-containing lamellar bodies in neighboring AT2 cells.[117] Torday and associates have demonstrated that triglycerides of fibroblast origin are used for surfactant phospholipid synthesis by AT2 cells in culture.[118]

Manipulation of the lipid-storing phenotype of fibroblasts during secondary septation when these fibroblasts are most abundant significantly alters alveolar formation. Peroxisome proliferator activated receptor-γ (PPARγ) agonists, the thiazolidinedione (rosiglitazone or pioglitazone) or the natural product prostaglandin-J2 (PGDJ2) promote fibroblast lipid-storage by increasing PPARγ and perilipin (also called adipocyte differentiation lipid-related protein, ADRP), which coats intracellular lipid droplets.[119,120] The administration of PPARγ-agonists concurrently increased epithelial surfactant proteins-B and -C and cytidyltransferase-α after both 1 and 7 days of treatment. Wang and associates also observed a small but significant decrease in mean linear intercept and increase in radial alveolar counts.[119] Varisco and associates recently demonstrated that only LF that express the Thy-1 antigen respond to PPARγ agonists.[121]

Characteristically, LIF display the surface antigen Thy-1 (CD90). This GPI-anchored glycoprotein originally was a marker of fibroblast phenotype, but we now have a much clearer picture of how Thy1 influences fibroblast phenotype and its significance for alveolar development and pulmonary fibrosis. Thy1+ lung fibroblasts are less abundant during secondary septation than they are in adult mice, although they increase approximately 4-fold between postnatal days 3 to 5, but then diminish by day 14. At P5 the Thy1 is most abundant in IF that contain lipid droplets, and forced expression of Thy1 promotes lipid-accumulation by increasing the expression of PPARγ and its downstream targets ADRP and fatty acid transporter solute carrier family 27 member-3.[121]

Absence of Thy-1 Promotes a Myofibroblastic Phenotype

Lung fibroblasts lacking Thy-1 exhibit characteristics of myofibroblasts including increased αSMA and collagen production. Secondary septation is impaired in Thy-1 null mice, which exhibit fewer alveoli at P21 with thickened septa, increased αSMA, collagen and elastin, and activation of TGFβ-signaling. Whereas the proliferation of endothelial and epithelial cells was reduced in Thy-1 null mice, fibroblast proliferation was increased.[122] PDGFRα-gene expression and the proliferative response to PDGF-A were both increased in IF lacking Thy-1.[123] Thy-1 inhibits myofibroblast differentiation through several mechanisms. Firstly, Thy-1 disrupts interactions between αVβ3 integrins and latent transforming growth factor binding protein-4 (LTBP4), which are required for activation of latent TGFβ through biochemical (thombospondin) and mechanical (stretch) means. Secondly, Thy-1 inhibits fibroblast migration by increasing Rho-GTPase activity and increased formation of focal adhesions. Finally, Thy-1 reduces LTBP4 gene-expression and suppresses the increase in LTBP4 in response to bleomycin. In summary, Thy-1 expression in IF confers the ability to accumulate lipids and restrain TGFβ-mediated fibrogenic responses.

Alveolar Myofibroblasts

The term alveolar myofibroblast (AMF) has generally been applied to pulmonary IF that express αSMA, an actin isoform that is most commonly associated with smooth muscle cells. The AMF is of functional importance during pulmonary development and disease. During lung development, it is present in primary septa at the site where the secondary septal buds form.[124,125] It is responsible for producing the elastin that localizes at the tip of the elongating septum and forms the alveolar contractile ring. During septal elongation, the AMF proliferates in a PDGF-A dependent process that is absolutely required for secondary septal formation.

Autocrine and paracrine factors stimulate αSMA production by cultured lung fibroblasts. These include TGFβs, PDGFs, IGFs, interleukin-4, and, in inflammatory states, interleukin-1β (IL-1β).[126] The effects of TGFβs, secreted by myofibroblasts themselves, have been most extensively studied.[126] TGFβ1 stimulates αSMA gene transcription through jun-kinase in lung myofibroblasts, with the effect mediated through a TGFβ responsive element in the proximal portion of the αSMA promoter.[127] The TGFβ-responsive elements in the αSMA promoter differ in myofibroblasts vs. smooth muscle and endothelial cells.[127] IL-1β stimulates inducible nitric oxide synthase in cultured lung fibroblasts, resulting in increased NO production, which reduces αSMA mRNA and protein.[128] IL-1β and other inflammatory mediators may be involved in interstitial pulmonary fibrosis as interstitial myofibroblasts, which in this context are also termed contractile interstitial cells, proliferate and produce more extracellular matrix proteins, most notably the interstitial collagens, in this disease.[129]

Structural Proteins in the Alveolar Wall

Birth drastically alters the mechanical stresses imposed on lung tissues, introducing phasic rather than tonic stretch, with new shear forces applied to airways and blood vessels at the onset of bidirectional airflow and an expanded circulation. This is accompanied by the production and deposition of a continuous, highly integrated, and efficient load-bearing network comprised of contractile cells (AMF and capillary pericytes) and extracellular matrix. The elastic fiber is the most important extracellular load-bearing element during normal tidal respiration and greatly improves mechanical efficiency.[130] The elastomeric protein, elastin, is located in the center of the elastic fiber, and is highly dependent on other surrounding microfibrillar components including fibrillins, fibulins, latent TGFβ-binding proteins, and lysyl oxidases for synthesis, targeted deposition, and cross-linking.[131] Much of what is known about elastin gene-regulation was covered in the prior version and most of the new information relates to elastic fiber assembly. Other proteins that are not contained in the elastic fiber also contribute to the mechanical properties of the alveolar interstitium. The structural collagens, types I, III, and IX, are also produced by IF and provide tensile strength to the alveolar wall. Interstitial fibroblasts produce proteoglycans that help to provide a hydrated environment in which the elastic fiber retains its elastomeric nature.[132] The glycosaminoglycans, hyaluronic acid, chondroitin sulfate, and heparin sulfate contribute to the stiffness of the postnatal alveolar walls and changes in their contents explain age-related differences in the mechanical properties of the distal lung.[133]

Elastin

The developmental regulation of elastin synthesis in the lung has been most extensively studied in mice and rats and, in these species, occurs primarily postnatally. Early morphologic studies established that extensive elastin accumulation in the alveoli occurs between postnatal days 4 and 20.[35,124,134] Amorphous elastic fibers are deposited adjacent to both LIF and AMF in the interstitium and are cross-linked extracellularly, enzymatically by lysyl oxidase and non-enzymatically through aldol condensation.[97,135] There is a close temporal correlation between an increase in elastic fiber length and the volume density of the interstitium of alveoli during postnatal days 4 through 20 in the rat. These morphologic findings have been corroborated by biochemical studies that demonstrate that tropoelastin (TE, the soluble elastin monomer) production is maximal during postnatal days 7 to 12, and that the peak in TE production precedes the maximal desmosine and cross-linked elastin accumulation during postnatal days 10 to 20.[136,137] Inhibition of elastin cross-linking by inhibiting lysyl oxidase (LOX, the rate limiting enzyme) during alveolarization markedly alters alveolar structure and decreases the gas-exchange surface area of the lung.[138] Together these observations indicate that critical regulatory events occur during the neonatal period to initiate and terminate elastin synthesis.

Tropoelastin is synthesized during the saccular phase of lung development and surrounds the terminal sacs.[139] Alveolar levels of TE mRNA reach a maximum around postnatal days 9 to 12 in the rat and then decline by postnatal day 15.[136,140] This postnatal increase in elastin gene expression is regulated both transcriptionally and post-transcriptionally.[141–143] Tropoelastin is rapidly exported from IF and associates with the microfibrils, which are comprised primarily of fibrillin along with microfibril-associated glycoproteins. This interaction is critically dependent on the association of basic amino acid residues in the carboxy terminus of the protein with microfibril-associated glycoproteins and with the amino terminal region of fibrillins-1 and -2.[144] TE monomers also associate with one another prior to cross-linking by a process termed coacervation. Coacervation is driven by hydrophobic interactions involving TE exon 26 and is promoted by sulfated glycosaminoglycans that interact with the positively-charged carboxy-terminal lysines on TE.[145,146] These protein–protein and protein–glycosaminoglycan interactions are accompanied by an enhancement in lysyl oxidase-mediated cross-linking.[147] From postnatal days 14 to 21, alveolar septal elastin undergoes cross-linking that confers increased chemical and proteolytic stability and contributes to the extraordinary longevity of elastin in the elastic fiber.[148]

Fibulins and Fibrillins

Fibulins-3, -4, and -5 are comprised of a series of calcium binding EGF (cbEGF) repeats that mediate interactions between elastin and lysyl oxidase.[149] Fibuluin-5 is located at the interface between elastin and the microfibrils, whereas fibulin-4 is found on the microfibrils themselves. Fibulin-5 regulates the extracellular deposition of elastin by preventing large aggregates and organizing its location and cross-linking.[149] Fibrillins are the most abundant constituents of the microfibrils and control the location and function of latent TGFβ-binding proteins, and indirectly the activation of latent TGFβs.[150,151] Fibrillins also influence the location and function of fibulins.[152] In fibulin-5 null mice, cross-linked elastin deposits are disorganized and fragmented whereas they are virtually absent in fibulin-4 null mice.[153] This difference may be attributed to two unique features of fibulin-4. In addition to binding tropoelastin, fibulin-4 increases elastin gene-transcription, leading to greater elastic fiber formation, both in fibroblast or smooth muscle cell cultures, and in the aorta.[154] Fibulin-4 also binds the propeptide region of lysyl oxidase and co-localizes lysyl oxidase to tropoelastin within nascent elastic fibers.[155]

Fibulins-4 and -5 likely work cooperatively to facilitate the deposition and cross-linking of tropoelastin (TE) along microfibrils. Fibulin-4 can independently bind to lysyl oxidase and facilitates the binding of this duplex to TE. The ternary complex can in turn interact with fibulin-5, which also may bind to TE. These complexes containing TE, one or both of the fibulins and lysyl oxidase can then interact with fibrillin-1, which interfaces with either the fibulin or TE components. However once the fibulins contact fibrillin their interaction with TE weakens, effectively releasing them from globules of TE. By inference, it appears that direct interactions between TE and fibrillin-1 must be sufficient to retain assembled TE monomers until they are cross-linked by lysyl oxidase.[155]

LTBP4 further improves the precision of the deposition of TE along microfibrils. LTBP4 distributes along microfibrils and leads to more uniform deposition of fibulin-TE complex. In the absence of LTBP4, fibulin-5 appears as discrete aggregates rather than as an even coating of the microfibrils, and the efficiency of TE-incorporation is reduced.[156]

The LTBPs also regulate elastic fiber and alveolar septal formation through their interactions with TGFβs. The availability of TGFβ-ligands and their down-stream signaling effects are regulated by TGFβ binding to, and release from, the extracellular matrix.[66,157] The LTBPs localize the TGFβs to the microfibrils in the extracellular matrix (LTBP3 and LTBP4 require fibrillin-1 to associate with microfibrils), and participate in the activation process, through which TGFβs are released from the small latency peptide. Deletion of the LTB4 gene increases TGFβ-signaling in the lung and is accompanied by diminished alveolar septal formation.[158] Deletion of only the LTBP4 gene produces disconnected, aggregated globules of elastin that are poorly integrated with the microfibrils. Using compound deletions of both LTBP3 and LTBP4, Dabovic and associates observed a further defect in alveolar septation with earlier mortality than in mice that only lacked functional LTBP4.[157] However, the elastic-fiber defects in mice with the compound gene deletion were qualitatively similar to those in mice lacking only LTBP4. Interestingly, in both the LTBP4 deletion and the LTBP4/LTBP3 compound deletions, the number of calponin-1 containing myofibroblasts was increased, despite a general increase in apoptosis of alveolar cells.

Paracrine Signaling to Mesenchymal Cells

Fibroblast Growth Factors

FGF-receptors (FGFR) on pulmonary mesenchymal cells engage FGF ligands (in particular FGF9 and FGF18) that originate from epithelial cells or within the mesenchyme. Epithelial derived FGF9 primarily regulates mesenchymal effects on airway branching, whereas mesenchymally derived FGF9 regulates mesenchymal proliferation during the pseudoglandular phase.[159] FGF9 has been most extensively studied during the pseudoglandular and canalicular stages when submesothelial ligands stimulate proliferation of the subjacent mesenchyme.[160] Mice bearing a generalized FGF9-gene deletion die at birth with markedly hypoplastic lungs, reduced mesenchyme, and shortened airways with fewer branch points. Although reduced in number at E18.5, the epithelial cells exhibited age-appropriate differentiation, normal vasculature, but reduced Wnt2a β-catenin canonical signaling, and diminished FGF10.[161] FGF9 acts upstream from Shh but also independently to increase subpleural mesenchymal proliferation.[162] Forced expression of FGF9 in epithelial cells during the pseudoglandular phase elevated proliferation and expanded the surrounding mesenchyme, where differentiation (indicated by αSMA) was reduced. Increased FGF9-mediated signaling in the mesenchyme leads to airway dilation, by augmenting the effects

of FGF7, FGF10, and BMP signaling on the epithelium.[162] Deletion of FGF9 reduced whereas over-expression increased mesenchymal Wnt2a and its downstream target lymphoid-enhancer factor-1 (Lef1), through canonical signaling via β-catenin. This resulted in diminished proliferation in both the mesenchymal and epithelial compartments during the pseudoglandular stage, and increased apoptosis during the saccular stage (E18.5). Wnt2a/β-catenin signaling also enhanced FGFR2 and FGFR1 gene expression, which further augmented FGF9-mediated signaling through a feed-forward loop.[161] Furthermore, FGF9 enhances its own signaling by increasing the expression of one of its receptors, FGFR2, in mesenchymal cells.[159]

Both FGFR3 and FGFR4 gene expression increase during secondary septation in rats and mice. Compound germ line deletion of FGFR3/R4 altered alveolar formation.[163,164] Although saccular structure was normal at birth, dilated airspaces appeared at P8 and to a greater extent at P28. This was accompanied by an increase in gene expression for multiple components of the elastic fibers including tropoelastin, lysyl oxidase, and fibrillin-1, among others. Elastic fibers and αSMA were more diffusely distributed whereas in control alveoli, the fibers were concentrated at septal tips and along the alveolar ducts.[165,166] These changes were dependent on the presence of the alveolar epithelium (potentially directed by epithelial-derived insulin-like growth factor-1) and were not observed in fibroblasts in the absence of epithelial cells.[165]

FGF18 increases during the first postnatal week in rats, stimulates lung fibroblast, but not epithelial cell proliferation and increases the expression of several genes characteristic of alveolar myofibroblasts (tropoelastin, lysyl oxidase, and αSMA, as well as FGF18 itself). FGF18 is a member of the FGF8 family and binds to FGFR2c, FGFR3, or FGFR4, and therefore is a potential ligand for requisite FGFR3, FGFR4 signaling during alveolarization.[167]

Notch Signaling

Notch signaling regulates alveolar development by orchestrating mesenchymal cell differentiation. Lunatic fringe encodes for a N-acetylglucosamine transferase, which is required for Notch receptor activation, particularly in response to Delta-like ligands (Dll). Using a germ line deletion of Lunatic fringe, Xu and associates showed that canonical signaling through Notch (with redundancy between Notch2 and Notch3) is required for differentiation of interstitial fibroblasts.[168] When Notch signaling was disrupted, fewer αSMA-expressing LF appeared along with diminished and mislocalized elastic fibers. Apparently this was not due to a reduction in mesenchymal progenitors, because they proliferated normally, apoptosis was not increased, and there was a normal complement of PDGFRα-expressing mesenchymal cells.[168] By inference, this argues for reduced mesenchymal differentiation. The alveolar surface area and the density of the pulmonary vasculature were diminished, without reducing the number or differentiation of epithelial cells. Lungs from mice in the late canalicular to early saccular stages were used to immunohistochemically localize cells that contained the Notch ligands Jagged or Dll. Notch ligands were present in epithelial and endothelial cells as well as in the mesenchyme. Others have shown that both epithelial and mesenchymal lung components are required to maintain a competent Notch signaling pathway including the Notch receptors, Dll and Jagged ligands, and Lunatic fringe.[169]

Additional Regulators of Alveolarization

Following the first edition of this chapter, other excellent reviews have described the regulation of alveolarization. This second edition emphasizes paracrine regulatory pathways, focusing on cell–cell interactions between mesenchymal and epithelial cell: in particular those involving PDGF-AA, TGFβs, and FGFs. This does not minimize the contributions of other important factors such as erythropoietin, glucocorticoids, epidermal growth factors, connective tissue growth factor (CCN2), VEGFs, hypoxia-inducible factors, and mechanical stimuli. The reader is referred to other reviews for more in-depth coverage of these signaling pathways during alveolarization.[12,57,170,171]

Mechanisms of Septal Thinning

Around postnatal day 14 in rats, following expansion of the IF population, movement of IF into the secondary septa, and prolific tropoelastin synthesis, there is a 20% decline in IF.[171] This loss results from IF apoptosis. Bruce and co-workers have shown that at postnatal day 16 in the rat, there is an abrupt increase in IF apoptosis that is characterized by an increase in chromatin condensation and DNA strand breaks, as well as an increase in the levels of Bax, a pro-apoptotic protein, and a decrease in the expression of the antiapoptotic protein Bcl-2.[172] Persistent expression of PDGFRα limits apoptosis and reduces premature alveolar thinning prior to P14.[173] This apoptotic phase is transient and ends after postnatal day 19. While the number of IF in the lung decreases, the mass of the ECM does not increase and assumes a larger proportion of the volume of the interstitium. The reduction in IF is accompanied by an increase in capillary surface area, which improves the efficiency of gas exchange.[1] During septal thinning, there is a fusion of the paired capillary loops in each septum by a process that most likely involves the formation of connecting pillars.[1] After breaching the interstitium between the two loops, these pillars increase in diameter and decrease in height. Thus the pillars assume the shape of disks that increase capillary surface area, while reducing the need for extensive

new endothelial cell proliferation (intussusceptive angiogenesis). As the alveolar walls thin, the pores of Kohn form and facilitate the merging of air sacs via the remodeling of the alveolar epithelial cells.[174] While the lung volume of humans and rats increases 23-fold between birth and adulthood, capillary volume increases about 35-fold, a finding that is consistent with an increase in the interconnecting capillary meshwork.

Developmental Defects in Alveolarization

Bronchopulmonary Dysplasia

Bronchopulmonary dysplasia (BPD) occurs in infants who are born prematurely with low birth weight. Although efforts to reduce the side effects of increased oxygen tension and mechanical ventilation have improved outcomes, infants continue to develop BPD, which is now viewed as disrupted and aberrant alveolarization. Infants dying of BPD have fewer and larger alveoli, with more attenuated alveolar walls and a paucity of alveolar capillaries.[8] It is hypothesized that this results from decreased secondary septation as well as injury to the existing primary septa. Multiple factors are thought to contribute to the defective alveolar formation that occurs in BPD, including prenatal or antenatal administration of glucocorticoids, nutritional deficiencies, inflammation, as well as the unavoidable effects of hyperoxia and mechanical ventilation.[77] Studies in prematurely delivered baboons and lambs that received exogenous surfactant and were ventilated using conditions that minimized barotrauma and hyperoxia, revealed that they had fewer and enlarged alveoli in their lungs, with a reduction in alveolar capillary surface area.[175,176] The pathobiology of BPD has recently been reviewed in detail.[57,170]

CONCLUSIONS

Although delicate and highly specialized, the alveolus is a focal hub for exchanging gases and for translating diaphragmatic and chest wall muscular forces into air-flow through the conducting airways. The last 10 years have provided a more integrated view of how epithelial and mesenchymal cells communicate and guide one another into the ideal locations, in the optimal proportions, and with the differentiated functions required to work cooperatively in their mission of gas-exchange. New information about how alveolar type 2 cells synthesize lipids, integrate them with surfactant proteins, and maintain the function pulmonary surfactant in the alveolar air-space has improved our understanding of respiratory distress in the infant and acute lung injury in adults. Protein interactions within the elastic fiber are offering new insights into how this important interstitial structural element is produced and organized. These combined insights bring us closer to developing strategies for promoting effective alveolarization in premature infants and possibly regenerating damaged alveoli in adults. However comparably little is known about how alveolarization is regulated by the altered mechanical forces of respiration and blood flow and increased oxygen tension that accompany birth. New information about oxygen- and mechano-sensing molecules in lung cells is laying a foundation for discoveries about how alveolar gene-networks are regulated by their dynamic environment.[177,178] Now appreciating these complexities, the next 10 years are an opportunity for expanding on and integrating these findings and applying them to the prevention and treatment of lung diseases.

ACKNOWLEDGEMENTS

The author would like to thank Jeanne Snyder Ph.D, who co-authored the prior version of this chapter. While preparing the second edition, he was reminded of Dr. Snyder's outstanding expertise in the areas of alveolar epithelial biology and how much he benefitted from their collaborations prior to her retirement. The author's work is supported by the Department of Veterans Affairs Research Service.

REFERENCES

1. Burri PH. Structural aspects of prenatal and postnatal development and growth of the lung. In: McDonald JA, editor. *Lung Growth and Development*. New York: M. Decker; 1997. pp. 1–35.
2. Zeltner TB, Carduff JH, Gehr P, Pfenninger J, Burri PH. The postnatal development and growth of the human lung. *I Morphometry Respir Physiol* 1987;**67**:247–67.
3. Wood JP, Kolassa JE, McBride JT. Changes in alveolar septal border lengths with postnatal lung growth. *Am J Physiol* 1998;**275**: L1157–63.
4. Crapo JD, Berry BE, Gehr P, Bachofen M, Weibel ER. Cell number and cell characteristics of the normal human lung. *Am Rev Respir Dis* 1982;**126**:332–7.
5. Evans MJ, Cabral JJ, Stephens RJ, Freeman G. Renewal of alveolar epithelium in the rat following exposure to NO_2. *Am J Pathol* 1973;**70**:175–98.
6. Perez-Gil J, Weaver TE. Pulmonary surfactant pathophysiology: current models and open questions. *Physiol (Bethesda)* 2010;**25**: 132–41.
7. Massaro GD, Massaro D. Formation of pulmonary alveoli and gas-exchange surface area: quantitation and regulation. *Annu Rev Physiol* 1996;**58**:73–92.
8. Jobe AH. The new bronchopulmonary dysplasia. *Curr Opin Pediatr* 2011;**23**:167–72.
9. Tuder RM, Petrache I. Pathogenesis of chronic obstructive pulmonary disease. *J Clin Invest* 2012;**122**:2749–55.
10. Agassandian M, Mallampalli RK. Surfactant phospholipid metabolism. *Biochim Biophys Acta* 2013;**1831**:612–25.
11. Whitsett JA, Wert SE, Weaver TE. Alveolar surfactant homeostasis and the pathogenesis of pulmonary disease. *Annu Rev Med* 2010;**61**:105–19.
12. Maeda Y, Davé V, Whitsett JA. Transcriptional control of lung morphogenesis. *Physiol Rev* 2007;**87**:219–44.
13. Hallman M. The surfactant system protects both fetus and newborn. *Neonatology* 2013;**103**:320–6.
14. Garbrecht MR, Klein JM, Schmidt TJ, Snyder JM. Glucocorticoid metabolism in the human fetal lung: implications for lung development and the pulmonary surfactant system. *Biol Neonate* 2006;**89**:109–19.
15. Hilgendorff A, Reiss I, Ehrhardt H, Eickelberg O, Alvira CM. Chronic lung disease in the preterm infant: lessons learned from animal models. *Am J Respir Cell Mol Biol* 2013;**50**:233–45. http://dx.doi.org/10.1165/rcmb.3013–0014TR.
16. Jobe AH. What is BPD in 2012 and what will BPD become? *Early Hum Dev* 2012;**88S2**:S27–8.
17. Kitaoka H, Burri PH, Weibel ER. Development of the human fetal airway tree: analysis of numerical density of airway endtips. *Anat Rec* 1996;**244**:207–13.
18. Rawlins E. The building blocks of lung development. *Dev Dyn* 2011;**240**:463–76.
19. Mallampalli RK, Acarregui MJ, Snyder JM. Differentiation of the alveolar epithelium in the fetal lung. In: McDonald JA, editor. *Lung Growth and Development*. New York, NY: Marcel Dekker, Inc.; 1997. pp. 119–62.
20. Stenmark KR, Abman SH. Lung vascular development: implications for the pathogenesis of bronchopulmonary dysplasia. *Annu Rev Physiol* 2005;**67**:623–61.
21. Mendelson CR. Role of transcription factors in fetal lung development and surfactant protein gene expression. *Annu Rev Physiol* 2000;**62**:875–915.
22. van Keymeulen A, Blanpain C. Tracing epithelial stem cells during development, homeostasis, and repair. *J Cell Biol* 2012;**197**:575–84.
23. Burri PH, Dbaly J, Weibel ER. The postnatal growth of the rat lung I: morphometry. *Anat Rec* 1974;**178**:711–30.
24. Kauffman SL, Burri PH, Weibel ER. The postnatal growth of the rat lung II. Autoradiography. *Anat Rec* 1974;**180**:63–76.
25. Adler K, Low RB, Leslie K, Mitchell J, Evans JN. Contractile cells in normal and fibrotic lung. *Lab Invest* 1989;**60**:473–85.
26. Mitchell JJ, Reynolds SE, Leslie KO, Low RB, Woodcock-Mitchell J. Smooth muscle cell markers in developing rat lung. *Am J Respir Cell Mol Biol* 1990;**3**:515–23.
27. Vidic B, Burri PH. Morphometric analysis of the remodeling of the rat pulmonary epithelium during early postnatal development. *Anat Rec* 1983;**207**:317–24.
28. Burri PH, Tarek MR. A novel mechanism of capillary growth in the rat pulmonary microcirculation. *Anat Rec* 1990;**228**:35–45.
29. Mund SI, Stampanoni M, Schittny JC. Developmental alveolarization of the mouse lung. *Dev Dynam* 2008;**237**:2108–16.
30. Butler JP, Loring SH, Patz S, Tsuda A, Yablonskiy DA, Mentzer SJ. Evidence for adult lung growth in humans. *N Engl J Med* 2012;**367**:244–7.
31. Mariani TJ, Pierce RA. Development of lung elastic matrix. In: Gaultier C, Bourbon JR, Post M, editors. *Lung Development*. New York: Oxford University Press; 2002. pp. 28–45.
32. Noguchi A, Samaha H. Developmental changes in tropoelastin gene expression in the rat lung studied by in situ hybridization. *Am J Respir Cell Mol Biol* 1991;**5**:571–8.
33. Nardell EA, Brody JS. Determinants of mechanical properties of rat lung during postnatal development. *J Appl Physiol: Respirat Environ Exercise Phsiol* 1982;**53**:140–8.
34. Cherukupalli K, Larson JE, Puterman M, Sekhon HS, Thurlbeck WM. Comparative biochemistry of gestational and postnatal lung growth and development in the rat and human. *Pediatr Pulmonol* 1997;**24**:12–21.

35. Goncalves C, Barros J, Honouio A, Rodrigues P, Bairos V. Quantification of elastin from the mouse lung during postnatal development. *Exper Lung Res* 2001;**27**:533–45.

36. Desai R, Wigglesworth JS, Aber V. Assessment of elastin maturation by radioimmunoassay of desmosine in the developing human lung. *Early Hum Dev* 1988;**16**:61–71.

37. Lin Y, Lechner AJ. Development of alveolar septa and cellular maturation within the perinatal lung. *Am J Respir Cell Mol Biol* 1991;**4**:59–64.

38. Hislop A, Howard S, Fairweather DVI. Morphometric studies on the structural development of the lung in Macaca fascicularis during fetal and postnatal life. *J Anat* 1984;**138**:95–112.

39. Schellenberg J-C, Liggins GC. Elastin and collagen in fetal sheeep lung. I. Ontogenesis. *Pediatr Res* 1987;**22**:335–8.

40. Willet KE, Jobe AH, Ikegami M, Newnham J, Brennan S, Sly PD. Antenatal endotoxin and glucocorticoid effects on lung morphometry in preterm lambs. *Pediatr Res* 2001;**48**:782–8.

41. Kaufmann SL. Cell proliferation in the mammalian lung. *Internat Rev Experi Pathol* 1980;**22**:131–91.

42. Evans MJ, Cabral LJ, Stephens RJ, Freeman G. Transformation of alveolar type 2 cells to type 1 cells following exposure to NO_2. *Experi Mol Pathol* 1975;**22**:142–50.

43. Rawlins E, Clark CP, Xue Y, Hogan BLM. The Id2+ distal tip lung epithelium contains individual multipotent embryonic progenitor cells. *Development* 2009;**136**:3741–5.

44. Okubo T, Knoepfler PS, Eisenman RN, Hogan BL. Nmyc plays an essential role during lung development as a dosage-sensitive regulator of progenitor cell proliferation and differentiation. *Development* 2005;**132**:1363–74.

45. Ramasamy SK, Mailleux AA, Gupte VV, Mata F, Sala FG, Veltmaat JM, Del Moral PM, De LS, Parsa S, Kelly LK, Kelly R, Shia W, Keshet E, Minoo P, Warburton D, Bellusci S. Fgf10 dosage is critical for the amplification of epithelial cell progenitors and for the formation of multiple mesenchymal lineages during lung development. *Dev Biol* 2007;**307**:237–47.

46. Guseh JS, Bores SA, Stanger BZ, Zhou Q, Anderson WJ, Melton DA, Rajagopal J. Notch signaling promotes airway mucous metaplasia and inhibits alveolar development. *Development* 2009;**136**: 1751–9.

47. Tsao PN, Vasconcelos M, Izvolsky KI, Qian J, Lu J, Cardoso WV. Notch signaling controls the balance of ciliated and secretory cell fates in developing airways. *Development* 2009;**136**:2297–307.

48. McQualter JL, Yuen K, Williams B, Bertoncello I. Evidence of an epithelial stem/progenitor cell hierarchy in the adult mouse lung. *Proc Natl Acad Sci USA* 2010;**107**:1416–9.

49. Vaughan AE, Chapman HA. Regenerative activity of the lung after epithelial injury. *Biochem Biophys Acta* 2013;**1832**:922–30.

50. Rawlins EL, Okubo T, Xue Y, Brass DM, Auten RL, Hasegawa H, Wang F, Hogan BL. The role of Scgb1a1+ Clara cells in the long-term maintenance and repair of lung airway, but not alveolar, epithelium. *Cell Stem Cell* 2009;**4**:525–34.

51. Barkauskas CE, Cronce MJ, Rackley CR, Bowie EJ, Keene DR, Randell SH, Noble PB, Hogan BLM. Type 2 alveolar cells are stem cells in adult lung. *J Clin Invest* 2013;**123**:325–36.

52. Rock JR, Barkauskas CE, Cronce MJ, Xue Y, Harris JR, Liang J, Noble PW, Hogan BL. Multiple stromal populations contribute to pulmonary fibrosis without evidence for epithelial to mesenchymal transition. *Proc Natl Acad Sci USA* 2011;**108**:E1475–83.

53. Chapman HA, Li X, Alexander JP, Brumwell A, Lorizio W, Tan K, Sonnenberg A, Wei Y, Vu TH. Integrin alpha6beta4 identifies an adult distal lung epithelial population with regenerative potential in mice. *J Clin Invest* 2011;**121**:2855–62.

54. Li X, Rossen N, Sinn PL, Hornick AL, Steines BR, Karp PH, Ernst SE, Adam RJ, Moninger TO, Levasseur DN, Zabner J. Integrin alpha6beta4 identifies human distal lung epithelial progenitor cells with potential as a cell-based therapy for cystic fibrosis lung disease. *PloS One* 2013;**8**:e83624.

55. Xu Y, Zhang M, Wang Y, Kadambi P, Dave V, Lu LJ, Whitsett JA. A systems approach to mapping transcriptional networks controlling surfactant homeostasis. *BMC Genomics* 2010;**11**:451.

56. Xu Y, Wang Y, Besnard V, Ikegami M, Wert SE, Heffner C, Murray SA, Donahue LR, Whitsett JA. Transcriptional programs controlling perinatal lung maturation. *PloS One* 2012;**7**:e37046.

57. Ahlfeld SK, Conway SJ. Aberrant signaling pathways of the lung mesenchyme and their contributions to the pathogenesis of bronchopulmonoary dysplasia. *Birth Defects Res (Part A)* 2011;**94**:3–15.

58. Rockich BE, Hrycaj SM, Shih HP, Nagy MS, Ferguson MA, Kopp JL, Sander M, Wellik DM, Spence JR. Sox9 plays multiple roles in the lung epithelium during branching morphogenesis. *Proc Natl Acad Sci USA* 2013;**110**:E4456–64.

59. Ding BS, Nolan DJ, Guo P, Babazadeh AO, Cao Z, Rosenwaks Z, Crystal RG, Simons M, Sato TN, Worgall S, Shido K, Rabbany SY, Rafii S. Endothelial-derived angiocrine signals induce and sustain regenerative lung alveolarization. *Cell* 2011;**147**:539–53.

60. Honda T, Isida K, Hayama M, Kubo K, Katsuyama T. Type II pneumocytes are preferentially located along thick elastic fibers forming the framework of human alveoli. *Anat Rec* 2000;**258**:34–8.

61. Barth K, Bläsche R, Kasper M. T1alpha/podoplanin shows raft-associatd distribution in mouse lung alveolar epithelial E10 cells. *Cell Physiol Biochem* 2010;**25**:103–12.

62. Williams MC, Cao Y, Hinds A, Rishi AK, Wetterwald A. T1 alpha protein is developmentally regulated and expressed by alveolar type I cells, choroid plexus, and ciliary epithelia of adult rats. *Am J Respir Cell Mol Biol* 1996;**14**:577–85.

63. Borok Z, Danto SI, Lubman RL, Cao Y, Williams MC, Crandall ED. Modulation of T1alpha expression with alveolar epithelial cell phenotype in vitro. *Am J Physiol* 1998;**275**:L155–64.

64. Kreda SM, Gynn MC, Fenstermacher DA, Boucher RC, Gabriel SE. Expression and localization of epithelial aquaporins in the adult human lung. *Am J Respir Cell Mol Biol* 2001;**24**:224–34.

65. Zhao L, Yee M, O'Reilly MA. Transdifferentiation of alveolar epithelial type II to type I cells is controlled by opposing TGF-β and BMP signaling. *Am J Physiol Lung Cell Mol Physiol* 2013;**305**:L409–18.

66. Massagué J. TGFβ signaling in context. *Nat Rev Mol Cell Biol* 2012;**13**:616–30.

67. Sountoulidis A, Stavropoulos A, Giaglis S, Apostolou E, Monteiro R, Chuva de Sousa Lopes SM, Chen H, Stripp BR, Mummery C, Andreakos E, Sideras P. Activation of the canonical bone morphogenetic protein (BMP) pathway during lung morphogenesis and adult lung tissue repair. *PloS One* 2012;**7**:e41460.

68. Abraham V, Chou ML, George P, Pooler P, Zaman A, Savani RC, Koval M. Heterocellular gap junctional communication between alveolar epithelial cells. *Am J Physiol Lung Cell Mol Physiol* 2001;**280**:L1085–93.

69. Isakson BE, Lubman RL, Seedorf GJ, Boitano S. Modulation of pulmonary alveolar type II cell phenotype and communication by extracellular matrix and KGF. *Am J Physiol Cell Physiol* 2001;**281**:C1291–9.

70. Newman GR, Campbell L, von RC, Jasani B, Gumbleton M. Caveolin and its cellular and subcellular immunolocalisation in lung alveolar epithelium: implications for alveolar epithelial type I cell function. *Cell Tissue Res* 1999;**295**:111–20.

71. Jung K, Schlenz H, Krasteva G, Muhlfeld C. Alveolar epithelial type II cells and their microenvironment in the caveolin-1-deficient mouse. *Anat Rec (Hoboken)* 2012;**295**:196–200.

72. Koval M. Claudin hetereogeneity and control of lung tight junctions. *Annu Rev Physiol* 2013;**75**:551–67.

73. Kaarteenaho R, Merikallio H, Lehtonen S, Harju T, Soini Y. Divergent expression of claudin -1, -3, -4, -5 and -7 in developing human lung. *Respir Res* 2010;**11**:59.

74. Werth M, Walentin K, Aue A, Schonheit J, Wuebken A, Pode-Shakked N, Vilianovitch L, Erdmann B, Dekel B, Bader M, Barasch J, Rosenbauer F, Luft FC, Schmidt-Ott KM. The transcription factor grainyhead-like 2 regulates the molecular composition of the epithelial apical junctional complex. *Development* 2010;**137**:3835–45.

75. Crapo JD, Young SL, Fram EK, Pinkerton KE, Barry BE, Crapo RO. Morphometric characteristics of cells in the alveolar region of mammalian lungs. *Am Rev Respir Dis* 1983;**128**:S42–6.

76. Avery ME, Mead J. Surface properties in relation to atelectasis and hyaline membrane disease. *Am J Dis Child* 1959;**97**:517–23.

77. Jobe AH, Ikegami M. Lung development and function in pre term infants in the surfactant treatment era. *Annu Rev Physiol* 2000;**62**:825–46.

78. Greenough A, Ahmed N. Perinatal prevention of bronchopulmonary dysplasia. *J Perinat Med* 2013;**41**:119–26.

79. Veldhuizen R, Nag K, Oreig S, Possmeyer F. The role of lipids in pulmonary surfactant. *Biochem Biophys Acta* 1998;**1408**:90–108.

80. Nunez JS, Torday JS. The developing rat lung fibroblast and alveolar type II cell actively recruit surfactant phospholipid substrate. *J Nutr* 1995;**125**:1639S–44S.

81. Cogo PE, Ori C, Simonato M, Verlato G, Isak I, Hamvas A, Carnielli VP. Metabolic precursors of surfactant disaturated-phosphatidylcholine in preterms with respiratory distress. *J Lipid Res* 2009;**50**:2324–31.

82. Matsumura Y, Ban N, Ueda K, Inagaki N. Characterization and classification of ATP-binding cassette transporter ABCA3 mutants in fatal surfactant deficiency. *J Biol Chem* 2006;**281**:34503–14.

83. von Nahmen A, Post A, Galla HJ, Sieber M. The phase behavior of lipid monolayers containing pulmonary surfactant protein C studied by fluorescence light microscopy. *Eur Biophys J* 1997;**26**:359–69.

84. Gower WA, Nogee LM. Surfactant dysfunction. *Pediatr Respir Rev* 2011;**12**:223–9.

85. Glasser JR, Mallampalli RK. Surfactant and its role in the pathobiology of pulmonary infection. *Microbes Infect* 2012;**14**:17–25.

86. Korfhagen TR, LeVine AM, Whitsett JA. Surfactant protein A (SP-A) gene targeted mice. *Biochim Biophys Acta* 1998;**1408**:296–302.

87. Botas C, Poulain F, Akiyama J, Brown C, Allen L, Goerke J, Clements J, Carlson E, Gillespie AM, Epstein C, Hawgood S. Altered surfactant homeostasis and alveolar type II cell morphology in mice lacking surfactant protein D. *Proc Natl Acad Sci USA* 1998;**95**:11869–74.

88. Korimilli A, Gonzalez LW, Guttentag SH. Intracellular localization and processing events in human surfactant protein B synthesis. *J Biol Chem* 2000;**275**:8672–9.

89. Clark JC, Wert SE, Bachurski CJ, Stahlman MT, Stripp BR, Weaver TE, Whitsett JA. Targeted disruption of the surfactant protein B gene disrupts surfactant homeostasis, causing respiratory failure in newborn mice. *Proc Natl Acad Sci USA* 1995;**92**:7794–8.

90. Glasser SW, Burhans MS, Korfhagen TR, Na CL, Sly PD, Ross GF, Ikegami M, Whitsett JA. Altered stability of pulmonary surfactant in SP-C-deficient mice. *Proc Natl Acad Sci USA* 2001;**98**:6366–71.

91. Volckaert T, Campbell A, Dill E, Li C, Minoo P, De Langhe S. Localized Fgf10 expression is not required for lung branching morphogenesis but prevents differentiation of epithelial progenitors. *Development* 2013;**140**:3731–42.

92. De Langhe SP, Carraro G, Warburton D, Hajihosseini MK, Bellusci S. Levels of mesenchymal FGFR2 signaling modulate smooth muscle progenitor cell commitment in the lung. *Dev Biol* 2006;**299**:52–62.

93. El Agha E, Herold S, Alam DA, Quantius J, Mackenzie B, Carraro G, Moiseenko A, Chao CM, Minoo P, Seeger W, Bellusci S. Fgf10-positive cells represent a progenitor cell population during lung development and postnatally. *Development* 2014;**141**:296–306.

94. Hokuto I, Perl A-KT, Whitsett JA. Prenatal, but not postnatal inhibition of fibroblast growth factor receptor signaling causes emphysema. *J Biol Chem* 2003;**278**:415–21.

95. Chen L, Acciani T, Le Cras T, Lutzko C, Perl A-KT. Dynamic regulation of platelet-derived growth factor receptor α expression in alveolar fibroblasts during realveolarization. *Am J Respir Cell Mol Biol* 2012;**47**:517 27.

96. Chang DR, Martinez AD, Miller RK, Ji H, Akiyama H, McCrea PD, Chen J. Lung epithelial branching program antagonizes alveolar differentiation. *Proc Natl Acad Sci USA* 2013;**110**:18042–51.

97. Brody JS, Kaplan NB. Proliferation of alveolar interstitial cells during postnatal lung growth. Evidence for two distinct populations of pulmonary fibroblasts. *Am Rev Respir Dis* 1983;**127**:763–70.

98. Vaccaro C, Brody JS. Ultrastructure of developing alveoli. I. The role of the interstitial fibroblast. *Anat Rec* 1978;**192**:467–80.

99. McQualter JL, Brouard N, Williams B, Baird BN, Sims-Lucas S, Yuen K, Nilsson SK, Simmons PJ, Bertoncello I. Endogenous fibroblastic progenitor cells in the adult mouse lung are highly enriched in the Sca-1 positive cell fraction. *Stem Cells* 2009;**27**:623–33.

100. McQualter JL, McCarty RC, Van der Velden J, O'Donoghue RJ, Asselin-Labat M-L, Bozinovski S, Bertoncello I. TGF-β signaling in stromal cells acts upstream of FGF-10 to regulate epithelial stem cell growth in adult lung. *Stem Cell Res* 2013;**11**:1222–33.

101. Boström H, Willetts K, Pekny M, Leveen P, Lindahl P, Hedstrand H, Pekna M, Hellstrom M, Gebre-Medin S, Schalling M, Nilsson M, Kurland S, Tornell J, Heath JK, Betsholtz C. PGDF-A signaling is a critical event in lung alveolar myofibroblast development and alveogenesis. *Cell* 1996;**85**:863–73.

102. Lindahl P, Karlsson L, Helstrom M, Gebre-Medhin S, Willetts K, Heath JK, Betsholtz C. Alveogenesis failure in PDGF-A-deficient mice is coupled to lack of distal spreading of alveolar smooth muscle cell progenitors during lung development. *Development* 1997;**124**:3943–53.

103. McGowan SE, McCoy DM. Platelet-derived growth factor-A and sonic hedgehog signaling direct lung fibroblast precursors during alveolar septal formation. *Am J Physiol Lung Cell Mol Physiol* 2013;**305**:L229–39.

104. Kimani PW, Holmes AJ, Grossmann RE, McGowan SE. PDGF-Rα gene expression predicts proliferation, but PDGF-A suppresses transdifferentiation of neonatal mouse lung myofibroblasts. *Respir Res* 2009;**10**:19.

105. McGowan SE, Grossmann RE, Kimani PW, Holmes AJ. Platelet-derived growth factor receptor-alpha-expressing cells localize to the alveolar entry ring and have characteristics of myofibroblasts during pulmonary alveolar septal formation. *Anat Rec* 2008;**291**:1649–61.

106. Tordet C, Marin L, Dameron F. Pulmonary di- and tri-glycerides during the perinatal development of the rat. *Experientia* 1981;**37**:333–4.

107. Maksvytis HJ, Vaccaro C, Brody JS. Isolation and characterization of the lipid-containing interstitial cell from the developing rat lung. *Lab Invest* 1981;**45**:248–59.

108. Chen H, Jackson S, Doro M, McGowan S. Perinatal expression of genes that may participate in lipid metabolism by lipid-laden lung fibroblasts. *J Lipids Res* 1998;**39**:2483–92.

109. Maksvytis HJ, Niles RM, Simanovsky L, Minassian IA, Richardson LL, Hamosh M, Hamosh P, Brody JS. In vitro characteristics of the lipid-filled interstitial cell associated with postnatal lung growth: evidence for fibroblast heterogeneity. *J Cell Physiol* 1984;**118**:113–23.

110. McGowan SE, Doro MM, Jackson S. Expression of lipoprotein receptor and apolipoprotein E genes by perinatal rat lipid-laden pulmonary fibroblasts. *Exper Lung Res* 2001;**27**:47–63.

111. McGowan SE, Harvey CS, Jackson SK. Retinoids, retinoic acid receptors, and cytoplasmic retinoid binding proteins in perinatal rat lung fibroblasts. *Am J Physiol (Lung Cell Mol Physiol)* 1995;**269**:L463–72.

112. Chytil F. The lungs and vitamin A. *Am J Physiol (Lung Cell Mol Physiol)* 1992;**262**:L517–27.

113. McGowan SE, Doro MM, Jackson SK. Endogenous retinoids increase perinatal elastin gene expression in rat lung fibroblasts and fetal explants. *Am J Physiol (Lung Cell Mol Physiol)* 1997;**273**:L410–6.

114. McGowan SE, Jackson SK, Jenkins-Moore M, Dai H-H, Chambon P, Snyder JM. Mice bearing deletions of retinoic acid receptors demonstrate reduced lung elastin and alveolar numbers. *Am J Respir Cell Mol Biol* 2000;**23**:162–7.

115. Deterding RR, Shannon JM. Proliferation and differentiation of fetal rat pulmonary epithelium in the absence of mesenchyme. *J Clin Invest* 1995;**95**:2963–72.

116. Warburton D, El-Hashash A, Carraro G, Tiozzo C, Sala F, Rogers O, De LS, Kemp PJ, Riccardi D, Torday J, Bellusci S, Shi W, Lubkin SR, Jesudason E. Lung organogenesis. *Curr Top Dev Biol* 2010;**90**:73–158.

117. Sorokin S, Padykula HA, Herman E. Comparative histochemical patterns in developing mammalian lungs. *Dev Biol* 1959;**1**:125–51.

118. Torday JS, Hua J, Slavin R. Metabolism and fate of neutral lipids of fetal lung fibroblast origin. *Biochem Biophys Acta* 1995;**1254**:198–206.

119. Wang Y, Santos R, Sakurai R, Shin E, Cerny L, Torday JS, Rehan VK. Peroxisome proliferator-activated receptor γ agonists enhance lung maturation in a neonatal rat model. *Pediatr Res* 2009;**65**:150–5.

120. McGowan SE, Jackson SK, Doro MM, Olson PJ. Peroxisome proliferators alter lipid acquisition and elastin gene expression in neonatal rat lung fibroblasts. *Am J Physiol (Lung Mol Cell Physiol)* 1997;**273**:L1249–57.

121. Varisco BM, Ambalavanan N, Whitsett JA, Hagood JS. Thy-1 signals through PPARγ to promote lipofibroblast differentiation in the developing lung. *Am J Respir Cell Mol Biol* 2012;**46**:765–72.

122. Nicola T, Hagood JS, James ML, MacEwen MW, Williams TA, Hewitt MM, Schwiebert L, Bulger A, Oparil S, Chen YF, Ambalavanan N. Loss of Thy-1 inhibits alveolar development in the newborn mouse lung. *Am J Physiol Lung Cell Mol Physiol* 2009;**296**:L738–50.

123. Hagood JS, Miller PJ, Lasky JA, Tousson A, Guo B, Fuller GM, McIntosh JC. Differential expression of platelet-derived growth factor-alpha receptor by Thy-1(-) and Thy-1(+) lung fibroblasts. *Am J Physiol* 1999;**277**:L218–24.

124. Burri PH. The postnatal growth of the rat lung: III morphology. *Anat Rec* 1974;**180**:77–98.

125. Wendel DP, Taylor DG, Albertine KH, Keating MT, Li DY. Impaired distal airway development in mice lacking elastin. *Am J Respir Cell Mol Biol* 2000;**23**:320–6.

126. Hinz B, Phan SH, Thannickal VJ, Prunotto M, Desmouliere A, Varga J, De Wever O, Mareel M, Gabbiani G. Recent developments in myofibroblast biology. Paradigms for connective tissue remodeling. *Am J Pathol* 2012;**180**:1340–55.

127. Roy SG, Nozaki Y, Phan SH. Regulation of alpha-smooth muscle actin gene expression in myofibroblast differentiation from rat lung fibroblasts. *Intl J Biochem Cell Biol* 2001;**33**:723–34.

128. Zhang HY, Phan SH. Inhibition of myofibroblast apoptosis by transforming growth factor beta(1). *Am J Respir Cell Mol Biol* 1999;**21**:658–65.

129. Hashimoto S, Gon Y, Takeshita I, Matsumoto K, Maruoka S, Horie T. Transforming growth factor-beta1 induces phenotypic modulation of human lung fibroblasts to myofibroblast through a c-Jun-NH2-terminal kinase-dependent pathway. *Am J Respir Crit Care Med* 2001;**163**:152–7.

130. Suki B, Stamenkovic I, Hubmayr R. Lung parenchymal mechanics. *Compr Physiol* 2011;**1**:1317–51.

131. Wagenseil JE, Mecham RP. New insights into elastic fiber assembly. *Birth Defects Res (Part C)* 2007;**81**:229–40.

132. Pasquali-Ronchetti I, Baccarani-Contri M. Elastic fiber during development and aging. *Microsc Res Tech* 1997;**38**:428–35.

133. Tanaka R, Al Jamal R, Ludwig MS. Maturational changes in extracellular matrix and lung tissue mechanics. *J Appl Physiol* 2001;**91**:2314–21.

134. Bruce MC, Lo PY. A morphometric quantitation of developmental changes in elastic fibers in rat lung parenchyma: variability with lung region and postnatal age. *J Lab Clin Med* 1991;**117**:226–33.

135. Kaplan NB, Grant MM, Brody JS. The lipid interstitial cell of the pulmonary alveolus. Age and species differences. *Am Rev Respir Dis* 1985;**132**:1307–12.

136. Bruce MC. Developmental changes in tropoelastin mRNA levels in rat lung: evaluation by in situ hybridization. *Am J Respir Cell Mol Biol* 1991;**5**:344–50.

137. Myers B, Dubick M, Last JA, Rucker RB. Elastin synthesis during perinatal lung development in the rat. *Biochem Biophys Acta* 1983;**761**:17–22.

138. Kida K, Thurlbeck WM. Lack of recovery of lung structure and function after the administration of beta-amino-proprionitrile in the postnatal period. *Am Rev Respir Dis* 1980;**122**:467–75.

139. Pierce RA, Mariencheck WI, Sandefur S, Crouch EC, Parks WC. Glucocorticoids upregulate tropoelastin expression during late stages of fetal lung development. *Am J Physiol Lung Cell Mol Physiol* 1995;**268**:L491–500.

140. Noguchi A, Firsching K, Kursar JD, Reddy R. Developmental changes of tropoelastin synthesis by rat pulmonary fibroblasts and effects of dexamethasone. *Pediatr Res* 1990;**28**:379–82.

141. Swee MH, Parks WC, Pierce RA. Developmental regulation of elastin production, expression of tropoelastin pre-mRNA persists after downregulaton of steady-state mRNA levels. *J Biol Chem* 1995;**270**:14899–906.

142. Bruce MC, Bruce EN, Janiga K, Chetty A. Hyperoxic exposure of developing rat lung decreases tropoelastin mRNA levels that rebound postexposure. *Am J Physiol (Lung Cell Mol Physiol)* 1993;**265**:L294–300.

143. Zhang M, Pierce RA, Wachi H, Mecham RP, Parks WC. An open reading frame element mediates posttranscriptional regulation of tropoelastin and responsiveness to transforming growth factor beta1. *Mol Cell Biol* 1999;**19**:7314–26.

144. Trask TM, Trask BC, Ritty TM, Abrams WR, Rosenbloom J, Mecham RP. Interaction of tropoelastin with the amino-terminal domains of fibrillin-1 and fibrillin-2 suggests a role for the fibrillins in elastic fiber assembly. *J Biol Chem* 2000;**275**:24400–6.

145. Jensen SA, Vrhovski B, Weiss AS. Domain 26 of tropoelastin plays a dominant role in association by coacervation. *J Biol Chem* 2000;**275**:28449–54.

146. Wu WJ, Vrhovski B, Weiss AS. Glycosaminoglycans mediate the coacervation of human tropoelastin through dominant charge interactions involving lysine side chains. *J Biol Chem* 1999;**274**:21719–24.

147. Stone PJ, Morris SM, Griffin S, Mithieux S, Weiss AS. Building elastin. Incorporation of recombinant human tropoelastin into extracellular matrices using nonelastogenic rat-1 fibroblasts as a source for lysyl oxidase. *Am J Resp Cell Mol Biol* 2001;**24**:733–9.

148. Shapiro SD, Endicott SK, Province MA, Pierce JA, Campbell EJ. Marked longevity of human lung parenchymal elastic fibers deduced from prevalence of D-aspartate and nuclear weapons-related radiocarbon. *J Clin Invest* 1991;**87**:1828–34.

149. Yanigisawa H, Davis EC. Unraveling the mechanisms of elastic fiber assembly: the roles of short fibulins. *Internat J Biochem Cell Biol* 2010;**42**:1084–93.

150. Davis MR, Summers KM. Structure and function of the mammalian fibrillin gene family: implications for human connective tissue diseases. *Mol Genet Metab* 2012;**107**:635–47.

151. Zhang H, Hu W, Ramirez F. Developmental expression of fibrillin genes suggests heterogeneity of extracellular microfibrils. *J Cell Biol* 1995;**129**:1165–76.

152. Lemaire R, Korn JH, Schiemann WP, Lafyatis R. Fibulin-2 and fibulin-5 alterations in tsk mice associated with disorganized hypodermal elastic fibers and skin tethering. *J Invest Dermatol* 2004;**123**:1063–9.

153. Nakamura T, Lozano PR, Ikeda Y, Iwanaga Y, Hinek A, Minamisawa S, Cheng CF, Kobuke K, Dalton N, Takada Y, Tashiro K, Ross JJ, Honjo T, Chien KR. Fibulin-5/DANCE is essential for elastogenesis in vivo. *Nature* 2002;**415**:171–5.

154. Chen Q, Zhang T, Roshetsky JF, Ouyang Z, Essers J, Fan C, Wang Q, Hinek A, Plow EF, Dicorleto PE. Fibulin-4 regulates expression of the tropoelastin gene and consequent elastic-fibre formation by human fibroblasts. *Biochem J* 2009;**423**:79–89.

155. Horiguchi M, Inoue T, Ohbayashi T, Hirai M, Noda K, Marmorstein LY, Yabe D, Takagi K, Akama TO, Kita T, Kimura T, Nakamura T. Fibulin-4 conducts proper elastogenesis via interaction with cross-linking enzyme lysyl oxidase. *Proc Natl Acad Sci USA* 2009;**106**:19029–34.

156. Noda K, Dabovic B, Takagi K, Inoue T, Horiguchi M, Hirai M, Fujikawa Y, Akama TO, Kusumoto K, Zilberberg L, Sakai LY, Koli K, Naitoh M, von Melcher H, Suzuki S, Rifkin DB, Nakamura T. Latent TGF-beta binding protein 4 promotes elastic fiber assembly by interacting with fibulin-5. *Proc Natl Acad Sci USA* 2013;**110**:2852–7.

157. Dabovic B, Chen Y, Choi J, Davis EC, Sakai LY, Todorovic V, Vassallo M, Zilberberg L, Singh A, Rifkin DB. Control of lung development by latent TGF-beta binding proteins. *J Cell Physiol* 2011;**226**:1499–509.

158. Dabovic B, Chen Y, Choi J, Vassallo M, Dietz HC, Ramirez F, von Melcher H, Davis EC, Rifkin DB. Dual functions for LTBP in lung development: LTBP-4 independently modulates elastogenesis and TGF-β activity. *J Cell Physiol* 2009;**219**:14–22.

159. Yin Y, Wang F, Ornitz DM. Mesothelial- and epithelial-derived FGF9 have distinct functions in the regulation of lung development. *Development* 2011;**138**:3169–77.

160. Colvin JS, White AC, Pratt SJ, Ornitz DM. Lung hypoplasia and neonatal death in Fgf9-null mice identify this gene as an essential regulator of lung mesenchyme. *Development* 2001;**128**:2095–106.

161. Yin Y, White AC, Huh SH, Hilton MJ, Kanazawa H, Long F, Ornitz DM. An FGF-WNT gene regulatory network controls lung mesenchyme development. *Dev Biol* 2008;**319**:426–36.

162. White AC, Xu J, Yin Y, Smith C, Schmid G, Ornitz DM. FGF9 and SHH signaling coordinate lung growth and development through regulation of distinct mesenchymal domains. *Development* 2006;**133**:1507–17.

163. Boucherat O, Franco-Montoya ML, Thibault C, Incitti R, Chailley-Heu B, Delacourt C, Bourbon JR. Gene expression profiling in lung fibroblasts reveals new players in alveolarization. *Physiol Genomics* 2007;**32**:128–41.

164. Park MS, Rieger-Fackeldey E, Schanbacher BL, Cook AC, Bauer JA, Rogers LK, Hansen TN, Welty SE, Smith CV. Altered expressions of fibroblast growth factor receptors and alveolarization in neonatal mice exposed to 85% oxygen. *Pediatr Res* 2007;**62**:652–7.

165. Srisuma S, Bhattacharya S, Simon DM, Solleti SK, Tyagi S, Starcher B, Mariani TJ. Fibroblast growth factor receptors control epithelial-mesenchymal interactions necessary for alveolar elastogenesis. *Am J Respir Crit Care Med* 2010;**181**:838–50.

166. Weinstein M, Xu X, Ohyama K, Deng CX. FGFR-3 and FGFR-4 function cooperatively to direct alveogenesis in the murine lung. *Development* 1998;**125**:3615–23.

167. Franco-Montoya M-L, Boucherat O, Thibault C, Chailley-Heu B, Incitti R, Delacourt C, Bourbon JR. Profiling target genes of FGF18 in the postnatal mouse lung: possible relevance for alveolar development. *Phsyiol Genomics* 2011;**32**:1226–40.

168. Xu K, Nieuwenhuis E, Cohen B, Wang W, Cantry A, Danska J, Coultas L, Rossant J, Wu MYJ, Piscione TD, Nagy A, Gossler A, Hicks GG, Hui C-C, Henkelman RN, Yu L, Sled JG, Gridley T, Egan SE. Lunatic fringe-mediated notch signaling is required for lung alveogenesis. *Am J Physiol Lung Cell Mol Physiol* 2009;**298**: L45–56.

169. Deimling J, Thompson K, Tseu I, Wang J, Keijzer R, Tanswell AK, Post M. Mesenchymal maintenance of distal epithelial cell phenotype during late fetal lung development. *Am J Physiol Lung Cell Mol Physiol* 2007;**292**:L725–41.

170. Madurga A, Mizikova I, Ruiz-Camp J, Morty RE. Recent advances in late lung development and the pathogenesis of bronchopulmonary dysplasia. *Am J Physiol Lung Cell Mol Physiol* 2013;**305**:L893–905.

171. Albertine KH. Progress in understanding the pathogenesis of BPD using the baboon and sheep models. *Semin Perinatol* 2013;**37**: 60–8.

172. Bruce MC, Honaker CE, Cross RJ. Lung fibroblasts undergo apoptosis following alveolarization. *Am J Respir Cell Mol Biol* 1999;**20**:228–36.

173. McGowan SE, McCoy DM. Fibroblasts expressing PDGF-receptor-alpha diminish during alveolar septal thinning in mice. *Pediatr Res* 2011;**70**:44–9.

174. Weiss MJ, Burri PH. Formation of interalveolar pores in the rat lung. *Anat Rec* 1996;**244**:481–9.

175. Coalson JJ, Winter V, deLemos RA. Decreased alveolarization in baboon survivors with bronchopulmonary dysplasia. *Am J Respir Crit Care Med* 1995;**152**:640–6.

176. Albertine KH, Jones GP, Starcher BC, Bohnsack JF, Davis PL, Cho SC, Carlton DP, Bland RD. Chronic lung injury in preterm lambs. Disordered respiratory tract development. *Am J Respir Crit Care Med* 1999;**159**:945–58.

177. Tschumperlin DJ, Liu F, Tager AM. Biomechanical regulation of mesenchymal cell function. *Curr Opin Rheumatol* 2013;**25**:92–100.

178. Janmey PA, Wells RG, Assoian RK, McCulloch CA. From tissue mechanics to transcription factors. *Differentiation* 2013;**86**:112–20.

Chapter 5

Pulmonary Vascular Development

Rosemary Jones*, Diane E. Capen† and Lynne Reid†
*Harvard Medical School and Department of Anesthesia, Critical Care and Pain Medicine, Massachusetts General Hospital, Boston, MA; †Department of Anesthesia, Critical Care and Pain Medicine, Massachusetts General Hospital, Charlestown, MA; ‡Department of Pathology, Harvard Medical School Children's Hospital, Boston, MA, USA

INTRODUCTION

The pulmonary and bronchial arterial systems, double venous systems, and lymphatics of the lung assemble into networks in close co-ordination with its forming airways and alveoli. Little is yet understood of the "patterning rules" that determine the spatial and fractal (tree-like) dimensions of developing units within the confines of mesenchymal or connective tissue spaces,[1] although familial patterns of branching in the human lung indicate that, at least for central vessels, there is a genetic component.[2] Regular and irregular vascular branching systems arise as networks evolve from expanding central arterial and venous structures, by secondary budding, branching and arborization, and by outgrowth of smaller offshoots.[3,4]

As the airways increasingly branch, the last (or terminal) bronchiole (TB) is reached: vessels proximal to the TB are termed pre-acinar and, immediately distal to the TB, the respiratory bronchiole (RB) marks the start of the lung's intra-acinar region in which vessels are associated with a RB or alveolar duct (AD), or lie more peripherally, i.e., in the alveolar-wall (AW). The pulmonary arteries supply capillaries within the acinus and pleura (except at the hilum) and drain to pulmonary veins. While arteries run centrally, the veins are distributed at the periphery of lung units, the most distal lying at the edge of acini. Venous tributaries arise from AWs, ADs, bronchial walls, pleura, and connective tissue sheaths, and drain to axial vessels that increase in size towards the hilum. The wall structure of vessels proximal to the acinus reflects their role as conduits of de-oxygenated or oxygenated blood, the structure of the small thin-walled intra-acinar vessels and capillaries of the alveolar-capillary membrane their role as gas-exchanging vessels. The main features of vascular beds forming in normal lung are vessel number, size, and wall structure.

The bronchial arteries supply the airway mucosa and peri-hilar structures; joining the main bronchi at the hilum before dividing with the bronchi, and sending one submucosal and one peri-bronchial branch along each airway wall to form communicating arcades.[5,6] When extra-pulmonary, they drain to pulmonary veins and to the right side of the heart via the azygos system; intrapulmonary, they form a network that anastomoses with the capillaries of the pulmonary arterial bed and drain via pulmonary veins to the left side of the heart. They provide nutrients to large airway and vascular structures.

Between the hilum and the start of the acinus, the simple endothelial channels of the lymphatics lie within connective tissue sheaths.[7,8] They extend deep within the pulmonary lobule in association with bronchovascular bundles, and are found in peri-vascular, peri-bronchiole, and inter-alveolar regions, as well as in the pleura and inter-lobular septae.[8,9] Liquid and plasma proteins move through the matrix to the lymphatic plexus surrounding terminal airways, and drain centrally through the lung's lymphatic channels to the systemic circulation. Abundant deep conduit and saccular lymphatics are present in the neonatal lung being about 3-fold more frequent than in the adult.[10] To maintain adequate function, both at rest and on exercise, each of these interconnecting systems must develop appropriately.

Gene knock-out and knock-in studies in mice continue to provide a wealth of data and increased understanding of lung development. Many genes initiating lung organogenesis and morphogenesis have been identified, especially ones regulating formation of the lung bud, the development of the trachea, right and left lung asymmetry, and the regulation of re-iterative airway branching patterns.[11,12] Genes regulating vasculoangiogenesis are also increasingly being identified,[13] and in the lung it is clear that the formation of pulmonary vascular units directs epithelial morphogenesis,[14–16] not only for normal branching morphogenesis but for alveolarization and maintenance of distal airspaces.[15] The importance of normal vascular development immediately before and after birth has emerged in relation to broncho-pulmonary dysplasia (BPD) in the human neonate,

The Lung. http://dx.doi.org/10.1016/B978-0-12-799941-8.00005-5

i.e., a critical period of change in the structural template of the distal lung.[17–19] Inhibition of vascular development results in fewer capillaries and large simplified airspaces in the postnatal (mouse) lung, reminiscent of the lung in BPD.[20]

As understanding of the role of instructive and permissive genes in the lung increases (genomics) so will our need to understand the modulation of cell phenotype by assembled proteins (proteomics) and the basis of vascular cell plasticity: as, for example, smooth muscle cells (SMCs), long considered cells with a defined phenotype, which have emerged as a continuum of cells expressing a range of genes and proteins,[21–23] and endothelial cell demonstration of inherent phenotypes that have a role in integrating signals from the circulation for vascular function.[24] Newer methodologies, based on an understanding of the regulatory role of non-coding microRNAs (miRNAs), at the interface between genes and proteins, perhaps offer a basis for the evident modulation of phenotype and function of lung cells. Acting via multiple signaling pathways to regulate transcription, mRNA processing, stability, and translation,[25] miRNAs influence lung airway and vascular development,[26] pulmonary artery SMC proliferation, differentiation and phenotype,[27,28] and angiogenesis.[29] A combination of these different approaches will continue to provide information essential to understanding developmental patterns of the lung.

CELLULAR BASIS OF VESSEL MORPHOGENESIS

Overview

The establishment of lung cell fate from the ventral foregut endoderm initiates embryonic lung development. As the lung anlage develops (as an out-pouch at the caudal end of the laryngotracheal groove from the foregut) it is vascularized by in-growth of a vascular plexus from the heart/aortic sac and dorsal aorta: and, as the trachea and dorsal esophagus separate, and the bronchial buds grow into the adjacent splanchnic mesoderm to give rise to airway generations, pulmonary vascularization starts with cells derived from the splanchnic plexus.

Developmental Stages of the Lung

The four developmental phases identified in the human and mouse lung (the latter included because of the focus of genetic regulation studies), are: (i) *pseudoglandular stage*, i.e., the formation of proximal airways and vessels (5–17 weeks human lung, embryonic day (E) 9.5–16.6 mouse); (ii) *canalicular stage*, i.e., the formation of distal airways and vessels (16–25 weeks human lung, E16.6–17.4 mouse); (iii) *saccular stage*, i.e., start of formation of the acinus by the development of immature saccules, increase in the

density of small vessels, and capillary reorganization to form a functional gas-exchange surface at birth (24 weeks to term in human lung; E17.4–E19 mouse); (iv) *alveolar stage*, i.e., septation of original saccules to form alveoli, and the formation of additional alveoli (late fetal period to childhood and young adult in human lung; and in the postnatal period through to young adulthood in rodents (P5 to P70).[30,31] Within this timetable the sequence of patterning is constant although the timing of each phase varies between species.[32]

Vascular Channel Formation in the Lung

Mesenchymal cells surrounding developing airways differentiate into angioblasts (cells committed to an endothelial lineage). As the angioblasts give rise to endothelial-like cells, capillary formation requires these cells to coalesce into a tube, their contiguous membranes to fuse, and junctional complexes to form.[33] Networks of large and small vessels evolve as initial vascular webs remodel into channels: the growth of small vessels from capillaries eventually giving rise to all large vessels in the lung. Further wall formation requires signaling between endothelial cells and from mural cell precursors that also develop from the mesenchymal cell population; and as mural cells develop, the formation of elastic laminae is critical for their organization into a media (SMCs), and adventitia (fibroblasts), to establish a fully mature vessel wall. While much data relating to signaling pathways regulating capillary and vessel assembly derives from studies in tumor growth,[34,35] the cell–cell interactions they identify highlight processes likely involved in normal lung vessel growth.

Homotypic and heterotypic contacts between cells and matrix are important in determining cell phenotype and in regulating tissue growth and organization.[36–43] Adhesion receptors at the endothelial cell surface (including members of the integrin, cadherin, immunoglobulin, selectin, and proteoglycan superfamilies) interact with those of adjacent cells or with matrix proteins such as collagens, fibronectins, laminins, and proteoglycans forming fibrils or other macromolecular arrays.[44–46] Cytoplasmic plaque proteins further link cell membrane receptors to the cytoskeleton transducing signals from the cell surface and regulating receptor function[44–46]; and nuclear DNA and its associated protein scaffold connect to matrix components through the cytoskeleton, contributing to communication between intracellular and extracellular environments.[47]

DEVELOPMENT (FORMATION AND GROWTH) OF ENDOTHELIAL CHANNELS

Overview

Angioblasts form capillary channels by vasculogenesis (de novo capillary formation) and by angiogenesis (the formation of new capillaries from existing ones). As new growth proceeds by cell proliferation and by capillary sub-division, vascular density is adjusted by release of endothe-lial and mural cell contacts and (as patterns of blood flow change) by the regression of un-perfused channels (vascu-lar pruning by cell loss via apoptotic and/or non-apoptotic mechanisms). Endothelial cell specification eventually determines the formation of arterial, venous, or lymphatic pathways, allowing the development of these systems in the lung to proceed in parallel.

Vasculogenesis

In vasculogenesis, local aggregates of progenitor cells form a primitive network,[48,49] as mesenchymal cells differenti-ating into angioblasts or hemangioblasts (blood cell pre-cursors) form sinusoidal nests of cells and spaces (blood islands) (Figure 1a,b). Loss of vesicles from the mesenchy-mal cell apical membranes gives rise to the lumen of the first channels. By thinning their cell processes and reform-ing their apical membrane, they become endothelial-like cells enclosing hemangioblasts. Mesenchymal cells release growth factors promoting their differentiation to endothelial cells.[50]

Angiogenesis (Sprouting)

In a widely recognized form of angiogenesis, pre-existing capillaries and small vessels form capillary-like sprouts (Figure 1c).[51–53] Focal degradation of the endothelial basement membrane and surrounding matrix by proteo-lytic enzymes leads to the extension of an endothelial pseudopodia through the gap to form a spur. Endothelial cell migration in the direction of the growth spur forms a sprout.[54,55] A single "tip" cell guides its direction. Rather than migrating singly, the migrating cells move as a sheet,[55] and proliferating cells behind the tip of the sprout increase its length.[54,56] Sprouts continue to elongate and to branch or fuse to an adjacent sprout at their blind-end to form a new loop (Figure 1c) and the development of a slit-like lumen, followed by the entry of plasma and blood cells, completes the formation of a continuous channel. The surrounding extracellular matrix influences the sig-naling pathways that modulate the pattern of endothelial cell organization into "tip" and trailing "stalk" cells lead-ing to angiogenesis and the branching pattern of vessel networks.[57]

Angiogenesis (Expansion)

Slow expansion of a forming vascular network, by increase in the diameter or length of existing units (Figure 1d), is achieved by endothelial and mural cell proliferation in the absence, or in excess, of cell loss.[55]

Angiogenesis (IMG, IAR, and IBR)

In further angiogenic processes, capillaries undergo inter-nal sub-division by intussusceptive microvascular growth (IMG)[58–61] and form into larger structures by intussuscep-tive arborization (IAR) and intussusceptive branch remodel-ing (IBR) (Figure 2 and Figure 3a-h).[62,63] In IMG, opposing endothelial cell membranes form an inter-endothelial bridge, and cell junctions re-align and the central region of the endothelial layer is perforated by an invading con-nective tissue post (1–2.5 mm in diameter) (Figure 2). The post is stabilized by inclusion of pericytes, myofibroblasts, and collagen fibrils,[59] and with growth evolves into a pillar (~2.5 μm in diameter) and forms a mature mesh.[59,61,64–66] IMG contributes to lung vascularization in the embryo, in septation after birth, and possibly to the continuing devel-opment of the gas-exchange surface in the young adult.[58,66] It also leads to the formation of complex networks in which vascular folds organize to establish compound loop systems.[67,68] Based on processes of IMG, IAR, and IBR (Figure 3), an overall concept of vascular morphogenesis, patterning and remodeling by intussusceptive angiogenesis (IA) has been proposed.[61,64,66,69,70] Mercox casts illustrate the complex pattern of vessel and capillaries in the devel-oping lung, and the extent to which they remodel within a remarkably short time frame (Figure 4a-d).

Endothelial Cell Fate (Arterial-Venous and Lymphatic Specification)

Our understanding of genetic factors specifying endothelial cell fate has expanded in recent years although how cells commit to arterial, venous, or lymphatic channels in the developing lung remains an intriguing question.[13,71]

Ephrin B2 and EphB4, a sub-class of receptor tyrosine kinases, were the first genes shown to be differentially expressed in arterial and venous endothelium. Mice lacking ephrin B2 develop large arterial and venous trunk vessels but have defects that prevent appropriate connections between the vessels.[71–73] Inactivation of EphB4 in EphB4 tau-lacZ mice results in vascular defects that parallel those in eph-rinB2-deficient mice. While EphrinB2 and EphB4 expres-sion is used to identify arterial and venous endothelial cells, expression is not required for specification of cell fate during vasculogenesis; but it is required to define and maintain arte-rio-venous boundaries to prevent mixing of cells marked for an arterial or venous fate (Figure 5). This important signaling system in embryonic development persists in the adult, indi-cating a need to maintain a boundary between the two vas-cular systems.[71,74] Interestingly, the signaling system (i.e., ephrin ligand to EphB4 receptor) is reversible (EphB4 recep-tor to ephrin ligand), changing the cellular response pathways initiated.[71] Furthermore, a degree of plasticity exists between ligand and receptor expression,[75] as in the response to excess

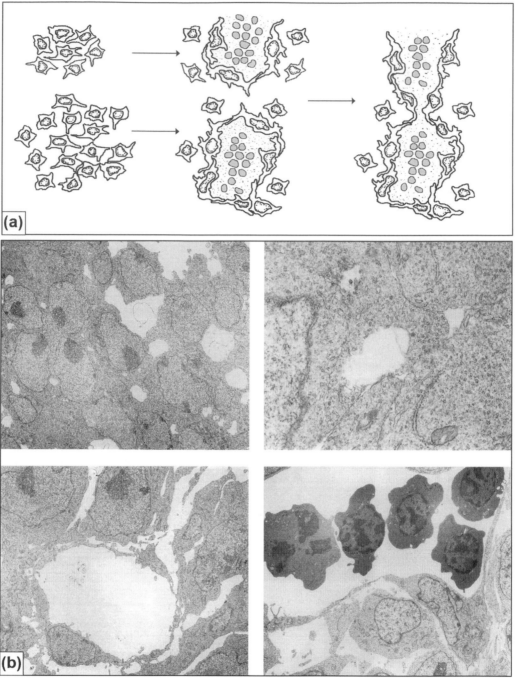

FIGURE 1 Capillary assembly by vasculogenesis, angiogenesis (sprouting) and expansion. (a) Vasculogenesis: aggregates of undifferentiated mesenchymal cells differentiate in situ into angioblasts, (i.e., cells committed to an endothelial lineage) and associate into blood islands; developing contacts, the angioblasts assemble into capillary-like structures that connect and retain hemangioblasts (hematopoietic precursor cells) within their lumen.[49] *Re-drawn and reproduced with permission.* (b) Vasculogenesis: Densely-packed mesenchymal cells containing cytoplasmic vesicles surround the developing lung bud of a mouse (E9, top left), and intercellular spaces form (which appear to result from discharge of intra-cytoplasmic vesicles, leaving a ruptured cell membrane) between plump, densely packed mesenchymal cells (top right). Thinning of lung mesenchymal cells adjacent to intercellular spaces, in a later fetus (E10), results in endothelial-like cells (lower left). Hematopoietic precursor cells are present in some luminal spaces (lower right). Original magnification: top left ×2,800; top right ×8,750; lower left and right ×5,250.[230] *Reproduced with permission from the American Thoracic Society.* (c) Angiogenesis: Endothelial sprouts form as the cell's processes are released from the constraint of a basement membrane, and in response to a signal triggering growth, invade the surrounding tissue (i.e., sprout in the direction of the growth signal) to form cell-lined channels. Wall destabilization, with focal degradation of the endothelial basement membrane and surrounding matrix, is followed by the formation of a spur as endothelial pseudopodia extend through the gap (top left) and endothelial cell migration in the direction of the growth spur forms a sprout (top right). While still connected at their origin, sprouts continue to elongate and to branch and fuse to an adjacent sprout at their blind-end to form a new loop and establish a contiguous network (lower left and right) and to branch or fuse to an adjacent sprout at their blind-end to form a new loop and establish a contiguous network. *Re-drawn from[55] and reproduced with permission.* (d) Growth of an existing vascular bed by simple expansion as wall cells are added and lumen diameter and segment length increases.[55] *Reproduced with permission from Wiley-Blackwell.*

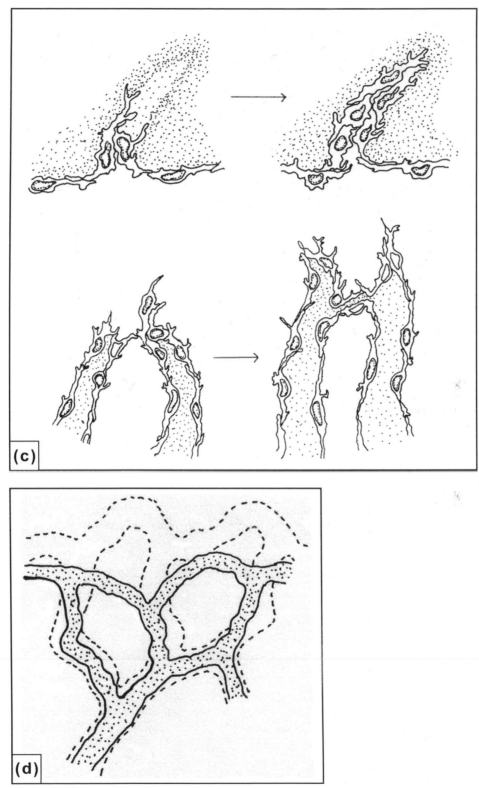

FIGURE 1 Cont'd

oxygen in the newborn retina where "venous" endothelial cells appear in segments previously consisting of "arterial" endothelial cells.[76] Cell specification is controlled upstream of the ephrin-Eph system by other signaling molecules that promote or inhibit endothelial cell fate. These include hedgehog (Hh), vascular endothelial growth factor (VEGF), Notch and chicken ovalbumin upstream-transcription factor (COUP-TFII), while epigenetic factors subsequently modify

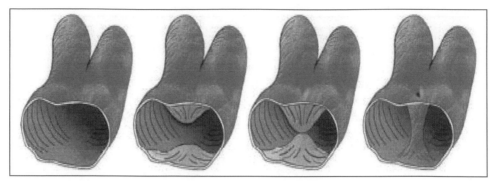

FIGURE 2 Capillary remodeling and growth by intussusceptive microvascular growth (IMG). The capillary lumen divides as the endothelial layer creates a contact zone between opposing endothelial cells. Following central perforation of the cell bilayer, the fused endothelial cells form a transluminal cuff, which is later invaded (and strengthened) by myofibroblasts or pericytes.[63] *Reproduced with permission.*

FIGURE 3 Capillary and vessel remodeling by intussusceptive microvascular growth (IMG), intussusceptive arborization (IAR), and intussusceptive branch remodeling (IBR). (a) Insertion of transluminal pillars (arrowheads) results in rapid expansion of the capillary plexus by IMG. (b–d) IAR generates feed vessels from the capillary plexus by vertical pillar formation in rows (arrows), which demarcate future vessels. Narrow tissue septae (arrow) formed by pillar reshaping and pillar fusion segregate the new vessel entity. Formation of horizontal pillars and folds (arrowheads) separate the newborn feeding vessels from the capillary plexus. (e–h) IBR finally adapts the deepness of the branching angle and the diameters of daughter vessels in the newly formed supplying and draining vessels by insertion of transluminal pillars at branching points (see e and f, arrows), and continues to lead to vascular pruning by repetitively eccentric formation, augmentation, and fusion of pillars (see f–h, arrowheads).[63] *Reproduced with permission.*

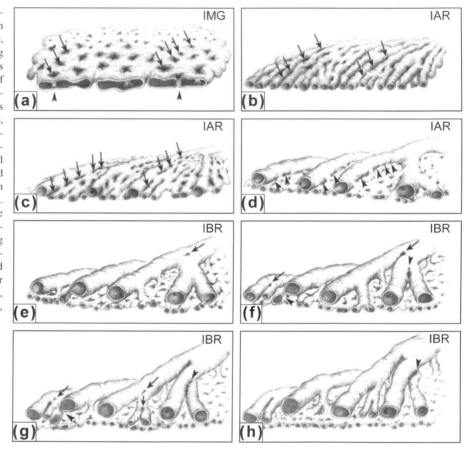

arterial-venous identity.[71] Other important factors (e.g., forkhead, sox, clcr, and snrk-1) also regulate endothelial fate.[71] In alveolar formation, ephrin-Eph signaling has a further role as a guidance cue in the outgrowth of secondary crests.[77]

Lymphatic (PROX-1) endothelial cells (LECs) arise by commitment of venous endothelial cells to a lymphatic fate.[75,78] In mice this occurs via ERK activation, leading to induction of SOX18 and PROX-1 expression, with PROX-1 required for initiation and maintenance of the LEC phenotype.[78] Although PROX-1 deficient endothelial cells can bud

from the cardinal vein (i.e., their normal source of origin), they display abnormal VEGF-induced migration and fail to express LEC markers. PROX-1 mediates lymphatic cell differentiation, including up-regulation of VEGF-R3 (via VEGF-C). VEGF-R3 is expressed by all endothelial cells during early development but is enhanced in cells committed to the LEC lineage, and becomes largely restricted to LECs as the lymphatic system develops.[78]

Relatively low endothelial proliferation rates in the young and adult lung raise the question of the source of new

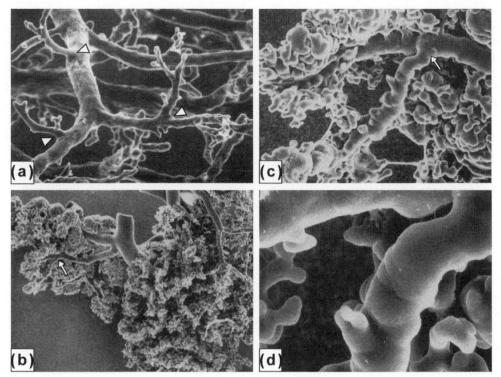

FIGURE 4 Scanning electron micrographs of Mercox casts illustrating normal pulmonary vascular growth patterns. (a) Vascular bed of a mouse at E14: note the presence of occasional small capillary buds and branches (arrowheads), intersecting at right angles with the lumen of larger adjacent capillary segments. (b–d) Vascular bed of a mouse at E15: extensive connections have now formed between the central and peripheral systems, revealing a complex peripheral network. (c–d) At higher magnification (see area indicated in b, arrow) blind-ending pre-capillary branches are seen to approach the future capillary network at right angles. Residual constrictions suggest sites of coalescence (fusion). It seems likely that the vessel upstream (from the arrow) is central, whereas the more irregular vessels below the arrow are derived from the peripheral system. Original magnification: (a) ×320; (b) ×69; (c) ×320; (d) ×1,200.[230] *Reproduced with permission from the American Thoracic Society.*

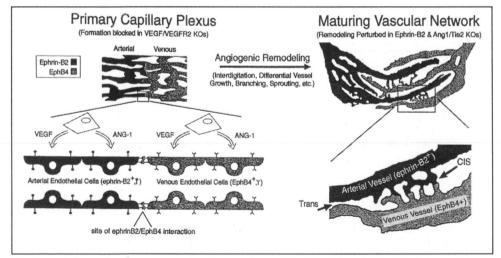

FIGURE 5 Arterial–venous specification. Distribution and sites of action of ephrin B2 and EphB4, shown in arteries and veins of a primary capillary plexus and in a maturing vascular network. They are presumed to interact between opposing arterial and venous endothelial cells in a *cis* manner (left). During later maturation and remodeling of the primary plexus (right) by interdigitation, branching, and differential growth of vascular segments, they remain localized to arterial and venous units (*cis* interactions) but may also interact at the interdigitating surfaces of large vessels (*trans* interactions).[73] *Reproduced with permission from Elsevier.*

cells for homeostasis and for vascular renewal/growth after injury.[79–83] Endothelial progenitor cells (EPCs) isolated from the bone marrow (circulating EPCs), and from the intima and media of vessels in different organs (resident EPCs), provide such a population. In addition, highly proliferative lung microvascular capillary endothelial cells offer a further resident cell population available for repair,[84,85] although it is unclear if these are truly EPCs or if they represent a proliferating endothelial cell population that exists among more numerous endothelial cells that do not proliferate.[86]

DEVELOPMENT OF VASCULAR MURAL CELLS

Overview

Mesenchymal cells recruited to endothelial channels as peri-endothelial cells differentiate into pericytes, SMCs, and fibroblasts, each regulating or modulating the response of the vessel wall to vaso-mediators.[34,35,87–91] Pericytes form a sub-intimal layer adjacent to endothelium, and SMCs and fibroblasts the vessel medial and adventitial layer. The presence of multipotent messenchymal stem cells (MSCs) and peri-vascular progenitor/stem cells in adult organs/tissues, including the lung, indicates a potential role for the cells in vascular homeostasis and repair.[92–96]

Pericytes

In capillaries, and especially small post-capillary venules, peri-endothelial cells investing vessel walls develop into pericytes (Figure 6a).[97–99] The cells derive from the mesoectoderm while endothelial cells develop from the mesodermal lateral plate.[100] Mesothelial cells undergoing epithelial to mesenchymal transition also differentiate into pericytes (or fibroblasts and SMCs).[101] Pericytes-endothelial cell communications determine the local characteristics of capillaries and microvessels in different organs.[102] Mature pericytes lack the extensive filament networks, dense bodies, and attachment plaques characteristic of SMCs.[98,99,103,104] Because these characteristics must be determined by transmission microscopy the cells are often described by their gene or protein expression,[101] although phenotyping the cells can prove difficult—especially during angiogenesis.[105] Pericytes and endothelial cells likely interact to assemble a basement membrane that is shared,[106,107] although commonly separating the two cells along much of their interface. Numerous pericyte processes extend between the membrane to contact the endothelial cell, and rare gap junctions allow nucleotides to pass between the cells.[103,104,108] Present as solitary cells, or forming a discontinuous layer enveloping capillaries, their processes usually extend

along the abluminal capillary surface, a single pericyte typically spanning several endothelial cells and sometimes capillary branches.[101] During angiogenesis and sprout formation, the cells prevent plasma from escaping from the vascular space into interstitial tissue (Figure 6b). Possessing cyclic GMP-dependent cyclase actin, desmin, vimentin, α-tropomyosin, and myosin,[109] their ability to express proteins typical of a contractile cell phenotype, such as smooth muscle myosin heavy chain (SM-MHC), α-smooth muscle actin (α-SMA), or desmin, and so the potential to contract or relax in response to vaso-mediators, and to regulate the diameter of capillaries, varies greatly in vascular beds.[110–113]

In adult organs, other peri-endothelial cells forming the sub-intimal layer of small lung vessels are intermediate cells rather than pericytes (Figure 6c). Midway between a pericyte and SMC in morphology, these cells do not share the endothelial cell basement membrane, and express filaments although not dense bodies or attachment plaques.[98,99] In response to injury, they typically acquire a SMC phenotype and separate from the adjacent endothelial layer by elastin synthesis and the assembly of an elastic lamina (Figure 6d,e).[114,115]

Smooth Muscle Cells and Elastic Laminae

In vessels where the transmural pressure is higher than in capillaries, the peri-endothelial cells typically acquire a SMC phenotype. Derived from mesenchymal cells in the developing lung (Figure 7a–c), the cells may develop from resident vascular progenitors in adult organs (see Figure 7d, and following text). Embryonic endothelial cells provide another source of SMCs[116–118]—the cells shifting to become "mesenchymal" cells expressing SM proteins.[119] SMC (or pericyte) investment of developing endothelial tubes is critical for vascular maturation. The network of contractile and cytoskeletal filaments occupying the cytoplasm (Figure 8a,b) of differentiated SMCs confers tensile strength and the ability to contract.[120,121]

SMCs exhibit a wide range of phenotypes at different stages of development, and even in adult organs retain a remarkable degree of plasticity, undergoing reversible changes in phenotype in response to local environmental changes, e.g., growth factors/inhibitors, mechanical influences, cell–cell and cell–matrix interactions, and inflammatory mediators.[21–23] In assembling vessels, they exhibit high rates of proliferation, migration, and production of extracellular matrix components (collagen, elastin, proteoglycans, cadherins, and integrins) while at the same time acquiring contractile capabilities, and the cells again switch to increase their proliferation and migration rates, and synthetic capacity, in response to vascular injury. Mechanisms controlling the expression of genes specific or selective for

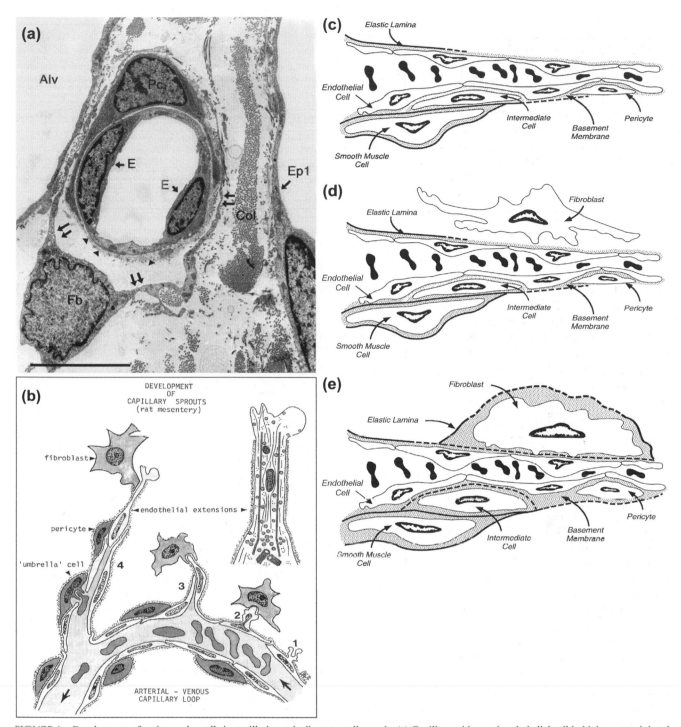

FIGURE 6   Development of peri-vascular cells in capillaries and adjacent small vessels: (a) Capillary with a peri-endothelial cell in high oxygen-injured adult rat lung. 80nm epon section stained with uranyl acetate and lead citrate. The capillary wall is formed of endothelial cells (E) and a pericyte (Pc), that shares the endothelial basal lamina (arrowheads). The processes of a nearby fibroblast (Fb) extend around the capillary (double arrows), and a basal lamina is absent from the cell. These fibroblasts can envelop capillary walls to become peri-endothelial cells (see text), as well as the walls of adjacent small vessels (see also Figure 6d,e). Collagen (Col), Alveolus (Alv), Epithelial type 1 cell (Ep1). Bar = 5μm.[149] *Reproduced with permission.* (b) Schematic drawing demonstrating the development of peri-vascular cells around capillary sprouts (based on intravital video recordings and electron micrographs of the mesenteric microcirculation of young rats). Small endothelial extensions or buds (1) penetrate the basal lamina (fine stippling) and evolve into cellular protrusions (2), develop into endothelial spurs and short sprouts (3), and gradually lengthen into full sprouts (4). Fibroblasts settle down on the adluminal walls of sprouts and transform into pericytes by sharing the endothelial cell basement membrane. Loss of plasma, blood cells, and platelets into the interstitial tissue during sprout formation is prevented by pericytes temporarily assuming an umbrella shape. Endothelial cells can contribute to the length of forming sprouts (regardless of proliferation) by organelle streaming and re-organization of their cytoplasm into long extensions.[103] *Reproduced with permission.* (c–e) Schematic illustration of peri-vascular cells in small lung vessels adjacent to capillaries. The location of mural cells in the sub-intima of a small vessel and an adjacent capillary in normal rat lung, are shown, i.e., an intermediate cell and pericyte respectively. In response to injury, fibroblasts present in the surrounding interstitium are recruited to the vessel wall where they align as peri-vascular cells. As the injured vessel wall remodels, the intermediate cells and fibroblasts express contractile proteins characteristic of SM, and the cells are separated from endothelium by the development of elastic lamina (see d and e).

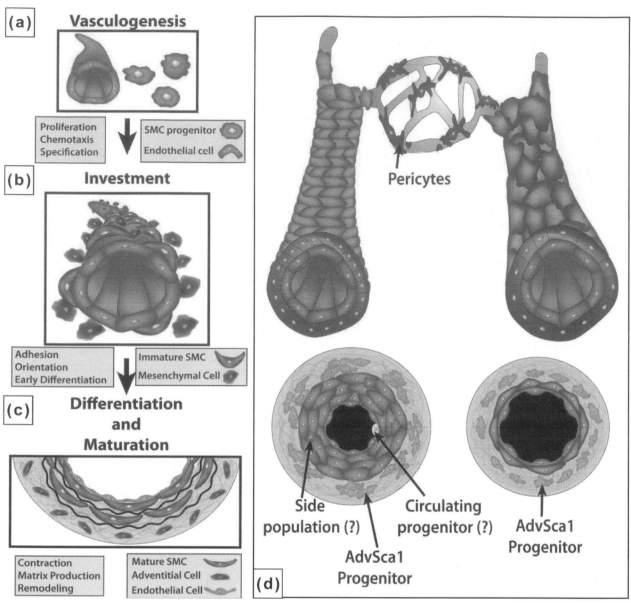

FIGURE 7 Vascular SMC development from embryonic progenitors. (a–c) Angioblasts differentiating into endothelial cells (red) self-assemble into a nascent capillary-like vascular network and become invested (b), as increasing cardiac output from the developing heart stimulates endothelial production of mesenchymal cell chemoattractants. SMC progenitors (brown) begin to invest the vessel wall around E10.5 in the mouse. Close contact with endothelial cells initiates SMC differentiation (blue cells). The position of the vessel within the embryo determines what type of SMC progenitor will be involved in producing the tunica media. Proliferation of SMC progenitor cells is required to supply SMCs in sufficient numbers for continued vascular development. As differentiation and maturation proceeds, layers of SMCs are added to the developing artery wall, and a cross-linked extracellular matrix forms, defining the structure of the mature tunica media. (d) Sources of vascular SMC progenitors in adults' large arteries (red) and veins (gray). Resident SMC progenitors have been identified in the adventitial layer (green cell clusters) and in the medial layer (blue cells). Some reports suggest that SMCs in atherosclerotic plaques and intimal masses can arise from circulating, bone marrow–derived progenitor cells (yellow), whereas others report finding no evidence to support that origin (see Reference [94]). In microvessels, pericytes (purple) have multi-lineage differentiation potentials and can act as SMC progenitor cells.[94] *Reproduced with permission from.[94]*

the SMC, and required for its differentiated function, continue to be understood.[21,22]

The number of SMCs investing vessel walls is typically proportional to blood flow and transmural pressure[122] and changes with vessel size. In the lung, however, the distribution of cells decreases along each pathway until small "resistance" arteries (located at the entrance to the acinus) are reached, where the wall thickness is high for lumen size. Additionally, most intra-acinar arteries (and veins) have no SMCs in their wall (Figure 9a). Notably, in disease, or in response to injury, SMCs develop in large numbers in the walls of many of

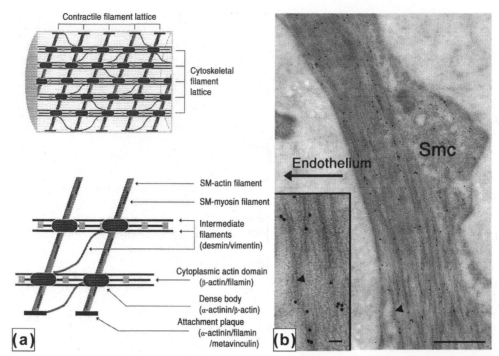

FIGURE 8   Arrangement of the SMC contractile and cytoskeletal filament lattice and organization of its structural components. (a) Oblique, face-polar, SM-myosin filaments (14–16 nm diameter) cross-bridge to α-SM-actin filaments (4–6nm) and anchor to the cytoskeleton at dense bodies—ovoid structures consisting of α-actinin and β-actin—to form the contractile apparatus. Longitudinal intermediate filaments (7–11 nm) of desmin (an SMC-specific protein) or vimentin, and a cytoplasmic domain of β-actin and filamin (an actin cross-linking protein), form the cell cytoskeleton. These filaments also anchor to the contractile apparatus at dense bodies, linking it to the cell's supporting structure to give the cell tensile strength; they also link the contractile apparatus to the plasmalemmal membrane and to elastic components of the extracellular matrix via peripherally located attachment plaques, i.e., submembranous structures (0.2–0.5 nm) containing α-actinin, filamin, metavinculin, or vinculin, which anchor at the cell membrane via proteins such as p-lectin. Other intermediate filaments traversing the network provide further support by anchoring to dense bodies and attachment plaques. Attachment plaques are separated by membrane regions rich in caveolae and characterized by transmembrane receptors or integrins that link components of the cytoskeleton to the extracellular matrix. Dense bodies and attachment plaques are considered the functional equivalent of Z-bands in striated muscle.[121] *Redrawn and reproduced with permission from Academic Press, San Diego, CA.* (b) Example of α-SMA immunoreactive sites (identified by 10nm gold particles) decorating the filaments of a SMC developing in the wall of an alveolar vessel (ED 31 µm) in high oxygen-injured adult rat lung. 80 nm Unicryl section stained with uranyl acetate and lead citrate. The cell lies surrounded by matrix with the vessel lumen and endothelium to the left. Filaments typically develop along the adluminal cell margin. The gold particles and filaments (at arrowhead) are shown at higher magnification in the inset. Bars = 1 µm and 0.1 µm.[270] *Reproduced with permission from Elsevier, London.*

these vessels, increasing their wall thickness ~10-fold (Figure 9b,c). These newly developing SMCs synthesize a myosin isoform that confers a high level of contractility, i.e., one typical of gut SMCs.[123]

In larger lung vessels, small bundles of collagen and collagen fibrils form between the SMCs and, in all but the smallest venules, they are surrounded by basement membrane and have extensive filaments, fusiform dense bodies, and attachment plaques. The signaling molecules/proteins required for their de-differentiation from a "contractile" to a "synthetic" phenotype, and to again revert to a contractile phenotype (see Figure 8a caption), appear in sequence in cells of developing vessels[91,124–131]; α-SMA expression is followed by calponin, h-caldesmon, α-tropomyosin, and metavinculin, while SM-MHC SM1 appears in immature cells during the fetal period and SM-MHC SM2, metavinculin, and α-tropomyosin appear in differentiated cells after birth.

The processes of SMCs penetrate their surrounding basal lamina to form lateral contacts and the basal lamina of adjacent endothelial cells to form myoendothelial junctions,[98]: and in a reverse pattern, endothelial cell contacts to SMCs increase in small vessels.[132] Differentiated SMCs establish a vessel media, separating from adjacent endothelial cells and from adventitial fibroblasts by the formation of an internal and external lamina; these form as elastin (synthesized by the cells) self-assembles into a lamina structure. In large vessels, additional lamina further divide the SMCs into multiple layers, while in small vessels, where SMCs are absent from the wall, a single elastic lamina separates endothelium from the surrounding connective tissue. Elastin regulation of SM growth is demonstrated by the development of obstructive intimal hyperplasia and death of mice that lack the elastin gene.[133] Endothelial cells, SMCs, and fibroblasts are each elastogenic but their relative contribution to

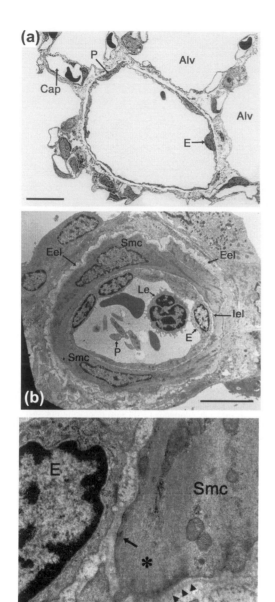

lamina formation at different levels of vascular pathways in the lung is unknown. Current understanding of lamina assembly derives from data of other sites.[134–146] Endostatin (an inhibitor of endothelial cell proliferation) present within matrix and elastic laminae of large vessels[147] may restrict sprouting from the wall.

In perinatal and adult vessels recent evidence suggests that SMC progenitors reside in a signaling domain, or niche environment, in the media/adventitia of vessels (see Figure 7d). Resident MSCs derived from the adult lung and other organs, including large and small blood vessels (and from capillaries in kidney), indicate cells in a peri-vascular cell niche,[92,95] providing a reservoir of "undifferentiated" cells in response to tissue demands.[92] The term peri-vascular stem cell has been suggested as more specific for these cells rather than MSC.[96] Similarly, endothelial precursor cells and stem cells in a distinct zone between the medial SM layer and adventitial fibroblast layer of large and middle-sized arteries and veins have been proposed to form a "vasculogenic" zone in blood vessel walls—a source of progenitor cells for postnatal vasculogenesis.[96] Such cells are thought to reside in an "adventitial cell niche"[94]; the niche essentially forming a signaling environment in which associated macrophages and T-cells control cell activity, preserving a group of cycling progenitor cells to sustain the population, and releasing others as needed into the vessel media/intima.[93] Their characterization is somewhat unclear as, unlike SMCs progenitors in embryonic development, which are characterized by the appearance of specific cytoskeletal and contractile protein isoforms, no markers currently are available to identify SMC progenitors in adult tissue.[94] These cells are unlikely to apply to SMC development in the smallest vessels of the adult lung in disease or in response to injury, because these vessels normally lack a defined layer of SMCs and adventitial cells, their wall consisting only of endothelial cells with or without an elastic lamina. In these vessels interstitial fibroblasts are recruited as peri-vascular cells that acquire a SMC phenotype (see following text).[148,149]

Fibroblasts

Adventitial fibroblasts form the outermost layer of vessel walls in which SMCs develop. Their number in the wall, and that of the SMCs, determines oxygen diffusion, which is restricted once the peri-vascular tissue cuff is

FIGURE 9   SMC development in injured distal lung vessel. (a) Alveolar wall vessel (ED ~35 μm) in normal adult rat lung. 80nm epon section stained with uranyl acetate and lead citrate. Typically, SMCs are absent in these vessels: endothelial cells (E) and the processes of peri-endothelial cells (P), including pericytes and intermediate cells (see text), form a thin wall. Alveolus (Alv), Capillary (Cap). Bar = 10 μm.[152] *Reproduced with permission from American Thoracic Society.* (b) SMCs have developed in an alveolar wall vessel (ED 20 μm) in high oxygen-injured adult rat lung. 80 nm epon section stained with uranyl acetate and lead citrate. Endothelial cells (E), and SMCs between an electron-lucent external (Eel) and internal elastic lamina (Iel) now form the vessel wall. Platelet (P), Leukocyte (Le). Bar = 10μm.[148] *Reproduced with permission.* (c) Higher magnification of SMCs forming the wall of the vessel in Figure 9b, showing the arrangement of intracellular filaments (*) and attachment plaques (arrows) characteristic of a contractile phenotype. Basement membrane (arrowheads) surrounds the cells. Bar = 1μm.[148] *Reproduced with permission.*

100μm-thick. The cells contain microfilaments (4–6nm), intermediate filaments, pinocytotic vesicles, and extensive RER; a basement membrane is absent, and collagen fibers and fibrils form along the cell's adluminal and abluminal margins. They typically express vimentin, actin isoforms and nonmuscle myosin but express SMC-specific proteins under certain conditions.[83,98,103,123,148–153]

The interstitial fibroblast, a resident cell of the lung's alveolar-capillary membrane, is a precursor SMC in small intra-acinar vessels, i.e., vessels without SMCs or adventitial fibroblasts in the normal lung. Evidence that, in small vessels and capillaries of mature vascular beds, they are recruited as peri-vascular cells, and differentiate into pericytes and SMCs, comes from studies of dorsal and mesenteric capillaries and of vascular remodeling in lung.[83,98,103,123,148–152] After migrating to align around adjacent vessels (see Figure 6d,e) they express filaments composed of contractile proteins (see Figure 8a,b; Figure 9b,c).[123,148,149,152] The recruited cells, likely with endothelial cells, synthesize the components of an internal elastic lamina. Similarly, the outermost layer of aligning fibroblasts produces an external lamina as they acquire a SM phenotype; this encloses the cells and defines the structure of the vessel wall from the surrounding interstitium (see Figure 6e).

In the developing lung, migrating fibroblasts termed myofibroblasts, i.e., cells with a phenotype between a fibroblast and SMC with many filaments, act as SM precursors in alveolar development, migrating into nascent septa and depositing elastin as a first step in the development of secondary septa.[154–156] In adult organs, myofibroblasts are specialized cells, characterized by dense filament networks; relatively sparse in the interstitium of the normal lung the cells increase in the lung in fibrosis. Like SMCs, myofibroblasts express α-SMA; although the transcriptional pathways for α-SMA in these two cell types differ.[157–159]

Depending on the species (being typical in the rat and mouse) an additional cuff of cardiac muscle surrounds extra-pulmonary venous segments, and the axial venous pathway and branches.[160–162] As the main branch of the pulmonary venous trunk grows out from the heart (to anastomose with the developing pulmonary plexus) and becomes established, the developing pulmonary veins are colonized by cardiac cells[162] aligning along the abluminal surface of the medial SMC layer.[161] Mapping the cells (identified in mice by cardiac troponin 1 gene and LacZ expression) demonstrates their accumulation along a caudo-cranial gradient and their spread into smaller branches.[162] The migrating cells are constrained by well-defined boundaries, eventually extending to the third bifurcation (of the venous trunk) but not to the capillary plexus. Venous endothelial cells or SMCs are a likely source of their recruitment[162] via a signaling system (e.g., ephrin/ephrin ligand) influencing the specification of arterial-venous segments and established before initiation of the circulation.[72,163]

CELL–CELL SIGNALING: ENDOTHELIAL/ MURAL CELL DEVELOPMENT

Overview

Vessel formation, and wall and network remodeling, are regulated by paracrine signals between receptor tyrosine kinases (RTKs) of endothelial and peri-endothelial cells and their ligands (Figure 10).[13,34,35,89,93,94] Ligands of interest in vessel formation include fibroblast growth factor (FGF), vascular endothelial growth factor (VEGF), platelet derived growth factor (PDGF), angiopoietins (ANG1 and ANG 2), and transforming growth factor β (TGFβ). Their effect is fine-tuned by the number and sub-type of receptors expressed by the target cell population, by the availability of ligand from producer cells, and by other signaling systems. Molecules such as extravasated plasma proteins, and inflammatory mediators such as prostaglandins, tumor necrosis factor-a, interleukins (ILs), and nitric oxide (NO), also induce angiogenesis in vivo,[34,35,164,165] but their role in regulating lung vascular development and growth is largely unknown.

Endothelial Channel Formation: VEGF/ VEGF-R

FGF-2 and VEGF secreted by mesenchymal cells (derived from the lungs of immortomice) promote "self" differentiation to endothelial cells.[50] The FGF/receptor family is widely expressed in the developing fetal lung[166,167]; and VEGF, a potent mitogen and survival factor, is required to maintain the differentiation and survival of angioblasts as endothelial cells. Alternative splicing produces homodimeric isoforms: $VEGF_{121}$, $VEGF_{165}$, $VEGF_{189}$, and $VEGF_{206}$. Of these, $VEGF_{121}$ and $VEGF_{165}$ act as survival factors for endothelial cells of immature vessels. VEGF signals controlling angiogenesis come from adjacent cells, stimulating endothelial cell RTKs that include VEGF-R1 (Flt-1), VEGF-R2 (Flk-1/KDR), and VEGF-R3 (Flt-4). Assembly of angioblasts into vascular channels requires expression of VEGF-R1[168–172]; and after differentiating to endothelial cells the cells express VEGF-R2. VEGF binding to VEGF-R2 results in proliferation while binding to VEGF-R1 results in tubule formation and sprouting. Nascent sprouts are guided by spatial cues, e.g., soluble VEGF-R1 produced by endothelial cells at the site of sprout emerging from a vessel wall.[173] VEGF-R1 further modulates VEGF-R2 signaling and blood vessel branching.[174,175] Immature vessels (that

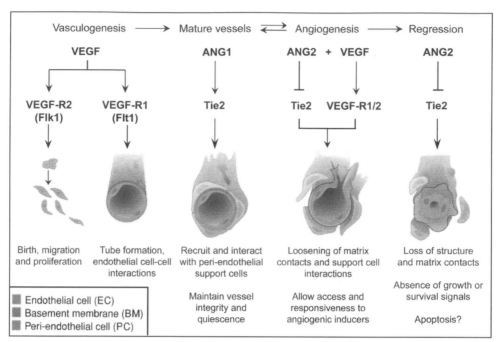

FIGURE 10   Regulation of vascular morphogenesis, maintenance, and remodeling by receptor tyrosine kinases (RTKs) and their ligands. A model for the regulation of vascular endothelium is demonstrated by the prototypical angiogenesis factor vascular endothelial growth factor (VEGF) and the class of angiogenic regulators angiopoietin-1 (ANG1) and angiopoietin-2 (ANG2). All three ligands bind to RTKs that have similar cytoplasmic signaling domains, yet their downstream signals elicit distinct cellular responses. Only VEGF binding to VEGF-R2 (Flk1) sends a classic proliferative signal. When first activated in embryogenesis, this interaction induces the birth and proliferation of endothelial cells. In contrast, VEGF binding to VEGF-R1 (Flt1) elicits endothelial cell–cell interactions and capillary tube formation, a process that closely follows proliferation and migration of endothelial cells. ANG1 binding to Tie-2 RTK recruits and likely maintains association of peri-endothelial support cells (pericytes, SMCs, myocardiocytes), thus solidifying and stabilizing newly formed blood vessels. ANG2, although highly homologous to ANG1, does not activate Tie-2 RTK; rather, it binds and blocks kinase activation in endothelial cells. The ANG2 negative signal causes vessel structures to become loosened, reducing endothelial cell contacts with matrix and disassociating peri-endothelial support cells. This loosening appears to render the endothelial cells more accessible and responsive to the angiogenic inducer VEGF (and likely to other inducers).[89] *Reproduced with permission from American Association Advancement of Science.*

lack peri-endothelial cells) need VEGF to prevent endothelial cells from detaching and undergoing apoptosis but during postnatal life endothelial cells become independent of paracrine VEGF signaling; autocrine signaling is required to maintain vascular homeostasis, however, although it is not needed for the angiogenic cascade triggered by vascular injury.[176]

Endothelial cells develop in VEGF null mice but the vessels are malformed; there are fewer small peripheral vessels and they are larger than normal.[177,178] VEGF-R1 null mice show similar changes, while VEGF-R2 null mice fail to develop endothelial cells, possibly because hemangioblasts fail to differentiate.[171,179] VEGF stabilizes developing vessel walls by promoting peri-endothelial cell development.[180–182] Endothelial cells lose VEGF dependence around the fourth postnatal week,[183] when most alveolarization is complete.

Epithelial cells are a paracrine source of VEGF isoforms (VEGF$_{120}$, VEGF$_{164}$, and VEGF$_{188}$), expression levels being highest in the canalicular stage of the developing lung (when most vessel growth is occurring) before levels decrease to those maintained postnatally and into adulthood; VEGF-R1 and VEGF-R2 each are expressed by endothelial cells closely apposed to the developing epithelium.[184–187] A link between alveolarization and angiogenesis is underscored by the abnormalities that occur in one process when the other is primarily affected via the VEGF system. Targeted exon deletion of the VEGF gene reveals that mice lacking the heparin-binding isoforms VEGF$_{164}$ and VEGF$_{188}$ have a variety of defects (Figure 11a-f), including fewer capillaries and distended and underdeveloped alveoli,[188] while VEGF overexpression greatly disrupts the overall architecture of the developing lung.[189,190]

Mural Cell Development: PDGF, ANG, and TGFβ

PDGF isoforms play an important role in lung vascular and alveolar development.[191–194] Of four identified genes (A, B, C, and D),[195] PDGF-A and PDGF-B are essential for mural cell support in developing vessels, while PDGF-C and PDGF-D are implicated in lymphatic vessel formation. Dimers of two homologous polypeptide PDGF chains, a secreted A-chain and cell associated B-chain (the *c-sis* homolog of the *v-sis* oncogene), dimerize to form functional

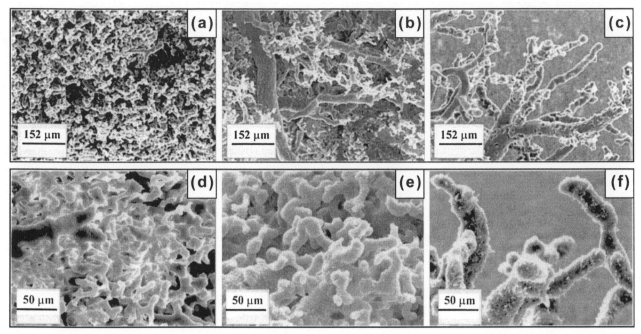

FIGURE 11 Scanning electron micrographs of Mercox casts illustrating disrupted pulmonary vascular growth patterns. (a–f) Different vascular growth patterns are shown in VEGF 120/120 littermate fetal mice (E17): wild-type (a, d), heterozygous (b, e), homozygous (c, f). In the homozygous fetus, there are fewer small peripheral vessels and they are of larger caliber than those in either the heterozygous or wild-type littermates. Bar (a–c) 152 μm and (d–f) 50 μm.[188] *Reproduced with permission from the American Thoracic Society.*

in vivo isoforms (PDGF-AA, -AB, or -BB), which in turn dimerize PDGF-Rα and PDGF-Rβ on the cell surface.

The absence of peri-endothelial cells, specifically pericytes, in PDGF-B gene deficient mice results in capillary dilatation, microaneurysms, and vascular leak and hemorrhage; the absence of the PDGF-Rβ produces a similar response. Absence of the PDGF-A gene results in an inability to form alveolar structures, in part because SMC progenitors fail to spread.[154,155,194,196]

Both PDGF-A and PDGF-B isoforms promote mesenchymal cell proliferation, PDGF-B initiating progression through the cell cycle and PDGF-A (a competence factor) requiring a further signal (e.g., IL-1) to induce mitogenesis.[197] PDGF-B alone induces chemotaxis.[198,199] It is suggested that SMCs use PDGF-B expression to enter the cell cycle and suppress cell differentiation and stimulate self-replication via synthesis of PDGF-A.[200,201] Actin re-organization and membrane ruffling, essential for cell migration, are induced by PDGF-AB and PDGF-B, and therefore regulated via PDGF-Rβ. PDGF-B further enhances wall stabilization by inducing VEGF expression in peri-endothelial cells.[202]

A shift in PDGF-A and PDGF-Rβ signaling occurs in interstitial fibroblasts recruited as peri-vascular cells, and in adjacent endothelial cells, in small intra-acinar vessels (Figure 12a,b). In both cell types, PDGF-AA and PDGF-Rβ are expressed constitutively in the lung; and in a growth response after injury, PDGF-Rβ expression increases at the expense of PDGF-AA (Figure 13a-h).[150,151] PDGF-BB and PDGF-Rα expression levels remain stable in the vascular

cell populations, although PDGF-BB levels significantly increase in alveolar macrophages (Figure 13i). The ability of VEGF to signal via the PDGF-Rβ highlights an alternative signaling pathway (VEGFA/PDGF-R) relevant to these findings in vascular wall remodeling.[203] More generally, PDGF-R signaling represents a critical vascular regulator and important therapeutic target in angiogenesis.[203,204]

Other angiogenic factors in addition to PDGF may be required to establish and maintain stable vessel networks, e.g., FGF-2. PDGF-AB, PDGF-Rβ, and PDGF-Rα each have binding patterns with FGF-2; FGF-2 up-regulates PDGF-Rα and PDGF-Rβ expression levels while PDGF-AB/FGF-2 (but not PDGF-AA/FGF-2) stabilizes vessels by recruiting pericytes.[205]

Studies implicating the complex interactions of PDGF-B with other signaling pathways that result in vascular maturation by pericyte/SMC differentiation and recruitment have been reviewed by Gaengel and colleagues[206], including the cross talk, cascade of events relating to ANG and TGFβ signaling, and to the sphingosine-1-phosphate (SIP)/endothelial cell differentiation gene (Edg1) ligand/receptor system (see Figure 14 and caption).

While the PDGF-PDGF-R system functions as a paracrine endothelial-to-mural cell loop, the ANG-Tie receptor axis represents a signaling loop mainly from mural cells to endothelium.[206] All ANGs are ligands for the Tie2 receptor: ANG-1 is expressed by mesenchymal cells and is the major ligand for Tie2 (see Figure 10), while ANG-2 is expressed

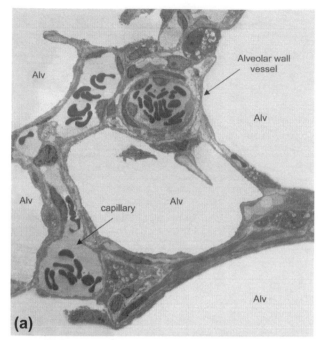

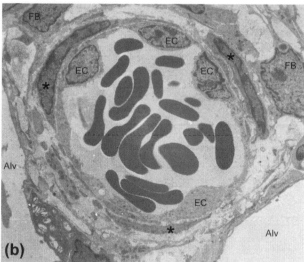

FIGURE 12   Interstitial fibroblasts align as peri-vascular cells in a distal lung vessel. (a) Alveolar wall vessel (ED 23 μm) in high oxygen-injured adult rat lung. Note the surrounding capillaries. Interstitial fibroblasts are in the process of migrating through the interstitium surrounding the vessel to establish contact and envelop the vessel wall. As the cells become peri-vascular they shift phenotype by expressing the contractile proteins and filaments of SMCs (see Figure 9b,c). Alv = alveolus. (b) Higher magnification of vessel shown in (a). Note the developing layer(s) of fibroblasts moving from an interstitial arrangement to become peri-vascular cells (*). As this occurs, signaling between adjacent endothelial and fibroblast populations changes (e.g., PDGF, see Figure 13). Alv = alveolus, EC = endothelial cell, FB = fibroblast.

by perivacular cells and is a negative Tie2 ligand (see Figure 10). ANG-1 and ANG-4 activate Tie1, its signal then being amplified by interaction with Tie2.[207] The roles of ANG-3 (in mouse) and ANG-4 (in human) are less well understood but are reported to have antagonistic (ANG-3) and agonistic (ANG-4)

effects.[88,89,206,208–214] Tie2 expression further modulates VEGF activity and is required for sprout formation.[212,213,215] ANG-1 together with VEGF enhances vascular density, whereas ANG-2 and VEGF result in longer sprouts.[39] In adult tissues, ANG-1 is widely expressed while ANG-2 is associated with sites of vascular remodeling.[206]

Tie2 null mice have normal numbers of endothelial cells and these assemble into immature channels that lack branching networks and large and small vessels. Tie2 is expressed throughout developing embryonic endothelium and in the quiescent vasculature during adulthood, while Tie1 is expressed later in embryonic endothelium and similarly persists into adulthood.[216,217] ANG-1 deficient mice die with similar vascular defects to Tie2 deficient ones.

TGFβ signaling also has a direct role in the development of peri-vascular cells, signaling through TGFβ type I receptors on vascular cells—activin receptor-like kinase (Alk)1 and Alk5—that trigger signaling pathways with opposing effects. Activation of Alk5 leads to phosphorylation of receptor-mediated Smads 2/3, which locate to the nucleus after association with Smad 4, and regulate transcription of genes inhibiting cell migration and proliferation and increasing SM differentiation. Conversely, activation of Alk1 leads to phosphorylation of Smad1/5 and the induction of genes triggering cell migration and proliferation, and decreasing SM differentiation. Moderate TGFβ levels lead to Alk1 signaling in endothelial cells, and so to increased proliferation, migration and tube formation; prolonged TGFβ signaling reduces Alk1 signaling, favoring the Alk5 pathway and matrix production, growth arrest, and differentiation of mural cells.[206] Mouse knockouts of TGFβ, Alk5 (and Smad 5) have similar phenotypes and die in utero between E9.5 and E11.5, the pattern of defects in these mice reflecting the complexity of the signaling systems involved.[206]

EMBRYONIC AND FETAL VASCULAR DEVELOPMENT

Overview

Pulmonary arteriograms demonstrate the changing pattern of the central large pulmonary arteries and smaller branches at different developmental ages in the human lung (Figure 15a,b). Distal vascular loops of small (pre-capillary) arteries, capillaries, and (post-capillary) veins serve as part of a gas-exchange surface for blood transiting the lung's complex arterial and venous networks (Figure 15b).

The vascular networks of the human embryo start to form in the 4th week of gestation with onset of organogenesis.[49] By the 16th week the adult pattern of central vascular and airway structures, consisting of lobar and segmental branches, is present. While the pre-acinar vessels develop

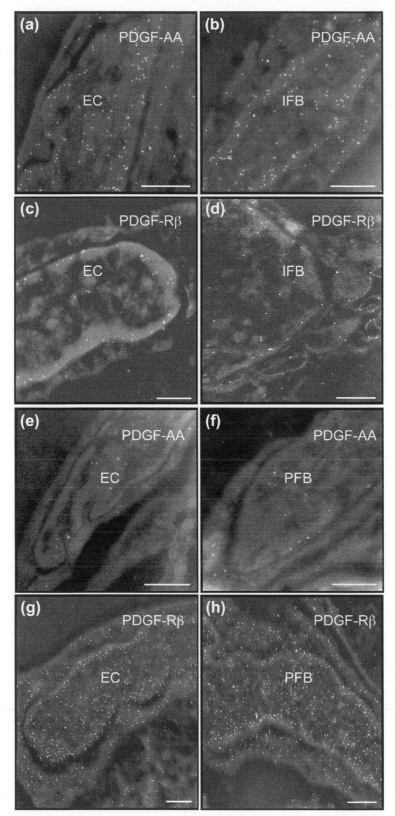

FIGURE 13 PDGF-AA/PDGF-Rβ signaling in distal vessels in normal and injured rat lung. (a–h) PDGF ligand (PDGF-AA) and receptor (PDGF-Rβ) signaling patterns change in endothelial cells, and in associated interstitial and perivascular fibroblasts, in alveolar vessels. In normal lung, endothelial cells and associated interstitial fibroblasts express: (a and b) high constitutive levels of PDGF-AA (EC, density $32.5\,\mu m^{-2}$; IFB, density $24.6\,\mu m^{-2}$) and (c and d) relatively low constitutive levels of PDGF-Rβ (EC, density $16.7\,\mu m^{-2}$; IFB, density $9.4\,\mu m^{-2}$). In the high oxygen-injured lung, in a reverse pattern, endothelial cells and peri-vascular fibroblasts (i.e., interstitial fibroblasts now aligned as peri-vascular cells) express: (e and f) low levels of PDGF-AA (EC, density $5.9\,\mu m^{-2}$; PFB, density $4.0\,\mu m^{-2}$) and (g and h) high levels of PDGF-Rβ (EC, density $76.0\,\mu m^{-2}$; PFB, density $89.6\,\mu m^{-2}$). (i) Expression levels of PDGF-BB increase in pulmonary alveolar macrophages (PAMs) of the injured lung—cells that were often closely apposed to the epithelial surface and therefore to the walls of small vessels—antigenic sites typically being distributed throughout the cytoplasm and localized in clusters in the PAM. Vessel diameters = $47\,\mu m$ (a and b), $30\,\mu m$ (c), $37\,\mu m$ (d), $29\,\mu m$ (e, f, and g) and $42\,\mu m$ (h). High resolution images are inverted to highlight the distribution in cells of (electron-dense) antigenic sites labeled with pA-AU (10nm). Bars = $0.5\,\mu m$ (a–i). (a, b, e, and f).[151] *Reproduced with permission. (c, d, g, h and i)*[150] *Reproduced with permission.*

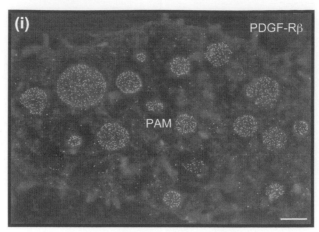

FIGURE 13 Cont'd

hand-in-hand with airways, intra-acinar ones and distal alveolar structures develop later.[218–227]

As the acinus develops, the distance between the TB and the pleura increases as RBs and ADs increase in number, and saccules appear (Figure 16). While the branching pattern and the size of large vessels can be assessed on angiograms,[220] additional details are obtained from serial reconstruction of vessels in tissue. For this, vessels are classed by their wall structure. The main pulmonary artery and its large branches, where SMCs are enveloped by many elastic lamina, are termed "elastic" arteries. More peripheral arteries with fewer lamina are termed "transitional", and as the arteries continue to decrease in size they are termed "muscular" when their wall consists simply of SMCs between an internal and external lamina; "partially muscular" when SMCs form a spiral in their wall and "nonmuscular" when SMCs are absent. A population count of distal arteries (or veins) assessed by their wall structure and location (see Introduction) is useful to assess the extent of muscularization during lung development and any change from the normal pattern.

By the 20th week, the full number of pre-acinar pulmonary vessels is present in each segment. During fetal life there is an increase in vessel size but little change in the density.[220,221,228] The vessels at the hilum grow faster than at the periphery so that the gradient of diameter against length increases with fetal age.

Early Vessel Development

Studies of early human lung development[118,229–234] reveal a main branching network evolving from the central arterial and venous trunks; as these structures expand, in diameter and length, offshoots grow out by irregular dichotomous branching. As the first vascular channels form, a complex pattern has been described in which the distal vasculature develops independently from the central vessels, the two connecting at some point in the mid-fetal stage.[229,231] In support of this, a similar pattern has emerged in the developing mouse lung.[230] In this model, vessels accompanying airways invest the lung at the hilum, and then form by sprouting angiogensis, while capillary development proceeds by vasculogenesis (see following text for details). After vascular channels form in the distal lung, and connect into a functioning system, however, further growth of the peripheral vasculature is considered likely to proceed by sprouting and nonsprouting forms of angiogenesis. Other models propose that the lung vasculature forms from distal vasculogenesis[118] or from distal angiogenesis.[232,233] As the developing channels increase in size the addition of mural cells supports their walls. While mesenchymal cells represent an important source of peri-endothelial precursors,[91,193,194] peri-endothelial cells can also develop from bronchial SMCs and endothelial cells.[118] These perivascular cells express the same smooth muscle proteins, however, despite their different origin.[118]

Serial sections of human embryos show blood lakes in the primitive mesenchyme surrounding the lung bud at the neck (at 32–44 days).[229,231] As the first (5–6) airway branches form at ~50.5 days, such lakes are abundant in the sub-pleural mesenchyme and, at this stage, the pulmonary artery accompanies airways only to the 3rd or 4th generation. Detailed reconstruction shows the first connection between developing distal and proximal networks to be between peripheral lakes and a thin-walled hilar vein, venous drainage thus being established before the pulmonary artery supply. At ~56.5 days, the branching of the pulmonary artery, a thick-walled blind-ended tube, lags behind airway branching by 2–3 generations. At 12–14 weeks, an extensive capillary network surrounds distal airway buds although well separated from them by the sub-pleural mesenchyme. By 22–23 weeks, the capillary network closely approaches airway epithelium and the pulmonary artery accompanies each airway branch.

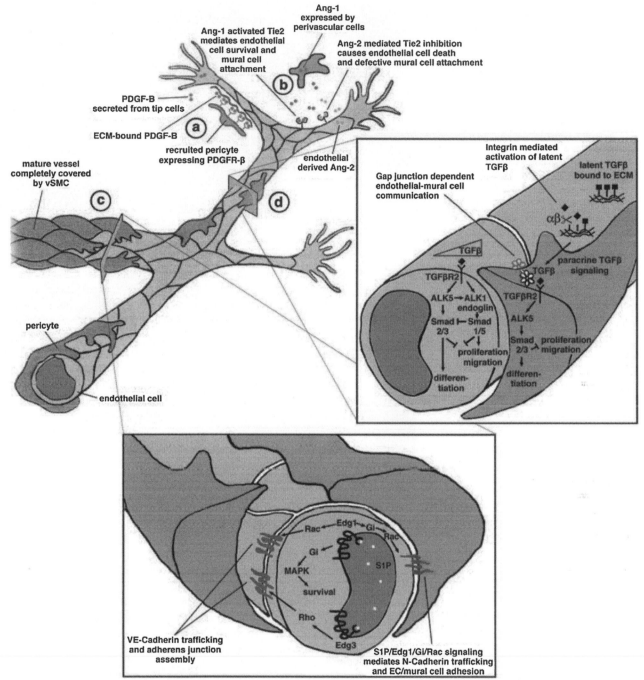

FIGURE 14 Signaling pathways in vascular smooth muscle cell/pericyte differentiation and recruitment and vascular maturation. (a) PDGFB secreted from tip cells is retained in close proximity to the growing endothelial vessel by heparan sulfate proteoglycans, where it serves as an attractant for co-migrating pericytes expressing PDGF-Rβ. (b) Ang-1, expressed by perivascular cells binds to, and activates, the Tie2 receptor thereby stimulating endothelial cell survival, angiogenesis, and subsequent mural cell attachment. Endothelial-derived Ang-2 inhibits Tie2 signaling. (c) S1P released from platelets or hematopoietic cells binds to and activates S1P/Edg receptors on endothelial cells. $S1P_1$/Edg1 signals via a G_i/MAPK cascade to stimulate cell survival, and via G_i/Rac to mediate N-Cadherin trafficking into adherens junctions and stabilize endothelial/mural cell adhesion. Additionally, S1Preceptors signal G_i independent via Rac ($S1P_1$/Edg1) and Rho ($S1P_3$/Edg3) to stimulate VE-Cadherin–based inter-endothelial cell adhesion. (d) TGF-β produced in endothelial cells induces SMC differentiation in adjacent peri-vascular cells via Alk5/Smad2/3. In endothelial cells, TGF-β can either signal via Alk5/Smad2/3 (to promote differentiation) or via Alk1/endoglin/Smad1/5 (to promote proliferation and migration). Which pathway prevails depends on the duration of TGF-β stimulation, and on TGF-β levels. Alk1 signaling is inhibitory to Alk5 signaling, whereas Alk5 signaling is required for full Alk1 activity. Integrin-mediated latent TGF-β activation as well as gap junction communication between endothelial and mural cells is critical for TGF-β paracrine signaling. ECM = extracellular matrix; vSMC = vascular smooth muscle cell; EC = endothelial cell.[206] *Reproduced with permission.*

FIGURE 15 Arteriograms illustrating the density of the pulmonary arterial system at different ages, and illustration of the density small vessels and capillaries in the distal adult lung. (a) Arteriograms showing the growth of pulmonary arteries in human lung, i.e., in a newborn (upper left), an 18-month-old infant (lower left), and an adult (right). The pre-acinar artery distribution (complete by 28 weeks in the fetus) is present at each time point shown, and the dense background haze, representing the growth of small intra-acinar arteries, increases with age. The arteriograms were prepared after injecting the pulmonary arterial bed with a barium sulfate–gelatin mixture.[227] *Reproduced by permission from the American Thoracic Society.* (b) Photomicrograph of an adult lung after injection with a silicone polymer to show the dense capillary networks that surrounds each alveolus. The polymer was perfused (at 50 mmHg infusion pressure), and fixed (2cm-thick) tissue slices, from a lung following acute myocardial infarction, were cleared by incubation in 50, 75, 85, and 100% glycerol (each for 24 h.) and examined with a Wild M8 stereomicroscope. Original magnification: (a) × 0.3. Note that the three lungs are photographed at the same magnification; (b) × 56.[271] *Reproduced with permission.*

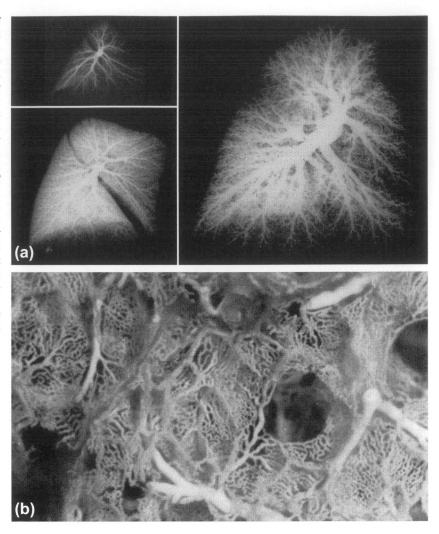

Conventional and Supernumerary Arteries and Veins

In the human fetal lung, as in the child and adult, the number of arterial branches exceeds that of airways.[220,221,228,235] Branches of the main pulmonary artery running alongside airways are termed "conventional" arteries, and the numerous additional branches, arising from the axial artery to the TB, i.e., pre-acinar and passing directly to adjacent respiratory tissue to supply the capillary bed, are termed "supernumerary" arteries (Figure 17). For example, in the posterior basal segment of the left lung of a 19-week fetus (length16mm),[220] the 25 bronchial branches present to the level of the TB are accompanied by 21 conventional arteries and 58 supernumerary arteries. The number of airways and conventional artery branches is similar from 18 weeks gestation onwards (Figure 18).

In the fetus, the veins arise from the saccular respiratory region, pleura, connective tissue septae, and airway walls. Conventional and supernumerary veins appear together, developing progressively from hilum to periphery. The axial veins running from the periphery to the hilum receive drainage from conventional and supernumerary tributaries. From 20 weeks of gestation the number of pre-acinar conventional veins is within the adult range for intra-segmental airway number. While the number of conventional veins in a segment equals that of the airways and conventional arteries, there are more supernumerary veins than supernumerary arteries. Many small arteries and veins develop in late fetal life and continue to increase with age, their density per unit area of lung increasing as RBs and saccules appear.

Vessel Wall Structure

Depending on gestational age, and their proximal to distal location, the wall structure of fetal pulmonary arteries is elastic, transitional, muscular, partially muscular, or non-muscular.[220,221] From 19 weeks of gestation an elastic structure is present to the same level in axial and conventional arteries, and in smaller vessels than in the adult. Because of high tensile strength and the ability to maintain vessel patency, these arteries can be regarded as supportive.

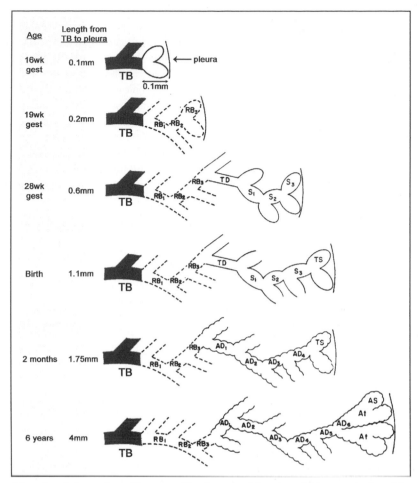

FIGURE 16   Development of the acinus in human lung. Six stages of development of the acinus: the increase in size of the acinus is shown by the length between the TB (the last airway segment that does not give rise to alveoli) and the pleura, i.e., a length 0.1mm in the fetus and 4 mm by 6 years of age. Note that changing length of the TB with age is not drawn to scale. The increase in length represents an increase in generations. A given generation may be traced vertically, permitting its structural remodeling to be followed. At 16 weeks gestation the airways end in a tubule close to the pleura. By 19 weeks the epithelium of the last airway generation has thinned to form the first respiratory bronchiolus (RB) and branching has led to second generation RB. By 28 weeks, further branching has given rise to a total of three generations of RB and one generation of transitional ducts (TDs), the latter giving rise to primitive saccules. By birth the number of saccules has increased, the last forming the terminal saccule. No true alveoli are present at this time, although a few indentations, representing future alveoli, appear just before birth. By 2 months after birth alveoli have developed in the walls of RB from TDs and from saccules, which transforms these last two structures into alveolar ducts (ADs): alveoli also open into the terminal saccules. By 6 years several changes have occurred with RB and ADs remodeling occurring in several ways: along some pathways a bronchiolus may transform into an extra generation of RB by centripetal alveolarization; distal RB may transform into ADs; further branching of another one or two AD generations may occur; or there may be no change. Certainty for some of the above, change is not possible because of variation between alveoli and cases. At the distal end, however, the terminal sac has probably transformed into the adult atrium, a short, wide, passage lined by alveoli, which has given rise to a number of alveolar sacs, formed by budding rather than branching, and each lined by alveoli. This pattern present at this time is similar to that seen in the adult: there is probably little further development save an increase in size.[222] *Reproduced with permission from BJM Publishing Group Ltd.*

Phenotypically distinct populations of SMCs in the wall of large developing pulmonary arteries raise the possibility of different lineages.[236] In small vessels, where present, the SMCs are immature.

The pulmonary veins are relatively free of muscle in the fetus. At 20 weeks gestation, for example, no muscle is present even in the largest veins. By 28 weeks, muscle fibers are seen within vein walls but a continuous muscle layer is present only at term.[221,228] As in arteries, the SMCs extend into veins as small as 80µm but unlike arteries the veins are thin-walled and no elastic lamina is present.[221,228] The walls of bronchial arteries and veins are typical of systemic vessels, being muscular and relatively thicker (for vessel size) than pulmonary arteries.[228]

Distribution of Intra-acinar Vessels

Pulmonary arteries with SMCs in their wall gradually extend into the intra-acinar region with lung growth and maturation; absent from the intra-acinar region in the fetus

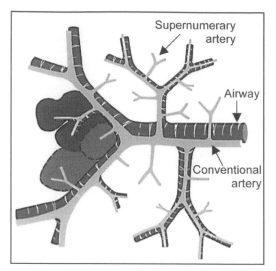

FIGURE 17 Schematic illustration of conventional and supernumerary artery branches arising from a broncho-arterial bundle in human lung. Conventional arteries arise at acute angles to the main axis and supply the respiratory region at the end of the axial pathway. Supernumerary arteries are short and have differing diameters. They arise at right angles to the main axis and supply adjacent air spaces, ultimately providing collateral circulation to a respiratory unit via the backdoor.[270] *Reproduced with permission from Elsevier, London.*

(Figure 19) they appear most distally only in the adult lung. A population count in the normal human lung at term shows that arteries greater than ED 150 μm are muscular and that below this they are partially muscular, the smallest being ~ED 25μm[220]; the largest non-muscular arteries are ~ED 90μm, and below ED 60μm all of the arteries are non-muscular (Figure 20). This relationship between vessel size and wall structure is similar at all fetal ages.[220] While the size of the smallest muscular artery, and the size range for partially muscular and non-muscular arteries, is the same in the fetal and adult lung, the wall thickness of an artery (of given diameter) is twice as thick in the fetus as in the adult.

Development of Bronchial Arteries and Veins

As cartilage plates form, bronchial arteries grow along the airway wall to supply the walls of vessels of the bronchopulmonary sheath.[237,238] Bronchial arteries and veins gradually increase in size and number and at term are found as far into lung as bronchioli.

The term pre-capillary anastomosis is used to describe a channel that is pre-capillary in position and larger than a capillary in size. A small number of these connections between

FIGURE 18 Distribution of conventional and supernumerary arteries by size in fetal lung. Conventional and supernumerary artery branches are shown by diameter, relative to the diameter of the axial arterial pathway from which they arise (see the black line), in a fetus (19 weeks gestation). In this (and subsequent images) each branch is seen as a vertical line whose height represents the diameter. From the lobar hilum to the terminal bronchiolus, the 25 airway generations are accompanied by 21 conventional arteries, and there are 58 supernumerary arteries. For clarity, the first eight conventional and 11 supernumerary arteries are shown separately (top image) from the remaining conventional and supernumerary ones (lower image). The asterisk indicates the point of overlap between the top and lower image. Note the change in scale of the vertical axis. At the hilum, the diameter of the main pathway is 1,300 μm. Conventional arteries have a mean diameter of 282.7 μm, and the supernumerary arteries are smaller, with a mean diameter of 62.6 μm.[220] *Reproduced with permission from Wiley-Blackwell.*

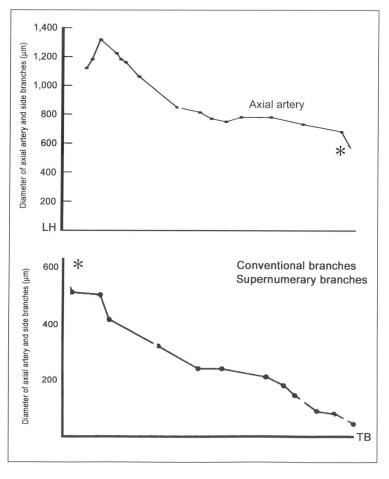

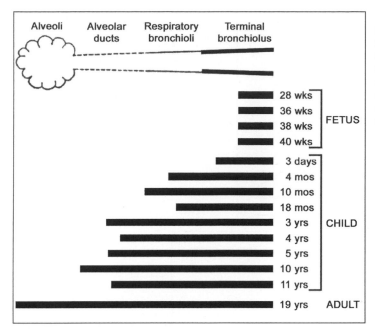

FIGURE 19 Development of pulmonary arteries with smooth muscle cells in the fetal, child, and adult lung. In the fetus pulmonary arteries with SMCs in their wall are not seen within the acinus (i.e., distal to the TB). Arteries with SMCs extend gradually into the acinus in childhood, but even by 11 years they are not yet present within the most distal region, i.e., the alveolar wall. Only in the adult (as shown at 19 years of age) are SMCs present in the most distal arteries.[239] *Reproduced with permission from BMJ Publishing Group Ltd.*

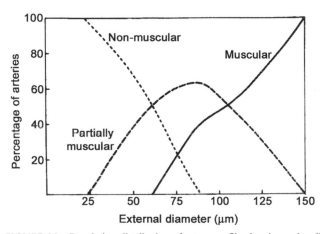

FIGURE 20 Population distribution of artery profiles by size and wall structure (muscular, partially muscular, and non-muscular) in the normal newborn: at this stage, muscular arteries range from 60 to > 150μm, partially muscular ones from 25–150μm and non-muscular ones from 25–90μm.[261] *Reproduced with permission from Wiley-Blackwell.*

bronchial and pulmonary arteries (ED 35–100μm) are present at all fetal ages. In the late fetal stage, pre-capillary anastomoses between pulmonary and bronchial arteries are found within the bronchial wall but none are larger than ED 15μm.

POSTNATAL VASCULAR DEVELOPMENT AND GROWTH

Overview

At 36 weeks to term of human gestation, the pre-acinar pattern of arteries and veins is complete. The lung now can support air breathing but structurally it is not the adult lung in miniature. It responds by a burst of vascular growth as

existing units expand in diameter and length, and many new intra-acinar units are added as the gas-exchange surface forms (see Figure 16).[30,87,221,222,228,239] Data from recent (helium-3 magnetic resonance) studies support the finding that, in addition to alveolar enlargement, new alveoli continue to develop in the human lung throughout childhood and adolescence.[240]

In the first 4 months after birth, as alveoli form and increase in size, the number of arteries per unit area of lung, and the density of capillary networks, increases.[59] These vessels are thought to form and grow by angiogenesis.[59] At birth, the respiratory saccules (primary septae) are supplied by small vessels and a double capillary system; within 2 weeks, as secondary septae form and enclose the interstitial tissue between the two capillary layers, these fuse into a single network.

In the first months, the diameter of proximal intra-acinar vessels increases more than distal ones, reflecting the burst of small vessels developing at the lung periphery. After 18 months the number of new vessels forming slows along with alveolar growth.[228,239] Between 4 months and 4 years, the number of arteries (up to ~ED 200μm) per unit area of lung increases greatly. While the ratio of intra-acinar arteries to alveoli remains similar throughout childhood their concentration per unit area of lung falls after 5 years of age as alveoli increase in size.[228,239]

Conventional and Supernumerary Arteries and Veins

The veins grow at the same time as airways and arteries. While the pre-acinar drainage pattern is complete half-way

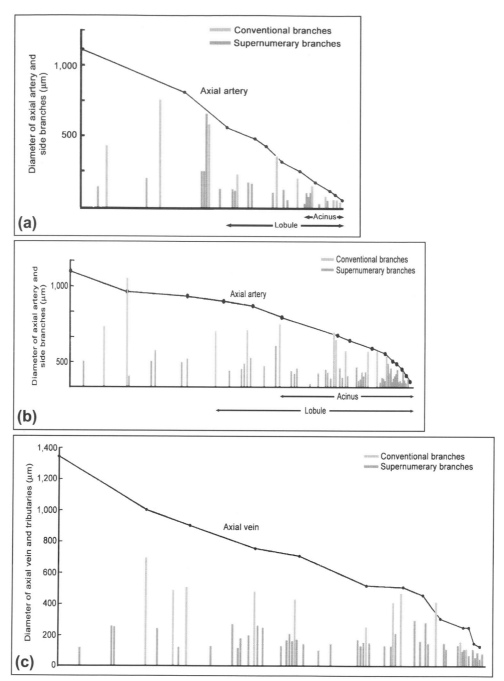

FIGURE 21 Distribution of conventional and supernumerary arteries and veins by size, relative to the diameter of the axial arterial pathway from which they arise (see the black line) in infant and child lung. (a) In the pre- and intra-lobular arteries of a 4-month-old child, the horizontal axis represents the last 5.1 mm of an axial pathway. The acinus is small with only five conventional and six supernumerary arteries. (b) In the pre- and intra-lobular arteries of a 5-year-old child, the horizontal axis represents the last 12.6 mm of an axial pathway. The length of the acinus and lobule has increased with age as have the number of conventional and supernumerary arteries (especially the latter). All vessels have also increased in size with age. (c) Distribution of veins (conventional and supernumerary) by size, relative to the diameter of the axial vein from which they arise (see the black line) in the lung of a 3-year-old child. The drainage pattern at the lung periphery is illustrated. The supernumerary veins include two types: type I and type II (the type II veins being larger on average than the type I veins). The actual length of the pathway is approximately 11 mm (a and b).[239] *Reproduced with permission from BJM Publishing Group Ltd. (c).*[221]

through fetal life, the intra-acinar pattern develops during childhood (Figure 21a–c). Both conventional and supernumerary vessels continue to develop in the postnatal lung, conventional arteries accompanying new airways appearing up to 18 months and new supernumerary arteries to 8 years of age.[228,239] Conventional veins, like the axial veins, run in their own connective tissue sheath. They enter the axial vein at an acute angle, are of similar size, and lie at some distance from the capillary bed they drain.[221] Supernumerary veins drain the lung tissue immediately around the axial vein. Some have no collagen sheath but pass directly through the main vein sheath to the axial vein: others receive post-capillary tributaries and are surrounded by a collagenous sheath continuous with the sheath of the axial vein.[221] Along the axial pathway they are equivalent in number to airway generations and the arteries accompanying them. Each type may be found along the length of the axial vein. Both conventional and supernumerary veins become more frequent toward the periphery.

Vessel Wall Structure

During childhood, the number of large arteries with an elastic or transitional wall remains constant. Between 4 and 10 months, vessels increase in size but muscle development lags. By 10–11 years, muscle extends further distally and is present in AD vessels but not AW ones (see Figure 19). The presence of a high population of thin-walled arteries within the acinus may provide children with an advantage, because arteries as large as ED 200μm that have virtually a capillary wall can contribute to oxygen transfer. The wall structure of veins is more developed in children; post-capillary veins consist only of endothelium but larger vessels have an internal and external lamina and an occasional SMC. In larger veins there is a continuous muscle coat, but even in the largest still no definite elastic lamina is present.[221]

Distribution of Intra-acinar Vessels

Based on the changes described earlier, the distribution of intra-acinar vessels shifts significantly in childhood with more partially muscular and non-muscular arteries than muscular ones present than in the fetal or adult lung (Figure 22).

Bronchial Artery to Pulmonary Artery Connections

By 10 weeks after birth pre-capillary anastomoses (formed in the fetal period) are obliterated by fibrin and muscle; pulmonary and bronchial arteries then communicate with the pulmonary veins only through their capillary bed. Pre-capillary vessels are present between the pulmonary and bronchial arteries but normally these are not functional; they have the potential, however, to open if a block occurs. In congenital heart defects, for example, these systems adapt to the altered hemodynamic state. The persistence of pre-capillary fetal anastomoses as well as selective opening of capillaries could both contribute to communication between the large vessels of different venous and arterial systems in the newborn and young lung.

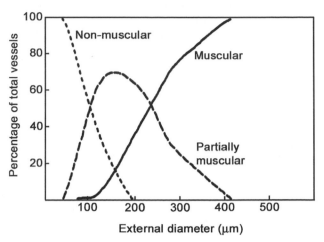

FIGURE 22 Population distribution of artery profiles by size and wall structure (muscular, partially muscular, and nonmuscular) in an 11-year-old child: the artery population is larger, and partially and non-muscular structures are now present in larger arteries than in the newborn (see Figure 20).[272] *Reproduced with permission from BJM Publishing Group Ltd.*

VASCULAR GROWTH AND REORGANIZATION IN THE ADULT

Overview

Angiograms demonstrate the branching pattern of central and peripheral pulmonary arteries and veins in the adult lung and density of peripheral vessels.[227] The branching pattern can be defined by counting the number of branches either as generations or by order.[235,241,242] A useful convention is to consider a segmental airway as the first generation; the trachea and lobar branches are counted separately. For example, the inferior lingular segment commonly has about 28 generations, the apical lower lobes may have as few as 15. Vessel branching parallels that of the airway in the above features. While endothelial cell turnover is extremely low in vessels of the adult lung, new vessels continue to form from pre-existing ones by angiogenesis in line with lung growth.

The lobule represents a group of (3–5) acini clumped at the end of an airway be it subpleural or deep in the lung. The acinus of the adult human lung is ~1cm in diameter or 1ml in volume, each acinus consisting of many alveoli. The extensive alveolar surface of the adult lung (~ 70m^2) is composed of ~ 300×10^6 alveoli each with many small vessel and capillary segments. The density of the capillary bed increases 4-fold between birth and adulthood, when the capillary endothelial surface area is ~126m.[59,243–246]

Conventional and Supernumerary Arteries and Veins

The supernumerary arteries of the adult lung remain more numerous than conventional ones both absolutely and relatively (Figure 23). They have a frequency ratio of 2.5–3.4:1

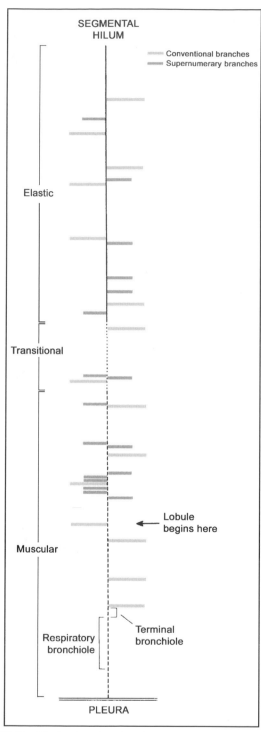

FIGURE 23 Distribution of pre-lobular and intra-lobular conventional and supernumerary arteries in adult lung. A pulmonary artery from the segmental hilum to the pleura, with conventional and supernumerary side branches included down to the start of the lobule is illustrated, and the conventional arteries thereafter. The 15 conventional arteries were numerically designated by reference to the bronchial generation. This part of the artery was 5 cm in length.[235] *Reproduced with permission from Elsevier, London.*

over the length of the preacinar segment and form 20–45% of the total cross-sectional area of side branches. At the origin of each supernumerary artery and parent conventional artery, the wall (as shown in cattle) is organized into a V-shaped musculo-elastic cushion: beginning as a funnel-shaped channel on the hilum side of each supernumerary artery it forms into a baffle that projects over the lumen to perhaps regulate blood flow.[247] Supernumerary arteries also appear to have their own vaso-regulatory pathways; their response to vasoconstrictors is greater than for conventional arteries, and the response to NO is different.[248]

Vascular Wall Structure

The wall of the main pulmonary artery has more than 7 elastic laminae in elastic segments and 4–7 laminae in transitional ones.[235] Elastic arteries are >3200 µm in diameter, transitional ones are 3200 µm–2000 µm, and muscular ones are <2000 µm. An elastic or transitional wall structure extends approximately halfway along an axial pathway (to about the 9th airway generation); the distal region of the pathway changes to a muscular structure in arteries about 2mm in diameter, and almost all branches from the axial pathway have a muscular wall. SMCs now extend along distal pathways and are present in AW vessels. Arteries as small as 30 µm in diameter may have a muscular wall and even smaller ones approaching the capillary bed can have a single SMC in their wall.

Distribution of Intra-acinar Vessels

All arteries greater than ED 150µm are muscular; partially muscular ones are most frequent in the ED 40–80µm size range, and non-muscular ones appear just below ED 130µm and increase in frequency to form the majority of the vessel population below 50 µm (Figure 24).

VESSELL WALL REORGANIZATION IN AGING

Overview

In general, age-related vascular wall remodeling includes changes to lumen size, wall thickening of the intima and media (but especially in the intima as migrating medial smooth muscle cells accumulate), and increased vascular stiffness.[249–251] Aging is associated with a reduced regenerative capacity of endothelium and by endothelial senescence as characterized by an increase in the rate of apoptosis.[252] Similar change in the response of smooth muscle cells to growth factors (e.g., TGF-β) adversely affects cells rates of proliferation and apoptosis. Reduced ability to maintain vascular wall structure may also be influenced by the depletion of adventitial precursor cells. Additional information

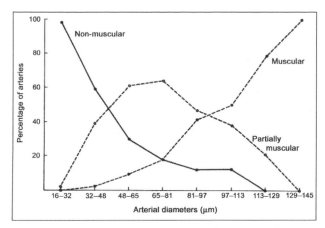

FIGURE 24 Population distribution of vessels by size and wall structure in the adult lung. In a typical pattern, as vessels increase in size the relative number of muscular artery profiles increases: the relative number of nonmuscular arteries decreases, and most partially muscular ones are found in the mid-size range.[273] *Reprinted with permission.*

about the natural fate of resident progenitor cells in aging arteries will allow a better understanding of the consequences of depleted progenitor pools for changes in the structure and function of large elastic arteries.[93] In addition to these structural changes in vessels of the aging lung, endothelial function declines. This results in the attenuation of endothelium-dependent dilator responses, which in turn change the balance of factors for relaxation and constriction released by the vascular cells.

Vascular Wall Structure

In the aged lung, the large elastic and transitional arteries have less elastic fibers than ones in the young adult.[250] This "age-related" loss in wall elasticity is associated with (a) vessel narrowing, caused by intimal and medial wall cells encroaching on the lumen,[250,253] as well as (b) alveolar wall thinning, (c) the presence of large and relatively simple alveolar structures, and (d) an overall fall in lung elastic recoil.[254] The intima of large elastic and transitional arteries is thickened by acellular deposits, although large muscular arteries appear free of this change. Hyaline deposits are present in the walls of small vessels less than ED 150μm.[250]

The medial thickness of vessels at all diameters is greater than in the normal young adult lung: most vessels are twice as thick, and the increase is nearly 3-fold in those of ~ ED 3000 μm.[250] The range for wall thickness, narrow in the young adult lung, is wide in the aged lung. The wall thickness of pulmonary arteries is greater than in the young adult as is the increase in the wall thickness of resistance arteries.

When the number of branches arising from the axial pulmonary artery was counted on arteriograms, and the diameter of successive branches to within 0.5 cm of the pleura and the distance between them measured, no difference was found between the aged and normal young control lung.

The axial pathways appear clearer in the aged lung, however, possibly because of a fall in the number of small blood vessels and capillaries.[255]

FAILURE TO DEVELOP THE NORMAL QUOTA OF VASCULAR UNITS AND A FUNCTIONALLY "NORMAL" LUNG

Overview

In many conditions there is disturbance in vascular development associated with major impairment of lung function, as in cystic fibrosis, congenital heart disease, and BPD.[256] In other conditions, adaptation to disturbance is associated with relatively normal lung function and good prognosis although under certain circumstances, as in environmental change or in disease, reduced pulmonary reserve becomes apparent. Examples of abnormal lung development that for a time at least are consistent with satisfactory function include the solitary lung in agenesis, the lung after repair of congenital diaphragmatic hernia (CDH) and the lung in certain musculo-skeletal disorders.

Single Lung in Agenesis

The single lung in agenesis provides an example of interference in the earliest stage of the development program so that a single airway branch gives rise to a single lung that typically fills both thoracic cavities. By 3 months of age, alveolar multiplication produces the number of alveoli and associated capillaries equivalent to that of two lungs[257]; quantitative analysis of the single lung in one infant at 3 months of age revealed that vascular, mesenchymal, and epithelial components of the single lung produce twice the normal alveolar quota and associated capillaries in response to the increased space available. This solitary lung is consistent with normal lung function. Studies in lambs have shown that the over distension of the lung before birth, or in the peri-natal period, stimulates alveolar multiplication.[258]

Repair of Congenital Diaphragmatic Hernia

In CDH, lung volume and airway branches are reduced in number.[259–261] Thoracic space is reduced owing to the upward movement of abdominal contents, suggesting that lung hypoplasia results predominantly from the reduced available space. The gestational age at which the hernia is produced determines the pattern of disturbance in lung and vascular growth.[262,263] The prognosis of infants delivered with CDH is strongly influenced by the deficiency in vascular development[260,264] and by the resulting development of pulmonary hypertension. Studies in a (nitrofen and bisdiamine-induced) mouse model indicate altered expression of the ANG-1/Tie2 signaling pathway, and in epithelial-endothelial crosstalk (i.e., ANG-1 being expressed in the developing lung bud and Tie2 being restricted to the vasculature in the surrounding mesenchyme), in the vasculopathy developing in CDH.[265] In a further study of the type and distribution of SMCs in the lungs of premature and term infants with CDH, more abundant contractile SMCs and the more distal distribution of these cells were present than in (age-matched) control lungs.[266]

After CDH repair, lung volume typically increases to fill the available thoracic space but airway, alveolar, and vascular multiplication do not catch-up.[267] Not uncommonly, the most hypoplastic lobe develops the radiographic features of pan-acinar emphysema. A gradient of hypoplasia is typically apparent, the ipsilateral lung, particularly the lower lobe, being most compressed. At least during adolescence, lung function is satisfactory although physiological studies detect reduced function.[268]

Musculoskeletal Disorders

Musculoskeletal disorders can impair lung growth metabolically or mechanically. In some of these disorders, the metabolic disturbance that impairs cartilage and bone development also affects tissues needed for lung growth; in other syndromes the resulting reduction of thoracic space also contributes to restrict lung growth and airway development.[256] For example, in osteogenesis imperfecta lung volume and airway branching are reduced because of reduced thoracic volume while the arteries are crowded and abnormally large. In camptomelic dwarfism the thorax is small, as is lung volume, but airway and arterial size and branching are more appropriate to the volume, suggesting that a metabolic effect is less important.[256] Dissociation between the size of the thorax and lung growth in the absence of metabolic impairment is illustrated by a case of Jeune syndrome (asphyxiating thoracic dystrophy), a rare inherited malformation characterized by disturbance in utero of bone formation, in which the thorax failed to grow, but the lungs grew to normal size by displacing the diaphragm.[269] At the time of death from hypoxia the small arteries of the lung had remodeled structurally, and although normal in volume, the lung was abnormal in shape as it conformed to the abnormal thoracic contours.

ACKNOWLEDGMENTS

Supported by NIH HL RO1 089252. We thank our colleagues for the permission granted to use illustrations from their publications.

REFERENCES

1. Dor Y, Djonov V, Keshet E. Making vascular networks in the adult: branching morphogenesis without a roadmap. *Trends Cell Biol* 2003;**13**:131–6.
2. Hislop A, Reid L. The similarity of the pulmonary artery branching system in siblings. *Forensic Sci* 1973;**2**:37–52.
3. Metzger RJ, Klein OD, Martin GR, Krasnow MA. The branching programme of mouse lung development. *Nature* 2008;**453**:745–50.
4. Warburton D. Developmental biology: order in the lung. *Nature* 2008;**453**:733–5.
5. Deffebach ME, Charan NB, Lakshminarayan S, Butler J. The bronchial circulation. Small, but a vital attribute of the lung. *Am Rev Respir Dis* 1987;**135**:463–81.
6. Paredi P, Barnes PJ. The airway vasculature: recent advances and clinical implications. *Thorax* 2009;**64**:444–50.
7. Leak LV, Ferrans VJ. Lymphatics and lymphoid tissue. In: Crystal RG, West JB, Barnes PJ, Cherniack NS, Weibel ER, editors. *The Lung: Scientific Foundations*. New York: Raven Press, Ltd.; 1991. pp. 779–86.
8. Sozio F, Rossi A, Weber E, Abraham DJ, Nicholson AG, Wells AU, et al. Morphometric analysis of intralobular, interlobular and pleural lymphatics in normal human lung. *J Anat* 2012;**220**:396–404.
9. Kretschmer S, Dethlefsen I, Hagner-Benes S, Marsh LM, Garn H, König P. Visualization of intrapulmonary lymph vessels in healthy and inflamed murine lung using CD90/Thy-1 as a marker. *PLoS One* 2013;**8**:e55201.
10. Dickie R, Cormack M, Semmler-Behnke M, Kreyling WG, Tsuda A. Deep pulmonary lymphatics in immature lungs. *J Appl Physiol* 2009;**107**:859–63.
11. Cornett B, Snowball J, Varisco BM, Lang R, Whitsett J, Sinner D. Wntless is required for peripheral lung differentiation and pulmonary vascular development. *Dev Biol* 2013;**379**:38–52.
12. Warburton D, El-Hashash A, Carraro G, Tiozzo C, Sala F, Rogers O, et al. Lung organogenesis. *Curr Top Dev Biol* 2010;**90**:73–158.
13. Heinke J, Patterson C, Moser M. Life is a pattern: vascular assembly within the embryo. *Front Biosci (Elite Ed)* 2012;**4**:2269–88.
14. Schwarz M, Lee M, Zhang F, Zhao J, Jin Y, Smith S, et al. EMAP II: a modulator of neovascularization in the developing lung. *Am J Physiol* 1999;**276**:L365–75.
15. van Tuyl M, Groenman F, Wang J, Kuliszewski M, Liu J, Tibboel D, et al. Angiogenic factors stimulate tubular branching morphogenesis of sonic hedgehog-deficient lungs. *Dev Biol* 2007;**303**:514–26.
16. Warburton D, Schwarz M, Tefft D, Flores-Delgado G, Anderson KD, Cardoso WV. The molecular basis of lung morphogenesis. *Mech Dev* 2000;**92**:55–81.
17. Abman SH. Impaired vascular endothelial growth factor signaling in the pathogenesis of neonatal pulmonary vascular disease. *Adv Exp Med Biol* 2010;**661**:323–35.
18. Meller S, Bhandari V. VEGF levels in humans and animal models with RDS and BPD: temporal relationships. *Exp Lung Res* 2012;**38**:192–203.
19. Thebaud B. Angiogenesis in lung development, injury and repair: implications for chronic lung disease of prematurity. *Neonatology* 2007;**91**:291–7.
20. Iosef C, Alastalo TP, Hou Y, Chen C, Adams ES, Lyu SC, et al. Inhibiting NF-kappaB in the developing lung disrupts angiogenesis and alveolarization. *Am J Physiol Lung Cell Mol Physiol* 2012;**302**:L1023–36.
21. Alexander MR, Owens GK. Epigenetic control of smooth muscle cell differentiation and phenotypic switching in vascular development and disease. *Annu Rev Physiol* 2012;**74**:13–40.
22. Fisher SA. Vascular smooth muscle phenotypic diversity and function. *Physiol Genomics* 2010;**42A**:169–87.
23. Owens GK, Kumar MS, Wamhoff BR. Molecular regulation of vascular smooth muscle cell differentiation in development and disease. *Physiol Rev* 2004;**84**:767–801.

24. Stevens T, Rosenberg R, Aird W, Quertermous T, Johnson FL, Garcia JC, et al. NHLBI workshop report: endothelial cell phenotypes in heart, lung, and blood diseases. *Am J Physiol Cell Physiol* 2001;**281**:C1422–33.

25. Joshi SR, McLendon JM, Comer BS, Gerthoffer WT. MicroRNAs-control of essential genes: implications for pulmonary vascular disease. *Pulm Circ* 2011;**1**:357–64.

26. Mujahid S, Nielsen HC, Volpe MV. MiR-221 and miR-130a regulate lung airway and vascular development. *PLoS One* 2013;**8**:e55911.

27. Davis BN, Hilyard AC, Nguyen PH, Lagna G, Hata A. Induction of microRNA-221 by platelet-derived growth factor signaling is critical for modulation of vascular smooth muscle phenotype. *J Biol Chem* 2009;**284**:3728–38.

28. Jalali S, Ramanathan GK, Parthasarathy PT, Aljubran S, Galam L, Yunus A, et al. Mir-206 regulates pulmonary artery smooth muscle cell proliferation and differentiation. *PLoS One* 2012;**7**:e46808.

29. Chan YC, Banerjee J, Choi SY, Sen CK. miR-210 the master hypoxamir. *Microcirculation* 2012;**19**:215–23.

30. Burri PH. Postnatal development and growth. In: Crystal RG, et al., editor. *The Lung: Scientific Foundations*. New York: Raven Press Ltd.; 1991. pp. 677–87.

31. Schittny JC, Mund SI, Stampanoni M. Evidence and structural mechanism for late lung alveolarization. *Am J Physiol Lung Cell Mol Physiol* 2008;**294**:L246–54.

32. Farrell PM. Morphologic aspects of lung maturation. In: Farrell PM, editor. *Lung Development: Biological and Clinical Perspectives: Neonatal Respiratory Distress*. New York, NY: Academic Press; 1982. pp. 13–25.

33. Dejana E, Corada M, Lampugnani MG. Endothelial cell-to-cell junctions. *FASEB J* 1995;**9**:910–8.

34. Carmeliet P. Developmental biology. One cell, two fates. *Nature* 2000;**408**:43–5.

35. Carmeliet P. Mechanisms of angiogenesis and arteriogenesis. *Nat Med* 2000;**6**:389–95.

36. Adams JC, Watt FM. Regulation of development and differentiation by the extracellular matrix. *Development* 1993;**117**:1183–98.

37. Carey DJ. Control of growth and differentiation of vascular cells by extracellular matrix proteins. *Annu Rev Physiol* 1991;**53**:161–77.

38. Lin CQ, Bissell MJ. Multi-faceted regulation of cell differentiation by extracellular matrix. *FASEB J* 1993;**7**:737–43.

39. McGowan SE. Extracellular matrix and the regulation of lung development and repair. *FASEB J* 1992;**6**:2895–904.

40. Risau W, Lemmon V. Changes in the vascular extracellular matrix during embryonic vasculogenesis and angiogenesis. *Dev Biol* 1988;**125**:441–50.

41. Roman J. Cell-cell and cell-matrix interactions in development of the lung vasculature. In: MacDonald JA, editor. *Lung Growth and Development*. New York: Marcel Dekker, Inc.; 1997. pp. 365–99.

42. Sannes PL, Burch KK, Khosla J, McCarthy KJ, Couchman JR, et al. Immunohistochemical localization of chondroitin sulfate, chondroitin sulfate proteoglycan, heparan sulfate proteoglycan, entactin, and laminin in basement membranes of postnatal developing and adult rat lungs. *Am J Respir Cell Mol Biol* 1993;**8**:245–51.

43. Taipale J, Keski-Oja J. Growth factors in the extracellular matrix. *FASEB J* 1997;**11**:51–9.

44. Gumbiner BM. Cell adhesion: the molecular basis of tissue architecture and morphogenesis. *Cell* 1996;**84**:345–57.

45. Polverini PJ. Cellular adhesion molecules. Newly identified mediators of angiogenesis. *Am J Pathol* 1996;**148**:1023–9.

46. Stromblad S, Cheresh DA. Cell adhesion and angiogenesis. *Trends Cell Biol* 1996;**6**:462–8.

47. Maniotis AJ, Bojanowski K, Ingber DE. Mechanical continuity and reversible chromosome disassembly within intact genomes removed from living cells. *J Cell Biochem* 1997;**65**:114–30.

48. Risau W, Flamme I. Vasculogenesis. *Annu Rev Cell Dev Biol* 1995;**11**:73–91.

49. Sadler TW. Embryonic period (third to eighth week). In: Gardner JN, editor. *Langman's Medical Embryology*. Sixth edition. Baltimore: Williams and Wilkins; 1990. pp. 61–84.

50. Yamamoto Y, Baldwin HS, Prince LS. Endothelial differentiation by multipotent fetal mouse lung mesenchymal cells. *Stem Cells Dev* 2012;**21**:1455–65.

51. Fantin A, Vieira JM, Plein A, Denti L, Fruttiger M, Pollard JW, et al. NRP1 acts cell autonomously in endothelium to promote tip cell function during sprouting angiogenesis. *Blood* 2013;**121**:2352–62.

52. Hayashi M, Majumdar A, Li X, Adler J, Sun Z, Vertuani S, et al. VE-PTP regulates VEGFR2 activity in stalk cells to establish endothelial cell polarity and lumen formation. *Nat Commun* 2013;**4**:1672.

53. Ribatti D, Crivellato E. "Sprouting angiogenesis", a reappraisal. *Dev Biol* 2012;**372**:157–65.

54. Ausprunk DH, Folkman J. Migration and proliferation of endothelial cells in preformed and newly formed blood vessels during tumor angiogenesis. *Microvasc Res* 1977;**14**:53–65.

55. Schoefl GI. Electron microscopic observations on the regeneration of blood vessels after injury. *Ann N Y Acad Sci* 1964;**116**:789–802.

56. Sholley MM, Ferguson GP, Seibel HR, Montour JL, Wilson JD. Mechanisms of neovascularization. Vascular sprouting can occur without proliferation of endothelial cells. *Lab Invest* 1984;**51**:624–34.

57. Mettouchi A. The role of extracellular matrix in vascular branching morphogenesis. *Cell Adh Migr* 2012;**6**:528–34.

58. Burri PH. Structural aspects of postnatal lung development - alveolar formation and growth. *Biol Neonate* 2006;**89**:313–22.

59. Burri PH, Tarek MR. A novel mechanism of capillary growth in the rat pulmonary microcirculation. *Anat Rec* 1990;**228**:35–45.

60. Caduff JH, Fischer LC, Burri PH. Scanning electron microscope study of the developing microvasculature in the postnatal rat lung. *Anat Rec* 1986;**216**:154–64.

61. De Spiegelaere W, Casteleyn C, Van den Broeck W, Plendl J, Bahramsoltani M, Simoens P, et al. Intussusceptive angiogenesis: a biologically relevant form of angiogenesis. *J Vasc Res* 2012;**49**:390–404.

62. Djonov VG, Kurz H, Burri PH. Optimality in the developing vascular system: branching remodeling by means of intussusception as an efficient adaptation mechanism. *Dev Dyn* 2002;**224**:391–402.

63. Kurz H, Burri PH, Djonov VG. Angiogenesis and vascular remodeling by intussusception: from form to function. *News Physiol Sci* 2003;**18**:65–70.

64. Burri PH, Hlushchuk R, Djonov V. Intussusceptive angiogenesis: its emergence, its characteristics, and its significance. *Dev Dyn* 2004;**231**:474–88.

65. Djonov V, Baum O, Burri PH. Vascular remodeling by intussusceptive angiogenesis. *Cell Tissue Res* 2003;**314**:107–17.

66. Djonov V, Schmid M, Tschanz SA, Burri PH. Intussusceptive angiogenesis: its role in embryonic vascular network formation. *Circ Res* 2000;**86**:286–92.

67. Augustin HG. Tubes branches, and pillars: the many ways of forming a new vasculature. *Circ Res* 2001;**89**:645–7.

68. Patan S, Munn LL, Tanda S, Roberge S, Jain RK, Jones RC. Vascular morphogenesis and remodeling in a model of tissue repair: blood vessel formation and growth in the ovarian pedicle after ovariectomy. *Circ Res* 2001;**89**:723–31.

69. Makanya AN, Hlushchuk R, Baum O, Velinov N, Ochs M, Djonov V. Microvascular endowment in the developing chicken embryo lung. *Am J Physiol Lung Cell Mol Physiol* 2007;**292**:L1136–46.

70. Makanya AN, Hlushchuk R, Djonov VG. Intussusceptive angiogenesis and its role in vascular morphogenesis, patterning, and remodeling. *Angiogenesis* 2009;**12**:113–23.

71. Swift MR, Weinstein BM. Arterial-venous specification during development. *Circ Res* 2009;**104**:576–88.

72. Wang HU, Chen ZF, Anderson DJ. Molecular distinction and angiogenic interaction between embryonic arteries and veins revealed by ephrin-B2 and its receptor Eph-B4. *Cell* 1998;**93**:741–53.

73. Yancopoulos GD, Klagsbrun M, Folkman J. Vasculogenesis, angiogenesis, and growth factors: ephrins enter the fray at the border. *Cell* 1998;**93**:661–4.

74. Aitsebaomo J, Portbury AL, Schisler JC, Patterson C. Brothers and sisters: molecular insights into arterial-venous heterogeneity. *Circ Res* 2008;**2008**:929–39.

75. Atkins GB, Jain MK, Hamik A. Endothelial differentiation: molecular mechanisms of specification and heterogeneity. *Arterioscler Thromb Vasc Biol* 2011;**31**:1476–84.

76. Claxton S, Fruttiger M. Oxygen modifies artery differentiation and network morphogenesis in the retinal vasculature. *Dev Dyn* 2005;**233**:822–8.

77. Vadivel A, van Haaften T, Alphonse RS, Rey-Parra GJ, Ionescu L, Haromy A, et al. Critical role of the axonal guidance cue EphrinB2 in lung growth, angiogenesis, and repair. *Am J Respir Crit Care Med* 2012;**185**:564–74.

78. Deng Y, Atri D, Eichmann A, Simons M. Endothelial ERK signaling controls lymphatic fate specification. *J Clin Invest* 2013;**123**:1202–15.

79. Howell K, Preston RJ, McLoughlin P. Chronic hypoxia causes angiogenesis in addition to remodelling in the adult rat pulmonary circulation. *J Physiol* 2003;**547**:133–45.

80. Jones R, Capen D. Pulmonary vascular remodeling by high oxygen, in Textbook of Pulmonary Vascular Disease. In: Yuan JX-J, et al., editor. *Springer Science & Business Media*. New York: LLC; 2011. pp. 733–58.

81. Jones R, Capen DE, Jacobson M, Cohen KS, Scadden DT, Duda DG. VEGFR2+PDGFRbeta+ circulating precursor cells participate in capillary restoration after hyperoxia acute lung injury (HALI). *J Cell Mol Med* 2009;**13**:3720–9.

82. Jones RC, Capen DE. A quantitative ultrastructural study of circulating (monocytic) cells interacting with endothelial cells in high oxygen-injured and spontaneously re-forming (FVB) mouse lung capillaries. *Ultrastruct Pathol* 2012;**36**:260–79.

83. Jones RC, Jacobson M. Angiogenesis in the hypertensive lung: response to ambient oxygen tension. *Cell Tissue Res* 2000;**300**:263–84.

84. Alvarez DF, Huang L, King JA, El Zarrad MK, Yoder MC, Stevens T. Lung microvascular endothelium is enriched with progenitor cells that exhibit vasculogenic capacity. *Am J Physiol Lung Cell Mol Physiol* 2008;**294**:L419–30.

85. Schniedermann J, Rennecke M, Buttler K, Richter G, Städtler AM, Norgall S. Mouse lung contains endothelial progenitors with high capacity to form blood and lymphatic vessels. *BMC Cell Biol* 2010;**11**:50.

86. Yoder MC. Is endothelium the origin of endothelial progenitor cells? *Arterioscler Thromb Vasc Biol* 2010;**30**:1094–103.

87. Burri PH, Moschopulos M. Structural analysis of fetal rat lung development. *Anat Rec* 1992;**234**:399–418.

88. Carmeliet P, Collen D. Genetic analysis of blood vessel formation role of endothelial versus smooth muscle cells. *Trends Cardiovasc Med* 1997;**7**:271–81.

89. Hanahan D. Signaling vascular morphogenesis and maintenance. *Science* 1997;**277**:48–50.

90. Le Lievre CS, Le Douarin NM. Mesenchymal derivatives of the neural crest: analysis of chimaeric quail and chick embryos. *J Embryol Exp Morphol* 1975;**34**:125–54.

91. Mitchell JJ, Reynolds SE, Leslie KO, Low RB, Woodcock-Mitchell J. Smooth muscle cell markers in developing rat lung. *Am J Respir Cell Mol Biol* 1990;**3**:515–23.

92. da Silva Meirelles L, Chagastelles PC, Nardi NB. Mesenchymal stem cells reside in virtually all post-natal organs and tissues. *J Cell Sci* 2006;**119**:2204–13.

93. Majesky MW, Dong XR, Hoglund V, Daum G, Mahoney WM, et al. The adventitia: a progenitor cell niche for the vessel wall. *Cells Tissues Organs* 2012;**195**:73–81.

94. Majesky MW, Dong XR, Regan JN, Hoglund VJ. Vascular smooth muscle progenitor cells: building and repairing blood vessels. *Circ Res* 2011;**108**:365–77.

95. Psaltis PJ, Harbuzariu A, Delacroix S, Holroyd EW, Simari RD. Resident vascular progenitor cells–diverse origins, phenotype, and function. *J Cardiovasc Transl Res* 2011;**4**:161–76.

96. Zengin E, Chalajour F, Gehling UM, Ito WD, Treede H, Lauke H, et al. Vascular wall resident progenitor cells: a source for postnatal vasculogenesis. *Development* 2006;**133**:1543–51.

97. Movat HZ, Fernando NV. The fine structure of the terminal vascular bed. IV. The venules and their perivascular cells (pericytes, adventitial cells). *Exp Mol Pathol* 1964;**34**:98–114.

98. Rhodin JA. Ultrastructure of mammalian venous capillaries, venules, and small collecting veins. *J Ultrastruct Res* 1968;**25**:452–500.

99. Sims DE. The pericyte–a review. *Tissue Cell* 1986;**18**:153–74.

100. Pardanaud L, Yassine F, Dieterlen-Lievre F. Relationship between vasculogenesis, angiogenesis and haemopoiesis during avian ontogeny. *Development* 1989;**105**:473–85.

101. Armulik A, Genove G, Betsholtz C. Pericytes: developmental, physiological, and pathological perspectives, problems, and promises. *Dev Cell* 2011;**21**:193–215.

102. Allt G, Lawrenson JG. Pericytes: cell biology and pathology. *Cells Tissues Organs* 2001;**169**:1–11.

103. Rhodin JA, Fujita H. Capillary growth in the mesentery of normal young rats. Intravital video and electron microscope analyses. *J Submicrosc Cytol Pathol* 1989;**21**:1–34.

104. Weibel ER. On pericytes, particularly their existence on lung capillaries. *Microvasc Res* 1974;**8**:218–35.

105. Krueger M, Bechmann I. CNS pericytes: concepts, misconceptions, and a way out. *Glia* 2010;**58**:1–10.

106. Stratman AN, Malotte KM, Mahan RD, Davis MJ, Davis GE. Pericyte recruitment during vasculogenic tube assembly stimulates endothelial basement membrane matrix formation. *Blood* 2009;**114**:5091–101.

107. Stratman AN, Schwindt AE, Malotte KM, Davis GE. Endothelial-derived PDGF-BB and HB-EGF coordinately regulate pericyte recruitment during vasculogenic tube assembly and stabilization. *Blood* 2010;**116**:4720–30.

108. Diaz-Flores L, Gutiérrez R, Madrid JF, Varela H, Valladares F, Acosta E, et al. Pericytes. Morphofunction, interactions and pathology in a quiescent and activated mesenchymal cell niche. *Histol Histopathol* 2009;**24**:909–69.

109. Shepro D, Morel NM. Pericyte physiology. *FASEB J* 1993;**7**:1031–8.

110. Attwell D, Buchan AM, Charpak S, Lauritzen M, Macvicar BA, Newman EA. Glial and neuronal control of brain blood flow. *Nature* 2010;**468**:232–43.

111. Kapanci Y, Ribaux C, Chaponnier C, Gabbiani G. Cytoskeletal features of alveolar myofibroblasts and pericytes in normal human and rat lung. *J Histochem Cytochem* 1992;**40**:1955–63.

112. Nehls V, Drenckhahn D. Heterogeneity of microvascular pericytes for smooth muscle type alpha-actin. *J Cell Biol* 1991;**113**:147–54.

113. Skalli O, Pelte MF, Peclet MC, Gabbiani G, Gugliotta P, Bussolati G, et al. Alpha-smooth muscle actin, a differentiation marker of smooth muscle cells, is present in microfilamentous bundles of pericytes. *J Histochem Cytochem* 1989;**37**:315–21.

114. Meyrick B, Fujiwara K, Reid L. Smooth muscle myosin in precursor and mature smooth muscle cells in normal pulmonary arteries and the effect of hypoxia. *Exp Lung Res* 1981;**2**:303–13.

115. Meyrick B, Reid L. The effect of continued hypoxia on rat pulmonary arterial circulation. An ultrastructural study. *Lab Invest* 1978;**38**:188–200.

116. Gittenberger-de Groot AC, DeRuiter MC, Bergwerff M, Poelmann RE. Smooth muscle cell origin and its relation to heterogeneity in development and disease. *Arterioscler Thromb Vasc Biol* 1999;**19**:1589–94.

117. Gittenberger-de Groot AC, Slomp J, DeRuiter MC, et al. Smooth muscle cell differentiation during early development and during intimal thickening formation in the ductus arterious. In: Schwartz SM, Mecham RP, editors. *The Vascular Smooth Muscle Cell. Molecular and Biological responses to the Extracellular Matrix*. San Diego, CA: Academic Press; 1995. pp. 17–36.

118. Hall SM, Hislop AA, Pierce CM, Haworth SG. Prenatal origins of human intrapulmonary arteries: formation and smooth muscle maturation. *Am J Respir Cell Mol Biol* 2000;**23**:194–203.

119. DeRuiter MC, Poelmann RE, VanMunsteren JC, Mironov V, Markwald RR, Gittenberger-de Groot AC. Embryonic endothelial cells transdifferentiate into mesenchymal cells expressing smooth muscle actins in vivo and in vitro. *Circ Res* 1997;**80**:444–51.

120. Small JV, Furst DO, Thornell LE. The cytoskeletal lattice of muscle cells. *Eur J Biochem* 1992;**208**:559–72.

121. Small JV, North AJ. Architecture of the smooth muscle cell. In: Schwartz SM, Mecham RP, editors. *The Vascular Smooth Muscle Cell. Molecular and Biological Responses to the Extracellular Matrix*. San Diego: Academic Press; 1995. pp. 169–88.

122. Langille BL. Remodeling of developing and mature arteries: endothelium, smooth muscle, and matrix. *J Cardiovasc Pharmacol* 1993;**21**:S11–7.

123. Jones, Steudel R, White S, Jacobson M, Low R. Microvessel precursor smooth muscle cells express head-inserted smooth muscle myosin heavy chain (SM-B) isoform in hyperoxic pulmonary hypertension. *Cell Tissue Res* 1999;**295**:453–65.

124. Allen KM, Haworth SG. Cytoskeletal features of immature pulmonary vascular smooth muscle cells: the influence of pulmonary hypertension on normal development. *J Pathol* 1989;**158**:311–7.

125. Borrione AC, Zanellato AM, Scannapieco G, Pauletto P, Sartore S. Myosin heavy-chain isoforms in adult and developing rabbit vascular smooth muscle. *Eur J Biochem* 1989;**183**:413–7.

126. Frid MG, Printesva OY, Chiavegato A, Faggin E, Scatena M, E V. Myosin heavy-chain isoform composition and distribution in developing and adult human aortic smooth muscle. *J Vasc Res* 1993;**30**:279–92.

127. Frid MG, Shekhonin BV, Koteliansky VE, Glukhova MA. Phenotypic changes of human smooth muscle cells during development: late expression of heavy caldesmon and calponin. *Dev Biol* 1992;**153**:185–93.

128. Giuriato L, Scatena M, Chiavegato A, Tonello M, Scannapieco G, Pauletto P, et al. Non-muscle myosin isoforms and cell heterogeneity in developing rabbit vascular smooth muscle. *J Cell Sci* 1992;**101**:233–46.

129. Kocher O, Skalli O, Cerutti D, Gabbiani F, Gabbiani G. Cytoskeletal features of rat aortic cells during development. An electron microscopic, immunohistochemical, and biochemical study. *Circ Res* 1985;**56**:829–38.

130. Owens GK. Regulation of differentiation of vascular smooth muscle cells. *Physiol Rev* 1995;**75**:487–517.

131. Sartore S, Scatena M, Chiavegato A, Faggin E, Giuriato L, Pauletto P. Myosin isoform expression in smooth muscle cells during physiological and pathological vascular remodeling. *J Vasc Res* 1994;**31**:61–81.

132. Rhodin JA. The ultrastructure of mammalian arterioles and precapillary sphincters. *J Ultrastruct Res* 1967;**18**:181–223.

133. Faury G. Role of elastin in the development of vascular function. Knock-out study of the elastin gene in mice. *J Soc Biol* 2001;**195**:151–6.

134. Brown-Augsburger P, Broekelmann T, Rosenbloom J, Mecham RP. Functional domains on elastin and microfibril-associated glycoprotein involved in elastic fibre assembly. *Biochem J* 1996;**318**:149–55.

135. Crouch EC, Noguchi A, Mecham RP, et al. Collagens and elastic fiber proteins in lung development. In: McDonald JA, editor. *Lung Growth and Development. Lung Biology in Health and Disease, executive ed. C. Lenfant*; vol. 100. New York: Marcel Dekker; 1997. pp. 327–63.

136. Davis EC, Mecham RP. Intracellular trafficking of tropoelastin. *Matrix Biol* 1998;**17**:245–54.

137. Faury G. Function-structure relationship of elastic arteries in evolution: from microfibrils to elastin and elastic fibres. *Pathol Biol (Paris)* 2001;**49**:310–25.

138. Hinek A, Wrenn DS, Mecham RP, Barondes SH. The elastin receptor: a galactoside-binding protein. *Science* 1988;**239**:1539–41.

139. Jaques A, Serafini-Fracassini A. Morphogenesis of the elastic fiber: an immunoelectronmicroscopy investigation. *J Ultrastruct Res* 1985;**92**:201–10.

140. Mariencheck MC, Davis EC, Zhang H, Ramirez F, Rosenbloom J, Gibson MA, et al. Fibrillin-1 and fibrillin-2 show temporal and tissue-specific regulation of expression in developing elastic tissues. *Connect Tissue Res* 1995;**31**:87–97.

141. Mecham RP, Davis EC. Elastic fiber structure and assembly. In: Yurchenco D, Birk DE, Mecham RP, editors. *Extracellular Matrix Assembly and Structure*. New York: Academic Press; 1994. pp. 281–314.

142. Mecham RP, Prosser I, Fukuda Y. Elastic fibers. In: Crystal RG, et al., editor. *The Lung: Scientific Foundations*. New York: Raven Press, Ltd.; 1991. pp. 389–98.

143. Robb BW, Wachi H, Schaub T, Mecham RP, Davis EC. Characterization of an in vitro model of elastic fiber assembly. *Mol Biol Cell* 1999;**10**:3595–605.

144. Rosenbloom J, Abrams WR, Mecham R. Extracellular matrix 4: the elastic fiber. *FASEB J* 1993;**7**:1208–18.

145. Sakai LY, Keene DR, Engvall E. Fibrillin, a new 350-kD glycoprotein, is a component of extracellular microfibrils. *J Cell Biol* 1986;**103**:2499–509.

146. Zhang H, Hu W, Ramirez F. Developmental expression of fibrillin genes suggests heterogeneity of extracellular microfibrils. *J Cell Biol* 1995;**129**:1165–76.

147. Miosge N, Sasaki T, Timpl R. Angiogenesis inhibitor endostatin is a distinct component of elastic fibers in vessel walls. *FASEB J* 1999;**13**:1743–50.

148. Jones R. Ultrastructural analysis of contractile cell development in lung microvessels in hyperoxic pulmonary hypertension. Fibroblasts and intermediate cells selectively reorganize nonmuscular segments. *Am J Pathol* 1992;**141**:1491–505.

149. Jones R. Role of interstitial fibroblasts and intermediate cells in microvascular wall remodelling in pulmonary hypertension. *Eur Resp Rev* 1993;**3**:569–75.

150. Jones R, Capen D, Jacobson M. PDGF and microvessel wall remodeling in adult lung: imaging PDGF-Rbeta and PDGF-BB molecules in progenitor smooth muscle cells developing in pulmonary hypertension. *Ultrastruct Pathol* 2006;**30**:267–81.

151. Jones R, Capen D, Jacobson M, Munn L. PDGF and microvessel wall remodeling in adult rat lung: imaging PDGF-AA and PDGF-Ralpha molecules in progenitor smooth muscle cells developing in experimental pulmonary hypertension. *Cell Tissue Res* 2006;**326**:759–69.

152. Jones R, Jacobson M, Steudel W. Alpha-smooth-muscle actin and microvascular precursor smooth-muscle cells in pulmonary hypertension. *Am J Respir Cell Mol Biol* 1999;**20**:582–94.

153. Desmouliere A, Gabbiani G. Smooth muscle cell and fibroblast biological and functional features: similarities and differences. In: Schwartz M, Mecham RP, editors. *The Vascular Smooth Muscle Cell. Molecular and Biological Responses to the Extracellular Matrix*. San Diego: Academic Press; 1995. pp. 329–59.

154. Bostrom H, Gritli-Linde A, Betsholtz C. PDGF-A/PDGF alpha-receptor signaling is required for lung growth and the formation of alveoli but not for early lung branching morphogenesis. *Dev Dyn* 2002;**223**:155–62.

155. Bostrom H, Willetts K, Pekny M, Levéen P, Lindahl P, Hedstrand H, et al. PDGF-A signaling is a critical event in lung alveolar myofibroblast development and alveogenesis. *Cell* 1996;**85**:863–73.

156. Lindahl P, Karlsson L, Hellström M, Gebre-Medhin S, Willetts K, Heath JK, et al. Alveogenesis failure in PDGF-A-deficient mice is coupled to lack of distal spreading of alveolar smooth muscle cell progenitors during lung development. *Development* 1997;**124**:3943–53.

157. Gan Q, Yoshida T, Li J, Owens GK. Smooth muscle cells and myofibroblasts use distinct transcriptional mechanisms for smooth muscle alpha-actin expression. *Circ Res* 2007;**101**:883–92.

158. Hinz B, Phan SH, Thannickal VJ, Prunotto M, Desmoulière A, Varga J, et al. Recent developments in myofibroblast biology: paradigms for connective tissue remodeling. *Am J Pathol* 2012;**180**:1340–55.

159. Phan SH. Genesis of the myofibroblast in lung injury and fibrosis. *Proc Am Thorac Soc* 2012;**9**:148–52.

160. Hosoyamada Y, Ichimura K, Koizumi K, Sakai T. Structural organization of pulmonary veins in the rat lung, with special emphasis on the musculature consisting of cardiac and smooth muscles. *Anat Sci Int* 2010;**85**:152–9.

161. Ludatscher RM. Fine structure of the muscular wall of rat pulmonary veins. *J Anat* 1968;**103**:345–57.

162. Millino C, Sarinella F, Tiveron C, Villa A, Sartore S, Ausoni S. Cardiac and smooth muscle cell contribution to the formation of the murine pulmonary veins. *Dev Dyn* 2000;**218**:414–25.

163. Gerety SS, Wang HU, Chen ZF, Anderson DJ. Symmetrical mutant phenotypes of the receptor EphB4 and its specific transmembrane ligand ephrin-B2 in cardiovascular development. *Mol Cell* 1999;**4**:403–14.

164. Jackson JR, Seed MP, Kircher CH, Willoughby DA, Winkler JD. The codependence of angiogenesis and chronic inflammation. *FASEB J* 1997;**11**:457–65.

165. Senger DR. Molecular framework for angiogenesis: a complex web of interactions between extravasated plasma proteins and endothelial cell proteins induced by angiogenic cytokines. *Am J Pathol* 1996;**149**:1–7.

166. Han RN, Buch S, Freeman BA, Post M, Tanswell AK. Platelet-derived growth factor and growth-related genes in rat lung. II. Effect of exposure to 85% O2. *Am J Physiol* 1992;**262**:L140–6.

167. Powell PP, Wang CC, Horinouchi H, Shepherd K, Jacobson M, Lipson M, et al. Differential expression of fibroblast growth factor receptors 1 to 4 and ligand genes in late fetal and early postnatal rat lung. *Am J Respir Cell Mol Biol* 1998;**19**:563–72.

168. Dumont DJ, Fong GH, Puri MC, Gradwohl G, Alitalo K, Breitman ML. Vascularization of the mouse embryo: a study of flk-1, tek, tie, and vascular endothelial growth factor expression during development. *Dev Dyn* 1995;**203**:80–92.

169. Gitay-Goren H, Cohen T, Tessler S, Soker S, Gengrinovitch S, Rockwell P, Klagsbrun M, et al. Selective binding of VEGF121 to one of the three vascular endothelial growth factor receptors of vascular endothelial cells. *J Biol Chem* 1996;**271**:5519–23.

170. Millauer B, Wizigmann-Voos S, Schnürch H, Martinez R, Møller NP, Risau W, et al. High affinity VEGF binding and developmental expression suggest Flk-1 as a major regulator of vasculogenesis and angiogenesis. *Cell* 1993;**72**:835–46.

171. Shalaby F, Rossant J, Yamaguchi TP, Gertsenstein M, Wu XF, Breitman M, et al. Failure of blood-island formation and vasculogenesis in Flk-1-deficient mice. *Nature* 1995;**376**:62–6.

172. Soker S, Takashima S, Miao HQ, Neufeld G, Klagsbrun M. Neuropilin-1 is expressed by endothelial and tumor cells as an isoform-specific receptor for vascular endothelial growth factor. *Cell* 1998;**92**:735–45.

173. Chappell JC, Taylor SM, Ferrara N, Bautch VL. Local guidance of emerging vessel sprouts requires soluble Flt-1. *Dev Cell* 2009;**17**:377–86.

174. Kappas NC, Zeng G, Chappell JC, Kearney JB, Hazarika S, Kallianos KG, et al. The VEGF receptor Flt-1 spatially modulates Flk-1 signaling and blood vessel branching. *J Cell Biol* 2008;**181**:847–58.

175. Kearney JB, Kappas NC, Ellerstrom C, DiPaola FW, Bautch VL. The VEGF receptor flt-1 (VEGFR-1) is a positive modulator of vascular sprout formation and branching morphogenesis. *Blood* 2004;**103**:4527–35.

176. Lee S, Chen TT, Barber CL, Jordan MC, Murdock J, Desai S, et al. Autocrine VEGF signaling is required for vascular homeostasis. *Cell* 2007;**130**:691–703.

177. Carmeliet P, Ferreira V, Breier G, Pollefeyt S, Kieckens L, Gertsenstein M, et al. Abnormal blood vessel development and lethality in embryos lacking a single VEGF allele. *Nature* 1996;**380**:435–9.

178. Ferrara N, Carver-Moore K, Chen H, Dowd M, Lu L, O'Shea KS, et al. Heterozygous embryonic lethality induced by targeted inactivation of the VEGF gene. *Nature* 1996;**380**:439–42.

179. Fong GH, Rossant J, Gertsenstein M, Breitman ML. Role of the Flt-1 receptor tyrosine kinase in regulating the assembly of vascular endothelium. *Nature* 1995;**376**:66–70.

180. Jakeman LB, Winer J, Bennett GL, Altar CA, Ferrara N. Binding sites for vascular endothelial growth factor are localized on endothelial cells in adult rat tissues. *J Clin Invest* 1992;**89**:244–53.

181. Lorquet S, Berndt S, Blacher S, Gengoux E, Peulen O, Maquoi E, et al. Soluble forms of VEGF receptor-1 and -2 promote vascular maturation via mural cell recruitment. *FASEB J* 2010;**24**:3782–95.

182. Tischer E, Gospodarowicz D, Mitchell R, Silva M, Schilling J, Lau K, et al. Vascular endothelial growth factor: a new member of the platelet-derived growth factor gene family. *Biochem Biophys Res Commun* 1989;**165**:1198–206.

183. Gerber HP, Hillan KJ, Ryan AM, Kowalski J, Keller GA, Rangell L, et al. VEGF is required for growth and survival in neonatal mice. *Development* 1999;**126**:1149–59.

184. Healy AM, Morgenthau L, Zhu X, Farber HW, Cardoso WV. VEGF is deposited in the subepithelial matrix at the leading edge of branching airways and stimulates neovascularization in the murine embryonic lung. *Dev Dyn* 2000;**219**:341–52.

185. Gebb SA, Shannon JM. Tissue interactions mediate early events in pulmonary vasculogenesis. *Dev Dyn* 2000;**217**:159–69.

186. Kalinichenko VV, Lim L, Stolz DB, Shin B, Rausa FM, Clark J, et al. Defects in pulmonary vasculature and perinatal lung hemorrhage in mice heterozygous null for the Forkhead Box f1 transcription factor. *Dev Biol* 2001;**235**:489–506.

187. Ng YS, Rohan R, Sunday ME, Demello DE, D'Amore PA. Differential expression of VEGF isoforms in mouse during development and in the adult. *Dev Dyn* 2001;**220**:112–21.

188. Galambos C, Ng YS, Ali A, Noguchi A, Lovejoy S, D'Amore PA, et al. Defective pulmonary development in the absence of heparin-binding vascular endothelial growth factor isoforms. *Am J Respir Cell Mol Biol* 2002;**27**:194–203.

189. Akeson AL, Cameron JE, Le Cras TD, Whitsett JA, Greenberg JM. Vascular endothelial growth factor A induces prenatal neovascularization and alters bronchial development in mice. *Pediatr Res* 2005;**57**:82–8.

190. Le Cras TD, Spitzmiller RE, Albertine K, Greenberg JM, Whitsett JA, Akeson AL. VEGF causes pulmonary hemorrhage, hemosiderosis, and air space enlargement in neonatal mice. *Am J Physiol Lung Cell Mol Physiol* 2004;**287**:L134–42.

191. Hellstrom M, Kalén M, Lindahl P, Abramsson A, Betsholtz C. Role of PDGF-B and PDGFR-beta in recruitment of vascular smooth muscle cells and pericytes during embryonic blood vessel formation in the mouse. *Development* 1999;**126**:3047–55.

192. Leveen P, Pekny M, Gebre-Medhin S, Swolin B, Larsson E, Betsholtz C. Mice deficient for PDGF B show renal, cardiovascular, and hematological abnormalities. *Genes Dev* 1994;**8**:1875–87.

193. Lindahl P, Hellström M, Kalén M, Betsholtz C. Endothelial-perivascular cell signaling in vascular development: lessons from knockout mice. *Curr Opin Lipidol* 1998;**9**:407–11.

194. Lindahl P, Johansson BR, Levéen P, Betsholtz C. Pericyte loss and microaneurysm formation in PDGF-B-deficient mice. *Science* 1997;**277**:242–5.

195. Fredriksson L, Li H, Eriksson U. The PDGF family: four gene products form five dimeric isoforms. *Cytokine Growth Factor Rev* 2004;**15**:197–204.

196. Lindahl P, Betsholtz C. Not all myofibroblasts are alike: revisiting the role of PDGF-A and PDGF-B using PDGF-targeted mice. *Curr Opin Nephrol Hypertens* 1998;**7**:21–6.

197. Simm A, Hoppe V, Tatje D, Schenzinger A, Hoppe J. PDGF-AA effectively stimulates early events but has no mitogenic activity in AKR-2B mouse fibroblasts. *Exp Cell Res* 1992;**201**:192–9.

198. Hirschi KK, Rohovsky SA, Beck LH, Smith SR, D'Amore PA. Endothelial cells modulate the proliferation of mural cell precursors via platelet-derived growth factor-BB and heterotypic cell contact. *Circ Res* 1999;**84**:298–305.

199. Hirschi KK, Rohovsky SA, D'Amore PA. PDGF, TGF-beta, and heterotypic cell-cell interactions mediate endothelial cell-induced recruitment of 10T1/2 cells and their differentiation to a smooth muscle fate. *J Cell Biol* 1998;**141**:805–14.

200. Holycross BJ, Blank RS, Thompson MM, Peach MJ, Owens GK. Platelet-derived growth factor-BB-induced suppression of smooth muscle cell differentiation. *Circ Res* 1992;**71**:1525–32.

201. Sjolund M, Rahm M, Claesson-Welsh L, Sejersen T, Heldin CH, Thyberg J. Expression of PDGF alpha- and beta-receptors in rat arterial smooth muscle cells is phenotype and growth state dependent. *Growth Factors* 1990;**3**:191–203.

202. Benjamin LE, Hemo I, Keshet E. A plasticity window for blood vessel remodelling is defined by pericyte coverage of the preformed endothelial network and is regulated by PDGF-B and VEGF. *Development* 1998;**125**:1591–8.

203. Ball SG, Shuttleworth CA, Kielty CM. Vascular endothelial growth factor can signal through platelet-derived growth factor receptors. *J Cell Biol* 2007;**177**:489–500.

204. Ball SG, Shuttleworth CA, Kielty CM. Platelet-derived growth factor receptors regulate mesenchymal stem cell fate: implications for neovascularization. *Expert Opin Biol Ther* 2010;**10**:57–71

205. Zhang J, Cao R, Zhang Y, Tanghong J, Cao Y, Wahlberg E. Differential roles of PDGFR-alpha and PDGFR-beta in angiogenesis and vessel stability. *FASEB J* 2009;**23**:153–63.

206. Gaengel K, Genové G, Armulik A, Betsholtz C. Endothelial mural cell signaling in vascular development and angiogenesis. *Arterioscler Thromb Vasc Biol* 2009;**29**:630–8.

207. Saharinen P, Kerkelä K, Ekman N, Marron M, Brindle N, Lee GM, et al. Multiple angiopoietin recombinant proteins activate the Tie1 receptor tyrosine kinase and promote its interaction with Tie2. *J Cell Biol* 2005;**169**:239–43.

208. Davis S, Aldrich TH, Jones PF, Acheson A, Compton DL, Jain V, et al. Isolation of angiopoietin-1, a ligand for the TIE2 receptor, by secretion-trap expression cloning. *Cell* 1996;**87**:1161–9.

209. Gale NW, Yancopoulos GD. Growth factors acting via endothelial cell-specific receptor tyrosine kinases: VEGFs, angiopoietins, and ephrins in vascular development. *Genes Dev* 1999;**13**:1055–66.

210. Koblizek TI, Weiss C, Yancopoulos GD, Deutsch U, Risau W. Angiopoietin-1 induces sprouting angiogenesis in vitro. *Curr Biol* 1998;**8**:529–32.

211. Papapetropoulos A, García-Cardeña G, Dengler TJ, Maisonpierre PC, Yancopoulos GD, Sessa WC. Direct actions of angiopoietin-1 on human endothelium: evidence for network stabilization, cell survival, and interaction with other angiogenic growth factors. *Lab Invest* 1999;**79**:213–23.

212. Sato TN, Tozawa Y, Deutsch U, Wolburg-Buchholz K, Fujiwara Y, Gendron-Maguire M, et al. Distinct roles of the receptor tyrosine kinases Tie-1 and Tie-2 in blood vessel formation. *Nature* 1995;**376**:70–4.

213. Suri C, Jones PF, Patan S, Bartunkova S, Maisonpierre PC, Davis S, et al. Requisite role of angiopoietin-1, a ligand for the TIE2 receptor, during embryonic angiogenesis. *Cell* 1996;**87**:1171–80.

214. Valenzuela DM, Griffiths JA, Rojas J, Aldrich TH, Jones PF, Zhou H, et al. Angiopoietins 3 and 4: diverging gene counterparts in mice and humans. *Proc Natl Acad Sci U S A* 1999;**96**:1904–9.

215. Asahara T, Chen D, Takahashi T, Fujikawa K, Kearney M, Magner M, et al. Tie2 receptor ligands, angiopoietin-1 and angiopoietin-2, modulate VEGF-induced postnatal neovascularization. *Circ Res* 1998;**83**:233–40.

216. Korhonen J, Polvi A, Partanen J, Alitalo K. The mouse tie receptor tyrosine kinase gene: expression during embryonic angiogenesis. *Oncogene* 1994;**9**:395–403.

217. Puri MC, Partanen J, Rossant J, Bernstein A. Interaction of the TEK and TIE receptor tyrosine kinases during cardiovascular development. *Development* 1999;**126**:4569–80.

218. Boyden EA, Tompsett DH. The changing patterns in the developing lungs of infants. *Acta Anat (Basel)* 1965;**61**:164–92.

219. Bucher U, Reid L. Development of the intrasegmental bronchial tree: the pattern of branching and development of cartilage at various stages of intra-uterine life. *Thorax* 1961;**16**:207–18.

220. Hislop A, Reid L. Intra-pulmonary arterial development during fetal life-branching pattern and structure. *J Anat* 1972;**113**:35–48.

221. Hislop A, Reid L. Fetal and childhood development of the intra-pulmonary veins in man–branching pattern and structure. *Thorax* 1973;**28**:313–9.

222. Hislop A, Reid L. Development of the acinus in the human lung. *Thorax* 1974;**29**:90–4.

223. Hislop A, Reid L. Growth and development of the respiratory system - Anatomical development. In: Davis JA, Dobbing J, editors. *Scientific Foundations of Pediatrics*. London, U.K: Heinemann Medical Publications, Ltd.; 1974. pp. 214–54.

224. Hislop A, Reid L. Formation of the pulmonary vasculature. In: Hodson WA, Lenfant C, editors. *Lung Biology in Health and Disease*. New York: Marcel Dekker Inc.; 1977. pp. 37–86. 1977.

225. Potter EL, Loosli CG. Prenatal development of the human lung. *AMA Am J Dis Child* 1951;**82**:226–8.

226. Reid L. 1976 Edward B.D. Neuhauser lecture: the lung: growth and remodeling in health and disease. *AJR Am J Roentgenol* 1977;**129**:777–88.

227. Reid LM. The pulmonary circulation: remodeling in growth and disease. The 1978 J. Burns Amberson lecture. *Am Rev Respir Dis* 1979;**119**:531–46.

228. Hislop A, Reid LM. Growth and development of the respiratory system: anatomical development. In: Davis JA, Dobbing J, editors. *Scientific Foundations of Pediatrics*. London: Heinemann Medical Publications; 1981. pp. 390–431.

229. deMello DE, Reid LM. Embryonic and early fetal development of human lung vasculature and its functional implications. *Pediatr Dev Pathol* 2000;**3**:439–49.

230. deMello DE, et al. Early fetal development of lung vasculature. *Am J Respir Cell Mol Biol* 1997;**16**:568–81.

231. Galambos C, deMello DE. Molecular mechanisms of pulmonary vascular development. *Pediatr Dev Pathol* 2007;**10**:1–17.

232. Parera MC, van Dooren M, van Kepen M, de Krijger R, Grosveld F, Tibboel D, et al. Distal angiogenesis: a new concept for lung vascular morphogenesis. *Am J Physiol Lung Cell Mol Physiol* 2005;**288**:L141–9.

233. Pereda J, Sulz L, San Martin S, Godoy-Guzmán C. The human lung during the embryonic period: vasculogenesis and primitive erythroblasts circulation. *J Anat* 2013;**222**:487–94.

234. Schachtner SK, Wang Y, Scott Baldwin H. Qualitative and quantitative analysis of embryonic pulmonary vessel formation. *Am J Respir Cell Mol Biol* 2000;**22**:157–65.

235. Elliott FM, Reid L. Some new facts about the pulmonary artery and its branching pattern. *Clin Radiol* 1965;**16**:193–8.

236. Frid MG, Moiseeva EP, Stenmark KR. Multiple phenotypically distinct smooth muscle cell populations exist in the adult and developing bovine pulmonary arterial media in vivo. *Circ Res* 1994;**75**:669–81.

237. Boyden EA. The developing bronchial arteries in a fetus of the twelfth week. *Am J Anat* 1970;**129**:357–68.

238. Boyden EA. The time lag in the development of bronchial arteries. *Anat Rec* 1970;**166**:611–4.

239. Hislop A, Reid L. Pulmonary arterial development during childhood: branching pattern and structure. *Thorax* 1973;**28**:129–35.

240. Narayanan M, Owers-Bradley J, Beardsmore CS, Mada M, Ball I, Garipov R, et al. Alveolarization continues during childhood and adolescence: new evidence from helium-3 magnetic resonance. *Am J Respir Crit Care Med* 2012;**185**:186–91.

241. Horsfield K, Gordon WI. Morphometry of pulmonary veins in man. *Lung* 1981;**159**:211–8.

242. Singhal S, Henderson R, Horsfield K, Harding K, Cumming G. Morphometry of the human pulmonary arterial tree. *Circ Res* 1973;**33**:190–7.

243. Gehr P, Bachofen M, Weibel ER. The normal human lung: ultrastructure and morphometric estimation of diffusion capacity. *Respir Physiol* 1978;**32**:121–40.

244. Warburton D, Lee MK. Current concepts on lung development. *Curr Opin Pediatr* 1999;**11**:188–92.

245. Weibel E. Lung cell biology. In: Fishman AP, Fisher A, Geiger S, editors. *Handbook of Physiology*. Bethesda, MD: American Physiological Society; 1985. pp. 47–91.

246. Weibel ER. Design and structure of the human lung. In: Fishman A, editor. *Pulmonary Diseases*. New York: McGraw-Hill; 1980. pp. 224–71.

247. Shaw AM, Bunton DC, Fisher A, McGrath JC, Montgomery I, Daly C, et al. V-shaped cushion at the origin of bovine pulmonary supernumerary arteries: structure and putative function. *J Appl Physiol* 1999;**87**:2348–56.

248. Bunton D, MacDonald A, Brown T, Tracey A, McGrath JC, Shaw AM, et al. 5-hydroxytryptamine- and U46619-mediated vasoconstriction in bovine pulmonary conventional and supernumerary arteries: effect of endogenous nitric oxide. *Clin Sci (Lond)* 2000;**98**:81–9.

249. Brunner EJ, Shipley MJ, Witte DR, Singh-Manoux A, Britton AR, Tabak AG, et al. Arterial stiffness, physical function, and functional limitation: the Whitehall II Study. *Hypertension* 2011;**57**:1003–9.

250. Semmens M. The pulmonary artery in the normal aged lung. *Br J Dis Chest* 1970;**64**:65–72.

251. Yildiz O. Vascular smooth muscle and endothelial functions in aging. *Ann N Y Acad Sci* 2007;**1100**:353–60.

252. Brandes RP, Fleming I, Busse R. Endothelial aging. *Cardiovasc Res* 2005;**66**:286–94.

253. Mackay EH, Banks J, Sykes B, Lee G. Structural basis for the changing physical properties of human pulmonary vessels with age. *Thorax* 1978;**33**:335–44.

254. Knudson RJ. Physiology of the aging lung. In: Crystal RG, et al., editors. *The Lung: Scientific Foundations*. New York: Raven Press, Ltd.; 1991. pp. 1749–59.

255. Reid L. *The aged lung, in The Pathology of Emphysema*. London: Lloyd-Luke (Medical Books), Ltd; 1967. pp. 22–28.

256. deMello DE, Reid L. Pre-and postnatal development of the pulmonary circulation. In: Haddad GG, Abman SH, Chernick V, editors. *Chernik-Mellins Basic Mechanisms of Pediatric Disease*. Hamilton and London: Marcel Dekker Inc.; 2002. pp. 77–101.

257. Ryland D, Reid L. Pulmonary aplasia–a quantitative analysis of the development of the single lung. *Thorax* 1971;**26**:602–9.

258. Nobuhara KK, Fauza DO, DiFiore JW, Hines MH, Fackler JC, Slavin R, et al. Continuous intrapulmonary distension with perfluorocarbon accelerates neonatal (but not adult) lung growth. *J Pediatr Surg* 1998;**33**:292–8.

259. Areechon W, Reid L. Hypoplasia of lung with congenital diaphragmatic hernia. *Br Med J* 1963;**1**:230–3.

260. Geggel RL, Murphy JD, Langleben D, Crone RK, Vacanti JP, Reid LM. Congenital diaphragmatic hernia: arterial structural changes and persistent pulmonary hypertension after surgical repair. *J Pediatr* 1985;**107**:457–64.

261. Kitagawa M, Hislop A, Boyden EA, Reid L. Lung hypoplasia in congenital diaphragmatic hernia. A quantitative study of airway, artery, and alveolar development. *Br J Surg* 1971;**58**:342–6.

262. DiFiore JW, Fauza DO, Slavin R, Peters CA, Fackler JC, Wilson JM. Experimental fetal tracheal ligation reverses the structural and physiological effects of pulmonary hypoplasia in congenital diaphragmatic hernia. *J Pediatr Surg* 1994;**29**:248–56. discussion 256–257.

263. DiFiore JW, Fauza DO, Slavin R, Wilson JM. Experimental fetal tracheal ligation and congenital diaphragmatic hernia: a pulmonary vascular morphometric analysis. *J Pediatr Surg* 1995;**30**:917–23. discussion 923–924.

264. Beals DA, Schloo BL, Vacanti JP, Reid LM, Wilson JM. Pulmonary growth and remodeling in infants with high-risk congenital diaphragmatic hernia. *J Pediatr Surg* 1992;**27**:997–1001. discussion 1001–1002.

265. Grzenda A, Shannon J, Fisher J, Arkovitz MS. Timing and expression of the angiopoietin-1-Tie-2 pathway in murine lung development and congenital diaphragmatic hernia. *Dis Model Mech* 2013;**6**:106–14.

266. Sluiter I, van der Horst I, van der Voorn P, Boerema-de Munck A, Buscop-van Kempen M, de Krijger R, et al. Premature differentiation of vascular smooth muscle cells in human congenital diaphragmatic hernia. *Exp Mol Pathol* 2013;**94**:195–202.

267. Adzick NS, Outwater KM, Harrison MR, Davies P, Glick PL, deLorimier AA, et al. Correction of congenital diaphragmatic hernia in utero. IV. An early gestational fetal lamb model for pulmonary vascular morphometric analysis. *J Pediatr Surg* 1985;**20**:673–80.

268. Wohl ME, Griscom NT, Strieder DJ, Schuster SR, Treves S, Zwerdling RG. The lung following repair of congenital diaphragmatic hernia. *J Pediatr* 1977;**90**:405–14.

269. Williams AJ, Vawter G, Reid LM. Lung structure in asphyxiating thoracic dystrophy. *Arch Pathol Lab Med* 1984;**108**:658–61.

270. Jones R, Reid L. Development of the pulmonary vasculature. In: Harding R, Pinkerton KE, Plopper CG, editors. *The Lung; Development, Aging and Environment*. 1st Edition; 2004. pp. 81–103.

271. Jones R, Zapol WM, Tomashefski JF, Kirton OC, Kobayashi K, Reid L. Pulmonary vascular pathology – human and experimental studies. In: Zapol WM, Falke KJ, editors. *Acute Respiratory Failure*. New York: Dekker; 1985. pp. 23–160.

272. Davies G, Reid L. Growth of alveoli and pulmonary arteries in childhood. *Thorax* 1970;**25**:669–81.

273. Reid L. *The Pathology of Emphysema*. London: Lloyd-Luke; 1967. 319–361.

Chapter 6

Developmental Physiology of the Pulmonary Circulation

Reuben B. Dodson*,†, Csaba Galambos*,‡ and Steven H. Abman*,§

*The Pediatric Heart Lung Center, Departments of †Surgery; ‡Pathology; §Pediatrics, University of Colorado Denver, Anschutz Medical Campus and Children's Hospital Colorado, Aurora, CO, USA

INTRODUCTION

Postnatal survival depends upon the ability of the pulmonary circulation to undergo rapid and dramatic vasodilation during the first minutes after birth. The resulting fall in pulmonary vascular resistance (PVR) allows for an 8–10-fold increase in pulmonary blood flow, which is essential for the lung to assume its postnatal role in gas exchange. Although the decrease in PVR during transition of the lung at birth is abrupt, the success of this critical event follows a lengthy series of carefully orchestrated events that characterize normal growth and maturation of the fetal pulmonary circulation. Normal development of the lung circulation is determined by the precise coordination of numerous signals from multiple cell types and the cell environment, which include diverse transcription factors, growth factors, chemokines, cytokines, vasoactive products, matrix proteins, and others.

In addition to developmental changes in lung vascular growth and structure, the pulmonary circulation also undergoes maturational changes in function at cellular, tissue, and organ levels. While maintaining high PVR in utero, the fetal lung must also acquire the ability to respond to vasodilator stimuli with advancing gestational age, prior to birth. Maturational changes in endothelial and smooth muscle cell function in the lung circulation are vital for enabling successful transition at birth. The vascular extracellular environment provides an important structural and mechanical component for vessel function as well as providing instruction signals that induce, define, and stabilize cellular phenotypes during maturation. Failure of the pulmonary circulation to successfully achieve or sustain this decrease in PVR causes severe hypoxemia in cardiopulmonary disorders that constitute the syndrome, persistent pulmonary hypertension of the newborn (PPHN). PPHN is a major clinical problem, contributing substantially to morbidity and mortality in both full-term and pre-term neonates. Mechanisms underlying

the pathogenesis of PPHN are unclear, but clinical and experimental data suggest that intrauterine events that impair vascular function can disrupt the normal maturational sequence of lung vascular development and growth. Therefore, understanding basic mechanisms of normal functional and structural development of the pulmonary circulation in utero and mechanisms that contribute to sustained pulmonary vasodilation at birth may provide insights into the syndrome of PPHN and its treatment.

This chapter reviews our current understanding of the physiologic signaling and growth regulation of the developing pulmonary circulation, including mechanisms that regulate vascular tone and reactivity in the fetal lung; mechanisms of extracellular and structural development of the vasculature; mechanisms that contribute to adaptive changes at birth and the early postnatal period and aging (see also Chapter 13 [Siew et al.]); and mechanisms that contribute to the failure of PVR to fall in neonates with PPHN.

LUNG VASCULAR GROWTH

Normal development of the human lung can be divided into five stages, namely the embryonic (3–7 weeks gestation), pseudoglandular (5–17 weeks), canalicular (16–26 weeks), saccular (24–38 weeks), and alveolar periods (up to 2–3 years of age).[1] Although branching morphogenesis, epithelial development and differentiation, and alveolarization have been extensively studied, relatively less is known about lung vascular growth and development (see also Chapter 5 [Jones et al.]). During each stage of development, the pulmonary circulation also undergoes marked changes in growth, structural remodeling, and maturation, as shown in Table 1. Lung vascular growth involves two basic processes: vasculogenesis, the formation of new blood vessels from endothelial cells within the immature mesenchyme; and angiogenesis, the formation of new blood vessels from sprouts of pre-existing vessels.[2] Embryonic stem cells

The Lung. http://dx.doi.org/10.1016/B978-0-12-799941-8.00006-7

TABLE 1 Stages of lung growth: vascular development

Stage	Events
Embryonic (<6 wk)	Vasculogenesis within immature mesenchyme; Pulmonary arteries branch form 6th aortic arches; veins as outgrowths from left atrium.
Pseudoglandular (<16 wk)	Parallel branching of large pulmonary arteries with central airways; lymphatics appear.
Canalicular (<24 wk)	Increased vessel proliferation and organization into capillary network around airspaces.
Saccular (<36 wk)	Marked vascular expansion with thinning and (<36 wk) condensation of mesenchyme; thin air-blood barrier; double capillary network in septae.
Alveolar (<2–3 yrs)	Accelerated vascular growth, fusion of the double capillary network with thinning of septae.
Postnatal (3 years to adulthood)	Marked vessel growth + remodeling, as surface area increases >20-fold.

become angioblasts, which are endothelial precursor cells that have not yet been incorporated into vessels. Relatively little is known concerning the exact origins of angioblasts, but endothelial differentiation occurs within the mesoderm. After the assembly of free angioblasts into cords, angioblasts differentiate into endothelial cells and form tubes (vasculogenesis). Angioblasts can migrate to distant sites and then organize into endothelial cells and capillaries. Endothelial cells were once considered homogeneous, but organ-related differences have now been found. Endothelial cell heterogeneity accounts for key functional differences in angiogenic responses and vascular functional development, and is likely to play important roles in key differences in postnatal function.[3] However, developmental mechanisms that contribute to basic functional differences between endothelial cells from different vascular beds, such as the divergent hypoxic responses of pulmonary and systemic arteries, remain uncertain.

Controversies exist regarding the relative roles of vasculogenesis and angiogenesis during early lung development, generating at least three alternate hypotheses.[4–6] First, deMello proposed concurrent roles for vasculogenesis and angiogenesis in which proximal pulmonary arteries and veins expand from endothelial proliferation from existing vessels, while the distal microvasculature develops from blood islands within the mesenchyme due to differentiation

of endothelial cells from immature precursors.[4] Alternatively, Hall et al suggest a more prominent role for vasculogenesis in which intrapulmonary vessels are derived from continuous expansion of the primary capillary plexus within the developing mesenchyme.[5] Finally, Parera et al suggest that distal angiogenesis plays a more prominent role, as capillary networks surrounding the terminal buds expand by formation of new capillaries from pre-existing vessels along with the expanding airway.[6] Recent studies have generated interest regarding the potential role for endothelial progenitor cells (EPC) during lung vascular development; however, data that clearly demonstrate the involvement of EPC in this process remain lacking.[7]

Multiple molecules contribute to angiogenesis in the developing lung.[8] Several growth factors have been shown to play important autocrine and paracrine roles in vascular development, including basic fibroblast growth factor (bFGF), transforming growth factor-beta (TGF-ß), vascular endothelial growth factor (VEGF), hepatocyte growth factor (HGF), and platelet derived growth factor (PDGF).[2,8,9] Characterization of their precise roles and the complex interactions between these signals in the regulation of endothelial cell differentiation, migration, and proliferation are unclear, but this field is rapidly expanding.[5–7,9,10] For example, bFGF induces early expression of the VEGF receptor, VEGF receptor-2 (VEGFR-2; or, KDR/flk-1), one of the earliest markers of the endothelial cell lineage.[4,11] Subsequent assembly of angioblasts within the mesoderm is likely due to stimulation of VEGFR-2 by its ligand, VEGF. The importance of paracrine signaling between VEGF, produced largely from airway epithelium, and its receptors (VEGFR-1 and VEGFR-2), as expressed by early endothelial cells, is best demonstrated by the marked vascular disruption and embryonic lethality in VEGF or VEGFR-deficient mice.[12–14] Signaling between airway epithelium and the immature mesenchyme provides important "crosstalk" that coordinates vascular development, with airspace growth. During branching morphogenesis in the embryo and early fetus, signals from the mesenchyme are critical for normal epithelial growth and function, such as type II cell differentiation, morphological development, and surfactant protein C expression.[15] In parallel, early vascular development within the rat embryonic lung mesenchyme is dependent on epithelial-derived products, such as VEGF, bFGF, TGF-ß, and others.[15–17]

Once the early endothelial tubes are formed, vessel growth is extended by angiogenesis due to sprouting and nonsprouting mechanisms (such as vessel fusion and "intussusception").[1,2,18] Vascular development involves overlapping mechanisms that are not mutually exclusive, and are clearly dependent upon timing. For example, another role of VEGF-VEGFR-2 signaling that occurs later in vascular development is to alter endothelial cell behavior in order to promote capillary fusion, which is distinct from its mitogenic effects.[19] In addition, VEGF continues to function throughout postnatal life as an endothelial "maintenance" or "survival factor," thereby sustaining normal vascular function through the expression of such key enzymes as endothelial nitric oxide synthase (eNOS) and prostacyclin synthase.[20] Disruption of VEGF signaling shortly before birth in the late gestation fetus impairs lung endothelial function, blunts vasodilation, and causes perinatal pulmonary hypertension.[21] Furthermore, inhibition of VEGF receptors during infancy inhibits vascular growth, causes pulmonary hypertension, and impairs alveolarization, which persists into adulthood.[22] Thus, growth factors such as VEGF, which are essential during embryonic lung development, continue to play an important role in preserving lung vascular function and maintaining lung architecture later in life.

Assembly and maturation of vessels require endothelial signals to recruit mesenchymal cells around the vessel, which later differentiate into mature smooth muscle cells.[5,23] Early signaling "loops" between endothelial cells and mesenchymal cells appear to involve the endothelial production of chemotactic factors, such as PDGF or heparin-binding epidermal growth factor, which cause migration of mesenchymal cells toward the developing vessel. This response is partly dependent upon the production of angiopoietin-1 by neighboring mesenchymal cells, which activates the endothelial TIE2 receptor and stimulates the release of migratory stimuli.[24] Subsequent maturation of the vessel wall involves ongoing production of growth factors and signaling molecules, which lead to smooth muscle cell proliferation, maturation of contractile proteins, and formation of extracellular matrix. As with endothelial cells, marked heterogeneity in smooth muscle cells exists within the developing vessel wall, and these may persist postnatally or with disease states.[25]

In addition to numerous growth factors and transcription factors, components of the extracellular matrix (ECM) are continually being found to play a critical role in endothelial growth, survival, migration, differentiation, and morphogenesis, as well as the regulation of growth factor release and activity.[26–29] In the embryo, early smooth muscle and fibroblast-mediated production of extracellular matrix proteins regulates vascular structure and growth.[27–30] The major structural ECM proteins are signaled through hemodynamic stress to provide vascular strength and compliance and are produced during a relatively narrow developmental window around birth. Extracellular matrix proteins contribute to: the initial assembly and organization of the vessel wall; cell proliferation; adventitial growth; smooth muscle cell phenotype; vascular compliance; and controlling the activities of diverse vasoactive mediators, growth factors, and cytokines.[28] Abnormal production or degradation of ECM components can markedly impair vessel growth, structure, and function leading to a wide range of pulmonary vascular diseases.[30]

Ongoing vascular growth and development continues during late gestation and early postnatal life. During the period of rapid alveolarization during infancy, the lung undergoes marked vascular growth, as reflected by the 20-fold increase in alveolar and capillary surface areas from birth to early childhood. Vascular growth and normal remodeling during infancy and childhood remains critical for the development and maintenance of normal lung architecture later in life. Based on strong experimental data, it has been speculated that early fetal events that disrupt angiogenesis may lead to sustained structural abnormalities of vascular and airspace growth, increasing susceptibility for late diseases such as pulmonary hypertension and emphysema.[31]

Although most studies examine growth of conventional pulmonary arteries and veins, little is known regarding the development of supernumerary vessels during fetal life, as well as anastomotic networks that connect pulmonary and bronchial vessels.[32,33] The existence of intrapulmonary arterial-venous connections (IAVC) has been established by past studies utilizing differential injection methods, serial histologic sectioning, and fluorescent bead injection in human subjects and animal models.[32,33] Physiologic demonstration of intrapulmonary shunting has been shown by echocardiography studies in fetal sheep, which are less prominent in the late neonatal period.[34] Late persistence or enhancement of IAVC has been previously reported in pathologic settings as neonatal heart disease requiring cavopulmonary shunt operations. More recently, the presence of striking IAVC, or "shunt" vessels, has been confirmed by histological 3-D reconstruction methods in lung autopsy tissue from preterm infants dying with severe chronic lung disease (bronchopulmonary dysplasia) and alveolar capillary dysplasia with misalignment of pulmonary veins.[35,36] As shown in Figure 1, prominent, thin walled, and centrally located abnormal intrapulmonary vessels connect small pulmonary arteries and veins with microvascular plexuses surrounding pulmonary arteries and airways. Vessels with such distinct appearance and course were not present in lungs of age-matched controls. We speculate that persistence of these "fetal shunt vessels" may contribute to the pathophysiology of neonatal lung diseases (Figure 1) and perhaps intrapulmonary shunt throughout adulthood, as previously noted.[37] Mechanisms that regulate the growth and development of IAVC during development and with disease remain unknown.

Physiology of the Fetal Pulmonary Circulation

Along with the remarkable progression of lung vascular growth and structure during development, the fetal pulmonary circulation also undergoes maturational changes in function. Pulmonary vascular resistance (PVR) is high throughout fetal life, especially in comparison with the low resistance of the systemic circulation. As a result, the

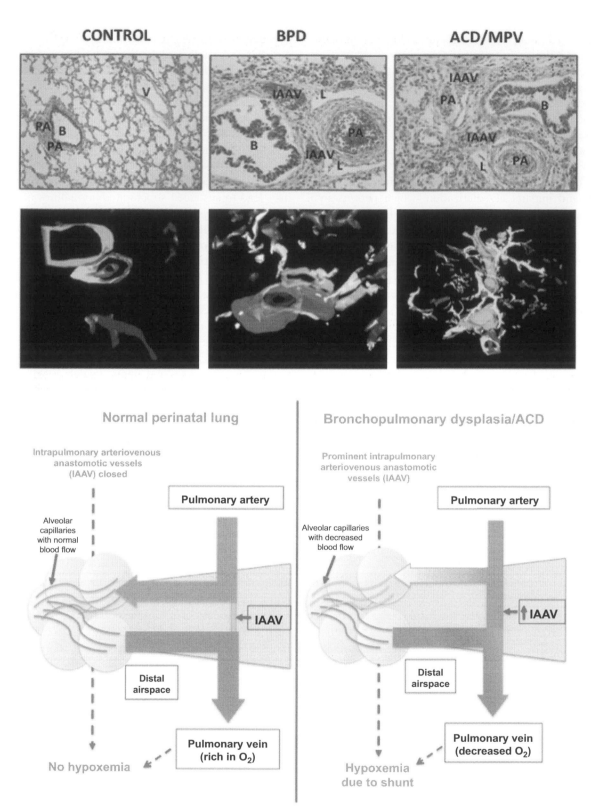

FIGURE 1 Intrapulmonary arteriovenous vessels (IAV) in the developing lung. Histologic sections show the presence of IAAV in severe BPD and ACD/MPV patients (upper panel). Three-dimensional reconstruction confirms that IAAV connects pulmonary veins with systemic microvessels surrounding airways and pulmonary arteries (endothelial cells; smooth muscle; lymphatics) in BPD and ACD lungs but not in age-matched newborn controls. Lower panel: a schematic illustration demonstrating the potential role of IAAV in shunt from BPD infants. (This figure is reproduced in color in the color plate section.) *(from Reference[35])*.

fetal sheep lung receives only 3–8% of combined ventricular output, with most of the right ventricular output crossing the ductus arteriosus to the aorta. Pulmonary artery hemodynamic stress (pressure and blood flow) progressively increases with advancing gestational age, along with increasing lung vascular growth.[38,39] Despite this increase in vascular surface area, PVR actually increases with gestational age when adjusted for lung or body weight; that is, pulmonary vascular tone increases during late gestation, especially prior to birth. Studies of the human fetus support these observations from fetal sheep.[40] Based on multiple Doppler ultrasound measurements that include assessments of human fetal left and distal pulmonary artery velocity waveforms, it has been demonstrated that pulmonary artery impedance progressively decreases during the second and early part of the third trimester. Pulmonary artery vascular impedance does not decrease further during the latter stage of the third trimester despite ongoing vascular growth.[40,41]

As shown in Table 2, several mechanisms contribute to high basal PVR in the fetus, including mechanical factors and chemical stimuli such as low O_2 tension, low basal production of vasodilator products (such as PgI_2 and NO), increased production of vasoconstrictors (including endothelin-1 (ET-1) or leukotrienes), and altered smooth muscle cell reactivity (such as enhanced myogenic tone).[42–51] The fluid-filled airspace in the fetal lung likely plays an important

role maintaining high lung volumes in utero, creating high vascular extra-luminal pressure, which further contributes to high PVR. Furthermore, cyclic stretch induced by breathing movements has been observed to reduce increased pulmonary blood flow, decreasing PVR and phasic reductions in intrapulmonary pressures, and are necessary for normal growth and structural maturation. Shear stress has been implicated as an important mechanical mechanism perhaps leading to endothelial cell maturation through stimulating eNOS increasing transcription activity of c-Jun mediate by protein kinase C (PKC).[52]

The fetal pulmonary circulation is characterized by progressive changes in responsiveness to vasoconstrictor and vasodilator stimuli (i.e., vasoreactivity). In the ovine fetus, the pulmonary circulation is initially poorly responsive to vasoactive stimuli during the early canalicular period, and responsiveness to several stimuli increases during late gestation. For example, the pulmonary vasoconstrictor response to hypoxia, and the vasodilator response to increased fetal PO_2 and acetylcholine increase with gestation (Figure 2).[53–55] As observed in the sheep fetus, human studies also demonstrate maturational changes in the fetal pulmonary vascular response to increased PaO_2.[40] Maternal hyperoxia does not increase fetal pulmonary blood flow between 20 and 26 weeks gestation, but increased PaO_2 caused pulmonary vasodilation in the 31–36 week fetus. These findings suggest that, in addition to structural maturation and growth of the fetal pulmonary circulation, the vessel wall also undergoes functional maturation, leading to enhanced vasoreactivity.

Mechanisms that contribute to progressive changes in pulmonary vasoreactivity during development are uncertain, but are likely due to maturational changes in endothelial cell function, especially with regard to NO production.[56–59] Figure 3 illustrates the NO-cGMP signaling cascade in the fetal lung. Lung endothelial NOS (eNOS, type III) mRNA and protein is present in the early fetus and increases with advancing

---

TABLE 2 Factors contributing to high pulmonary vascular resistance in the normal fetus

1. Mechanical (presence of fetal lung liquid, lack of air-liquid interface)

2. Structural (e.g., decreased surface area, vessel wall compliance)

3. Functional:
   a. Decreased dilator activity:
      - Nitric oxide
      - Adenosine
      - Prostacyclin
      - Adrenomedullin
      - Atrial natriuretic peptide
      - Endothelium-derived hyperpolarization factor (EDHF)
   b. Increased constrictor stimuli:
      - Endothelin-1
      - Serotonin
      - Leukotrienes C4, D4
      - Endothelium-derived constricting factors (EDCF)
   c. Altered smooth muscle cell signaling:
      - High myogenic tone/responses
      - High cGMP PDE5 activity
      - Increased rho kinase activity
      - Altered calcium handling
      - Impaired K^+ – channel expression, function

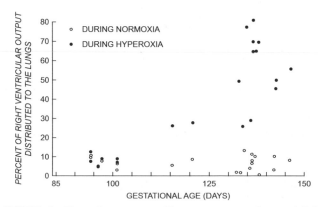

FIGURE 2 The pulmonary vasodilator response to increased PO_2 increases with gestational age in fetal sheep. Despite achieving similar elevations in arterial PO_2, the proportion of right ventricular output to the lung remains low until late in gestation. *(from Reference[54]).*

FIGURE 3 Schematic diagram illustrating the NO-cGMP signaling cascade in the fetal lung.

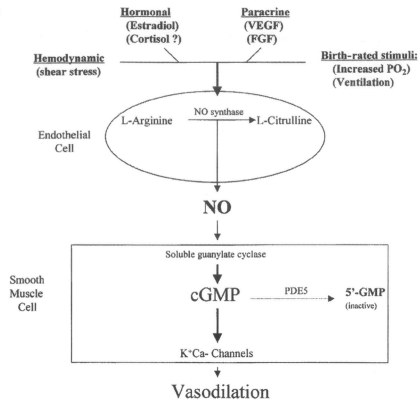

gestation in utero and during the early postnatal period in rats and sheep.[60–62] The timing of this increase in lung eNOS content immediately precedes and parallels changes in the capacity to respond to endothelium-dependent vasodilators, as shown by in vivo and in vitro studies (Figure 4).[58,63] This increase in lung endothelial NOS content coincides with the capacity to respond to endothelium-dependent vasodilator stimuli, such as oxygen and acetylcholine.[53,54] In contrast, fetal pulmonary arteries are already quite responsive to exogenous NO much earlier in gestation[53–68] These findings suggest that the ability of the endothelium to produce or sustain production of NO in response to specific stimuli during maturation lags behind the capacity of fetal pulmonary smooth muscle to relax to NO. This may account for clinical observations that extremely preterm neonates are highly responsive to inhaled NO.[64]

Although most studies of the perinatal lung have focused on the role of eNOS in vasoregulation, the other NOS isoforms, including neuronal NOS (nNOS; type I) and inducible NOS (iNOS; type II), have been identified by immunostaining in the rat, sheep, and human fetal lung.[66–68] Lung nNOS mRNA and protein increase in parallel with eNOS expression during development in the fetal rat.[61] Inducible (Type II) NOS has also been detected in the ovine fetal lung, and is predominantly expressed in airway epithelium and vascular smooth muscle, with little expression in vascular endothelium.[67] Whether the "non-endothelial" (types I and II) isoforms contribute to the physiologic responses of NO-dependent modulation of fetal pulmonary vascular tone has been controversial. Treatment of pregnant rats with an iNOS-selective antagonist caused constriction of the great vessels

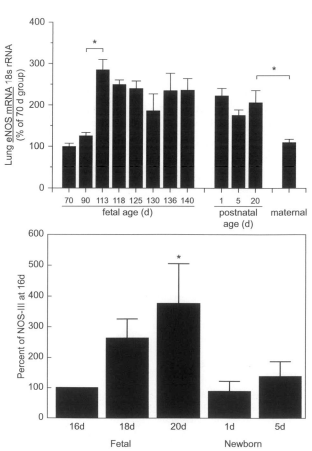

FIGURE 4 Maturational changes in endothelial NO synthase (type III) expression in the ovine fetal lung (*panel a; from Reference*[61]) and the rat lung. (*panel b; from Reference*[62]).

(main pulmonary artery and thoracic aorta) and ductus arteriosus in fetal rats.[65] Selective iNOS and nNOS antagonists increase fetal PVR and inhibit shear stress-induced vasodilation at doses that do not inhibit acetylcholine-induced pulmonary vasodilation.[65,67,69,70] These findings support the speculation that iNOS and nNOS may also modulate pulmonary vascular tone in utero and at birth (see following text).[71]

NOS expression and activity are affected by multiple factors including oxygen tension, hemodynamic forces, hormonal stimuli (e.g., estradiol), paracrine factors (including VEGF), substrate and cofactor availability, superoxide production (which inactivates NO), and others.[72–76] Past studies have shown that estradiol acutely releases NO and upregulates eNOS expression in fetal pulmonary artery endothelial cells.[75] Although estradiol does not cause acute pulmonary vasodilation in vivo, prolonged estradiol treatment causes late vasodilation that is sustained despite cessation of estradiol infusion.[73,75] In contrast, VEGF acutely releases NO and causes abrupt pulmonary vasodilation in vivo.[77] Chronic inhibition of VEGF receptors downregulates eNOS and induces pulmonary hypertension in the late gestation fetus.[21] Thus, diverse hormonal and paracrine factors can regulate NOS expression and activity during development.

Vascular responsiveness to endogenous or exogenous NO is also dependent upon several smooth muscle cell enzymes, including soluble guanylate cyclase, cGMP-specific (types V) phosphodiesterase (PDE5), and cGMP kinase.[78–81] Several studies have shown that soluble guanylate cyclase, which produces cGMP in response to NO activation, is abundant in the late fetal and early neonatal rat lung, with markedly reduced levels detected in the adult lung. Similarly, PDE5, which limits cGMP-mediated vasodilation by hydrolysis and inactivation of cGMP, is also normally active in utero[82,83] In the fetal lung, PDE5 expression has been localized to vascular smooth muscle, and PDE5 activity is high in comparison with the postnatal lung.[81] Infusions of selective PDE5 antagonists cause potent fetal pulmonary vasodilation. Thus, PDE5 activity appears to play a critical role in pulmonary vasoregulation during the perinatal period, and must be taken into account when assessing responsiveness to endogenous NO and related vasodilator stimuli.

Functionally, the NO-cGMP cascade (Figure 3) plays several important physiologic roles in vasoregulation of the fetal pulmonary circulation.[72] These include: (1) modulation of basal PVR[42]; (2) mediating the vasodilator response to specific physiologic and pharmacologic stimuli[42–44]; and (3) opposing strong myogenic tone in the normal fetal lung.[50] Studies in fetal sheep have demonstrated that intrapulmonary infusions of NO synthase inhibitors increase basal PVR by 35%.[42] Because inhibition of NO synthase increases basal PVR at least as early as 0.75 gestation (112 days) in fetal sheep, endogenous NOS activity appears to contribute to vasoregulation throughout late gestation.[68] NOS inhibition also selectively blocks pulmonary vasodilation to such stimuli as acetylcholine, oxygen, and shear stress in the normal fetus.[42,44,50,84]

In addition to high PVR and altered vasoreactivity during development, the fetal pulmonary circulation is further characterized by its ability to oppose sustained pulmonary vasodilation during prolonged exposure to vasodilator stimuli. For example, increased PaO_2 increases fetal pulmonary blood flow during the first hour of treatment; however, blood flow returns toward baseline values over time despite maintaining high PaO_2 (Figure 5).[59] Similar responses are observed during acute hemodynamic stress

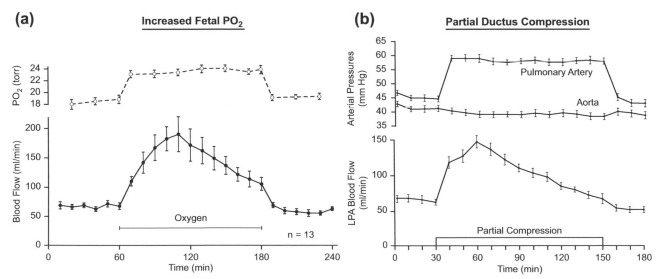

FIGURE 5   (a) Effect of increased PO_2 on pulmonary blood flow in the sheep fetus. As shown, pulmonary blood flow initially increases with the rise in PO_2, but subsequently falls towards baseline values. (b) Time-dependent response of pulmonary blood flow during partial compression of the ductus arteriosus. As shown, ductus compression increases pulmonary artery pressure (PAP) and blood flow, but despite constant PAP, blood flow returns to basal levels over time. *(from References*[56,59]*).*

(shear stress) caused by partial compression of the ductus arteriosus[56] or with infusions of several pharmacologic agents.[57] These findings suggest that unique mechanisms exist in the fetal pulmonary circulation that oppose vasodilation and maintain high PVR in utero. We have speculated that this response reflects the presence of an augmented myogenic response within the fetal pulmonary circulation.

Recent studies have suggested that NO release plays an additional role in modulating high intrinsic or myogenic tone in the fetal pulmonary circulation. The myogenic response is commonly defined by the presence of increased vasoconstriction caused by acute elevation of intravascular pressure or "stretch stress."[85] Previous in vitro studies demonstrated the presence of a myogenic response in sheep pulmonary arteries, and that fetal pulmonary arteries have greater myogenic activity than neonatal or adult arteries.[86,87] More recent studies of intact fetal sheep have demonstrated that high myogenic tone is normally operative in the fetus and contributes to maintaining high PVR in utero.[56–59] These studies demonstrate that NOS inhibition not only blocks vasodilation to several physiologic stimuli, but acute inhibition of NO production unmasks a potent myogenic response (Figure 5).[50] Thus, NOS inhibition unmasks a potent myogenic response that maintains high PVR in the normal fetus. Down-regulation of NO activity, as observed in experimental pulmonary hypertension, augments myogenic activity and increases the risk for unopposed vasoconstriction in response to stretch stress at birth.

Because eNOS protein is present at a stage of lung development when blood flow is absent or minimal, it has been hypothesized that NO may potentially contribute to angiogenesis during early lung development.[72] Whether early eNOS expression implies a role in promoting vascular growth or is merely a marker of growing endothelial cells is unknown. Recent studies report conflicting data regarding the effects of eNOS activity in promoting new vessel formation in different experimental models of angiogenesis. Although NO can inhibit endothelial cell mitogenesis and proliferation, it has also been shown to mediate the angiogenic effects of substance P and vascular endothelial growth factor in vitro.[88,89] In culture, growing bovine aortic endothelial cells express more eNOS mRNA and protein than confluent cells, but NOS inhibition does not affect their rate of proliferation in vitro.[90] NO has also been shown to decrease smooth muscle proliferation in vitro,[91,92] but NO may have biphasic, dose-dependent effects on the growth of fetal pulmonary artery smooth muscle cells. High doses of NO donors inhibit smooth muscle cell growth, but low doses cause paradoxical stimulation in vitro. Whether NO modulates smooth muscle cell growth in vivo remains controversial; one study reported the failure of chronic NOS inhibition to alter pulmonary vascular structure during late gestation in the fetal sheep.[93] Thus, although multiple studies have examined the role of NO in vascular growth and

remodeling, its effects vary between studies, and the effects of NO on angiogenesis and structure of the vessel wall remain controversial.

Although other vasodilator products, including prostacyclin (PgI_2), are released upon stimulation of the fetal lung (e.g., by increased shear stress), basal prostaglandin release appears to play a less important role than NO in fetal pulmonary vasoregulation. For example, cyclooxygenase inhibition has minimal effect on basal PVR and does not increase myogenic tone in the fetal lamb.[94] The physiologic roles of other dilators, including adrenomedullin, adenosine, and endothelium-derived hyperpolarizing factor (EDHF), are uncertain. EDHF is a short-lived product of cytochrome P450 activity that is produced by vascular endothelium, and has been found to cause vasodilation through activation of calcium-activated K^+ channels in vascular smooth muscle in vitro.[95] K^+-channel activation appears to modulate basal PVR and vasodilator responses to shear stress and increased oxygen tension in the fetal lung, but whether this is partly related to EDHF activity is unknown.[45,96]

Carbon monoxide (CO) is a gaseous molecule produced by heme-oxygenase and has been shown to have several vascular effects, including vasodilation in the adult systemic and pulmonary vascular beds.[97] CO may act in part through activation of soluble guanylate cyclase, increasing cGMP content in vascular smooth muscle, and causing vasodilation, as described for NO.[98] Despite several studies that suggest an important role in vasoregulation, CO has yet to be shown to modulate vascular tone or growth in the perinatal lung. For example, inhaled CO treatment of late-gestation fetal sheep had no effect on PVR, and infusions of a heme-oxygenase inhibitor did not alter basal pulmonary vascular tone.[99] Further studies are needed to demonstrate its physiologic importance in the developing lung circulation.

Vasoconstrictors have long been considered as potentially maintaining high PVR in utero. Several candidates, including lipid mediators (thromboxane A_2, leukotrienes C_4 and D_4, and platelet-activating factor) and endothelin-1 (ET-1), have been extensively studied. Thromboxane A_2, a potent pulmonary vasoconstrictor that has been implicated in animal models of Group B Streptococcal sepsis, does not appear to influence PVR in the normal fetus. In contrast, inhibition of leukotriene production causes fetal pulmonary vasodilation[49]; however, questions have been raised regarding the specificity of the antagonists used in these studies.

Endothelin-1 (ET-1), a potent vasoconstrictor and comitogen that is produced by vascular endothelium, has been demonstrated to play a key role in fetal pulmonary vasoregulation (Figure 6).[46,100–102] In the human lung, ET-1 expression is greater during the saccular and alveolar stages than in the earlier canalicular stage. Although ET-1 causes an intense vasoconstrictor response in vitro, its effects in the intact pulmonary circulation are complex. Brief infusions of ET-1 cause transient vasodilation, but PVR progressively

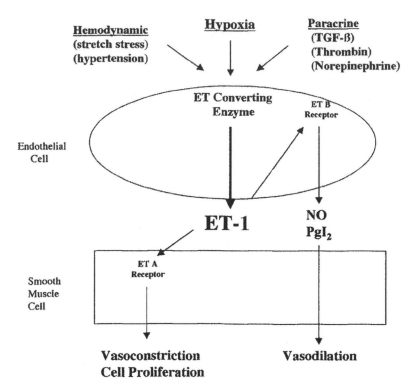

FIGURE 6   Schematic diagram illustrating the Endothelin-1 (ET-1) signaling cascade in the developing lung circulation.

increases during prolonged treatment.[103] The biphasic pulmonary vascular effects during pharmacologic infusions of ET-1 are explained by the presence of at least two different ET receptors, ET A and ET B. The ET B receptor, localized to the endothelium in the sheep fetus, mediates the ET-1 vasodilator response through the release of NO.[46,101,103–105] The ET A receptor is located on vascular smooth muscle, and when activated, causes marked constriction. Although capable of both vasodilator and constrictor responses, ET-1 is more likely to play an important role as a pulmonary vasoconstrictor in the normal fetus. This is suggested in extensive fetal studies showing that inhibition of the ET A receptor decreases basal PVR and augments the vasodilator response to shear stress-induced pulmonary vasodilation.[104,105] Thus, ET-1 modulates PVR through the ET A and B receptors, but its predominant role is as a vasoconstrictor through stimulation of the ET A receptor.

CONTROL OF THE DUCTUS ARTERIOSUS

As most of the right ventricular output crosses the ductus arteriosus (DA) in utero, patency of the DA is vital for fetal survival and well-being. Premature DA closure in utero causes severe pulmonary hypertension, congestive heart failure, hydrops fetalis, or severe hypoxemia. In contrast, an inability of the DA to close after birth may complicate lung disease in the premature newborn with respiratory distress syndrome or cause high flow pulmonary vascular injury during postnatal life. In addition, maintaining DA patency can be critical for survival in newborns and infants with ductus-dependent

cyanotic congenital heart disease. Finally, insights into the unique nature of regulation of the DA, especially with regard to smooth muscle cell tone, proliferation, and synthetic functions, provide important lessons in vascular biology. For example, changes in PO_2 have striking effects on DA smooth muscle that are unique from its neighboring smooth muscle cells in systemic (aortic) and pulmonary circulations. Low PO_2 constricts pulmonary vessels but dilates the DA; conversely, the increase in PO_2 at birth contributes to the fall in PVR but paradoxically constricts the DA.

Past studies have shown the important role of intramural prostaglandin production on DA tone, as evidenced by the potent constrictor effects of cyclooxygenase inhibitors. However, recent studies have shown that regulation of the DA is complex and involves multiple signals, such as ET-1, carbon monoxide (CO), the cytochrome P450 system, and K^+ channel activities.[106–108] Recent in vitro data suggest that a cytochrome P450-based mono-oxygenase reaction transduces the signal for DA closure, and that DA constriction is mediated through increased ET-1 production and stimulation of the ET A receptor. Furthermore, CO can relax the DA, but its effects vary according to ambient oxygen tension, and CO may act in part by inhibiting ET-1 production. Other studies have demonstrated that ion channels mediate early changes in DA tone, primarily through voltage-gated K^+ channels. Decreased oxygen tension inhibits voltage-gated K^+ channels, opening Ca^{2+} channels and causing constriction in pulmonary artery smooth muscle cells; in fetal DA, oxygen activates a calcium-activated K^+ channel and causes vasodilation. ET may act by inhibiting voltage-activated K^+ channels.[107,109]

Mechanisms of Pulmonary Vasodilation at Birth

Within minutes of birth, pulmonary artery pressure falls and blood flow increases in response to birth-related stimuli. Mechanisms contributing to the fall in PVR at birth include establishment of an air–liquid interface, rhythmic lung distension, increased oxygen tension, and altered production of vasoactive substances. Physical mechanical stimuli, such as increased shear stress, ventilation, and increased oxygen, cause pulmonary vasodilation in part by increasing production of vasodilators, NO, and PgI_2.[42–45,48–51] Pretreatment with the arginine analogue, nitro-L-arginine, blocks NOS activity, and attenuates the decline in PVR after delivery of near term fetal sheep (Figure 7).[42] These findings suggest that about 50% of the rise in pulmonary blood flow at birth may be directly related to the acute release of NO. Specific mechanisms that cause NO release at birth include the marked rise in shear stress, increased oxygen, and ventilation.[44] Increased PaO_2 triggers NO release, which augments vasodilation through cGMP kinase-mediated stimulation of K^+ channels.[45,109–111] Although the endothelial isoform of NO synthase (type III) has been presumed to be the major contributor of NO at birth, recent studies suggest that other isoforms (inducible (type II) and neuronal (type I)) may be important sources of NO release in utero and at birth as well.[65,68–71] Although early studies were performed in fetuses at term, NO also contributes to the rapid decrease in PVR at birth in preterm lambs, at least as early as 112–115 days (0.7 term).[63]

Other vasodilator products, including PgI_2, also modulate changes in pulmonary vascular tone at birth.[48,51] Rhythmic lung distension and shear stress stimulate both PgI_2 and NO production in the late gestation fetus, but increased PO_2 triggers NO activity and overcomes the effects of prostaglandin inhibition at birth. In addition, the vasodilator effects of exogenous PgI_2 are blocked by NO synthase inhibitors, suggesting that NO modulates PgI_2 activity in the perinatal lung.[94] Adenosine release may also contribute to the fall in PVR at birth, but its actions may be partly through enhanced production of NO.[112]

Thus, although NO does not account for the entire fall in PVR at birth, NOS activity appears important in achieving postnatal adaptation of the lung circulation. Transgenic eNOS knock-out mice successfully make the transition at birth without evidence of PPHN.[113,114] This finding suggests that eNOS -/- mice may have adaptive mechanisms, such as a compensatory vasodilator mechanisms (such as upregulation of other NOS isoforms or dilator prostaglandins) or less constrictor tone. Interestingly, these animals are more sensitive to the development of pulmonary hypertension at relatively mild decreases in PaO_2[113,114] and have higher neonatal mortality when exposed to hypoxia after birth. We speculate that isolated eNOS deficiency alone may not be sufficient for the failure of postnatal adaptation, but that decreased ability to produce NO in the setting of a perinatal stress (such as hypoxia, inflammation, hypertension, or upregulation of vasoconstrictors) may cause PPHN.

Although these studies were performed in term animals, similar mechanisms also contribute to the rapid decrease in PVR at birth in preterm lambs.[63,115] The pulmonary vasodilator responses to ventilation with hypoxic gas mixtures (or rhythmic distension) of the lung or increased PaO_2 are partly due to stimulation of NO release in preterm lambs at least as early as 112–115 days (0.7 term).[63] Other vasodilator products, including PgI_2, also modulate changes in pulmonary vascular tone at birth.[47–51] Rhythmic lung distension and shear stress stimulate both PgI_2 and NO production in the late gestation fetus; increased PO_2 triggers NO activity but does not appear to alter PgI_2 production in vivo.

Mechanisms That Cause Failure of Pulmonary Vasodilation at Birth

Some newborns fail to achieve or sustain the normal decline in PVR after birth, which constitutes the clinical syndrome known as PPHN.[116,117] As a clinical syndrome, PPHN includes diverse cardiac and pulmonary disorders, or occurs as an idiopathic disorder, in the absence of significant cardiac or pulmonary disease. Although these diverse diseases have features that are distinct from each other, they are generally included within this clinical syndrome because they share a common pathophysiologic feature, i.e., high PVR leading to right-to-left shunting of blood across the ductus arteriosus or foramen ovale and marked hypoxemia. Despite multiple therapeutic strategies, morbidity and mortality in neonates with severe PPHN remain high. Although cardiac and lung dysfunctions may contribute to the clinical course

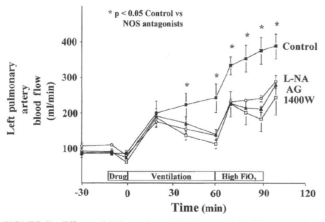

FIGURE 7   Effects of NO synthase inhibition at birth. Near term fetal lambs were delivered by cesarean section and ventilated with low (10%) and high (100%) oxygen tensions. As shown, the non-selective NOS antagonist, L-NA, impaired the rise in blood flow during delivery, but the effects were not different from the type II antagonists, aminoguanidine (AG) and 1400W. *(from Reference[71])*.

of PPHN, abnormalities of the pulmonary circulation are its critical features. PPHN is characterized by altered pulmonary vascular reactivity, structure, and in some cases, growth.[118]

Autopsy studies of the lungs of newborns with fatal PPHN have revealed severe hypertensive structural remodeling even in newborns who die shortly after birth, suggesting that many cases of severe disease are associated with chronic intrauterine stress (Figure 8).[118–120] However, the exact intrauterine events that alter pulmonary vascular reactivity and structure are poorly understood. Epidemiologic studies have demonstrated strong associations between PPHN and maternal smoking and ingestion of cold remedies that include aspirin or other non-steroidal anti-inflammatory products.[121] Because these agents can induce partial constriction of the ductus arteriosus (DA), it is possible that pulmonary hypertension due to DA narrowing contributes to PPHN (see following text).[122–124] Other perinatal stresses, including chorioamnionitis, placental vascular lesions, and intrauterine growth restriction, are associated with PPHN[125]; however, most neonates who are exposed to these prenatal stresses do not develop PPHN. Recent epidemiologic studies suggest that maternal use of antidepressants may increase the risk for PPHN, which has been further supported by animal studies.[126]

Circulating levels of L-arginine, the substrate for NO, are decreased in some newborns with PPHN, suggesting that impaired NO production may contribute to the pathophysiology of PPHN, as observed in experimental studies.[127–129] It is possible that genetic factors increase susceptibility for pulmonary hypertension. A recent study reported strong links between PPHN and polymorphisms

of the carbamoyl phosphate synthase gene[130]; however, the importance of this finding is uncertain. Studies of adults with idiopathic primary pulmonary hypertension have identified abnormalities of bone morphogenetic protein (BMP) receptor genes[131]; whether polymorphisms of genes for the BMP or TGF-ß receptors, other critical growth factors, vasoactive substances or other products increase the risk for some newborns to develop PPHN is unknown.

Several experimental models have been used to explore the pathogenesis and pathophysiology of PPHN.[131,132] Such models have included exposure to acute or chronic hypoxia after birth, chronic hypoxia in utero, placement of meconium into the airways of neonatal animals, sepsis, and others. Although each model demonstrates interesting physiologic responses that may be relevant to particular clinical settings, most studies examine only brief changes in the pulmonary circulation, and mechanisms underlying altered lung vascular structure and function of PPHN remain poorly understood. Clinical observations that neonates with severe PPHN who die during the first days after birth already have pathologic signs of chronic pulmonary vascular disease suggesting that intrauterine events may play an important role in this syndrome.[118–120] Adverse intrauterine stimuli during late gestation, such as abnormal hemodynamics, changes in substrate or hormone delivery to the lung, hypoxia, inflammation, or others, may potentially alter lung vascular function and structure, contributing to abnormalities of postnatal adaptation. Several investigators have used animal models to examine the effects of chronic intrauterine stresses, such as hypoxia or hypertension, in an attempt to mimic PPHN. Whether chronic hypoxia alone can cause PPHN is controversial. Although a rodent study

Pathogenesis of PPHN

FIGURE 8 Schematic illustration of pathogenetic mechanisms underlying PPHN.

Prenatal Factors

- Maternal NSAID, SSRI use
- Premature closure of the DA
- C-section delivery
- Post-term (> 41 weeks)
- Large for gestational age
- Abnormal placenta
- Altered lung development
- Cardiovascular abnormalities

Postnatal Factors

- Hyperoxia/oxidative stress
- Ventilator-induced injury
- Asphyxia
- Inflammation/Infection

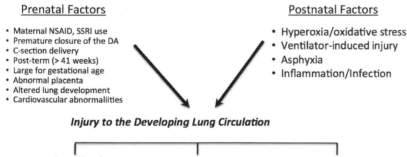

Injury to the Developing Lung Circulation

Impaired Vasoreactivity   Decreased Angiogenesis   Altered Vascular Structure

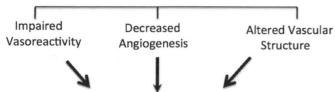

Persistent Pulmonary Hypertension of the Newborn
- Failure to decrease PVR at birth
- Extra-pulmonary shunting across DA, PFO
- Severe hypoxemia

showed that maternal hypoxia increases pulmonary vascular smooth muscle thickening in newborns,[133] this observation has not been reproduced by more extensive studies in maternal rats or guinea pigs.[134]

Animal studies suggest that, in contrast to hypoxia, intrauterine hypertension due to either renal artery ligation or constriction of the ductus arteriosus can cause structural and physiologic changes that resemble features of clinical PPHN.[122-124] Pulmonary hypertension induced experimentally by early closure of the DA in fetal sheep alters lung vascular reactivity and structure, causing the failure of postnatal adaptation at delivery and providing a model of PPHN. Over days, pulmonary artery pressure and PVR progressively increase, but flow remains low and PaO_2 is unchanged.[123] Marked right ventricular hypertrophy and structural remodeling of small pulmonary arteries develop after 8 days of hypertension. After delivery, these lambs have persistent elevation of PVR despite mechanical ventilation with high oxygen concentrations.

Studies with this model show that chronic hypertension without high blood flow impairs fetal lung vascular structure and function. This model is characterized by endothelial cell dysfunction and altered smooth muscle cell reactivity and growth, including decreased NO production and activity due to down-regulation of lung endothelial NO synthase mRNA and protein expression.[134-138] Fetal pulmonary hypertension also decreases soluble guanylate cyclase activity and up-regulated cGMP-specific phosphodiesterase (type 5; PDE5) activities, suggesting further impairments in the NO-cGMP cascade.[139-142] Recent studies further suggest that impaired soluble guanylate cyclase (sGC) activity contributes to high PVR and decreased responsiveness to NO.[139-140] Novel sGC activators and stimulators cause potent pulmonary vasodilation in experimental PPHN, suggesting a potential therapeutic role for these agents in infants with PPHN who are refractory to other therapies, including inhaled NO.[139-140] An elegant series of studies has shown that chronic intrauterine hypertension impairs NO-cGMP signaling through the generation of reactive oxygen species, especially superoxide.[143] Thus, disruption of the NO-cGMP cascade appears to play an essential role in the pathogenesis and pathophysiology of experimental PPHN (Figure 9). Abnormalities of NO production and responsiveness contribute to altered structure and function of the developing lung circulation, leading to failure of postnatal cardiorespiratory adaptation.

Upregulation of endothelin-1 (ET-1) may also contribute to the pathophysiology of PPHN. Circulating levels of ET-1, a potent vasoconstrictor and co-mitogen for vascular smooth muscle cell hyperplasia, are increased in human newborns with severe PPHN.[144] In the experimental model of PPHN induced by compression of the DA in fetal sheep, lung ET-1 mRNA and protein content is markedly increased, and the balance of ET receptors is altered, favoring vasoconstriction.[145] Chronic inhibition of the ET A receptor attenuates the severity of pulmonary hypertension, decreases pulmonary artery wall thickening, and improves the fall in PVR at birth in this model.[105] Thus, experimental studies have shown the important role of the NO-cGMP cascade and the ET-1 system in the regulation of vascular tone and reactivity of the fetal and transitional pulmonary circulation.

CONCLUSIONS

Physiologically, the fetal pulmonary circulation is characterized by the presence of high vascular resistance and the ability to oppose vasodilation and maintain low blood flow in utero; the ability to respond to vasoactive stimuli and progressive acquisition increases with maturation. Experimental studies have clearly shown the important role of the NO-cGMP cascade in the regulation of vascular tone and reactivity of

FIGURE 9 Schematic diagram illustrating the potential role of abnormalities of NO-cGMP cascade in the pathophysiology of persistent pulmonary hypertension of the newborn (PPHN).

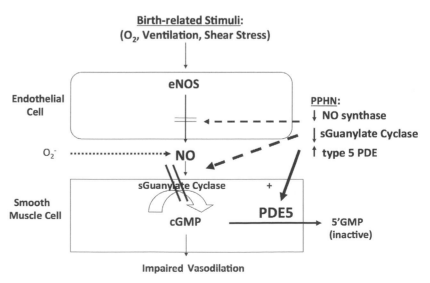

the fetal and transitional pulmonary circulation, and that abnormalities in this system contribute to abnormal pulmonary vascular tone and reactivity in an experimental model of PPHN. Further studies of the NO-cGMP cascade may provide helpful insights into new clinical strategies for more successful treatment of neonatal pulmonary vascular disease. In addition, because studies of vascular growth suggest important functions of NO in angiogenesis, we speculate that fetal NO production may contribute to normal lung vascular development. Insights into mechanisms that regulate changes in pulmonary vascular function are critical for understanding the pathogenesis and pathophysiology of neonatal diseases.

REFERENCES

1. Burri PH. Structural aspects of prenatal and postnatal development and growth of the lung. In: McDonald JA, editor. *Lung Growth and Development*. New York: M. Decker; 1997. pp. 1–35.

2. Schachtner S, Taichman D, Baldwin HS. Mechanisms of lung vascular development. In: Haddad GG, Chernick V, Abman SH, editors. *Basic Mechanisms of Pediatric Respiratory Disease*. Hamilton: BC Decker; 2002. pp. 49–67.

3. Stevens T, Rosenberg R, Aird W, Quertermous T, Johnson FL, Garcia JGN, et al. NHLBI workshop report: endothelial cell phenotypes in heart, lung and blood diseases. *Am J Physiol* 2001;**282**:C1422–33.

4. deMello DE, Reid LM. Embryonic and early fetal development of human lung vasculature and its functional implications. *Pediatr Deve Pathol* 2000;**3**:439–49.

5. Hall SM, Hislop AA, Pierce CM, Haworth SG. Prenatal origins of human intrapulmonary arteries: formation and smooth muscle maturation. *Am J Respir Cell Mol Biol* 2000;**23**:194–203.

6. Parera MC, van Dooren M, van Kempen M, de Krijger R, Grosveld F, Tibboel D, et al. Distal angiogenesis: a new concept for lung vascular morphogenesis. *Am J Physiol LCMP* 2005;**288**:L141–9.

7. O'Reilly M, Thebaud B. The promise of stem cells in BPD. *Semin Perinatol* 2013;**37**:79–84.

8. Risau W. Mechanisms of angiogenesis. *Nature* 1997;**386**:671–4.

9. Poole TJ, Finkelstein EB, Cox CM. The role of FGF and VEGF in angioblast induction and migration during vascular development. *Dev Dyn* 2001;**220**:1–17.

10. Morrell NW, Weiser MCM, Stenmark KR. Development of the pulmonary vasculature. In: Gaultier C, Bourbon JR, Post M, editors. *Lung Development*. NY: Oxford; 1999. pp. 152–95.

11. Gebb SA, Shannon JM. Tissue interactions mediate early events in pulmonary vasculogenesis. *Dev Dyn* 2000;**217**:159–69.

12. Carmeliet P, Ferreira V, Breier G, Pollefeyt S, Kieckens L, Gertsenstein M, et al. Abnormal blood vessel development and lethality in embryos lacking a single VEGF allele. *Nature* 1996;**380**:435–9.

13. Ferrara N, Carver-Moore K, Chen H, Dowd M, Lu L, O'Shea KS, et al. Heterozygous embryonic lethality induced by targeted inactivation of the VEGF gene. *Nature* 1996;**380**:439–42.

14. Shalaby F, Rossant J, Yamaguchi TP, Gertsenstein M, Wu XF, Breltmn ML, et al. Failure of blood-island formation and vasculogenesis in flk1 deficient mice. *Nature* 1995;**376**:62–6.

15. Shannon JM, Deterding RR. Epithelial-mesenchymal interactions in lung development. In: McDonald JA, editor. *Lung Growth and Development*. NY: Marcel-Dekker; 1997. pp. 81–118.

16. Acarregui M, Penisten ST, Goss KL, Ramirez K, Snyder JM. Vascular endothelial growth factor gene expression in human fetal lung in vitro. *Am J Repir Cell Mol Biol* 1999;**20**:14–23.

17. Hanahan D. Signaling vascular morphogenesis and maintenance. *Science* 1997;**277**:48–50.

18. Drake CJ, Little CD. VEGF and vascular fusion: implications for normal and pathological vessels. *J Histochem Cytochem* 1999;**47**:1351–5.

19. Drake CJ, Little CD. The de novo formation of blood vessels in the early embryo and in the developing lung. In: Weir EK, Sl Archer, Reeves JT, editors. *Fetal and Neonatal Pulmonary Circulation*. NY: Futura; 1999. pp. 19–30.

20. Hood JD, Meininger CJ, Ziche M, Granger HJ. VEGF upregulates ecNOS message, protein and NO production in human endothelial cells. *Am J Physiol* 1998;**274**:H1054–8.

21. Grover TR, Parker TA, Zenge JP, Markham NE, Kinsella JP, Abman SH. Intrauterine hypertension decreases lung VEGF expression and VEGF inhibition causes pulmonary hypertension in the ovine fetus. *Am J Phyiol Lung Cell Mol Biol* 2003;**284**:L508–17.

22. Le Cras TD, Markham NE, Tuder RM, Voelkel NF, Abman SH. Treatment of newborn rats with a VEGF receptor inhibitor causes pulmonary hypertension and abnormal lung structure. *Am J Phyiol Lung Cell Mol Biol* 2002;**283**:L555–62.

23. Folkman J, D'Amore PA. Blood vessel formation: what is its molecular basis. *Cell* 1996;**87**:1153–5.

24. Suri C, Jones PF, Patan S, Bartunkova S, Maisonpierre PC, Davis S, et al. Requisite role of angiopoietin-1 and ligand for the TIE2 receptor, during embryonic angiogenesis. *Cell* 1996;**87**:1171–80.

25. Frid MG, Moiseeva EP, Stenmark KR. Multiple phenotypically distinct smooth muscle cell populations exist in the adult and developing bovine pulmonary arterial media in vivo. *Circ Res* 1994;**75**:669–81.

26. Carey DJ. Control of growth and differentiation of vascular cells by extracellular matrix. *Annu Rev Physiol* 1991;**53**:161–77.

27. Roman J. Cell-cell and cell-matrix interactions in development of the lung vasculature. In: McDonald JA, editor. *Lung Growth and Development*. NY: Marcel Dekker; 1997. pp. 365–400.

28. Rupp PA, Little CD. Integrins in vascular development. *Circ Res* 2001;**89**:566–72.

29. Wagenseil JE, Mecham RP. Vascular extracellular matrix and arterial mechanisms. *Physiol Rev* 2009;**89**:957–89.

30. Stenmark KR, Mecham RP, Durmowicz AG, Parks WC. Persistence of the fetal pattern of tropoelastin gene expression in severe neonatal pulmonary hypertension. *J Clin Invest* 1994;**93**:1234–42.

31. Robbins IM, Moore TM, Blaisdell CJ, Abman SH. Improving outcomes for pulmonary vascular disease. *Am J Respir Crit Care Med* 2012;**185**:1015–20.

32. Robertson B. Anastomoses in the human lung: postnatal formation and obliteration of arterial anastomoses in the human lung: a microangiographic and histologic study. *Pediatrics* 1969;**43**:971.

33. Wilkinson MJ, Fagan DG. Postmortem demonstration of intrapulmonary arteriovenous shunting. *Arch Dis Child* 1990;**65**:435–7.

34. McMullan DM, Hanley FL, Cohen GA, Portman MA, Piemer RK. Pulmonary arteriovenous shunting in the normal fetal lung. *J Am Coll Cardiol* 2004;**44**:1497–500.

35. Galambos C, Lucas Sims, Abman SH. Histologic evidence of intrapulmonary anastomoses by three-dimensional reconstruction in severe BPD. *Ann Am Thorac Soc* 2013;**10**:474–81.

36. Galambos C, Lucas Sims, Abman SH. Three-dimensional reconstruction identifies misaligned pulmonary veins as intrapulmonary shunt vessels in alveolar capillary dysplasia. *J Pediatr* 2014;**164**:192–5.

37. Lovering AT, Romer LM, Haverkamp HC, Pegelow DF, Hokanson JS, Eldridge MW. Intrapulmonary shunting and pulmonary gas exchange during normoxic and hypoxic exercise in healthy humans. *J Appl Physiol* 2008;**104**:1418–25.

38. Heymann MA, Soifer SJ. Control of fetal and neonatal pulmonary circulation. In: Weir EK, Reeves JT, editors. *Pulmonary vascular Physiology and Pathophysiology*. NY: Marcel-Dekker; 1989. pp. 33–50.

39. Rudolph AM. Fetal and neonatal pulmonary circulation. *Ann Rev Physiol* 1979;**41**:383–95.

40. Rasanen J, Huhta JC, Weiner S, Ludomirski A. Role of the pulmonary circulation in the distribution of human fetal cardiac output during the second half of pregnancy. *Circulation* 1996;**94**:1068–73.

41. Rasanen J, Wood DC, Debbs RH, Cohen J, Weiner S, Huhta JC. Reactivity of the human fetal pulmonary circulation to maternal hyperoxygenation increases during the second half of pregnancy. A randomized study. *Circulation* 1998;**97**:257–62.

42. Abman SH, Chatfield BA, Hall SL, McMurtry IF. Role of endothelium-derived relaxing factor during transition of pulmonary circulation at birth. *Am J Physiol* 1990;**259**:H1921–7.

43. Cassin S. Role of prostaglandins, thromboxanes and leukotrienes in the control of the pulmonary circulation in the fetus and newborn. *Semin Perinatol* 1987;**11**:53–63.

44. Cornfield DN, Chatfield BA, McQueston JA, McMurty IF, Abman SH. Effects of birth-related stimuli on L-arginine -dependent pulmonary vasodilation in the ovine fetus. *Am J Physiol* 1992;**262**:H1474–81.

45. Cornfield DN, Reeves HL, Tolarova S, Weir EK, Archer S. Oxygen causes fetal pulmonary vasodilation through activation of a calcium-dependent potassium channel. *Proc Natl Acad Sci* 1996;**93**:8089–94.

46. Ivy DD, Kinsella JP, Abman SH. Physiologic characterization of endothelin A and B receptor activity in the ovine fetal lung. *J Clin Inves* 1994;**93**:2141–8.

47. Leffler CW, Hessler JR, Green RS. Mechanism of stimulation of pulmonary prostacyclin synthesis at birth. *Prostaglandins* 1984;**28**:877–87.

48. Leffler CW, Tyler TL, Cassin S. Effect of indomethacin on pulmonary vascular response to ventilation of fetal goats. *Am J Physiol* 1978;**234**:H346–51.

49. Soifer SJ, Loitz RD, Roman C, Heymann MA. Leukotriene end organ antagonists increase pulmonary blood flow in fetal lambs. *Am J Physiol* 1985;**249**:570.

50. Storme L, Rairhig RL, Abman SH. In vivo evidence for a myogenic response in the ovine fetal pulmonary circulation. *Pediatr Res* 1999;**45**:425–31.

51. Velvis H, Moore P, Heymann MA. Prostaglandin inhibition prevents the fall in pulmonary vascular resistance as the result of rhythmic distension of the lungs in fetal lambs. *Pediatr Res* 1991;**30**:62–7.

52. Wedgwood S, Bekker JM, Black SM. Shear stress regulation of endothelial NO synthase in fetal pulmonary arterial endothelial cells involves PKC. *Am J Physiol LCMP* 2001;**281**:L490–8.

53. Lewis AB, Heymann MA, Rudolph AM. Gestational changes in pulmonary vascular responses in fetal lambs in utero. *Circ Res* 1976;**39**:536–41.

54. Morin FC, Egan EA, Ferguson W, Lundgren CEG. Development of pulmonary vascular response to oxygen. *Am J Physiol* 1988;**254**:H542–6.

55. Rudolph AM, Heymann MA, Lewis AB. Physiology and pharmacology of the pulmonary circulation in the fetus and newborn. In: Hodson W, editor. *Development of the Lung*. NY: Marcel Dekker; 1977. pp. 497–523.

56. Abman SH, Accurso FJ. Acute effects of partial compression of the ductus arteriosus on the fetal pulmonary circulation. *Am J Physiol* 1989;**257**:H626–34.

57. Abman SH, Accurso FJ. Sustained fetal pulmonary vasodilation during prolonged infusion of atrial natriuretic factor and 8-bromo-guanosine monophosphate. *Am J Physiol* 1991;**260**:H183–92.

58. Abman SH, Chatfield BA, Rodman DM, Hall SL, McMurtry IF. Maturation-related changes in endothelium-dependent relaxation of ovine pulmonary arteries. *Am J Physiol* 1991;**260**:L280–5.

59. Accurso FJ, Alpert B, Wilkening RB, Petersen RG, Meschia G. Time-dependent response of fetal pulmonary blood flow to an increase in fetal oxygen tension. *Respir Physiol* 1986;**63**:43–52.

60. Halbower AC, Tuder RM, Franklin WA, Pollock JS, Forstermann U, Abman SH. Maturation-related changes in endothelial NO synthase immunolocalization in the developing ovine lung. *Am J Physiol* 1994;**267**:L585–91.

61. North AJ, Star RA, Brannon TS, Ujiie K, Wells LB, Lowenstien CJ, Snyder SH, Shaul PW. NO synthase type I and type III gene expression are developmentally regulated in rat lung. *Am J Physiol* 1994;**266**:L635–41.

62. Parker TA, Le Cras TD, Kinsella JP, Abman SH. Developmental changes in endothelial NO synthase expression in the ovine fetal lung. *Am J Physiol* 2000;**278**:L202–8.

63. Kinsella JP, Ivy DD, Abman SH. Ontogeny of NO activity and response to inhaled NO in the developing ovine pulmonary circulation. *Am J Physiol* 1994;**267**:H1955–61.

64. Abman SH, Kinsella JP, Schaffer MS, Wilkening RB. Inhaled nitric oxide therapy in a premature newborn with severe respiratory distress and pulmonary hypertension. *Pediatrics* 1993;**92**:606–9.

65. Bustamante SA, Pang Y, Romero S, Pierce MR, Voelker CA, Thompson JH, et al. Inducible NOS and the regulation of central vessel caliber in the fetal rat. *Circulation* 1996;**94**:1948–53.

66. Kobzik L, Bredt DS, Lowenstein CJ, Drazen J, Gaston B, Sugarbaker D, et al. NOS in human and rat lung: immunocytochemical and histochemical localization. *Am J Respir Cell Mol Biol* 1993;**9**:371–7.

67. Rairigh RL, Parker TA, Ivy DD, Kinsella JP, Fan I, Abman SH. Role of inducible nitric oxide synthase in the transition of the pulmonary circulation at birth. *Circ Res* 2001;**88**:721–6.

68. Sherman TS, Chen Z, Yuhanna IS, Lau KS, Margraf LR, Shaul PW. NO synthase isoform expression in the developing lung epithelium. *Am J Physiol* 1999;**276**:L383–90.

69. Rairhig R, Le Cras TD, Ivy DD, Kinsella JP, Richter G, Horan MP, et al. Role of inducible nitric oxide synthase in regulation of pulmonary vascular tone in the late gestation ovine fetus. *J Clin Invest* 1998;**101**:15–21.

70. Rairigh RL, Storme L, Parker TA, Le Cras TD, Markham N, Jakkula M, et al. Role of neuronal nitric oxide synthase in regulation of vascular and ductus arteriosus tone in the ovine fetus. *Am J Physiol Lung Cell Mol Physiol* 2000;**278**:L105–10.

71. Rairigh RL, Parker TA, Ivy DD, Kinsella JP, et al. Role of inducible nitric oxide synthase in the pulmonary vascular response to birth-related stimuli in the ovine fetus. *Circulation Res* 2001;**88**:721–6. http://dx.doi.org/10.1161/hh0701.088683.

72. Abman SH, Kinsella JP, Parker TA, Storme L, Le Cras TD. Physiologic roles of NO in the perinatal pulmonary circulation. In: Weir EK, Archer SL, Reeves JT, editors. *Fetal and Neonatal Pulmonary Circulation*. NY: Futura; 1999. pp. 239–60.

73. MacRitchie AN, Jun SS, Chen Z, German Z, Yuhanna IS, Sherman TS, et al. Estrogen upregulates endothelial NO synthase gene expression in fetal pulmonary artery endothelium. *Circ Res* 1997;**81**:355–62.

74. Parker TA, Kinsella JP, Galan HL, Richter G, Abman SH. Prolonged infusions of estradiol dilate the ovine fetal pulmonary circulation. *Pediatr Res* 2000;**47**:89–96.

75. Shaul PW. NO in the developing lung. *Adv Pediatr* 1995;**42**: 367–414.

76. Parker TA, Afshar S, Kinsella JP, Ivy DD, Shaul PW, Abman SH. Effects of chronic estrogen receptor blockade on the pulmonary circulation in the late gestation ovine fetus. *Am J Physiol Heart Circ Physiol* 2001;**281**:H1005–14.

77. Grover TR, Zenge JP, Parker TA, Abman SH. Vascular endothelial growth factor causes pulmonary vasodilation through activation of the phosphatidylinositol-3-kinase-nitric oxide pathway in the late-gestation ovine fetus. *Pediatr Res* 2002;**52**:907–12.

78. Beavo JA, Reifsnyder DH. Primary sequence of cyclic nucleotide phosphodiesterase isozymes and the design of selective inhibitors. *Trends Pharmacol Sci* 1990;**11**:150–5.

79. Braner DA, Fineman JR, Chang R, Soifer SJ. M and B 22948, a cGMP phosphodiesterase inhibitor, is a pulmonary vasodilator in lambs. *Am J Physiol* 1993;**264**:H252–8.

80. Cohen AH, Hanson K, Morris K, Fouty B, McMurtry IF, Clarke W, et al. Inhibition of cGMP-specific phosphodiesterase selectively vasodilates the pulmonary circulation in chronically hypoxic rats. *J Clin Invest* 1996;**97**:172–9.

81. Hanson KA, Burns F, Rybalkin SD, Miller J, Beavo J, Clarke WR. Developmental changes in lung cGMP phosphodiesterase-5 activity, protein and message. *Am J Resp Crit Care Med* 1995;**158**:279–88.

82. Thusu KG, Morin FC, Russell JA, Stein horn RH. The cGMP phosphodiesterase inhibitor zaprinast enhances the effect of NO. *Am J Resp Crit Care Med* 1995;**152**:1605–10.

83. Ziegler JW, Ivy DD, Fox JJ, Kinsella JP, Clarke WR, Abman SH. Dipyridamole, a cGMP phosphodiesterase inhibitor, causes pulmonary vasodilation in the ovine fetus. *Am J Physiol* 1995;**269**:H473–9.

84. McQueston JA, Cornfield DN, McMurtry IF, Abman SH. Effects of oxygen and exogenous L-arginine on endothelium-derived relaxing factor activity in the fetal pulmonary circulation. *Am J Physiol* 1993;**264**:H865–71.

85. Meininger GA, Davis MJ. Cellular mechanisms involved in the vascular myogenic response. *Am J Physiol* 1992;**263**:H647–59.

86. Belik J, Stephens NL. Developmental differences in vascular smooth muscle mechanics in pulmonary and systemic circulations. *J Appl Physiol* 1993;**74**:682–7.

87. Kulik TJ, Evans JN, Gamble WJ. Stretch-induced contraction in pulmonary arteries. *Am J Physiol* 1988;**255**:H1191–8.

88. Morbidelli L, Chang C-H, Douglas JG, Granger HJ, Maggi CA, Geppetti P, et al. NO mediates mitogenic effect of VEGF on coronary venular endothelium. *Am J Physiol* 1996;**270**:H411–5.

89. Ziche M, Morbidelli L, Masini E, Amerini S, Granger HJ, Maggi CA, et al. NO mediates angiogenesis in vivo and endothelial cell growth and migration in vitro promoted by substance P. *J Clin Invest* 1994;**94**:2036–44.

90. Arnal J-F, Yamin J, Dockery S, Harrison DG. Regulation of endothelial NO synthase mRNA, protein, and activity during cell growth. *Am J Physiol* 1994;**267**:C1381–8.

91. Garg UC, Hassid A. NO-generating vasodilators and 8-bromo-cGMP inhibit mitogenesis and proliferation of cultured rat vascular smooth muscle cells. *J Clin Invest* 1989;**83**:17744–7.

92. Thomae KR, Nakayama DK, Billiar TR, Simmons RL, Pitt BR, Davies P. Effect of NO on fetal pulmonary artery smooth muscle growth. *J Surg Res* 1996;**270**:H411–5.

93. Fineman JR, Wong J, Morin FC, Wild LM, Soifer SJ. Chronic NO inhibition in utero produces persistent pulmonary hypertension in newborn lambs. *J Clin Invest* 1994;**93**:2675–83.

94. Zenge JP, Rairigh RL, Grover TR, Storme L, Parker TA, Abman SH. NO and prostaglandins modulate the pulmonary vascular response to hemodynamic stress in the late gestation fetus. *Am J Physiol Lung Cell Mol Physiol* 2001;**281**:L1157–63.

95. Campbell WB, Harder DR. Prologue: EDHF what is it? *Am J Physiol* 2001;**280**:H2413–6.

96. Storme L, Rairigh RL, Parker TP, Cornfield DN, Kinsella JP, Abman SH. Potassium channel blockade attenuates shear stress-induced pulmonary vasodilation in the ovine fetus. *Am J Physiol* 1999;**276**:L220–8.

97. Lin H, McGrath JJ. Vasodilating effects of carbon monoxide. *Life Sci* 1988;**43**:1813.

98. Kharitonov VG. Basis of guanylate cyclase activation of carbon monoxide. *Proc Natl Acad Sci USA* 1995;**92**:2568.

99. Grover TR, Rairigh RL, Zenge JP, Abman SH, Kinsella JP. Inhaled carbon monoxide does not alter pulmonary vascular tone in the ovine fetus. *Am J Physiol. Lung Cell Mol Physiol* 2000;**278**:L779–84.

100. Boulanger C, Luscher TF. Release of endothelin from the porcine aorta. Inhibition by endothelium-derived nitric oxide. *J Clin Invest* 1990;**85**:587–90.

101. Ivy DD, Abman SH. Role of endothelin in perinatal pulmonary vaso-regulation. In: Weir EK, Archer SL, Reeves JT, editors. *Fetal and Neonatal Pulmonary Circulation*. NY: Futura; 1999. pp. 279–302.

102. Yanagisawa M, Kurihara H, Kimura S, Tomobe Y, Kobayashi M, Mitsui Y, et al. A novel potent vasoconstrictor peptide produced by vascular endothelial cells. *Nature* 1988;**332**:411–5.

103. Chatfield BA, McMurtry IF, Hall SL, Abman SH. Hemodynamic effects of endothelin-1 on the ovine fetal pulmonary circulation. *Am J Physiol* 1991;**261**:R182–7.

104. Ivy DD, Kinsella JP, Abman SH. Physiologic characterization of endothelin A and B receptor activity in the ovine fetal lung. *J Clin Inves* 1996;**93**:2141–8.

105. Ivy DD, Parker TA, Abman SH. Prolonged endothelin B receptor blockade causes pulmonary hypertension in the ovine fetus. *Am J Physiol* 2000:L758–65.

106. Coceani F, Kelsey L, Seiditz F, Marks GS, McLaughlin BE, Vreman HJ, et al. Carbon monoxide formation in the ductus arteriosus in the lamb: implications for the regulation of muscle tone. *Br J Pharmacol* 1997;**120**:599–603.

107. Reeve HL, Weir EK. Regulation of ion channels in the ductus arteriosus. In: Weir EK, Archer SL, Reeves JT, editors. *Fetal and Neonatal Pulmonary Circulation*. NY: Futura; 1999. pp. 319–30.

108. Tristani-Firouzi M, Reeve HL, Tolarova S, Weir EK, Archer SL. Oxygen-induced constriction of rabbit ductus arteriosus occurs via inhibition of 4-aminopyridine, voltage-sensitive K channels. *J Clin Invest* 1996;**98**:1959–65.

109. Archer SL, Huang JMC, Hampl V, Nelson DP, Shultz PJ, Weir EK. NO and cGMP cause vasorelaxation by activation of a charybdo-toxin-sensitive K channel by cGMP-dependent protein kinase. *Proc Natl Acad Sci USA* 1994;**91**:7583–7.

110. Rhodes MT, Porter VA, Saqueton CB, Herron JM, Resnik ER, Cornfield DN. Pulmonary vascular response to normoxia and Kca channel activity is developmentally regulated. *Am J Physiol* 2001;**280**:L1250–7.

111. Tristani-Firouzi M, Martin EB, Tolarova S, Weir EK, Archer SL, Cornfield DN. Ventilation-induced pulmonary vasodilation at birth is modulated by potassium channel activity. *Am J Physiol* 1996;**271**:H2353–9.

112. Konduri GG, Mital S, Gervasio CT, Rotta AT, Forman K. Purine nucleotides contribute to pulmonary vasodilation caused by birth-related stimuli in the ovine fetus. *Am J Physiol* 1997;**272**:H2377–84.

113. Fagan KA, Fouty BW, Tyler RC, Morris KG, Helper LK, Sato K, et al. The pulmonary circulation of mice with either homozygous or heterozygous disruption of endothelial NO synthase is hyper-responsive to chronic hypoxia. *J Clin Invest* 1999;**103**:291–9.

114. Steudel W, Scherrer-Crosbie M, Bloch KD, Weiman J, Huang PL, Jones RC, et al. Sustained pulmonary hypertension and right ventricular hypertrophy after chronic hypoxia in mice with congenital deficiency of NOS III. *J Clin Invest* 1998;**101**:2468–77.

115. Kinsella JP, McQueston JA, Rosenberg AA, Abman SH. Hemodynamic effects of exogenous nitric oxide in ovine transitional pulmonary circulation. *Am J Physiol* 1992;**263**:H875–80.

116. Kinsella JP, Abman SH. Recent developments in the pathophysiology and treatment of persistent pulmonary hypertension of the newborn. *J Pediatrics* 1995;**126**:853–64.

117. Levin DL, Heymann MA, Kitterman JA, Gregory GA, Phibbs RH, Rudolph AM. Persistent pulmonary hypertension of the newborn. *J Pediatr* 1976;**89**:626–33.

118. Geggel RL, Reid LM. The structural basis of persistent pulmonary hypertension of the newborn. *Clin Perinatol* 1984;**3**:525–49.

119. Murphy JD, Rabinovitch M, Goldstein JD, Reid LM. The structural basis for PPHN infant. *J Pediatr* 1981;**98**:962–7.

120. Murphy JD, Vawter G, Reid LM. Pulmonary vascular disease in fatal meconium aspiration. *J Pediatr* 1984;**104**:758–62.

121. Van Marter LJ, Leviton A, Allred EN, Pagano M, Sullivan KF, Cohen A, et al. PPHN and smoking and aspirin and nonsteroidal antiinflammatory drug consumption during pregnancy. *Pediatrics* 1996;**97**:658–63.

122. Abman SH, Shanley PF, Accurso FJ. Failure of postnatal adaptation of the pulmonary circulation after chronic intrauterine pulmonary hypertension in fetal lambs. *J Clin Invest* 1989;**83**:1849–58.

123. Levin DL, Hyman AI, Heymann MA, Rudolph AM. Fetal hypertension and the development of increased pulmonary vascular smooth muscle: a possible mechanism for persistent pulmonary hypertension of the newborn infant. *J Pediatr* 1978;**92**:265–9.

124. Morin FC. Ligating the ductus arteriosus before birth causes persistent pulmonary hypertension in the newborn lamb. *Pediatr Res* 1989;**25**:245–50.

125. Williams MC, Wyble LE, O'Brien WF, Nelson RM, Schwenke JR, Casanova C. PPHN and asymmetric growth restriction. *Obstet Gynecol* 1998;**91**:336–41.

126. Delaney C, Gien J, Roe G, Isenberg N, Kailey J, Abman SH. Serotonin contributes to high pulmonary vascular tone in a sheep model of persistent pulmonary hypertension of the newborn. *Am J Physiol* 2013;**304**:L894–9001.

127. Castillo L, DeRojas-Walker T, Yu YM, Sanchez M, Chapman TE, Shannon D, et al. Whole body arginine metabolism and NO synthesis in newborns with persistent pulmonary hypertension. *Pediatr Res* 1995;**38**:17–24.

128. Dollberg S, Warner BW, Myatt L. Urinary nitrite and nitrate concentrations in patients with idiopathic PPHN and effect of ECMO. *Pediatr Res* 1994;**37**:31–4.

129. Pearson DL, Dawling S, Walsh WF, Haines JL, Chritman BW, Bazyk A, et al. Neonatal pulmonary hypertension: urea cycle intermediates, NO production and carbamoyl phosphate synthetase function. *N Engl J Med* 2001;**344**:1932–8.

130. Newman JH, Wheeler L, Lane KB, Loyd E, Gaddipati R, Phillips JA, et al. Mutation in the gene for bone morphogenetic protein receptor II as a cause of primary pulmonary hypertension in a large kindred. *N Engl J Med* 2001;**345**:367–71.

131. Haworth SG, Reid LM. Persistent fetal circulation. Newly recognized structural features. *J Pediatr* 1976;**88**:614–20.

132. Stenmark KR, Abman SH, Accurso FJ. Etiologic mechanisms of persistent pulmonary hypertension of the newborn. In: Weir EK, Reeves JT, editors. *Pulmonary Vascular Physiology and Pathophysiology*. NY: Marcel-Dekker; 1989. p. 335.

133. Goldberg SJ, Levy RA, Siassi B. Effects of maternal hypoxia and hyperoxia upon the neonatal pulmonary vasculature. *Pediatrics* 1971;**48**:528.

134. McQueston JA, Kinsella JP, Ivy DD, McMurtry IF, Abman SH. Chronic pulmonary hypertension *in utero* impairs endothelium-dependent vasodilation. *Am J Physiol* 1995;**268**:H288–94.

135. Shaul PW, Yuhanna IS, German Z, Chen Z, Steinhorn RH, Morin FC. Pulmonary endothelial NO synthase gene expression is decreased in fetal lambs with pulmonary hypertension. *Am J Physiol* 1997;**272**:L1005–12.

136. Storme L, Rairigh RL, Parker TA, Kinsella JP, Abman SH. Acute intrauterine pulmonary hypertension impairs endothelium-dependent vasodilation in the ovine fetus. *Pediatr Res* 1999;**45**:575–81.

137. Storme L, Rairhig RL, Parker TA, Kinsella JP, Abman SH. Chronic pulmonary hypertension abolishes flow-induced vasodilation and increases the myogenic response in the ovine fetal pulmonary circulation. *Am J Physiol* 2001.

138. Villamor E, Le Cras TD, Horan M, Halbower AC, Tuder RM, Abman SH. Chronic hypertension impairs endothelial NO sythase in the ovine fetus. *Am J Physiol* 1997;**16**:L1013–20.

139. Chester M, Gien J, Tourneux P, Seedorf G, Grover TR, Stasch H-P, et al. Cinaciguat, a soluble guanylate cyclase activator, augments cGMP production and causes potent pulmonary vasodilation in experimental persistent pulmonary hypertension of the newborn. *Am J Physiol Lung* 2011;**301**(5):L755–64.

140. Deruelle P, Grover TR, Storme L, Abman SH. Effects of BAY 41-2272, a direct soluble guanylate cyclase activator, on pulmonary vasoreactivity in the ovine fetus. *Am J Physol LCMP* 2005;**288**: 727–33.

141. Hanson KA, Beavo JA, Abman SH, Clarke WR. Chronic pulmonary hypertension increases fetal lung cGMP activity. *Am J Physiol* 1998;**275**:L931–41.

142. Steinhorn RH, Russell JA, Morin FC. Disruption of cGMP production in pulmonary arteries isolated from fetal lambs with pulmonary hypertension. *Am J Physiol* 1995;**268**:H1483–9.

143. Farrow KN, Lakshminrusimha S, Reda WJ, Wedgwood S, Czech L, Gugino S, et al. Superoxide dismutase restores eNOS expression and function in resistance pulmonary arteries from neonatal lambs with persistent pulmonary hypertension. *Am J Physiol LCMP* 2008;**295**:979–89.

144. Rosenberg AA, Kennaugh J, Koppenhafer SL, Loomis M, Chatfield BA, Abman SH. Increased immunoreactive endothelin-1 levels in persistent pulmonary hypertension of the newborn. *J Pediatr* 1993;**123**:109–14.

145. Ivy DD, LeCras TD, Horan MP, Abman SH. Increased lung prepro-endothelin-1 and decreased endothelin B receptor gene expression after chronic pulmonary hypertension in the ovine fetus. *Am J Physiol* 1998;**274**:L535–41.

Chapter 7

Development of Salt and Water Transport across Airway and Alveolar Epithelia

Jonathan H. Widdicombe

Department of Physiology & Membrane Biology, University of California – Davis, Davis, CA, USA

TRANSPORT PROCESSES UNDERLYING SECRETION AND ABSORPTION ACROSS PULMONARY EPITHELIA

The fetal lung is filled with liquid, secreted across alveolar and airway epithelia as a consequence of active secretion of Cl^-. This process generates a lumen-negative transepithelial voltage (V_{te}) that draws cations (mainly Na^+) into the lumen predominantly via the paracellular pathway (i.e., through the tight junctions). The transfer of Na^+ and Cl^- results in local osmotic gradients across the epithelium that drive fluid secretion by osmosis.[1] The main route of fluid flow across leaky epithelia is thought to be predominantly through the cells (transcellular).[2,3] A high osmotic permeability of this transcellular pathway ensures that the secreted liquid is virtually iso-osmolar with interstitial fluid.[4,5] The presence of aquaporins in the plasma membranes of both airway and alveolar epithelia contributes to this high permeability.[6]

Chloride secretion across airway epithelium has been studied in great detail,[7,8] and appears to be by a mechanism widespread in vertebrate epithelia.[9] Though studied in less detail, all available evidence suggests that Cl^- secretion by alveolar epithelium is by the same mechanism.[10–12] In the original version of this model, Cl^- entry across the basolateral membrane was by cotransport with Na^+ and K^+ in the ratio $1Na^+{:}1K^+{:}2Cl^-$. This carrier (NKCC) is inhibited by loop diuretics, such as bumetanide. However, even at supramaximal doses, bumetanide inhibits only ~60% of Cl^- secretion across dog tracheal epithelium,[13] and a similar failure to abolish Cl^- secretion has been reported for alveolar epithelium.[14] Most of the Cl^- secretion that is independent of NKCC probably involves entry of Cl^- (as NaCl) across the basolateral membrane by the parallel operation of Na^+/H^+ and Cl^-/HCO_3^- exchangers.[15] Net basolateral uptake of Cl^- by either mechanism requires a transmembrane Na^+ gradient. This is generated by the Na^+-K^+-ATPase, which is restricted to the basolateral membrane[16] and maintains intracellular $[Na^+]$ at a level about one-tenth that of extracellular $[Na^+]$.[17,18]

Net exit of Cl^- across the apical membrane occurs via anion channels down a favorable electrochemical gradient. One of the more important of these channels in adult lungs is the cystic fibrosis transmembrane conductance regulator (CFTR), the protein that is defective in cystic fibrosis (CF). It is regulated primarily by cAMP-dependent phosphorylation. In addition, the airway apical membrane contains Ca-activated Cl channels (CaCC).[19] Of these, TME16a predominates, though others may also be present in significant amounts.[8] Fetal lung development requires the active secretion of Cl^- (and water), yet lung development is normal in CF both of humans and transgenic animals.[20–23] Thus CFTR does not play an important role in secretion of fetal lung liquid. Instead, the Cl^- channel primarily responsible is most probably CLC-2,[24] though other anion channels may contribute (see later).

The basolateral membrane is K^+-selective, and the K^+ that enters on the Na^+-K^+-ATPase and NKCC recycles through the basolateral K^+ channels.[25] By hyperpolarizing the cell, the basolateral K^+ channels also serve to maintain the driving force for Cl^- exit across the apical membrane.[26] A plethora of K^+ channels have been detected in pulmonary epithelia and epithelial cell lines,[27] but only two, Kv7.1 and KCa3.1, have been established as being both basolateral in location and important in active Cl^- secretion by airway epithelium.[15,28] Type II cells have the same two channels in their basolateral membrane as well as ATP-activated K^+ channels.[27,29,30]

The main process mediating fluid absorption is electrogenic absorption of Na^+ by the mechanism first proposed by Koefoed-Johnsen and Ussing.[31] Net Na^+ entry across the apical membrane is conductive and driven by both electrical and chemical gradients. The basolateral Na^+-K^+-ATPase

The Lung. http://dx.doi.org/10.1016/B978-0-12-799941-8.00007-9

actively extrudes the Na^+ entering across the apical membrane. The K^+ pumped in recycles through basolateral K^+ channels.

Most of the Na^+ entering across the apical membrane usually does so via epithelial Na^+ channels (ENaC), all forms of which are inhibited by amiloride. Until recently this channel was believed to have only three subunits (α, β, γ) that assembled as homo- or hetero-trimers. Alpha subunits were thought essential for function,[32] and knock-out of α-subunits in mice leads to early death due to incomplete absorption of fetal lung liquid.[33] In the Xenopus oocyte expression system, the $\alpha\beta\gamma$ trimer shows the greatest channel activity,[32] and was generally regarded as responsible for the great majority of ENaC function in mammalian epithelia. However, there is evidence in alveolar Type I and Type II cells that several different trimers are present with different channel properties.[34] Trimers of all three subunits produce channels that are highly selective for Na^+ over K^+. Alpha-homotrimers, by contrast, are essentially non-selective for Na^+ over K^+. In addition, there are small numbers of moderately selective channels composed of some combination of α with either β or γ. The initial evidence for several forms of ENaC came from patch-clamp studies of isolated Type II cells,[35] that were allowed to attach for a few hours before patching. Such cells often lose their polarity, and this can lead to expression of channels not normally present, or scarce, in native epithelium.[36-38] However, this earlier work on isolated single cells was largely confirmed by studies on lung slices, in which both Type I and Type II cells can be identified and patch-clamped while in their natural polarized state.[34,39]

The relative frequencies of the different forms of ENaC in *airway* epithelium are unknown, though all three subunits are expressed.[40] Significant levels of the non-selective α_3 form of the channel seem unlikely. Being non-selective this would result in K^+ secretion, but active K^+ secretion across airway epithelium is only about 1/100th the level of active Na^+ absorption.[8]

Recently, a fourth subunit of ENac (δ) has been described that can substitute for α-ENaC with little or no loss of function.[41] It is expressed in both alveolar and airway epithelium.[41-43] In human nasal epithelium it occurs at the same frequency as α-ENaC (i.e., ~0.15 μm^{-2} of apical membrane), where it has been proposed to account for ~50% of Na^+ absorption.[44] Currently, it has no known role in Na^+ absorption across alveolar epithelium.[41]

Depending on species, stage of development, and lung region, amiloride-insensitive Na^+ channels may mediate up to about 50% of active Na^+ absorption.[45] In rat Type I cells, this amiloride-insensitive Na^+ absorption has been ascribed to cyclic nucleotide-gated channels; the specific channel involved being CNHGA1.[46] In some cases Na^+-glucose cotransport may also contribute to active transepithelial Na^+ absorption.[47,48]

Both $\alpha 1$ and $\alpha 2$ subunits of the Na^+-K^+-ATPase are present in alveolar Type I cells, but dose-response curves to ouabain suggest that $\alpha 1$ is most important in alveolar fluid absorption.[49]

Pulmonary epithelia show a number of active transepithelial ion transport processes in addition to Na^+ absorption and Cl^- secretion. These include secretion of K^+, H^+, and HCO_3^-.[8,50-52] However, these solute flows are small and unlikely to contribute much to transepithelial volume flows. The very low $[HCO_3^-]$ (~2 mEq/l) of fetal sheep lung liquid is notable.[53-55] It may reflect Na^+/H^+ exchange in the apical membrane of alveolar epithelium,[56] but there is no reason to think this process would much influence transepithelial volume secretion. In contrast to sheep, the $[Cl^-]$ in lung liquid of fetal dogs is only slightly less than in plasma.[57]

Apical membrane Cl^- channels are generally thought of as being involved in secretion of Cl^- and water, but they may also enhance *absorption*. In tissues that actively accumulate Cl^- across the basolateral membrane, E_{Cl} (the reversal potential for Cl^-) may be depolarized relative to V_a (the apical membrane voltage), and the electrochemical potential gradient for Cl^- will be outward across the apical membrane. If the apical membrane Cl^- channels are open, such tissues will secrete Cl^-. However, if Cl^- is not actively accumulated, E_{Cl} will lie between V_a and V_b (the basolateral membrane potential). The electrochemical potential gradient for Cl^- will be inward across the apical membrane and outward across the basolateral. If Cl^- transport pathways are present in both membranes, then in the presence of active transepithelial Na^+ absorption the tissue will show net transepithelial Cl^- absorption.

In airway epithelium, when V_{te} is less hyperpolarized than about −30 mV, the electrochemical gradient for Cl^- is outward, and opening of apical membrane Cl^- channels results in active secretion of Cl^-, with Na^+ and water following.[58] This net secretion of Cl^- is sustained if Cl^- uptake across the basolateral membrane is simultaneously activated.[8] However, if V_{te} is more hyperpolarized than about −30 mV, opening of CFTR enhances fluid absorption, with Cl^- flowing inward through CFTR or CaCC in the apical membrane and then outward across the basolateral membrane via cAMP- or swelling-activated Cl^- channels or the K^+-Cl^- cotransporter.[29,36,58,59]

Microelectrode measurements of V_a and aCl_i have been used to determine the electrochemical driving forces for Cl^- across the membranes of airway epithelium under various conditions. Unfortunately, such information is lacking for alveolar epithelium. However, fetal lung explants actively secrete Cl^- (see later), so it seems likely that fetal alveolar epithelium actively accumulates intracellular Cl^-, and that aCl_i is higher than predictable for passive distribution according to V_a. Absorption of liquid at birth, however, is stimulated by activation of CFTR,[60] suggesting that aCl_i is now lower than for equilibrium with V_a, though whether this is due to changes in V_a or aCl_i (or both) is unknown. In cultures of adult alveolar epithelium, activation of CFTR also enhances liquid absorption.[60,61]

Not only will apical membrane Cl^- channels influence fluid flows by enhancing net transcellular movement of Cl^- in either direction, but they will also influence active Na^+ absorption by changes in V_a. For instance, an inward electrochemical gradient for Cl^- means that E_{Cl} is hyperpolarized relative to V_a. Under these circumstances, opening Cl^- channels will hyperpolarize V_a and stimulate Na^+ influx via ENaC and other cation channels.[62]

CELLULAR BASIS OF SECRETION AND ABSORPTION

What are the relative amounts of liquid shifted by airway vs. alveolar epithelium? In general, the greater the height of its cells, the greater the amount of vectorial ion transport shown by an epithelium; columnar cells have a greater surface area of transporting membrane per unit area of epithelium than do squamous cells. Across airway and alveolar epithelia, the short-circuit current (I_{sc}), defined in detail later, is approximately equal to the sum of active Cl^- secretion and active Na^+ absorption.[63] Native tracheal epithelia of the larger mammalian species (dog, human, cow, horse, sheep, etc.) have cells that average ~50 μm in height and baseline I_{sc}'s of ~100 μA/cm^2. Well-differentiated cultures of airway epithelium have cell heights and I_{sc}'s that are similar to native epithelium.[64–66] By contrast alveolar epithelia are much thinner (height ≤ 2 μm). The I_{sc} across alveolar epithelia of amphibians is ~2 μA per cm^2 of transporting area (the epithelium is thrown into ridges so the transporting area is about 6 times the apparent area).[67–69] Primary cultures of mammalian Type II cells are also highly squamous, and show I_{sc}'s of 2–10 μA/cm^2.[11,60,70] Type I cells of native epithelium are flatter than Type II cells, and may be expected to show even lower rates of transepithelial active ion transport. V_{te} in rat alveoli is ~5 mV,[71] and in rabbits is ~1 mV. Assuming the same resistance as in amphibian lung (i.e., 6000 kΩ.cm^2),[67] this corresponds to I_{sc}'s of ~1 or ~0.2 μA/cm^2, respectively. However, the total area of alveolar epithelium is much greater than that of the airway, the ratio being ~100:1 in humans.[12] Thus the airway epithelium provides a small area of epithelium with high levels of transepithelial transport, whereas the alveolar epithelium comprises a much greater area but probably has much lower transport rates. However, the area of epithelium of the alveoli is so much greater than that of the airways that the former would seem the more important in generating liquid movements into or out of the lung.

What cell types are responsible for fluid movements across the airway epithelium? In early fetal development, airway epithelium consists of undifferentiated columnar cells. Depending on species, ciliated cells appear from 30% to 90% of the way through gestation. In adults, they are the most frequent epithelial cell type in all airway generations. At least in the larger airways, goblet cells appear shortly after the ciliated.[72] Both goblet and ciliated cells are present before birth in the large airways of all species studied (monkey, human rabbit, rat, mouse, ferret), though their relative numbers may change quite dramatically postnatally (see Chapter ??). In respiratory bronchioles, goblet cells are absent being replaced by Clara cells. The ion transport processes of the undifferentiated airway epithelium of the early fetus have not, as far as I know, been studied. However, in mucociliary epithelium, it is likely that most of both active Na^+ absorption and active Cl^- secretion is performed by ciliated cells. They are the predominant columnar cell type of the epithelium, and contain all the necessary ion channels.[8] There is no evidence for significant levels of either Na^+ absorption or Cl^- secretion by goblet cells. In fact, one study provides clear evidence against it. Thus, when highly ciliated cultures of guinea-pig tracheal epithelium were exposed to IL-13, 60% of the ciliated cells were converted into goblet cells, and the I_{sc} declined by almost exactly the same amount.[73] In the epithelium of the smaller airways, Clara cells contribute to both active Na^+ absorption and Cl^- secretion.[74–76]

Fully differentiated Type I and Type II appear much later in gestation than ciliated and goblet cells of the airway epithelium. In most species they are first found after ~60% of gestation, but are not fully functional until after about 80–85% (see Chapter ??). The early, undifferentiated epithelium of the alveoli expresses CFTR and secretes Cl^- and liquid.[77] In differentiated alveolar epithelium, early immunocytochemical studies showed Na^+-K^+-ATPase to be localized in Type II but not Type I cells,[78] and cultures of Type II cells showed active Na^+ absorption and Cl^- secretion.[11,60] This led to the idea that they were primarily responsible for transepithelial ion transport. Type I cells, by contrast, were comparatively rich in aquaporins, and isolated Type I cells had an osmotic permeability about 5× that of Type II cells.[4] Thus it was proposed that Type I cells allowed water to follow the ions actively transported by the Type II cells. Later work, however, showed that Type I cells had ENaC, amiloride-sensitive Na^+ uptake and Na^+-K^+-ATPase.[79,80] They thus contained the machinery necessary for active transepithelial Na^+ absorption. It was further shown that isolated cells of either type showed similar levels of amiloride-sensitive, ^{22}Na uptake.[80] Therefore, given that Type I cells constitute ~95% of the alveolar epithelial surface, it was proposed that they are mainly responsible for active Na^+ absorption by the alveolar epithelium.[81] However, in these studies the amiloride-sensitive ^{22}Na uptakes were normalized to total cell protein. But, isolated Type II cells are approximately smooth spheres, whereas isolated type I cells are characterized by multiple long, thin cytoplasmic extensions.[4] When expressed per unit area of membrane, the ^{22}Na uptake of Type I cells would therefore be much less than for Type II. This conclusion contrasts with recent patch-clamp data suggesting that ENaC is present at the same density in the apical membranes of Type I and Type II cells.[81] Both cell types express β-adrenergic receptors[82,83]; as described later, β-adrenergic agents are potent stimulators of lung liquid absorption in both adult and neonatal lungs.

Type II cells contain functional CFTR and can actively secrete Cl⁻ and water.[60,84,85] Interestingly, they lack CaCC.[84] Type I cells also contain CFTR and show ^{36}Cl⁻ influx that is blocked by CFTR inhibitors.[81,86] Both cell types may therefore be involved in active secretion of Cl⁻ and water.

INTACT ADULT LUNG

Transport of ions and water has been studied in several preparations of intact adult lung: isolated perfused lungs, lungs perfused in situ, nonperfused lungs in vitro, and lungs in vivo.[10] The health of non-perfused lungs in vitro declines within a couple of hours, as manifested by a generalized increase in endothelial permeability. However, even without ventilation or perfusion, active clearance of fluid from the alveoli may persist for several hours,[87] and this preparation is obviously the only one suitable for studying intact human lungs.[88] In comparison, the in situ perfused lung has the advantages of preserving the lymphatic drainage and of avoiding the trauma associated with removing the lungs from the thorax. Viability is maintained for about twice as long as in isolated lungs. Finally, liquid can be introduced into lung lobes of unanesthetized animals. This preparation has the disadvantage that liquid removal can only be determined once: at the end of the experiment, when the animal is anesthetized and the lung liquid removed. A disadvantage of all intact lung preparations is that it is not possible to distinguish between the contributions of the epithelia of the alveoli and distal airways.

In intact lungs, liquid absorption is stimulated by instillation of β-adrenergic agents or cAMP (~2 to 3-fold increase).[10,89,90] A recent patch-clamp study on rat lung slices showed that β₁-adrenegic agents activated non-selective ENaC on both Type I and Type II cells, whereas β₂-agents activated highly selective ENaC also on both cell types.[90] Other agents known to stimulate alveolar liquid clearance in intact lungs include endotoxin,[91] TGF-α,[92] acetylcholine,[93] LTD₄,[94] norepinephrine (through both α- and β-receptors),[95] T₃,[96] lipoxin,[97] and dopamine.[98] Some have found TNF-α to stimulate lung liquid absorption (by a cAMP-independent mechanism).[48,99,100] Others have found it to reduce mRNA levels for all subunits of ENaC and

result in alveolar flooding.[101] TGF-β inhibits fluid absorption,[102] as does thrombin (by promoting internalization of Na⁺-K⁺-ATPase).[103]

In perfused lungs, addition of ouabain to the perfusate inhibits liquid clearance generally by ≥90%.[88,104] In the mouse, amiloride causes the lungs to switch from absorption to secretion.[105] In other species, however, multiple studies have shown that liquid absorption is inhibited by 40–70% by luminal amiloride.[10] Conclusive evidence for an involvement of ENaC in the amiloride-sensitive fluid absorption was provided by studies in which α-ENaC levels were reduced by the use of small interfering RNA (siRNA).[106] In control lungs, removal of liquid was stimulated 165% by the β-adrenergic agent, terbutaline. Treatment of rats intratracheally with a plasmid that generated siRNA to α-ENaC caused this protein to become undetectable in Type II cells. This treatment also inhibited baseline fluid absorption by 30%, but completely eliminated the stimulatory effect of terbutaline. Thus, ENaC involvement in liquid absorption may be less under baseline than stimulated conditions.

Most of the amiloride-insensitive absorption probably reflects the activity of cyclicGMP-mediated Na⁺ channels (CGNC). For instance, in rat lungs in vivo, terbutaline increased liquid clearance by 85%. Amiloride and 1-cis-diltiazem (an inhibitor of cyclic-nucleotide gated channels) inhibited nearly equal fractions of the terbutaline-stimulated liquid absorption, and their actions were additive.[107] A permeable analogue of cyclic GMP also induced both amiloride-sensitive and insensitive fluid absorption.[107] A recent review by O'Brodovich et al.[45] provides more information on amiloride-insensitive absorption of lung liquid.

Apical membrane processes other than Na⁺ channels may contribute to liquid absorption. In one study, for instance, amiloride inhibited the absorption of luminal liquid from isolated perfused rat lungs by ~60%. However, amiloride in combination with either phloridzin (an inhibitor of Na⁺-glucose cotransport) or glucose removal abolished fluid absorption.[48] A feature of these experiments was that the lung was perfused with blood from a second rat. Thus, the presence or absence of Na⁺-glucose cotransport in various preparations may depend on hormonal influences.

As discussed above, β-adrenergic agents enhance fluid absorption from intact lungs. However, in perfused rat lungs, when Na⁺ (and fluid) absorption was first blocked with a combination of amiloride and phloridzin, cAMP or a β-adrenergic agent stimulated liquid secretion.[89] Thus, the transport apparatus for Cl⁻ secretion is present in adult pulmonary epithelium, and liquid secretion can occur under certain conditions. In this particular case hyperpolarization of the apical membrane by amiloride was probably required for Cl⁻ secretion. However, Cl⁻ secretion can be induced without first blocking Na⁺ channels. Thus, it was found that elevation of left atrial pressure in mice to15 cm H₂O (which raises pulmonary capillary pressures and mimics heart failure) caused the lung to switch from being absorptive to secretory. Further, the liquid secretion was abolished by bumetanide or blockers of CFTR indicating that it was driven by transepithelial Cl^- secretion.[105] Thus, active Cl^- secretion across the pulmonary epithelium may contribute to the pulmonary edema associated with cardiac failure.

CULTURES OF ADULT TYPE II CELLS

Cell culture allows tubular epithelia to be converted into cell sheets. It also allows epithelial sheets to be created from individual types of isolated cells. If tight junctions form, and the cultures are confluent, the sheets develop a finite transepithelial electrical resistance (R_{te}). Further, if the active transport processes operating across the sheet are electrogenic (i.e., generate a V_{te}) then the summed magnitude of these processes can be determined with Ussing's short-circuit current technique. In this method, a sheet of epithelium is mounted between plastic half-chambers and current passed across it to clamp V_{te} to zero (the sheet is short-circuited). Provided certain conditions are met, this current, the short-circuit current or I_{sc}, is equal to the sum of all the active ion transport processes across the tissue.[108] The specific processes that account for the I_{sc} can be determined with radioisotopes,[109] or by the use of blockers of specific transport proteins.

In 1982, Mason et al. reported that adult rat Type II cells grown on permeable supports, when studied in Ussing chambers, showed V_{te} of ~1 mV that was enhanced by terbutaline and abolished by amiloride.[110] These results indicated that active amiloride-sensitive absorption of Na^+ is quantitatively the most important active ion transport process operating across adult Type II cells; there was no evidence for active Cl^- secretion. This basic conclusion has been confirmed in several subsequent studies.[11] Most compellingly, the use of radioisotopes has shown that the I_{sc} across rat Type II cell cultures is entirely accounted for by net absorption of Na^+; there was no active secretion of Cl^- under any experimental condition.[111]

Evidence for active absorption of salt and water by Type II cell cultures comes not only from Ussing chamber studies but also from the development of "domes" when the cultures are grown on solid supports. Domes are transient blisters where absorbed liquid accumulates between the epithelium and the growth support (generally a plastic petri dish). They grow until ruptured by hydrostatic pressure, and then subside. Both amiloride and ouabain reduce their numbers.[112]

In addition to β-adrenergic agents, all the other factors that stimulate liquid absorption across the wall of the intact lung stimulate Na^+ and liquid absorption across Type II cell cultures.[11,39,93,94,100,113–115]

In the above studies the Type II cells were grown in submersion culture (i.e., with medium several millimeters deep on their mucosal surfaces). Under these conditions Type II cells lose their typical markers (e.g., lamellar bodies) and start expressing features and proteins more characteristic of Type I cells.[115–120] Airway epithelial cells have long been grown with an "air-liquid interface."[121] In this approach, cells are grown on porous-bottomed inserts, and culture medium is added to the basolateral side of the tissue only. Once the cultures become confluent, form tight junctions, and develop a significant R_{te}, only a film of liquid ~15 μm deep remains on the mucosal surface.[122] Such air-interface feeding promotes a vast improvement in structural and functional differentiation.[64,66,121] Interestingly, adult Type II cells grown in this way show cAMP-dependent secretion of Cl^- and liquid.[60,84]

Ussing chamber results from airway epithelial cultures and native epithelium can be compared. Unfortunately the same is not true of Type II cells, and the applicability of results on cultures to native cells is unclear. The low I_{sc} (~2 μA/cm^2) of Type II cell cultures could, therefore, be due to dedifferentiation in culture. In this regard, it is worth noting that with improvement in culture conditions, baseline I_{sc} of airway epithelial cell cultures increased over the years from ~2 to ~50 μA/cm^2.[64,66,123,124] However, though Type II cells cultured on a collagen gel with an air-liquid interface showed a considerably improved ultrastructure compared to cells grown in submersion culture,[125] there was little increase in I_{sc}.[60,84,126]

INTACT FETAL LUNG

The first evidence for liquid secretion by the fetal lung was that ligation of tracheas of fetal rabbits led to lungs abnormally distended with liquid.[127,128] Similar findings had earlier been reported for congenital obstructions of the trachea in human fetuses.[129] In near-term fetal lambs, use of tracheal cannulas has shown that fetal lung liquid is produced at almost 500 ml per day.[53] An arrangement of cartilage and muscle at the laryngeal outlet acts as a sphincter that prevents entry of amniotic fluid but opens periodically to allow lung fluid to be discharged.[130] Secretion of lung liquid generates an intraluminal pressure of 1.5 to 3 mm Hg,[131] and either this pressure or the elevated volume is necessary for normal lung development; hypoplastic lungs result when lung liquid is drained out through a tracheal cannula.[132,133] The secretion of lung liquid is blocked by bumetanide[134,135] and stimulated by prolactin.[136]

Tracheal cannulae can be used to replace fetal lung liquid with experimental media.[137] When both impermeant and permeant tracers are present in the introduced medium, the concentration of the former will decline solely due to liquid secretion into the pulmonary lumen. By contrast, the concentration of permeant markers will decline both by dilution and by diffusion across the pulmonary epithelium. By comparing the rates of change in concentration of the two types of marker, transepithelial flux of the permeant marker from lumen to blood can be estimated. Flux from blood to lung lumen can be measured simply by injecting tracer into the blood and sampling pulmonary liquid at timed intervals. Unidirectional fluxes of Na+, K+, and Cl− were measured in this way. If they were passive (i.e., entirely diffusional) then they should have obeyed Ussing's flux ratio equation,[138] which predicts that in the absence of solvent drag:

$$J_{PL}/J_{LP} = a_P/a_L \bullet e^{zVteF/RT} .$$

Where J_{PL} and J_{LP} are the unidirectional fluxes from plasma-to-lumen and lumen-to-plasma, a_P and a_L are the ion's activities in plasma and lung lumen, and z, F, R, and T are, respectively, the ion's valency, Faraday's constant, the Universal gas constant, and the temperature in °A. A V_{te} of ~4 mV (lung lumen negative) was measured, and after the possibility of significant solvent drag had been excluded, the predicted and measured flux ratios were compared.[139] It was concluded that Na+ was passively distributed, but that K+ and Cl− were actively secreted. The initial studies were done on fetuses at 123–144 days if gestation (term = 147 days), but similar results were later obtained as early as 84 days gestation.[140]

FETAL LUNG EXPLANTS

When pieces of fetal lung are placed in submersion culture, their open ends seal and they form cysts.[141] Liquid transport into these cysts can be estimated by their change in size. Also, they can be pierced with a microelectrode, and the types of ion transport present determined from the sensitivity of their V_{te} to pharmacological agents.

Pieces of lung from 14- or 16-days gestation rats (term = 22 days) were maintained in culture for up to 3 weeks. The cells lining the cysts contained typical lamellar bodies, and cyst volume increased ~3-fold from 1 to 2-wk in culture.[142] At 1 wk in culture the [Cl−] inside the cysts was 42% greater than in the bathing medium, even though the cysts had a lumen negative V_{te} of ~3 mV. In both 1 wk and 2 wk old cysts, V_{te} was essentially abolished by ouabain or bumetanide and increased 50% by terbutaline. By contrast, amiloride microinjected into the lumen had no effect on V_{te} of 2-wk cysts, and only a small inhibition on 1-wk cysts (−20%). All these results are consistent with the presence of cAMP-mediated Cl− secretion, and with active Na+ absorption being absent or of comparatively trivial magnitude. Similar results were shortly afterwards published by others.[143]

Isolated dispersed Type II cells from 18-day gestation fetal rats were plated on collagen gels, and formed cysts with the apical membranes facing the cyst lumen.[144] A V_{te} of ~2 mV, lumen negative, developed. A 24-h. exposure to

bumetanide decreased the total number of cysts and the volume of individual cysts. Both parameters were increased by 24-h. exposure to cAMP analogues. Neither bumetanide nor cAMP altered 3H-thymidine incorporation.

Explant cultures of human fetal lung behaved similarly to those from rat.[77] Thus, addition of isoproterenol or a combination of CPT-cAMP/IBMX caused cysts to swell more rapidly than untreated controls. Further, the lumen negative V_{te} (~4 mV) was doubled or trebled by these agents, an effect inhibited by loop diuretics. These explants were shown to contain CFTR, and the volume and V_{te} of explants from fetuses with CF did not increase on elevation of cAMP. Thus, CFTR is presumably involved in the cAMP-dependent stimulation of Cl− secretion in these alveolar cysts. This is puzzling as lung development is normal in CF, whether in humans or transgenic animals.[20–23] However, baseline fluid secretion was normal in the CF explants suggesting that baseline Cl− secretion via anion channels other than CFTR is sufficient for normal lung development. There are several possible candidates for these alternative channels. CLC-2, -3, and -5 are present in fetal lungs of humans and various other species.[145–149] The most likely candidate, though, is CLC-2. It is found in the apical membrane of fetal airway epithelium.[146] It is activated by extracellular acidity, as is Cl− secretion across rat fetal distal lung epithelial cells.[150] Further, the pH of fetal lung liquid (at least in sheep) is ~6.3.[55] When levels of CLC-2 in rat fetal lung cysts were halved with antisense oligonucleotides to CLC-2 gene, both V_{te} and liquid secretion were halved.[24] Keratinocyte growth factor stimulates fluid secretion into fetal lung cysts,[151,152] and also upregulates expression of CLC-2.[153] It has the same effects on lung cysts from wild-type or CFTR knock-out mice.[152] In a human CF airway cell line (IB-3), overexpression of CLC-2 increases Cl− conductance.[154] The promoter region of the CLC-2 gene contains binding sites for the Sp1 and Sp3 transcription factors that are involved in lung development and are downregulated postnatally (as also is the CLC-3 gene).[155]

In addition to β-adrenergic agents, cAMP analogues and KGF, other agents shown to stimulate Cl−/fluid secretion by fetal lung explants include adenosine, ATP, and UTP,[156] prostaglandins,[157] and atrial natriuretic factor.[158]

CULTURES OF FETAL ALVEOLAR TYPE II CELLS

Surprisingly, several studies using primary cultures of fetal Type II cells have all shown amiloride-sensitive I_{sc}, and have often failed to detect Cl− secretion. When Cl− secretion has been demonstrated it is sometimes only under permissive conditions (e.g., with Cl−-free solution in the mucosal bath, or after the apical membrane has been hyperpolarized by amiloride). V_{te} is generally small (≤5 mV) and R_{te}

high ($100–1000$ $\Omega.cm^2$). The significance of these values is uncertain, given that properties of cultured cells are very dependent on growth conditions, and that most of the cultures used were somewhat dedifferentiated (they had lamellar bodies, but not as many or as large as in native Type II cells).[11,159] Also, of necessity, most cultures were initiated from cells derived from fetuses close to term, and it is possible that they attained the adult phenotype during the several days they were kept in culture.

In the first cultures of fetal alveolar epithelium (from rats at 18–20 days gestation), I_{sc} was potently inhibited by amiloride, but bumetanide and DPAC (diphenylamine-2-carboxylate, a blocker of CFTR and other Cl^- channels), were without effect.[70] Terbutaline stimulated I_{sc}, but sensitivity to amiloride showed that this was due to stimulation of active Na^+ absorption rather than Cl^- secretion. These cultures were derived from fetuses within 2–4 days of term, and their ion transport may have switched from Cl^- secretion to Na^+ absorption during the culture period of 2–6 days. Though there may well be changes in the balance of the transport processes in culture, the younger the fetus, the greater is the level of Cl^- secretion relative to Na^+ absorption.[160] For instance, in one series of experiments, amiloride inhibited I_{sc} across cells from 18-day fetuses by 36%, and bumetanide (in the continued presence of amiloride) then inhibited I_{sc} by 33%. By contrast, in cells from 21-day fetuses, amiloride inhibited I_{sc} by 68%, and bumetanide increased I_{sc} by 4%.[160] Other work[161] also suggested that cultures from 18-day gestation rats showed significant levels of Cl^- secretion: amiloride inhibited I_{sc} by 15% and DPAC by 50%, though the dose of DPAC (3 mM) may have been toxic. Later studies on midgestation human epithelial cells or on 137–142-day fetal sheep cells (term = 145 days), showed clear-cut evidence for active cAMP-dependent Cl^- secretion that was at least as great as the levels of amiloride-sensitive Na absorption.[85,162,163] Interestingly, in both rat and human cells there is evidence for considerable levels of bumetanide-insensitive Cl^- secretion. For instance, in cultures of fetal human alveolar epithelium (18–24 wk gestation), amiloride reduced I_{sc} by 20%, ouabain by ~80%, and DPAC by 26%, but bumetanide was without effect.[162] Terbutaline, ATP, and ionomycin all stimulated I_{sc} in cultures pretreated with amiloride, consistent with the presence of both cAMP- and Ca^{2+}-activated Cl channels.[162] Later studies,[84] however, showed that Type II cultures lacked the likely candidates for CaCC, and that increases in I_{sc} in response to elevation of $[Ca^{2+}]_i$ were inhibited by blockers of CFTR.[84] Perhaps Ca^{2+}, by opening basolateral K^+ channels, hyperpolarizes V_a and drives Cl^- through constitutively open CFTR. On the other hand, there is accumulating evidence for a number of signaling pathways directly linking elevation of $[Ca^{2+}]_i$ to activation of CFTR.[164]

ADULT AIRWAY EPITHELIUM

As reviewed in detail elsewhere,[8,63] the I_{sc} across most adult airway epithelia can be accounted for by active absorption of Na^+; there is generally no net movement of Cl^-. Under "open-circuit" conditions (i.e., with the tissue at its spontaneous V_{te}), this active absorption of Na^+ results in essentially electrically neutral absorption of Na^+ and Cl^- with water following at a rate of ~5 $\mu l.cm^{-2}.h^{-1}$.[8] Active absorption of Na^+ by airway epithelia is little affected by neurohumoral agents.[8,165] Some airway epithelia, especially those of the trachea, when short-circuited, may also show active secretion of Cl^-, which can be stimulated by a wide range of neurohumoral agents acting through either intracellular cAMP or Ca^{2+}.[8] However, it is questionable whether these Cl^--secreting tissues secrete much fluid under open-circuit conditions. Thus, the presence of a luminal negative V_{te} of ~30 mV causes an equivalent depolarization of the apical membrane, reducing the driving force for Cl^- exit or even creating a driving force for entry.[58] In fact, most of the adult airway epithelia that show net Cl^- secretion under short-circuit conditions show net Cl^- absorption or no net Cl^- movement under open-circuit.[8]

The larger bronchi have been studied in sheep and dogs.[57,166] They tend to have lower V_{te} and lower R_{te} than tracheas, though I_{sc} is similar. They are exclusively Na^+-absorbing even in the dog, in which the trachea shows active Cl^- secretion.

Small bronchi and bronchioles are hard to study and prone to dissection damage. Miniature Ussing chambers have been used,[167,168] but generally the airways are perfused intact, and the effects of transport blockers on V_{te} assessed. From the effects of amiloride, bumetanide or Cl^- channel blockers, it appears that they are predominantly Na^+ absorbing with levels of Cl^- secretion that vary from zero to about the same as the active Na^+ absorption.[167–172] Cultures of human bronchiolar epithelium are also predominantly Na^+-absorbing.[173]

FETAL AND NEWBORN AIRWAY EPITHELIUM

Three studies have compared the transport properties of fetal and adult tracheas in Ussing chambers. In the first,[57] fetal sheep tracheas, studied at 140–145 days (term = 145 days), showed net secretion of Cl^- that was not significantly different from the I_{sc}. There was no net transepithelial Na^+ movement. Both net Cl^- secretion and I_{sc} were stimulated by isoproterenol, and to approximately the same extent. In the tracheas of the mothers, by contrast, the I_{sc} was accounted for by net Na^+ absorption; there was no net Cl^- movement. Isoproterenol had no effect on any of the unidirectional or net fluxes of Na^+ or Cl^- across fetal epithelium. Under open-circuit, fetal tracheas secreted and adult tracheas absorbed

NaCl. In direct contrast, a later study found net Na^+ absorption, but no net Cl^- movement across both fetal and adult sheep tracheas.[174] However, studies with pharmacological agents were consistent with a shift from fetal Cl^- secretion to adult Na^+ absorption. Thus, isoproterenol increased net Cl^- flux towards the lumen in both fetuses and adults, but the change was much greater in fetuses. Conversely, amiloride had negligible effects on Na^+ fluxes in the fetus, but reduced net Na^+ absorption by 60% in the adult. In the third study,[175] in short-circuited fetal dog tracheal epithelium (1–3 days before term), I_{sc} was entirely accounted for by net Cl^- secretion. Second generation fetal bronchi from dogs are also Cl^--secreting.[175]

Compared to cultures from adult animals, cultures of fetal rabbit tracheal epithelium showed a smaller inhibitory effect of amiloride on I_{sc} (-21% vs. -35%), and a larger effect of furosemide (-74% in fetal cells, -53% in adult).[176] This is consistent with the shift from secretion to absorption after birth. However, the large effect of furosemide on I_{sc} across adult cultures is surprising; I_{sc} of adult rabbit tracheal epithelium is entirely accounted for by net Na^+ absorption.[177,178]

Fetal rat tracheas in submersion culture seal to form cysts that accumulate fluid over 2 weeks in culture.[142] They develop higher V_{te}'s (8–21 mV) than alveolar cysts grown under the same conditions (1–4 mV). The V_{te} is stimulated by β-adrenergic agents, abolished by ouabain and inhibited 70% by bumetanide. The effects of amiloride on V_{te} are comparatively small (−6% at 1 week and −17% at 2 weeks).

In airways, the switch from Cl^- secretion to Na^+ absorption occurs at varying times after birth depending on species. Sheep tracheas studied within the first two days after birth were predominantly Cl^--secreting. However, at 2, 4, and 8 weeks after birth, and in adults, there was no net Cl^- secretion, and Na^+ absorption accounted for the I_{sc}.[166] In dog tracheas, by contrast, for up to 46 days after birth, the predominant ion transport process was Cl^- secretion.[175]

PERINATAL ABSORPTION OF LIQUID

Overview

For about twenty years from the mid-1980s, it was generally accepted that liquid absorption from the lung began shortly before birth, and was due mainly to a hormonally induced switch from Cl^- secretion to amiloride-sensitive Na^+ absorption. Recently, however, prenatal absorption of liquid has been questioned. Further, it has been suggested that other factors are of more importance than solute transport in liquid absorption immediately after birth. One of these is transmural hydrostatic pressure gradients generated during early inspirations. Another is contractions of fetal thoracic and abdominal muscles. The change in fetal blood oxygen tensions favors active Na^+ absorption over active

Cl^- secretion. However, the effects of increased O_2 tension on ion transport are slow, and may be more important in keeping adult lungs free of luminal liquid than in absorbing fetal lung liquid.

Time Course of Liquid Removal

Early work indicated that the rate of liquid secretion into the fetal lung declines shortly before birth,[179,180] but the rate at which liquid is lost from the lungs via the trachea does not change.[180] The end result is a decline in lung liquid volume. In the sheep, for instance, secretory rates of 7.4, 16.8, and 7.1 ml/h were reported at 119, 135, and 142 days gestation, and the corresponding volumes of lung liquid were 51, 105, and 70 ml.[179] However, later studies on sheep have found lung liquid volume to increase continuously up to birth.[181] Prenatal removal of liquid is potentially important: in lambs, respiratory performance after birth is inversely related to the volume of liquid in the lungs at the first breath.[182]

Some lung liquid is lost via the mouth during labor. Possibly because of this, rabbits delivered by Caesarian had lungs that contained more water than those of rabbits delivered vaginally.[183,184] In 1–7-day-old human newborns, the average thoracic gas volume of babies delivered vaginally was 32.7 ± 1.7 ml/kg, but in those delivered by Caesarian section it was significantly lower (19.7 ± 1.4 ml/kg).[185] This is possibly part of the reason that human babies delivered by Cesarean surgery show a small increase in the likelihood of later complications (transient tachypnea of the newborn, respiratory distress syndrome) compared to those delivered naturally.[186,187] The mechanism by which labor stimulates removal of lung liquid could be as simple as compression of the chest wall during delivery,[188] or could reflect transient increases in levels of epinephrine or other hormones that stimulate transepithelial solute transport (see the following text). After birth, liquid is removed equally efficiently from the lungs of rabbits born vaginally or by Cesarean operation.[184]

How rapidly is liquid removed at birth, and what is the corresponding flow of liquid across the epithelium? The volume of the human lung at birth is ~100 ml and the alveolar surface area is ~4 m^2.[189] As shown in Figure 1, the lungs of human newborns fill with air at an initial rate of ~75 ml/h,[185] and the half-time for liquid removal is ~1-h. This corresponds to a rate of liquid removal of ~2 $\mu l.cm^{-2}.h^{-1}$.

This is consistent with considerations of individual alveoli; at 2 $\mu l.cm^{-2}.h^{-1}$, it will take an alveolus of 100 μm radius ~2-h to clear all its luminal liquid. The slowness of lung liquid removal means that, compared to adults, newborns must be significantly hypoxic for at least an hour or so after birth. However, they are used to it; the normal fetal arterial O_2 tension is ~ 20 mm Hg.[190] Removal of liquid from the sheep lung is considerably slower than from the human. A total lung volume of ~250 ml is removed at an initial rate of ~25 ml/h.[191]

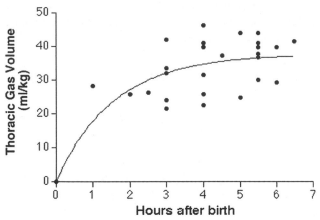

FIGURE 1 Filling of human lungs with air after birth. The linear regression of lung volume against time is statistically significant. However, the line fitted to the data is the best least-squares exponential approach to an asymptote, with an initial increase in volume of 75 ml/h. *From Milner et al.*[185]

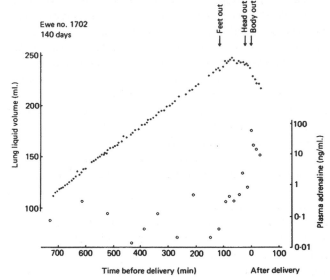

FIGURE 2 Changes in lung liquid movement and maternal plasma epinephrine levels at birth in the sheep. *From Brown et al.*[191] *See text for details.*

Humoral Regulation

In fetal sheep, the removal of lung liquid immediately after birth is associated with a surge in plasma epinephrine content (Figure 2).[191] Further, there is a progressive increase in sensitivity to epinephrine between 122 and 142 days of gestation. Thus, at 122 days, infusion of epinephrine merely caused a slight inhibition of liquid secretion. At 133 days, epinephrine induced a small level of absorption with the critical concentration for epinephrine being ~0.5 ng/ml. Close to term, at 142 days, the magnitude of the switch from secretion to absorption was considerably greater than at 133 days, and the critical concentration of epinephrine had declined to 0.08 ng/ml.[191] Isoproterenol had similar effects on lung liquid movement as epinephrine, and the actions of both agents were inhibited by propranolol (a β-adrenergic blocker); norepinephrine was without effect on lung liquid secretion.[192]

To respond to epinephrine the pulmonary epithelium must first be primed with triiodothyronine (T_3) and hydrocortisone (HC), circulating concentrations of which progressively increase during ovine gestation.[193,194] Thyroidectomy of fetal sheep completely abolishes the actions of epinephrine on lung liquid absorption,[195] and infusion of T_3 restores responsiveness.[196] Fetal infusion of T_3 did not change the age at which responsiveness to epinephrine first developed (i.e., ~122 days). However, if both T_3 and hydrocortisone were infused (to produce plasma levels in the upper part of the range seen just before full term), then responsiveness to epinephrine could be induced in younger fetuses (116–120 days) than normal.[197]

There is abundant evidence that glucocorticoids upregulate ENaC and Na^+-K^+-ATPase in the fetal (and adult) lung of all species studied (human, rabbit, rat, mouse).[198–204] In newborn humans, levels of α-ENaC mRNA in nasal epithelial cells correlated positively with cord plasma cortisol levels.[205] Chronic exposure to epinephrine also upregulates Na^+-K^+-ATPase,[206] though the short-term effects of this agent are on the opening probability of ENaC.[34] T_3 stimulates Na^+-K^+-ATPase activity in cultures of fetal (and adult) distal lung epithelium.[207,208]

There is evidence that epinephrine may not be required for lung liquid removal. In rabbits, an irreversible β-adrenergic agonist inhibited surfactant production but had no effect on lung water content at or after birth.[209] In sheep propranolol does not prevent absorption of liquid at birth,[210] though it is possible that occupancy of only a small fraction of the total number of receptors is needed for the effects of epinephrine. If not epinephrine, then what is stimulating liquid absorption? AVP, at concentrations within the range seen during labor, inhibited liquid secretion in >140 day fetal sheep by ~80%, an effect that occurred without change in circulating epinephrine levels.[211] EGF slows lung liquid secretion in fetal sheep, even in the presence of β-blockade.[212] PGE_2 and atrial natriuretic factor have also been shown to reduce tracheal liquid production in fetal lambs.[158,213]

Work with explants of fetal rat lung or trachea indicated that the effects of hormones on lung liquid absorption were mediated predominantly by alveolar rather than airway epithelium.[214] Cultures of rat lung and trachea were initiated at 14 or 16 days of gestation, respectively. Both tracheas and lung buds showed active Cl^- secretion. The V_{te} in both was stimulated by terbutaline and inhibited by bumetanide. Further, pretreatment with bumetanide abolished the effects of terbutaline. However, pretreatment of lung buds for 2 days with HC + T_3 produced results consistent with the presence of active Na^+ absorption. Thus, terbutaline now increased V_{te} in the presence of bumetanide, and this increase was blocked by amiloride. By contrast, in tracheal buds exposed to HC + T_3, the terbutaline-induced increase in V_{te} was blocked only by bumetanide, not amiloride. Thus, only in the lung buds did the combination of T_3, HC, and β-adrenergic agent convert Cl^- secretion to Na^+ absorption. A combination of T_3, HC, and terbutaline reduced the wet weights (a measure of liquid secretion) of lung buds relative to control, but was without effect on tracheal buds.

In support of a relatively greater role for solute transport across alveolar epithelium vs. airway epithelium in perinatal liquid removal are the findings (described earlier) that the epithelium of the larger airways of sheep and dogs does not become absorptive until some days after birth. Also the airways have a highly unfavorable surface area to volume ratio compared to alveoli, and it would take ~ 15-h to remove all the liquid from a newborn human trachea of 5 cm length and 3 mm diameter if absorption were at the adult rate of 5 $\mu l.cm^{-2}.h^{-1}$. However, following birth, luminal liquid is drawn down along the airways towards the alveoli.[215] Ultimately, therefore it is the alveolar epithelium that is responsible for clearance of most airway liquid. In addition, as discussed later, some of the liquid in the large airways may be pushed out through the mouth by contraction of fetal abdominal and thoracic muscles.[216]

Fetal and adult blood O_2 tensions are very different: 20 and 100 mm Hg, respectively. Carbon dioxide tensions also differ: 60 mm Hg in the fetus and 40 mm Hg in the adult. To determine whether the change from fetal to adult gas tensions contributed to neonatal lung liquid absorption, lung explants from 14-, 20-, 22-day fetal and 2-day neonatal rats were exposed to fetal or adult P_{O2} and P_{CO2}.[217] The wet-to-dry weight ratios of 14-day fetal or 2-day neonatal explants were independent of gas tensions. However, the water contents of 20-day and 22-day explants were markedly increased by fetal as opposed to adult gas tensions. Thus, in the perinatal period, fetal gas tensions favored liquid secretion.[217] The effects of gas tension have also been investigated in primary cultures of fetal distal lung epithelium. When O_2 was switched from 3 to 21%, amiloride-sensitive I_{sc} increased by ~2-fold after 18 h and ~8-fold after 48 h. Others have reported similarly slow changes in Na^+ absorption following changes in P_{O2}.[218,219] The increase in Na^+ absorption produced by elevated P_{O2} is associated with increased numbers of both ENaC and basolateral Na^+-K^+-ATPase.[202,218–220] The slowness of the above changes in Na^+ transport suggest that PO_2 may be more important at maintaining liquid absorption in the adult than in removing lung liquid at birth.

Some of the developmental changes in ion and water transport across pulmonary epithelium may be brought about by changes in the extracellular matrix. Thus, when isolated fetal pneumocytes were grown on matrix derived from mixed lung cells at the canalicular phase, they reverted to

an immature phenotype with less amiloride-sensitive Na^+ absorption and more Cl^- secretion than when grown on a variety of other substrates.[218]

Role of Active Solute Transport

Multiple studies have indicated that ENaC is central to the hormonally mediated perinatal absorption of lung liquid. In catheterized fetal sheep in utero, epinephrine, given shortly before term, increased whole-lung V_{te}. Further, as revealed by use of the Ussing flux ratio equation, Cl^- transport went from active to passive and Na^+ transport from passive to active. These changes were abolished by luminal amiloride.[221] Permeable analogues of cAMP also induce a change from secretion to absorption in near-term sheep fetuses, and this too is blocked by amiloride.[222] Immediately at birth, and before the first breath, amiloride was instilled into guinea-pig lungs. Compared to saline-controls, it led to a dose-dependent increase in the amount of extravascular lung water (EVLW) 4-h after birth. It had no effects on EVLW in 9-day old animals.[223] Similar results were described for fetal rabbits removed near term and artificially ventilated.[224] Transgenic mice lacking α-ENaC fail to remove lung liquid and die of respiratory distress within forty hours of birth.[33] Removal of γ-ENaC or β-ENac slows but does not prevent neonatal lung liquid clearance.[225,226] Thus, both α_3 and αβγ-ENaC may be involved. In newborn rat lung, blockers of Na^+/H^+ exchange or Na^+-glucose cotransport have no effect on liquid removal.[227]

Developmental Changes in Transport Proteins

Developmental changes in transport proteins are consistent with a neonatal switch from active Cl^- secretion to active Na^+ absorption. Levels of NKCC in rats increase ~2.5-fold between gestational days 18 and 21 (E18 & E22), but two days after birth (P2) had dropped to ~one-third of the E18 level.[228]

Developmental changes in a number of pulmonary epithelial anion channels have been described. Of these, the expression pattern of CLC-2 most closely parallels the changes in Cl^- secretion. CLC-5 protein levels in rat lung are highest in the fetus and during the first two weeks after birth. There is a decline by 21 days after birth, and the protein is virtually undetectable in adult lungs.[147] CLC-3 protein levels in rat lung are trivial in the fetus, but increase markedly after birth.[147] CLC-2 mRNA and protein expression are most abundant in fetal rat lung and diminished shortly after birth.[146] In human lung, the mRNA for both CLC-2 and CLC-3 peaks at 22 weeks gestation, declining to much lower levels in the adult.[145] In situ hybridization showed CFTR transcript in the fetal primordial epithelium of human lung at the pseudoglandular stage. During development, expression then decreased in the alveolar spaces and became limited to the epithelium of the small airways. After birth, message became undetectable in the alveoli and decreased in the small airways.[229] CFTR protein levels progressively increase between10 and 20 weeks of gestation, but thereafter decrease.[230] In sheep, mRNA levels for CFTR are highest early in fetal live (E52-E80), and then decline progressively to term, with no increase after birth.[149,231]

Several studies on expression of ENaC in rat lungs[199,232,233] have produced similar results. At all times mRNA for α-ENaC was considerably more abundant than mRNA for the other subunits. Alpha-ENaC mRNA was first detectable in fetal rat lung at E18, which is the time when alveoli first start to form and when alveolar α-ENaC is first detectable by immunocytochemistry.[234] Transcripts for β- and γ-ENaC became detectable a couple of days later. Between these times and term (E22), levels of all three subunits progressively increased. Levels of mRNA for all three subunits decreased shortly after birth,[233] returning to levels in the adult that were about the same as at term (α), twice term (β), or quadruple term (γ).[199] In mice, levels of mRNA for all three subunits of ENaC increase sharply to near adult levels immediately before birth.[235] In sheep lungs, the relative levels of mRNA were β-ENaC >> α-ENaC > γ-ENaC. All increased markedly from E80 to term, and then declined (by ~50%) after birth. For α-ENaC protein in whole cell homogenates, levels were a little greater at term than at E80, and declined by a further 50% after birth. Beta-ENaC levels were fairly constant throughout gestation, but again declined after birth.[149] In human nasal epithelium, levels of mRNA for α- and β-ENaC were greater than for γ-ENaC. They all increased throughout gestation and dropped within a day after birth.[236]

Aggregate ion channel activity in the plasmalemma can be altered by changing the probability of individual channels being open or by insertion or removal of channels. One of the ways of removing active channels from the membrane is to sequester them in caveolae. These organelles are enriched in caveolin, and vesicles prepared from guinea-pig lungs were separated into caveolin-rich and caveolin-poor fractions.[237] At 61 days of gestation, ENaC was found in the caveolin-rich section and CFTR in the caveolin-poor. By contrast, at 68 days of gestation (i.e., term), the distribution had reversed. Thus, it appears that the balance between Cl⁻ secretion and Na⁺ absorption can be regulated by exchange of channels between caveolae (where they are inactive) and the rest of the apical membrane (where they are active). Pharmacological disruption of caveolin-rich regions at day 61 caused ENaC to appear in the caveolin-poor areas and switched the direction of fluid movement from secretory to absorptive.

There is a large literature on changes in airway epithelial Na⁺-K⁺-ATPase expression in lung disease,[238,239] or under the influence of hormones.[206,240] Comparatively little is known about how the numbers and turnover rates of Na⁺-K⁺-ATPase change throughout development. In dispersed rabbit Type II cells, levels of ouabain-sensitive ⁸⁶Rb⁺ uptake in fetal, neonatal, and adult pneumocytes were in the ratio 1:3:12, respectively.[241] Studies in which ouabain-sensitive Rb⁺ uptakes were compared to pump numbers (determined from [³H]-ouabain binding) showed that during the transition from fetus to newborn there was no change in the numbers of pump sites, but that turnover of individual pumps increased ~4-fold.[242] By contrast, in moving from newborn to adult, there was no change in turnover rate, but a 5-fold increase in pump density (from ~250 to ~1000 pumps per μm² of basolateral membrane). Thus there is a dramatic increase in the rate of Na⁺ entry at birth (mainly through ENaC) that increases the rate of Na⁺-K⁺-ATPase turnover presumably by elevating [Na⁺]ᵢ. The increase in pump numbers between neonatal and adult cells may reflect the effects of chronically elevated [Na⁺]ᵢ, a condition that leads to synthesis and membrane insertion of further Na-K-ATPase units in cells in general.[243] Rats also show changes in lung Na⁺-K⁺-ATPase levels in the neonatal period, but the timing is a little different from the rabbit. Thus, Na⁺-K⁺-ATPase α₁ subunit increased from fetal day 17 to fetal days 20–22 and then declined in the early postnatal period.[244] Direct measurement of Na⁺-K⁺-ATPase activity showed a 2.6-fold increase between days 17 and 20–22.

Absorption of lung liquid at birth could potentially involve changes in permeability to water as well as changes in ion transport. In fact, in the alveolar epithelium of rats, there are marked increases in the expression of all the major lung aquaporins (AQP-1, -4, and -5) in the perinatal period.[245,246] Further, switching from 4% to 21% O₂, a change known to participate in the neonatal switch from Cl⁻ secretion to Na⁺ absorption, also increased AQP-4 expression.[245] However, mice deficient in all three of the above aquaporins show no alteration in clearance of lung water at birth.[247]

Role of Starling Forces

It has long been appreciated that changes in Starling forces and epithelial permeability could contribute to lung liquid absorption at birth. During early inspirations, interstitial hydrostatic pressures are highly negative.[248,249] Also, the characteristic end-expiratory pauses (expiratory grunting) that occur between the newborn's first few breaths are caused by expiration against a closed glottis, and will create positive alveolar pressures[250] that should also promote liquid absorption. The first breaths are associated with an increase in permeability to non-electrolytes that could correspond to an increase in hydraulic conductivity.[251] An increase in plasma protein concentration in the last few days before birth increases the oncotic driving force for absorption,[248,249] and there is an increase in the diffusional (osmotic) water permeability of rabbit lung epithelium immediately after birth.[252]

Two quantitative considerations point to these forces being important in alveolar liquid clearance at birth. First, as discussed earlier, the rate of fluid absorption from postnatal human lung corresponds to ~2μl.cm^{-2}.h^{-1}. However, cultured rat alveolar Type II cells absorb liquid at 0.84 μl. cm^{-2}.h^{-1}.[253] Grown with an air-liquid interface, the cells used in these studies showed excellent levels of differentiation with abundant lamellar bodies. The absorption rate across the much thinner Type I cells is likely to be much less. Second, it is true that mice lacking α-ENaC have waterlogged airspaces at death. However, their median age of survival is ~20-h.[33] Would they survive that long if no lung liquid were being removed?

Recent studies utilizing new approaches to lung imaging suggest that Starling forces, at least in rabbits, are in fact considerably more important than active transepithelial Na$^+$ absorption in the removal of fetal lung liquid at birth.[254] Aeration of newborn rabbit lungs was viewed by phase contrast X-ray imaging. The air-liquid interface moved distally only during inspiration. Between breaths there was either no significant movement of the interface or a slight proximal movement.[215] In a follow-up study, newly born pups were sedated and subjected to positive pressure ventilation.[255] Again, the air-liquid interface moved distally only during inspiration. Further, the rate of this movement was unaffected by either amiloride or epinephrine. Between breaths, however, there was a small proximal movement of the air-liquid interface that was interpreted as being due to back leak of liquid driven by the positive interstitial pressure created by the presence of absorbate. This was blocked by amiloride. Another effect of amiloride was to increase (by ~15%) the positive inspiratory pressures needed to achieve a given degree of ventilation. The conclusions were (1) that aeration of the lungs did not require active Na$^+$ absorption, (2) even with Na$^+$ absorption present, the primary factor driving lung fluid absorption at birth was a transpulmonary pressure gradient generated during inspiration,[255] and (3) the main role of active solute transport was to prevent ingress of liquid during expiration.

Other recent work points to a role for fetal muscle contractions in expulsion of fluid from the sheep lung. Thus, during birth, contractions of the fetal abdominal and thoracic muscles, mostly associated with uterine contractions, generated fluid flows along the trachea. Most of these expelled a small volume of fluid, and approximately 250 such fetal contractions resulted over ~6-h in the expulsion of ~150 ml of liquid, a volume that is approximately equal to all the fetal lung liquid.[216]

CONCLUSIONS

There is no doubt that active secretion of Cl$^-$ is responsible for filling the fetal lungs with fluid and expanding their volume during development. Further, the turgor pressure generated by this liquid secretion is thought necessary for normal lung development. Interestingly, Cl$^-$ channels other than CFTR seem to mediate the secretion of Cl$^-$; lung development is apparently normal in CF either of humans or of transgenic mice, pigs, and ferrets. At birth, there is a rapid absorption of lung liquid. Initially, this was thought to be driven predominantly by Starling forces or by simple compression of the chest during birth. However, after about 1980 it became generally accepted that fluid was removed from the lungs when, under the influence of epinephrine and other factors, active secretion of Cl$^-$ was replaced by active absorption of Na$^+$ as the primary transepithelial solute transport across alveolar epithelium. Though several types of Na$^+$ absorption may contribute, most studies suggest that conductive flow of Na$^+$ across the apical membrane through ENaC is of primary importance. More recently, however, improvements in imaging of lungs in vivo have led to the conclusion that transepithelial hydrostatic pressures generated during inspiration are of more importance than active Na$^+$ absorption in removal of lung liquid at birth. Contraction of fetal abdominal and thoracic muscles may also play an important role in fluid expulsion, at least in the sheep. The relative importance of these various mechanisms for removing lung liquid is currently uncertain, and probably varies significantly between species. Active absorption of Na$^+$ across both alveolar and airway epithelia persists into the adult, where it guards against alveolar flooding in inflammatory lung disease.

REFERENCES

1. Spring KR. Routes and mechanism of fluid transport by epithelia. *Annu Rev Physiol* 1998;**60**:105–19.
2. Kovbasnjuk O, Leader JP, Weinstein AM, Spring KR. Water does not flow across the tight junctions of MDCK cell epithelium. *Proc Natl Acad Sci USA* 1998;**95**:6526–30.
3. Schnermann J, Chou C-L, Ma T, Traynor T, Knepper MA, Verkman AS. Defective proximal tubular fluid reabsorption in transgenic aquaporin-1 null mice. *Proc Natl Acad Sci USA* 1998;**95**:9660–4.
4. Dobbs LG, Gonzalez R, Matthay MA, Carter EP, Allen L, Verkman AS. Highly water-permeable type I alveolar epithelial cells confer high water permeability between the airspace and vasculature in rat lung. *Proc Natl Acad Sci USA* 1998;**95**:2991–6.
5. Folkesson HG, Matthay MA, Frigeri A, Verkman AS. Transepithelial water permeability in microperfused distal airways. Evidence for channel-mediated water transport. *J Clin Invest* 1996;**97**:664–71.
6. Verkman AS. Role of aquaporins in lung liquid physiology. *Respir Physiolo Neurobiol* 2007;**159**:324–30.
7. Welsh MJ. Electrolyte transport by airway epithelia. *Physiol Rev* 1987;**67**:1143–84.
8. Widdicombe JH. Airway Epithelium. *Morgan Claypool Life Sci* 2012.
9. Frizzell RA, Field M, Schultz SG. Sodium-coupled chloride transport by epithelial tissues. *Am J Phys* 1979;**236**:F1–8.
10. Matthay MA, Folkesson HG, Verkman AS. Salt and water transport across alveolar and distal airway epithelia in the adult lung. *Am J Phys* 1996;**270**:L487–503.

11. Matalon S. Mechanisms and regulation of ion transport in adult mammalian alveolar type II pneumocytes. *Am J Phys* 1991;**261**: C727–38.
12. Wilson SM, Olver RE, Walters DV. Developmental regulation of lumenal lung fluid and electrolyte transport. *Respir Physiolo Neurobiol* 2007;**159**:247–55.
13. Widdicombe JH, Nathanson IT, Highland E. Effects of "loop" diuretics on ion transport by dog tracheal epithelium. *Am J Phys* 1983;**245**:C388–96.
14. McCray PB, Bettencourt JD, Bastacky J. Developing bronchopulmonary epithelium of the human fetus secretes fluid. *Am J Phys* 1992;**262**:L270–9.
15. Frizzell RA, Hanrahan JW. Physiology of epithelial chloride and fluid secretion. *Cold Spring Harb Perspect Med* 2012;**2**:1–19. a009563.
16. Widdicombe JH, Basbaum CB, Yee JY. Localization of Na pumps in the tracheal epithelium of the dog. *J Cell Biol* 1979;**82**:380–90.
17. Widdicombe JH, Basbaum CB, Highland E. Ion contents and other properties of isolated cells from dog tracheal epithelium. *Am J Phys* 1981;**241**:C184–92.
18. Shorofsky SR, Field M, Fozzard HA. Changes in intracellular sodium with chloride secretion in dog tracheal epithelium. *Am J Phys* 1986;**250**:C646–50.
19. Anderson M, Welsh M. Calcium and cAMP activate different chloride channels in the apical membrane of normal and cystic fibrosis epithelia. *Proc Natl Acad Sci USA* 1991;**88**:6003–7.
20. Sturgess J, Imrie J. Quantitative evaluation of the development of tracheal submucosal glands in infants with cystic fibrosis and control infants. *Am J Pathol* 1982;**106**:303–11.
21. Rogers CS, Stoltz DA, Meyerholz DK, Ostedgaard LS, Rokhlina T, Taft PJ, et al. Disruption of the CFTR gene produces a model of cystic fibrosis in newborn pigs. *Science* 2008;**321**:1837–41.
22. Sun X, Sui H, Fisher JT, Yan Z, Liu X, Cho H-J, et al. Disease phenotype of a ferret CFTR-knockout model of cystic fibrosis. *J Clin Invest* 2010;**120**:3149 60.
23. Guilbault C, Saeed Z, Downey GP, Radzioch D. Cystic fibrosis mouse models. *Am J Respir Cell Mol Biol* 2007;**36**:1–7.
24. Blaisdell CJ, Morales MM, Andrade ACO, Bamford P, Wasicko M, Welling P. Inhibition of CLC-2 chloride channel expression interrupts expansion of fetal lung cysts. *Am J Phys* 2004;**286**:L420–6.
25. Welsh MJ. Evidence for a basolateral membrane potassium conductance in canine tracheal epithelium. *Am J Phys* 1983;**244**:C377–84.
26. Welsh MJ, Smith PL, Frizzell RA. Chloride secretion by canine tracheal epithelium II. The cellular electrical potential profile. *J Membr Biol* 1982;**70**:227–38.
27. Bardou O, Trinh NT, Brochiero E. Molecular diversity and function of K$^+$ channels in airway and alveolar epithelial cells. *Am J Physiol Lung Cell Mol Physiol* 2009;**296**:L145–55.
28. Namkung W, Song Y, Mills AD, Padmawar P, Finkbeiner WE, Verkman AS. In situ measurement of airway surface liquid [K$^+$] using a ratioable K+-sensitive fluorescent dye. *J Biol Chem* 2009;**284**:15916–26.
29. Lee SY, Maniak PJ, Rhodes R, Ingbar DH, O'Grady SM. Basolateral Cl$^-$ transport is stimulated by terbutaline in adult rat alveolar epithelial cells. *J Membr Biol* 2003;**191**:133–9.
30. Leroy C, Dagenais A, Berthiaume Y, Brochiero E. Molecular identity and function in transepithelial transport of K(ATP) channels in alveolar epithelial cells. *Am J Physiol Lung Cell Mol Physiol* 2004;**286**:L1027–37.
31. Koefoed-Johnsen V, Ussing HH. The nature of the frog skin potential. *Acta Physiol Scand* 1958;**42**:298–308.
32. Canessa CM, Schild L, Buell G, Thorens B, Gautschi I, Horisberger JD, et al. Amiloride-sensitive epithelial Na$^+$ channel is made of three homologous subunits. *Nature* 1994;**367**:463–7.
33. Hummler E, Barker P, Gatzy J, Beermann F, Verdumo C, Schmidt A, et al. Early death due to defective neonatal lung liquid clearance in alpha-ENaC-deficient mice. *Nat Genet* 1996;**12**:325–8.
34. Eaton DC, Helms MN, Koval M, Bao HF, Jain L. The contribution of epithelial sodium channels to alveolar function in health and disease. *Annu Rev Physiol* 2009;**71**:403–23.
35. Matalon S, O'Brodovich H. Sodium channels in alveolar epithelial cells: molecular characterization, biophysical properties, and physiological significance. *Annu Rev Physiol* 1999;**61**:627–61.
36. Fischer H, Illek B, Finkbeiner WE, Widdicombe JH. Basolateral Cl channels in primary airway epithelial cultures. *Am J Physiol Lung Cell Mol Physiol* 2007;**292**:L1432–43.
37. Anderson MP, Sheppard DN, Berger HA, Welsh MJ. Chloride channels in the apical membrane of normal and cystic fibrosis airway and intestinal epithelia. *Am J Phys* 1992;**263**:L1–14.
38. Solc CK, Wine JJ. Swelling-induced and depolarization-induced Cl$^-$ channels in normal and cystic fibrosis epithelial cells. *Am J Phys* 1991;**261**:C658–74.
39. Helms MN, Self J, Bao HF, Job LC, Jain L, Eaton DC. Dopamine activates amiloride-sensitive sodium channels in alveolar type I cells in lung slice preparations. *Am J Physiol Lung Cell Mol Physiol* 2006;**291**:L610–8.
40. Burch LH, Talbot CR, Knowles MR, Canessa CM, Rossier BC, Boucher RC. Relative expression of the human epithelial Na$^+$ channel subunits in normal and cystic fibrosis airways. *Am J Phys* 1995;**269**:C511–8.
41. Ji HL, Zhao RZ, Chen ZX, Shetty S, Idell S, Matalon S. delta ENaC: a novel divergent amiloride-inhibitable sodium channel. *Am J Physiol Lung Cell Mol Physiol* 2012;**303**:L1013–26.
42. Nie HG, Chen L, Han DY, Li J, Song WF, Wei SP, et al. Regulation of epithelial sodium channels by cGMP/PKGII. *J Physiol* 2009;**587**:2663–76.
43. Zhao RZ, Nie HG, Su XF, Han DY, Lee A, Huang Y, et al. Characterization of a novel splice variant of delta ENaC subunit in human lungs. *Am J Physiol Lung Cell Mol Physiol* 2012;**302**:L1262–72.
44. Bangel-Ruland N, Sobczak K, Christmann T, Kentrup D, Langhorst H, Kusche-Vihrog K, et al. Characterization of the epithelial sodium channel delta-subunit in human nasal epithelium. *Am J Respir Cell Mol Biol* 2010;**42**:498–505.
45. O'Brodovich H, Yang P, Gandhi S, Otulakowski G. Amiloride-insensitive Na$^+$ and fluid absorption in the mammalian distal lung. *Am J Physiol Lung Cell Mol Physiol* 2008;**294**:L401–8.
46. Wilkinson WJ, Benjamin AR, De Proost I, Orogo-Wenn MC, Yamazaki Y, Staub O, et al. Alveolar epithelial CNGA1 channels mediate cGMP-stimulated, amiloride-insensitive, lung liquid absorption. *Pflugers Arch* 2011;**462**:267–79.
47. Joris L, Quinton PM. Evidence for electrogenic Na-glucose cotransport in tracheal epithelium. *Pflugers Arch* 1989;**415**:118–20.
48. Basset G, Crone C, Saumon G. Fluid absorption by rat lung in situ: pathways for sodium entry in the luminal membrane of alveolar epithelium. *J Physiol* 1987;**384**:325–45.
49. Ridge KM, Olivera WG, Saldias F, Azzam Z, Horowitz S, Rutschman DH, et al. Alveolar type 1 cells express the alpha2 Na,K-ATPase, which contributes to lung liquid clearance. *Circ Res* 2003;**92**:453–60.

50. Saumon G, Basset G, Bouchonnet F, Crone C. Cellular effects of beta-adrenergic and of cAMP stimulation on potassium transport in rat alveolar epithelium. *Pflugers Arch* 1989;**414**:340–5.

51. Fischer H, Widdicombe JH. Mechanisms of acid and base secretion by the airway epithelium. *J Membr Biol* 2006;**211**:139–50.

52. Lubman RL, Danto SI, Crandall ED. Evidence for active H⁺ secretion by rat alveolar epithelial cells. *Am J Phys* 1989;**257**:L438–45.

53. Mescher EJ, Platzker AC, Ballard PL, Kitterman JA, Clements JA, Tooley WH. Ontogeny of tracheal fluid, pulmonary surfactant, and plasma corticoids in the fetal lamb. *J Appl Phys* 1975;**39**:1017–21.

54. Adams FH, Fujiwara T, Rowsham G. The nature and origin of the fluid in fetal lamb lung. *J Pediatr* 1963;**63**:881–8.

55. Adamson TM, Boyd RD, Platt HS, Strang LB. Composition of alveolar liquid in the foetal lamb. *J Physiol* 1969;**204**:159–68.

56. Shaw AM, Steele LW, Butcher PA, Ward MR, Olver RE. Sodium-proton exchange across the apical membrane of the alveolar type II cell of the fetal sheep. *Biochim Biophys Acta* 1990;**1028**:9–13.

57. Cotton CU, Lawson EE, Boucher RC, Gatzy JT. Bioelectric properties and ion transport of airways excised from adult and fetal sheep. *J Appl Phys* 1983;**55**:1542–9.

58. Uyekubo SN, Fischer H, Maminishkis A, Illek B, Miller SS, Widdicombe JH. cAMP-dependent absorption of chloride across airway epithelium. *Am J Phys* 1998;**275**:L1219–27.

59. Itani OA, Lamb FS, Melvin JE, Welsh MJ. Basolateral chloride current in human airway epithelia. *Am J Physiol Lung Cell Mol Physiol* 2007;**293**:L991–9.

60. Li X, Comellas AP, Karp PH, Ernst SE, Moninger TO, Gansemer ND, et al. CFTR is required for maximal transepithelial liquid transport in pig alveolar epithelia. *Am J Physiol Lung Cell Mol Physiol* 2012;**303**:L152–60.

61. Fang X, Song Y, Hirsch J, Galietta LJ, Pedemonte N, Zemans RL, et al. Contribution of CFTR to apical-basolateral fluid transport in cultured human alveolar epithelial type II cells. *Am J Physiol Lung Cell Mol Physiol* 2006;**290**:L242–9.

62. Widdicombe JH. How does cAMP increase active Na absorption across alveolar epithelium? *Am J Phys* 2000;**276**:L231–2.

63. Finkbeiner WE, Widdicombe JH. Control of nasal airway secretions, ion transport, and water movement. In: Parent RA, editor. *Comparative Biology of the Normal Lung. Treatise on Pulmonary Toxicology*, **vol. 1**. Boca Raton, FL: CRC Press; 1992. pp. 633–57.

64. Yamaya M, Finkbeiner WE, Chun SY, Widdicombe JH. Differentiated structure and function of cultures from human tracheal epithelium. *Am J Phys* 1992;**262**:L713–24.

65. Kondo M, Finkbeiner WE, Widdicombe JH. Cultures of bovine tracheal epithelium with differentiated ultrastructure and ion transport. *Vitro Cell Dev Biol* 1993;**29A**:19–24.

66. Kondo M, Finkbeiner WE, Widdicombe JH. Simple technique for culture of highly differentiated cells from dog tracheal epithelium. *Am J Phys* 1991;**261**:L106–17.

67. Kim KJ. Active Na⁺ transport across Xenopus lung alveolar epithelium. *Respir Physiol* 1990;**81**:29–39.

68. Gatzy JT. Ion transport across the excised bullfrog lung. *Am J Phys* 1975;**228**:1162–71.

69. Fischer H, Van Driessche W, Clauss W. Evidence for apical sodium channels in frog lung epithelial cells. *Am J Phys* 1989;**256**:C764–71.

70. O'Brodovich H, Rafii B, Post M. Bioelectric properties of fetal alveolar epithelial monolayers. *Am J Phys* 1990;**258**:L201–6.

71. Ballard ST, Gatzy JT. Alveolar transepithelial potential difference and ion transport in adult rat lung. *J Appl Phys* 1991;**70**:63–9.

72. Plopper CG, Alley JL, Weir AJ. Differentiation of tracheal epithelium during fetal lung maturation in the rhesus monkey macaca mulata. *Am J Anat* 1986;**175**:59–71.

73. Kondo M, Tamaoki J, Takeyama K, Nakata J, Nagai A. Interleukin-13 induces goblet cell differentiation in primary cell culture from Guinea pig tracheal epithelium. *Am J Respir Cell Mol Biol* 2002;**27**:536–41.

74. Van Scott MR, Davis CW, Boucher RC. Na⁺ and Cl⁻ transport across rabbit nonciliated bronchiolar epithelial (Clara) cells. *Am J Phys* 1989;**256**:C893–901.

75. Van Scott MR, Penland CM, Welch CA, Lazarowski E. Beta-adrenergic regulation of Cl⁻ and HCO₃⁻ secretion by Clara cells. *Am J Respir Cell Mol Biol* 1995;**13**:344–51.

76. Chinet TC, Gabriel SE, Penland CM, Sato M, Stutts MJ, Boucher RC, et al. CFTR-like chloride channels in non-ciliated bronchiolar epithelial (Clara) cells. *Biochem Biophys Res Commun* 1997;**230**:470–5.

77. McCray Jr PB, Reenstra WW, Louie E, Johnson J, Bettencourt JD, Bastacky J. Expression of CFTR and presence of cAMP-mediated fluid secretion in human fetal lung. *Am J Phys* 1992;**262**:L472–81.

78. Schneeberger EE, McCarthy KM. Cytochemical localization of Na⁺-K⁺-ATPase in rat type II pneumocytes. *J Appl Phys* 1986;**60**:1584–9.

79. Borok Z, Liebler JM, Lubman RL, Foster MJ, Zhou B, Li X, et al. Na transport proteins are expressed by rat alveolar epithelial type I cells. *Am J Physiol Lung Cell Mol Physiol* 2002;**282**:L599–608.

80. Johnson MD, Widdicombe JH, Allen L, Barbry P, Dobbs LG. Alveolar epithelial type I cells actively transport sodium and are likely to play a role in mediating lung liquid homeostasis. *Proc Natl Acad Sci* 2002;**99**:1966–71.

81. Johnson MD, Bao HF, Helms MN, Chen XJ, Tigue Z, Jain L, et al. Functional ion channels in pulmonary alveolar type I cells support a role for type I cells in lung ion transport. *Proc Natl Acad Sci USA* 2006;**103**:4964–9.

82. Liebler JM, Borok Z, Li X, Zhou B, Sandoval AJ, Kim KJ, et al. Alveolar epithelial type I cells express beta2-adrenergic receptors and G-protein receptor kinase 2. *J Histochem Cytochem* 2004;**52**:759–67.

83. Fabisiak JP, Vesell ES, Rannels DE. Interactions of beta adrenergic antagonists with isolated rat alveolar type II pneumocytes. I. Analysis, characterization and regulation of specific beta adrenergic receptors. *J Pharmacol Exp Ther* 1987;**241**:722–7.

84. Bove PF, Grubb BR, Okada SF, Ribeiro CM, Rogers TD, Randell SH, et al. Human alveolar type II cells secrete and absorb liquid in response to local nucleotide signaling. *J Biol Chem* 2010;**285**:34939–49.

85. McCray Jr PB, Bettencourt JD, Bastacky J, Denning GM, Welsh MJ. Expression of CFTR and a cAMP-stimulated chloride secretory current in cultured human fetal alveolar epithelial cells. *Am J Respir Cell Mol Biol* 1993;**9**:578–85.

86. Johnson M, Allen L, Dobbs L. Characteristics of Cl⁻ uptake in rat alveolar type I cells. *Am J Physiol Lung Cell Mol Physiol* 2009;**297**:L816–27.

87. Sakuma T, Pittet JF, Jayr C, Matthay MA. Alveolar liquid and protein clearance in the absence of blood flow or ventilation in sheep. *J Appl Phys* 1993;**74**:176–85.

88. Sakuma T, Okaniwa G, Nakada T, Nishimura T, Fujimura S, Matthay MA. Alveolar fluid clearance in the resected human lung. *Am J Respir Crit Care Med* 1994;**150**:305–10.

89. Saumon G, Basset G, Bouchonnet F, Crone C. cAMP and beta-adrenergic stimulation of rat alveolar epithelium. Effects on fluid absorption and paracellular permeability. *Pflugers Arch* 1987;**410**:464–70.

90. Downs CA, Kriener LH, Yu L, Eaton DC, Jain L, Helms MN. beta-Adrenergic agonists differentially regulate highly selective and nonselective epithelial sodium channels to promote alveolar fluid clearance in vivo. *Am J Physiol Lung Cell Mol Physiol* 2012;**302**:L1167–78.

91. Garat C, Rezaiguia S, Meignan M, D'Ortho MP, Harf A, Matthay MA, et al. Alveolar endotoxin increases alveolar liquid clearance in rats. *J Appl Phys* 1995;**79**:2021–8.

92. Folkesson HG, Pittet JF, Nitenberg G, Matthay MA. Transforming growth factor-alpha increases alveolar liquid clearance in anesthetized ventilated rats. *Am J Phys* 1996;**271**:L236–44.

93. Takemura Y, Helms MN, Eaton AF, Self J, Ramosevac S, Jain L, et al. Cholinergic regulation of epithelial sodium channels in rat alveolar type 2 epithelial cells. *Am J Physiol Lung Cell Mol Physiol* 2013;**304**:L428–37.

94. Sloniewsky DE, Ridge KM, Adir Y, Fries FP, Briva A, Sznajder JI, et al. Leukotriene D4 activates alveolar epithelial Na,K-ATPase and increases alveolar fluid clearance. *Am J Respir Crit Care Med* 2004;**169**:407–12.

95. Azzam ZS, Adir Y, Crespo A, Comellas A, Lecuona E, Dada LA, et al. Norepinephrine increases alveolar fluid reabsorption and Na,K-ATPase activity. *Am J Respir Crit Care Med* 2004;**170**:730–6.

96. Bhargava M, Runyon MR, Smirnov D, Lei J, Groppoli TJ, Mariash CN, et al. Triiodo-L-thyronine rapidly stimulates alveolar fluid clearance in normal and hyperoxia-injured lungs. *Am J Respir Crit Care Med* 2008;**178**:506–12.

97. Wang Q, Lian QQ, Li R, Ying BY, He Q, Chen F, et al. Lipoxin A(4) activates alveolar epithelial sodium channel, Na,K ATPase, and increases alveolar fluid clearance. *Am J Respir Cell Mol Biol* 2013;**48**:610–8.

98. Barnard ML, Olivera WG, Rutschman DM, Bertorello AM, Katz AI, Sznajder JI. Dopamine stimulates sodium transport and liquid clearance in rat lung epithelium. *Am J Respir Crit Care Med* 1997;**156**:709–14.

99. Rezaiguia S, Garat C, Delclaux C, Meignan M, Fleury J, Legrand P, et al. Acute bacterial pneumonia in rats increases alveolar epithelial fluid clearance by a tumor necrosis factor-alpha-dependent mechanism. *J Clin Invest* 1997;**99**:325–35.

100. Fukuda N, Jayr C, Lazrak A, Wang Y, Lucas R, Matalon S, et al. Mechanisms of TNF-alpha stimulation of amiloride-sensitive sodium transport across alveolar epithelium. *Am J Physiol Lung Cell Mol Physiol* 2001;**280**:L1258–65.

101. Yamagata T, Yamagata Y, Nishimoto T, Hirano T, Nakanishi M, Minakata Y, et al. The regulation of amiloride-sensitive epithelial sodium channels by tumor necrosis factor-alpha in injured lungs and alveolar type II cells. *Respir Physiolo Neurobiol* 2009;**166**:16–23.

102. Frank J, Roux J, Kawakatsu H, Su G, Dagenais A, Berthiaume Y, et al. Transforming growth factor-beta1 decreases expression of the epithelial sodium channel alphaENaC and alveolar epithelial vectorial sodium and fluid transport via an ERK1/2-dependent mechanism. *J Biol Chem* 2003;**278**:43939–50.

103. Vadasz I, Morty RE, Olschewski A, Konigshoff M, Kohstall MG, Ghofrani HA, et al. Thrombin impairs alveolar fluid clearance by promoting endocytosis of Na^+,K^+-ATPase. *Am J Respir Cell Mol Biol* 2005;**33**:343–54.

104. Basset G, Crone C, Saumon G. Significance of active ion transport in transalveolar water absorption: a study on isolated rat lung. *J Physiol* 1987;**384**:311–24.

105. Solymosi EA, Kaestle-Gembardt SM, Vadasz I, Wang L, Neye N, Chupin CJ, et al. Chloride transport-driven alveolar fluid secretion is a major contributor to cardiogenic lung edema. *Proc Natl Acad Sci USA* 2013;**110**:E2308–16.

106. Li T, Folkesson HG. RNA interference for alpha-ENaC inhibits rat lung fluid absorption in vivo. *Am J Phys* 2006;**290**:L649–60.

107. Norlin A, Lu LN, Guggino SE, Matthay MA, Folkesson HG. Contribution of amiloride-insensitive pathways to alveolar fluid clearance in adult rats. *J Appl Phys* 2001;**90**:1489–96.

108. Ussing HH, Zerahn K. Active transport of sodium as the source of electric current in short-circuited isolated frog skin. *Acta Physiol Scand* 1951;**23**:110–27.

109. Widdicombe JH. Electrical methods for studying ion and fluid transport across airway epithelia. In: Allegra L, Braga P, editors. *Methods in Bronchial Mucology*. New York: Raven Press; 1988. pp. 335–45.

110. Mason RJ, Williams MC, Widdicombe JH, Sanders MJ, Misfeldt DS, Berry LC. Transepithelial transport by pulmonary alveolar type II cells in primary culture. *Proc Natl Acad Sci USA* 1982;**79**:6033–7.

111. Kim KJ, Cheek JM, Crandall ED. Contribution of active Na^+ and Cl^- fluxes to net ion transport by alveolar epithelium. *Respir Physiol* 1991;**85**:245–56.

112. Goodman BE, Fleischer RD, Crandall ED. Evidence for active Na^+ transport by cultured monolayers of pulmonary alveolar epithelial cells. *Am J Phys* 1983;**245**:C78–83.

113. Roux J, Carles M, Koh H, Goolaerts A, Ganter MT, Chesebro BB, et al. Transforming growth factor beta1 inhibits cystic fibrosis transmembrane conductance regulator-dependent cAMP-stimulated alveolar epithelial fluid transport via a phosphatidylinositol 3-kinase-dependent mechanism. *J Biol Chem* 2010;**285**:4278–90.

114. Mutlu GM, Koch WJ, Factor P. Alveolar epithelial beta 2-adrenergic receptors: their role in regulation of alveolar active sodium transport. *Am J Respir Crit Care Med* 2004;**170**:1270–5.

115. Willis BC, Kim KJ, Li X, Liebler J, Crandall ED, Borok Z. Modulation of ion conductance and active transport by TGF beta 1 in alveolar epithelial cell monolayers. *Am J Physiol Lung Cell Mol Physiol* 2003;**285**:L1192–200.

116. Takahashi K, Mitsui M, Takeuchi K, Uwabe Y, Kobayashi K, Sawasaki Y, et al. Preservation of the characteristics of the cultured human type II alveolar epithelial cells. *Lung* 2004;**182**:213–26.

117. Sakamoto T, Hirano K, Morishima Y, Masuyama K, Ishii Y, Nomura A, et al. Maintenance of the differentiated type II cell characteristics by culture on an acellular human amnion membrane. *Vitro Cell Dev Biol Anim* 2001;**37**:471–9.

118. Elbert KJ, Schafer UF, Schafers HJ, Kim KJ, Lee VH, Lehr CM. Monolayers of human alveolar epithelial cells in primary culture for pulmonary absorption and transport studies. *Pharm Res* 1999;**16**:601–8.

119. Filippatos GS, Hughes WF, Qiao R, Sznajder JI, Uhal BD. Mechanisms of liquid flux across pulmonary alveolar epithelial cell monolayers. *Vitro Cell Dev Biol Anim* 1997;**33**:195–200.

120. Williams MC. Alveolar type I cells: molecular phenotype and development. *Annu Rev Physiol* 2003;**65**:669–95.

121. Whitcutt MJ, Adler KB, Wu R. A biphasic chamber system for maintaining polarity of differentiation of cultured respiratory tract epithelial cells. *In Vitro* 1988;**24**:420–8.

122. Johnson LG, Dickman KG, Moore KL, Mandel LJ, Boucher RC. Enhanced Na^+ transport in an air-liquid interface culture system. *Am J Phys* 1993;**264**:L560–5.

123. Coleman DL, Tuet IK, Widdicombe JH. Electrical properties of dog tracheal epithelial cells grown in monolayer culture. *Am J Phys* 1984;**246**:C355–9.

124. Widdicombe JH, Coleman DL, Finkbeiner WE, Tuet IK. Electrical properties of monolayers cultured from cells of human tracheal mucosa. *J Appl Phys* 1985;**58**:1729–35.

125. Dobbs LG, Pian MS, Maglio M, Dumars S, Allen L. Maintenance of the differentiated type II cell phenotype by culture with an apical air surface. *Am J Phys* 1997;**273**:L347–54.

126. Jiang X, Ingbar DH, O'Grady SM. Adrenergic regulation of ion transport across adult alveolar epithelial cells: effects on Cl⁻ channel activation and transport function in cultures with an apical air interface. *J Membr Biol* 2001;**181**:195–204.

127. Jost A, Policard A. Contribution experimental a l'etude du development prenatal du poumon chez le lapin. *Arch Anat Microscop Morpol Exp* 1948;**37**:323–32.

128. Carmel JA, Friedman F, Adams FH. Tracheal ligation and lung development. *Am J Dis Child* 1965;**109**:452–6.

129. Potter EL, Bohlender GP. Intrauterine respiration in relation to development of the fetal lung. *Am J Obstet Gynecol* 1941;**42**:14–22.

130. Adams FH, Desilets DT, Towers B. Control of flow of fetal lung fluid at the laryngeal outlet. *Respir Physiol* 1967;**2**:302–9.

131. Fewell JE, Johnson P. Upper airway dynamics during breathing and during apnoea in fetal lambs. *J Physiol* 1983;**339**:495–504.

132. Alcorn D, Adamson TM, Lambert TF, Maloney JE, Ritchie BC, Robinson PM. Morphological effects of chronic tracheal ligation and drainage in the fetal lamb lung. *J Anat* 1977;**123**:649–60.

133. Moessinger AC, Harding R, Adamson TM, Singh M, Kiu GT. Role of lung fluid volume in growth and maturation of the fetal sheep lung. *J Clin Invest* 1990;**86**:1270–7.

134. Cassin S, Gause G, Perks AM. The effects of bumetanide and furosemide on lung liquid secretion in fetal sheep. *Exp Biol Med* 1986;**181**:427–31.

135. Thom J, Perks AM. The effects of furosemide and bumetanide on lung liquid production by in vitro lungs from fetal guinea pigs. *Can J Physiol Pharmacol* 1990;**68**:1131–5.

136. Cassin S, Perks AM. Studies of factors which stimulate lung fluid secretion in fetal goats. *J Physiol* 1982;**4**:311–25.

137. Strang LB. Fetal lung liquid: secretion and reabsorption. *Physiol Rev* 1991;**71**:991–1016.

138. Ussing HH. The distinction by means of tracers between active transport and diffusion. The transfer of iodide across the isolated frog skin. *Acta Physiol Scand* 1949;**19**:43–56.

139. Olver RE, Strang LB. Ion fluxes across the pulmonary epithelium and the secretion of lung liquid in the foetal lamb. *J Physiol* 1974;**241**:327–57.

140. Olver RE, Schneeberger EE, Walters DV. Epithelial solute permeability, ion transport and tight junction morphology in the developing lung of the fetal lamb. *J Physiol* 1981;**315**:395–412.

141. McAteer JA, Cavanagh TJ, Evan AP. Submersion culture of the intact fetal lung. *Vitro* 1983;**19**:210–8.

142. Krochmal EM, Ballard ST, Yankaskas JR, Boucher RC, Gatzy JT. Volume and ion transport by fetal rat alveolar and tracheal epithelia in submersion culture. *Am J Phys* 1989;**256**:F397–407.

143. McCray Jr PB, Bettencourt JD, Bastacky J. Secretion of lung fluid by the developing fetal rat alveolar epithelium in organ culture. *Am J Respir Cell Mol Biol* 1992;**6**:609–16.

144. McCray Jr PB, Welsh MJ. Developing fetal alveolar epithelial cells secrete fluid in primary culture. *Am J Phys* 1991;**260**:L494–500.

145. Lamb FS, Graeff RW, Clayton GH, Smith RL, Schutte BC, McCray Jr PB. Ontogeny of CLCN3 chloride channel gene expression in human pulmonary epithelium. *Am J Respir Cell Mol Biol* 2001;**24**:376–81.

146. Murray CB, Morales MM, Flotte TR, McGrath-Morrow SA, Guggino WB, Zeitlin PL. CIC-2: a developmentally dependent chloride channel expressed in the fetal lung and downregulated after birth. *Am J Respir Cell Mol Biol* 1995;**12**:597–604.

147. Edmonds RD, Silva IV, Guggino WB, Butler RB, Zeitlin PL, Blaisdell CJ. ClC-5: ontogeny of an alternative chloride channel in respiratory epithelium. *Am J Phys* 2002;**282**:L501–7.

148. Ringman Uggla A, Zelenina M, Eklof AC, Aperia A, Frenckner B. Expression of chloride channels in trachea-occluded hyperplastic lungs and nitrofen-induced hypoplastic lungs in rats. *Pediatr Surg Int* 2009;**25**:799–806.

149. Jesse NM, McCartney J, Feng X, Richards EM, Wood CE, Keller-Wood M. Expression of ENaC subunits, chloride channels, and aquaporins in ovine fetal lung: ontogeny of expression and effects of altered fetal cortisol concentrations. *Am J Physiol Regul Integr Comp Physiol* 2009;**297**:R453–61.

150. Blaisdell CJ, Edmonds RD, Wang X-T, Guggino S, Zeitlin PL. pH-related chloride secretion in fetal lung epithelia. *Am J Phys* 2000;**278**:L1248–55.

151. Graeff RW, Wang G, McCray Jr PB. KGF and FGF-10 stimulate liquid secretion in human fetal lung. *Pediatr Res* 1999;**46**:523–9.

152. Zhou L, Graeff RW, McCray Jr PB, Simonet WS, Whitsett JA. Keratinocyte growth factor stimulates CFTR-independent fluid secretion in the fetal lung in vitro. *Am J Phys* 1996;**271**:L987–94.

153. Blaisdell CJ, Pellettieri JP, Loughlin CE, Chu S, Zeitlin PL. Keratinocyte growth factor stimulates CLC-2 expression in primary fetal rat distal lung epithelial cells. *Am J Respir Cell Mol Biol* 1999;**20**:842–7.

154. Schwiebert EM, Cid-Soto LP, Stafford D, Carter M, Blaisdell CJ, Zeitlin PL, et al. Analysis of ClC-2 channels as an alternative pathway for chloride conductance in cystic fibrosis airway cells. *Proc Natl Acad Sci USA* 1998;**95**:3879–84.

155. Chu S, Blaisdell CJ, Liu MZ, Zeitlin PL. Perinatal regulation of the ClC-2 chloride channel in lung is mediated by Sp1 and Sp3. *Am J Phys* 1999;**276**:L614–24.

156. Barker PM, Gatzy JT. Effects of adenosine, ATP, and UTP on chloride secretion by epithelia explanted from fetal rat lung. *Pediatr Res* 1998;**43**:652–9.

157. McCray Jr PB, Bettencourt JD. Prostaglandins stimulate fluid secretion in human fetal lung. *J Dev Physiol* 1993;**19**:29–36.

158. Castro R, Ervin MG, Ross MG, Sherman DJ, Leake RD, Fisher DA. Ovine fetal lung fluid response to atrial natriuretic factor. *Am J Obstet Gynecol* 1989;**161**:1337–43.

159. Dobbs LG. Isolation and culture of alveolar type II cells. *Am J Phys* 1990;**258**:L134–47.

160. Rao AK, Cott GR. Ontogeny of ion transport across fetal pulmonary epithelial cells in monolayer culture. *Am J Phys* 1991;**261**:L178–87.

161. Barker PM, Stiles AD, Boucher RC, Gatzy JT. Bioelectric properties of cultured epithelial monolayers from distal lung of 18-day fetal rat. *Am J Phys* 1992;**262**:L628–36.

162. Barker PM, Boucher RC, Yankaskas JR. Bioelectric properties of cultured monolayers from epithelium of distal human fetal lung. *Am J Phys* 1995;**268**:L270–77.

163. Tessier GJ, Lester GD, Langham MR, Cassin S. Ion transport properties of fetal sheep alveolar epithelial cells in monolayer culture. *Am J Phys* 1996;**270**:L1008–16.

164. Billet A, Hanrahan JW. The secret life of CFTR as a calcium-activated chloride channel. *J Physiol* 2013;**591**:5273–8.

165. Cullen JJ, Welsh MJ. Regulation of sodium absorption by canine tracheal epithelium. *J Clin Invest* 1987;**79**:73–9.

166. Phipps RJ, Abraham WM, Mariassy AT, Torrealba PJ, Sielczak MW, Ahmed A, et al. Developmental changes in the tracheal mucociliary system in neonatal sheep. *J Appl Phys* 1989;**67**:824–32.

167. Al-Bazzaz FJ, Gailey C. Ion transport by sheep distal airways in a miniature chamber. *Am J Physiol Lung Cell Mol Physiol* 2001;**281**:L1028–34.

168. Shamsuddin AK, Quinton PM. Surface fluid absorption and secretion in small airways. *J Physiol* 2012;**590**:3561–74.

169. Ballard ST, Taylor AE. Bioelectric properties of proximal bronchiolar epithelium. *Am J Phys* 1994;**267**:L79–84.

170. Ballard ST, Fountain JD, Inglis SK, Corboz MR, Taylor AE. Chloride secretion across distal airway epithelium: relationship to submucosal gland distribution. *Am J Phys* 1995;**268**:L526–31.

171. Inglis SK, Corboz MR, Taylor AE, Ballard ST. Regulation of ion transport across porcine distal bronchi. *Am J Phys* 1996;**270**: L289–97.

172. Wang X, Lytle C, Quinton PM. Predominant constitutive CFTR conductance in small airways. *Respir Res* 2005;**6**:7.

173. Blouquit-Laye S, Chinet T. Ion and liquid transport across the bronchiolar epithelium. *Respir Physiolo Neurobiol* 2007;**159**:278–82.

174. Olver RE, Robinson EJ. Sodium and chloride transport by the tracheal epithelium of fetal, new-born and adult sheep. *J Physiol* 1986;**375**:377–90.

175. Cotton CU, Boucher RC, Gatzy JT. Bioelectric properties and ion transport across excised canine fetal and neonatal airways. *J Appl Phys* 1988;**65**:2367–75.

176. Zeitlin PL, Loughlin GM, Guggino WB. Ion transport in cultured fetal and adult rabbit epithelia. *Am J Phys* 1988;**254**:C691–8.

177. Jarnigan F, Davis JD, Bromberg PA, Gatzy JT, Boucher RC. Bioelectric properties and ion transport of excised rabbit trachea. *J Appl Physiol: Respirat Environ Exercise Physiol* 1983;**55**:1884–92.

178. Melon J. Activite secretoire de la muquese nasale. *Acta Otorhinolaryngol Belg* 1968;**22**:11–244.

179. Kitterman JA, Ballard PL, Clements JA, Mescher EJ, Tooley WH. Tracheal fluid in fetal lambs: spontaneous decrease prior to birth. *J Appl Phys* 1979;**47**:985–9.

180. Dickson KA, Maloney JE, Berger PJ. Decline in lung liquid volume before labor in fetal lambs. *J Appl Phys* 1986;**61**:2266–72.

181. Lines A, Hooper SB, Harding R. Lung liquid production rates and volumes do not decrease before labor in healthy fetal sheep. *J Appl Phys* 1997;**82**:927–32.

182. Berger PJ, Smolich JJ, Ramsden CA, Walker AM. Effect of lung liquid volume on respiratory performance after caesarean delivery in the lamb. *J Physiol* 1996;**492**(Pt 3):905–12.

183. Bland RD, Bressack MA, McMillan DD. Labor decreases the lung water content of newborn rabbits. *Am J Obstet Gynecol* 1979;**135**:364–7.

184. Bland RD, McMillan DD, Bressack MA, Dong L. Clearance of liquid from lungs of newborn rabbits. *J Appl Phys* 1980;**49**:171–7.

185. Milner AD, Saunders RA, Hopkin IE. Effects of delivery by caesarean section on lung mechanics and lung volume in the human neonate. *Arch Dis Child* 1978;**53**:545–8.

186. Jain NJ, Kruse LK, Demissie K, Khandelwal M. Impact of mode of delivery on neonatal complications: trends between 1997 and 2005. *J Matern Fetal Neonatal Med* 2009;**22**:491–500.

187. Jain L, Eaton DC. Physiology of fetal lung fluid clearance and the effect of labor. *Semin Perinatol* 2006;**30**:34–43.

188. Karlberg P, Adams FH, Beubelle F, Wallgren G. Alteration of the infant's thorax during vaginal delivery. *Acta Obstet Gynecol Scand* 1962;**41**:223–9.

189. Hislop AA, Wigglesworth JS, Desai R. Alveolar development in the human fetus and infant. *Early Hum Dev* 1986;**13**:1–11.

190. Armstrong L, Stenson BJ. Use of umbilical cord blood gas analysis in the assessment of the newborn. *Arch Dis Child Fetal Neonatal Ed* 2007;**92**:F430–34.

191. Brown MJ, Olver RE, Ramsden CA, Strang LB, Walters DV. Effects of adrenaline and of spontaneous labour on the secretion and absorption of lung liquid in the fetal lamb. *J Physiol* 1983;**344**:137–52.

192. Walters DV, Olver RE. The role of catecholamines in lung liquid absorption at birth. *Pediatr Res* 1978;**12**:239–42.

193. Nathanielsz PW, Comline RS, Silver M, Paisey RB. Cortisol metabolism in the fetal and neonatal sheep. *J Reprod Suppl* 1972;**16**(Suppl. 16):39–59.

194. Fraser M, Liggins GC. Thyroid hormone kinetics during late pregnancy in the ovine fetus. *J Dev Physiol* 1988;**10**:461–71.

195. Barker PM, Brown MJ, Ramsden CA, Strang LB, Walters DV. The effect of thyroidectomy in the fetal sheep on lung liquid reabsorption induced by adrenaline or cyclic AMP. *J Physiol* 1988;**407**:373–83.

196. Barker PM, Markiewicz M, Parker KA, Walters DV, Strang LB. Synergistic action of triiodothyronine and hydrocortisone on epinephrine-induced reabsorption of fetal lung liquid. *Pediatr Res* 1990;**27**:588–91.

197. Barker PM, Walters DV, Markiewicz M, Strang LB. Development of the lung liquid reabsorptive mechanism in fetal sheep: synergism of triiodothyronine and hydrocortisone. *J Physiol* 1991;**433**:435–49.

198. Dagenais A, Denis C, Vives MF, Girouard S, Masse C, Nguyen T, et al. Modulation of alpha-ENaC and alpha1-Na$^+$-K$^+$-ATPase by cAMP and dexamethasone in alveolar epithelial cells. *Am J Phys* 2001;**281**:L217–30.

199. Tchepichev S, Ueda J, Canessa C, Rossier BC, O'Brodovich H. Lung epithelial Na channel subunits are differentially regulated during development and by steroids. *Am J Phys* 1995;**269**:C805–12.

200. Mustafa SB, DiGeronimo RJ, Petershack JA, Alcorn JL, Seidner SR. Postnatal glucocorticoids induce alpha-ENaC formation and regulate glucocorticoid receptors in the preterm rabbit lung. *Am J Physiol Lung Cell Mol Physiol* 2004;**286**:L73–80.

201. Venkatesh VC, Katzberg HD. Glucocorticoid regulation of epithelial sodium channel genes in human fetal lung. *Am J Phys* 1997;**273**:L227–33.

202. Thome UH, Davis IC, Nguyen SV, Shelton BJ, Matalon S. Modulation of sodium transport in fetal alveolar epithelial cells by oxygen and corticosterone. *Am J Phys* 2003;**284**:L376–85.

203. Hao H, Wendt CH, Sandhu G, Ingbar DH. Dexamethasone stimulates transcription of the Na$^+$-K$^+$-ATPase beta1 gene in adult rat lung epithelial cells. *Am J Physiol Lung Cell Mol Physiol* 2003;**285**: L593–601.

204. Guney S, Schuler A, Ott A, Hoschele S, Zugel S, Baloglu E, et al. Dexamethasone prevents transport inhibition by hypoxia in rat lung and alveolar epithelial cells by stimulating activity and expression of Na$^+$-K$^+$-ATPase and epithelial Na$^+$ channels. *Am J Physiol Lung Cell Mol Physiol* 2007;**293**:L1332–8.

205. Janer C, Pitkanen OM, Helve O, Andersson S. Airway expression of the epithelial sodium channel a-subunit correlates with cortisol in term newborns. *Pediatrics* 2011;**128**:e414–21.

206. Rahman MS, Gandhi S, Otulakowski G, Duan W, Sarangapani A, O'Brodovich H. Long-term terbutaline exposure stimulates alpha1-Na+-K+-ATPase expression at posttranscriptional level in rat fetal distal lung epithelial cells. *Am J Physiol Lung Cell Mol Physiol* 2010;**298**:L96–104.

207. Lei J, Wendt CH, Fan D, Mariash CN, Ingbar DH. Developmental acquisition of T3-sensitive Na-K-ATPase stimulation by rat alveolar epithelial cells. *Am J Physiol Lung Cell Mol Physiol* 2007;**292**:L6–14.

208. Lei J, Mariash CN, Bhargava M, Wattenberg EV, Ingbar DH. T3 increases Na-K-ATPase activity via a MAPK/ERK1/2-dependent pathway in rat adult alveolar epithelial cells. *Am J Physiol Lung Cell Mol Physiol* 2008;**294**:L749–54.

209. McDonald Jr JV, Gonzales LW, Ballard PL, Pitha J, Roberts JM. Lung beta-adrenoreceptor blockade affects perinatal surfactant release but not lung water. *J Appl Phys* 1986;**60**:1727–33.

210. Bland RD, Nielson DW. Developmental changes in lung epithelial ion transport and liquid movement. *Annu Rev Physiol* 1992;**54**:373–94.

211. Wallace MJ, Hooper SB, Harding R. Regulation of lung liquid secretion by arginine vasopressin in fetal sheep. *Am J Phys* 1990;**258**:R104–11.

212. Kennedy KA, Wilton P, Mellander M, Rojas J, Sundell H. Effect of epidermal growth factor on lung liquid secretion in fetal sheep. *J Dev Physiol* 1986;**8**:421–33.

213. Kitterman JA. Fetal lung development. *J Dev Physiol* 1984;**6**:67–82.

214. Krochmal-Mokrzan EM, Barker PM, Gatzy JT. Effects of hormones on potential difference and liquid balance across explants from proximal and distal fetal rat lung. *J Physiol* 1993;**463**:647–65.

215. Hooper SB, Kitchen MJ, Wallace MJ, Yagi N, Uesugi K, Morgan MJ, et al. Imaging lung aeration and lung liquid clearance at birth. *FASEB J* 2007;**21**:3329–37.

216. Stockx EM, Pfister RE, Kyriakides MA, Brodecky V, Berger PJ. Expulsion of liquid from the fetal lung during labour in sheep. *Respir Physiolo Neurobiol* 2007;**157**:403–10.

217. Barker PM, Gatzy JT. Effect of gas composition on liquid secretion by explants of distal lung of fetal rat in submersion culture. *Am J Phys* 1993;**265**:L512–17.

218. Baines DL, Ramminger SJ, Collett A, Haddad JJ, Best OG, Land SC, et al. Oxygen-evoked Na+ transport in rat fetal distal lung epithelial cells. *J Physiol* 2001;**532**:105–13.

219. Ramminger SJ, Baines DL, Olver RE, Wilson SM. The effects of PO2 upon transepithelial ion transport in fetal rat distal lung epithelial cells. *J Physiol* 2000;**524**:539–47.

220. Jain M, Sznajder JI. Effects of hypoxia on the alveolar epithelium. *Proc Am Thorac Soc* 2005;**2**:202–5.

221. Olver RE, Ramsden CA, Strang LB, Walters DV. The role of amiloride-blockable sodium transport in adrenaline-induced lung liquid reabsorption in the fetal lamb. *J Physiol* 1986;**376**:321–40.

222. Walters DV, Ramsden CA, Olver RE. Dibutyryl cAMP induces a gestation-dependent absorption of fetal lung liquid. *J Appl Phys* 1990;**68**:2054–9.

223. O'Brodovich H, Hannam V, Seear M, Mullen JB. Amiloride impairs lung water clearance in newborn guinea pigs. *J Appl Phys* 1990;**68**:1758–62.

224. Song G-W, Sun B, Curstedt T, Grossmann G, Robertson B. Effect of amiloride and surfactant on lung liquid clearance in newborn rabbits. *Respir Physiol* 1992;**88**:233–46.

225. Barker PM, Nguyen MS, Gatzy JT, Grubb B, Norman H, Hummler E, et al. Role of gammaENaC subunit in lung liquid clearance and electrolyte balance in newborn mice. Insights into perinatal adaptation and pseudohypoaldosteronism. *J Clin Invest* 1998;**102**:1634–40.

226. McDonald FJ, Yang B, Hrstka RF, Drummond HA, Tarr DE, McCray Jr PB, et al. Disruption of the beta subunit of the epithelial Na+ channel in mice: hyperkalemia and neonatal death associated with a pseudohypoaldosteronism phenotype. *Proc Natl Acad Sci USA* 1999;**96**:1727–31.

227. O'Brodovich H, Hannam V, Rafii B. Sodium channel but neither Na(+)-H+ nor Na-glucose symport inhibitors slow neonatal lung water clearance. *Am J Respir Cell Mol Biol* 1991;**5**:377–84.

228. Ringman A, Zelenina M, Eklof A-C, Aperia A, Frencker B. NKCC-1 and ENaC are down-regulated in nitrofen-induced hypoplastic lungs with congenital diaphragmatic hernia. *Pediatr Surg Int* 2008;**24**:993–1000.

229. Tizzano EF, O'Brodovich H, Chitayat D, Benichou JC, Buchwald M. Regional expression of CFTR in developing human respiratory tissues. *Am J Respir Cell Mol Biol* 1994;**10**:355–62.

230. Marcorelles P, Montier T, Gillet D, Lagarde N, Ferec C. Evolution of CFTR protein distribution in lung tissue from normal and CF human fetuses. *Pediatr Pulmonol* 2007;**42**:1032–40.

231. Broackes-Carter FC, Mouchel N, Gill D, Hyde S, Bassett J, Harris H. Temporal regulation of CFTR expression during ovine lung development: implications for CF gene therapy. *Hum Mol Genet* 2002;**11**:125–31.

232. O'Brodovich H, Canessa C, Ueda J, Rafii B, Rossier BC, Edelson J. Expression of the epithelial Na+ channel in the developing rat lung. *Am J Phys* 1993;**265**:C491–6.

233. Watanabe S, Matsushita K, Stokes JB, McCray PB. Developmental regulation of epithelial sodium channel subunit mRNA expression in rat colon and lung. *Am J Phys* 1998;**275**:G1227–35.

234. Smith DE, Otulakowski G, Yeger H, Post M, Cutz E, O'Brodovich HM. Epithelial Na+ channel (ENaC) expression in the developing normal and abnormal human perinatal lung. *Am J Respir Crit Care Med* 2000;**161**:1322–31.

235. Talbot CL, Bosworth DG, Briley EL, Fenstermacher DA, Boucher RC, Gabriel SE, et al. Quantitation and localization of ENaC subunit expression in fetal, newborn, and adult mouse lung. *Am J Respir Cell Mol Biol* 1999;**20**:398–406.

236. Helve O, Janer C, Pitkanen O, Andersson S. Expression of the epithelial sodium channel in airway epithelium of newborn infants depends on gestational age. *Pediatrics* 2007;**120**:1311–6.

237. Beard LL, Li T, Hu Y, Folkesson HG. Fetal lung epithelial ion channels relocate in the cell membrane during late gestation. *Anat Rec (Hoboken)* 2011;**294**:1461–71.

238. Zhou G, Dada LA, Sznajder JI. Regulation of alveolar epithelial function by hypoxia. *Eur Respir J* 2008;**31**:1107–13.

239. Helenius IT, Dada LA, Sznajder JI. Role of ubiquitination in Na,K-ATPase regulation during lung injury. *Proc Am Thorac Soc* 2010;**7**:65–70.

240. Bhargava M, Lei J, Mariash CN, Ingbar DH. Thyroid hormone rapidly stimulates alveolar Na,K-ATPase by activation of phosphatidylinositol 3-kinase. *Curr Opin Endocrinol Diabetes Obes* 2007;**14**:416–20.

241. Bland RD, Boyd CA. Cation transport in lung epithelial cells derived from fetal, newborn, and adult rabbits. *J Appl Phys* 1986;**61**:507–15.

242. Chapman DC, Widdicombe JH, Bland RD. Developmental differences in rabbit lung epithelial Na+-K+-ATPase. *Am J Phys* 1990;**259**:L481–7.

243. Lamb JF. Regulation of the abundance of sodium pumps in isolated animal cells. *Int J Biochem* 1990;**22**:1365–70.

244. Ingbar DH, Weeks CB, Gilmore-Hebert M, Jacobsen E, Duvick S, Dowin R, et al. Developmental regulation of Na, K-ATPase in rat lung. *Am J Phys* 1996;**270**:L619–29.

245. Ruddy MK, Drazen JM, Pitkanen OM, Rafii B, O'Brodovich HM, Harris HW. Modulation of aquaporin 4 and the amiloride-inhibitable sodium channel in perinatal rat lung epithelial cells. *Am J Phys* 1998;**274**:L1066–72.

246. Umenishi F, Carter EP, Yang B, Oliver B, Matthay MA, Verkman AS. Sharp increase in rat lung water channel expression in the perinatal period. *Am J Respir Cell Mol Biol* 1996;**15**:673–9.

247. Song Y, Fukuda N, Bai C, Ma T, Matthay MA, Verkman AS. Role of aquaporins in alveolar fluid clearance in neonatal and adult lung, and in oedema formation following acute lung injury: studies in transgenic aquaporin null mice. *J Physiol* 2000;**525**:771–9.

248. Fike CD, Lai-Fook SJ, Bland RD. Alveolar liquid pressures in newborn and adult rabbit lungs. *J Appl Phys* 1988;**64**:1629–35.

249. Raj JU. Alveolar liquid pressure measured by micropuncture in isolated lungs of mature and immature fetal rabbits. *J Clin Invest* 1987;**79**:1579–88.

250. Mortola JP. Dynamics of breathing in newborn mammals. *Physiol Rev* 1987;**67**:187–243.

251. Egan EA, Olver RE, Strang LB. Changes in non-electrolyte permeability of alveoli and the absorption of lung liquid at the start of breathing in the lamb. *J Physiol* 1975;**244**:161–79.

252. Carter EP, Umenishi F, Matthay MA, Verkman AS. Developmental changes in water permeability across the alveolar barrier in perinatal rabbit lung. *J Clin Invest* 1997;**100**:1071–8.

253. Fang X, Song Y, Zemans R, Hirsch J, Matthay MA. Fluid transport across cultured rat alveolar epithelial cells: a novel in vitro system. *Am J Physiol Lung Cell Mol Physiol* 2004;**287**:L104–10.

254. te Pas AB. Davis PG, Hooper SB, Morley CJ. From liquid to air: breathing after birth. *J Ped* 2008;**152**:607–11.

255. Siew ML, Wallace MJ, Allison BJ, Kitchen MJ, te Pas AB, Islam MS, et al. The role of lung inflation and sodium transport in airway liquid clearance during lung aeration in newborn rabbits. *Pediatr Res* 2013;**73**:443–9.

Chapter 8

Physical, Endocrine, and Growth Factors in Lung Development

Megan J. Wallace, Stuart B. Hooper and Annie R.A. McDougall

The Ritchie Centre, MIMR-PHI Institute of Medical Research, and The Department of Obstetrics and Gynaecology, Monash University, Clayton, VIC, Australia

INTRODUCTION

Before birth, the lung is filled with liquid and plays no role in gas exchange. But when the umbilical cord is cut, the lung must be sufficiently developed to take over the role of gas exchange, a role that it is has never played before. Therefore, to survive the transition from prenatal to postnatal life, humans, like other mammals, must have a functionally mature respiratory system by the time of birth. The lung must develop an intricate branched structure for the conduction of gases to and from the gas-exchanging regions of the lung. The gas-exchanging regions of the lung must develop a large surface area that is closely apposed to a rich vascular network to enable efficient exchange of gases. The tissue separating the air spaces from the vascular network must be thin enough for rapid gas exchange, elastic enough to allow ease of expansion, yet strong enough to prevent damage to the delicate tissue structures. The cell types lining the gas-exchanging region of the lung must have differentiated into type I alveolar epithelial cells (AECs) that have long, thin cytoplasmic extensions that reduce the distance for gas exchange, and type-II AECs that produce and secrete surfactant that reduces surface tension at the air-liquid interface.

The lungs of most neonates born at term take over the role of gas exchange smoothly and effectively. However, infants that are born preterm, before the lungs have had sufficient time to develop, or who have suffered some compromise to their lung development, may suffer respiratory insufficiency at birth; this is a major cause of neonatal morbidity and mortality and can lead to long-term deficits in respiratory health. Because appropriate lung development during fetal life is essential for the successful transition to extra-uterine life, it is important to understand the factors that control prenatal lung development. The factors that regulate lung development include physical factors, circulating factors, and locally produced factors.

ROLE OF PHYSICAL FACTORS IN REGULATING FETAL LUNG DEVELOPMENT

During fetal life, the future airways are filled with a liquid (fetal lung liquid) that is secreted by the lungs (see Chapter 7 [Widdicombe]). This luminal liquid maintains the lungs in a constantly distended state, stimulating their growth and maturation.[1] The secretion of fetal lung liquid is driven by an osmotic gradient that is created by the net movement of Cl^- ions into the lung lumen.[2,3] This Cl^- gradient is driven by Na^+/K^+-ATPase pumps located on the basolateral surface of pulmonary epithelial cells. The exit of Na^+ from the cells via Na^+/K^+-ATPase pumps creates a gradient for Na^+ re-entry, via $Na^+K^+2Cl^-$ co-transporters that are also located on the basolateral surface of the cells. This causes a build-up of Cl^- within the cells and creates an electrochemical gradient that favors the movement of Cl^- across the apical surface of the epithelium into the lung lumen, via specific Cl^- channels. The net movement of Cl^- ions into the lung lumen establishes an osmotic gradient that favors the movement of water into the lung lumen.[2,3] Much of our knowledge about lung liquid secretion and the role of lung expansion in lung growth is derived from studies in fetal sheep. In late gestation fetal sheep, lung liquid is secreted at 2–5 mL/h/kg.[4–7] Lung liquid leaves the lung by flowing out of the trachea where it is either swallowed or enters the amniotic sac.[8]

It is often stated that the fetal lung is either collapsed or maintained at the same degree of expansion before and after birth. However, it is now clear that the healthy fetal lung is expanded to a greater degree before birth than after birth, and that the basal degree of lung expansion is a critical regulator of lung growth and maturation.

The Lung. http://dx.doi.org/10.1016/B978-0-12-799941-8.00008-0

Regulation of the Basal Degree of Lung Expansion in the Fetus

Although there has been some controversy in the literature as to the exact volume of lung liquid retained in the lung during fetal life, data from fetal sheep show that the resting lung liquid volume (35–45 mL/kg)[6,9–11] is considerably higher than the resting lung volume (functional residual capacity) in the neonate (25–30 mL/kg) (Figure 1).[6] The fetus actively participates in maintaining its lung liquid volume; however, the discrepancies reported in the literature are probably because the physical environment of the fetus also influences lung liquid volume. Values that have been reported from dead, anesthetized (and often exteriorized), or paralyzed fetuses will underestimate lung liquid volume because lung liquid is rapidly lost following death, anesthesia, and paralysis.[12] Similarly, even measurements in chronically catheterized fetuses in utero are questionable unless it was verified at the time of measurement that there was an adequate volume of amniotic liquid and that the animals were not in labor.[10,13]

The basal degree of lung expansion before birth is regulated by the resting volume of lung liquid, which in turn is regulated by the balance between the rate of lung liquid secretion and the rate at which it flows out of the lungs via the trachea. However, in practice (in the absence of labor), it is the rate of efflux that is the principal mechanism determining lung liquid volume, as alterations in lung liquid secretion rate are usually associated with corresponding alterations in lung liquid efflux, with the net result being very little change in lung liquid volume.[14]

Role of the Upper Airway and the Trans-Pulmonary Pressure Gradient in Regulating Fetal Lung Liquid Volume

The rate of lung liquid efflux via the trachea is regulated by the resistance to lung liquid flow through the glottis, as well as the pressure gradient between the lung lumen and amniotic sac (the trans-pulmonary pressure gradient).[8,15] During apnea, the glottis is adducted, which provides a high degree of resistance to liquid flow along the trachea, causing lung liquid to accumulate within the lungs. This accumulation of lung liquid, together with the inherent recoil of elastic lung tissue, raises intra-luminal pressure to 1–2 mmHg above the pressure within the amniotic sac.[8,16] However, during episodes of fetal breathing movements (FBM), the glottis dilates, which reduces resistance to lung liquid efflux and enables lung liquid to flow along the trachea, following its pressure gradient (Figure 2).[8,16] Thus, during episodes of FBM there is a net loss of lung liquid via the trachea. However, because the diaphragm contracts at the same time as the

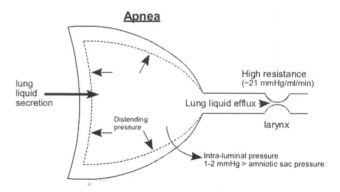

Apnea

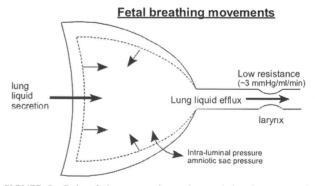

Fetal breathing movements

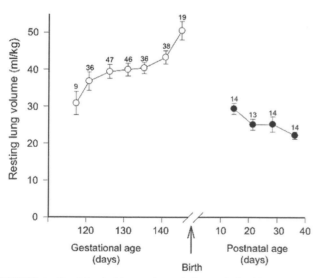

FIGURE 1 Basal luminal lung volumes before and after birth. The fetal values (open circles) represent fetal lung liquid volumes measured in healthy, anesthetized fetal sheep in utero using a dye dilution technique. The postnatal values (closed circles) represent measurements of the volume of air remaining in the lung at end expiration (functional residual capacity), measured in lambs using a helium dilution technique. Symbols are mean ± SEM and the numbers above each symbol represent the number of animals at each age. *Data from Harding and Hooper (1996).*[6]

FIGURE 2 Role of the upper airway in regulating lung expansion. Schematic diagram showing the function of the upper airway in regulating the efflux of lung liquid during fetal "apnea" (upper diagram) and episodes of fetal breathing movements (FBM; lower diagram). During apnea (non-FBM periods), the glottis is constricted, providing a high resistance to the efflux of lung liquid. The continued secretion of lung liquid therefore provides a distending force on the lung, raising intra-luminal pressure 1–2 mmHg above the pressure in the amniotic sac. This pressure gradient is referred to as the trans-pulmonary pressure gradient. During FBM, the glottis dilates, allowing lung liquid to flow along the trachea, following its pressure gradient, from the lung lumen to the mouth where it is either swallowed or it enters the amniotic sac. *Modified from Harding and Hooper (1996).*[6]

glottis dilates, the intrapulmonary pressure is reduced during FBM, which reduces the intrapulmonary pressure gradient and limits the amount of lung liquid that flows along the trachea, compared to that which would occur if the glottis dilated without simultaneous contraction of the diaphragm.[12]

The trans-pulmonary pressure gradient in the fetus can also be influenced by external factors, particularly as the chest wall is compliant, which limits its ability to resist external influences. For example, increases in fetal trunk flexion due to fetal body movements or compression by the uterus (e.g., during labor or in the presence of oligohydramnios) can cause increased abdominal and intra-thoracic pressure. This increases the trans-pulmonary pressure gradient, increasing lung liquid loss via the trachea.[10,13,17]

Role of the Chest Wall in Maintaining Lung Volumes

After birth, chest wall stiffness plays an important role in maintaining end-expiratory lung volume (functional residual capacity), by opposing lung recoil and preventing lung collapse. The tendency of the lung to collapse away from the chest wall generates a sub-atmospheric intra-pleural pressure of ~5 cmH$_2$O. The mechanical strain placed on the chest wall by the development of this sub-atmospheric pressure after birth is thought to play an important role in the increase in chest wall stiffness that occurs within weeks after birth. In contrast, before birth the fetal lung is maintained in a distended state by the retention of lung liquid, producing a 1–2 mmHg distending pressure. As a result, at rest, the intra-pleural pressure during fetal life is essentially zero,[16] suggesting that the intact chest wall plays little, if any, role in maintaining fetal lung liquid volume at rest. However, the diaphragm of the fetus does dome up into the chest suggesting that lung recoil, and/or abdominal pressure, plays a role in the position of the diaphragm. As described earlier, diaphragmatic contractions during FBM limit the loss of fetal lung liquid and are therefore important for maintaining basal lung expansion.

Evidence for the Role of Basal Lung Expansion in Lung Development

A wealth of clinical and experimental evidence indicates that lung growth and maturation during fetal life are critically dependent on the degree to which the lungs are expanded by liquid. Sustained increases in lung liquid volume accelerate most aspects of lung growth and maturation while sustained reductions in lung liquid volume retard lung development.

Clinical Evidence

A wide variety of developmental disorders cause pulmonary hypoplasia, including oligohydramnios, congenital diaphragmatic hernia (CDH), space-occupying lesions like tumors or pleural effusions, and musculo-skeletal disorders. Despite the diversity of these disorders, they all cause pulmonary hypoplasia via a common mechanism: a prolonged reduction in fetal lung expansion.

Oligohydramnios, a reduction in amniotic fluid volume, occurs in 5–10% of all pregnancies. It is often caused by premature rupture of the fetal membranes but can also be caused by a variety of congenital conditions, particularly those that result in inadequate production of fetal urine (e.g., kidney agenesis or dyplasia) or that prevent the flow of urine into the amniotic sac (e.g., agenesis, dysplasia, or stenosis of the ureters). The severity of lung hypoplasia varies, depending on the gestational age at onset and the duration of exposure to oligohydramnios; but when severe, the resulting lung hypoplasia can be lethal.[18] When the volume of amniotic fluid is reduced, the uterus compresses the fetus increasing flexion of the fetal body (Figure 3).[13] The increase in fetal trunk flexion increases abdominal and intra-thoracic pressures, which increases the trans-pulmonary pressure gradient causing increased efflux of lung liquid along the trachea and a reduction in lung expansion (Figure 4).[13]

CDH occurs in about one in 5,000 live births and is caused by a failure of the diaphragm to close during embryonic development.[19] The defect is usually unilateral with left-sided hernias occurring more frequently (70–95%) than right-sided hernias.[19,20] The defect allows the abdominal contents to migrate into the chest, thereby resulting in lung hypoplasia. It is not clear whether the abdominal contents compress the lungs, or whether they simply occupy space, in a similar manner to that which is thought to occur for cysts, pleural effusions and tumors, thus reducing the space available for lung growth. In ~40% of cases CDH occurs with multiple other defects, in which case the risk of neonatal

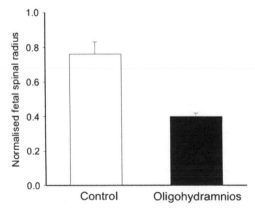

FIGURE 3   The effect of oligohydramnios on fetal spinal flexion. The effect of oligohydramnios, induced by drainage of amniotic fluid, on the degree of spinal flexion (measured as normalized spinal radius) in fetal sheep. A smaller normalized spinal radius is indicative of a greater curvature of the spine. *Data from Harding et al. (1990).*[13]

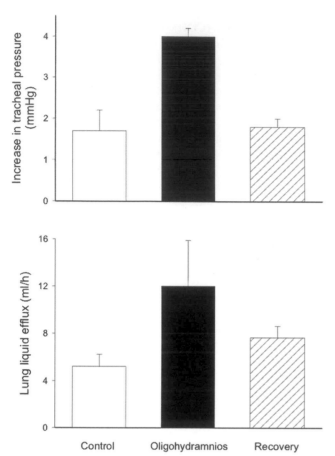

FIGURE 4   Fetal tracheal pressure and lung liquid efflux before, during, and after a 48-hour period of oligohydramnios. Tracheal pressures (upper panel) and lung liquid efflux along the trachea (lower panel) during a control period (open bars), a 48-hour period of oligohydramnios (induced by draining amniotic fluid; solid bars) and during a recovery period (cross-hatched bars), in fetal sheep. The values in each period are taken during non-labor uterine contractions, demonstrating that the fetus is compressed, causing increased abdominal and tracheal pressures and resulting in increased efflux of lung liquid, during uterine contractions when amniotic fluid is insufficient. *Data taken from Harding et al. (1990).*[13]

death is very high (~85%).[21–23] In ~60% of cases, CDH is the sole defect, and in recent years overall survival in this group has improved dramatically to 50%–80%.[24–26] The large variability in reported occurrence and survival rates is likely due to a number of factors, including improvements in neonatal care and prenatal interventions, but also may be skewed by the most severe cases undergoing termination of pregnancy.[19,27,28]

Musculo-skeletal disorders in the fetus can also result in severe pulmonary hypoplasia. The lung hypoplasia in these cases is also likely mediated by a chronic reduction in lung expansion. As diaphragmatic contractions and contraction of the glottis[8,12,15,29] are both important mechanisms for maintaining lung expansion, deficits in these muscles or alterations in the compliance of the chest wall and diaphragm are likely to have major effects on lung expansion.

Experimental Evidence

Most of the evidence demonstrating that the basal degree of fetal lung expansion is critical for regulating lung growth, and maturation is derived from experiments in which lung expansion was reduced (by draining luminal liquid from the lungs), or by obstruction or ligation of the trachea, which causes lung liquid to accumulate within the lungs, increasing lung expansion. The first experiment to demonstrate this relationship showed that ligation of the fetal rabbit trachea in utero caused the pulmonary "airspaces" to be more dilated with liquid; this study demonstrated that the liquid must be a product of the lung itself and not inhaled amniotic fluid.[30] If the trachea was sectioned without ligation, allowing free passage of liquid between the lungs and amniotic cavity, the bronchioles and alveoli appeared to be collapsed.[30] Numerous subsequent studies in fetal sheep,[31–34] rabbits,[35–37] rats,[38] and mice[39] have shown that chronically increased lung expansion, induced by tracheal obstruction or ligation, causes an increase in lung growth and alters lung structure. Conversely, prolonged reductions in lung expansion, induced experimentally by draining liquid out of the lungs in fetal sheep, reduce lung growth. Indeed, a 25% reduction in fetal lung expansion causes a 25% reduction in lung DNA content,[32] whereas total lung deflation causes lung growth to cease altogether.[31,33,40] These studies indicate that a basal degree of lung expansion is a fundamental requirement for lung growth. A large body of literature has subsequently been devoted to understanding the mechanisms by which lung expansion regulates lung growth, in the hope that this understanding may lead to novel treatments to accelerate lung growth and development in infants with lung hypoplasia, or that are born preterm.

Increases in fetal lung growth in response to increases in lung expansion follow a specific time-course[34,36,39] that closely parallels the increase in lung volume,[34] and that is gestational age dependent.[35,41] In sheep, tracheal obstruction during the early alveolar stage of lung development causes an 80% increase in lung liquid volume within the first day, followed by a plateau at 1–2 days after obstruction. This initial increase in lung expansion is associated with lung hypertrophy as indicated by an increase in lung tissue protein content that exceeds the increase in DNA content.[34] After 2 days of tracheal obstruction, lung liquid volume increases linearly to reach a maximum at day 7[32,34,41] and this is associated with lung hyperplasia with all cell types in the distal lung proliferating.[42] The percentage increase in lung growth (as determined by the increase in DNA content) also increased linearly between 2 and 7 days after tracheal obstruction and reached a maximum on day 7.[32,34] The increase in lung DNA content was highly correlated to the increase in lung volume, but not to the increase in intra-luminal pressure, which increases during the first day after tracheal obstruction then remains at a similar level for at least 10 days, indicating that it is the increase in lung liquid volume,[34] rather than intra-luminal pressure, that regulates lung growth during fetal life. Although DNA accumulates gradually between days 2–7, the rate of cell proliferation is greatest 36–48 hours after tracheal obstruction (~800% above control),[34,41,43] which indicates that the mechanisms that mediate the effect of lung expansion on lung growth are likely to be most active within 1–2 days of an increase in lung expansion. The finding that lung liquid volume did not increase between 1 and 2 days indicates that lung volumes must reach an initial structural limit imposed by the lung within the first day of tracheal obstruction and suggests that the further accumulation of lung liquid must depend on structural remodeling of the extracellular matrix of the lung and increased cell proliferation; this is supported by the observed increases in protein content and DNA synthesis during the first 1–2 days after tracheal obstruction.[34] The lack of a continued increase in lung liquid volume after 7 days of tracheal obstruction is likely due to the structural limits imposed by the chest wall, as any further increases would be limited by the rate of chest wall and diaphragm growth. Indeed, we and others commonly observe that following prolonged tracheal obstruction in late gestation fetal sheep, the diaphragm has everted[32,33,44] and that the lungs have distinct indentations visible on their surface that align with individual ribs (unpublished observations).

Tracheal obstruction in the fetus is usually also associated with increased structural maturation of the lung with an increase in collagen deposition,[34] elastin deposition,[45] thinning of interstitial lung tissue,[31,36,37,39] and an increase in the formation of capillaries[46] and alveoli.[36,39,42] These features are all consistent with structural maturation of the lung.

Increases in lung expansion induced by tracheal ligation or obstruction in fetal sheep and rabbits during the late pseudoglandular/early canalicular stage of lung development induce a much slower acceleration in lung growth[36,41] than during the saccular/alveolar stage,[34] which is consistent with the low compliance of the lung at that stage. Despite the slow onset, increases in lung expansion at this stage of development allow lung growth to proceed to a much greater extent,[36,47] likely because the rib cage is more compliant at this age, imposing less of a restriction on lung expansion and lung growth. However, tracheal ligation during this stage of development in sheep was characterized by a marked increase in mesenchymal cell proliferation that resulted in increased, rather than decreased, interstitial tissue thickness.[47] In contrast, in rabbits[36] and mice,[39]

tracheal obstruction at this stage of development was associated with a thinning of mesenchymal lung tissue. The differences between species are likely due to the duration of tracheal ligation relative to the stages of lung development. In sheep, the trachea was ligated from ~75–90 days of gestation, which is still during the canalicular stage of development.[47] In contrast, the trachea was ligated for 1–5 days in rabbits[36] and for 24–36 hours in mice[39], but tissue thinning only occurred once the lungs reached the saccular stage of lung development. This suggests that the beneficial effects of tracheal ligation are greater in lungs at a later stage of development. This is consistent with a study that demonstrated that corticosteroid pretreatment (which matures the fetal lung), accelerates the response of the lung to tracheal obstruction.[48]

The degree of basal lung expansion is also a potent regulator of alveolar epithelial cell (AEC) differentiation. Reductions in fetal lung expansion decrease the proportion of type-I AECs and increase the proportion of type-II AECs, while increases in lung expansion are associated with an increase in the proportion of type-I AECs and a reduction in the proportion of type-II AECs.[31,38,49–51] The effects of changes in lung expansion on AECs are both potent and rapidly reversible,[52] indicating that type-I and type-II cells can trans-differentiate from one cell type to the other, likely via an intermediate cell type.[52] In vitro studies have also indicated that type-I and type-II cells are capable of trans-differentiation when exposed to alterations in static stretch.[53] Together these studies also indicate that type-I cells are not terminally differentiated as had previously been believed. Importantly, glucocorticoids, which are commonly thought to promote the type-II cell phenotype, cannot prevent the trans-differentiation of type-II to type-I cells in response to an increase in basal distension.[54–56] This indicates that basal lung expansion is a much more important regulator of the type-II AEC than glucocorticoids. The differentiation of type-II to type-I AECs that is induced by static stretch[56] can be partially prevented in vitro by a Rho Kinase inhibitor[55] suggesting that the Rho-GTPase/ROCK pathway may mediate this effect, likely via its critical role in regulating the actin cytoskeleton, which must undergo substantial re-organization as cells trans-differentiate away from the cuboidal shape of type-II AECs, towards the large flattened shape typical of type-I AECs.

The profound increases in fetal lung growth and structural maturation induced by tracheal obstruction in animal studies has led to its application in animal models of lung hypoplasia and in human fetuses with congenital diaphragmatic hernia to promote lung development and improve postnatal survival, as discussed below.

At the time of birth, the basal level of lung expansion is dramatically reduced from 35–45 mL/kg[6,9–11] in fetal sheep to a resting lung volume (functional residual capacity) of 25–30 mL/kg in newborn lambs (Figure 1). This large reduction in lung expansion is due to the loss of the high distending pressure of lung liquid before birth and the development of an air–liquid interface after birth, creating surface tension within the lung, which increases lung recoil. Consistent with the changes that occur during fetal life, this reduction in basal lung expansion at birth is associated with a reduction in the rate of lung growth after birth (Figure 5),[57] an alteration in AEC phenotype (Figure 6),[58] and a reduction in the rate at which new alveoli are formed (Figure 7).[59]

Lung growth in neonates, adolescents, and adults also appears to be regulated by the degree of basal lung expansion.[60–62] One of the most common models used to study lung growth in the adolescent and adult is compensatory lung growth following pneumonectomy. However, the physical forces to which the lung is exposed after birth are much more complex than during fetal life so the response to lung growth stimuli, particularly in adults, can vary widely depending on species and the stimulus used.[63] For example, because rodent epiphyses never close, thoracic growth and alveolarization continue throughout life in rodents enabling robust compensatory lung growth at all ages,[63] whereas in dogs whose chest wall and lung growth ceases early in life, >50% of the lung must be resected before compensatory lung growth occurs.[61] In contrast, the effects on alveolar epithelial cell differentiation after birth do not appear to be related to basal distension as newborn lambs exposed to CPAP or PEEP have no alterations in surfactant protein expression or in the relative proportions of type-I and type-II AECs.[60,64] This may reflect the much more complex nature of the physical forces to which AECs are exposed to in the air-filled postnatal lung, than before birth. The role of mechanical forces in compensatory lung growth will be explored in more detail in Chapter 12 (Hsia et al.).

Evidence for the Role of Phasic Lung Expansion in Lung Development

Fetal breathing movements (FBMs) are thought to influence lung growth by both influencing basal lung expansion as described earlier, and by phasic effects on lung distension. In healthy fetuses, FBMs occur in discrete episodes 40–50% of the time towards the end of gestation. Individual FBMs lower intra-thoracic pressure by up to 5 mmHg as the diaphragm contracts and they cause small oscillations in the flow of lung liquid along the trachea as the glottis dilates.[8]

Numerous studies have used cell-stretching devices to determine the effects of phasic tissue stress on lung cells in culture. Some studies aimed to identify the mechanisms that underlie ventilation-induced lung injury in preterm infants and adults, and those studies generally exposed lung cells to 10–25% phasic distension to mimic volutrauma.[65,66] Other studies aimed to identify the effects of phasic FBM on fetal lung cells and those studies usually exposed cells to 5% phasic distension, either constantly at a rate of ~60 cycles/min or intermittently (e.g., 15 min/h), in either 2-dimensional, or 3-dimensional cultures.[67–70] The 5% distension regimen of fetal lung cells caused an increase in the rate of lung cell proliferation[71,72] and differentiation of epithelial cells,[73–75] implying that the phasic distension as a result of FBM may be important for lung development in

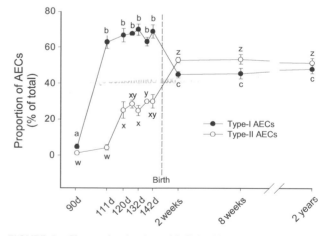

FIGURE 6 Changes in alveolar epithelial cell proportions before and after birth in sheep. Undifferentiated epithelial cells differentiate into type-I (solid circles) and type-II (open circles) alveolar epithelial cells (AECs) during the canalicular stage of lung development, between 90 and 120 d of gestation age (GA) in fetal sheep. The type-I cells appear first, between 90 and 111 d of GA, and then remain stable until birth, accounting for 60–70% of all AECs. Type-II AECs differentiate slightly later, between 111 and 120 d of GA and then remain stable until birth, accounting for ~30% of all AECs. After birth, when there is a reduction in lung expansion, due to the loss of the distending influence of lung liquid and the development of an air–liquid interface, the relative proportions of each AEC type change. The proportion of type-I AECs decreases to ~45% of all AECs, while the proportion of type-II AECs increases to ~55% of all AECs. For each cell type, values that do not share a common letter are significantly different from each other. *Figure reproduced from Flecknoe et al. (2003).*[58]

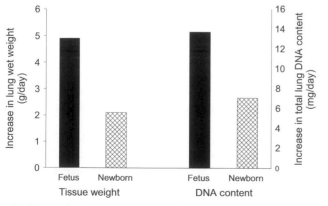

FIGURE 5 The rate of lung growth before and after birth in sheep. The rate of lung growth was calculated from the increase in wet lung weight (g/day; left panel) and total lung DNA content (right panel), collected from fetuses at 114 and 138 days of gestation (closed bars), and from lambs at 1 and 46 days postnatal age (cross-hatched bars) (*n* = 5 at each age). *Figure reproduced from Hooper and Wallace (2006).*[57]

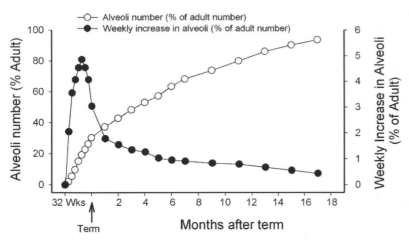

FIGURE 7   Alveolar number and the weekly rate of alveolar formation before and after birth in humans. The accumulation of alveoli (open circles) and the weekly rate of alveolar formation (closed circles) before and after birth in humans, expressed as a percentage of the adult number of alveoli. *Modified from Jobe 2011.*[59] *Idealized curves data from Langston et al. 1984.*[290,291]

a similar manner to basal distension. Phasic distension of cells in culture also induces the expression of many growth factors, including PDGF, VEGF, EGF, and TGF[68,75,76] and other components of intracellular signaling pathways such as IP$_3$, protein kinase C, calcium, cAMP, and MAPK[71,73,77,78] that may mediate the effects of cell stretch/distension on lung cell proliferation and differentiation. Those studies are described in more detail later (mechanotransduction and growth factor).

It remains controversial, however, as to whether the phasic component of FBMs is important for lung development in vivo, because most studies that have attempted to eliminate FBMs in vivo, either by fetal paralysis, replacement of part of the chest wall with a compliant membrane, spinal cord transection, or by reversible blockade or transection of the phrenic nerve[12,79–83] have been complicated by a reduction in basal lung expansion that could have been responsible for the resulting lung hypoplasia. Studies of knockout mice that lack skeletal muscle and that develop lung hypoplasia[84] are also likely to be confounded by reductions in the basal degree of lung expansion. It has been suggested that the cyclic 5% distension stimulus, as used in vitro, greatly exceeds that which occurs as a result of individual FBMs in vivo. Individual FBMs in vivo are essentially isovolumic, causing less than a 0.5% (10-fold less than used in vitro) change in lung volume with each breathing movement.[85,86] This is because lung liquid is much more viscous than air and because the fetal chest wall is very compliant. As a result the chest wall is drawn in with each contraction of the diaphragm.[86]

Regardless of whether the regimens aimed at mimicking FBMs replicate the true effect of FBMs in vivo, they are nevertheless likely to provide important information about the mechanisms by which lung cells detect and respond to changes in their mechanical environment. It is important to understand the mechanisms that underlie distension-induced lung growth and maturation, as these mechanisms likely reflect the normal growth stimuli during development

and could be manipulated to promote lung development in infants born with a lung growth deficit. In a seminal study performed by Moessinger and colleagues in 1990 the left main bronchus of fetal sheep was ligated, while liquid was drained continuously from the right lung for 25 days.[33] They found that the left lung underwent hyperplasia while the right lung was hypoplastic, thereby demonstrating that the local physical environment, rather than circulating factors, must be the primary driving force for expansion-mediated lung growth.[33]

Another interesting feature of fetal lung development is that fetal airway smooth muscle exhibits rhythmic waves of spontaneous contractile activity, similar to gastrointestinal smooth muscle, while postnatal airway smooth muscle exhibits tonic contractions.[87,88] These peristaltic-like waves of airway smooth muscle contraction in the fetus occur at a rate of 2–12/min and cause a 30–50% reduction in luminal diameter of the airways in excised lungs of a variety of species at different stages of lung development.[87] The waves propagate from the larger airways toward the smaller airways and it is thought that they may act to propel lung liquid toward the distal buds of the developing airways, promoting phasic local lengthening and expansion of the terminal buds and potentially providing a local stimulus for lung bud outgrowth or differentiation of the surrounding mesenchymal cells into smooth muscle cell precursors.[87]

MECHANOTRANSDUCTION MECHANISMS

During lung development, lung cells are exposed to a range of different physiological stimuli; of these, physical forces are thought to be a primary growth stimulus. The transmission of mechanical stimuli, such as distension, into biochemical responses is referred to as mechanotransduction. Mechanotransduction plays an important role in cellular growth and differentiation and is a critical regulator of three-dimensional tissue structure.[89,90] In general, the response of a cell to a mechanical stimulus is to reduce the impact of force on that cell. Cells exposed to increased strain will align their intracellular structural fibres along the direction of strain and eventually recruit, synthesize, and align new intracellular structural fibres along this plane. In addition, a cell exposed to strain may synthesize new extracellular matrix (ECM) components, altering the extracellular framework to resist the load.[89,90] As described earlier, responses to strain include cellular proliferation, differentiation, migration, or other alteration in cell function that can lead to marked changes in the size, structure, and function of the lung. Many of these adaptive changes occur as a result of the activation of signaling pathways that are mechanically linked to the ECM and cytoskeleton, by the opening of stretch-activated calcium channels, or by the synthesis or release of growth factors, cytokines, and other factors acting in an autocrine or paracrine manner (Figure 8).

The Role of the ECM and "Outside-in" Cell Signaling

Cells are attached to the ECM through adhesion molecules such as integrins, a family of trans-membrane proteins that cluster together to form focal adhesion complexes (FACs). FACs include cytoskeletal-associated proteins, such as paxillin, talin, α-actinin, and vinculin, that link the integrins to the actin bundles of the cytoskeleton. This forms a structural continuum and mechanical coupling between the ECM and the cell's inner scaffold, via ECM receptors like integrins. It is via this coupling that mechanical forces can be detected and transmitted into intracellular signals. The model of tensegrity[91] describes how the 3-D structure of cells is determined by this mechanical coupling between the intracellular cytoskeletal filaments and the surrounding ECM. At rest, the cytoskeletal filaments exert an isometric tension (termed pre-stress) on the ECM, which in turn influences the shape of the cell. In cultured alveolar epithelial cells this tension is 0.1–0.2 kPa.[92] External mechanical strain results in changes in forces within the ECM that are transmitted to the cytoskeleton via integrins,[93,94] consequently changing the shape of the cell, the alignment of the cytoskeleton and the interaction between proteins that bind to the cytoskeleton.[91,95,96] Integrin signaling also alters gene transcription in a stress-dependent manner.[97] The role of the cytoskeleton in mediating the effects of stretch on cell function is supported by the observation that: (1) mechanical stress of integrin receptors causes the anchoring of phosphorylated proteins (which are major intracellular signaling molecules) to the cytoskeleton[98] and (2) that an intact cytoskeleton is required for stress-induced cell proliferation and cyclin D1 expression in fibroblasts.[99] Mechanical forces may also modulate post-transcriptional gene expression, as binding of cells to the ECM and mechanical tension applied to integrins leads to the recruitment of pre-existing mRNA and ribosomes to FACs through tension-dependent restructuring of the cytoskeleton.[100]

This structural continuum from the ECM via cell surface ECM receptors continues to the nucleus. The cytoskeleton is physically coupled to the nucleoskeleton, via the LINC complex that contains nesprin, sun, and lamin proteins.[90] Nesprin proteins on the outer nuclear membrane couple the micro- and intermediate filaments of the cytoskeleton to the SUN proteins on the inner nuclear membrane. SUN proteins are then bound to the internal nuclear scaffold and nuclear proteins via A-type lamins,[95] providing a mechanism for the effects of physical stretch on gene transcription via the transfer of mechanical signals from the ECM to the cytoskeleton and then to the nucleus (Figure 8).[101] The concept that mechanical signals can be transferred from the ECM to the nucleus is supported by the finding that mechanical stimuli exerted on the ECM can deform the nuclei of endothelial cells[93] and lung epithelial cells.[102] Changes in nuclear shape are associated with altered cell function and phenotype,[95] as the binding of cell surface receptors to the ECM controls chromatin structure and acetylation of gene promoters in a way that regulates the access of transcription factors to DNA, thus controlling epigenetic gene expression.[101] Maintenance of the internal tension (pre-stress) of cells regulates chromatin organisation, as disruption of actin filaments has been shown to cause histone hypoacetylation,[103] and changes in intracellular tension have been correlated with changes in histone acetylation,[104] possibly due to the recruitment of histone deacetylase to the cytoskeleton and translocation to the nucleus following stress.[103]

The importance of integrin signaling for normal lung growth and development has been demonstrated in a number of integrin-null mice generated in recent years. Integrin α3-null mice have abnormal branching morphogenesis[105]; however, double α3/α6 integrin null-mice have a complete lack of the left lung and severe right lung hypoplasia.[106] The lungs of integrin α8-null mouse have large dilated airspaces with defects in secondary septation, abnormal elastin deposition, and fusion of the medial and caudal lobes of the right lung.[107] These defects appear due to the inability of the mesenchymal cells to form focal contacts and control cell migration.[107] Lung epithelial cell formation and differentiation also appears to require integrin signaling as β1−integrin is essential for the development of lung epithelium[108]; furthermore, increases in SP-C induced by 5% phasic stretch of fetal lung epithelial cells in culture are reduced by blocking antibodies against β1, α3, and α6-integrins.[78]

Stretch-Induced Activation of Intracellular Signaling Pathways

Focal adhesion sites also include numerous other signaling molecules that are activated or inactivated following activation of integrins and other ECM receptors. Mechanical stress is transmitted to the cell via integrins, causing activation of downstream pathways such as the cAMP

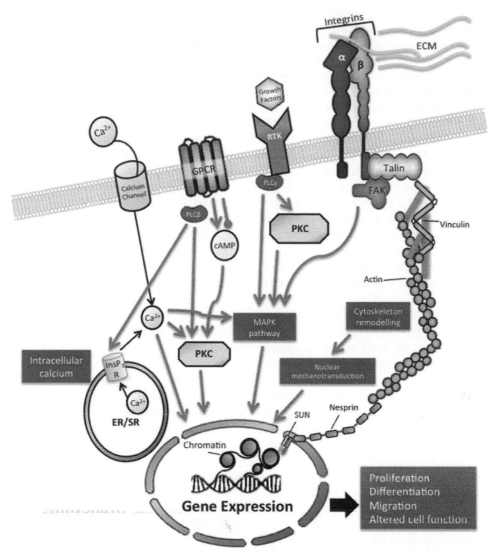

FIGURE 8 A schematic diagram summarizing some of the major mechanotransduction and cell signaling pathways involved in regulating fetal lung growth and maturation. These pathways include increases in intracellular calcium,[110] signaling through G-protein coupled receptors (GPCR),[179] growth factors binding to their tyrosine kinase receptors (RTK)[68,76,138] or integrin signaling.[91] Calcium enters the cell via opening of stretch-activated calcium channels, causing influx of calcium ions from the extracellular space. Calcium can also be released from intracellular stores, including the endoplasmic reticulum (ER) or the sarcoplasmic reticulum (SR), when inositol 1,4,5-triphosphate (InsP3) binds to the InsP3 receptor (InsP3R). Calcium can activate calcium associated signaling molecules, such as calmodulin, which then activates other cell signaling proteins and transcription factors. GPCR are activated by a wide variety of factors known to influence lung development including cytokines, PTHrP, and prostaglandins. This causes the release of intracellular G-proteins, which in turn activate PLCβ. PLCβ catalyzes the hydrolysis of PIP2 into InsP3, activating calcium signaling pathways and DAG, which activates the PKC pathway. GPCR can also activate or inhibit cAMP signaling. Growth factors can initiate intracellular signaling by binding to RTKs, which activate PLCγ and/or the Ras/Raf/MAPK pathway. Calcium, cAMP, PLCβ, and PLCγ can all activate the PKC pathway, which in turn activates the mitogen-activated protein kinase (MAPK) pathway activating ERK1/2, which can enter the nucleus and form transcriptional complexes or phosphorylate other transcription factors in the nucleus or targets in the cytoplasm. Force changes in the extracellular matrix (ECM) are detected by integrins. There are at least 18 alpha and 8 beta integrin subunits that form heterodimers between an alpha and beta subunit. Force is transmitted from the ECM through the integrins to focal adhesions. Focal adhesions are large protein complexes that connect the ECM to the filamentous (F)-actin, via integrins and other connecting proteins, including talin, focal adhesion kinase (FAK), and vinculin. Force transmission through focal adhesions results in remodeling of F-actin. F-actin is further connected to the nucleo-skeleton via nesprin binding to the intra-nuclear protein SUN. SUN binds to the nuclear lamin, which in turn connects to chromatin. Force can therefore be transmitted from the ECM to the chromatin to alter gene transcription. Alteration of gene transcription via any of these pathways can lead to alterations in cell proliferation, differentiation, migration, and other cell functions, which coordinate to promote lung growth and development. (This figure is reproduced in color in the color plate section.)

and mitogen activated protein kinase (MAPKs)[97,98] pathways. The signaling molecules associated with focal adhesion sites include focal adhesion kinase (FAK), which is an intracellular tyrosine kinase that binds to a number of

other signaling, adaptor, and structural proteins, including PI-3-kinase, pp60*src*, GRB2, and p130Cas. Exposing fetal lung cells to cyclic stretch leads to increased pp60*src* activity and its translocation to the cytoskeleton via actin filament-associated protein.[77] pp60[src] increases activation and tyrosine phosphorylation of phospholipase C (PLC-γ1), leading to increases in diacylgycerol (DAG) and inositol 1,4,5-triphosphate (InsP3), all of which are increased by stretch of fetal lung cells.[72] InsP3 activates the release of calcium from intracellular stores, via binding to the InsP3 receptor (InsP3R) that, via DAG, leads to activation of PKC and its downstream mitogenic signaling pathway.[109] PLC-γ1 and PKC activation is required for cyclic stretch-induced proliferation of fetal lung cells in culture.[72,77]

Mechanical forces can also be converted into biochemical signals by the opening and closing of stretch-activated ion channels causing changes in intracellular ion concentrations, such as calcium, which is involved in many signaling pathways.[110] Blocking either strain-induced calcium channels with gadolinium or chelating intracellular calcium stores with BAPTA/AM abolishes the proliferation of lung cells in culture,[71] suggesting that calcium entry into cells, via stretch activated ion channels is vital for fetal lung growth. Blockade of voltage-dependent calcium channels using Nifedipine also causes hypoplasia in lung explants.[111] mRNA levels of calmodulin, the main calcium-signaling molecule, increase in response to an increase in fetal lung expansion[112] when lung cells are proliferating at rates ~800% above those in control fetuses. Inhibition of calmodulin activity in type-II AECs of transgenic mice results in disrupted lung development[113] while postnatal rats treated with a potent calmodulin antagonist, trifluoperazine, had impaired lung growth and structural development.[114] Hemipneumonectomy in young rats induces compensatory lung growth in the remaining lung, which is thought to be expansion mediated.[115,116] The time-course for the increase in lung growth following postnatal hemipneumonectomy in rats is very similar to that induced by tracheal obstruction in fetal sheep[34] with a maximal increase in both cell proliferation and in calmodulin expression occurring within two days of hemipneumonectomy.[116] Trifluoperazine reduces both calmodulin activity and the increase in lung growth normally induced by hemipneumonectomy.[116] These studies indicate that calcium influx and calcium signaling are critical for enabling lung growth, both before and after birth, and that it is critical for mediating expansion-induced lung growth.

Role of Locally Produced Factors in Promoting Lung Growth and Maturation

Recently, studies have begun to focus on the genes that are involved in sensing or responding to mechanical forces during accelerated fetal lung growth with the aim of identifying the factors that regulate lung development. This has been achieved by either targeting specific genes of interest or by identifying genes that are differentially expressed following an increase in lung expansion.[117–119] In fetal sheep,[118] many

of the most highly up-regulated genes, such as connective tissue growth factor (Ctgf), cysteine-rich 61 (Cyr61), early growth response 1 (Egr1), thrombospondin-1 (Tsp1), and trophoblast antigen 2 (Trop2), have established roles in cell proliferation and growth. Tsp1, Trop2, and Egr1 are highly correlated with lung cell proliferation during normal and altered lung growth,[43,120] and we have recently shown that Trop2, a calcium signaling molecule, regulates the proliferation and migration of fetal lung myofibroblasts in culture.[121] Myofibroblasts secrete elastin and are necessary for alveolar formation.[122,123] Increases and decreases in fetal lung expansion cause corresponding increases and decreases in alveolar formation, an effect which is likely mediated by corresponding alterations in mRNA levels of the elastin precursor, tropoelastin, and altered deposition of elastin.[45,124] Another factor, Vitamin D3 up-regulated protein 1 (VDUP1) that inhibits cell proliferation[125] and promotes cell differentiation, was inversely related to lung cell proliferation and positively related to SP-B expression in models of normal and altered lung development.[118,126] These studies support the theory that expansion of the lungs induces changes in the expression of many genes that are likely to coordinate lung growth and alveolar development.

In fetal mice, serial analysis of gene expression (SAGE) was used to identify genes with altered transcription levels following increased lung expansion.[119] The majority of genes that were up-regulated were involved in cytoskeletal regulation or glycolysis, while the majority of down-regulated genes were associated with lipid metabolism. Increased lung expansion also led to a decrease in lipid-laden cells and an increase in alpha smooth muscle actin (α-SMA) containing cells, consistent with the differentiation of lipofibroblasts into myofibroblasts, which are important for driving alveolar development.[119] Increases in α-SMA and altered cytoskeletal organization of mesenchymal cells following increased lung expansion may be explained by increased ROCK expression. ROCK is a downstream target of RhoA, a regulator of focal adhesions and cytoskeletal organization.[127] ROCK expression is increased in mice with lung hyperplasia induced by tracheal occlusion and decreased in mice with lung hypoplasia induced by oligohydramnios.[128] Inhibition of RhoA also inhibited the growth response to tracheal occlusion,[128] supporting the role of RhoA and ROCK in mediating expansion-induced lung growth. In rabbits[117] and sheep,[45,118] many extracellular matrix genes, or genes that regulate the extracellular matrix, are also up-regulated following increased lung expansion; these include tenascin, fibulins, collagens, matrix metalloproteins (MMPs), tissue inhibitors of MMPs (TIMPS), and lysyl oxidase, which regulates the cross-linking of elastin. Genes involved in regulating the arrangement of the actin-cytoskeleton and cell migration (tubulin, Arp2/3, sITGB1, ITGA6, and DBN1) were also upregulated.[117,118] The activation of these genes suggests a coordinated response leading to increased lung growth, alveolar development, and remodeling of the lung to attenuate tissue and increase "air" space.

ROLE OF GROWTH FACTORS IN LUNG DEVELOPMENT

The role of growth factors in regulating embryonic lung development has been discussed in Chapter 1. Here, we briefly describe the role of growth factors in later stages of lung development, particularly factors that may be involved in the transduction of mechanical stimuli like lung expansion, and the coordination of alveolarization, airway growth, ECM remodeling, cell differentiation, and angiogenesis.

The levels of PDGF-A and PDGF-B and their receptors PDGFR-α and PDGFR-β are correlated with rates of cell proliferation during the later stages of lung development, decreasing between the canalicular and saccular stages.[129,130] In lung explants from mid-gestation fetal lungs, cell proliferation is decreased by inhibition of either PDGF-A and -B translation or protein function.[131,132] PDGF-A is vital in regulating normal lung mesenchymal cell development as PDGF-A knockout mice show a severe failure of alveolar septation due to a failure in the maturation of alveolar myofibroblasts and an associated lack of elastin.[122,123] In contrast, over-expression of PDGF-A causes perinatal lethality due to marked mesenchymal cell proliferation and a failure to progress to the saccular stage of development.[133] These studies demonstrate that PDGF-A signaling must be very tightly regulated during the saccular and alveolar stages of lung development. In contrast to their roles in mesenchymal cell development, exogenous PDGF-A or -B have no effect on the proliferation of neonatal type-II AECS in culture.[134] PDGF-A and -B contain shear-stress response elements in their promoters,[135,136] suggesting they could mediate the effect of lung expansion on lung development. Indeed PDGF-B and PDGFR-β expression are increased by cyclic stretch of fetal lung cells[68] and antisense oligonucleotides for PDGF-B and PDGFR-β reduce cell proliferation in unstretched and cyclically stretched fetal lung cells.[68,137] Surprisingly, however, PDGF-B expression decreased by ~25% in fetal sheep exposed to tracheal obstruction[138] when lung cells were proliferating most rapidly, suggesting that PDGF-B is unlikely to mediate the effect of increased lung expansion on lung growth in vivo.

Vascular endothelial growth factor (VEGF) is a potent endothelial cell mitogen that signals via the VEGF receptor (VEGF-R) and is critical for normal lung development. Pulmonary VEGF and VEGF-R gene expression increase throughout gestation,[139] coinciding with increased cross-sectional area of the vascular bed.[140,141] The mitogenic effect of VEGF is not limited to endothelial cells as exogenous VEGF also increases the proliferation of pulmonary mesenchymal cells and type-II AECs in culture, whereas VEGF-A blockade decreases the proliferation of mesenchymal

cells.[142,143] Exposing cultured lung cells to cyclic stretch increases VEGF expression.[76,144] VEGF expression is also increased following tracheal obstruction after 24 h in mice[46] and after 36 h and 2 d in fetal sheep,[138] coinciding with the peak increase in cell proliferation.[34] VEGF is decreased in rabbits with lung hypoplasia induced by diaphragmatic hernia, and its expression is increased in these rabbits following subsequent tracheal obstruction[145]; this suggests that VEGF may be an important mediator of the increase in cell proliferation induced by an increase in fetal lung expansion. When VEGF signaling was inhibited in vivo using VEGF-R neutralizing antibodies or the VEGF-R inhibitor SU-5416, both pulmonary vascular development and alveolarization were impaired.[146,147] VEGF therapy, on the other hand, has been shown to restore capillary development and alveolarization in neonatal rats exposed to hyperoxia.[148,149]

Insulin-like growth factors (IGFs) I and II are potent mitogens that signal via the type I IGF receptor (IGF1R). IGF1 and IGF1R are most highly expressed during the saccular and alveolar stages of lung development, whereas IGF-II is predominately expressed during the pseudoglandular and canalicular stages.[150,151] IGF-II knockout mice have delayed lung development with increased cellularity and decreased septation, most likely due to an inhibition of cell differentiation.[152] Mutations of the IGF-I receptor (IGF1R), alone or in combination with IGF-I or IGF-II mutations, are lethal after birth due to respiratory failure.[153] Lung-specific deletion of IGF1R results in severely hypoplastic lungs with thickened mesenchyme, increased cell proliferation and apoptosis, and delayed differentiation of epithelial and endothelial cells.[154] Exogenous IGF-I, however, has no effect on neonatal type-II AEC proliferation in vitro.[134] IGF-I and -II are up-regulated by increases in lung expansion[32,155] and down-regulated by reductions in lung expansion induced by diaphragmatic hernia[155] or the abolition of FBM.[29] However, the increase in IGF-II expression only occurs after 7 d of TO[32,138] when cell proliferation is almost back to control levels.[34] Furthermore, mRNA levels for the IGF-II receptor (IGF1R) are reduced from 2 to 10 d of TO.[138] This suggests that IGFs may not be involved in expansion-induced lung growth, but may be involved in other aspects of lung development.

Transforming growth factor β (TGFβ) is co-localized with collagen I, collagen III, and fibronectin in the fetal lung.[156] Aberrant expression of TGFβ in distal respiratory epithelial cells, driven by the SPC promoter, delays lung development[157,158] with fewer, narrower acinar buds, thickened mesenchyme between airways, reduced vessel numbers and reduced invasion of vessels around distal airsacs. There was also a decrease in SPB and SPC levels and a reduced number of lamellar bodies in the type-II AECs[157,158] indicative of reduced proliferation or differentiation of type-II AECs. In support of this, exogenous TGFβ decreases the proliferation of cultured neonatal type-II AECs.[134] Fetal lung expansion induces the trans-differentiation of type-II AECs to type-I AECs and decreases surfactant protein expression,[10,31,34,49,159] which may be mediated by an increase in TGFβ. However, TGFβ1 mRNA levels are only increased after 10 d of increased lung expansion in fetal sheep, while at the same time TGFβ1 protein levels decrease,[138] which is well after type-II AECs have trans-differentiated.[34] In contrast to TGFβ, TGFα, and epidermal growth factor (EGF) increase the proliferation of cultured type-II AECs from rabbits,[134] and EGF treatment accelerates lung development in fetal monkeys and rats.[160–162] Mechanical strain of fetal type-II AECs in culture also induces the release of EGF and TGFα, promoting the differentiation of type-II AECs.[70,75] The effects of both EGF and TGFα are mediated by the EGF receptor (EGF-R) and this signaling pathway is critical in promoting lung development in vivo; this is evident because EGF-R null mice die of respiratory failure at birth, due to hypercellular mesenchyme between the airways and a reduction in surfactant levels.[163,164]

A number of fibroblast growth factors (FGFs 1–4, 7, 9, 10, and 18) and their receptors (FGF-R1-5) are expressed in the developing lung.[165–167] FGF10 and FGF-R2 are critical for the initiation of embryonic lung development as knockout mice fail to develop lungs past the trachea.[168,169] FGF-R2 is also involved in development of the distal lung as inhibition of its activity in the pseudoglandular-to-saccular stage of development resulted in emphysematous postnatal lungs with thinner alveolar walls and larger airspaces.[170] However there was no effect when FGF-R2 activity was blocked only during the alveolar stage.[170] FGF-R3/FGF-R4 double mutant mice also have thinner mesenchyme and larger airspaces.[166] FGF9 knockout mice die at birth due to respiratory failure, apparently due to reduced mesenchymal tissue and airway branching, as there were no changes in the alveolar regions of the lungs.[171] In contrast, FGF7 (also known as keratinocyte growth factor; KGF) and FGF18 are important for alveolarization. Selective inhibition of FGF7 in the alveolar stage of lung development impairs secondary septation and reduces alveolar number.[172] FGF7 and FGF2 are increased following cyclic stretch of postnatal lungs[144] but are also up-regulated in the distal epithelium and mesenchyme in the hypoplastic lungs of rats with nitrofen-induced CDH,[173] so their roles are as yet unclear. FGF-18 knockout mice have small lungs with reduced alveolar space, thicker interstitial tissue, and altered capillary development in the alveolar region.[174] Interestingly, over-expression of FGF-18 in the developing lungs also alters alveolarization with a complete lack of alveolar structures,[175] indicating that FGF18 levels must be tightly regulated for alveolarization to progress normally. FGF18 levels are also decreased in the hypoplastic lungs of rats with nitrofen-induced CDH, suggesting FGF18 may be involved in regulating expansion-induced lung growth.[176]

Parathyroid hormone-related protein (PTHrP) is secreted by lung epithelial cells and its receptor, PTHrP-R, is localized to the mesenchymal cells of the lung, forming a feedback loop that regulates surfactant production.[177–180] Deletion of either parathyroid hormone-related protein (PTHrP) or the PTH/PTHrP receptor results in respiratory failure at birth with severe lung hypoplasia, a non-distensible rib-cage, and reduced lung liquid clearance.[181,182] PTHrP plays a role in surfactant production, increasing phospholipid biosynthesis in fetal mouse lung explants.[180] In addition, PTHrP expression and receptor binding is increased by stretch of lung cells in vitro and by lung expansion in vivo,[179,183] whereas their expression is decreased in rats with nitrofen-induced CDH[184]; this suggests that PTHrP signaling is likely involved in expansion-induced lung growth and development.

MicroRNAs Involved in Fetal Lung Growth and Development

Recently, a new field has emerged, investigating the potential role of microRNAs (miRNA), in fetal lung development. MicroRNAs are small (~22nt) non-coding RNA fragments that regulate gene expression at the post-transcriptional level (Figure 9). They regulate many biological processes such as cell proliferation, differentiation, apoptosis, developmental timing, and immune function.[185] The majority of miRNA genes that have been characterized are located in the regions between protein-encoding genes, although some are located within introns. They are synthesized by RNA polymerase II, initially as longer primary (pri)-miRNA sequences that are then cleaved into shorter precursor (pre)-miRNA sequences by the RNase Drosha and the co-factor DiGeorge syndrome critical region gene 8 (DGCR8). Pre-miRNA is exported out of the nucleus and is cleaved by the RNase Dicer into the ~22nt double-stranded miRNA, which is then separated into single strands and one strand is recruited by the Argonaute (Ago) proteins to form a protein complex called RNA-induced silencing complex (RISC). RISC guides the miRNA to its target mRNA, leading to mRNA degradation or the repression of mRNA translation. The critical role of miRNAs in regulating fetal lung growth was demonstrated by the selective deletion of the DICER gene in lung epithelial cells, resulting in inhibition of all miRNA processing and the arrest of airway branching and abnormal airway growth.[186]

The miRNA cluster miR-17-92, which includes 6 miRNA genes (miR-17, -18a, -19a, -20a, -19b-1, and -92-1) is highly expressed in E11.5 mouse lungs with decreasing expression as gestation progresses.[187] Over-expression of miR-17-92 in lung epithelial cells caused increased epithelial progenitor cell proliferation and decreased epithelial differentiation,[187] whereas deletion of miR-17-92 in mice caused neonatal death due to severe lung hypoplasia.[188] The miR-17-92 cluster may regulate lung growth by inhibiting FGF10 and E-cadherin signaling.[189] Another miRNA family, let-7 shows increased expression in the lung with increasing gestational age. Let-7 miRNAs inhibit cell proliferation in lung cancer,[190] therefore the low levels of let-7 early in

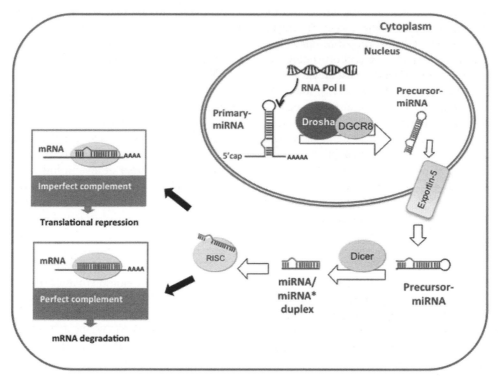

FIGURE 9   Schematic diagram of the mechanism by which microRNA (miRNA) regulate gene expression. MicroRNAs are short, non-coding RNA sequences, about 22 nucleotides (nt) in length, that play an important role in the regulation of mRNA levels at the post-transcriptional level by targeting mRNA for degradation and by repressing mRNA translation. RNA polymerase II (RNA Pol II) transcribes miRNA genes to produce long primary-miRNA strands (usually hundreds of nucleotides in length) that form a hairpin loop. Each primary miRNA strand may contain 1–6 miRNA precursors. Primary miRNA strands then undergo two cleavage steps. The first cleavage occurs in the nucleus by the RNA endonuclease (RNase) Drosha together with the cofactor DiGeorge syndrome critical region 8 (DGCR8), to yield 70–90 nt oligonucleotides called precursor miRNA. Precursor miRNA is then exported to the cytoplasm by the exportin-5 protein where it is cleaved by the RNase Dicer, to generate an imperfect ~22nt double-stranded miRNA sequence. The RNA-induced silencing complex (RISC) then interacts with one strand of the miRNA duplex via a member of the Argonaute (Ago) protein family. The other strand of the miRNA duplex called the passenger strand (denoted miRNA*) is usually degraded but can also target separate mRNA molecules. The mechanism by which RISC leads to mRNA degradation or translational repression remains controversial,[288] but may be related to the Ago protein member and the degree of complementarity between the miRNA and the mRNA targets. The miRNAs generally bind to the 3′ untranslated region of mRNA target and if the miRNA sequence is a perfect complement for the mRNA strand, the RNase activity of AGO 2 can de-adenylate the 3′ poly A tail of mRNA and cleave the 5′ cap leading to mRNA degradation. Interactions of AGO1 with imperfect complementation between miRNA and mRNA sequences are thought to repress translation of the mRNA strand either by preventing the initiation of translation or by interfering with the elongation of the protein sequence.[185,288,289](This figure is reproduced in color in the color plate section.)

gestation are likely to permit rapid lung growth while its high abundance at E17.5[187] may inhibit lung cell proliferation. One of Let-7's known gene targets is the Ras gene, which is associated with tyrosine kinase signaling pathways, including those of FGFs.[185] Various other miRNAs also have increased expression during the saccular and alveolar stage of lung development, including miR-127, miR-154, miR-323, miR-335, miR-337, miR-351, and miR-370[191,192]; however no functional role has yet been established for these miRNAs.

CIRCULATING FACTORS AND METABOLIC INFLUENCES ON LUNG DEVELOPMENT

The Role of Corticosteroids in Fetal Lung Growth and Development

The pre-parturient increase in circulating glucocorticoid levels facilitates survival of the newborn at birth. Infants born prior to the pre-parturient increase in glucocorticoids are likely to suffer respiratory insufficiency at birth. To mature the fetal lungs and increase the chance of survival, synthetic glucocorticoids are now routinely administered to women in preterm labor.[193] The reported effects of glucocorticoids on lung development include altered growth, tissue remodeling to thin the interstitial tissue,[194–196] maturation of the mechanisms that enable lung liquid reabsorption at birth[7,197] (see following text) and differentiation of type-II AECs and surfactant synthesis.[196,198] Although it is often assumed that the principal beneficial effect of glucocorticoids is on maturation of type-II AECs and the surfactant system, it has now become clear that the principal effect of glucocorticoids is on remodeling of the lung structure to promote more efficient gas exchange.[199,200]

Glucocorticoids increase the expression of markers of the type-II AEC phenotype, namely surfactant protein (SP)-A, -B, -C, and -D both in vitro[201–203] and in vivo.[204–206] However, glucocorticoid regulation of surfactant and surfactant protein gene expression is complex with a dose-dependent effect observed in vitro.[207] Glucocorticoid-receptor null mice, as well as mice that have had the glucocorticoid-receptor deleted from the mesenchyme only, almost all die at birth due to respiratory failure owing to structurally immature lungs. However, these mice have normal surfactant protein levels[208,209] and a greater proportion of type-II AECs than controls,[200] suggesting that glucocorticoids are not necessary for maturation of type-II AECs and surfactant production.

Glucocorticoid-induced remodeling of lung tissue primarily occurs by thinning of the interstitial lung tissue, which decreases the distance for gas diffusion and improves lung compliance.[194,210–213] Indeed, airway development in glucocorticoid receptor knockout (GRKO) mice appears to cease during the pseudoglandular stage.[209] The lungs of fetal GRKO mice are hypercellular with increased proliferation of interstitial and epithelial cells, increased septal wall thickness, and increased airway-to-capillary diffusion distances.[199,200] Glucocorticoid-induced thinning of pulmonary interstitial tissue appears to be primarily mediated via mesenchymal cells as mesenchymal cell-specific GRKO mice have a virtually identical phenotype to the complete GRKO mice.[208,214] One study has also shown 50% perinatal lethality and impaired lung development in an epithelial cell-specific GR knockout[215]; however, two other laboratories have shown that both epithelial- and endothelial-cell-specific GR deletions do not affect survival at birth and result in normal lung morphology.[208,214] The differences between the three doxycycline-inducible epithelial GR knockout studies may relate to the duration of GR knockout, or the duration of exposure to doxycycline, which may have deleterious effects on its own. These studies suggest that glucocorticoid signaling primarily matures the lungs via its actions on the mesenchymal cells.[208]

There are contradictory data on the effects of glucocorticoids on lung growth with the prevalent assumption being that glucocorticoids increase lung maturation at the expense of lung growth. Maternal betamethasone treatment causes decreases in both fetal body weight and lung growth.[216–219] However, when administered directly to the fetus, betamethasone had no effect on fetal body or lung growth,[220] suggesting that reported effects of maternal betamethasone administration on fetal and lung growth are mediated via the placenta. Infusion of physiological doses of cortisol directly to the fetus, to mimic the prenatal cortisol surge, also did not affect fetal body weight, lung weight, or lung DNA content, although it almost doubled lung protein content.[197] There is no difference in the fetal body weight between glucocorticoid receptor-null mice and wildtype controls at E18.5; however, the GR null mice have increased lung wet weight, cell proliferation, and DNA content,[200] and histologically the lungs are hypercellular. They also have reduced expression of p21[CIP1],[199] an inhibitor of cell proliferation, which

is consistent with high rates of proliferation of lung cells in these mice. The apparent contradictions between studies are likely to be due to differences in species, dose, number of doses, and the route of administration.

Glucocorticoids also affect alveolarization, although this appears to be dependent upon whether synthetic glucocorticoids are administered using doses similar to those used clinically for women in preterm labour, or whether physiological levels of endogenous glucocorticoids are used. Studies in sheep[221] and rats[222] have demonstrated a decrease in the number of alveoli following treatment with betamethasone or dexamethasone, suggesting a reduction in secondary septation. However, continuous infusion of physiological levels of cortisol (the endogenous glucocorticoid in sheep), directly to the fetus, caused an increase in the number of alveoli.[196] A glucocorticoid-induced increase in lung compliance and alveolarization is also likely responsible for the high lung liquid volumes observed in late gestation fetal sheep,[1,6,197] (Figure 1) and for the greater increase in lung expansion and subsequent lung growth in cortisol-infused fetuses following tracheal obstruction.[223] Similarly, lung liquid volume and the rate of lung liquid secretion are substantially reduced in fetal sheep following bilateral fetal adrenalectomy,[7] which prevents the preparturient increase in glucocorticoids.

An additional, often overlooked beneficial effect of glucocorticoids on the fetal lung is an increase in the lung's ability to reabsorb lung liquid. Lung liquid reabsorption is activated by elevated levels of adrenaline (epinephrine) and arginine vasopressin (AVP) during labor and is responsible for clearing significant volumes of liquid from the lungs before birth. Adrenaline and AVP stimulate the opening of amiloride-blockable Na^+-channels located on the apical surface of AECs, allowing increased Na^+ entry, reversing the osmotic gradient across the pulmonary epithelium and favoring the movement of water from the lumen into the interstitium.[3,224–226] This mechanism, however, only develops late in gestation[227] during the preparturient increase in glucocorticoids.[197,7,224] Glucocorticoids increase the expression of Na^+-K^+-ATPase and epithelial sodium channels (ENaC) in lung, in vitro[228–231] and in vivo,[232–235] whereas GR-null mice have reduced mRNA levels of the ENaCγ subunit.[200] Thus, the lungs of preterm infants who have not been exposed to the cortisol surge prior to birth, are likely to have a reduced ability to absorb lung liquid that likely impairs lung aeration and gas exchange immediately after birth.

The Role of Retinoids in Fetal Lung Growth and Development

All-trans-retinoic acid (RA), the active form of vitamin A, is the ligand for retinoic acid receptors (RARs) and retinoid X receptors (RXRs), which act as transcription factors. Both the ligand and its receptors have been implicated in the regulation of alveolarization. In the lung, retinoids are stored in lipid-laden fibroblasts as retinyl esters. As gestation progresses, the retinyl ester content of these cells decreases, but the content of the metabolically active retinoids, retinol, and retinoic acid increases.[236] Retinoic acid can then enter the nucleus, altering gene activity by binding to the RARs (α, β, and γ) or RXRs (α, β, and γ).[237,238] RARs and RXRs form heterodimers and affect gene transcription by binding to the retinoic acid response element consensus sequences in retinoic acid responsive genes. RARs are expressed in the majority of epithelial cells in the fetal lung, as well as some mesenchymal cells.[239] In humans, expression of RAR-α and RAR-γ significantly increases throughout gestation, whereas expression of RAR-β decreases.[239] In contrast, in rats, expression of RAR-α and RAR-β both decrease late in gestation.[240]

Alveolarization occurs normally in RAR-α null mice,[241] although overexpression of RAR-α results in fewer and larger alveoli, an increase in airspace, a decrease in surface area for gas exchange, and a decrease in differentiated type-I and type-II AECs.[242] RAR-β null mice have increased septation and form alveoli twice as fast as wild-type mice, while RAR-β agonist treatment of newborn rats impairs septation.[243] However, in later life, the RAR-β null mice have impaired formation of the distal alveolar region and impaired lung function.[244] RAR-γ deletion leads to reduced alveolar number, surface area, wall density, and elastin content.[245] These changes were more pronounced when RAR-γ deletion was coupled with a single copy of the RXR-α gene and this also led to a reduction in the number of lipid-laden fibroblasts in the lungs.[245] These studies suggest that retinoic acid regulates alveolarization and that it involves complex interactions between very tightly regulated levels of the different receptor subtypes.

It has been suggested that retinoic acid could be used to induce alveolar development. The administration of retinoic acid to neonatal mice can reverse a glucocorticoid-induced impairment of septation.[246] Similarly, maternal retinoic acid administration can reverse fetal lung hypoplasia induced in a nitrofen-induced model of CDH.[247,248] However, maternal retinoic acid treatment could not reverse lung hypoplasia induced by a surgical model of CDH in rabbits[249] or by oligohydramnios in fetal rats[250] and it did not accelerate structural or functional maturation of the lungs in preterm fetal sheep.[251] Thus, further studies are required to determine the complex role of retinoic acid and its receptors in fetal lung development and alveolarization.

Other Circulating Factors Involved in Fetal Lung Growth and Development

Growth hormone, thyroid hormones, sex hormones, and a variety of other circulating factors have been implicated in fetal lung growth; however, much less is known about their role in development. Although growth hormone levels are high in the fetus and decrease just prior to birth, fetal growth is thought to be independent of GH. Many fetal tissues contain GH receptors (GHR),[252] particularly the lung,[253] but these receptors are thought to be immature, making fetal GH largely inactive.[253] GHR expression is complex; alternate leader exon usage of the 5′-UTR confers tissue-specific regulation,[254] whereas alternate mRNA splicing can produce a short inactive form of the GHR.[255] Near to term, cortisol is thought to increase GHR expression and activate the adult alternately spliced version of the receptor, which may explain how postnatal growth becomes GH-dependent. However, while overall fetal growth is thought to be independent of GH, a recent study has shown that the lung growth response to an increase in fetal lung expansion is GH-dependent. In the absence of GH (due to hypophysectomy), the initial growth response to tracheal obstruction was abolished, whereas the re-infusion of GH restored the growth response.[256] These data indicate that GH may play a permissive role in enabling lung cell proliferation during fetal life in response to alterations in lung expansion. GH is thought to play a similar role after birth in regulating compensatory lung growth following hemipneumonectomy, which is also thought to be expansion-dependent.[257]

Thyroid hormones can influence septation in the fetal lung, increasing the number of alveoli and the surface area for gas exchange. Triiodothyronine (T3) treatment increases septation in newborn rats,[258] while blocking the conversion of thyroxine to T3 with propylthiouracil inhibits septation.[258,259] Basal levels of thyroid hormones are also required for maturation of the lung liquid reabsorption mechanism and may act in a synergistic mechanism with glucocorticoids.[260,261] Androgens and estrogens have also been linked to fetal lung development.[262] There is a higher incidence of respiratory distress syndrome (RDS) in male infants compared to females.[263] This is thought to be due to androgens, which delay surfactant synthesis,[264] reduce lung structural maturation,[265] and inhibit the maturational effects of glucocorticoids on the lung.[266] Conversely, estrogens are believed to increase fetal lung maturation, by increasing surfactant production[267] and alveolarization.[268,269]

Metabolic Influences on Lung Development: Hypoxia and Nutrition

Impairments in the delivery of oxygen and nutrients to the developing fetus can cause intra-uterine growth restriction (IUGR) and impair fetal lung development, leading to respiratory deficits in infants, children, and adults.[270–273] The most common cause of IUGR is placental insufficiency, an umbrella term for a broad range of deficits in placental growth or function that limit the availability of oxygen and nutrients for fetal development.

Placental insufficiency in sheep, during the saccular and alveolar stages of fetal lung development, induces IUGR and impairs development of the bronchial walls, reduces alveolar number, increases inter-alveolar wall thickness, and increases the blood-gas barrier diffusion distances; many of these effects persist into adult life.[274–276] These alterations are likely responsible for the observed impairments in gas exchange and respiratory function following IUGR.[277–279] Mice exposed to hypoxia prenatally also had delayed lung maturation with reduced surfactant protein expression that may have been mediated by reductions in VEGF and VEGF-R.[280] In contrast, IUGR induced by uterine artery ligation in pregnant rats only transiently inhibited alveolar development with increases in septal wall thickness and reduced elastin expression in females; this effect may have been mediated by retinoids as RARβ, which inhibits alveolar development was increased in female IUGR rats. In that study, lung development was normal in male IUGR rats.[281] The differences between these studies may be related to the timing of the insult relative to lung development or to the duration of oxygen and nutrient restriction.

IUGR infants are often born preterm,[282] and lung immaturity can exacerbate IUGR-related deficits in respiratory function. These infants often require respiratory support, which can injure their lungs and lead to chronic lung disease and abnormal lung development known as bronchopulmonary dysplasia (BPD).[283] However, even mild preterm birth in sheep, without the need for respiratory support, is sufficient to cause abnormal lung development with an increase in protein concentration of the lung and an increase in blood-gas-barrier thickness.[284] This suggests that even slightly premature exposure of the lung to the altered physical, metabolic, and endocrine environment that occurs after birth may be sufficient to cause permanent changes in lung development. The effect of altered nutrition on lung development is explored in more detail in Chapter 18 [Harding & De Matteo].

Treatments for Infants with Inappropriate Lung Development

More than 10% of all infants are born either preterm, before the lungs have fully developed, or following an adverse intra-uterine environment, which compromises lung development, causing lung hypoplasia or other alterations in lung development. Improvements in perinatal care over the last 30–40 years, including the routine administration of glucocorticoids to women in preterm labor, and surfactant administration for respiratory distress, have dramatically improved the survival of these infants. However, glucocorticoids have also been associated with adverse side-effects[217,222] and the increase in survival rates of the most vulnerable infants has come at the expense of higher rates of severe morbidities like bronchopulmonary dysplasia (BPD). There remains,

therefore, a pressing need to develop improved therapeutic options to promote lung development. To achieve this, it is necessary to understand the mechanisms that underlie normal and abnormal lung development, so that those mechanisms can be manipulated.

The finding that tracheal obstruction can be used to accelerate lung growth in animal studies[40,223,285] has prompted its use in human fetuses with CDH, in order to reverse lung hypoplasia before the lungs are required for gas exchange at birth. Success in human fetuses was initially limited and complicated by an increased risk of preterm birth.[286] However, improved surgical techniques and better prediction of which fetuses will most benefit from obstructing the trachea have resulted in vastly improved survival rates for infants that previously would have died at birth or who would have been candidates for elective termination.[19,20]

The most severely affected preterm and growth restricted infants, and infants with lung hypoplasia, still require respiratory support at birth to survive. Respiratory support is usually provided in terms of supplemental oxygen and/or continuous positive airway pressure (CPAP) or mechanical ventilation. However, respiratory support itself is associated with BPD[283] and poor neurodevelopmental outcomes.[287] In order to reduce lung and brain injury at birth it is important that we increase our understanding of how to most effectively improve the transition of the liquid-filled fetal lung to an air-filled lung that can sustain adequate gas exchange. The most recent advances in our understanding of the transition at birth in term and preterm infants is described in Chapter 13 [Siew et al.] and the pulmonary consequences of preterm birth and the administration of hyperoxic gases are described in more detail in Chapters 17 [Albertine et al.] and 28 [Sozo and O'Reilly], respectively.

The advent of transgenic mouse approaches, microarrays and large-scale gene sequencing approaches, some of which have been described earlier, are also yielding very important information on the mechanisms likely to underlie normal and abnormal lung development. Taking these findings forward to produce effective treatments will depend on our ability to decipher, interrogate, and integrate the information from those studies and on our ability to test new therapies in appropriate animal models, ensuring that both short- and long-term beneficial and deleterious effects have been fully investigated. In the next few years we should see similar advances in our understanding of the role of miRNAs and epigenetics (discussed in Chapter 16 [Joss-Moore et al.]) in the regulation of lung development.

CONCLUSION

Lung growth and maturation before and after birth are dependent upon a complex interplay between physical forces, circulating factors, and locally produced factors that act together to produce a highly integrated and tightly controlled process of lung cell growth, proliferation, differentiation, and maturation. Alterations to any of those factors can have dramatic and long-lasting effects on the lung. Having a clear understanding of how cells detect and respond to alterations in their environment will therefore be critical for the development of new therapeutic strategies to improve lung development and the long-term outcomes for infants born with lung growth deficits.

REFERENCES

1. Hooper SB, Harding R. Fetal lung liquid: a major determinant of the growth and functional development of the fetal lung. *Clin Exp Pharmacol Physiol* 1995;**22**:235–47.
2. Olver RE, Strang LB. Ion fluxes across the pulmonary epithelium and the secretion of lung liquid in the foetal lamb. *J Physiol* 1974;**241**:327–57.
3. Olver RE, Ramsden CA, Strang LB, Walters DV. The role of amiloride-blockable sodium transport in adrenaline-induced lung liquid reabsorption in the fetal lamb. *J Physiol* 1986;**376**:321–40.
4. Olver RE, Schneeberger EE, Walters DV. Epithelial solute permeability, ion transport and tight junction morphology in the developing lung of the fetal lamb. *J Physiol* 1981;**315**:395–412.
5. Mescher EJ, Platzker AC, Ballard PL, Kitterman JA, Clements JA, Tooley WH. Ontogeny of tracheal fluid, pulmonary surfactant, and plasma corticoids in the fetal lamb. *J Appl Phys* 1975;**39**:1017–21.
6. Harding R, Hooper SB. Regulation of lung expansion and lung growth before birth. *J Mol Endocrinol* 1996;**81**:209–24.
7. Wallace MJ, Hooper SB, Harding R. Role of the adrenal glands in the maturation of lung liquid secretory mechanisms in fetal sheep. *Am J Phys* 1996;**270**:R33–40.
8. Harding R, Bocking AD, Sigger JN. Influence of upper respiratory tract on liquid flow to and from fetal lungs. *J Appl Phys* 1986;**61**:68–74.
9. Pfister RE, Ramsden CA, Neil HL, Kyriakides MA, Berger PJ. Volume and secretion rate of lung liquid in the final days of gestation and labour in the fetal sheep. *J Physiol* 2001;**535**:889–99.
10. Lines A, Hooper SB, Harding R. Lung liquid production rates and volumes do not decrease before labor in healthy fetal sheep. *J Appl Phys* 1997;**82**:927–32.
11. Cassin S, Perks AM. Estimation of lung liquid production in fetal sheep with blue dye dextran and radioiodinated serum albumin. *J Appl Phys* 2002;**92**:1531–8.
12. Miller AA, Hooper SB, Harding R. Role of fetal breathing movements in control of fetal lung distension. *J Appl Phys* 1993;**75**:2711–7.
13. Harding R, Hooper SB, Dickson KA. A mechanism leading to reduced lung expansion and lung hypoplasia in fetal sheep during oligohydramnios. *Am J Obstet Gynecol* 1990;**163**:1904–13.
14. Dickson KA, Harding R. Restoration of lung liquid volume following its acute alteration in fetal sheep. *J Physiol* 1987;**385**:531–43.
15. Harding R, Bocking AD, Sigger JN. Upper airway resistances in fetal sheep: the influence of breathing activity. *J Appl Phys* 1986;**60**:160–5.
16. Vilos GA, Liggins GC. Intrathoracic pressures in fetal sheep. *J Dev Physiol* 1982;**4**:247–56.
17. Dickson KA, Harding R. Fetal breathing and pressures in the trachea and amniotic sac during oligohydramnios in sheep. *J Appl Phys* 1991;**70**:293–9.
18. Moessinger AC, Collins MH, Blanc WA, Rey HR, James LS. Oligohydramnios-induced lung hypoplasia: the influence of timing and duration in gestation. *Pediatr Res* 1986;**20**:951–4.
19. Jani JC, Nicolaides KH. Fetal surgery for severe congenital diaphragmatic hernia? *Ultrasound Obstet Gynecol* 2012;**39**:7–9.
20. Jani JC, Benachi A, Nicolaides KH, Allegaert K, Gratacos E, Mazkereth R, et al. Prenatal prediction of neonatal morbidity in survivors with congenital diaphragmatic hernia: a multicenter study. *Ultrasound Obstet Gynecol* 2009;**33**:64–9.
21. Skari H, Bjornland K, Haugen G, Egeland T, Emblem R. Congenital diaphragmatic hernia: a meta-analysis of mortality factors. *J Pediatr Surg* 2000;**35**:1187–97.
22. Stege G, Fenton A, Jaffray B. Nihilism in the 1990s: the true mortality of congenital diaphragmatic hernia. *Pediatrics* 2003;**112**:532–5.
23. Witters I, Legius E, Moerman P, Deprest J, Van Schoubroeck D, Timmerman D, et al. Associated malformations and chromosomal anomalies in 42 cases of prenatally diagnosed diaphragmatic hernia. *Am J Med Genet* 2001;**103**:278–82.
24. Deprest J, Jani J, Gratacos E, Vandecruys H, Naulaers G, Delgado J, et al. Fetal intervention for congenital diaphragmatic hernia: the European experience. *Semin Perinatol* 2005;**29**:94–103.
25. Ruano R, Takashi E, da Silva MM, Campos JA, Tannuri U, Zugaib M. Prediction and probability of neonatal outcome in isolated congenital diaphragmatic hernia using multiple ultrasound parameters. *Ultrasound Obstet Gynecol* 2012;**39**:42–9.
26. Harrison MR, Keller RL, Hawgood SB, Kitterman JA, Sandberg PL, Farmer DL, et al. A randomized trial of fetal endoscopic tracheal occlusion for severe fetal congenital diaphragmatic hernia. *N Engl J Med* 2003;**349**:1916–24.
27. Colvin J, Bower C, Dickinson JE, Sokol J. Outcomes of congenital diaphragmatic hernia: a population-based study in Western Australia. *Pediatrics* 2005;**116**:e356–63.
28. Gallot D, Boda C, Ughetto S, Perthus I, Robert-Gnansia E, Francannet C, et al. Prenatal detection and outcome of congenital diaphragmatic hernia: a French registry-based study. *Ultrasound Obstet Gynecol* 2007;**29**:276–83.
29. Harding R, Hooper SB, Han VK. Abolition of fetal breathing movements by spinal cord transection leads to reductions in fetal lung liquid volume, lung growth, and IGF-II gene expression. *Pediatr Res* 1993;**34**:148–53.
30. Jost A, Policard A. Contribution experimentale a l'etude du developpement prenatal du poumon chez le lapin. *Archives D'Anatomie Microscopique* 1948;**37**:323–32.
31. Alcorn D, Adamson TM, Lambert TF, Maloney JE, Ritchie BC, Robinson PM. Morphological effects of chronic tracheal ligation and drainage in the fetal lamb lung. *J Anat* 1977;**123**:649–60.
32. Hooper SB, Han VK, Harding R. Changes in lung expansion alter pulmonary DNA synthesis and IGF-II gene expression in fetal sheep. *Am J Phys* 1993;**265**:L403–9.
33. Moessinger AC, Harding R, Adamson TM, Singh M, Kiu GT. Role of lung fluid volume in growth and maturation of the fetal sheep lung. *J Clin Invest* 1990;**86**:1270–7.
34. Nardo L, Hooper SB, Harding R. Stimulation of lung growth by tracheal obstruction in fetal sheep: relation to luminal pressure and lung liquid volume. *Pediatr Res* 1998;**43**:184–90.
35. M.E. De Paepe, Johnson BD, Papadakis K, Luks FI. Lung growth response after tracheal occlusion in fetal rabbits is gestational age-dependent. *Am J Respir Cell Mol Biol* 1999;**21**:65–76.

36. De Paepe ME, Johnson BD, Papadakis K, Sueishi K, Luks FI. Temporal pattern of accelerated lung growth after tracheal occlusion in the fetal rabbit. *Am J Pathol* 1998;**152**:179–90.

37. Carmel JA, Friedman F, Adams FH. Fetal tracheal ligation and lung development. *Am J Dis Child* 1965;**109**:452–6.

38. Yoshizawa J, Chapin CJ, Sbragia L, Ertsey R, Gutierrez JA, Albanese CT, et al. Tracheal occlusion stimulates cell cycle progression and type I cell differentiation in lungs of fetal rats. *Am J Physiol Lung Cell Mol Physiol* 2003;**285**:L344–53.

39. Maltais F, Seaborn T, Guay S, Piedboeuf B. In vivo tracheal occlusion in fetal mice induces rapid lung development without affecting surfactant protein C expression. *Am J Physiol Lung Cell Mol Physiol* 2003;**284**:L622–32.

40. Nardo L, Hooper SB, Harding R. Lung hypoplasia can be reversed by short-term obstruction of the trachea in fetal sheep. *Pediatr Res* 1995;**38**:690–6.

41. Keramidaris E, Hooper SB, Harding R. Effect of gestational age on the increase in fetal lung growth following tracheal obstruction. *Exp Lung Res* 1996;**22**:283–98.

42. Nardo L, Maritz G, Harding R, Hooper SB. Changes in lung structure and cellular division induced by tracheal obstruction in fetal sheep. *Exp Lung Res* 2000;**26**:105–19.

43. McDougall AR, Hooper SB, Zahra VA, Sozo F, Lo CY, Cole TJ, et al. The oncogene Trop2 regulates fetal lung cell proliferation. *Am J Physiol Lung Cell Mol Physiol* 2011;**301**:L478–89.

44. Polglase GR, Wallace MJ, Morgan DL, Hooper SB. Increases in lung expansion alter pulmonary hemodynamics in fetal sheep. *J Appl Phys* 2006;**101**:273–82.

45. Joyce BJ, Wallace MJ, Pierce RA, Harding R, Hooper SB. Sustained changes in lung expansion alter tropoelastin mRNA levels and elastin content in fetal sheep lungs. *Am J Physiol Lung Cell Mol Physiol* 2003;**284**:L643–9.

46. Cloutier M, Maltais F, Piedboeuf B. Increased distension stimulates distal capillary growth as well as expression of specific angiogenesis genes in fetal mouse lungs. *Exp Lung Res* 2008;**34**:101–13.

47. Probyn ME, Wallace MJ, Hooper SB. Effect of increased lung expansion on lung growth and development near midgestation in fetal sheep. *Pediatr Res* 2000;**47**:806–12.

48. Boland RE, Nardo L, Hooper SB. Cortisol pretreatment enhances the lung growth response to tracheal obstruction in fetal sheep. *Am J Phys* 1997;**273**:L1126–31.

49. Flecknoe S, Harding R, Maritz G, Hooper SB. Increased lung expansion alters the proportions of type I and type II alveolar epithelial cells in fetal sheep. *Am J Physiol Lung Cell Mol Physiol* 2000;**278**:L1180–85.

50. Joe P, Wallen LD, Chapin CJ, Lee CH, Allen L, Han VK, et al. Effects of mechanical factors on growth and maturation of the lung in fetal sheep. *Am J Phys* 1997;**272**:L95–105.

51. Lines A, Nardo L, Phillips ID, Possmayer F, Hooper SB. Alterations in lung expansion affect surfactant protein A, B, and C mRNA levels in fetal sheep. *Am J Phys* 1999;**276**:L239–245.

52. Flecknoe SJ, Wallace MJ, Harding R, Hooper SB. Determination of alveolar epithelial cell phenotypes in fetal sheep: evidence for the involvement of basal lung expansion. *J Physiol* 2002;**542**:245–53.

53. Shannon JM, Jennings SD, Nielsen LD. Modulation of alveolar type II cell differentiated function in vitro. *Am J Phys* 1992;**262**:L427–436.

54. Flecknoe SJ, Boland RE, Wallace MJ, Harding R, Hooper SB. Regulation of alveolar epithelial cell phenotypes in fetal sheep: roles of cortisol and lung expansion. *Am J Physiol Lung Cell Mol Physiol* 2004;**287**:L1207–14.

55. Foster CD, Varghese LS, Gonzales LW, Margulies SS, Guttentag SH. The Rho pathway mediates transition to an alveolar type I cell phenotype during static stretch of alveolar type II cells. *Pediatr Res* 2010;**67**:585–90.

56. Foster CD, Varghese LS, Skalina RB, Gonzales LW, Guttentag SH. In vitro transdifferentiation of human fetal type II cells toward a type I-like cell. *Pediatr Res* 2007;**61**:404–9.

57. Hooper SB, Wallace MJ. Role of the physicochemical environment in lung development. *Clin Exp Pharmacol Physiol* 2006;**33**:273–9.

58. Flecknoe SJ, Wallace MJ, Cock ML, Harding R, Hooper SB. Changes in alveolar epithelial cell proportions during fetal and postnatal development in sheep. *Am J Physiol Lung Cell Mol Physiol* 2003;**285**:L664–70.

59. Jobe A. Lung development and maturation. In: Martin RJ, Fanaroff AA, Walsh MC, editors. *Neonatal-Perinatal Medicine*. Elsevier; 2011. pp. 1075–206.

60. Flecknoe SJ, Crossley KJ, Zuccala GM, Searle JE, Allison BJ, Wallace MJ, et al. Increased lung expansion alters lung growth but not alveolar epithelial cell differentiation in newborn lambs. *Am J Physiol Lung Cell Mol Physiol* 2007;**292**:L454–61.

61. Hsia CC. Signals and mechanisms of compensatory lung growth. *J Appl Phys* 2004;**97**:1992–8.

62. Zhang S, Garbutt V, McBride JT. Strain-induced growth of the immature lung. *J Appl Phys* 1996;**81**:1471–6.

63. Ad Hoc Statement Committee. A. T. S. Mechanisms and limits of induced postnatal lung growth. *Am J Respir Crit Care Med* 2004;**170**:319–43.

64. Lines AL, Davey MG, Harding R, Hooper SB. Effect of increased lung expansion on surfactant protein mRNA levels in lambs. *Pediatr Res* 2001;**50**:720–5.

65. Sanchez-Esteban J, Wang Y, Cicchiello LA, Rubin LP. Cyclic mechanical stretch inhibits cell proliferation and induces apoptosis in fetal rat lung fibroblasts. *Am J Physiol Lung Cell Mol Physiol* 2002;**282**:L448–56.

66. Lee HS, Wang Y, Maciejewski BS, Esho K, Fulton C, Sharma S, et al. Interleukin-10 protects cultured fetal rat type II epithelial cells from injury induced by mechanical stretch. *Am J Physiol Lung Cell Mol Physiol* 2008;**294**:L225–32.

67. Liu M, Xu J, Souza P, Tanswell B, Tanswell AK, Post M. The effect of mechanical strain on fetal rat lung cell proliferation: comparison of two- and three-dimensional culture systems. *In Vitro Cell Dev Biol Anim* 1995;**31**:858–66.

68. Liu M, Liu J, Buch S, Tanswell AK, Post M. Antisense oligonucleotides for PDGF-B and its receptor inhibit mechanical strain-induced fetal lung cell growth. *Am J Phys* 1995;**269**:L178–84.

69. Sanchez-Esteban J, Tsai SW, Sang J, Qin J, Torday JS, Rubin LP. Effects of mechanical forces on lung-specific gene expression. *Am J Med Sci* 1998;**316**:200–4.

70. Wang Y, Maciejewski BS, Soto-Reyes D, Lee HS, Warburton D, Sanchez-Esteban J. Mechanical stretch promotes fetal type II epithelial cell differentiation via shedding of HB-EGF and TGF-alpha. *J Physiol* 2009;**587**:1739–53.

71. Liu M, Xu J, Tanswell AK, Post M. Inhibition of mechanical strain-induced fetal rat lung cell proliferation by gadolinium, a stretch-activated channel blocker. *J Cell Physiol* 1994;**161**:501–7.

72. Liu M, Xu J, Liu J, Kraw ME, Tanswell AK, Post M. Mechanical strain-enhanced fetal lung cell proliferation is mediated by phospholipase C and D and protein kinase C. *Am J Phys* 1995;**268**:L729–38.

73. Wang Y, Maciejewski BS, Lee N, Silbert O, McKnight NL, Frangos JA, et al. Strain-induced fetal type II epithelial cell differentiation is mediated via cAMP-PKA-dependent signaling pathway. *Am J Physiol Lung Cell Mol Physiol* 2006;**291**:L820–7.

74. Silbert O, Wang Y, Maciejewski BS, Lee HS, Shaw SK, Sanchez-Esteban J. Roles of RhoA and Rac1 on actin remodeling and cell alignment and differentiation in fetal type II epithelial cells exposed to cyclic mechanical stretch. *Exp Lung Res* 2008;**34**:663–80.

75. Wang Y, Huang Z, Nayak PS, Matthews BD, Warburton D, Shi W, et al. Strain-induced differentiation of fetal type II epithelial cells is mediated via the integrin alpha6beta1-ADAM17/tumor necrosis factor-alpha-converting enzyme (TACE) signaling pathway. *J Biol Chem* 2013;**288**:25646–57.

76. Muratore CS, Nguyen HT, Ziegler MM, Wilson JM. Stretch-induced upregulation of VEGF gene expression in murine pulmonary culture: a role for angiogenesis in lung development. *J Pediatr Surg* 2000;**35**:906–12. discussion 912-903.

77. Liu M, Qin Y, Liu J, Tanswell AK, Post M. Mechanical strain induces pp60src activation and translocation to cytoskeleton in fetal rat lung cells. *J Biol Chem* 1996;**271**:7066–71.

78. Sanchez-Esteban J, Wang Y, Filardo EJ, Rubin LP, Ingber DE. Integrins beta1, alpha6, and alpha3 contribute to mechanical strain-induced differentiation of fetal lung type II epithelial cells via distinct mechanisms. *Am J Physiol Lung Cell Mol Physiol* 2006;**290**:L343–50.

79. Alcorn D, Adamson TM, Maloney JE, Robinson PM. Morphological effects of chronic bilateral phrenectomy or vagotomy in the fetal lamb lung. *J Anat* 1980;**130**:683–95.

80. Fewell JE, Lee CC, Kitterman JA. Effects of phrenic nerve section on the respiratory system of fetal lambs. *J Appl Physiol Respir Environ Exerc Physiol* 1981;**51**:293–7.

81. Liggins GC, Vilos GA, Campos GA, Kitterman JA, Lee CH. The effect of bilateral thoracoplasty on lung development in fetal sheep. *J Dev Physiol* 1981;**3**:275–82.

82. Liggins GC, Vilos GA, Campos GA, Kitterman JA, Lee CH. The effect of spinal cord transection on lung development in fetal sheep. *J Dev Physiol* 1981;**3**:267–74.

83. Wigglesworth JS, Desai R. Effect on lung growth of cervical cord section in the rabbit fetus. *Early Hum Dev* 1979;**3**:51–65.

84. Baguma-Nibasheka M, Gugic D, Saraga-Babic M, Kablar B. Role of skeletal muscle in lung development. *Histol Histopathol* 2012;**27**:817–26.

85. Dawes GS, Fox HE, Leduc BM, Liggins GC, Richards RT. Respiratory movements and rapid eye movement sleep in the foetal lamb. *J Physiol* 1972;**220**:119–43.

86. Harding R, Liggins GC. Changes in thoracic dimensions induced by breathing movements in fetal sheep. *Reprod Fertil Dev* 1996;**8**:117–24.

87. Schittny JC, Miserocchi G, Sparrow MP. Spontaneous peristaltic airway contractions propel lung liquid through the bronchial tree of intact and fetal lung explants. *Am J Respir Cell Mol Biol* 2000;**23**:11–8.

88. Sparrow MP, Lamb JP. Ontogeny of airway smooth muscle: structure, innervation, myogenesis and function in the fetal lung. *Respir Physiol Neurobiol* 2003;**137**:361–72.

89. Ingber DE. Cellular mechanotransduction: putting all the pieces together again. *Faseb J* 2006;**20**:811–27.

90. Wang N, Tytell JD, Ingber DE. Mechanotransduction at a distance: mechanically coupling the extracellular matrix with the nucleus. *Nat Rev Mol Cell Biol* 2009;**10**:75–82.

91. Alenghat FJ, Ingber DE. Mechanotransduction: all signals point to cytoskeleton, matrix, and integrins. *Sci STKE* 2002;**2002**. pe6.

92. Laurent VM, Fodil R, Canadas P, Fereol S, Louis B, Planus E, et al. Partitioning of cortical and deep cytoskeleton responses from transient magnetic bead twisting. *Ann Biomed Eng* 2003;**31**:1263–78.

93. Maniotis AJ, Chen CS, Ingber DE. Demonstration of mechanical connections between integrins, cytoskeletal filaments, and nucleoplasm that stabilize nuclear structure. *Proc Natl Acad Sci U S A* 1997;**94**:849–54.

94. Sims JR, Karp S, Ingber DE. Altering the cellular mechanical force balance results in integrated changes in cell, cytoskeletal and nuclear shape. *J Cell Sci* 1992;**103**:1215–22.

95. Dahl KN, Ribeiro AJ, Lammerding J. Nuclear shape, mechanics, and mechanotransduction. *Circ Res* 2008;**102**:1307–18.

96. Ingber DE. Mechanobiology and diseases of mechanotransduction. *An Med* 2003;**35**:564–77.

97. Meyer CJ, Alenghat FJ, Rim P, Fong JH, Fabry B, Ingber DE. Mechanical control of cyclic AMP signalling and gene transcription through integrins. *Nat Cell Biol* 2000;**2**:666–8.

98. Schmidt C, Pommerenke H, Durr F, Nebe B, Rychly J. Mechanical stressing of integrin receptors induces enhanced tyrosine phosphorylation of cytoskeletally anchored proteins. *J Biol Chem* 1998;**273**:5081–5.

99. Bohmer RM, Scharf E, Assoian RK. Cytoskeletal integrity is required throughout the mitogen stimulation phase of the cell cycle and mediates the anchorage-dependent expression of cyclin D1. *Mol Biol Cell* 1996;**7**:101–11.

100. Chicurel ME, Singer RH, Meyer CJ, Ingber DE. Integrin binding and mechanical tension induce movement of mRNA and ribosomes to focal adhesions. *Nature* 1998;**392**:730–3.

101. Lelievre SA. Contributions of extracellular matrix signaling and tissue architecture to nuclear mechanisms and spatial organization of gene expression control. *Biochim Biophys Acta* 2009;**1790**:925–35.

102. Pajerowski JD, Dahl KN, Zhong FL, Sammak PJ, Discher DE. Physical plasticity of the nucleus in stem cell differentiation. *Proc Natl Acad Sci U S A* 2007;**104**:15619–24.

103. Le Beyec J, Xu R, Lee SY, Nelson CM, Rizki A, Alcaraz J, et al. Cell shape regulates global histone acetylation in human mammary epithelial cells. *Exp Cell Res* 2007;**313**:3066–75.

104. Kim YB, Yu J, Lee SY, Lee MS, Ko SG, Ye SK, et al. Cell adhesion status-dependent histone acetylation is regulated through intracellular contractility-related signaling activities. *J Biol Chem* 2005;**280**:28357–64.

105. Kreidberg JA, Donovan MJ, Goldstein SL, Rennke H, Shepherd K, Jones RC, et al. Alpha 3 beta 1 integrin has a crucial role in kidney and lung organogenesis. *Development* 1996;**122**:3537–47.

106. De Arcangelis A, Mark M, Kreidberg J, Sorokin L, Georges-Labouesse E. Synergistic activities of alpha3 and alpha6 integrins are required during apical ectodermal ridge formation and organogenesis in the mouse. *Development* 1999;**126**:3957–68.

107. Benjamin JT, Gaston DC, Halloran BA, Schnapp LM, Zent R, Prince LS. The role of integrin alpha8beta1 in fetal lung morphogenesis and injury. *Dev Biol* 2009;**335**:407–17.

108. Berger TM, Hirsch E, Djonov V, Schittny JC. Loss of beta1-integrin-deficient cells during the development of endoderm-derived epithelia. *Anat Embryol* 2003;**207**:283–8.

109. Berridge MJ. Calcium signalling and cell proliferation. BioEssays : News and Reviews in Molecular. *Cell Dev Biol* 1995;**17**:491–500.

110. Wirtz HR, Dobbs LG. The effects of mechanical forces on lung functions. *Respir Physiol* 2000;**119**:1–17.

111. Roman J. Effects of calcium channel blockade on mammalian lung branching morphogenesis. *Exp Lung Res* 1995;**21**:489–502.

112. Gillett AM, Wallace MJ, Gillespie MT, Hooper SB. Increased expansion of the lung stimulates calmodulin 2 expression in fetal sheep. *Am J Physiol Lung Cell Mol Physiol* 2002;**282**:L440–7.

113. Wang J, Campos B, Kaetzel MA, Dedman JR. Expression of a calmodulin inhibitor peptide in progenitor alveolar type II cells disrupts lung development. *Am J Phys* 1996;**271**:L245–50.

114. Ofulue AF, Sekhon H, Cherukupalli K, Khadempour H, Thurlbeck WM. Morphometric and biochemical changes in lungs of growing rats treated with a calmodulin antagonist. *Pediatr Pulmonol* 1991;**10**:46–51.

115. Rannels DE. Role of physical forces in compensatory growth of the lung. *Am J Phys* 1989;**257**:L179–89.

116. Ofulue AF, Matsui R, Thurlbeck WM. Role of calmodulin as an endogenous initiatory factor in compensatory lung growth after pneumonectomy. *Pediatr Pulmonol* 1993;**15**:145–50.

117. Vuckovic A, Herber-Jonat S, Flemmer AW, Roubliova XI, Jani JC. Alveolarization genes modulated by fetal tracheal occlusion in the rabbit model for congenital diaphragmatic hernia: a randomized study. *PLoS One* 2013;**8**. e69210.

118. Sozo F, Wallace MJ, Zahra VA, Filby CE, Hooper SB. Gene expression profiling during increased fetal lung expansion identifies genes likely to regulate development of the distal airways. *Physiol Genomics* 2006;**24**:105–13.

119. Seaborn T, St-Amand J, Cloutier M, Tremblay MG, Maltais F, Dinel S, et al. Identification of cellular processes that are rapidly modulated in response to tracheal occlusion within mice lungs. *Pediatr Res* 2008;**63**:124–30.

120. Sozo F, Hooper SB, Wallace MJ. Thrombospondin-1 expression and localization in the developing ovine lung. *J Physiol* 2007;**584**:625–35.

121. McDougall AR, Hooper SB, Zahra VA, Cole TJ, Lo CY, Doran T, et al. Trop2 regulates motility and lamellipodia formation in cultured fetal lung fibroblasts. *Am J Physiol Lung Cell Mol Physiol* 2013;**305**:L508–21.

122. Lindahl P, Karlsson L, Hellstrom M, Gebre-Medhin S, Willetts K, Heath JK, et al. Alveogenesis failure in PDGF-A-deficient mice is coupled to lack of distal spreading of alveolar smooth muscle cell progenitors during lung development. *Development* 1997;**124**:3943–53.

123. Bostrom H, Willetts K, Pekny M, Leveen P, Lindahl P, Hedstrand H, et al. PDGF-A signaling is a critical event in lung alveolar myofibroblast development and alveogenesis. *Cell* 1996;**85**:863–73.

124. Cock ML, Joyce BJ, Hooper SB, Wallace MJ, Gagnon R, Brace RA, et al. Pulmonary elastin synthesis and deposition in developing and mature sheep: effects of intrauterine growth restriction. *Exp Lung Res* 2004;**30**:405–18.

125. Schulze PC, De Keulenaer GW, Yoshioka J, Kassik KA, Lee RT. Vitamin D3-upregulated protein-1 (VDUP-1) regulates redox-dependent vascular smooth muscle cell proliferation through interaction with thioredoxin. *Circ Res* 2002;**91**:689–95.

126. Filby CE, Hooper SB, Sozo F, Zahra VA, Flecknoe SJ, Wallace MJ. VDUP1: a potential mediator of expansion-induced lung growth and epithelial cell differentiation in the ovine fetus. *Am J Physiol Lung Cell Mol Physiol* 2006;**290**:L250–8.

127. Riveline D, Zamir E, Balaban NQ, Schwarz US, Ishizaki T, Narumiya S, et al. Focal contacts as mechanosensors: externally applied local mechanical force induces growth of focal contacts by an mDia1-dependent and ROCK-independent mechanism. *J Cell Biol* 2001;**153**:1175–86.

128. Cloutier M, Tremblay M, Piedboeuf B. ROCK2 is involved in accelerated fetal lung development induced by in vivo lung distension. *Pediatr Pulmonol* 2010;**45**:966–76.

129. Han RN, Liu J, Tanswell AK, Post M. Ontogeny of platelet-derived growth factor receptor in fetal rat lung. *Microse Res Tech* 1993;**26**:381–8.

130. Buch S, Jassal D, Cannigia I, Edelson J, Han R, Liu J, et al. Ontogeny and regulation of platelet-derived growth factor gene expression in distal fetal rat lung epithelial cells. *Am J Respir Cell Mol Biol* 1994;**11**:251–61.

131. Souza P, Sedlackova L, Kuliszewski M, Wang J, Liu J, Tseu I, et al. Antisense oligodeoxynucleotides targeting PDGF-B mRNA inhibit cell proliferation during embryonic rat lung development. *Development* 1994;**120**:2163–73.

132. Souza P, Kuliszewski M, Wang J, Tseu I, Tanswell AK, Post M. PDGF-AA and its receptor influence early lung branching via an epithelial-mesenchymal interaction. *Development* 1995;**121**:2559–67.

133. Li J, Hoyle GW. Overexpression of PDGF-A in the lung epithelium of transgenic mice produces a lethal phenotype associated with hyperplasia of mesenchymal cells. *Dev Biol* 2001;**239**:338–49.

134. Ryan RM, Mineo-Kuhn MM, Kramer CM, Finkelstein JN. Growth factors alter neonatal type II alveolar epithelial cell proliferation. *Am J Phys* 1994;**266**:L17–22.

135. Resnick N. Collins T, Atkinson W, Bonthron DT, Dewey CF, Jr., Gimbron MA, Jr. Platelet-derived growth factor B chain promoter contains a cis-acting fluid shear-stress-responsive element. *Proc Natl Acad Sci U S A* 1993;**90**:7908.

136. Khachigian LM, Anderson KR, Halnon NJ, Gimbrone MA. Jr., Resnick N, Collins T. Egr-1 is activated in endothelial cells exposed to fluid shear stress and interacts with a novel shear-stress-response element in the PDGF A-chain promoter. *Arterioscler Thromb Vasc Biol* 1997;**17**:2280–6.

137. Buch S, Jones C, Liu J, Han RN, Tanswell AK, Post M. Differential regulation of platelet-derived growth factor genes in fetal rat lung fibroblasts. *Exp Cell Res* 1994;**211**:142–9.

138. Wallace MJ, Thiel AM, Lines AM, Polglase GR, Sozo F, Hooper SB. Role of platelet-derived growth factor-B, vascular endothelial growth factor, insulin-like growth factor-II, mitogen-activated protein kinase and transforming growth factor-beta1 in expansion-induced lung growth in fetal sheep. *Reprod Fertil Dev* 2006;**18**:655–65.

139. Bhatt AJ, Amin SB, Chess PR, Watkins RH, Maniscalco WM. Expression of vascular endothelial growth factor and Flk-1 in developing and glucocorticoid-treated mouse lung. *Pediatr Res* 2000;**47**:606–13.

140. Schachtner SK, Wang Y, Scott Baldwin H. Qualitative and quantitative analysis of embryonic pulmonary vessel formation. *Am J Respir Cell Mol Biol* 2000;**22**:157–65.

141. Levin DL, Rudolph AM, Heymann MA, Phibbs RH. Morphological development of the pulmonary vascular bed in fetal lambs. *Circulation* 1976;**53**:144–51.

142. Majka S. Fox K, McGuire B, Crossno J, Jr., McGuire P, Izzo A. Pleiotropic role of VEGF-A in regulating fetal pulmonary mesenchymal cell turnover. *Am J Physiol Lung Cell Mol Physiol* 2006;**290**:L1183–92.

143. Brown KR, England KM, Goss KL, Snyder JM, Acarregui MJ. VEGF induces airway epithelial cell proliferation in human fetal lung in vitro. *Am J Physiol Lung Cell Mol Physiol* 2001;**281**:L1001–10.

144. Quinn TP, Schlueter M, Soifer SJ, Gutierrez JA. Cyclic mechanical stretch induces VEGF and FGF-2 expression in pulmonary vascular smooth muscle cells. *Am J Physiol Lung Cell Mol Physiol* 2002;**282**:L897–903.

145. Sanz-Lopez E, Maderuelo E, Pelaez D, Chimenti P, Lorente R, Munoz MA, et al. Changes in the expression of vascular endothelial growth factor after fetal tracheal occlusion in an experimental model of congenital diaphragmatic hernia. *Critical Care Res Pract* 2013;**2013**:958078.

146. McGrath-Morrow SA, Cho C, Cho C, Zhen L, Hicklin DJ, Tuder RM. Vascular endothelial growth factor receptor 2 blockade disrupts postnatal lung development. *Am J Respir Cell Mol Biol* 2005;**32**:420–7.

147. Jakkula M, Le Cras TD, Gebb S, Hirth KP, Tuder RM, Voelkel NF, et al. Inhibition of angiogenesis decreases alveolarization in the developing rat lung. *Am J Physiol Lung Cell Mol Physiol* 2000;**279**:L600–7.

148. Thebaud B, Ladha F, Michelakis ED, Sawicka M, Thurston G, Eaton F, et al. Vascular endothelial growth factor gene therapy increases survival, promotes lung angiogenesis, and prevents alveolar damage in hyperoxia-induced lung injury: evidence that angiogenesis participates in alveolarization. *Circulation* 2005;**112**:2477–86.

149. Kunig AM, Balasubramaniam V, Markham NE, Morgan D, Montgomery G, Grover TR, et al. Recombinant human VEGF treatment enhances alveolarization after hyperoxic lung injury in neonatal rats. *Am J Physiol Lung Cell Mol Physiol* 2005;**289**:L529–35.

150. Clemmons DR. Insulin-like growth factor binding proteins and their role in controlling IGF actions. *Cytokine Growth Factor Rev* 1997;**8**:45–62.

151. Batchelor DC, Hutchins AM, Klempt M, Skinner SJ. Developmental changes in the expression patterns of IGFs, type 1 IGF receptor and IGF-binding proteins-2 and -4 in perinatal rat lung. *J Mol Endocrinol* 1995;**15**:105–15.

152. Silva D, Venihaki M, Guo WH, Lopez MF. Igf2 deficiency results in delayed lung development at the end of gestation. *Endocrinology* 2006;**147**:5584–91.

153. Liu JP, Baker J, Perkins AS, Robertson EJ, Efstratiadis A. Mice carrying null mutations of the genes encoding insulin-like growth factor I (Igf-1) and type 1 IGF receptor (Igf1r). *Cell* 1993;**75**:59–72.

154. Epaud R, Aubey F, Xu J, Chaker Z, Clemessy M, Dautin A, et al. Knockout of insulin-like growth factor-1 receptor impairs distal lung morphogenesis. *PLoS One* 2012;**7**:e48071.

155. Nobuhara KK, DiFiore JW, Ibla JC, Siddiqui AM, Ferretti ML, Fauza DO, et al. Insulin-like growth factor-I gene expression in three models of accelerated lung growth. *J Pediatr Surg* 1998;**33**:1057–60. discussion 1061.

156. Heine UI, Wahl SM, Munoz EF, Allen JB, Ellingsworth LR, Flanders KC, et al. Transforming growth factor-beta 1 specifically localizes in elastin during synovial inflammation: an immunoelectron microscopic study. *Archiv fur Geschwulstforschung* 1990;**60**:289–94.

157. Zhou L, Dey CR, Wert SE, Whitsett JA. Arrested lung morphogenesis in transgenic mice bearing an SP-C-TGF-beta 1 chimeric gene. *Dev Biol* 1996;**175**:227–38.

158. Zeng X, Gray M, Stahlman MT, Whitsett JA. TGF-beta1 perturbs vascular development and inhibits epithelial differentiation in fetal lung in vivo. *Dev Dyn* 2001;**221**:289–301.

159. De Paepe ME, Papadakis K, Johnson BD, Luks FI. Fate of the type II pneumocyte following tracheal occlusion in utero: a time-course study in fetal sheep. Virchows Archiv. *Virchows Arch* 1998;**432**:7–16.

160. Plopper CG, St George JA, Read LC, Nishio SJ, Weir AJ, Edwards L, et al. Acceleration of alveolar type II cell differentiation in fetal rhesus monkey lung by administration of EGF. *Am J Phys* 1992;**262**:L313–21.

161. Goetzman BW, Read LC, Plopper CG, Tarantal AF, George-Nascimento C, Merritt TA, et al. Prenatal exposure to epidermal growth factor attenuates respiratory distress syndrome in rhesus infants. *Pediatr Res* 1994;**35**:30–6.

162. Ma L, Wang AH, Frieda L, He HY, Ma GX, Wang HW, et al. Effect of epidermal growth factor and dexamethasone on fetal rat lung development. *Chin Med J* 2009;**122**:2013–6.

163. Miettinen PJ, Berger JE, Meneses J, Phung Y, Pedersen RA, Werb Z, et al. Epithelial immaturity and multiorgan failure in mice lacking epidermal growth factor receptor. *Nature* 1995;**376**:337–41.

164. Sibilia M, Wagner EF. Strain-dependent epithelial defects in mice lacking the EGF receptor. *Science* 1995;**269**:234–8.

165. Powell PP, Wang CC, Horinouchi H, Shepherd K, Jacobson M, Lipson M, et al. Differential expression of fibroblast growth factor receptors 1 to 4 and ligand genes in late fetal and early postnatal rat lung. *Am J Respir Cell Mol Biol* 1998;**19**:563–72.

166. Weinstein M, Xu X, Ohyama K, Deng CX. FGFR-3 and FGFR-4 function cooperatively to direct alveogenesis in the murine lung. *Development* 1998;**125**:3615–23.

167. Shannon JM, Hyatt BA. Epithelial-mesenchymal interactions in the developing lung. *Annu Rev Physiol* 2004;**66**:625–45.

168. De Moerlooze L, Spencer-Dene B, Revest JM, Hajihosseini M, Rosewell I, Dickson C. An important role for the IIIb isoform of fibroblast growth factor receptor 2 (FGFR2) in mesenchymal-epithelial signalling during mouse organogenesis. *Development* 2000;**127**:483–92.

169. Min H, Danilenko DM, Scully SA, Bolon B, Ring BD, Tarpley JE, et al. Fgf-10 is required for both limb and lung development and exhibits striking functional similarity to Drosophila branchless. *Genes Dev* 1998;**12**:3156–61.

170. Hokuto I, Perl AK, Whitsett JA. FGF signaling is required for pulmonary homeostasis following hyperoxia. *Am J Physiol Lung Cell Mol Physiol* 2004;**286**:L580–87.

171. Colvin JS, White AC, Pratt SJ, Ornitz DM. Lung hypoplasia and neonatal death in Fgf9-null mice identify this gene as an essential regulator of lung mesenchyme. *Development* 2001;**128**:2095–106.

172. Padela S, Yi M, Cabacungan J, Shek S, Belcastro R, Masood A, et al. A critical role for fibroblast growth factor-7 during early alveolar formation in the neonatal rat. *Pediatr Res* 2008;**63**:232–8.

173. Friedmacher F, Doi T, Gosemann JH, Fujiwara N, Kutasy B, Puri P. Upregulation of fibroblast growth factor receptor 2 and 3 in the late stages of fetal lung development in the nitrofen rat model. *Pediatr Surg Int* 2012;**28a**:195–9.

174. Usui H, Shibayama M, Ohbayashi N, Konishi M, Takada S, Itoh N. Fgf18 is required for embryonic lung alveolar development. *Biochem Biophys Res Commun* 2004;**322**:887–92.

175. Whitsett JA, Clark JC, Picard L, Tichelaar JW, Wert SE, Itoh N, et al. Fibroblast growth factor 18 influences proximal programming during lung morphogenesis. *J Biol Chem* 2002;**277**:22743–9.

176. Takahashi H, Friedmacher F, Fujiwara N, Hofmann A, Kutasy B, Gosemann JH, et al. Pulmonary FGF-18 gene expression is down-regulated during the canalicular-saccular stages in nitrofen-induced hypoplastic lungs. *Pediatr Surg Int* 2013;**29**:1199–203.

177. Hastings RH, Duong H, Burton DW, Deftos LJ. Alveolar epithelial cells express and secrete parathyroid hormone-related protein. *Am J Respir Cell Mol Biol* 1994;**11**:701–6.

178. Lee K, Deeds JD, Segre GV. Expression of parathyroid hormone-related peptide and its receptor messenger ribonucleic acids during fetal development of rats. *Endocrinology* 1995;**136**:453–63.

179. Torday JS, Rehan VK. Stretch-stimulated surfactant synthesis is coordinated by the paracrine actions of PTHrP and leptin. *Am J Physiol Lung Cell Mol Physiol* 2002;**283**:L130–5.

180. Torday JS, Sun H, Wang L, Torres E, Sunday ME, Rubin LP. Leptin mediates the parathyroid hormone-related protein paracrine stimulation of fetal lung maturation. *Am J Physiol Lung Cell Mol Physiol* 2002;**282**:L405–10.

181. Karaplis AC, Luz A, Glowacki J, Bronson RT, Tybulewicz VL, Kronenberg HM, et al. Lethal skeletal dysplasia from targeted disruption of the parathyroid hormone-related peptide gene. *Genes Dev* 1994;**8**:277–89.

182. Ramirez MI, Chung UI, Williams MC. Aquaporin-5 expression, but not other peripheral lung marker genes, is reduced in PTH/PTHrP receptor null mutant fetal mice. *Am J Respir Cell Mol Biol* 2000;**22**:367–72.

183. Cilley RE, Zgleszewski SE, Chinoy MR. Fetal lung development: airway pressure enhances the expression of developmental genes. *J Pediatr Surg* 2000;**35**:113–8. discussion 119.

184. Doi T, Lukosiute A, Ruttenstock E, Dingemann J, Puri P. Disturbance of parathyroid hormone-related protein signaling in the nitrofen-induced hypoplastic lung. *Pediatr Surg Int* 2010;**26**:45–50.

185. Khoshgoo N, Kholdebarin R, Iwasiow BM, Keijzer R. MicroRNAs and lung development. *Pediatr Pulmonol* 2013;**48**:317–23.

186. Harris KS, Zhang Z, McManus MT, Harfe BD, Sun X. Dicer function is essential for lung epithelium morphogenesis. *Proc Natl Acad Sci U S A* 2006;**103**:2208–13.

187. Lu Y, Thomson JM, Wong HY, Hammond SM, Hogan BL. Transgenic over-expression of the microRNA miR-17-92 cluster promotes proliferation and inhibits differentiation of lung epithelial progenitor cells. *Dev Biol* 2007;**310**:442–53.

188. Ventura A, Young AG, Winslow MM, Lintault L, Meissner A, Erkeland SJ, et al. Targeted deletion reveals essential and overlapping functions of the miR-17 through 92 family of miRNA clusters. *Cell* 2008;**132**:875–86.

189. Carraro G, El-Hashash A, Guidolin D, Tiozzo C, Turcatel G, Young BM, et al. miR-17 family of microRNAs controls FGF10-mediated embryonic lung epithelial branching morphogenesis through MAPK14 and STAT3 regulation of E-Cadherin distribution. *Dev Biol* 2009;**333**:238–50.

190. Zhong Z, Dong Z, Yang L, Chen X, Gong Z. Inhibition of proliferation of human lung cancer cells by green tea catechins is mediated by upregulation of let-7. *Exp Ther Med* 2012;**4**:267–72.

191. Bhaskaran M, Wang Y, Zhang H, Weng T, Baviskar P, Guo Y, et al. MicroRNA-127 modulates fetal lung development. *Physiol Genomics* 2009;**37**:268–78.

192. Williams AE, Moschos SA, Perry MM, Barnes PJ, Lindsay MA. Maternally imprinted microRNAs are differentially expressed during mouse and human lung development. *Dev Dyn* 2007;**236**:572–80.

193. Liggins GC, Howie RN. A controlled trial of antepartum glucocorticoid treatment for prevention of the respiratory distress syndrome in premature infants. *Pediatrics* 1972;**50**:515–25.

194. Kitterman JA, Liggins GC, Campos GA, Clements JA, Forster CS, Lee CH, et al. Prepartum maturation of the lung in fetal sheep: relation to cortisol. *J Appl Phys* 1981;**51**:384–90.

195. Bolt RJ, van Weissenbruch MM, Lafeber HN, Delemarre-van de Waal HA. Glucocorticoids and lung development in the fetus and preterm infant. *Pediatr Pulmonol* 2001;**32**:76–91.

196. Boland R, Joyce BJ, Wallace MJ, Stanton H, Fosang AJ, Pierce RA, et al. Cortisol enhances structural maturation of the hypoplastic fetal lung in sheep. *J Physiol* 2004;**554**:505–17.

197. Wallace MJ, Hooper SB, Harding R. Effects of elevated fetal cortisol concentrations on the volume, secretion, and reabsorption of lung liquid. *Am J Phys* 1995;**269**:R881–7.

198. Liggins GC. Premature delivery of foetal lambs infused with glucocorticoids. *J Endocrinol* 1969;**45**:515–23.

199. Bird AD, Tan KH, Olsson PF, Zieba M, Flecknoe SJ, Liddicoat DR, et al. Identification of glucocorticoid-regulated genes that control cell proliferation during murine respiratory development. *J Physiol* 2007;**585**:187–201.

200. Cole TJ, Solomon NM, Van Driel R, Monk JA, Bird D, Richardson SJ, et al. Altered epithelial cell proportions in the fetal lung of glucocorticoid receptor null mice. *Am J Respir Cell Mol Biol* 2004;**30**:613–9.

201. Boggaram V, Smith ME, Mendelson CR. Posttranscriptional regulation of surfactant protein-A messenger RNA in human fetal lung in vitro by glucocorticoids. *Mol Endocrinol* 1991;**5**:414–23.

202. Deterding RR, Shimizu H, Fisher JH, Shannon JM. Regulation of surfactant protein D expression by glucocorticoids in vitro and in vivo. *Am J Respir Cell Mol Biol* 1994;**10**:30–7.

203. Nichols KV, Floros J, Dynia DW, Veletza SV, Wilson CM, Gross I. Regulation of surfactant protein A mRNA by hormones and butyrate in cultured fetal rat lung. *Am J Phys* 1990;**259**:L488–95.

204. Mariencheck W, Crouch E. Modulation of surfactant protein D expression by glucocorticoids in fetal rat lung. *Am J Respir Cell Mol Biol* 1994;**10**:419–29.

205. Schellhase DE, Shannon JM. Effects of maternal dexamethasone on expression of SP-A, SP-B, and SP-C in the fetal rat lung. *Am J Respir Cell Mol Biol* 1991;**4**:304–12.

206. Tan RC, Ikegami M, Jobe AH, Yao LY, Possmayer F, Ballard PL. Developmental and glucocorticoid regulation of surfactant protein mRNAs in preterm lambs. *Am J Phys* 1999;**277**:L1142–8.

207. Whitsett JA, Pilot T, Clark JC, Weaver TE. Induction of surfactant protein in fetal lung. Effects of cAMP and dexamethasone on SAP-35 RNA and synthesis. *J Biol Chem* 1987;**262**:5256–61.

208. Bird AD, Choo YL, Hooper SB, McDougall AR, Cole TJ. Mesenchymal glucocorticoid receptor regulates development of multiple cell layers of the mouse lung. *Am J Respir Cell Mol Biol* 2014;**50**:419–28.

209. Cole TJ, Blendy JA, Monaghan AP, Schmid W, Aguzzi A, Schutz G. Molecular genetic analysis of glucocorticoid signaling during mouse development. *Steroids* 1995;**60**:93–6.

210. Crone RK, Davies P, Liggins GC, Reid L. The effects of hypophysectomy, thyroidectomy, and postoperative infusion of cortisol or adrenocorticotrophin on the structure of the ovine fetal lung. *J Dev Physiol* 1983;**5**:281–8.

211. Ikegami M, Polk D, Tabor B, Lewis J, Yamada T, Jobe A. Corticosteroid and thyrotropin-releasing hormone effects on preterm sheep lung function. *J Appl Phys* 1991;**70**:2268–78.

212. Kendall JZ, Lakritz J, Plopper CG, Richards GE, Randall GC, Nagamani M, et al. The effects of hydrocortisone on lung structure in fetal lambs. *J Dev Physiol* 1990;**13**:165–72.

213. Warburton D, Parton L, Buckley S, Cosico L, Enns G, Saluna T. Combined effects of corticosteroid, thyroid hormones, and beta-agonist on surfactant, pulmonary mechanics, and beta-receptor binding in fetal lamb lung. *Pediatr Res* 1988;**24**:166–70.

214. Habermehl D, Parkitna JR, Kaden S, Brugger B, Wieland F, Grone HJ, et al. Glucocorticoid activity during lung maturation is essential in mesenchymal and less in alveolar epithelial cells. *Mol Endocrinol* 2011;**25**:1280–8.

215. Manwani N, Gagnon S, Post M, Joza S, Muglia L, Cornejo S, et al. Reduced viability of mice with lung epithelial-specific knockout of glucocorticoid receptor. *Am J Respir Cell Mol Biol* 2010;**43**:599–606.

216. Schellenberg JC, Liggins GC, Stewart AW. Growth, elastin concentration, and collagen concentration of perinatal rat lung: effects of dexamethasone. *Pediatr Res* 1987;**21**:603–7.

217. Ikegami M, Jobe AH, Newnham J, Polk DH, Willet KE, Sly P. Repetitive prenatal glucocorticoids improve lung function and decrease growth in preterm lambs. *Am J Respir Crit Care Med* 1997;**156**:178–84.

218. French NP, Hagan R, Evans SF, Godfrey M, Newnham JP. Repeated antenatal corticosteroids: size at birth and subsequent development. *Am J Obstet Gynecol* 1999;**180**:114–21.

219. Adamson IY, King GM. Postnatal development of rat lung following retarded fetal lung growth. *Pediatr Pulmonol* 1988;**4**:230–6.

220. Jobe AH, Newnham J, Willet K, Sly P, Ikegami M. Fetal versus maternal and gestational age effects of repetitive antenatal glucocorticoids. *Pediatrics* 1998;**102**:1116–25.

221. Willet KE, Jobe AH, Ikegami M, Newnham J, Brennan S, Sly PD. Antenatal endotoxin and glucocorticoid effects on lung morphometry in preterm lambs. *Pediatr Res* 2000;**48**:782–83.

222. Massaro GD, Massaro D. Formation of alveoli in rats: postnatal effect of prenatal dexamethasone. *Am J Phys* 1992;**263**:L37–41.

223. Davey MG, Hooper SB, Cock ML, Harding R. Stimulation of lung growth in fetuses with lung hypoplasia leads to altered postnatal lung structure in sheep. *Pediatr Pulmonol* 2001;**32**:267–76.

224. Brown MJ, Olver RE, Ramsden CA, Strang LB, Walters DV. Effects of adrenaline and of spontaneous labour on the secretion and absorption of lung liquid in the fetal lamb. *J Physiol* 1983;**344**:137–52.

225. Hooper SB, Wallace MJ, Harding R. Amiloride blocks the inhibition of fetal lung liquid secretion caused by AVP but not by asphyxia. *J Appl Phys* 1993;**74**:111–5.

226. Wallace MJ, Hooper SB, Harding R. Regulation of lung liquid secretion by arginine vasopressin in fetal sheep. *Am J Phys* 1990;**258**:R104–11.

227. Walters DV, Olver RE. The role of catecholamines in lung liquid absorption at birth. *Pediatr Res* 1978;**12**:239–42.

228. Barquin N, Ciccolella DE, Ridge KM, Sznajder JI. Dexamethasone upregulates the Na-K-ATPase in rat alveolar epithelial cells. *Am J Phys* 1997;**273**:L825–30.

229. Chalaka S, Ingbar DH, Sharma R, Zhau Z, Wendt CH. Na(+)-K(+)-ATPase gene regulation by glucocorticoids in a fetal lung epithelial cell line. *Am J Phys* 1999;**277**:L197–203.

230. Itani OA, Auerbach SD, Husted RF, Volk KA, Ageloff S, Knepper MA, et al. Glucocorticoid-stimulated lung epithelial Na(+) transport is associated with regulated ENaC and sgk1 expression. *Am J Physiol Lung Cell Mol Physiol* 2002;**282**:L631–41.

231. Lazrak A, Samanta A, Venetsanou K, Barbry P, Matalon S. Modification of biophysical properties of lung epithelial Na(+) channels by dexamethasone. *Am J Physiol Cell Physiol* 2000;**279**:C762–70.

232. Ingbar DH, Duvick S, Savick SK, Schellhase DE, Detterding R, Jamieson JD, et al. Developmental changes of fetal rat lung Na-K-ATPase after maternal treatment with dexamethasone. *Am J Phys* 1997;**272**:L665–72.

233. Mustafa SB, DiGeronimo RJ, Petershack JA, Alcorn JL, Seidner SR. Postnatal glucocorticoids induce alpha-ENaC formation and regulate glucocorticoid receptors in the preterm rabbit lung. *Am J Physiol Lung Cell Mol Physiol* 2004;**286**:L73–80.

234. Noda M, Suzuki S, Tsubochi H, Sugita M, Maeda S, Kobayashi S, et al. Single dexamethasone injection increases alveolar fluid clearance in adult rats. *Crit Care Med* 2003;**31**:1183–9.

235. Tchepichev S, Ueda J, Canessa C, Rossier BC, O'Brodovich H. Lung epithelial Na channel subunits are differentially regulated during development and by steroids. *Am J Phys* 1995;**269**:C805–12.

236. McGowan SE, Harvey CS, Jackson SK. Retinoids, retinoic acid receptors, and cytoplasmic retinoid binding proteins in perinatal rat lung fibroblasts. *Am J Phys* 1995;**269**:L463–72.

237. Kastner P, Grondona JM, Mark M, Gansmuller A, LeMeur M, Decimo D, et al. Genetic analysis of RXR alpha developmental function: convergence of RXR and RAR signaling pathways in heart and eye morphogenesis. *Cell* 1994;**78**:987–1003.

238. Kliewer SA, Umesono K, Mangelsdorf DJ, Evans RM. Retinoid X receptor interacts with nuclear receptors in retinoic acid, thyroid hormone and vitamin D3 signalling. *Nature* 1992;**355**:446–9.

239. Rajatapiti P, Kester MH, de Krijger RR, Rottier R, Visser TJ, Tibbocl D. Expression of glucocorticoid, retinoid, and thyroid hormone receptors during human lung development. *J Clin Endocrinol Metab* 2005;**90**:4309–14.

240. Grummer MA, Thet LA, Zachman RD. Expression of retinoic acid receptor genes in fetal and newborn rat lung. *Pediatr Pulmonol* 1994;**17**:234–8.

241. Massaro GD, Massaro D, Chambon P. Retinoic acid receptor-alpha regulates pulmonary alveolus formation in mice after, but not during, perinatal period. *Am J Physiol Lung Cell Mol Physiol* 2003;**284**:L431–3.

242. Yang L, Naltner A, Yan C. Overexpression of dominant negative retinoic acid receptor alpha causes alveolar abnormality in transgenic neonatal lungs. *Endocrinology* 2003;**144**:3004–11.

243. Massaro GD, Massaro D, Chan WY, Clerch LB, Ghyselinck N, Chambon P, et al. Retinoic acid receptor-beta: an endogenous inhibitor of the perinatal formation of pulmonary alveoli. *Physiol Genomics* 2000;**4**:51–7.

244. Snyder JM, Jenkins-Moore M, Jackson SK, Goss KL, Dai HH, Bangsund PJ, et al. Alveolarization in retinoic acid receptor-beta-deficient mice. *Pediatr Res* 2005;**57**:384–91.

245. McGowan S, Jackson SK, Jenkins-Moore M, Dai HH, Chambon P, Snyder JM. Mice bearing deletions of retinoic acid receptors demonstrate reduced lung elastin and alveolar numbers. *Am J Respir Cell Mol Biol* 2000;**23**:162–7.

246. Massaro GD, Massaro D. Retinoic acid treatment partially rescues failed septation in rats and in mice. *Am J Physiol Lung Cell Mol Physiol* 2000;**278**:L955–60.

247. Montedonico S, Nakazawa N, Puri P. Retinoic acid rescues lung hypoplasia in nitrofen-induced hypoplastic foetal rat lung explants. *Pediatr Surg Int* 2006;**22**:2–8.

248. Montedonico S, Sugimoto K, Felle P, Bannigan J, Puri P. Prenatal treatment with retinoic acid promotes pulmonary alveologenesis in the nitrofen model of congenital diaphragmatic hernia. *J Pediatr Surg* 2008;**43**:500–7.

249. Gallot D, Coste K, Jani J, Roubliova X, Marceau G, Velemir L, et al. Effects of maternal retinoic acid administration in a congenital diaphragmatic hernia rabbit model. *Pediatr Pulmonol* 2008;**43**:594–603.

250. Chen CM, Chou HC, Wang LF, Lang YD, Yeh CY. Retinoic acid fails to reverse oligohydramnios-induced pulmonary hypoplasia in fetal rats. *Pediatr Res* 2007;**62**:553–8.

251. Willet KE, Jobe AH, Ikegami M, Newnham J, Sly PD. Antenatal retinoic acid does not alter alveolization or postnatal lung function in preterm sheep. *Eur Respir J* 2000;**16**:101–7.

252. Garcia-Aragon J, Lobie PE, Muscat GE, Gobius KS, Norstedt G, Waters MJ. Prenatal expression of the growth hormone (GH) receptor/binding protein in the rat: a role for GH in embryonic and fetal development? *Development* 1992;**114**:869–76.

253. Batchelor DC, Lewis RM, Breier BH, Gluckman PD, Skinner SJ. Fetal rat lung epithelium has a functional growth hormone receptor coupled to tyrosine kinase activity and insulin-like growth factor binding protein-2 production. *J Mol Endocrinol* 1998;**21**:73–84.

254. Li J, Gilmour RS, Saunders JC, Dauncey MJ, Fowden AL. Activation of the adult mode of ovine growth hormone receptor gene expression by cortisol during late fetal development. *Faseb J* 1999;**13**:545–52.

255. Finidori J. Regulators of growth hormone signaling. *Vitam Horm* 2000;**59**:71–97.

256. Nardo L, Young IR, Hooper SB. Influence of growth hormone on the lung growth response to tracheal obstruction in fetal sheep. *Am J Physiol Lung Cell Mol Physiol* 2000;**278**:L453–9.

257. Brody JS, Buhain WJ. Hormonal influence on post-pneumonectomy lung growth in the rat. *Respir Physiol* 1973;**19**:344–55.

258. Massaro D, Teich N, Massaro GD. Postnatal development of pulmonary alveoli: modulation in rats by thyroid hormones. *Am J Phys* 1986;**250**:R51–5.

259. Steele RE, Wekstein DR. Influence of thyroid hormone on homeothermic development of the rat. *Am J Phys* 1972;**222**:1528–33.

260. Barker PM, Markiewicz M, Parker KA, Walters DV, Strang LB. Synergistic action of triiodothyronine and hydrocortisone on epinephrine-induced reabsorption of fetal lung liquid. *Pediatr Res* 1990;**27**:588–91.

261. Barker PM, Strang LB, Walters DV. The role of thyroid hormones in maturation of the adrenaline-sensitive lung liquid reabsorptive mechanism in fetal sheep. *J Physiol* 1990;**424**:473–85.

262. Seaborn T, Simard M, Provost PR, Piedboeuf B, Tremblay Y. Sex hormone metabolism in lung development and maturation. *Trends Endocrinol Metab* 2010;**21**:729–38.

263. Farrell PM, Avery ME. Hyaline membrane disease. *Am Rev Respir Dis* 1975;**111**:657–88.

264. Fleisher B, Kulovich MV, Hallman M, Gluck L. Lung profile: sex differences in normal pregnancy. *Obstet Gynecol* 1985;**66**:327–30.

265. Dammann CE, Ramadurai SM, McCants DD, Pham LD, Nielsen HC. Androgen regulation of signaling pathways in late fetal mouse lung development. *Endocrinology* 2000;**141**:2923–9.

266. Willet KE, Jobe AH, Ikegami M, Polk D, Newnham J, Kohan R, et al. Postnatal lung function after prenatal steroid treatment in sheep: effect of gender. *Pediatr Res* 1997;**42**:885–92.

267. Carey MA, Card JW, Voltz JW, Germolec DR, Korach KS, Zeldin DC. The impact of sex and sex hormones on lung physiology and disease: lessons from animal studies. *Am J Physiol Lung Cell Mol Physiol* 2007;**293**:L272–8.

268. Massaro D, Clerch LB, Massaro GD. Estrogen receptor-alpha regulates pulmonary alveolar loss and regeneration in female mice: morphometric and gene expression studies. *Am J Physiol Lung Cell Mol Physiol* 2007;**293**:L222–8.

269. Morani A, Barros RP, Imamov O, Hultenby K, Arner A, Warner M, et al. Lung dysfunction causes systemic hypoxia in estrogen receptor beta knockout (ERbeta-/-) mice. *Proc Natl Acad Sci U S A* 2006;**103**:7165–9.

270. Edwards CA, Osman LM, Godden DJ, Campbell DM, Douglas JG. Relationship between birth weight and adult lung function: controlling for maternal factors. *Thorax* 2003;**58**:1061–5.

271. Hoo AF, Stocks J, Lum S, Wade AM, Castle RA, Costeloe KL, et al. Development of lung function in early life: influence of birth weight in infants of nonsmokers. *Am J Respir Crit Care Med* 2004;**170**:527–33.

272. Lucas JS, Inskip HM, Godfrey KM, Foreman CT, Warner JO, Gregson RK, et al. Small size at birth and greater postnatal weight gain: relationships to diminished infant lung function. *Am J Respir Crit Care Med* 2004;**170**:534–40.

273. Tyson JE, Kennedy K, Broyles S, Rosenfeld CR. The small for gestational age infant: accelerated or delayed pulmonary maturation? Increased or decreased survival? *Pediatrics* 1995;**95**:534–8.

274. Maritz GS, Cock ML, Louey S, Joyce BJ, Albuquerque CA, Harding R. Effects of fetal growth restriction on lung development before and after birth: a morphometric analysis. *Pediatr Pulmonol* 2001;**32**:201–10.

275. Maritz GS, Cock ML, Louey S, Suzuki K, Harding R. Fetal growth restriction has long-term effects on postnatal lung structure in sheep. *Pediatr Res* 2004;**55**:287–95.

276. Wignarajah D, Cock ML, Pinkerton KE, Harding R. Influence of intrauterine growth restriction on airway development in fetal and postnatal sheep. *Pediatr Res* 2002;**51**:681–8.

277. Cock ML, Camm EJ, Louey S, Joyce BJ, Harding R. Postnatal outcomes in term and preterm lambs following fetal growth restriction. *Clin Exp Pharmacol Physiol* 2001;**28**:931–7.

278. Joyce BJ, Louey S, Davey MG, Cock ML, Hooper SB, Harding R. Compromised respiratory function in postnatal lambs after placental insufficiency and intrauterine growth restriction. *Pediatr Res* 2001;**50**:641–9.

279. Moss TJ, Harding R. Ventilatory and arousal responses to respiratory stimuli of full term, intrauterine growth restricted lambs. *Respir Physiol* 2001;**124**:195–204.

280. Tsao PN, Wei SC. Prenatal hypoxia downregulates the expression of pulmonary vascular endothelial growth factor and its receptors in fetal mice. *Neonatology* 2013;**103**:300–7.

281. Joss-Moore L, Carroll T, Yang Y, Fitzhugh M, Metcalfe D, Oman J, et al. Intrauterine growth restriction transiently delays alveolar formation and disrupts retinoic acid receptor expression in the lung of female rat pups. *Pediatr Res* 2013;**73**:612–20.

282. Gilbert WM, Danielsen B. Pregnancy outcomes associated with intrauterine growth restriction. *Am J Obstet Gynecol* 2003;**188**:1596–9. discussion 1599-1601.

283. Jobe AH, Bancalari E. Bronchopulmonary dysplasia. *Am J Respir Crit Care Med* 2001;**163**:1723–9.

284. Cock M, Hanna M, Sozo F, Wallace M, Yawno T, Suzuki K, et al. Pulmonary function and structure following mild preterm birth in lambs. *Pediatr Pulmonol* 2005;**40**:336–48.

285. Jani JC, Flemmer AW, Bergmann F, Gallot D, Roubliova X, Muensterer OJ, et al. The effect of fetal tracheal occlusion on lung tissue mechanics and tissue composition. *Pediatr Pulmonol* 2009;**44**:112–21.

286. Deprest JA, Nicolaides K, Gratacos E. Fetal surgery for congenital diaphragmatic hernia is back from never gone. *Fetal Diagn Ther* 2011;**29**:6–17.

287. Short EJ, Klein NK, Lewis BA, Fulton S, Eisengart S, Kercsmar C, et al. Cognitive and academic consequences of bronchopulmonary dysplasia and very low birth weight: 8-year-old outcomes. *Pediatrics* 2003;**112**:e359.

288. Fabian MR, Sonenberg N. The mechanics of miRNA-mediated gene silencing: a look under the hood of miRISC. *Nat Struct Mol Biol* 2012;**19**:586–93.

289. Sessa R, Hata A. Role of microRNAs in lung development and pulmonary diseases. *PulmonaryCirculation* 2013;**3**:315–28.

290. Langston C, Kida K, Reed M, Thurlbeck WM. Human lung growth in late gestation and in the neonate. *Am Rev Respir Dis* 1984;**129**:607–13.

291. Hislop AA, et al. Alveolar development in the human fetus and infant. *Early Hum Dev* 1986;**13**:1–11.

Chapter 9

The Development of the Pulmonary Surfactant System

Sandra Orgeig*, Janna L. Morrison*, Lucy C. Sullivan† and Christopher B. Daniels‡

*School of Pharmacy & Medical Sciences, Sansom Institute for Health Research, University of South Australia, Adelaide, SA; †Department of Microbiology and Immunology, The University of Melbourne, Melbourne, VIC; ‡Barbara Hardy Institute, University of South Australia, Adelaide, SA, Australia

INTRODUCTION

At birth, the lung must rapidly transition from a fluid-filled to an air-filled organ capable of gas exchange. Birth presents both a physical challenge, associated with formation of an air–liquid interface, as well as an environmental challenge associated with exposure to a vast array of foreign pathogens. Essential to successful transition is a complex lipid-protein mixture, the pulmonary surfactant system. Surfactant forms a film at the alveolar air–liquid interface that reduces and varies the surface tension of the lung lining to maintain alveolar stability, thereby reducing the risk of atelectasis (alveolar collapse) and pulmonary edema, and reducing the work of breathing. In addition, the surfactant system is actively involved in protecting against foreign pathogens. As the pulmonary surfactant system develops during late gestation, infants born preterm are at an increased risk of developing neonatal respiratory distress syndrome. In this chapter we first consider the normal composition, function and development of the pulmonary surfactant system, and then discuss its dysfunctional development in response to preterm birth and genetic factors. The role of environmental factors in shaping surfactant development is discussed in Chapter 27.

ASSEMBLY AND RELEASE OF SURFACTANT

The surfactant lipids and the specific surfactant-associated proteins (SP-A, -B, -C, and -D) are synthesized in the lung by alveolar epithelial type II cells (AECII) that contain lamellar bodies. Lamellar bodies are storage organelles for surfactant and consist of a dense proteinaceous core and lipid bilayers arranged in parallel, stacked lamellae. The phospholipids are synthesized within the endoplasmic reticulum but the exact transport mechanisms to the lamellar bodies are unknown.[1] One possibility includes vesicular transport involving the Golgi apparatus, although this may

not be a major pathway for phospholipid transfer.[2,3] Another possible mechanism involves specific phospholipid transfer proteins, such as ABCA3, member A3 of the adenosine triphosphate binding cassette (ABC) family, which is located on the limiting membrane of lamellar bodies.[1,4] A final possible pathway involves the direct transfer of surfactant lipids from the endoplasmic reticulum to lamellar bodies at specific membrane contact sites.[1,4] Unlike the lipids, the low molecular weight hydrophobic proteins, SP-B and SP-C, are synthesized as immature proproteins and undergo extensive post-translational processing during trafficking involving vesicular transport from the endoplasmic reticulum via the Golgi apparatus and multivesicular bodies to the lamellar bodies.[3,5] Unlike the coordinated secretion of the co-packaged phospholipids, SP-B and -C, the large multimeric hydrophilic surfactant proteins, SP-A and SP-D, are not controlled by adrenergic and purinergic secretagogues (see following text). Instead, following translation, SP-A and -D undergo significant modifications in the endoplasmic reticulum and Golgi apparatus,[6] but then appear to bypass the lamellar bodies and are constitutively secreted.[3,7,8] After the lamellar bodies are released into the hypophase (the fluid lining the epithelium), they swell and unravel into tubular myelin, another form of surfactant. The surface film is derived from tubular myelin, and it is this surface film that regulates surface tension at the air–liquid interface (reviewed in[9]; Figure 1).

COMPOSITION OF PULMONARY SURFACTANT

In humans, pulmonary surfactant is comprised of ~80–85% phospholipids, 5–10% neutral lipids and 8–10% protein, with 5–6% consisting of the four specific surfactant proteins.[9] The components and their main functions are summarized in Table 1.

The Lung. http://dx.doi.org/10.1016/B978-0-12-799941-8.00009-2

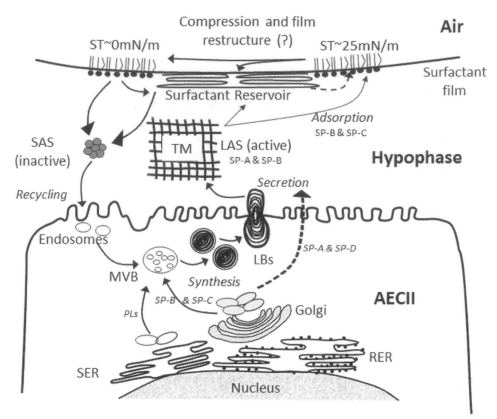

FIGURE 1   Schematic diagram of the life cycle of pulmonary surfactant. Surfactant proteins and lipids are synthesized in rough endoplasmic reticulum (RER) and smooth ER (SER), respectively, and transported to the Golgi Apparatus (Golgi). SP-B and -C and the phospholipids are transported via vesicles from the trans-Golgi network to multivesicular bodies (MVB) before being packaged into lamellar bodies (LBs). LB secretion into the liquid lining of the alveoli (hypophase) occurs via exocytosis across the alveolar epithelial type II cell (AECII) plasma membrane, and is stimulated by various secretagogues. SP-A and -D are secreted constitutively via a non-LB pathway. Within the hypophase, LBs swell and unravel, forming tubular myelin (TM), consisting of lipids and proteins (particularly SP-A and SP-B); TM is contained within the large aggregate surfactant (LAS) fraction that can be isolated from lung lavage. This fraction is surface active and supplies the lipids to the air–liquid interface as well as the surfactant reservoir, which is a multilayer structure associated with the surface film (also known as the surface-associated phase). The adsorption of lipids to the air–liquid interface is mediated by the hydrophobic surfactant proteins, SP-B and -C. As the mixed molecular film is compressed, some of the lipid is squeezed out into the multilayer reservoir and the film undergoes a restructuring, rendering it capable of reducing surface tension (ST) to near 0 mN/m. Upon re-expansion, some lipids from the reservoir re-enter the surface film. Lipids from the surface film and the reservoir become inactive, forming part of the small aggregate surfactant (SAS) fraction and are eventually taken back up by the AECII via endocytosis. These lipids are then recycled via the endocytic pathway to MVBs and combined with new lipids and proteins from the Golgi to form new LBs. *Figure reproduced with modifications and permission from.*[323]

Lipids

The phospholipids consist predominantly of phosphatidylcholine (PC, ~80%) with smaller amounts of the anionic phospholipids, phosphatidylinositol (PI) and phosphatidylglycerol (PG), as well as lysophosphatidylcholine (LPC), phosphatidylserine (PS), phosphatidylethanolamine (PE), and sphingomyelin.[10] Although there are exceptions,[11] the most abundant molecular species (30–60%) in most mammalian surfactants is the disaturated phospholipid, dipalmitoylphosphatidylcholine (DPPC), also denoted as PC16:0/16:0, indicating a choline headgroup and two identical fatty acids in the sn-1 and sn-2 positions, consisting of 16 carbon atoms and no double bonds.[11] Most species also contain significant levels of palmitoyl–myristoyl phosphatidylcholine (PMPC) (PC16:0/14:0) and numerous unsaturated

molecular species such as palmitoyl–palmitoleoyl phosphatidylcholine (PPPC) (PC16:0/16:1), palmitoyl–oleoyl phosphatidylcholine (POPC) (PC16:0/18:1), and palmitoyl–linoleoyl phosphatidylcholine (PC16:0/18:2).[11,12]

While the disaturated phospholipid DPPC is largely responsible for reducing surface tension[13] (see following text), the functions of the so-called "minor" phospholipids remain obscure. Due to their fully saturated palmitic acid chains, DPPC molecules can be tightly compressed, eliminating water molecules from the interface and thus markedly reducing surface tension.[14] The unsaturated phospholipids, with their kinked side chains, are unable to sustain a similar packing density and are normally unable to reduce surface tension as efficiently. Among the "minor" phospholipids, the acidic phospholipids, PG and/or PI, are the most abundant. While in humans, PG contributes ~10% by weight, it is one

TABLE 1 Summary of the functions of the lipid and protein components of pulmonary surfactant

Component	Relative Composition	Functions
Phospholipids (PL)	~80–85% of surfactant by mass	
Phosphatidylcholine (PC)	~70–80% of total PL and ~50% of PC is dipalmitoyl PC (DPPC)	Principal component responsible for lowering surface tension; capable of generating surface tensions of near 0 mN/m under compression.
Acidic Phospholipids: Phosphatidylglycerol (PG) and Phosphatidylinositol (PI)	~8–15% show inverse relationship; PG high in adults; low in fetus	Precise roles unknown; potential role in surface tension reduction; possible role for PG in DPPC film enrichment through interaction with SP-B, and in distribution of SP-B in disordered regions; possible role in suppressing viral infection and inflammatory responses.
Other Phospholipids: Phosphatidylethanolamine (PE) – Phosphatidylserine (PS), Lysophosphatidylcholine (LPC), Sphingomyelin	3–5% for most species, 12% in humans; trace PLs <3%	Functions unknown; may support the formation of structures such as tubular myelin or be involved in signaling events in surfactant metabolism; high levels of LPC occur during lung injury and are associated with increased membrane permeability.
Neutral Lipids	~5–10% of surfactant by mass	
Cholesterol	80–90% of neutral lipids; second most abundant molecule of surfactant	Interactions with PLs and proteins complex, dependent on in vitro system; generally enhances film adsorption, but increases minimum surface tension; varies dynamically with temperature and ventilation, possibly to regulate phase transition of phospholipids.
Cholesteryl Esters		May represent storage form of cholesterol.
Free Fatty Acids; predominantly palmitate (16:0) Mono-, di-, and triacylglycerides		May improve surface activity and/or adsorption rate when in combination with DPPC, PG and proteins.
Surfactant Proteins (SP)	~5–6% of surfactant by mass	
Hydrophilic Surfactant Proteins: SP-A and SP-D		Role in the innate immune system of the lung; recognise a broad spectrum of pathogens; SP-A is responsible for the structural integrity of tubular myelin; SP-D may have a role in surfactant homeostasis.
Hydrophobic Surfactant Proteins: SP-B and SP-C		Enhance the adsorption of phospholipids to, and influence the molecular sorting and stability of, the surfactant film; SP-B required for the formation of tubular myelin; may have a role as an anti-bacterial agent; deletion of SP-B is fatal; SP-C may be important in maintaining tight packing of PLs in lamellar bodies or in keeping surfactant lipids in close proximity to the surfactant film.

For further information, see the text or recent review articles of surfactant lipids[4,10] or surfactant proteins.[18,26,322]

of the most variable phospholipids among species. In most cases the proportion of the combination of both phospholipids (i.e., PG + PI) is relatively constant, suggesting that it is the content of the anionic phospholipids, as opposed to the specific phospholipid, that is most important.[1] Through their anionic charge, PG and PI may enhance the adsorption of DPPC to the air–liquid interface[15] and may direct the preferential distribution of SP-B in disordered regions of surfactant membranes and interfacial films through ionic interactions with the positive charges of SP-B.[16–18] Moreover, PG appears to have a role in the innate immune functions of surfactant because it can suppress viral infection and inflammatory responses in the lung.[19,20] The roles of the other minor

phospholipids are unclear, but they may support the formation of structures such as tubular myelin or be involved in signaling events in surfactant metabolism.[10,21]

The neutral lipid, cholesterol, is the second most abundant lipid in mammalian surfactant. Cholesterol accounts for 80–90% of the neutral lipid fraction, which also includes free fatty acids and mono-, di-, and triacylglycerides.[10,22,23] Cholesterol maintains fluidity of surfactant and promotes surface film respreading[22]; this is achieved by disruption of the cohesive forces between the phospholipids, thereby reducing the phase transition temperature of the surface film.[24] The synthetic origin of surfactant cholesterol is not clear. While the internal membranes of multivesicular bodies, which are involved in the biosynthesis of lamellar bodies, are enriched in cholesterol, it is likely that the endocytic recycling pathway is an important source of cholesterol.[1] However, labeling experiments have shown that the majority of cholesterol present in lamellar bodies is not released into the alveolar hypophase. Furthermore, cholesterol levels appear to be independently regulated from surfactant phospholipids.[22] The origin, function, and regulation of surfactant cholesterol require further investigation.

Proteins

Surfactant proteins A and D are large hydrophilic proteins that are members of a family of collagenous carbohydrate binding proteins, known as collectins, or calcium-dependent (C-type) lectins. Collectins consist of oligomers of trimeric subunits that can recognize, inhibit, and inactivate a broad spectrum of foreign pathogens.[25,26] Hence, both SP-A and SP-D are important effector molecules of the innate immune system.[27-29] SP-A knockout mice lack tubular myelin and do not experience respiratory difficulties,[30] but they are more susceptible to pathogen infection.[31] SP-D knockout mice, on the other hand, develop marked disturbances in surfactant homeostasis leading to alveolar lipoproteinosis and an increase of the intracellular surfactant pool.[6,32] Furthermore, the lungs of these mice develop an emphysema-like phenotype[33] and are in a chronic inflammatory state.[25]

The low molecular weight hydrophobic surfactant proteins, SP-B and SP-C, strongly interact with surfactant lipids and promote the formation and adsorption of the surface-active film to the air–liquid interface.[26] They are integral to the regulation of the movement of the surfactant lipids between the surface associated phase, the multilayer structure in the hypophase associated with the surfactant film, and the interfacial surfactant film itself (Figure 1). Hence, they are intricately involved in lowering surface tension. While SP-B deficiency results in death in mice[34] and humans,[35] SP-C deficient mice are viable at birth and grow normally to adulthood without apparent pulmonary abnormalities.[36] Hence, SP-B is the only surfactant protein that is essential for lung function.[37]

FUNCTIONS OF THE SURFACTANT FILM

Following secretion of surfactant from the AECII, the surface active surfactant film is formed at the air–liquid interface, associated with a multilayer surfactant phospholipid reservoir, known as the "surface associated phase" (Figure 1). The film is essential for lung function as it reduces and varies surface tension at the air–liquid interface. Due to hydrogen bonding, water molecules experience a high force of attraction to each other. Hence, water molecules at an air–liquid interface experience a greater force of attraction to the other water molecules in the bulk phase, as opposed to the air above. This causes water to have a high surface tension of

~70 mN/m at 37°C.[13] Usually, surface tension is quantified as the force required to stretch the rectangular surface fluid by a known length (usually 1cm), and is expressed either as dyn/cm or mN/m (where 1dyn/cm = 10^{-5} N/cm or 1 mN/m). Any molecule at the air–liquid interface that displaces a water molecule from the interface will reduce the surface tension. This ability to interfere with the interaction of the surface water molecules, and therefore vary the surface tension, is termed surface-activity.[38] Due to the amphipathic nature of phospholipid molecules (i.e., possession of both a hydrophobic and a hydrophilic end), they adsorb to the air–liquid interface forming a surface-active film that lowers the surface tension (γ) of water. A PC film is capable of reducing surface tension to ~23–25 mN/m, defined as the equilibrium surface tension (γ_{eq}).[13] As phospholipids are insoluble, when the film is dynamically compressed, the interfacial surface area is reduced and the molecules are packed closer together, eliminating more water molecules from the interface, thereby further reducing surface tension.[13] The lowest surface tension that can be measured at the lowest surface area is the minimum surface tension (γ_{min}). Under dynamic compression in vitro, a "good" surfactant is generally able to reach a minimum surface tension (γ_{min}) of < 5mN/m, and an excellent surfactant can achieve < 1mN/m.[39] The extent of surface area compression required to reach γ_{min} is known as the % surface area compression (%SAcomp).[40] It is this ability of surfactant to lower and vary surface tension that underlies the physical functions of surfactant to promote alveolar stability, maintain airway patency and pulmonary fluid balance, and prevent adherence of epithelial tissues.[12]

However, the precise mechanism by which the surfactant film reduces and varies surface tension on a breath-by-breath basis remains controversial. Surfactant purportedly regulates surface tension at the air–liquid interface by dynamically regulating the surface concentration of the chief tension lowering constituent (DPPC) in relation to fluidizing components such as unsaturated phospholipid molecular species and cholesterol. However, some recent studies refute this classical model of surface tension reduction.

The Classical Model of Surfactant Film Function

The classical model for surfactant function relied on DPPC, at 40–50% by weight, being the only significant component capable of reducing surface tension to the low values required to stabilize alveoli.[41] Films enriched in DPPC are required to stabilize alveoli and maintain lung function. Compression results in squeeze-out of the fluid unsaturated phospholipid components from the interface, leading to an increased surface concentration of DPPC. As DPPC only undergoes a thermal chain melting phase transition (Tc) at 41°C, which is above the body temperature of most homeothermic mammals (35–37°C), a film enriched in DPPC could attain a tightly packed state, known as the gel-like condensed phase during alveolar compression. This state is characterized by tight packing of the phospholipid molecules with exclusion of water molecules and lowering of surface tension to near zero. As alveoli expand, the more fluid components such as unsaturated phospholipids and cholesterol would be retrieved from the bulk phase into the surface film. Neutral lipids and unsaturated phospholipids lower the thermal chain melting phase transition of the film, increasing its fluidity and enabling re-spreading over the expanding alveolar surface.[18] The hydrophobic surfactant proteins, SP-B & -C, are critical for regulating the dynamic behavior of lipids.[18]

Limitations of the Classical Model

In the past 10 years, research has shifted thinking regarding the composition and function of the surfactant film. While the concept that DPPC reduces surface tension and that other lipids act as fluidizers remain relevant, we now know that surface tension is not reduced by a compositional modification of the surface film on a breath by breath basis during ventilation. Rather, this is due to a structural rearrangement of the whole film occurring under dynamic compression.[18,42,43] Central to this understanding are the discoveries that a condensed film composed of almost pure DPPC is not required for reaching low surface tensions[43] and that films of fluid phospholipids can reach low tensions if compressed rapidly enough.[42,44] This debate was fuelled by findings that the molar composition of phospholipid molecular species is highly variable, with molecular alterations occurring with age and respiratory rate[45] as well as diet,[46] such that the minor, more fluid phospholipids such as PPPC (PC16:0/16:1) and PMPC (PC16:0/14:0) can assume significant proportions. In addition, our own work with heterothermic mammals that can enter a state of reduced body temperature and metabolism (e.g., torpor or hibernation) has impacted upon this debate, with the finding that in many of these species DPPC is not the major surfactant phospholipid species, but that PPPC is much more prevalent.[11] Moreover, the lipid composition is altered within an individual in response to changes in body temperature[47–50] and in turn this results in changes in surfactant activity and film structure.[48,51–55] Collectively these findings indicate that surface activity is primarily determined by the packing propensity of the specific lipid mixture, which is dependent not only on the acyl chain composition of the lipids, but also on temperature, the presence of other lipids such as cholesterol and the state of compression of the film at the air–liquid interface.

FUNCTIONS OF THE PULMONARY SURFACTANT SYSTEM

Due to the complex composition of the lipo-protein mixture, pulmonary surfactant is involved in a suite of pulmonary functions associated with its surface tension lowering properties, but it also carries out an important host defence function.

Lung Stability

In the mammalian lung, a γ_{min} of near zero at end-expiration is a necessary condition for alveolar stability. Alveoli (5–10 μm diameter) are lined with a fluid layer approximately 0.2 μm thick.[56] The surfactant film lining the interface between this fluid and the air space generates a surface tension-related elastic retractile force. According to the law of Young and LaPlace, $P = -2\gamma/r$ (where γ is surface tension measured in mN/m and r is the radius of curvature measured in m), very small alveoli will experience a very large collapse pressure unless the surface tension was very low. Pulmonary surfactant reduces the surface tension to near zero at very low lung volumes, such that the retractile forces are similar both within separate regions of an alveolus as well as between respiratory units.[57,58] This, together with the structural interdependence of the alveoli, provides the lung with alveolar stability and maintains a relatively large alveolar surface area necessary for efficient gas exchange even at low lung volumes.[59]

Alveolar Interdependence and Anti-Adherence

The elastic recoil of alveoli is responsible for about one-third of lung compliance. As alveoli are inter-connected, any alveolus tending to collapse will be held open, because it will be supported by the walls of adjoining alveoli; this interaction between alveoli is termed interdependence. A model for the structure–function relationship of the lung parenchyma, the alveoli and surface tension is shown in Figure 2.[58–60] Furthermore, as alveoli are not perfect spheres, but rather polyhedral in shape, their walls may come into contact upon expiration. The work required to separate contiguous alveolar walls is directly proportional to the surface tension of the fluid lining.[61] By reducing the surface tension of the alveolar hypophase, pulmonary surfactant greatly reduces the work required to initiate lung inflation. This function of surfactant is termed anti-adhesive (or "anti-glue") and is believed to be particularly important in non-mammals,[62] but might be significant in aiding the reinflation of a partially collapsed or atelectatic lung.

Prevention of Pulmonary Edema and Maintenance of Airway Patency

The luminal surface of alveoli and small conducting airways is lined with a fluid similar in composition to interstitial fluid. Moreover, lungs have a large surface area,

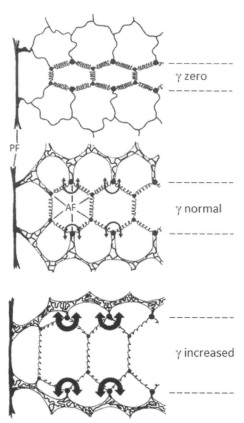

FIGURE 2   Schematic diagram of the alveolar interdependence model, illustrating the structure-function behavior of lung parenchyma in response to alterations in surface tension (γ). The functional unit is the alveolar duct (or a set of ducts forming an acinus) embraced by peripheral connective tissue fibres. The peripheral fibres (PF) are connected to the pleura, are the main force-bearing element, and are largely independent of changes in surface tension (γ). The axial fibres (AF) are rings of tissue forming the entrance of alveoli; they are influenced by the surface tension of the air–liquid interface, which is continuous along the alveolar wall. The 2-dimensional alveolar walls represent a negligible mechanical component. Low surface tensions allow a large alveolar surface area between slightly stretched axial fibres. However, when surface tension is abnormally high, the axial fibres become more stretched resulting in duct enlargement, flattening of alveoli, and a decreased alveolar surface area. *Figure reproduced with permission from.*[59]

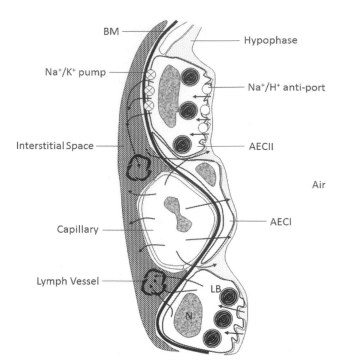

FIGURE 3   Schematic diagram of an alveolar wall illustrating the movement of fluid (arrows) between the fluid lining the air spaces (hypophase) and the interstitial space. The small radius of curvature of corners and crevices leads to a large negative fluid pressure in the alveolus, which tends to draw fluid into the alveolus from the interstitium. Furthermore, under hydrostatic pressure, net fluid movement occurs out of the capillaries into the surrounding tissue and the alveolus. Sodium pumps in the alveolar membrane of epithelial type II cells remove sodium from the hypophase and transport it into the interstitium, causing a net passive fluid movement out of the alveolus, thereby preventing fluid build-up. Excess fluid in the interstitium is removed by the lymphatic circulation. BM, basement membrane; LB, lamellar bodies; N, nucleus.

high blood flow and leaky capillary endothelial cells and are therefore susceptible to fluid disturbances. Areas with a high surface tension would tend to draw fluid from the intersitium.[63] Fluid pressure is large and negative in the alveolar corners (because the radius of curvature is very small), but small in the flatter regions of the alveolus. Normally fluid probably flows from the interstitial space into the alveolar hypophase via alveolar seams, crevices, and corners (Figure 3). Fluid flow may occur back into the interstitium via flatter parts or AECII.[64] In this latter case, fluid movement is aided by Na+ pumps on the AECII membrane that set up a concentration gradient, causing fluid to flow passively down an osmotic gradient. The pumping of Na+ from the hypophase into the interstitium may also promote the reabsorption of fluid from the small airways. Fluid build-up within the alveolus (pulmonary edema) increases the gas diffusion distance and compromises gas exchange. By lowering surface tension, surfactant reduces the amount of fluid entering the alveolar space; it also decreases surface tension to values lower than the surrounding interstitium, and therefore encourages the movement of fluid from the alveolus into the tissue[65] (Figure 3).

Pulmonary surfactant also aids in airway patency. As the terminal airways leading to alveoli lack cartilage, they are prone to narrowing. During expiration, these airways become narrower and fluid may accumulate in the narrowest section.[66] Because narrow portions of the airway experience greater pressures than wide portions, this could result in fluid extravasation into the narrower regions. By lowering surface tension during expiration, pulmonary surfactant lowers the pressure in narrow sections to less than that in wider portions, thereby maintaining airway patency.[66]

Host Defence

The lung is exposed to air-borne bacteria, fungi, viruses, and other foreign material. Surfactant contributes to lung immunity by aiding the mucociliary escalator and by directly inhibiting and inactivating infectious agents or particles. Surfactant may directly influence the transport of particles from the peripheral to the upper airways, as demonstrated in vitro by the regular compression and expansion of a monomolecular film, resulting in the unidirectional transport of particles that had been deposited on the film.[67] In the airways, surfactant is present in the periciliary fluid of the mucus lining of the bronchiolar surface[68,69] and is associated with the cilial tracts.[70] Surfactant may aid the mucus layer and the cilial tract to remove particles from the airways in one of two ways. Firstly, surfactant in the periciliary fluid may lower the surface tension and viscosity of the fluid bathing the cilia, thereby facilitating more efficient beating.[70,71] Secondly, surfactant may improve mucociliary transport, by coating both the mucus droplets and the periciliary-gel interface along the airways, resulting in an anti-adhesive effect, thereby allowing greater cilia-mucus coupling.[70,72] Using tracheal rings in vitro, it was shown that surfactant increases cilial beat frequency in a dose-dependent manner.[73] This effect would facilitate the upward removal of particles from the bronchial tree.

Pulmonary surfactant also plays a significant role in innate pulmonary host defence by virtue of the hydrophilic surfactant proteins, SP-A and SP-D, which are capable of recognizing, inhibiting, and inactivating a broad spectrum of foreign pathogens.[26] These functions are achieved by the C-terminal carbohydrate recognition domain that is present in both proteins. The formation of multimers enables clustering of the carbohydrate recognition domains that in turn enables multivalent interactions between the carbohydrate recognition domain and the carbohydrates on the surface of various pathogens. Both SP-A and SP-D are capable of directly killing micro-organisms or indirectly killing pathogens by enhancing their uptake by phagocytes.[12] Interestingly, SP-A is capable of mounting both a pro- and an anti-inflammatory response, depending on whether a pro-inflammatory response is needed to combat infection or whether an anti-inflammatory action is required to limit inflammation and avoid tissue damage (reviewed in[25]). On the other hand, SP-D is predominantly anti-inflammatory, as it is capable of directly modulating macrophage and dendritic cell function as well as T-cell dependent inflammatory events (reviewed in[25]).

REGULATION OF SURFACTANT SECRETION

The release of surfactant from adult AECII, either in vitro or in the intact lung, is mediated by numerous biochemical and mechanical stimuli.[8,74–77] A working model for the signal

transduction pathways that mediate surfactant release was first proposed by Rooney et al.,[78,79] and later refined.[8,74,80] Here we briefly outline this model (Figure 4) and focus on the most important factors that stimulate surfactant secretion. Information on other biochemical agonists that stimulate secretion, as well as substrates, downstream targets, receptor subtypes, and protein subfamilies is provided in detailed reviews.[4,74,80,81]

Signaling Mechanisms of Surfactant Secretion

Stimuli may activate one or more of three potential pathways involving three classes of protein kinases (PK), i.e., PKA, PKC, and Ca^{2+}/calmodulin-dependent protein kinase (CaMK) (Figure 4). The first involves activation of adenylate cyclase (AC), generation of cyclic AMP (cAMP) and subsequent activation of cAMP-dependent protein kinase, i.e., PKA. This pathway is activated by ß-adrenergic and adenosine A_{2B} receptors that are coupled to adenylate cyclase via the heterotrimeric GTP-binding protein (G-protein), G_s. The second pathway involves direct or indirect activation of PKC. Phorbol esters and cell permeable diacylglycerols (DAGs) directly activate PKC. ATP and UTP bind to purinergic receptors that are coupled to phospholipase C (PLC)-ß3 via a different G-protein, G_q. PLC-ß3 hydrolyzes phosphatidylinositol bisphosphate into DAG and inositol trisphosphate (IP3). While DAG activates PKC, IP3 promotes mobilization of Ca^{2+} from intracellular stores, which activates the third pathway involving CaMK. Stimulation of CaMK can also be accomplished by ionophores and mechanical strain, both of which promote influx of extracellular Ca^{2+} into the cell.[8,74,80,82] Activation of PKC, particularly by ATP, is the most potent pathway for surfactant secretion (~5-fold greater than basal, unstimulated secretion), whereas stimuli that activate PKA or CaMK increase secretion by 2- to 3-fold. Simultaneous activation of different pathways stimulates secretion up to 12- to 15-fold.[74] The downstream mechanisms regulating lamellar body movement to, and fusion with, the plasma membrane are not fully elucidated, but likely involve Ca^{2+}-dependent phosphorylated annexin A7 trafficking to the plasma membrane and interacting with the soluble N-ethylmaleimide-sensitive fusion protein attachment receptors (SNARE) protein, SNAP23[83–85] (Figure 4).

Autonomic Agonists

Regulation of surfactant secretion by β-adrenergic agonists occurs in vivo and in isolated AECII. The order of potency on isolated AECII is isoproterenol > adrenaline > noradrenaline, indicating that the β_2 receptor mediates the response.[76] Regulation of surfactant phospholipid secretion by β-adrenergic agonists occurs in vivo, for example, in response to labor[86] and in isolated AECII of rat[87–89] and mouse.[90] Moreover, β-adrenergic agonists also stimulate

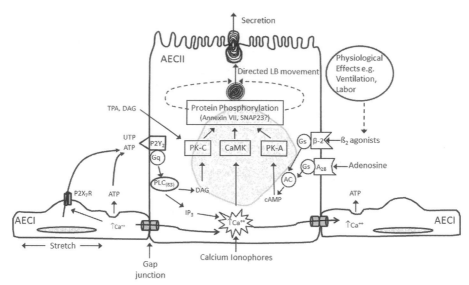

FIGURE 4 Schematic diagram summarizing the major factors and signaling pathways that stimulate surfactant secretion from alveolar epithelial type II cells (AECII). Several beta-2 adrenergic agonists, including isoproterenol, adrenaline, and noradrenaline, stimulate the β-2 adrenergic receptor. The receptor is coupled to a heterotrimeric G protein (Gs), which stimulates adenylate cyclases (AC) to produce cyclic AMP (cAMP), which in turn stimulates the cAMP dependent protein kinase A (PK-A). Similarly the adenosine A_{2B} receptor stimulates PK-A via the same intermediates. Another pathway involves the direct or indirect stimulation of protein kinase C (PK-C). The synthetic surfactant secretagogue tetra-decanoylphorbol acetate (TPA) and cell-permeable diacylglyerols (DAGs) are potent direct stimulators of PK-C. ATP and UTP bind to the purinergic receptor (P2Y$_2$) which is coupled to a heterotrimeric G protein Gq, which stimulates phospholipase C (PLC)-ß3. Activation of PLC-ß3 leads to the formation of inositol triphosphate (IP$_3$) and DAG. The latter stimulates PK-C, while IP$_3$ feeds into the third secretory mechanism, which involves the elevation of intracellular Ca^{2+} levels by IP$_3$ by calcium ionophores or by stretch of the basement membrane. Calcium in turn stimulates calmodulin-dependent protein kinase (CaMK). The stimulatory effect of stretch is mediated indirectly by the P2Y$_2$ receptor, and therefore the PK-C and/or CaMK signalling pathways. Specifically stretch is mediated by the alveolar epithelial type I cells (AECI) that are connected to AECII via gap junctions. Cytoplasmic Ca^{2+} increases in either cell can spread via gap junctions, and the intercellular transmission of Ca^{2+} triggers the secretion of ATP from AECI into the extracellular spaces via stimulation of the specific P2 purinergic receptor (P2X$_7$R), which in turn activates Ca^{2+} signalling pathways in AECII via P2Y$_2$ purinergic receptors to induce surfactant secretion from AECII in a paracrine manner. However, physiological stimulation of surfactant secretion by ventilation or labor may be mediated via the β-2 receptor. The exact subsequent mechanisms leading to lamellar body exocytosis are not well understood, but are thought to involve protein kinase-stimulated protein phosphorylation, which presumably activates contractile proteins to move lamellar bodies to the apical surface to fuse with the plasma membrane. *Figure modified from: Mason and Voelker (Figures 1 and 2)[76]; Rooney (Figure 2)[8] and Andreeva et al. (Figure 3)[74] with permission obtained from Elsevier.*

SP-B and -C secretion from isolated AECII of rat[91] and mouse.[90] In contrast, SP-A and SP-D secretion are largely constitutive and thus unregulated.[8,76] Adrenergic agonists decrease lamellar body volume density within 30 min. of injection,[92] demonstrating that the response is rapid.

While AECII possess muscarinic cholinergic receptors,[93] cholinergic agonists fail to stimulate surfactant secretion from isolated AECII of rats[87,94] or humans.[95] However, cholinergic agonists do enhance PC secretion in the adult isolated perfused rat[96] and rabbit[97] lung as well as in vivo in the neonatal rabbit.[98] It is possible that cholinergic agonists may affect surfactant secretion indirectly, either by causing contraction of smooth muscle cells, resulting in distortion of AECII,[96] or via activation of adrenergic receptors in response to catecholamines released from the adrenal medulla.[8] On the other hand, cholinergic agonists do act directly on isolated AECII of a heterothermic marsupial to release surfactant lipids.[99,100] The parasympathetic nervous system may have evolved in ectotherms to stimulate surfactant secretion at lower temperatures without elevating metabolic rate,[101] whereas adrenergic influences on the surfactant system may have developed subsequent to the radiation of tetrapods.[99]

Ventilation

Hyperventilation results in stimulation of surfactant secretion within minutes,[102] and a single deep breath increases the amount of alveolar phospholipid, possibly through increased secretion.[103] Ventilation-induced increases in surfactant secretion are likely to occur via direct mechanical stimulation of AECII, which are preferentially located in corners of alveoli where they are exposed to maximum distortion.[104] Physical stretch of isolated AECII results in an increase in surfactant secretion equivalent to that induced by a combination of agonists.[105,106] The stretch-induced surfactant secretion is accompanied by an increase in cytoplasmic Ca^{2+} concentration, which appears to be an important intracellular messenger, both in the isolated AECII[105] and in the intact alveolus.[107] However, the concept that the AECII is the alveolar stretch sensor was recently queried: in that study, fluorimetric measurements of cytoplasmic Ca^{2+} concentration indicate synchronous oscillations between adjacent alveolar epithelial type I cells (AECI) and AECII following a 15 sec. alveolar stretch.[107] When gap junctions were blocked with heptanol, only AECI responded asynchronously to the stretch with an increase in intracellular Ca^{2+}, whereas AECII failed to respond. This suggests firstly that Ca^{2+} signals can spread from AECI to AECII, presumably via gap junctions, and secondly that during lung inflation AECI act as alveolar mechanotransducers that regulate AECII secretion.[107] This suggestion was recently confirmed with the demonstration that mechanical stimulation of AECI elicits an intracellular calcium wave that is transmitted to neighboring AECI and AECII specifically via gap junctions (Figure 4), as determined by specific gap junction blockers in a heterocellular culture system of AECI and II cells,[108] allowing for the coordinated regulation of the alveolar epithelium. Moreover, the intercellular transmission of Ca^{2+} triggers the secretion of ATP from AECI into the extracellular spaces via stimulation of the specific P2 purinergic receptor ($P2X_7R$), which in turn activates Ca^{2+} signaling pathways in AECII via $P2Y_2$ purinergic receptors to induce surfactant secretion from AECII in a paracrine manner.[109,110]

DEVELOPMENT OF THE PULMONARY SURFACTANT SYSTEM

The pulmonary surfactant system generally develops during late gestation, with the initial appearance of surfactant lipids and proteins in lung tissue in the form of lamellar bodies, before their appearance in amniotic fluid or alveolar lavage.[111] However, the ontogenic pattern of lung development and surfactant maturation differs between species (Figure 5). When examining these differences it is important to consider relative stage of lung maturation in different species (Figure 5). In humans, lamellar bodies are detected at ~24 weeks of gestation (term ~40 weeks), corresponding to the mid-canalicular stage of lung development. Thereafter low levels of DPPC are detected in amniotic fluid or alveolar lavage between 24 and 28 weeks gestation in the mid-late canalicular phase and these increase steadily thereafter until term.[111] Consequently some premature infants (born 26–32wks) have sufficient quantities of surfactant in the airways and do not develop respiratory distress syndrome (RDS),[111] although surfactant maturation between individuals is highly variable.[112] In contrast, in most "laboratory" mammals, surfactant matures relatively later during gestation.[111] For example, mature lamellar bodies appear only at 80–90% of gestation in sheep, monkey, guinea pig, rabbit, rat, and mouse[12,113] compared to the human at ~60%. However, more importantly, if viewed in terms of stage of lung development, lamellar bodies appear only marginally later in sheep, monkey and rabbit (late canalicular stage) and in fact appear earlier during lung development

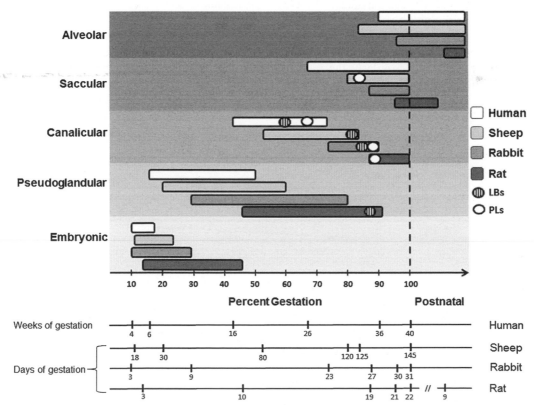

FIGURE 5 Comparison of the development of the human, sheep, rat, and rabbit lung indicating the relative gestational period for each of the five phases of lung development, both in terms of the percent of gestation and the weeks/days of gestation. Also indicated are the approximate appearance of mature lamellar bodies (LBs, hatched circles) and phospholipids (PLs) in the lavage or amniotic fluid (white circle). *Figure adapted from*[324,325] *with additional data from.*[111,113]

in rodents (late pseudoglandular stage)[12,113] (Figure 5). On the other hand, in guinea pigs, lamellar bodies do appear relatively later in the saccular stage.[113] Consequently, in all these species, the relatively later appearance of DPPC in amniotic fluid or alveolar lavage does not occur until closer to term, which corresponds to a variable range of lung developmental stages among species: early canalicular in rat, late canalicular in rabbit, and early saccular in sheep[113] compared to mid-late canalicular in humans (Figure 5). Surfactant accumulation in the alveolar compartment thereafter occurs exponentially.[111,113] For example, total PC is low in fetal sheep 3–4 weeks prior to birth (term ~145 days), but reaches adult levels 3–4 days prior to parturition.[12,114] Similarly, phospholipid in lung homogenates of fetal rats increases 10- to 20-fold between gestational days 19 and 21 (term ~22 days).[115] Furthermore, in rabbits,[86] rats,[116] and monkeys,[113] there is significant secretion of surfactant phospholipids during and immediately after birth. The delayed maturation means that many of these animals, if delivered prematurely, more so than humans, will die at birth or develop RDS.[12,111]

The composition and saturation of the surfactant lipids also changes during gestation. For example, the disaturated phospholipids appear late in gestation, with the relative saturation of PC increasing 4- to 5-fold in the lavage of the rabbit fetus between the 24th and 30th day of gestation (term, ~31days) in the late canalicular stage.[12,117] The proportions of the two acidic phospholipids (PG and PI) also change during development with the relative proportion of PG increasing, while that of PI decreases, leading to an increase in the PG/PI ratio. Similarly the ratio of PC to sphingomyelin in amniotic fluid, the so-called lecithin/sphingomyelin (L/S) ratio increases with gestation.[118]

Both of these ratios are used as an index of fetal lung maturity.[12,118,119]

The presence of surfactant proteins in amniotic fluid occurs late in human gestation. At 16–20 weeks of gestation SP-A mRNA is undetectable in fetal lung tissues[120]; after ~20 weeks, both SP-A mRNA and faint immunohistochemical signal for SP-A protein are detected in cells of the terminal airways[121] (Figure 6). Differentiated AECII containing few lamellar bodies can be observed in human fetal lung tissue as early as 22 weeks gestation; however, active secretion of surfactant occurs only after 30 weeks[122] at which time SP-A can be detected in the amniotic fluid.[123] Similarly, SP-B appears in amniotic fluid after 31weeks.[124] Both SP-A and SP-B increase exponentially after 30-31weeks until term.[124] Both SP-B and SP-C proprotein and their mRNA appear in proximal airways of the lungs as determined by immunohistochemistry and in situ hybridization much earlier, by the 15th week of pregnancy (Figure 6) and in the distal lung and AECII by the 25th week.[125] SP-D protein and mRNA in human lung tissues is first detected in the second trimester as early as 16 weeks.[126]

In many mammalian species, the prenatal maturation of surfactant proteins occurs relatively later than in humans. In rat lung tissue, SP-A protein is very low up to day 18, increases from day 19 and reaches its maximum on the day after birth.[127,128] But between the day of birth and postnatal day 7, there is a 35–40% reduction in SP-A content.[129] Expression of SP-A mRNA closely mirrors that of SP-A protein, as it is undetectable on day 17 of gestation, barely detectable on day 18, increases 3- to 4-fold on day 19, when there is a transition from the pseudoglandular to the canalicular state, and a further 6- to 9-fold between day 19 and day 21[129] (Figure 6). In contrast, the SP-B content in fetal rat lungs is low throughout gestation and increases only after

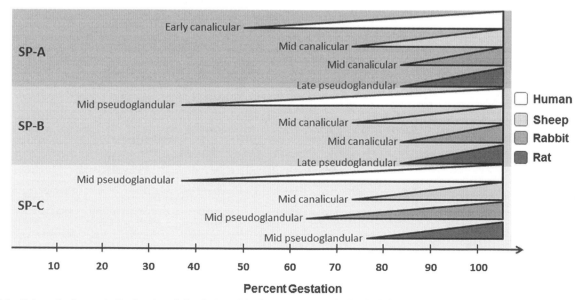

FIGURE 6   Schematic diagram indicating the relative timing of the first appearance of mRNA of the surfactant proteins, SP-A, -B, and -C, in lung tissue of human, sheep, rabbit, and rat, both in terms of the percent gestation as well as the lung developmental period. *Data from.*[111,113,120,121,125]

birth, reaching its maximum on postnatal day 4.[128] SP-B mRNA content is first detectable on day 18 and increases to adult levels by day 20.[129] On the other hand, SP-C mRNA levels are detectable earlier at day 17 and increase to adult levels by day 20–21[129] (Figure 6). SP-D content is very low until day 18, with a 4-fold increase on day 19 and a further 2-fold increase between day 21 and the day of birth.[127] The adult level of SP-D is about 25% of the adult level of SP-A. Unlike SP-A there is no decrease in SP-D content after birth.[127] The expression of SPs very late in gestation in rat is consistent with the relative immaturity of several organ systems in rats, such that survival is low after preterm birth.

In rabbit lung tissue, SP-A and SP-B mRNAs remain relatively low until 26 days gestation when they increase dramatically to a maximum on day 30.[130] However, SP-C mRNA appears relatively earlier from day 19[131] to day 22[130] and is present in maximal concentration at term.[132] In situ hybridization localizes SP-C mRNA in all epithelial cells of the pre-alveolar region at day 19, i.e., about 7 days before the appearance of differentiated AECII. By day 27 of gestation, SP-C mRNA is restricted to epithelial cells with the morphologic characteristics of AECII.[132]

In preterm sheep, SP-A, -B, and -C mRNA are not detectable in lung tissue at ~100 days (term is 145–150 days), begin to appear at ~115–120 days and then increase rapidly thereafter to term.[133]

Regulation of Surfactant Development

Regulation of surfactant maturation in the fetus is under complex multifactorial control with significant redundancy. In addition to a vast array of neurohormonal, growth and transcription factors, mechanical and physical factors are also involved in regulating surfactant maturation. The system is further complicated by the fact that the different components (e.g., lipids versus proteins) are regulated by different factors at different times during development. Moreover, different species demonstrate different developmental trajectories, as outlined earlier.

Biochemical Factors

There are numerous neurohormonal, growth, and transcription factors that stimulate and regulate the maturation of surfactant lipids and proteins in both fetal AECII in vitro or in the intact fetus.[12] The most important of these are glucocorticoids, thyroid hormones and autonomic neurotransmitters,[80] which are reviewed in more detail in the following text and illustrated in Figure 7. Other relevant hormones and

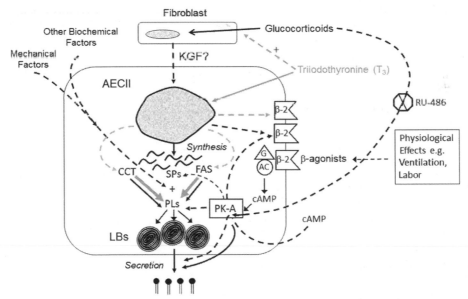

FIGURE 7 Schematic diagram summarizing neurohormonal regulation of surfactant maturation in fetal alveolar epithelial type II (AECII) cells. Glucocorticoids (black arrows) and thyroid hormones (grey arrows) increase surfactant production as well as the number of adrenergic receptors, rendering AECIIs more responsive to surfactant secretagogues. Glucocorticoids stimulate surfactant synthesis by AECII indirectly, possibly via keratinocyte growth factor (KGF) produced by fibroblasts, a pathway that is enhanced by T_3.[179] Thyroid hormones act directly on nuclear receptors of AECII to increase surfactant synthesis, but the pathway is unclear. While glucocorticoids increase gene expression of surfactant proteins (SPs) as well as that of the enzymes cytidine triphosphate phosphocholine cytidylyltransferase (CCT) and fatty acid synthase (FAS) to increase phospholipid (PL) synthesis, thyroid hormones increase the activity of these enzymes (indicated by thick grey arrows). Together, increased synthesis of SPs and PLs leads to an increase in lamellar bodies (LBs) and hence surfactant secretion. In addition, glucocorticoids stimulate surfactant secretion via downstream signaling factors of the PK-A signaling pathway and via a glucocorticoid receptor-dependent pathway (blocked by the glucocorticoid receptor antagonist RU-4860[164]). Several beta adrenergic agonists, including iso-proterenol, adrenaline, and noradrenaline stimulate the β-2 receptor to increase surfactant secretion. The receptor is coupled to adenylate cyclase (AC), which produces cyclic AMP (cAMP) via a trimeric GTP-binding protein (G), in order to stimulate cAMP dependent protein kinase A (PK-A). cAMP administered to isolated AECII also acts via PK-A to increase β-2 receptor number.[196] Other factors including hormones, growth factors, and physical factors are described in the text. Dashed lines indicate pathways that are not fully elucidated. *Figure reproduced with modifications and permission from.*[12]

growth factors include thyrotropin-releasing hormone,[134,135] adrenocorticotropin hormone (ACTH),[136] corticotropin-releasing hormone,[137] prolactin,[138] prostaglandins,[139] insulin,[140,141] estrogen,[142] parathyroid hormone,[143] leptin,[144] keratinocyte growth factor (KGF),[145] epidermal growth factor, insulin-like growth factor 1, retinoic acid, and gastrin releasing peptide.[146] The last two decades have seen a very large increase in our understanding of growth and transcription factors and the associated signal transduction pathways that regulate gene expression of the surfactant proteins during development (reviewed by[147–149]). Some relevant aspects are mentioned in the following text, but a detailed discussion of the molecular regulation is beyond the scope of this chapter.

Glucocorticoids

Glucocorticoids are the most studied hormones that accelerate surfactant maturation. During late gestation, activation of the hypothalamic-pituitary-adrenal axis (HPA) leads to an increase in plasma concentration of ACTH and cortisol, which facilitates lung maturation.[150] Liggins et al first observed that fetal exposure to glucocorticoids led to increased survival of preterm lambs,[151,152] which indicated that glucocorticoids specifically induce surfactant synthesis.[152] Since then there has been a plethora of studies investigating the timing, dosage, and mechanisms by which glucocorticoids promote lung development, specifically surfactant synthesis and thinning of parenchymal tissue (reviewed by[153]). Although there have been many conflicting results owing to differences in timing, dosage, species and whether it is the fetus or the mother that is treated,[154] it is clear that antenatal glucocorticoid treatment reduces the incidence of RDS by ~35–40%,[155] improves fetal lung mechanics after as little as 8–15 h,[156] and changes the surfactant system, but only after more prolonged treatment.[12,153,157]

The glucocorticoid-induced changes to the surfactant system include, for example, an increase in phospholipid synthesis and the increased appearance of phospholipid in the alveolar compartment (reviewed in[12]). Specifically, glucocorticoids increase the amount[158] and activity[159] of cytidine triphosphate (CTP) phosphocholine cytidylyltransferase (CCT), an enzyme that catalyzes the regulatory step in PC synthesis. Glucocorticoids also increase the expression of fatty acid synthase (FAS) that catalyzes the de novo synthesis of fatty acids and is thought to be a significant requirement for phospholipid synthesis in the fetus[80] (Figure 7). Glucocorticoids have a minimal effect on isolated AECII[160–162]; however, it is thought that the glucocorticoid effect is mediated by mesenchymal factors secreted by fibroblasts[161] (Figure 7). A likely candidate is keratinocyte growth factor (KGF), which is also known as fibroblast growth factor 7,[80,145] and has been shown to increase PL synthesis, CCT activity, and

FAS expression in isolated AECII.[145] Hence, it is likely that glucocorticoid stimulation of PL synthesis is mediated by KGF.[12,80,163] Although the synthetic action of glucocorticoids on AECII may be indirect, it appears that the effect on secretion may be direct, as the secretory response was blocked by the glucocorticoid receptor antagonist RU-486[164] (Figure 7).

Glucocorticoids also appear to regulate the surfactant proteins, by stimulating their gene expression (summarized in[12]) (Figure 7). However, the effect of glucocorticoids on surfactant protein gene expression is not straightforward, as it is species, dose, and gestational age dependent.[165] For example, dexamethasone has a biphasic effect on SP-A gene expression in human fetal lung explants with low doses being stimulatory and high doses being inhibitory.[166,167] Moreover, the stimulatory effect is likely mediated indirectly rather than by direct binding of the glucocorticoid receptor (GR) to the SP genes,[168] because no glucocorticoid response elements (GRE) have been identified in the promoter regions of the surfactant protein genes.[149] This is supported by the fact that GR-/- mice demonstrate normal levels of SP gene expression in the perinatal and postnatal periods.[149,168] It is likely that the actions of glucocorticoids on surfactant protein production are mediated by thyroid transcription factor-1 (TTF-1 or TITF1, also known as Nkx2.1 or T/EBP [thyroid-specific-enhancer binding protein]), which is stimulated by glucocorticoids.[169,170] However, this stimulation also appears to be indirect, as the TTF-1 promoter lacks a GRE.[148] However, the genes for SP-A, -B, and -C possess a TTF-1 binding element (TBE)[171–173] and TTF-1 has been shown to be critical for activation of SP-A, -B, and -C gene expression.[148,149,174] SP-D is also activated by TTF-1, but only in combination with the transcription factor, nuclear factor of activated T cells (NFATc3)[174] (Figure 8).

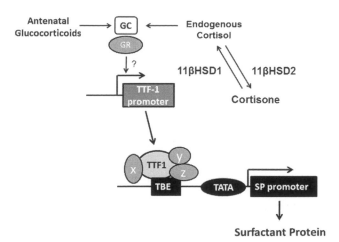

FIGURE 8 Diagrammatic representation of the mechanism by which endogenous (circulating or locally produced) cortisol and antenatal glucocorticoids act in the lung to increase the gene and protein expression of surfactant protein. GC: glucocorticoid; GR: GC, receptor; TTF-1: thyroid transcription factor-1; TBE: TTF-1 binding element; x, y, and z indicate cofactors for different SP genes. *Figure reproduced with permission from.*[326]

Thyroid Hormones

The thyroid hormones, triiodothyronine (T_3) and thyroxine (T_4), enhance surfactant development, particularly the lipid component, through stimulatory effects on secretion and synthesis (summarized in[12]) (Figure 7). The surge in fetal plasma cortisol concentrations also causes an increase in plasma T_3, but not T_4 concentrations[175,176] that results in an increase in pulmonary surfactant content in late mammalian gestation (reviewed in[177]). Although T_3 is the most potent of the thyroid hormones, it does not readily cross the placenta. Hence, thyrotropin-releasing hormone (TRH) has been maternally administered to increase fetal T_3 and hence surfactant lipid concentrations.[135,178] Moreover, thyroid hormones administered alone may not increase surfactant synthesis[179]; rather, glucocorticoids and thyroid hormones act synergistically to enhance surfactant PC synthesis[138,179–182] (Figure 7).

The effects of thyroid hormones are mediated via receptors in the nuclei of AECII[183] (Figure 7), which appear in the human between 12 and 19 weeks of gestation. Throughout this time, binding affinity increases.[184] However, the exact maturation mechanism remains unclear.[185] The activities of the enzymes, which control the synthesis of DPPC, are increased by administration of T_3.[158,186] The thyroid hormones may also act to enhance surfactant lipid metabolism.[187] Fetal administration of thyroid hormones also increases the number of β-adrenergic receptors in the lungs of rabbits[188] and rats.[189]

However, apart from one report of increased SP-A protein in human fetal lung culture in response to T_3[190] it does not appear that thyroid hormones stimulate surfactant protein abundance or gene expression,[185] and may in fact decrease surfactant protein gene expression.[191,192]

Autonomic Stimulation

Adrenergic agonists also contribute to surfactant maturation, presumably by the same signaling mechanisms as in the adult[193,194] (see section "Signaling mechanisms of surfactant secretion" and Figure 4). Stimulation of fetal lung tissues from a range of mammalian species with adrenergic agonists or directly with the signaling intermediate, cAMP, stimulates both phospholipid secretion and synthesis as well as surfactant protein secretion and synthesis via increased gene transcription (summarized in[12]) (Figure 7).

Adrenergic stimulation is very important physiologically during development and at parturition.[12] Specifically, towards the end of gestation, AECII respond increasingly to β-adrenergic agonists and other secretagogues.[194,195] This maturation of the receptor pathways[194] may be facilitated by cAMP, which has been shown to increase β-adrenergic receptor concentration in human fetal lung and specifically in AECII through a PKA-dependent mechanism[196] (Figure 7).

Glucocorticoids transiently increase the expression of β-adrenergic receptor in human airway epithelial cells.[197] Furthermore, administration of dexamethasone to newborn rat AECII increases the secretory response to a range of secretagogues.[164] As plasma glucocorticoid concentrations increase towards term, this mechanism may serve to mature the lungs in preparation for birth.[164] Together these effects likely contribute to the increase in surfactant content during late gestation. During birth, there is a massive surge in circulating catecholamine levels,[198] which provide profound adrenergic stimulation. However, sympathetic nerve activity may also contribute.[199] Interestingly, the β-receptors mature more rapidly in females than males,[200] which may confer an advantage upon females.

The influence of cholinergic agonists on surfactant development during the perinatal period is less clear and more contradictory. Pilocarpine, a muscarinic receptor agonist, stimulates PC release in fetal lung explants in the rat,[201] but does not stimulate PC surfactant release in newborn rabbit lung slices.[139] Furthermore, carbachol does not stimulate PC secretion from fetal hamster lung slices.[202] A recent in vivo study showed that vagal denervation of fetal sheep lung did not alter alveolar architecture, including morphology of AECII and their lamellar bodies, SP-A and -B, or total phospholipids in either lung tissue or bronchoalveolar lavage fluid. Hence, vagal innervation at mid-late gestation is not required for maturation of the pulmonary surfactant system.[203] Hence, it appears that the parasympathetic nervous system does not play a significant role in the maturation of the surfactant system.

Mechanical Factors

As discussed in detail in Chapter 8 both tonic and cyclic distension of the lung, produced respectively by the hydrostatic pressure of fetal lung liquid and fetal breathing movements (FBM), are important for normal lung growth, cellular differentiation, and maturation.[204] As these processes are responsible for regulating the relative proportions of AECI and II, they also have an important role in regulating surfactant maturation. For example, studies have shown that artificially induced lung expansion by tracheal occlusion for periods ranging from 2 days to 4 weeks in the sheep fetus during either mid or late gestation promotes a decrease in AECII number[205–207] and a concomitant increase in AECI number[205] by promoting differentiation. These changes are associated with a decrease in SP-A, -B, and -C mRNA,[206,208,209] SP-A protein,[207,208] and saturated PC[207] in lung tissue of the sheep fetus. Conversely, fetal interventions leading to a decrease in fetal lung fluid either through lung liquid drainage[208] or the rapid release of the tracheal occlusion[210] can fully or partially restore surfactant protein gene expression. However, short-term tracheal occlusion (<7 days) late in gestation in rat and rabbit fetuses, induces rapid lung growth without the deleterious effect on AECII.[211,212] While these results may indicate that the duration and timing of tracheal occlusion are important in balancing the effects of tracheal occlusion on lung growth versus lung maturation,[212] it is also possible that there are species differences, likely related to the much shorter gestation period of rats (term, 22 days) and rabbits (term, 31 days) compared with sheep (term, 147 days). This possibility is supported by the fact that a short (2 days) period of tracheal occlusion late in ovine gestation (126 days) was sufficient to reduce both AECII number[205] and SP-A, -B, and -C gene expression.[213] Interestingly, lung distension in postnatal lambs induced by continuous positive airway pressure (CPAP) for 12 h. does not alter SP mRNA levels or DNA synthesis, suggesting that surfactant maturation and lung growth responses to tonic mechanical deformation are very different before and after birth.[213]

A series of ex vivo and in vitro studies has shown that tonic versus cyclical stretching or deformation have predominantly opposite effects on alveolar epithelial cells and hence the surfactant system. Based on early studies, Dobbs and Gutierrez proposed that "mechanical distension promotes expression of the type I cell phenotype and inhibits expression of the type II cell phenotype; contraction has the opposite effects."[214] Extrapolating from this and taking the above animal studies into account, it appears that continuous static stretch favors the formation of AECI,[205] whereas cyclic stretch favors the formation of AECII.[215] This is borne out by the demonstration that static distension of excised rat lungs for as little as 4 h decreases steady-state SP-A and SP-B mRNA levels.[216] In contrast, cyclical stretching of isolated cultured AECII, or in co-culture with fibroblasts[215] or with other lung cells[217] greatly stimulates gene expression of SP-B and -C,[215,217] expression of SP-B and -C protein as well as saturated PC.[215] Cyclic stretching of cells from the human pulmonary epithelial cell line, H441, in culture for 24 h. also increases SP-A and SP-B expression and enhances PC synthesis.[216]

Collectively, these data indicate that it is the dynamic mechano-deformation that occurs during FBM that is the likely critical stimulus for regulating surfactant protein and gene expression.[216] Although the mechanisms by which stretch is translated into molecular signals to modify gene expression in the alveolar epithelium are still relatively poorly understood, a number of molecular candidates are beginning to emerge. While a detailed discussion of these complex interactions is beyond the scope of this review, the reader is referred to several recent studies that show how the physical forces are mediated intrinsically by the cytoskeleton[218,219] and extrinsically by cell–cell and cell–matrix interactions.[220,221] Also significant is mesenchyme-epithelial signaling with various paracrine mediators such as cytokines,[222] enzymes,[223] lipids,[224] growth factors,[225,226] and hormones[227] released by neighboring cells such as fibroblasts and the extracellular matrix that stimulate alveolar epithelial cells.[228]

SURFACTANT DEFICIENCY LEADING TO NEONATAL RESPIRATORY DISTRESS SYNDROME

Much of the interest in pulmonary surfactant has focused on the consequences of the failure of the system, particularly in the preterm infant. Von Neergaard[229] first suggested that the alveolar collapse, observed in some newborns, may be a result of high surface tension in the lungs. Avery and Mead[230] performed autopsies on infants who had died with hyaline membrane disease; they defined the condition as infiltration of alveoli by cellular and fibrotic material, noted an absence of foam in the airways after death and concluded that a lack of surface-active film was partly responsible for the regions of alveolar collapse seen in respiratory distress syndrome (RDS). The link between impaired pulmonary surfactant function due to a deficiency and RDS was later confirmed independently by Clements et al.[231] and Pattle et al.,[232] and sparked worldwide interest from researchers in medicine, biology, and physics.[233]

Surfactant Deficiency Caused by Prematurity

If birth occurs before the complete maturation of the surfactant system, the pathological consequences are that the reabsorption of fetal lung fluid and the subsequent maintenance of a low work of breathing are compromised. If fetal lung fluid remains in the alveoli, the diffusion distance increases, which may cause hypoxemia, with or without hypercapnia. The low arteriolar PO_2 reduces the extent of arteriolar dilation and hence pulmonary hypertension may remain. In addition, the high work of inspiration and elevated elastic recoil of the lung results in a very high frequency, low volume breathing pattern, accompanied by a collapsed sternum, elevated diaphragm, and reduced lung volume. In the adult, the stiffness and natural tendency of the chest wall to spring outwards reduces the extent of the lung collapse. However, in the neonate, the ribs are highly compliant and cannot oppose the decreased lung compliance and hence atelectasis will result.[234,235]

In general, surfactant from infants with RDS has reduced levels of total phospholipid,[236,237] especially PC[238–240] and PG[241–244]; SP-A is also greatly reduced.[236,237] These changes result in reduced surface-activity in the lungs.[244,245] If the surfactant system is not activated or augmented soon after birth in very preterm infants, fibroblasts may invade the alveolar interstitium and form large quantities of hyaline cartilage and elastin, which thicken and stiffen the membrane. This alteration in the membranes can exacerbate RDS and lead to death.[246] Oxygen therapy and ventilation are used to treat RDS but do not address the primary cause, i.e., a lack of surfactant. Antenatal synthetic glucocorticoids such as dexamethasone or betamethasone reverse many of the symptoms, leading to improved surface activity, a reduction in the amount of inhibitory proteins and an increase in the level of SP-A and SP-D.[247,248] Numerous clinical trials have shown that the combination of antenatal glucocorticoid therapy in women with threatened preterm birth and the postnatal administration of intra-tracheal surfactants after birth are remarkably effective, leading to dramatic reductions in mortality.[155,249,250]

Genetic Causes of Surfactant Deficiency During Development

Although surfactant deficiency due to immaturity is the principal cause of RDS in preterm infants, the last 20 years have seen an ever-increasing understanding of the complex genetic basis to RDS. Despite advances in neonatal care that have markedly improved outcomes for preterm infants with RDS, there is a subset of full-term or near term infants that suffer a lethal form of RDS that responds poorly to therapy.[251,252] Moreover, different racial groups demonstrate differential improvement in mortality rates following the establishment of routine surfactant therapy.[253] These pathological and epidemiological findings contributed to the realization of a genetic basis to RDS. The identification in 1993 of an inherited deficiency of SP-B in a term newborn with lethal respiratory distress was the first demonstration of a specific genetic mechanism for acute inherited RDS.[35,251] Since that initial report, mutations in other surfactant-specific or surfactant-regulatory genes have been described,[254] which has allowed the identification of genetic causes for previously classified "idiopathic" interstitial lung disease.[252,255] Hence, it is the advances in therapy that have led to the realization of new disease mechanisms and hence have led to an improved understanding of basic surfactant biology.[250,254]

SP-B Deficiency

Inherited SP-B deficiency, which was also originally known as congenital alveolar proteinosis, is an autosomal recessive disorder that leads to lethal respiratory failure immediately after birth.[251] Heterozygous carriers are normally non-symptomatic. Although more than 40 separate mutations have been identified,[256] most specific to a certain family,[257] the most common null mutation, which accounts for 60–70% of mutant alleles, is due to a 2 base pair insertion in codon 121 of the SP-B gene, resulting in a frameshift mutation.[35,252,256,258] Generally, these mutations lead to a complete absence or loss of function of SP-B, which leads to a severe acute form of RDS in the term neonate that clinically and radiologically resembles RDS of prematurity.[252,259] Surfactant lipid metabolism is also affected, in that phospholipid to protein and PG/PI ratios of bronchoalveolar lavage fluid are decreased.[260–262] Lung ultrastructural abnormalities are also present with a reduced number of lamellar bodies and absence of tubular myelin.[263] While

affected infants can be supported temporarily with assisted ventilation and supplemental O_2,[261,262] exogenous surfactant replacement does not reverse the surfactant dysfunction.[264] Hence, without a lung transplant, infants die within the first few months of life.[252,256] However, there have been some cases of inherited SP-B deficiency that do not match either the genetic or pathological pattern of inherited SP-B deficiency; these include partial,[265] transient,[266] or milder SP-B deficiency leading to prolonged survival.[267] The outcomes for infants with SP-B deficiency who have received a lung transplant are similar to those seen in infants receiving transplants for other reasons.[262,268]

The SP-B gene targeted mouse also suffers from lethal respiratory failure and all homozygous null mutants (SP-B -/-) die within 30 min. of birth,[34] which is evidence of the crucial role of this protein in the successful transition to air breathing. AECII of SP-B -/- mice lack lamellar bodies and secreted tubular myelin, demonstrating that SP-B is crucial for these complexes.[269] Occasionally, a few loosely arranged lamellae are observed in their lamellar bodies; it is likely that these represent disorganized lamellar bodies, and that SP-B is responsible for the packaging of surfactant phospholipids into concentric lamellae.[269] The failure of proper lamellar body assembly due to SP-B deficiency has been reproduced in vitro using adenoviral-mediated antisense SP-B expression in a differentiating isolated AECII culture system.[270]

The abnormal cellular processing also appears to affect other surfactant components; for example, SP-C processing is abnormal in SP-B deficient humans and mice with AECII exhibiting an accumulation of intermediate pro-SP-C peptides.[261,269,271] This suggests that there is a block in normal SP-C processing as a result of SP-B deficiency. As the latter stages of normal SP-C processing have been localized to the lamellar bodies,[272,273] this disruption is readily understood. Impaired lamellar body processing could also result in abnormal phospholipid secretion, which may lead to delayed accumulation of phospholipid in amniotic fluid.[264] Both lavage and lung tissue of SP-B deficient infants have a decreased phospholipid to protein ratio, possibly a result of increased protein, as surfactant PL content is normal in SP-B -/- mice.[34] Furthermore, both lung tissue and lavage exhibit elevated PI and decreased PG concentrations, whereas disaturated PC is elevated in lung tissue.[261] However, phospholipid synthetic rates are normal.[34,261]

The neonatal phenotype of hereditary SP-B deficiency in humans and SP-B -/- mice has been confirmed using an adult compound transgenic mouse model in which the mouse SP-B cDNA is conditionally expressed under control of exogenous doxycycline, so that SP-B expression can be temporarily turned off and on.[274] In adult SP-B-/- mice, decreased SP-B was associated with low alveolar content of PG, accumulation of misprocessed SP-C proprotein in the air spaces, increased protein content in alveolar lavage fluid, and altered surfactant activity in vitro.[274] Consistent with surfactant dysfunction, hysteresis, maximal tidal volumes, and end-expiratory volumes were decreased. Reduction of alveolar SP-B content causes surfactant dysfunction and respiratory failure, which confirms the importance of SP-B for postnatal lung function.[274]

In SP-B -/- mice, both residual and total lung volumes and compliance are markedly decreased. Pressure-volume curves exhibit no hysteresis and significant atelectasis is evident in the lungs. In heterozygous mice (SP-B +/-), lung compliance is only slightly decreased.[34] As surfactant replacement therapy in SP-B deficient infants is rarely successful, perfluorocarbon ventilation has been trialled and has prolonged survival by improving lung expansion, lung compliance, and oxygenation.[34] However, lung function is not completely restored, as lungs virtually collapse at end-expiration, presumably because the perfluorocarbons only reduce surface tension to ~15mN/m[275] and not to <1 mN/m required for alveolar stability.[39] Hence, perfluorocarbons that have been used in RDS[276] may also be useful for the treatment of hereditary SP-B deficiency. However, perfluorocarbon also causes structural damage to epithelial, vascular, and alveolar structures in SP-B -/- and SP-B +/- mice, and may not be successful if used long term.[34]

Another therapeutic possibility receiving attention is gene therapy, an approach particularly suited to recessive single gene defect inherited disorders.[277] A gene therapy regimen for treating SP-B deficiency could overcome the limitations of surfactant replacement therapies or lung transplantation. Among the various viral and non-viral gene delivery tools available, only adenoviral vectors have been tested for delivering SP-B cDNA to the lungs of animal models. While this has been successful in the short term, the technique is hampered by host defence issues, resulting in clearance of the virus or virus-infected cells or by promoter inactivation in surviving cells. Furthermore, subsequent vector administration is affected by the host antibody response (reviewed by[277]).

ABCA3 Mutations

Following the discovery of hereditary SP-B deficiency, there were reports of familial cases of fatal respiratory disease in full-term infants with symptoms of surfactant deficiency, but in whom SP-B was present; this suggested the presence of additional genetic mechanisms.[278,279] One such mechanism includes mutations in member A3 of the ATP-binding cassette (ABCA3) lipid transport proteins, which cause a relatively rare autosomal recessive disorder that is presently the most common genetic cause of respiratory failure in full-term infants.[255]

Newborn infants with an ABCA3 mutation present with severe RDS requiring ventilatory support; the clinical findings are consistent with surfactant deficiency[279] and radiographic findings demonstrate diffuse alveolar disease and atelectasis.[255,279] Some infants also present with persistent pulmonary hypertension that is unresponsive to conventional therapies.[280] Respiratory failure occurs despite surfactant replacement and ventilatory support, and most infants die in the first months of life.[255]

The ABCA3 gene encodes a 1704 amino acid protein that belongs to the superfamily of ABC transporters,[281] which are highly conserved transmembrane proteins that use the hydrolysis of ATP to energise diverse biological systems and to transport a multitude of substrates across all cellular membranes.[282,283] Fourteen ABC transporter genes have been associated with distinct genetic diseases in humans.[279,282] More than 150 distinct ABCA3 mutations linked with severe neonatal lung disease have been identified. Of these, the vast majority cause neonatal lung disease and death within the first months of life, but there are also less frequently occurring and milder ABCA3 mutations that are associated with chronic lung disease that present in childhood.[255,284,285]

In the lung, the gene for ABCA3 is exclusively and highly expressed in AECII,[281,286] and the protein is localised to the limiting membrane of lamellar bodies,[286–288] which first pointed to the protein having an important role in surfactant metabolism.[279] ABCA3 selectively transports PC, sphingomyelin, and cholesterol to lamellar bodies. Transfection studies have shown that lipid trafficking by ABCA3 across lamellar body membranes is necessary for lamellar body biogenesis as a key step in the assembly of surfactant in AECII.[289,290] Using different site-directed mutant GFP-tagged proteins, it was confirmed that there are two categories of mutant ABCA3 proteins leading to type I and II surfactant deficiencies. Type I mutants are unable to sort appropriately into the limiting membrane of lamellar bodies, instead remaining within the endoplasmic reticulum; i.e., they have disturbed intracellular trafficking, whereas type II mutants have impaired ATP hydrolysis.[291]

ABCA3 mutations result in histological features that are characteristic of congenital alveolar proteinosis.[279,285] The respiratory epithelium undergoes cuboidal hyperplasia with remodelling and loss of alveoli, as well as accumulation of lipid-rich macrophages in the airspaces.[255] Bronchoalveolar lavage fluid from patients with ABCA3-related lung disease is deficient in both PG and PC and has greatly reduced surface activity.[260] The majority of lamellar bodies appear abnormally small in lungs of patients with ABCA3-related disease.[279,285] They also frequently contain eccentrically placed, electron-dense inclusions, and tightly packed phospholipid lamellae, likely representing lamellar bodies that lack appropriate lipid constituents. This morphological feature can be used in the tentative diagnosis of ABCA3-related disease.[255]

SP-C Mutations

The clinical manifestations of lung disease resulting from mutations in the gene coding for SP-C (SFTPC) are similar to those observed in ABCA3 deficient patients,[292,293] presenting as interstitial lung disease with onset usually in older children and adults.[252] Histological features include diffuse alveolar damage, interstitial thickening with mild lymphocytic inflammation, thickening of the alveolar septa, accumulation of alveolar macrophages and evidence of eosinophilic alveolar proteinosis, and regenerating alveolar epithelium lined by hyperplastic AECII.[255,256] Unlike lung disease due to SP-B or ABCA3 deficiencies, mutant SFTPC-induced disease is inherited in an autosomal dominant fashion.[252] However, it also frequently results from de novo mutations causing apparent sporadic disease[294,295]; a mutation on one allele is sufficient to cause disease.[296,297]

Approximately 40 different missense, splice, or frameshift SFTPC mutations have been documented as leading to either familial or sporadic interstitial lung disease.[255] They are all predicted to alter the coding sequence of proSP-C as opposed to being null mutations,[252] which leads to two distinct protein expression phenotypes (mistargeted vs. aggregating), each capable of triggering a characteristic subset of cellular responses.[255,298] The mistargeted proSP-C mutants disrupt the endosome/lysosome system through abnormal accumulation, producing both organellar dysfunction and cytotoxicity.[298,299] In contrast, missense and splicing mutations reside predominantly within the SFTPC BRICHOS domain of proSP-C that is found in the carboxy-terminal and has homology to a group of proteins that are linked to familial neurodegenerative disease. These mutations produce proprotein misfolding or aberrant processing and intracellular aggregate formation.[298,300] Hence, lung disease is not necessarily caused by a deficiency of SP-C, but is rather the result of a toxicity mechanism due to the accumulating misfolded or mistargeted protein.[252,301]

The accumulation of misfolded proSP-C within the endoplasmic reticulum and Golgi apparatus in the AECII leads to a condition known as endoplasmic reticulum or ER stress, which induces an ER-to-nucleus signaling pathway called unfolded protein response (UPR), which can eventually trigger cell death pathways.[255,302] For example, in vitro expression of the SP-C BRICHOS mutant, SP-C(Delta exon4), in both transfected A549 and HEK293 cells stimulates a broad ER stress response and hence apoptosis through multiple UPR signalling pathways.[298] Targeted expression of the same SP-C mutant in AECII of transgenic mice results in a dose-dependent cytotoxicity leading to disruption of lung morphogenesis and respiratory failure at birth.[303] Moreover, the infection of mutant transfected cells with respiratory syncytial virus exacerbates the cytotoxicity and results in pronounced accumulation of mutant proprotein followed by extensive cell death associated with proteasome inhibition.[304]

Hence, transfected cells that have become adapted to chronic ER stress imposed by misfolded SP-C demonstrate increased susceptibility to an environmental stress, specifically, viral infection.[304] SP-C deficient mice are also more susceptible to respiratory syncytial virus infection,[305] as well as bacterial infection,[306] demonstrating an enhanced pulmonary pro-inflammatory response,[305,306] and reduced phagocytic activity of alveolar macrophages.[306] Collectively, these results may explain the wide variability in the age of onset of interstitial lung disease in patients with SFTPC mutations, suggesting a secondary environmental insult that may ultimately overwhelm the homeostatic cytoprotective response[304] and possibly the innate immune response of the lung.

SFTPC mutations also lead to abnormalities of lamellar bodies with a mixture of normal lamellar bodies containing electron-dense vesicles, intracellular membranous aggregates, or disorganized lamellar bodies.[256] Occasionally larger composite bodies containing two or more smaller lamellar bodies are found.[256] This phenotype is also reflected in SP-C deficient mice that demonstrate large intact lamellar bodies and tubular myelin.[36,256,307]

Perturbations of surfactant composition and function may also play a role in the pathogenesis of lung disease in patients with SFTPC mutations,[256] due to SP-C's role in reuptake and catabolism of secreted surfactant. Different mutations show a combination of typical histopathology patterns consisting of both non-specific interstitial pneumonia and pulmonary alveolar proteinosis.[294,308] Patients demonstrate increased alveolar total phospholipid, but a lack of PG,[308] as well as an intra-alveolar accumulation of SP-A,[294,308] precursors of SP-B, mature SP-B, aberrantly processed proSP-C, as well as mono- and dimeric SP-C.[294] In a different mutation, airway surfactant demonstrates little or no mature SP-B or SP-C, but an increase in SP-A content. These changes are associated with an increased minimum surface tension of 20mN/m, relative to the normal value of <5mN/m.[309] Cultured MLE-12 alveolar epithelial cells stably expressing the mutant SP-C(I73T) demonstrate decreased intracellular PC level and increased lyso-PC level without appreciable changes in other phospholipids.[310]

While SFTPC mutations are not generally lethal, unlike those of SP-B in hereditary SP-B deficiency, they do lead to a complex suite of lung diseases in neonates, infants, and adults. Moreover, they are associated with more variable, and hence, unpredictable short- and long-term outcomes that present significant clinical and diagnostic challenges.[309] An increased understanding of the molecular mechanisms in this suite of sublethal interstitial lung diseases provides hope for future therapeutic tools.

NKX2.1 Mutations

Defects of the NKX2.1 gene, encoding thyroid transcription factor-1, cause brain-thyroid-lung syndrome,[311,312] reflecting the tissue expression sites of this gene. The syndrome is characterized by benign hereditary chorea, congenital hypothyroidism, and respiratory disease.[313] However, the classical triad of organ system involvement is not always present and the severity of the chorea, respiratory symptoms, and hypothyroidism vary widely.[312,314,315] Although the majority of affected patients display neurological and/or thyroidal problems, lung disease is less frequent, but is responsible for a considerable mortality.[315]

The respiratory phenotypes resulting from deletions or complete loss-of-function mutations on one copy (haplo-insufficiency) of NKX2.1 include severe RDS and interstitial lung disease.[252,314,316,317] The incidence and prevalence of lung disease due to NKX2.1 haplo-insufficiency are unknown.[252] The majority of reported mutations apparently occur de novo causing sporadic disease. Rare familial cases of pulmonary disease have also been reported.[252] In addition to complete deletions of NKX2.1,[317] the range of mutations includes splice site and missense mutations within the homeodomain.[315] Little is known about the natural history of the lung disease due to NKX2.1 haplo-insufficiency.[252] Fatal lung disease in infancy has been reported,[313] but older patients with relatively milder or no lung diseases are also recognized.[315] It is unknown whether some mutations may result in milder disease.[252]

NKX2.1 expression is evident in human fetal lung epithelium from 11 weeks gestation,[315,318,319] reflecting its importance in early lung development and differentiation. NKX2.1 is an important transcription factor, regulating the expression of multiple surfactant related genes, including those for ABCA3 and SP-A to -D in the lung epithelium.[171,320,321] Surfactant abnormalities documented in one particular NKX2.1 mutation[313] included an abnormally low amount of SP-C in relation to SP-B in bronchoalveolar lavage fluid as well as low levels of surfactant phospholipids. Hence, this disease results in a disturbance of both surfactant protein and lipid homeostasis as a consequence of NKX2.1 haplo-insufficiency. Not surprisingly, the majority of NKX2.1 mutations result in congenital surfactant deficiency syndrome,[315] and lead to lethal respiratory failure of the newborn due to disruption of surfactant homeostasis.[313] A minority of patients develop recurrent pulmonary infections of mild-to-severe degree without RDS at term.[315]

In summary, due to the similarity in symptoms among all of the genetic diseases described earlier, only genetic analysis will reveal the definitive cause.[256] Disease that persists after the first week of life or is especially severe should be considered for early testing. With neonatal onset of disease, ABCA3 and SFTPB analyses should be considered first. If there is evidence for hypothyroidism, analysis for NKX2.1 is indicated. If testing for these genes is negative, and the lung disease does not resolve, or if there is a family history of lung disease inherited in a dominant pattern, then SFTPC mutational analysis should be pursued.[252]

CONCLUSIONS

The pulmonary surfactant system is complex in terms of its composition and function, and neither the specific functions nor the exact biophysical interactions of its different components are fully understood. During development, numerous regulatory elements interact to affect surfactant function and maturity. The surfactant system develops in utero during late gestation, and it is crucial that this system is established and operational at birth. Dysfunction of the surfactant system at this time, either through preterm birth or genetic causes, may lead to RDS.

Although the use of antenatal glucocorticoids and the administration of exogenous surfactant to the neonate have been remarkably successful in the treatment of RDS, some cases are resistant to therapy. This has led to the understanding that there is a genetic basis to many cases of acute neonatal RDS. The identification of a number of specific gene defects relating to the pulmonary surfactant system has indicated a specific cause for many lung diseases of infancy, childhood, or adulthood that were previously classified as idiopathic. These discoveries have opened up new areas of research, including the search for genetic markers of risk for RDS in infancy and the search for new therapeutic targets for interstitial lung disease, including the development of gene therapy strategies.

ACKNOWLEDGEMENTS

This work was supported by the Australian Research Council, National Health & Medical Research Council and the Heart Foundation of Australia.

REFERENCES

1. Perez-Gil J, Weaver TE. Pulmonary surfactant pathophysiology: current models and open questions. *Physiology* 2010;**25**:132–41.
2. Osanai K, Mason RJ, Voelker DR. Pulmonary surfactant phosphatidylcholine transport bypasses the brefeldin A sensitive compartment of alveolar type II cells. *Biochim Biophys Acta Mol Cell Biol Lipids* 2001;**1531**:222–9.
3. Osanai K, Tsuchihara C, Hatta R, Oikawa T, Tsuchihara K, Iguchi M, Seki T, Takahashi M, Huang J, Toga H. Pulmonary surfactant transport in alveolar type II cells. *Respirology* 2006;**11**:S70–3.
4. Goss V, Hunt AN, Postle AD. Regulation of lung surfactant phospholipid synthesis and metabolism. *Biochim Biophys Acta-Mol Cell Biol Lipids* 2013;**1831**:448–58.
5. Weaver TE. Synthesis, processsing and secretion of surfactant proteins B and C. *Biochim Biophys Acta* 1998;**1408**:173–9.
6. Hawgood S, Poulain FR. The pulmonary collectins and surfactant metabolism. *Annu Rev Physiol* 2001;**63**:495–519.
7. Voorhout WF, Veenendaal T, Kuroki Y, Ogasawara Y, Vangolde LMG, Geuze HJ. Immunocytochemical localization of surfactant protein-D (SP-D) in type-II cells, Clara cells, and alveolar macrophages of rat lung. *J Histochem Cytochem* 1992;**40**:1589–97.

8. Rooney SA. Regulation of surfactant secretion. *Comp Biochem Physiol A* 2001;**129**:233–43.

9. Goerke J. Pulmonary surfactant: Functions and molecular composition. *Biochim Biophys Acta* 1998;**1408**:79–89.

10. Veldhuizen RAW, Nag K, Orgeig S, Possmayer F. The role of lipids in pulmonary surfactant. *Biochim Biophys Acta Mol Basis Dis* 1998;**1408**:90–108.

11. Lang CJ, Postle AD, Orgeig S, Possmayer F, Bernhard W, Panda AK, et al. Dipalmitoylphosphatidylcholine is not the major surfactant phospholipid species in all mammals. *Am J Physiol Regul Integr Comp Physiol* 2005;**289**:R1426–39.

12. Orgeig S, Morrison JL, Daniels CB. Prenatal development of the pulmonary surfactant system and the influence of hypoxia. *Respir Physiol Neurobiol* 2011;**178**:129–45.

13. Possmayer F. Physicochemical aspects of pulmonary surfactant. In: Polin RA, Fox WW, Abman SH, editors. *Fetal and Neonatal Physiology*. Philadelphia: W. B. Saunders Company; 2004. pp. 1014–34.

14. Possmayer F. Biophysical activities of pulmonary surfactant. In: Polin RA, Fox WW, editors. *Fetal and Neonatal Physiology*. Philadelphia: WB Saunders Co.; 1991. pp. 459–962.

15. Meban C. Effect of lipids and other substances on the adsorption of dipalmitoylphosphatidylcholine. *Proc R Soc Med* 1981;**15**:1029–31.

16. Perez-Gil J, Casals C, Marsh D. Interactions of hydrophobic lung surfactant proteins SP-B and SP-C with dipalmitoylphosphatidylcholine and dipalmitoylphosphatidylglycerol bilayers studied by electron spin resonance spectroscopy. *Biochemistry* 1995;**34**:3964–71.

17. Nag K, Taneva SG, Perez-Gil J, Cruz A, Keough KM. Combinations of fluorescently labeled pulmonary surfactant proteins SP-B and SP-C in phospholipid films. *Biophys J* 1997;**72**:2638–50.

18. Perez-Gil J. Structure of pulmonary surfactant membranes and films: the role of proteins and lipid-protein interactions. *Biochim Biophys Acta Biomembr* 2008;**1778**:1676–95.

19. Kuronuma K, Mitsuzawa H, Takeda K, Nishitani C, Chan ED, Kuroki Y, et al. Anionic pulmonary surfactant phospholipids inhibit inflammatory responses from alveolar macrophages and U937 cells by binding the lipopolysaccharide-interacting proteins CD14 and MD-2. *J Biol Chem* 2009;**284**:25488–500.

20. Numata M, Chu HW, Dakhama A, Voelker DR. Pulmonary surfactant phosphatidylglycerol inhibits respiratory syncytial virus–induced inflammation and infection. *Proc Natl Acad Sci* 2010;**107**:320–5.

21. Veldhuizen EJ, Haagsman HP. Role of pulmonary surfactant components in surface film formation and dynamics. *Biochim Biophys Acta* 2000;**1467**:255–70.

22. Orgeig S, Daniels CB. The roles of cholesterol in pulmonary surfactant: insights from comparative and evolutionary studies. *Comp Biochem Physiol A Mol Integr Physiol* 2001;**129**:75–89.

23. Batenburg JJ. Surfactant phospholipids: synthesis and storage. *Am J Phys* 1992;**262**:L367–85.

24. Presti FT, Pace RJ, Chan SI. Cholesterol phospholipid interaction in membranes. 2. Stoichiometry and molecular packing of cholesterol-rich domains. *Biochemistry* 1982;**21**:3831–5.

25. Orgeig S, Hiemstra PS, Veldhuizen EJA, Casals C, Clark HW, Haczku A, et al. Recent advances in alveolar biology: evolution and function of alveolar proteins. *Respir Physiol Neurobiol* 2010;**173**:S43–54.

26. Haagsman HP, Diemel RV. Surfactant-associated proteins: functions and structural variation. *Comp Biochem Physiol A Mol Integr Physiol* 2001;**129**:91–108.

27. Wright JR, Wager RE, Hawgood S, Dobbs LG, Clements JA. Surfactant apoprotein Mr=26000-36000 enhances uptake of liposomes by type II cells. *J Biol Chem* 1987;**262**:2888–94.

28. Tenner AJ, Robinson SL, Borchelt J, Wright JR. Human pulmonary surfactant protein (SP-A), a protein structurally homologous to C1q, can enhance FcR-and CR1- mediated phagocytosis. *J Biol Chem* 1989;**264**:13923–8.

29. Casals C. Role of surfactant protein A (SP-A)/lipid interactions for SP-A functions in the lung. *Pediatr Pathol Mol Med* 2001;**20**: 249–68.

30. Korfhagen TR, Bruno MD, Ross GF, Huelsman KM, Ikegami M, Jobe AH, et al. Altered surfactant function and structure in SP-A gene targeted mice. *Proc Natl Acad Sci U S A* 1996;**93**:9594–9.

31. LeVine AM, Kurak KE, Bruno MD, Stark JM, Whitsett JA, Korfhagen TR. Surfactant protein-A-deficient mice are susceptible to Pseudomonas aeruginosa infection. *Am J Respir Cell Mol Biol* 1998;**19**:700–8.

32. Botas C, Poulain F, Akiyama J, Brown C, Allen L, Goerke J, et al. Altered surfactant homeostasis and alveolar type II cell morphology in mice lacking surfactant protein D. *Proc Natl Acad Sci U S A* 1998;**95**:11869–74.

33. Ochs M, Knudsen L, Allen L, Stumbaugh A, Levitt S, Nyengaard JR, et al. GM-CSF mediates alveolar epithelial type II cell changes, but not emphysema-like pathology, in SP-D-deficient mice. *Am J Physiol Lung Cell Mol Physiol* 2004;**287**:L1333–41.

34. Tokieda K, Whitsett JA, Clark JC, Weaver TE, Ikeda K, McConnell KB, et al. Pulmonary dysfunction in neonatal SP-B deficient mice. *Am J Phys* 1997;**17**:L875–82.

35. Nogee LM, Garnier G, Dietz HC, Singer L, Murphy AM, deMello DE, et al. A mutation in the surfactant protein B gene responsible for fatal neonatal respiratory disease in multiple kindreds. *J Clin Invest* 1994;**93**:1860–3.

36. Glasser SW, Burhans MS, Korfhagen TR, Na CL, Sly PD, Ross GF, et al. Altered stability of pulmonary surfactant in SP-C-deficient mice. *Proc Natl Acad Sci U S A* 2001;**98**:6366–71.

37. Weaver TE, Conkright JJ. Functions of surfactant proteins B and C. *Annu Rev Physiol* 2001;**63**:555–78.

38. Daniels CB, Lopatko OV, Orgeig S. Evolution of surface activity related functions of vertebrate pulmonary surfactant. *Dev Pharmacol Ther* 1998;**25**:716–21.

39. Schürch S, Bachofen H, Goerke J, Green F. Surface properties of rat pulmonary surfactant studied with the captive bubble method: adsorption, hysteresis, stability. *Biochim Biophys Acta* 1992;**1103**:127–36.

40. Schürch S, Bachofen H, Possmayer F. Surface activity in situ, in vivo, and in the captive bubble surfactometer. *Comp Biochem Physiol A* 2001;**129**:195–207.

41. Goerke J, Clements JA. Alveolar surface tension and lung surfactant. In: Macklem PT, Mead J, editors. *Handbook of Physiology, Section 3: The Respiratory System. Vol. III; Mechanics of Breathing, Part I*, Washington, D.C: American Physiological Society; 1985. pp. 247–60.

42. Crane JM, Hall SB. Rapid compression transforms interfacial monolayers of pulmonary surfactant. *Biophys J* 2001;**80**:1863–72.

43. Zuo YY, Veldhuizen RAW, Neumann AW, Petersen NO, Possmayer F. Current perspectives in pulmonary surfactant – Inhibition, enhancement and evaluation. *Biochim Biophys Acta Biomembr* 2008;**1778**:1947–77.

44. Smith EC, Crane JM, Laderas TG, Hall SB. Metastability of a super-compressed fluid monolayer. *Biophys J* 2003;**85**:3048–57.

45. Bernhard W, Hoffmann S, Dombrowsky H, Rau GA, Kamlage A, Kappler M, et al. Phosphatidylcholine molecular species in lung surfactant: composition in relation to respiratory rate and lung development. *Am J Respir Cell Mol Biol* 2001;**25**:725–31.

46. Pynn CJ, Picardi MV, Nicholson T, Wistuba D, Poets CF, Schleicher E, et al. Myristate is selectively incorporated into surfactant and decreases dipalmitoylphosphatidylcholine without functional impairment. *Am J Physiol Regul Integr Comp Physiol* 2010;**299**:R1306–16.

47. Slocombe NC, Codd JR, Wood PG, Orgeig S, Daniels CB. The effect of alterations in activity and body temperature on the pulmonary surfactant system in the lesser long-eared bat Nyctophilus geoffroyi. *J Exp Biol* 2000;**203**:2429–35.

48. Lopatko OV, Orgeig S, Palmer D, Schürch S, Daniels CB. Alterations in pulmonary surfactant after rapid arousal from torpor in the marsupial Sminthopsis crassicaudata. *J Appl Phys* 1999;**86**:1959–70.

49. Orgeig S, Daniels CB, Lopatko OV, Langman C. Effect of torpor on the composition and function of pulmonary surfactant in the heterothermic mammal (Sminthopsis crassicaudata). In: Geiser F, Hulbert AJ, Nicol SC, editors. *Adaptations to the Cold: Tenth International Hibernation Symposium*. Armidale: University of New England Press; 1996. pp. 223–32.

50. Langman C, Orgeig S, Daniels CB. Alterations in composition and function of surfactant associated with torpor in Sminthopsis crassicaudata. *Am J Phys* 1996;**271**:R437–45.

51. Suri LNM, Cruz A, Veldhuizen RAW, Staples JF, Possmayer F, Orgeig S, et al. Adaptations to hibernation in lung surfactant composition of 13-lined ground squirrels influence surfactant lipid phase segregation properties. *Biochim Biophys Acta Biomembr* 2013;**1828**:1707–14.

52. Suri LNM, McCaig L, Picardi MV, Ospina OL, Veldhuizen RAW, Staples JF, et al. Adaptation to low body temperature influences pulmonary surfactant composition thereby increasing fluidity while maintaining appropriately ordered membrane structure and surface activity. *Biochim Biophys Acta Biomembr* 2012;**1818**:1581–9.

53. Orgeig S, Bernhard W, Biswas SC, Daniels CB, Hall SB, Hetz SK, et al. The anatomy, physics, and physiology of gas exchange surfaces: is there a universal function for pulmonary surfactant in animal respiratory structures? *Integr Comp Biol* 2007;**47**:610–27.

54. Codd JR, Orgeig S, Daniels CB, Schurch S. Alterations in surface activity of pulmonary surfactant in Gould's wattled bat during rapid arousal from torpor. *Biochim Biophys Res Commun* 2003;**308**:463–8.

55. Lopatko OV, Orgeig S, Daniels CB, Palmer D. Alterations in the surface properties of lung surfactant in the torpid marsupial Sminthopsis crassicaudata. *J Appl Phys* 1998;**84**:146–56.

56. Bastacky J, Lee CYC, Goerke J, Koushafar H, Yager D, Kenaga L, et al. Alveolar lining layer is thin and continuous: low temperature scanning electron microscopy of rat lung. *J Appl Phys* 1995;**79**:1615–28.

57. Bachofen H, Hildebrandt J, Bachofen M. Pressure-volume curves of air- and liquid-filled excised lungs - surface tension in situ. *J Appl Physiol* 1970;**29**:422–31.

58. Schürch S. Surface tension at low lung volumes: dependence on time and alveolar size. *Respir Physiol* 1982;**48**:339–55.

59. Bachofen H, Schürch S, Urbinelli M, Weibel ER. Relations among alveolar surface tension, surface area, volume and recoil pressure. *J Appl Physiol* 1987;**62**:1878–87.

60. Bachofen H, Schürch S. Alveolar surface forces and lung architecture. *Comp Biochem Physiol A* 2001;**129**:183–93.

61. Sanderson RJ, Paul GW, Vatter AE, Filley GF. Morphological and physical basis for lung surfactant action. *Proc R Soc Med* 1976;**27**:379–92.

62. Daniels CB, Orgeig S. The comparative biology of pulmonary surfactant: past, present and future. *Comp Biochem Physiol A* 2001;**129**:9–36.

63. Crapo JD. New concepts in the formation of pulmonary edema. *Am Rev Respir Dis* 1993;**147**:790–2.

64. Gil J, Bachofen H, Gehr P, Weibel ER. Alveolar volume-surface area relation in air- and saline-filled lungs fixed by vascular perfusion. *J Appl Physiol* 1979;**47**:990–1001.

65. Guyton AC, Moffat DS, Adair TH. Role of alveolar surface tension in transepithelial movement of fluid. In: Robertson B, Van Golde LMG, Batenburg JJ, editors. *Pulmonary Surfactant*. Amsterdam: Elsevier Science Publishers; 1984. pp. 171–85.

66. Enhorning G. Pulmonary surfactant function studied with the pulsating bubble surfactometer (PBS) and the capillary surfactometer (CS). *Comp Biochem Physiol A* 2001;**129**:221–6.

67. Rensch H, von Seefeld H, Gebhardt KF, Renzow D, Sell PJ. Stop and go particle transport in the peripheral airways? *Respiration* 1983;**44**:346–50.

68. Gil J, Weibel ER. Extracellular lining of bronchioles after perfusion-fixation of rat lungs for electron microscopy. *Anat Rec* 1971;**169**:185–200.

69. Yoneda K. Mucous blanket of rat bronchus. An ultrastructural study. *Am Rev Respir Dis* 1976;**114**:837–42.

70. Morgenroth K, Bolz J. Morphological features of the interaction between mucus and surfactant on the bronchial mucosa. *Respiration* 1985;**47**:225–31.

71. Kilburn KH. Alveolar clearance of particles. A bullfrog lung model. *Arch Environ Health* 1969;**18**:556–63.

72. Allegra L, Bossi R, Braga P. Influence of surfactant on mucociliary transport. *Eur J Respir Dis* 1985;**142**:71–6.

73. Kakuta Y, Sasaki H, Takishima T. Effect of artificial surfactant on ciliary beat frequency in guinea pig trachea. *Respir Physiol* 1991;**83**:313–22.

74. Andreeva AV, Kutuzov MA, Voyno-Yasenetskaya TA. Regulation of surfactant secretion in alveolar type II cells. *Am J Physiol Lung Cell Mol Physiol* 2007;**293**:L259–71.

75. Chander A, Fisher AB. Regulation of lung surfactant secretion. *Am J Phys* 1990;**258**:L241–53.

76. Mason RJ, Voelker DR. Regulatory mechanisms of surfactant secretion. *Biochim Biophys Acta* 1998;**1408**:226–40.

77. Wright JR, Dobbs LG. Regulation of pulmonary surfactant secretion and clearance. *Annu Rev Physiol* 1991;**53**:395–414.

78. Rooney SE, Young SL, Mendelson CR. Molecular and cellular processing of lung surfactant. *FASEB J* 1994;**8**:957–67.

79. Griese M, Gobran LI, Rooney SA. Signal-transduction mechanisms of ATP-stimulated phosphatidylcholine secretion in rat type-II pneumocytes - interactions between ATP and other surfactant secretagogues. *Biochim Biophys Acta* 1993;**1167**:85–93.

80. Rooney SA. Regulation of surfactant-associated phospholipid synthesis and secretion. In: Polin RA, Fox WW, Abman SH, editors. *Fetal and Neonatal Physiology*. Philadelphia: Saunders; 2004. pp. 1042–54.

81. Dietl P, Haller T. Exocytosis of lung surfactant: from the secretory vesicle to the air-liquid interface. *Annu Rev Physiol* 2005;**67**:595–621.

82. Frick M, Bertocchi C, Jennings P, Haller T, Mair N, Singer W, et al. Ca2+ entry is essential for cell strain-induced lamellar body fusion in isolated rat type II pneumocytes. *Am J Obstet Gynecol* 2004;**286**:L210–20.

83. Chander A, Gerelsaikhan T, Vasa PK, Holbrook K. Annexin A7 trafficking to alveolar type II cell surface: possible roles for protein insertion into membranes and lamellar body secretion. *Biochim Biophys Acta Mol Cell Res* 2013;**1833**:1244–55.

84. Gerelsaikhan T, Chen XL, Chander A. Secretagogues of lung surfactant increase annexin A7 localization with ABCA3 in alveolar type II cells. *Biochim Biophys Acta Mol Cell Res* 2011;**1813**:2017–25.

85. Gerelsaikhan T, Vasa PK, Chander A. Annexin A7 and SNAP23 interactions in alveolar type II cells and in vitro: a role for Ca2+ and PKC. *Biochim Biophys Acta Mol Cell Res* 2012;**1823**:1796–806.

86. Marino P, Rooney S. The effect of labor on surfactant secretion in newborn rabbit lung slices. *Biochim Biophys Acta* 1981;**664**:389–96.

87. Dobbs LG, Mason RJ. Pulmonary alveolar type II cells isolated from rats. *J Clin Invest* 1979;**63**:378–87.

88. Brown LA, Longmore WJ. Adrenergic and cholinergic regulation of lung surfactant secretion in the isolated perfused rat lung and in the alveolar type II cell in culture. *J Biol Chem* 1981;**256**:66–72.

89. Mettler NR, Gray ME, Schuffman S, Lequire VS. Beta-adrenergic induced synthesis and secretion of phosphatidylcholine by isolated pulmonary alveolar type II cells. *Lab Invest* 1981;**45**:575–86.

90. Gobran LI, Rooney SA. Pulmonary surfactant secretion in briefly cultured mouse type II cells. *Am J Physiol Lung Cell Molr Physiol* 2004;**286**:L331–6.

91. Gobran LI, Rooney SA. Regulation of SP-B and SP-C secretion in rat type II cells in primary culture. *Am J Physiol Lung Cell Molr Physiol* 2001;**281**:L1413–9.

92. Smith DM, Griffin LS. Stereologic analysis of the in vivo alveolar type II cell response to isoproterenol or saline administration. *Histol Histopathol* 1987;**2**:95–106.

93. Keeney SK, Oelberg DG. Alpha-adrenergic and muscarinic receptors in adult and neonatal rat type II pneumocytes. *Lung* 1993;**171**:355–66.

94. Brown LS, Longmore WJ. Adrenergic and cholinergic regulation of lung surfactant secretion in the isolated perfused rat lung and in the alveolar type II cell in culture. *J Biol Chem* 1981;**256**:66–72.

95. Robinson PC, Voelker DR, Mason RJ. Isolation and culture of human alveolar type II epithelial cells. Characterization of their phospholipid secretion. *Am Rev Respir Dis* 1984;**130**:1156–60.

96. Massaro D, Clerch L, Massaro GD. Surfactant secretion: evidence that cholinergic stimulation of secretion is indirect. *Am J Phys* 1982;**243**:C39 45.

97. Oyarzun MJ, Clements JA. Ventilatory and cholinergic control of pulmonary surfactant in the rabbit. *J Appl Physiol* 1977;**43**:39–45.

98. Abdellatif MM, Hollingsworth M. Effect of oxotremorine and epinephrine on lung surfactant secretion in neonatal rabbits. *Pediatr Resh* 1980;**14**:916–20.

99. Wood PG, Lopatko OV, Orgeig S, Joss JM, Smits AW, Daniels CB. Control of pulmonary surfactant secretion: an evolutionary perspective. *Am J Physiol* 2000;**278**:R611–9.

100. Ormond CJ, Orgeig S, Daniels CB. Neurochemical and thermal control of surfactant secretion by alveolar type II cells isolated from the marsupial Sminthopsis crassicaudata. *J Comp Physiol B* 2001;**171**:223–30.

101. Wood PG, Lopatko OV, Orgeig S, Codd JR, Daniels CB. Control of pulmonary surfactant secretion from type II pneumocytes isolated from the lizard, Pogona vitticeps. *Am J Phys* 1999;**277**:R1705–11.

102. Nicholas TE, Power JHT, Barr HA. Surfactant homeostasis in the rat lung during swimming exercise. *J Appl Phys* 1982;**53**:1521–8.

103. Nicholas TE, Power JHT, Barr HA. The pulmonary consequences of a deep breath. *Respir Physiol* 1982;**49**:315–24.

104. Wirtz H, Schmidt M. Ventilation and secretion of pulmonary surfactant. *Clin Invest* 1992;**70**:3–13.

105. Wirtz HRW, Dobbs LG. Calcium mobilization and exocytosis after one mechanical stretch of lung epithelial cells. *Science* 1990;**250**:1266–9.

106. Edwards YS, Sutherland LM, Power JH, Nicholas TE, Murray AW. Cyclic stretch induces both apoptosis and secretion in rat alveolar type II cells. *FEBS Lett* 1999;**448**:127–30.

107. Ashino Y, Ying X, Dobbs LG, Bhattacharya J. [Ca2+]i oscillations regulate type II cell exocytosis in the pulmonary alveolus. *Am J Physiol Lung Cell Mol Physiol* 2000;**279**:L5–13.

108. Isakson BE, Seedorf GJ, Lubman RL, Evans WH, Boitano S. Cell–cell communication in heterocellular cultures of alveolar epithelial cells. *Am J Respir Cell Mol Biol* 2003;**29**:552–61.

109. Mishra A. New insights of P2X7 receptor signaling pathway in alveolar functions. *Hum Mutat* 2013;**20**.

110. Mishra A, Chintagari NR, Guo YJ, Weng TT, Su LJ, Liu L. Purinergic P2X(7) receptor regulates lung surfactant secretion in a paracrine manner. *J Cell Sci* 2011;**124**:657–68.

111. Hallman M, Glumoff V, Ramet M. Surfactant in respiratory distress syndrome and lung injury. *Comp Biochem Physiol A Mol Integr Physiol* 2001;**129**:287–94.

112. Hallman M, Bry K, Hoppu K, Lappi M, Pohjavuori M. Inositol supplementation in premature-infants with respiratory-distress syndrome. *N Engl J Med* 1992;**326**:1233–9.

113. Cockshutt AM, Possmayer F. Metabolism of surfactant lipids and proteins in the developing lung. In: Robertson B, van Golde LMG, Batenburg JJ, editors. *Pulmonary Surfactant: From Molecular Biology to Clinical Practice*. Amsterdam: Elsevier Science Publishers; 1992. pp. 339–77.

114. Benson BJ, Kitterman JA, Clements JA, Mescher EJ, Tooley WH. Changes in phospholipid composition of lung surfactant during development in the fetal lamb. *Biochim Biophys Acta* 1983;**753**:83–8.

115. Katyal SL, Estes LW, Lombardi B. Method for the isolation of surfactant from homogenates and lavages of lung of adult, newborn, and fetal rats. *Lab Invest* 1977;**36**:585–92.

116. Maniscalco WM, Wilson CM, Gross I, Gobran L, Rooney SA, Warshaw JB. Development of glycogen and phospholipid metabolism in fetal and newborn rat lung. *Biochim Biophys Acta* 1978;**530**:333–46.

117. Torday JS, Nielson HC. Surfactant phospholipid ontogeny in fetal rabbit lung lavage and amniotic fluid. *Biol Neonate* 1981;**39**:266–71.

118. Gluck L, Kulovich M. Lecithin-sphingomyelin ratios in amniotic fluid in normal and abnormal pregnancy. *Am J Obstet Gynecol* 1973;**115**:539–46.

119. Hallman M, Kulovich M, Kirkpatrick E, Sugarman RG, Gluck L. Phosphatidylinositol and phosphatidylglycerol in amniotic fluid: indices of lung maturity. *Am J Obstet Gynecol* 1976;**125**:613–7.

120. Ballard PL, Hawgood S, Liley HG, Wellenstein G, Gonzales LW, Benson B, et al. Regulation of pulmonary surfactant apoprotein SP 28-36 gene in fetal human lung. *Proc Natl Acad Sci U S A* 1986;**83**:9527–31.

121. Khoor A, Gray ME, Hull WM, Whitsett JA, Stahlman MT. Developmental expression of SP-A and SP-A mRNA in the proximal and distal respiratory epithelium in the human fetus and newborn. *J Clin Endocrinol Metab* 1993;**41**:1311–9.

122. Mendelson CR, Boggaram V. Hormonal control of the surfactant system in fetal lung. *Annu Rev Physiol* 1991;**53**:415–40.

123. Snyder JM, Kwun JE, Obrien JA, Rosenfeld CR, Odom MJ. The concentration of the 35-kDA surfactant apoprotein in amniotic-fluid from normal and diabetic pregnancies. *Pediatr Res* 1988;**24**:728–34.

124. Pryhuber GS, Hull WM, Fink I, McMahon MJ, Whitsett JA. Ontogeny of surfactant proteins A and B in human amniotic fluid as indices of fetal lung maturity. *Pediatr Res* 1991;**30**:597–605.

125. Khoor A, Stahlman MT, Gray ME, Whitsett JA. Temporal-spatial distribution of SP-B and SP-C proteins and mRNAs in developing respiratory epithelium of human lung. *Lab Invest* 1994;**42**(9):1187–99.

126. Dulkerian SJ, Gonzales LW, Ning Y, Ballard PL. Regulation of surfactant protein D in human fetal lung. *Am J Respir Cell Mol Biol* 1996;**15**:781–6.

127. Ogasawara Y, Kuroki Y, Shiratoari M, Shimizu H, Miyamura K, Akino T. Ontogeny of surfactant apoprotein D, SP-D, in the rat lung. *Biochim Biophys Acta* 1991;**1083**:252–6.

128. Shimizu H, Miyamura K, Kuroki Y. Appearance of surfactant proteins, SP-A and SP-B, in developing rat lung and the effects of in vivo dexamethasone treatment. *Biochim Biophys Acta* 1991;**1081**:53–60.

129. Schellhase DE, Emrie PA, Fisher JH, Shannon JM. Ontogeny of surfactant apoproteins in the rat. *Pediatr Res* 1989;**26**:167–74.

130. Xu J, Yao L, Possmayer F. Regulation of mRNA levels for pulmonary surfactant-associated proteins in developing rabbit lungs. *Biochim Biophys Acta* 1995;**1254**:302–10.

131. Ohashi T, Polk D, Ikegami M, Ueda T, Jobe A. Ontogeny and effects of exogenous surfactant treatment on SP-A, SP-B, and SP-C messenger-RNA expression in rabbit lungs. *Am J Phys* 1994;**267**:L46–51.

132. Wohlford-Lenane CL, Durham PL, Snyder JM. Localization of surfactant-associated protein-C (SP-C) messenger-RNA in fetal rabbit lung-tissue by insitu hybridization. *Am J Respir Cell Mol Biol* 1992;**6**:225–34.

133. Tan RC, Ikegami M, Jobe AH, Yao LY, Possmayer F, Ballard PL. Developmental and glucocorticoid regulation of surfactant protein mRNAs in preterm lambs. *Am J Phys* 1999;**277**:L1142–8.

134. Oulton M, Rasmusson M, Yoon R, Fraser M. Gestation-dependent effects of the combined treatment of glucocorticoids and thyrotropin-releasing hormone on surfactant production by fetal rabbit lung. *Am J Obstet Gynecol* 1989;**160**:961–7.

135. Rooney S, Marino P, Gobran L, Gross I, Warshaw J. Thyrotropin-releasing hormone increases the amount of surfactant in lung lavage from fetal rabbits. *Pediatr Res* 1979;**13**:623–5.

136. Rooney SA. The surfactant system and lung phospholipid biochemistry. *Am Rev Respir Dis* 1985;**131**:439–60.

137. Emanuel RL, Torday JS, Asokananthan N, Sunday ME. Direct effects of corticotropin-releasing hormone and thyrotropin-releasing hormone on fetal lung explants. *Peptides* 2000;**21**:1819–29.

138. Schellenberg JC, Liggins GC, Manzai M, Kitterman JA, Lee CC. Synergistic hormonal effects on lung maturation in fetal sheep. *J Appl Physiol* 1988;**65**:94–100.

139. Marino PA, Rooney SA. Surfactant secretion in a newborn rabbit lung slice model. *Biochim Biophys Acta* 1980;**620**:509–19.

140. Mendelson CR, Johnston JM, MacDonald PC, Snyder JM. Multihormonal regulation of surfactant synthesis by human fetal lung in vitro. *J Clin Endocrinol Metab* 1981;**53**:307–17.

141. Hallman M, Wermer D, Epstein BL, Gluck L. Effects of maternal insulin or glucose infusion on the fetus: study on lung surfactant phospholipids, plasma myoinositol, and fetal growth in the rabbit. *Am J Obstet Gynecol* 1982;**142**:877–82.

142. Gross I, Wilson CM, Ingleson LD, Brehier A, Rooney SA. The influence of hormones on the biochemical development of fetal rat lung in organ culture. I. Estrogen. *Biochim Biophys Acta* 1979;**575**:375–83.

143. Rubin LP, Torday JS. Parathyroid Hormone-related Protein (PTHrP) biology in fetal lung development. In: Mendelson CR, editor. *Endocrinology of the Lung.* Totowa, NJ: Humana Press Inc.; 2000. pp. 269–97.

144. Torday JS, Sun H, Wang L, Torres E, Sunday ME, Rubin LP. Leptin mediates the parathyroid hormone-related protein paracrine stimulation of fetal lung maturation. *Am J Physiol Lung Cell Mol Physiol* 2002;**282**:L405–10.

145. Chelly N, Mouhieddine-Gueddiche OB, Barlier-Mur AM, Chailley-Heu B, Bourbon JR. Keratinocyte growth factor enhances maturation of fetal rat lung type II cells. *Am J Respir Cell Mol Biol* 1999;**20**:423–32.

146. Fraslon C, Bourbon JR. Comparison of effects of epidermal and insulin-like growth-factors, gastrin releasing peptide and retinoic acid on fetal lung-cell growth and maturation in vitro. *Biochim Biophys Acta* 1992;**1123**:65–75.

147. Boggaram V. Regulation of lung sufactant protein gene expression. *Front Biosci* 2003;**8**:D751–67.

148. Boggaram V. Thyroid transcription factor-1 (TTF-1/Nkx2.1/TITF1) gene regulation in the lung. *Clin Sci* 2009;**116**:27–35.

149. Mendelson CR. Role of transcription factors in fetal lung development and surfactant protein gene expression. *Annu Rev Physiol* 2000;**62**:875–915.

150. Gross I, Ballard PL. Hormonal therapy for prevention of respiratory distress syndrome. In: Polin RA, Fox WW, Abman SH, editors. *Fetal and Neonatal Physiology.* Philadelphia: W.B. Saunders & Company; 2004.

151. Liggins GC. Premature parturition after infusion of corticotrophin or cortisol into foetal lambs. *J Endocrinol* 1968;**42**:323–9.

152. Liggins GC. Premature delivery of foetal lambs infused with glucocorticoids. *J Endocrinol* 1969;**45**:515–23.

153. Jobe AH, Ikegami M. Lung development and function in preterm infants in the surfactant treatment era. *Annu Rev Physiol* 2000;**62**:825–46.

154. Morrison JL, Botting KJ, Soo PS, McGillick EV, Hiscock J, Zhang S, McMillen IC, Orgeig S. Antenatal steroids and the IUGR fetus: are exposure and physiological effects on the lung and cardiovascular system the same as in normally grown fetuses? *J Pregnancy* 2012:839656.

155. Roberts D, Dalziel SR. Antenatal corticosteroids for accelerating fetal lung maturation for women at risk of preterm birth. Review by 'The Cochrane Collaboration'. In: *The Cochrane Library.* John Wiley & Sons, Ltd; 2008. pp. 1–167.

156. Ikegami M, Polk D, Jobe A. Minimum interval from fetal betamethasone treatment to postnatal lung responses in preterm lambs. *Am J Obstet Gynecol* 1996;**174**:1408–13.

157. Ballard PL, Ning Y, Polk D, Ikegami M, Jobe AH. Glucocorticoid regulation of surfactant components in immature lambs. *Am J Phys* 1997;**17**:L1048–57.

158. Sharma A, Gonzales LW, Ballard PL. Hormonal regulation of cholinephosphate cytidylyltransferase in human fetal lung. *Biochim Biophys Acta* 1993;**1170**:237–44.

159. Viscardi RM, Weinhold PA, Beals TM, Simon RH. Cholinephosphate cytidylyltransferase in fetal rat lung cells: activity and subcellular distribution in response to dexamethasone, triiodothyronine, and fibroblast-conditioned medium. *Exp Lung Res* 1989;**15**:223–37.

160. Smith BT. Lung maturation in the fetal rat. Acceleration by injection of fibroblast-pneumocyte factor. *Science* 1979;**204**:1094–5.

161. Post M, Barsoumian A, Smith BT. The cellular mechanism of glucocorticoid acceleration of fetal lung maturation. Fibroblast-pneumonocyte factor stimulates choline-phosphate cytidylyltransferase activity. *J Biol Chem* 1986;**261**:2179–84.

162. Suzuki R, Ichikawa F, Endo C, Hoshi K, Sato A. A method of measuring dipalmitoylphosphatidylcholine (DPPC) by high-performance liquid chromatography (HPLC): effect of dexamethasone on DPPC secretion in fetal rat alveolar type II cell culture. *J Clin Endocrinol Metab* 1996;**22**:481–8.

163. Chelly N, Henrion A, Pinteur C, Chailley-Heu B, Bourbon JR. Role of keratinocyte growth factor in the control of surfactant synthesis by fetal lung mesenchyme. *Endocrinology* 2001;**142**:1814–9.

164. Isohama Y, Rooney SA. Glucocorticoid enhances the response of type II cells from newborn rats to surfactant secretagogues. *Biochim Biophys Acta* 2001;**1531**:241–50.

165. Orgeig S, Crittenden TA, Marchant C, McMillen IC, Morrison JL. Intrauterine growth restriction delays surfactant protein maturation in the sheep fetus. *Am J Physiol Lung Cell Mol Physiol* 2010;**298**:L575–83.

166. Boggaram V, Smith ME, Mendelson CR. Regulation of expression of the gene encoding the major surfactant protein (SP-A) in human fetal lung in vitro. Disparate effects of glucocorticoids on transcription and on mRNA stability. *J Biol Chem* 1989;**264**:11421–7.

167. Mendelson CR, Gao E, Li J, Young PP, Michael LF, Alcorn JL. Regulation of expression of surfactant protein-A. *Biochim Biophys Acta* 1998;**1408**:132–49.

168. Whitsett JA, Glasser SW. Regulation of surfactant protein gene transcription. *Biochim Biophys Acta* 1998;**1408**:303–11.

169. Gonzales LW, Guttentag SH, Wade KC, Postle AD, Ballard PL. Differentiation of human pulmonary type II cells in vitro by glucocorticoid plus cAMP. *Am J Physiol Lung Cell Mol Physiol* 2002;**283**:L940–51.

170. Losada A, Tovar JA, Xia HM, Diez-Pardo JA, Santisteban P. Down-regulation of thyroid transcription factor-1 gene expression in fetal lung hypoplasia is restored by glucocorticoids. *Endocrinology* 2000;**141**:2166–73.

171. Bohinski RJ, Di Lauro R, Whitsett JA. The lung specific surfactant protein B gene promotor is a target for thyroid transcription factor 1 and hepatocyte nuclear factor 3, indicating common factors for organ specific gene expression along the foregut axis. *Mol Cell Biol* 1994;**14**:5671–81.

172. Bruno MD, Bohinski RJ, Huelsman KM, Whitsett JA, Korfhagen TR. Lung cell-specific expression of the murine surfactant protein A (SP-A) gene is mediated by interactions between the SP-A promoter and thyroid transcription factor-1. *J Biol Chem* 1995;**270**:6531–6.

173. Kelly SE, Bachurski CJ, Burhans MS, Glasser SW. Transcription of the lung-specific surfactant protein C gene is mediated by thyroid transcription factor 1. *J Biol Chem* 1996;**271**:6881–8.

174. Davé V, Childs T, Whitsett JA. Nuclear factor of activated T cells regulates transcription of the surfactant protein D gene (sftpd) via direct interaction with thyroid transcription factor-1 in lung epithelial cells. *J Biol Chem* 2004;**279**:34578–88.

175. Nwosu UC, Kaplan MM, Utiger RD, Delivoria-Papadopulos M. Surge of fetal plasma triiodothyronine before birth in sheep. *Am J Obstet Gynecol* 1978;**132**:489–94.

176. Thomas AL, Krane EJ, Nathanielsz PW. Changes in the fetal thyroid axis after induction of premature parturition by low dose continuous intravascular cortisol infusion to the fetal sheep at 130 days of gestation. *Endocrinology* 1978;**103**:17–23.

177. Post M, Smith BT. Hormonal control of surfactant metabolism. In: Robertson B, Van Golde LMG, Batenburg JJ, editors. *Pulmonary Surfactant: From Molecular Biology to Clinical Practice.* Amsterdam: Elsevier Science Publishers B.V.; 1992. pp. 379–424.

178. Liggins GC. Thyrotrophin-releasing hormone (TRH) and lung maturation. *Reprod Fertil Dev* 1995;**7**:443–50.

179. Smith BT, Sabry K. Glucocorticoid-thyroid synergism in lung maturation: A mechanism involving epithelial-mesenchymal interaction. *Proc Natl Acad Sci U S A* 1983;**80**:1951–4.

180. Ballard PL, Hovey ML, Gonzales LK. Thyroid hormone stimulation of phosphatidylcholine synthesis in cultured fetal rabbit lung. *J Clin Invest* 1984;**74**:898–905.

181. Torday JS, Dow KE. Synergistic effect of triiodothyronine and dexamethasone on male and female fetal-rat lung surfactant synthesis. *Dev Pharmacol Ther* 1984;**7**:133–9.

182. Gonzales LW, Ballard PL, Ertsey R, Williams MC. Glucocorticoids and thyroid hormones stimulate biochemical and morphological differentiation of human fetal lung in organ culture. *J Clin Endocrinol Metab* 1986;**62**:678–91.

183. Wilson M, Hitchcock KR, Douglas WH, DeLellis RA. Hormones and the lung. II. Immunohistochemical localization of thyroid hormone binding in type II pulmonary epithelial cells clonally-derived from adult rat lung. *Anat Rec* 1979;**195**:611–9.

184. Gonzales LW, Ballard PL. Identification and characterization of nuclear 3,5,3'- triiodothyronine-binding dites in fetal human lung. *J Clin Endocrinol Metab* 1981;**53**:21–8.

185. Gross I. Regulation of fetal lung maturation. *Am J Phys* 1990;**259**:L337–44.

186. Das DK. Hormonal-control of pulmonary surfactant synthesis. *Arch Environ Health* 1983;**224**:1–12.

187. Redding RA, Douglas WH, Stein M. Thyroid hormone influence upon lung surfactant metabolism. *Science* 1972;**175**:994–6.

188. Das DK, Ayromlooi J, Bandyopadhyay D, Bandyopadhyay S, Neogi A, Steinberg H. Potentiation of surfactant release in fetal lung by thyroid hormone action. *J Appl Physiol* 1984;**56**:1621–6.

189. Whitsett JA, Darovec-Beckerman C, Pollinger J, Moore JJJ. Ontogeny of [beta] adrenergic receptors in the rat lung: effects of hypothyroidism. *Pediatr Res* 1982;**16**:381–7.

190. Whitsett JA, Weaver TE, Lieberman MA, Clark JC, Daugherty C. Differential-effects of epidermal growth-factor and transforming growth-factor-beta on synthesis of MR = 35,000 surfactant-associated protein in fetal lung. *J Biol Chem* 1987;**262**:7908–13.

191. Nichols KV, Floros J, Dynia DW, Veletza SV, Wilson CM, Gross I. Regulation of surfactant protein A mRNA by hormones and butyrate in cultured fetal rat lung. *Am J Physiol Lung Cell Mol Physiol* 1990;**259**:L488–95.

192. Floros J, Gross I, Nichols KV, Veletza SV, Dynia D, Lu HW, et al. Hormonal effects on the surfactant protein-B (SP-B) messenger-RNA in cultured fetal-rat lung. *Am J Respir Cell Mol Biol* 1991;**4**:449–54.

193. Rooney SA. Lung surfactant. *Environ Health Perspect* 1984;**55**:205–26.

194. Griese M, Gobran LI, Rooney SA. Ontogeny of surfactant secretion in type II pneumocytes from fetal, newborn and adult rats. *Am J Phys* 1992;**262**:L337–43.

195. Gobran LI, Rooney SA. Adenylate cyclase-coupled ATP receptor and surfactant secretion in type II pneumocytes from newborn rats. *Am J Physiol Lung Cell Mol Physiol* 1997;**272**:L187–96.

196. Duffy DM, Ballard PL, Goldfien A, Roberts JM. Cyclic adenosine-3',5'-monophosphate increases beta-adrenergic-receptor concentration in cultured human fetal lung explants and type-II cells. *Endocrinology* 1992;**131**:841–6.

197. Aksoy MO, Mardini IA, Yang Y, Bin W, Zhou S, Kelsen SG. Glucocorticoid effects on the beta-adrenergic receptor-adenylyl cyclase system of human airway epithelium. *J Allergy Clin Immunol* 2002;**109**:491–7.

198. Padbury J, Roberman B, Oddie T, Hobel C, Fisher D. Fetal catecholamine release in response to labor and delivery. *Obstetrics Gynecol* 1982;**60**:607–11.

199. Crittenden DJ, Alexander LA, Beckman DL. Sympathetic nerve influence on alveolar type II cell ultrastructure. *Life Sci* 1994;**55**:1229–35.

200. Warburton D, Parton L, Buckley S, Cosico L, Saluna T. Beta-receptors and surface active material flux in fetal lamb lung: female advantage. *J Appl Phys* 1987;**63**:828–33.

201. Pysher TJ, Konrad KD, Reed GB. Effects of hydrocortisone and pilocarpine on fetal rat lung explants. *Lab Invest* 1977;**37**:588–94.

202. Delahunty TJ, Johnston JM. Neurohumoral control of pulmonary surfactant secretion. *Lung* 1979;**157**:45–51.

203. Gahlot L, Green FHY, Rigaux A, Schneider JM, Hasan SU. Role of vagal innervation on pulmonary surfactant system during fetal development. *J Appl Phys* 2009;**106**:1641–9.

204. Khan PA, Cloutier M, Piedboeuf B. Tracheal occlusion: a review of obstructing fetal lungs to make them grow and mature. *Am J Med Genet C Semin Med Genet* 2007;**145C**:125–38.

205. Flecknoe S, Harding R, Maritz G, Hooper SB. Increased lung expansion alters the proportions of type I and type II alveolar epithelial cells in fetal sheep. *Am J Phys* 2000;**278**:L1180–5.

206. Piedboeuf B, Laberge JM, Ghitulescu G, Gamache M, Petrov P, Belanger S, et al. Deleterious effect of tracheal obstruction on type II pneumocytes in fetal sheep. *Pediatr Pulmonol* 1997;**41**:473–9.

207. Joe P, Wallen LD, Chapin CJ, Lee CH, Allen L, Han VK, et al. Effects of mechanical factors on growth and maturation of the lung in fetal sheep. *Am J Physiol Lung Cell Mol Physiol* 1997;**272**:L95–105.

208. Lines A, Nardo L, Phillips ID, Possmayer F, Hooper SB. Alterations in lung expansion affect surfactant protein A, B and C mRNA levels in fetal sheep. *Am J Physiol Lung Cell Mol Physiol* 1999;**276**:L239–45.

209. Islam S, Donahoe PK, Schnitzer JJ. Tracheal ligation increases mitogen-activated protein kinase activity and attenuates surfactant protein B mRNA in fetal sheep lungs. *J Surg Res* 1999;**84**:19–23.

210. Lines A, Gillett AM, Phillips ID, Wallace MJ, Hooper SB. Re-expression of pulmonary surfactant proteins following tracheal obstruction in fetal sheep. *Exp Physiol* 2001;**86**:55–63.

211. De Paepe ME, Johnson BD, Papadakis K, Luks FI. Lung growth response after tracheal occlusion in fetal rabbits is gestational age-dependent. *Am J Respir Cell Mol Biol* 1999;**21**:65–76.

212. Danzer E, Robinson LE, Davey MG, Schwarz U, Volpe M, Adzick NS, et al. Tracheal occlusion in fetal rats alters expression of mesenchymal nuclear transcription factors without affecting surfactant protein expression. *J Pediatr Surg* 2006;**41**:774–80.

213. Lines AL, Davey MG, Harding R, Hooper SB. Effect of increased lung expansion on surfactant protein mRNA levels in lambs. *Pediatr Res* 2001;**50**:720–5.

214. Dobbs LG, Gutierrez JA. Mechanical forces modulate alveolar epithelial phenotypic expression. *Comp Biochem Physiol A* 2001;**129**:261–6.

215. Sanchez-Esteban J, Cicchiello LA, Wang Y, Tsai SW, Williams LK, Torday JS, et al. Mechanical stretch promotes alveolar epithelial type II cell differentiation. *J Appl Phys* 2001;**91**:589–95.

216. Sanchez-Esteban J, Tsai SW, Sang J, Qin J, Torday JS, Rubin LP. Effects of mechanical forces on lung-specific gene expression. *Am J Med Sci* 1998;**316**:200–4.

217. Nakamura T, Liu M, Mourgeon E, Slutsky A, Post M. Mechanical strain and dexamethasone selectively increase surfactant protein C and tropoelastin gene expression. *Am J Physiol Lung Cell Mol Physiol* 2000;**278**:L974–80.

218. Foster CD, Varghese LS, Gonzales LW, Margulies SS, Guttentag SH. The Rho pathway mediates transition to an alveolar type I cell phenotype during static stretch of alveolar type II cells. *Pediatr Res* 2010;**67**:585–90.

219. Silbert O, Wang Y, Maciejewski BS, Lee HS, Shaw SK, Sanchez–Esteban J. Roles of RhoA and Rac1 on actin remodeling and cell alignment and differentiation in fetal type II epithelial cells exposed to cyclic mechanical stretch. *Exp Lung Res* 2008;**34**:663–80.

220. Liu X-Y, Chen X-F, Ren Y-H, Zhan Q-Y, Wang C, Yang C. Alveolar type II cells escape stress failure caused by tonic stretch through transient focal adhesion disassembly. *Int J Biol Sci* 2011;**7**: 588–99.

221. Sanchez-Esteban J, Wang YL, Filardo EJ, Rubin LP, Ingber DE. Integrins beta(1), alpha(6), and alpha(3) contribute to mechanical strain-induced differentiation of fetal lung type II epithelial cells via distinct mechanisms. *Am J Physiol Lung Cell Mol Physiol* 2006;**290**:L343–50.

222. Hokenson MA, Wang YL, Hawwa RL, Huang ZP, Sharma S, Sanchez-Esteban J. Reduced IL-10 production in fetal type II epithelial cells exposed to mechanical stretch is mediated via activation of IL-6-SOCS3 signaling pathway. *Plos One* 2013;**8**.

223. Hawwa RL, Hokenson MA, Wang YL, Huang ZP, Sharma S, Sanchez-Esteban J. Differential expression of MMP-2 and -9 and their inhibitors in fetal lung cells exxposed to mechanical stretch: regulation by IL-10. *Lung* 2011;**189**:341–9.

224. Torday JS, Sun H, Qin J. Prostaglandin E2 integrates the effects of fluid distension and glucocorticoid on lung maturation. *Am J Phys* 1998;**274**:L106–11.

225. Huang Z, Wang Y, Nayak PS, Dammann CE, Sanchez-Esteban J. Stretch-induced fetal type II cell differentiation is mediated via ErbB1-ErbB4 interactions. *J Biol Chem* 2012;**287**:18091–102.

226. Sanchez-Esteban J, Wang Y, Gruppuso PA, Rubin LP. Mechanical stretch induces fetal type II cell differentiation via an epidermal growth factor receptor-extracellular-regulated protein kinase signaling pathway. *Am J Respir Cell Mol Biol* 2004;**30**:76–83.

227. Torday JS, Rehan VK. Stretch-stimulated surfactant synthesis is coordinated by the paracrine actions of PTHrP and leptin. *Am J Physiol Lung Cell Mol Physiol* 2002;**283**:L130–5.

228. Torday JS, Sanchez-Esteban J, Rubin LP. Paracrine mediators of mechanotransduction in lung development. *Am J Med Sci* 1998;**316**:205–8.

229. von Neergaard KV. New notions on a fundamental principle of respiratory mechanics: the retractile force of the lung, dependent on the surface tension in the alveoli. *Zeitschrift für Gesundheit und Experimentelle Medizin* 1929;**66**:373–94.

230. Avery ME, Mead J. Surface properties in relation to atelectasis and hyaline membrane disease. *Am Med Assoc J Dis Child* 1959;**97**: 517–23.

231. Clements JA, Hustead RF, Johnson RP, Gribetz I. Pulmonary surface tension and alveolar stability. *J Appl Phys* 1961;**16**(3):444–50.

232. Pattle RE, Claireaux AE, Davies PA, Cameron AH. Inability to form a lung - lining film as a cause of the respiratory - distress syndrome in the newborn. *The Lancet* 1962;**2**:469–73.

233. Clements JA, Avery ME. Lung surfactant and neonatal respiratory distress syndrome. *Am J Respir Crit Care Med* 1998;**157**:S59–66.

234. Welty S, Hansen TN, Corbet A. Respiratory distress in the perterm infant. In: Taeusch HW, et al., editor. *Avery's Diseases of the Newborn*. Philadelphia, USA: Elsevier Saunders; 2005. p. 687.

235. Jobe AH. Pathophysiology of respiratory distress syndrome. In: Polin RA, Fox WW, editors. *Fetal and Neonatal Physiology*. Philadelphia: Saunders; 1991. pp. 949–62.

236. Stevens PA, Schadow B, Bartholain S, Segerer H, Obladen M. Surfactant protein A in the course of respiratory distress syndrome. *Eur J Pediatr* 1992;**151**:596–600.

237. Taieb J, Francoual J, Magny JF, Fraslon C, Messaoudi C, Lindenbaum A, et al. Surfactant associated protein A determination using a chemiluminescence system–application to tracheal aspirates from newborns. *Clin Chim Acta* 1995;**235**:229–34.

238. Shelley SA, Kovacevic M, Paciga JE, Balis JU. Sequential changes of surfactant phosphatidylcholine in hyaline-membrane disease of the newborn. *N Engl J Med* 1979;**300**:112–6.

239. Motoyama EK, Namba Y, Rooney SA. Phosphatidylcholine content and fatty acid composition of tracheal and gastric liquids from premature and full-term newborn infants. *Clin Chim Acta* 1976;**70**:449–54.

240. Ashton MR, Postle AD, Hall MA, Smith SL, Kelly FJ, Normand IC. Phosphatidylcholine composition of endotracheal tube aspirates of neonates and subsequent respiratory disease. *Arch Dis Child* 1992;**67**:378–82.

241. Hallman M, Merritt TA, Pohjavuori M, Gluck L. Effect of surfactant substitution on lung effluent phospholipids in respiratory distress syndrome: evaluation of surfactant phospholipid turnover, pool size, and the relationship to severity of respiratory failure. *Pediatr Res* 1986;**20**:1228–35.

242. Griese M, Dietrich P, Reinhardt D. Pharmacokinetics of bovine surfactant in neonatal respiratory distress syndrome. *Am J Respir Crit Care Med* 1995;**152**:1050–4.

243. Morley CJ, Brown BD, Hill CM, Barson AJ, Davis JA. Surfactant abnormalities in babies dying from sudden infant death syndrome. *The Lancet* 1982;**i**:1320–2.

244. Ikegami M, Jacobs H, Jobe A. Surfactant function in respiratory distress syndrome. *J Pediatr* 1983;**102**:443–7.

245. Griese M, Westerburg B. Surfactant function in neonates with respiratory distress syndrome. *Respiration* 1998;**65**:136–42.

246. Wigglesworth JS. Pathology of neonatal respiratory distress. *Proc R Soc Med* 1977;**70**:861–3.

247. Hallman M, Merritt TA, Kari A, Bry K. Factors affecting surfactant responsiveness. *Annu Mediaev* 1991;**23**:693–8.

248. Kari MA, Raivio KO, Venge P, Hallman M. Dexamethasone treatment of infants at risk for chronic lung disease: surfactant components and inflammatory parameters in airway specimens. *Pediatr Res* 1994;**36**:387–93.

249. Soll R, Ozek E. Multiple versus single doses of exogenous surfactant for the prevention or treatment of neonatal respiratory distress syndrome. Review by 'The Cochrane Collaboration'. In: *The Cochrane Library*. John Wiley & Sons, Ltd; 2009. pp. 1–22.

250. Whitsett JA, Stahlman MT. Impact of advances in physiology, biochemistry, and molecular biology on pulmonary disease in neonates. *Am J Respir Crit Care Med* 1998;**157**:S67–71.

251. Nogee LM, Demello DE, Colton HR. Deficiency of pulmonary surfactant protein B in congenital alveolar proteinosis. *N Engl J Med* 1993;**328**:406–10.

252. Gower WA, Nogee LM. Surfactant Dysfunction. *Paediatr Respir Rev* 2011;**12**:223–9.

253. Hamvas A, Wise PH, Yang RK, Wampler NS, Noguchi A, Maurer MM, et al. The influence of the wider use of surfactant therapy on neonatal mortality among blacks and whites. *N Engl J Med* 1996;**334**:1635–40.

254. Hamvas A, Cole FS, Nogee LM. Genetic disorders of surfactant proteins. *Neonatology* 2007;**91**:311–7.

255. Whitsett JA, Wert SE, Weaver TE. Alveolar surfactant homeostasis and the pathogenesis of pulmonary disease. *Annu Rev Med* 2010;**61**:105–19.

256. Wert SE, Whitsett JA, Nogee LM. Genetic disorders of surfactant dysfunction. *Pediatr Dev Pathol* 2009;**12**:253–74.

257. Nogee LM. Genetic mechanisms of surfactant deficiency. *Biol Neonate* 2004;**85**:314–8.

258. Hamvas A, Nogee LM, deMello DE, Cole FS. Pathophysiology and treatment of surfactant protein-B deficiency. *Biol Neonate* 1995;**67**:18–31.

259. Herman TE, Nogee LM, McAlister WH, Dehner LP. Surfactant protein-B deficiency - radiographic manifestations. *Pediatr Radiol* 1993;**23**:373–5.

260. Garmany TH, Moxley MA, White FV, Dean M, Hull WM, Whitsett JA, et al. Surfactant composition and function in patients with ABCA3 mutations. *Pediatr Res* 2006;**59**:801–5.

261. Beers MF, Hamvas A, Moxley MA, Gonzales LW, Guttentag SH, Solarin KO, et al. Pulmonary surfactant metabolism in infants lacking surfactant protein B. *Am J Respir Cell Mol Biol* 2000;**22**:380–91.

262. Hamvas A, Nogee LM, Mallory GB, Spray TL, Huddleston CB, August A, et al. Lung transplantation for treatment of infants with surfactant protein B deficiency. *J Pediatr* 1997;**130**:231–9.

263. Demello DE, Nogee LM, Heyman S, Krous HF, Hussain M, Merritt A, et al. Molecular and phenotypic variability in the congenital alveolar proteinosis syndrome-associated with inherited surfactant protein-B deficiency. *J Pediatr* 1994;**125**:43–50.

264. Hamvas A, Cole FS, Demello DE, Moxley M, Whitsett JA, Colten HR, et al. Surfactant protein-B deficiency - antenatal diagnosis and prospective treatment with surfactant replacement. *J Pediatr* 1994;**125**:356–61.

265. Ballard PL, Nogee LM, Beers MF, Ballard RA, Planer BC, Polk L, et al. Partial deficiency of surfactant protein-B in an infant with chronic lung-disease. *Pediatrics* 1995;**96**:1046–52.

266. Klein JM, Thompson MW, Snyder JM, George TN, Whitsett JA, Bell EF, et al. Transient surfactant protein B deficiency in a term infant with severe respiratory failure. *J Pediatr* 1998;**132**:244–8.

267. Dunbar AE, Wert SE, Ikegami M, Whitsett JA, Hamvas A, White FV, et al. Prolonged survival in hereditary surfactant protein B (SP-B) deficiency associated with a novel splicing mutation. *Pediatr Res* 2000;**48**:275–82.

268. Palomar LM, Nogee LM, Sweet SC, Huddleston CB, Cole FS, Hamvas A. Long-term outcomes after infant lung transplantation for surfactant protein B deficiency related to other causes of respiratory failure. *J Pediatr* 2006;**149**:548–53.

269. Stahlman MT, Gray MP, Falconieri MW, Whitsett JA, Weaver TE. Lamellar body formation in normal and surfactant protein B-deficient fetal mice. *Lab Invest* 2000;**80**:395–403.

270. Foster CD, Zhang PX, Gonzales LW, Guttentag SH. In vitro surfactant protein B deficiency inhibits lamellar body formation. *Am J Respir Cell Mol Biol* 2003;**29**:259–66.

271. Vorbroker DK, Profitt SA, Nogee LM, Whitsett JA. Aberrant processing of surfactant protein C in hereditary SP-B deficiency. *Am J Phys* 1995;**268**:L647–56.

272. Voorhout WF, Weaver TE, Haagsman HP, Geuze HJ, Van Golde LM. Biosynthetic routing of pulmonary surfactant proteins in alveolar type II cells. *Microsc Res Tech* 1993;**26**:366–73.

273. Beal SM. Sudden infant death syndrome: epidemiological comparisons between South Australia and communities with a different incidence. *Aust Paediatr J* 1986;(Suppl.):13–6.

274. Melton KR, Nesslein LL, Ikegami M, Tichelaar JW, Clark JC, Whitsett JA, et al. SP-B deficiency causes respiratory failure in adult mice. *Am J Physiol Lung Cell Mol Physiol* 2003;**285**:L543–549.

275. Kylstra JA, Schoenfisch WH. Alveolar surface tension in fluorocarbon-filled lungs. *J Appl Phys* 1972;**33**:32–5.

276. Curley AE, Halliday HL. The present status of exogenous surfactant for the newborn. *Early Hum Dev* 2001;**61**:67–83.

277. Aneja MK, Rudolph C. Gene therapy of surfactant protein B deficiency. *Curr Opin Mol Ther* 2006;**8**:432–8.

278. Tredano M, Griese M, de Blic J, Lorant T, Houdayer C, Schumacher S, et al. Analysis of 40 sporadic or familial neonatal and pediatric cases with severe unexplained respiratory distress: relationship to SFTPB. *AmJ Med Genet A* 2003;**119A**:324–39.

279. Shulenin S, Nogee LM, Annilo T, Wert SE, Whitsett JA, Dean M. ABCA3 gene mutations in newborns with fatal surfactant deficiency. *N Engl J Med* 2004;**350**:1296–303.

280. Kunig AM, Parker TA, Nogee LM, Abman SH, Kinsella JP. ABCA3 deficiency presenting as persistent pulmonary hypertension of the newborn. *J Pediatr* 2007;**151**:322–4.

281. Hallman M, Haataja R. Genetic basis of respiratory distress syndrome. *Front Biosci* 2007;**12**:2670–82.

282. Dean M, Rzhetsky A, Allikmets R. The human ATP-binding cassette (ABC) transporter superfamily. *Genome Res* 2001;**11**:1156–66.

283. Higgins CF. ABC transporters - from microorganisms to man. *Annu Rev Cell Biol* 1992;**8**:67–113.

284. Bullard JE, Wert SE, Nogee LM. ABCA3 deficiency: neonatal respiratory failure and interstitial lung disease. *Semin Perinatol* 2006;**30**:327–34.

285. Doan ML, Guillerman RP, Dishop MK, Nogee LM, Langston C, Mallory GB, et al. Clinical, radiological and pathological features of ABCA3 mutations in children. *Thorax* 2008;**63**:366–73.

286. Mulugeta S, Gray JM, Notarfrancesco KL, Gonzales LW, Koval M, Feinstein SI, et al. Identification of LBM180, a lamellar body limiting membrane protein of alveolar type II cells, as the ABC transporter protein ABCA3. *J Biol Chem* 2002;**277**:22147–55.

287. Yamano G, Funahashi H, Kawanami O, Zhao LX, Ban N, Uchida Y, et al. ABCA3 is a lamellar body membrane protein in human lung alveolar type II cells. *Febs Letters* 2001;**508**:221–5.

288. Langmann T, Mauerer R, Zahn A, Moehle C, Probst M, Stremmel W, et al. Real-time reverse transcription-PCR expression profiling of the complete human ATP-binding cassette transporter superfamily in various tissues. *Clin Chem* 2003;**49**:230–8.

289. Nagata K, Yamamoto A, Ban N, Tanaka AR, Matsuo M, Kioka N, et al. Human ABCA3, a product of a responsible gene for *abca3* for fatal surfactant deficiency in newborns, exhibit unique ATP hydrolysis activity and generates intracellular multilamellar vesicles. *Arch Environ Health* 2004;**324**:262–8.

290. Cheong N, Madesh M, Gonzales LW, Zhao M, Yu K, Ballard PL, et al. Functional and trafficking defects in ATP binding cassette A3 mutants associated with respiratory distress syndrome. *J Biol Chem* 2006;**281**:9791–800.

291. Matsumura Y, Ban N, Ueda K, Inagaki N. Characterization and classification of ATP-binding cassette transporter ABCA3 mutants in fatal surfactant deficiency. *J Biol Chem* 2006;**281**:34503–14.

292. Thouvenin G, Abou Taam R, Flamein F, Guillot L, Le Bourgeois M, Reix P, et al. Characteristics of disorders associated with genetic mutations of surfactant protein C. *Arch Dis Child* 2010;**95**:449–54.

293. Mechri M, Epaud R, Emond S, Coulomb A, Jaubert F, Tarrant A, et al. Surfactant protein C gene (SFTPC) mutation-associated lung disease: high-resolution computed tomography (HRCT) findings and its relation to histological analysis. *Pediatr Pulmonol* 2010;**45**:1021–9.

294. Brasch F, Griese M, Tredano M, Johnen G, Ochs M, Rieger C, Mulugeta S, et al. Interstitial lung disease in a baby with a de novo mutation in the SFTPC gene. *Eur Respir J* 2004;**24**:30–9.

295. Cameron HS, Somaschini M, Carrera P, Hamvas A, Whitsett JA, Wert SE, et al. A common mutation in the surfactant protein C gene associated with lung disease. *J Pediatr* 2005;**146**:370–5.

296. Nogee LM, Dunbar 3rd AE, Wert SE, Askin F, Hamvas A, Whitsett JA. A mutation in the surfactant protein C gene associated with familial interstitial lung disease. *N Engl J Med* 2001;**344**:573–9.

297. Thomas AQ, Lane K, Phillips 3rd J, Prince M, Markin C, Speer M, et al. Heterozygosity for a surfactant protein C gene mutation associated with usual interstitial pneumonitis and cellular nonspecific interstitial pneumonitis in one kindred. *Am J Respir Crit Care Med* 2002;**165**:1322–8.

298. Maguire JA, Mulugeta S, Beers MF. Multiple ways to die: delineation of the unfolded protein response and apoptosis induced by Surfactant Protein C BRICHOS mutants. *Int J Biochem Cell Biol* 2012;**44**:101–12.

299. Hawkins AE, Zhao M, Beers MF, Mulugeta S. Impairment of epithelial cell endocytosis by mistargeted SP-C I73T mutant protein. *Faseb Journal* 2011;**25**:865–912.

300. Sanchez-Pulido L, Devos D, Valencia A. BRICHOS: a conserved domain in proteins associated with dementia, respiratory distress and cancer. *Trends Biochem Sci* 2002;**27**:329–32.

301. Stevens PA, Pettenazzo A, Brasch F, Mulugeta S, Baritussio A, Ochs M, et al. Nonspecific interstitial pneumonia, alveolar proteinosis, and abnormal proprotein trafficking resulting from a spontaneous mutation in the surfactant protein C gene. *Pediatr Res* 2005;**57**:89–98.

302. Hamvas A. Inherited surfactant protein-B deficiency and surfactant protein-C associated disease: clinical features and evaluation. *Semin Perinatol* 2006;**30**:316–26.

303. Bridges JP, Wert SE, Nogee LM, Weaver TE. Expression of a human surfactant protein C mutation associated with interstitial lung disease disrupts lung development in transgenic mice. *J Biol Chem* 2003;**278**:52739–46.

304. Bridges JP, Xu Y, Na CL, Wong HR, Weaver TE. Adaptation and increased susceptibility to infection associated with constitutive expression of misfolded SP-C. *J Cell Biol* 2006;**172**:395–407.

305. Glasser SW, Witt TL, Senft AP, Baatz JE, Folger D, Maxfield MD, et al. Surfactant protein C-deficient mice are susceptible to respiratory syncytial virus infection. *Am J Physiol Lung Cell Mol Physiol* 2009;**297**:L64–72.

306. Glasser SW, Senft AP, Whitsett JA, Maxfield MD, Ross GF, Richardson TR, et al. Macrophage dysfunction and susceptibility to pulmonary Pseudomonas aeruginosa infection in surfactant protein C-deficient mice. *J Immunol* 2008;**181**:621–8.

307. Glasser SW, Detmer EA, Ikegami M, Na CL, Stahlman MT, Whitsett JA. Pneumonitis and emphysema in SP-C gene targeted mice. *J Biol Chem* 2003;**278**:14291–8.

308. Stevens PA, Pettenazzo A, Brasch F, Mulugeta S, Baritussio A, Ochs M, et al. Nonspecific interstitial pneumonia, alveolar proteinosis, and abnormal proprotein trafficking resulting from a spontaneous mutation in the surfactant protein C gene. *Pediatr Res* 2005;**57**:89–98.

309. Hamvas A, Nogee LM, White FV, Schuler P, Hackett BP, Huddleston CB, et al. Progressive lung disease and surfactant dysfunction with a deletion in surfactant protein C gene. *Am J Respir Cell Mol Biol* 2004;**30**:771–6.

310. Woischnik M, Sparr C, Kern S, Thurm T, Hector A, Hartl D, et al. A non-BRICHOS surfactant protein c mutation disrupts epithelial cell function and intercellular signaling. *Bmc Cell Biology* 2010;**11**:88.

311. Guillot L, Carre A, Szinnai G, Castanet M, Tron E, Jaubert F, et al. NKX2-1 mutations leading to surfactant protein promoter dysregulation cause interstitial lung disease in "Brain-Lung-Thyroid Syndrome". *Hum Mutat* 2010;**31**:E1146–62.

312. Willemsen MA, Breedveld GJ, Wouda S, Otten BJ, Yntema JL, Lammens M, et al. Brain-Thyroid-Lung syndrome: a patient with a severe multi-system disorder due to a de novo mutation in the thyroid transcription factor 1 gene. *Eur J Pediatr* 2005;**164**:28–30.

313. Kleinlein B, Griese M, Liebisch G, Krude H, Lohse P, Aslanidis C, et al. Fatal neonatal respiratory failure in an infant with congenital hypothyroidism due to haploinsufficiency of the NKX2-1 gene: alteration of pulmonary surfactant homeostasis. *Annu Rev Physiol* 2011;**96**:F453–6.

314. Krude H, Schutz B, Biebermann H, von Moers A, Schnabel D, Neitzel H, et al. Choreoathetosis, hypothyroidism, and pulmonary alterations due to human NKX2-1 haploinsufficiency. *J Clin Invest* 2002;**109**:475–80.

315. Carre A, Szinnai G, Castanet M, Sura-Trueba S, Tron E, Broutin-L'Hermite I, et al. Five new TTF1/NKX2.1 mutations in brain-lung-thyroid syndrome: rescue by PAX8 synergism in one case. *Hum Mol Genet* 2009;**18**:2266–76.

316. Pohlenz J, Dumitrescu A, Zundel D, Martine U, Schonberger W, Koo E, et al. Partial deficiency of thyroid transcription factor 1 produces predominantly neurological defects in humans and mice. *J Clin Invest* 2002;**109**:469–73.

317. Iwatani N, Mabe H, Devriendt K, Kodama M, Miike T. Deletion of NKX2.1 gene encoding thyroid transcription factor-1 in two siblings with hypothyroidism and respiratory failure. *J Pediatr* 2000;**137**:272–6.

318. Szinnai G, Lacroix L, Carre A, Guimiot F, Talbot M, Martinovic J, et al. Sodium/iodide symporter (NIS) gene expression is the limiting step for the onset of thyroid function in the human fetus. *J Clin Endocrinol Metab* 2007;**92**:70–6.

319. Trueba SS, Auge J, Mattei G, Etchevers H, Martinovic J, Czernichow P, et al. PAX8, TITF1, and FOXE1 gene expression patterns during human development: new insights into human thyroid development and thyroid dysgenesis-associated malformations. *J Clin Endocrinol Metab* 2005;**90**:455–62.

320. Liu C, Glasser SW, Wan H, Whitsett JA. GATA-6 and thyroid transcription factor-1 directly interact and regulate surfactant protein-C gene expression. *J Biol Chem* 2002;**277**:4519–25.

321. Yi M, Tong GX, Murry B, Mendelson CR. Role of CBP/p300 and SRC-1 in transcriptional regulation of the pulmonary surfactant protein-A (SP-A) gene by thyroid transcription factor-1 (TTF-1). *J Biol Chem* 2002;**277**:2997–3005.

322. Haagsman HP, Hogenkamp A, van Eijk M, Veldhuizen EJA. Surfactant collectins and innate immunity. *Neonatology* 2008;**93**:288–94.

323. Foot NJ, Orgeig S, Daniels CB. The evolution of a physiological system: the pulmonary surfactant system in diving mammals. Special Issue: Frontiers in Comparative Physiology II: Respiratory Rhythm, Pattern and Responses to Environmental Change. *Respir Physiol Neurobiol* 2006;**154**:118–38.

324. Pringle KC. Human-fetal lung development and related animal models. *Clin Sci* 1986;**29**:502–13.

325. Roubliova XI, Biard JM, Ophalvens L, Gallot D, Jani JC, Verbeken EK, et al. Morphology of the developing fetal lung – the rabbit experimental model. In: Mendez-Vilas A, Diaz J, editors. *Modern Research and Educational Topics in Microscopy*. Formatex; 2007. pp. 417–25.

326. Morrison JL, Botting KJ, Soo PS, McGillick EV, Hiscock J, Zhang S, et al. Antenatal steroids and the IUGR fetus: Are exposure and physiological effects on the lung and cardiovascular system the same as in normally grown fetuses? *J Pregnancy* 2012;**2012**:15.

Chapter 10

Ontogeny of the Pulmonary Immune System

Candace M. Crowley and Lisa A. Miller

Department of Anatomy, Physiology, & Cell Biology, School of Veterinary Medicine, University of California – Davis, Davis, CA, USA

INTRODUCTION

Despite undergoing a comprehensive regime of vaccinations starting at birth, human infants have a limited ability to mount an effective immune response against pathogens. The cellular basis for the immunological restrictions of the fetus and neonate are multifaceted. Based on data obtained from human and laboratory animal studies, developmental immaturity of immune cell phenotypes in conjunction with developmental immaturity of organ systems are contributing factors towards microbial susceptibility in the infant. However, it is important to appreciate that the immunological progression to adult functional status does not necessarily operate on a linear trajectory. Further, while geriatric subjects also exhibit enhanced susceptibility towards pathogens, the mechanisms for compromised immunity in the elderly differ from that of an infant. This chapter will discuss current knowledge in the field of developmental immunology, with an emphasis on the pulmonary system. Establishment of mucosal immunity in the immature lung is an area of investigation that continues to suffer from sparse information in the literature, particularly in humans. Postnatal development of rodent airways and immunity is significantly different from that in humans; further, there are species differences in the release of cytokines and other mediators by leukocytes. Because of these important distinctions between human and rodent pulmonary/immune system development, it is imperative to expand our basic knowledge of the maturing mucosal immune system in the lung within the context of primates (either human or monkey). As such, this chapter will focus on data obtained from human/non-human primate studies, with supportive data from other species included where appropriate.

POSTNATAL MATURATION OF SYSTEMIC IMMUNITY

Innate Immunity

Because lymphocyte development takes place during the prenatal period in humans, the newborn immune system contains functional T and B lymphocytes, albeit in a naïve state. Given the time required to establish robust effector lymphocyte responses (hence the term "adaptive" immunity) after birth, cells of the innate immune system are of particular importance in protecting the neonate from infectious disease. Based upon numerous cord blood studies, it is apparent that the first line of immune defense mediated by granulocytes and monocytes is not as functionally robust in comparison with adult counterparts.

At birth, cord blood neutrophils from healthy newborns exhibit an attenuated chemotactic response to stimuli such as leukotriene B4 in vitro, as compared with peripheral blood neutrophils from adults.[1] Interestingly, the chemotactic ability of newborn neutrophils to leukotriene B4 is enhanced in vitro by a larger pore size chemotaxis chamber, suggesting that limited deformability plays a role in attenuation of chemotaxis. In a related study, it was shown that newborn neutrophils had a diminished capacity to undergo cellular orientation in response to a chemotactic gradient, which correlated with a reduced capacity for cytoplasmic microtubule complex assembly.[2] Impairment of migratory behavior in neutrophils has also been attributed to decreased receptor binding sites for formyl-methionyl-leucyl-phenylalanine, reduced cell surface expression of Mac-1 (CD11b/CD18), and diminished actin polymerization.[2–4] Trafficking limitations of human neonatal neutrophils were recently confirmed using microflow chambers, demonstrating

The Lung. http://dx.doi.org/10.1016/B978-0-12-799941-8.00010-9

that the capacity to roll and adhere directly correlated with gestational age of leukocytes as well as endothelial cells.[5] Measures of neonatal neutrophil function also suggest developmental impairment of phagocytosis, levels of bactericidal permeability-increasing protein, and reactive-oxygen species formation as compared to adult counterparts.[6–8] The formation of neutrophil extracellular traps, or "NETs", a mechanism that involves extrusion of antimicrobial proteins and nuclear material that serve to trap and kill bacteria, is significantly impaired in neonates and independent of reactive-oxygen species formation; this impairment may be related to reduced caspase-3 expression and activity, as well as resistance to apoptosis.[9,10] Despite functional deficiencies in host pathogen defenses within the neutrophil compartment, newborns are at significant health risk for severe inflammatory responses in the lung (bronchopulmonary dysplasia) and the gut (necrotizing enterocolitis). It has been proposed that the mechanism for enhanced local tissue injury in neonates is apoptotic resistance and subsequent prolonged survival of neutrophils due to reduced FasL responsiveness.[11]

When considering immunological strategies to enhance neonatal host defense, cord blood and infant monocytes are of intense interest, as these cells appear to be the principal responders to microbial ligand stimulation and are at least partly responsible for the enhanced susceptibility towards infectious disease during early childhood. Further, in the presence of appropriate cytokine or adhesive stimuli, monocytes may also differentiate into tissue macrophage or dendritic cells, which have varying degrees of overlapping functions with their originating cells. In contrast with neutrophils, chemotactic migration and phagocytic capability of cord blood monocytes does not appear to significantly differ from adult counterparts.[12,13] Earlier studies have suggested that cord blood monocytes can be simply distinguished from adult counterparts by surface expression of CD14, but recent work suggests that the umbilical cord monocyte population is actually deficient in CD14+CD16+ cells.[13,14] Expression of CD80 and IILA-DR does not reach adult levels in monocyte-derived dendritic cells until the first 3 months of life, and in monocytes until 6–9 months of life, indicating a progressive increase in antigen presenting responsiveness with age.[15]

While the initial characterization of neonatal monocyte phenotype focused on cellular markers, recent work has emphasized age-dependent functional differences. Pathogen recognition receptors (PPR), which are essential for the initial recognition of microbes in monocytes and dendritic cells, have been an area of intense investigation in innate immunity. The identification of a mammalian Toll-like receptor (TLR) in 1998 as a membrane-bound cellular sensor for lipolysaccharide has expanded to the discovery of other related PPR family members, including nucleotide-binding

oligomerization domain (NOD)-like receptors that respond at the intracellular level.[16,17] A broad survey of TLR1-9 expression by flow cytometry indicates that monocyte expression of TLRs is comparable between neonates and adults,[18] although it has been reported that CD14+CD16+ TLR4 high cells are much reduced in cord blood.[13] Despite having what appears to be levels of TLR protein comparable with adults, stimulation of neonatal monocytes with microbial ligands results in profound differences in cytokine production that often correlate with chronological age. Comprehensive analysis of monocyte responses to TLR2, TLR3, TLR4, and TLR9 ligands in multiple birth cohorts has shown that anti-viral cytokines are generated in a linear trajectory with age, whereas proinflammatory cytokine synthesis is varied and in some cases (such as IL-10) progressively declines with age.[19–21] Collectively, these findings indicate that development of TLR responses during early life does not follow a linear trajectory, but is differentially regulated with age.

Whereas cytokine synthesis by monocytes may be compromised in the newborn, gene array analysis of neonatal mononuclear cells stimulated with lipopolysaccharide suggests that neonates exhibit a higher level of gene expression for IL-1 receptor/TLR signaling molecules as compared to 12 month old infants; it is speculated that increased signaling through Myd88-dependent pathways may serve as a protective mechanism for the newborn while other innate pathways mature.[22] Other PPRs that have been evaluated during infancy include NOD1 and NOD2 via stimulation with γ-D-Glu-mDAP and muramyl dipeptide, respectively; responsiveness of NOD-associated pathways in peripheral blood appears to be stable during the first year of life.[21] In contrast, aluminum potassium sulfate (alum) stimulation of IL-1β and CXCL8 production through the NALP3 inflammasome pathway in peripheral blood declines with increasing infant age, suggesting that efficacy of alum-containing vaccines may progressively wane with maturity.[21]

Adaptive Immunity

The cellular and antibody-mediated effector responses of the adaptive immune system are crucial, both in defense against invading pathogens and in mediating allergic and inflammatory diseases. The number and functionality of T and B cell subsets significantly impact protective and pathological responses in the periphery and mucosa. Therefore, it is vital to understand their developmental regulation in order to fully realize the complexity of neonatal immunity. Longitudinal studies using human cord and neonatal blood have described the maturation trajectory of adaptive immune responses, as well as identified unique functional features that may contribute to enhanced neonatal disease susceptibility.

Fetal and adult hematopoietic stem cells (HSCs) possess several differences, including tissue localization and functional status. Fetal HSC are mainly located in the liver, while adult HSC are maintained in the bone marrow. In addition, fetal HSC are skewed towards proliferation and display a unique set of transcription factors as compared to adult HSC. This has been demonstrated in elegant studies by J. Mold and colleagues, which suggest that maturity of lymphocytes during infancy may in fact be a layering process of overlapping fetal and adult populations.[23] Fetal lymphocytes display autoreactive antigen receptors, a feature that has been linked with reduced antigen receptor diversity and lower expression of terminal deoxynucleotidyl transferase, a key enzyme in generating TCR and BCR diversity. Despite this limitation, studies in mice have suggested that fetal lymphocyte receptors may possess enhanced cross-reactivity, with fetal T cells capable of responding to a larger range of encountered antigens and pathogens. While it remains to be seen whether fetal or adult lymphocytes separately contribute to susceptibility to infection in the infant, it is intriguing to speculate that the shift in function and decline in lymphocytes with maturity is in fact due to waning of neonatal populations and overtaking of adult populations.[24]

Human neonatal blood lymphocytes differ from their adult counterparts both in expression of cell surface markers and in overall number within the circulation, emphasizing the need to evaluate lymphocyte subsets both by frequency and absolute counts.[25,26] Total leukocyte counts in humans are highest at birth and decline with maturation.[25,27–29] Total blood lymphocyte counts are elevated at 1 week of age and remain high, relative to adult blood values, until approximately 6 years of age.[25,28] A peak of blood T lymphocyte counts and frequency occurs at 1 week of age, followed by a peak of blood B lymphocyte counts and frequency at 2–3 months of age (FIGURE 1).[28] CD4+ T cell numbers increase from birth to 3 months, while CD8+ T cells increase from birth to 11 months of age.[30] Following the early rise in circulating numbers after birth, peripheral blood T lymphocyte populations show a modest, but progressive, reduction in frequency and number. In contrast, B lymphocyte populations remain relatively stable both in frequency and number until 2 years of age.[28,31] For both T and B peripheral blood lymphocytes, a dramatic drop in absolute counts occurs between the ages of 7 and 17 years.[25] Although B lymphocyte frequency continues to fall with the onset of adulthood, the frequency of T lymphocytes appears to slightly increase at this age.[25,29] Based on these findings alone, it is clear that the first 6 years of life are an important period of immune system growth and maturation. The pronounced changes in T and B lymphocyte populations observed during the first few months of life suggest that this is a particularly significant developmental window within which environmental stimuli may impact normal development.

Although studies demonstrate that most cord blood T cells are in a naïve state and express CD45RA,[32] several studies on neonatal CD4+ and CD8+ T cells demonstrate a fraction that in fact do express activation markers (CD69) and an effector/memory phenotype (CD45RA- and CCR7-).[33] Thus, it has been suggested that during the prenatal to postnatal transition, a new wave of T cells with a unique functionality replaces the prenatal pool.[24] Neonatal T cells are generally believed to display an altered functional capacity. A high proportion of neonatal T cells display surface CD1 and CD38, as well as co-expression of CD4 and CD8, indicating that these cells possess an immature functional status.[26] The relative frequencies of activation marker positive (CD25 and HLA-DR) subpopulations of CD3+ T lymphocytes are considerably higher in adults versus infants, with CD25 expression peaking at 1 week of age and HLA-DR demonstrating little change in expression over the first year.[28]

Neonatal T cells may respond to challenge, but the magnitude or speed of the response can vary compared to adult

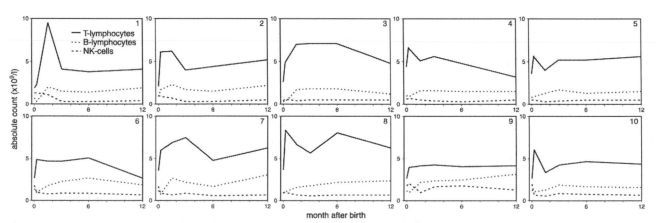

FIGURE 1 Numbers of T-lymphocytes, B-lymphocytes, and NK-cells in blood (cells ×/10^9/litre) from birth to 1 year of age. Each graph represents one individual infant (infants 1–10). *Reproduced from de Vries et al.*[28]

T cells.[34] The cytokine profile of neonates is biased towards a Th2 response as a result of deficient IL-12 production by neonatal dendritic cells, hypomethylation of the Th2 locus, and hyper-responsiveness to IL-4 by T cells.[26,34] Correspondingly, because IL-12 plays a key role in the differentiation of Th1 cells, interferon gamma expression is reduced in neonatal lymphocytes.[35] In addition, IL-4, a key cytokine in Th2 cell differentiation, is deficient during early life,[36] and cord blood T cells produce less IL-5 and GM-CSF in the presence of dendritic cells compared to adult cells.[37] In contrast, expression of IL-1, IL-6, IL-13, TNF-alpha, and interferon-alpha in neonatal cells is comparable to that of adults.[35,38–41]

In addition, neonatal T cells display a reduced ability for TCR engagement and activation as well as an enhanced susceptibility to anergy post stimulation.[26] Recent studies have demonstrated differential expression of microRNAs that regulate cytokine gene transcription between infants and adults, which may further contribute to age-dependent differences in cytokine production.[34] A deficiency of cytokines that is central to the differentiation of memory/effector T helper cells at birth may represent a key regulatory step in the subsequent structuring of a unique immune cell repertoire that is personalized to the newborn environment. Interestingly, proportionally more neonatal than adult T cells undergo homeostatic proliferation and begin secreting cytokines in vitro in response to IL-7 and IL-15 alone. Increased cytokine receptor expression might make the neonatal T cells more prone to clonal expansion and acquisition of an effector-like phenotype without exposure to cognate antigen.[42]

CD4+ T cell immunity, particularly Th1 responses, are reduced during infancy, which may contribute to enhanced susceptibility to HSV and MTB infections for which Th1 responses are key for protection. It has been proposed that reduced CD154 expression contributes to limited antigen-specific induction of interferon gamma synthesis. In addition, decreased IL-2 production by naïve CD4+T cells in infants is linked to anergic status post activation. Despite functional limitations, it appears that neonatal CD4+ T cells have enhanced calcium signaling and Erk phosphorylation post activation compared to adult cells, which has been linked to higher levels of the microRNA MiR-181a in neonates. However, this enhanced upstream signaling is decoupled from AP-1 dependent transcription, thereby preventing robust T cell activation.[43] Despite deficiencies in neonatal adaptive immunity, strong, adult level T cell responses can develop in human infants under certain conditions, as is seen in BCG vaccination (Th1 response) and in congenital cytomegalovirus infection (CTL response).[42] However, the mechanistic explanation for these conditionally robust responses is still not well understood.

Lymphoproliferation and cytokine studies on postnatally collected lymphocytes indicate that a newborn does come in contact with antigens during gestation, most likely via transplacental transfer.[44–51] Sources of such antigens include common environmental allergens such as house dust mite and milk protein, as well as parasites. At present, it is not known how this apparent development of immune responses to antigens in utero can affect the maturation of immunity and alter subsequent development of allergic disease.

Th17 cells are involved in protection against bacterial and fungal infections, particularly at mucosal sites. Interestingly, naïve CD4+ T cells from preterm and term infants were found to have a higher capacity to develop into Th17 cells in vitro compared to adults, expressed higher levels of Th17-cell signaling components and transcription factors favoring Th17 differentiation and produced higher levels of Th17 cytokines. However, the differentiation capacity, functionality, and mucosal homing ability of preterm neonatal Th17 cells may be limited in vivo, which may contribute to enhanced susceptibility to disease in early life.[52] Further research into in vivo ontogeny of Th17 cells is needed to elucidate phenotypic and functional differences between neonatal and adult populations.

Recent studies have shown that the fetal immune response is geared towards a tolerogenic phenotype, with CD4+CD25hiFoxP3+ regulatory T cells prominent in circulation. This is most likely responsible for a controlled, tolerant response to a vast range of maternal antigens during gestation. Although no significant difference has been observed in the frequency of CD4+CD25hiFoxP3+ Tregs between fetal and neonatal thymocytes, higher levels of TGFβ in fetal peripheral lymph nodes has been suggested to skew towards peripheral Treg generation. CD4+CD25+ Treg cells begin to appear in peripheral blood from 4 to 10 days after birth.[24] Frequency of naïve Tregs in peripheral blood within the CD4+T cell compartment decreases with age, with percentages around 4–10% in cord blood decreasing to around 1% in adulthood.[53] In comparison, effector Tregs follow the opposite trajectory. A very low percentage of effector Tregs is found in cord blood (0–0.5%), while the percentage in young adults and in older adults are 1–2.5% and 1–4%, respectively.[53] This reciprocal pattern for naïve and effector Tregs may be reflective of increased exposure to environmental factors and the necessity for regulatory cells to control an increasing pool of effector lymphocytes.

Compared to adults, infant naïve B cells have decreased levels of CD21, CD40, CD80, and CD86 molecules involved in B cell activation and maturation. There is a higher frequency and cell count for CD1c+ and CD5+ B lymphocytes within the first year of life, as compared with adult values.[28] Absolute counts of B cells expressing CD1c, an MHC-like molecule that presents lipids to CD1-restricted T cells, continue to rise up until 3 months of age, yet the frequency of this population declines following the first week of life. The functional importance of this trajectory is at present unknown. In contrast, both absolute counts and frequency of CD5+ cells peak at 6 weeks of age.[28] CD5+ B lymphocytes, prominent in neonatal circulation, have now been described as B-1 cells: a distinct, self-renewing subset of IgM secreting B lymphocytes that exhibit a restricted V-region repertoire. The secretion of highly cross-reactive IgM by this B lymphocyte subset may be an important early defense mechanism for the immature infant immune system against a broad range of antigens. In addition, recent studies have suggested roles for B-1 cells in tissue homeostasis by mediating clearance of apoptotic cells and by tuning cell survival and inflammation.[24,54]

In addition, germinal center responses are impaired during the first 4 months of life, and antibody responses are delayed, short-lived, and have impaired affinity as compared to adults[34] Production of significant levels of immunoglobulin isotypes other than IgM in response to antigenic challenge is poor.[55] Neonatal B lymphocytes can secrete IgE in vitro provided there are high levels of exogenous IL-4: this suggests that the lack of appropriate cytokines in the environment may play a role in this apparent deficiency of immunoglobulin production.[56] Cord blood mononuclear cells express very low levels of CD40 ligand (CD40L).[57–59] As the interaction of CD40L on activated T lymphocytes with CD40 on B lymphocytes is required for isotype switching, the apparent lack of sufficient CD40L on newborn mononuclear cells and lower expression of CD40 on neonatal B cells[34] suggests a potential mechanism whereby secretion of IgA, IgG, and IgE is suppressed.

During gestation, NK cells increase in number in the periphery. Following birth, absolute numbers and frequency of circulating NK cells are reduced and continue to fall during the first year of life, reaching adult levels near age 5.[28,60] CD56^{bright} NK cells, which are known to secrete high amounts of cytokine and do not possess robust cytotoxic capabilities, are slightly increased in newborns as compared to adults. In addition, neonatal NK cells have a higher frequency of CD94/NKG2A, an inhibitory functional receptor, suggesting that neonatal NK cells are more skewed towards inhibition versus activation. Corresponding to these observations, neonatal NK cells have impaired cytolytic abilities, which may be linked to lower levels of perforin and granzyme, lower numbers of cytoplasmic granules, and/or decreased granulation. Lower levels of NK cell activating cytokines, including IL-2 and IL-12, and increased susceptibility to apoptosis during early life are also suggested to limit NK cell functionality. As NK cells are important in protection against intracellular pathogens and have demonstrated involvement in inflammatory diseases, this developmental defect may contribute to increased susceptibility in the infant. For example, lower levels of circulating NK cells are associated with severe respiratory infections with influenza or RSV during early life.[60]

Myeloid-derived suppressor cells (MDSCs), innate immune cells that are produced during inflammation and infection, have been demonstrated to suppress T and NK cell responses. Recently, neutrophilic/granulocytic MDSCs (Gr-MDSCs) were found to be enriched in neonatal cord blood as compared to adult levels. In addition, these cells suppressed Th1, Th2, and Th17 responses, including IFN-γ, IL-5, and IL-17 production, and reduced NK cell function and cytotoxicity.[61] This suggests that Gr-MDSCs may also play a role in increased susceptibility to infection during neonatal life.

As the adaptive immune system is key in mediating protection against invading pathogens and in driving pathological conditions, unique features of the neonatal adaptive immune system likely enhance susceptibility to disease in early life. Understanding the neonatal system is a prerequisite for developing effective vaccinations and therapeutics targeted at infants and children, and in understanding persistent health effects of environmental exposures during development.

POSTNATAL MATURATION OF PULMONARY MUCOSAL IMMUNITY

Resident Immune/Inflammatory Cells of Developing Airways

Airways of newborn infants without pulmonary lesions contain few leukocytes. In contrast, histological examination of lungs from stillborn infants diagnosed with congenital pneumonia show the presence of alveolar macrophages as early as 20 weeks' gestation.[62] Alveolar macrophages are observed histologically within alveoli of healthy preterm *Macaca nemestrina* monkeys at 140 days, albeit at very low numbers.[63] Macrophages have also been identified within the airway compartment by lavage of healthy fetal monkeys at 162–165 days of gestation (term = 168 days).[64] In addition to pulmonary pathology during gestation, alveolar macrophage number within airways correlates with duration of postnatal life for both monkey and man. Alveolar macrophages have been found within airways of preterm human infants that had no apparent pulmonary lesions, but survived for at least 48 hours.[62] Alveolar macrophages in lavage of healthy newborn monkeys are increased 33-fold at 2 days of age (versus fetal) and 4-fold at 3–4 weeks of age (versus 2 days).[64] Based on these data, it has been speculated that in healthy infants, "seeding" of leukocytes within the airways must be initiated following birth and exposure to environmental stimuli; this process likely continues during the first 2 years of life.[65]

A multi-center study with standardized methods has reported that bronchoalveolar lavage from healthy adults contains an average of 85.2% alveolar macrophages, 11.8% lymphocytes, 1.6% neutrophils, and 0.2% eosinophils.[66]

Reference values for bronchoalveolar lavage (BAL) from healthy children do exist and are generally similar to those reported for adults, yet the range in ages sampled was wide (4 months to 16 years), thus limiting the ability to determine age-related effects.[67-72] A single study of 16 children aged 2-32 months without parenchymal lung disease reported that neutrophil frequency was higher in younger children, particularly those less than 12 months of age.[69] There are no significant changes in lavage cell profiles of children between the age groups of 3-8 and 8-14 years, suggesting that accumulation of resident immune/inflammatory cells within airways is essentially complete after the age of three.[71] It is important to note that methods for obtaining airway lavage samples can produce varying results. For example, in preterm infants, leukocyte populations from tracheal aspirates versus deep pulmonary lavage yielded pronounced differences in cellular phenotype.[73] Although this study had too few subjects to draw significant conclusions with regard to correlation of cell profiles with airway levels, the findings suggest that the region of sampling should be taken into consideration for interpretation of results.

In a small study of 18 healthy children (3 months to 10 years), it has been reported that age significantly correlates with the frequency of lymphocytes within lavage; however, there was no other correlation with other lavage cell phenotypes.[67] A direct comparison of airway lavage from children less than 2 years versus 2-6 years of age does show a significant decrease in frequency of alveolar macrophages with age, with a concomitant increase in lymphocyte frequency.[74] Available data on the immunophenotype of lavage lymphocyte subsets in normal children are limited. It has been reported that the overall frequency of B lymphocytes, T lymphocytes, and NK cells in lavage from children (3-16 years of age) is similar to that obtained from adults, although the limited number of subjects and wide range of age groups in this study may mask significant changes that could occur in younger subjects.[75] Several groups have found striking differences in the CD4/CD8 ratio of lavage lymphocytes from children (3 months to 16 years of age); CD4/CD8 ratios ranged from 0.6 to 0.8 for children, in contrast with 1.8-2.7 for adults.[66,67,72,75-78] The reduction in CD4/CD8 values for BAL from children appears to be due to an overall increase in CD8 cell number.[75] Comparative studies in normal infant rhesus macaques have shown similar lymphocyte frequency and CD4/CD8 ratios of 0.5 or less in lung lavage, with a predominant memory T helper cell phenotype that is CD45RA-/CD62L-/CD11a^{lo}.[79]

In a histological analysis of human fetal airway tissues (with no apparent lung abnormalities), it has been reported that T lymphocytes, mast cells, and macrophages were observed within lung interstitium as early as the pseudoglandular stage of development (9-18 weeks).[80] From the pseudoglandular to saccular/alveolar stage, numbers of T lymphocytes and mast cells progressively increased in number within lung interstitium, whereas numbers of tissue macrophages decreased with time in utero. In this same study, neutrophils and B lymphocytes were rarely observed within lung parenchyma up until the saccular/alveolar stage (27-41 weeks); a few B lymphocytes were localized within lymphoid structures of the trachea and small bronchi. Comparatively, a postpartum study conducted in infants (2 weeks-6 months), many of which succumbed to sudden infant death syndrome, reported finding no lymphoid follicular structures or plasma cells within the bronchus at birth.[81] The investigators of the aforementioned study observed small lymphoid nodules within the airway wall as early as 12 hours of life; the structures continued to increase in number after 8 weeks of age. A few B cells with surface IgM were also found in bronchial tissue immediately following birth, with IgA and IgG plasma cells appearing at 4 and 13 weeks, respectively. The ability for plasma cells to undergo isotype switching is an important indicator of developmental immune maturation, as this step is dependent upon T and B lymphocyte interactions via CD40/C40L.

A more recently published study of immune cell profiles in a group of 10 infants (11-840 days) who had died of natural causes unrelated to sudden infant death syndrome reported a prevalence of CD56+, CD8+, Granzyme B+, and CD68+ cells in airways and parenchyma, consistent with a predominant contribution of innate immunity in the very young.[82] In comparison with CD8+ cells, CD4+ were less abundant in conducting airways of human infants, which is consistent with reported BAL CD4/CD8 ratio profiles for young children that are CD8 dominant (ratio less than 1).[66,67,72,75-78,82] While findings in postmortem pediatric lungs are invaluable, it should be noted that inconsistent tissue processing and morphometric methods make global interpretation cautious at best. In light of these challenges, unbiased sampling and quantitation strategies for CD3+CD4+ T cells indicate that the overall volume of this cell type within infant rhesus macaque monkey conducting airways is not dramatically different from adult rhesus macaque monkey conducting airways, suggesting that it is the volume of CD8+ cells within the lung that progressively declines with postnatal development.[83,84]

Bronchus-associated lymphoid tissue (BALT) was first formally described in rabbit lung in the early 1970s.[85,86] Structurally, BALT consists of a follicular aggregation of lymphocytes within the airway lamina propria, containing a central, germinal center-like region of B lymphocytes surrounded by T lymphocytes (reviewed in[87]). There are no afferent venules; endothelial venules within the peripheral regions of BALT allow for trafficking of lymphocytes into the BALT structure. The most distinguishing feature of BALT is a dome-like protrusion into the bronchial epithelium. In the human adult, BALT is not normally found in the lung but can be found under disease conditions.[88,89]

In contrast, BALT is often observed in children and adolescents under apparently healthy conditions, likely indicating the antigenic stimulation and expansion of memory lymphocyte populations that this age group undergoes. In the fetal lung, BALT has been observed as early as 20 weeks and most often correlates with chorioamnionitis or intrauterine pneumonia; BALT is only found in 10% of "normal" fetal lungs.[90] BALT has also been observed within infant airways, with the caveat that post-mortem specimens were obtained from infants that had died from sudden infant death syndrome.[91] No evidence exists to indicate that sudden infant death syndrome is related to the presence of BALT, but because of the limited ability to obtain information from healthy human infants, it is possible that both of these phenomena are physiologically linked. More recent postmortem investigation of BALT in a very limited sample size of school-age children (ages 2–15 years) appears to support the notion that the presence of prominent lymphoid aggregates in the airways is a normal feature, suggesting a role in the establishment of mucosal immunity throughout childhood development.[92] As with lymphoid nodules within the conducting airways, peripheral lymphoreticular aggregates have also been evaluated within alveolar walls of the newborn infant lung. Peripheral lymphoreticular aggregates are collections of lymphocytes but not considered BALT due to the lack of lymph node architecture and domed intrusion into the bronchial epithelium. In the human infant, lymphoreticular aggregates are observed following the first week of life; like bronchial lymphoid follicles, aggregates progressively increase in number during the first year of life.[93]

Developmental Expression of Pulmonary Cytokines and Chemokines

Very little data are available on the expression of cytokines during infancy, particularly with respect to in situ production in the lungs. In the fetal lung, protein and mRNA expression of colony stimulating factor-1 (CSF-1), a hematopoietic growth factor that regulates the development of tissue macrophages, is positively correlated with gestational age and is present as early as 10 weeks of age.[94] In addition, IL-6, CXCL8 (IL-8), and IL-10 protein are detectable throughout the pseudoglandular, canalicular, and alveolar stages of development by standard immunostaining methods; expression is primarily localized to epithelial cells and observed as early as 9–18 weeks gestation.[80] Correspondingly, mRNA and protein for IL-6 receptor and CXCR8 (IL-8 receptor) is detectable in the 8–9 week gestational age lung.[95] CXCR8 mRNA and protein expression is maintained in the 16–17 week gestational lung but in contrast, IL-6 receptor protein is not detectable during this period of development.[95] CCL11 (eotaxin), CCL24 (eotaxin-2), and CCL26 (eotaxin-3) mRNA are detectable in the pseudoglandular stage of human fetal lung development, but only

CCL11 is increased with gestational age.[96] Despite the constitutive presence of eosinophilic chemokines in the fetal lung, there is no corresponding evidence of localized eosinophil accumulation in the normal prenatal lung, suggesting that eotaxins have an alternative developmental function. Indeed, there is some evidence to suggest the eotaxins may contribute to airway epithelial cell proliferation in the perinatal lung.[96–98]

Messenger RNA for IL-1 alpha, IL-1 beta, IL-6, CXCL8, and TNF-alpha is detectable in cells obtained by tracheal aspirates and deep pulmonary lavage from premature infants at 1, 7, and 28 days of age.[73] Infants that eventually develop chronic lung disease such as bronchopulmonary dysplasia show elevated levels of IL-1 beta, IL-6, and CXCL8 in lavage at different time points within the first month of life, correlating with inflammatory cell influx into the airways.[99,100] In animal models, exposure to hyperoxic conditions mimics many of the features of bronchopulmonary dysplasia.[101,102] Interestingly, adult animals are more susceptible to the toxic effects of hyperoxia than newborn animals, suggesting that supplementary mechanisms exist in the young that are no longer present in the adult.[101,103] In newborn rats, CINC-1 and MIP-2 (homologues for human CXCL8 and CXCL2) expression is detectable in whole lung homogenates; expression of these chemokines is increased in response to hyperoxia and occurs in parallel with neutrophil accumulation within airways.[104] In comparison with adult mice, exposure of newborn mice to hyperoxic conditions results in elevated expression of TNF-alpha, IL-1 beta, and IL-6.[105] As an extension of this approach, chemokine expression was evaluated in lung homogenates of mice; newborn mice exposed to hyperoxic conditions showed higher elevations of CCL2 (MCP-1), MIP-2, and CXCL10 (IP-10) mRNA within airways as compared with adult counterparts.[106] Further, chemokine induction in the hyperoxic newborn mouse occurred prior to the onset of significant airway inflammation, whereas in the adult, chemokine expression was concurrent with ongoing airway inflammation. Although cytokines and chemokines have a primary role in the induction and regulation of airways inflammation, it has been suggested that elevation of these proteins in newborn airways in response to hyperoxic conditions also functions as a protective mechanism.[106]

Developmental Regulation of Pulmonary Immune/Inflammatory Cell Trafficking

In the adult, lymphocytes may accumulate within the lung via two different mechanisms: (1) expansion of precursors that are already present within the lung parenchyma or (2) expansion of lymphocytes within regional lymph nodes, followed by trafficking via the circulation into the lung.[107] Because there are limited numbers of precursor or "memory" lymphocytes within neonatal airways, it is likely that the infant is most dependent upon immune cell expansion in regional lymph nodes to respond to environmental exposure to antigens and infectious agents. Trafficking patterns of circulating lymphocytes have been best described in the sheep, whereby cannulation of lymph nodes allows for access to lymphocytes that are draining from specific sites; radioactive labeling and subsequent reinfusion into animals enables one to follow recirculation patterns of lymphocytes obtained from regional lymph nodes. A direct comparison of lymphocyte migration patterns in fetal versus adult sheep suggests a significant incapacity of fetal lymphocytes to be transported to mucosal sites such as the small intestine or lung.[108]

A related finding in a rodent model of infectious disease supports the notion that trafficking of lymphocytes to mucosal organs during the fetal/neonatal period is somewhat restricted. It has been shown that, unlike adult mice, neonatal mice exhibit a 3-week delay in the initiation of a pulmonary inflammatory response to *Pneumocystis carinii* infection.[109] Adoptive transfer of normal splenocytes from adult mice into neonatal SCID mice does not eliminate the delay in pulmonary inflammatory response to *Pneumocystis carinii* infection. In reverse, adoptive transfer of splenocytes from neonatal mice into adult SCID mice does permit resolution of *Pneumocystis carinii* infection.[110] These findings clearly demonstrate that newborn lymphocytes are functional and capable of responding to infectious agents such as *Pneumocystis carinii*. Importantly, the data suggest that the neonatal lung environment is the limiting factor in resolving infection. Further analysis showed fewer antigen presenting cells and reduced efficiency of phagocytic activity within neonatal mouse airways; both factors could contribute to the apparent ineffectiveness of the newborn lung to promote robust lymphocyte recruitment into the airways in response to a challenge with an infectious organism.[109]

Pulmonary Defense Mechanisms During Infancy

Alveolar Macrophage

As discussed earlier, the alveolar macrophage is the predominant leukocyte phenotype in the lung for all age groups (greater than 89%). It has been reported that the frequency of alveolar macrophages in BAL is significantly higher in children less than 2 years of age (98% frequency), as compared with children and teenagers up to 17 years of age (91–92% frequency).[74] The predominance of macrophages within airways of infants is likely to be representative of the first level of innate immunity established for the "naïve" lung; as the infant matures and comes in contact with environmental antigens, cells from the adaptive arm of the immune system will contribute to the resident leukocyte population of the airways.

Functionally, the alveolar macrophage of the infant lung is not identical to that of the adult. As compared with their more mature counterparts, alveolar macrophages from children under 2 years of age express less HLA-DR, are impaired with regards to respiratory burst (as measured by the NBT assay), and produce less IL-1 and TNF-alpha in response to LPS stimulation.[74] In the newborn monkey, the ability of alveolar macrophages to migrate in response to endotoxin-activated plasma is significantly inhibited as compared with adults; chemotactic ability is increased to nearly adult levels as early as 6 days of age.[111] In a similar fashion, intracellular killing of *Candida albicans* by newborn monkey alveolar macrophages is much reduced, but functions at nearly adult capacity by 6 days of age. The ability to phagocytose *Candida albicans* is also much reduced in the newborn monkey alveolar macrophage; however, this function does not reach adult capacity until after 6 months of age. In contrast with the monkey, newborn rat macrophages have greater phagocytic capacity for certain gram-negative and positive bacteria as compared with adults and may represent a compensatory mechanism for the immature neonatal immune system.[112] These findings emphasize the distinctive immunological differences between primate and other laboratory animal species.

Dendritic Cells

Dendritic cells are the primary antigen-presenting cells of the lung and are critical for the immunosurveillance of inhaled antigens. By the 13th week of gestation, HLA-DR positive dendritic-like cells are observed in the human fetal lung.[113] HLA-DR positive cells within the fetal lung are dispersed as single cells within the interstitium, do not stain for markers of macrophage/monocyte lineage, and increase in frequency with gestational age (to 21 weeks). Studies performed in the rat demonstrate that MHC Class II positive dendritic cells are very few in number within neonatal airways.[114,115] Isolated MHC Class II positive dendritic cells from the fetal rat lung are not as effective as adult-derived cells in stimulating cell proliferation within the context of an autologous mixed lymphocyte reaction, yet, at birth, dendritic cells of the newborn rat function nearly as well as in the adult.[115] A conflicting study reported that MHC class II positive dendritic cells obtained from airways of very young rats respond poorly to maturation signals via GM-CSF and provide an attenuated response to microbial challenge,[116] which is now supported by a more recent study in the near-term fetal baboon showing a reduced ability to phagocytose *E. coli* in comparison with adult dendritic cells.[117] Dendritic cells isolated from lung tissues of near-term fetal baboons also exhibit lower levels of MHC Class II, CD11c, and CD86 as compared with adult counterparts.[117] Cumulatively, these studies suggest that lung-derived dendritic cells from the newborn have limited functional competency, as compared with their adult counterparts.

Epithelium

While originally thought to be a cell type that contributes exclusively towards lung function, the airway epithelium is now appreciated as an important immune entity within the pulmonary mucosa (reviewed in[118]). Both conducting epithelial cells as well as type II cells of the gas exchange areas of the lung have the ability to respond to PAMPS and proinflammatory cytokines. Moreover, all lung epithelial cells can produce cytokines that promote and exacerbate immune responses. With the change in paradigm associated with function of lung epithelia, additional consideration should be given to the potential role of neonatal airway epithelial cells as critical mediators of environmental signaling to the immune system.

At present, the capacity for airway epithelial cells to function as effective antigen presenting cells remains poorly defined. MHC class II expression by airway epithelial cells has been reported in fetuses greater than 21 weeks of age, primarily in conjunction with lung inflammation.[119] Most infants at birth express MHC class II on airway epithelium regardless of the presence of an inflammatory response. Secretory component (receptor for IgM and dimeric IgA) is also expressed by airway epithelial cells at birth, suggesting a role in transport of immunoglobulins across the epithelial barrier. Despite the apparent ability to contribute to adaptive immunity, the airway epithelium at infancy is limited in responsiveness to microbial challenge, as evidenced by recent in vitro studies directly comparing airway epithelium from very young rhesus monkeys with that of adults. Synthesis of CXCL8 and IL-6 protein in response to lipopolysaccharide treatment is significantly attenuated in infant monkey airway epithelial cells in comparison with adult airway epithelium.[120] Constitutively, adult monkey airway epithelial cells express higher levels of TLR10, IL-6, and IL-1 alpha as compared with infant monkey airway epithelial cells. Surprisingly, infant monkey airway epithelial cells express higher levels of petidoglycan recognition protein 2(PGLYRP2) and PGLYRP3; relatively little is known about this unique class of PPR, but these molecules may serve as important enhancers of innate immunity within the lung while adaptive immunity is established in the first year of life.

CONCLUSIONS

Epidemiological studies suggest that early childhood immune-related events can have a profound impact on the development of airways disease later in life (reviewed by Holt).[121] Experimental data to support this notion in the very young human population does not exist and cannot be obtained due to ethical reasons. Limited descriptive findings from post-mortem human specimens and peripheral blood have provided important insights into the dramatic changes that occur within the pulmonary/systemic immune system during the first year of life. Yet, these data provide a restricted view of the developmental dynamics that occur within the airways during this important period of development. In addition, data obtained from adult populations are not comparable to infants owing to immune changes associated with introduction of antigens from the environment. If the infant immune system (pulmonary or otherwise) is plastic and can be "molded," then the prudent introduction of therapeutics or antigenic stimuli must occur at this time. The caveat to immune modulation of neonates for medicinal purposes is that this approach may in fact have a detrimental effect on protective immunity towards pathogens or could ultimately promote autoimmune reactivity. The continued use of relevant animal models to address mechanistic questions with regard to immune events during postnatal development will begin to fill this large gap in knowledge and ultimately support the safety and efficacy of clinical trials for long-term immune protection in humans.

ACKNOWLEDGMENTS

Supported by NIH ES011617, NIH ES000628, NIH HL081286, NIH HL097087, NIH OD011107, EPA STAR Grant 832947, and NIH T32 HL007013.

REFERENCES

1. Dos Santos C, Davidson D. Neutrophil chemotaxis to leukotriene B4 in vitro is decreased for the human neonate. *Pediatr Res* 1993;**33**:242–6.
2. Anderson DC, Hughes BJ, Wible LJ, Perry GJ, Smith CW, Brinkley BR. Impaired motility of neonatal PMN leukocytes: relationship of abnormalities of cell orientation and assembly of microtubules in chemotactic gradients. *J Leukoc Biol* 1984;**36**:1–15.
3. Bortolussi R, Howlett S, Rajaraman K, Halperin S. Deficient priming activity of newborn cord blood-derived polymorphonuclear neutrophilic granulocytes with lipoppolysaccharide and tumor necrosis factor-alph triggered with formyl-methionyl-leucyl-phenylalanine. *Pediatr Res* 1993;**34**:243–8.
4. Harris MC, Shalit M, Southwick FS. Diminished actin polymerization by neutrophils from newborn infants. *Pediatr Res* 1993;**33**:27–31.
5. Nussbaum C, Gloning A, Pruenster M, Frommhold D, Bierschenk S, Genzel-Boroviczeny O, et al. Neutrophil and endothelial adhesive function during human fetal ontogeny. *J Leukoc Biol* 2013;**93**:175–84.
6. Filias A, Theodorou GL, Mouzopoulou S, Varvarigou AA, Mantagos S, Karakantza M. Phagocytic ability of neutrophils and monocytes in neonates. *BMC Pediatr* 2011;**11**:29.
7. Levy O, Martin S, Eichenwald E, Ganz T, Valore E, Carroll SF, et al. Impaired innate immunity in the newborn: newborn neutrophils are deficient in bactericidal/permeability-increasing protein. *Pediatrics* 1999;**104**:1327–33.
8. Moriguchi N, Yamamoto S, Isokawa S, Andou A, Miyata H. Granulocyte functions and changes in ability with age in newborns; Report no. 1: flow cytometric analysis of granulocyte functions in whole blood. *Pediatr Int* 2006;**48**:17–21.

9. Yost CC, Cody MJ, Harris ES, Thornton NL, McInturff AM, Martinez ML, et al. Impaired neutrophil extracellular trap (NET) formation: a novel innate immune deficiency of human neonates. *Blood* 2009;**113**:6419–27.

10. Song C, Wang C, Huang L. Human neonatal neutrophils are resistant to apoptosis with lower caspase-3 activity. *Tohoku J Exp Med* 2011;**225**:59–63.

11. Hanna N, Vasquez P, Pham P, Heck DE, Laskin JD, Laskin DL, et al. Mechanisms underlying reduced apoptosis in neonatal neutrophils. *Pediatr Res* 2005;**57**:56–62.

12. Pahwa SG, Pahwa R, Grimes E, Smithwick E. Cellular and humoral components of monocyte and neutrophil chemotaxis in cord blood. *Pediatr Res* 1977;**11**:677–80.

13. Pedraza-Sanchez S, Hise AG, Ramachandra L, Arechavaleta-Velasco F, King CL. Reduced frequency of a CD14+ CD16+ monocyte subset with high toll-like receptor 4 expression in cord blood compared to adult blood contributes to lipopolysaccharide hyporesponsiveness in newborns. *Clin Vaccine Immunol* 2013;**20**:962–71.

14. Murphy FJ, Reen DJ. Differential expression of function-related antigens on newborn and adult monocyte subpopulations. *Immunology* 1996;**89**:587–91.

15. Nguyen M, Leuridan E, Zhang T, De Wit D, Willems F, Van Damme P, et al. Acquisition of adult-like TLR4 and TLR9 responses during the first year of life. *PLoS One* 2010;**5**:e10407.

16. Poltorak A, He X, Smirnova I, Liu MY, Van Huffel C, Du X, et al. Defective LPS signaling in C3H/HeJ and C57BL/10ScCr mice: mutations in Tlr4 gene. *Science* 1998;**282**:2085–8.

17. Broz P, Monack DM. Newly described pattern recognition receptors team up against intracellular pathogens. *Nat Rev Immunol* 2013;**13**:551–65.

18. Dasari P, Zola H, Nicholson IC. Expression of Toll-like receptors by neonatal leukocytes. *Pediatr Allergy Immunol* 2011;**22**:221–8.

19. Corbett NP, Blimkie D, Ho KC, Cai B, Sutherland DP, Kallos A, et al. Ontogeny of Toll-like receptor mediated cytokine responses of human blood mononuclear cells. *PLoS One* 2010;**5**:e15041.

20. Burl S, Townend J, Njie-Jobe J, Cox M, Adetifa UJ, Touray E, et al. Age-dependent maturation of Toll-like receptor-mediated cytokine responses in Gambian infants. *PLoS One* 2011;**6**:e18185.

21. Lisciandro JG, Prescott SL, Nadal-Sims MG, Devitt CJ, Pomat W, Siba PM, et al. Ontogeny of Toll-like and NOD-like receptor-mediated innate immune responses in Papua New Guinean infants. *PLoS One* 2012;**7**:e36793.

22. Martino D, Holt P, Prescott S. A novel role for interleukin-1 receptor signaling in the developmental regulation of immune responses to endotoxin. *Pediatr Allergy Immunol* 2012;**23**:567–72.

23. Mold JE, Venkatasubrahmanyam S, Burt TD, Michaelsson J, Rivera JM, Galkina SA, et al. Fetal and adult hematopoietic stem cells give rise to distinct T cell lineages in humans. *Science* 2010;**330**:1695–9.

24. Mold JE, McCune JM. Immunological tolerance during fetal development: from mouse to man. *Adv Immunol* 2012;**115**:73–111.

25. Erkeller-Yuksel FM, Deneys V, Yuksel B, Hannet I, Hulstaert F, Hamilton C, et al. Age-related changes in human blood lymphocyte subpopulations. *J Pediatr* 1992;**120**:216–22.

26. Holt PG, Upham JW, Sly PD. Contemporaneous maturation of immunologic and respiratory functions during early childhood: implications for development of asthma prevention strategies. *J Allergy Clin Immunol* 2005;**116**:16–24. quiz 5.

27. Kato I. Leukocytes in infancy and childhood. *J Pediatr* 1935;**7**:7–15.

28. de Vries E, de Bruin-Versteeg S, Comans-Bitter WM, de Groot R, Hop WC, Boerma GJ, et al. Longitudinal survey of lymphocyte subpopulations in the first year of life. *Pediatr Res* 2000;**47**:528–37.

29. Schultz C, Reiss I, Bucsky P, Gopel W, Gembruch U, Ziesenitz S, et al. Maturational changes of lymphocyte surface antigens in human blood: comparison between fetuses, neonates and adults. *Biol Neonate* 2000;**78**:77–82.

30. Tsao PN, Chiang BL, Yang YH, Tsai MJ, Lu FL, Chou HC, et al. Longitudinal follow-up of lymphocyte subsets during the first year of life. *Asian Pac J Allergy Immunol* 2002;**20**:147–53.

31. Comans-Bitter WM, de Groot R, van den Beemd R, Neijens HJ, Hop WC, Groeneveld K, et al. Immunophenotyping of blood lymphocytes in childhood. Reference values for lymphocyte subpopulations. *J Pediatr* 1997;**130**:388–93.

32. Beck R, Lam-Po-Tang PR. Comparison of cord blood and adult blood lymphocyte normal ranges: a possible explanation for decreased severity of graft versus host disease after cord blood transplantation. *Immunol cell biol* 1994;**72**:440–4.

33. Michaelsson J, Mold JE, McCune JM, Nixon DF. Regulation of T cell responses in the developing human fetus. *J Immunol* 2006;**176**:5741–8.

34. PrabhuDas M, Adkins B, Gans H, King C, Levy O, Ramilo O, et al. Challenges in infant immunity: implications for responses to infection and vaccines. *Nat Immunol* 2011;**12**:189–94.

35. Wilson CB, Westall J, Johnston L, Lewis DB, Dower SK, Alpert AR. Decreased production of interferon-gamma by human neonatal cells. Intrinsic and regulatory deficiencies. *J Clin Invest* 1986;**77**:860–7.

36. Tang ML, Kemp AS. Ontogeny of IL-4 production. *Pediatr Allergy Immunol* 1995;**6**:11–9.

37. Adkins B. T-cell function in newborn mice and humans. *Immunology today* 1999;**20**:330–5.

38. Weatherstone KB, Rich EA. Tumor necrosis factor/cachectin and interleukin-1 secretion by cord blood monocytes from premature and term neonates. *Pediatr Res* 1989;**25**:342–6.

39. Yachie A, Takano N, Yokoi T, Kato K, Kasahara Y, Miyawaki T, et al. The capability of neonatal leukocytes to produce IL-6 on stimulation assessed by whole blood culture. *Pediatr Res* 1990;**27**:227–33.

40. Cederblad B, Riesenfeld T, Alm GV. Deficient herpes simplex virus induced interferon-alpha production by blood leukocytes of preterm and term newborn infants. *Pediatr Res* 1990;**27**:7–10.

41. Williams TJ, Jones CA, Miles EA, Warner JO, Warner JA. Fetal and neonatal IL-13 production during pregnancy and at birth and subsequent development of atopic symptoms. *J Allergy Clin Immunol* 2000;**105**:951–9.

42. Adkins B, Leclerc C, Marshall-Clarke S. Neonatal adaptive immunity comes of age. *Nat Rev Immunol* 2004;**4**:553–64.

43. Palin AC, Ramachandran V, Acharya S, Lewis DB. Human neonatal naive CD4+ T cells have enhanced activation-dependent signaling regulated by the microRNA miR-181a. *J Immunol* 2013;**190**:2682–91.

44. Thornton CA, Vance GH. The placenta: a portal of fetal allergen exposure. *Clin Exp Allergy* 2002;**32**:1537–9.

45. Jones AC, Miles EA, Warner JO, Colwell BM, Bryant TN, Warner JA. Fetal peripheral blood mononuclear cell proliferative responses to mitogenic and allergenic stimuli during gestation. *Pediatr Allergy Immunol* 1996;**7**:109–16.

46. Van Duren-Schmidt K, Pichler J, Ebner C, Bartmann P, Forster E, Urbanek R, et al. Prenatal contact with inhalant allergens. *Pediatr Res* 1997;**41**:128–31.

47. Szepfalusi Z, Huber WD, Ebner C, Granditsch G, Urbanek R. Early sensitization to airborne allergens. *Int Arch Allergy Immunol* 1995;**107**:595–8.

48. Szepfalusi Z, Nentwich I, Gerstmayr M, Jost E, Todoran L, Gratzl R, et al. Prenatal allergen contact with milk proteins. *Clin Exp Allergy* 1997;**27**:28–35.

49. Prescott SL, Macaubas C, Holt BJ, Smallacombe TB, Loh R, Sly PD, et al. Transplacental priming of the human immune system to environmental allergens: universal skewing of initial T cell responses toward the Th2 cytokine profile. *J Immunol* 1998;**160**:4730–7.

50. Novato-Silva E, Gazzinelli G, Colley DG. Immune responses during human schistosomiasis mansoni. XVIII. Immunologic status of pregnant women and their neonates. *Scand J Immunol* 1992;**35**:429–37.

51. Fievet N, Ringwald P, Bickii J, Dubois B, Maubert B, Le Hesran JY, et al. Malaria cellular immune responses in neonates from Cameroon. *Parasite Immunol* 1996;**18**:483–90.

52. Black A, Bhaumik S, Kirkman RL, Weaver CT, Randolph DA. Developmental regulation of Th17-cell capacity in human neonates. *Eur J Immunol* 2012;**42**:311–9.

53. Sakaguchi S, Miyara M, Costantino CM, Hafler DA. FOXP3+ regulatory T cells in the human immune system. *Nat Rev Immunol* 2010;**10**:490–500.

54. Baumgarth N. The double life of a B-1 cell: self-reactivity selects for protective effector functions. *Nat Rev Immunol* 2011;**11**:34–46.

55. Gathings WE, Kubagawa H, Cooper MD. A distinctive pattern of B cell immaturity in perinatal humans. *Immunol Rev* 1981;**57**:107–26.

56. Pastorelli G, Rousset F, Pene J, Peronne C, Roncarolo MG, Tovo PA, et al. Cord blood B cells are mature in their capacity to switch to IgE-producing cells in response to interleukin-4 in vitro. *Clin Exp Immunol* 1990;**82**:114–9.

57. Durandy A, De Saint Basile G, Lisowska-Grospierre B, Gauchat J-F, Forveille M, Kroczek RA, et al. Undetectable CD40 ligand expression on T cells and low B cell responses to CD40 binding agonists in human newborns. *J Immunol* 1995;**154**:1560–8.

58. Brugnoni D, Airao P, Graf D, Marconi M, Lebowitz M, Plebani A, et al. Ineffective expression of CD40 ligand on cord blood T cells may contribute to poor immunoglublin production in the newborn. *Eur J Immunol* 1994;**24**:1919–24.

59. Nonoyama S, Penix LA, Edwards CP, Lewis DB, Ito S, Aruffo A, et al. Diminished expression of CD40 ligand by activated neonatal T cells. *J Clin Invest* 1995;**95**:66–75.

60. Guilmot A, Hermann E, Braud VM, Carlier Y, Truyens C. Natural killer cell responses to infections in early life. *J Innate Immun* 2011;**3**:280–8.

61. Rieber N, Gille C, Kostlin N, Schafer I, Spring B, Ost M, et al. Neutrophilic myeloid-derived suppressor cells in cord blood modulate innate and adaptive immune responses. *Clin Exp Immunol* 2013;**174**:45–52.

62. Alenghat E, Esterly JR. Alveolar macrophages in perinatal infants. *Pediatrics* 1984;**74**:221–3.

63. Jackson JC, Chi EY, Wilson CB, Truog WE, Teh EC, Hodson WA. Sequence of inflammatory cell migration into lung during recovery from hyaline membrane disease in premature newborn monkeys. *Am Rev Respir Dis* 1987;**135**:937–40.

64. Jackson JC, Palmer S, Wilson CB, Standaert TA, Truog WE, Murphy JH, et al. Postnatal changes in lung phospholipids and alveolar macrophages in term newborn monkeys. *Respir Physiol* 1988;**73**:289–300.

65. Grigg J, Riedler J. Developmental airway cell biology. The "normal" young child. *Am J Respir Crit Care Med* 2000;**162**:S52–5.

66. The BAL. Cooperative Group Steering Committee. Bronchoalveolar lavage constituents in healthy individuals, idiopathic pulmonary fibrosis, and selected comparison groups. *Am Rev Respir Dis* 1990;**141**:S169–202.

67. Riedler J, Grigg J, Stone C, Tauro G, Robertson CF. Bronchoalveolar lavage cellularity in healthy children. *Am J Respir Crit Care Med* 1995;**152**:163–8.

68. Ratjen F, Bredendiek M, Brendel M, Meltzer J, Costabel U. Differential cytology of bronchoalveolar lavage fluid in normal children. *Eur Respir J* 1994;**7**:1865–70.

69. Midulla F, Villani A, Merolla R, Bjermer L, Sandstrom T, Ronchetti R. Bronchoalveolar lavage studies in children without parenchymal lung disease: cellular constituents and protein levels. *Pediatr Pulmonol* 1995;**20**:112–8.

70. Khan TZ, Wagener JS, Bost T, Martinez J, Accurso FJ, Riches DW. Early pulmonary inflammation in infants with cystic fibrosis. *Am J Respir Crit Care Med* 1995;**151**:1075–82.

71. Heaney LG, Stevenson EC, Turner G, Cadden IS, Taylor R, Shields MD, et al. Investigating paediatric airways by non-bronchoscopic lavage: normal cellular data. *Clin Exp Allergy* 1996;**26**:799–806.

72. de Mendonca Picinin IF, Camargos P, Mascarenhas RF, Santos SM, Marguet C. Cell count and lymphocyte immunophenotyping of bronchoalveolar lavage fluid in healthy Brazilian children. *Eur Respir J* 2011;**38**:738–9.

73. LoMonaco MB, Barber CM, Sinkin RA. Differential cytokine mRNA expression by neonatal pulmonary cells. *Pediatr Res* 1996;**39**:248–52.

74. Grigg J, Riedler J, Robertson CF, Boyle W, Uren S. Alveolar macrophage immaturity in infants and young children. *Eur Respir J* 1999;**14**:1198–205.

75. Ratjen F, Bredendiek M, Zheng L, Brendel M, Costabel U. Lymphocyte subsets in bronchoalveolar lavage fluid of children without bronchopulmonary disease. *Am J Respir Crit Care Med* 1995;**152**:174–8.

76. Clement A, Chadelat K, Masliah J, Housset B, Sardet A, Grimfeld A, et al. A controlled study of oxygen metabolite release by alveolar macrophages from children with interstitial lung disease. *Am Rev Respir Dis* 1987;**136**:1424–8.

77. Costabel U, Bross KJ, Ruhle KH, Lohr GW, Matthys H. Ia-like antigens on T-cells and their subpopulations in pulmonary sarcoidosis and in hypersensitivity pneumonitis. Analysis of bronchoalveolar and blood lymphocytes. *Am Rev Respir Dis* 1985;**131**:337–42.

78. Hunninghake GW, Crystal RG. Pulmonary sarcoidosis: a disorder mediated by excess helper T-lymphocyte activity at sites of disease activity. *N Engl J Med* 1981;**305**:429–34.

79. Miller LA, Plopper CG, Hyde DM, Gerriets JE, Pieczarka EM, Tyler NK, et al. Immune and airway effects of house dust mite aeroallergen exposures during postnatal development of the infant rhesus monkey. *Clin Exp Allergy* 2003;**33**:1686–94.

80. Hubeau C, Puchelle E, Gaillard D. Distinct pattern of immune cell population in the lung of human fetuses with cystic fibrosis. *J Allergy Clin Immunol* 2001;**108**:524–9.

81. El Kaissouni J, Bene MC, Thionnois S, Monin P, Vidailhet M, Faure GC. Maturation of B cells in the lamina propria of human gut and bronchi in the first months of human life. *Dev Immunol* 1998;**5**:153–9.

82. Dos Santos AB, Binoki D, Silva LF, de Araujo BB, Otter ID, Annoni R, et al. Immune cell profile in infants' lung tissue. *Ann Anat* 2013;**195**:596–604.

83. Miller LA, Gerriets JE, Tyler NK, Abel K, Schelegle ES, Plopper CG, et al. Ozone and allergen exposure during postnatal development alters the frequency and airway distribution of CD25+ cells in infant rhesus monkeys. *Toxicol Appl Pharmacol* 2009;**236**:39–48.

84. Miller LA, Hurst SD, Coffman RL, Tyler NK, Stovall MY, Chou DL, et al. Airway generation-specific differences in the spatial distribution of immune cells and cytokines in allergen-challenged rhesus monkeys. *Clin Exp Allergy* 2005;**35**:894–906.

85. Bienenstock J, Johnston N, Perey DYE. Bronchial lymphoid tissue. I. Morphologic characteristics. *Lab Invest* 1973;**28**:686–92.

86. Bienenstock J, Johnston N, Perey DYE. Bronchial lymphoid tissue. II. Functional characteristics. *Lab Invest* 1973;**28**:693–8.

87. Tschernig T, Pabst R. Bronchus-associated lymphoid tissue (BALT) is not present in the normal adult lung but in different diseases. *Pathobiology* 2000;**68**:1–8.

88. Hiller AS, Tschernig T, Kleemann WJ, Pabst R. Bronchus-associated lymphoid tissue (BALT) and larynx-associated lymphoid tissue (LALT) are found at different frequencies in children, adolescents and adults. *Scand J Immunol* 1998;**47**:159–62.

89. Pabst R, Gehrke I. Is the bronchus-associated lymphoid tissue (BALT) an integral structure of the lung in normal mammals, including humans? *Am J Respir Cell Mol Biol* 1990;**3**:131–5.

90. Gould SJ, Isaacson PG. Bronchus-associated lymphoid tissue (BALT) in human fetal and infant lung. *J Pathol* 1993;**169**:229–34.

91. Tschernig T, Kleemann WJ, Pabst R. Bronchus-associated lymphoid tissue (BALT) in the lungs of children who had died from sudden infant death syndrome and other causes. *Thorax* 1995;**50**:658–60.

92. Heier I, Malmstrom K, Sajantila A, Lohi J, Makela M, Jahnsen FL. Characterisation of bronchus-associated lymphoid tissue and antigen-presenting cells in central airway mucosa of children. *Thorax* 2011;**66**:151–6.

93. Emery JL, Dinsdale F. The postnatal development of lymphoreticular aggregates and lymph nodes in infants' lungs. *J Clin Pathol* 1973;**26**:539–45.

94. Roth P, Stanley ER. Colony-stimulating factor-1 expression in the human fetus and newborn. *J Leukoc Biol* 1995;**58**:432–7.

95. Dame JB, Juul SE. The distribution of receptors for the pro-inflammatory cytokines interleukin (IL)-6 and IL-8 in the developing human fetus. *Early Hum Dev* 2000;**58**:25–39.

96. Haley KJ, Sunday ME, Porrata Y, Kelley C, Twomey A, Shahsafaei A, et al. Ontogeny of the eotaxins in human lung. *Am J Physiol Lung Cell Mol Physiol* 2008;**294**:L214–24.

97. Cook EB, Stahl JL, Lilly CM, Haley KJ, Sanchez H, Luster AD, et al. Epithelial cells are a major cellular source of the chemokine eotaxin in the guinea pig lung. *Allergy Asthma Proc* 1998;**19**:15–22.

98. Chou DL, Gerriets JE, Schelegle ES, Hyde DM, Miller LA. Increased CCL24/eotaxin-2 with postnatal ozone exposure in allergen-sensitized infant monkeys is not associated with recruitment of eosinophils to airway mucosa. *Toxicol Appl Pharmacol* 2011;**257**:309–18.

99. Kotecha S, Wilson L, Wangoo A, Silverman M, Shaw RJ. Increase in interleukin (IL)-1 beta and IL-6 in bronchoalveolar lavage fluid obtained from infants with chronic lung disease of prematurity. *Pediatr Res* 1996;**40**:250–6.

100. Munshi UK, Niu JO, Siddiq MM, Parton LA. Elevation of interleukin-8 and interleukin-6 precedes the influx of neutrophils in tracheal aspirates from preterm infants who develop bronchopulmonary dysplasia. *Pediatr Pulmonol* 1997;**24**:331–6.

101. Frank L, Bucher JR, Roberts RJ. Oxygen toxicity in neonatal and adult animals of various species. *J Appl Phys* 1978;**45**:699–704.

102. Holm BA, Matalon S, Finkelstein JN, Notter RH. Type II pneumocyte changes during hyperoxic lung injury and recovery. *J Appl Physiol* 1988;**65**:2672–8.

103. Bonikos DS, Bensch KG, Northway Jr WH. Oxygen toxicity in the newborn. The effect of chronic continuous 100 percent oxygen exposure on the lungs of newborn mice. *Am J Pathol* 1976;**85**:623–50.

104. Deng H, Mason SN, Auten RL. Lung inflammation in hyperoxia can be prevented by antichemokine treatment in newborn rats. *Am J Respir Crit Care Med* 2000;**162**:2316–23.

105. Johnston CJ, Wright TW, Reed CK, Finkelstein JN. Comparison of adult and newborn pulmonary cytokine mRNA expression after hyperoxia. *Exp Lung Res* 1997;**23**:537–52.

106. D'Angio CT, Johnston CJ, Wright TW, Reed CK, Finkelstein JN. Chemokine mRNA alterations in newborn and adult mouse lung during acute hyperoxia. *Exp Lung Res* 1998;**24**:685–702.

107. Joel DD, Chanana AD. Comparison of pulmonary and intestinal lymphocyte migrational patterns in sheep. *Ann N Y Acad Sci* 1985;**459**:56–66.

108. Cahill RN, Poskitt DC, Hay JB, Heron I, Trnka Z. The migration of lymphocytes in the fetal lamb. *Eur J Immunol* 1979;**9**:251–3.

109. Garvy BA, Qureshi MH. Delayed inflammatory response to Pneumocystis carinii infection in neonatal mice is due to an inadequate lung environment. *J Immunol* 2000;**165**:6480–6.

110. Qureshi MH, Garvy BA. Neonatal T cells in an adult lung environment are competent to resolve Pneumocystis carinii pneumonia. *J Immunol* 2001;**166**:5704–11.

111. Kurland G, Cheung ATW, Miller ME, Ayin SA, Cho MM, Ford EW. The ontogeny of pulmonary defenses: alveolar macrophage function in neonatal and juvenile rhesus monkeys. *Pediatr Res* 1988;**23**:293–7.

112. Lee PT, Holt PG, McWilliam AS. Role of alveolar macrophages in innate immunity in neonates: evidence for selective lipopolysaccharide binding protein production by rat neonatal alveolar macrophages. *Am J Respir Cell Mol Biol* 2000;**23**:652–61.

113. Hofman FM, Danilovs JA, Taylor CR. HLA-DR (Ia)-positive dendritic-like cells in human fetal nonlymphoid tissues. *Transplantation* 1984;**37**:590–4.

114. Nelson DJ, McMenamin C, McWilliam AS, Brenan M, Holt PG. Development of the airway intraepithelial dendritic cell network in the rat from class II major histocompatibility (Ia)-negative precursors: differential regulation of Ia expression at different levels of the respiratory tract. *J Exp Med* 1994;**179**:203–12.

115. McCarthy KM, Gong JL, Telford JR, Schneeberger EE. Ontogeny of Ia+ accessory cells in fetal and newborn rat lung. *Am J Respir Cell Mol Biol* 1992;**6**:349–56.

116. Nelson DJ, Holt PG. Defective regional immunity in the respiratory tract of neonates is attributable to hyporesponsiveness of local dendritic cells to activation signals. *J Immunol* 1995;**155**:3517–24.

117. Awasthi S, Wolf R, White G. Ontogeny and phagocytic function of baboon lung dendritic cells. *Immunol Cell Biol* 2009;**87**:419–27.

118. Proud D, Leigh R. Epithelial cells and airway diseases. *Immunol Rev* 2011;**242**:186–204.

119. Peters U, Papadopoulos T, Muller-Hermelink HK. MHC class II antigens on lung epithelial of human fetuses and neonates. Ontogeny and expression in lungs with histologic evidence of infection. *Lab Invest* 1990;**63**:38–43.

120. Maniar-Hew K, Clay CC, Postlethwait EM, Evans MJ, Fontaine JH, Miller LA. The innate immune response to LPS in airway epithelium is dependent upon chronological age and antecedent exposures. *Am J Respir Cell Mol Biol* 2013;**49**:710–20.

121. Holt PG. Programming for responsiveness to environmental antigens that trigger allergic respiratory disease in adulthood is initiated during the perinatal period. *Environ Health Perspect* 1998;**106**(Suppl. 3):795–800.

Chapter 11

Development of Antioxidant and Xenobiotic Metabolizing Enzyme Systems

Michelle V. Fanucchi

Department of Environmental Health Sciences, School of Public Health, University of Alabama at Birmingham, Birmingham, AL, USA

IMPORTANCE OF ANTIOXIDANTS AND XENOBIOTIC METABOLIZING ENZYMES

When considering the potential pulmonary toxicity of a compound, a backdrop of developmental issues should be kept in mind. These issues have been addressed in more detail previously[1] and will be only briefly mentioned here. First, one must remember that lung development is a multievent process and occurs during both prenatal and postnatal periods. The initial evagination of the tracheobronchial bud occurs early in gestation and a significant amount of lung growth and development continues after birth. Only a few maturational events need to be complete at birth for survival. Second, overall growth, branching morphogenesis, and cellular differentiation are events common to all stages of lung development. Third, all of these developmental events occur in combination with a steadily increasing total cell mass.

The interplay and balance between the xenobiotic activating enzymes and detoxifying enzyme systems have been shown to be critical factors in dictating the toxic response of the respiratory system to bioactivated compounds in adults.[2–5] This interplay is also important in perinatal animals during pulmonary morphogenetic and differentiation processes. An obvious role of antioxidants in the lung is protecting the lung from oxidative stress at birth, when the lung switches from a relatively hypoxic state to a relatively hyperoxic state. However, the antioxidant enzymes (superoxide dismutase, catalase, and glutathione peroxidase), nonenzymatic antioxidants (glutathione) as well as the xenobiotic metabolizing enzymes (cytochrome P450 monooxygenases, glutathione *S*-transferases, epoxide hydrolases, and glucuronyl transferases) all have important roles in modulating cellular interaction with environmental toxicants. Antioxidants and antioxidant enzymes protect the lung from oxidant pollutants such as ozone and nitrogen dioxide. Xenobiotic metabolizing enzymes act on compounds to make them more water-soluble and increase their rate of elimination. Many lung-targeted toxicants, such as furans,[6] chlorinated hydrocarbons,[7,8] aromatic hydrocarbons,[3,9] indoles,[10,11] and pyrrolizdine alkaloids[12] require bioactivation in order to produce their toxicity. The reactive metabolites of these bioactivated compounds are then detoxified by a number of pathways. Other compounds, such as organometallic[13] and amphiphilic agents,[14] are toxic when introduced to the body and also require metabolic detoxification. This section defines the pattern of differentiation for antioxidant enzyme pathways as well as bioactivation and detoxification enzyme systems during pre- and postnatal lung development.

DEVELOPMENT OF ANTIOXIDANT ENZYME SYSTEMS

During late gestation, changes in the fetal lung include the development of the antioxidant enzymes. This group of enzymes includes superoxide dismutase (SOD), catalase, and glutathione peroxidase (GPx). These enzymes play an important role in the detoxification of highly reactive oxygen metabolites that are produced during normal aerobic cellular respiration as well as during oxidant injury. In general, the pulmonary antioxidant enzyme system develops during the last 10–15% of gestation in humans[15] as well as in laboratory animals such as rats, hamsters, guinea pigs, rabbits, and lambs.[16–24]

Superoxide Dismutase

The antioxidant enzyme SOD rapidly catalyzes the conversion of superoxide anion to hydrogen peroxide and oxygen. Manganese SOD (MnSOD), an inducible form of superoxide dismutase, is predominantly located in the

The Lung. http://dx.doi.org/10.1016/B978-0-12-799941-8.00011-0

mitochondria,[25,26] while the non-inducible copper-zinc SOD (Cu,ZnSOD) is located in the cytosol.[27] There is also an extracellular form of SOD (EC-SOD) localized in extracellular matrix and extracellular fluid.[28,29] EC-SOD is thought to play a role in modulating nitric oxide concentrations by controlling the amount of superoxide anion available to react with it,[30] but very little is known about this form in the developing lung.[31] In most experimental animals studied, SOD activity increases throughout the pre- and postnatal periods, however, there are some species differences. A comparison of the timecourse of expression of pulmonary antioxidant enzymes during lung development is represented in Figure 1.

In rats, pulmonary Cu,ZnSOD activity and total enzyme content peak during the late gestational period (last 10–15% of gestation) and activity peaks again early in the postnatal period, finally reaching adult levels at 4 weeks.[19,22,32–34] Temporal expression of mRNA does not correlate with the activity levels.[34,35] In contrast, pulmonary MnSOD activity and mRNA content remain steady throughout late gestation and the early postnatal period in the rat,[16,32,35] even though total enzyme content increases 6.9-fold.[19] In the mouse, Cu,ZnSOD mRNA expression peaks at 15 days gestation (early pseudoglandular stage of lung development), is low during the later stages of fetal lung development, and peaks again at birth.[35,36] In rabbits, pulmonary Cu,ZnSOD and MnSOD activities also peak in late gestation and in the postnatal period.[21] The only species in which pulmonary EC-SOD protein expression has been evaluated is the rabbit. Expression is low and contained within the intracellular compartment during the fetal period. Expression increases and shifts to the extracellular compartment with increasing age.[31] In contrast to experimental animals, human Cu,ZnSOD and MnSOD activity remains constant during fetal and neonatal lung development, and no surge in expression is seen.[15,37]

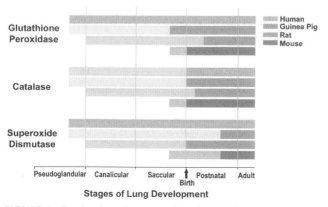

FIGURE 1 Species-dependent expression of antioxidant enzyme protein and/or activity during stages of lung development. Pale portions of bars represent transitional expression, dark portions of bars represent mature expression for human,[16,38] guinea pig,[22] rat,[18,20,23,34,35,101] and mouse.[39]

Glutathione Peroxidase

Glutathione Peroxidase (GPx) is a cytosolic enzyme that catalyzes the reduction of hydrogen peroxide to water and oxygen as well as catalyzing the reduction of peroxide radicals to alcohols and oxygen.[18] As with SOD, GPx activity and total enzyme content increase during late gestation in two of the experimental animals studied: guinea pig and rat (Figure 1).[19,21,22,32,35] However, mRNA expression steadily decreases during fetal development and does not correlate with activity levels.[16,35] In the mouse, however, both GPx activity and mRNA levels peak in late gestation and then activity steadily decreases throughout the postnatal period.[36,38] In humans, GPx activity is lowest at 10–20 weeks gestation, highest during the early postnatal period (41–50 weeks postconception), and drops at 3–5 months of age.[15,39]

Catalase

Like GPx, catalase accelerates the combination of two hydrogen peroxide molecules to produce water and oxygen. Catalase activity steadily increases during both the fetal and postnatal periods of lung development in humans[15] and in all experimental animals studied, including guinea pigs, rats, and mice (Figure 1).[19,22,32,35,38] In contrast to GPx, though, mRNA expression during this time is consistent with the increase in activity.[35] In humans, catalase activity increases 3.5-fold between 10 weeks of gestation and 3 months postnatal age, unlike SOD and GPx, which remain constant throughout human lung development.[15] The increase in catalase activity during the late gestational period has been suggested by McElroy and colleagues to be linked to maturation of the surfactant system and has been correlated temporally to increases in lung dipalmitoylphosphatidylcholine (DPPC) content.[15]

Glutathione

Glutathione (γ-glutatamylcysteinylglycine, GSH), a tripeptide, acts as an antioxidant by directly scavenging free radicals through the donation of a hydrogen atom. In the lung, GSH is found in both intra- and extracellular compartments. In the cell, 85–90% of the GSH resides in the cytosol, 10–15% in the mitochondria, and small percentages in the endoplasmic reticulum and nucleus.[40] The synthesis of glutathione is homeostatically controlled, both inside the cell and outside.[41–44] Glutathione is not taken up by pulmonary cells in its intact form. It is synthesized and utilized through the gamma-glutamyl cycle, a multistep, ATP-dependent process. Very simplistically, it begins with the cleavage of extracellular glutathione and other glutamate-containing peptides by γ-glutamyl transpeptidase, a membrane-bound enzyme. The resulting amino acids can then be moved by specific transporters into the intracellular environment.[45] The rate-limiting step in the synthesis of glutathione, however, is the formation of γ-glutamylcysteine by γ -glutamylcysteine synthetase (γ -GCS). The formation of glutathione is complete when glutathione synthetase catalyzes the addition of glycine to γ -glutamylcysteine to form γ -glutamylcysteinylglycine (glutathione).[45] Even though glutathione plays an important role in protection against both oxidant and reactive metabolite injury, there is very little information available on glutathione levels or gamma-glutamyl cycle enzyme activity during fetal and postnatal lung development.

Evidence from the few rodent studies available suggests that neonates have similar glutathione levels as adults. In rats, pulmonary glutathione levels are at mature levels right before birth, decrease 30% at birth and then increase back to mature levels between 10 and 15 days.[46] The activities of two important enzymes of the gamma-glutamyl cycle, γ-GCS and glutathione synthetase, are similar in the lungs of 1–2 week old mice and adult mice.[38] γ-Glutamyl transpeptidase mRNA is detectable by polymerase chain reaction in late fetal and early postnatal rat lung, and is located in the type II cell.[47] However, pulmonary γ-glutamyl transpeptidase activity is very low in newborn rat, and only gradually increases to mature levels.[47,48] In the late postnatal and adult rat lung, γ-glutamyl transpeptidase is found in higher abundance in Clara cells than in type II cells.[47] Pulmonary glutathione synthetic activity is affected by several factors. One factor, maternal nutrition, has been shown to be very important in regulating the perinatal activity of glutathione cycle enzymes. Pulmonary γ-GCS activity is lower in rat pups from dams that were fed low protein diets than in pups from dams fed diets supplemented with casein.[48] Surgical procedures in newborn guinea pigs also affect glutathione synthetic activity and result in increased pulmonary glutathione content.[49]

Miscellaneous Antioxidants

There is little information regarding the expression of other antioxidants during lung development. Ceruloplasmin, a free radical scavenger, is a major extracellular antioxidant in human lung epithelial fluid.[50] Ceruloplasmin mRNA has been detected in fetal lung of species as varied as mouse and baboon.[51] The mRNA first appears in baboon lung during the pseudoglandular stage (60 days gestation) in bronchial epithelium. By the saccular stage of development (140 days gestation), ceruloplasmin mRNA is detected in all airway epithelium. In mice, ceruloplasmin mRNA is detected during the saccular stage and is expressed in epithelium of all airways. Peroxiredoxin, an antioxidant that reduces hydrogen peroxide to molecular oxygen, has also been described in lungs of newborn baboons.[52] Both pulmonary peroxiredoxin mRNA and activity are expressed at low levels in fetal baboons, and are upregulated by oxygen treatment. Urate and ascorbate (vitamin C) levels in bronchoalveolar lavage fluid have been described in preterm human infants.[53] Urate

levels drop during the first 2 weeks of life, while ascorbate levels drop during the first week then increase during the second week of life.

DEVELOPMENT OF XENOBIOTIC METABOLIZING ENZYME SYSTEMS

Cytochrome P450 Monooxygenases

The cytochrome P450 (CYP) monooxygenases are very important enzymes in both bioactivating and detoxifying compounds. In adults, immunoreactive CYP enzymes have been detected in four pulmonary cell types: the nonciliated bronchiolar (Clara) cell, the type II epithelial cell, the endothelial cell, and the alveolar macrophage. The developmental expression of CYP monooxygenases is the most extensively studied of all the pulmonary xenobiotic metabolizing enzymes. Even so, the expression of pulmonary CYP monooxygenases in the differentiating lung has been evaluated in only five species: rabbit, hamster, mouse, rat, and goat.[54–63] Some of the CYP isozymes (*and representative substrates*) that have been evaluated during lung development include CYP1A1 (*ethoxyresorufin*), CYP2B (*O,O,S-trimethylphosphorothioate*), CYP2F (*naphthalene*), and CYP4B (*2-aminofluorene*). In addition, NADPH CYP reductase, important for electron transport during the catalytic cycle of CYP monooxygenases, has also been evaluated. In general, the first detectable expression of CYP monooxygenases occurs after the development of the smooth endoplasmic reticulum (SER). Protein for NADPH CYP reductase is detected before the monooxygenase isozymes in all species evaluated (Figure 2). The youngest age at which the intracellular protein can be detected immunohistochemically varies

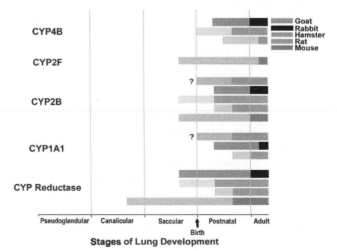

FIGURE 2 Species-dependent expression of CYP monooxygenase protein and/or activity during stages of lung development. Pale portions of bars represent transitional expression, dark portions of bars represent mature expression for goat,[63] rabbit,[55] hamster,[57] rat,[59,66] and mouse.[62] Question mark (?) indicates a lack of available information during the fetal period.

substantially within the five species evaluated. Isozymes CYP2B and CYP2F are expressed earliest: they are detectable in the late fetal stage in mice and hamsters,[56,61] in the early postnatal period of rabbits and rats,[54,58,64,65] and around 6 weeks of age in goats,[62,63] CYP4B is detected 2–3 days of age later. Activity for these proteins is first detected approximately 2–3 days after the protein is immunologically detectable. The temporal and spatial distribution of immunoreactive pulmonary CYP protein has been described in detail for the rabbit.[54] Initially, immunoreactive protein is detected only in the most apical border of a small percentage of the nonciliated cell population. During the period of time in which the amount of detectable protein increases, the distribution changes in two ways. First, an immunologically detectable protein is found in an increasing proportion of the nonciliated cells. Second, the distribution of detectable protein within an individual cell moves from the apex to the base until it is evenly distributed throughout the cell. The timeframe between when the protein is first detectable and when it reaches mature distribution and intensity also varies among species. Mature expression of CYP proteins can take up to 4 weeks in rabbits and mice,[54,61] 3 weeks in hamsters,[56] and as little as 1 week in rats.[58,65] Mature expression of the immunoreactive protein, however, does not necessarily indicate mature enzyme activity, especially in rats and rabbits.[58,65] This suggests that the activity of these enzymes continues to increase after the protein density and organelle composition have reached adult levels. It is critical to understand the temporal and spatial development of CYP monooxygenase activity in order to be able to extrapolate among species.

Glutathione S-Transferase

It is essential to understand the temporal and spatial development of the detoxifying enzymes, including glutathione S-transferase, in order to appreciate the role they play in protecting the lung during development. Glutathione S-transferases (GSTs) are a group of dimeric cytosolic enzymes that catalyze the conjugation of GSH to electrophilic compounds. In human tissues, GSTs are grouped into four classes based on isoelectric points: alpha, mu, pi, and theta.[66–68] All four classes have been described in adult lung, although only the expression of GST alpha, mu, and pi have been evaluated in developing lung of mice,[69,70] rabbits,[71] goats,[62] guinea pigs[71] and humans.[72–75] Unlike CYP monooxygenases, which are expressed in only four cell types in the lung, pulmonary GSTs are expressed in multiple cell types including both ciliated and nonciliated bronchiolar epithelial cells, alveolar epithelial cells, endothelial cells, and smooth muscle cells. The ubiquitous distribution of the transferase enzymes suggests a role in protecting cells from reactive intermediates from both exogenous and endogenous metabolism. In goats, rabbits and guinea pigs, information regarding the expression of GSTs is available for only the postnatal period of

lung development (Figure 3).[62,71] Rabbits and mice at birth have pulmonary GST protein expression and activity levels similar to adults, while GST expression is not mature in goats until later in the postnatal period. In mice, the only laboratory animal in which GST expression has been described during fetal lung development, GST isozymes are present very early in development, however, the isozymes do not reach mature expression levels until early in the postnatal period.[70] In contrast to mice, human fetal lung has been shown to contain more overall GST activity than adult lung.[74] During human lung development, the total activity of pulmonary GST decreases 5-fold between 13 weeks gestation and birth and then remains constant.[75] Pulmonary GST activity is due to a combination of all the isozymes, and the contribution of each isozyme changes during development. In humans,[72,76,77] GST1 (mu) increases as a percentage of total GST activity over development, from 5% at 10 weeks gestation (pseudoglandular stage of lung development) to 25% at birth and 60% in adult tissue. GST2 (alpha) contributes 50% of the total activity consistently. GST3 (pi) decreases as a percentage of total activity over development, from 50% of the total GST activity at 10 weeks gestation to 15% at birth, and is almost undetectable in adult tissue. These activity data are supported by immunohistochemical data,[73] GST2 is expressed consistently throughout development: strongly in proximal airway epithelium and weakly in more distal airways in fetus, newborns, and adults. GST3 decreases in expression during development. It is expressed strongly in all epithelial cells in early gestation, and then expression is lost in the distal airways by 24 weeks of gestation.[73,74,78]

Epoxide Hydrolase

There are three general forms of epoxide hydrolase (EH) which all create 1,2-dihydrodiols from epoxides, although

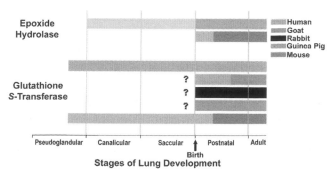

FIGURE 3   Species-dependent expression of glutathione S-transferase (alpha, mu, and pi combined) and epoxide hydrolase (mEH and cEH combined) protein and/or activity during stages of lung development. Pale portions of bars represent transitional expression, dark portions of bars represent mature expression for goat,[63] rabbit,[72,102,103] guinea pig,[72] mouse,[71] and human.[73–76,84–86] Question mark (?) indicates a lack of available information during the fetal period. (This figure is reproduced in color in the color plate section.)

each form has different substrate specificities and tissue distribution.[79] The first form, microsomal EH (mEH), is involved in the conversion of cyclic epoxides and is found in high levels in the smooth endoplasmic reticulum (SER) and in lower levels in other membranous organelles. The second form, cytosolic EH (cEH), hydrates aliphatic epoxides and is found in the cytosol and peroxisomes. The third form, cholesterol EH, catalyzes the hydration of cholesterol epoxides and is located in the microsomal fraction. Substrate specificity may also differ immensely among species and tissues (reviewed in Wixtrom and Hammock[79]). As with most metabolic enzymes, the highest levels of both mEH and cEH activity are found in the liver, and this is true in adults of a wide variety of species: rabbits,[80] humans,[78] rats,[81] and mice.[70,82] The ratio of cytosolic to microsomal EH activity, though, varies by species and by organ. In humans, cEH activity is higher than mEH activity in both the liver and the lung.[78,83–85] In mice, while cEH activity is higher than mEH activity in the liver, the reverse is true in the lung.[86] EH activity also varies within the lung. EH activity is reported to be present in microdissected airways of dogs with the highest activity in distal airway generations and lower activities in proximal airway generations.[87]

There are few studies on the development of epoxide hydrolase (Figure 3). In mice, the predominant pulmonary form of epoxide hydrolase is microsomal, with the cytosolic form found only in vascular smooth muscle.[70] Immunoreactive mEH protein is detectable in airway epithelial cells soon after birth, and reaches mature density and distribution near the age of weaning. However, specific activity for mEH in mouse lung is detectable at mature levels at 7 days postnatal age.[70] In contrast to the mouse, the predominant pulmonary form of epoxide hydrolase in humans is cytosolic and EH mRNA expression gradually increases to adult levels by 65 days postnatal.[55] Increases in fetal human pulmonary EH activity do not correlate with increases in mRNA expression.[85]

Glucuronyl Transferase

Glucuronidation is also a major pathway for the metabolic elimination of parent compounds (such as 4-ipomeanol) or primary metabolites (such as 1-naphthol) and as such can provide important means of protecting extrahepatic tissues from toxicants.[88] As with the other Phase II enzymes discussed in this chapter, glucuronyl transferases are widely distributed throughout tissues, with the highest activity found in the liver. These enzymes have been described in the lungs of adult rabbits,[89] dogs,[87,90] humans,[84] and sheep.[91] Similar to CYP monooxygenases, the distribution of uridine diphosphate-glucuronyl transferase is restricted to bronchial epithelial cells, Clara cells and type II pneumocytes in rat lung. There are reported species-specific differences in activity levels. UDP-glucuronyl-transferase activity has been shown to be evenly distributed throughout

the respiratory tract of the dog and is similar to activity levels found in the liver.[87] In the human,[92] pulmonary UDP-glucuronyl transferase activity is considerably lower than hepatic activity. In contrast to the detailed studies of the development of hepatic glucuronyl transferases, there is only one report on the developmental expression of this enzyme in the lung. In goats, pulmonary UDP-glucuronyl transferase activity has been reported to be lowest at birth and highest in adults. However, the pattern of enzyme expression also varies by substrate used.[62]

Miscellaneous Enzymes

In addition to the four enzyme systems already mentioned, there is a small amount of information available concerning the expression of two other enzymes during lung development: flavin-containing monooxygenases and sulfotransferases.[64,93] Flavin-containing monooxygenases are important oxidative metabolizing enzymes and have much in common with the cytochrome P450 monooxygenases. They have similar molecular weights, are localized in SER, have the highest expression in liver, require NADPH, and have multiple isozymes.[94,95] To date, flavin-containing monooxygenases have only been studied during lung development in rabbits.[64] Activity, protein, and mRNA are all expressed as early as 25 days gestation (canalicular stage). Flavin-containing monooxygenase expression is high prenatally (except for an unexplained decrease at 28 days gestation), drops immediately after birth, and then steadily increases throughout postnatal lung development. This expression pattern matches that of CYP2B4 and CYP4B1 in the rabbit lung.[64] Sulfation is a major detoxification pathway, resulting in a highly water-soluble product.[96,97] There are two major subfamilies: phenol sulfotransferase and hydroxysteroid sulfotransferase, each of which has a different substrate specificity.[98] Human pulmonary hydroxysteroid sulfotransferase, which sulfates steroids and cholesterol, is found in low levels early in gestation (around 56 days gestation) and expression peaks at 1 year after birth.[93] Hydroxysteroid sulfotransferase is expressed in most ciliated, nonciliated, and basal airway cells, but not in mucus-secreting cells. Human phenol sulfotransferase is highly expressed and widely distributed in fetal lung.[93,99] After birth, expression decreases and the distribution is restricted to the proximal airways.

CONCLUSIONS

Humans are exposed to multiple compounds early in life, yet most toxicological studies focus on the effects in adults. In addition, decisions regarding acceptable levels of environmental contaminants are based on adult data, possibly missing the most susceptible portion of the population, our children. Indeed there is growing evidence that exposure to bioactivated pollutants produces much higher pulmonary toxicity in neonates than in adults. Although the human lung needs to be sufficiently formed at birth in order to perform its primary function of gas-exchange, lung development continues for approximately 8–12 years.[100] During both the pre- and postnatal period of lung development, there are rapid rates of cellular differentiation, cell division and alveolarization occurring, making the early postnatal lung uniquely susceptible to injury by environmental oxidant and toxic air pollutants.[1] The enzyme systems responsible for protection from oxidant injury and toxicant-induced injury differentiate during the perinatal period, with the majority of differentiation activity occurring for an extended period of time after birth. Each of these enzyme systems is expressed in a different pattern during pre- and postnatal lung development. Postnatal patterns for expression of these enzyme systems suggest that their distribution and activity in postnatal lung are not reflective of the situation in adults. In the absence of clear epidemiologic evidence suggesting an association between chemical or oxidant exposure and an adverse human health consequence, we still do not, in many instances, understand how results obtained in animal models extrapolate to the human and whether or not there is an age-related susceptibility.

As shown in Figure 1, the pulmonary antioxidant enzyme system in humans appears to mature earlier than the corresponding systems in laboratory animals, although most species follow a similar timecourse of expression. Variability between species becomes much greater when evaluating the Phase I and II enzyme systems (Figures 2 and 3). The few data available indicate that Phase II enzymes are expressed early in lung development, suggesting a role in protecting the lung during the period of rapid cellular differentiation and cell division. CYP monooxygenases are expressed later during lung development, suggestive of a role in protecting the lung from exogenous, rather than endogenous, compounds.

It is clear that there is a lack of information regarding the expression of antioxidants and Phase I and Phase II enzyme systems during lung development in laboratory animals as well as in humans. It is imperative that we obtain more information regarding the temporal and spatial expression of these enzymes in order to understand the mechanisms of susceptibility to environmental insults and how we may extrapolate among species.

REFERENCES

1. Fanucchi M, Plopper C. Pulmonary developmental responses to toxicants. In: Roth RA, editor. *Toxicology of the Respiratory System.* 1st ed. New York: Elsevier Science, Inc.; 1997. pp. 203–20.

2. Buckpitt A, Chang AM, Weir A, Van Winkle L, Duan X, Philpot R, et al. Relationship of cytochrome p450 activity to clara cell cytotoxicity. Iv. Metabolism of naphthalene and naphthalene oxide in microdissected airways from mice, rats, and hamsters. *Mol Pharmacol* 1995;**47**(1):74–81.

3. Verschoyle RD, Carthew P, Wolf CR, Dinsdale D. 1-nitronaphthalene toxicity in rat lung and liver: Effects of inhibiting and inducing cytochrome p450 activity. *Toxicol Appl Pharmacol* 1993;**122**(2):208–13.

4. Buckpitt AR, Franklin RB. Relationship of naphthalene and 2-methylnaphthalene metabolism to pulmonary bronchiolar epithelial cell necrosis. *Pharmacol Ther* 1989;**41**(1–2):393–410.

5. Statham CN, Boyd MR. Distribution and metabolism of the pulmonary alkylating agent and cytotoxin, 4-ipomeanol, in control and diethylmaleate-treated rats. *Biochem Pharmacol* 1982;**31**(8):1585–9.

6. Dutcher JS, Boyd MR. Species and strain differences in target organ alkylation and toxicity by 4-ipomeanol. Predictive value of covalent binding in studies of target organ toxicities by reactive metabolites. *Biochem Pharmacol* 1979;**28**(23):3367–72.

7. Boyd MR, Statham CN, Longo NS. The pulmonary clara cell as a target for toxic chemicals requiring metabolic activation; studies with carbon tetrachloride. *J Pharmacol Exp Ther* 1980;**212**(1):109–14.

8. Moussa M, Forkert PG. 1,1-dichloroethylene-induced alterations in glutathione and covalent binding in murine lung: Morphological, histochemical, and biochemical studies. *J Pathol* 1992;**166**(2):199–207.

9. Reid WD, Ilett KF, Glick JM, Krishna G. Metabolism and binding of aromatic hydrocarbons in the lung. Relationship to experimental bronchiolar necrosis. *Am Rev Respir Dis* 1973;**107**(4):539–51.

10. Yost GS, Kuntz DJ, McGill LD. Organ-selective switching of 3-methylindole toxicity by glutathione depletion. *Toxicol Appl Pharmacol* 1990;**103**(1):40–51.

11. Nocerini MR, Carlson JR, Yost GS. Adducts of 3-methylindole and glutathione: Species differences in organ-selective bioactivation. *Toxicol Lett* 1985;**28**(2–3):79–87.

12. Teicher BA, Crawford JM, Holden SA, Lin Y, Cathcart KN, Luchette CA, Flatow J. Glutathione monoethyl ester can selectively protect liver from high dose bcnu or cyclophosphamide. *Cancer* 1988;**62**(7):1275–81.

13. Hanzlik RP, Stitt R, Traiger GJ. Toxic effects of methylcyclopentadienyl manganese tricarbonyl (mmt) in rats: Role of metabolism. *Toxicol Appl Pharmacol* 1980;**56**(3):353–60.

14. Kacew S, Parulekar MR, Narbaitz R, Ruddick JA, Villeneuve DC. Modification by phenobarbital of chlorphentermine-induced changes in lung morphology and drug-metabolizing enzymes in newborn rats. *J Toxicol Environ Health* 1981;**8**(5–6):873–84.

15. McElroy MC, Postle AD, Kelly FJ. Catalase, superoxide dismutase and glutathione peroxidase activities of lung and liver during human development. *Biochim Biophys Acta* 1992;**1117**(2):153–8.

16. Clerch LB, Massaro D. Rat lung antioxidant enzymes: Differences in perinatal gene expression and regulation. *Am J Phys* 1992;**263** (4 Pt 1):L466–70.

17. Asayama K, Hayashibe H, Dobashi K, Uchida N, Kobayashi M, Kawaoi A, Kato K. Immunohistochemical study on perinatal development of rat superoxide dismutases in lungs and kidneys. *Pediatr Res* 1991;**29**(5):487–91.

18. Frank L, Sosenko IR. Development of lung antioxidant enzyme system in late gestation: Possible implications for the prematurely born infant. *J Pediatr* 1987;**110**(1):9–14.

19. Hayashibe H, Asayama K, Dobashi K, Kato K. Prenatal development of antioxidant enzymes in rat lung, kidney, and heart: Marked increase in immunoreactive superoxide dismutases, glutathione peroxidase, and catalase in the kidney. *Pediatr Res* 1990;**27**(5):472–5.

20. Frank L, Price LT, Whitney PL. Possible mechanism for late gestational development of the antioxidant enzymes in the fetal rat lung. *Biol Neonate* 1996;**70**(2):116–27.

21. Rickett GM, Kelly FJ. Developmental expression of antioxidant enzymes in guinea pig lung and liver. *Development* 1990;**108**(2):331–6.
22. Sosenko IR, Frank L. Thyroid inhibition and developmental increases in fetal rat lung antioxidant enzymes. *Am J Physiol* 1989;**257** (2 Pt 1):L94–9.
23. Tanswell AK, Tzaki MG, Byrne PJ. Hormonal and local factors influence antioxidant enzyme activity of rat fetal lung cells in vitro. *Exp Lung Res* 1986;**11**(1):49–59.
24. Walther FJ, Ikegami M, Warburton D, Polk DH. Corticosteroids, thyrotropin-releasing hormone, and antioxidant enzymes in preterm lamb lungs. *Pediatr Res* 1991;**30**(6):518–21.
25. Weisiger RA, Fridovich I. Superoxide dismutase. Organelle specificity. *J Biol Chem* 1973;**248**(10):3582–92.
26. Weisiger RA, Fridovich I. Mitochondrial superoxide simutase. Site of synthesis and intramitochondrial localization. *J Biol Chem* 1973;**248**(13):4793–6.
27. Crapo JD, Oury T, Rabouille C, Slot JW, Chang LY. Copper,zinc superoxide dismutase is primarily a cytosolic protein in human cells. *Proc Natl Acad Sci USA* 1992;**89**(21):10405–9.
28. Marklund SL. Extracellular superoxide dismutase in human tissues and human cell lines. *J Clin Invest* 1984;**74**(4):1398–403.
29. Oury TD, Crapo JD, Valnickova Z, Enghild JJ. Human extracellular superoxide dismutase is a tetramer composed of two disulphide-linked dimers: A simplified, high-yield purification of extracellular superoxide dismutase. *Biochem J* 1996;**317**(Pt 1):51–7.
30. Huie RE, Padmaja S. The reaction of no with superoxide. *Free Radic Res Commun* 1993;**18**(4):195–9.
31. Nozik-Grayck E, Dieterle CS, Piantadosi CA, Enghild JJ, Oury TD. Secretion of extracellular superoxide dismutase in neonatal lungs. *Am J Physiol Lung Cell Mol Physiol* 2000;**279**(5):L977–84.
32. Tanswell AK, Freeman BA. Differentiation-arrested rat fetal lung in primary monolayer cell culture. III. Antioxidant enzyme activity. *Exp Lung Res* 1984;**6**(2):149–58.
33. Hass MA, Massaro D. Developmental regulation of rat lung cu,zn-superoxide dismutase. *Biochem J* 1987;**246**(3):697 703.
34. Kakkar P, Jaffery FN, Viswanathan PN. Neonatal developmental pattern of superoxide dismutase and aniline hydroxylase in rat lung. *Environ Res* 1986;**41**(1):302–8.
35. Chen Y, Frank L. Differential gene expression of antioxidant enzymes in the perinatal rat lung. *Pediatr Res* 1993;**34**(1):27–31.
36. de Haan JB, Tymms MJ, Cristiano F, Kola I. Expression of copper/zinc superoxide dismutase and glutathione peroxidase in organs of developing mouse embryos, fetuses, and neonates. *Pediatr Res* 1994;**35**(2):188–96.
37. Strange RC, Cotton W, Fryer AA, Drew R, Bradwell AR, Marshall T, et al. Studies on the expression of cu,zn superoxide dismutase in human tissues during development. *Biochim Biophys Acta* 1988;**964**(2):260–5.
38. Harman AW, McKenna M, Adamson GM. Postnatal development of enzyme activities associated with protection against oxidative stress in the mouse. *Biol Neonate* 1990;**57**(3–4):187–93.
39. Frank L. Developmental aspects of experimental pulmonary oxygen toxicity. *Free Radic Biol Med* 1991;**11**(5):463–94.
40. Raza H. Dual localization of glutathione s-transferase in the cytosol and mitochondria: Implications in oxidative stress, toxicity and disease. *FEBS J* 2011;**278**(22):4243–51.
41. Melikian AA, Bagheri K, Hecht SS, Hoffmann D. Metabolism of benzo[a]pyrene and 7 beta,8 alpha-dihydroxy-9 alpha, 10 alpha-epoxy-7,8,9,10-tetrahydrobenzo[a pyrene in lung and liver of newborn mice. *Chem Biol Interact* 1989;**69**(2–3):245–57.
42. Meister A, Anderson ME. Glutathione. *Annu Rev Biochem* 1983;**52**:711–60.
43. Meister A. Selective modification of glutathione metabolism. *Science* 1983;**220**(4596):472–7.
44. Meister A. Glutathione, metabolism and function via the gamma-glutamyl cycle. *Life Sci* 1974;**15**(2):177–90.
45. Deneke SM, Fanburg BL. Regulation of cellular glutathione. *Am J Physiol* 1989;**257**(4 Pt 1):L163–73.
46. Martensson J, Jain A, Stole E, Frayer W, Auld PA, Meister A. Inhibition of glutathione synthesis in the newborn rat: A model for endogenously produced oxidative stress. *Proc Natl Acad Sci USA* 1991;**88**(20):9360–4.
47. Oakes SM, Takahashi Y, Williams MC, Joyce-Brady M. Ontogeny of gamma-glutamyltransferase in the rat lung. *Am J Physiol* 1997;**272** (4 Pt 1):L739–44.
48. Langley-Evans SC, Wood S, Jackson AA. Enzymes of the gamma-glutamyl cycle are programmed in utero by maternal nutrition. *Ann Nutr Metab* 1995;**39**(1):28–35.
49. Lavoie JC, Spalinger M, Chessex P. Glutathione synthetic activity in the lungs in newborn guinea pigs. *Lung* 1999;**177**(1):1–7.
50. Pacht ER, Davis WB. Role of transferrin and ceruloplasmin in antioxidant activity of lung epithelial lining fluid. *J Appl Physiol* 1988;**64**(5):2092–9.
51. Yang F, Friedrichs WE, deGraffenried L, Herbert DC, Weaker FJ, Bowman BH, et al. Cellular expression of ceruloplasmin in baboon and mouse lung during development and inflammation. *Am J Respir Cell Mol Biol* 1996;**14**(2):161–9.
52. Das KC, Pahl PM, Guo XL, White CW. Induction of peroxiredoxin gene expression by oxygen in lungs of newborn primates. *Am J Respir Cell Mol Biol* 2001;**25**(2):226–32.
53. Vyas JR, Currie A, Dunster C, Kelly FJ, Kotecha S. Ascorbate acid concentration in airways lining fluid from infants who develop chronic lung disease of prematurity. *Eur J Pediatr* 2001;**160**(3):177–84.
54. Plopper CG, Weir AJ, Morin D, Chang A, Philpot RM, Buckpitt AR. Postnatal changes in the expression and distribution of pulmonary cytochrome p450 monooxygenases during clara cell differentiation in rabbits. *Mol Pharmacol* 1993;**44**(1):51–61.
55. Simmons DL, Kasper CB. Quantitation of mrnas specific for the mixed-function oxidase system in rat liver and extrahepatic tissues during development. *Arch Biochem Biophys* 1989;**271**(1):10–20.
56. Strum JM, Ito T, Philpot RM, DeSanti AM, McDowell EM. The immunocytochemical detection of cytochrome p-450 monooxygenase in the lungs of fetal, neonatal, and adult hamsters. *Am J Respir Cell Mol Biol* 1990;**2**(6):493–501.
57. Parandoosh Z, Franklin MR. Developmental changes in rabbit pulmonary cytochrome p-450 subpopulations. *Life Sci* 1983;**33**(13): 1255–60.
58. Ji CM, Cardoso WV, Gebremichael A, Philpot RM, Buckpitt AR, Plopper CG, et al. Pulmonary cytochrome p-450 monooxygenase system and clara cell differentiation in rats. *Am J Physiol* 1995;**269**(3 Pt 1):L394–402.
59. Ji CM, Plopper CG, Witschi HP, Pinkerton KE. Exposure to sidestream cigarette smoke alters bronchiolar epithelial cell differentiation in the postnatal rat lung. *Am J Respir Cell Mol Biol* 1994;**11**(3):312–20.
60. Fouts JR, Devereux TR. Developmental aspects of hepatic and extrahepatic drug-metabolizing enzyme systems: Microsomal enzymes and components in rabbit liver and lung during the first month of life. *J Pharmacol Exp Ther* 1972;**183**(2):458–68.

61. Fanucchi MV, Murphy ME, Buckpitt AR, Philpot RM, Plopper CG. Pulmonary cytochrome p450 monooxygenase and clara cell differentiation in mice. *Am J Respir Cell Mol Biol* 1997;**17**(3): 302–14.

62. Eltom SE, Babish JG, Schwark WS. The postnatal development of drug-metabolizing enzymes in hepatic, pulmonary and renal tissues of the goat. *J Vet Pharmacol Ther* 1993;**16**(2):152–63.

63. Eltom SE, Schwark WS. Cyp1a1 and cyp1b1, two hydrocarbon-inducible cytochromes p450, are constitutively expressed in neonate and adult goat liver, lung and kidney. *Pharmacol Toxicol* 1999;**85**(2):65–73.

64. Larsen-Su S, Krueger SK, Yueh MF, Lee MY, Shehin SE, Hines RN, et al. Flavin-containing monooxygenase isoform 2: Developmental expression in fetal and neonatal rabbit lung. *J Biochem Mol Toxicol* 1999;**13**(3-4):187–93.

65. Gebremichael A, Chang AM, Buckpitt AR, Plopper CG, Pinkerton KE. Postnatal development of cytochrome p4501a1 and 2b1 in rat lung and liver: Effect of aged and diluted sidestream cigarette smoke. *Toxicol Appl Pharmacol* 1995;**135**(2):246–53.

66. Landi S. Mammalian class theta gst and differential susceptibility to carcinogens: A review. *Mutat Res* 2000;**463**(3):247–83.

67. Mannervik B. The isoenzymes of glutathione transferase. *Adv Enzymol Relat Areas Mol Biol* 1985;**57**:357–417.

68. Mannervik B, Danielson UH. Glutathione transferases–structure and catalytic activity. *Crc Crit Rev Biochem* 1988;**23**(3):283–337.

69. Buetler TM, Eaton DL. Complementary DNA cloning, messenger rna expression, and induction of alpha-class glutathione s-transferases in mouse tissues. *Cancer Res* 1992;**52**(2):314–8.

70. Fanucchi MV, Buckpitt AR, Murphy ME, Storms DH, Hammock BD, Plopper CG. Development of phase ii xenobiotic metabolizing enzymes in differentiating murine clara cells. *Toxicol Appl Pharmacol* 2000;**168**(3):253–67.

71. James MO, Foureman GL, Law FC, Bend JR. The perinatal development of epoxide-metabolizing enzyme activities in liver and extrahepatic organs of guinea pig and rabbit. *Drug Metab Dispos* 1977;**5**(1):19–28.

72. Beckett GJ, Howie AF, Hume R, Matharoo B, Hiley C, Jones P, et al. Human glutathione s-transferases: Radioimmunoassay studies on the expression of alpha-, mu- and pi-class isoenzymes in developing lung and kidney. *Biochim Biophys Acta* 1990;**1036**(3):176–82.

73. Cossar D, Bell J, Strange R, Jones M, Sandison A, Hume R. The alpha and pi isoenzymes of glutathione s-transferase in human fetal lung: In utero ontogeny compared with differentiation in lung organ culture. *Biochim Biophys Acta* 1990;**1037**(2):221–6.

74. Pacifici GM, Franchi M, Colizzi C, Giuliani L, Rane A. Glutathione s-transferase in humans: Development and tissue distribution. *Arch Toxicol* 1988;**61**(4):265–9.

75. Fryer AA, Hume R, Strange RC. The development of glutathione s-transferase and glutathione peroxidase activities in human lung. *Biochim Biophys Acta* 1986;**883**(3):448–53.

76. Datta K, Roy SK, Mitra AK, Kulkarni AP. Glutathione s-transferase mediated detoxification and bioactivation of xenobiotics during early human pregnancy. *Early Hum Dev* 1994;**37**(3):167–74.

77. Strange RC, Davis BA, Faulder CG, Cotton W, Bain AD, Hopkinson DA, et al. The human glutathione s-transferases: Developmental aspects of the gst1, gst2, and gst3 loci. *Biochem Genet* 1985;**23** (11–12):1011–28.

78. Pacifici GM, Franchi M, Bencini C, Repetti F, Di Lascio N, Muraro GB. Tissue distribution of drug-metabolizing enzymes in humans. *Xenobiotica* 1988;**18**(7):849–56.

79. Wixtrom RN, Hammock BD. Membrane-bound and soluble-fraction epoxide hydrolases: Methodological aspects. In: Zakim D, Vessey DA, editors. *Biochemical pharmacology and toxicology.* New York: Wiley; 1985. pp. 3–93.

80. Hassett C, Turnblom SM, DeAngeles A, Omiecinski CJ. Rabbit microsomal epoxide hydrolase: Isolation and characterization of the xenobiotic metabolizing enzyme cdna. *Arch Biochem Biophys* 1989;**271**(2):380–9.

81. de Waziers I, Cugnenc PH, Yang CS, Leroux JP, Beaune PH. Cytochrome p 450 isoenzymes, epoxide hydrolase and glutathione transferases in rat and human hepatic and extrahepatic tissues. *J Pharmacol Exp Ther* 1990;**253**(1):387–94.

82. Waechter F, Bentley P, Bieri F, Muakkassah-Kelly S, Staubli W, Villermain M. Organ distribution of epoxide hydrolases in cytosolic and microsomal fractions of normal and nafenopin-treated male dba/2 mice. *Biochem Pharmacol* 1988;**37**(20):3897–903.

83. Pacifici GM, Rane A. Metabolism of styrene oxide in different human fetal tissues. *Drug Metab Dispos* 1982;**10**(4):302–5.

84. Pacifici GM, Temellini A, Giuliani L, Rane A, Thomas H, Oesch F. Cytosolic epoxide hydrolase in humans: Development and tissue distribution. *Arch Toxicol* 1988;**62**(4):254–7.

85. Omiecinski CJ, Aicher L, Swenson L. Developmental expression of human microsomal epoxide hydrolase. *J Pharmacol Exp Ther* 1994;**269**(1):417–23.

86. Rouet P, Dansette P, Frayssinet C. Ontogeny of benzo(a)pyrene hydroxylase, epoxide hydrolase and glutathione-s transferase in the brain, lung and liver of c57bl/6 mice. *Dev Pharmacol Ther* 1984;**7**(4):245–58.

87. Bond JA, Harkema JR, Russell VI. Regional distribution of xenobiotic metabolizing enzymes in respiratory airways of dogs. *Drug Metab Dispos* 1988;**16**(1):116–24.

88. Statham CN, Dutcher JS, Kim SH, Boyd MR. Ipomeanol 4-glucuronide, a major urinary metabolite of 4-ipomeanol in the rat. *Drug Metab Dispos* 1982;**10**(3):264–7.

89. Baron J, Voigt JM. Localization, distribution, and induction of xenobiotic-metabolizing enzymes and aryl hydrocarbon hydroxylase activity within lung. *Pharmacol Ther* 1990;**47**(3):419–45.

90. Kawalek JC, el Said KR. Maturational development of drug-metabolizing enzymes in dogs. *Am J Vet Res* 1990;**51**(11):1742–5.

91. Kawalek JC, el Said KR. Maturational development of drug-metabolizing enzymes in sheep. *Am J Vet Res* 1990;**51**(11):1736–41.

92. Pacifici GM, Kubrich M, Giuliani L, de Vries M, Rane A. Sulphation and glucuronidation of ritodrine in human foetal and adult tissues. *Eur J Clin Pharmacol* 1993;**44**(3):259–64.

93. Hume R, Barker EV, Coughtrie MW. Differential expression and immunohistochemical localisation of the phenol and hydroxysteroid sulphotransferase enzyme families in the developing lung. *Histochem Cell Biol* 1996;**105**(2):147–52.

94. Ziegler DM. Flavin-containing monooxygenases: Enzymes adapted for multisubstrate specificity. *Trends Pharmacol Sci* 1990;**11**(8):321–4.

95. Cashman JR. Human flavin-containing monooxygenase: Substrate specificity and role in drug metabolism. *Curr Drug Metab* 2000;**1**(2):181–91.

96. Falany CN. Enzymology of human cytosolic sulfotransferases. *Faseb J* 1997;**11**(4):206–16.

97. Coughtrie MW, Sharp S, Maxwell K, Innes NP. Biology and function of the reversible sulfation pathway catalysed by human sulfotransferases and sulfatases. *Chem Biol Interact* 1998;**109**(1–3):3–27.

98. Glatt H, Boeing H, Engelke CE, Ma L, Kuhlow A, Pabel U, et al. Human cytosolic sulphotransferases: Genetics, characteristics, toxicological aspects. *Mutat Res* 2001;**482**(1–2):27–40.

99. Richard K, Hume R, Kaptein E, Stanley EL, Visser TJ, Coughtrie MW. Sulfation of thyroid hormone and dopamine during human development: Ontogeny of phenol sulfotransferases and arylsulfatase in liver, lung, and brain. *J Clin Endocrinol Metab* 2001;**86**(6):2734–42.

100. Burri PH. Postnatal development and growth. In: Crystal RG, editor. *The lung: Scientific foundations*. 2nd ed. Philadelphia: Lippencott-Raven Publishers; 1997. pp. 1013–26.

101. Bucher JR, Roberts RJ. The development of the newborn rat lung in hyperoxia: A dose-response study of lung growth, maturation, and changes in antioxidant enzyme activities. *Pediatr Res* 1981;**15**(7):999–1008.

102. Serabjit-Singh CJ, Bend JR. Purification and biochemical characterization of the rabbit pulmonary glutathione s-transferase: Stereoselectivity and activity toward pyrene 4,5-oxide. *Arch Biochem Biophys* 1988;**267**(1):184–94.

103. Gardlik S, Gasser R, Philpot RM, Serabjit-Singh CJ. The major alpha-class glutathione s-transferases of rabbit lung and liver. Primary sequences, expression, and regulation. *J Biol Chem* 1991;**266**(29):19681–7.

Chapter 12

Stretch and Grow: Mechanical Forces in Compensatory Lung Growth

Connie C.W. Hsia*, Cuneyt Yilmaz* and Ewald R. Weibel†

*Department of Internal Medicine, University of Texas Southwestern Medical Center, Dallas, TX, USA; †Institute of Anatomy, University of Bern, Bern, Switzerland

INTRODUCTION

How does the lung grow? The answer to this question remains fragmentary despite the recent explosion of information concerning the cellular and molecular events that influence developmental and compensatory lung growth. In particular, we lack an adequate understanding of how the escalating number of molecular signals and mediators known to act on individual components of the lung are coordinated to form the complex three-dimensional architecture that fully meets functional requirements.[1] We understand even less about the altered coordination of growth and remodeling events in adaptation to a loss of functioning lung units caused by injury or disease. While normal postnatal lung maturation reaches its limit at somatic maturity, it is apparent that in the face of destruction of some lung units, growth and remodeling may be re-initiated in the remaining units by the appropriate institution of mechanical stimuli leading to partial or complete restoration of gas exchange capacity. What signals trigger growth re-initiation? How do the signals act on the constitutive components to effect balanced alveolar septal growth while minimizing distortion and maximizing functional benefit? The objective of this chapter is to compare the characteristics of developmental and post-pneumonectomy (PNX) growth in order to (a) illustrate key principles in the regulation of growth and remodeling in relation to mechanical tissue stress and strain, and (b) interpret the innate potential for lung re-growth in the context of structure–function relationships with respect to the whole organ.

TISSUE AND MECHANICAL FORCES IN LUNG DEVELOPMENT

The lung's basic constructional problem is that a very large gas exchanger must be linked to two systems for coordinated air and blood supply from central sources to each of the 400 million gas exchange units. This is achieved during lung development by stepwise branching and growth of the airway tube associated with parallel branching of pulmonary arteries and veins connected to the primitive microvascular network that enwraps the branching airway tubes. As a final step the gas exchanging alveolo-capillary complex results from the transformation of the last generations of airways and vessels with drastic reduction of tissue separating air and blood.[2] This process is governed by the architectural principle of complexity, based on fractal geometry,[3] coupled with the principle of correlativity between airways and blood vessels throughout.[1,4]

This development occurs under two very different conditions: in the first pre-natal phase lung development and growth are essentially driven by tissue forces because the lung is not yet ventilated and poorly perfused, whereas in the postnatal phase the growing lung is subjected to various mechanical forces, the ventilator activity of thorax and diaphragm, as well as the now elevated perfusion with blood. Morphogenesis of lung architecture is governed by complex but systematic branching of the airways, blood vessels with mesenchyme following suit. This branching process is described as dichotomy or bifurcation, but it does not need to be symmetric, particularly in the major airways where variations in length of airways are mostly related to evenly filling the irregular space of the chest cavity with peripheral lung units. While airway branching occurs only at the tips, all airway segments grow in length and diameter about in proportion to overall lung growth. The result is a space-filling branching tree with a quite even distribution of the terminal tips in the lung volume. This growth process has all the characteristics of a fractal tree[3,5]: the branching process is self-similar, that is, the proportions between parent and daughter branch dimensions are about the same from one

The Lung. http://dx.doi.org/10.1016/B978-0-12-799941-8.00012-2

generation to the next as a result of proportional growth, and the tips are space-filling.

While the airway tree develops in complexity by growth and branching, the mesenchyme and the vasculature undergo related changes. The terminal airway buds, where further branching can occur, are in close contact with a cell-rich mesenchyme, a decisive factor because the formation of a new pair of branches is induced by interaction of epithelial cells with the mesenchyme through a complex cascade of growth factors and receptors.[6–9]

In the lung, the fractal branching of the broncho-vasculo-alveolar network serves to maximize gas exchange surface area within a confined thoracic cavity defined by the diaphragm and the relatively rigid rib cage. Because form follows function and vice versa, physiologic variables such as ventilation, perfusion, and diffusion must also follow fractal principles,[10] and growth and function of the lung are intrinsically linked to that of the thorax. During development the enlarging rib cage exerts a negative intrathoracic pressure resulting in chronic mechanical stress and strain on the lung (Figure 1). Mechanical stress activates a cascade of cellular events leading to cell proliferation, growth and remodeling of lung parenchyma in addition to inducing a host of metabolic alterations. Lung growth and remodeling in turn relieves septal stress and strain, creating a negative feedback loop that continues until somatic maturity is reached, at which time the epiphyses close and the thorax reaches its maximum size and final shape.[11] The lungs, which must conform to the size and shape of their container, also stop growing as the mechanical signals diminish. During the period of active growth, recoil of the lung exerts traction on the thorax and modulates thoracic growth in a reciprocal manner.

Mechanical stretch of alveolar septa in vitro stimulates alveolar epithelial cell proliferation,[12,13] and induces apoptosis,[14] signal transduction pathways[13,15] ion channel flux,[16,17] turnover of matrix proteins,[16,18] cytoskeletal proteins,[19] cytokine growth factors,[20,21] as well as gene expression of

surfactant-associated proteins.[22] Any process that interferes with mechanical lung–thorax interdependence can alter lung growth and development. For example, increasing luminal pressure by tracheal occlusion in utero accelerates branching morphogenesis of the embryonic lung; the process is modulated via signaling of the fibroblast growth factor-10 pathway.[23,24] Fetal breathing movements and chest expansion are critical for pre-natal lung development.[25] Even the lack of in utero skeletal muscle movements, as in transgenic mice without the muscle-specific transcription factor myogenin, secondarily impairs lung organogenesis.[26] Congenital diaphragmatic hernia (CDH) is a well-known example of reduced lung stress/strain causing lung hypoplasia.[27] Following restoration of mechanical lung–thorax interactions by surgical repair of the diaphragmatic defect, "catch-up" alveolar growth occurs associated with vascular remodeling and amelioration of pulmonary arterial hypertension,[28] Long-term pulmonary gas exchange in survivors is surprisingly well-preserved,[29,30] highlighting the innate potential for compensatory alveolar development. In a fetus with CDH, temporary antenatal tracheal occlusion has been employed as an adjunct modality in addition to definitive surgical repair to improve lung growth and function.[31] Another condition of deranged lung–thorax mechanical coupling, severe childhood kyphoscoliosis, also blunts alveolar development[32–35] resulting in fewer alveoli and enlarged alveolar airspaces. Conversely, selective inhibition of lung growth may deform the growing rib cage in this reciprocal relationship.[36]

Mechanical lung–thorax interdependence continues after birth. Post-natal lung stretch induced by continuous positive airway pressure in weanling ferrets increases total lung capacity, lung weight, protein, and DNA content without changing lung recoil.[37] Sustained segmental lung distension with perfluorocarbon in neonatal lambs also accelerates lung growth in an age-dependent fashion.[38] New alveolar tissue is laid down preferentially at the lung periphery,[39,40] where mechanical forces are largely borne by the septa due

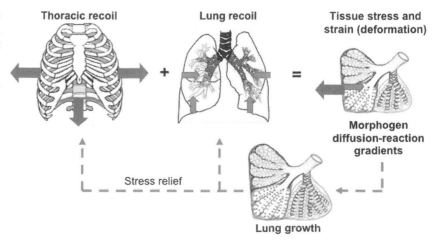

FIGURE 1 Mechanical interactions between the lung and thorax. Recoil of the rib cage exerts an outward force (red arrows) while recoil of the lung exerts an inward force (blue arrows). The resulting stress and strain on the lung parenchyma create diffusion-reaction gradients of morphogens that transduce lung growth to relieve tissue stress. This feedback loop continues during post-natal development until the bony epiphyses close upon reaching somatic maturation. *Adapted from.*[11]

to a relative absence of physical support from the more rigid airway and vascular scaffold.

MECHANICAL FORCES FOLLOWING PNEUMONECTOMY

The anatomical effects of removing one lung by pneumonectomy (PNX) is illustrated by magnetic resonance imaging (MRI) in a human subject (Figure 2, top panels) and by computerized tomography (CT) in a dog (Figure 2, lower panels) compared to their normal counterparts. Normally the right and left lungs comprise 55–58% and 42–45% of total volume and diffusing capacity, respectively, and receive the corresponding proportions of total ventilation and blood flow.[41,42] Gross mediastinal shift develops following unilateral PNX. The remaining lung expands across the midline and its volume increases to ~90% of that of two normal lungs. Blood flow to the remaining lung at a given cardiac output increases by a factor corresponding to the reciprocal of the remaining lung fraction as the entire cardiac output is now directed through the remaining lung. The ipsilateral rib cage and hemidiaphragm are distorted. The mediastinal structures are displaced and rotated. The cardiac fossa, a potential space in which the heart resides, loses its compliance as the heart becomes surrounded by the rigid rib cage, the elevated right hemidiaphragm and a stiffer remaining lung. The magnitude of post-PNX mechanical strain is readily appreciated from three-dimensional CT reconstruction of lung lobes, conducting airways and blood vessels (Figure 3). Following right PNX, the trachea deviates to the right side, and the lobar airways are displaced, rotated, and splayed. Each airway generation becomes elongated and dilated as a result of chronic asymmetric traction[43] caused by the increased airflow to the enlarged parenchymal units.

In the early post-operative period, acute microcirculatory hyperperfusion may stretch the capillary walls leading to increased endothelial permeability, recruit macrophages and circulating cells,[44] and may cause pulmonary edema.[45] Acute mediastinal shift can cause hemodynamic instability as a result of cardiac herniation and/or torsion, kinking of major blood vessels and leakage of lymph.[46–48] Sustained mechanical strain of intrathoracic structures can cause the "post-pneumonectomy syndrome," particularly after right PNX and in pediatric patients, where a distorted main stem or lobar bronchus gradually becomes obstructed, necessitating surgical reposition of the mediastinum.[49,50] In addition, disrupting the normal mechanical coupling between lung and thorax may cause asymmetric thoracic deformities.[51] Dilatation of the esophagus may develop and lead to reflux esophagitis.[52,53] Earlier measures that have been used clinically to prevent or correct mediastinal shift after PNX include thoracoplasty, pneumothorax, oleothorax, and plastic sponge plombage.[54] More recently tracheobronchial

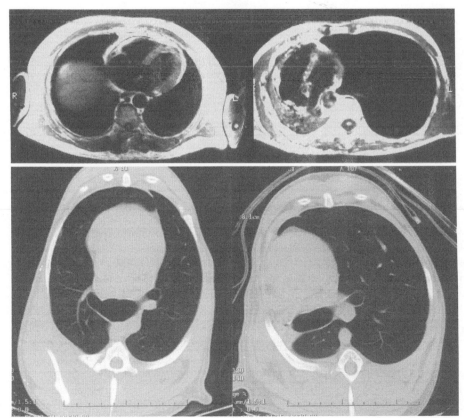

FIGURE 2 Upper panels: magnetic resonance images from a normal subject (left) and a patient after right PNX (right). Lower panels: CT images from a normal dog (left) and a dog after right PNX (right). *See text for discussion.*

FIGURE 3 Three-dimensional HRCT reconstructions of conducting airways, pulmonary blood vessels, and individual lobes in adult canine lungs are shown. Left panels: normal; middle panels: the left lung following 58% lung resection by right pneumonectomy; right panels: the remaining right and left cranial lobes and the often incompletely separated left middle lobe (also termed the inferior segment of the left cranial lobe) following 70% lung resection. (This figure is reproduced in color in the color plate section.)

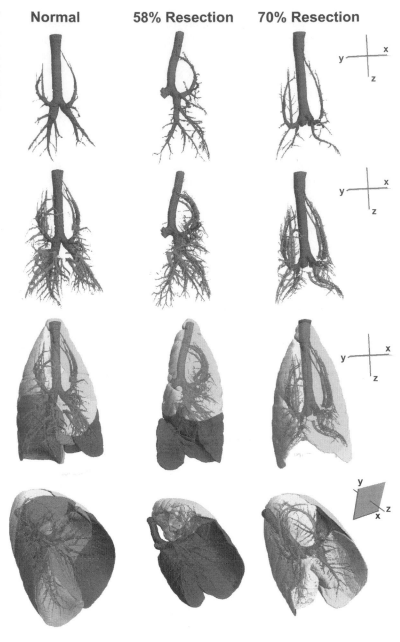

Normal   **58% Resection**   **70% Resection**

stent, injections of sulfur hexafluoride into the pleural space, and implantation of fixed volume or expandable prosthesis,[50,53,55,56] have been employed on the side of PNX. Surgical intervention may relieve symptomatic airway obstruction and stabilize hemodynamics, but does not always confer long-term benefit in gas exchange as it impairs compensatory response to the loss of alveolar tissue.[57,58]

POST-PNEUMONECTOMY COMPENSATORY RESPONSE

In general, the remaining lung adapts to the loss of functioning tissue through: (a) greater utilization or recruitment of remaining physiological reserves; (b) remodeling of remaining lung structure; and (c) compensatory growth of new lung tissue. All three mechanisms are invoked following PNX, detailed in the following sections. In this regard PNX provides a useful model for understanding the consequences and the adaptive mechanisms in response to a known and reproducible severity of alveolar destruction irrespective of the specific etiology of disease. Early studies of PNX, summarized by Schilling in 1965,[41] often regarded post-PNX lung expansion as detrimental, causing "emphysema-like" changes of the remaining parenchyma leading to gas exchange and mechanical dysfunction. However, later studies suggest that sustained mechanical strain on the remaining lung is in fact a major signal initiating beneficial adaptation,[11,43,59,60] and long-term compensation, evidenced by the maintenance of a near normal exercise capacity in adult dogs after removal of 55–58% of lung by right PNX.[61]

Recruitment of Physiologic Reserves

In any transport system, "load" is the flux through the system and "capacity" is the maximum flux that can be handled, essentially determined by the structural design of the system. Most of the time, capacity far exceeds flux by several folds, unless the system is maximally stressed, for example, in the respiratory system when the organism demands maximal O_2 consumption to be achieved in heavy exercise. The difference between load and capacity is termed "safety factor" in engineering or "functional reserves" in physiology.[62] The preservation of adequate structural and functional reserves that are not utilized at basal state, or even in limited activity, but can be readily recruited under metabolic stress is an essential organ feature that ensures survival and adaptation of the individual as well as the species.[63,64]

Physiologic reserves for pulmonary gas exchange are huge, attributed to both structural and non-structural variables. Structural determinants of diffusive oxygen transport include the effective alveolar-capillary surface area at the tissue–air interface, and effective diffusion path length from the air–tissue interface to the capillary erythrocytes. Non-structural determinants of diffusive oxygen transport include the magnitude as well as the distribution of lung inflation, pulmonary blood flow and erythrocyte mass (hematocrit) within alveolar capillaries. At rest, erythrocytes are unevenly distributed within the lung; many capillaries particularly at the apex are perfused with plasma only and devoid of erythrocytes.[65] Studies using in vivo videomicroscopy in dog lung have shown that as blood flow or pressure increases, the number of subpleural capillary segments perfused with erythrocytes progressively increase,[66] that is, effective surface area for gas exchange increases without intrinsic structural change. The result of alveolar-capillary recruitment is a linear increase in diffusive gas transport, measured as lung diffusing capacity (DL) for carbon monoxide (DL_{CO}), oxygen (DL_{O2}), or nitric oxide (DL_{NO}), as cardiac output increases.[67-69] From rest to

peak exercise, DL can increase two-fold without evidence of reaching an upper limit as a result of recruitment of microvascular reserves.[67,70,71] The ability to increase DL with respect to cardiac output is critical for maintaining adequate oxygenation of the end-capillary blood leaving the lung. If the ratio of DL to blood flow (DL/Q) falls below a critical level end-capillary O_2 saturation will fall precipitously and arterial hypoxemia will develop.[72]

Although the measurement of DL has been performed for many decades, its importance as a source of physiologic compensation was not widely understood until recently. Following PNX as in all destructive lung diseases, the entire ventilation and blood flow are directed to the remaining functioning lung units, causing the remaining alveolar surfaces to unfold, and the remaining capillaries to open and distend. The effective surface for gas exchange and therefore DL per unit of lung is higher than that expected based on anatomical lung destruction. Through cumulative studies in animals and human subjects from rest up to heavy exercise, we have validated a conceptual framework for interpreting structural and functional compensation in diffusive gas exchange from the DL vs. blood flow (DL–$\dot{Q}$) relationship (Figure 4).[72] Normally DL increases linearly with $\dot{Q}$. When some alveolar-capillary units are destroyed, DL is reduced at a given blood flow, but as long as the remaining units can be recruited DL should continue to increase along a lower parallel relationship (i.e., lower intercept and normal slope of the DL-$\dot{Q}$ curve). If alveolar-capillary destruction is diffuse and the remaining lung units cannot be recruited, the slope of the DL–Q relationship will also be reduced. On the other hand, formation of new gas exchange tissue, by whatever mechanism, would increase DL at a given blood flow and return the entire relationship towards normal. Accordingly, the slope of the DL–$\dot{Q}$ relationship provides an index of the integrity of microvascular recruitment, while the absolute magnitude of DL at a given $\dot{Q}$ provides an index of the effective gas exchange surface area and/or effective diffusing capacity. An example of compensation by recruitment is seen in dogs after 42–45% lung resection, where DL at a given workload declines by only 25% because the remaining capillary bed is normal and diffusive reserves

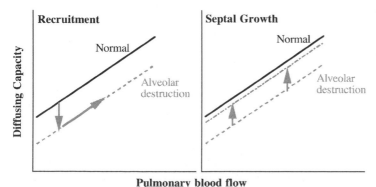

FIGURE 4 Compensation in gas exchange achieved by microvascular recruitment (left panel) or septal regrowth (right panel). Alveolar destruction reduces DL at a given pulmonary blood flow. Blood flow and ventilation, redirected to the remaining functioning units, cause those units to distend, thereby recruiting gas exchange surface area and DL along a parallel lower relationship. As a result, apparent DL is higher than expected from anatomical alveolar destruction. Growth of new alveolar-capillary tissue elevates the entire relationship between DL and pulmonary blood flow back to normal. Arrows indicate direction of change. *Adapted from*.[72]

can be recruited.[73] Similarly, in patients after PNX where the remaining lung is relatively normal, recruitment of DL is preserved and the decline in DL at a given workload is considerably less than expected from the amount of lung removed.[74,75] In contrast, for the same resting DL, patients with idiopathic pulmonary fibrosis who are unable to recruit microvascular reserves demonstrate earlier and more severe reductions in DL and arterial hypoxemia upon exercise.[74,76]

Acinar and Septal Remodeling

Considerable enhancement of gas exchange can be achieved through remodeling of air space and septal architecture to maximize the efficiency of oxygen uptake in the remaining lung without significant addition of new septal tissue. Burri et al.[77,78] proposed that initially after PNX, alveolar duct airspace expands more than alveolar airspace followed by rearrangement of existing septal tissue that ultimately restores the original volume proportions between alveolar ducts and alveoli. The net result is lengthening of interalveolar septa and increased gas exchange surface even in the absence of alveolar multiplication. In adult dogs following 42–45% lung resection by left PNX,[59] structural response is heterogeneous among the remaining lobes; the most caudal lobe shows the largest response, as it also experiences the largest volume expansion. Total volume of alveolar type-2 (AT2) epithelial cells, the putative resident progenitor cell type, in the remaining lung is selectively increased but without overt growth of the other major septal tissue components.[79,80] In the air-ventilated lung, part of the alveolar septal surface is in fact, folded up beneath the surfactant film, and expands fully only at total lung capacity.[81] Maintaining inflation at a relatively higher level after PNX recruits some of this "reserve" surface. Thus, the remaining alveolar surfaces unfold, the alveolar airspaces enlarge and the septa become thinner in response to lung expansion. Septal unfolding increases the effective surface area for gas exchange, while thinning of the septum reduces the effective barrier thickness. Rearrangement of cell-matrix components within the septa can also optimize the harmonic mean barrier thickness, an index of resistance to diffusion, in the absence of a net change in septal tissue volume. Along with microvascular recruitment, architectural remodeling is sufficient to mitigate the post-PNX reduction in DL at a given blood flow and maintain a near-normal arterial oxygen saturation during exercise.[82] Both alveolar-capillary recruitment and architectural remodeling are ubiquitous adaptive mechanisms that are invoked regardless of whether tissue re-growth is initiated. In progressive lung diseases, alveolar-capillary recruitment and remodeling in the remaining normal lung units are responsible for the clinical observation that overt arterial hypoxemia at rest and escalation of exertional dyspnea often do not develop until lung destruction is extensive (≥50 to 60% of total).[83]

Re-Initiation of Alveolar Septal Growth

In contrast to 42–45% resection by left PNX, the response to 55–58% resection by right PNX in adult dogs is characterized by balanced though limited growth of new septal tissue and capillaries in addition to microvascular recruitment and remodeling.[60] In a graded threshold-dependent alveolar structural response, the AT2 cells are activated first.[79] As more lung units are lost, other alveolar cells also proliferate or hypertrophy, leading to an overt increase in septal tissue volume. Intra-acinar epithelial, smooth muscle, and capillary endothelial cells and fibroblasts multiply.[84,85] Post-PNX capillary endothelial cells also produce angiocrine growth factors that induce proliferation of epithelial progenitor cells.[86] Peripheral alveolar tissue grows more vigorously than central alveolar tissue,[40] as peripheral tissue can expand by adding new septa, increasing the complexity of existing septa, or perhaps branching to form another generation of alveolar ducts. Respiratory bronchioles also increase in number after PNX,[87] presumably via (a) transformation of the first generation of alveolar ducts into respiratory bronchioles or (b) alveolization of the terminal bronchiole. Chronic volume expansion and increased perfusion to the remaining lung increase alveolar strain and endothelial distention and shear, respectively; when these signals exceed a critical threshold septal cell proliferation, matrix protein synthesis, and angiogenesis are stimulated.[18,88–91]

The induction of compensatory alveolar growth is dependent on the species, maturity, and ability of the remaining lung to expand. There is extensive literature detailing the vigorous post-PNX lung growth in mice,[85,92–97] rats,[98–107] ferrets,[108–110] and rabbits.[111–113] In rodents, PNX activates a host of early response[92–105] and pro-inflammatory[114] genes, associated with rapid induction of septal tissue growth and restoration of lung volume, weight, DNA, and protein content to that of two lungs within 2 weeks.[94,104] There is strong evidence from large,[71,79] and small,[115] animals that AT2 cells actively participate in compensatory lung growth with a minor contribution of resident bronchoalveolar stem cells.[116] Conflicting evidence exists on whether circulating bone marrow derived progenitor cells contribute significantly.[117,118]

Compared to larger mammals, in rodents lung structure is less stratified, alveolar-capillary reserves are limited, and the thorax is more compliant. Somatic growth continues throughout life, and the epiphyses never fully close; thus, post-natal alveolar growth is easily stimulated, for example, following resection of only one or two lobes.[77,98,119,120] Although mice normally lack a bronchial circulation, a functioning bronchial circulation can be generated de novo within 1 week after unilateral pulmonary artery ligation.[121] In addition, the less rigid rib cage and bronchovascular support in rodent lungs means that a given mechanical force results in greater alveolar tissue deformation, which could contribute to the more vigorous growth response. The brisk post-PNX proliferative activities in rodent lungs are associated with increased susceptibility to experimental carcinogenesis and tumor metastasis.[94,95]

In contrast to rodents, compensatory lung growth in large adult animals is limited by a relatively rigid thorax, and exhibits a protracted course where functional compensation lags behind tissue growth owing to progressive remodeling of the bronchovascular and the alveolo-capillary scaffold.[59,60] During the first few weeks following right PNX in adult dogs, when cell proliferation and/or hypertrophy is most pronounced, total septal tissue volume increases more than 2-fold compared to the control lung following sham PNX.[60] However, aerobic capacity remains markedly reduced; both the magnitude and the recruitment of DL are impaired.[122] The alveolar septa are thickened due to a disproportional increase in the volume of interstitial cells and matrix by more than 3-fold above control levels, leading to a longer mean alveolar epithelial-to-erythrocyte barrier distance and increased resistance to gas diffusion that offsets the functional benefit of septal tissue growth. Subsequently, tissue remodeling becomes evident consisting of thinning of the septum and increased complexity of the gas exchange surfaces, resulting in normalization of the mean barrier distance for diffusion and a larger effective alveolar-capillary surface area. As a result, DL at a given blood flow and the slope of DL recruitment improves progressively, although DL never completely normalizes. Upon exercise one year post-surgery,[70,73] DL_{CO} and arterial O_2 saturation at a given blood flow per unit of remaining lung is higher following right PNX than left PNX while DL_{CO} and arterial O_2 saturation in their respective control groups are similar, that is, higher intensity of lung destruction elicits greater intensity of compensatory response (Figure 5). Thus, diffusion impairment is the major limitation following 45–55% lung resection; within this range, compensation occurs primarily via physiological recruitment and remodeling of the remaining lung, until a threshold of resection (~50%) is exceeded after which septal tissue growth is initiated. However, active septal tissue growth does not translate immediately into functional enhancement of DL until after the normal alveolar morphology is reconstituted. The determinants of the growth threshold remain unclear, but it is important to grasp the concept that optimal functional enhancement requires both tissue-capillary growth as well as architectural remodeling at all levels, down to the alveolar-capillary barrier.

Pneumonectomy During Somatic Maturation

In both large and small animals, compensatory lung growth is most vigorous during somatic maturation but diminishes with age[60,71,123,124]; functional restoration is more complete in young than adult animals. In dogs undergoing PNX as puppies, resting DL and in vivo estimates of lung air and tissue volume return within 8 weeks to control values for two lungs.[125,126] When raised to maturity one year later, aerobic capacity and DL are normal at rest and up to peak exercise compared to simultaneously raised SHAM control animals.[71] Alveolar architecture normalizes such that DL estimated from morphometric measurements of surface area and harmonic mean diffusion path in the postmortem fixed lung is also normal compared to both lungs of control animals.[71] The relative volume increases of most septal cell volumes and surface areas in the remaining lung are greater in dogs undergoing right PNX as puppies than as adults,[71] with the exception that AT2 cell volume increases to a similar extent irrespective of maturity; this latter finding reflects the pleiotropic functions of AT2 cells. As resident progenitors,[127] AT2 cells transdifferentiate into alveolar type 1 (AT1) cells[128,129] to repair injured epithelium or extend the epithelial surface in growing lungs. In addition, AT2 cells synthesize and process surfactant components to reduce surface tension and facilitate innate immunity as well as produce and/or metabolize cytokines and growth factors. Both AT1 cells and AT2 cells maintain barrier defenses, regulate fluid

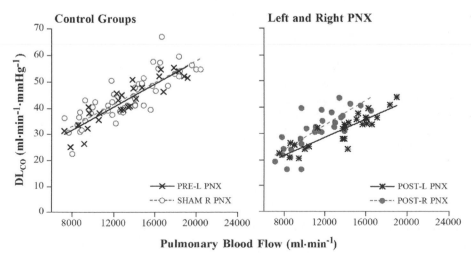

FIGURE 5   At a given blood flow, DL_{CO} is higher in adult dogs after right PNX (55–58% resection) than after left PNX (42–45% resection, right panel). In comparison, DL_{CO} at a given blood flow is similar in the respective control groups (pre-L-PNX and sham R-PNX, left panel). *Adapted from.*[70]

fluxes at the air–tissue interface, and participate in epithelial–mesenchymal interactions.[130,131] Therefore, regardless of the extent of active cell proliferation, the demands on all epithelial cell functions intensify following PNX.

The maturity-dependent compensatory response suggests that developmental and post-PNX stimuli are synergistic. Somatic growth is not affected by PNX; hence normal progressive enlargement of the thorax further intensifies the already accelerated growth of the remaining lung. In patients followed for more than 30 years after PNX, there is an age-related gradual decrease in ventilatory capacity and reserve depending on whether PNX was performed during infancy, early childhood, adolescence or after, although adaptation is evident at all ages.[75,132,133]

The effective surface area, tissue-plasma-erythrocyte diffusion barrier, and physical features of erythrocytes are key determinants of the functional usefulness of alveolar septal growth. These variables are independent of the anatomical alveolar distortion that occurs after PNX or in disease. Estimates of DL based on these morphometric variables correlate strongly with physiologic measurements of DL during heavy exercise in the same animals[71]; an indication that structural variables accurately index physiologic reserves of pulmonary gas exchange. In contrast, another frequently used parameter,

the number of alveoli, does not take into account surface complexities at the air–tissue or tissue–capillary interface, and may not accurately reflect gas exchange capacity.

MANIPULATING MECHANICAL SIGNALS IN COMPENSATORY LUNG GROWTH

Experimental manipulation of mechanical lung stress and strain on post-PNX adaptation dates to the 1930s. Cohn[119] reported that wax plombage packed into the chest of rats after lobectomy prevents expansion of the remaining lung, an observation replicated in adult ferrets when one lung was replaced by an oil-filled silicone balloon.[134] Others[96,134–136] have found that plombage alters the shape of the remaining lung,[136] and only delays but does not eliminate the post-PNX increase in mitotic index,[135] DNA synthesis,[96] or lung volume.[136] These early studies were of short duration and the corresponding structural or functional changes were not examined.

To avoid the unintended adverse effects caused by the weight, rigidity, and unnatural shape of plombage material used in previous studies, we used MRI to construct a physical model of the normal right lung of an adult dog,[137] and manufacture an inflatable silicone prosthesis (Figure 6A). Following right PNX, the prosthesis was placed in the

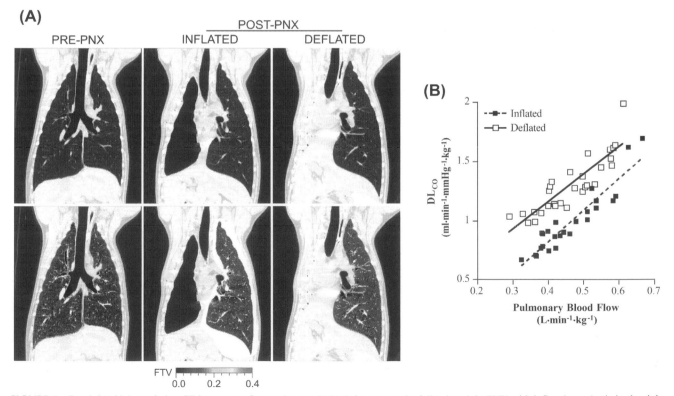

FIGURE 6 Panel 6A: high resolution CT images are from a dog pre-PNX (left), ~4 months following right PNX with inflated prosthesis in the right hemithorax to prevent lateral expansion of the remaining lung (middle), and ~4 months following deflation of the prosthesis allowing lung expansion and mediastinal shift (right). The corresponding color maps show regional fractional tissue volume (FTV). *Adapted from.*[141,142] Panel 6B: at a given blood flow, DL_{CO} is markedly lower in adult dogs with inflated prosthesis to prevent mediastinal shift after right PNX compared to pneumonectomized animals with deflated prosthesis. *Adapted from.*[137]

empty hemithorax and inflated via a subcutaneous injection port buried at the neck. In some animals the prosthesis was kept inflated with an SF_6-air mixture to a volume ~20% above resting supine functional residual capacity–sufficient to keep the mediastinum at the midline. In other animals the prosthesis was deflated except for a minimum amount of air to prevent pleating, resulting in marked mediastinal shift. Gas volume in the prosthesis was checked by helium dilution and refilled, and a normal mediastinal position verified by chest X-rays at regular intervals. After surgical recovery, the animals were trained to run on a treadmill; exercise studies and thoracic high resolution CT (HRCT) were performed at selected time points, followed by postmortem studies. Results were compared to adult dogs post-PNX without prosthesis and those following SHAM PNX.

The presence of silicone prosthesis even when deflated modestly reduced maximal oxygen uptake compared to post-PNX animals without prosthesis. Inflation of the prosthesis, however, further reduced maximal oxygen uptake by ~30% with earlier development of arterial hypoxemia during exercise compared to animals with deflated prosthesis. Ventilation-perfusion distributions and maximal cardiac output were not different between inflated and deflated prosthesis groups. At a given pulmonary blood flow, DL_{CO}[137] and DL_{O2},[138] were reduced in animals with inflated prosthesis compared to deflated prosthesis (Figure 6B); the reductions were due to a low membrane diffusing capacity, that is, a lower conductance by diffusion across the tissue-plasma barrier; the findings suggest that minimizing lung strain impairs structural adaptation of gas exchange units, an interpretation confirmed by postmortem lung morphometry.[139] With the inflated prosthesis, the remaining lung showed septal crowding with an elevated capillary blood volume. The volumes of septal tissue and cell components and the alveolar-capillary surface areas were significantly lower than in animals with deflated prosthesis or in both lungs of SHAM controls, but higher than in the left lung of SHAM animals.[139] The adaptive response to mechanical signals is preserved even when signal institution is delayed. In pneumonectomized ferrets when the resected lung was replaced with an oil-filled silicone balloon, lung volume increased following balloon deflation 3–13 weeks later.[134] In the canine model, delayed deflation of the initially inflated prosthesis 3 weeks post-PNX led to upregulated signaling via hypoxia-inducible factor-1-alpha (HIF-1α) and its effectors including paracrine erythropoietin receptor (EPOR) and VEGF.[140] Delayed deflation of the prosthesis more than 4 months post-PNX led to progressive increases in lung tissue volume, DL_{CO} and exercise capacity, and stimulated alveolar septal growth.[137,139,141,142]

Even when post-PNX lateral expansion is prevented, the remaining lung continued to increase 20–30% in air and tissue volume via caudal displacement of the diaphragm and modest outward expansion of the thorax. This observation suggests that the remaining lung does not simply expand passively to fill an empty space, but responds to intrinsic signals that induce the lung to grow and expand even when space is not readily available. Because mechanical signals associated with lung expansion accounts for on average ~50% of the observed functional and structural compensation depending on the parameter being assessed, additional signals must also play a significant role.[134,136,137,139,141,142] Neither plombage nor prosthesis eliminates completely the post-PNX tidal stretch of the remaining lung, or the endothelial distention and shear forces. These additional mechanical factors could contribute to the residual compensation. Lung growth may also be modulated by non-mechanical signals, such as exercise-induced alveolar hypoxia,[104] endocrine or paracrine growth mediators such as adrenocorticosteroid hormone,[143] epidermal growth factor,[144] hepatocyte growth factor,[145] platelet-derived growth factor,[146] retinoic acid,[107] and nitric oxide,[147] among others. These modulating influences interact with mechanical signals in ways that remain incompletely understood.

Recent applications of non-invasive in vivo imaging by HRCT advance the serial monitoring of lung growth.[137,142,148–152] Quantitative CT image analysis of lung parenchyma using in vivo calibration of attenuation and the segmentation of the lobes permits measurement of the magnitude and regional gradients of air and tissue volumes. The HRCT-derived lung tissue volume correlates strongly with alveolar septal tissue-blood volume measured by morphometry in the postmortem fixed lung, but the former is systematically higher owing to the inclusion of blood volume in microvessels larger than capillaries.[148] When transpulmonary pressure is known and imaging is repeated at different lung inflation levels, regional static lung compliance could be estimated. Applying state-of-the-art non-rigid registration algorithms to match landmarks on paired images obtained at different transpulmonary pressures, regional parenchyma deformation—linear displacement and strain along reference axes, the direction and magnitude of principal strain, and the shear distortion along reference planes, could be derived.[151] The resulting 3D topographical maps of parenchyma deformation visually represent the in vivo distribution of mechanical signals (Figure 7). These studies have shown that post-natal lung growth is relatively homogeneous[148,149]; mechanical strain during passive inflation is most pronounced in the lung periphery, corresponding to the region of most rapid cell proliferation.[39] In contrast, following extensive resection, mechanical strain and shear increase heterogeneously in the remaining lobes, particularly in the regions closest to the removed lobes and in regions that expand around the mediastinum or surrounding the heart. Longitudinal monitoring by

FIGURE 7 Visualization of mechanical deformation in adult canine lung during passive inflation (from 15 to 30 cmH_2O of transpulmonary pressure) in a normal dog (left column, PRE, 7 lobes), and in the same animal 3 and 15 months following resection of 70% of lung units (middle and right columns, POST3 and POST15, respectively) by removing both caudal lobes, the right middle and the infra-cardiac lobes. The remaining 3 lobes are: left and right cranial lobes and the often incompletely separated left middle lobe (also termed inferior segment of the left cranial lobe).[151] Upper row: the distribution of fractional tissue volume (FTV) is shown in coronal section. White lines indicate lobar fissures. FTV increases heterogeneously from PRE to POST3, then diminishes by POST15. The remaining lobes expand 2- to 3-fold in volume. Second row: vector field maps show the direction and magnitude of regional parenchyma displacement, which is normally highest adjacent to the diaphragm (PRE), diminishes in the stiffer remaining lobes at POST3, and then increases by POST15 especially at the caudal end of the remaining lobes. Lower 2 rows: 3D vector field maps show the distribution of principal lung strain in two orientations. Principal strain is normally highest in the peripheral and caudal regions (PRE), increases heterogeneously at POST3, especially in the peripheral and caudal regions of the remaining lobes compared to the corresponding lobes PRE-resection. The early increases in lobar volume, FTV, and strain at POST3 coincide with the period of active alveolar-capillary growth. From POST3 to POST15, principal strain in the remaining lobes declines in the absence of active alveolar-capillary growth, consistent with gradual tissue remodeling and relaxation. *Adapted from.*[151] (This figure is reproduced in color in the color plate section.)

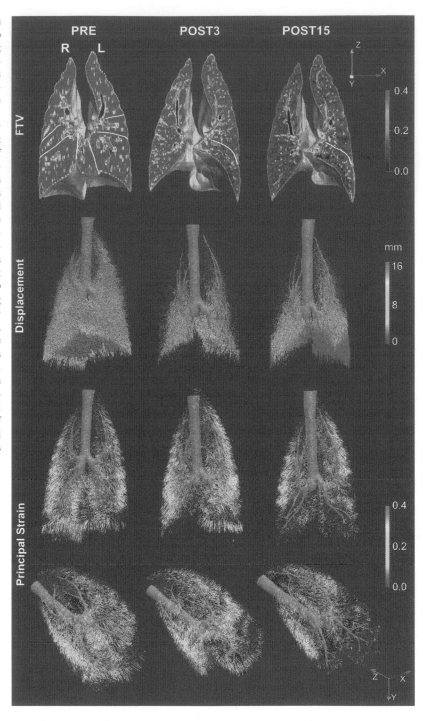

serial HRCT detects the early post-PNX increase in regional strain as well as in air and tissue volumes followed by a later increase in regional compliance due to architectural remodeling that progresses even after the cessation of active tissue growth.[151,152] Although less well established than HRCT, other emergent techniques using positron emission tomography,[153] and MRI,[154,155] also show promising applications in the non-invasive monitoring of lung growth.

COMPENSATORY AIRWAY GROWTH, REMODELING, AND FUNCTION

The generation of new alveolar tissue and capillaries must be accompanied by appropriate architectural remodeling in order to achieve functional improvement. While the intra-acinar airways could increase their size and complexity by adding new septa to split an existing alveolus, perhaps by branching to form another generation of alveolar ducts either out of alveolar complexes or from terminal alveolar sacs, or by alveolizing the terminal bronchiole, the branching patterns of extra-acinar conducting airways are set in utero prior to transformation of the peripheral generations of the pseudoglandular airway tubes into saccules and, eventually, alveolar ducts and sacs[2,156]; accordingly the number of acini as ventilator units are also set at this stage. For conducting airways, lengthening and dilatation are the only options for post-natal adaptation to match the ventilatory requirements of the enlarging parenchymal units. In larger bronchi, the rigid cartilaginous rings or plates reduce mechanical strain experienced by airway cells at a given airway stress or lung stretch; hence strain-related responses are also blunted. At a given total ventilation, post-PNX airway flow resistance (R_{aw}) in the remaining lung is elevated due to increased air flow through the expanding residual lung in relation to a diminished total airway cross-sectional area. Anatomical airway distortion as a result of asymmetric lung expansion is an additional factor that may limit efficient ventilation. In response to the flow and traction forces, the remaining airways lengthen[108,111,157] and dilate.[108] Airway lengthening increases R_{aw} while airway dilatation decreases R_{aw}; hence the net functional compensation depends on a balance of the two processes. Since R_{aw} is inversely proportional to the 4th power of airway diameter and directly related to airway length, only a minimal increase in airway diameter is needed to offset the effect of airway lengthening.

To determine the time course of airway adaptation and its contribution to flow resistance, we measured in vivo airway dimensions by HRCT at normal distending pressures in immature dogs 4 and 10 months after right PNX.[43] By 4 months after PNX the remaining airway lengthened with minimal dilatation without a significant net change of R_{aw}. Post-PNX airway dilatation became apparent later. By 10 months after PNX, the average cross-sectional area of a given lobar bronchus was 24% greater than in the control airway, which can be expected to attenuate lobar R_{aw} by 50% compared to that of a normal lobe at the same volume flow. However, since total airway cross-sectional area still remains below that in two normal lungs, the net reduction of estimated work of breathing against R_{aw} in the whole animal is only ~30%. Thus, airway remodeling after PNX progresses slowly; the delayed increase in total airway volume and cross-sectional area lags behind compensatory parenchymal growth,[126] and only partially mitigates the increased work of breathing in the whole lung. The HRCT-derived airway dimensions predict a 3-fold higher work of breathing against R_{aw} 10 months after PNX,[43] consistent with the 2.5-fold actual increase in work of breathing done against the whole lung measured at rest and during exercise in pneumonectomized immature dogs raised to maturity.[158] Post-PNX airway remodeling amplifies the normal pattern of declining R_{aw} during post-natal maturation and requires the addition of new airway tissue and/or a reduction in compliance of airway wall.

Long-term elevation in work of breathing is greater following right than left PNX,[159] suggesting that the intensity of airway growth and remodeling does not increase with more extensive lung resection. The increase in work of breathing is also similar regardless of the maturity of the animal at the time of lung resection,[158] in contrast to the maturity-dependent compensation in gas exchange function. Thus, parallel physiologic and structural analysis reached similar conclusions that compensatory growth and remodeling of the parenchyma are more vigorous and rapid than that of conducting airways. After PNX the rate and magnitude of increase in airway volume or cross-sectional area at all generations are less than expected from the increase in lung air or septal tissue volume. This unequal potential for compensatory growth among different structural components, termed "dysanaptic growth," was first applied to explain the highly variable expiratory flow rates with respect to lung volume observed in normal subjects.[160] The term was later used to explain gender differences in the expiratory flow–volume relationships of teenagers,[161] and the lower maximal airflow rates relative to a larger increase in lung volume observed in natives of high altitude,[162] in children after lobectomy,[163] and in post-PNX animals.[43,164,165]

A similar "dysanaptic" adaptive pattern is observed in the response of pulmonary blood vessels to PNX. In immature dogs raised to maturity after PNX, pulmonary arterial (PA) pressure and total pulmonary resistance during exercise are persistently elevated even after normal alveolar oxygen transport has been restored.[71,158] Although pulmonary vascular resistance is invariably elevated post-PNX, exercise-induced pulmonary arterial hypertension as well as structural changes in pulmonary arteries are more pronounced in dogs pneumonectomized as adults than as puppies.[158,166] A corresponding age-related increase in PA pressure has also been observed in patients after PNX at exercise and sometimes at rest.[167] In adult humans as well as dogs, long-term mechanical and hemodynamic impairment become disproportionately further accentuated after extensive bilateral resection removing up to 70% of lung and impose major ventilatory and cardiac limitations on exercise capacity.[83,168] While dysfunction of alveolar-capillary gas exchange is the physiological bottleneck following resection of up to 55% of lung, with more extensive destruction (up to 70% of lung units removed) the adaptive response of the more plastic alveolar septa outstrips that of the less plastic conducting structures, and the physiological bottleneck shifts to airway and hemodynamic dysfunction.[168] Consequently, during post-PNX adaptation the extent of concurrent growth and remodeling of conducting airway and blood vessels determines the upper limit of functional compensation that can be achieved regardless of the strength or completeness of alveolar septal regeneration.

The dysanapsis in compensatory growth between the gas exchanging lung parenchyma and the conducting airways and blood vessels has significant effects on the architecture of the gas exchange units that may have functional consequences. The compensatory growth of acinar structures enlarges the volume and the surface area of acinar airways as well as alveolar capillaries. But since the conducting airways cannot proliferate, the number of transitional bronchioles that lead into the acini as ventilatory units[169] is unchanged with the consequence that acinar volume about doubles when the residual lung expands to fill the chest cavity, irrespective of whether there is tissue growth or not. This causes the total path length of acinar airways, that is, the distance from the transitional bronchiole to the end alveolar sacs, to lengthen by a factor of ~1.3. Because gas transport within acinar airspaces occurs mainly by diffusion, the longer path length slows the rate at which inspired O_2 reaches the alveolar tissue interface. This phenomenon, termed gas phase resistance or stratified inhomogeneity, has been detected in post-PNX lungs.[122,170]

A longer path length may also become important because the capillary units supplying different alveoli along the acinar airways are perfused in parallel but ventilated in series.[171] It has been postulated that this arrangement may reduce the O_2 tension in alveolar air of the most peripheral ducts and sacs via the process of screening where O_2 is absorbed by blood in the proximal alveoli leading to a PO_2 gradient along the duct sequence[171,172]; this gradient can become significant because the last generation of acinar airways, the alveolar sacs, contains half the gas exchange surface so that the PO_2 reigning at this level will significantly affect O_2 transfer into capillary blood. This screening effect is largely determined by the total length of the acinar airways; it appears that in the normal human lung the size of acini is such that an ill-effect of screening is avoided,[171,173,174] but if acinar length increases by 30% such an effect cannot be excluded, and could contribute to the limitation of O_2 upon exercise. Screening could be particularly important if acinar expansion is uneven. In such instances, a low PO_2 in the largest alveolar sacs may cause venous admixture to end-capillary blood leaving the lung, unless hypoxic vasoconstriction restricts blood flow to these under-ventilated areas.

REGULATORY PATTERNS DURING DEVELOPMENTAL AND COMPENSATORY GROWTH

Post-PNX signals broadly upregulate genes involved in cellular growth and differentiation, cellular respiration, cytoprotection, synthesis of structural proteins, angiogenesis, immune function, signal transduction, and metabolism.[92,114,140,175-177] It has been suggested that compensatory lung growth might simply recapitulate normal growth.[84] We tested this hypothesis directly in dogs undergoing right PNX as puppies or as adults compared to their age-matched controls and observed age-related differential expression patterns of several growth modulators. In normal growing lungs, cell proliferation markers are elevated nearly 24-fold compared to the adult lung; the vigorous proliferation is further enhanced by 80% 3 weeks post-PNX in puppies followed by subsequent decline to baseline.[40] Whereas the expression of epidermal growth factor (EGF) as well as its receptor (EGFR) and the surfactant protein system are modestly but significantly elevated in the maturing lung compared to adult lung, the expression of both proteins are modestly but significantly lower post-PNX compared to age-matched controls. The surfactant protein expression pattern also differs between post-natal and compensatory lung growth. In the normal growing lung, surfactant protein-A (SP-A) and -C (pro SP-C) are 60–80% lower while SP-D and pro SP-B levels were slightly higher compared to the adult lung. In contrast, in the post-PNX lung SP-A level increases nearly 5-fold and SP-D increases modestly compared to sham controls, while pro SP-B and pro SP-C levels remain unchanged. The selective and prominent upregulation in SP-A after PNX is consistent with a large differentiated and active population of AT2 and bronchiolar club cells. The cytoprotective pathway HIF-1α and its signaling via paracrine erythropoietin (EPO) and its receptor (EPOR) are upregulated during both post-natal and post-PNX lung growth.[140,175,178] While EPOR mRNA and protein expression in lung is upregulated during both types of growth, the expression of antisense EPOR transcripts is selectively downregulated during development but upregulated post-PNX.[175,176] While the actions of antisense EPOR transcripts and the mechanisms underlying developmental vs. non-developmental responses remain to be fully understood, the divergent directions of response suggest the invocation of distinct non-developmental mechanisms during compensatory growth.

What factors might contribute to the different regulatory patterns between developmental and compensatory lung growth? Obviously the metabolic and hormonal milieu in the post-PNX lung differs from that in the normal lung. Mechanical strain on the remaining lung after PNX is imposed suddenly at a larger magnitude than during maturation and as space becomes available to accommodate the expanding alveolar-capillary units. After PNX, minute ventilation is unchanged but volumetric flow through the remaining lung units with each breath is doubled. It is possible that a mismatch between static lung strain due to lung expansion and cyclic strain due to tidal breathing elicits different responses than either stimulus alone. In addition to tissue-specific growth factors, the early influx of inflammatory and immune cells and mediators and the invocation of cytoprotective pathways following PNX suggest the need to balance growth and remodeling against possible tissue injury (Figure 8). During development the increase in tidal volume matches the ventilatory requirement of the animal.

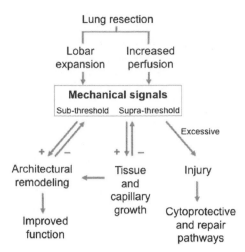

FIGURE 8 A summary diagram shows the mechanical signals and their effects during structure-function adaptation following major lung resection.

PNX markedly alters the distribution and uniformity of mechanical stress and strain experienced by the thorax and the remaining lung; mechanical heterogeneity could cause excessive increases in strain to develop in some regions leading to cell injury, thereby necessitating the recruitment of non-developmental cytoprotective mechanisms. In developing lungs the increase in pulmonary blood flow matches growth of the heart and microvascular bed. In contrast, the rapid and disproportionate post-PNX increase in perfusion to the remaining lung may selectively exaggerate endothelium-derived response to developmental signals. These possibilities remain speculative at present.

AMPLIFYING COMPENSATORY LUNG GROWTH

Post-PNX lung growth is highly sensitive to pharmacological manipulation[107,144,145,179-183]; yet, no growth factor has been convincingly shown to enhance alveolar function. In the canine model, supplementation with oral all-trans retinoic acid (RA) amplifies active compensatory lung growth following right PNX,[179,181] leading to larger alveolar tissue-capillary volumes and surface areas, capillary neo-formation, and persistent deposition of connective tissue-fiber-matrix in the remaining lung. However, growth of AT2 cells lags behind that of other septal components, and supplementation fails to enhance alveolar remodeling such as the slow increase in gas exchange surface area, the gradual thinning of the diffusion barrier, or the transition of newly formed double capillary morphology into the mature single capillary morphology. As a result, alveolar-capillary basal lamina remains thickened; the conductance of diffusion barrier and overall gas exchange and mechanical function are not enhanced compared to placebo treatment.[181] Furthermore, RA supplementation fails to re-initiate lung growth following left PNX where mechanical signals fall below the

threshold for induction of whole-lung growth,[184] The dissociation between structural and functional responses illustrates several general principles: (1) Pharmacological agents modulate mechanosensitive signals of growth but may not be able to re-initiate growth de novo. (2) Stimulation of a single or a few signaling pathways may skew the overall tissue response. (3) Optimal functional gain requires not only the balanced induction of cell proliferation and/or hypertrophy but also the appropriate remodeling to restore and maintain normal architectural relationships at both micro- and macro-levels. Furthermore, in contrast to short-term intense pharmacological stimulation, natural remodeling of the post-PNX lung scaffold is gradual and sustained and associated with only modest upregulation of key growth mediators,[40,176] suggesting that an approach of less intensive, more sustained, and balanced pharmacological stimulation may be more effective in enhancing functionally useful growth.

CONCLUSIONS

This review utilizes the PNX model of compensatory lung growth to illustrate important structure–function relationships during adaptation to the loss of lung units and during attempts to re-initiate lung growth. We first note that compensatory growth leads to full restoration of lung diffusing capacity only if PNX is performed during the period of active somatic growth and maturation whereas PNX in adults leads to effective compensatory growth only if more than ~50% of the lungs are removed. This implies that post-PNX linear tissue stress of more than a factor of 1.3 is required for triggering the processes that lead to tissue neo-formation or actual growth. Post PNX compensation is characterized by: (a) rapid induction of alveolar cellular proliferation and hypertrophy in a threshold- and maturity-dependent manner, (b) progressive acinar remodeling to minimize distortion and optimize function, leading to (c) partial-to-complete restoration of gas exchange capacity. A thorough assessment of lung growth must take into account the physiological consequences of structural responses. Recent evidence directly supports a major role of mechanical signals in the induction and perpetuation of post-PNX lung growth, although non-mechanical signals also play a role in modulating mechanosensitive responses. Compensatory lung growth not only recapitulates normal developmental events but also invokes non-developmental regulatory mechanisms. As the magnitude of mechanical stress and strain increases with greater loss of lung units, a disparate pattern of adaptation emerges between distal lung parenchyma and conducting bronchovasculature, reflecting differential plasticity of the respective components. The extent of disparity or "mismatch" between the adaptive potential of these components effectively limits the functional compensation that could be derived from alveolar septal growth and regeneration. Post-PNX administration of exogenous growth promoters has thus far been able to modestly augment alveolar structural dimensions but not significantly enhance lung function. Emergent approaches involving gene delivery, stem cell-based therapy, repopulation of decellularized scaffolds, and other bioengineered lungs may face a similar challenge. This structure-function gap constitutes a major obstacle to meaningful translational progress in the field of lung repair and regeneration. Overcoming this gap will require thorough understanding not only of the individual cells, fibers, and matrix but also the physico-chemical integration among diverse components at both microscopic and macroscopic levels of architectural organization.

ACKNOWLEDGMENTS

Supported by National Heart Lung and Blood Institute Grants R01 HL40700 and UO1 HL111146.

REFERENCES

1. Weibel ER. It takes more than cells to make a good lung. *Am J Respir Crit Care Med* 2013;**187**:342–6.
2. Burri PH. Structural aspects of prenatal and postnatal development and growth of the lung. In: McDonald JA, editor. *Lung Growth and Development*. New York: Marcel Dekker; 1997. pp. 1–35.
3. Mandelbrot BB. *The Fractal Geometry of Nature*. New York: Freeman; 1983.
4. Weibel ER. *The Pathway for Oxygen: Structure and Function in the Mammalian Respiratory System*. Cambridge (MA): Harvard University Press; 1984.
5. Weibel ER. Fractal geometry: a design principle for living organisms. *Am J Physiol* 1991;**261**:L361–9.
6. Alescio T, Cassini A. Induction in vitro of tracheal buds by pulmonary mesenchyme grafted on tracheal epithelium. *J Exp Zool* 1962;**150**:83–94.
7. Spooner BS, Wessells NK. Mammalian lung development: interactions in primordium formation and bronchial morphogenesis. *J Exp Zool* 1970;**175**:445–54.
8. Cardoso WV, Lu J. Regulation of early lung morphogenesis: questions, facts and controversies. *Development* 2006;**133**:1611–24.
9. Arora R, Metzger RJ, Papaioannou VE. Multiple roles and interactions of Tbx4 and Tbx5 in development of the respiratory system. *PLoS Genet* 2012;**8**:e1002866.
10. Robertson HT, Altemeier WA, Glenny RW. Physiological implications of the fractal distribution of ventilation and perfusion in the lung. *Ann Biomed Eng* 2000;**28**:1028–31.
11. Hsia CC. Signals and mechanisms of compensatory lung growth. *J Appl Physiol* 2004;**97**:1992–8.
12. Liu M, Skinner SJ, Xu J, Han RN, Tanswell AK, Post M. Stimulation of fetal rat lung cell proliferation in vitro by mechanical stretch. *Am J Physiol* 1992;**263**:L376–83.
13. Chess PR, Toia L, Finkelstein JN. Mechanical strain-induced proliferation and signaling in pulmonary epithelial H441 cells. *Am J Physiol Lung Cell Mol Physiol* 2000;**279**:L43–51.
14. Sanchez-Esteban J, Wang Y, Cicchiello LA, Rubin LP. Cyclic mechanical stretch inhibits cell proliferation and induces apoptosis in fetal rat lung fibroblasts. *Am J Physiol Lung Cell Mol Physiol* 2002;**282**:L448–56.
15. Liu M, Qin Y, Liu J, Tanswell AK, Post M. Mechanical strain induces pp60src activation and translocation to cytoskeleton in fetal rat lung cells. *J Biol Chem* 1996;**271**:7066–71.
16. Xu J, Liu M, Liu J, Caniggia I, Post M. Mechanical strain induces constitutive and regulated secretion of glycosaminoglycans and proteoglycans in fetal lung cells. *J Cell Sci* 1996;**109**:1605–13.
17. Liu M, Xu J, Tanswell AK, Post M. Inhibition of mechanical strain-induced fetal rat lung cell proliferation by gadolinium, a stretch-activated channel blocker. *J Cell Physiol* 1994;**161**:501–7.
18. Breen EC. Mechanical strain increases type I collagen expression in pulmonary fibroblasts in vitro. *J Appl Physiol* 2000;**88**:203–9.
19. Smith PG, Moreno R, Ikebe M. Strain increases airway smooth muscle contractile and cytoskeletal proteins in vitro. *Am J Physiol* 1997;**272**:L20–7.
20. Liu M, Liu J, Buch S, Tanswell AK, Post M. Antisense oligonucleotides for PDGF-B and its receptor inhibit mechanical strain-induced fetal lung cell growth. *Am J Physiol* 1995;**269**:L178–84.
21. Waters CM, Chang JY, Glucksberg MR, DePaola N, Grotberg JB. Mechanical forces alter growth factor release by pleural mesothelial cells. *Am J Physiol* 1997;**272**:L552–7.
22. Sanchez-Esteban J, Tsai SW, Sang J, Qin J, Torday JS, Rubin LP. Effects of mechanical forces on lung-specific gene expression. *Am J Med Sci* 1998;**316**:200–4.
23. Unbekandt M, del Moral PM, Sala FG, Bellusci S, Warburton D, Fleury V. Tracheal occlusion increases the rate of epithelial branching of embryonic mouse lung via the FGF10-FGFR2b-Sprouty2 pathway. *Mech Dev* 2008;**125**:314–24.
24. Papadakis K, Luks FI, De Paepe ME, Piasecki GJ, Wesselhoeft Jr CW. Fetal lung growth after tracheal ligation is not solely a pressure phenomenon. *J Pediatr Surg* 1997;**32**:347–51.
25. Harding R, Hooper SB, Han VK. Abolition of fetal breathing movements by spinal cord transection leads to reductions in fetal lung liquid volume, lung growth, and IGF-II gene expression. *Pediatr Res* 1993;**34**:148–53.
26. Tseng BS, Cavin ST, Booth FW, Olson EN, Marin MC, McDonnell TJ, et al. Pulmonary hypoplasia in the myogenin null mouse embryo. *Am J Respir Cell Mol Biol* 2000;**22**:304–15.
27. Nagaya M, Akatsuka H, Kato J, Niimi N, Ishiguro Y. Development in lung function of the affected side after repair of congenital diaphragmatic hernia. *J Pediatr Surg* 1996;**31**:349–56.
28. Beals DA, Schloo BL, Vacanti JP, Reid LM, Wilson JM. Pulmonary growth and remodeling in infants with high-risk congenital diaphragmatic hernia. *J Pediatr Surg* 1992;**27**:997–1001. discussion 1001–1002.
29. Ijsselstijn H, Tibboel D, Hop WJ, Molenaar JC, de Jongste JC. Long-term pulmonary sequelae in children with congenital diaphragmatic hernia. *Am J Respir Crit Care Med* 1997;**155**:174–80.
30. Marven SS, Smith CM, Claxton D, Chapman J, Davies HA, Primhak RA, Powell CV. Pulmonary function, exercise performance, and growth in survivors of congenital diaphragmatic hernia. *Arch Dis Child* 1998;**78**:137–42.
31. McHugh K, Afaq A, Broderick N, Gabra HO, Roebuck DJ, Elliott MJ. Tracheomegaly: a complication of fetal endoscopic tracheal occlusion in the treatment of congenital diaphragmatic hernia. *Pediatr Radiol* 2010;**40**:674–80.
32. Berend N, Marlin GE. Arrest of alveolar multiplication in kyphoscoliosis. *Pathology (Phila)* 1979;**11**:485–91.
33. Davies G, Reid L. Effect of scoliosis on growth of alveoli and pulmonary arteries and on right ventricle. *Arch Dis Child* 1971;**46**:623–32.

34. Olgiati R, Levine D, Smith JP, Briscoe WA, King TK. Diffusing capacity in idiopathic scoliosis and its interpretation regarding alveolar development. *Am Rev Respir Dis* 1982;**126**:229–34.

35. Boffa P, Stovin P, Shneerson J. Lung developmental abnormalities in severe scoliosis. *Thorax* 1984;**39**:681–2.

36. Lund DP, Mitchell J, Kharasch V, Quigley S, Kuehn M, Wilson JM. Congenital diaphragmatic hernia: the hidden morbidity. *J Pediatr Surg* 1994;**29**:258–62. discussion 262–254.

37. Zhang S, Garbutt V, McBride JT. Strain-induced growth of the immature lung. *J Appl Physiol* 1996;**81**:1471–6.

38. Nobuhara KK, Fauza DO, DiFiore JW, Hines MH, Fackler JC, Slavin R, et al. Continuous intrapulmonary distension with perfluorocarbon accelerates neonatal (but not adult) lung growth. *J Pediatr Surg* 1998;**33**:292–8.

39. Massaro GD, Massaro D. Postnatal lung growth: evidence that the gas-exchange region grows fastest at the periphery. *Am J Physiol* 1993;**265**:L319–22.

40. Foster DJ, Yan X, Bellotto DJ, Moe OW, Hagler HK, Estrera AS, et al. Expression of epidermal growth factor and surfactant proteins during postnatal and compensatory lung growth. *Am J Physiol Lung Cell Mol Physiol* 2002;**283**:L981–90.

41. Schilling JA. Pulmonary resection and sequelae of thoracic surgery. In: Fenn WO, Rahn H, editors. *Handbook of Physiology. Section 3: Respiration.* Washington, D.C.: American Physiological Society; 1965. pp. 1531–63.

42. Johnson Jr RL, Cassidy SS, Haynes M, Reynolds RL, Schulz W. Microvascular injury distal to unilateral pulmonary artery occlusion. *J Appl Physiol* 1981;**51**:845–51.

43. Dane DM, Johnson Jr RL, Hsia CCW. Dysanaptic growth of conducting airways after pneumonectomy assessed by CT scan. *J Appl Physiol* 2002;**93**:1235–42.

44. Chamoto K, Gibney BC, Ackermann M, Lee GS, Lin M, Konerding MA, et al. Alveolar macrophage dynamics in murine lung regeneration. *J Cell Physiol* 2012;**227**:3208–15.

45. Shapira OM, Shahian DM. Postpneumonectomy pulmonary edema. *Ann Thorac Surg* 1993;**56**:190–5.

46. Patel DR, Shrivastav R, Sabety AM. Cardiac torsion following intrapericardial pneumonectomy. *J Thorac Cardiovasc Surg* 1973;**65**:626–8.

47. Baaijens PF, Hasenbos MA, Lacquet LK, Dekhuijzen PN. Cardiac herniation after pneumonectomy. *Acta Anaesthesiol Scand* 1992;**36**:842–5.

48. Vallieres E, Shamji FM, Todd TR. Postpneumonectomy chylothorax. *Ann Thorac Surg* 1993;**55**:1006–8.

49. Powell RW, Luck SR, Raffensperger JG. Pneumonectomy in infants and children: the use of a prosthesis to prevent mediastinal shift and its complications. *J Pediatr Surg* 1979;**14**:231–7.

50. Audry G, Balquet P, Vazquez MP, Dejerine ES, Baculard A, Boule M, et al. Expandable prosthesis in right postpneumonectomy syndrome in childhood and adolescence. *Ann Thorac Surg* 1993;**56**:323–7.

51. Jacobsen S, Rosenklint A, Halkier E. Post-pneumonectomy scoliosis. *Acta Orthop Scand* 1974;**45**:867–72.

52. Weekley LB, Read R, Wu EY, Takeda S, Hsia CCW, Johnson Jr RL. Gastroesophageal intussusception associated with pneumonectomy in a dog. *Contemp Top Lab Anim Sci* 1997;**36**:91–3.

53. Soll C, Hahnloser D, Frauenfelder T, Russi EW, Weder W, Kestenholz PB. The postpneumonectomy syndrome: clinical presentation and treatment. *Eur J Cardiothorac Surg* 2009;**35**:319–24.

54. Gaensler EA, Strieder JW. Progressive changes in pulmonary function after pneumonectomy; the influence of thoracoplasty, pneumothorax, oleothorax, and plastic sponge plombage on the side of pneumonectomy. *J Thoracic Surg* 1951;**22**:1–34.

55. Shen KR, Wain JC, Wright CD, Grillo HC, Mathisen DJ. Postpneumonectomy syndrome: surgical management and long-term results. *J Thorac Cardiovasc Surg* 2008;**135**:1210–6. discussion 1216–1219.

56. Harada K, Hamaguchi N, Shimada Y, Saoyama N, Minamimoto T, Inoue K. Use of sulfur hexafluoride, SF6, in the management of the postpneumonectomy pleural space. *Respiration* 1984;**46**:210–8.

57. Grillo HC, Shepard JA, Mathisen DJ, Kanarek DJ. Postpneumonectomy syndrome: diagnosis, management and results. *Ann Thorac Surg* 1992;**54**:638–50.

58. Birath G, Malmberg R, Simonsson BG. Lung function after pneumonectomy in man. *Clin Sci* 1965;**29**:59–72.

59. Hsia CCW, Fryder-Doffey F, Stalder-Navarro V, Johnson Jr RL, Weibel ER. Structural changes underlying compensatory increase of diffusing capacity after left pneumonectomy in adult dogs [Published erratum in J. Clin. Invest., 93(2):913, 1994.]. *J Clin Invest* 1993;**92**:758–64.

60. Hsia CCW, Herazo LF, Fryder-Doffey F, Weibel ER. Compensatory lung growth occurs in adult dogs after right pneumonectomy. *J Clin Invest* 1994;**94**:405–12.

61. Hsia CCW, Herazo LF, Johnson Jr RL. Cardiopulmonary adaptations to pneumonectomy in dogs. I. Maximal exercise performance. *J Appl Physiol* 1992;**73**:362–7.

62. Weibel ER, Taylor CR, Bolis L, editors. *Principles of Animal Design.* Cambridge, U.K: Cambridge University Press; 1998.

63. Weibel ER. *Symmorphosis.* Cambridge, MA: Harvard University Press; 2000.

64. Hsia CC. Coordinated adaptation of oxygen transport in cardiopulmonary disease. *Circulation* 2001;**104**:963–9.

65. König MF, Lucocq JM, Weibel ER. Demonstration of pulmonary vascular perfusion by electron and light microscopy. *J Appl Physiol* 1993;**75**:1877–83.

66. Okada O, Presson Jr RG, Kirk KR, Godgey PS, Capen RL, Wagner WW. Capillary perfusion patterns in single alveolar walls. *J Appl Physiol* 1992;**72**:1838–44.

67. Hsia CC, McBrayer DG, Ramanathan M. Reference values of pulmonary diffusing capacity during exercise by a rebreathing technique. *Am J Respir Crit Care Med* 1995;**152**:658–65.

68. Johnson Jr RL, Heigenhauser GJF, Hsia CCW, Jones NL, Wagner PD. Determinants of gas exchange and acid-base balance during exercise. In: Rowell LB, Shepherd JT, editors. *Handbook of Physiology; Section 12, Exercise Regulation and Integration of Multiple Systems.* New York: American Physiological Society & Oxford University Press; 1996. pp. 515–84.

69. Tamhane RM, Johnson Jr RL, Hsia CC. Pulmonary membrane diffusing capacity and capillary blood volume measured during exercise from nitric oxide uptake. *Chest* 2001;**120**:1850–6.

70. Hsia CCW, Herazo LF, Ramanathan M, Johnson Jr RL. Cardiopulmonary adaptations to pneumonectomy in dogs. IV. Membrane diffusing capacity and capillary blood volume. *J Appl Physiol* 1994;**77**:998–1005.

71. Takeda S, Hsia CCW, Wagner E, Ramanathan M, Estrera AS, Weibel ER. Compensatory alveolar growth normalizes gas exchange function in immature dogs after pneumonectomy. *J Appl Physiol* 1999;**86**:1301–10.

72. Hsia CC. Recruitment of lung diffusing capacity: update of concept and application. *Chest* 2002;**122**:1774–83.

73. Carlin JI, Hsia CCW, Cassidy SS, Ramanathan M, Clifford PS, Johnson Jr RL. Recruitment of lung diffusing capacity with exercise before and after pneumonectomy in dogs. *J Appl Physiol* 1991;**70**:135–42.

74. Hsia CCW, Ramanathan M, Estrera AS. Recruitment of diffusing capacity with exercise in patients after pneumonectomy. *Am Rev Respir Dis* 1992;**145**:811–6.

75. Giammona ST, Mandelbaum I, Battersby JS, Daly WJ. The late cardiopulmonary effects of childhood pneumonectomy. *Pediatrics* 1966;**37**:79–88.

76. Hughes JMB, Lockwood DNA, Jones HA, Clark RJ. DL_{CO}/Q and diffusion limitation at rest and on exercise in patients with interstitial fibrosis. *Respir Physiol* 1991;**83**:155–66.

77. Burri PH, Pfrunder HB, Berger LC. Reactive changes in pulmonary parenchyma after bilobectomy: a scanning electron microscopic investigation. *Exp Lung Res* 1982;**4**:11–28.

78. Berger LC, Burri PH. Timing of the quantitative recovery in the regenerating rat lung. *Am Rev Respir Dis* 1985;**132**:777–83.

79. Hsia CC, Johnson Jr RL. Further examination of alveolar septal adaptation to left pneumonectomy in the adult lung. *Respir Physiol Neurobiol* 2006;**151**:167–77.

80. Ravikumar P, Yilmaz C, Dane DM, Bellotto DJ, Estrera AS, Hsia CC. Defining a stimuli-response relationship in compensatory lung growth following major resection. *J Appl Physiol* 2014;**116**:816–24.

81. Bachofen H, Schürch S, Urbinelli M, Weibel ER. Relations among alveolar surface tension, surface area, volume, and recoil pressure. *J Appl Physiol* 1987;**62**:1878–87.

82. Hsia CCW, Carlin JI, Wagner PD, Cassidy SS, Johnson Jr RL. Gas exchange abnormalities after pneumonectomy in conditioned foxhounds. *J Appl Physiol* 1990;**68**:94–104.

83. DeGraff Jr AC, Taylor HF, Ord JW, Chuang TH, Johnson Jr RL. Exercise limitation following extensive pulmonary resection. *J Clin Invest* 1965;**44**:1514–22.

84. Cagle PT, Thurlbeck WM. Postpneumonectomy compensatory lung growth. *Am Rev Respir Dis* 1988;**138**:1314–26.

85. Voswinckel R, Motejl V, Fehrenbach A, Wegmann M, Mehling T, Fehrenbach H, et al. Characterisation of post-pneumonectomy lung growth in adult mice. *Eur Respir J* 2004;**24**:524–32.

86. Ding BS, Nolan DJ, Guo P, Babazadeh AO, Cao Z, Rosenwaks Z, et al. Endothelial-derived angiocrine signals induce and sustain regenerative lung alveolarization. *Cell* 2011;**147**:539–53.

87. Hsia CCW, Zhou XS, Bellotto DJ, Hagler HK. Regenerative growth of respiratory bronchioles in dogs. *Am J Physiol Lung Cell Mol Physiol* 2000;**279**:L136–42.

88. Rannels DE. Role of physical forces in compensatory growth of the lung. *Am J Physiol* 1989;**257**:L179–89.

89. Mourgeon E, Xu J, Tanswell AK, Liu M, Post M. Mechanical strain-induced posttranscriptional regulation of fibronectin production in fetal lung cells. *Am J Physiol* 1999;**277**:L142–9.

90. Dobbs LG, Gutierrez JA. Mechanical forces modulate alveolar epithelial phenotypic expression. *Comp Biochem Physiol A Mol Integr Physiol* 2001;**129**:261–6.

91. Muratore CS, Nguyen HT, Ziegler MM, Wilson JM. Stretch-induced upregulation of VEGF gene expression in murine pulmonary culture: a role for angiogenesis in lung development. *J Pediatr Surg* 2000;**35**:906–12. discussion 912–903.

92. Landesberg LJ, Ramalingam R, Lee K, Rosengart TK, Crystal RG. Upregulation of transcription factors in lung in the early phase of postpneumonectomy lung growth. *Am J Physiol Lung Cell Mol Physiol* 2001;**281**:L1138–49.

93. Bardocz S, Tatar-Kiss S, Kertai P. The effect of alpha-difluoromethylornithine on ornithine decarboxylase activity in compensatory growth of mouse lung. *Acta Biochim Biophys Hung* 1986;**21**:59–65.

94. Brown LM, Malkinson AM, Rannels DE, Rannels SR. Compensatory lung growth after partial pneumonectomy enhances lung tumorigenesis induced by 3-methylcholanthrene. *Cancer Res* 1999;**59**:5089–92.

95. Brown LM, Welch DR, Rannels DE, Rannels SR. Partial pneumonectomy enhances melanoma metastasis to mouse lungs. *Chest* 2002;**121**:28S–9S.

96. Brody JS, Burki R, Kaplan N. Deoxyribonucleic acid synthesis in lung cells during compensatory lung growth after pneumonectomy. *Am Rev Respir Dis* 1978;**117**:307–16.

97. Fehrenbach H, Voswinckel R, Michl V, Mehling T, Fehrenbach A, Seeger W, et al. Neoalveolarisation contributes to compensatory lung growth following pneumonectomy in mice. *Eur Respir J* 2008;**31**:515–22.

98. Wandel G, Berger LC, Burri PH. Morphometric analysis of adult rat lung after bilobectomy. *Am Rev Respir Dis* 1983;**128**:968–72.

99. Rannels DE, Stockstill B, Mercer RR, Crapo JD. Cellular changes in the lungs of adrenalectomized rats following left pneumonectomy. *Am J Respir Cell Mol Biol* 1991;**5**:351–62.

100. Thet LA, Law DJ. Changes in cell number and lung morphology during early postpneumonectomy lung growth. *J Appl Physiol* 1984;**56**:975–8.

101. Nattie EE, Wiley CW, Bartlett Jr D. Adaptive growth of the lung following pneumonectomy in rats. *J Appl Physiol* 1974;**37**:491–5.

102. McAnulty RJ, Guerreiro D, Cambrey AD, Laurent GJ. Growth factor activity in the lung during compensatory growth after pneumonectomy: evidence of a role for IGF-1. *Eur Respir J* 1992;**5**:739–47.

103. Faridy EE, Sanii MR, Thliveris JA. Influence of maternal pneumonectomy on fetal lung growth. *Respir Physiol* 1988;**72**:195–209.

104. Sekhon HS, Smith C, Thurlbeck WM. Effect of hypoxia and hyperoxia on postpneumonectomy compensatory lung growth. *Exp Lung Res* 1993;**19**:519–32.

105. Gilbert KA, Rannels DE. Increased lung inflation induces gene expression after pneumonectomy. *Am J Physiol* 1998;**275**:L21–9.

106. Kaza AK, Kron IL, Long SM, Fiser SM, Stevens PM, Kern JA, et al. Epidermal growth factor receptor up-regulation is associated with lung growth after lobectomy. *Ann Thorac Surg* 2001;**72**:380–5.

107. Kaza AK, Kron IL, Kern JA, Long SM, Fiser SM, Nguyen RP, et al. Retinoic acid enhances lung growth after pneumonectomy. *Ann Thorac Surg* 2001;**71**:1645–50.

108. McBride JT. Postpneumonectomy airway growth in the ferret. *J Appl Physiol* 1985;**58**:1010–4.

109. Kirchner KK, McBride JT. Changes in airway length after unilateral pneumonectomy in weanling ferrets. *J Appl Physiol* 1990;**68**:187–92.

110. McBride JT, Kirchner KK, Russ G, Finkelstein J. Role of pulmonary blood flow in postpneumonectomy lung growth. *J Appl Physiol* 1992;**73**:2448–51.

111. Boatman ES. A morphometric and morphological study of the lungs of rabbits after unilateral pneumonectomy. *Thorax* 1977;**32**:406–17.

112. Das RM, Thurlbeck WM. The events in the contralateral lung following pneumonectomy in the rabbit. *Lung* 1979;**156**:165–72.

113. Langston C, Sachdeva P, Cowan MJ, Haines J, Crystal RG, Thurlbeck WM. Alveolar multiplication in the contralateral lung after unilateral pneumonectomy in the rabbit. *Am Rev Respir Dis* 1977;**115**:7–13.

114. Paxson JA, Parkin CD, Iyer LK, Mazan MR, Ingenito EP, Hoffman AM. Global gene expression patterns in the post-pneumonectomy lung of adult mice. *Respir Res* 2009;**10**:92.

115. Chamoto K, Gibney BC, Ackermann M, Lee GS, Konerding MA, Tsuda A, et al. Alveolar epithelial dynamics in postpneumonectomy lung growth. *Anat Rec (Hoboken)* 2013;**296**:495–503.

116. Nolen-Walston RD, Kim CF, Mazan MR, Ingenito EP, Gruntman AM, Tsai L, et al. Cellular kinetics and modeling of bronchioalveolar stem cell response during lung regeneration. *Am J Physiol Lung Cell Mol Physiol* 2008;**294**:L1158–65.

117. Voswinckel R, Ziegelhoeffer T, Heil M, Kostin S, Breier G, Mehling T, et al. Circulating vascular progenitor cells do not contribute to compensatory lung growth. *Circ Res* 2003;**93**:372–9.

118. Suga A, Ueda K, Takemoto Y, Nishimoto A, Hosoyama T, Li TS, et al. Significant role of bone marrow-derived cells in compensatory regenerative lung growth. *J Surg Res* 2013;**183**:84–90.

119. Cohn R. Factors affecting the postnatal growth of the lung. *Anat Rec* 1939;**75**:195–205.

120. Burri PH, Sehovic S. The adaptive response of the rat lung after bilobectomy. *Am Rev Respir Dis* 1979;**119**:769–77.

121. Mitzner W, Lee W, Georgakopoulos D, Wagner E. Angiogenesis in the mouse lung. *Am J Pathol* 2000;**157**:93–101.

122. Hsia CC, Herazo LF, Ramanathan M, Johnson Jr RL, Wagner PD. Cardiopulmonary adaptations to pneumonectomy in dogs. II. VA/Q relationships and microvascular recruitment. *J Appl Physiol* 1993;**74**:1299–309.

123. Johnson Jr RL, Cassidy SS, Grover R, Ramanathan M, Estrera A, Reynolds RC, et al. Effect of pneumonectomy on the remaining lung in dogs. *J Appl Physiol* 1991;**70**:849–58.

124. Holmes C, Thurlbeck WM. Normal lung growth and response after pneumonectomy in rats at various ages. *Am Rev Respir Dis* 1979;**120**:1125–36.

125. Takeda S, Ramanathan M, Wu EY, Estrera AS, Hsia CCW. Temporal course of gas exchange and mechanical compensation after right pneumonectomy in immature dogs [Published corrigenda in J. Appl. Physiol. 80(6): after Table of Contents, 1996.]. *J Appl Physiol* 1996;**80**:1304–12.

126. Takeda S, Wu EY, Epstein RH, Estrera AS, Hsia CCW. In vivo assessment of changes in air and tissue volumes after pneumonectomy. *J Appl Physiol* 1997;**82**:1340–8.

127. Reddy R, Buckley S, Doerken M, Barsky L, Weinberg K, Anderson KD, et al. Isolation of a putative progenitor subpopulation of alveolar epithelial type 2 cells. *Am J Physiol Lung Cell Mol Physiol* 2004;**286**:L658–667.

128. Danto SI, Shannon JM, Borok Z, Zabski SM, Crandall ED. Reversible transdifferentiation of alveolar epithelial cells. *Am J Respir Cell Mol Biol* 1995;**12**:497–502.

129. Flecknoe S, Harding R, Maritz G, Hooper SB. Increased lung expansion alters the proportions of type I and type II alveolar epithelial cells in fetal sheep. *Am J Physiol Lung Cell Mol Physiol* 2000;**278**:L1180–5.

130. Fehrenbach H. Alveolar epithelial type II cell: defender of the alveolus revisited. *Respir Res* 2001;**2**:33–46.

131. Williams MC. Alveolar type I cells: molecular phenotype and development. *Annu Rev Physiol* 2003;**65**:669–95.

132. Laros CD, Westermann CJJ. Dilatation, compensatory growth, or both after pneumonectomy during childhood and adolescence. A thirty-year follow-up study. *J Thorac Cardiovasc Surg* 1987;**93**:570–6.

133. Butler JP, Loring SH, Patz S, Tsuda A, Yablonskiy DA, Mentzer SJ. Evidence for adult lung growth in humans. *N Engl J Med* 2012;**367**:244–7.

134. McBride JT. Lung volumes after an increase in lung distension in pneumonectomized ferrets. *J Appl Physiol* 1989;**67**:1418–21.

135. Fisher JM, Simnett JD. Morphogenetic and proliferative changes in the regenerating lung of the rat. *Anat Rec* 1973;**176**:389–96.

136. Olson LE, Hoffman EA. Lung volumes and distribution of regional air content determined by cine x-ray CT of pneumonectomized rabbits. *J Appl Physiol* 1994;**76**:1774–85.

137. Wu EY, Hsia CC, Estrera AS, Epstein RH, Ramanathan M, Johnson Jr RL. Preventing mediastinal shift after pneumonectomy does not abolish physiologic compensation. *J Appl Physiol* 2000;**89**:182–91.

138. Hsia CC, Johnson Jr RL, Wu EY, Estrera AS, Wagner H, Wagner PD. Reducing lung strain after pneumonectomy impairs oxygen diffusing capacity but not ventilation-perfusion matching. *J Appl Physiol* 2003;**95**:1370–8.

139. Hsia CC, Wu EY, Wagner E, Weibel ER. Preventing mediastinal shift after pneumonectomy impairs regenerative alveolar tissue growth. *Am J Physiol Lung Cell Mol Physiol* 2001;**281**:L1279–87.

140. Zhang Q, Bellotto DJ, Ravikumar P, Moe OW, Hogg RT, Hogg DC, et al. Postpneumonectomy lung expansion elicits hypoxia-inducible factor-1alpha signaling. *Am J Physiol Lung Cell Mol Physiol* 2007;**293**:L497–504.

141. Dane DM, Yilmaz C, Estrera AS, Hsia CC. Separating in vivo mechanical stimuli for postpneumonectomy compensation: physiological assessment. *J Appl Physiol* 2013;**114**:99–106.

142. Ravikumar P, Yilmaz C, Bellotto DJ, Dane DM, Estrera AS, Hsia CC. Separating in vivo mechanical stimuli for postpneumonectomy compensation: imaging and ultrastructural assessment. *J Appl Physiol* 2013;**114**:961–70.

143. Bennett RA, Colony PC, Addison JL, Rannels DE. Effects of prior adrenalectomy on postpneumonectomy lung growth in the rat. *Am J Physiol* 1985;**248**:E70–74.

144. Kaza AK, Laubach VE, Kern JA, Long SM, Fiser SM, Tepper JA, et al. Epidermal growth factor augments postpneumonectomy lung growth. *J Thorac Cardiovasc Surg* 2000;**120**:916–21.

145. Sakamaki Y, Matsumoto K, Mizuno S, Miyoshi S, Matsuda H, Nakamura T. Hepatocyte growth factor stimulates proliferation of respiratory epithelial cells during postpneumonectomy compensatory lung growth in mice. *Am J Respir Cell Mol Biol* 2002;**26**:525–33.

146. Yuan S, Hannam V, Belcastro R, Cartel N, Cabacungan J, Wang J, et al. A role for platelet-derived growth factor-BB in rat postpneumonectomy compensatory lung growth. *Pediatr Res* 2002;**52**:25–33.

147. Leuwerke SM, Kaza AK, Tribble CG, Kron IL, Laubach VE. Inhibition of compensatory lung growth in endothelial nitric oxide synthase-deficient mice. *Am J Physiol Lung Cell Mol Physiol* 2002;**282**:L1272–8.

148. Ravikumar P, Yilmaz C, Dane DM, Johnson Jr RL, Estrera AS, Hsia CC. Regional lung growth following pneumonectomy assessed by computed tomography. *J Appl Physiol* 2004;**97**:1567–74. discussion 1549.

149. Ravikumar P, Yilmaz C, Dane DM, Johnson Jr RL, Estrera AS, Hsia CC. Developmental signals do not further accentuate nonuniform postpneumonectomy compensatory lung growth. *J Appl Physiol* 2007;**102**:1170–7.

150. Yilmaz C, Ravikumar P, Dane DM, Bellotto DJ, Johnson Jr RL, Hsia CC. Noninvasive quantification of heterogeneous lung growth following extensive lung resection by high-resolution computed tomography. *J Appl Physiol* 2009;**107**:1569–78.

151. Yilmaz C, Tustison NJ, Dane DM, Ravikumar P, Takahashi M, Gee JC, et al. Progressive adaptation in regional parenchyma mechanics following extensive lung resection assessed by functional computed tomography. *J Appl Physiol* 2011;**111**:1150–8.

152. Hsia CC, Tawhai MH. What can imaging tell us about physiology? Lung growth and regional mechanical strain. *J Appl Physiol* 2012;**113**:937–46.

153. Gibney BC, Park MA, Chamoto K, Ysasi A, Konerding MA, Tsuda A, et al. Detection of murine post-pneumonectomy lung regeneration by 18FDG PET imaging. *EJNMMI Res* 2012;**2**:48.

154. Wang W, Nguyen NM, Guo J, Woods JC. Longitudinal, non-invasive monitoring of compensatory lung growth in mice after pneumonectomy via He and H MRI. *Am J Respir Cell Mol Biol* 2013;**49**:697–703.

155. Butler JP, Loring SH, Patz S, Tsua A, Yablonskiy DA, Mentzer SJ. Evidence for adult lung growth in humans. *N Engl J Med* 2012;**367**:244–7.

156. Kitaoka H, Burri PH, Weibel ER. Development of the human fetal airway tree: analysis of the numerical density of airway endtips. *Anat Rec* 1996;**244**:207–13.

157. Yee NW, Hyatt RE. Effect of left pneumonectomy on lung mechanics in rabbits. *J Appl Physiol* 1983;**54**:1612–7.

158. Takeda S, Ramanathan M, Estrera AS, Hsia CCW. Postpneumonectomy alveolar growth does not normalize hemodynamic and mechanical function. *J Appl Physiol* 1999;**87**:491–7.

159. Hsia CCW, Herazo LF, Ramanathan M, Claassen H, Fryder-Doffey F, Hoppeler H, et al. Cardiopulmonary adaptations to pneumonectomy in dogs. III. Ventilatory power requirements and muscle structure. *J Appl Physiol* 1994;**76**:2191–8.

160. Green M, Mead J, Turner JM. Variability of maximum expiratory flow-volume curves. *J Appl Physiol* 1974;**37**:67–74.

161. Merkus PJ, Borsboom GJ, Van Pelt W, Schrader PC, Van Houwelingen HC, Kerrebijn KF, et al. Growth of airways and air spaces in teenagers is related to sex but not to symptoms. *J Appl Physiol* 1993;**75**:2045–53.

162. Brody JS, Lahiri S, Simpser M, Motoyama EK, Velasquez T. Lung elasticity and airway dynamics in Peruvian natives to high altitude. *J Appl Physiol* 1977;**42**:245–51.

163. McBride JT, Wohl ME, Strieder DJ, Jackson AC, Morton JR, Zwerdling RG, et al. Lung growth and airway function after lobectomy in infancy for congenital lobar emphysema. *J Clin Invest* 1980;**66**:962–70.

164. Arnup ME, Greville HW, Oppenheimer L, Mink SN, Anthonisen NR. Dynamic lung function in dogs with compensatory lung growth. *J Appl Physiol* 1984;**57**:1569–76.

165. Greville HW, Arnup ME, Mink SN, Oppenheimer L, Anthonisen NR. Mechanism of reduced maximum expiratory flow in dogs with compensatory lung growth. *J Appl Physiol* 1986;**60**:441–8.

166. Davies P, McBride J, Murray GF, Wilcox BR, Shallal JA, Reid L. Structural changes in the canine lung and pulmonary arteries after pneumonectomy. *J Appl Physiol* 1982;**53**:859–64.

167. Staněk V, Widimsky J, Hurych J, Petrikova J. Pressure, flow and volume changes during exercise within pulmonary vascular bed in patients after pneumonectomy. *Clin Sci* 1969;**37**:11–22.

168. Hsia CC, Dane DM, Estrera AS, Wagner HE, Wagner PD, Johnson Jr RL. Shifting sources of functional limitation following extensive (70%) lung resection. *J Appl Physiol* 2008;**104**:1069–79.

169. Haefeli BB, Weibel ER. Morphometry of the human pulmonary acinus. *Anat Rec* 1988;**220**:401–14.

170. Hsia CC, Yan X, Dane DM, Johnson Jr RL. Density-dependent reduction of nitric oxide diffusing capacity after pneumonectomy. *J Appl Physiol* 2003;**94**:1926–32.

171. Sapoval B, Filoche M, Weibel ER. Smaller is better — but not too small: a physical scale for the design of the mammalian pulmonary acinus. *Proc Natl Acad Sci U S A* 2002;**99**:10411–6.

172. Weibel ER, Sapoval B, Filoche M. Design of peripheral airways for efficient gas exchange. *Respir Physiol Neurobiol* 2005;**148**:3–21.

173. Swan AJ, Tawhai MH. Evidence for minimal oxygen heterogeneity in the healthy human pulmonary acinus. *J Appl Physiol* 2011;**110**:528–37.

174. Foucquier A, Filoche M, Moreira AA, Andrade Jr JS, Arbia G, Sapoval B. A first principles calculation of the oxygen uptake in the human pulmonary acinus at maximal exercise. *Respir Physiol Neurobiol* 2013;**185**:625–38.

175. Zhang Q, Zhang J, Moe OW, Hsia CC. Synergistic upregulation of erythropoietin receptor (EPO-R) expression by sense and antisense EPO-R transcripts in the canine lung. *Proc Natl Acad Sci USA* 2008;**105**:7612–7.

176. Foster DJ, Moe OW, Hsia CC. Upregulation of erythropoietin receptor during postnatal and postpneumonectomy lung growth. *Am J Physiol Lung Cell Mol Physiol* 2004;**287**:L1107–15.

177. Wolff JC, Wilhelm J, Fink L, Seeger W, Voswinckel R. Comparative gene expression profiling of post-natal and post-pneumonectomy lung growth. *Eur Respir J* 2010;**35**:655–66.

178. Zhang Q, Moe OW, Garcia JA, Hsia CC. Regulated expression of hypoxia-inducible factors during postnatal and postpneumonectomy lung growth. *Am J Physiol Lung Cell Mol Physiol* 2006;**290**:L880–9.

179. Ravikumar P, Dane DM, McDonough P, Yilmaz C, Estrera AS, Hsia CC. Long-term post-pneumonectomy pulmonary adaptation following all-trans-retinoic acid supplementation. *J Appl Physiol* 2011;**110**:764–73.

180. Yan X, Bellotto DJ, Foster DJ, Johnson Jr RL, Hagler HH, Estrera AS, et al. Retinoic acid induces nonuniform alveolar septal growth after right pneumonectomy. *J Appl Physiol* 2004;**96**:1080–9.

181. Dane DM, Yan X, Tamhane RM, Johnson Jr RL, Estrera AS, Hogg DC, et al. Retinoic acid-induced alveolar cellular growth does not improve function after right pneumonectomy. *J Appl Physiol* 2004;**96**:1090–6.

182. Kaza AK, Kron IL, Leuwerke SM, Tribble CG, Laubach VE. Keratinocyte growth factor enhances post-pneumonectomy lung growth by alveolar proliferation. *Circulation* 2002;**106**:I120–4.

183. Li D, Fernandez LG, Dodd-o J, Langer J, Wang D, Laubach VE. Upregulation of hypoxia-induced mitogenic factor in compensatory lung growth after pneumonectomy. *Am J Respir Cell Mol Biol* 2005;**32**:185–91.

184. Yan X, Bellotto DJ, Dane DM, Elmore RG, Johnson Jr RL, Estrera AS, et al. Lack of response to all-trans retinoic acid supplementation in adult dogs following left pneumonectomy. *J Appl Physiol* 2005;**99**:1681–8.

Chapter 13

Pulmonary Transition at Birth

Melissa L. Siew*, Marcus J. Kitchen†, Arjan B. te Pas‡, Richard Harding§ and Stuart B. Hooper*

*The Ritchie Centre, MIMR-PHI Institute of Medical Research,; †School of Physics, Monash University, Clayton, VIC, Australia; §Department of Anatomy and Developmental Biology, Monash University, Clayton, VIC, Australia,; ‡Division of Neonatology, Department of Pediatrics, Leiden University Medical Centre, Leiden, The Netherlands

INTRODUCTION

The transition from fetal to neonatal life is dependent upon major changes within the cardiovascular and respiratory systems that allow the infant to initiate pulmonary gas exchange, a function that is performed by the placenta during fetal life. For this to occur, a number of important adaptive events must take place. First and foremost, the liquid that occupies the lung lumen throughout gestation must be cleared to allow the entry of air into the distal gas-exchange regions. This in turn triggers the second event, dilation of the pulmonary vascular bed, which results in a marked increase in pulmonary blood flow (PBF). This not only facilitates pulmonary gas exchange in the newborn infant, but also provides vital venous return and preload for the left ventricle that is lost once the umbilical cord is clamped. This is because during fetal life, the lungs receive only a small fraction of right ventricular output, as most (~90%) flows from the main pulmonary artery, through the ductus arteriosus and into the descending aorta. Consequently, as PBF is low in the fetus, pulmonary venous return is unable to provide much preload for the left ventricle, which instead comes from the umbilical circulation (see the following text).

Although it is known that lung aeration is the major precipitating event that triggers the physiological changes underpinning the transition to newborn life, the mechanisms involved are not fully understood. The entry of air not only increases oxygen levels within distal lung tissue, it also creates an air–liquid interface across the large internal surface of the lung that is not present in the liquid-filled fetal lung. The resulting surface tension that forms increases the recoil pressure of the lungs, which not only opposes lung expansion and increases the lung's tendency to collapse, but also alters the distribution of force within the distal airways and surrounding tissue. In addition, as the entry of air into the distal airways displaces lung liquid from the airways into the surrounding tissue, the accumulation of liquid within the distal lung tissue causes an increase in interstitial tissue pressure. It is likely that one or all of these mechanisms are responsible for triggering the physiological changes that underpin the transition to newborn life.

Many of the changes in lung physiology that occur late in gestation to facilitate the transition to air-breathing are closely linked to the processes that initiate labor. Indeed, early evidence that the increase in fetal circulating corticosteroid concentrations preceding birth plays an important role in maturing the lungs[1] has gained universal acceptance and has led to the widespread use of antenatal corticosteroids in women at risk of delivering before term.[2,3] Although the success of antenatal corticosteroids in reducing respiratory insufficiency in the preterm infant is incontrovertible,[4] the mechanisms by which corticosteroids act on the immature lung to facilitate gas exchange after birth are not well understood.[5] The aim of this chapter is to review the physiological changes that occur in the lung around the time of birth, and how these changes are regulated.

FETAL LUNG MATURATION, GLUCOCORTICOIDS, AND BIRTH

In most mammalian species, particularly those giving birth to precocial offspring, birth is preceded by an exponential-like increase in circulating corticosteroid concentrations. In some species (e.g., sheep), the increase in circulating fetal cortisol levels not only initiates parturition,[6] but also stimulates maturation of a variety of fetal organ systems, including the lung.[7–9] From a teleological standpoint, it is appropriate that the same processes that initiate parturition also contribute to the maturation of fetal organ systems that, at birth, must rapidly assume a role that is vital for postnatal survival. Preterm birth represents an example of the failure of this link between parturition and maturation, resulting in the birth of an infant before vital organs such as the lungs have been able to mature to the point that they can sustain independent life.

To determine the mechanisms by which fetal corticosteroids accelerate lung maturation, most attention has focussed on surfactant production and type-II alveolar epithelial cell (AEC) maturation; the role of corticosteroids in fetal lung development has been the subject of a number of reviews.[10,5]

The Lung. http://dx.doi.org/10.1016/B978-0-12-799941-8.00013-4

As described in Chapter 9 (Orgeig et al.), the phospholipid component of surfactant plays a vital role in lung function by forming a stable monolayer at the air–liquid interface, thereby reducing surface tension and hence lung recoil. In particular, the recruitment and expulsion of phospholipids from this monolayer during the respiratory cycle help to stabilize the lung, particularly at end-expiration. It is well established that exogenous corticosteroids increase surfactant synthesis by inducing many of the synthetic enzymes and by inducing surfactant protein expression both in vivo and in vitro.[10,5] However, some discrepancies persist in the literature, particularly relating to the induction of surfactant protein gene expression, which may reflect differences in the dose and duration of corticosteroid exposure and differences between in vitro and in vivo studies.[10] Indeed, most experimental studies have used high doses of synthetic glucocorticoids (e.g., dexamethasone or betamethasone) that have a ~30-fold greater bioactivity than natural cortisol.

The role of endogenous corticosteroids in fetal lung maturation has been examined using glucocorticoid receptor (GR) deficient mice.[11] Homozygous GR (-/-) deficient mice die of respiratory insufficiency at birth, confirming that endogenous corticosteroids play a vital role in fetal lung maturation.[11] However, the deficiency primarily causes an immature tissue architecture rather than type-II alveolar epithelial cell (AEC) dysfunction or surfactant deficiency (Figure 1). Indeed, GR deficient mice have increased type-II AEC numbers, reduced type-I AEC numbers and a largely unaltered surfactant protein gene expression.[11] Instead, GR knockout mice have hyperplastic lungs with thick interalveolar walls caused by high rates of mesenchymal cell proliferation that persist into late gestation. More recently, using cell-specific (epithelial, endothelial, and fibroblast

cell specific) GR knockouts, it has been found that the associated lung phenotype and neonatal mortality only persists when the knockout is specific to lung fibroblasts.[12] These findings indicate that corticosteroids primarily affect the developing architecture of the lung by selectively reducing proliferation of distal lung fibroblasts. This allows the distal lung tissue to thin, the airways to expand, and the surrounding capillaries to come into closer apposition with the distal airways. Furthermore, as the fibroblasts are largely responsible for extracellular matrix production, altering the fibroblast population may substantially modify the extracellular matrix components of the distal airway tissue and therefore its compliance.[5]

AIRWAY LIQUID CLEARANCE BEFORE BIRTH

In the fetus, the pulmonary airways contain a liquid that is secreted by the epithelium (see Chapter 7 [Widdicombe]) and, during the latter part of gestation, the volume of this liquid is greater than the equivalent volume of air in the neonate, which is essential for lung development.[13] However, before effective air-breathing can commence after birth, the airways must be cleared of this liquid and the mechanisms driving this process have been the subject of much investigation. Although numerous mechanisms have been identified, the relative contributions of each must vary depending on the timing and mode of delivery.[14–23] Initially, it was proposed that lung liquid clearance begins days, even as long as a week, before labor onset, based on studies that found reductions in lung liquid volume over the last week of gestation in fetal sheep.[24,25] However, there has been considerable debate as to how this was initiated and whether the

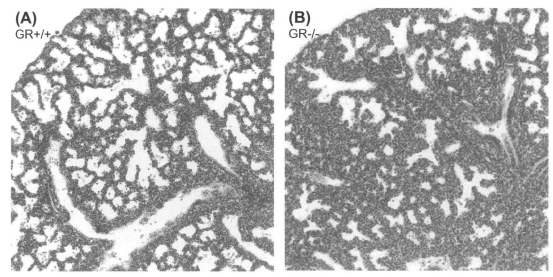

FIGURE 1   Lung histology of wildtype (GR +/+; A) and glucocorticoid receptor deficient (GR -/-; B) mice at E18.5. Note the significantly thickened lung tissue and reduced gas-exchange area in the lungs of the GR (-/-) mice (B) compared to normal mice (A). (This figure is reproduced in color in the color plate section.) *Images used with permission from T. Cole, Monash University.*

observed reductions were an experimental artefact resulting from reduced amniotic fluid volume.[13] Indeed, more recent studies in fetal sheep have failed to show a reduction in lung liquid volume in healthy fetuses near term with normal amniotic fluid volumes.[26]

As outlined in Chapter 8 (Wallace et al.), the fetal respiratory system is highly compliant such that even very small changes in trans-pulmonary pressure can cause large changes in lung liquid volume. Thus, one would expect considerable natural variability in lung liquid volume late in gestation, depending upon the factors influencing the trans-pulmonary pressure gradient. For example, reductions in intra-uterine volume (perhaps due to amniotic fluid loss) are known to increase flexion of the fetal trunk, which increases abdominal pressure and elevates the diaphragm.[27] Thus even minor changes in fetal posture could increase the trans-pulmonary pressure gradient and increase lung liquid loss.[27] It is unlikely that reduced lung liquid production rates could account for a reduction in lung liquid volume late in gestation. This is because a reduction in fetal lung liquid secretion rates simply leads to a simultaneous reduction in liquid loss via the trachea, resulting in no net change in lung liquid volume.[28,29]

Some evidence indicates that lung liquid volumes may decrease after labor onset, particularly during delivery. It is often assumed that the passage of the fetus through the birth canal "squeezes" liquid from the fetal lung. However, as the chest offers little resistance, compared to the head and shoulders, to the infant's passage through the birth canal, this assumption may be incorrect.[30] Instead, large amounts of liquid can be lost in response to a reduction in intra-uterine volume due to membrane rupture and amniotic fluid loss and/ or to the contractions and shortening of the myometrium. As discussed earlier, both of these factors can impose marked changes in fetal posture, as has been shown in both sheep[27] and humans,[31] leading to increased transpulmonary pressure gradients and lung liquid loss.[27] Indeed, in the absence of amniotic fluid, even mild non-labour uterine contractions cause phasic compression of the fetus that increases the transpulmonary pressure gradient and the loss of lung liquid.[27] Consequently, strong coordinated uterine contractions associated with labor may increase lung liquid loss due to increased trans-pulmonary pressure gradients, particularly after the membranes have ruptured, which may explain the large "gushes" of liquid that have been observed following delivery of the infant's head. Similarly, marked reductions in airway liquid volumes have been observed shortly after labor onset in sheep (as indicated by uterine electromyography), many hours before the second stage of labor commences.[26]

LUNG LIQUID CLEARANCE AT BIRTH

Until recently, the primary mechanism driving airway liquid clearance at birth was thought to involve the reversal of an osmotic gradient across the pulmonary epithelium that normally drives fetal lung liquid secretion. A detailed account of the ionic basis for this proposed mechanism is presented in Chapter 7 (Widdicombe). In brief, the osmotic gradient that promotes lung liquid secretion during fetal life is thought to be reversed at birth due to the activation of amiloride-inhibitable epithelial Na^+ channels (ENaCs) located on the apical surface of pulmonary epithelial cells.[32,33] It is proposed that active labour stimulates a large increase in fetal circulating concentrations of adrenaline (epinephrine) and arginine vasopressin (AVP) that act via β-adrenergic and AVP V_2 receptors, respectively, to induce a cAMP-mediated activation of ENaCs.[32] This, in turn, leads to an increase in Na^+ and Na^+-linked flux of chloride ions across the pulmonary epithelium, from lumen into the interstitium, resulting in a reversal of the osmotic gradient and liquid movement across the epithelium.[34,33]

There is much experimental evidence consistent with this proposal. Indeed, the ability of adrenaline and AVP to inhibit fetal lung liquid secretion and initiate liquid reabsorption matures late in gestation, increasing in an exponential-like manner close to term.[34–36] This maturational increase is dependent upon the actions of both cortisol and triiodothyronine (T3), as thyroidectomy[37] and adrenalectomy[38] abolish the gestational age-related increase in lung liquid reabsorption induced by adrenaline. Similarly, infusions of either cortisol alone or cortisol and T3 together (T3 alone had no effect) precociously mature the Na reabsorptive response to adrenaline.[39,40] This mechanism is therefore unlikely to be active at the time of preterm birth, which is consistent with the finding that RNA transcripts for the ENaC subunits are virtually undetectable in the lung of preterm infants.[17] However, although this may partially explain why preterm infants commonly suffer from airway liquid retention, the majority of preterm infants are able to aerate their lungs and begin pulmonary ventilation within minutes of birth, presumably in the absence of this mechanism.[41]

More recent studies have questioned whether adrenaline-induced ENaC activation is the primary mechanism driving airway liquid clearance at birth.[42,43] For instance, as most normal healthy infants clear their airways of liquid and establish effective gas exchange within seconds to minutes of birth,[44–46] the primary mechanism of airway liquid clearance must be rapid. This is supported by recent X-ray imaging experiments (see the following text) demonstrating that lung aeration occurs at a rate of ~3 ml/kg/sec during inspiration,[42,43] which is considerably greater than the maximum reabsorption rates that can be achieved with adrenaline (~10 mL/kg/h).[35,33,47] Indeed, pharmacological doses of adrenaline are required to achieve these reabsorption rates (~10 mL/kg/h) and adrenaline would need to be elevated for hours to clear the airways of all liquid.[35,48,33,47] However, if this were the case, other physiological parameters such as an increase in heart rate would also be evident. Although studies detailing the heart rate changes in infants immediately after birth show an increase (from ~100 to 160 bpm) in the first few minutes[49] this increase is from a relatively low value (~100) and there is no evidence of a sustained tachycardia.[49]

ENaC knockout studies in mice have provided some of the most compelling evidence supporting the role of adrenaline-induced ENaC activation in airway liquid clearance at birth.[17,19] ENaC is encoded by 3 different genes (α-, β-, and γ-), with the mature protein comprising of 3 subunits.[17] Deletion of the α-ENaC gene in mice results in neonatal mortality within 40 hours of birth due to respiratory failure, indicating that α-ENaC is critical for postnatal survival.[50,17] As lung wet weights were increased, it was assumed that adrenaline-induced activation of Na reabsorption was disrupted by deletion of the α-ENaC gene.[50] However, these newborn mice also did not feed well and had poor costal retractions indicating that energy supply and inspiratory activity were likely reduced.[17] As inspiratory activity is a major determinant of lung aeration (see the following text), poor inspiratory efforts may have contributed to the failure to clear airway liquid in these mice.[17] Indeed, deletion of β- and γ-ENaC subunits did not cause respiratory failure in newborn pups, despite a 6-fold reduction in ENaC activity.[17] Similarly, infants that have gene mutations that markedly reduce ENaC activity (pseudoaldosteronism) do not exhibit neonatal respiratory failure at birth.[51]

Studies have shown that ENaC inhibition with amiloride can reverse the inhibitory effects of adrenaline and vasopressin on lung liquid reabsorption[52,53,36] and delays, but does not block airway liquid clearance at birth.[20] More recently, an imaging approach has been used (see the following text) to determine the effect of ENaC inhibition on lung aeration at birth in ventilated term newborn rabbits.[54] Amiloride had no effect on airway liquid clearance or functional residual capacity (FRC) accumulation,[54] which consistently occurred exclusively during lung inflations, as previously reported.[42,43,55,56] However, amiloride did increase the rate of liquid re-entry into the airways, as determined by the rate of FRC decrease, between inflations. This suggests that Na+ reabsorption after birth helps to keep the airways cleared of liquid and to maintain FRC after the liquid has been cleared.[54]

AIRWAY LIQUID CLEARANCE AFTER BIRTH

Other potential mechanisms that may contribute to airway liquid clearance at birth include changes in pulmonary epithelial pore sizes, increases in oncotic pressures, as well as increases in the trans-pulmonary hydrostatic pressure gradient generated during inspiration. Recently, a series of studies have used phase-contrast X-ray imaging to observe the air–liquid interface as it moves through the airways and into the distal air spaces after birth.[42,57,54,43] Phase-contrast X-ray imaging uses both absorption and refractive index

differences between air and water to produce contrast, allowing air–water boundaries to be visualized.[57–59,60] As the lung is 80% air-filled at FRC this technique is ideal for imaging the lung, allowing the small airways (including alveoli) to be resolved with a high degree of spatial resolution (Figure 2).

Phase-contrast X-ray videoimaging sequences of spontaneously breathing, term newborn rabbits at birth clearly demonstrate that lung liquid clearance occurs during inspiration.[42,43] The air–liquid interface moves distally towards the terminal airways only during inspiration and, although some proximal movement can occur during expiration, little or no distal movement occurs between breaths.[42,43] This indicates that liquid moves from the airways into the

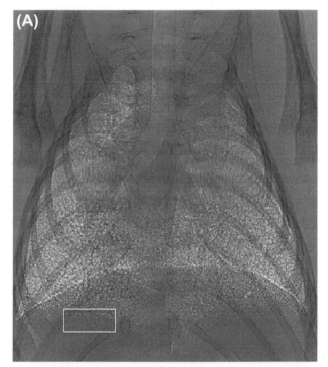

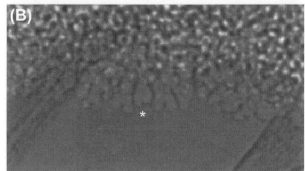

FIGURE 2   Phase-contrast X-ray image of the aerated lung of a newborn rabbit (A). Bright, speckled regions indicate the presence of air in the lungs. Note that the trachea, major bronchi and the shadow of the diaphragm are clearly visible. The area within the white rectangle in (A) is magnified in (B). Note that individual alveoli can be observed at the periphery of the lung (*).

surrounding tissue during inspiration and that little or no liquid re-enters during expiration. As a result, FRC progressively increases with each breath, and the increase in FRC equals the volume of liquid leaving the airways.[42,43] Simultaneous plethysmography confirmed that FRC accumulation only occurred during inspiration, resulting in a "step-like" increase in FRC with each breath (Figure 3).[42,43] Based on these findings, it was concluded that trans-epithelial pressures generated during inspiration provide the pressure gradient for liquid to leave the airways and enter the surrounding lung tissue.[43]

Although it is not well recognized, the association between spontaneous breathing and FRC accumulation after birth is well established and has been reported using a variety of different animal models.[19,45,53] Recent phase-contrast imaging studies now confirm that these two processes are causally linked and progress simultaneously.[61,42,43] Understanding how inspiration (during a spontaneous breath) could drive airway liquid clearance is premised on the knowledge that inspiration causes expansion-induced pressure reductions in both the intra-pleural space and peri-alveolar interstitial tissue.[43] This generates a pressure gradient between the airways and surrounding tissue (across the airway wall), and between the lower and upper airways that drives liquid movement distally through the airways and into the tissue. Within the peri-alveolar tissue, the liquid forms into perivascular fluid cuffs, from where it is gradually cleared via the pulmonary vasculature and lymphatics.[62]

Although liquid movement into the tissue occurs rapidly, within 3–5 breaths (Figure 2), liquid clearance from the tissue can take hours.[62,63] As a result, pulmonary interstitial tissue pressure transiently (~4 h) increases[63] and the chest wall expands[42] immediately after birth. Chest wall expansion is required to accommodate both the increase in gas volume as well as the volume of liquid within the interstitial tissue that had resided within the airways before lung aeration.[42] This understanding provides a rational basis for why infants need a compliant chest wall at birth. That is, a compliant chest wall will allow an increase in intra-thoracic volume without substantially increasing the chest's recoil pressure, which would otherwise increase interstitial tissue pressures and force liquid re-entry into the airways at FRC.

Small transient (~4 h) increases in interstitial tissue pressure normally occur at birth,[63] which creates a small pressure gradient that facilitates liquid re-entry into the airways at FRC. This is consistent with a gradual decline in FRC that was noted between breaths,[54,43] although the rate of liquid re-entry is considerably slower than the rate it leaves the airways during inspiration.[54,43] This is likely because of the quantitative differences in trans-epithelial pressure gradients that occur in the lung during inspiration and at FRC.

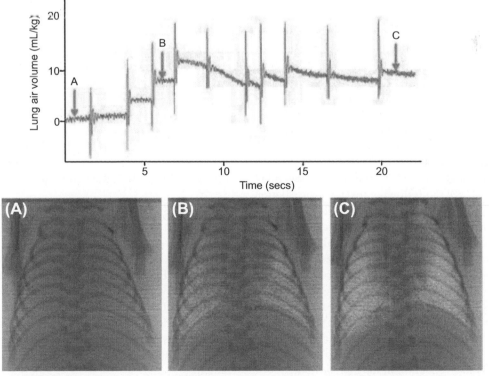

FIGURE 3 Increase in lung aeration that occurs with breathing onset in newborn rabbits. A, B, and C are phase-contrast X-ray images of the lungs at volumes indicated on the lung volume trace. Note the absence of air in the lungs at A, before the onset of breathing. Within 3 breaths, there is an increase in lung aeration (B). After 10 breaths, lungs are very well aerated (C). *Figure reproduced from[109] with permission.*

Furthermore, as much of the airway liquid present at birth must be accommodated within the interstitial tissue following lung aeration when airway liquid volumes are large it is likely that the resulting increase in interstitial tissue pressures will be higher. These higher pressures must increase the potential for liquid to re-enter the airways at FRC and may also explain why infants born by caesarean section are more likely to suffer "wet lung" or transient tachypnoea of the newborn.[19] That is, in the absence of the intra-partum mechanisms for airway liquid clearance (see earlier), greater volumes of airway liquid will result in higher interstitial tissue pressures that increase the likelihood of liquid re-entry into the airways at FRC. This concept is consistent with the tachypnoea and increased grunting and expiratory braking that are characteristic of infants with wet lung and why continuous positive airway pressure (CPAP) is an effective treatment for these infants.

THE PHYSIOLOGICAL CONSEQUENCES OF LUNG AERATION

The consequences of replacing airway (lung) liquid with air at birth have long been overlooked as important contributors to the changes in cardiorespiratory physiology that occur at this time. In particular, the effects of increasing lung recoil and decreasing resting lung expansion are not widely acknowledged or understood. Before birth, the volume of liquid retained within the future airways is considerably greater than the end-expiratory volume of the air-filled lung after birth.[13,32] Although the lung is commonly thought to "expand at birth," the entry of air must cause resting lung volumes to decrease within a few hours after birth when compared to the liquid-filled fetal lung before labor onset. Before birth, fetal lung expansion is maintained by active liquid retention due to fetal skeletal muscle activity in conjunction with continued lung liquid secretion.[13] During fetal apnea, glottic adduction restricts liquid efflux, which results in an internal hydrostatic distending pressure of 1–2mmHg and resting intra-pleural pressures that are close to zero.[64,65] During fetal breathing movements (FBM; see the following text), phasic dilation of the glottis reduces its resistance to lung liquid efflux,[66] resulting in increased lung liquid loss. However, this liquid loss is minimized by diaphragmatic contractions that oppose liquid efflux via the trachea.[13,32] After birth, the replacement of liquid with air causes an air–liquid interface to form across the large internal surface of the lung, which increases lung recoil, despite the presence of surfactant. As air is compressible it is also less able to oppose the increase in lung recoil caused by surface tension and combined with the absence of the distending influence of lung liquid, leads to a reduction in lung expansion.[13,32]

The reduction in lung luminal expansion caused by the entry of air helps to explain a number of physiological changes that occur after birth. For example, the increase in lung recoil and the partial collapse of the lung away from the chest wall explains why intra-pleural pressures become subatmospheric after birth.[64] Before birth, intra-pleural pressure is similar to ambient (amniotic sac) pressure,[64,65] but within hours of birth, it decreases to 2–4 cmH$_2$O below atmospheric pressure.[64,63] This indicates that the mechanical load experienced by the chest wall has increased with the increase in lung recoil, which likely plays an important role in stiffening the chest wall after birth.[67] Similarly, as the liquid is cleared from the interstitial tissue space after birth, interstitial tissue pressure decreases and becomes subatmospheric and similar to intra-pleural pressure.[63] As a result, the interstitial tissue/capillary wall transmural pressures must increase, thereby facilitating capillary recruitment and expansion that would help sustain the reduction in pulmonary vascular resistance after birth (see the following text).

A predictable consequence of the decrease in lung expansion associated with lung aeration at birth is its effect on the alveolar epithelium. As the degree of mechanical strain imposed on AECs determines their differentiated state in vitro,[68–70] altering the degree of lung expansion has profound effects on AEC populations[71,72]; type-I AECs provide the large surface area for gas exchange whereas type-II produce and secrete surfactant. Increased fetal lung expansion increases the proportions of type-I AECs,[71,72] due to type-II to type-I AEC trans-differentiation via an intermediate cell type,[72] thereby reducing the type-II AEC population from ~30% to ~2%, within 10 days.[72] As a subsequent sustained (~7 days) reduction in fetal lung expansion can restore the proportion of type-II AECs, it appears that type-I AECs can trans-differentiate into type-II cells[73]; similar results have been obtained in vitro[68,70] It is not surprising therefore that the proportion of type-I AECs decreases from 60–65% in the fetus to ~40% after birth, whereas the proportion of type-II cells increases from ~30% to 50–55% after birth (Figure 4).[74] These studies provide compelling evidence that the change in the local mechanical environment of the lung caused by lung aeration alters the resident populations of type-I and type-II AECs within the epithelium.

CHANGES IN PULMONARY BLOOD FLOW AT BIRTH

Development of the pulmonary vascular bed is described in detail in Chapter 5 (Jones et al.), and Chapter 6 (Dodson et al.) describes the cellular processes involved with maintaining a high fetal pulmonary vascular resistance (PVR) and how PVR is reduced at birth. Here we focus on the consequences of the decrease in PVR at birth, particularly for cardiac output and how it interacts with umbilical cord clamping.

Control of the pulmonary circulation and pulmonary vascular resistance (PVR) in adults has few similarities with control of the systemic circulation. In adults, the pulmonary circulation can be characterized as a high flow, low pressure circuit with a low mean arterial pressure (~15 mmHg) and high flow that equals cardiac output. As a result, PVR is low at rest, but has the capacity to decrease further in response to increases in cardiac output,[75] due to recruitment and distension of capillary beds.[75] Furthermore, as the pulmonary circulation is connected in series between the right and left sides of the heart in adults, the entire output of the right ventricle (RV) passes through the lungs and pulmonary venous return provides the sole source of preload for the left ventricle (LV). Thus, the output of both ventricles must be equal to

avoid blood pooling in either the systemic or pulmonary circulations. In contrast, in the fetus, the presence of two shunts allows both ventricles to work independently to supply output to the systemic circulation. These shunts are the foramen ovale (FO), which allows venous return to bypass the right side of the heart and directly enter the left atrium, and the ductus arteriosus (DA), which shunts blood from the main pulmonary artery into the descending aorta (Figure 5). As a result, in fetal sheep, the RV provides 66% of combined ventricular output, with only 33% coming from the LV.[76]

In the fetus, PVR is high, pulmonary blood flow (PBF) is low and mean pulmonary arterial pressure is ~5 mmHg

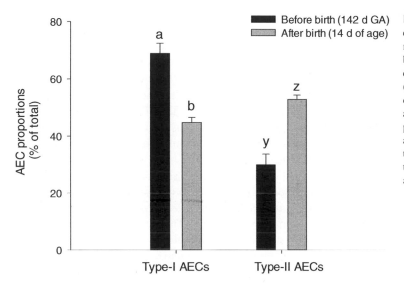

FIGURE 4 The proportions of type-I and type-II alveolar epithelial cells (AECs), expressed as a percentage of the total number of alveolar epithelial cells counted, before (black bars) and after (grey bars) birth. Within each cell type, different letters indicate the values that are significantly different (*p*<0.05). Note that type-I AECs predominate before birth and decrease in proportion after birth. Type-II AECs, in contrast, are of a low proportion before birth but increase to become the predominant cell type after birth. The decrease in type-I cells and increase in type-II cells may result from the reduction in the basal degree of lung expansion that occurs at birth due to the increase in lung recoil associated with the formation of an air-liquid interface. *Data taken from.*[110]

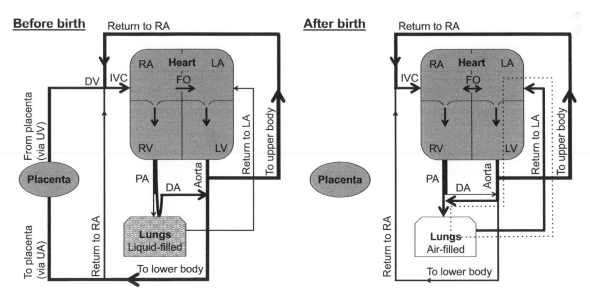

FIGURE 5 Blood circulation before and after birth. Following lung aeration and clamping of the umbilical cord, pulmonary vascular resistance decreases, pulmonary blood flow increases and blood flow through the ductus arteriosus (DA) changes from right-to-left to predominantly left-to-right; this establishes a LV-lung-LV short circuit (dotted lines) that helps to maintain LV output. UA, umbilical artery; UV, umbilical vein; DV, ductus venosus; IVC, inferior vena cava; RA, right atrium; LA, left atrium; RV, right ventricle; FO, foramen ovale; LV, left ventricle; PA, pulmonary artery; DA, ductus arteriosus.

greater than systemic arterial pressure.[77,78] Thus, only a small proportion (~10%) of RV output flows through the lungs, with the majority by-passing the lungs and flowing through the DA directly into the descending aorta.[79,80,81,76] This flow is referred to as right-to-left shunting. Within the pulmonary arteries, blood only flows towards the lungs briefly during systole, whereas throughout most of diastole, blood flows retrogradely away from the lungs and exits the pulmonary circulation by flowing across the DA (right-to-left) and into the aorta.[79,81] As a result, blood flow through the DA proceeds continuously throughout the cardiac cycle with the high retrograde PBF contributing to both the high DA flow during diastole as well as the "low" mean PBF (Figure 5).[79,81] Although it is commonly assumed that fetal PBF is uniformly low, this assumption is incorrect as PBF is very variable and can increase 10-fold depending upon fetal activity, particularly late in gestation.[82] For example, fetal breathing movements (FBM) significantly increase PBF due to a decrease in PVR, which is thought to result from an increase in the capillary/interstitial tissue transmural pressure achieved during inspiration.[82] This causes capillary distension and recruitment, which increases PBF in close association with increasing inspiratory effort. Each individual FBM causes a large change in the PBF waveform, although the predominant effect is a marked reduction in the amount of retrograde flow during diastole, which is indicative of reduced PVR.[82]

The consequence of a high PVR and a low PBF in the fetus is that pulmonary venous return is unable to provide sufficient preload to sustain left ventricular (LV) output. Instead, in the fetus, preload for the LV is primarily derived from umbilical venous return, which flows via the ductus venous, inferior vena cava and FO directly into the left atrium (Figure 5).[76] Despite being structurally different, this circulatory arrangement is analogous to the adult, whereby oxygenated blood preferentially returns to the left side of the heart.

As the umbilical circulation is a low resistance vascular bed, it receives ~30–50% of the combined ventricular output of the fetus.[83] As a consequence, umbilical venous return is large (30–50% of total) and high enough to supply the majority of LV preload in the fetus. Thus, clamping the umbilical cord at birth represents a major disturbance to the fetal circulation, particularly to afterload and the supply of venous return and preload for both ventricles.[84,85]

Specifically, removal of the low resistance umbilical circulation with cord clamping at birth markedly increases downstream peripheral resistance.[84,85] This results in ~30% increase in arterial blood pressure over the first four heart beats after cord clamping, which in turn causes a transient, pressure driven, increase in cerebral blood flow (Figure 6).[84] However, both RV and LV output then markedly decrease (by ~50%), mostly due to the loss of umbilical venous return caused by umbilical cord clamping.[84,79] This causes a severe reduction in preload, which decreases cardiac output and causes a transient decrease in blood pressure.[84,79] Cardiac output remains low, due to the low preload, until pulmonary ventilation commences and PBF increases.[84] At this time, the rapid and large increase in PBF restores preload to the LV by increasing pulmonary venous return, thereby markedly increasing cardiac output. The increase in PBF also likely restores some preload to the RV, presumably via left-to-right flow through the foramen ovale, as RV output also rapidly increases (Figure 5).[84]

As umbilical cord clamping increases systemic vascular resistance, whereas lung aeration decreases PVR, PVR decreases below systemic vascular resistance, resulting in a reversal in blood flow (from right-to-left to left-to-right) through the DA.[79] Within ~10–20 mins of ventilation onset, the extent of the left-to-right shunting through the DA is so large (Figure 7) that the LV contributes to ~50% of PBF, resulting in a substantial LV-lung-LV short circuit within the systemic circulation.[79] The consequence of this is an increase in cardiac output, which in turn causes a rebound increase in arterial pressure and carotid blood flow.[84] Thus, following umbilical cord occlusion the loss of umbilical venous return and preload causes substantial reductions in cardiac output (CO), which cannot be restored until the lung aerates and PBF increases. This represents a major upheaval within the fetal circulatory system, resulting in large swings in CO that cause large swings in blood pressures and flows. To avoid these large swings in CO, logically, umbilical cord clamping should be delayed until after ventilation has commenced and PBF has increased. This allows the source of preload for the ventricles to immediately switch from the umbilical circulation to pulmonary venous return without significant interruption.[84]

Numerous clinical trials have examined the benefits of delayed umbilical cord clamping and have found that,

FIGURE 6 Recordings of carotid arterial pressure (CAP) and carotid artery blood flow (CABF) in unventilated lambs at the time of umbilical cord clamping (dotted line). Note the immediate increase in mean CAP and CABF after the clamp. *Figure adapted from.*[84]

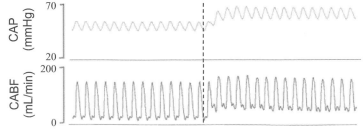

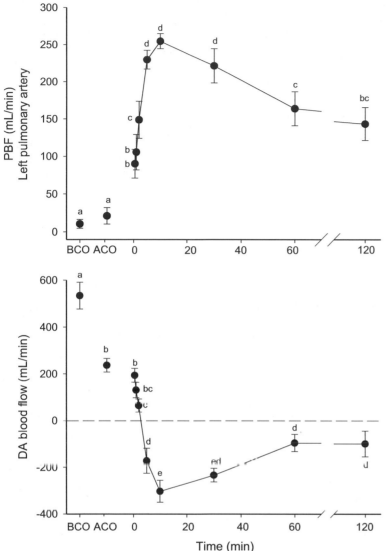

FIGURE 7 Mean left pulmonary blood flow (PBF; top panel) and ductus arteriosus blood flow (DA; bottom panel) in lambs before umbilical cord occlusion (BCO), after umbilical cord occlusion (ACO) and at ventilation onset (time 0) for 120 min. Within each panel, different letters indicate that values are significantly different from each other (p<0.05). Values above zero indicate blood flow from the pulmonary artery into the aorta (right-to-left) whereas values below zero indicate blood flow from the aorta into the pulmonary circulation (left-to-right). Note the large decrease in DA blood flow once the umbilical cord has been occluded that significantly reverses the flow of blood within 10 min. PBF only significantly increases at ventilation onset, but not in response to umbilical cord occlusion. *Reproduced from[79] with permission.*

although variable, delays of ~1 min provide some benefits for the infant.[85–89] The underlying rationale for the reported benefits is that delayed cord clamping increases placenta-to-infant blood transfer, resulting in increased neonatal blood volume and haematocrit.[90] While this may in part explain some benefits, a recent study has shown that it also greatly stabilizes the circulatory transition at birth and mitigates all of the large swings in CO, arterial pressure, and blood flow that occur in response to cord clamping.[84] This is because, as explained earlier, if ventilation commences and PBF increases before the cord is clamped, then pulmonary venous return can immediately replace umbilical venous return as the main source of preload for the LV. As a result, CO does not decrease following cord clamping and heart rates remain much higher.[84] Interestingly, the 30% increase (over 4 heart beats) in arterial blood pressure caused by cord clamping is also greatly reduced when it follows lung aeration, indicating that the decrease in PVR is

largely sufficient to compensate for the increase in systemic vascular resistance caused by cord clamping.[84] This suggestion is consistent with the finding that, following lung aeration, despite a decrease in PVR, shunting through the DA remains right-to-left while the umbilical cord is open (unclamped).[84] This indicates that downstream resistance in the systemic circulation remains lower than the pulmonary circulation. However, immediately (within seconds) after the cord is clamped, flow through the DA switches and becomes left-to-right, indicating that after removal of the umbilical circulation, downstream resistance is lower in the pulmonary circulation than in the systemic circulation.[84]

In summary, the increase in PBF at birth is not only essential for pulmonary gas exchange in the newborn, but is also essential to replace umbilical venous return as the primary source of preload for the LV (and initially RV). Thus, any significant delay between umbilical cord clamping and the onset of pulmonary ventilation will result in a prolonged period of markedly reduced cardiac output. However, if umbilical cord clamping is delayed until after the lungs have aerated and PBF has increased, then pulmonary venous return can immediately replace umbilical venous return as the major source of preload for the left ventricle upon cord clamping after birth.

DYNAMIC CHANGES IN THE DUCTUS ARTERIOSUS AT BIRTH

The ductus arteriosus (DA) is a large vascular shunt interposed between the main pulmonary artery and descending aorta, allowing the majority of RV output to enter the systemic circulation. Soon after birth, the DA must close to separate the pulmonary and systemic circulations, particularly to protect the pulmonary circulation from the high pressure of the systemic circulation. Failure of DA closure (i.e., patent ductus arteriosus) can result in high PBF, due to sustained left-to-right shunting, and exposure of the pulmonary vascular bed to high systemic arterial pressures, which increases the risk of pulmonary haemorrhage.[91] The mechanisms underlying closure of the DA at birth are complex and involve both chemical and physical factors; the chemical control of ductal closure at birth has been the focus of much attention and is discussed in Chapter 6 [Dodson et al.]. Here we focus on the physical relationship between PVR and blood flows through the pulmonary circulation and DA after birth.

During fetal life, the DA shunts blood from the pulmonary circulation to the systemic circulation due to the pressure gradient (of 4–5 mmHg) between these circulations.[76,78,92] After birth, due to the large decrease in PVR and increase in systemic vascular resistance, the pressure gradient across the DA reverses and blood predominantly shunts from left-to-right (from the systemic to the pulmonary circulation) across the DA (Figure 7).[79] As a result, retrograde flow along the left and right pulmonary arteries quickly diminishes (within minutes of birth) resulting in only forward flow through the pulmonary arteries even during diastole (Figure 5).[79] However, the DA flow profile is complex, with small amounts of right-to-left shunting persisting during peak systole, whereas for the remainder of the cardiac cycle blood flows left-to-right.[79] This complex flow profile likely results from differences in the timing of when the peak systolic pressures, emanating from the RV and LV, reach either end of the DA. As the pressure peak derived from the RV is likely to enter the DA slightly ahead of the pressure peak from the LV, which has further to travel, then the pressure gradient will favor right-to-left flow. However, this quickly reverses once the pressure peak from the LV reaches the DA-aorta junction, thereby facilitating left-to-right flow.

While the DA remains open, blood will pass between the pulmonary and systemic circulations, depending upon the pressure gradient across this vessel, thereby preventing any substantial decrease in pulmonary arterial pressure.

Functional closure of the DA begins within hours of birth in lambs and is clearly shown by the substantial decrease in both absolute DA flow, but most particularly by the pulsatile DA flow caused by ventricular contraction and relaxation.[79] Although mediators responsible for DA closure have been identified, it is interesting to speculate whether the anatomic relationship between the pulmonary artery, DA and the descending aorta could be involved. Indeed, considerable turbulence must occur at this site when blood flow through the DA switches from right-to-left to left-to-right, which could elicit the release of endothelial-derived vasoactive factors (see Chapter 6 [Dodson et al.]) that contribute to constriction of the DA.

FETAL BREATHING AND THE ONSET OF CONTINUOUS BREATHING AT BIRTH

Although it is often stated that an infant takes its first breath at birth, respiratory movements begin during fetal life, long before birth. Like postnatal breathing, FBM involve rhythmical contractions of the diaphragm and other inspiratory muscles, such as dilator muscles of the pharynx and larynx, which are driven by brainstem centres that are stimulated by CO_2.[93] The major differences between fetal breathing and postnatal breathing are that in the fetus: (1) the lungs are liquid-filled and hence tidal volume is very small (see Chapter 8 [Wallace et al.]), (2) breathing movements are episodic, (3) breathing movements are inhibited, rather than stimulated, by hypoxia, and (4) breathing movements play no role in gas exchange, but instead result in a net oxygen consumption.

Breathing movements in healthy human and ovine fetuses can be detected before mid-gestation and continue until the onset of labor.[93] Characteristically, they occur in episodes, which become more organized later in gestation as fetal behavioral states become organized. During late gestation, FBM in humans and sheep occur 40–50% of the time[93] and are primarily associated with a state resembling rapid eye movement sleep; during episodes of "quiet sleep" fetuses are largely apneic.[94]

The incidence of FBM is reduced during active labour, but the mechanisms underlying this inhibition are not well understood.[93] Although it is well established that prostaglandins can inhibit FBM,[95] it is thought that prostaglandins such as PGE_2 are not involved in the labour-related reduction of FBM.[96] Alternatively, adenosine is known to inhibit FBM via cerebral adenosine receptors and can be released into the fetal circulation from the placenta and fetal liver.[97]

During parturition and when the umbilical cord is cut at birth, the fetus-neonate may become hypoxemic, hypercapnic and acidemic, as well as being exposed to a lower environmental temperature resulting in increased heat loss. It will also be exposed to a greatly increased degree of external sensory stimuli; in addition, its behavioural state may change to one of arousal.[98] What triggers continuous breathing at the time of birth is the subject of much debate, although it is apparent that many factors, such as those listed earlier, are likely to be involved. A major factor is thought to be increased CO_2 production, perhaps resulting from an increased rate of metabolism at birth, and/or an increased sensitivity to CO_2. Circulating levels of catecholamines are elevated at birth.[99] which would be expected to stimulate metabolic activity and hence CO_2 production. A reduction at birth in circulating concentrations of adenosine, which is produced by the placenta, can stimulate thermogenesis[100] leading to an increased CO_2 production. A decrease in the concentrations of one or more circulating factors of placental origin (e.g., prostaglandin E_2 (PGE_2), adenosine, progesterone metabolites) that are known to inhibit FBM may also facilitate continuous breathing after birth.[95,101,102] Removal of the placenta from the fetal circulation results in an increased ventilatory sensitivity to CO_2 suggesting that an inhibitory factor may act at the level of the fetal brainstem.[98] Central and peripheral chemoreceptors are active in the fetus,[103] although it has been suggested that their sensitivity may be tonically suppressed by factors released from the placenta into the fetal circulation. For instance, PGE_2 is released by the placenta[102] and its receptors have been identified in the fetal brainstem.[104]

Studies of fetal sheep maintained ex utero by extracorporeal oxygenation with the umbilical cord occluded, thereby eliminating potential effects of placental factors, have shown that continuous breathing is dependent upon blood CO_2 levels,[105] supporting the notion that CO_2 plays a crucial role in the maintenance of continuous breathing after birth.

The integrity of the vagus nerves are essential for the onset of adequate breathing at birth[106] and although the critical pathways have not yet been identified, it is likely that volume receptive feedback from the lungs is involved. Studies in unanesthetised neonatal lambs have shown that the application of negative airway pressures or creation of a tracheostomy, both of which would reduce end-expiratory lung volumes (FRC) and the amount of vagal neural traffic from lung volume receptors (pulmonary stretch receptors), results in profound hypoventilation, periodic breathing and active glottic adduction during periods of apnea.[107,108] This indicates that volume receptive vagal feedback at end-expiration, which is normally maintained by an adequate FRC, is essential for continuous breathing in the newborn, and explains, at least in part, the benefits of positive end-expiratory pressure (PEEP) in the treatment of infantile apnea.

CONCLUSIONS

Following normal gestation, labor, and term delivery, the fetal lung is well prepared for its critical role of gas exchange after birth. Both endocrine and physical factors play a major role in preparing the lung for its function after birth and breathing activity is largely responsible for clearing the airways of liquid. This process of lung aeration triggers the cardiovascular transition at birth by decreasing PVR, resulting in an increase in pulmonary venous return and an increase in preload for both ventricles. As preload for the LV is primarily derived from the umbilical circulation during fetal life, clamping of the umbilical cord causes a large reduction in preload for the LV until the lung aerates and PBF increases. However, aerating the lung and increasing PVF before the umbilical cord is clamped allows the source of preload to immediately switch from the umbilical circulation to the pulmonary circulation. As a result, the changes in CO associated with birth and cord clamping are greatly minimized.

If gestation is shortened as a result of preterm birth, maturation of the lung may not have occurred to a sufficient degree, resulting in respiratory compromise. In particular, the lung may not have developed structurally and alveolar epithelial type-ll cells may not produce sufficient quantities of surfactant, resulting in respiratory distress. If birth occurs as a result of caesarean section without labor, the absence of the intra-partum mechanisms for airway liquid clearance will likely increase the volume of liquid that must be cleared into the lung tissue. The resulting increase in interstitial pressures must increase the tendency for liquid to re-enter the airways, potentially resulting in transient tachypnoea of the newborn. Although much is already known, there remain many unanswered questions relating to physiological and molecular mechanisms underlying the preparation of the lung for birth and its postnatal adaptation to air breathing.

REFERENCES

1. Liggins GC. Premature delivery of foetal lambs infused with glucocorticoids. *J Endocrinol* 1969;**45**:515–23.
2. Crowley P, Chalmers I, Keirse MJNC. The effects of corticosteroid administration before preterm delivery: an overview of the evidence from controlled trials. *Br J Obstet Gynaecol* 1990;**97**:11–25.
3. Liggins GC, Howie RN. A controlled trial of antepartum glucocorticoid treatment for prevention of the respiratory distress syndrome in premature infants. *Pediatrics* 1972;**50**:515–25.
4. Crowley P. Prophylactic corticosteroids for preterm birth. *Cochrane Database Syst Rev* 2000. CD000065.
5. Jobe AH, Ikegami M. Lung development and function in preterm infants in the surfactant treatment era. *Annu Rev Physiol* 2000;**62**:825–46.
6. Liggins GC. The foetal role in the initiation of parturition in the ewe. In: Wolstenholme GEW, O'Connor M, editors. *Foetal Autonomy (Ciba Foundation Symposium)*. London: Churchill; 1990. pp. 218.
7. Crone RK, Davies P, Liggins GC, Reid L. The effects of hypophysectomy, thyroidectomy, and postoperative infusion of cortisol or adrenocorticotrophin on the structure of the ovine fetal lung. *J Dev Physiol* 1983;**5**:281–8.

8. Kitterman JA, Liggins GC, Campos GA, Clements JA, Forster CS, Lee CH, et al. Prepartum maturation of the lung in fetal sheep: relation to cortisol. *J Appl Physiol* 1981;**51**:384–90.

9. Liggins GC, Schellenberg JC, Finberg K, Kitterman JA, Lee CH. The effects of ACTH1-24 or cortisol on pulmonary maturation in the adrenalectomized ovine fetus. *J Dev Physiol* 1985;**7**:105–11.

10. Ballard PL. The glucocorticoid domain in the lung and mechanisms of action. In: Mendelson CR, editor. *Endocrinology of the Lung.* Totowa: Humana Press Inc.; 2000. pp. 1–44.

11. Cole TJ, Blendy JA, Monaghan AP, Krieglstein K, Schmid W, Aguzzi A, et al. Targeted disruption of the glucocorticoid receptor gene blocks adrenergic chromaffin cell development and severely retards lung maturation. *Genes Dev* 1995;**9**:1608–21.

12. Bird AD, Choo YL, Hooper SB, McDougall ARA, Cole TJ. Mesenchymal glucocorticoid receptor regulates development of multiple cell layers of the mouse lung. *Am J Respir Cell Mol Biol* 2014;**50**:419–28.

13. Harding R, Hooper SB. Regulation of lung expansion and lung growth before birth. *J Appl Physiol* 1996;**81**:209–24.

14. Barker PM, Olver RE. Invited review: Clearance of lung liquid during the perinatal period. *J Appl Physiol* 2002;**93**:1542–8.

15. Bland RD. Lung liquid clearance before and after birth. *Semin Perinatol* 1988;**12**:124–33.

16. Bland RD, Nielson DW. Developmental changes in lung epithelial ion transport and liquid movement. *Annu Rev Physiol* 1992;**54**: 373–94.

17. Hummler E, Planes C. Importance of ENaC-mediated sodium transport in alveolar fluid clearance using genetically-engineered mice. *Cell Physiol Biochem* 2010;**25**:63–70.

18. Jain L, Dudell GG. Respiratory transition in infants delivered by cesarean section. *Semin Perinatol* 2006;**30**:296–304.

19. Jain L, Eaton DC. Physiology of fetal lung fluid clearance and the effect of labor. *Semin Perinatol* 2006;**30**:34–43.

20. O'Brodovich H, Hannam V, Seear M, Mullen JBM. Amiloride impairs lung water clearance in newborn guinea pigs. *J Appl Physiol* 1990;**68**:1758–62.

21. Probyn ME, Hooper SB, Dargaville PA, McCallion N, Crossley K, Harding R, et al. Positive end expiratory pressure during resuscitation of premature lambs rapidly improves blood gases without adversely affecting arterial pressure. *Pediatric Res* 2004;**56**:198–204.

22. Siew ML, te Pas AB, Wallace MJ, Kitchen MJ, Lewis RA, Fouras A, et al. Positive end expiratory pressure enhances development of a functional residual capacity in preterm rabbits ventilated from birth. *J Appl Physiol* 2009;**106**:1487–93.

23. te Pas AB, Davis PG, Hooper SB, Morley CJ. From liquid to air: breathing after birth. *J Pediatr* 2008;**152**:607–11.

24. Dickson KA, Maloney JE, Berger PJ. Decline in lung liquid volume before labor in fetal lambs. *J Appl Physiol* 1986;**61**:2266–72.

25. Kitterman JA, Ballard PL, Clements JA, Mescher EJ, Tooley WH. Tracheal fluid in fetal lambs: spontaneous decrease prior to birth. *J Appl Physiol* 1979;**47**:985–9.

26. Lines A, Hooper SB, Harding R. Lung liquid production rates and volumes do not decrease before labor in healthy fetal sheep. *J Appl Physiol* 1997;**82**:927–32.

27. Harding R, Hooper SB, Dickson KA. A mechanism leading to reduced lung expansion and lung hypoplasia in fetal sheep during oligohydramnios. *Am J Obstet Gynecol* 1990;**163**:1904–13.

28. Dickson KA, Harding R. Restoration of lung liquid volume following its acute alteration in fetal sheep. *J Physiol* 1987;**385**:531–43.

29. Hooper SB, Dickson KA, Harding R. Lung liquid secretion, flow and volume in response to moderate asphyxia in fetal sheep. *J Dev Physiol* 1988;**10**:473–85.

30. Bland RD. Loss of liquid from the lung lumen in labor: more than a simple "squeeze". *Am J Physiol Lung Cell Mol Physiol* 2001;**280**:L602–5.

31. Albuquerque CA, Smith KR, Saywers TE, Johnson C, Cock ML, Harding R. Relation between oligohydramnios and spinal flexion in the human fetus. *Early Hum Dev* 2002;**68**:119–26.

32. Hooper SB, Harding R. Fetal lung liquid: a major determinant of the growth and functional development of the fetal lung. *Clin Exp Pharmacol Physiol* 1995;**22**:235–47.

33. Olver RE, Ramsden CA, Strang LB, Walters DV. The role of amiloride-blockable sodium transport in adrenaline-induced lung liquid reabsorption in the fetal lamb. *J Physiol* 1986;**376**:321–40.

34a. Brown MJ, Olver RE, Ramsden CA, Strang LB, Walters DV. Effects of adrenaline and of spontaneous labour on the secretion and absorption of lung liquid in the fetal lamb. *J Physiol* 1983;**344**:137–52.

34b. Matalon S, O'Brodovich H. Sodium channels in alveolar epithelial cells: molecular characterization, biophysical properties, and physiological significance. *Annu Rev Physiol* 1999;**61**:627–61.

35. Hooper SB, Harding R. Effects of beta-adrenergic blockade on lung liquid secretion during fetal asphyxia. *Am J Physiol* 1989;**257**:R705–10.

36. Wallace MJ, Hooper SB, Harding R. Regulation of lung liquid secretion by arginine vasopressin in fetal sheep. *Am J Physiol* 1990;**258**:R104–111.

37. Barker PM, Brown MJ, Ramsden CA, Strang LB, Walters DV. The effect of thyroidectomy in the fetal sheep on lung liquid reabsorption induced by adrenaline or cyclic AMP. *J Physiol* 1988;**407**:373–83.

38. Wallace MJ, Hooper SB, Harding R. Role of the adrenal glands in the maturation of lung liquid secretory mechanisms in fetal sheep. *Am J Physiol* 1996;**270**:R1–8.

39. Barker PM, Markiewicz M, Parker KA, Walters DV, Strang LB. Synergistic action of triiodothyronine and hydrocortisone on epinephrine-induced reabsorption of fetal lung liquid. *Pediatric Res* 1990;**27**:588–91.

40. Wallace MJ, Hooper SB, Harding R. Effects of elevated fetal cortisol concentrations on the volume, secretion and reabsorption of lung liquid. *Am J Physiol* 1995;**269**:R881–887.

41. O'Donnell CP, Kamlin CO, Davis PG, Morley CJ. Crying and breathing by extremely preterm infants immediately after birth. *J Pediatr* 2010;**156**:846–7.

42. Hooper SB, Kitchen MJ, Wallace MJ, Yagi N, Uesugi K, Morgan MJ, et al. Imaging lung aeration and lung liquid clearance at birth. *FASEB J* 2007;**21**:3329–37.

43. Siew ML, Wallace MJ, Kitchen MJ, Lewis RA, Fouras A, te Pas AB, et al. Inspiration regulates the rate and temporal pattern of lung liquid clearance and lung aeration at birth. *J Appl Physiol* 2009;**106**:1888–95.

44. Karlberg P, Cherry RB, Escardo FE, Koch G. Respiratory studies in newborn infants. II. Pulmonary ventilation and mechanics of breathing in first minutes of life, including onset of respiration. *Acta Paediatrica* 1962;**51**:121–36.

45. Mortola JP. Dynamics of breathing in newborn mammals. *Physiol Rev* 1987;**67**:187–243.

46. te Pas AB, Davis PG, Kamlin CO, Dawson J, O'Donnell CP, Morley CJ. Spontaneous breathing patterns of very preterm infants treated with continuous positive airway pressure at birth. *Pediatric Res* 2008;**64**:281–5.

47. Walters DV, Olver RE. The role of catecholamines in lung liquid absorption at birth. *Pediatric Res* 1978;**12**:239–42.
48. Lagercrantz H, Bistoletti P. Catecholamine release in newborn-infant at birth. *Pediatric Res* 1973;**11**:889–93.
49. Dawson JA, Kamlin CO, Wong C, te Pas AB, Vento M, Cole TJ, et al. Changes in heart rate in the first minutes after birth. *Arch Dis Child Fetal Neonatal Ed* 2010;**95**:F177–181.
50. Hummler E, Barker P, Gatzy J, Beermann F, Verdumo C, Schmidt A, et al. Early death due to defective neonatal lung liquid clearance in alpha ENaC-deficient mice. *Nat Genet* 1996;**12**:325–8.
51. Bonny O, Rossier BC. Disturbances of Na/K balance: pseudohypoaldosteronism revisited. Journal of the American Society of Nephrology. *JASN* 2002;**13**:2399–414.
52. Olver RE, Robinson EJ. Sodium and chloride transport by the tracheal epithelium of fetal, new-born and adult sheep. *J Physiol* 1986;**375**:377–90.
53. Olver RE, Walters DV, Wilson M. Developmental regulation of lung liquid transport. *Annu Rev Physiol* 2004;**66**:77–101.
54. Siew ML, Wallace MJ, Allison BJ, Kitchen MJ, te Pas AB, Islam MS, et al. The role of lung inflation and sodium transport in airway liquid clearance during lung aeration in newborn rabbits. *Pediatr Res* 2013;**73**:443–9.
55. te Pas AB, Siew M, Wallace MJ, Kitchen MJ, Fouras A, Lewis RA, et al. Effect of sustained inflation length on establishing functional residual capacity at birth in ventilated premature rabbits. *Pediatric Res* 2009;**66**:295–300.
56. te Pas AB, Siew M, Wallace MJ, Kitchen MJ, Fouras A, Lewis RA, et al. Establishing functional residual capacity at birth: the effect of sustained inflation and positive end expiratory pressure in a preterm rabbit model. *Pediatr Res* 2009;**65**:537–41.
57. Kitchen MJ, Lewis RA, Morgan MJ, Wallace MJ, Siew MLL, Siu KKW, et al. Dynamic measures of lung air volume using phase contrast X-ray imaging. *Phys Med Biol* 2008;**53**:6065–77.
58. Kitchen MJ, Lewis RA, Hooper SB, Wallace MJ, Siu KKW, Williams I, et al. Dynamic studies of lung fluid clearance with phase contrast imaging. *Am Inst Phys* 2007;**879**:1903–7.
59. Kitchen MJ, Lewis RA, Yagi N, Uesugi K, Paganin D, Hooper SB, et al. Phase contrast X-ray imaging of mice and rabbit lungs: a comparative study. *Br J Radiol* 2005;**78**:1018–27.
60. Lewis RA, Yagi N, Kitchen MJ, Morgan MJ, Paganin D, Siu KK, et al. Dynamic imaging of the lungs using x-ray phase contrast. *Phys Med Biol* 2005;**50**:5031–40.
61. Hooper SB, Kitchen MJ, Siew ML, Lewis RA, Fouras A, te Pas AB, et al. Imaging lung aeration and lung liquid clearance at birth using phase contrast X-ray imaging. *Clin Exp Pharmacol Physiol* 2009;**36**:117–25.
62. Bland RD, McMillan DD, Bressack MA, Dong L. Clearance of liquid from lungs of newborn rabbits. *J Appl Physiol* 1980;**49**:171–7.
63. Miserocchi G, Poskurica BH, Del Fabbro M. Pulmonary interstitial pressure in anesthetized paralyzed newborn rabbits. *J Appl Physiol* 1994;**77**:2260–8.
64. Avery ME, Cook CD. Volume-pressure relationships of lungs and thorax in fetal, newborn, and adult goats. *J Appl Physiol* 1961;**16**:1034–8.
65. Vilos GA, Liggins GC. Intrathoracic pressures in fetal sheep. *J Dev Physiol* 1982;**4**:247–56.
66. Harding R, Bocking AD, Sigger JN. Upper airway resistances in fetal sheep: the influence of breathing activity. *J Appl Physiol* 1986;**60**:160–5.
67. Davey MG, Johns DP, Harding R. Postnatal development of respiratory function in lambs studied serially between birth and 8 weeks. *Respir Physiol* 1998;**113**:83–93.
68. Danto SI, Shannon JM, Borok Z, Zabski SM, Crandall ED. Reversible transdifferentiation of alveolar epithelial cells. *Am J Respir Cell Mol Biol* 1995;**12**:497–502.
69. Gutierrez JA, Gonzalez RF, Dobbs LG. Mechanical distension modulates pulmonary alveolar epithelial phenotypic expression in vitro. *Am J Physiol* 1998;**274**:L196–202.
70. Shannon JM, Jennings SD, Nielsen LD. Modulation of alveolar type II cell differentiated function in vitro. *Am J Physiol* 1992;**262**:L427–36.
71. Alcorn D, Adamson TM, Lambert TF, Maloney JE, Ritchie BC, Robinson PM. Morphological effects of chronic tracheal ligation and drainage in the fetal lamb lung. *J Anat* 1977;**123**:649–60.
72. Flecknoe S. Harding R, Maritz G, and Hooper SB. Increased lung expansion alters the proportions of type I and type II alveolar epithelial cells in fetal sheep. *Am J Physiol* 2000;**278**:L1180–5.
73. Flecknoe SJ, Wallace MJ, Harding R, Hooper SB. Determination of alveolar epithelial cell phenotypes in fetal sheep: evidence for the involvement of basal lung expansion. *J Physiol* 2002;**542**:245–53.
74. Flecknoe SJ, Wallace MJ, Harding R, Hooper SB. Changes in alveolar epithelial cell proportions before and after birth. *Am J Respir Crit Care Med* 2002;**165**. A643.
75. West JB. Pulmonary blood flow and metabolism. In: West JB, editor. *Physiological Basis of Medical Pratice*. Baltimore: Williams & Wilkins; 1989. pp. 529–36.
76. Rudolph AM. Fetal and neonatal pulmonary circulation. *Annu Rev Physiol* 1979;**41**:383–95.
77. Heymann MA. Control of the pulmonary circulation in the perinatal period. *J Dev Physiol* 1984;**6**:281–90.
78. Hooper SB. Role of luminal volume changes in the increase in pulmonary blood flow at birth in sheep. *Exp Physiol* 1998;**83**:833–42.
79. Crossley KJ, Allison BJ, Polglase GR, Morley CJ, Davis PG, Hooper SB. Dynamic changes in the direction of blood flow through the ductus arteriosus at birth. *J Physiol Aug* 2009.
80. Dawes GS, Mott JC, Widdicombe JG. The foetal circulation in the lamb. *J Physiol* 1954;**126**:563–87.
81. Polglase GR, Hooper SB. Role of intra-luminal pressure in regulating PBF in the fetus and after birth. *Curr Pediatr Rev* 2006;**2**:287–99.
82. Polglase GR, Wallace MJ, Grant DA, Hooper SB. Influence of fetal breathing movements on pulmonary hemodynamics in fetal sheep. *Pediatr Res* 2004;**56**:932–8.
83. Rudolph AM, Heymann MA. Circulatory changes during growth in the fetal lamb. *Circ Res* 1970;**26**:289–99.
84. Bhatt S, Alison BJ, Wallace EM, Crossley KJ, Gill AW, Kluckow M, et al. Delaying cord clamping until ventilation onset improves cardiovascular function at birth in preterm lambs. *J Physiol Lond* 2013;**591**:2113–26.
85a. Cernadas JMC, Carroli G, Pellegrini L, Otano L, Ferreira M, Ricci C, et al. The effect of timing of cord clamping on neonatal venous hematocrit values and clinical outcome at term: A randomized, controlled trial. *Pediatrics* 2006;**117**:E779–86.
85b. Dawes GS, Mott JC, Widdicombe JG. Closure of the foramen ovale in newborn lambs. *J Physiol* 1955;**128**:384–95.
86. Chaparro CM, Neufeld LM, Alavez GT, Cedilla REL, Dewey KG. Effect of timing of umbilical cord clamping on iron status in Mexican infants: a randomised controlled trial. *Lancet* 2006;**367**:1997–2004.

87. Emhamed MO, van Rheenen P, Brabin BJ. The early effects of delayed cord clamping in term infants born to Libyan mothers. *Trop Doct* 2004;**34**:218–22.

88. Gupta R, Ramji S. Effect of delayed cord clamping on iron stores in infants born to anemic mothers: a randomized controlled trial. *Indian pediatr* 2002;**39**:130–5.

89. Saigal S, Surainde.Y Usher R, Chua LB, Oneill A. Placental transfusion and hyperbilirubinemia in the premature. *Pediatrics* 1972;**49**:406–19.

90. Mercer JS. Current best evidence: A review of the literature on umbilical cord clamping. *J Midwifery Womens Health* 2001;**46**:402–14.

91. West JB. Invited review: pulmonary capillary stress failure. *J Appl Physiol* 2000;**89**:2483–9.

92. Reid DL, Thornburg KL. Pulmonary pressure-flow relationships in the fetal lamb during in utero ventilation. *J Appl Physiol* 1990;**69**:1630–6.

93. Harding R. Fetal breathing movements. In: Crystal RG, West JB, Weibel ER, Barnes PJ, editors. *The Lung: Scientific Foundations.* New York: Lippincott-Raven; 1997. p. 2093.

94. Dawes GS, Fox HE, Leduc BM, Liggins GC, Richards RT. Respiratory movements and rapid eye movement sleep in the foetal lamb. *J Physiol* 1972;**220**:119–43.

95. Kitterman JA. Arachidonic acid metabolites and control of breathing in the fetus and newborn. *Semin Perinatol* 1987;**11**:43–52.

96. Wallen LD, Murai DT, Clyman RI, Lee CH, Mauray FE, Kitterman JA. Effects of meclofenamate on breathing movements in fetal sheep before delivery. *J Appl Physiol* 1988;**64**:759–66.

97. Koos BJ, Maeda T, Jan C. Adenosine A(1) and A(2A) receptors modulate sleep state and breathing in fetal sheep. *J Appl Physiol* 2001;**91**:343–50.

98. Adamson SL. Regulation of breathing at birth. *JDevPhysiol* 1991;**15**:45–52.

99. Hagnevik K, Faxelius G, Irestedt L, Lagercrantz H, Lundell B, Persson B. Catecholamine surge and metabolic adaptation in the newborn after vaginal delivery and caesarean section. *Acta Paediatr Scand* 1984;**73**:602–9.

100. Sawa R, Asakura H, Power GG. Changes in plasma adenosine during simulated birth of fetal sheep. *J Appl Physiol* 1991;**70**:1524–8.

101. Crossley KJ, Nicol MB, Hirst JJ, Walker DW, Thorburn GD. Suppression of arousal by progesterone in fetal sheep. *Reprod Fertil Dev* 1997;**9**:767–73.

102. Thorburn GD. The placenta and the control of fetal breathing movements. *Reprod Fertil Dev* 1995;**7**:577–94.

103. Jansen AH, Chernick V. Onset of breathing and control of respiration. *Semin Perinatol* 1988;**12**:104–12.

104. Tai TC, MacLusky NJ, Adamson SL. Ontogenesis of prostaglandin E2 binding sites in the brainstem of the sheep. *Brain Res* 1994;**652**:28–39.

105. Kuipers IM, Maertzdorf WJ, De Jong DS, Hanson MA, Blanco CE. Initiation and maintenance of continuous breathing at birth. *Pediatr Res* 1997;**42**:163–8.

106. Wong KA, Bano A, Rigaux A, Wang B, Bharadwaj B, Schurch S, et al. Pulmonary vagal innervation is required to establish adequate alveolar ventilation in the newborn lamb. *J Appl Physiol* 1998;**85**:849–59.

107. Harding R. State-related and developmental changes in laryngeal function. *Sleep* 1980;**3**:307–22.

108. Johnson P. Physiological aspects of regular, periodic and irregular breathing in adults and in the perinatal period. In: Von Euler C, Lagercrantz H, editors. *Central Nervous Control Mechanisms in Breathing.* Oxford: Pergamon Press; 1979. pp. 337–51.

109. Hooper SB, Siew ML, Kitchen MJ, te Pas AB. Establishing functional residual capacity in the non-breathing infant. *Semin Fetal Neonatal Med* 2013;**18**:336–43.

110. Flecknoe SJ, Wallace MJ, Cock ML, Harding R, Hooper SB. Changes in alveolar epithelial cell proportions during fetal and postnatal development in sheep. *Am J Physiol Lung Cell Mol Physiol* 2003;**285**:L664–70.

Chapter 14

Normal Aging of the Lung

Kent E. Pinkerton*, Matt J. Herring*, Dallas M. Hyde* and Francis H.Y. Green[†]

*Department of Anatomy, Physiology and Cell Biology, School of Veterinary Medicine, Center for Health and the Environment, California National Primate Research Center, University of California – Davis, Davis, CA, USA; [†]Department of Pathology & Laboratory Medicine, University of Calgary, Calgary, Alberta, Canada

INTRODUCTION

Development of the respiratory system for a variety of mammalian species has been analyzed both anatomically, as well as physiologically, during early postnatal life.[1–3] However, few studies have quantitatively examined structural changes that occur with aging. Descriptions of age-related alterations in the respiratory system that do exist,[4–6] have focused almost exclusively on the gas exchange portions of the lungs. Less information is available for detailed description of cellular and structural changes in the aging process of the tracheobronchial tree. The majority of data related to lung aging is typically confined to brief descriptive observations of control animals through lifetime toxicity studies.[7,8]

This chapter will cover topics of normal aging within a number of mammalian species. These will focus primarily on the mouse, rat, dog, and rhesus monkey. Aging of the human lung will be covered in the final chapter (Chapter 27). Life span characteristics for each of the species are different, with the mouse having a median life span of 29–30 months; the rat, 30–34 months; the dog, 12–14 years; the rhesus monkey, 27–30 years; and the human, 72–78 years. These striking differences in life span are likely to have a significant impact on the resultant aging process within the respiratory system of each species. However, there also exists a multitude of morphologic features in aging that cross boundaries for all species. These include postnatal alveolarization of the lungs during early childhood development that is complete in small laboratory species within 4–6 weeks of birth, in dogs within the first year of life, and in humans within the first 8 years of life. Thinning of alveolar septa within the gas exchange portions of the lung is a common characteristic across all species. Pores or fenestrations within alveolar walls are also present in all species. These anatomical structures are thought to serve in cross-collateral ventilation between adjacent alveoli. They may also serve to facilitate the migration of cells in the airspace such as macrophages in a short circuit path from one alveolus to another. These pores appear to enlarge during the aging process and may play an important role in the development of an emphysematous condition in the lungs with larger and greater numbers of fenestrations and loss of alveolar septal walls with age. These characteristics have been studied in detail in the mouse, dog, and human. Airspace enlargement is a characteristic feature seen in all aging species, however, it is unclear whether such enlargement is always accompanied by the destruction or partial loss of alveolar wall structures. Nevertheless, these general characteristics present in each species described in this chapter can serve to better understand the process of lung aging over the normal life span.

The most vital function of the lungs is gas exchange. Oxygen is transferred from the inspired air to the blood, while carbon dioxide, a byproduct of cellular metabolism, is released from the blood into the airspace and expired. To allow for the most efficient exchange of these gases, air must be brought into close proximity to the vascular supply as it passes through the lungs. That portion of the lungs involved in gas exchange represents 80–90% of the total lung volume. Billions of cells in the lungs are arranged to form a delicate, yet sturdy and highly vascular, air/tissue interface creating a surface area 25 times greater than that of the external surface of the body. This tissue barrier separates the inspired air from the blood by less than one micrometer. It is easily deformed by the passage of blood through the underlying capillary bed, causing bulging of the walls into the air space with the outline of red blood cells easily visible through the thin alveolar partitions (Figure 1). This highly efficient surface design for gas exchange is limited to the small space enclosed by the bony thorax and muscular, dome-shaped diaphragm. A highly ordered airway branching beginning at the carina of trachea gives rise to millions of alveoli lining several generations of alveolar ducts and sacs to bring about this exponential increase in area within a tightly packaged volume. The

The Lung. http://dx.doi.org/10.1016/B978-0-12-799941-8.00014-6

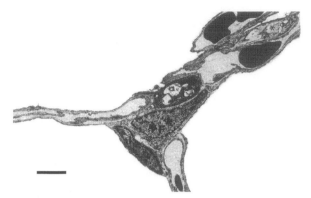

FIGURE 1   An alveolar septum from the lungs of a 5-month-old Fischer 344 rat. A fibroblast (F) can be seen at the junction of the septa. A pocket (*) of interstitial matrix is present, but most of the alveolar septum is composed of a thin air-to-blood tissue barrier (bar is equal to 3 μm).

process of aging during early life leads to the formation of approximately 80% of all alveoli postnatally. Later in life the number of alveoli may be reduced through a destructive process with loss of alveolar septal wall surface area and/or the rearrangement of essential extracellular matrix components, such as collagen and elastin, causing the alveoli to become stretched and shallow and leading to distention of alveolar ducts, referred to alveolar ductasia. These phenomena will be discussed in this chapter as well as in Chapter 27 (Environmental Determinants of Lung Aging).

AGING, BODY MASS, AND THE LUNGS IN MAMMALS

For the mammalian lung, a strong allometric relationship exists between body mass and the following features: lung volume, lung capillary volume, alveolar surface area, and pulmonary diffusing capacity.[9] The postnatal process during development is associated with significant increases in body mass and lung volume. For the mouse, the most dramatic increases in body mass occur during the first month of life. Subsequent weight changes beyond 2 months of age

occur at a slow rate, but continue a slight increase up to 19 months of age, followed by a usual decrease beyond 28 months of age (Table 1). For rats, the two most commonly used strains for lifetime studies are Sprague Dawley and Fischer 344 rats. The popularity of this species for lifetime studies is attributed to their small body size, simple housing requirements, and relatively short life span of 30–36 months of age. The Fischer 344 rat has been used extensively because of a slow increase in body mass with age and resistance to pulmonary disease (Figure 2).

Dogs cover a vast span in body size and therefore have pulmonary size characteristics that cover a wide range of values compared with mice and rats. Lung characteristics, such as airspace volume or alveolar surface area, may span at least two orders of magnitude in dogs. However, the lungs of dogs possess many characteristics similar to the human lung, including the presence of respiratory bronchioles, which are absent in the lungs of mice and rats. Pulmonary physiology and lung tissue structure in dogs also change in a similar fashion to humans during the process of lung growth and development.[10]

The adult rhesus monkey varies in body size and subsequent body characteristics, but not to the same degree as a dog. Female adult rhesus monkeys can have the most variable body weight, ranging from 4 to 13 kg, but have small lung volumes (360–675 cm³).[11] Similar to canines, respiratory bronchioles are present in nonhuman primate lungs making them a suitable model for human lung morphology and airway disease. Morphological comparisons between humans and macaques indicate similarities in alveolar segmental arrangement, alveolar number in adults, structure, branching pattern or airways, arterial structure and arterial changes after birth.[12–14] Being a rather long-lived species, the rhesus monkey can be difficult to study regarding aging properties of the lung and other organs.

For the human, lung development continues beyond birth, with over 80% of alveoli formed postnatally in the lung. This process of alveolarization is thought to be complete by 8 years of age. A recent study using Helium-3 magnetic resonance imaging has shown neo-alveolarization continues

TABLE 1 Body weight for various strains of aging mice

	1 month	2 months	9 months	17–19 months	25–28 months
Body weight (g)					
Strain					
BALB/cNNia[21,†]	19.98 ± 0.72	23.98 ± 0.77*	28.46 ± 0.83*	31.27 ± 0.46*	28.42 ± 0.65*
SAM-P/1[26,‡]	21.60 ± 3.20	29.70 ± 1.20*	29.00 ± 3.10	28.90 ± 3.10	—
SAM-R/1[26,‡]	23.80 ± 0.80	28.70 ± 1.00	29.80 ± 3.50	32.70 ± 2.50	29.70 ± 2.60

*p<0.05 for comparison to the immediately younger age group.
†All values are mean ± SEM.
‡All values are mean ± SD.

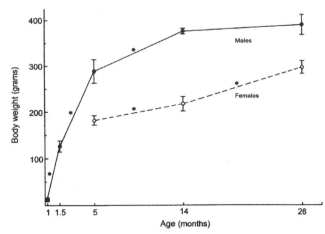

FIGURE 2 Changes in body weight with increasing age in male and female Fischer 344 rats. Each datum point represents the mean ± SD of four animals. The asterisk on the line connecting two different age groups denotes a significant change in body weight ($p < 0.05$).

from ages 7 to 21 in human lungs. This study is based on alveolar size being maintained but lung volume steadily increasing during this time. A stereological approach to alveolar growth in humans that eliminates this bias would be ideal to better understand the growth curve of the human lung.[15] During adolescence, the lungs continue to expand to fill the dimensions of the thoracic cavity and increase physiological capacity. By age 18, lung growth is considered complete.

The following sections will cover normal aging in the mouse, rat, dog, and rhesus monkey, with an emphasis on lung anatomy.

LIFE SPAN CHARACTERISTICS OF THE MOUSE

The white-footed mouse maintained under barrier conditions can survive to 8 years of age.[16,17] The median life span for this mouse strain is 4.5–6 years.[18] Mouse strains most commonly used in research, including BALB/c, Swiss-Webster, and C57BL/6, have a shorter median life span of 29–30 months. The life span of the laboratory mouse is dependent on the strain, nutritional status, and environment in which they are maintained. Special mouse strains have been established, including the senescence-accelerated mouse (SAM), a murine model of rapid aging.[19] Both prone (SAM-P/1) and resistant (SAM-R/1) strains have been developed and have median life spans of 11.9 and 17.5 months, respectively.[19] Homozygous mutant klotho gene (KL$^{-/-}$) mice also exhibit a shortened life span (12–14 weeks), infertility, arteriosclereosis, skin atrophy, osteoporosis, and emphysema.[20]

The Aging Mouse Lung

The majority of studies using mice begin with animals at 1–2 months of age. Experimental studies on the process of pulmonary aging rarely extend beyond 12 months of age. Therefore, the amount of information available on the effects of aging on the structure and function of the lungs in mice, including age-related changes of the airways, vasculature, nerves, lymph vessels, or immune system is limited. Selected changes in the pulmonary parenchyma of the mouse from 1 to 28 months of age have been established.[5,21,22] Studies with SAM have prompted a growing interest in using this mouse strain as a potential model of the aging process. Senescence in this mouse strain is marked by behavioral changes, hair loss,[23] senile amyloidosis,[24] cataracts,[23] and osteoporosis.[25] Pulmonary studies in SAM have noted accelerated changes in the degree of pulmonary hyperinflation similar to that noted in other mouse strains at older ages.[26,27] Therefore, SAM may provide insights of the aging process in mouse lungs over a shorter period of time compared with other conventional mouse strains.

Changes in Lung Volume during Aging Mice

Changes in total lung volume with increasing age in mice of differing strains are given in Table 2. Kawakami and colleagues[21] found that the volume of the lungs continued

TABLE 2 Lung volumes in various strains of aging mice

	1 month	2 months	9 months	17–19 months	25–28 months
Lung volume (mL)					
Strain					
BALB/cNNia[21],†	0.92 ± 0.03	1.14 ± 0.03*	1.47 ± 0.07	1.68 ± 0.03	1.87 ± 0.10*
SAM-P/1[26],‡	0.80 ± 0.17	0.97 ± 0.07	1.34 ± 0.11	1.47 ± 0.10	-
SAM-R/1[26],‡	0.70 ± 0.13	0.77 ± 0.10**	1.22 ± 0.12**	1.47 ± 0.10	1.57 ± 0.10

*p<0.05 for comparison to the immediately younger age group.
**p<0.05 for comparison to SAM/P1 of the same age.
†All values are mean ± SEM.
‡All values are mean ± SD.

to increase in a significant age-dependent manner to age 25–28 months. These increases occur as a process of aging in approximately a linear manner throughout the life span of the mouse (Table 2). Lung volume in female C57BL/6J mice has been shown to decrease between 6 months and 24 months of age.[28] Measurement of lung changes in mice during aging is based on a simple technique that utilizes intratracheal instillation of fixative at a standard pressure until the lungs become rigid in an inflated state. For those measurements listed under Table 2, the lungs were removed from the thoracic cavity and degassed before instillation of the fixative at 25–30 cm water pressure. Once the lungs were fixed, the volume was measured by the immersion method of liquid displacement.[29]

Fixed lung volume can serve as a simple index for measurements of size, volume, and surface area of those structural components present in the lungs by applying morphometric techniques. Intratracheal instillation of fixative at a standard pressure is one of the most accurate and reproducible means of inflating excised lungs. However, without the retaining forces of the chest wall, the increased compliance of pulmonary tissues in excised old mouse lungs could cause distention beyond those volumes permitted within the thoracic limits of the animal. However, for those measurements given in Table 2, special care was taken to ensure that intrapulmonary pressures not exceed 15 cm after an initial inflating pressure of 25 cm. Therefore, subsequent measures of parenchymal airspace volume and surface area, as determined using morphometric techniques on these same tissues, can be considered as a true reflection of changes due to aging in the lungs of each mouse strain.

Tracheobronchial Airways of Aging Mice

In the trachea and proximal bronchi of the mouse, the epithelial cells form a pseudostratified layer, with a staggered arrangement of nuclei located above the basal lamina. The airway epithelium is composed of three primary cell types: basal cells, ciliated cells, and secretory cells. In the mouse, secretory cells present in the trachea form the nonciliated bronchiolar epithelial cells or Clara cells. In the mouse, the location of these cells within the trachea is unique because Clara cells are typically found only in more distal bronchioles of the airways in most species.

Little information is available on epithelial cell populations of the tracheobronchial tree during the aging process in the mouse. Only studies of mice aged 2–3 months have reported the number of epithelial cells per millimeter of basal lamina in the trachea to be approximately 215.[30] The epithelial cell types in the trachea at this age consist of 10% basal cells, 39% ciliated cells, 49% Clara cells, and 2% unknown cells. The density of epithelial cells does not change in second and third airway generations compared with the trachea. The proportion of epithelial cell types

also remains constant. Within the mainstem bronchi of the mouse, basal cells constitute 4% of the total population, ciliated cells 47%, Clara cells 46%, and unknown cells 3%.[30] Epithelial density in airway generations four through six in the mouse is slightly reduced, with 199 cells per millimeter of basal lamina. The proportion of epithelial cell types shifts, with basal cells accounting for 1%, ciliated cells 36%, Clara cells 61%, and unknown cells 2%. Epithelial cells of the bronchioles, including the terminal bronchiole, consist of a simple cuboidal to columnar epithelial layer formed by ciliated cells and Clara cells. The relative proportion of Clara cells in the bronchiole is as high as 80%, and ciliated cells constitute the remaining cells of the epithelial airway lining. Clara cells within the terminal bronchiole of the mouse have dome-shaped apical surfaces protruding into the lumen of the airway and numerous secretory granules and mitochondria found within the apical cytoplasm. Little-to-nothing is known about the effects of aging on the structure or function of Clara cells in the mouse.

Parenchymal Lung Structure in Aging Mice

The parenchyma of the lungs is composed of alveoli, alveolar ducts, and alveolar sacs. These structures form approximately 90% of the total lung volume in the mouse.[21] The alveoli represent the smallest anatomical unit involved in gas exchange and are composed of the airspace bounded by the alveolar wall and its opening into the alveolar duct. The alveolar duct is formed by the airspace shared in common with alveoli opening along a common channel created by the tissue ridges forming the mouth opening of individual alveoli. The branching of alveolar ducts to form discrete alveolar duct generations begins at the bronchiole-alveolar duct junction (BADJ) and ends three to five generations away as a blind alveolar sac formed by several alveoli. The total alveolar surface area formed by these structures in the aging mouse is depicted in Figure 3 for BALB/c mice and two strains of senescent mice (SAM).

The proportion of air volume found in ducts and alveoli is given in Table 3. Total volume of alveolar air is approximately 0.5 ml at 1 month of age and steadily increases until 28 months of age. The age-related increase in air volume is more dramatic in young animals compared with older animals. The total volume of air within alveolar ducts and sacs increases in a consistent and significant degree with increasing age in mice (Table 3). A three-fold increase in the air volume of the ducts and sacs occurred from age 1 to 28 months. The total volume of the alveolar wall (Table 3) was not significantly changed from age 1 to 9 months. In older animals of 19 months of age, alveolar wall thickness was significantly increased compared with younger age groups.

An increase in total alveolar surface area (Figure 3) in combination with a continual increase in the total air volumes of the alveoli, ducts, and sacs (Table 3) strongly suggests that

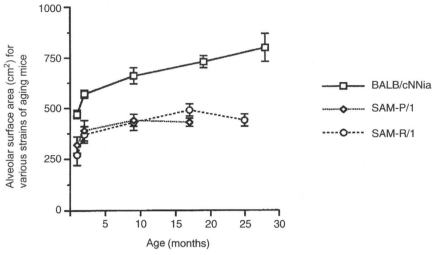

FIGURE 3   Alveolar surface area in aging strains of mice.

TABLE 3 Absolute air and tissue volumes in aging BALB/cNNia mice[21] †

	1 month	2 months	9 months	19 months	28 months
Total volume of compartment (mL)					
Air in ducts	0.126 ± 0.008	0.178 ± 0.060*	0.303 ± 0.015*	0.341 ± 0.020*	0.387 ± 0.010*
Bronchial and bronchiolar air	0.075 ± 0.008	0.105 ± 0.009	0.127 ± 0.010	0.106 ± 0.013	0.143 ± 0.020
Alveolar air	0.497 ± 0.019	0.621 ± 0.018*	0.786 ± 0.036	0.890 ± 0.029	0.969 ± 0.074
Alveolar wall	0.110 ± 0.006	0.112 ± 0.009	0.121 ± 0.009	0.176 ± 0.010*	0.184 ± 0.008

*$p < 0.05$ for comparison to the immediately younger age group.
†All values are mean ± SEM.

the aging process in the mouse lung results in a hyperinflated lung with significant airspace distention. Although there is some evidence for the loss of alveolar surface area in SAM mice with increasing age, there is no loss in alveolar surface area in the BALB/c mouse during the aging process. Numbers of alveoli in C57BL/6J female mice do not significantly decrease with age.[28] Klotho deficient mice display a homogenenous enlargement of airspaces and increased lung compliance.[31] Abnormal activation of Vitamin D appears to play a role in the alveolar wall destruction in Klotho mice.[32]

Interalveolar pores form a common communication between adjacent alveoli. Changes in the size and frequency of these alveolar pores are best detected using scanning electron microscopy on critical point dried lung tissues. Alveolar pores or wall fenestrations increase in number and frequency as the lungs of the mouse age. At 1 month of age, these pores are relatively sparse within the walls of alveoli but increase in size with age. The frequency of interalveolar pores per alveolus varies depending on location within the lung parenchyma.[22] Subpleural and peribronchiolar regions

appear to have alveoli that contain higher numbers of interalveolar pores compared with parenchymal tissues in other regions of the lungs. The number of pores per alveolus at 1 month of age more than doubles by the time these animals reach 28 months of age (Figure 4). The total area of interalveolar pores in the aging BALB/cNNia mouse increases throughout life. From 1 month to 28 months of age, the total area of alveolar pores increases more than four-fold (Table 4).

A rapid increase in the number of interalveolar pores early in life has been described in both mice[22] and dogs.[33] Interalveolar pores are extremely rare during the first 10 days of life in mice, but rapidly increase in number and size after day 14.[34]

The increase in the total area of interalveolar pores as well as the number of interalveolar pores per alveolus corresponds to the increase in total alveolar surface area during the life span of the mouse. Rapid expansion of the lungs postnatally may account in part for the formation of some pores; however, degenerative processes during aging may also be responsible for an increase in the total interalveolar pore surface area as well as the frequency of interalveolar pores per alveolus. Pore enlargement may occur as a result of the rupture of tissue strands between adjacent pores, especially in the senescent animal. Pump[35] described a stage of fenestration in the lungs that would allow for these tissues to attenuate and rupture between intervening capillaries. This process could also occur in the aging mouse lungs. The proportion of alveolar wall formed by pores is about 3.5% at 1 month, 5.9% at 2 months, and 8.5 to 9% in mice older than 9 months.

A decrease in pulmonary elasticity with aging has been verified physiologically. The morphological and chemical basis for changes with age in elastic tissue and the organization of the elastic fiber network throughout the lungs and alveoli is not well understood. A number of investigators have examined elastin content in aging mouse lungs by morphology through measurement of total fiber length.[5,21,26] They found aging in mouse lungs occurs in the absence of statistically significant changes in total elastic fiber length,

despite a significant increase in pulmonary volume (Tables 2 and 3). However, if elastic fiber length is normalized to pulmonary volume, an age-related decrease in elastic fibers was noted in BALB/c mice.[21] This decrease in elastic fibers was associated with an increase in the static compliance of excised lungs of aged mice, due to a progressive loss of elastic recoil pressure. If an increase in pulmonary compliance could occur without the destruction of elastic fibers, an increase in the total elastic fiber length would be expected.[13] However, biochemical analysis of elastic content demonstrated a loss of elastic fibers in aging lungs. Huang et al. demonstrated a delay between the reduction in elastin of lung parenchyma in C57BL/6J mice and the increase in collagen that happened later in senescence.[36] Special care must be taken to separate pseudoelastin, a form of elastin in human lungs that increases with age, and elastin during biochemical analysis or the actual amount of elastin will be overestimated. Pseudoelastin has not been found in mice. Therefore, Ranga and colleagues[5] have suggested this absence of pseudoelastin fibers accounts for the decrease in elastin content in the aging BALB/c mouse lung. Twenty-four-month-old C57BL/6 mice do exhibit a profibrotic phenotype regarding mRNA expression with increased expression of MMP-2, MMP-9, TGF-β1, and TGF-β receptor 1.[37]

Studies to measure histological changes in lung elastic fibers in SAM mice have found no evidence of destruction of the alveolar wall or elastic fibers during aging in the lungs. Physiological studies have also demonstrated lung compliance in these mice at age 10 months to be significantly greater than that noted at 2 months of age.[26] Therefore, age-related changes in lung distension in SAM occur in a manner similar to that noted for other strains of mice. In contrast, for mice of the SAM-P/I strain, changes in lung hyperdistention occur in an accelerated manner compared with the resistant strain (SAM-R/1) or other mouse strains. Therefore, SAM-P/I strain mice could prove useful in the study of senile-related lung hyperinflation.

Macrophages and Lung Aging (Mouse)

Aging is thought to be associated with a decline in immune function. Changes in macrophage function can be an important component in compromised lung defense. Phagocytic cells collected by bronchoalveolar lavage are

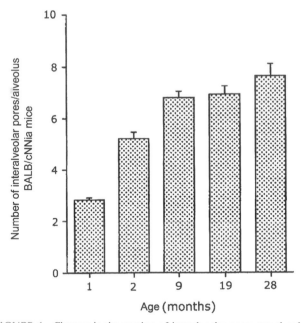

FIGURE 4   Changes in the number of interalveolar pores per alveolus with aging in the BALB/cNNia mouse.

TABLE 4 Total area of interalveolar pores in aging BALB/cNNia mice[21] †					
	1 month	2 months	9 months	19 months	28 months
Total area of pores (10^{-2}m^2)	0.083 ± 0.009	0.170 ± 0.015*	0.281 ± 0.025*	0.338 ± 0.028	0.357 ± 0.022

*p<0.05 for comparison to the immediately younger age group.
†All values are mean ± SEM.

TABLE 5 Bronchoalveolar lavage recovery in aging C57BL/6, SPF mice[38] [†]

	1.5 months	26 months
Recovery ratio[‡]	0.93 ± 0.02	0.98 ± 0.01
Total cell count (× 10^5)	4.88 ± 0.47	7.49 ± 2.04
Differential cell count (%)		
Macrophage	97.0 ± 0.60	94.8 ± 1.30
Neutrophil	0.10 ± 0.60	1.90 ± 0.96
Lymphocyte	1.82 ± 0.60	2.83 ± 0.61

[†]All values are mean ± SEM.
[‡]Recovery ratio = recovery fluid/injected fluid of bronchoalveolar lavage.

useful to study the characteristics of these cells. Higashimoto and colleagues[38] found in C57BL/ 6 mice of differing ages that the number of cells recovered by bronchoalveolar lavage was greater in 26-month-old animals compared with 1.5-month-old animals (Table 5). The cell differentials at both ages were similar, with 95–96% of macrophages recovered from the lungs. Neutrophils represented a small fraction (0.1%) in the youngest animals and a slightly higher fraction in older animals (1.8%). Lymphocytes in both young and old animals were approximately 2%. This slight increase in the number of neutrophils and lymphocytes within the lungs of aged animals compared with young animals may reflect a subtle change in the immune system of these animals or the need for greater numbers of cells in old versus young animals to maintain the proper sterility of the lungs. The number of cells recovered by bronchoalveolar lavage in old mice compared with young mice may also be a simple reflection of the aging process with greater numbers of phagocytic cells present in the lungs of old mice. The ability of cells to phagocytize particles has been tested to demonstrate cells recovered from 26-month-old mice had a proportionally greater percentage that were not able to phagocytize latex spheres as easily as those cells recovered from the lungs of mice 1.5 months of age.[38] Compared to 12-week-old Balb/c mice, 24-month-old Balb/c mice showed increased expression of eight inflammation-related genes (including CD20, CD72, and IL-8RB) along with increased numbers of CD4, CD8, B cells, and macrophages.[39] This may indicate a proinflammatory shift occurring in the lungs of aging mice along with a decrease in immune function. Age-dependent macrophage dysfunction has been characterized in several studies based on exposure to bacterial products and may be the result of weak NF-κB and MAPK activation.[40–42] Activation of macrophages may also be reduced via heme oxygenase-1 (HO-1). In response to LPS, upregulation of HO-1 was significantly impaired in the lungs of 65–66-week-old mice compared to 9–11-week-old mice.[43]

Antioxidant Enzyme Activity in the Aging Mouse

Antioxidant enzymes represent an important means of protecting cells from damage due to gases and particles that have the oxidizing capacity to alter vital cellular components. Superoxide dismutase (SOD), catalase (Cat), and the glutathione (GSH) system all serve to protect against the toxic effects of oxidants. GSH in sufficient concentrations can adequately detoxify oxidants through conjugation. However, if glutathione concentrations become depleted, toxic intermediates can form to cause pulmonary cell injury. Pulmonary glutathione concentrations have been determined over the life span of male C57BL/6 mice. Studies used mice at age 3, 6, 12, 26, and 31 months. GSH concentrations decreased by 30% in the lungs of aging mice, whereas GSSG cysteine and cystine concentrations remained unchanged. Depletion of pulmonary GSH by injection of acetaminophen demonstrated that young (age 3—6 months) and mature (age 12 months) mice recovered hepatic GSH levels more efficiently than senescent mice (age 31 months), but no differences were noted in the lungs.[44] These findings suggest that detoxification capacity decreases as age increases in the mouse. However, no information is known regarding cell numbers or cell types responsible for maintaining GSH concentrations in the lungs of mice during the aging process.

Conclusions for Lung Aging in the Mouse

Little information is available to adequately describe the effects of aging in mouse lungs. Although changes in air space and tissue volumes and alveolar surface area with aging are known, less is known for cell populations lining the lung airways or forming the alveolar structures of the pulmonary parenchyma over the life span of the mouse. A number of studies have implicated that lung aging in mice is associated with decreases in specific functional and structural parameters. These include increases in the phagocytic cell populations present in the lung airspaces, but decreased ability to engulf foreign particles. Decreases in antioxidant defense systems have also been noted in the lungs of aging mice. From a structural perspective, hyperinflation of the lungs and increases in interalveolar pore size and number are key in the lung aging process in mice. Future knowledge of changes in pulmonary cell number, type, distribution, and function with aging would greatly increase our understanding of their impact on pulmonary physiology, metabolism, and immunity in the mouse. Age-related changes could significantly affect the normal function of the lungs as well as greatly increase host susceptibility to injury.

LIFE SPAN CHARACTERISTICS OF THE RAT

As previously mentioned, aging has been examined in a number of rat strains, but the two most commonly used for lifetime studies are Sprague-Dawley and Fischer 344 rats. The life span characteristics of the Fischer 344 rat have been well documented.[45–52] The majority of rats and mice maintained under barrier conditions show no gross pathological changes of the lungs. A review by Boorman and colleagues,[53] however, does discuss the pathology of nasal, laryngeal, tracheal, and respiratory systems that may be associated with aging in the rat.

The Aging Rat Lung

Development of the respiratory system in the rat has been analyzed morphometrically during early postnatal life,[1,2,3] but few studies have quantitatively examined structural changes, which occur with aging. Descriptions of age-related alterations in the respiratory system which do exist[9] have focused almost exclusively on the gas exchange portions of the lungs. Only one study has examined morphometry of terminal and respiratory bronchioles in aged (27-month-old) Fischer 344 rats showing dilation of both bronchioles.[54] Less information is available that provides detailed descriptions of cellular and structural changes in the tracheobronchial airway tree with age with the exception of the postnatal changes in the nonciliated bronchiolar epithelial (Clara) cell.[2,3] The majority of data related to lung aging in the rat are confined to brief descriptive observations of control animals in lifetime toxicity studies.[7,8]

The Tracheobronchial Tree and Epithelium of the Aging Rat

The tracheobronchial epithelium of adult rats varies in terms of the types of cells present and the relative proportions of specific cell types throughout the conducting airway tree. In the trachea, four cell types have been identified: basal cells, serous cells, ciliated cells, and mucous goblet cells.[55,56] In contrast, the epithelium of bronchi and bronchioles consists of ciliated cells and nonciliated bronchiolar epithelial (Clara) cells.[55–57] The composition of epithelial cells also varies from proximal to distal airway generations with postnatal development.[55,57]

The transformation of tracheal epithelial cells in the rat beginning in the perinatal period through adult age appears to be continual. At the time of birth, the rat trachea contains some mature ciliated cells, obvious secretory cells, and the beginning of basal cell differentiation.[55,58,59] Ciliogenesis begins at about 80% gestation in the rat. Nonciliated secretory cells are obvious in the tracheal epithelium at about 90–95% gestation.

One study evaluated the general characteristics of the airway epithelium in the aging rat. Pinkerton and colleagues[60]

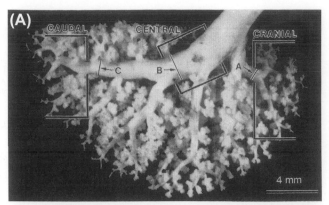

FIGURE 5 Location of tissue samples taken from the rat lung. (A) Silicone cast of the tracheobronchial airway tree. Lung casts served to standardize sampling. Samples of terminal BADJs were taken from the three regions (cranial, central, and caudal) indicated with letters and arrows within the figure. (B) Mediastinal half of a fixed, microdissected rat lung. Note how closely the pathways and sampling regions match the silicone cast shown here (bar is equal to 4 μm).

examined epithelial cells in three different airway generations of the left lung for male Fischer 344 rats aged 5 and 22 months. Illustrated in Figure 5 are three airway levels (A, B, and C) designated as cranial, central, and caudal bronchi, respectively. At each level, ciliated, non-ciliated, and basal cells were identified. The average volume density (cell volume/ basal lamina surface area) of each cell type was determined and expressed in Table 6. No significant differences in the abundance of each cell type were noted from 5 to 22 months of age. Total epithelial cell volume also remained constant from 5 to 22 months of age. No appreciable differences were noted between each of the three bronchial regions examined. These findings suggest few changes occur during aging with a lack of significant shifts in the proportion of epithelial cell types in the tracheobronchial tree of the aging rat. However, far less is known regarding the molecular and biochemical vitality of cells as the organism ages.

Parenchymal Lung Structure in the Aging Rat

Total alveolar air space volume of the lungs shows a progressive increase over a 2-year period of growth in Fischer 344 rats (Figure 6). Air space volume was measured in lungs

TABLE 6 Volume density of epithelial cells of rat tracheobronchial airways of male Fischer 344 rats

Epithelial cell type	Age (months)	Bronchus		
		Cranial	Central	Caudal
Ciliated cells	5	4.50 ± 0.88	2.74 ± 0.33	4.73 ± 0.33
	22	3.80 ± 0.56	3.33 ± 0.29	5.17 ± 0.73
Nonciliated cells	5	1.73 ± 0.30	2.07 ± 0.28	1.83 ± 0.40
	22	2.20 ± 0.52	1.89 ± 0.22	2.65 ± 0.45
Basal cells	5	0.00 ± 0.00	0.08 ± 0.03	0.02 ± 0.02
	22	0.00 ± 0.00	0.09 ± 0.06	0.00 ± 0.00
Total epithelial volume	5	6.23 ± 0.67	4.89 ± 0.43	6.58 ± 0.31
	22	6.01 ± 0.66	5.30 ± 0.55	7.82 ± 1.16

Values are presented as means ± SEM expressed as $\mu m^3/\mu m^2$; $n = 4$ for each age group.

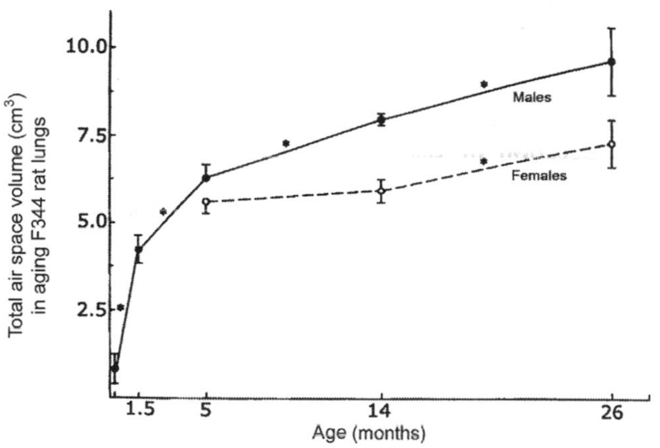

FIGURE 6   Total air space volume (cm^3) in glutaraldehyde-fixed lungs of male and female Fischer 344 rats. Each point represents the mean (± SD) of four animals. Asterisks indicate significant changes between consecutive age groups ($p < 0.05$).[4]

fixed following airway instillation of 2% glutaraldehyde at a hydrostatic pressure of 20 cm.[12,61] Pulmonary physiology performed on male rats prior to aldehyde instillation demonstrated that the lungs were fixed at 75% of total lung capacity.[12,62,63] The marked increase in air space volume from 1 week to 5 months of age was associated with significant increase in alveolar number. Randell and associates estimated the peak rate of formation for new alveoli in the neonatal rat to be 1000 per minute.[64] Light microscopic examination of large airways, terminal bronchioles, large blood vessels, alveoli, and alveolar septa demonstrated no detectable differences from 5 to 26 months of age. Animals 26 months of age appeared to have slightly enlarged alveolar ducts compared with those in younger age groups, but otherwise were indistinguishable from younger animals (Figure 7). The proportion of the lungs that forms the lung parenchyma was 0.81 to 0.82 in young and old animals, respectively.[12,65] Fischer 344/N rats were noted to have enlarged alveolar ducts and alveoli in 24-month-old animals. These alveoli were wider and shallower compared to younger animals.[54]

Epithelial and capillary surface areas of the gas exchange regions of the lungs are given in Table 7. From 1 to 6 weeks of age the surface area formed by epithelial type 1 cells increased six-fold, while epithelial type II cell surface area increased three-fold. The squamous type I cell covered more than 95% of the total alveolar surface, while the cuboidal type II cell covered the remaining 5%. The alveolar type III cell is a rather rare epithelial cell type within the alveoli and contributes very little to the total alveolar surface area. Type III cells are found most frequently within alveolar regions near bronchiole-alveolar duct junctions.[57] From 6 weeks to 5 months of age alveolar surface area almost doubled. From 5 to 26 months of age total alveolar surface area remained unchanged. Since air space size increased by 50% from 5 to 26 months of age in the absence of any significant change in alveolar surface area, alveolar duct enlargement in older animals could explain, in part, how air space volume increased without a loss of surface area within the lung parenchyma. Enlargement of alveolar ducts was not quantitatively confirmed, but emphysematous changes (alveolar wall destruction) appeared to be absent in the aging Fischer 344 rat.[49,53,66,67]

Changes in capillary surface area with age were proportional to changes in alveolar surface area (Table 7). The capillaries within the alveolar septa of the neonatal rat undergo a fascinating transformation from a double capillary system to a single capillary system associated with the growth of secondary alveolar septa to form new alveoli. This reorganization of the pulmonary vasculature is complete by 3 weeks of age.[1,12] From 5 to 26 months of age the total surface area of the capillary bed remained unchanged in both male and female rat lungs (Table 7).

FIGURE 7 Photomicrographs of the lung parenchyma from a Fischer 344 rat 5 months of age (A) and 26 months of age (B). A slight increase in alveolar size can be noted in the lungs of 26-month-old rats compared with the lungs of 5-month-old rats (bar is equal to 100 μm).

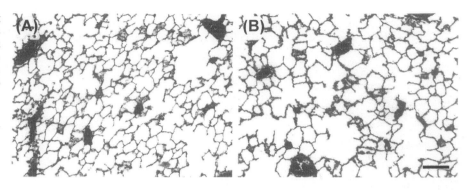

TABLE 7 Alveolar epithelial and capillary surface area (cm²/both lungs) in the aging Fischer 344 rat [4] [†]

	1 week	6 weeks	5 months	14 months	26 months**
Males					
No.	4	4	4	4	4
Alveolar epithelium					
Type I	320 ± 27	2085 ± 113*	3914 ± 390*	4501 ± 324	4628 ± 442
Type II	32 ± 3	107 ± 27	147 ± 47	66 ± 23	66 ± 40
Capillary endothelium	320 ± 27	2510 ± 162*	3832 ± 394*	4543 ± 117	4491 ± 484
Females					
No.	—	—	4	4	4
Alveolar epithelium					
Type I	—	—	3422 ± 127	3268 ± 297	4022 ± 24
Type II	—	—	85 ± 20	58 ± 24	56 ± 25
Capillary endothelium			3259 ± 184	3146 ± 305	3569 ± 163

*$p < 0.05$ for comparison to the immediately younger age group.
**None of the differences between the 26-month group and the 5-month group were statistically significant.
[†]All values are mean ± SEM.

Alveolar Tissue Compartments in the Aging Rat

During the first months of life in the rat, alveolar tissue volumes increase dramatically. However, little change is noted in tissue volumes from 5 to 26 months of life, with the exception of the noncellular component of the interstitium (Table 8). The volume of the capillary bed also increases dramatically during the first 5 months of life, but remains relatively unchanged from 5 to 26 months of age. Alveolar tissue volumes for epithelial, interstitial, and endothelial compartments of the lung parenchyma are presented in Table 8. Cell number for each major alveolar cell type is given in Table 9. Morphometric characteristics of each of these cell types are presented in Table 10. The lung parenchyma consists of three tissue compartments, the epithelium, the interstitium, and the endothelium. Alveolar macrophages form a unique tissue compartment of individual cells that freely migrate along the surfaces of alveoli and airways and are also present within the pulmonary connective tissues as interstitial macrophages.

The composition of the alveolar epithelium in the rat changes dramatically during postnatal development (Table 8). The volume of type I epithelium increased more than six-fold from 1 week to 6 weeks of age, while the type II epithelium displayed a more modest three-fold increase during the same period. The total surface area of type I and type II cells also increased more than five-fold from 1 to 6 weeks of age (Table 8). The ratio of type II to type I cells in the lungs at 6 weeks of age was 1.8, but it decreased to 1.0 by 26 months of age (Figure 8).[12] The significance of the reduction in type II to type I cell ratio is unknown, but may contribute to an altered secretory response in type II cells of old rats since a greater surface area must be served per type II cell to form the surfactant lining layer of the lungs compared with that in younger animals.

TABLE 8 Alveolar tissue volumes (mm³/both lungs) in the aging Fischer 344 rat [4] †

	1 week	6 weeks	5 months	14 months	26 months**
Males					
No.	4	4	4	4	4
Total alveolar tissue	116 ± 14	296 ± 5*	427 ± 46*	454 ± 20	464 ± 43
Epithelium					
Type I	11 ± 2	61 ± 4*	82 ± 6*	84 ± 3	84 ± 8
Type II	7 ± 1	22 ± 3	37 ± 9	30 ± 4	28 ± 9
Interstitium					
Cellular	61 ± 6	77 ± 7	68 ± 5	57 ± 8	57 ± 7
Noncellular	13 ± 1	46 ± 3	128 ± 16*	151 ± 13	178 ± 29**
Endothelium	23 ± 4	84 ± 4*	94 ± 9	111 ± 6	96 ± 9
Macrophages	1 ± 0.4	6 ± 1	19 ± 7	21 ± 7	21 ± 7
Females					
No.	—	—	4	4	4
Total alveolar tissue	—	—	359 ± 15	313 ± 23	370 ± 22
Epithelium					
Type I	—	—	79 ± 7	58 ± 7*	66 ± 6**
Type II	—	—	25 ± 1	20 ± 3	18 ± 5
Interstitium					
Cellular	—	—	60 ± 7	49 ± 4	56 ± 6
Noncellular	—	—	81 ± 10	110 ± 11	153 ± 8*,**
Endothelium	—	—	93 ± 5	70 ± 4*	67 ± 5**
Macrophages	—	—	22 ± 9	6 ± 3	9 ± 3

*p<0.05 for comparison to the immediately younger age group.
**p<0.05 for comparison of the 26-month group to the 5-month group.
†All values are mean ± SEM.

The type II cell undergoes dramatic changes during the perinatal period. Just prior to birth, the type II cell of the rat is heavily laden with glycogen and first acquires its characteristic lamellar bodies 48–72 h prior to parturition. From the terminal day of gestation to a few hours later, the cellular content of glycogen drops from about 10% to zero. On the day of birth, the cells have polarized their secretory granules, the lamellar bodies, toward the basal pole. The putative immediate precursor of the lamellar body contains eccentrically placed vesicles in addition to slips of phospholipid lamellae and is termed a composite body. These are polarized toward the basal side of type II cells throughout the life span of the animal, but the mature lamellar bodies become randomly distributed between the ages of 2 and 6 weeks.[68] In contrast, the Clara cell has its secretory granules polarized toward the apical region like many other secretory cells in the body. Another interesting polarization of type II cell intracellular organelles is that of the light and dark multivesicular bodies. Light multivesicular bodies are not rich in lysosomal enzyme and accumulate endocytosed membrane markers most quickly, like endosomes do in other cells. Dark multivesicular bodies are rich in lysosomal enzymes and are basally polarized-like composite bodies. There could be functional differences between the two multivesicular body types, but more direct experimental evidence is needed to define them.

Type II cellular composition between birth and adulthood differs mainly by a doubling of the volume density of the lamellar bodies and a 50% increase in mitochondrial volume density. Neither the appearance nor the size distribution of lamellar bodies changes during postnatal life, but dramatic changes do occur in mitochondria. They shift from an isolated spheroid form to a highly branched

TABLE 9 Alveolar cell number per both lungs (× 10⁶) in the aging Fischer 344 rat[4] [†]

	Total Number of Cells/Both Lungs	Alveolar Type I	Alveolar Type II	Interstitial	Endothelial	Macrophages
Males						
1 week	224 ± 19	15 ± 1	18 ± 1	118 ± 9	71 ± 8	2 ± 0.5
6 weeks	460 ± 33	30 ± 4*	54 ± 3*	114 ± 7	253 ± 22*	9 ± 1
5 months	668 ± 16*	54 ± 1*	81 ± 7*	163 ± 10*	341 ± 9*	29 ± 7*
14 months	554 ± 27*	63 ± 6	59 ± 7	134 ± 10	284 ± 16*	14 ± 3*
26 months	527 ± 56**	59 ± 7	57 ± 8	130 ± 15	263 ± 26	17 ± 6
Females						
5 months	466 ± 34	39 ± 5	50 ± 7	133 ± 16	223 ± 11	20 ± 6
14 months	437 ± 12	48 ± 4	57 ± 5	120 ± 4	205 ± 8	8 ± 1
26 months	476 ± 12	49 ± 5	51 ± 4	137 ± 3	223 ± 8	17 ± 3

*p<0.05 for comparison to the immediately younger age group.
**p<0.05 for comparison of the 26-month group to the 5-month group.
[†]All values are mean ± SEM.

interconnected web in the adult cell (Figure 9). The meaning of such a shift is unknown but very similar changes occur in other phyla (such as insect flight muscle) and may accompany cell division. Although the notion is untested, it is easy to see how division of the mitochondria would be simplified by the fetal disconnected form. The adult mitochondria have only a small surface area-volume ratio advantage over the spheroid form, and it seems unlikely to be the entire advantage of that shape. The extended shape may allow an entire mitochondrion to respond to very focal cellular changes in high energy phosphates.

One more key change occurs in type II cell morphology during postnatal development. At birth, each type II pneumocyte has as many as 200 cytoplasmic extensions, which perforate the basement membrane, and these are termed foot processes. A fraction of these have close apposition with interstitial lipofibroblasts. Although there is an electron-dense cytoplasmic condensation, there is no intragap substructure on tilted sections to suggest a true gap junction. Occasional profiles from serial sections show cytoplasmic bridges between the epithelial type II cell and the mesenchymal lipofibroblast. The function of these three structures is unknown but is widely speculated to be related to epithelial mesenchymal interactions. They can be induced with corticosteroid administration in the perinatal period, and they decrease by 20-fold in number with maturation of the lung. The foot processes proliferate during reaction to injury resulting in interstitial pneumonitis in humans.

Over the life span of the Fischer 344 rat, the absolute volume of the cellular interstitium of the lung parenchyma does not change (Table 8). Interstitial cell number (Table 9) and cell size (Table 10) were found similar at 1 week

and 26 months of age, although the types and ratios of interstitial cell types forming the parenchyma cell pool were markedly different in neonatal pups compared with adult rats (Figure 10).[69] In contrast, the noncellular components of the interstitium demonstrated dramatic volume changes from 1 week through 5 months of age, increasing approximately 10-fold in total volume. From 5 to 26 months of age, interstitial volume continued to change at a lower, but significant rate (Table 8). Compared to 5-month-old rats, the interstitial matrix increased 18% by 14 months of age and 39% by 26 months of age (Figure 11). Females also demonstrated a significant increase in interstitial matrix volume from 81 mm at 5 months to 110 mm at 14 months, a 36% increase. By 26 months of age a further 89% increase in matrix volume was noted compared to 5 months of age. These changes could be a result of two different conditions: (a) pulmonary edema and (b) an increase in the noncellular matrix components of the lung, such as collagen or elastin.

In a study by Vincent and colleagues,[70] changes in the constituents of the interstitial matrix at the level of the proximal alveolar region (PAR) were examined in Fischer 344 rats. The PAR consists of alveolar tissue sampled in a perpendicular orientation relative to the once axis of the terminal bronchiole approximately 300–400 μm down the alveolar ducts. The overall characteristics of the alveolar septum in this region were similar to those of the more distal parenchyma. Morphometric analysis of the PAR revealed that the increase in interstitial matrix volume in the parenchyma could be attributed almost entirely to a thickening of basement membranes and the deposition of collagen fibers. Basement membrane thickness went from 40–45 nm at 4–6 months of age to 75–80 nm at 20–24 months.

TABLE 10 Morphometric characteristics of the major parenchymal cells in the aging Fischer 344 rat[4][†]

	Males					Females		
	1 week	6 weeks	5 months	14 months	26 months	5 months	14 months	26 months
Alveolar type I cells								
Average volume (μm³)	709 ± 120	2106 ± 121*	1530 ± 121*	1371 ± 151	1462 ± 156	2170 ± 508	1214 ± 125*	1385 ± 120
Average surface area (μm²)	2343 ± 145	8608 ± 998*	7612 ± 667	8457 ± 1122	8699 ± 685	10074 ± 1615	7431 ± 109	9523 ± 863
Average basement membrane surface area/cell (μm²)	2147 ± 101	7146 ± 679*	7287 ± 755	7478 ± 1270	7976 ± 702	9146 ± 1350	6779 ± 163	8528 ± 861
% of total alveolar cells	6.7	6.5	8.1	11.4	11.2	8.4	11.0	10.3
% of alveolar surface covered	91.3	96.1	96.4	98.6	98.6	97.6	98.3	98.6
Alveolar type II cells								
Average volume (μm³)	369 ± 60	401 ± 42	455 ± 108	526 ± 95	519 ± 178	541 ± 93	346 ± 27	352 ± 73
Average surface area (μm²)								
With microvilli	188 ± 20	292 ± 79	230 ± 41	223 ± 74	287 ± 218	347 ± 127	217 ± 73	217 ± 75
Without microvilli	183 ± 24	200 ± 55	185 ± 56	107 ± 33	124 ± 73	187 ± 64	97 ± 38	105 ± 38
% of total alveolar cells	8.0	11.7	12.1	10.6	10.8	10.7	13.0	10.7
% of alveolar surface covered	8.7	3.9	3.6	1.4	1.4	2.4	1.7	1.4
Interstitial cells								
Average volume (μm³)	513 ± 39	684 ± 67	427 ± 55*	452 ± 106	447 ± 51	456 ± 28	414 ± 40	411 ± 42
% of total alveolar cells	52.7	24.8	24.4	24.2	24.7	28.5	27.5	28.8
Endothelial cells								
Average volume (μm³)	321 ± 44	334 ± 13	275 ± 25	394 ± 31	377 ± 55	421 ± 37	344 ± 28	303 ± 25
Average surface area (μm²)	459 ± 32	1000 ± 32*	1121 ± 95	1618 ± 102*	1725 ± 142**	1467 ± 97	1560 ± 198	1604 ± 80
% of total alveolar cells	31.7	55.0	51.0	51.3	49.9	47.9	46.9	46.8
Macrophages								
Average volume	863 ± 143	672 ± 182	639 ± 131	1488 ± 427	1398 ± 513	1091 ± 196	821 ± 444	551 ± 107
% of total alveolar cells	0.9	2.0	4.3	2.5	3.2	4.3	1.8	3.6

*$p < 0.05$ for comparison to the immediately younger age group.
**$p < 0.05$ for comparison of the 26-month group to the 5-month group.
†All values are mean ± SEM.

Similarly, collagen fiber volume, normalized to the surface of the alveolar epithelium, increased by more than 100% during this same period. Basement membrane volume and collagen fiber volume were similar in the proximal alveolar regions and accounted for 50% of the noncellular matrix at 4–6 months of age and 80% at 20–24 months of age. By comparison, no change was noted in the relative volume of elastin and remaining acellular space. The volume ratio of collagen fibers to elastin fibers shifted from 3 to 5 in the young adults (4–6 months) to 10 in the older animals (20–24 months). Mercer and Crapo[71] found the same collagen to elastin fiber ratio in Sprague-Dawley rats as was seen in young Fischer 344 rats. Yamamoto et al.[54] found no significant change in collagen or elastin distribution and density in Fischer 344 rats.

It appears that the volume of the ground substance, relative to epithelial surface, is not significantly modified in the lungs of the aging Fischer 344 rats. Therefore, it is unlikely that edema contributes significantly to the increase in interstitial matrix volume with age. The absence of detectable

changes in the volume of elastin is consistent with its low rate of synthesis and slow turnover in adult animals (reviewed by[72,73]). A progressive thickening of basement membranes may be explained by epithelial and endothelial cell turnover if the cells lay down basement membrane material during each cell cycle, and possibly also during interphase. The increase in the volume of collagen fibers documented by morphometric approaches in adult rats confirms our previous morphologic observations,[12] and is in agreement with biochemical studies of collagen done in the lungs of Lewis rats[73,74] and Fischer 344 rats.[76]

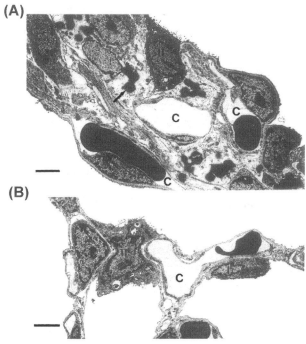

FIGURE 10   (A) An alveolar septum from the lungs of a 1-week-old Fischer 344 rat. The alveolar surface is covered by type I (I) and type II (II) epithelial cells. Note the presence of capillaries (C) on both sides of the septum and numerous lipid-containing cells (arrow) in the interstitium (bar is equal to 3 μm). (B) An alveolar septum from the lungs of a 6-week-old Fischer 344 rat (bar is equal to 3 μm).

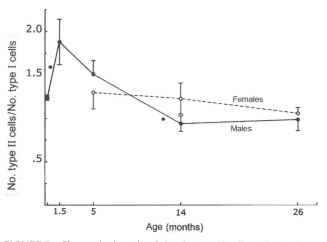

FIGURE 8   Changes in the ratio of alveolar type II cell number to alveolar type I cell number in the lungs of aging male and female Fischer 344 rats. Each datum point represents the mean ± SEM of four animals. An asterisk along the line connecting two age groups denotes a significant change in this ratio ($p < 0.05$).

FIGURE 9   Computer-assisted three-dimensional reconstruction of the mitochondria from an alveolar type II cell within the lungs of a 1-day old neonate (A) and an adult (B) Sprague-Dawley rat. The globular appearance of individual mitochondria in the neonate is in striking contrast to the filamentous branching mitochondria of the adult.[68]

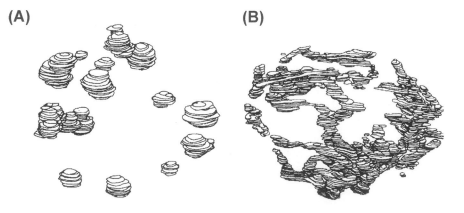

(A)

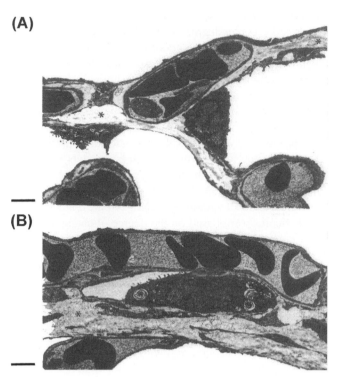

(B)

FIGURE 11   The alveolar septum from the lungs of male (A) and female (B) Fischer 344 rats 26 months of age. A prominent interstitial matrix space (*) is present in the septum. Some cytoplasmic extensions of interstitial cells are also present. Collagen is present within the matrix. In spite of the increase in the interstitial matrix, a thin air-to-blood tissue barrier is still maintained (bars are equal to 3 μm).

Morphologic and biochemical evidence supports the notion that there is a continuous deposition of mature cross-linked extracellular collagen in the lung parenchyma of aging rats that can be interpreted as an age-related excess of collagen (or fibrosis). However, it should be noted that qualitative examination revealed no visible change in the overall gross architecture of the matrix in older animals compared to younger rats. Although speculative at present, it is conceivable that the dominant stimulus for collagen fiber deposition, through life, is the stress and strain exerted on fibers and translated to fibroblasts. Following this concept, fibroblasts may add fibrils to the existing network while maintaining the general spatial relationship between collagen fibers and other tissue components and result in a net enrichment in collagen. It is not clear how this increase in collagen mass and the shift in the ratio of collagen to elastin with age impact on the micro mechanical behavior of the lungs. A logical assumption would be stiffer lungs, but physiologic measurements suggest otherwise.[12,67] Three-dimensional reconstructions in concert with physiologic measurements[71,77] could specifically address this issue in the senescent rat lung.

An interesting feature of senescent lungs in the Lewis rat observed by Mays et al.[75] was an abrupt four-fold decrease in the measured rate of gross collagen synthesis between 15 months and 24 months of age. More than 80% of newly synthesized collagen was apparently degraded intracellularly in 15-month-old animals and 60% degraded intracellularly at 24 months. The net result is only a small fraction of product being deposited as extracellular cross-linked collagen. Mays and associates suggest that maintenance of a high rate of collagen synthesis provides an adaptive capacity that allows the fibroblast to rapidly redirect the procollagen from intracellular degradation pathways toward secretory pathways in response to injury or pathologic insult. In the Fischer 344 rat there was no decline in fibroblast populations in aging lungs.[4] If the observations of Mays et al.[75] can be extended to other strains of rats, then a drop in the rate of collagen synthesis may constitute an intrinsic characteristic of the senescent lung fibroblast in vivo resulting in a compromised ability to repair collagen fibers.

If these assumed dynamics of the lung interstitium prove to be true, there are specific and profound metabolic changes that accompany aging that need to be understood to more fully appreciate the nature of the morphologic and functional alterations seen in normal aging and as a consequence of exposure to toxicants by the aged lung.[78] Acute toxicity studies are seldom performed in senescent animals, while chronic studies are usually not extended through the last one-third of the life span due to low survival rates. Therefore, our present understanding of age-related interstitial changes is severely limited.

The endothelium forms the third major tissue compartment of the alveolar septum. The volume of the endothelium in the lungs of Fischer 344 rats increased more than three-fold from 1 to 6 weeks of age. Beyond 6 weeks of age, total endothelial cell volume did not increase significantly through 26 months of age (Table 8). The total surface area of the capillary endothelium, like the epithelial surface, demonstrated nearly a 10-fold increase from 1 week to 6 weeks of age. From 6 weeks to 5 months of age capillary surface area increased by an additional 20%. From 5 months to 26 months of age the surface area of the capillary endothelium did not change significantly. Endothelial cell volume and surface area were relatively unchanged in male and female rats from 5 months to 26 months of age (Tables 7 and 8).

Alveolar Macrophages in the Aging Rat

Alveolar macrophages form an important defense against inhaled particulates and pathogens in the lungs. Without these cells, the sterility of the lungs would be severely compromised. Alveolar macrophages at 1 week of age numbered approximately 2 million cells and increased to 9 million cells by 6 weeks of age. Numbers were increased to 20 million cells by 5 months of age. Although no further increases in alveolar macrophage number were evident at 26 months of age (Table 10), this population of cells is highly dynamic, and numbers can change rapidly through the recruitment of monocytes and macrophages from the interstitium and the blood[79] and by the in situ proliferation of cells within the lung air spaces.[80]

Conclusions for Lung Aging in the Rat

The dynamics of lung growth, development, and aging in the rat are a continuous process that involves every tissue compartment of the lungs. Significant changes in aging adult rats are primarily within alveolar type II cells and the noncellular portions of the interstitium. Because such changes may influence the response of the lungs to inhaled chemical agents and dusts, the age of the rat should be considered in the evaluation of any experimental study. Different responses of the lungs to inhaled pollutants have been noted in young versus old rats.[77,81] Such differences may be due to changes within target cell populations and/or alterations in the functional status of cells through the aging process. Although cellular changes are most prominent during postnatal growth and development, modifications in cells continue through advanced age. Changes in cell number, size, and function associated with aging are likely to impact on lung physiology, metabolism, and immunity. Such changes could significantly alter the normal functions of the lung and its susceptibility to injury. Therefore, an understanding of the aging process in the rat is essential to the evaluation and interpretation of chronic (lifetime) exposures to a variety of substances that present a potential health risk to all mammalian species.

GENERAL CHARACTERISTICS OF THE LUNGS IN AGING DOGS

Morphological changes in the lungs during aging have been studied most extensively in the dog. As discussed earlier in this chapter, a strong allometric relationship exists between body mass and several features of the respiratory tract. Table 11 illustrates a number of these morphometric characteristics of the lungs for dogs of differing body mass and age.

Changes in alveolar size, enlargement of respiratory bronchioles and alveolar ducts, and the accumulation of anthracotic pigment are all common features of the aging process in the dog lung. Robinson and Gillespie[82] studied the aging process in the lungs of 20 beagle dogs, ranging from less than 1 year of age to 10 years of age to determine if a similar pattern could be noted in this species. They found that the primary lesion in older dogs was large accumulations of pulmonary macrophages containing dust and golden-brown pigment in the walls of respiratory bronchioles and at the mouth openings of alveoli into alveolar ducts. They observed that the dust-laden macrophages became more prominent with increasing age. Focal pneumonitis was also frequently associated with the accumulation of macrophages in the lungs. They further noted an increase in the volume of the lungs occupied by alveolar ducts. In

the airways, aging was associated with larger submucosal glands and a greater degree of calcification of bronchial cartilage. Due to the typical outdoor habitat of dogs, the accumulation of macrophages loaded with dusts and anthracotic pigment is to be expected with age. Although not studied, impaired clearance of particles may also occur with increasing age, leading to greater retention of pigment with age.

The relationship of each of these lung features with age were further analyzed by Robinson and Gillespie[82] using linear regression to determine which parameters were significantly associated with the aging process. Tissue sections taken from a total of 33 blocks from both lungs of each animal were examined without knowledge of the dog's identity or age. Scores were based on the degree of change relative to the youngest animals observed. A level of statistical significance was chosen as $p < 0.05$. Significant age-related changes were identified as increases in pigment surrounding respiratory bronchioles and alveolar duct mouth openings, enlargement of lumenal size of respiratory bronchioles and alveolar ducts, increased abundance of airway submucosal glands, and bronchial calcification of cartilage with age. Two features not changed with age were subpleural air space size and the degree of focal pneumonitis.[82]

The Tracheobronchial Tree of the Aging Dog

Little information is available to describe or quantify the aging process of the trachea or conducting airways in the lungs of dogs. It is known that body mass is proportional to the dimensions of the trachea in terms of its length to diameter ratio. In most mammalian species, this ratio is approximately 8:1. Compared to other species, the trachea and central airways of the dog are slightly larger due to their need of dissipating body heat by panting. The relatively large central airways also minimize impedance produced by rapid shallow breathing. Age-related alterations of the bronchi and the nonrespiratory bronchiole are unknown with the exception of age-related calcification of cartilaginous plates within the bronchi as well as hypertrophic changes of the submucosal glands.[82] No changes have been described for the nonrespiratory bronchioles of dogs with aging. The most peripheral airways in the lungs of dogs are formed by respiratory bronchioles and alveolar ducts.

Mauderly and Hahn[83] studied mucous velocity in the trachea of dogs 1 to 15 years of age. They found the velocity of mucous flow to increase in the trachea during the first few years of life, followed by a gradual age-related reduction in flow rates beginning around 4 years of age (Figure 11). With extrapolation of equivalent dog years to human years, a close correlation has been found between dogs and humans with similar patterns of decline in tracheal mucous velocity rates with increasing years during adult life. No information is available from this study to correlate changes in flow rates with cellular and/or functional parameters in epithelial cells lining the trachea.

Although extensive studies have been done to examine the branching pattern of the tracheobronchial tree in dogs,[84,85] only one study focused on postnatal growth. Horsfield examined resin casts of the bronchial tree from four Labrador dogs weighing 0.5, 3.4, 7.5, and 30.0 kg, respectively. The youngest dog was one week of age, while the ages of the remaining dogs were not given. Horsfield ordered the branching pattern of the lungs by number, mean diameter, and mean length of branches in each generation. These were plotted by order and expressed according to diameter and length. It was found that the diameter and length of similar airway orders ran in a parallel fashion for dogs of varying size. If these measurements were normalized to body weight, they were found to be identical. This correction factor was equal to the cube root of body weight. These observations by Horsfield were strong evidence that airway branching morphogenesis of the bronchial tree of the dog is complete at birth and postnatal growth reflects a simple expansion of the length and diameter of each airway generation in the absence of any further airway branching.

Parenchymal Lung Structure in the Aging Dog

To define morphological changes of the aging canine lung, Hyde and colleagues[6] studied the parenchymal characteristics of the lungs of 14 beagle dogs using morphometric analysis. These dogs were the same beagles studied by Robinson and Gillespie,[82] ranging from 1 to 10 years of age. Parenchymal tissues were analyzed to measure the volume and surface density of tissues and air. The volume of alveoli, alveolar ducts, and alveolar sacs (i.e., endings of the alveolar duct), as well as the number of alveoli per unit volume of lung parenchyma were analyzed. They found that the volume density of alveoli decreased with age, while the volume density of alveolar ducts and sacs increased with age. The number of alveoli per unit volume also decreased with age. No significant correlation was observed between the number density of alveolar ducts and alveolar sacs with age. Alveolar tissue density also decreased with age.

Multiple regression analysis was also used by Hyde et al.[6] to determine the relationship between stereological parameters of the lung parenchyma, body weight, age, and diffusion capacity. Analyses were repeated on the male and female populations as well as the total population. There were no significant multiple correlations among stereological lung parameters, diffusing capacity, body weight, and age. No sex-related differences in the slope of the regression lines relating to morphometric parenchymal lung values and age were noted.

Robinson and Gillespie[82] observed a significant increase in alveolar duct profile area. They observed minimal emphysema with only occasional signs of fibrosis associated with focal pneumonitis. These observations were confirmed by the morphometric measurements of Hyde and co-workers,[6] showing an increased volumetric density of alveolar ducts associated with decreases in the volumetric density of alveoli and parenchymal tissue, as well as decreases in the numerical density of alveoli, and the surface density of parenchymal tissues. These observations strongly correlate with a prevalence of lung hyperdistention or ductasia in the aging process for dogs. Similar findings have also been noted in the human lung, without a significant decrease in the number of alveoli. In ductasia, those alveoli adjacent to enlarged alveolar ducts in respiratory bronchioles decreased in depth, thus maintaining a constant number of alveoli within the lung.[6]

The observed decrease in surface density in the lungs of dogs with age can be explained by an increase in lung volume, a loss of interalveolar septa, a rearrangement of the geometry of the lung by alveolar flattening and duct enlargement, or a combination of any of these anatomical changes. Similar changes have also been observed in aged human lungs with a geometric rearrangement of alveoli in alveolar ducts, resulting in an increase in the average interalveolar septal distance. The total alveolar surface area for the lungs decreased as a result of an increase in the average interalveolar septal distance rather than an increase in lung volume. In these dogs, no significant change in total lung capacity with age was observed.[82] The reduction in volume density of alveolar parenchymal tissue could be interpreted as a loss of interalveolar septa. This type of reduction has also been observed in human lungs.

Cross-collateral ventilation of the lungs within the parenchymal regions is accomplished in part through communications formed by alveolar pores discussed earlier in the lungs of mice. A number of studies have examined the alveolar pores during aging in dogs.[6,33] The pores appear to be absent at birth, but become prominent within the first year of life.[33] The average number and size of the pores remain constant through the first 10 years of life (Figure 12). Prolonged exposure to a number of environmental air pollutants has been shown to dramatically increase the number and size of the pores,[6] with extensive fenestration of alveolar walls.

Conclusions for the Aging Dog Lung

Aging of canine lungs has many similarities to that seen in human lungs. The aging process in dogs occurs over a shorter time frame than that for humans. The accumulations of dust-laden macrophages as well as lumenal enlargement of alveolar ducts are hallmarks of this aging process. Although some tissue thinning and loss of alveolar septal tissues may occur, the increase in lumenal size of alveolar ducts is in large measure due to the stretching and shallowing of alveolar outpocketings within these regions of

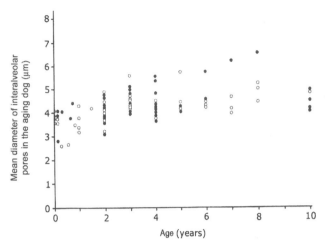

FIGURE 12   Average diameter of pores plotted against age. Each point represents one dog. Closed circles are males and open circles are females.

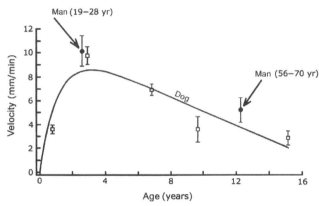

FIGURE 13   Tracheal mucous velocity (mean and SEM) for beagle dogs versus age. The fitted function that describes this relation is $V(t) = 11\ [1/2 - \exp(-0.9t)] - 0.6t$. The available data for humans are also shown after transforming for age, as described by Mauderly and Hahn.[83]

the lung parenchyma. Changes in the relative velocity of mucous flow in the trachea with aging also parallel changes observed in the aging human trachea (Figure 13). Little is known regarding the canine aging process at the cellular level for either the airways or gas exchange portions of the lungs. However, a primary advantage of studying dogs is the extensive similarity of this species with that of the human lung in composition and structure. These similarities may offer unique opportunities to better understand the aging process of the lungs for both dogs and humans.

GENERAL CHARACTERISTICS OF THE LUNGS IN AGING RHESUS MONKEY

Despite ongoing longitudinal studies of aging being conducted at the National Institute of Aging, the Wisconsin National Primate Research Center, and the California National Primate Research Center, very little is known

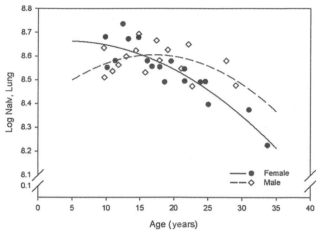

FIGURE 14   Number of alveoli in female (solid dots) versus male (open diamonds) Rhesus macaque monkeys with increasing age. Females demonstrate a significant decrease during the aging process compared with males. The lines are plotted as a log quadratic function of the log number of alveoli in the lung (Nalv, lung) vs. age (yr) for females (solid line) and males (dotted line). Reproduced by permission, the American Physiological Society (AJP:Lung).

about changes in the rhesus monkey lung regarding senescence. Rhesus monkeys are potentially an excellent model for lung senescence in humans because they have similar segmental arrangement, structure and branching of airways, arterial structure, and arterial changes after birth.[14] However, they are a long-lived species with a median life span of 25 years, which makes them an investment in time, resources, and effort for studying senescence.[86]

Morphometric analysis on postnatal lung growth was performed up to 7 years of age in both males and females of rhesus monkeys by Hyde et al.[14] Gender differences were noted in number of alveoli, alveolar surface area, and size of alveoli. Based on body weight, females were found to have more and greater alveolar surface area, which is also seen in rodents, although lung volume was the best indicator of alveolar number.[14]

Lung Structure in the Aging Rhesus

The number of alveoli in the aging rhesus monkey decreases at approximately 10–15 years of age and continues to decrease until death (Figure 14). Females have been shown to decrease initially at a faster rate than males and have a more significant loss of alveoli during senescence.[11] The loss of alveoli did not correlate with any changes in lung volume; subsequently, there was an increase in alveolar duct air and a decrease in alveolar air, indicating shallower and wider alveoli during senescence.

Collagen and elastin content had an inverse relationship during senescence in the rhesus monkey. Collagen volume in interalveolar septal tissue increased while the volume of elastin decreased. Volume of arteries and veins in the lung also increased during the aging process in the lungs of these rhesus monkeys. It is unclear if the volume increases were due primarily to the wall or the lumen increasing.[11]

Conclusions of the Aging Rhesus Lung

A considerable amount of research remains to be done regarding morphological changes in the rhesus monkey lung during the aging process. Sex differences in aging set this species apart from rodents, which may shed light on the sexual disparities of certain human lung diseases associated with aging, such as chronic obstructive pulmonary disease (COPD). With so few centers offering the capacity for longitudinal studies of aging nonhuman primate lungs, this research remains at a disadvantage compared to other shorter-lived species. However, the recompense for the research is that it is potentially a better model for human lung aging than other species.[86]

OVERALL CONCLUSIONS

Aging is a natural process in the respiratory system. A number of similarities have been noted in all species associated with aging. An increase in the total alveolar airspace volume is a natural consequence of aging with enlargement of alveolar ducts immediately beyond the terminal and respiratory bronchioles. This enlargement is typically seen in the form of ductasia with stretching and shallowing of alveoli in affected regions. Although some destruction of alveolar tissue walls may be evident in the aging lung, such changes appear to be due to an increase in the size and frequency of alveolar pores connecting adjacent alveoli. An increase in the number of phagocytic cells in the lung airspace is also a common finding, with less efficient phagocytic properties. Increased collagen deposition with focal increases in the interstitium also appears to be a consequence of the aging process. Metabolic functions of cells are likely to be compromised in the lungs as aging advances. A number of environmental factors may accentuate all these changes during aging, with the greatest consequences being asthma, emphysema, and COPD. These topics will be more fully covered in Chapter 27.

REFERENCES

1. Burri PH, Dbaly J, Weibel ER. The postnatal growth of the rat lung: I. Morphometry. *Anat Rec* 1974;**178**:711–30.
2. Massaro GD, Davis L, Massaro D. Postnatal development of the bronchiolar Clara cell in rats. *Am J Phys* 1984;**247**:C197–203.
3. Massaro GD, Massaro D. Development of bronchiolar epithelium in rats. *Am J Phys* 1986;**250**:R783–8.
4. Pinkerton KE, Barry BE, O'Neil JJ, Raub JA, Pratt PC, Crapo JD. Morphologic changes in the lung during the life span of Fischer 344 rats. *J Anat* 1982;**164**:155–74.
5. Ranga V, Kleinerman J, Sorensen J. Age-related changes in elastic fibers and elastin of lung. *Am Rev Respir Dis* 1979;**119**:369–76.
6. Hyde D, Orthoefer J, Dungworth D, Tyler W, Carter R, Lum H. Morphometric and morphologic evaluation of pulmonary lesions in beagle dogs chronically exposed to high ambient levels of air pollutants. *Lab Invest* 1978;**4**:455–69.
7. Coleman GL, Barthold SW, Osbaldistan GW, Foster SJ, Jonas AM. Pathological changes during aging in barrier-reared Fischer 344 rats. *J Gerontol* 1977;**32**:258–78.
8. Goodman DG, Ward JM, Squire RA, Chu KC, Linhart MS. Neoplastic and nonneoplastic lesions in aging F344 rats. *Toxicol Appl Pharmacol* 1979;**48**:237–48.
9. Pinkerton KE, Gehr P, Crapo JD. Architecture and cellular composition of the air-blood barrier. In: Parent RA, editor. *Treatise on Pulmonary Toxicology, Comparative Biology of the Normal Lung*, vol I. Boca Raton, FL: CRC Press; 1992. pp. 121–8.
10. Boyden EA, Thompsett DH. The postnatal growth of the lung in the dog. *Acta Anat* 1961;**47**:185–215.
11. Herring MJ, Avdalovic MV, Quesenberry CL, Putney LF, Tyler NK, Ventimiglia FF, et al. Accelerated structural decrements in the aging female rhesus macaque lung compared with males. *Am J Physiol Lung Cell Mol Physiol* 2013;**304**:L125–34.
12. Tyler WS, Julian MD, editors. *Gross and Subgross Anatomy of Lungs, Pleura, Connective Tissue Septa, Distal Airways, and Structural Units*. Boca Raton, FL: CRC Press; 1991.
13. McBride JT. Architecture of the tracheobronchial tree. In: *Treatise on Pulmonary Toxicology: Comparative Biology of the Normal Lung*. Boca Raton, FL: CRC Press; 1992. pp. 49–61.
14. Hyde DM, Blozis SA, Avdalovic MV, Putney LF, Dettorre R, Quesenberry NJ, et al. Alveoli increase in number but not size from birth to adulthood in rhesus monkeys. *Am J Physiol Lung Cell Mol Physiol* 2007;**293**:L570–9.
15. Narayanan M, Owers-Bradley J, Beardsmore CS, Mada M, Ball I, Garipov R, et al. Alveolarization continues during childhood and adolescence: new evidence from helium-3 magnetic resonance. *Am J Respir Crit Care Med* 2012;**185**:186.
16. Sacher GA, Hart RW. Longevity, aging and comparative cellular and molecular biology of the house mouse (Mus musculus) and the white-footed mouse (Peromyscus leucepus) birth defects. *Original article series* 1978;**14**:71–96. March of Dimes Foundation, New York.
17. Banfield AWF. *The Mammals of Canada*. Toronto: University of Toronto Press; 1974. p. 438.
18. Burger J, Gochfield M. Survival and reproduction in Peromyscus leucopus in the laboratory: viable model for aging studies. *Growth Dev Aging* 1992;**56**:17–22.
19. Takeda T, Hosokawa M, Takeshita S, Irino M, Higuchi K, Matsushita T, et al. A new murine model of accelerated senescence. *Mech Ageing Dev* 1981;**17**:183–94.
20. Kuro-o M, Matsumura Y, Aizawa H, Kawaguchi H, Suga T, Utsugi T, et al. Mutation of the mouse klotho gene leads to a syndrome resembling ageing. *Nature* 1997;**390**:45–51.
21. Kawakami M, Paul JL, Thurlbeck WM. The effect of age on lung structure in male BALB/cNNia inbred mice. *Am J Anat* 1984:170–1.
22. Ranga V, Kleinerman J. Interalveolar pores in mouse lungs. *Am Rev Respir Dis* 1980;**122**:477–81.
23. Hosokawa M, Takeshita S, Higuchi K, Shimizu K, Irino M, Toda K, et al. Cataract and other ophthalmic lesions in senescence accelerated mouse (SAM): morphology and incidence of senescence accelerated ophthalmic changes in mice. *Exp Eye Res* 1984;**38**:105–14.

24. Takeshita S, Hosokawa M, Irino M, Higuchi K, Shimizu K, Yasuhira K, et al. Spontaneous age-associated amyloidosis in senescence accelerated mouse (SAM). *Mech Ageing Dev* 1982;**26**:91–102.

25. Matsushita M, Tsuboyama T, Kasai R, Okumura H, Yamamuro T, Higuchi K, et al. Age-related changes in bone mass in the senescence accelerated mouse (SAM): SAM-R/3 and SAM-P/6 as new murine models for senile osteoporosis. *Am J Pathol* 1986;**125**: 276–83.

26. Kurozumi M, Matsushita T, Hosokawa M, Takeda T. Age-related changes in lung structure and function in the senescence-accelerated mouse (SAM): SAM-P/I as a new murine model of senile hyperinflation of lung. *Am J Respir Crit Care Med* 1994;**149**: 776–82.

27. Uejima Y, Fukuchi Y, Nagase T, Tabata R, Orimo H. A new murine model of aging lung: the senescence accelerated mouse (SAM)-P. *Mech Ageing Dev* 1991;**61**:223–36.

28. Glassberg MK, Choi R, Manzoli V, Shahzeidi S, Rauschkolb P, Voswinckel R, et al. 17β-estradiol replacement reverses age-related lung disease in estrogen-deficient C57BL/6J mice. *Endocrinology* 2013;**155**:441–8.

29. Scherle WA. A simple method for volumetry of organs in quantitative stereology. *Mikroskopie* 1970;**26**:57–60.

30. Pack RJ, Al-Ugaily LH, Morris G. The cells of the tracheobronchial epithelium of the mouse: quantitative light and electron microscope study. *J Anat* 1981;**132**:71.

31. Ishii M, Yamaguchi Y, Yamamoto H, Hanaoka Y, Ouchi Y. Airspace enlargement with airway cell apoptosis in klotho mice: a model of aging lung. *J Gerontol A Biol Sci Med Sci* 2008;**63**:1289–98.

32. Sato A, Hirai T, Imura A, Kita N, Iwano A, Muro S, et al. Morphological mechanism of the development of pulmonary emphysema in klotho mice. *Proc Natl Acad Sci USA* 2007;**104**:2361–5.

33. Martin H. The effect of aging on the alveolar pores of Kohn in the dog. *Am Rev Respir Dis* 1963;**88**:773–8.

34. Amy RWM, Bowes D, Burri PH, Haines J, Thurlbeck WM. Postnatal growth of the mouse lung. *J Anat* 1977;**124**:131–51.

35. Pump K. Emphysema and its relation to age. *Am Rev Respir Dis* 1976;**114**:5–13.

36. Huang K, Rabold R, Schofield B, Mitzner W, Tankersley CG. Age-dependent changes of airway and lung parenchyma in C57BL/6J mice. *J Appl Phys* 2006;**102**:200–6.

37. Sueblinvong V, Neujahr DC, Mills ST, Roser-Page S, Ritzenthaler JD, Guidot D, et al. Predisposition for disrepair in the aged lung. *Am J Med Sci* 2012;**344**:41.

38. Higashimoto Y, Fukuchi Y, Shimada Y, Ishida K, Ohata M, Furuse T, et al. The effects of aging on the function of alveolar macrophages in mice. *Mech Aging Dev* 1991;**69**:207–17.

39. Aoshiba K, Nagai A. Chronic lung inflammation in aging mice. *FEBS Lett* 2007;**581**:3512–6.

40. Renshaw M, Rockwell J, Engleman C, Gewirtz A, Katz J, Sambhara S. Cutting edge: impaired Toll-like receptor expression and function in aging. *J Immunol* 2002;**169**:4697–701.

41. Boehmer ED, Goral J, Faunce DE, Kovacs EJ. Age-dependent decrease in Toll-like receptor 4-mediated proinflammatory cytokine production and mitogen-activated protein kinase expression. *J Leukoc Biol* 2004;**75**:342–9.

42. Boyd AR, Shivshankar P, Jiang S, Berton MT, Orihuela CJ. Age-related defects in TLR2 signaling diminish the cytokine response by alveolar macrophages during murine pneumococcal pneumonia. *Exp Gerontol* 2012;**47**:507–18.

43. Ito Y, Betsuyaku T, Moriyama C, Nasuhara Y, Nishimura M. Aging affects lipopolysaccharide-induced upregulation of heme oxygenase-1 in the lungs and alveolar macrophages. *Biogerontology* 2009;**10**:173–80.

44. Chen TS, Richie Jr JP, Lang CA. Life span profiles of glutathione and acetaminophen detoxification. *Drug Metab Dispos Biol Fate Chem* 1990;**18**:882–7.

45. Boorman GA, Eustis SL. Lung. In: Boorman GA, Eustis SL, Elwell MR, Montgomery CA, MacKenzie WF, editors. *Pathology of the Fischer Rat.* New York: Academic; 1990. pp. 339–67.

46. Chesky JA, Rockstein M. Life span characteristics in the male Fischer rat. *Exp Aging Res* 1976;**2**:399–407.

47. Jacobs BB, Huseby RA. Neoplasms occurring in aged Fischer rats with special reference to testicular, uterine, and thyroid tumors. *J Natl Cancer Inst* 1967;**39**:303–9.

48. Massaro EJ. Mortality and growth characteristics of rat strains commonly used in aging research. *Exp Aging Res* 1980;**6**:219–33.

49. Mauderly JL, Likens SA. Relationships of age and sex to function of Fischer 344 rats. *Fed Proc* 1980;**39**:10–91.

50. Rockstein M, Chesky JA, Sussman ML. Comparative biology and evolution of aging. In: Finch CE, Hayflick L, editors. *Handbook of the Biology of Aging.* New York: Van Nostrand Reinhold; 1977. pp. 3–34.

51. Sass B, Rabstein LS, Madison R, Nims RM, Peters RL, Kelloff GJ. Incidence of spontaneous neoplasms in F344 rats throughout the natural life-span. *J Natl Cancer Inst* 1975;**54**:1449–56.

52. Snell KC. Spontaneous lesions of the rat. In: Ribelin WE, McCoy JR, editors. *The Pathology of Laboratory Animals.* Springfield: Thomas; 1965. pp. 211–302.

53. Boorman GA, Morgan KT, Uriah LC. Nose, larynx and trachea. In: Boorman GA, Eustis SL, Elwell MR, Montgomery CA, MacKenzie WF, editors. *Pathology of the Fischer Rat.* New York: Academic; 1990. pp. 315–37.

54. Yamamoto Y, Tanaka A, Kanamaru A, Tanaka S, Tsubone H, Atoji Y, et al. Morphology of aging lung in F344/N rat: alveolar size, connective tissue, and smooth muscle cell markers. *Anat Rec A Discov Mol Cell Evol Biol* 2003;**272**:538–47.

55. Jeffery PK, Reid LM. Ultrastructure of airway epithelium and submucosal glands during development. In: Hodson, editor. *Development of the Lung.* New York: Dekker; 1977. pp. 87–134.

56. Plopper CG, Mariassy AT, Wilson DW, Alley JL, Nishio SJ, Nettesheim P. Comparison of nonciliated tracheal epithelial cells in six mammalian species: ultrastructure and population densities. *Exp Lung Res* 1983;**5**:281–94.

57. Chang L, Mercer RR, Crapo JD. Differential distribution of brush cells in rat lung. *Anat Rec* 1986;**216**:49–54.

58. Cireli E. Elektronenmikroskopische Anaivse der priiund postnatalen Differenzierung des Epithels der oberen Luftwege der Ratte. *Z Mikrosk Anat Forsch* 1966;**41**:132–78.

59. Kober HJ. Die lumenseitige Oberfl@che der Rattentrachea wiihrend der Ontogenese. *Z Mikrosk Anat Forsch* 1975;**89**:399–409.

60. Pinkerton KE, Weller BL, Menaché MG, Plopper CG. Part XIII. A comparison of changes in the tracheobronchial epithelium and pulmonary acinus in male rats at 3 and 20 months. *Health Effects Inst Res Rep Number* 1998;**85**.

61. Crapo JD, Peters-Golden M, Marsh-Salin J, Shelburne JS. Pathologic changes in the lungs of oxygen-adapted rats. A morphometric analysis. *Lab Invest* 1978;**39**:640–53.

62. Takezawa J, Miller FJ, O'Neil JJ. Single-breath diffusing capacity and lung volumes in small laboratory mammals. *J Appl Phys* 1980;**48**:1052–9.

63. Hayatdavoudi G, Crapo JD, Miller FJ, O'Neil JJ. Factors determining degree of inflation in intratracheally fixed rat lungs. *J Appl Phys* 1980;**48**:389–93.

64. Randell SH, Mercer RR, Young SL. Postnatal growth of pulmonary acini and alveoli in normal and oxygen-exposed rats studied by serial section reconstructions. *Am J Anat* 1989;**186**:55–68.

65. Burri PH. The postnatal growth of the rat lung, III: Morphology. *Anat Reconstructionist* 1974;**180**:77–98.

66. Liebow AA. Summary: biochemical and structural changes in the aging lung. In: Cander L, Moyer JH, editors. *Aging of the Lung*. New York: Grune and Stratton; 1964. pp. L113–22.

67. Mauderly JL. Effect of age on pulmonary structure and function of immature and adult animals and man. *Fed Proc* 1979;**38**:173–7.

68. Young SL, Spain CL, Fram EK, Larson E. Development of type II pneumocytes in the rat lung. *Am J Physiol Lung Cell Mol Physiol* 1991;**260**:L113–22.

69. Brody JS, Kaplan NB. Proliferation of alveolar intersitial cells during postnatal lung growth. Evidence for two distinct populations of pulmonary fibroblasts. *Am Rev Respir Dis* 1983;**127**:763–70.

70. Vincent R, Mercer RR, Chang LY, Pinkerton KE, Crapo JD. Morphometric study of interstitial matrix in the lungs of the aging rat. *FASEB J* 1990;**4**:AI915 (abstr).

71. Mercer RR, Crapo JD. Spatial distribution of collagen and elastin fibers in the lungs. *J Appi Physiol* 1990;**69**:756–65.

72. Rucker RB, Dubick MA. Elastin metabolism and chemistry: potential roles in lung development and structure. *Environ Health Perspect* 1984;**55**:179–91.

73. Foster JA, Curtiss SW. The regulation of lung elastin synthesis. *Am J Physiol* 1990;**259**:L13–23.

74. Mays PK, Bishop JE, Laurent GJ. Age-related changes in the proportion of types I and III coliagen. *Mech Aging Dev* 1988;**45**:203–12.

75. Mays PK, McAnulty RJ, Laurent GJ. Age-related changes in lung collagen metabolism. A role for degradation in regulating lung coliagen production. *Am Rev Respir Dis* 1989;**140**:410–6.

76. Sahebjami H. Lung tissue elasticity during the lifespan of Fischer 344 rats. *Exp Lung Res* 1991;**17**:887–902.

77. Mercer RR, Crapo JD. Three-dimensional reconstruction of alveoli in the rat lung for pressure-volume relationships. *J Appi Physiol* 1987;**62**:1480–7.

78. Stiles J, Tyler WS. Age-related morphometric differences in responses of rat lungs to ozone. *Toxicol Appl Pharmacol* 1988;**92**:274–85.

79. Brain JD, Sorokin S, Godieski IJ. Quantification, origin, and fate of pulmonary macrophages. In: Brain JD, Proctor DF, Reid LM, editors. *Lung Biol Health Dis Respir Defense Mech*, vol. 5. New York: Dekker; 1977. p. 849. part 11.

80. Shellito J, Esparza C, Armstrong C. Maintenance of the normal rat alveolar macrophage cell population. The roles of monocyte influx and alveolar macroph- age proliferation in situ. *Am Rev Respir Dis* 1987;**135**:78–82.

81. Tyler WS, Tyler NK, Last JA, Barstow TI, Magliano DJ, Hinds DM. Effects of ozone on lung and somatic growth. Pair fed rats after ozone exposure and recovery periods. *Toxicology* 1987;**46**:1–20.

82. Robinson NE, Gillespie JR. Morphologic features of the lungs of aging dogs. *Am Rev Respir Dis* 1973;**108**:1192–9.

83. Mauderly JL, Hahn FF. The effects of age on lung function and structure of adult animals. *Adv Vet Med* 1982;**26**:35–77.

84. Horsefield K, Cumming G. Morphology of the bronchial tree in the dog. *Respir Physiol* 1976;**26**:173–82.

85. Raabe OG, Yeh HC, Schum GM, Phalen RM. *Tracheobronchial Geometry: Human, Dog, Rat, Hamster*. Albuquerque, NM: Lovelace Foundation; 1976.

86. Roth GS, Mattison JA, Ottinger MA, Chachich ME, Lane MA, Ingram DK. Aging in rhesus monkeys: relevance to human health interventions. *Science* 2004;**305**:1423–6.

Chapter 15

Cell-Based Strategies for the Treatment of Injury to the Developing Lung

Megan O'Reilly* and Bernard Thébaud†,‡

*Department of Pediatrics and Women and Children's Health Research Institute, University of Alberta, Edmonton, AB; †Sprott Centre for Stem Cell Research, Ottawa Hospital Research Institute, Ottawa, ON; ‡Division of Neonatology, Department of Pediatrics, Children's Hospital of Eastern Ontario (CHEO) and CHEO Research Institute, Ottawa, ON, Canada

INTRODUCTION

Advances in perinatal and neonatal care have led to improved survival following preterm birth, with infants born as early as 23–24 weeks of gestation now being capable of survival. However, most very and extremely preterm infants require prolonged respiratory support to ensure survival, and this can result in lung injury and the development of chronic lung disease of prematurity known as bronchopulmonary dysplasia (BPD). Therefore, the task of protecting the immature lung from injury has become increasingly challenging. Recent evidence demonstrates that BPD has life-long respiratory complications. Ex-preterm children, adolescents, and adults who developed BPD commonly present with evidence of pulmonary complications such as cough, wheeze, poor lung function, and reduced exercise capacity. Furthermore, survivors of moderate–severe BPD are likely to present with emphysema in early adulthood (17–33 years of age). However, currently there is no effective treatment for BPD. Such a treatment should not only target the repair of lung injury, but should also promote normal lung growth and function, leading to improved lung function and respiratory health in ex-preterm children and adults. Exciting discoveries in stem cell biology over recent years offer new insights into the pathogenesis of BPD and, more importantly, open new therapeutic avenues.

Stem Cell Biology: Understanding the Basics

Stem cells are primitive cells capable of extensive self-renewal and have the potential to give rise to multiple differentiated cellular phenotypes.[1] Not only are they critical for organogenesis and growth during the early stages of development, but they also contribute to organ repair and regeneration throughout life. Stem cells exhibit varying differentiation potencies and are typically categorized into embryonic and adult stem cells. The potency of a stem cell refers to the range of possible fates open to cells during differentiation.

Embryonic Stem Cells (ESCs)

ESCs are derived from the early blastocyst and represent the most potent of stem cells due to their pluripotency (i.e., their ability to differentiate into cell types derived from all three germinal lineages: endoderm, mesoderm, or ectoderm) and their ability for indefinite self renewal.

Adult Stem Cells

Adult stem cells, or ASCs, are also called somatic stem cells. In contrast to ESCs, ASCs have increasing degrees of fate restriction and are either multipotent (i.e., differentiate into a limited range of cell types) or unipotent (i.e., can generate only one cell type).[2] Residual pools of multi- or unipotent ASCs are thought to reside in almost all adult organs and are considered important for tissue repair and maintenance throughout adulthood. ASCs have the ability to contribute to tissue repair and regeneration by repopulation during growth, injury, or disease. Highly proliferative tissues, such as the intestinal epithelium or the hematopoietic compartment of the bone marrow, rely on a pool of ASCs that are organized in a classical hierarchy to maintain homeostasis. In a classical stem cell hierarchy, ASCs give rise to transit-amplifying cells, which then produce terminally differentiated cells.[3] In contrast, anatomically complex tissues that have a slow turnover (i.e., lung, brain, heart, kidney) do not appear to support a classical stem cell hierarchy, but rather are maintained by cell populations organized in a non-classical hierarchy. The ASCs in a non-classical hierarchy do not typically participate in normal tissue maintenance, but can be activated to participate in repair after progenitor cell depletion. The

The Lung. http://dx.doi.org/10.1016/B978-0-12-799941-8.00015-8

ASCs give rise to transit-amplifying cells, which then produce terminally differentiated cells or facultative progenitor cells (i.e., cells with proliferative capacity but functional properties of differentiated cell types in a quiescent state).[3]

ENDOGENOUS LUNG STEM/PROGENITOR CELLS

In the lung, several local epithelial cell types function as both differentiated functional cells and transit-amplifying progenitors that proliferate in response to airway or alveolar injuries.[4] Currently, the localization and properties of lung stem cell niches and the type of cells within each niche are of major interest, yet also present controversy. The complexity of the lung architecture combined with an extensive diversity of lung cell types and niches has hindered the identification of true lung stem cells. Recent insights into the basic concepts of lung stem cell biology have been provided by several investigators.[5–18] These studies have demonstrated that the adult lung harbors rare populations of multipotent endogenous stem cells that are tightly regulated by specific micro-environmental cellular niches and can be putatively recruited to repopulate the injured lung (summarized in Table 1). Although technological advances have improved isolation and characterization methods for the identification of putative resident lung stem cells in animals and humans, uncertainty remain as to whether the stem cells identified by different studies are the same. Techniques that have been utilized by hematologists for many years have also been applied to identify lung stem cells.[7,8] In contrast, others have utilized genetic techniques, such as lineage tracing, to identify similar lung stem cells.[9,10] Combining these two approaches could enhance identification of distal lung progenitor(s).[18]

Distal Lung Epithelial Stem/Progenitor Cells

In the distal lung it is speculated that multiple niches support different cell populations and their progenitors.

Clara Cells

A subset of the Clara cell that exhibits stem cell characteristics, termed the variant Clara cell (Clara[v]), has been located within two different niches: in the neuroendocrine bodies (NEBs) and at the bronchoalveolar duct junctions (BADJs).[13,14] A third niche, the parabronchial smooth muscle cells (PSMCs) has also been proposed to harbor Clara[v] cells.[15]

Bronchioalveolar Stem Cells (BASCs)

Multipotent stem cells in the distal lung capable of differentiating into epithelial cells specific to the bronchioles and the alveoli have been identified. The dual-lineage BASCs located at the BADJ niche express both bronchiolar

(CCSP) and alveolar (SP-C) markers and proliferate in response to airway and alveolar injury.[6] However, based on the employed techniques, there has been some ambiguity regarding the lineage potential[19,20] and contribution of these cells to alveolar repair.[10]

Alveolar Type II (AT2) Pneumocytes

Cuboidal AT2 pneumocytes have long been considered progenitors of the alveolar epithelium based on their capacity to replenish themselves and generate terminally differentiated AT1 pneumocytes.[21,22] Since then, AT2 pneumocytes have been speculated to contain a subpopulation of progenitors cells that can undergo reactivation into a progenitor-like state in response to injury cues.[16,23–26]

Lung Endothelial Progenitor Cells (EPCs)

EPCs are a population of circulating and resident vascular precursor cells. Assessment of the contribution of endogenous lung EPCs in lung vascular repair and lung regeneration and remodeling is often hampered by their rarity, lack of distinguishing markers and the inability to discriminate circulating EPCs and resident EPCs.[27] However, Alvarez et al. showed that the lung microvasculature is enriched with a population of EPCs, termed resident microvascular endothelial progenitor cells, that were shown to be highly proliferative and capable of renewing the entire hierarchy of endothelial cell growth potentials.[17]

Mesenchymal Stromal Cells (MSCs)

MSCs are multipotent cells derived from mesenchymal tissues, and are often referred to by their tissue of origin. MSCs can be easily sourced from bone marrow (BM), adipose tissue, muscle, peripheral blood, umbilical cord blood (UCB), umbilical cord (UC), Wharton's jelly, and placenta. These cells possess the ability to differentiate and form various mesenchymal cell types, including bone, cartilage, and adipose cells. For these reasons, MSCs appear as an attractive candidate for regenerative tissue applications, and have thus been widely utilized in pre-clinical studies of neonatal lung injury (as discussed in the subsequent sections).

Characterization of MSCs

Identification of MSCs in vitro requires compliance with a specific set of characteristics.[28] First and foremost, MSCs are plastic-adherent, meaning that when cultured they possess the ability to readily adhere to plastic culture dishes and form fibroblast-like colonies. Secondly MSCs can be induced into various specialized cell lineages, including adipogenic, osteogenic, chondrogenic, myogenic, and neurogenic-like lineages. Finally, MSCs can be characterized by their expression of defined cell surface marker profiles.

TABLE 1 Stem/progenitor cells in the lung

Category	Stem Cell	Attributed Differentiated Phenotype	Niche/location	Defining Characteristics	References
Distal lung epithelial stem cells	Clara^V cells	Distal airway epithelium	BADJs, NEBs, and PSMCs	Express CCSP; survives and repopulates distal airway epithelium following naphthalene injury; dependent on paracrine signaling of Fgf10	13–15
	BASCs	Bronchioalveolar epithelial cells	BADJs	Resistant to naphthalene injury and proliferate in response; co-express CCSP and SP-C	6
	p63+ bronchiolar progenitor cells	Alveolar epithelium	Bronchioles	Krt5+ pods, proliferative	25
	Id2+ distal epithelial cells	All epithelial cell lineages, including neuroendocrine cells	Distal tip of the embryonic lung	Express T1α, β-tubulin, Scgb1a1, CGRP, and pro-SP-C, contribute to all epithelial lineages in pseudoglandular stage but only contribute to alveoli in later lung developmental stages	9
	Multipotent lung epithelial progenitors	Airway and alveolar epithelium	No specific location	EpCAM^{high}, CD49f+, CD104+, CD24^{low}, Sca-1-, CD45-, CD31- lung epithelial cfus, form colonies in Matrigel, serially passaged and retain multipotent potential	8
	Alveolar type-II pneumocytes	Alveolar type-I pneumocytes	Alveolar surface	All alveolar type-II pneumocytes	21
	A subset of alveolar type-II pneumocytes	Alveolar type-I and mature type-II pneumocytes	Alveolar surface	E-cadherin negative subset of alveolar type-II cells, proliferative, high telomerase activity, resistant to oxygen-induced injury, laminin receptor α6β4+	16,23,24
Lung MSCs	SP cells	Epithelial, vascular, endothelial, hematopoietic, mesenchymal potential	No specific location	Efflux of Hoechst dye, CD45-, c-kit-, CD11b-, CD34-, CD14-, CD44+, CD90+, CD105+, CD106+, CD73+, Sca-1+, high telomerase expression	30,31
	Multipotent MSCs	Osteocytes, chondrocytes, adipocytes, fibroblasts, smooth muscle cells	No specific location	Express vimentin, collagen IV and laminin, CD44+, CD73+, CD90+, CD105+, CD106+, express ICAM-1	74
	Fibroblastic progenitor cells	Fibroblasts, chondrocytes, osteocytes, lipofibroblasts (Thy-1^{high} cells)	No specific location	CD45-, CD31-, Sca-1+, CD34+, coexpress immunophenotypic markers (Thy-1, platelet-derived growth factor)	7
Lung EPCs	Resident microvascular EPCs	Endothelial cells	Pulmonary microcirculation	CD31+, CD144+, CD34+, CD309+, CD45-, express endothelial nitric oxide synthase and von Willebrand factor, form vascular networks in Matrigel	17

Acronyms: CCSP: Clara cell secretory protein; Fgf: Fibroblast growth factor; SP-C: Surfactant protein C; Krt: Keratin; T1α: Alveolar Type I cell marker; Scgb1a1: Secretoglobin family 1A member 1(also known as CCSP); CGRP: Calcitonin gene related peptide; EpCAM: Epithelial cell adhesion molecule; cfus: colony forming units.

Positive expression of the markers CD73, CD90, and CD105 are a requirement for MSCs. Furthermore, MSCs must be negative for the markers CD11b and CD14 (marker of monocytes and macrophages), CD34 (marker of primitive hematopoietic stem cells and endothelial cells), CD45 (marker for pan-leukocytes), CD79α or CD19 (marker for B cells), and HLA-DR (marker for MHC class II cell surface receptor).[28]

MSCs in the Lung

Current knowledge on lung mesenchymal precursors is limited, however, there is evidence that small populations of resident lung cells expressing certain phenotypic characteristics of MSCs with progenitor capacity exist within the lung. For example, resident lung "side population" (SP) cells have been isolated and shown to exhibit both mesenchymal and epithelial potential.[29–31] These SP cells have been found at all levels of the airway tree and appear to exhibit a relatively uniform phenotype regardless of which section they were isolated from.[31] Although it has been demonstrated that these SP cells are a source of adult lung MSCs,[30] the role of SP cells in endogenous lung repair is not completely understood. Furthermore, a population of endogenous fibroblastic progenitor cells with clonogenic potential in the adult lung has been described, which are predominantly representative of mesenchymal cell lineages.[7]

Benefits of MSCs as Therapeutic Cells

The ethical and social acceptability of MSCs, as well as their exhibited immune privilege, are key benefits for their use in cell transplantation and tissue regenerative medicine. Both differentiated and undifferentiated MSCs elicit no lymphocyte alloreactivity, thus eliminating the concern of graft-versus-host disease following allogeneic transplantation.[28] Furthermore, MSCs have the ability to regulate hematopoietic cells and to secrete multiple regulatory molecules, such as growth factors, and anti-inflammatory cytokines that can modulate immune responses.[32] Although the anti-inflammatory and immune-modulator mechanisms of actions are not well understood, it is thought that they involve secretion of soluble mediators as well as cell-to-cell contact.

THERAPEUTIC POTENTIAL OF STEM CELLS FOR NEONATAL LUNG INJURY

The recent surge in our knowledge of stem cell biology and the availability of advanced research tools in this field have enabled researchers to explore the role of stem cells in neonatal lung injury. Several studies in experimental animal models provide compelling evidence for the beneficial effects of stem cell therapy for neonatal lung injury (summarized in Table 2). These studies have not only focused on the therapeutic benefit of stem cells in experimental models, but have also focused on resident lung stem cells in health and disease.

Perturbation of Stem Cells in Neonatal Lung Injury

Recent animal and human studies suggest that damage or depletion of stem cells in the developing lung likely contributes to the pathogenesis of BPD.

Evidence in Animal Models

Exposure of neonatal rodents to high levels of oxygen is extensively utilized as an injury model to investigate experimental BPD. Several recent studies have demonstrated how lung epithelial cells, MSCs, and EPCs are perturbed in oxygen-challenged neonatal rodent models of experimental BPD.[33–36] Significantly reduced numbers of circulating and resident lung MSCs have been reported in rodents in experimental BPD.[34] SP cells also demonstrate comparable characteristics in experimental BPD; reduced numbers of endothelial cells and decreased endothelial differentiation potential have been demonstrated in rodents.[33] Depleted numbers of putative resident lung EPCs have also been demonstrated in experimental BPD in rodents.[35]

Evidence in Preterm Infants

The presence of MSCs in tracheal aspirates of preterm infants has been shown to be indicative of an increased risk of developing BPD.[37] These reported MSCs isolated from tracheal aspirates of preterm infants display a pattern of lung-specific gene expression and secrete pro-inflammatory cytokines, and are distinct from lung fibroblasts.[38]

Analysis of endothelial-colony-forming cells (ECFCs, highly proliferative and self-renewing late-outgrowth EPCs) from UCB samples of preterm infants show altered characteristics, including increased susceptibility to in vitro oxygen-exposure[39] and reduced numbers among BPD compared to non-BPD preterm infants.[40,41] In contrast to the above-mentioned reports of depleted numbers of ECFCs, it has recently been reported that no association exists between the number of EPCs at birth and the subsequent development of BPD.[42] Although these studies highlight the need for a better understanding of resident lung stem/progenitor cells during disease, they do provide a strong rationale for stem cell supplementation for the prevention or repair of lung injury.

Effect of Environment and Aging on Stem Cells

It is well known that aging is associated with progressive decline in lung function, a result of many age-induced structural alterations to the lung.[43,44] Increases in the airspace volume of the alveolar region along with reductions in the supporting parenchymal tissue and elastic recoil are common features of an aging lung.[45] Also, aging is

TABLE 2 Studies examining the therapeutic effect of stem/progenitor cells in experimental models of neonatal lung injury

Experimental Model	Therapeutic Cell or Product	Outcomes	Suggested Mechanism	References
Hyperoxia-induced neonatal lung injury	BM-derived MSCs (i.t.)	Improved survival Improved alveolar structure/prevented alveolar arrest Prevented vascular growth arrest Improved exercise capacity Reduced pulmonary hypertension	Engraftment as AT2 Paracrine mechanisms	34
	BM-derived MSCs or CdM (i.v.)	Improved alveolar structure / prevented alveolar arrest Attenuated inflammation Prevented vascular growth arrest Prevented pulmonary hypertension	Paracrine mechanisms Immunomodulatory effects	50
	BM-derived MSCs or CdM (i.v.)	Increased number of BASCs Improved alveolar structure/prevented alveolar arrest	Stimulation of BASCs Paracrine mechanisms	57
	BM-derived MSCs (i.p.)	Improved survival Improved alveolar structure/prevented alveolar arrest Attenuated inflammation Inhibited lung fibrosis	Engraftment as AT2 Reduction in ECM remodeling and fibrosis gene expression (TGF-β1, collagen 1α, TIMP-1) Anti-inflammatory effects	59
	BM-derived MSC-CdM (i.v.)	Improved alveolar structure Attenuated myofibroblast infiltration Improved lung function Reversed pulmonary hypertention and RV hypertrophy Attenuated pulmonary arterial remodeling Rescued loss of pulmonary blood vessels	Paracrine mechanisms Cytoprotective effects Activation of BASCs	54
	BM-derived MSCs or CdM BM-derived MSCs or CdM (i.t.)	Improved alveolar structure Improved vascular development Improved alveolar structure Improved vascular development Reduced lung inflammation Up-regulated angiogenic factors VEGF and angiopoeitin-1 Attenuated down-regulation of TTF-1	Modulation of TTF 1, an important factor in lung Modulation of TTF-1, an important factor in lung morphogenesis Increase in vasculogenic growth factors	56
	hUCB-derived MSCs (i.t.; i.p.)	Improved survival and growth restriction Improved alveolar structure Attenuated lung fibrosis, inflammation and ROS activity Up-regulated growth factors VEGF and HGF i.t. administration better than i.p. Better protection when administered in early-phase of injury, rather than late-phase	Paracrine anti-inflammatory, anti-fibrotic and anti-oxidative effects	51–53
	hUCB-derived MSCs and MSC-CdM (i.t.) hUC-derived PCs and PC-CdM (i.t.)	Prevented and restored impaired alveolar growth Improved lung function and exercise capacity Prevented impaired lung angiogenesis Prevented pulmonary arterial wall remodeling and RV hypertrophy Persistent benefit on lung architecture and exercise capacity at 6 months No adverse effects on lung structure in treated control animals at 6 months	Paracrine mechanisms	55

Continued

TABLE 2 Studies examining the therapeutic effect of stem/progenitor cells in experimental models of neonatal lung injury—cont'd

Experimental Model	Therapeutic Cell or Product	Outcomes	Suggested Mechanism	References
	BM-derived MSC-CdM (i.p.) and preconditioned CdM (O₂-exposed)	Improved alveolar structure Reduced pulmonary hypertension O₂-preconditioning of CdM enhanced expression of anti-oxidant STC-1	Paracrine mechanisms and anti-oxidative effects	58
	BM-derived ACs (i.v.)	Improved alveolar structure Improved vascular growth	Paracrine mechanisms	65
	hAECs (i.p.)	Prevented postnatal growth restriction Improved alveolar structure Moderately improved lung inflammation	Immune modulation	70
Ventilation-induced neonatal lung injury	hAECs (i.t.; i.v.)	Improved alveolar structure / prevented alveolar arrest Attenuated lung fibrosis Attenuated lung inflammation	Differentiation into AT1 and AT2 cells Immune modulation	68
LPS-induced (i.a.) neonatal lung injury	hAECs (i.t.; i.v.)	Improved alveolar structure Increased surfactant protein expression Attenuated inflammation	Immunomodulatory effects	69
Bleomycin-induced (i.p.) neonatal lung injury	hUCB-derived ECFCs and CdM (i.v.; i.p.)	Decreased RV hypertrophy No beneficial effects on alveolar structure nor vessel density or wall thickness	Paracrine mechanisms	66

Acronyms: AC: Angiogenic cell; AT1: Alveolar epithelial type 1; AT2: Alveolar epithelial type 2; BASC: Bronchioalveolar stem cell; BM: Bone marrow; CdM: Conditioned media; ECFC: Endothelial colony forming cell; ECM: Extracellular matrix; hAEC: Human amnion epithelial cell; HGF: Hepatocyte growth factor; hUC: Human umbilical cord; hUCB: Human umbilical cord blood; i.a.: Intraamniotic, i.p.: Intraperitoneal; i.t.: Intratracheal; i.v.: Intravenous; LPS: Lipopolysaccharide; MSC: Mesenchymal stem cell; PC: Perivascular cell; ROS: Reactive oxygen species; RV: Right ventricle; STC-1: Stanniocalcin 1; TGF-β1: Transforming growth factor-β1; TIMP1: Tissue inhibitor of metalloproteinase 1; TTF-1: Thyroid transcription factor; VEGF: Vascular endothelial growth factor.

thought to be associated with a decline in pulmonary immune function. The capacity of self-renewal by stem cells invariably declines with increasing age, and eventually results in the inability of tissues to be repaired. However, the exact mechanism of this decline in stem cell self-renewal capacity remains incompletely understood. It is possible that senescence-related alterations to resident lung stem cells increase pulomonary susceptibility to injury and disrepair.[36] Furthermore, impaired recruitment of BM-MSCs into the lung, or recruitment of BM-MSCs that promote fibrosis rather than repair, are likely situations that may also be present in the aging lung.[46] Therefore, not only can early life insults such as preterm birth and the development of BPD lead to stem cell perturbation, but the eventual onset of aging and cell senescence also plays a role. This is concerning when considering that recent reports show increased oxidative stress in the airways of ex-preterm non-BPD and BPD adolescents[47] in conjunction with reports of early-onset emphysema in ex-preterm BPD adults,[48,49] which suggests that preterm birth and the subsequent development of BPD may accelerate and/or exacerbate the normal age-induced pulmonary

changes. Not only does this intensify the rationale for cell-based therapies in neonatal lung injury, but it also highlights the need for further research in our aging ex-preterm population.

Pre-Clinical Experimental Studies Using Stem Cells in Neonatal Lung Injury

In response to the findings of stem cell perturbation in the BPD lung, some studies have focused on the stimulation of endogenous stem cell pools or their therapeutic replacement with exogenous-derived stem cells (overview in Figure 1). Various different stem cell types have received much attention and have shown similar beneficial effects in various studies utilizing different experimental models of BPD.

MSCs in Pre-Clinical Experimental Studies

Of the many different types of stem cell therapies that have been used in experimental models, MSCs are the most extensively examined cell type. Numerous aspects of neonatal lung injury have been ameliorated by the administration of

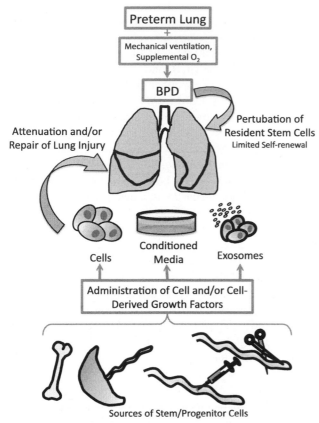

FIGURE 1 Current sources of stem/progenitor cells for lung regeneration in experimental models of neonatal lung injury. Many studies have investigated the effects of cells and cell-derived growth factors, such as conditioned media and exosomes, to promote lung regeneration following injury. Stem/progenitor cells can be sourced from the bone marrow, placenta amnion, umbilical cord blood, and umbilical cord.

either bone marrow-derived, umbilical cord blood-derived, or umbilical cord-derived MSCs.[34,50–59] In those experimental models of BPD, the MSCs exerted their therapeutic effects by mitigating lung inflammation, preventing lung vascular damage and alveolar growth impairment, inhibiting lung fibrosis, and improving exercise tolerance.

A common observation reported when treating the neonatally injured lung with MSCs is their low engraftment and differentiation into the lung (less than 5% engraftment), suggesting a paracrine-mediated therapeutic effect. These speculations are supported by in vitro and in vivo studies in which the use of conditioned media (CdM) from MSCs has been shown to protect alveolar epithelial and lung microvasculature endothelial cells from oxidative stress, prevent oxygen-induced alveolar growth impairment, and stimulate a subset of stem cells, known as bronchoalveolar stem cells, to aid in lung repair.[34,50,54,57] Furthermore, the therapeutic benefits of CdM from MSCs may surpass those of MSCs, with in vivo findings indicating a more profound therapeutic effect of MSC-CdM in preventing/repairing lung injury than that of MSCs.[50] In

addition, it was recently shown that preconditioning the CdM, by exposing the BM-derived MSCs to high levels of oxygen prior to their collection, enhanced its paracrine effect as shown by increased expression of the naturally occurring anti-oxidant stanniocalcin-1.[58] These findings support the notion that the potential mechanisms through which MSCs exert their actions are paracrine-mediated, and that the therapeutic benefits of MSC-CdM may in fact surpass those of MSCs.

Recently, it has been demonstrated that the major paracrine anti-inflammatory and therapeutic mediators by which MSCs exert their action may be driven by the membrane microvesicles secreted from MSCs called exosomes.[60] Exosomes that were fractionated from mouse and human cord MSC-CdM were used in an adult rodent model of hypoxia-induced pulmonary hypertension. They were found to exert a pleiotropic protective effect on the lung, attenuated lung macrophage influx, decreased proinflammatoy cytokine levels in bronchoalveolar lavage fluid, and also inhibited vascular remodeling and pulmonary hypertension. Further supporting the importance of exosomes in the paracrine effect of MSCs, it has been demonstrated that MSC-derived exosome-depleted media exerted no effect, nor did exosomes fractionated from fibroblasts. To date there are no published studies that have utilized exosomes in experimental models of neonatal lung injury. However given their appeal as ideal therapeutic agents, it is likely that new studies employing exosomes will emerge in the near future as possible treatments for BPD.

MSCs have commonly been obtained from bone marrow; however a major limitation of this source is availability. The harvest of bone marrow itself is invasive, and is only obtained by an elective surgical procedure usually undertaken in young individuals. Furthermore, bone marrow contains numerous cell types of which only 1 in 100,000 are MSCs, and the number and lifespan of BM-derived MSCs declines with increasing age of the donor.[61] Therefore, supply of BM-derived MSCs for therapeutic use may be somewhat limited. Thus, compared to BM as a source of MSCs, the umbilical cord itself (i.e., Wharton's jelly) represents a very appealing source of MSCs for therapeutic use in the newborn due to its clinically relevant, easily accessible, ethically viable, and readily available source of stem/progenitor cells. As the umbilical cord is usually discarded after the birth of a baby, obtaining the necessary tissue does not require invasive surgical procedures and the supply is not limited. Furthermore, the frequency of MSCs amongst other cells within the umbilical cord following isolation is approximately 1 in 300, a much greater ratio compared to that in BM and further accentuating the appeal of MSC isolation from umbilical cord.[62] UCB is also another appealing source of MSCs compared to BM; however in terms of the ratio of MSCs to other cells in the source, cord blood does not rate as high as the cord itself, with only 1 in 500 million cells being MSCs.[61]

MSCs isolated from human cord blood have been shown to prevent hyperoxia-induced alveolar growth arrest and alleviate fibrotic changes in the neonatal rat lung.[51–53] The focus has been on the differing therapeutic effects according to the route of administration,[53] dose of MSCs,[52] and the timing of the dose.[51] More favorable outcomes arose from intra-tracheal administration with a minimum of 5×10^4 cells and optimal protective effects with 5×10^5 cells. Furthermore, it was recently shown that treatment in the early stages of lung injury (pre-alveolar stage of lung development, i.e., ~28–34 weeks of gestation in humans) exhibited more significant protection, rather than treatment in the late stages of lung injury and inflammation (post-alveolar stage of lung development, i.e., >37 weeks of gestation in humans).[51] More recently, data from our own laboratory has demonstrated the beneficial long-term effects of UC-derived MSCs in an oxygen-challenged rat model of BPD.[55] Long-term assessment of 6-month-old rats showed no adverse effects of either MSC or MSC-CdM therapy, and also demonstrated persistent improvements in adult exercise capacity and lung structure.[55]

Umbilical cord-derived PCs have also been shown to exhibit similar reparative potential to UCB- and UC-derived MSCs in a rat model of neonatal lung injury.[55] UC-derived PCs, as a whole cell therapy or as growth factor producers (i.e., CdM), rescued oxygen-induced arrested alveolar growth and improved long-term lung function.[55] The low engraftment into the lungs indicates that these cells act via immune modulation, rather than by cell engraftment and differentiation. More detailed assessment of the therapeutic potential of these cells in other models of neonatal and adult lung injury will be of interest.

A recent study has shown the potential benefits of a particular population of BM-derived cells that express the marker c-kit, a tyrosine kinase receptor that regulates angiogenesis.[63] These c-kit[pos] cells were used in a rat model of neonatal lung injury and were administered once during the hyperoxia-exposure period. BM-derived c-kit[pos] cells were shown to improve alveolarization, as evidenced by increased lung septation and reduced MLI, as well as increased lung vascular density, reduced lung apoptosis, and increased secretion of pro-angiogenic factors.[63] The engraftment of these cells into the lung was also very low (<2% of total lung cells), supporting their paracrine-mediated action.

Other Repair Cells in Pre-Clinical Experimental Studies

Given the importance of lung angiogenesis and vascular growth factors during lung growth and repair, EPCs and ECFCs are appealing candidate cell types.[64] Treatment of hyperoxia-challenged neonatal mice with intravenously administered BM-derived angiogenic cells (myeloid-like precursor cells) showed restoration of alveolar structure and vessel density compared to the control (room air-exposed) mice.[65] In a different experimental model of BPD, newborn rats exposed to bleomycin received CdM from human UCB-derived ECFCs.[66] The ECFCs were isolated from both preterm and term UCB, and then CdM was produced from the ECFCs that were either grown in room air or mild hyperoxic conditions. CdM from term and preterm ECFCs grown in room air were found to promote cell growth and angiogenesis in vitro; however, the in vitro benefit of preterm ECFC-CdM was not evident when collected during mild hyperoxia. Interestingly, administration of term and preterm ECFC-CdM (from both room air and hyperoxic conditions) did not appear to attenuate bleomycin-induced alveolar simplification or improve the reduced vessel density.[66] Data from our own laboratory supports the beneficial role of ECFC administration in a neonatal rodent oxygen-challenged model of BPD. UCB-derived ECFCs administered to neonatal mice after established hyeroxia-induced alveolar growth were found to arrest attenuate lung injury; mice presented with improved alveolar structure and lung vascular growth, as well as improved lung compliance and attenuation of pulmonary hypertension.[67] Furthermore, in compliance with the engraftment rates of MSCs, ECFCs also showed a low rate of engraftment into recipient lungs, suggesting a paracrine-mediated effect in this model.

Recently, the therapeutic potential of human amnion epithelial cells (hAECs) has been investigated in two different sheep models of neonatal lung injury[68,69] as well as the more commonly utilized oxygen-challenged rodent model of neonatal lung injury.[70] Since hAECs are sourced from placentae, which are normally discarded after birth, they present an easily accessible and ethically viable cell therapy candidate. In a model of lung injury induced by lipopolysaccharide (LPS) administration in fetal sheep, intravenous delivery of hAECs attenuated pulmonary inflammation and improved lung function and structure.[69] Lung injury induced by in utero ventilation of fetal sheep was also mitigated by treatment with hAECs, and these cells localized within the fetal lung and differentiated into AT1 and AT2 cells.[68] More recently, hAECs have been reported to partially reduce hyperoxia-induced lung inflammation and structural damage in neonatal mice.[70] Administration of hAECs to neonatal mice for 3 consecutive days normalized the reduction in postnatal growth, as well as attenuating some structural damage and lung inflammation. However, the hAECs did not succeed in completely restoring the lung architecture; hyperoxia-induced alterations in alveolar airspace volume, septal tissue volume, tissue-to-airspace ratio, and parenchymal collagen content all remained unchanged by the administration of hAECs.[70] It is possible that the relatively modest reparative effects of hAECs shown in this particular model are attributed to the relatively young age of assessment combined with no "recovery" in room air following hyperoxia-exposure.

Findings from the above-mentioned studies have broadened our understanding of stem cells in neonatal lung injury and indicate that a variety of stem cells can prevent and/or repair neonatal lung injury in various experimental models. The global aim of the various studies that investigate the use of stem cells in such settings is to generate evidence to create a strong rationale for translating this potential breakthrough into the clinic. Excitingly, the first few clinical trials have recently commenced.

FROM BENCH TO BEDSIDE: CLINICAL TRIALS IN PRETERM INFANTS WITH BPD

Based on the promising evidence from the pre-clinical studies in experimental models of BPD, phase I clinical trials are currently underway in preterm infants with BPD. Recently, the results of the first phase I clinical trial were presented at the Annual Pediatric Academic Societies Meeting.[71] The authors assessed the safety and efficacy of human UCB-derived MSCs administered as a prophylactic therapy in 9 preterm infants born between 23 and 29 weeks of gestational age (500–1250 g) and were continuing to receive invasive mechanical ventilation. Two different doses of cells were investigated: a low dose of 10 million cells per kg of body weight, and a high dose of 20 million cells per kg. The cells were administered within the first two weeks after birth by a single intratracheal injection. The authors reported that none of the patients showed any serious adverse effects or dose limiting toxicity related with human UCB-MSCs. The inflammatory markers matrix metalloproteinase (MMP)-9 and interleukin (IL)-8 were reduced in the tracheal aspirate 7 days after MSC transplantation. Furthermore, it was shown that MSC treatment reduced the severity of BPD and the length of hospital stay. The findings from this phase I clinical trial indicate that intratracheal transplantation of human UCB-derived MSCs in preterm infants was feasible, safe and may be effective in attenuating the severity of BPD. Also being conducted in the same center is a phase I clinical trial investigating the long-term follow up of these infants who received the MSC therapy. Furthermore, a phase II clinical trial has commenced in multiple centers to evaluate the efficacy and safety of human UCB-derived MSCs for treatment of BPD in preterm infants versus a control group. There are also reports of another phase I clinical trial that will evaluate the effects of hAECs in preterm infants at risk of developing BPD. It is anticipated that the findings from these clinical trials will pave the way for the future treatment of BPD in preterm babies.

CONCLUSIONS

Half a century after the landmark discovery of stem cells by the Canadian researchers Till and McCulloch in 1961,[72] their therapeutic potential in regenerative medicine is now being harnessed for treatment of neonatal lung injury.

Much more needs to be learned about the characteristics and mechanism of action of these various repair cells. For some however, the time is ripe to conduct carefully planned phase I and II studies to assess the feasibility and safety of cell-based therapies in humans. These studies will instruct us about the design of further animal and clinical studies in order to determine the optimal cell-based strategy. Almost half a century since the first description of BPD,[73] there may be a therapy on the horizon.

REFERENCES

1. Blau HM, Brazelton TR, Weimann JM. The evolving concept of a stem cell: Entity or function? *Cellule* 2001;**105**:829–41.
2. Stevenson K, McGlynn L, Shiels PG. Stem cells: Outstanding potential and outstanding questions. *Scott Med J* 2009;**54**:35–7.
3. Stripp BR. Hierarchical organization of lung progenitor cells: Is there an adult lung tissue stem cell? *Proc Am Thorac Soc* 2008;**5**:695–8.
4. Rawlins EL, Hogan BL. Epithelial stem cells of the lung: Privileged few or opportunities for many? *Development* 2006;**133**:2455–65.
5. Giangreco A, Reynolds SD, Stripp BR. Terminal bronchioles harbor a unique airway stem cell population that localizes to the bronchoalveolar duct junction. *Am J Pathol* 2002;**161**:173–82.
6. Kim CF, Jackson EL, Woolfenden AE, Lawrence S, Babar I, Vogel S, Crowley D, Bronson RT, Jacks T. Identification of bronchioalveolar stem cells in normal lung and lung cancer. *Cell* 2005;**121**:823–35.
7. McQualter JL, Brouard N, Williams B, Baird BN, Sims-Lucas S, Yuen K, et al. Endogenous fibroblastic progenitor cells in the adult mouse lung are highly enriched in the sca-1 positive cell fraction. *Stem Cells* 2009;**27**:623–33.
8. McQualter JL, Yuen K, Williams B, Bertoncello I. Evidence of an epithelial stem/progenitor cell hierarchy in the adult mouse lung. *Proc Natl Acad Sci U S A* 2010;**107**:1414–9.
9. Rawlins EL, Clark CP, Xue Y, Hogan BL. The id2+ distal tip lung epithelium contains individual multipotent embryonic progenitor cells. *Development* 2009;**136**:3741–5.
10. Rawlins EL, Okubo T, Xue Y, Brass DM, Auten RL, Hasegawa H, et al. The role of scgb1a1+ clara cells in the long-term maintenance and repair of lung airway, but not alveolar, epithelium. *Cell Stem Cell* 2009;**4**:525–34.
11. Hegab AE, Ha VL, Gilbert JL, Zhang KX, Malkoski SP, Chon AT, et al. Novel stem/progenitor cell population from murine tracheal submucosal gland ducts with multipotent regenerative potential. *Stem Cells* 2011;**29**:1283–93.
12. Hong KU, Reynolds SD, Watkins S, Fuchs E, Stripp BR. Basal cells are a multipotent progenitor capable of renewing the bronchial epithelium. *Am J Pathol* 2004;**164**:577–88.
13. Giangreco A, Shen H, Reynolds SD, Stripp BR. Molecular phenotype of airway side population cells. *Am J Physiol Lung Cell Mol Physiol* 2004;**286**:L624–30.
14. Hong KU, Reynolds SD, Giangreco A, Hurley CM, Stripp BR. Clara cell secretory protein-expressing cells of the airway neuroepithelial body microenvironment include a label-retaining subset and are critical for epithelial renewal after progenitor cell depletion. *Am J Respir Cell Mol Biol* 2001;**24**:671–81.
15. Volckaert T, Dill E, Campbell A, Tiozzo C, Majka S, Bellusci S, et al. Parabronchial smooth muscle constitutes an airway epithelial stem cell niche in the mouse lung after injury. *J Clin Invest* 2011;**121**:4409–19.

16. Reddy R, Buckley S, Doerken M, Barsky L, Weinberg K, Anderson KD, et al. Isolation of a putative progenitor subpopulation of alveolar epithelial type 2 cells. *Am J Physiol Lung Cell Mol Physiol* 2004;**286**:L658–67.

17. Alvarez DF, Huang L, King JA, ElZarrad MK, Yoder MC, Stevens T. Lung microvascular endothelium is enriched with progenitor cells that exhibit vasculogenic capacity. *Am J Physiol Lung Cell Mol Physiol* 2008;**294**:L419–30.

18. Barkauskas CE, Cronce MJ, Rackley CR, Bowie EJ, Keene DR, Stripp BR, et al. Type 2 alveolar cells are stem cells in adult lung. *J Clin Invest* 2013;**123**:3025–36.

19. Bertoncello I, McQualter JL. Endogenous lung stem cells: What is their potential for use in regenerative medicine? *Expert Rev Respir Med* 2010;**4**:349–62.

20. Snyder JC, Teisanu RM, Stripp BR. Endogenous lung stem cells and contribution to disease. *J Pathol* 2009;**217**:254–64.

21. Adamson IY, Bowden DH. The type 2 cell as progenitor of alveolar epithelial regeneration. A cytodynamic study in mice after exposure to oxygen. *Lab Invest* 1974;**30**:35–42.

22. Brody JS, Williams MC. Pulmonary alveolar epithelial cell differentiation. *Annu Rev Physiol* 1992;**54**:351–71.

23. Driscoll B, Buckley S, Bui KC, Anderson KD, Warburton D. Telomerase in alveolar epithelial development and repair. *Am J Physiol Lung Cell Mol Physiol* 2000;**279**:L1191–1198.

24. Chapman HA, Li X, Alexander JP, Brumwell A, Lorizio W, Tan K, et al. Integrin alpha6beta4 identifies an adult distal lung epithelial population with regenerative potential in mice. *J Clin Invest* 2011;**121**:2855–62.

25. Kumar PA, Hu Y, Yamamoto Y, Hoe NB, Wei TS, Mu D, et al. Distal airway stem cells yield alveoli in vitro and during lung regeneration following h1n1 influenza infection. *Cell* 2011;**147**:525–38.

26. Kajstura J, Rota M, Hall SR, Hosoda T, D'Amario D, Sanada F, et al. Evidence for human lung stem cells. *N Engl J Med* 2011;**364**:1795–806.

27. McQualter JL, Bertoncello I. Concise review: Deconstructing the lung to reveal its regenerative potential. *Stem Cells* 2012;**30**:811–6.

28. Spencer ND, Gimble JM, Lopez MJ. Mesenchymal stromal cells: Past, present, and future. *Vet Surg* 2011;**40**:129–39.

29. Majka SM, Beutz MA, Hagen M, Izzo AA, Voelkel N, Helm KM. Identification of novel resident pulmonary stem cells: Form and function of the lung side population. *Stem Cells* 2005;**23**:1073–81.

30. Martin J, Helm K, Ruegg P, Varella-Garcia M, Burnham E, Majka S. Adult lung side population cells have mesenchymal stem cell potential. *Cytotherapy* 2008;**10**:140–51.

31. Reynolds SD, Shen H, Reynolds PR, Betsuyaku T, Pilewski JM, Gambelli F, et al. Molecular and functional properties of lung sp cells. *Am J Physiol Lung Cell Mol Physiol* 2007;**292**:L972–83.

32. Sueblinvong V, Weiss DJ. Stem cells and cell therapy approaches in lung biology and diseases. *Transl Res* 2010;**156**:188–205.

33. Irwin D, Helm K, Campbell N, Imamura M, Fagan K, Harral J, et al. Neonatal lung side population cells demonstrate endothelial potential and are altered in response to hyperoxia-induced lung simplification. *Am J Physiol Lung Cell Mol Physiol* 2007;**293**:L941–51.

34. van Haaften T, Byrne R, Bonnet S, Rochefort GY, Akabutu J, Bouchentouf M, et al. Airway delivery of mesenchymal stem cells prevents arrested alveolar growth in neonatal lung injury in rats. *Am J Respir Crit Care Med* 2009;**180**:1131–42.

35. Balasubramaniam V, Mervis CF, Maxey AM, Markham NE, Abman SH. Hyperoxia reduces bone marrow, circulating, and lung endothelial progenitor cells in the developing lung: Implications for the pathogenesis of bronchopulmonary dysplasia. *Am J Physiol Lung Cell Mol Physiol* 2007;**292**:L1073–84.

36. Yee M, Vitiello PF, Roper JM, Staversky RJ, Wright TW, McGrath-Morrow SA, et al. Type ii epithelial cells are critical target for hyperoxia-mediated impairment of postnatal lung development. *Am J Physiol Lung Cell Mol Physiol* 2006;**291**:L1101–11.

37. Popova AP, Bozyk PD, Bentley JK, Linn MJ, Goldsmith AM, Schumacher RE, et al. Isolation of tracheal aspirate mesenchymal stromal cells predicts bronchopulmonary dysplasia. *Pediatrics* 2010;**126**:e1127–33.

38. Bozyk PD, Popova AP, Bentley JK, Goldsmith AM, Linn MJ, Weiss DJ, et al. Mesenchymal stromal cells from neonatal tracheal aspirates demonstrate a pattern of lung-specific gene expression. *Stem Cells Dev* 2011;**20**:1995–2007.

39. Baker CD, Ryan SL, Ingram DA, Seedorf GJ, Abman SH, Balasubramaniam V. Endothelial colony-forming cells from preterm infants are increased and more susceptible to hyperoxia. *Am J Respir Crit Care Med* 2009;**180**:454–61.

40. Borghesi A, Massa M, Campanelli R, Bollani L, Tzialla C, Figar TA, et al. Circulating endothelial progenitor cells in preterm infants with bronchopulmonary dysplasia. *Am J Respir Crit Care Med* 2009;**180**:540–6.

41. Baker CD, Balasubramaniam V, Mourani PM, Sontag MK, Black CP, Ryan SL, et al. Cord blood angiogenic progenitor cells are decreased in bronchopulmonary dysplasia. *Eur Respir J* 2012;**40**:1516–22.

42. Paviotti G, Fadini GP, Boscaro E, Agostini C, Avogaro A, Chiandetti L, et al. Endothelial progenitor cells, bronchopulmonary dysplasia and other short-term outcomes of extremely preterm birth. *Early Hum Dev* 2011;**87**:461–5.

43. Sharma G, Goodwin J. Effect of aging on respiratory system physiology and immunology. *Clin Interv Aging* 2006;**1**:253–60.

44. Wang L, Green FHY, Smiley-Jewell SM, Pinkerton KE. Susceptibility of the aging lung to environmental injury. *Semin Respir Crit Care Med* 2010;**31**:539–53.

45. Pinkerton KE, Green FHY. Normal aging of the lung. In: Harding R, Pinkerton KE, Plopper CG, editors. *The Lung: Development, Aging and the Environment*. London: Elsevier; 2004. pp. 213–33.

46. Mora AL, Rojas M. Aging and lung injury repair: A role for bone marrow derived mesenchymal stem cells. *J Cell Biochem* 2008;**105**:641–7.

47. Filippone M, Bonetto G, Corradi M, Frigo AC, Baraldi E. Evidence of unexpected oxidative stress in airways of adolescents born very preterm. *Eur Respir J* 2012;**40**:1253–9.

48. Aukland SM, Rosendahl K, Owens CM, Fosse KR, Eide GE, Halvorsen T. Neonatal bronchopulmonary dysplasia predicts abnormal pulmonary hrct scans in long-term survivors of extreme preterm birth. *Thorax* 2009;**64**:405–10.

49. Wong PM, Lees AN, Louw J, Lee FY, French N, Gain K, et al. Emphysema in young adult survivors of moderate-to-severe bronchopulmonary dysplasia. *Eur Respir J* 2008;**32**:321–8.

50. Aslam M, Baveja R, Liang OD, Fernandez-Gonzalez A, Lee C, Mitsialis SA, et al. Bone marrow stromal cells attenuate lung injury in a murine model of neonatal chronic lung disease. *Am J Respir Crit Care Med* 2009;**180**:1122–30.

51. Chang YS, Choi SJ, Ahn SY, Sung DK, Sung SI, Yoo HS, et al. Timing of umbilical cord blood derived mesenchymal stem cells transplantation determines therapeutic efficacy in the neonatal hyperoxic lung injury. *PLoS One* 2013;**8**:e52419.

52. Chang YS, Choi SJ, Sung DK, Kim SY, Oh W, Yang YS, et al. Intratracheal transplantation of human umbilical cord blood derived mesenchymal stem cells dose-dependently attenuates hyperoxia-induced lung injury in neonatal rats. *Cell Transplant* 2011;**20**:1843–54.

53. Chang YS, Oh W, Choi SJ, Sung DK, Kim SY, Choi EY, et al. Human umbilical cord blood-derived mesenchymal stem cells attenuate hyperoxia-induced lung injury in neonatal rats. *Cell Transplant* 2009;**18**:869–86.

54. Hansmann G, Fernandez-Gonzalez A, Aslam M, Vitali SH, Martin T, Mitsialis SA, et al. Mesenchymal stem cell-mediated reversal of bronchopulmonary dysplasia and associated pulmonary hypertension. *Pulm Circ* 2012;**2**:170–81.

55. Pierro M, Ionescu L, Montemurro T, Vadivel A, Weissmann G, Oudit G, et al. Short-term, long-term and paracrine effect of human umbilical cord-derived stem cells in lung injury prevention and repair in experimental bronchopulmonary dysplasia. *Thorax* 2012. http://dx.doi.org/10.1136/thoraxjnl-2012-202323.

56. Sutsko RP, Young KC, Ribeiro A, Torres E, Rodriguez M, Hehre D, et al. Long term reparative effects of mesenchymal stem cell therapy following neonatal hyperoxia-induced lung injury. *Pediatr Res* 2012;**73**:46–53.

57. Tropea KA, Leder E, Aslam M, Lau AN, Raiser DM, Lee JH, et al. Bronchioalveolar stem cells increase after mesenchymal stromal cell treatment in a mouse model of bronchopulmonary dysplasia. *Am J Physiol Lung Cell Mol Physiol* 2012;**302**:L829–37.

58. Waszak P, Alphonse R, Vadivel A, Ionescu L, Eaton F, Thebaud B. Preconditioning enhances the paracrine effect of mesenchymal stem cells in preventing oxygen-induced neonatal lung injury in rats. *Stem Cells Dev* 2012;**21**:2789–97.

59. Zhang X, Wang H, Shi Y, Peng W, Zhang S, Zhang W, et al. The role of bone marrow-derived mesenchymal stem cells in the prevention of hyperoxia-induced lung injury in newborn mice. *Cell Biol Int* 2012;**36**:589–94.

60. Lee C, Mitsialis SA, Aslam M, Vitali SH, Vergadi E, Konstantinou G, et al. Exosomes mediate the cytoprotective action of mesenchymal stromal cells on hypoxia-induced pulmonary hypertension. *Circulation* 2012;**126**:2601–11.

61. Sarugaser R, Ennis J, Stanford WL, Davies JE. Isolation, propagation, and characterization of human umbilical cord perivascular cells (hucpvcs). *Methods Mol Biol* 2009;**482**:269–79.

62. Sarugaser R, Lickorish D, Baksh D, Hosseini MM, Davies JE. Human umbilical cord perivascular (hucpv) cells: A source of mesenchymal progenitors. *Stem Cells* 2005;**23**:220–9.

63. Ramachandran S, Suguihara C, Drummond S, Chatzistergos K, Klim J, Torres E, et al. Bone marrow-derived c-kitpositive cells attenuate neonatal hyperoxia-induced lung injury. *Cell Transplant* 2013. http://dx.doi.org/10.3727/096368913X667736.

64. Thebaud B, Abman SH. Bronchopulmonary dysplasia - where have all the vessels gone? Roles of angiogenic growth factors in chronic lung disease. *Am J Resp Crit Care* 2007;**175**:978–85.

65. Balasubramaniam V, Ryan SL, Seedorf GJ, Roth EV, Heumann TR, Yoder MC, et al. Bone marrow-derived angiogenic cells restore lung alveolar and vascular structure after neonatal hyperoxia in infant mice. *Am J Physiol Lung Cell Mol Physiol* 2010;**298**: L315–23.

66. Baker CD, Seedorf GJ, Wisniewski BL, Black CP, Ryan SL, Balasubramaniam V, et al. Endothelial colony-forming cell conditioned media promotes angiogenesis in vitro and prevents pulmonary hypertension in experimental bronchopulmonary dysplasia. *Am J Physiol Lung Cell Mol Physiol* 2013;**305**:L73–81.

67. Alphonse R, Vadivel A, Waszak P, Coltan L, Fung M, Eaton F, et al. Existence, functional impairment and therapeutic potential of endothelial colony forming cells (ecfcs) in oxygen-induced arrested alveolar growth. *Am J Respir Crit Care Med* 2011;**183**:A1237.

68. Hodges RJ, Jenkin G, Hooper SB, Allison B, Lim R, Dickinson H, et al. Human amnion epithelial cells reduce ventilation-induced preterm lung injury in fetal sheep. *Am J Obstet Gynecol* 2012;**206**(448): e448–15.

69. Vosdoganes P, Hodges RJ, Lim R, Westover AJ, Acharya RY, Wallace EM, et al. Human amnion epithelial cells as a treatment for inflammation-induced fetal lung injury in sheep. *Am J Obstet Gynecol* 2011;**205**(156):e126–33.

70. Vosdoganes P, Lim R, Koulaeva E, Chan ST, Acharya R, Moss TJ, et al. Human amnion epithelial cells modulate hyperoxia-induced neonatal lung injury in mice. *Cytotherapy* 2013;**15**:1021–9.

71. Ahn SY, Chang YC, Kim ES, Yoo HS, Sung SI, Choi SJ, et al. *Human umbilical cord blood derived mensenchymal stem cells transplantation for bronchopulmonary dysplasia: Results of a phase I dose escalation clinical study.* Washington DC: Pediatric Academic Societies Annual Meeting; 2013.

72. Till JE, Mc CE. A direct measurement of the radiation sensitivity of normal mouse bone marrow cells. *Radiat Res* 1961;**14**:213–22.

73. Northway Jr WH, Rosan RC, Porter DY. Pulmonary disease following respirator therapy of hyaline-membrane disease. Bronchopulmonary dysplasia. *N Engl J Med* 1967;**276**:357–68.

Chapter 16

Epigenetics and the Developmental Origins of Lung Disease

Lisa A. Joss-Moore*, Robert H. Lane† and Kurt H. Albertine*

*Department of Pediatrics, University of Utah, Salt Lake City, UT; †Department of Pediatrics, Medical College of Wisconsin, WI, USA

INTRODUCTION

Organs and cells undergo a preset developmental course that anticipates a normal environment. Environmental disruptions to this preset developmental course produce long-term changes in organ structure and function that predispose towards adult disease. The premise that early-life events that disrupt a developmental course can lead to long-term biological changes has been coined the "developmental origins of disease hypothesis" following seminal observations by physician and researcher, the late David Barker.[1] In his original studies, Barker linked low birth weight (a surrogate for poor in utero conditions) to adult cardiovascular disease, disruptions in glucose homeostasis, and early death.[2–4] Barker's early epidemiologic studies spawned a research direction, now 25 years strong. The "developmental origins of lung disease" is relatively recent addition to the field and is now an active area of investigation.

Researchers studying the developmental origins of disease are beginning to unravel some of the molecular mechanisms by which early-life disruptions lead to the long-term development of disease. Experimental research using animal models is at the forefront of understanding the molecular underpinnings of the developmental origins of lung disease. Recent studies are identifying key molecular players that contribute to multiple aspects of lung dysfunction, including impaired lung development, lung inflammatory response, and injury repair. Identification of key molecular players is revealing the occurrence of persistent, subtle changes in gene expression profiles resulting from early-life insults. The subtle changes in gene expression profiles are accompanied by aberrations in epigenetic modifications, leading to the notion that epigenetics participates in the dysfunction.

Epigenetic modifications are part of the molecular toolbox regulating expression of genetic material, and can be thought of as the blueprint for how a particular cell should use its genomic repertoire. Epigenetic modifications are patterned during development yet are dynamic and responsive to the environment. Traits such as these facilitate an epigenetic contribution to the developmental origins of lung disease.

HUMAN EVIDENCE FOR THE DEVELOPMENTAL ORIGINS OF LUNG DISEASE

Seminal reports linking early life events to long-term lung disease have characterized the development of chronic airflow obstruction in men of low birth weight.[5–7] A unique set of child health records of men born in the English counties of Hertfordshire and Derbyshire between 1911 and 1930 made the studies possible. The records contained details of birth weight, weight at 1 year of age, and occurrence of lower respiratory tract infection in the first 5 years of life. Follow-up assessment of these men 60 to 70 years later demonstrated that low birth weight and a lower respiratory tract infection in the first 2 years of life predicted impaired adult lung function.[7] Strikingly, those with the combination of low birth weight and low weight at 1 year of age were more likely to have died from chronic obstructive pulmonary disease.[5]

Low birth weight is often associated with preterm birth. Preterm birth is often accompanied by respiratory failure and necessitates postnatal management with mechanical ventilation (MV) and oxygen-rich gas. Preterm infants subjected to MV and oxygen frequently develop chronic lung disease of early infancy (bronchopulmonary dysplasia, BPD).[8,9] Survivors of BPD have impaired lung function and increased susceptibility to lung infection during adolescence and adulthood.[10–14]

Low birth weight also results from intrauterine growth restriction (IUGR). IUGR has multiple etiologies. In developed countries, the two most common causes of IUGR are uteroplacental insufficiency secondary to maternal vascular disorders and fetal exposure to maternal tobacco smoke (MTS). IUGR in MTS exposed offspring may also be

The Lung. http://dx.doi.org/10.1016/B978-0-12-799941-8.00016-X

related to uteroplacental insufficiency or may be related to additional factors associated with MTS.[15–17] Interestingly, male infants are more severely affected by IUGR than female infants for reasons that remain unclear.[18–23] In term infants, IUGR increases the need for respiratory support in the neonatal period.[24–28]

An important consideration is that IUGR often accompanies preterm birth, due to either maternal or fetal factors. The combination of IUGR and preterm birth is important because in an IUGR infant that is also preterm, the administration of MV and oxygen acts as a "second hit," increasing the likelihood of subsequent disease. When an IUGR infant is also born preterm, the incidence and severity of BPD are greater.[29–32] In one study examining risk factors for BDP in infants born before the 28th week of gestation, IUGR functioned as an independent predictor of BPD, after adjusting for other risk factors including gestational age.[29]

IUGR impacts lung function in infants and adolescents. Several studies show that after adjusting for gestational age and maternal age, weight, and tobacco use, lower birth weight is associated with lower than expected forced expiratory volume in 1 second (FEV$_1$) and forced vital capacity (FVC).[33–35]

MTS, regardless of IUGR, has long-term effects on lung disease and lung function later in life. The development of asthma is causally linked to MTS exposure.[36–40] Other studies demonstrate that MTS contributes to the development of airway hyper-responsiveness and long-term declines in lung function.[37,41–44] MTS exposure may also increase the risk of developing chronic obstructive pulmonary disease.[39]

LESSONS FROM ANIMAL STUDIES

Mechanisms underlying the developmental origins of lung disease are being uncovered using animal studies. Studies in different animal species provide insight into physiology as well as molecular paradigms. While overlap exists in the utility of various animal models, large animal models (e.g., sheep) provide valuable physiologic understanding, while rodent models facilitate molecular manipulation and assessment. An important aspect of different animal models is differences in the developmental timing of the lung.

Lung development transitions through five distinct phases: embryonic, pseudoglandular, canalicular, saccular, and alveolar. The majority of perinatal insults that produce long-term changes in the lung occur during the second half of gestation and the early postnatal period; therefore, the saccular and alveolar phases are most relevant. The human lung is in the saccular phase of development at approximately 28–32 weeks gestation. During the saccular stage, the lung parenchyma is composed of smooth-walled sacs that are lined by cuboidal epithelial cells. These cells contribute to surfactant production. The alveolar stage begins at approximately 32 weeks human gestation and continues

into postnatal life. Structural transformation of saccules to alveoli occurs by formation and elongation of secondary septa. Secondary septa subdivide the saccules into anatomic alveoli. At the same time, the thick walls of the saccules become thin, while capillaries grow in number and proximity to the thinning epithelial cells. Concurrently, alveolar type II epithelial cells synthesize and secrete surfactant and surfactant apoproteins.[45,46]

Relative developmental timing in the sheep lung is similar to that of humans. In sheep, the alveolar epithelium differentiates, and a large increase in pulmonary capillary surface area occurs before birth at term.[47] Developmental lung timing in the sheep and human is different to that of mice and rats, in which the alveolar stage occurs after birth at term (Figure 1).[48,49]

Numerous animal models of perinatal insults, including IUGR, preterm birth with prolonged MV and oxygen exposure, or a combination of both, have shown impaired alveolarization. The degree and specific alveolar effect depends upon the severity and timing of the perinatal insults.[50–52] Impaired alveolar formation, decreased alveolar number, and decreased internal surface area are evident in sheep and baboons born preterm and managed with MV.[9,53,54]

IUGR in the sheep impairs alveolar formation, with IUGR producing fewer, larger alveoli, thicker alveolar septa and a thicker blood–air barrier.[55–57] Similar findings are seen in the IUGR rat, with lungs characterized by thickened alveolar walls, decreased septation, and decreased alveolar number.[58,59]

Fetal exposure to MTS, independent of IUGR, has similar effects on the structural development of the lung. In rats and non-human primates, MTS results in fewer, larger alveoli, as well as airway remodeling.[60–63] Of the numerous chemicals found in tobacco smoke, nicotine is arguably the most detrimental for lung development. Effects of MTS on lung outcomes are mirrored in studies examining the effects of maternal nicotine in isolation. Maternal nicotine administration alters lung growth and causes alveolar impairment in sheep, rats, and non-human primates.[64–67]

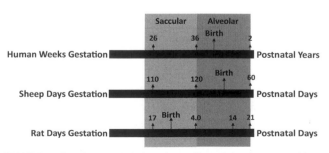

FIGURE 1   Developmental timing of saccular and alveolar stages of lung development in human, sheep, and rat. Timing of lung maturation relative to birth is similar in human and sheep lungs, with alveolar formation underway at the time of birth at term. In contrast, the rat lung begins alveolar formation after birth at term.

Of note, detrimental effects on lung development following maternal nicotine administration are also observed in the second-generation offspring of rats.[68,69]

Disruption to alveolar formation caused by preterm birth with MV, IUGR, or maternal tobacco/nicotine administration provides an opportunity to examine epigenetic mechanisms contributing to alveolar disruption. For alveoli to develop normally, gene expression patterns in the cells of the lung need to be precisely regulated. Precise regulation of gene expression ensures temporal and spatial gene expression patterns that support the mesenchymal–epithelial interactions necessary for alveolar formation. The timing and magnitude of gene expression is, in part, determined by epigenetic modifications. Disruption of normal epigenetic modifications during a developmentally sensitive time can alter gene expression and may contribute to changes in alveolar formation.

EPIGENETICS IN THE DEVELOPMENTAL ORIGINS OF LUNG DISEASE

Perinatal insults alter the lung epigenome and therefore alter the transcription of key genes during lung development.[50,70] The long-term implications of developmentally disrupted epigenetics and gene transcription are twofold. First, alterations in gene transcription can change the final structure and function of an organ, for example, via changes in levels of apoptotic, proliferative, or extracellular matrix genes. Second, when the epigenetic determinants of a gene's expression are altered by an insult during development, the altered epigenetic determinants disrupt the preset developmental course by becoming the new epigenetic platform upon which future modifications are built. Implications of this second outcome may be anticipated to change the way the lung responds to subsequent stressors, even after development is complete.

Epigenetic Basics

Epigenetic mechanisms form the basis of the regulation of gene expression. Epigenetic modifications dictate developmentally-specific gene expression as well as cell-specific and organ-specific gene expression. Regulation of developmentally and location-specific gene expression tends to involve epigenetically regulated gene activation or silencing. However, epigenetic modifications do not just dictate an "on or off" effect, epigenetic modifications also determine the magnitude of expression of a particular gene. Short and long-term changes to the magnitude of gene expression in the lung contribute to the developmental origins of lung disease.

Epigenetic mechanisms regulate gene expression by directing the interactions among the transcription machinery, transcription factors and specific regions of the DNA. In eukaryotic cells, DNA is packaged in the nucleus in combination with nuclear proteins of the histone family. The DNA and protein combination is known as chromatin. Chromatin consists of repeating units of 147 base pairs of DNA wrapped twice around a core of 8 histone proteins, forming a unit called a nucleosome.[71] The protein core consists of two copies each of 4 histone proteins H2A, HB2, H3, and H4. Adjacent nucleosomes are connected by short pieces of linker DNA (Figure 2). Epigenetic regulating mechanisms include DNA methylation, covalent histone modifications, and microRNAs.

DNA Methylation

One of the most studied epigenetic modifications is DNA methylation. DNA methylation occurs primarily on the cytosine (C) of a CpG dinucleotide. Overall CpG density in the mammalian genome is relatively low. Despite overall low levels, however, CpGs tend to be clustered in CpG "islands," consisting of more than a 200 base pair region with a CG content of at least 50%.[72] CpG islands are commonly found in the promoter region of mammalian genes and are usually methylated, which is associated with gene silencing. At promoters, DNA methylation prevents transcription directly by blocking the binding of transcriptional activators. Alternatively, DNA methylation at promoter regions prevents transcription indirectly by recruiting methyl-binding complexes that contain histone deacetylase (HDAC) activity.[73,74]

In contrast, low density CpG regions are found in the coding and other regions of genes, as well as between genes, and are more frequently methylated.[75–77] Low density CpGs located in the coding region tend to be associated with exons more than with introns.[78] Methylation of low density CpG regions may have a role in alternative exon usage. Methylation of DNA is significantly enhanced in regions

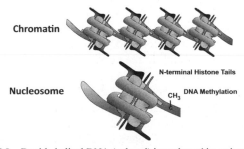

FIGURE 2 Double helical DNA (colored) is packaged in an increasingly complex protein scaffold, collectively known as chromatin. At its lowest level, the DNA is wrapped twice around a protein core, forming a unit called a nucleosome (bottom). The protein core consists of two copies each of four different histone proteins H2A, HB2, H3, and H4. Epigenetic modifications include methylation of the DNA as well as covalent modifications to the N-terminal tails of the histone proteins. Adapted from[118]: reproduced with permission from NeoReviews, Vol. 12, pages e498-e505, Copyright 2011 by the AAP.

of exons that are included in the final transcript compared with exons that are spliced out.[77] The mechanisms by which exonic DNA methylation affects alternative splicing are still being elucidated and may involve recruitment of methyl CpG binding protein 2 (MeCP2) and subsequent HDAC activities, or physical slowing of RNA polymerase II transit, allowing splicing to take place.[77,79]

Histone Modifications

The nucleosome core is occupied by the globular portion of the histone proteins, while the unstructured, N-terminal "tails" of the histone proteins extend freely from the nucleosome.[71] Post-translational, covalent modifications to histone proteins occur largely, but not exclusively, on these tails (reviewed in[80]). The modifications (marks) include acetylation, mono-, di- and tri-methylation, phosphorylation, and ubiquitylation.

The complexity of potential combinations of histone modifications along a gene is high because of the potential number of modifications, modifiable amino acids, and the number of nucleosomes along the length of a gene. However, evidence supports a limited number of combinations of histone modifications. Genome-wide mapping of global patterns of histone modifications with techniques, such as chromatin immunoprecipitation with parallel DNA sequencing (ChIP-seq), have revealed that distinct patterns of histone modifications are associated with distinct elements within the DNA of a genome.[80,81]

Specific gene regions can be associated with particular modification. For example, promoters tend to have high levels of histone 3 (H3), lysine 4 (K4) trimethylation (me^3), while putative enhancers are characterized by enriched H3K4me^1 alone or with H3K27 (acetylation) ac or H3K27me^3.[82,83] Gene bodies tend to be enriched with H3K36me^2 (or me^3) in association with transcriptional activation. H3K4me^2 (or^3) and K36me^2 (or me^3) may contribute to regulating stability of the nucleosome during RNA polymerase II transit.[80,84] Also enriched in the promoter and body of the gene is H4K20me^1.[85] H4K20me^1 may function in transcriptional initiation and promoter clearance as well as nucleosome stability in the body of the gene.[86]

The dynamic nature of histone modifications is important in regulating gene expression. Genes that are being actively transcribed, or genes that are poised for transcription pending an activating signal (such as transcription factor binding to an enhancer), are characterized by rapid acetylation and deacetylation.[87–89] Acetyl groups are placed on histones by histone acetyltransferase (HAT) enzymes and removed by histone deacetylatase (HDAC) enzymes. Under basal conditions, chromatin that is being rapidly acetylated and deacetylated is enriched in HAT activity as well as HDAC activity. Rapid local changes in histone acetylation occur on nucleosomes at or near promoters at the time of activation

of transcription from poised genes.[90] While mechanisms are still being elucidated, evidence suggests that rapid acetylation and deacetylation may facilitate nucleosome mobilization during polymerase transit, thus physically facilitating transcriptional activation and elongation.[90–92]

MicroRNA

MicroRNAs (miRNA) are small non-coding RNAs that function to reduce protein translation (reviewed in[93]). MicroRNAs may be transcribed from independent genes or from within introns of coding genes. The transcription of miRNAs is controlled by the same epigenetic mechanisms as coding genes. Pre-miRNAs undergo processing by Drosha and Dicer enzymes to produce miRNAs approximately 21 base-pairs long. The processed miRNAs bind to complementary sequences on the 3′ untranslated region (3′UTR) of target mRNAs and prevent translation by either accelerating mRNA degradation, or by blocking the passage of the ribosome (Figure 3).

Loss-of-function and gain-of-function studies indicate that miRNA may have a limited role in normal cellular states. However, during times of cellular stress, such as that invoked by perinatal insults, miRNAs direct translation to optimize stress response pathways.[94] Interestingly, a subset of miRNAs also controls the expression of chromatin modifying enzymes, including DNA methyltransferases and histone deacetylases.[95] Given that the transcription, processing, and turnover of miRNAs is particularly impacted by cell stress, an epigenetic-miRNA regulatory circuit may be important in the transcriptional response to cellular stressors such as those induced by perinatal insults.[94,95]

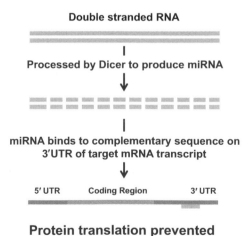

Protein translation prevented

FIGURE 3   Double stranded RNA is transcribed from an independent gene, or from within an intron of a coding gene. The RNA is processed by the Dicer complex. Processed microRNA (miRNA) bind to complementary sequences in the 3′ untranslated region (3′UTR) of target mRNA transcript. Binding of the miRNA prevents translation of the mRNA into protein by either accelerating mRNA degradation, or by blocking passage of the ribosome.

Understanding the complex nature of epigenetic modifications and how they regulate eukaryotic gene transcription is far from complete; however, several concepts have emerged. First, epigenetic modifications need to be considered in the context of one another because no single modification dictates the transcriptional fate of a gene. Second, the effects of epigenetic modifications must be considered along the entire length of a gene, not just the 5′ regulatory region. Finally, genes that have complex architecture (e.g., multiple promoters, alternative exon usage, multiple 3′ termination sites) are generally characterized by high levels of epigenetic complexity.

EPIGENETICS IN THE DEVELOPMENTAL ORIGINS OF LUNG DISEASE

Environmental factors, including perinatal insults, lead to epigenetic changes and subsequent changes in gene expression. While we understand that epigenetic changes occur in the context of perinatal insults, and we know that concomitant changes in gene expression and subsequent phenotype occur, the causal relationships and mechanistic details are poorly understood. Research into the developmental origins of lung disease is just beginning to scratch the surface of this complex and important topic.

The identification of relevant candidate genes is important in the developmental origins of lung disease. The high complexity of epigenetic regulation and the myriad ways that perinatal insults can affect lung epigenetics imply that many genes may be affected. In order to ultimately develop targeted interventions, it is necessary to focus on genes that are causative and amenable to manipulation. Several characteristics characterize candidate genes in the developmental origins of lung disease. Candidate genes include those that are crucial for lung development and/or subsequent response to injury. Candidate genes are also likely to have complex gene architecture, such as being transcribed from multiple promoters, having alternative exon usage and alternative termination sites. Expression of genes with complex gene architecture requires more epigenetic regulation than expression of genes with simple architecture. Genes requiring higher levels of epigenetic regulation to ensure proper expression are more likely to be disrupted in the face of environmentally-induced epigenetic changes.

Candidate genes such as peroxisome proliferator activated receptor gamma (PPARγ) and insulin-like growth factor 1 (IGF1) are epigenetically altered by perinatal insults. Both genes have complex architecture, with transcription from multiple promoters and alternative exon usage.[96,97] Crucial to the developmental origins of lung disease, both PPARγ and IGF1 have integral roles in lung development and disease. Integral roles are also shared with histone modifying enzymes, because these enzymes add or remove the histone marks that regulate gene expression (e.g., HATs, HDACs).

PPARγ

PPARγ, a key player in lung development and lung repair, is a member of the nuclear receptor family of transcription factors.[98–100] Roles for PPARγ in the lung are varied and include epithelial–mesenchymal interactions, lipid homeostasis, and inflammatory control.[101–103] However, perhaps the most interesting role of PPARγ is the direct transcriptional regulation of epigenetic modifying enzymes. Several chromatin modifying enzymes have PPAR response elements (PPRE) in their promoters and are bona fide transcriptional targets of PPARγ.[104] One of these PPARγ responsive genes is the set domain containing histone methyltransferase, Setd8, which places the H4K20me[1] mark in the promoter region and gene body of target genes.[104]

Setd8 promotes cellular differentiation by methylating multiple target genes at the same time, thus allowing a rapid complex response to a single stimulus.[104] Recent evidence demonstrates that H4K20me[1] and Setd8 mediate target gene activation in wingless (Wnt) signaling.[105] Wnt signaling is important in the context of the developmental origins of lung disease because it is essential for both lung development as well as repair after lung injury.[106–108]

IUGR alters the expression and epigenetics of PPARγ and Setd8 in the rat lung.[100] IUGR decreases levels of PPARγ expression in parallel with reductions in Setd8. In the setting of reduced Setd8, levels of genome-wide H4K20me[1] are reduced. Gene-specific levels of H4K20me[1] along the PPARγ gene are also reduced, most notably at Exon 4 within the body of the PPARγ gene.[100] The PPARγ-Setd8-H4K20me[1] axis can be enhanced with activation of PPARγ. Supplementation of IUGR rats with the PPARγ agonist, docosahexanoic acid (DHA), restores PPARγ levels, Setd8 levels as well as global and gene-specific H4K20me[1].[100] These results suggest a potential way to treat perturbations in regulation of gene expression after a perinatal insult.

Other epigenetic effects of IUGR on the PPARγ gene include alterations in H3 methylation along the PPARγ gene, in a sex-specific manner.[109] Specifically, IUGR decreases H3K9me[3] along the PPARγ gene in male neonatal rats. In contrast, IUGR increases H3K9me[3] along the PPARγ gene in female neonatal rats.[109] Sex-divergence in H3K9me[3] is interesting because it is not apparent control (non-IUGR) rats. Sex-divergent changes in PPARγ epigenetics following IUGR demonstrate that males respond differently than females to the insult of IUGR.

IUGR also has sex-divergent effects on chromatin interactions with proteins that regulate the histone modifications. The methyl binding protein MeCP2 bridges DNA methylation and histone modifications. MeCp2 binding to methylated DNA associates with histone methyltransferase activity resulting in H3K9me^3, and transcriptional repression.[110] In the rat lung, IUGR alters both MeCP2 expression and MeCP2 occupancy of the PPARγ promoters in a sex-divergent manner.[109] IUGR does not affect MeCP2 levels or occupancy at the PPARγ promoters in male lung. In contrast, IUGR increases MeCP2 expression and MeCP2 occupancy of the PPARγ promoters in female lung.[109] Thus, epigenetic mechanisms may participate in sex-specific responses to perinatal insults.

IGF1

Another gene that is important for alveolar formation is IGF1. Lung IGF1 expression is increased in preterm infants who died from respiratory distress syndrome or BPD.[111] Increased IGF1 expression is also present in the lung of preterm lambs following prolonged MV.

In lambs, both prematurity and MV affect the histone code defining the IGF1 regulatory sequences and gene body. Compared to term lambs, prematurity alters histone modifications within the two promoter regions of IGF1. For both promoter regions, prematurity decreases H3K14ac, and increases H3K4me^3. MV of preterm lambs increases H3K4me^3 within promoter 1 and H3K36me^3 within the body of IGF1 (exons 4 and 6) compared to nasal HFNV.[112] The mechanism by which the mode of ventilation changes the IGF-1 histone code is currently unknown. The implication of changes in the IGF1 histone code with MV is that MV during lung development may change the way IGF1 gene transcription is regulated both in the short-term and in the long-term. Increased IGF1 transcription may contribute to the lung pathologies observed in the preterm lambs managed by MV.

Histone Deacetylation

The HDAC family of enzymes removes acetyl groups from lysine residues of histone proteins, and facilitates the rapid deacetylation of transcriptionally active genes. HDAC1 and 2 are both Class I deacetylases that target similar substrates. However, HDAC1 and HDAC2 also have unique substrates. HDAC1 specific substrates include the androgen receptor and p53, while HDAC2 specific substrates include the glucocorticoid receptor (GR).[113] Substrate specificity of HDAC enzymes allows for fine-tuning of deacetylase activity in the regulation of gene expression.

HDAC1 expression is increased in the lungs of preterm lambs supported by MV. Examination of acetylation levels in the lungs of MV preterm lambs reveals genome-wide histone *hypo*acetylation. In contrast, genome-wide histone *hyper*acetylation is associated with non-invasive high-frequency nasal ventilation, the ventilation mode less likely to be associated with lung injury. Histone modifications H3K14ac, H3K18ac, H3K27ac, H3K36me^3 are lower in lungs of preterm lambs managed by MV than high frequency nasal ventilation. Taken together, these data suggest that MV upsets the acetylation–deacetylation equilibrium in the lung by increasing deacetylation. Notably, when preterm lambs are treated with the HDAC inhibitors, valproic acid or trichostatin A, all histone modifications increase in MV group. Measures of alveolar formation are improved with increased histone acetylation following inhibition of HDACs in preterm lambs managed by MV.[114]

HDAC2 is also important in the development and management of lung disease. Corticosteroids suppress inflammation by recruiting HDAC2 to NF-kappa B driven proinflammatory gene promoters and inhibiting transcription.[115] HDAC2 also deacetylates GR, thus enabling GR to bind to the NF-kappa B complex, again leading to the inhibition of NF-kappa B-dependent proinflammatory gene transcription.[116] Levels and activity of HDAC2 are reduced in the lungs of patients with chronic obstructive pulmonary disease, as well as in mice exposed to tobacco smoke.[117] The result of decreased HDAC2 level and activity is that corticosteroids are not effective mediators of inflammation in chronic obstructive pulmonary disease, and steroid resistance ensues.

FUTURE PERSPECTIVES

While the complex mechanistic role of epigenetics in the developmental origins of lung disease is beginning to be revealed, several important questions remain unanswered. Given that the long-term goals of the field are to improve the lung health outcomes of individuals exposed to perinatal insults, the development of targeted interventions is vital. To develop targeted interventions, a complete understanding of the developmental origins of lung disease is needed. For instance, greater mechanistic understanding of the means by which perinatal insults cause later lung disease will facilitate development of interventions. Because of the fundamental nature of epigenetics in regulating gene expression, interventions will need to be as specific as possible and delivered as closely as possible to the lung. Fortunately, accessibility of the lung via the airways (nebulization) brings promise to the possibility of targeted delivery of treatment compounds to the lung.

Another bright spot in rectifying epigenetic defects lies in the concept that epigenetic changes result from altered environmental cues and thus stand to be corrected by other environmental cues. For example, epigenetic changes stemming from deficient signaling through PPARγ may be corrected by dietary normalization of PPARγ signaling.

CONCLUSIONS

Perinatal insults such as preterm birth and MV with oxygen support, IUGR, or MTS exposure during critical periods of lung development predispose to impaired lung function and lung disease in later life. A molecular response to perinatal insults involves altered epigenetic mechanisms as well as altered expression of key lung genes and pathways. A detailed understanding of the molecular mechanisms driving the developmental origins of lung disease has the exciting potential to lead to effective treatments and improve lung health in susceptible individuals.

ACKNOWLEDGMENTS

This work was supported by National Institute of Heath grants HL062875(KHA), HL110002(KHA), HL07744(KHA), and DK084036(LJM) and the Department of Pediatrics at the University of Utah.

REFERENCES

1. Barker DJ, Osmond C. Infant mortality, childhood nutrition, and ischaemic heart disease in England and Wales. *Lancet* 1986;**1**:1077–81.
2. Barker DJ, Osmond C, Golding J, Kuh D, Wadsworth ME. Growth in utero, blood pressure in childhood and adult life, and mortality from cardiovascular disease. *BMJ* 1989;**298**:564–7.
3. Hales CN, Barker DJ, Clark PM, Cox LJ, Fall C, Osmond C, et al. Fetal and infant growth and impaired glucose tolerance at age 64. *BMJ* 1991;**303**:1019–22.
4. Phipps K, Barker DJ, Hales CN, Fall CH, Osmond C, Clark PM. Fetal growth and impaired glucose tolerance in men and women. *Diabetologia* 1993;**36**:225–8.
5. Barker DJ, Godfrey KM, Fall C, Osmond C, Winter PD, Shaheen SO. Relation of birth weight and childhood respiratory infection to adult lung function and death from chronic obstructive airways disease. *BMJ* 1991;**303**:671–5.
6. Shaheen S, Barker DJ. Early lung growth and chronic airflow obstruction. *Thorax* 1994;**49**:533–6.
7. Shaheen SO, Barker DJ, Shiell AW, Crocker FJ, Wield GA, Holgate ST. The relationship between pneumonia in early childhood and impaired lung function in late adult life. *Am J Respir Crit Care Med* 1994;**149**:616–9.
8. Jobe AH. The new bronchopulmonary dysplasia. *Curr Opin Pediatr* 2011;**23**:167–72.
9. Albertine KH. Progress in understanding the pathogenesis of BPD using the baboon and sheep models. *Semin Perinatol* 2013;**37**:60–8.
10. Doyle LW, Faber B, Callanan C, Freezer N, Ford GW, Davis NM. Bronchopulmonary dysplasia in very low birth weight subjects and lung function in late adolescence. *Pediatrics* 2006;**118**:108–13.
11. Greenough A. Long-term pulmonary outcome in the preterm infant. *Neonatology* 2008;**93**:324–7.
12. Kairamkonda VR, Richardson J, Subhedar N, Bridge PD, Shaw NJ. Lung function measurement in prematurely born preschool children with and without chronic lung disease. *J Perinatol* 2008;**28**:199–204.
13. Pei L, Chen G, Mi J, Zhang T, Song X, Chen J, et al. Low birth weight and lung function in adulthood: retrospective cohort study in China, 1948–1996. *Pediatrics* 2010;**125**:e899–905.
14. Wong PM, Lees AN, Louw J, Lee FY, French N, Gain K, et al. Emphysema in young adult survivors of moderate-to-severe bronchopulmonary dysplasia. *Eur Respir J* 2008;**32**:321–8.
15. Kinzler WL, Vintzileos AM. Fetal growth restriction: a modern approach. *Curr Opin Obstet Gynecol* 2008;**20**:125–31.
16. Cetin I, Alvino G. Intrauterine growth restriction: implications for placental metabolism and transport. A review. *Placenta* 2009;**30**(Suppl. A):S77–82.
17. Rosenberg A. The IUGR newborn. *Semin Perinatol* 2008;**32**:219–24.
18. Naeye RL, Burt LS, Wright DL, Blanc WA, Tatter D. Neonatal mortality, the male disadvantage. *Pediatrics* 1971;**48**:902–6.
19. Khoury MJ, Marks JS, McCarthy BJ, Zaro SM. Factors affecting the sex differential in neonatal mortality: the role of respiratory distress syndrome. *Am J Obstet Gynecol* 1985;**151**:777–82.
20. Chen SJ, Vohr BR, Oh W. Effects of birth order, gender, and intrauterine growth retardation on the outcome of very low birth weight in twins. *J Pediatr* 1993;**123**:132–6.
21. Copper RL, Goldenberg RL, Creasy RK, DuBard MB, Davis RO, Entman SS, et al. A multicenter study of preterm birth weight and gestational age-specific neonatal mortality. *Am J Obstet Gynecol* 1993;**168**:78–84.
22. Fanaroff AA, Wright LL, Stevenson DK, Shankaran S, Donovan EF, Ehrenkranz RA, et al. Very-low-birth-weight outcomes of the National Institute of Child Health and Human Development Neonatal Research Network, May 1991 through December 1992. *Am J Obstet Gynecol* 1995;**173**:1423–31.
23. Jennische M, Sedin G. Gender differences in outcome after neonatal intensive care: speech and language skills are less influenced in boys than in girls at 6.5 years. *Acta Paediatr* 2003;**92**:364–78.
24. McIntire DD, Bloom SL, Casey BM, Leveno KJ. Birth weight in relation to morbidity and mortality among newborn infants. *N Engl J Med* 1999;**340**:1234–8.
25. Minior VK, Divon MY. Fetal growth restriction at term: myth or reality? *Obstet Gynecol* 1998;**92**:57–60.
26. Hoo AF, Stocks J, Lum S, Wade AM, Castle RA, Costeloe KL. Development of lung function in early life: influence of birth weight in infants of nonsmokers. *Am J Respir Crit Care Med* 2004;**170**:527–33.
27. Tyson JE, Kennedy K, Broyles S, Rosenfeld CR. The small for gestational age infant: accelerated or delayed pulmonary maturation? Increased or decreased survival? *Pediatrics* 1995;**95**:534–8.
28. Lucas JS, Inskip HM, Godfrey KM, Foreman CT, Warner JO, Gregson RK, et al. Small size at birth and greater postnatal weight gain: relationships to diminished infant lung function. *Am J Respir Crit Care Med* 2004;**170**:534–40.
29. Bose C, Van Marter LJ, Laughon M, O'Shea TM, Allred EN, Karna P, et al. Fetal growth restriction and chronic lung disease among infants born before the 28th week of gestation. *Pediatrics* 2009;**124**:e450–8.
30. Regev RH, Lusky A, Dolfin T, Litmanovitz I, Arnon S, Reichman B. Excess mortality and morbidity among small-for-gestational-age premature infants: a population-based study. *J Pediatr* 2003;**143**:186–91.
31. Reiss I, Landmann E, Heckmann M, Misselwitz B, Gortner L. Increased risk of bronchopulmonary dysplasia and increased mortality in very preterm infants being small for gestational age. *Arch Gynecol Obstet* 2003;**269**:40–4.
32. Torrance HL, Mulder EJ, Brouwers HA, van Bel F, Visser GH. Respiratory outcome in preterm small for gestational age fetuses with or without abnormal umbilical artery Doppler and/or maternal hypertension. *J Matern Fetal Neonatal Med* 2007;**20**:613–21.
33. Edwards CA, Osman LM, Godden DJ, Campbell DM, Douglas JG. Relationship between birth weight and adult lung function: controlling for maternal factors. *Thorax* 2003;**58**:1061–5.

34. Lawlor DA, Ebrahim S, Davey Smith G. Association of birth weight with adult lung function: findings from the British Women's Heart and Health Study and a meta-analysis. *Thorax* 2005;**60**:851–8.

35. Rona RJ, Gulliford MC, Chinn S. Effects of prematurity and intra-uterine growth on respiratory health and lung function in childhood. *BMJ* 1993;**306**:817–20.

36. Alati R, Al Mamun A, O'Callaghan M, Najman JM, Williams GM. In utero and postnatal maternal smoking and asthma in adolescence. *Epidemiology* 2006;**17**:138–44.

37. Gilliland FD, Berhane K, Li YF, Rappaport EB, Peters JM. Effects of early onset asthma and in utero exposure to maternal smoking on childhood lung function. *Am J Respir Crit Care Med* 2003;**167**:917–24.

38. Jaakkola JJ, Gissler M. Maternal smoking in pregnancy, fetal development, and childhood asthma. *Am J Public Health* 2004;**94**:136–40.

39. Hylkema MN, Blacquiere MJ. Intrauterine effects of maternal smoking on sensitization, asthma, and chronic obstructive pulmonary disease. *Proc Am Thorac Soc* 2009;**6**:660–2.

40. Henderson AJ, Newson RB, Rose-Zerilli M, Ring SM, Holloway JW, Shaheen SO. Maternal Nrf2 and gluthathione-S-transferase polymorphisms do not modify associations of prenatal tobacco smoke exposure with asthma and lung function in school-aged children. *Thorax* 2010;**65**:897–902.

41. Singh SP, Barrett EG, Kalra R, Razani-Boroujerdi S, Langley RJ, Kurup V, et al. Prenatal cigarette smoke decreases lung cAMP and increases airway hyperresponsiveness. *Am J Respir Crit Care Med* 2003;**168**:342–7.

42. Cunningham J, Dockery DW, Speizer FE. Maternal smoking during pregnancy as a predictor of lung function in children. *Am J Epidemiol* 1994;**139**:1139–52.

43. Gilliland FD, Berhane K, McConnell R, Gauderman WJ, Vora H, Rappaport EB, et al. Maternal smoking during pregnancy, environmental tobacco smoke exposure and childhood lung function. *Thorax* 2000;**55**:271–6.

44. Tager IB, Ngo L, Hanrahan JP. Maternal smoking during pregnancy. Effects on lung function during the first 18 months of life. *Am J Respir Crit Care Med* 1995;**152**:977–83.

45. Burri PH. Fetal and postnatal development of the lung. *Annu Rev Physiol* 1984;**46**:617–28.

46. Burri PH. Structural aspects of postnatal lung development - alveolar formation and growth. *Biol Neonate* 2006;**89**:313–22.

47. Alcorn DG, Adamson TM, Maloney JE, Robinson PM. A morphologic and morphometric analysis of fetal lung development in the sheep. *Anat Rec* 1981;**201**:655–67.

48. Burri PH, Dbaly J, Weibel ER. The postnatal growth of the rat lung. I. Morphometry. *Anat Rec* 1974;**178**:711–30.

49. Burri PH, Moschopulos M. Structural analysis of fetal rat lung development. *Anat Rec* 1992;**234**:399–418.

50. Joss-Moore LA, Lane RH. The developmental origins of adult disease. *Curr Opin Pediatr* 2009;**21**:230–4.

51. Harding R, Maritz G. Maternal and fetal origins of lung disease in adulthood. *Semin Fetal Neonatal Med* 2012;**17**:67–72.

52. Briana DD, Malamitsi-Puchner A. Small for gestational age birth weight: impact on lung structure and function. *Paediatr Respir Rev* 2013;**14**:256–62.

53. Albertine KH, Jones GP, Starcher BC, Bohnsack JF, Davis PL, Cho SC, et al. Chronic lung injury in preterm lambs. Disordered respiratory tract development. *Am J Respir Crit Care Med* 1999;**159**:945–58.

54. Coalson JJ, Winter V, deLemos RA. Decreased alveolarization in baboon survivors with bronchopulmonary dysplasia. *Am J Respir Crit Care Med* 1995;**152**:640–6.

55. Harding R, Cock ML, Louey S, Joyce BJ, Davey MG, Albuquerque CA, et al. The compromised intra-uterine environment: implications for future lung health. *Clin Exp Pharmacol Physiol* 2000;**27**:965–74.

56. Maritz GS, Cock ML, Louey S, Joyce BJ, Albuquerque CA, Harding R. Effects of fetal growth restriction on lung development before and after birth: a morphometric analysis. *Pediatr Pulmonol* 2001;**32**:201–10.

57. Maritz GS, Cock ML, Louey S, Suzuki K, Harding R. Fetal growth restriction has long-term effects on postnatal lung structure in sheep. *Pediatr Res* 2004;**55**:287–95.

58. O'Brien EA, Barnes V, Zhao L, McKnight RA, Yu X, Callaway CW, et al. Uteroplacental insufficiency decreases p53 serine-15 phosphorylation in term IUGR rat lungs. *Am J Physiol Regul Integr Comp Physiol* 2007;**293**:R314–22.

59. Karadag A, Sakurai R, Wang Y, Guo P, Desai M, Ross MG, et al. Effect of maternal food restriction on fetal rat lung lipid differentiation program. *Pediatr Pulmonol* 2009;**44**:635–44.

60. Blacquiere MJ, Timens W, Melgert BN, Geerlings M, Postma DS, Hylkema MN. Maternal smoking during pregnancy induces airway remodelling in mice offspring. *Eur Respir J* 2009;**33**:1133–40.

61. Collins MH, Moessinger AC, Kleinerman J, Bassi J, Rosso P, Collins AM, et al. Fetal lung hypoplasia associated with maternal smoking: a morphometric analysis. *Pediatr Res* 1985;**19**:408–12.

62. Avdalovic M, Putney, Tyler N, Finkbeiner W, Pinkerton K, Hyde D. In utero and postnatal exposure to environmental tobacco smoke (ETS) alters alveolar and respiratory bronchiole (RB) growth and development in infant monkeys. *Toxicol Pathol* 2009;**37**:256–63.

63. Manoli SE, Smith LA, Vyhlidal CA, An CH, Porrata Y, Cardoso WV, et al. Maternal smoking and the retinoid pathway in the developing lung. *Respir Res* 2012;**13**:42.

64. Sandberg K, Poole SD, Hamdan A, Arbogast P, Sundell HW. Altered lung development after prenatal nicotine exposure in young lambs. *Pediatr Res* 2004;**56**:432–9.

65. Sekhon HS, Jia Y, Raab R, Kuryatov A, Pankow JF, Whitsett JA, et al. Prenatal nicotine increases pulmonary alpha7 nicotinic receptor expression and alters fetal lung development in monkeys. *J Clin Invest* 1999;**103**:637–47.

66. Rehan VK, Sakurai R, Wang Y, Santos J, Huynh K, Torday JS. Reversal of nicotine-induced alveolar lipofibroblast-to-myofibroblast transdifferentiation by stimulants of parathyroid hormone-related protein signaling. *Lung* 2007;**185**:151–9.

67. Maritz GS, Dennis H. Maternal nicotine exposure during gestation and lactation interferes with alveolar development in the neonatal lung. *Reprod Fertil Dev* 1998;**10**:255–61.

68. Leslie FM. Multigenerational epigenetic effects of nicotine on lung function. *BMC Med* 2013;**11**:27.

69. Rehan VK, Liu J, Naeem E, Tian J, Sakurai R, Kwong K, et al. Perinatal nicotine exposure induces asthma in second generation offspring. *BMC Med* 2012;**10**:129.

70. Joss-Moore LA, Metcalfe DB, Albertine KH, McKnight RA, Lane RH. Epigenetics and fetal adaptation to perinatal events: diversity through fidelity. *J Anim Sci* 2010;**88**(13 Suppl.):E216–22.

71. Luger K, Mader AW, Richmond RK, Sargent DF, Richmond TJ. Crystal structure of the nucleosome core particle at 2.8 A resolution. *Nature* 1997;**389**:251–60.

72. Gardiner-Garden M, Frommer M. CpG islands in vertebrate genomes. *J Mol Biol* 1987;**196**:261–82.

73. Huh I, Zeng J, Park T, Yi SV. DNA methylation and transcriptional noise. *Epigenetics Chromatin* 2013;**6**:9.

74. Klose RJ, Bird AP. Genomic DNA methylation: the mark and its mediators. *Trends Biochem Sci* 2006;**31**:89–97.

75. Illingworth R, Kerr A, Desousa D, Jorgensen H, Ellis P, Stalker J, et al. A novel CpG island set identifies tissue-specific methylation at developmental gene loci. *PLoS Biol* 2008:6. e22.

76. Liang P, Song F, Ghosh S, Morien E, Qin M, Mahmood S, et al. Genome-wide survey reveals dynamic widespread tissue-specific changes in DNA methylation during development. *BMC Genomics* 2011;**12**:231.

77. Maunakea AK, Chepelev I, Cui K, Zhao K. Intragenic DNA methylation modulates alternative splicing by recruiting MeCP2 to promote exon recognition. *Cell Res* 2013;**23**:1256–69.

78. Choi JK. Contrasting chromatin organization of CpG islands and exons in the human genome. *Genome Biol* 2010;**11**:R70.

79. Shukla S, Kavak E, Gregory M, Imashimizu M, Shutinoski B, Kashlev M, et al. CTCF-promoted RNA polymerase II pausing links DNA methylation to splicing. *Nature* 2011;**479**:74–9.

80. Zentner GE, Henikoff S. Regulation of nucleosome dynamics by histone modifications. *Nat Struct Mol Biol* 2013;**20**:259–66.

81. Huff JT, Plocik AM, Guthrie C, Yamamoto KR. Reciprocal intronic and exonic histone modification regions in humans. *Nat Struct Mol Biol* 2010;**17**:1495–9.

82. Rada-Iglesias A, Bajpai R, Swigut T, Brugmann SA, Flynn RA, Wysocka J. A unique chromatin signature uncovers early developmental enhancers in humans. *Nature* 2011;**470**:279–83.

83. Zentner GE, Tesar PJ, Scacheri PC. Epigenetic signatures distinguish multiple classes of enhancers with distinct cellular functions. *Genome Res* 2011;**21**:1273–83.

84. Wagner EJ, Carpenter PB. Understanding the language of Lys36 methylation at histone H3. *Nat Rev Mol Cell Biol* 2012;**13**:115–26.

85. Smolle M, Workman JL. Transcription-associated histone modifications and cryptic transcription. *Biochim Biophys Acta* 2013;**1829**:84–97.

86. Karlic R, Chung HR, Lasserre J, Vlahovicek K, Vingron M. Histone modification levels are predictive for gene expression. *Proc Natl Acad Sci U S A* 2010;**107**:2926–31.

87. Barth TK, Imhof A. Fast signals and slow marks: the dynamics of histone modifications. *Trends Biochem Sci* 2010;**35**:618–26.

88. Zhang DE, Nelson DA. Histone acetylation in chicken erythrocytes. Rates of acetylation and evidence that histones in both active and potentially active chromatin are rapidly modified. *Biochem J* 1988;**250**:233–40.

89. Spencer VA, Davie JR. Dynamically acetylated histone association with transcriptionally active and competent genes in the avian adult beta-globin gene domain. *J Biol Chem* 2001;**276**:34810–5.

90. Waterborg JH. Dynamics of histone acetylation in vivo. A function for acetylation turnover? *Biochem Cell Biol* 2002;**80**:363–78.

91. Reinke H, Gregory PD, Horz W. A transient histone hyperacetylation signal marks nucleosomes for remodeling at the PHO8 promoter in vivo. *Mol Cell* 2001;**7**:529–38.

92. Walia H, Chen HY, Sun JM, Holth LT, Davie JR. Histone acetylation is required to maintain the unfolded nucleosome structure associated with transcribing DNA. *J Biol Chem* 1998;**273**:14516–22.

93. Krol J, Loedige I, Filipowicz W. The widespread regulation of microRNA biogenesis, function and decay. *Nat Rev Genet* 2010;**11**:597–610.

94. Leung AK, Sharp PA. MicroRNA functions in stress responses. *Mol Cell* 2010;**40**:205–15.

95. Sato F, Tsuchiya S, Meltzer SJ, Shimizu K. MicroRNAs and epigenetics. *FEBS J* 2011;**278**:1598–609.

96. Fajas L, Auboeuf D, Raspe E, Schoonjans K, Lefebvre AM, Saladin R, et al. The organization, promoter analysis, and expression of the human PPARgamma gene. *J Biol Chem* 1997;**272**:18779–89.

97. Fu Q, Yu X, Callaway CW, Lane RH, McKnight RA. Epigenetics: intrauterine growth retardation (IUGR) modifies the histone code along the rat hepatic IGF-1 gene. *FASEB J* 2009;**23**:2438–49.

98. Simon DM, Arikan MC, Srisuma S, Bhattacharya S, Tsai LW, Ingenito EP, et al. Epithelial cell PPAR[gamma] contributes to normal lung maturation. *Faseb J* 2006;**20**:1507–9.

99. Cerny L, Torday JS, Rehan VK. Prevention and treatment of bronchopulmonary dysplasia: contemporary status and future outlook. *Lung* 2008;**186**:75–89.

100. Joss-Moore LA, Wang Y, Baack ML, Yao J, Norris AW, Yu X, et al. IUGR decreases PPARgamma and SETD8 Expression in neonatal rat lung and these effects are ameliorated by maternal DHA supplementation. *Early Hum Dev* 2010;**86**:785–91.

101. Wang Y, Santos J, Sakurai R, Shin E, Cerny L, Torday JS, et al. Peroxisome proliferator-activated receptor gamma agonists enhance lung maturation in a neonatal rat model. *Pediatr Res* 2009;**65**:150–5.

102. Simon DM, Arikan MC, Srisuma S, Bhattacharya S, Andalcio T, Shapiro SD, et al. Epithelial cell PPARgamma is an endogenous regulator of normal lung maturation and maintenance. *Proc Am Thorac Soc* 2006;**3**:510–1.

103. Lian X, Yan C, Qin Y, Knox L, Li T, Du H. Neutral lipids and peroxisome proliferator-activated receptor-{gamma} control pulmonary gene expression and inflammation-triggered pathogenesis in lysosomal acid lipase knockout mice. *Am J Pathol* 2005;**167**:813–21.

104. Wakabayashi K, Okamura M, Tsutsumi S, Nishikawa NS, Tanaka T, Sakakibara I, et al. The peroxisome proliferator-activated receptor gamma/retinoid X receptor alpha heterodimer targets the histone modification enzyme PR-Set7/Setd8 gene and regulates adipogenesis through a positive feedback loop. *Mol Cell Biol* 2009;**29**:3544–55.

105. Li Z, Nie F, Wang S, Li L. Histone H4 Lys 20 monomethylation by histone methylase SET8 mediates Wnt target gene activation. *Proc Natl Acad Sci U S A* 2011;**108**:3116–23.

106. Dasgupta C, Sakurai R, Wang Y, Guo P, Ambalavanan N, Torday JS, et al. Hyperoxia-induced neonatal rat lung injury involves activation of TGF-{beta} and Wnt signaling and is protected by rosiglitazone. *Am J Physiol Lung Cell Mol Physiol* 2009;**296**:L1031–41.

107. Crosby LM, Waters CM. Epithelial repair mechanisms in the lung. *Am J Physiol Lung Cell Mol Physiol* 2010;**298**:L715–31.

108. Villar J, Cabrera NE, Valladares F, Casula M, Flores C, Blanch L, et al. Activation of the Wnt/beta-catenin signaling pathway by mechanical ventilation is associated with ventilator-induced pulmonary fibrosis in healthy lungs. *PLoS One* 2011;**6**:e23914.

109. Joss-Moore LA, Wang Y, Ogata EM, Sainz AJ, Yu X, Callaway CW, et al. IUGR differentially alters MeCP2 expression and H3K9Me3 of the PPARgamma gene in male and female rat lungs during alveolarization. *Birth Defects Res A Clin Mol Teratol* 2011;**91**:672–81.

110. Fuks F, Hurd PJ, Wolf D, Nan X, Bird AP, Kouzarides T. The methyl-CpG-binding protein MeCP2 links DNA methylation to histone methylation. *J Biol Chem* 2003;**278**:4035–40.

111. Chetty A, Andersson S, Lassus P, Nielsen HC. Insulin-like growth factor-1 (IGF-1) and IGF-1 receptor (IGF-1R) expression in human lung in RDS and BPD. *Pediatr Pulmonol* 2004;**37**:128–36.

112. McCoy M, Metcalfe D, Metcalfe B, Beck B, Fu Q, McKnight R, et al. *Ventilation Mode Affects Ovine Pulmonary IGF-1 Epigenetic Characteristics*; 2009. E-PAS2009:3705.1.

113. Dokmanovic M, Clarke C, Marks PA. Histone deacetylase inhibitors: overview and perspectives. *Mol Cancer Res* 2007;**5**:981–9.

114. Hamvas A, Deterding R, Balch WE, Schwartz DA, Albertine KH, Whitsett JA, et al. Diffuse lung disease in children: summary of a scientific conference. *Pediatr Pulmonol* 2013;**49**:400–9.

115. Shakespear MR, Halili MA, Irvine KM, Fairlie DP, Sweet MJ. Histone deacetylases as regulators of inflammation and immunity. *Trends Immunol* 2011;**32**:335–43.

116. Ito K, Yamamura S, Essilfie-Quaye S, Cosio B, Ito M, Barnes PJ, et al. Histone deacetylase 2-mediated deacetylation of the glucocorticoid receptor enables NF-kappaB suppression. *J Exp Med* 2006;**203**:7–13.

117. Yao H, Rahman I. Role of histone deacetylase 2 in epigenetics and cellular senescence: implications in lung inflammaging and COPD. *Am J Physiol Lung Cell Mol Physiol* 2012;**303**:L557–66.

118. Joss-Moore L, Lane R. Perinatal Nutrition, Epigenetics, and Disease. *Neoreviews* 2011;**12**:e498–505.

Environmental Influences on Lung Development and Aging

Chapter 17

Pulmonary Consequences of Preterm Birth

Kurt H. Albertine[*,†] and Bradley A. Yoder[*]

Departments of Pediatrics; †Medicine, and Neurobiology & Anatomy, University of Utah School of Medicine, Salt Lake City, UT, USA

INTRODUCTION

Today, infants born as early as 22 weeks postconceptional age (term is 40 weeks) may survive if supported by antenatal steroids, postnatal surfactant replacement, mechanical ventilation, supplemental oxygen, antibiotics, and appropriate nutrition. These, and other, supports are required because of the sudden event of preterm birth, which causes an abrupt change in environment. Two questions are addressed in this chapter. First, what are the causes and adverse outcomes of preterm birth? Second, how does the abrupt change in environment dysregulate molecular pathways that are necessary for normal lung development during the canalicular stage through the alveolar stage?

CAUSES AND ADVERSE OUTCOMES OF PRETERM BIRTH

Definition of Preterm Birth

Preterm birth is defined as any birth, regardless of birth weight, that occurs before 37 completed weeks from the first day of menstrual cycle.[1] Pregnancies that end prior to 20 completed weeks of gestation are commonly termed miscarriages. Thus, a reasonable definition of preterm birth is any delivery that occurs between 20 and 37 weeks of gestation.

Incidence of Preterm Birth

In the USA, the overall rate of prematurity is 1 in 8 live births, or approximately 12%. Preterm birth rates increased from 9% in the 1980s to a peak of 13% in 2006, an increase of over 30%.[2] Since 2006, the preterm birth rate has steadily declined to the most recent rate of 12% in 2012.[3]

Despite this decline, the incidence of preterm birth remains markedly higher among non-Hispanic African-American women, nearly twice that of non-Hispanic Caucasian women.[3,4] A reduction in iatrogenic late-preterm births is the most important factor contributing to both the dramatic increase and the recent decline in preterm birth rates.[5,6]

Factors Leading to Preterm Birth

Preterm birth is a complex clinical problem with multiple causative factors. The most common etiologies are divided into three major groups: spontaneous preterm labor, preterm rupture of membranes, and medically indicated for either fetal or maternal indications.[7,8] The relative distribution among these three groups varies among different studies, but approximately one-third of preterm births can be attributed to each of these groups. However, spontaneous labor and preterm rupture of membranes are much more common etiologies for births occurring before 32 weeks' gestation, while a medical/fetal indication is more common among late-preterm births.[7,8]

Although a wide range of factors may lead to iatrogenic preterm birth, the most common are related to maternal hypertensive disorders (primarily pre-eclampsia, which is, by far, the most common medical indication for preterm birth), diabetes, poor fetal growth, and placental problems (previa, accrete, abruption).[9] The mechanisms behind spontaneous preterm labor and preterm rupture of membranes remain elusive; however, infection is considered a common contributing factor. Maternal infection is clinically identified in 30–40% of pregnancies complicated by preterm labor, and is much more common as gestational age decreases.[10,11] Infection in the maternal cervix or vagina may involve extra-embryonic fetal tissue (i.e., infection of the fetal membranes, called chorioamnionitis) or uterine decidua. Alternatively, low-grade systemic infection in the mother may cross the placenta and inflame the fetus and/or placental villi (called villitis) (Figure 1).

Preterm Birth and Perinatal Mortality

Between the 1980s and 1990s, death of extremely-low-birth-weight (ELBW) preterm infants (defined as weighing <1,000 g at birth) declined.[12] This improved survival was accompanied by increased rates of significant neonatal morbidity.[13] Subsequently, from the 1990s to late 2000s, further improvement in survival of extremely preterm infants has not occurred (Figure 2).[14,15] An important point is that major differences occur between centers in survival of these high-risk infants.[16] Such differences

The Lung. http://dx.doi.org/10.1016/B978-0-12-799941-8.00017-1

appear more related to center-specific approaches in care than to differences in demographic characteristics.[17,18] As an example, associated with specific changes in perinatal practices within our center, we have seen a 50% increase in survival of preterm infants over the past 4 years (Figure 3). In other words, overall mortality among extremely-low-gestational-age neonates decreased from 21% to 9% at our center (Yoder; unpublished observation).

Perinatal mortality associated with preterm birth is also related to sex, plurality (twins, triplets, etc.), birth defects, intrauterine growth rate, late-onset sepsis, and neonatal respiratory failure.[12,19–21] Specific examples include a mortality rate of male preterm infants born before 29 completed weeks of gestation that is approximately double that of females; mortality rates 2–3 times higher for twin versus singleton preterm infants; and mortality rates nearly four times greater in the presence of significant birth defects.

Preterm Birth and Adverse Outcomes

Among the survivors of preterm birth, morbidities such as poor in-hospital growth, intraventricular hemorrhage, late-onset sepsis, neonatal respiratory failure, necrotizing enterocolitis, and later neurodevelopmental impairments are frequent. The most commonly reported morbidities include postnatal growth failure,[12] late-onset infections, bronchopulmonary dysplasia (BPD), and intraventricular hemorrhage.[22,23] Though initial improvements in survival of preterm infants were accompanied by increasing rates of serious morbidities,[13] more recent studies report decreased rates of morbidity.[14,15] Nonetheless, amongst the smallest and youngest preterm infants, morbidities remain major concerns. Late-onset infections are commonly reported at rates greater than 50%, BPD is diagnosed in approximately 40% of ELBW infants, and rates for significant intraventricular hemorrhage, necrotizing enterocolitis, and retinopathy of prematurity are reported in the 10–15% range.[24,25]

Some VLBW infant survivors of newborn intensive care are now older than 20 years of age.[26] Studies of these infants are providing both good and bad news about adverse outcomes later in life. Encouraging observations are that 51% of the survivors had IQ scores within the normal range, 74% had completed high school, and 41% were continuing their education beyond high school. In addition, alcohol use, illicit-drug use, and criminal behavior occurred at a frequency that was not different from those of peers born at normal birth weight (>2,500 gm). On the other hand, these infants had more

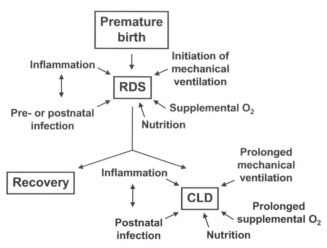

FIGURE 1   Factors contributing to the development of respiratory distress syndrome (RDS) and its evolution to chronic lung disease (CLD) following preterm birth.

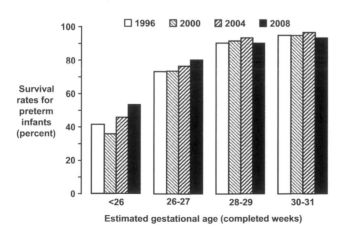

FIGURE 2   Histogram summarizing the results of a clinical study that assessed survival rates among extremely-low-gestational-age-neonates (defined as gestation <30 weeks at birth) stratified for completed weeks of gestation. Improvement in survival has not occurred from the 1990s to late 2000s.[15]

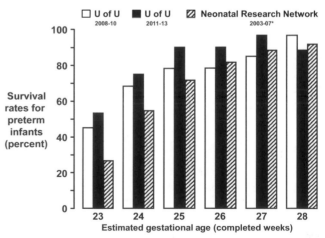

FIGURE 3   Histogram summarizing results that make the important point that major differences occur between centers in survival of high-risk preterm infants. For example, we have seen an increase in survival of extremely-low-gestational-age-neonates over the past 4 years associated with specific changes in perinatal practices within our University of Utah (U of U) center. Our center's data are juxtaposed to results reported by the Neonatal Research Network.[23]

chronic health problems, such as cerebral palsy, blindness, and deafness.[26] As a group, they also were shorter, had lower IQ scores, and lower scores on academic achievement tests compared to peer young adults who had normal birth weight. The VLBW infants were also less likely to have graduated from high school or enrolled in post secondary education compared to peer young adults who had normal birth weight. Whether these differences will diminish as preterm survivors reach maturity, or as survivors who had the benefit of recent advances in perinatal care reach adulthood, remains to be seen.

In summary, the incidence of preterm birth remains relatively high, the size and gestation of preterm infants that can be successfully supported has decreased over time, and chronic lung disease remains a common problem among the smallest preterm infants. How does the abrupt change in environment dysregulate molecular pathways and thereby change the trajectory of lung development during the canalicular stage through the alveolar stage? This question will be addressed in the remainder of this chapter.

PRETERM BIRTH AS AN ENVIRONMENTAL INFLUENCE ON LUNG DEVELOPMENT

Before birth, respiratory gas exchange is a major function of the placenta, which is interposed between the maternal and fetal circulations. The immature lung in utero does not participate in this vital function. Indeed the morphology of the human fetal lung in utero is not sufficiently developed to permit the exchange of oxygen (O_2) and carbon dioxide (CO_2) until 37 weeks of gestation. To emphasize the structural immaturity of the human preterm lung, a description is provided of the architectural organization of the mature lung, with the focus on efficient respiratory gas exchange.

Efficient exchange of O_2 and CO_2 across the parenchyma of the mature lung occurs in relatively large anatomic units that are referred to as terminal respiratory units. Each terminal respiratory unit stems from the most proximal respiratory bronchiole and includes all alveolar ducts together with their accompanying alveoli.[27] In the adult human lung, such units contain approximately 100 alveolar ducts and 2000 anatomic alveoli. A terminal respiratory unit is approximately 5 mm in diameter and has a volume of about 0.02 milliliters at functional residual capacity. Together, the two lungs of the adult human contain about 150,000 terminal respiratory units.[28] Clusters of 10–12 terminal respiratory units comprise the acinus, an anatomic unit recognized by pathologists.[29]

The function of the terminal respiratory unit is such that diffusion of O_2 and CO_2 is so rapid that the partial pressures of each gas are uniform throughout the unit. From each inspired breath, O_2 that reaches the respiratory bronchiole diffuses in alveolar duct gas and alveolar gas because the incoming, fresh air has a higher O_2 concentration than the alveolar gas. Oxygen subsequently diffuses

through the air-blood barrier into the red blood cells, where it combines with hemoglobin as the red blood cells flow along the capillaries. Carbon dioxide diffuses in the opposite direction.

Gas diffusion across the air-blood barrier of the normal adult human lung is facilitated by an exceedingly thin air-blood barrier. The average thickness of this barrier is about 1.5 µm.[28,30] For O_2, the air-blood barrier's anatomic elements consist, in order, of alveolar epithelium and its subjacent basal lamina, alveolar wall interstitium, basal lamina of the capillary endothelium, the capillary endothelium, plasma, and the membrane of red blood cells. These anatomic elements are traversed in the opposite direction by CO_2.

The terminal respiratory units of the preterm infant are incompletely developed. Developmental immaturity is greater with earlier preterm birth because thickness of the gas-exchange barrier is inversely related to gestational age. This structural-temporal relationship is best illustrated by reviewing the stages of human lung development (Table 1 and Figure 4).[31] Branching of the airways is complete by 16 weeks of gestation. However, the future bronchi and bronchioles branch into relatively undifferentiated mesenchyme that contains no terminal respiratory units or capillaries (Figure 4a). This period of lung development is known as the pseudoglandular stage. The lung up to this gestational age is more or less a solid organ. The next stage of fetal lung development is the canalicular stage.

The canalicular stage occurs from weeks 17 to 28 of gestation in humans. Lung architecture at this stage is characterized by thick partitions of mesenchyme between the air canals (future airways). To this thick connective tissue impediment for gas exchange is another impediment; namely, cuboidal or even columnar epithelial cells that line the air canals. Although capillary proliferation is robust during the canalicular stage compared to the pseudoglandular stage, the forming and growing capillaries are far removed from the tall epithelium that lines the primitive air canals (Figure 4b). Thus, the diffusion distance from the airspaces to flowing red blood cells is large, which impedes efficient respiratory gas exchange.

The next stage of lung development is the saccular stage, which in humans occurs from about week 24 to week 36 of gestation. The saccular stage is structurally characterized by transformation of the primitive air canals into distal air sacs, thinning of their lining epithelial cells, thinning of the mesenchyme, and proliferation and growth of capillaries (Figure 4c). Combination of these structural transformations creates a larger airspace surface area, thinner connective tissue, thinner epithelial cells that line the future respiratory bronchioles, alveolar ducts, and alveoli, and larger surface area of capillaries. These developmental transformations lead to a shorter distance between the expanding surface areas of the gas-exchange airspaces and capillaries.

TABLE 1 Stages of lung development in humans

Gestation Age	Lung Developmental Stage	Principal Process
4–6 weeks	Embryonic	Development of major airways and vessels
6–16 weeks	Pseudoglandular	Morphogenesis of the conducting airways and related large blood vessels
17–28 weeks	Canalicular	Formation of respiratory bronchioles and alveolar ducts, differentiation of alveolar type II epithelial cells, and canalization of microvessels into the alveolar wall mesenchyme
~24–36 weeks (term is 40 weeks)	Saccular	Division of the distal airspaces into smaller units
~32 weeks of gestation and continuing through the first 18–24 months of postnatal life*	Alveolar	Completion of alveolar formation

*Controversial. The developmental range is week 32 of gestation to 20 years of postnatal life.[31,185–189]

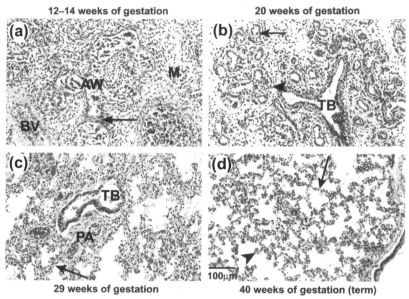

12–14 weeks of gestation **20 weeks of gestation**

29 weeks of gestation **40 weeks of gestation (term)**

FIGURE 4 Normal lung development in humans. (a) Although the lung of this infant stillborn at 12–14 weeks of gestation shows considerable autolysis (postmortem degeneration), a smooth muscle wall (arrow) is visible around airways (AW) that are lined by (largely desquamated) columnar epithelium. There is no development in the intervening mesenchyme (M), which is devoid of blood vessels (BV), except at the periphery of the developing lung tissue. (b) By 20 weeks of gestation, smooth-walled respiratory bronchioles and alveolar ducts lined by low cuboidal epithelium (arrow) are visible beyond the larger Y-shaped terminal bronchiole (TB). Numerous capillaries (arrowhead) are visible deep within the developing interstitium (mesenchyme). (c) At 30 weeks of gestation, the distal airspaces are subdivided, and many capillaries are present in the thick primary septa (arrow). A terminal bronchiole (TB) and neighboring pulmonary artery (PA) are visible in the center of the panel. (d) By 40 weeks of gestation, alveolar development has progressed to a point where the primary septa are thinner (arrow) and secondary septa (crests; arrowhead) protrude into the developing alveoli (A). The alveolar walls are replete with capillaries. Thus, at term, the infant lung clearly resembles its adult form, although considerable growth and development are yet to occur. All four panels show tissue sections that were stained with hematoxylin and eosin and are the same magnification (the scale bar is 100 μm in length). (This figure is reproduced in color in the color plate section.)

However, intimate structural association between the respiratory gas exchange airspaces and capillaries is not accomplished during the saccular stage.

Intimate structural association between the respiratory gas exchange airspaces and blood capillaries occurs during the alveolar stage of lung development. This stage occurs from about 32 weeks of human gestation until the first 18–24 months of postnatal life. This stage is when anatomic (definitive) alveoli are formed by subdivision of the saccules by secondary septa (also called secondary crests) (Figure 4d).

The outcome is vastly increased surface areas for alveoli and alveolar capillaries. Thinning of the intervening mesenchyme in the developing alveolar walls reduces the diffusion distance for O_2 and CO_2 between the alveolar air and the subjacent capillaries through which red blood cells flow. At the same time, the epithelial cells that line the developing alveoli differentiate into two cell populations.[32] One population, known as alveolar type II epithelial cells, remains cuboidal. Arising from this population of epithelial cells are alveolar type I (squamous) epithelial cells. These terminally differentiated cells, although fewer in number than their cuboidal counterpart, cover about 90–95% of the alveolar surface area of the peripheral lung.[33] Both structural attributes of alveolar type I epithelial cells provide a large, thin cellular barrier specialized for efficient respiratory gas exchange.

Alveolar type II epithelial cells and the surfactant phospholipids and apoproteins that they synthesize, secrete, and recycle are important for efficient lung function postnatally. In the human lung, type II cells contain lamellar bodies that become histologically recognizable at the transition from the canalicular to the saccular stages of lung development (weeks 20–24 of gestation). Surfactant synthesis begins later, during the saccular stage (about 30 weeks of gestation).[34] From the saccular stage to the end of term gestation, the concentrations of dipalmitoyl phosphatidylcholine and other surfactant phospholipids increase in fetal lung tissue, lung liquid, and amniotic liquid. Surfactant apoproteins (SP) A, B, and C, expressed only in lung tissue, also are developmentally regulated but not concordantly. Human SP-A mRNA and protein in lung tissue are not detectable until the saccular stage of lung development (about 30 weeks of gestation).[34–37] In the human fetus, SP-B and SP-C mRNAs are detectable in lung tissue at very low levels earlier than SP-A, during the canalicular stage (about 24 weeks of gestation).[38] The first detection of SP-B and SP-C proteins, however, occurs later in development, during the saccular stage (about 30 weeks of gestation).[36] To our knowledge, human SP-D protein has only been detected in amniotic liquid during the saccular stage of lung development (about 24 weeks of gestation).[39,40] Thus, the cellular and biochemical machinery to reduce surface tension at air–liquid interfaces in the future airspaces develops during the third trimester in humans. Detection of these secretory products in amniotic fluid provides a means of assessing lung development prior to birth.[41] For these reasons, the earlier that preterm birth occurs, the greater the risk for respiratory distress and subsequent respiratory failure (Figure 1).

Before birth at term gestation, the fetal lungs are filled with liquid that is secreted by the epithelial cells that line the potential airspaces, as well as inhaled amniotic liquid. Adequate production of lung luminal liquid is required for normal intrauterine expansion and growth of the immature lung.[42] The produced liquid flows centrally along the conducting airways towards the oropharynx, where it is either swallowed or expelled into the amniotic sac. When drainage exceeds production, the fetal lung is not exposed to continuous liquid distending pressure; as a consequence, lung growth is inhibited. This occurs in fetuses with prolonged oligohydramnios due to premature rupture of amniotic membranes.[43] Other conditions that inhibit lung growth include diaphragmatic hernia,[44] and pulmonary artery occlusion.[45]

Liquid flows into the potential airspaces of the fetal lung because chloride is secreted across the apical surface of fetal respiratory epithelial cells into the potential airspaces.[46] The resultant elevated concentration of chloride pulls water into the potential airspaces. Near the time of term gestation, the secretory activity of the respiratory epithelium switches from a predominantly chloride-secreting membrane to a predominantly sodium-absorbing membrane along the basolateral surface of alveolar epithelial cells.[47–49] This molecular switch is necessary to remove water from the developing respiratory gas exchanges airspaces around the time of birth.

Pumping of water from the airspaces is important because a physiologic impediment to respiratory gas exchange is liquid (water), especially for O_2. In the developing lung in utero, liquid is present in the potential airspaces, the surrounding thick mesenchyme, and the tall epithelial cells. Water is an impediment because gas diffusion is much slower in water than in a gas. For example, the solubility of O_2 in water is low (0.03 ml/L × mm Hg P_aO_2) compared to CO_2, the solubility of which is about 20 times higher (0.7 ml/L × mm Hg P_aCO_2). The advantage for CO_2 persists even though the driving pressure for CO_2 diffusion is only one-tenth that for O_2 entering the blood. Therefore, oxygen diffusion is much slower in the relatively over-hydrated environment of the preterm lung. To this physiological impediment is added the extensive diffusion distance between the lumen of the potential airspaces and the sparse and widely spaced capillaries.

Other structural and functional characteristics of the immature lung at preterm birth confound efficient respiratory gas exchange. One characteristic is that the immature airways have little smooth muscle in their wall, the epithelium is immature, and epithelial cell-cell adhesion is weak. Physiologic studies of human tracheobronchial segments ex vivo indicate that pressure–volume relations are affected by maturity,[50,51] such that airway compliance is inversely related to maturity. The greater compliance of the airways of preterm infants may contribute to the need for higher airway pressures, and the associated increase in lung volume needed to inflate the collapsed gas exchange regions compared to the airways of term infants.[52] The dearth of airway smooth muscle and weak adhesion between epithelial cells of the immature airways may make them susceptible to injury by the increased airway pressure and lung volume that are required to attain adequate oxygenation and ventilation. Evidence of injury is manifest as sloughed airway epithelial cells in tracheal aspirates and lung lavage liquid, as well as development of air leaks (interstitial emphysema and pneumothorax).

Acute Lung Injury (Respiratory Distress Syndrome, RDS)

When an infant is born before term, and its lungs are structurally immature and deficient in surfactant and its associated apoproteins, surface tension forces are high and the immature airspaces are unstable.[53] Airspace instability results in airspace collapse (atelectasis). Atelectasis contributes to ventilation-perfusion mismatch. Such mismatch leads to intrapulmonary shunt and thus contributes to poor oxygenation.[54]

Opening the collapsed airspaces requires the recruitment of collapsed airspaces, as well as airspaces yet to be recruited. High inspiratory pressures may be used in this effort. However, the danger is that only the ventilated airspaces will fill with the supplied volume of gas, leading to over-inflation (distension) of the already recruited airspaces; the mechanism leading to the resultant lung injury is identified below. In effect, the extreme effort required to expand the lungs with the first breath must be repeated with each breath because surfactant is not present to prevent collapse of the airspaces. For these reasons, preterm infants who have these characteristics usually develop acute lung injury (RDS) (Figure 1).

RDS affects about 50,000 preterm infants annually in the United States. Not surprisingly, the incidence of RDS increases with decreasing gestational age[12,19,20,24] and exceeds 50% of preterm infants born before 30 weeks of gestation.[55] Because the total number of late-preterm infants is so much larger than those born at earlier gestations, nearly one-half of all infants diagnosed with RDS are older than 33 weeks' gestation at birth.[56,57] RDS is more prevalent and severe in male compared to female preterm infants, for reasons that remain unclear.

RDS is characterized by tachypnea, chest retractions, and cyanosis (hypoxemia) soon after birth. Chest radiographs demonstrate low lung volumes, air bronchograms, and diffuse opacification (i.e., reticular-granular infiltrates). Lung function studies have shown increased airway resistance[58] and decreased pulmonary compliance.[59] Pulmonary artery pressure and pulmonary vascular resistance are elevated, particularly in preterm infants with severe RDS.[60]

The pathologic findings in RDS are similar to those in adults with acute respiratory distress syndrome (ARDS),[61] superimposed on the immature lung.[62] Invasive mechanical ventilation with O_2-rich gas is frequently necessary very early in the course of this disease in preterm infants, particularly the tiniest among them ($\leq$750 gm). Therefore, the lung injury reflects both the consequences of lung immaturity and its treatment. Within the first 3–4 hours of ventilatory support, the lungs may only show uneven ventilation of distal airspaces, due to surfactant deficiency and excess fetal lung liquid in the airspaces and interstitium (Figure 5a). By 12–24 hours of ventilatory support, necrosis of airway and airspace epithelial cells develops. Also, plasma proteins that leak from the underlying microvessels blanket the airspace walls, forming hyaline membranes, which are brightly eosinophilic transudates of plasma proteins admixed with necrotic epithelial cells and fibrin (Figure 5b).[63] Hyaline membranes accumulate, especially at branch points. Accumulation of hyaline membranes is the histopathologic basis of the coined term of hyaline membrane disease (HMD). Hyaline membranes are similar to the exudative phase of ARDS in adults, but the alveolar type II epithelial cell hyperplasia that constitutes the proliferative phase of ARDS in adults is not prominent in autopsy slides of the lung of preterm infants who died with RDS (HMD).

Invasive mechanical ventilation of preterm infants is frequently associated with ventilator-induced lung injury (Figure 1),[64] even following repeated doses of exogenous surfactant. The injurious effects of invasive mechanical ventilation depend on a number of factors, among which are the magnitudes of airway pressures (barotrauma) and lung volume (volutrauma),[65,66] the concentration of inspired O_2, and the duration of ventilation support. Several experimental studies compared indices of ventilator-induced lung injury between conditions that raised airway pressure and increased lung volume. One study distinguished between the effects of airway pressures and volume. Distinction was accomplished by subjecting rats to identical peak airway pressure and either large or small tidal volume ventilation.[67] Small tidal volume ventilation with high peak airway pressure was accomplished by strapping the thorax and abdomen. Rats subjected to high airway pressure and large tidal volume ventilation developed increased permeability pulmonary edema and ultrastructural evidence of cellular damage. Strikingly, the rats that were subjected to high peak airway pressure, by strapping the abdomen, and small tidal volume had neither pulmonary edema nor ultrastructural evidence of cellular damage. Results of this study, and those of similar studies that used rabbit kits[68] and young lambs,[69] suggest that large lung volumes, rather than high peak airway pressures per se, are important in the pathogenesis of ventilator-induced lung injury. For this reason, volutrauma is identified as the source of ventilator-induced lung injury.

How does volutrauma induce lung injury, especially following preterm birth? Evidence suggests that some of the circumstances that have already been described are exacerbated. For example, regional over-inflation increases microvascular permeability, either directly or indirectly (through release of inflammatory mediators from sequestered leukocytes in the lung), which leads to airspace flooding. Airspace flooding, in turn, is associated with inactivation of surfactant, which in turn leads to less compliant lungs and atelectasis. If these cycles are not broken, lung injury recurs, necessitating higher airway pressures, larger tidal volume, and higher O_2 concentration in the inhaled gas.

25 weeks + 6 hours ventilation   **29 weeks + 2 days ventilation**

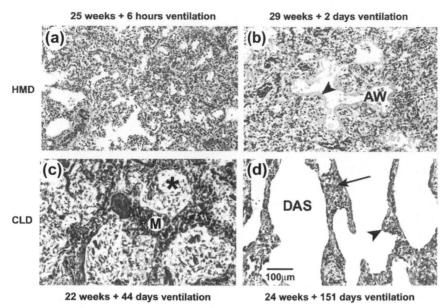

22 weeks + 44 days ventilation   **24 weeks + 151 days ventilation**

FIGURE 5   Hyaline membrane disease (HMD; panels a and b) and chronic lung disease of prematurity (CLD; panels c and d) in humans. (a) HMD, birth at 25 weeks of gestation followed by 6 hours of mechanical ventilation. In the first few hours of life, the preterm, mechanically ventilated lung shows uneven expansion of the distal airspaces, vascular congestion, and interstitial edema. Some patchy hemorrhage also may be present. (b) HMD, birth at 29 weeks of gestation followed by 2 days of mechanical ventilation. Hyaline membranes (arrowhead) are uniformly present by 12–24 hours, but may appear within 3–4 hours of preterm birth. Hyaline membranes are typically most evident at the branch points of distal airways (AW). The distribution of ventilation is uneven, with centriacinar expansion and peripheral collapse. (c) and (d) Infants who progress to CLD show marked simplification of the distal airspaces (compare panels c and d with the expected appearance of the lung at 30 and 40 weeks of gestation in Figure 4, panels c and d, respectively; all four panels are at the same magnification). (c) CLD, birth at 22 weeks of gestation followed by 44 days of mechanical ventilation (28–29 weeks post conceptual age). The distal airspaces appear as distended circles that are filled, in this specimen, with cellular debris (*). Thick, cellular mesenchyme (M) separates the adjacent airspaces. (d) CLD, birth at 24 weeks of gestation followed by 151 days of mechanical ventilation (45 weeks post conceptual age). Simplification and over-distension of the distal airspaces (DAS) is clearly evident. The primary septa (arrow) are thick and cellular. Secondary septa (crests; arrowhead) are infrequently visible; those that are visible are short, thick, and devoid of capillaries near their tip. Panels a and b show tissue sections that were stained with hematoxylin and eosin. Panels c and d show tissue sections that were stained with Mason's trichrome. All four panels are the same magnification (the scale bar is 100 μm in length). (This figure is reproduced in color in the color plate section.)

As the extent and duration of mechanical ventilation with O_2-rich gas increases, the acute lung injury transforms to chronic lung disease (CLD) of prematurity.

An additional predisposing factor for the development of RDS and subsequent neonatal CLD may be O_2 toxicity and overwhelmed endogenous anti-oxidants. Oxygen toxicity occurs when prolonged use of increased concentrations of inspired oxygen is needed for preterm infants with RDS.[66,70–73] Evidence of oxidant stress in preterm infants is detected as shifts in the plasma concentrations of reduced glutathione and glutathione disulfide.[74] Oxidant stress is exacerbated by endogenous release of toxic metabolites of O_2 (e.g., superoxide anion, hydroxyl radical) from activated neutrophils and macrophages that accumulate in the lung during the acute inflammatory response characteristic of RDS.[75–78]

Endogenous anti-oxidant enzymes, such as catalase, copper-zinc superoxide dismutase or manganese superoxide dismutase, may be overwhelmed during the evolution of RDS and subsequent neonatal CLD. Imbalances in proteinases (elastase and collagenase) and inactivation of proteinases by O_2 (α_1-anti-proteinase) may also play a role.[79]

Such imbalances may result in the destruction of elastin and collagen, two prominent extracellular matrix molecules in the lung that are necessary for alveolar formation.[80–82]

Acute lung injury is evident as increased permeability pulmonary edema that is associated with accumulation of leukocytes (Figure 1), particularly neutrophils, in the vascular, interstitial, and airspace compartments of the lung.[77,78,83] We showed that when preterm lambs were ventilated for 8 hours with 100% O_2, they had a significant decrease in the number of circulating neutrophils in the first 30–90 min of life. The decrease correlated with increased number of neutrophils that accumulated in the lung's airspaces at the end of the 8-hour study.[77] A complimentary, retrospective chart review study in human preterm infants subsequently showed that a low concentration of mature neutrophils in the systemic circulation within 24 hours of birth (ascribed to egress of neutrophils from the circulation coupled with an inability of the immature bone marrow to replenish the circulating pool) is associated with more severe respiratory distress during the first postnatal week of life.[84] Such preterm infants were more likely to need mechanical ventilation with supplemental O_2.

Because lung immaturity is a consistent feature of RDS, strategies have been developed to accelerate lung development before preterm delivery. One strategy is administration of hormones, such as glucocorticoids.[85,86] The seminal studies of Liggins demonstrated that glucocorticoids accelerated lung maturation of fetal sheep and decreased the incidence of RDS in human infants after antenatal glucocorticoids therapy.[87,88] The maturational effect of glucocorticoid treatment on lung structure was first described by Kikkawa and colleagues.[89] Among the structural effects, glucocorticoids (betamethasone or dexamethasone) accelerate the maturation of alveolar type II epithelial cells, which are the source of pulmonary surfactant. For these reasons, mothers who are in preterm labor, and are at risk of giving birth to a preterm infant between 24 and 35 weeks of gestation, are treated antenatally with glucocorticoids.[90]

A treatment strategy to reduce lung stiffness (increase lung compliance) after preterm birth is to instill surfactant into the airways.[91] Surfactant replacement therapy is beneficial because once the exogenous surfactant becomes widely and thinly distributed in the lung, it reduces surface tension at air–liquid interfaces, thereby stabilizing airspaces when they are deflated. After surfactant replacement, oxygenation improves swiftly, followed for several hours by progressive improvement in respiratory gas exchange and lung mechanics.[92,93] Other improvements, at least in preterm lambs, are reduced vascular injury and edema.[94] Thus, surfactant replacement therapy reduces the incidence and severity of RDS following preterm birth.

Neonatal CLD (Bronchopulmonary Dysplasia, BPD)

Prematurely born infants who develop neonatal CLD today have demographic characteristics that are different from those who developed BPD over 40 years ago.[95] Preterm infants who develop neonatal CLD today are younger (22–26 weeks of gestation compared to 31–34 weeks of gestation), smaller (<1000 gm versus >2000 gm at birth), and the radiological and pathological features of their lung disease are less severe and more peripheral.[96–98]

The radiological findings include lung hyperinflation, emphysema, and interstitial densities. The principal pathological findings, at autopsy, are dilated and simplified air canals and air sacs, leading to the histopathological term of alveolar simplification (Figure 5c and 5d).[96,97] Subsequently studies showed that the number of alveoli was not significantly different than the number expected to be present at term gestation,[98] that is, alveolarization had apparently not progressed after birth. Generally missing are the pathological changes in the central airways. Recognition of these distinctions between neonatal CLD that is seen today and BPD that was described by Northway et al. in 1967 prompted clinicians and investigators to call today's neonatal CLD "new" BPD. Whether the disease that is seen today is "new" or a less severe variant of BPD is debated.

Survival of preterm infants with RDS has significantly increased since the introduction of antenatal glucocorticoid and postnatal surfactant replacement therapies, and gentler ventilation strategies.[12,99–101] As described earlier, survival rates at some centers are better than 90% for preterm infants weighing 500–1500 g at birth (Table 2).[102,103] Nonetheless, over 40%, or 10,000–15,000, of these preterm infants go on to develop neonatal CLD despite high rates of antenatal glucocorticoids and postnatal surfactant replacement therapy (Figure 1). As expected, the smallest infants (<750 g at birth) are at greater risk for neonatal CLD than larger preterm infants.[24,25]

Risk factors for neonatal CLD among older preterm and term neonates include meconium aspiration pneumonia,[104] neonatal pneumonia,[105] congestive heart failure,[106] and congenital diaphragmatic hernia.[107,108] Other risk factors for neonatal CLD include male sex, "white race," greater severity of RDS, and a low Apgar score 1 min after birth. After adjusting for these risk factors, the development of neonatal CLD is primarily associated most with very low birth weight, gestation sepsis, and a need for sustained mechanical ventilation.[109,110]

The histopathological characteristic of neonatal CLD is alveolar simplification (Figure 1). Alveolar simplification is evident as distended distal airspaces that have few secondary septa, alveolar walls that are thick and cellular, and capillaries that are separated from the airspaces. These features are recapitulated in chronically ventilated preterm

TABLE 2 Time-related decrease in mortality is accompanied by increase in BPD

Outcome	Year Intervals (# ≤1500 gm)					
	1976–80 (n=665)	1981–85 (n=834)	1986–90 (n=875)	1991–95 (n=1046)	1996–97 (n=417)	P
Died	28%	20%	17%	11%	6%	<0.001
BPD (28d)	11%	22%	33%	30%	27%	<0.001
Either	35%	37%	45%	38%	31%	0.723

Adapted from.[102]

baboons[111–117] and preterm lambs.[118–125] Simplification is related to failure of secondary septa (crests) to sprout into the distal air sacs. A consequence of failed secondary septation is reduced airspace surface area for respiratory gas exchange.

A molecular player in secondary septation is elastin. This role is revealed by studies using knockout mice for PDGF-AA[126] or elastin[127] suggest that elastin synthesis and secretion at the site of initial outgrowth of secondary septa may be involved. Our studies using chronically ventilated preterm lambs with evolving neonatal CLD indicate that the elastin gene is excessively and continuously upregulated.[118] Upregulation of elastin gene expression is not limited, however, to the points of sprouting secondary septa. Upregulation is also evident in the thick walls of air sacs, as well as in the walls of pulmonary arterioles and bronchioles. Although the results from knockout mice and preterm lambs seem inconsistent, together they suggest that elastin gene expression has to be tightly regulated for normal secondary septation. Too little (knockout mice) or too much (chronically ventilated preterm lambs) expression of elastin disrupts normal secondary septation.

Alveolar simplification also includes reduced capillary growth in the lung. This was quantified in studies that used chronically ventilated preterm baboons[128] and preterm lambs.[119,121,125] Our studies using chronically ventilated preterm lambs also show that the stunted alveolar secondary septa do not have capillaries.[125] Therefore, another consequence of alveolar simplification is reduced capillary surface area for juxtaposing flowing red blood cells next to ventilated airspaces. The physiologic significance of this observation is that the secondary septa cannot contribute to respiratory gas exchange because the capillary surface area is inadequate.

Extra-alveolar microvessels are also reduced in number in the lung in neonatal CLD. Postmortem histology and angiography studies of preterm infants who had neonatal CLD for 1–7 months had lower arterial density compared to other preterm infants who had neonatal CLD for less than one month.[129] This observation is corroborated in chronically ventilated preterm lambs; our results show that the number of pre- and post capillary pulmonary arterioles and venules is significantly less compared to term lambs.[121]

The structure and function of intrapulmonary arterial vessels are also affected in preterm infants who develop neonatal CLD. Autopsy studies show structural remodeling of arterial vessels. The most characteristic changes are greater muscularization of small arteries and extension of smooth muscle into the smallest arterioles that normally are partially muscular or non-muscular.[130,131] Studies of lung tissue from chronically ventilated preterm lambs demonstrate that 3 weeks of mechanical ventilation leads to failed regression of vascular smooth muscle.[121]

Nitric Oxide Signaling and Inhaled Nitric Oxide in Chronically Ventilated Preterm Neonates

Diminished blood vessel formation and persistent muscularization may contribute to elevated pulmonary arterial pressure (pulmonary hypertension) and pulmonary vascular resistance that typically accompany neonatal CLD.[121,132,133] A molecular contribution to the functional and structural consequences of prolonged mechanical ventilation may be diminished abundance of endothelial nitric oxide synthase (eNOS), an enzyme that catalyzes endogenous generation of nitric oxide (NO) by vascular endothelial cells.[134] Three other links to NO are that (1) it acts as a signaling molecule to inhibit growth of vascular smooth muscle cells in culture,[135] (2) blocking endogenous NO production increases vascular smooth muscle growth in response to injury in vivo,[136] and (3) providing exogenous NO inhibits vascular smooth muscle growth after mechanical injury.[137–139]

We assessed the abundance and location of eNOS protein in third to fifth generation intrapulmonary arteries and pulmonary arterioles adjacent to terminal bronchioles, respectively.[120,122] Our results showed less eNOS protein in the endothelium of small intrapulmonary arterial vessels of preterm lambs that were mechanically ventilated for 3 weeks compared to reference lambs that were born at term gestation (reference term lambs). By comparison, neither protein abundance nor localization of iNOS was different compared to reference term lambs.

Nitric oxide signaling downstream of eNOS is effected by soluble guanylate cyclase (sGC). This enzyme is the intermediary enzyme through which NO induces increased cyclic guanosine monophosphate (cGMP) in vascular smooth muscle cells, leading to vasodilation. Like eNOS protein, sGC protein is less in the smooth muscle cells of the intrapulmonary arteries of chronically ventilated preterm lambs than in reference term lambs. These results suggest a molecular basis, in addition to a structural basis, for elevated pulmonary arterial pressure (pulmonary hypertension) that occurs in chronically ventilated preterm lambs.

Reduced endogenous eNOS and sGC in the pulmonary arterial vessels led us to treat chronically ventilated preterm lambs with inhaled nitric oxide (iNO).[140,141] A physiologic rationale for such treatment is that iNO decreases pulmonary vascular resistance (lowers pulmonary arterial pressure), as well as increases arterial oxygenation in newborn animals and humans with pulmonary hypertension.[142,143] For one study, we gave iNO (5–15 ppm) for 1 hour to preterm lambs after 1, 2, and 3 weeks of mechanical ventilation.[122] Pulmonary vascular resistance decreased acutely at 1 and 2 weeks of life; however, it did not decrease acutely at week 3.

We ruled out that the dysfunction was at the level of pulmonary vascular smooth muscle cells not relaxing. This was demonstrated by infusing 8-bromo-guanosine 3′,5′-cyclic monophosphate (8-bromo-cGMP) for 15–30 min in four of the same preterm lambs at the end of week 3 of invasive mechanical ventilation. Pulmonary vascular resistance decreased consistently in the four preterm lambs, indicating that pulmonary vascular smooth muscle cells remained capable of relaxing at a time when pulmonary vascular responsiveness to iNO was lost.

Our next study provided iNO (5–15 ppm) continuously for 3 weeks.[141] We anticipated that continuous iNO would decrease pulmonary vascular resistance and smooth muscle abundance in the wall of pulmonary arterial vessels. Instead, continuous iNO did not decrease pulmonary vascular resistance or muscularization of pulmonary arterial vessels. Surprisingly, eNOS protein abundance did not decrease over time in pulmonary arterial vessels, an observation that will be returned to. Finding that eNOS protein abundance remained at control levels is consistent with results of other studies that NO donors increased NOS expression in cultured pulmonary artery endothelial cells isolated from fetal sheep.[144] The utility of our finding that eNOS protein abundance did not decrease over time is that it may help provide a basis for developing new therapeutic or preventive interventions to reduce pulmonary vascular resistance.

Mechanical ventilation for 3 weeks also led to greater airway resistance in preterm lambs compared to reference term lambs.[122] Associated with greater airway resistance was thicker airway smooth muscle. By comparison, eNOS and sGC protein abundance was less compared to reference term lambs.[120] These responses in third to fifth generation airways and terminal bronchioles mirrored the responses of the neighboring pulmonary arterial vessels.

A surprising result when iNO was given continuously was that it had functional and structural effects on the airways.[141] For example, continuous iNO for 3 weeks led to significantly (~40%) lower airway resistance in preterm lambs that received iNO compared to other preterm lambs that received vehicle. Airways in the lungs of the iNO-treated preterm lambs had significantly (~50%) less smooth muscle compared to vehicle-treated preterm lambs. In this regard, our results are consistent with studies that showed that NO acts as a signaling molecule to inhibit growth of airway smooth muscle cells in culture.[145] The airways also had an abundance of eNOS and sGC proteins that were comparable to those unventilated reference term lambs. These results suggest that the effect of iNO may be diffusion-distance limited because effects were not detected in the neighboring pulmonary arterioles.

Nutrition and Vitamin A Replacement Therapy in Chronically Ventilated Preterm Neonates

Nutrition is important in the pathogenesis of neonatal CLD from several standpoints. One of these is energy expenditure related to rapid growth in the face of limited nutritional reserves, both of which apply to preterm infants. To that energy expenditure is added, the additional energy needed for breathing. If these compounded energy needs are not met, a catabolic state ensues that may contribute to the pathogenesis of neonatal CLD.[146] In addition, most anti-oxidant enzymes require trace amounts of metals, such as copper, zinc, and selenium, for optimal activity.[147,148] Deficiencies in such trace metals, therefore, may contribute to the imbalance between oxidants and anti-oxidants and participate in the pathogenesis of neonatal CLD.[149] Vitamin E (tocopherol) is an anti-oxidant that prevents peroxidation of lipid membranes. However, a number of studies showed that vitamin E supplementation does not alter the pathogenesis of neonatal CLD.[150–154]

Vitamin A (retinol) concentrations may be low in the plasma of some preterm infants who are less than 36 weeks of gestation.[155–158] This observation prompted several vitamin A supplementation trials. However, the results have been inconsistent, with some trials showing improvement,[159–161] while others showed no benefit.[162]

Experimental animal studies are providing intriguing results. In one study, pregnant rats were repeatedly treated with dexamethasone to inhibit fetal lung development, and the newborn pups were then treated with all-trans-retinoic acid.[163] Alveolar formation in the treated rat pups was greater than that in the untreated control rat pups.

Results from chronically ventilated preterm baboons and preterm lambs given supplemental vitamin A daily are not consistent. Supplemental vitamin A (Aquasol; 5000 U/Kg/d) given to preterm baboons that were supported by mechanical ventilation for 14 days did not improve ventilation or oxygenation, alveolar formation, capillary growth, or expression of vascular growth factors compared to historical controls.[164] Nor did vitamin A treatment reduce elastin expression. In contrast, our studies using preterm lambs that were given supplemental vitamin A (Aquasol; 5000 U/Kg/d) during mechanical ventilation for 21 days demonstrated improved alveolar formation, capillary growth, and expression of vascular growth factors compared to untreated preterm lambs.[125] Coincidently, vitamin A supplementation reduced both expression of elastin and accumulation of elastic fibers compared to untreated preterm lambs. These results led to new, ongoing studies that are determining the efficacy of supplementation with a mixture of vitamin A and all-trans retinoic acid added daily to colostrum and milk.

Non-Invasive Ventilation Mode Provides Opportunities to Identify Molecular Mechanisms

Until recently, discovering molecular mechanisms that contribute to alveolar simplification was hindered by lack of a positive-outcome gold-standard control for alveolar formation in chronically ventilated preterm neonates. Previously, unventilated, gestation age-matched reference fetuses, or unventilated or ventilated reference term neonates served as references. However, these reference groups provide only developmental references. They do not meet the clinical setting of preterm delivery, respiratory failure, and prolonged mechanical ventilation with O_2-rich gas.

Gentler non-invasive ventilation modes, such as nasal continuous positive airway pressure (nasal CPAP), are being used today that may lower incidence of neonatal CLD in preterm infants.[165] Animal studies are providing data that suggest that nasal CPAP leads to less inflammation,[166] more surfactant production,[167] and improved alveolar formation and capillary growth compared to invasive mechanical ventilation.[168,169] In this regard, our group reported that non-invasive high-frequency nasal ventilation (HFNV) in spontaneously breathing preterm lambs is associated with adequate arterial blood gas values for 3 days[124] or 21 days.[170] Our results show that significantly lower fractional inspired oxygen fraction (FIO_2) and lower respiratory system pressures (peak inspiratory pressure and mean airway pressure) are necessary to maintain target physiologic levels values compared to invasive mechanical ventilation. Our results also show that mean intratracheal pressure is very low during non-invasive HFNV: ~1/10th that during invasive mechanical ventilation. The latter result is exciting because it raises the possibility that long-term use of non-invasive HFNV may be an alternative approach to sustain acceptable respiratory support for preterm infants and reduce the severity of neonatal CLD.[171]

We use the two ventilation modes to identify molecular participants in the pathogenesis of thick and cellular distal airspace walls in preterm lambs supported by invasive mechanical ventilation. For normal formation of alveolar walls, the thick and cellular walls of the saccules have to be reduced. We found that thinning and reduced cellularity occur with non-invasive HFNV support for 3 days[124] or 21 days[170] compared to invasive mechanical ventilation. Both thinning and reduced cellularity are related to more apoptosis, and less proliferation, among mesenchymal cells in the saccular walls to attain thinner and less cellular alveolar walls.[124]

Another necessary developmental process is growth of capillaries in the thinning walls and sprouting alveolar secondary septa. Our studies show that vascular endothelial growth factor (VEGF), its functional receptor 2 (VEGF-R2), and midkine (neurite growth-promoting factor 2) have more mRNA and protein abundance in the developing alveolar parenchyma when non-invasive HFNV is used compared to invasive mechanical ventilation. These positive structural and molecular outcomes are accompanied by adequate respiratory gas exchange across the developing lung compared to invasive mechanical ventilation for 3 days[124] or 21 days.[170]

The results of our recent studies suggest that long-term use of non-invasive HFNV as we use it may be more effective than nasal CPAP or nasal intermittent mandatory ventilation.[172–175] Our suggestion is supported by a clinical trial that used the same type of ventilator that we used.[176] That clinical trial used non-invasive HFNV in a prospective, unmasked, randomized, controlled study of 46 eligible newborn infants who were hospitalized for transient tachypnea. The results showed that non-invasive HFNV shortened the period of transient tachypnea by 50%. However, that clinical trial did not test efficacy when non-invasive HFNV is used for longer periods. This test will be important to do in preterm infants at risk of developing neonatal CLD.

Finally, a molecular mechanism that we are focusing on is epigenetic regulation of gene expression because epigenetics adjusts gene expression in response to environmental changes.[177] Regulation of gene expression by epigenetics occurs through modifications to histone tails, methylation of DNA, types of microRNAs, and positioning of nucleosomes.[178,179] Modifications of histone tails participate in gene regulation through influences on transcription initiation, elongation, and/or termination of transcription along a gene locus. Methylation of DNA occurs at sites called cytosine, linked by phosphate, to guanine (CpG sites). Like modifications to histone tails, methylation of DNA has varied functions on transcriptional regulation. The sites of action along DNA are at transcriptional start sites, at regulatory elements, along a gene locus, and at repeat sequences.[180] MicroRNAs are RNA molecules that do not code for a protein. Instead, microRNAs silence transcription when they bind to the 3′ untranslated region of target genes.[181] Positioning of nucleosomes plays a role in exposing or not exposing transcription start sites to transcription complexes and RNA polymerase. Our studies are beginning to test the roles of these four determinants of epigenetic regulation.

For neonatal CLD, the initial environmental change is preterm birth, followed by invasive mechanical ventilation and exposure to oxygen-rich gas, each of which may trigger epigenetic regulatory responses. We are evaluating the impact of ventilation mode on modifications to histone tails. Our results are revealing that invasive mechanical ventilation leads to genome-wide *hypo*acetylation in the lung compared to non-invasive HFNV of preterm lambs. More often than not, genome-wide *hypo*acetylation is also less in the lung during invasive mechanical ventilation compared to reference unventilated fetal lambs (matched for postconceptional age) and reference term lambs (matched for postnatal age).

Our genome-wide epigenetic observations are intriguing; however, they do not identify epigenetic effects on a specific gene that is dysregulated during the evolution of neonatal CLD. To this end, we are analyzing insulin-like growth factor-1 (IGF-1) as a prototypic epigenetically regulated gene[177,182] that is involved in both lung development normally[183] and evolving neonatal CLD.[184] Our recent results show that IGF-1 mRNA and protein levels are higher in the lung of preterm lambs supported by invasive mechanical ventilation compared to non-invasive HFNV. Furthermore, our results show that the pattern of modifications of histone tails along the length of the IGF-1 gene (the histone code) is scrambled in the lung when preterm lambs are supported by invasive mechanical ventilation compared to non-invasive HFNV. We do not know, at this time, the meaning of the scrambled histone code on regulation of IGF-1 expression in the lung during invasive mechanical ventilation. Positioning of nucleosomes may be involved because we have found that ventilation of preterm lambs is associated with more transcription start sites for IGF-1 compared to reference unventilated fetal lambs. However, we are yet to test DNA methylation or microRNAs in the pathway leading to increased mRNA and protein levels in the lungs of preterm lambs that are supported by invasive mechanical ventilation.

CONCLUSION

Preterm birth is an environmental stress that disrupts lung development. Structural and functional immaturity of the preterm infant's lungs necessitates use of antenatal steroids, perinatal surfactant replacement therapy, ventilation and O_2 support, and other supportive measures (Figure 1). Acute lung injury often occurs because of lung immaturity and the treatments that it necessitates. If rapid recovery does not follow, prolonged invasive mechanical ventilation with O_2-rich gas is often the only alternative. The affected preterm infants frequently progress to neonatal CLD. Challenges remain to better understand lung developmental biology and alveolar formation, and how both processes are impaired by preterm birth, invasive mechanical ventilation with O_2-rich gas, infection and inflammation, inadequate nutrition, and poor growth. This chapter addressed two questions: first, what are the causes and adverse outcomes of preterm birth and, second, how does the abrupt change in environment dysregulate molecular pathways that are necessary for normal lung development during the canalicular stage through to the alveolar stage. Results of new mechanistic studies using chronically ventilated preterm lambs suggest that invasive mechanical ventilation with O_2-rich gas may dysregulate epigenetic determinants of regulation of expression of genes that direct normal lung development.

REFERENCES

1. AAP Committee on Fetus and Newborn and ACOG Committee on Obstetric Practice. *Guidelines for Perinatal Care.* 7th ed. Elk Grove Village, IL: Am Acad Pediatr and Am Coll Obstet Gynecol; 2013.
2. Martin JA, Hamilton BE, Osterman MJ, Curtin SC, M.A. M, Mathews TJ. Final Date for 2012. *Natl Vital Stat Rep* 2013a;62.
3. Martin JA, Osterman MJ. Centers for Disease C, Prevention. Preterm births – United States, 2006 and 2010. *MMWR Surveill Summ* 2013b;**62**(Suppl. 3):136–8.
4. Goldenberg RL, Cliver SP, Mulvihill FX, et al. Medical, psychosocial, and behavioral risk factors do not explain the increased risk for low birth weight among black women. *Am J Obstet Gynecol* 1996;**175**:1317–24.
5. Davidoff MJ, Dias T, Damus K, et al. Changes in the gestational age distribution among U.S. singleton births: impact on rates of late preterm birth, 1992 to 2002. *Semin Perinatol* 2006;**30**:8–15.
6. Oshiro BT, Kowalewski L, Sappenfield W, et al. A multistate quality improvement program to decrease elective deliveries before 39 weeks of gestation. *Obstet Gynecol* 2013;**121**:1025–31.
7. Goldenberg RL, Culhane J, Iams JD, Romero R. Preterm Birth 1- Epidemiology and causes of preterm birth. *Lancet* 2008;**371**: 75–84.
8. Henderson JJ, McWilliam OA, Newnham JP, Pennell CE. Preterm birth aetiology 2004–2008. Maternal factors associated with three phenotypes: spontaneous preterm labour, preterm pre-labour rupture of membranes and medically indicated preterm birth. *J Mat-Fet Neo Med* 2012;**25**:642–7.
9. Wong AE, Grobman WA. Medically indicated-iatrogenic prematurity. *Clin Perinatol* 2011;**38**:423–39.
10. Romero R, Gomez R, Chaiworapongsa T, Conoscenti G, Kim JC, Kim YM. The role of infection in preterm labour and delivery. *Paediatr Perinat Epidemiol* 2001;**15**(Suppl. 2):41–56.
11. DiGiulio DB, Romero R, Amogan HP, et al. Microbial prevalence, diversity and abundance in amniotic fluid during preterm labor: a molecular and culture-based investigation. *PLoS One* 2008;**3**: e3056.
12. Lemons JA, Bauer CR, Oh W, et al. Very low birth weight outcomes of the National Institute of Child health and human development neonatal research network, January 1995 through December 1996. NICHD Neonatal Research Network. *Pediatrics* 2001;107. E1.
13. Wilson-Costello D, Friedman H, Minich N, Fanaroff AA, Hack M. Improved survival rates with increased neurodevelopmental disability for extremely low birth weight infants in the 1990s. *Pediatrics* 2005;**115**:997–1003.
14. Doyle LW, Roberts G, Anderson PJ. Victorian infant collaborative study group. outcomes at age 2 years of infants <28 weeks' gestational age born in Victoria in 2005. *J Pediatr* 2010;**156**:49–53.
15. Rüegger C, Hegglin M, Adams M, Bucher HU. For the Swiss Neonatal Network. Population based trends in mortality, morbidity and treatment for very preterm and very low birth weight infants over 12 years. *BMC Pediatr* 2012;**12**:17.
16. Alleman BW, Bell EF, Li L, et al. Individual and center-level factors affecting mortality among extremely low birth weight infants. *Pediatrics* 2013;**132**:e175–184.

17. Smith PB, Ambalavanan N, Li L, et al. Approach to infants born at 22 to 24 weeks' gestation: relationship to outcomes of more-mature infants. *Pediatrics* 2012;**129**:e1508–16.

18. Mehler K, Grimme J, Abele J, Huenseler C, Roth B, Kribs A. Outcome of extremely low gestational age newborns after introduction of a revised protocol to assist preterm infants in their transition to extrauterine life. *Acta Paediatr* 2012;**101**:1232–9.

19. Stoll BJ, Hansen N, Fanaroff AA, et al. Late-onset sepsis in very low birth weight neonates: the experience of the NICHD neonatal research network. *Pediatrics* 2002;**110**:285–91.

20. Angus DC, Linde-Zwirble WT, Clermont G, Griffin MF, Clark RH. Epidemiology of neonatal respiratory failure in the United States: projections from California and New York. *Am J Respir Crit Care Med* 2001;**164**:1154–60.

21. Adams-Chapman I, Hansen NI, Shankaran S, Bell EF, Boghossian NS, Murray JC. Ten-year review of major birth defects in VLBW infants. *Pediatrics* 2013;**132**:49–61.

22. Costeloe K, Hennessy E, Gibson AT, Marlow N, Wilkinson AR. for the EPICure study group. The EPICure study: outcomes to discharge from hospital for infants Born at the threshold of viability. *Pediatrics* 2000;**106**:659–71.

23. Stoll BJ, Hansen NI, Bell EF, et al. Neonatal outcomes of extremely preterm infants from the NICHD Neonatal Research Network. *Pediatrics* 2010;**126**:443–56.

24. Finer NN, Carlo WA, Walsh MC, et al. Early CPAP versus surfactant in extremely preterm infants. *NEJM* 2010;**362**:1970–9.

25. Kirpalani H, Millar D, Lemyre B, Yoder B, Chiu A, Roberts R. Nasal intermitte positive pressure ventilation or nasal continuous positive airway pressure (nCPAP) for extremely low birth weight infants- the NIPPV international randomized controlled trial. *NEJM* 2013;**369**:611–20.

26. Hack M, Flannery DJ, Schluchter M, Cartar L, Borawski E, Klein N. Outcomes in young adulthood for very-low-birth-weight infants. *NEJM* 2002;**346**:149–57.

27. Hayek H. *The Human Lung*. New York: Hafner; 1960.

28. Weibel ER. *Morphometry of the Lung*. New York: Academic Press; 1963.

29. Schreider JP, Raabe OG. Structure of the human respiratory acinus. *Am J Anat* 1981;**162**:221–32.

30. Crapo JD, Young SL, Fram EK, Pinkerton KE, Barry BE, Crapo RO. Morphometric characteristics of cells in the alveolar region of mammalian lungs. *Am Rev Respir Dis* 1983;**128**:S42–6.

31. Burri PH. Structural aspects of prenatal and postnatal development and growth of the lung. In: McDonald JA, editor. *Lung Growth and Development*, vol. 100. New York: Marcel Dekker; 1997. pp. 1–35.

32. Albertine KH, Pysher TJ. Impaired lung growth after injury in premature lung. In: Polin RA, Fox WW, Abman S, editors. *Fetal and Neonatal Physiology*, vol. 1. 4th ed. New York: Elsevier Science; 2011. pp. 1039–47.

33. Crapo JD, Barry BE, Gehr P, Bachofen M, Weibel ER. Cell number and cell characteristics of the normal human lung. *Am Rev Respir Dis* 1982;**126**:332–7.

34. King RJ, Ruch J, Gikas EG, Platzker AC, Creasy RK. Appearance of apoproteins of pulmonary surfactant in human amniotic fluid. *J Appl Phys* 1975;**39**:735–41.

35. Ballard PL, Hawgood S, Liley H, et al. Regulation of pulmonary surfactant apoprotein SP 28-36 gene in fetal human lung. *Proc Natl Acad Sci USA* 1986;**83**:9527–31.

36. Pryhuber GS, Hull WM, Fink I, McMahan MJ, Whitsett JA. Ontogeny of surfactant proteins A and B in human amniotic fluid as indices of fetal lung maturity. *Pediatr Res* 1991;**30**:597–605.

37. Liley HG, Hawgood S, Wellenstein GA, Benson B, White RT, Ballard PL. Surfactant protein of molecular weight 28,000–36,000 in cultured human fetal lung: cellular localization and effect of dexamethasone. *Mol Endocrinol* 1987;**1**:205–15.

38. Liley HG, White RT, Warr RG, Benson BJ, Hawgood S, Ballard PL. Regulation of messenger RNAs for the hydrophobic surfactant proteins in human lung. *J Clin Invest* 1989;**83**:1191–7.

39. Miyamura K, Malhotra R, Hoppe HJ, et al. Surfactant proteins A (SP-A) and D (SP-D): levels in human amniotic fluid and localization in the fetal membranes. *Biochim Biophys Acta* 1994;**1210**:303–7.

40. Inoue T, Matsuura E, Nagata A, et al. Enzyme-linked immunosorbent assay for human pulmonary surfactant protein D. *J Immunol Methods* 1994;**173**:157–64.

41. Clements JA, Platzker AC, Tierney DF, et al. Assessment of the risk of the respiratory-distress syndrome by a rapid test for surfactant in amniotic fluid. *NEJM* 1972;**286**:1077–81.

42. Harding R, Hooper SB. Regulation of lung expansion and lung growth before birth. *J Appl Phys* 1996;**81**:209–24.

43. Adzick NS, Harrison MR, Glick PL, Villa RL, Finkbeiner W. Experimental pulmonary hypoplasia and oligohydramnios: relative contributions of lung fluid and fetal breathing movements. *J Pediatr Surg* 1984;**19**:658–65.

44. Harrison MR, Bressack MA, Chug AM, de Lorimier AA. Correction of congenital diaphragmatic hernia in utero. II. Simulated correction permits fetal lung growth with survival at birth. *Surgery* 1980;**88**:260–8.

45. Wallen LD, Perry SF, Alston JT, Maloney JE. Morphometric study of the role of pulmonary arterial flow in fetal lung growth in sheep. *Pediatr Res* 1990;**27**:122–7.

46. Olver RE, Strang LB. Ion fluxes across the pulmonary epithelium and the secretion of lung liquid in the foetal lamb. *J Physiol* 1974;**241**:327–57.

47. Chapman DL, Carlton DP, Nielson DW, Cummings JJ, Poulain FR, Bland RD. Changes in lung lipid during spontaneous labor in fetal sheep. *J Appl Phys* 1994;**76**:523–30.

48. O'Brodovich H, Hannam V, Seear M, Mullen JB. Amiloride impairs lung water clearance in newborn guinea pigs. *J Appl Phys* 1990;**68**:1758–62.

49. Olver RE, Robinson EJ. Sodium and chloride transport by the tracheal epithelium of fetal, new-born and adult sheep. *J Physiol* 1986;**375**:377–90.

50. Croteau JR, Cook CD. Volume-pressure and length-tension measurements in human tracheal and bronchial segments. *J Appl Phys* 1961;**16**:170–2.

51. Burnard ED, Grauaug A. Dyspnoea and apnoea in the newborn: some results of investigation. *Med J Aust* 1965;**1**:445–55.

52. Frank L. Pathophysiology of lung injury and repair: special features of the immature lung. In: Polin RA, Fox WW, editors. *Fetal and Neonatal Physiology*. Philadelphia: WB Saunders; 1992. pp. 914–26.

53. Reynolds EO, Roberton NR, Wigglesworth JS. Hyaline membrane disease, respiratory distress, and surfactant deficiency. *Pediatrics* 1968;**42**:758–68.

54. Avery ME, Mead J. Surface properties in relation to atelectasis and hyaline membrane disease. *AMA J Dis Child* 1959;**97**:517–23.

55. Kendig JW, Notter RH, Cox C, et al. A comparison of surfactant as immediate prophylaxis and as rescue therapy in newborns of less than 30 weeks' gestation. *NEJM* 1991;**324**:865–71.

56. Yoder BA, Gordon MC, Barth WH. Late-preterm birth: How does the changing obstetric paradigm affect the epidemiology of significant respiratory and neurological complications? *Obstet Gynecol* 2008;**111**:814–22.

57. Consortium on Safe L, Hibbard JU, Wilkins I, et al. Respiratory morbidity in late preterm births. *JAMA* 2010;**304**:419–25.

58. Goldman SL, Gerhardt T, Sonni R, et al. Early prediction of chronic lung disease by pulmonary function testing. *J Pediatr* 1983;**102**:613–7.

59. Gerhardt T, Hehre D, Feller R, Reifenberg L, Bancalari E. Serial determination of pulmonary function in infants with chronic lung disease. *J Pediatr* 1987;**110**:448–56.

60. Golan A, Zalzstein E, Zmora E, Shinwell ES. Pulmonary hypertension in respiratory distress syndrome. *Pediatr Pulmonol* 1995;**19**:221–5.

61. Albertine KH. Histopathology of pulmonary edema and the acute respiratory distress syndrome. In: Matthay MA, Ingbar DH, editors. *Pulmonary Edema. Lung Biology in Health and Disease*, vol. 116. New York: Marcel Dekker; 1998. pp. 37–84.

62. Wigglesworth JS. Pathology of neonatal respiratory distress. *Proc Royal Soc Med London B* 1977;**70**:861–3.

63. Stocker JT. The respiratory tract. In: Dehner LP, Stocker JT, editors. *Pediatric Pathology*. 2nd ed. Philadelphia: Lippiincott Williams; 2001. pp. 445–517.

64. Dreyfuss D, Saumon G. Ventilator-induced lung injury: lessons from experimental studies. *Am J Respir Crit Care Med* 1998;**157**:294–323.

65. DeLemos RA, Coalson JJ, Gerstmann DR, et al. Ventilatory management of infant baboons with hyaline membrane disease: the use of high frequency ventilation. *Pediatr Res* 1987;**21**:594–602.

66. Davis JM, Dickerson B, Metlay L, Penney DP. Differential effects of oxygen and barotrauma on lung injury in the neonatal piglet. *Pediatr Pulmonol* 1991;**10**:157–63.

67. Dreyfuss D, Soler P, Basset G, Saumon G. High inflation pressure pulmonary edema. Respective effects of high airway pressure, high tidal volume, and positive end-expiratory pressure. *Am Rev Respir Dis* 1988;**137**:1159–64.

68. Adkins WK, Hernandez LA, Coker PJ, Buchanan B, Parker JC. Age effects susceptibility to pulmonary barotrauma in rabbits. *Crit Care Med* 1991;**19**:390–3.

69. Carlton DP, Cummings JJ, Scheerer RG, Poulain FR, Bland RD. Lung overexpansion increases pulmonary microvascular protein permeability in young lambs. *J Appl Phys* 1990;**69**:577–83.

70. Northway Jr WH, Rosan RC, Porter DY. Pulmonary disease following respirator therapy of hyaline-membrane disease. Bronchopulmonary dysplasia. *N Engl J Med* 1967;**276**:357–68.

71. DeLemos RA, Coalson JJ, Gerstmann DR, Kuehl TJ, Null Jr DM. Oxygen toxicity in the premature baboon with hyaline membrane disease. *Am Rev Respir Dis* 1987;**136**:677–82.

72. Rinaldo JE, Borovetz H. Deterioration of oxygenation and abnormal lung microvascular permeability during resolution of leukopenia in patients with diffuse lung injury. *Am Rev Respir Dis* 1985;**131**:579–83.

73. Crapo JD, Barry BE, Foscue HA, Shelburne J. Structural and biochemical changes in rat lungs occurring during exposures to lethal and adaptive doses of oxygen. *Am Rev Respir Dis* 1980;**122**:123–43.

74. Smith CV, Hansen TN, Martin NE, McMicken HW, Elliott SJ. Oxidant stress responses in premature infants during exposure to hyperoxia. *Pediatr Res* 1993;**34**:360–5.

75. Rinaldo JE, English D, Levine J, Stiller R, Henson J. Increased intrapulmonary retention of radiolabeled neutrophils in early O_2 toxicity. *Am Rev Respir Dis* 1988;**137**:345–52.

76. Bagchi A, Viscardi RM, Taciak V, Ensor JE, McCrea KA, Hasday JD. Increased activity of interleukin-6 but not tumor necrosis factor-alpha in lung lavage of premature infants is associated with the development of bronchopulmonary dysplasia. *Pediatr Res* 1994;**36**:244–52.

77. Carlton DP, Albertine KH, Cho SC, Lont M, Bland RD. Role of neutrophils in lung vascular injury and edema after premature birth in lambs. *J Appl Phys* 1997;**83**:1307–17.

78. Lorant DE, Albertine KH, Bohnsack JF. Chronic lung disease of early infancy: role of neutrophils. In: Bland RD, Coalson JJ, editors. *Chronic Lung Disease of Early Infancy. Lung Biology in Health and Disease*, vol. 137. New York: Marcel Dekker; 2000. pp. 793–811.

79. Ossanna PJ, Test ST, Matheson NR, Regiani S, Weiss SJ. Oxidative regulation of neutrophil elastase-alpha-1-proteinase inhibitor interactions. *J Clin Invest* 1986;**77**:1939–51.

80. Merritt TA, Cochrane CG, Holcomb K, et al. Elastase and alpha 1-proteinase inhibitor activity in tracheal aspirates during respiratory distress syndrome. Role of inflammation in the pathogenesis of bronchopulmonary dysplasia. *J Clin Invest* 1983;**72**:656–66.

81. Bruce MC, Schuyler M, Martin RJ, Starcher BC, Tomashefski Jr JF, Wedig KE. Risk factors for the degradation of lung elastic fibers in the ventilated neonate. Implications for impaired lung development in bronchopulmonary dysplasia. *Am Rev Respir Dis* 1992;**146**:204–12.

82. Thibault DW, Mabry SM, Ekekezie II, Truog WE. Lung elastic tissue maturation and perturbations during the evolution of chronic lung disease. *Pediatrics* 2000;**106**:1452–9.

83. Jackson JC, Chi EY, Wilson CB, Truog WE, Teh EC, Hodson WA. Sequence of inflammatory cell migration into lung during recovery from hyaline membrane disease in premature newborn monkeys. *Am Rev Respir Dis* 1987;**135**:937–40.

84. Ferreira PJ, Bunch TJ, Albertine KH, Carlton DP. Circulating neutrophil concentration and respiratory distress in premature infants. *J Pediatr* 2000;**136**:466–72.

85. Doyle LW, Kitchen WH, Ford GW, Rickards AL, Lissenden JV, Ryan MM. Effects of antenatal steroid therapy on mortality and morbidity in very low birth weight infants. *J Pediatr* 1986;**108**:287–92.

86. Van Marter LJ, Leviton A, Kuban KC, Pagano M, Allred EN. Maternal glucocorticoid therapy and reduced risk of bronchopulmonary dysplasia. *Pediatrics* 1990;**86**:331–6.

87. Liggins GC. Premature delivery of foetal lambs infused with glucocorticoids. *J Endocrinol* 1969;**45**:515–23.

88. Liggins GC, Howie RN. A controlled trial of antepartum glucocorticoid treatment for prevention of the respiratory distress syndrome in premature infants. *Pediatrics* 1972;**50**:515–20.

89. Kikkawa Y, Kaibara M, Motoyama EK, Orzalesi MM, Cook CD. Morphologic development of fetal rabbit lung and its acceleration with cortisol. *Am J Pathol* 1971;**64**:423–42.

90. Conference NIoHC. Effects of corticosteroids for fetal maturation. *JAMA* 1995;**273**:412–8.

91. Jobe AH. Pulmonary surfactant therapy. *NEJM* 1993;**328**:861–8.

92. Kwong MS, Egan EA, Notter RH, Shapiro DL. Double-blind clinical trial of calf lung surfactant extract for the prevention of hyaline membrane disease in extremely premature infants. *Pediatrics* 1985;**76**:585–92.

93. Enhorning G, Shennan A, Possmayer F, Dunn M, Chen CP, Milligan J. Prevention of neonatal respiratory distress syndrome by tracheal instillation of surfactant: a randomized clinical trial. *Pediatrics* 1985;**76**:145–53.

94. Carlton DP, Bland RD. Surfactant and lung fluid balance. In: Robertson B, Taeusch HW, editors. *Surfactant Therapy for Lung Disease.* New York: Marcel Dekker; 1995. pp. 33–46.

95. Northway Jr WH, Moss RB, Carlisle KB, et al. Late pulmonary sequelae of bronchopulmonary dysplasia. *NEJM* 1990;**323**:1793–9.

96. Hislop AA, Haworth SG. Pulmonary vascular damage and the development of cor pulmonale following hyaline membrane disease. *Pediatr Pulmonol* 1990;**9**:152–61.

97. Margraf LR, Tomashefski Jr JF, Bruce MC, Dahms BB. Morphometric analysis of the lung in bronchopulmonary dysplasia. *Am Rev Respir Dis* 1991;**143**:391–400.

98. Husain AN, Siddiqui NH, Stocker JT. Pathology of arrested acinar development in postsurfactant bronchopulmonary dysplasia. *Hum Pathol* 1998;**29**:710–7.

99. Merritt TA, Hallman M, Berry C, et al. Randomized, placebo-controlled trial of human surfactant given at birth versus rescue administration in very low birth weight infants with lung immaturity. *J Pediatr* 1991;**118**:581–94.

100. Hodson WA. Ventilation strategies and bronchopulmonary dysplasia. In: Bland RD, Coalson JJ, editors. *Chronic Lung Disease in Early Infancy. Lung Biology in Health and Disease*, vol. 137. New York: Marcel Dekker; 2000. pp. 173–208.

101. LeFlore JL, Salhab WA, Broyles RS, Engle WD. Association of antenatal and postnatal dexamethasone exposure with outcomes in extremely low birth weight neonates. *Pediatrics* 2002;**110**:275–9.

102. Byrne BJ, Mellen BG, Lindstrom DP, Cotton RB. Is the BPD epidemic diminishing? *Semin Perinatol* 2002;**26**:461–6.

103. Smith VC, Zupancic JA, McCormick MC, et al. Trends in severe bronchopulmonary dysplasia rates between 1994 and 2002. *J Pediatr* 2005;**146**:469–73.

104. Rhodes PG, Hall RT, Leonides JC. Chronic pulmonary disease in neonates with assisted ventilation. *Pediatrics* 1975;**55**:788–96.

105. Campognone P, Singer DB. Neonatal sepsis due to nontypable Haemophilus influenzae. *Am J Dis Child* 1986;**140**:117–21.

106. Mayes L, Perkett E, Stahlman MT. Severe bronchopulmonary dysplasia: a retrospective review. *Acta Paediatr Scand* 1983;**72**:225–9.

107. Bos AP, Hussain SM, Hazebroek FW, Tibboel D, Meradji M, Molenaar JC. Radiographic evidence of bronchopulmonary dysplasia in high-risk congenital diaphragmatic hernia survivors. *Pediatr Pulmonol* 1993;**15**:231–4.

108. Lagatta JM, Clark RH, Brousseau DC, Hoffmann RG, Spitzer AR. Varying patterns of home oxygen use in infants at 23-43 weeks' gestation discharged from United States neonatal intensive care units. *J Pediatr* 2013;**163**:976–82. e972.

109. Laughon MM, Langer JC, Bose CL, et al. Prediction of bronchopulmonary dysplasia by postnatal age in extremely premature infants. *Am J Respir Crit Care Med* 2011;**183**:1715–22.

110. Klinger G, Sokolover N, Boyko V, et al. Perinatal risk factors for bronchopulmonary dysplasia in a national cohort of very-low-birthweight infants. *Am J Obstet Gynecol* 2013;**208**(115):e1–9.

111. Escobedo MB, Hilliard JL, Smith F, et al. A baboon model of bronchopulmonary dysplasia. I. Clinical features. *Exp Mol Pathol* 1982;**37**:323–34.

112. Coalson JJ, Kuehl TJ, Escobedo MB, et al. A baboon model of bronchopulmonary dysplasia. II. Pathologic features. *Exp Mol Pathol* 1982;**37**:335–50.

113. Coalson JJ, Winter VT, Gerstmann DR, Idell S, King RJ, Delemos RA. Pathophysiologic, morphometric, and biochemical studies of the premature baboon with bronchopulmonary dysplasia. *Am Rev Respir Dis* 1992;**145**:872–81.

114. DeLemos RA, Coalson JJ. The contribution of experimental models to our understanding of the pathogenesis and treatment of bronchopulmonary dysplasia. *Clin Perinatol* 1992;**19**:521–39.

115. Coalson JJ, Winter VT, Siler-Khodr T, Yoder BA. Neonatal chronic lung disease in extremely immature baboons. *Am J Respir Crit Care Med* 1999;**160**:1333–46.

116. Yoder BA, Siler-Khodr T, Winter VT, Coalson JJ. High-frequency oscillatory ventilation: effects on lung function, mechanics, and airway cytokines in the immature baboon model for neonatal chronic lung disease. *Am J Respir Crit Care Med* 2000;**162**:1867–76.

117. McGreal EP, Chakraborty M, Winter VT, Jones SA, Coalson JJ, Kotecha S. Dynamic expression of IL-6 trans-signalling molecules in the lungs of preterm baboons undergoing mechanical ventilation. *Neonatology* 2011;**100**:130–8.

118. Pierce RA, Albertine KH, Starcher BC, Bohnsack JF, Carlton DP, Bland RD. Chronic lung injury in preterm lambs: disordered pulmonary elastin deposition. *Am J Physiol Lung Cell Mol Physiol* 1997;**272**:L452–60.

119. Albertine KH, Jones GP, Starcher BC, et al. Chronic lung injury in preterm lambs. Disordered respiratory tract development. *Am J Respir Crit Care Med* 1999;**159**:945–58.

120. MacRitchie AN, Albertine KH, Sun J, et al. Reduced endothelial nitric oxide synthase in lungs of chronically ventilated preterm lambs. *Am J Physiol Lung Cell Mol Physiol* 2001;**281**:L1011–20.

121. Bland RD, Albertine KH, Carlton DP, et al. Chronic lung injury in preterm lambs: abnormalities of the pulmonary circulation and lung fluid balance. *Pediatr Res* 2000;**48**:64–74.

122. Bland RD, Ling CY, Albertine KH, et al. Pulmonary vascular dysfunction in preterm lambs with chronic lung disease. *Am J Physiol Lung Cell Mol Physiol* 2003;**285**:L76–85.

123. Bland RD, Xu L, Ertsey R, et al. Dysregulation of pulmonary elastin synthesis and assembly in preterm lambs with chronic lung disease. *Am J Physiol Lung Cell Mol Physiol* 2007;**292**:L1370–84.

124. Reyburn B, Li M, Metcalfe DB, et al. Nasal ventilation alters mesenchymal cell turnover and improves alveolarization in preterm lambs. *Am J Respir Crit Care Med* 2008;**178**:407–18.

125. Albertine KH, Dahl MJ, Gonzales LW, et al. Chronic lung disease in preterm lambs: effect of daily vitamin A treatment on alveolarization. *Am J Physiol Lung Cell Mol Physiol* 2010;**299**:L59–72.

126. Lindahl P, Karlsson L, Hellstrom M, et al. Alveogenesis failure in PDGF-A-deficient mice is coupled to lack of distal spreading of alveolar smooth muscle cell progenitors during lung development. *Development* 1997;**124**:3943–53.

127. Wendel DP, Taylor DG, Albertine KH, Keating MT, Li DY. Impaired distal airway development in mice lacking elastin. *Am J Respir Cell Mol Biol* 2000;**23**:320–6.

128. Maniscalco WM, Watkins RH, Pryhuber GS, Bhatt A, Shea C, Huyck H. Angiogenic factors and alveolar vasculature: development and alterations by injury in very premature baboons. *Am J Respir Cell Mol Biol* 2002;**282**:L811–23.

129. Gorenflo M, Vogel M, Obladen M. Pulmonary vascular changes in bronchopulmonary dysplasia: a clinicopathologic correlation in short- and long-term survivors. *Pediatr Pathol* 1991;**11**:851–66.

130. Tomashefski Jr JF, Oppermann HC, Vawter GF, Reid LM. Bronchopulmonary dysplasia: a morphometric study with emphasis on the pulmonary vasculature. *Pediatr Pathol* 1984;**2**:469–87.

131. Bush A, Busst CM, Knight WB, Hislop AA, Haworth SG, Shinebourne EA. Changes in pulmonary circulation in severe bronchopulmonary dysplasia. *Arch Dis Child* 1990;**65**:739–45.

132. Benatar A, Clarke J, Silverman M. Pulmonary hypertension in infants with chronic lung disease: non-invasive evaluation and short term effect of oxygen treatment. *Arch Dis Child Fetal Neonatal.* 1995;**72**:F14–19.

133. Gill AB, Weindling AM. Pulmonary artery pressure changes in the very low birthweight infant developing chronic lung disease. *Arch Dis Child* 1993;**68**:303–7.

134. Sherman TS, Chen Z, Yuhanna IS, Lau KS, Margraf LR, Shaul PW. Nitric oxide synthase isoform expression in the developing lung epithelium. *Am J Physiol Lung Cell Mol Physiol* 1999;**276**:L383–90.

135. Thomae KR, Nakayama DK, Billiar TR, Simmons RL, Pitt BR, Davies P. The effect of nitric oxide on fetal pulmonary artery smooth muscle growth. *J Surg Res* 1995;**59**:337–43.

136. Rudic RD, Shesely EG, Maeda N, Smithies O, Segal SS, Sessa WC. Direct evidence for the importance of endothelium-derived nitric oxide in vascular remodeling. *J Clin Invest* 1998;**101**:731–6.

137. Guo JP, Panday MM, Consigny PM, Lefer AM. Mechanisms of vascular preservation by a novel NO donor following rat carotid artery intimal injury. *Am J Phys* 1995;**269**:H1122–1131.

138. Lee JS, Adrie C, Jacob HJ, Roberts Jr JD, Zapol WM, Bloch KD. Chronic inhalation of nitric oxide inhibits neointimal formation after balloon-induced arterial injury. *Circ Res* 1996;**78**:337–42.

139. Seki J, Nishio M, Kato Y, Motoyama Y, Yoshida K. FK409, a new nitric-oxide donor, suppresses smooth muscle proliferation in the rat model of balloon angioplasty. *Atherosclerosis* 1995;**117**:97–106.

140. Bland RD, Kullama LK, Day RW, Carlton DP, MacRitchie AN, Albertine KH. Nitric oxide inhalation decreases pulmonary vascular resistance in preterm lambs with evolving chronic lung. *Pediatr Res* 1997;**40**:247A.

141. Bland RD, Albertine KH, Carlton DP, MacRitchie AJ. Inhaled nitric oxide effects on lung structure and function in chronically ventilated preterm lambs. *Am J Respir Crit Care Med* 2005;**172**:899–906.

142. McCurnin DC, Pierce RA, Chang LY, et al. Inhaled NO improves early pulmonary function and modifies lung growth and elastin deposition in a baboon model of neonatal chronic lung disease. *Am J Physiol Lung Cell Mol Physiol* 2005;**288**:L450–9.

143. Steinhorn RH. Neonatal pulmonary hypertension. *Pediatr Crit Care Med* 2010;**11**:S79–84.

144. Yuhanna IS, MacRitchie AN, Lantin-Hermoso RL, Wells LB, Shaul PW. Nitric oxide (NO) upregulates NO synthase expression in fetal intrapulmonary artery endothelial cells. *Am J Respir Cell Mol Biol* 1999;**21**:629–36.

145. Hamad AM, Johnson SR, Knox AJ. Antiproliferative effects of NO and ANP in cultured human airway smooth muscle. *Am J Physiol Lung Cell Mol Physiol* 1999;**277**:L910–8.

146. Frank L, Groseclose E. Oxygen toxicity in newborn rats: the adverse effects of undernutrition. *J Appl Phys* 1982;**53**:1248–55.

147. Forman HJ, Rotman EI, Fisher AB. Roles of selenium and sulfur-containing amino acids in protection against oxygen toxicity. *Lab Invest* 1983;**49**:148–53.

148. Hawker FH, Ward HE, Stewart PM, Wynne LA, Snitch PJ. Selenium deficiency augments the pulmonary toxic effects of oxygen exposure in the rat. *Eur Respir J* 1993;**6**:1317–23.

149. Thibeault DW. The precarious antioxidant defenses of the preterm infant. *Am J Perinatol* 2000;**17**:167–81.

150. Ehrenkranz RA, Bonta BW, Ablow RC, Warshaw JB. Amelioration of bronchopulmonary dysplasia after vitamin E administration. A preliminary report. *NEJM* 1978;**299**:564–9.

151. Hittner HM, Godio LB, Rudolph AJ, et al. Retrolental fibroplasia: efficacy of vitamin E in a double-blind clinical study of preterm infants. *NEJM* 1981;**305**:1365–71.

152. Ehrenkranz RA, Ablow RC, Warshaw JB. Effect of vitamin E on the development of oxygen-induced lung injury in neonates. *Ann NY Acad Sci* 1982;**393**:452–66.

153. Saldanha RL, Cepeda EE, Poland RL. The effect of vitamin E prophylaxis on the incidence and severity of bronchopulmonary dysplasia. *J Pediatr* 1982;**101**:89–93.

154. Hansen TN, Hazinski TA, Bland RD. Vitamin E does not prevent oxygen-induced lung injury in newborn lambs. *Pediatr Res* 1982;**16**:583–7.

155. Brandt RB, Mueller DG, Schroeder JR, et al. Serum vitamin A in premature and term neonates. *J Pediatr* 1978;**92**:101–4.

156. Hustead VA, Gutcher GR, Anderson SA, Zachman RD. Relationship of vitamin A (retinol) status to lung disease in the preterm infant. *J Pediatr* 1984;**105**:610–5.

157. Shenai JP, Chytil F, Stahlman MT. Vitamin A status of neonates with bronchopulmonary dysplasia. *Pediatr Res* 1985;**19**:185–8.

158. Shenai JP, Rush MG, Stahlman MT, Chytil F. Plasma retinol-binding protein response to vitamin A administration in infants susceptible to bronchopulmonary dysplasia. *J Pediatr* 1990;**116**:607–14.

159. Shenai JP, Kennedy KA, Chytil F, Stahlman MT. Clinical trial of vitamin A supplementation in infants susceptible to bronchopulmonary dysplasia. *J Pediatr* 1987;**111**:269–77.

160. Tyson JE, Ehrenkranz RA, Stoll BJ, et al. Vitamin (Vit.) A supplementation to increase survival without chronic lung disease (CLD) in extremely low birth weight (ELBW) infants: a 14-center randomized trial. *Pediatr Res* 1998;**43**:199A.

161. Tyson JE, Wright LL, Oh W, et al. Vitamin A supplementation for extremely-low-birth-weight infants. National Institute of Child Health and Human Development Neonatal Research Network. *NEJM* 1999;**340**:1962–8.

162. Pearson E, Bose C, Snidow T, et al. Trial of vitamin A supplementation in very low birth weight infants at risk for bronchopulmonary dysplasia. *J Pediatr* 1992;**121**:420–7.

163. Massaro GD, Massaro D. Postnatal treatment with retinoic acid increases the number of pulmonary alveoli in rats. *Am J Physiol Lung Cell Mol Physiol* 1996;**270**:L305–10.

164. Pierce RA, Joyce B, Officer S, et al. Retinoids increase lung elastin expression but fail to alter morphology or angiogenesis genes in premature ventilated baboons. *Pediatr Res* 2007;**61**:703–9.

165. Van Marter LJ, Allred EN, Pagano M, et al. Do clinical markers of barotrauma and oxygen toxicity explain interhospital variation in rates of chronic lung disease? The Neonatology Committee for the Developmental Network. *Pediatrics* 2000;**105**:1194–201.

166. Jobe AH, Kramer BW, Moss TJ, Newnham JP, Ikegami M. Decreased indicators of lung injury with continuous positive expiratory pressure in preterm lambs. *Pediatr Res* 2002;**52**:387–92.

167. Mulrooney N, Champion Z, Moss TJ, Nitsos I, Ikegami M, Jobe AH. Surfactant and physiologic responses of preterm lambs to continuous positive airway pressure. *Am J Respir Crit Care Med* 2005;**171**:488–93.

168. Thomson MA, Yoder BA, Winter VT, et al. Treatment of immature baboons for 28 days with early nasal continuous positive airway pressure. *Am J Respir Crit Care Med* 2004;**169**:1054–62.

169. Thomson MA, Yoder BA, Winter VT, Giavedoni L, Chang LY, Coalson JJ. Delayed extubation to nasal continuous positive airway pressure in the immature baboon model of bronchopulmonary dysplasia: lung clinical and pathological findings. *Pediatrics* 2006;**118**:2038–50.

170. Null DM, Alvord J, Leavitt W, et al. High-frequency nasal ventilation for 21 d maintains gas exchange with lower respiratory pressures and promotes alveolarization in preterm lambs. *Pediatr Res* 2014;**75**:507–16.

171. Carlo WA. Gentle ventilation: the new evidence from the SUPPORT, COIN, VON, CURPAP, Colombian Network, and Neocosur Network trials. *Early Hum Dev* 2012;**88**(Suppl. 2):S81–3.

172. Friedlich P, Lecart C, Posen R, Ramicone E, Chan L, Ramanathan R. A randomized trial of nasopharyngeal-synchronized intermittent mandatory ventilation versus nasopharyngeal continuous positive airway pressure in very low birth weight infants after extubation. *J Perinatol* 1999;**19**:413–8.

173. Khalaf MN, Brodsky N, Hurley J, Bhandari V. A prospective randomized, controlled trial comparing synchronized nasal intermittent positive pressure ventilation versus nasal continuous positive airway pressure as modes of extubation. *Pediatrics* 2001;**108**:13–7.

174. Barrington KJ, Bull D, Finer NN. Randomized trial of nasal synchronized intermittent mandatory ventilation compared with continuous positive airway pressure after extubation of very low birth weight infants. *Pediatrics* 2001;**107**:638–41.

175. De Paoli AG, Davis PG, Lemyre B. Nasal continuous positive airway pressure versus nasal intermittent positive pressure ventilation for preterm neonates: a systematic review and meta-analysis. *Acta Paediatr* 2003;**92**:70–5.

176. Dumas De La Roque E, Bertrand C, Tandonnet O, et al. Nasal high frequency percussive ventilation versus nasal continuous positive airway pressure in transient tachypnea of the newborn: A pilot randomized controlled trial (NCT00556738). *Pediatr Pulmonol* 2011;**46**:218–23.

177. Joss-Moore LA, Albertine KH, Lane RH. Epigenetics and the developmental origins of lung disease. *Mol Genet Metab* 2011;**104**:61–6.

178. Kouzarides T. Chromatin modifications and their function. *Cell* 2007;**128**:693–705.

179. Roth DM, Balch WE. Modeling general proteostasis: proteome balance in health and disease. *Curr Opin Cell Biol* 2011;**23**:126–34.

180. Jones PA. Functions of DNA methylation: islands, start sites, gene bodies and beyond. *Nat Rev Genet* 2012;**13**:484–92.

181. Sato F, Tsuchiya S, Meltzer SJ, Shimizu K. MicroRNAs and epigenetics. *FEBS J* 2011;**278**:1598–609.

182. Fu Q, Yu X, Callaway CW, Lane RH, McKnight RA. Epigenetics: intrauterine growth retardation (IUGR) modifies the histone code along the rat hepatic IGF-1 gene. *FASEB J* 2009;**23**:2438–49.

183. Lallemand AV, Ruocco SM, Joly PM, Gaillard DA. In vivo localization of the insulin-like growth factors I and II (IGF I and IGF II) gene expression during human lung development. *Int J Dev Biol* 1995;**39**:529–37.

184. Chetty A, Andersson S, Lassus P, Nielsen HC. Insulin-like growth factor-1 (IGF-1) and IGF-1 receptor (IGF-1R) expression in human lung in RDS and BPD. *Pediatr Pulmonol* 2004;**37**:128–36.

185. Dunnill MS. Postnatal growth of the lung. *Thorax* 1962;**17**:329–33.

186. Emery JL, Wilcock PF. The post-natal development of the lung. *Acta Anat (Basel)* 1966;**65**:10–29.

187. Davies G, Reid L. Growth of the alveoli and pulmonary arteries in childhood. *Thorax* 1970;**25**:669–81.

188. Boyden EA. Development and growth of the airways. In: Hodson WA, editor. *Development of the Lung. Lung Biology in Health and Disease*, vol. 6. New York: Marcel Dekker; 1977. pp. 3–35.

189. Langston C, Kida K, Reed M, Thurlbeck WM. Human lung growth in late gestation and in the neonate. *Am Rev Respir Dis* 1984;**129**:607–13.

Chapter 18

The Effects of Neonatal Hyperoxia on Lung Development

Foula Sozo* and Megan O'Reilly[†]

*Department of Anatomy and Developmental Biology, Monash University, Melbourne, VIC, Australia; [†]Department of Pediatrics and Women and Children's Health Research Institute, University of Alberta, Edmonton, AB, Canada

INTRODUCTION

Owing to lung immaturity, babies born before term often experience respiratory insufficiency and require respiratory support. Improvements in prenatal and neonatal care have resulted in the majority of these infants, even those born as early as 22–24 weeks of gestation, surviving. Despite improved survival, however, these infants often develop bronchopulmonary dysplasia (BPD), which is characterized by altered lung development, leading to an increased risk of poor lung function, asthma, and susceptibility to pulmonary infection later in life. The adverse long-term outcomes for babies born preterm have been attributed to factors associated with their respiratory support, including prolonged mechanical ventilation and supplemental oxygen therapy, as well as poor nutrition and postnatal infection. As supplemental oxygen is an unavoidable component of the care of most preterm infants, this chapter will focus on the effects of neonatal exposure to high oxygen concentrations (hyperoxia) on development of the conducting airways and gas-exchanging regions of the lung; we also consider later effects on lung function and susceptibility to infection. The potential mechanisms involved in hyperoxia-induced lung injury and potential treatments to prevent or repair alterations in lung development will also be explored.

PRETERM BIRTH

Preterm birth (defined as birth prior to 37 completed weeks of gestation) affects approximately 10% of pregnancies worldwide, and can be further categorized into very preterm birth (<32 weeks of gestation) and extremely preterm birth (<28 weeks of gestation). Very preterm infants are born during the saccular stage of lung development (25 weeks gestation to term), which is prior to the development of definitive alveoli when pulmonary surfactant production is limited. Extremely preterm infants are born during the early saccular stage or canalicular stage of lung development (16–26 weeks gestation), which precedes the formation of terminal saccules, prior to the lung interstitium thinning sufficiently to allow adequate gas exchange, and before the onset of surfactant production. Owing to the immaturity of the surfactant system and gas-exchanging tissue, infants born very and extremely preterm usually develop respiratory distress (RDS) and hence require prolonged respiratory support to ensure their survival. The degree of respiratory support required and the length of stay in hospital are proportional to the degree of prematurity, with extremely preterm infants having a hospital stay more than six times longer than infants born after 32 weeks of gestation.[1]

Oxygen therapy is the most common form of respiratory support given to preterm infants, with the requirement for prolonged supplemental oxygen increasing with each week of decreasing gestational age at birth; approximately 20% of babies born very preterm and around 84% of babies born extremely preterm require oxygen therapy at 28 days.[2] Oxygen is administered in order to achieve adequate oxygen saturation in the blood, with the aim of preventing tissue hypoxia. In the past, the clinical use of oxygen was quite liberal, with 100% O_2 frequently being administered to preterm infants for prolonged periods.[3] However, it was soon observed that although oxygen treatment improved survival of these infants, there was an increase in morbidity, in particular retinopathy of prematurity (ROP) and BPD.[3–5] Initially BPD was characterized histologically by regions of atelectasis and overinflation with the presence of emphysematous changes in the alveoli, hyaline membranes, alveolar wall thickening, fibrosis, airway epithelial lesions including hyperplasia and metaplasia, necrosis of the bronchiolar and alveolar epithelium, bronchiolar obstruction by inflammatory cell exudate, bronchiolar smooth muscle hypertrophy, and vascular lesions (Figure 1).[3]

The majority of infants born preterm now survive due to improved neonatal care, including the introduction of exogenous surfactant administration and gentler ventilation

The Lung. http://dx.doi.org/10.1016/B978-0-12-799941-8.00018-3

329

FIGURE 1   Image showing human BPD lung compared to experimental BPD lung. Panels A and B depict the alveolar region of a term human lung (A) and a preterm human lung with BPD (B). Panels C and D depict the alveolar region of a neonatal mouse lung that was exposed to normal room air (C) and hyperoxic gas (D). The structural characteristics of the neonatal mouse lung exposed to hyperoxic gas are similar to the preterm human BPD lung, which makes it an appealing animal model for use in studies investigating various aspects associated with preterm birth, BPD, and pulmonary-related outcomes. *Panels A and B were modified from the original source: Agrons et al. RadioGraphics 2005;25:1047–1073[147] with permission from the Radiological Society of North America.*

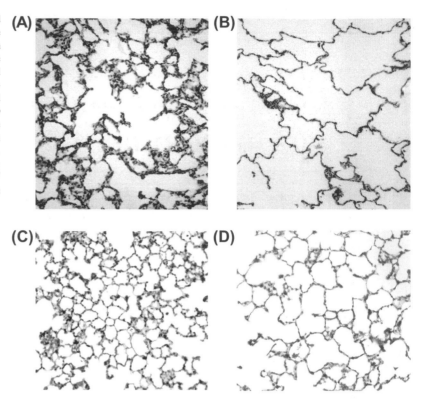

techniques; however, the respiratory support of very and extremely preterm infants unavoidably injures the lungs. This injury includes inflammation, inappropriate repair mechanisms, and arrested alveolar development (collectively termed BPD or chronic lung disease (CLD)). Histopathologically, the "new BPD" (i.e., the current form) is characterized by decreased numbers of alveoli (i.e., larger and simplified), variable interstitial fibrosis and fewer alveolar capillaries; in the conducting airways there is an increased amount of airway smooth muscle (ASM).[6,7] However, due to the increased survival of these preterm infants, the pathology of "new BPD" is based on the findings of only one autopsy study of 14 infants who were treated with surfactant.[7] Altered lung structure following preterm birth and BPD is thought to be the major factor contributing to the life-long risk of illness that is observed in preterm survivors. The long-term pulmonary outcomes for survivors of preterm birth will be described in the following sections.

With improvements in neonatal intensive care, the administration of oxygen has also changed; babies are normally resuscitated in room air (21% O_2) and then the fraction of inspired oxygen (FiO_2) is increased until adequate blood oxygenation levels are achieved.[8] It is thought that this change in practice results in less injury to immature lungs and better long-term outcomes. Recent clinical trials have attempted to determine the optimal blood oxygenation level (O_2 saturation) in preterm infants that improves survival but reduces the harmful effects of excessive oxygen concentrations, including ROP. The Supplemental

Therapeutic Oxygen for Prethreshold Retinopathy of Prematurity (STOP-ROP) trial found that although there was no difference in ROP progression when infants were randomized to saturation targets of 89–94% or 96–99%, infants targeted to the higher O_2 saturation had higher rates of pneumonia and BPD, and were more likely to require oxygen therapy and be hospitalized following discharge.[9] In the Benefits Of Oxygen Saturation Targeting (BOOST) trial, a high O_2 saturation range of 95–98% did not alter the frequency of deaths or major developmental abnormalities (e.g., blindness or cerebral palsy) at 12 months of age compared to a standard target of 91–94%, although the higher saturation group had a greater oxygen dependency and rehospitalization rate.[10] In contrast, the Surfactant, Positive Pressure, and Pulse Oximetry Randomized Trial (SUPPORT) found that the incidence of ROP was reduced in infants who were targeted to arterial O_2 saturations of 85–89% compared to 91–95%; mortality, however, was increased.[11] Similarly, the recent BOOST II trial, which targeted the same O_2 saturation ranges as the SUPPORT trial, found that the lower target O_2 saturation group had a reduced rate of ROP, but an increased rate of necrotizing enterocolitis and death; there was no difference in the incidence of BPD.[12] In contrast, the results of the Canadian Oxygen Trial (COT) suggest that targeting a lower O_2 saturation of 85–89% may be beneficial as it reduced the gestational age at last use of oxygen therapy and had no significant effects on the rates of ROP, necrotizing enterocolitis, severe BPD, hospital readmissions, and death or disability at 18 months of age compared to targeting a high O_2 saturation of 91–95%.[13] Therefore, although there have now been several clinical trials investigating the optimal blood oxygenation level in preterm infants, the levels that are most safe and lead to better long-term outcomes are currently not clear.

What are the Known Effects of Neonatal Hyperoxia on Functional Outcomes?

Due to the multi-factorial etiology of BPD, it is difficult to separate the many contributing factors to determine their individual effect on respiratory functional outcomes in children and adults who were born preterm. Therefore, in the following sections that outline the long-term pulmonary outcomes of preterm birth and BPD, the role of neonatal hyperoxia per se is unclear.

Long-Term Pulmonary Outcomes of Preterm Birth

Recent evidence suggests that infants born very and extremely preterm, especially if they develop BPD, have long-term respiratory complications that reach beyond infancy into childhood and adulthood. At present, however, few studies have investigated the long-term pulmonary outcomes in adults beyond their early twenties. This is particularly important because large numbers of very preterm subjects are now approaching mid-adulthood (i.e., 35–45 years of age) and may be at risk of persistent respiratory morbidity.

A recent meta-analysis that included studies in children and adults born preterm (age range: 5–23 years old) in the pre- and post-surfactant eras (1964–2000) showed that the forced expiratory volume in 1 second (FEV_1) is decreased in preterm-born survivors, even if they do not develop BPD.[14] This suggests that all survivors of preterm birth are at risk of long-term deficits in pulmonary function. It also emphasizes the need to follow these cohorts of ex-preterm subjects into later life to determine whether functional deficits translate into higher rates of chronic obstructive lung disease.

Pulmonary Outcomes during Infancy

Numerous studies have documented abnormal pulmonary function during infancy following very preterm birth and the development of BPD.[15–21] These studies indicate that infants born very preterm, both with and without BPD, experience reduced lung function in the first few months after birth. Compared to term-born infants, preterm infants with BPD are more likely to be symptomatic with recurrent wheezing[19] and require re-hospitalization during the first 2 years after birth due to respiratory illness.[18,20] Functional tests show significantly reduced airway function (i.e., lower FEVs and flow rates) and increased residual lung volume in

infants born extremely–very preterm (range: 23–34 weeks of gestation) compared to term-born controls.[15–17,19–21] Reduced pulmonary diffusing capacity of carbon monoxide (DL_{CO}), a measure of alveolar gas exchange, has also been demonstrated in infants and young children with chronic lung disease of prematurity, suggesting impairment of alveolar development, albeit alveolar volume appeared normal.[22] Furthermore, respiratory variables are significantly different to those of term-born infants, indicated by a faster breathing frequency,[16,20] greater tidal volume (V_T),[16] increased dead space,[16] and greater minute ventilation (V_E).[16] There are also considerable differences in respiratory system compliance and resistance, with preterm infants typically presenting with lower compliance and higher resistance than term-born controls.[16] Of particular concern is the reported reduction in respiratory compliance and increase in respiratory resistance in healthy late preterm infants (33–36 weeks of gestation) compared to term-born controls.[23]

Pulmonary Outcomes during Childhood

There are now abundant data on pulmonary outcomes during childhood following preterm birth, with the majority of follow-up studies assessing respiratory symptoms, pulmonary function, and exercise capacity in children aged between 7 and 12 years.[24] Such studies indicate that children born preterm have an increased risk of respiratory symptoms including cough,[18,25] wheeze,[18,25,26] and asthma.[25,27,28] Assessment of pulmonary function throughout childhood shows that preterm birth increases the risk of lower FEVs, flow rates, and lung capacities, and increased residual volume; these impairments indicate airflow obstruction and are particularly common in individuals who developed BPD.[18,27–34] Importantly, even children born extremely preterm during the surfactant era are at increased risk of ventilation inhomogeneity, gas trapping, airway hyper-responsiveness, and reductions in forced expiratory flows, and asthma.[25,26,30,35,36] Interestingly, a recent study that reported a significantly reduced FEV_1 in children born extremely preterm in the surfactant era also showed that alveolar size and number were normal, suggesting catch-up alveolarization.[36] However, even if alveolar structure is normalized in childhood, measures of alveolar function, such as DL_{CO}, can remain abnormal. Reduced DL_{CO} has been commonly reported among prematurely born children in both the pre-[37] and post-surfactant eras.[26,31,34,38] In addition to the numerous functional abnormalities associated with preterm birth (with and without BPD), inflammation of the lower airways has also been reported. Sputum analyses in children born preterm in the pre- and post-surfactant eras were reported to have found significantly higher proportions of neutrophils and interleukin-8 (IL-8) values, in combination with lower FEV_1 and forced expiratory flows.[39]

Exercise capacity is also compromised in children born preterm (both with and without BPD) compared to term-born controls; this is indicative of reduced aerobic capacity, as evidenced by a lower maximal heart rate,[18] higher breathing frequency,[31] smaller V_T,[31] reduced maximal oxygen consumption (VO_{2max}),[30-32] reduced maximal minute ventilation (V_{Emax}),[32] reduced post-exercise FEV_1 and exercised-induced bronchoconstriction,[28] a shorter exercise time,[32] and a shorter distance that could be run.[30] Interestingly, oxygen supplementation during the neonatal period in infants born very preterm has been identified as an independent risk factor for asthma in childhood.[40]

Pulmonary Outcomes during Adolescence and Early Adulthood

Poor pulmonary function remains common in adolescence and young adulthood following preterm birth.[24] Recent studies of subjects aged 17–22 years (both males and females) born very preterm show that respiratory symptoms persist into late adolescence and early adulthood, with significantly more ex-preterm subjects reporting cough, wheeze, or asthma compared to term-born controls.[41-43] Adolescents born preterm also had a significantly greater risk of being hospitalized for respiratory illness.[42] As with pulmonary function outcomes in childhood, FEVs, flow rates, and lung capacities all remained significantly lower in late adolescence and young adulthood following preterm birth compared to control subjects.[44-47] Pulmonary diffusing capacity also remained lower in ex-preterm subjects during late-adolescence, with the reported DL_{CO} being significantly lower than those born at term.[38] Furthermore, exercise capacity remained lower than that of controls, with significantly lower maximal heart rate, faster breathing frequency, reduced VO_{2max}, and reduced V_{Emax} in response to exercise tests.[45,47] The presence of ongoing airway inflammation is supported by recent evidence demonstrating higher exhaled breath condensate 8-isoprostane levels, a biomarker of oxidative stress, in ex-preterm non-BPD and BPD adolescents.[44] Alarmingly, CT studies reveal apparently permanently disrupted alveolar development and emphysema at adult age in ex-preterm infants with BPD.[48,49] Functional data from such follow-up studies suggest that very preterm birth, or associated factors, can permanently affect the conducting airways and gas-exchanging tissue of the lungs.

It is important to recognize that the majority of studies of adolescent and young adult survivors of very preterm birth described above were undertaken with individuals born during the pre-surfactant era (years of birth ranged from 1977 to 1985). Although adverse pulmonary outcomes have been reported in infants and children born following the introduction of exogenous surfactant,[35] further longitudinal follow-up studies are required from current survivors of preterm birth who were treated with surfactant to gain a thorough understanding of their long-term pulmonary outcomes. To date there are limited reports of functional studies in adolescent and young adult subjects from the surfactant era.[44,46] It is anticipated that data on lung function, respiratory symptoms, and exercise capacity, as well as data from imaging studies, will become available as survivors of preterm birth who were the first to be treated with exogenous surfactant are now entering this age range.

EFFECTS OF NEONATAL HYPEROXIA ON LUNG DEVELOPMENT

Alterations in the architecture of both the alveoli and the conducting airways are likely to affect lung function. Stimulated by the adverse long-term pulmonary outcomes observed in survivors of preterm birth, numerous experimental studies have investigated the potential involvement of neonatal hyperoxia. Specifically, studies have investigated the effects of neonatal hyperoxia on the development of the gas-exchanging region of the lung, although the effects on the conducting airways of the lung have received little attention. The most common experimental model used for addressing these issues is the newborn rodent; newborn mice or rats are exposed to hyperoxia, ranging from 40 to 100% O_2, but most commonly >80% O_2, for different periods of time after birth. Newborn rodents are born during the saccular stage of lung development and develop alveoli after birth. Therefore, their alveoli develop in the extra-uterine environment under hyperoxic conditions, as occurs with preterm infants. Most studies have focused on the immediate alterations to lung architecture caused by neonatal hyperoxia, although few studies have examined whether these alterations persist later in life and whether lung function is also perturbed. The following sections outline these findings and Figure 2 illustrates potential pathways leading to altered lung structure as a result of hyperoxia exposure.

Effects of Hyperoxia on Development of Alveoli

Effects on Alveolar Number and Structure

Numerous studies have demonstrated that exposure of the immature lung to hyperoxia (>90% O_2) results in enlarged air spaces[50,51] and larger alveoli, as indicated by an increase in mean linear intercept (MLI),[52-54] thinned interstitia,[51] and decreased secondary septal crests[51,55]; there is also a significant reduction in septal thickness[51,52] and alveolar number, as indicated by a decrease in radial alveolar count (RAC) (Table 1).[50-53] These findings are evident after as little as 24 hours of exposure to 95% oxygen.[51] Mean alveolar surface area is also decreased.[55] In addition, after 2 weeks of exposure to 100% O_2, there was an increase in alveolar epithelial cell proliferation, thickened alveolar septa, and focal alveolar

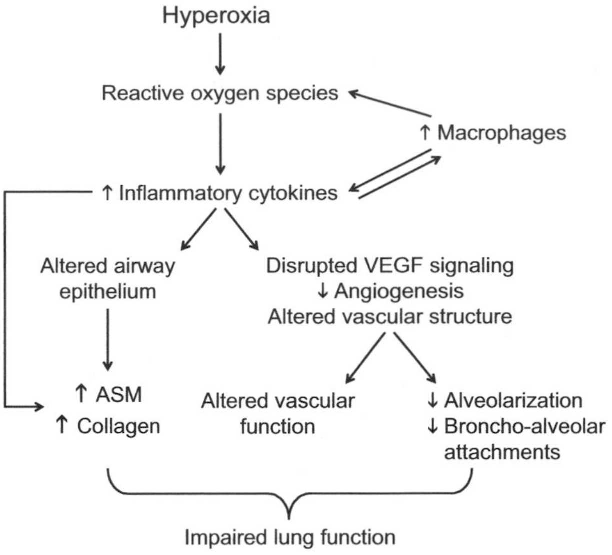

FIGURE 2 Diagram showing potential pathways whereby exposure of the immature lung to hyperoxic gas leads to impaired alveolarization, disrupted pulmonary vascular development and function, and alterations in the structure of the airways, all of which lead to altered lung function later in life. ASM, airway smooth muscle; VEGF, vascular endothelial growth factor.

atelectasis.[56] As the duration of exposure increased, alveoli became more distorted and thrombi appeared within the alveolar capillary lumina.[56] Similarly, exposure to 85% O_2 resulted in greatly enlarged airspaces following 7 days of exposure and this became more pronounced with an increasing duration of exposure.[57] In contrast, exposure to a lower oxygen concentration of 60% for 14 days did not increase MLI but increased the variance in average MLI measurements between fields within each lung section.[58] Furthermore, there were no alterations in the estimated alveolar surface area per unit lung volume or the estimated total number of alveoli in the lung despite a decrease in the secondary crest/tissue ratio.[58] Together, these studies show that the oxygen concentration and duration of oxygen exposure are important in the development of alveolar injury.

Alveolar simplification, characterized by larger and fewer air spaces and decreased septation, increased MLI, and decreased RAC were significantly attenuated in mice deficient in matrix metalloproteinase (MMP)-9, suggesting that MMP-9 also plays an important role in hyperoxia-induced alterations in alveolar architecture.[59] Furthermore, appropriate pulmonary vascular development has been shown to be necessary for normal alveolar formation, as discussed further in the following sections. Studies have also indicated that hyperoxia can induce epigenetic changes, resulting in a state of cell senescence and growth inhibition and consequently altered alveolar development. For example, exposure of the immature lung to 80% O_2 for 14 days decreased histone deacetylase (HDAC) activity and HDAC1 and HDAC2 expression.[60] Since HDACs normally silence gene transcription, these changes were associated

with increased expression of the cell cycle arrest regulators p53 and p21, which was further associated with a decrease in cell proliferation.[60]

Studies have investigated whether alterations in alveolar formation persist following the period of hyperoxia, or whether the lung is able to undergo spontaneous repair following recovery in room air; some studies have only investigated effects following 1–2 weeks of recovery in room air,[61–64] whilst others have reported effects into early adulthood[65,66,85] and with aging.[65,67,68,87] Following 1–2 weeks of recovery, exposure to moderate to high O_2 concentrations (65–95% O_2) after birth resulted in enlarged alveoli,[64] as indicated by an increase in MLI,[61–63] and reduced septation,[64] as indicated by decreased RAC,[61–63] internal surface area (lung volume/MLI), and nodal point density.[61,62] The effects observed following 1–2 weeks of recovery persisted into early adulthood, with 4 days of hyperoxia causing a persistent increase in MLI and mean chord length at 8 weeks of age, indicating that neonatal hyperoxia increases alveolar size through to early adulthood; these effects were dose-dependent with 40% oxygen having no effect and 100% oxygen having a greater effect than 60% and 80% oxygen.[66] Furthermore, studies have demonstrated that neonatal hyperoxia results in decreased RAC and number of alveoli per unit surface area of parenchyma, suggesting impaired alveolarization at 7.5 months of age, which is mid-adulthood in mice.[67] Similarly, the parenchymal airspace fraction and MLI increased and the internal surface area, density of alveolar septa, and number of alveoli decreased in 2-month-old and 22-month-old rats exposed to 100% O_2 for 8 days after birth.[65] In contrast, in 67-week-old mice, prior exposure to 100% O_2 for 4 days after birth did not cause a persistence in the alveolar simplification observed in early adulthood, with mean chord length significantly reduced following hyperoxia-exposure rather than being increased.[68] These studies suggest that although the effects of neonatal exposure to hyperoxia can persist into old age, the effects appear to be dependent on both the concentration of oxygen and the duration of oxygen exposure.

Effects on Pulmonary Angiogenesis

In addition to reduced alveolarization, BPD is characterized by reduced vascularization of the lung. Consistent with this effect, infants who develop BPD are at risk of pulmonary hypertension and right ventricular hypertrophy as a result of vascular remodeling and impaired vascular growth. The role of neonatal hyperoxia in altering the pulmonary vasculature of the immature lung has received considerable interest, in particular because functionally, angiogenesis and alveolarization are intimately linked; alveolar formation can be disrupted when pulmonary angiogenesis is inhibited.[69]

Studies investigating the effects of neonatal hyperoxia on the pulmonary vasculature have shown that exposure

TABLE 1 Effects of hyperoxia on development of alveoli

Species	O$_2$	Duration	Lung Development Stage Coinciding with Hyperoxia	Observed Effects	Reference
Rat; Sprague Dawley	>96%	Birth to P8, RA to P60 or 22 mo.	Late-saccular to early-alveolar	Increased MLI and parenchymal airspace fraction at P60 and 22 mo. Reduced alveolar number and alveolar septa at P60 and 22 mo.	65
	>95%	P0.5–14	Late-saccular to mid-alveolar	Increased MLI Reduced RAC	53
	95%	Birth to 24 h.	Late-saccular	Enlarged airspaces Thinned interstitium Decreased secondary septal crests and septal thickness	51
		Birth to P14	Late-saccular to mid-alveolar	Increased MLI Reduced RAC	50,54
		Birth to P14, RA to P21 or P45	Late-saccular to mid-alveolar	Increased MLI at P21 and P45	64
		Birth to P7	Late-saccular to early-alveolar	Increased MLI Reduced secondary septation, septal thickness, RAC, and tissue density	52
	>95% (at Denver altitude)	P1-7, RA to P21	Late-saccular to early-alveolar	Increased MLI Reduced RAC	63
	75%	P2-14, RA to P21	Late-saccular to mid-alveolar	Increased MLI Reduced secondary septation and RAC Distal airspace enlargement	62
	60%	Birth to P14	Late-saccular to mid-alveolar	Decreased secondary crest/tissue ratio No increase in MLI, however increased variance in average MLI observed No reduction in alveolar surface area No reduction in alveolar number	58
Rat; Wistar	>90%	P3–13	Late-saccular to early-alveolar	Reduced alveolar surface area Reduced secondary crest number	55
Mouse; FVB/N	85%	P1–28	Late-saccular to late-alveolar	Simplified acinar structure Reduced alveolar number Enlarged terminal airspaces	57
	80%	P1–15	Late-saccular to mid-alveolar	Reduced RAC	60
	65% (80% at Denver altitude)	P1-11, RA to P21	Late-saccular to early-alveolar	Increased MLI Reduced radial alveolar count and internal surface area	61
Mouse; CD1	90%	P5-13	Early-alveolar	Increased MLI Reduced alveolar counts	59
Mouse; C57Bl/Ka	100%	P1 to P14, P21, P28, or 6 wks	Late-saccular to mid- and late-alveolar	Increased alveolar epithelial cell proliferation Thickened alveolar septa and focal alveolar atelectasis	146

TABLE 1 Effects of hyperoxia on development of alveoli—cont'd

Species	O_2	Duration	Lung Development Stage Coinciding with Hyperoxia	Observed Effects	Reference
Mouse; C57Bl/6J	100%	P1–4, RA to 67 wks	Late-saccular	Reduced mean alveolar cord length	[68]
	40%, 60%, 80%, 100%	P1–4, RA to 8 wks	Late-saccular	Increased MLI and mean cord length (not 40% O_2)	[66]
Mouse; Swiss	65%	Birth to P28	Late-saccular to late-alveolar	Fewer and larger alveoli	[67]

Abbreviations: h., hours; MLI, mean linear intercept; mo., months; P, postnatal day; RA, room air/normoxia; RAC, radial alveolar count; wks, weeks.

to elevated oxygen concentrations can induce features of pulmonary hypertension immediately following exposure. For example, exposure of neonatal rat pups to hyperoxia (95% O_2) for 14 days decreased pulmonary capillary density and increased medial wall thickness of small pulmonary arteries (Table 2).[50,54] These structural changes were associated with *decreased vascular endothelial growth factor (VEGF) and VEGF receptor 2 (VEGFR-2) expression*[54] and a reduced number of circulating and resident mesenchymal stem cells (MSCs).[64] Furthermore, pulmonary arterial acceleration time, measured by echo-doppler, was decreased and right ventricular hypertrophy was evident, both of which are suggestive of pulmonary hypertension.[50] In contrast, exposure to >90% oxygen from day 3 to day 13 after birth did not decrease the number of pulmonary microvessels despite an increase in smooth muscle content; however, pups exposed to hyperoxia had reduced body weight in this study.[55] Exposure to a lower degree of hyperoxia (75% O_2) for 14 days also stimulated arterial remodeling, indicated by an increase in medial wall thickness (muscularization) of pulmonary arterioles (25–75 μm diameter), a decreased number of small (25–50 μm diameter) and intermediate (50–100 μm diameter) blood vessels in the lung, reduced blood vessel density, and right ventricular hypertrophy.[70] Similarly, exposure to 60% O_2 for 14 days resulted in medial wall thickening in pulmonary arterioles (20–65 μm diameter) and a reduced number of small vessels in the lung periphery. These effects are potentially due to a reduction in the expression of VEGF-A and angiopoietin-1 (Ang-1).[58] Reductions in VEGF, VEGFR-2, endothelial nitric oxide synthase (NOS), and erythropoietin receptor expression and a reduction in pulmonary vessel density following exposure to moderate hyperoxia (65% oxygen) were associated with a reduction in endothelial progenitor cells (EPCs) in the circulation, bone marrow, and lungs.[71]

The effects of neonatal hyperoxia on the pulmonary vasculature have been shown to persist following recovery in room air, with a reduction in vessel density observed at 3 weeks of age after 10–14 days of hyperoxia.[54,61,62] In the presence of reduced body weight, however, vessel density and the wall thickness of small pulmonary arteries (20–60 μm diameter) were not different and right ventricular hypertrophy was not evident in hyperoxia-exposed animals; however, this lack of difference may be due to a shorter duration of hyperoxia.[63] In contrast, the decrease in microvessel density following exposure to 70% O_2 for 10 days persisted to 41 days of age; however, right ventricular systolic pressure was unaltered, suggesting that pulmonary hypertension was absent.[72] Reductions in microvessel density occurred in the presence of an increase in CD45-positive lung cell populations, suggesting that an increase in resident lung precursor cells with endothelial potential was insufficient to rescue the observed defects in pulmonary vascular development.[72] Studies have also investigated the persistence of effects of hyperoxia on the pulmonary vasculature into old age. At 67 weeks of age, mice exposed to 100% O_2 for 4 days after birth had right ventricular hypertrophy, a reduction in the number of pulmonary microvessels, no change in vessel wall thickness, and an increased number of dilated arterioles; the changes were associated with a reduction in platelet endothelial cell adhesion molecule (PECAM) but not VEGF, and alterations in bone morphogenetic protein (BMP) signaling.[68] Furthermore, the persistent pulmonary hypertension observed in mice exposed to neonatal hyperoxia likely resulted in cardiac failure, causing an increase in mortality with age.[68] Other studies, however, have not shown an increase in mortality in hyperoxia-exposed rats with aging, despite a persistence in the reduction in pulmonary vessel density (arteries of 25–200 μm diameter) and the increase in right ventricular weight and right ventricular systolic pressure observed at 2 months of age to 22 months of age.[65] In addition, there was a significant increase in the wall thickness of vessels (25–100 μm diameter) and muscularization of small arterioles at both 2 and 22 months of age in rats that had been exposed to neonatal hyperoxia; however, the wall thickness in vessels 25–75 μm diameter and

TABLE 2 Effects of hyperoxia on pulmonary angiogenesis

Species	O$_2$	Duration	Lung Development Stage Coinciding with Hyperoxia	Observed Effects	Reference
Rat; Sprague Dawley	>96%	Birth to P8, RA to P60 or 22 mo.	Late-saccular to early-alveolar	Reduced pulmonary vessel density (25–200μm diameter) Increased vessel wall thickness (25–100μm diameter) and small arteriole muscularization	65
	95%	Birth to P14	Late-saccular to mid-alveolar	Reduced pulmonary capillary density Increased medial wall thickness of small pulmonary arteries Reduced expression of VEGF and VEGFR-2	50,54,64
	>95% (at Denver altitude)	P1–7, RA to P21	Late-saccular to early-alveolar	No reduction in pulmonary vessel density No increase in wall thickness of small pulmonary arteries (20–60μm diameter)	63
	60%	Birth to P14	Late-saccular to mid-alveolar	Increased medial wall thickness of pulmonary arterioles (20–65μm diameter) Reduced number of small vessels in lung periphery Reduced expression of VEGF-A and Ang-1	58
	75%	P2–14, RA to P21	Late-saccular to mid-alveolar	Reduced pulmonary vessel density	62
Rat; Wistar	>90%	P3–13	Late-saccular to early-alveolar	No reduction in pulmonary microvessel number	55
Mouse; C57Bl/6J	100%	P1–4, RA to 67 wks	Late-saccular	Reduced number of microvessels Increased number of dilated arterioles No increase in vessel wall thickness Reduced expression of PECAM No reduction in VEGF expression	68
Mouse; HO-1 TG	75%	Birth to P14	Late-saccular to mid-alveolar	Increased medial wall thickness of pulmonary arterioles (25–75μm diameter) Reduced blood vessel number: both small (20–50μm) and intermediate (50–100μm) Reduced pulmonary vessel density	70
Mouse; C57Bl/6-ROSA	70% (at Denver altitude)	P1–10, RA to P41	Late-saccular to early-alveolar	Reduced pulmonary vessel density	72
Mouse; C57Bl/6J	65% (80% at Denver altitude)	P1–10	Late-saccular to early-alveolar	Reduced pulmonary vessel density Reduced expression of VEGF, VEGFR-2, endothelial NOS, and erythropoietin receptor Reduced EPCs in circulation, BM, and lungs	71
Mouse; FVB/N	65% (80% at Denver altitude)	P1–11, RA to P21	Late-saccular to early-alveolar	Reduced pulmonary vessel density	61

Abbreviations: Ang-1, angiopoietin-1; BM, bone marrow; EPCs, endothelial progenitor cells; HO-1, heme oxygenase-1; mo., months; NOS, nitric oxide synthase; P, postnatal day; PECAM, platelet endothelial cell adhesion molecule; RA, room air/normoxia; TG, transgenic; VEGF, vascular endothelial growth factor; VEGFR-2, vascular endothelial growth factor receptor-2; wks, weeks.

extent of muscularization decreased with aging.[65] Overall, neonatal hyperoxia appears to cause permanent alterations in the pulmonary vasculature, which could affect cardiopulmonary function and lead to increased disease risk.

Effects on Pulmonary Extracellular Matrix (ECM)

Neonatal hyperoxia results in increased pulmonary fibrosis[53,55,57] and increased interstitial thickness,[57,58] which is associated with an increase in tissue fraction.[58] After 2 weeks of neonatal exposure to 85% O_2, increased collagen deposition within the parenchymal region was evident, and this increased progressively with continued exposure to oxygen[57]; in contrast, parenchymal elastin deposition was not altered.[57] After 2 weeks of 100% O_2 exposure, however, alveolar septa contained an increased amount of elastic fibers; increasing the duration of hyperoxia continued to result in more severe fibrosis at the expense of alveoli.[56] Thickened bundles of elastin fibers lining alveolar walls were also evident in adult mice exposed to 60%, 80%, or 100% O_2 after birth, but not 40% O_2, suggesting neonatal hyperoxia causes persistent dose-dependent remodeling of the ECM.[66] Myofibroblasts in the lung parenchyma deposit elastin, and it has been shown that neonatal hyperoxia significantly decreases the expression of lipogenic markers whilst increasing the expression of myogenic markers; this suggests that hyperoxia induces transdifferentiation of alveolar interstitial lipofibroblasts to myofibroblasts.[51] There are limited studies investigating the signaling pathways by which hyperoxia alters the ECM. Exposure to hyperoxia increases MMP-2 and MMP-9, with an increase in MMP-9 evident in both the mesenchyme and alveolar epithelium; MMPs are proteolytic enzymes that degrade ECM.[59] The increase in protein levels of type 1 collagen, tropoelastin and alpha-smooth muscle actin (a marker for myofibroblasts), and aberrant elastin deposition in the alveolar walls observed following hyperoxia-exposure were ameliorated in hyperoxia-exposed MMP-9 deficient mice, suggesting a role for MMPs in hyperoxia-induced remodeling of the extracellular matrix.[59] A role for growth factors in the dysplastic growth caused by hyperoxia has also been suggested. For example, exposure to 60% O_2 for 2 weeks resulted in patchy areas of interstitial thickening that were not due to increases in collagen deposition but were associated with increases in DNA synthesis, insulin-like growth factor (IGF)-1 and IGF receptor type 1.[73]

Effects on Pulmonary Surfactant

Exposure of the immature lung to hyperoxia can change the composition of pulmonary surfactant; this is important because surfactant has an essential role in stabilizing alveoli and in pulmonary host defense. Exposure of premature baboons, maintained on ventilatory support, to 100% O_2 for 6 days led to a two-fold increase in *SP-B* mRNA levels and a four-fold increase in *SP-C* mRNA levels, without changes in *SP-A* mRNA levels.[74] Interestingly, after 10 days of oxygen exposure, there were no longer any significant differences in *SP-B* and *SP-C* mRNA levels, suggesting the effects of hyperoxia are dependent on the length of exposure.[74] Additionally, hyperoxia did not alter surfactant phospholipid composition, despite impairment in the ability of the surfactant to reduce surface tension.[74] In preterm lambs at approximately 0.8 of term, ventilation with 100% O_2 had no effect on *SP-A*, *SP-B*, and *SP-C* mRNA levels; however, at 0.9 of term mechanical ventilation increased *SP-A* mRNA levels and close to term it increased *SP-A* and *SP-B* mRNA levels; these findings suggest that the effects of mechanical ventilation and/or hyperoxia are dependent on gestational age or lung maturity.[75] In rodents, neonatal hyperoxia progressively increased mRNA expression of *SP-A* and *SP-B*, whilst *SP-C* and *SP-D* mRNA expression was increased earlier during the exposure period and then declined, although levels continued to be elevated with respect to normoxic controls; SP-A and SP-D protein levels were also increased in lung lavage (SP-B and SP-C were not studied).[76] In contrast, in another rodent study, hyperoxia progressively increased *SP-A*, *SP-B*, and *SP-D* mRNA levels but did not change *SP-C* mRNA levels.[77] Combined, these data suggest differential regulation of specific SPs by hyperoxia at a pre- and post-translational level in a time-dependent manner.

In addition to effects on SP expression, hyperoxia may affect alveolar type II cell differentiation. Exposure to 95% O_2 for 24 hours induced a significant decrease in the pulmonary expression of lipogenic markers and a significant increase in the expression of myogenic markers; these findings indicate that hyperoxia causes differentiation of lipofibroblasts to myofibroblasts, which in turn may affect lipofibroblast-induced differentiation of type II cells.[51] In support of these observations, culturing fetal rat type II cells[64] and human fetal type II cells[78] under hyperoxic conditions decreases cell viability[64,78] and mature SP-B protein levels despite no change in *SP-B* mRNA levels.[78]

Further studies on the long-term effects of neonatal hyperoxia on the surfactant system have shown that 8-week-old mice previously exposed to 100% O_2 for 4 days after birth had 70% fewer type II cells, potentially due to an inhibition of type II cell proliferation; however, levels of T1a, a marker of type I cells, was also increased, indicating epithelial cell differentiation rather than increased cell death.[79] Additionally, the reduction in alveolar type II cells following neonatal hyperoxia is dose-dependent, with a decline in type II cells following exposure to 60%, 80%, and 100% O_2, whilst the increase in T1a occurs following exposure to 40–100% O_2.[66] Although alveolar epithelial cell differentiation appears to be persistently affected by neonatal hyperoxia at early adulthood, surfactant phospholipid composition, and surface activity of large surfactant aggregates from bronchoalveolar lavage were not altered.[66]

The effect of neonatal hyperoxia on the reduction in type II cell number, however, does not appear to persist in aged mice, with the expression of proSP-C increased and T1a reduced in hyperoxia-exposed mice at 22 months of age.[68]

Effects of Hyperoxia on Development of Conducting Airways

Effects on the Large Airways

Few studies have investigated the effects of neonatal hyperoxia on the development of the conducting airways of lung; in those that have, the focus has been on the smooth muscle of the trachea or large bronchi. For example, exposure of newborn rats to 50% O_2 for 15 days increased the smooth muscle area in isolated tracheal rings.[80] Similarly, exposure of 21-day-old rats to >95% O_2 for 8 days increased the thickness of the smooth muscle layer of the large airways (1000–3000 μm in diameter).[81] Airway smooth muscle cells have been isolated from human fetal trachea at the canalicular stage of lung development; exposure of these cells to hyperoxia for 48 hours resulted in increased proliferation when exposed to <60% O_2 (30%, 40%, and 50%) but increased apoptosis at >60% O_2 (60%, 70%, 90%), suggesting that effects of hyperoxia on smooth muscle are dose-dependent.[82]

Effects on the Small Conducting Airways

Few studies have examined effects of hyperoxia on the small conducting airways (bronchioles), which are important determinants of lung function. Exposure of newborn rats to 95% O_2 for 7 days decreased the number and increased the size of Clara cells and ciliated cells, the major cell types within the rodent bronchiolar epithelium.[83,84] The changes in epithelial cell numbers and size, however, did not persist following 48 hours of recovery in room air.[84] In contrast, recent studies examining the effects of neonatal exposure to 65% O_2 for 7 days have shown that bronchiolar epithelial cell composition is affected in early adulthood[85]; in early adulthood the proportion of Clara cells in the bronchiolar epithelium was decreased, whilst the proportion of ciliated cells was increased. The discrepancy between studies may be due to the difference in measurements made, or to the concomitant reduction in body weight and lung volume during the period of hyperoxia in one of the studies.[84] Previous studies have shown that early postnatal under-nutrition can diminish the conversion of Clara to ciliated cells.[86] In addition, neonatal hyperoxia in the presence of reduced body growth does not result in alterations in bronchiolar epithelial cell composition.[85] Therefore, hyperoxia in the absence of postnatal growth restriction appears to persistently alter the epithelium of the bronchioles. As Clara cells are progenitor cells of the bronchiolar epithelium, a reduced proportion may have implications for epithelial recovery from injury.

Neonatal hyperoxia can also alter the composition of the outer airway wall, resulting in a decrease in the amount of collagen in early adulthood, which may decrease airway wall rigidity and potentially increase the risk for airway narrowing during expiration.[85] The combination of neonatal hyperoxia and growth restriction, however, results in an increase in bronchiolar collagen that may increase airway wall stiffness.[85] Regardless, neonatal hyperoxia is likely to affect the function of the small airways. In addition to collagen, ASM is also affected. Immediately following exposure of 21-day-old rats to >95% O_2 for 8 days, the thickness of the ASM in airways <1000 μm in diameter was increased.[81] Similarly, in mice at mid-adulthood, the area of ASM was increased by neonatal hyperoxia.[87] Because hyperoxia was combined with impaired growth in both studies, the effects of hyperoxia-exposure in isolation on smooth muscle of the bronchioles are unknown. As ASM regulates airway contractility and increased ASM is a feature of obstructive lung diseases such as asthma and chronic obstructive pulmonary disease, exposure of the immature lung to hyperoxia has the potential to persistently affect airway function.

Effects of Neonatal Hyperoxia on Later Lung Function

Airway Function

Exposing neonatal mice to 100% O_2 for 4 days after birth decreases airway resistance and airway elastance at 2 months of age, which is early adulthood in rodents[66]; in contrast, exposure to lower concentrations of oxygen appears to have no effect on baseline airway function.[66,85] Exposure of older Sprague-Dawley rats (21 days old) to hyperoxia (>95% O_2), however, does not alter baseline airway resistance, suggesting that the developing airways are most susceptible to the effects of hyperoxia.[81] Earlier studies suggested that neonatal hyperoxia in rats increases airway reactivity due to increased ASM contraction in the larger airways, including trachea and intralobar bronchi.[80,88] An increase in respiratory system resistance in response to methacholine, indicating airway responsiveness, is also observed in newborn rats exposed to 60% O_2 for 14 days[89] and in 21-day-old rats exposed to >95% O_2.[81]

Numerous studies have investigated the potential mechanisms by which neonatal hyperoxia alters the responsiveness of ASM in later life. For example, the bronchial smooth muscle relaxation response to epithelium-dependent and -independent stimulation was not altered after hyperoxia.[88] However, other studies have suggested that hyperoxia causes airway hyper-reactivity via a disruption in nitric oxide (NO) signaling, which normally induces airway relaxation.[90,91] In support of this, hyperoxia impairs relaxation of the trachea via attenuation of cAMP production and prostaglandin E_2 release.[91] In addition, hyperoxia may further impair smooth muscle relaxation downstream of cAMP signaling by increasing phosphorylation of the myosin-binding subunit of smooth muscle myosin phosphatase, thus inhibiting its phosphatase activity and altering contractile properties.[92]

Respiratory System Mechanics

Respiratory system compliance is decreased immediately following 14 days exposure of neonatal rats to 60% O_2, although this occurred in the presence of growth restriction.[89] In contrast, previous studies have shown that hyperoxia for 4 days after birth increased lung compliance, tissue elastance and tissue damping in early adulthood (2 months); however, these effects were dose-dependent with only O_2 concentrations greater than 60% affecting lung compliance and tissue damping.[66] In a study in mice, prolonged exposure to 65% O_2 (for 28 days) in the absence of growth restriction led to increased lung compliance at 9.5 months of age, suggesting that neonatal hyperoxia persistently affects lung function up until mid-adulthood.[67]

MECHANISMS OF ALTERED LUNG DEVELOPMENT

Oxidative Stress

Antioxidant Defence Systems

Under normal conditions, there is a balance between the production of reactive oxygen species (ROS) and the antioxidant defences that protect cells from oxidative stress in vivo.[93] Therefore, an imbalance between the generation of ROS and the elimination of ROS via antioxidant defence systems can lead to oxidative stress. ROS is a collective term that includes both oxygen radicals such as superoxide ($O_2^{\bullet-}$) and hydroxyl ($OH^{\bullet}$), and non-radical oxidizing agents such as hydrogen peroxide (H_2O_2) that can easily be converted into radicals.[94] Although ROS are produced during normal metabolism and are involved in the modulation of several physiological functions, if they are produced in excessive amounts they can overwhelm or inactivate the antioxidant defence system, and damage major cellular components via oxidative stress, such as membrane lipids, proteins, carbohydrates, and DNA.[95]

The reducing environment within cells assists in preventing damage caused by free radicals, and is maintained through the action of antioxidant enzymes and substances.[94] The antioxidant defences can be divided into two main groups: enzymatic defences and non-enzymatic defences. Enzymatic defences within the cells rely on a number of enzymes that act to reduce the generation of ROS; examples of such enzymes include superoxide dismutase (SOD), glutathione peroxidase (Gpx), and catalase (CAT).[93,94] SOD catalyses the conversion of $O_2^{\bullet-}$ to H_2O_2. However, H_2O_2 has the ability to form a highly reactive ROS, $OH^{\bullet}$, which is produced when H_2O_2 proceeds through the Fenton reaction; the resultant $OH^{\bullet}$ can induce cellular injury.[94] Conversion of H_2O_2 to $OH^{\bullet}$ through the Fenton reaction is prevented by the conversion of H_2O_2 to water (H_2O) via the enzymes Gpx and CAT. Through these important enzymatic defence systems, ROS levels are maintained in balance. It is therefore important that SOD, Gpx, and CAT work in equilibrium to scavenge ROS and reduce oxidative stress. Non-enzymatic defences are generally in the form of biological compounds such as glutathione, thioredoxin, vitamins A, C, and E, melatonin, and polyphenols.[93]

The enzymatic antioxidant defence system undergoes significant maturation in the last 15% of gestation (i.e., from about 32 weeks in humans), when there is an approximately 150% increase in antioxidant enzymes. Interestingly, the later gestational increase in antioxidant enzymes parallels the maturational pattern of the pulmonary surfactant system, and during this timeframe the non-enzymatic antioxidants reach the fetus via the placenta in increasing amounts.[93] Normally, pulmonary surfactant contains substantial amounts of SOD and CAT that likely prevent oxidative degradation of the lipids in surfactant and subsequent surfactant inactivation by ROS.[93]

Preterm Birth and Oxidative Stress

Infants born preterm have an immature antioxidant defence system and this can increase their vulnerability to the damaging effects of oxygen toxicity on the developing lung.[96,97] It has been shown that preterm infants have a significantly higher level of oxidative stress (as indicated by increased urinary 8-hydroxyl-2'-deoxyguanosine; a sensitive marker of oxidative stress) over the first 100 days of life than term-born controls. This significant increase in oxidative stress is likely the result of lower Gpx activity and reduced SOD levels in preterm infants. Interestingly, the preterm infants in that study[98] had not undergone mechanical ventilation, oxygen supplementation, or any other intensive care procedure, indicating that the reduction in antioxidant defences and increased oxidative stress were due to prematurity per se. Furthermore, it is known that the levels of non-enzymatic defences are low in preterm infants.[99–101] Not only do vitamins A, C, and E have important roles in normal physiology, but they also exhibit antioxidant properties and are essential in non-enzymatic antioxidant defence pathways.[93] Vitamin A concentration in preterm infants was found to be significantly lower in term-born infants[99]; another study investigating vitamin A status in preterm infants showed that vitamin A concentrations were further reduced in the presence of BPD.[100] A study of cord blood from preterm infants showed that they have lower vitamin A, E, and C levels compared to term infants.[101] As most strategies for the respiratory support of very preterm infants include supplemental oxygen, it is probable that the combination of hyperoxia and an immature antioxidant defence system further augments oxidative stress and may play a causal role in altering lung development and later lung function.

Effects of Hyperoxia on Oxidative Stress and Antioxidant Defences

Prolonged exposure to hyperoxic gas can induce the generation of excessive ROS within the lung, which can cause an acute inflammatory response leading to tissue injury.[96,102] Newborn mice exposed to hyperoxia (95% O_2, 7 days) have significantly elevated ROS concentrations in lung homogenates.[103] The importance of antioxidant defence systems has been demonstrated by many studies that have utilized transgenic mice to over-express particular enzymatic defences.[103–105] Newborn transgenic mice over-expressing extracellular SOD (EC-SOD) were exposed to 95% O_2 for 7 days; they were shown to exhibit significant preservation of alveolar surface and volume density, as well as reduced pulmonary neutrophil influx compared to wild-type mice exposed to hyperoxia.[104] EC-SOD over-expression was also shown to preserve bronchiolar and alveolar epithelial proliferation (in particular type II cells), which can be impaired by hyperoxia.[105] Furthermore, a recent study showed that over-expression of EC-SOD preserved angiogenesis following exposure to hyperoxia.[103] Together these studies show that although the antioxidant defence pathways are compromised following neonatal hyperoxia, treatments that restore the imbalance are likely to provide a beneficial effect on lung development.

Inflammation

Preterm Birth and Pulmonary Inflammation

Inflammation can alter the structure of the lung via release of pro-inflammatory cytokines and induction of an inflammatory cascade.[106] Macrophages and neutrophils are major sources of inflammatory cytokines and chemokines in the lung and play a crucial role in pulmonary inflammation.[107] Inflammation is strongly associated with the development of BPD, and numerous inflammatory mediators including chemokines, cytokines, and growth factors are present in tracheal aspirate and bronchoalveolar lavage (BAL) fluid of infants with BPD.[108] Furthermore, preterm infants with BPD have much higher and persisting numbers of macrophages and neutrophils in their BAL fluid.[106] Increased concentrations of IL-8 have also been found in tracheal aspirate and BAL fluid of preterm infants, and are strongly correlated with the development of BPD.[109,110] An increase in macrophage recruitment to lung tissue can arise in response to the production of monocyte chemoattractant protein (MCP)-1, which is a monocyte-selective chemokine that can attract monocytes from the circulation into the lungs; these monocytes can then differentiate into macrophages.

Preterm infants who developed BPD have been shown to have significantly increased concentrations of MCP-1 in tracheal aspirate in the first weeks of life,[111,112] and up-regulated gene expression of *chemokine ligand 2* (which encodes the MCP-1 protein) in lung autopsy tissue.[113] Furthermore, the magnitude of MCP-1 concentration has been correlated with the duration of oxygen support; MCP-1 concentrations were significantly higher in preterm infants who were oxygen-dependent at 28 postnatal days compared to preterm infants who were not.[111,112]

Hyperoxia and Pulmonary Inflammation

Increased numbers of immune cells, including greater numbers of macrophages and neutrophils, are common features in experimental studies of neonatal hyperoxia.[57,114,115] Increased levels of pro-inflammatory mediators, such as IL-1α, IL-1β, IL-6, IL-8/cytokine-induced neutrophil chemoattractant (CINC)-1, and MIP-1 are also characteristic of hyperoxia-induced neonatal lung injury.[96] Increased expression of IL-1α and MIP-1α has been shown to precede the neutrophil influx, suggesting that they may be important in modulating subsequent lung inflammation.[57] Other studies have investigated MCP-1 in experimental models of neonatal hyperoxia[116–118]; neonatal hyperoxia was shown to result in a significant increase in lung *MCP-1* mRNA expression,[116] increased lung MCP-1 protein levels,[117] and increased MCP-1 concentration in BAL fluid[118] when compared to room air exposure. The increase in MCP-1 concentration was also accompanied by significant increases in macrophage number in BAL fluid.[118] The relationship between increased MCP-1 concentration and increased macrophage number is further demonstrated when MCP-1 is blocked with neutralizing antibodies that prevent macrophage accumulation in hyperoxia-exposed lungs.[118] This finding indicates that macrophage recruitment is under the control of macrophage chemokines, such as MCP-1. Increased CINC-1 has also been reported in the lung following neonatal hyperoxia; this was accompanied by significantly increased myeloperoxidase activity.[119] The neutrophil chemokine CINC-1 signals through the neutrophil CXC chemokine receptor-2 to induce an influx of neutrophils into the lung following hyperoxia-exposure. The role of CINC-1 in hyperoxia-induced neonatal lung injury was further demonstrated in a study showing that blockade of the CXC receptor-2 in neonatal rat pups exposed to hyperoxia; neutrophil accumulation was prevented, ROS production was attenuated, and increased alveolar formation was observed.[119]

Mast cells are also implicated in hyperoxia-induced lung injury. They contribute to allergy-induced airway hyper-responsiveness by release of histamine, cytokines, and proteases that result in ASM contraction.[89] In tracheal rings isolated from neonatal rats, type-I mast cells and macrophages (located in the submucosa and connective tissue), and granulocytes (located in the connective tissue) were increased by 15 days of exposure to 50% O_2.[80] In a different study, exposure of neonatal rats to 60% O_2 for 14 days resulted in the development of airway hyper-reactivity, as well as significant accumulation of mast cells.[89] The role of mast cells in mediating airway hyper-reactivity in hyperoxia-induced lung injury was demonstrated by normalization of lung function after inhibition of mast cells by systematic treatment with cromolyn and imatinib mesylate.[89]

INFLUENCE OF HYPEROXIA ON SUSCEPTIBILITY TO INFECTION

Given that preterm infants are at increased risk of respiratory infections and have high re-hospitalization rates,[120–122] several studies have investigated whether neonatal hyperoxia plays a contributory role. Studies in baboons delivered preterm and mechanically ventilated have indicated that exposure to 80–100% O_2 for 21 days causes significant increases in coagulase-negative staphylococci infection of the oropharynx and more severe histological and clinical pneumonia than baboons treated with lower O_2.[123] However, even baboons treated with lower O_2 concentrations than those used in previous studies, with a median FiO_2 at 28 days of 0.32 (range 0.21–0.50), developed pneumonia, which was correlated with increased IL-8 levels in tracheal aspirates.[124] In more recent studies, neonatal hyperoxia in rodents has been associated with an increased susceptibility to influenza A virus in adulthood.[125] Neonatal hyperoxia increased the number of macrophages, neutrophils, and lymphocytes in the airways following infection with influenza A, and was associated with increased levels of MCP-1 (CCL2) in lavage fluid.[125] Furthermore, adult mice exposed to neonatal hyperoxia experienced a decrease in body weight and an increase in mortality in response to infection compared to mice that had not been exposed to hyperoxia, despite levels of virus-specific antibodies in the blood being equivalent and the function of cytotoxic T cells not disrupted.[125,126] Further studies using this model suggest that the altered response to infection with influenza A virus by hyperoxia is dose-dependent, with 80% and 100% O_2 eliciting an exacerbated response whilst 40% or 60% O_2 had no effect.[127] The exact mechanisms by which hyperoxia alters the immune response are not clear, although epithelial immune function may be disrupted.[128] In this regard, when eosinophil-associated RNase 1 (Ear1), which is reduced in the airway epithelium following hyperoxia, is delivered to the lung before infection with influenza virus, viral replication and leukocyte recruitment is reduced during infection and survival is improved.[128] Recent studies suggest a role for MCP-1 in the elevated recruitment of leukocytes in response to infection with influenza A virus in adult mice

exposed to hyperoxia after birth, although MCP-1 does not appear to contribute to the alveolar simplification observed following hyperoxia.[129] A role for the antioxidant SOD has been proposed in the exacerbated fibrotic response to influenza A virus infection in adult mice exposed to hyperoxia after birth.[129] Furthermore, enhancing the expression of EC-SOD in type-II alveolar epithelial cells appears to preserve alveolar development in adult mice exposed to hyperoxia after birth, and blunts the reduction in body weight and improves survival in response to infection.[129] Therefore, there appears to be multiple pathways or signaling events involved in the altered response to infection following hyperoxia-exposure. Although the effects of neonatal hyperoxia on susceptibility to respiratory viral infections, in particular influenza, have recently been studied, it will be important to determine whether effects on the immune system are more widespread and whether susceptibility to other infections to which preterm survivors are vulnerable can also be attributed to neonatal hyperoxia.

POTENTIAL THERAPIES

Following the establishment and characterization of neonatal hyperoxia-induced lung injury as a model of experimental BPD, many studies have used it to investigate an array of potential treatments. These treatments range from factors that promote angiogenesis, prevent oxidative stress and pulmonary inflammation, as well as the introduction of stem/progenitor cells to repair the lung. The following sections briefly summarize major advances in these areas of potential treatments.

Angiogenic Factors

The normal progress of alveolarization throughout lung development is highly dependent on angiogenesis. The requirement of angiogenic growth factors, in particular VEGF, for the development of the vasculature has been demonstrated by inactivation of VEGF alleles and knockout of the VEGF receptors; these studies show that inactivation results in lethal phenotypes characterized by deficient organization of endothelial cells (reviewed in[69]). Furthermore, decreased lung VEGF and its receptor combined with impaired pulmonary vascularization are commonly reported features of neonatal hyperoxia-induced lung injury.[54,62] Adenovirus-mediated VEGF gene therapy[54] or treatment with recombinant human VEGF (rhVEGF)[62] has been demonstrated to enhance lung vessel growth and preserve alveolarization in rodent models of neonatal hyperoxia, as well as improve survival. VEGF-induced angiogenesis is mediated, in part, by nitric oxide (NO). It is thought that increased production of NO in the developing lung promotes angiogenesis. NO stimulates the production of guanosine 3′,5′-cyclic monophosphate (cGMP), which is inactivated by phosphodiesterase enzymes (PDE). Sildenafil is a specific inhibitor of PDE5, and protects the activity of cGMP and increases NO. It has been shown that treatment with sildenafil in a rodent model of hyperoxia-induced lung injury can preserve alveolar growth and lung angiogenesis, as well as reducing pulmonary hypertension associated with oxygen-induced lung injury.[50] An additional factor that is involved in the production of NO is L-citrulline. Endogenous NO is produced from the metabolism of L-arginine to L-citrulline. Treatment with L-citrulline in a neonatal rodent model of hyperoxia-exposure has been shown to increase plasma L-arginine and L-citrulline concentrations, as well as resulting in preserved alveolar and vascular growth.[130] Treatment with a different angiogenic factor, erythropoietin (EPO), has also been investigated.[55] EPO stimulates endothelial cells to increase cell migration and proliferation, increases the production and release of endothelin-1 (required for vasoconstriction and local cellular growth), and induces angiogenesis. Not only is EPO an angiogenic factor, but it also possesses antioxidant properties. Administration of rhEPO in a rodent model of hyperoxia-induced lung injury resulted in improved alveolar structure, enhanced vascularity, and decreased fibrosis.[55] Recently, the effect of connective tissue growth factor (CTGF) in hyperoxia-induced neonatal lung injury has been investigated.[131] CTGF is a matricellular protein that plays an important role in tissue development and remodeling, and its expression is up-regulated following hyperoxia. Using a CTGF-neutralizing antibody, the effects of neonatal hyperoxia exposure were prevented, as shown by improved alveolarization and vascular development.[131] Therefore, improving pulmonary angiogenesis, which can be achieved by targeting a variety of factors, can ameliorate the detrimental effects of hyperoxia-exposure on alveolar development.

Antioxidant and Anti-Inflammatory Factors

As described earlier, hyperoxia can induce the generation of excessive ROS within the lung, which can lead to oxidative stress and an acute inflammatory response leading to tissue injury.[96,97] Several studies have investigated the effects of attenuating excessive oxidative stress production and reducing the degree of inflammation within the lung to prevent lung injury induced by neonatal hyperoxia. Recently, the effects of pentoxifylline (PTX) treatment in a rodent model of hyperoxia exposure has been shown to reduce lung edema, decrease macrophage infiltration, and increase the activities of the antioxidant enzymes SOD, CAT, and Gpx.[132] Furthermore, treatment with PTX was shown to improve survival, increase gene, and protein expression of VEGF as well as improve pulmonary vascularization.[132] Like sildenafil, PTX is an inhibitor of PDE and has immunomodulatory and anti-fibrotic properties. Interestingly, although PTX exerted

therapeutic potential in relation to oxidative stress and inflammation, it did not improve alveolar structure or attenuate pulmonary fibrosis.[132] Non-enzymatic antioxidants, such as vitamins A, C, and E, are important in protection from oxidative stress.[93] Vitamin A is essential for normal lung development and maturation, and its effects are mediated through its action on retinol-binding protein and the retinoic acid receptor.[93] A recent study demonstrated that a combination therapy of vitamin A and retinoic acid (VARA) increased lung retinol stores in neonatal rodents exposed to hyperoxia.[133] This was accompanied by attenuation of hyperoxia-induced increases in mediators of oxidative stress and mRNA and protein expression of key pro-inflammatory mediators. VARA therapy also attenuated hyperoxia-induced alveolar simplification and alterations in lung function.[133] Curcumin is a lipophilic polyphenol compound that is an active ingredient of the Indian spice turmeric, and also has potent antioxidant and anti-inflammatory properties. When administered to neonatal rats exposed to hyperoxia, it inhibited oxidative stress and effectively blocked activation of transforming growth factor beta (TGF-β) and subsequent lung injury.[134] Caffeine is widely used to treat apnoea of prematurity, but it has also been shown to reduce the rate of BPD.[135] However, it is unclear exactly how caffeine exerts its beneficial effects. Therefore, a recent study has investigated the effects of caffeine therapy in the presence of hyperoxia using a rodent model of hyperoxia-induced lung injury.[135] Caffeine was shown to attenuate the perturbation in alveolar development, diminish leukocyte infiltration into the lung, and decrease the pulmonary mRNA expression of chemokines and pro-inflammatory cytokines, suggesting that caffeine can have an anti-inflammatory role.[135] At the sites of tissue injury following hyperoxia-exposure, cells can undergo death via apoptosis or necrosis,[96] which if exaggerated may disrupt normal alveolar development. Therefore, it has been speculated that protection against pulmonary cell death via activation of the pro-survival Akt pathway could prevent arrested alveolar development in hyperoxia-induced lung injury. This speculation was supported by a recent study that inhibited the pro-survival Akt pathway during a critical period of alveolar development and demonstrated impaired alveolarization.[136] Furthermore, when adenovirus-mediated Akt gene transfer was introduced into a neonatal rodent model of hyperoxia-exposure, it preserved alveolar architecture.[136] These recent studies highlight the therapeutic potential of targeting oxidative stress and inflammation pathways for the treatment of hyperoxia-induced lung injury.

Stem/Progenitor Cells

Recent studies show that stem/progenitor cells are perturbed in neonatal rodent models of hyperoxia-induced lung injury (reviewed in[137]), which provided a strong rationale for stem/progenitor cell supplementation to prevent or repair lung injury. Over recent years, many different types of stem cell therapies have been investigated in experimental models of BPD. Here we refer to some of the most recent studies that have used the neonatal hyperoxia-induced lung injury model; for further detailed information regarding the use of stem/progenitor cells in BPD, refer to Chapter 15 (O'Reilly and Thébaud). MSCs have been extensively examined as a therapeutic cell; they can be sourced from the bone marrow, umbilical cord blood, Wharton's jelly, placenta, and adipose tissue. One of the first studies to identify a therapeutic potential of stem/progenitor cells in hyperoxia-induced neonatal lung injury utilized bone marrow-derived MSCs.[64] Treatment with MSCs was shown to improve survival, improve exercise tolerance, and attenuate alveolar and lung vascular injury.[64] Several other studies have confirmed the therapeutic benefit of MSCs, with common outcomes including improved survival, prevention of alveolar arrest and improved alveolar structure, prevention in vascular growth arrest, and attenuation of lung fibrosis and inflammation.[138–142] Several of these effects have also been observed in hyperoxia-induced lung injury following administration of a different type of stem/progenitor cell, bone marrow-derived angiogenic cells.[61] Recently, studies have utilized the cell-free conditioned media obtained from MSCs.[143–145] Administration of conditioned media to hyperoxia-exposed neonatal rodents has shown beneficial effects as evidenced by reversal of parenchymal fibrosis and peripheral pulmonary artery devascularization, partial reversal of alveolar injury, normalized lung function, and attenuation of peripheral pulmonary artery muscularization.[143–145] Therefore, as yet unidentified factors secreted by these cells in the media apparently have the potential to prevent or repair hyperoxia-induced lung injury.

CONCLUSIONS

Advances in neonatal care have enabled infants to survive birth as early as 22–23 weeks of gestation; with extreme prematurity the immature and surfactant-deficient lung is commonly exposed to hyperoxic gas for extended periods. This can lead to life-long alterations in the lungs as a result of injury and altered development induced in the first weeks after birth; oxidative stress and inflammation are thought to be the underlying causes of these changes. The conducting airways, gas-exchanging tissue, and the pulmonary vasculature are all likely to be affected by hyperoxia. These structural changes to the lung can result in an increased risk of poor pulmonary function that can persist into adult life. Ongoing inflammation may also increase the risk of respiratory infection and asthma later in life. Improvements in long-term outcomes are likely to result from changes in clinical practice that target minimizing the exposure to hyperoxic gas, as well as therapies aimed at preventing and/or repairing neonatal lung injury. As supplemental oxygen is likely to be unavoidable in the care of very preterm infants, future studies should aim to determine the most effective methods for reducing lung injury in infancy and for identifying long-term effects on the lungs.

ACKNOWLEDGEMENTS

We acknowledge our funding support from a Program Grant from the National Health and Medical Research Council of Australia (ID 606789) and Monash University.

REFERENCES

1. Hintz SR, Bann CM, Ambalavanan N, Cotten M, Das A, Higgins RD. Predicting time to hospital discharge for extremely preterm infants. *Pediatrics* 2010;**125**:E146–54.
2. Chow SSW. *Report of the Australian and New Zealand neonatal network 2010*. Sydney: ANZNN; 2013.
3. Northway Jr WH, Rosan RC, Porter DY. Pulmonary disease following respirator therapy of hyaline-membrane disease. Bronchopulmonary dysplasia. *N Engl J Med* 1967;**276**:357–68.
4. Campbell K. Intensive oxygen therapy as a possible cause of retrolental fibroplasia: A clinical approach. *Med J Aust* 1951;**2**:48–50.
5. Ryan H. Retrolental fibroplasia: A clinicopathologic study. *Am J Ophthalmol* 1952;**35**:329–42.
6. Coalson JJ. Pathology of new bronchopulmonary dysplasia. *Semin Neonatol* 2003;**8**:73–81.
7. Husain AN, Siddiqui NH, Stocker JT. Pathology of arrested acinar development in postsurfactant bronchopulmonary dysplasia. *Hum Pathol* 1998;**29**:710–7.
8. Richmond S, Wyllie J. European resuscitation council guidelines for resuscitation 2010 section 7. Resuscitation of babies at birth. *Resuscitation* 2010;**81**:1389–99.
9. STOP-ROP Multicenter Study Group. Supplemental therapeutic oxygen for prethreshold retinopathy of prematurity (STOP-ROP), a randomized, controlled trial. I: Primary outcomes. *Pediatrics* 2000;**105**:295–310.
10. Askie LM, Henderson-Smart DJ, Irwig L, Simpson JM. Oxygen-saturation targets and outcomes in extremely preterm infants. *N Engl J Med* 2003;**349**:959–67.
11. Carlo WA, Finer NN, Walsh MC, Rich W, Gantz MG, Laptook AR, et al. Target ranges of oxygen saturation in extremely preterm infants. *N Engl J Med* 2010;**362**:1959–69.
12. Stenson BJ, Tarnow-Mordi WO, Darlow BA, Simes J, Juszczak E, Askie L, et al. Oxygen saturation and outcomes in preterm infants. *N Engl J Med* 2013;**368**:2094–104.
13. Schmidt B, Whyte RK, Asztalos EV, Moddemann D, Poets C, Rabi Y, et al. Effects of targeting higher vs lower arterial oxygen saturations on death or disability in extremely preterm infants: A randomized clinical trial. *JAMA* 2013;**309**:2111–20.
14. Kotecha SJ, Edwards MO, Watkins WJ, Henderson AJ, Paranjothy S, Dunstan FD, et al. Effect of preterm birth on later FEV1: A systematic review and meta-analysis. *Thorax* 2013;**68**:760–6.
15. Friedrich L, Pitrez PM, Stein RT, Goldani M, Tepper R, Jones MH. Growth rate of lung function in healthy preterm infants. *Am J Respir Crit Care Med* 2007;**176**:1269–73.
16. Hjalmarson O, Sandberg K. Abnormal lung function in healthy preterm infants. *Am J Respir Crit Care Med* 2002;**165**:83–7.

17. Friedrich L, Stein RT, Pitrez PM, Corso AL, Jones MH. Reduced lung function in healthy preterm infants in the first months of life. *Am J Respir Crit Care Med* 2006;**173**:442–7.

18. Gross SJ, Iannuzzi D, Kveselis DA, Anbar RD. Effect of preterm birth on pulmonary function at school age: A prospective controlled study. *J Pediatr* 1998;**133**:188–92.

19. Robin B, Kim YJ, Huth J, Klocksieben J, Torres M, Tepper RS, et al. Pulmonary function in bronchopulmonary dysplasia. *Pediatr Pulmonol* 2004;**37**:236–42.

20. Tepper RS, Morgan WJ, Cota K, Taussig LM. Expiratory flow limitation in infants with bronchopulmonary dysplasia. *J Pediatr* 1986;**109**:1040–6.

21. Sanchez-Solis M, Garcia-Marcos L, Bosch-Gimenez V, Perez-Fernandez V, Pastor-Vivero MD, Mondejar-Lopez P. Lung function among infants born preterm, with or without bronchopulmonary dysplasia. *Pediatr Pulmonol* 2012;**47**:674–81.

22. Balinotti JE, Chakr VC, Tiller C, Kimmel R, Coates C, Kisling J, et al. Growth of lung parenchyma in infants and toddlers with chronic lung disease of infancy. *Am J Resp Crit Care* 2010;**181**:1093–105.

23. McEvoy C, Venigalla S, Schilling D, Clay N, Spitale P, Nguyen T. Respiratory function in healthy late preterm infants delivered at 33-36 weeks of gestation. *J Pediatr* 2013;**162**:464–9.

24. Narang I. Review series: What goes around, comes around: Childhood influences on later lung health? Long-term follow-up of infants with lung disease of prematurity. *Chron Respir Dis* 2010;**7**:259–69.

25. Fawke J, Lum S, Kirkby J, Hennessy E, Marlow N, Rowell V, et al. Lung function and respiratory symptoms at 11 years in children born extremely preterm: The epicure study. *Am J Respir Crit Care Med* 2010;**182**:237–45.

26. Lum S, Kirkby J, Welsh L, Marlow N, Hennessy E, Stocks J. Nature and severity of lung function abnormalities in extremely pre-term children at 11 years of age. *Eur Respir J* 2011;**37**:1199–207.

27. Doyle LW. VICS-Group: Respiratory function at age 8-9 years in extremely low birthweight/very preterm children born in Victoria in 1991-1992. *Pediatr Pulmonol* 2006;**41**:570–6.

28. Joshi S, Powell T, Watkins WJ, Drayton M, Williams EM, Kotecha S. Exercise-induced bronchoconstriction in school-aged children who had chronic lung disease in infancy. *J Pediatr* 2013;**162**:813–8: e811.

29. Brostrom EB, Thunqvist P, Adenfelt G, Borling E, Katz-Salamon M. Obstructive lung disease in children with mild to severe BPD. *Respir Med* 2010;**104**:362–70.

30. Smith LJ, van Asperen PP, McKay KO, Selvadurai H, Fitzgerald DA. Reduced exercise capacity in children born very preterm. *Pediatrics* 2008;**122**:E287–93.

31. Welsh L, Kirkby J, Lum S, Odendaal D, Marlow N, Derrick G, et al. The epicure study: Maximal exercise and physical activity in school children born extremely preterm. *Thorax* 2010;**65**:165–72.

32. Santuz P, Baraldi E, Zaramella P, Filippone M, Zacchello F. Factors limiting exercise performance in long-term survivors of bronchopulmonary dysplasia. *Am J Respir Crit Care Med* 1995;**152**:1284–9.

33. Kulasekaran K, Gray PH, Masters B. Chronic lung disease of prematurity and respiratory outcome at eight years of age. *J Paediatr Child Health* 2007;**43**:44–8.

34. Cazzato S, Ridolfi L, Bernardi F, Faldella G, Bertelli L. Lung function outcome at school age in very low birth weight children. *Pediatr Pulmonol* 2012.

35. Hacking DF, Gibson AM, Robertson C, Doyle LW. Respiratory function at age 8-9 after extremely low birthweight or preterm birth in Victoria in 1997. *Pediatr Pulmonol* 2012.

36. Narayanan M, Beardsmore CS, Owers-Bradley J, Dogaru CM, Mada M, Ball I, et al. Catch-up alveolarization in ex-preterm children. Evidence from (3)He magnetic resonance. *Am J Respir Crit Care Med* 2013;**187**:1104–9.

37. Hakulinen AL, Jarvenpaa AL, Turpeinen M, Sovijarvi A. Diffusing capacity of the lung in school-aged children born very preterm, with and without bronchopulmonary dysplasia. *Pediatr Pulmonol* 1996;**21**:353–60.

38. Satrell E, Roksund O, Thorsen E, Halvorsen T. Pulmonary gas transfer in children and adolescents born extremely preterm. *Eur Respir J* 2013;**42**:1536–44.

39. Teig N, Allali M, Rieger C, Hamelmann E. Inflammatory markers in induced sputum of school children born before 32 completed weeks of gestation. *J Pediatr* 2012;**161**:1085–90.

40. Mai XM, Gaddlin PO, Nilsson L, Finnstrom O, Bjorksten B, Jenmalm MC, et al. Asthma, lung function and allergy in 12-year-old children with very low birth weight: A prospective study. *Pediatr Allergy Immunol* 2003;**14**:184–92.

41. Doyle LW, Faber B, Callanan C, Freezer N, Ford GW, Davis NM. Bronchopulmonary dysplasia in very low birth weight subjects and lung function in late adolescence. *Pediatrics* 2006;**118**:108–13.

42. Halvorsen T, Skadberg BT, Eide GE, Roksund OD, Carlsen KH, Bakke P. Pulmonary outcome in adolescents of extreme preterm birth: A regional cohort study. *Acta Paediatrica* 2004;**93**:1294–300.

43. Narang I, Rosenthal M, Cremonesini D, Silverman M, Bush A. Longitudinal evaluation of airway function 21 years after preterm birth. *Am J Respir Crit Care Med* 2008;**178**:74–80.

44. Filippone M, Bonetto G, Corradi M, Frigo AC, Baraldi E. Evidence of unexpected oxidative stress in airways of adolescents born very preterm. *Eur Respir J* 2012;**40**:1253–9.

45. Clemm H, Roksund O, Thorsen E, Eide GE, Markestad T, Halvorsen T. Aerobic capacity and exercise performance in young people born extremely preterm. *Pediatrics* 2012;**129**:e97–105.

46. Kotecha SJ, Watkins WJ, Paranjothy S, Dunstan FD, Henderson AJ, Kotecha S. Effect of late preterm birth on longitudinal lung spirometry in school age children and adolescents. *Thorax* 2012;**67**:54–61.

47. Vrijlandt EJLE, Gerritsen J, Boezen HM, Grevink RG, Duiverman EJ. Lung function and exercise capacity in young adults born prematurely. *Am J Respir Crit Care Med* 2006;**173**:890–6.

48. Aukland SM, Rosendahl K, Owens CM, Fosse KR, Eide GE, Halvorsen T. Neonatal bronchopulmonary dysplasia predicts abnormal pulmonary hrct scans in long-term survivors of extreme preterm birth. *Thorax* 2009;**64**:405–10.

49. Wong PM, Lees AN, Louw J, Lee FY, French N, Gain K, et al. Emphysema in young adult survivors of moderate-to-severe bronchopulmonary dysplasia. *Eur Respir J* 2008;**32**:321–8.

50. Ladha F, Bonnet S, Eaton F, Hashimoto K, Korbutt G, Thebaud B. Sildenafil improves alveolar growth and pulmonary hypertension in hyperoxia-induced lung injury. *Am J Respir Crit Care Med* 2005;**172**:750–6.

51. Rehan VK, Wang Y, Patel S, Santos J, Torday JS. Rosiglitazone, a peroxisome proliferator-activated receptor-gamma agonist, prevents hyperoxia-induced neonatal rat lung injury in vivo. *Pediatr Pulmonol* 2006;**41**:558–69.

52. Dasgupta C, Sakurai R, Wang Y, Guo P, Ambalavanan N, Torday JS, et al. Hyperoxia-induced neonatal rat lung injury involves activation of TGF-{beta} and Wnt signaling and is protected by rosiglitazone. *Am J Physiol Lung Cell Mol Physiol* 2009;**296**:L1031–41.

53. Lee JH, Sung DK, Koo SH, Shin BK, Hong YS, Son CS, et al. Erythropoietin attenuates hyperoxia-induced lung injury by down-modulating inflammation in neonatal rats. *J Korean Med Sci* 2007;**22**:1042–7.

54. Thebaud B, Ladha F, Michelakis ED, Sawicka M, Thurston G, Eaton F, et al. Vascular endothelial growth factor gene therapy increases survival, promotes lung angiogenesis, and prevents alveolar damage in hyperoxia-induced lung injury: Evidence that angiogenesis participates in alveolarization. *Circulation* 2005;**112**:2477–86.

55. Ozer EA, Kumral A, Ozer E, Yilmaz O, Duman N, Ozkal S, et al. Effects of erythropoietin on hyperoxic lung injury in neonatal rats. *Pediatr Res* 2005;**58**:38–41.

56. Bonikos DS, Bensch KG, Northway Jr WH. Oxygen toxicity in the newborn. The effect of chronic continuous 100 percent oxygen exposure on the lungs of newborn mice. *Am J Pathol* 1976;**85**:623–50.

57. Warner BB, Stuart LA, Papes RA, Wispe JR. Functional and pathological effects of prolonged hyperoxia in neonatal mice. *Am J Physiol Lung Cell Mol Physiol* 1998;**19**:L110–7.

58. Masood A, Yi M, Lau M, Belcastro R, Shek S, Pan J, et al. Therapeutic effects of hypercapnia on chronic lung injury and vascular remodeling in neonatal rats. *Am J Physiol Lung Cell Mol Physiol* 2009;**297**:L920–30.

59. Chetty A, Cao GJ, Severgnini M, Simon A, Warburton R, Nielsen HC. Role of matrix metalloprotease-9 in hyperoxic injury in developing lung. *Am J Physiol Lung Cell Mol Physiol* 2008;**295**:L584–92.

60. Londhe VA, Sundar IK, Lopez B, Maisonet TM, Yu Y, Aghai ZH, et al. Hyperoxia impairs alveolar formation and induces senescence through decreased histone deacetylase activity and up-regulation of p21 in neonatal mouse lung. *Pediatr Res* 2011;**69**:371–7.

61. Balasubramaniam V, Ryan SL, Seedorf GJ, Roth EV, Heumann TR, Yoder MC, et al. Bone marrow-derived angiogenic cells restore lung alveolar and vascular structure after neonatal hyperoxia in infant mice. *Am J Physiol Lung Cell Mol Physiol* 2010;**298**:L315–23.

62. Kunig AM, Balasubramaniam V, Markham NE, Morgan D, Montgomery G, Grover TR, et al. Recombinant human VEGF treatment enhances alveolarization after hyperoxic lung injury in neonatal rats. *Am J Physiol Lung Cell Mol Physiol* 2005;**289**:L529–35.

63. Lin YJ, Markham NE, Balasubramaniam V, Tang JR, Maxey A, Kinsella JP, et al. Inhaled nitric oxide enhances distal lung growth after exposure to hyperoxia in neonatal rats. *Pediatr Res* 2005;**58**:22–9.

64. van Haaften T, Byrne R, Bonnet S, Rochefort GY, Akabutu J, Bouchentouf M, et al. Airway delivery of mesenchymal stem cells prevents arrested alveolar growth in neonatal lung injury in rats. *Am J Respir Crit Care Med* 2009;**180**:1131–42.

65. Thibeault DW, Mabry S, Rezaiekhaligh M. Neonatal pulmonary oxygen toxicity in the rat and lung changes with aging. *Pediatr Pulmonol* 1990;**9**:96–108.

66. Yee M, Chess PR, McGrath-Morrow SA, Wang Z, Gelein R, Zhou R, et al. Neonatal oxygen adversely affects lung function in adult mice without altering surfactant composition or activity. *Am J Physiol Lung Cell Mol Physiol* 2009;**297**:L641–9.

67. Dauger S, Ferkdadji L, Saumon G, Vardon G, Peuchmaur M, Gaultier C, et al. Neonatal exposure to 65% oxygen durably impairs lung architecture and breathing pattern in adult mice. *Chest* 2003;**123**:530–8.

68. Yee M, White RJ, Awad HA, Bates WA, McGrath-Morrow SA, O'Reilly MA. Neonatal hyperoxia causes pulmonary vascular disease and shortens life span in aging mice. *Am J Pathol* 2011;**178**:2601–10.

69. Thebaud B. Angiogenesis in lung development, injury and repair: Implications for chronic lung disease of prematurity. *Neonatology* 2007;**91**:291–7.

70. Fernandez-Gonzalez A, Alex Mitsialis S, Liu X. Kourembanas S. Vasculoprotective effects of heme oxygenase-1 in a murine model of hyperoxia-induced bronchopulmonary dysplasia. *Am J Physiol Lung Cell Mol Physiol* 2012;**302**:L775–84.

71. Balasubramaniam V, Mervis CF, Maxey AM, Markham NE, Abman SH. Hyperoxia reduces bone marrow, circulating, and lung endothelial progenitor cells in the developing lung: Implications for the pathogenesis of bronchopulmonary dysplasia. *Am J Physiol Lung Cell Mol Physiol* 2007;**292**:L1073–84.

72. Irwin D, Helm K, Campbell N, Imamura M, Fagan K, Harral J, et al. Neonatal lung side population cells demonstrate endothelial potential and are altered in response to hyperoxia-induced lung simplification. *Am J Physiol Lung Cell Mol Physiol* 2007;**293**:L941–51.

73. Han RN, Buch S, Tseu I, Young J, Christie NA, Frndova H, et al. Changes in structure, mechanics, and insulin-like growth factor-related gene expression in the lungs of newborn rats exposed to air or 60% oxygen. *Pediatr Res* 1996;**39**:921–9.

74. Minoo P, Segura L, Coalson JJ, King RJ, DeLemos RA. Alterations in surfactant protein gene expression associated with premature birth and exposure to hyperoxia. *Am J Physiol* 1991;**261**:L386–92.

75. Woods E, Ohashi T, Polk D, Ikegami M, Ueda T, Jobe AH. Surfactant treatment and ventilation effects on surfactant SP A, SP B, and SP C mRNA levels in preterm lamb lungs. *Am J Physiol* 1995;**269**:L209–14.

76. White CW, Greene KE, Allen CB, Shannon JM. Elevated expression of surfactant proteins in newborn rats during adaptation to hyperoxia. *Am J Respir Cell Mol Biol* 2001;**25**:51–9.

77. ter Horst SA, Fijlstra M, Sengupta S, Walther FJ, Wagenaar GT. Spatial and temporal expression of surfactant proteins in hyperoxia-induced neonatal rat lung injury. *BMC Pulm Med* 2006;**6**:8.

78. Johnston LC, Gonzales LW, Lightfoot RT, Guttentag SH, Ischiropoulos H. Opposing regulation of human alveolar type ii cell differentiation by nitric oxide and hyperoxia. *Pediatr Res* 2010;**67**:521–5.

79. Yee M, Vitiello PF, Roper JM, Staversky RJ, Wright TW, McGrath-Morrow SA, et al. Type ii epithelial cells are critical target for hyperoxia-mediated impairment of postnatal lung development. *Am J Physiol Lung Cell Mol Physiol* 2006;**291**:L1101–11.

80. Denis D, Fayon MJ, Berger P, Molimard M, De Lara MT, Roux E, et al. Prolonged moderate hyperoxia induces hyperresponsiveness and airway inflammation in newborn rats. *Pediatr Res* 2001;**50**:515–9.

81. Hershenson MB, Aghili S, Punjabi N, Hernandez C, Ray DW, Garland A, et al. Hyperoxia-induced airway hyperresponsiveness and remodeling in immature rats. *Am J Physiol* 1992;**262**:L263–9.

82. Hartman WR, Smelter DF, Sathish V, Karass M, Kim S, Aravamudan B, et al. Oxygen dose responsiveness of human fetal airway smooth muscle cells. *Am J Physiol Lung Cell Mol Physiol* 2012;**303**:L711–9.

83. Massaro GD, McCoy L, Massaro D. Hyperoxia reversibly suppresses development of bronchiolar epithelium. *Am J Physiol* 1986;**251**:R1045–50.

84. Massaro GD, McCoy L, Massaro D. Development of bronchiolar epithelium: Time course of response to oxygen and recovery. *Am J Physiol* 1988a;**254**:R755–60.

85. O'Reilly M, Hansbro PM, Horvat JC, Beckett EL, Harding R, Sozo F. Bronchiolar remodeling in adult mice following neonatal exposure to hyperoxia: relation to growth. *Anat Rec* 2014;**297**:758–69.

86. Massaro GD, McCoy L, Massaro D. Postnatal undernutrition slows development of bronchiolar epithelium in rats. *Am J Physiol* 1988b;**255**:R521–6.

87. O'Reilly M, Harding R, Sozo F. Altered small airways in aged mice following neonatal exposure to hyperoxic gas. *Neonatology* 2014;**105**:39–45.

88. Belik J, Jankov RP, Pan J, Tanswell AK. Chronic O_2 exposure enhances vascular and airway smooth muscle contraction in the newborn but not adult rat. *J Appl Physiol* 2003;**94**:2303–12.

89. Schultz ED, Potts EN, Mason SN, Foster WM, Auten RL. Mast cells mediate hyperoxia-induced airway hyper-reactivity in newborn rats. *Pediatr Res* 2010;**68**:70–4.

90. Iben SC, Dreshaj IA, Farver CF, Haxhiu MA, Martin RJ. Role of endogenous nitric oxide in hyperoxia-induced airway hyperreactivity in maturing rats. *J Appl Physiol* 2000;**89**:1205–12.

91. Mhanna MJ, Haxhiu MA, Jaber MA, Walenga RW, Chang CH, Liu S, et al. Hyperoxia impairs airway relaxation in immature rats via a camp-mediated mechanism. *J Appl Physiol* 2004;**96**:1854–60.

92. Smith PG, Dreshaj A, Chaudhuri S, Onder BM, Mhanna MJ, Martin RJ. Hyperoxic conditions inhibit airway smooth muscle myosin phosphatase in rat pups. *Am J Physiol Lung Cell Mol Physiol* 2007;**292**:L68–73.

93. Davis JM, Auten RL. Maturation of the antioxidant system and the effects on preterm birth. *Semin Fetal Neonatal Med* 2010;**15**:191–5.

94. Bayir H. Reactive oxygen species. *Crit Care Med* 2005;**33**:S498–501.

95. Comhair SA, Erzurum SC. Antioxidant responses to oxidant-mediated lung diseases. *Am J Physiol Lung Cell Mol Physiol* 2002;**283**:L246–55.

96. Bhandari V. Hyperoxia-derived lung damage in preterm infants. *Semin Fetal Neonatal Med* 2010;**15**:223–9.

97. Saugstad OD. Oxygen and oxidative stress in bronchopulmonary dysplasia. *J Perinat Med* 2010;**38**:571–7.

98. Nassi N, Ponziani V, Becatti M, Galvan P, Donzelli G. Anti-oxidant enzymes and related elements in term and preterm newborns. *Pediatr Int* 2009;**51**:183–7.

99. Brandt RB, Mueller DG, Schroeder JR, Guyer KE, Kirkpatrick BV, Hutcher NE, et al. Serum vitamin A in premature and term neonates. *J Pediatr* 1978;**92**:101–4.

100. Shenai JP, Chytil F, Stahlman MT. Vitamin A status of neonates with bronchopulmonary dysplasia. *Pediatr Res* 1985;**19**:185–8.

101. Baydas G, Karatas F, Gursu MF, Bozkurt HA, Ilhan N, Yasar A, et al. Antioxidant vitamin levels in term and preterm infants and their relation to maternal vitamin status. *Arch Med Res* 2002;**33**:276–80.

102. Zaher TE, Miller EJ, Morrow DM, Javdan M, Mantell LL. Hyperoxia-induced signal transduction pathways in pulmonary epithelial cells. *Free Radic Biol Med* 2007;**42**:897–908.

103. Perveen S, Patel H, Arif A, Younis S, Codipilly CN, Ahmed M. Role of EC-SOD overexpression in preserving pulmonary angiogenesis inhibited by oxidative stress. *PLoS One* 2012;**7**. e51945.

104. Ahmed MN, Suliman HB, Folz RJ, Nozik-Grayck E, Golson ML, Mason SN, et al. Extracellular superoxide dismutase protects lung development in hyperoxia-exposed newborn mice. *Am J Respir Crit Care Med* 2003;**167**:400–5.

105. Auten RL, O'Reilly MA, Oury TD, Nozik-Grayck E, Whorton MH. Transgenic extracellular superoxide dismutase protects postnatal alveolar epithelial proliferation and development during hyperoxia. *Am J Physiol Lung Cell Mol Physiol* 2006;**290**:L32–40.

106. Speer CP. Inflammation and bronchopulmonary dysplasia: A continuing story. *Semin Fetal Neonatal Med* 2006;**11**:354–62.

107. Rozycki HJ, Comber PG, Huff TF. Cytokines and oxygen radicals after hyperoxia in preterm and term alveolar macrophages. *Am J Physiol Lung Cell Mol Physiol* 2002;**282**:L1222–8.

108. Ryan RM, Ahmed Q, Lakshminrusimha S. Inflammatory mediators in the immunobiology of bronchopulmonary dysplasia. *Clin Rev Allergy Immunol* 2008;**34**:174–90.

109. D'Angio CT, Basavegowda K, Avissar NE, Finkelstein JN, Sinkin RA. Comparison of tracheal aspirate and bronchoalveolar lavage specimens from premature infants. *Biol Neonate* 2002;**82**:145–9.

110. Kotecha S, Chan B, Azam N, Silverman M, Shaw RJ. Increase in interleukin-8 and soluble intercellular adhesion molecule-1 in bronchoalveolar lavage fluid from premature infants who develop chronic lung disease. *Arch Dis Child Fetal Neonatal Ed* 1995;**72**:F90–6.

111. Baier RJ, Majid A, Parupia H, Loggins J, Kruger TE. Cc chemokine concentrations increase in respiratory distress syndrome and correlate with development of bronchopulmonary dysplasia. *Pediatr Pulmonol* 2004;**37**:137–48.

112. Baier RJ, Loggins J, Kruger TE. Monocyte chemoattractant protein-1 and interleukin-8 are increased in bronchopulmonary dysplasia: Relation to isolation of Ureaplasma urealyticum. *J Investig Med* 2001;**49**:362–9.

113. De Paepe ME, Greco D, Mao Q. Angiogenesis-related gene expression profiling in ventilated preterm human lungs. *Exp Lung Res* 2010;**36**:399–410.

114. Auten RL, Mason SN, Auten KM, Brahmajothi M. Hyperoxia impairs postnatal alveolar epithelial development via NADPH oxidase in newborn mice. *Am J Physiol Lung Cell Mol Physiol* 2009;**297**:L134–42.

115. Bhandari V, Choo-Wing R, Homer RJ, Elias JA. Increased hyperoxia-induced mortality and acute lung injury in IL-13 null mice. *J Immunol* 2007;**178**:4993–5000.

116. ter Horst SA, Wagenaar GT, de Boer E, van Gastelen MA, Meijers JC, Biemond BJ, et al. Pentoxifylline reduces fibrin deposition and prolongs survival in neonatal hyperoxic lung injury. *J Appl Physiol* 2004;**97**:2014–9.

117. Stenger MR, Rose MJ, Joshi MS, Rogers LK, Chicoine LG, Bauer JA, et al. Inhaled nitric oxide prevents 3-nitrotyrosine formation in the lungs of neonatal mice exposed to >95% oxygen. *Lung* 2010;**188**:217–27.

118. Vozzelli MA, Mason SN, Whorton MH, Auten Jr RL. Antimacrophage chemokine treatment prevents neutrophil and macrophage influx in hyperoxia-exposed newborn rat lung. *Am J Physiol Lung Cell Mol Physiol* 2004;**286**:L488–93.

119. Yi M, Jankov RP, Belcastro R, Humes D, Copland I, Shek S, et al. Opposing effects of 60% oxygen and neutrophil influx on alveologenesis in the neonatal rat. *Am J Respir Crit Care Med* 2004;**170**:1188–96.

120. Doyle LW, Anderson PJ. Long-term outcomes of bronchopulmonary dysplasia. *Semin Fetal Neonatal Med* 2009;**14**:391–5.

121. Stoll BJ, Gordon T, Korones SB, Shankaran S, Tyson JE, Bauer CR, et al. Late-onset sepsis in very low birth weight neonates: A report from the National Institute of Child Health and Human Development Neonatal Research Network. *J Pediatr* 1996;**129**:63–71.

122. Webber S, Wilkinson AR, Lindsell D, Hope PL, Dobson SR, Isaacs D. Neonatal pneumonia. *Arch Dis Child* 1990;**65**:207–11.

123. Coalson JJ, Gerstmann DR, Winter VT, Delemos RA. Bacterial colonization and infection studies in the premature baboon with bronchopulmonary dysplasia. *Am Rev Respir Dis* 1991;**144**:1140–6.

124. Coalson JJ, Winter VT, Siler-Khodr T, Yoder BA. Neonatal chronic lung disease in extremely immature baboons. *Am J Respir Crit Care Med* 1999;**160**:1333–46.

125. O'Reilly MA, Marr SH, Yee M, McGrath-Morrow SA, Lawrence BP. Neonatal hyperoxia enhances the inflammatory response in adult mice infected with influenza A virus. *Am J Respir Crit Care Med* 2008;**177**:1103–10.

126. Giannandrea M, Yee M, O'Reilly MA, Lawrence BP. Memory CD8+ T cells are sufficient to alleviate impaired host resistance to influenza A virus infection caused by neonatal oxygen supplementation. *Clin Vaccine Immunol* 2012;**19**:1432–41.

127. Buczynski BW, Yee M, Lawrence BP, O'Reilly MA. Lung development and the host response to influenza A virus are altered by different doses of neonatal oxygen in mice. *Am J Physiol Lung Cell Mol Physiol* 2012;**302**:L1078–87.

128. O'Reilly MA, Yee M, Buczynski BW, Vitiello PF, Keng PC, Welle SL, et al. Neonatal oxygen increases sensitivity to influenza A virus infection in adult mice by suppressing epithelial expression of ear1. *Am J Pathol* 2012;**181**:441–51.

129. Buczynski BW, Yee M, Martin KC, Lawrence BP, O'Reilly MA. Neonatal hyperoxia alters the host response to influenza A virus infection in adult mice through multiple pathways. *Am J Physiol Lung Cell Mol Physiol* 2013;**305**. L282–L90.

130. Vadivel A, Aschner JL, Rey-Parra GJ, Magarik J, Zeng H, Summar M, et al. L-citrulline attenuates arrested alveolar growth and pulmonary hypertension in oxygen-induced lung injury in newborn rats. *Pediatr Res* 2010;**68**:519–25.

131. Alapati D, Rong M, Chen S, Hehre D, Rodriguez MM, Lipson KE, et al. Connective tissue growth factor antibody therapy attenuates hyperoxia-induced lung injury in neonatal rats. *Am J Respir Cell Mol Biol* 2011;**45**:1169–77.

132. Almario B, Wu S, Peng J, Alapati D, Chen S, Sosenko IR. Pentoxifylline and prevention of hyperoxia-induced lung injury in neonatal rats. *Pediatr Res* 2012;**71**:583–9.

133. James ML, Ross AC, Nicola T, Steele C, Ambalavanan N. VARA attenuates hyperoxia-induced impaired alveolar development and lung function in newborn mice. *Am J Physiol Lung Cell Mol Physiol* 2013;**304**:L803–12.

134. Sakurai R, Li Y, Torday JS, Rehan VK. Curcumin augments lung maturation, preventing neonatal lung injury by inhibiting TGF-beta signaling. *Am J Physiol Lung Cell Mol Physiol* 2011;**301**:L721–30.

135. Weichelt U, Cay R, Schmitz T, Strauss E, Sifringer M, Buhrer C, et al. Prevention of hyperoxia-mediated pulmonary inflammation in neonatal rats by caffeine. *Eur Respir J* 2013;**41**:966–73.

136. Alphonse RS, Vadivel A, Coltan L, Eaton F, Barr AJ, Dyck JR, et al. Activation of Akt protects alveoli from neonatal oxygen-induced lung injury. *Am J Respir Cell Mol Biol* 2011;**44**:146–54.

137. O'Reilly M, Thebaud B. The promise of stem cells in bronchopulmonary dysplasia. *Semin Perinatol* 2013;**37**:79–84.

138. Aslam M, Baveja R, Liang OD, Fernandez-Gonzalez A, Lee C, Mitsialis SA, et al. Bone marrow stromal cells attenuate lung injury in a murine model of neonatal chronic lung disease. *Am J Respir Crit Care Med* 2009;**180**:1122–30.

139. Chang YS, Choi SJ, Sung DK, Kim SY, Oh W, Yang YS, et al. Intratracheal transplantation of human umbilical cord blood derived mesenchymal stem cells dose-dependently attenuates hyperoxia-induced lung injury in neonatal rats. *Cell Transplant* 2011.

140. Chang YS, Oh W, Choi SJ, Sung DK, Kim SY, Choi EY, et al. Human umbilical cord blood-derived mesenchymal stem cells attenuate hyperoxia-induced lung injury in neonatal rats. *Cell Transplant* 2009;**18**:869–86.

141. Tropea KA, Leder E, Aslam M, Lau AN, Raiser DM, Lee JH, et al. Bronchioalveolar stem cells increase after mesenchymal stromal cell treatment in a mouse model of bronchopulmonary dysplasia. *Am J Physiol Lung Cell Mol Physiol* 2012;**302**:L829–37.

142. Zhang X, Wang H, Shi Y, Peng W, Zhang S, Zhang W, et al. Role of bone marrow-derived mesenchymal stem cells in the prevention of hyperoxia-induced lung injury in newborn mice. *Cell Biol Int* 2012;**36**:589–94.

143. Hansmann G, Fernandez-Gonzalez A, Aslam M, Vitali SH, Martin T, Mitsialis SA, et al. Mesenchymal stem cell-mediated reversal of bronchopulmonary dysplasia and associated pulmonary hypertension. *Pulm Circ* 2012;**2**:170–81.

144. Pierro M, Ionescu L, Montemurro T, Vadivel A, Weissmann G, Oudit G, et al. Short-term, long-term and paracrine effect of human umbilical cord-derived stem cells in lung injury prevention and repair in experimental bronchopulmonary dysplasia. *Thorax* 2013;**68**:475–84.

145. Waszak P, Alphonse R, Vadivel A, Ionescu L, Eaton F, Thebaud B. Preconditioning enhances the paracrine effect of mesenchymal stem cells in preventing oxygen-induced neonatal lung injury in rats. *Stem Cells Dev* 2012;**21**:2789–97.

146. Bonikos DS, Bensch KG, Ludwin SK, Northway Jr WH. Oxygen toxicity in the newborn. The effect of prolonged 100 percent O_2 exposure on the lungs of newborn mice. *Lab Invest* 1975;**32**:619–35.

147. Agrons GA, Courtney SE, Stocker JT, Markowitz RI. From the archives of the afip: Lung disease in premature neonates; Radiologic-pathologic correlation. *Radiographics* 2005;**25**:1047–73.

Chapter 19

The Influence of Nutrition on Lung Development before and after Birth

Richard Harding and Robert De Matteo

Department of Anatomy and Developmental Biology, Monash University, Clayton Campus, VIC, Australia

INTRODUCTION

In common with many other body organs, the lung is affected by the prevailing nutritional environment during its development. Both epidemiological and experimental studies show that nutritional and oxygen status during development can induce alterations in the structure of the developing lung, and that some of these can persist such that later lung structure is altered and lung function is affected. Significant associations between the early nutritional environment and lung development, as well as lung function and respiratory symptoms or illnesses during childhood or adulthood, have been the subject of numerous reviews.[1–10] The persistence of altered respiratory function following early life alterations in lung development has been referred to as "tracking"[11] or "developmental programming."[12] Although epidemiological and clinical studies have indicated that low birthweight is associated with later respiratory illness, it is not yet clear which of the many factors associated with low birthweight and restricted postnatal growth are responsible for altered lung development; in particular, many studies of the later effects of low birthweight (LBW) fail to distinguish between LBW caused by preterm birth and intrauterine growth restriction (IUGR). In this chapter, we have attempted to summarize current understanding as to how the early nutritional environment impacts upon lung development and respiratory function; hypoxia is included, as oxygen is essential for nutrient utilization in the developing organism and hypoxia can contribute to growth restriction. As well as influencing lung development, it is increasingly apparent that nutritional impairments during development can influence the rate of "pulmonary aging." Indeed, it may be reasonably concluded that the early nutritional environment is important for the entire lifespan of the lung. The lung is particularly vulnerable to early life "programming" owing to its apparently limited potential for regeneration and recovery from alterations induced during development.

Firstly we consider causes of restricted fetal nutrition and growth.

CAUSES OF RESTRICTED FETAL NUTRITION AND GROWTH

The underlying causes of intrauterine growth restriction (IUGR) or being small for gestational age (SGA) are varied and can include maternal, fetal, or placental factors (Table 1); however, a common underlying cause is reduced placental transport of substrates including oxygen from mother to fetus. Blood samples taken from the umbilical cord show that IUGR arising from a range of causes is associated with fetal hypoxemia, hypoglycemia, hyperlactinemia, acidemia, elevated blood lactate, and altered endocrine status indicative of intrauterine stress.[13] IUGR or SGA are usually defined as fetal body weight being below the 10th percentile for gestational age (Figure 1).

Maternal Causes of IUGR

These include chronic disease states of the mother, drug use (e.g., cigarette smoking, alcohol abuse), poor maternal nutritional status or diet, and exposure to environmental factors. These factors act to alter nutrient availability to the placenta and fetus and may modify placental and fetal gene expression and function[14,15] leading to fetal growth abnormalities and long-term adverse effects on the health outcome of offspring.

Maternal vascular disease, whether chronic hypertension, pre-eclampsia, diabetes mellitus, or collagen vascular disease, accounts for 25–30% of IUGR cases due to impairments of utero-placental perfusion in non-anomalous fetuses.[16] The reduction in utero-placental perfusion is thought to result from the combination of hypoxia, an imbalance of angiogenic and anti-angiogenic factors,

The Lung. http://dx.doi.org/10.1016/B978-0-12-799941-8.00019-5

349

TABLE 1 Common causes of fetal growth restriction in human pregnancy

Maternal

Chronic Disease States	Drug Use	Diet	Genetic Predisposition	Disease	Hypoxemia
• Chronic hypertension • Diabetes mellitus • Renal insufficiency • Collagen vascular disease	• Nicotine • Ethanol • Steroids • Narcotics	• Undernutrition • Obesity	• Acquired thrombophilias	• Malaria • Pneumonia	• High altitude • Anemia • Cyanotic heart disease

Placental

Placental Anomalies	Placental Dysfunction	Chronic Inflammatory Conditions
• Abnormal cord insertion • Circumvallate placenta • Chorioangioma	• Placental abruption • Placenta praevia	• Placentitis • Villitis • Chorioamnionitis

Fetal

Chromosomal Disorders	Congenital Malformations	Infection	Multiple Gestation
• Sex chromosome disorders • Trisomy 18, 13, 21	• Congenital heart disease • Anencephaly	• Viral • Bacterial	• Twin-to-twin transfusion syndrome

inflammation, and altered immunity.[17] The incidence of IUGR is increased in the presence of chronic maternal hypertension and is directly correlated with the severity of the hypertension.[18] Maternal diabetes is a risk factor for IUGR due to damage of the placental microcirculation. Maternal renal insufficiency may also be accompanied by IUGR.[19]

Maternal hypoxemia has multiple causes, including heart disease, chronic anemia, sickle cell anemia, and high altitude. Human pregnancies that have experienced chronic mild to moderate hypoxia due to high altitude have shown that fetal oxygen delivery and fetal oxygen consumption remain constant; however, glucose availability for fetal growth is limited due to a preferential anaerobic consumption of glucose by the placenta.[20] This hypoxia-induced metabolic reprogramming of the placenta[21] may be an initiating step in the progression of fetal growth restriction.

The single most common cause of restricted fetal growth in Western societies is maternal cigarette smoking, and a dose-response relationship has been demonstrated.[22] Smoking by the mother's domestic partner also increases the risk of IUGR.[23] The smoking-related impairment of fetal growth likely results from a combination of carbon monoxide exposure, which decreases the O_2 carrying capacity of fetal hemoglobin, and nicotine, which induces the release of maternal catecholamines. The repeated release of maternal catecholamines by cigarette smoking can reduce maternal perfusion of the placenta, and hence

nutrient and oxygen delivery to the fetus. Maternal cigarette smoking is associated with chronically increased resistance in the uterine and umbilical arteries.[24] Maternal smoking during pregnancy is associated with a decrease in placental transport function resulting in restricted nutrient and oxygen transfer to the fetus, thereby affecting fetal development.

Excessive maternal alcohol (ethanol) consumption can cause IUGR and the fetal alcohol spectrum disorder syndrome (FASD).[25] Reduced birthweight can result from maternal alcohol ingestion of only 1–2 standard drinks per day.[26] The maternal use of drugs such as steroids, dilantin, coumadin, cocaine, and heroin has also been implicated in causing IUGR.

Maternal infectious diseases account for 5–10% of IUGR cases. Worldwide, *Plasmodium falciparum* (malaria) accounts for most cases of infection-related IUGR[27] and is thought to induce changes in placental structure and function, impairing placental blood flow, vascular development and/or nutrient transport.[28] Rubella and cytomegalovirus infections are associated with IUGR; varicella zoster virus, human immunodeficiency, and a first episode of herpes simplex virus infection may also be associated with impaired fetal growth.[29] Pneumonia has been associated with placental abruption and IUGR.[30]

Recent evidence suggests that hereditary thrombophilia may not be associated with IUGR[31,32]; however, antiphospholipid antibody syndrome (an acquired

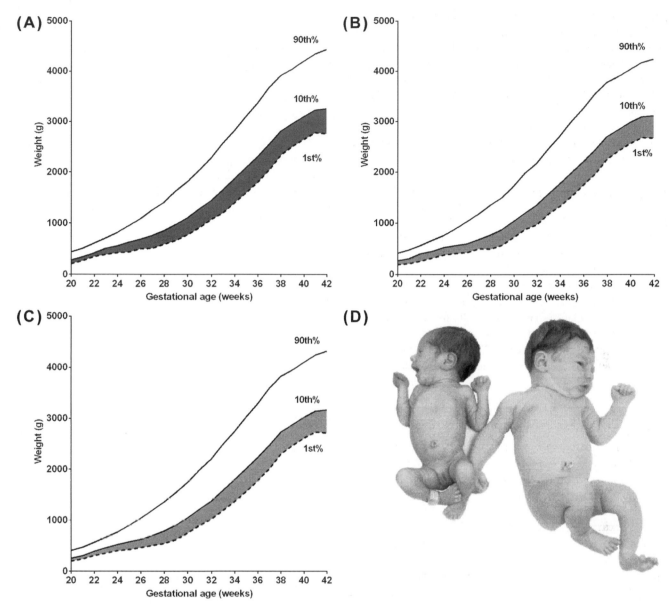

FIGURE 1 Birthweight percentiles for live singleton infants born to Australian women (1998–2007) for (A) males, (B) females, and (C) males and females combined, showing the 1st, 10th, and 90th percentiles. Note the rapid increase in fetal weight during the last third of gestation. By definition intrauterine growth restriction (IUGR) is present when estimated fetal weight falls below the 10th percentile for gestational age (depicted as shaded area). Data obtained from Dobbins et al.[209] (D) Comparison of an average sized baby at term on the right to a baby with fetal growth restriction on the left. *Image obtained from Science Photo Library.* (This figure is reproduced in the color plate section.)

immune-mediated thrombophilic condition), anticardio-lipin antibody syndrome or lupus erythematosus can all lead to IUGR.[33,34]

A sub-optimal maternal diet and nutritional status during gestation have been shown to reduce fetal growth in both humans and experimental animals.[35–37] Maternal undernutrition or overnutrition may alter fetal growth by altering placental growth and vascularity, impairing placental function, increasing oxidative stress in the placenta and fetus, or by producing an endocrine or hormonal imbalance.[36,38] A significant proportion of women of reproductive age and

in pregnancy are overweight or obese. Obesity induces a chronic low grade inflammation and an intrauterine inflammatory environment may alter fetal growth and development.[35]

Placental Causes of IUGR

The placenta is fundamental for an adequate exchange of nutrients and gases between mother and fetus; thus placental dysfunction (insufficiency) is a major cause of IUGR, accounting for 25–40% of cases of IUGR. A decrease in placental mass or impaired umbilical cord development can restrict the rate of substrate delivery to the fetus and thus can contribute to IUGR.[39] In particular, placental anomalies such as circumvallate placenta and chorioangioma of the placenta have been associated with restricted fetal growth.[40] Placental abruption and placenta praevia can chronically restrict placental function, and occur more frequently among cigarette smokers.[41] Chronic inflammatory conditions such as placentitis, villitis, and chorioamnionitis have been associated with placental dysfunction and IUGR.[42]

Fetal Causes of IUGR

Fetal anomalies arising from chromosomal disorders (e.g., trisomy 18, 13, 21, and sex chromosome disorders) and congenital malformations (both chromosomal and non-chromosomal) account for 15 to 20% of IUGR cases.[16,43] Multiple gestation is another cause of restricted fetal growth.[44] IUGR can occur in any multiple gestation pregnancy, but is severe if there is a shared fetal circulation. Twinning results in a decreased placental mass in relation to fetal mass, thereby restricting fetal nutrient supply. The incidence of IUGR in twins is 15–30% and the figure is increased with greater numbers of fetuses.[16,45] Fetal viral infections such as parvovirus, rubella, and cytomegalovirus can cause up to 5% of IUGR cases, particularly if they occur in early gestation.[29]

Fetal nutrient restriction is likely to exert its greatest effects on organ development during mid-late gestation, when the fetus is growing rapidly (Figure 1). Measurements of fetal growth throughout gestation indicate that the greatest divergence from normal growth profiles occurs during the second half of gestation when fetal organs are growing in complexity. Hence it is likely that IUGR primarily affects developmental processes that occur during the late stages of in utero development; in the lung this is an important time for saccular, alveolar, and vascular development.

ASSOCIATION BETWEEN IUGR, GENES, AND LONG-TERM HEALTH OUTCOMES

Modifications of DNA methylation, or other epigenetic alteration, in the fetus and/or placenta are thought to be an underlying mechanism leading to IUGR and long-term disease.[35,46] Although alterations in nutrient availability,[47] and exposure to toxic substances,[48] play a major role in modifying fetal growth and organ structure through the programming of specific genes, a reduction in placental blood flow may also have permanent effects on the fetus and placenta.[14] In placentas of IUGR infants nearly 500 differentially methylated regions have been reported,[49] while altered choline intake in human pregnancies can modulate the epigenetic state of fetal cortisol-regulating genes in humans as well as the epigenomic status of fetal derived tissues.[50] These adaptions may ensure fetal survival in the face of short-term stress; however, a contrasting

environment in extrauterine life (e.g., over supply of nutrients) can increase the risk of adult onset diseases such as diabetes, obesity, and coronary heart disease,[51–53] providing evidence for the "developmental origins of health and disease" hypothesis.

PROGRAMMING EFFECTS OF GROWTH RESTRICTION ON LUNG FUNCTION AND RESPIRATORY HEALTH: HUMAN DATA

An increasing number of studies show that IUGR or being SGA, which are markers of impaired fetal nutrition and/or oxygenation, are associated with adverse health outcomes, including alterations in respiratory function at all stages of postnatal life. As low birthweight can also be caused by preterm birth, birthweight per se may not be a reliable indicator of fetal nutrition; therefore, in this chapter we have made the distinction between low birthweight (LBW) and IUGR. As outlined above, IUGR can be caused by multiple factors, and this has led to difficulties, in human studies, in identifying individual factors responsible for altered lung development.

Respiratory Consequences of IUGR that are Observed in Infancy

Recent studies controlling for gestational age show that being exposed to IUGR increases the risk of adverse metabolic and respiratory complications in the early postnatal period.[54,55] Although IUGR may increase lung maturity by accelerating the maturation of pulmonary surfactant,[56] this is likely to be at the expense of altered lung development. In neonates born either preterm or at term, IUGR increases the risk of respiratory insufficiency and the need for respiratory support after birth.[57,58] Although few studies of lung function have been conducted in infants, it has been shown that IUGR infants of non-smoking mothers have impaired lung function, as indicated by reduced forced vital capacity (FVC) and forced expiratory flow (FEF$_{75}$).[59,60] The reasons for poorer gas exchange in infants exposed to IUGR are unclear, but they could relate to delayed clearance of lung liquid from the airways, reduced pulmonary perfusion, structural alterations in the walls of the conducting airways, impaired alveolarization, or altered surfactant release and/or composition. Experimental studies have reported a range of developmental alterations in the lungs following IUGR or maternal nutrient restriction during pregnancy (see the following text).

Preterm birth occurs in 8–12% of live births and a high proportion of preterm infants have experienced IUGR.[61] In the neonatal period, preterm infants who are also growth restricted have a greater risk of morbidity and mortality,[62] and a greater requirement for ventilatory support than non-IUGR preterm infants.[63] IUGR increases the risk of chronic lung disease in preterm neonates born before 28 weeks.[64] Thus, low birthweight infants may suffer from the combined effects of preterm birth and IUGR.

Respiratory Effects of IUGR That Are Observed during Childhood

Several studies have demonstrated adverse effects of IUGR on the lung function of children. A study of more than 2000 children aged 6–11 years showed that IUGR was associated with a reduction in FEV_1; importantly, this effect was seen in children born both at term and preterm.[65] Similarly, a study of twins aged 7–15 years showed that smaller twins had lower forced expiratory flow (FEF) rates than their larger siblings.[66] More recently, spirometric studies of school-age children (8–9 years of age) revealed impaired lung function in subjects who experienced IUGR.[67] Together, these studies suggest that factors that restrict fetal growth alter the development of conducting airways, although alterations in the lung parenchyma, chest wall, or respiratory muscle function may contribute to the reduced pulmonary flows and volumes. Airway reactivity may not be affected by IUGR. This was tested in children and adolescents from multiple pregnancies, and it was found that IUGR was not associated with altered bronchoreactivity,[68] however, respiratory infections in the post-neonatal period were associated with later hyper-responsiveness.

As seen in infants, there is evidence that prior IUGR exacerbates the effects of preterm birth on lung function in children. In 6 10 year old survivors of preterm birth, being subjected to IUGR led to a greater reduction in FEF_{25-75} than in preterm children (born later than 26 weeks) who were not growth restricted.[69]

Respiratory Effects of IUGR That Can Be Observed in Adulthood

Several studies have indicated that low birthweight is associated with altered lung function in adult life. In some of these studies the gestational age at birth of the subjects was unknown, so it is possible that the low birthweight could have been due, at least in part, to preterm birth in those studies. However, given that many individuals in these studies were born at a time when few severely preterm infants survived the neonatal period, it is likely that low birthweight was mainly attributable to IUGR and/or moderately preterm birth. The first study to show an association between birthweight and adult respiratory health was a large study of British men aged 59–70 years. This study showed that low birthweight was associated with reduced FEV_1 (after adjustment for height, age, and smoking status), and an increased risk of death from respiratory causes.[12] A similar study of men and women living in southern India also showed that FEV_1 was positively correlated with birthweight.[40] An analysis of lung function in adults who experienced the Dutch winter famine of World War 2 as fetuses showed that they had a tendency towards impaired lung function.[41] Since these studies were performed, a number of other studies

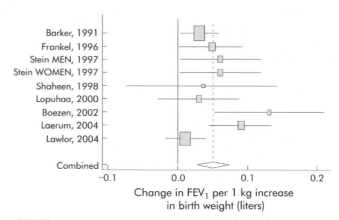

FIGURE 2 Meta-analysis of nine studies examining the relationship between birthweight and forced expiratory volume in one second (FEV_1, litres) assessed in adult men and women. For each of the 9 studies, the regression coefficient (shown as change in FEV_1 per one kg increase in birthweight) is shown by a box. The size of the box indicates the weight given to that study. The horizontal lines show 95% confidence intervals. The overall effect (diamond) reflects the combination of regression coefficients in each study: vertical points of the diamond indicate pooled regression coefficient and horizontal points indicate the 95% confidence limits (upper and lower). *Taken with permission from Lawlor et al.*[74]

have confirmed a significant relationship between birthweight at term and adult respiratory illness (see[70-73]). A meta-analysis of 9 such studies of adult subjects, including males and females, has shown a significant reduction in FEV_1 with decreasing birthweight (Figure 2).[74] Together, these studies provide good evidence that restricted nutrition during fetal life can alter lung development such that lung function is impaired through to adult life; specifically, it is likely that development of the conducting airways and/or lung parenchyma had been altered. Unfortunately there are no histological examinations of human lungs from adults born with IUGR.

Chronic obstructive pulmonary disease (COPD) has been identified as a pulmonary disease with origins in early life.[5,6] However it is still not clear which early life factors increases the risk of COPD: potential risk factors include low birthweight (i.e., IUGR and/or preterm birth), poor growth in infancy, preterm birth and associated chronic lung disease, maternal smoking or early respiratory illness in infancy and childhood.[75]

Postnatal Growth Restriction: Effects on Lung Function

Inadequate nutrition during infancy and childhood can arise from a variety of causes including preterm birth, restricted nutrient availability, infections, and gastro-intestinal illness. As the final stages of lung development continue into infancy and early childhood, nutrient restriction after birth has the potential to alter lung structure and later lung function.[76] Numerous studies using laboratory animals indicate that impaired nutrition after birth, like prenatal nutrient restriction, can alter lung development (see below), but there are limited data from humans, especially in those born at term. One such study showed a significant relation between weight at one year (an index of early postnatal growth) and adult deaths from respiratory disease.[12] Severe protein deficiency may occur in infancy and childhood in developing countries in situations of poverty and during times of famine. Effects of malnutrition on lung function have been documented in undernourished children. In Indian children aged 6–12 years, evidence of current malnutrition and body wasting, evidenced by low weight for height, was associated with lower than expected peak expiratory flow rates.[77] Similarly, children who were wasted and stunted had lower lung volumes and reduced forced inspiratory flow rates,[78] indicating that poor nutrition influences aspects of lung or respiratory muscle development in children.[79]

A group that is particularly vulnerable to inadequate nutrition during early postnatal life is preterm infants, who represent 8–12% of all live births.[76,80] Impaired nutrition can occur in these infants as a result of reduced fat deposition during fetal life, hypoxia, parenteral feeding, poor feeding ability, gut immaturity, or unavailability of breast milk. Owing to the recognition that early life undernutrition likely contributes to the etiology of neonatal chronic lung disease or BPD[81] it is clear that preterm infants, whether or not they were prenatally exposed to IUGR, require particular nutritional management.[81] As peripheral lung development is ongoing in preterm infants, undernutrition, or a lack of necessary minerals or micronutrients, could have detrimental effects on lung anti-oxidant and defense mechanisms, surfactant production, and the process of alveolar formation.

Undernutrition during any stage of life may impair the development and function of skeletal muscles of the thoracic "pump" (diaphragm, intercostal, abdominal), contributing to reduced inspiratory or expiratory flow rates. It is now considered likely that nutritional factors resulting in weight loss contribute to respiratory illnesses such as COPD,[82,83] although it is recognized that increased energy expenditure due to airway obstruction may also be involved. A study in rats found that prolonged undernutrition (40% of estimated requirements) led to reduced oxidative capacity in the diaphragm, secondary to a reduced production of NADH in the Krebs cycle.[84] Biopsies of adult human intercostal muscles indicate that nutritional status, as well as gender, was related to muscle fiber morphometry, although no relation was found between muscle structure and lung function or respiratory muscle strength.[85]

Undernutrition may also affect other aspects of the lungs, such as immune defence. As in animal studies, data from humans indicates that malnutrition causes a reduction in macrophage function, mucociliary clearance, and specific B and T cell responses to infection.[86] Such effects may explain the elevated incidence of respiratory infections in undernourished infants and children.

EFFECTS OF NUTRIENT RESTRICTION ON THE DEVELOPING LUNG: EXPERIMENTAL FINDINGS

The relation between nutrition and lung development has been the subject of a number of comprehensive reviews.[2,9,87,88] In this section we provide a concise review of studies examining the effects of nutrient restriction on lung development, although interpretation is made difficult owing to differences in species used, the developmental stage at which nutrition was manipulated, and the nature and severity of the dietary restriction (e.g., protein or caloric restriction, acute or chronic). As different species reach term at differing stages of lung development, care must be taken to make comparisons taking these stages into account; this is important, for example, when comparing rodent models, in which alveolar development is largely a postnatal event, with long gestation species (e.g., larger mammals such as primates and ruminants), in which alveolar formation begins before birth.[89] Another major problem with studies on the effects of nutrient, and oxygen, restriction on lung development is that these interventions can induce a range of endocrine adjustments that could in themselves affect lung development; such changes include elevated circulating levels of corticosteroids, catecholamines, and prostaglandins[90,91]; furthermore, maternal undernutrition can alter the endocrine status of both mother and fetus.[92]

Fetal Lung Expansion

During lung development, the future airways and air-spaces are not collapsed but contain a unique liquid (fetal lung liquid/fluid) thought to be secreted by pulmonary epithelial cells.[93] This liquid plays a crucial role in prenatal lung development as it maintains the lungs in an expanded state, which is now recognized as being essential for normal lung growth and structural development.[94] As fetal lung liquid production depends upon metabolic activity of the airway epithelium, this process is likely affected by restricted nutrient and oxygen availability. Indeed it has been shown that both acute,[95,96] and chronic hypoxia and hypoglycemia,[97,98] inhibit the secretion of lung liquid. Of most importance to lung development is the volume of liquid retained within the lung lumen[94]; it is apparent that this volume is not significantly altered, when adjusted for lung or body weight, as a result of prolonged hypoxemia and hypoglycemia.[97] The likely reason is that the volume of lung liquid is regulated independent of production, and is determined principally by physical factors, the most important being the transpulmonary pressure gradient and the resistance of the upper respiratory tract.[94]

Fetal Breathing Movements

Fetal breathing movements (FBM) play an important role in lung development by retarding the efflux of liquid from the future airspaces of the lungs, thereby maintaining an adequate level of lung expansion.[94] FBM may also stimulate lung development via phasic alterations in lung tissue stress.[99,100] Few studies, however, have examined the impact of IUGR on FBM, although one such study in ovine IUGR fetuses found that the incidence of FBM was decreased and the proportion of time spent "apneic" was increased.[101] It is well established that acute hypoxemia in the fetus inhibits FBM; however, in the presence of chronic hypoxemia, FBM return to a frequency and amplitude similar to those of a normoxic fetus.[102,103] In the human growth restricted fetus, FBM are reduced in incidence, and this is associated with elevated adenosine concentrations in fetal blood.[104] Thus it appears that IUGR, probably due to the associated chronic hypoxemia and hypoglycemia, may have an inhibitory effect on FBM; this in turn may affect lung development, although this effect is likely not due to a reduced volume of lung liquid.

Fetal Lung Growth

Numerous experimental studies have shown that restricted fetal nutrient (and oxygen) supply can alter lung development, but the effects vary according to the form and gestational timing of the intervention. In sheep, IUGR was found to either reduce,[101] or have no effect,[105] on fetal lung weight, relative to fetal body weight, near to term. Maternal undernutrition for 10 days during late gestation led to reduced fetal lung weight (wet) in the absence of IUGR.[106] More prolonged periods of fetal undernutrition, induced by chronic placental insufficiency, restrict fetal growth, but lung growth relative to body growth is not impaired.[97,107] Similarly, in the rat, 4 days of late gestational undernutrition did not affect fetal lung weight or DNA content.[108] Fetal hypoxemia, in contrast, reduced the rate of DNA synthesis in the fetal lung, which may contribute to altered lung tissue remodeling during development.[109]

Alveolarization

Studies in a number of species have shown that undernutrition during the later stages of lung development can interfere with alveolarization. Most studies have been performed in the rat, in which alveoli are formed after birth.[110] Intermittent starvation during the first postnatal week resulted in enlarged (and hence fewer) alveoli, thicker septa, with an apparent reduction in elastin deposition at 2 weeks of age.[111] Protein deficiency during postnatal development initially inhibited lung growth and altered lung morphometry, but normal lung parenchymal maturity was apparent at weaning.[112] Thus it appears that the degree and type of

nutrient restriction during development may determine the form of structural alteration in the lungs.

The effects of IUGR resulting from fetal hypoxemia and nutritional restriction during late gestation have been studied in the near-term ovine fetus, 8-week-old lamb and 2-year-old (young adult) sheep.[113–115] In these studies the induced IUGR coincided with the saccular and alveolar stages of lung development. At 8 weeks after term-birth, but not in the near-term fetus, there was a thickening of the inter-alveolar septa and a reduction in the number of alveoli per respiratory unit (Figure 3); at 8 weeks after birth there was also an increase in the mean alveolar diameter.[114] Importantly, this reduction in alveolar number was still evident in adult sheep at 2 years of age[116]; this reduction in alveolar number was associated with a reduction in the number of bronchiolar-alveolar attachments, which could contribute to impaired lung function later in life.[117] Thus it appears that adequate fetal growth and nutrition is necessary for normal alveolar formation and that the effects of IUGR can persist into adulthood. Similar effects on alveolar formation have been made in the undernourished postnatal rat,[111] and in the guinea pig following undernutrition late in gestation and during the neonatal period.[118] Reductions in the number of alveoli and bronchiolar-alveolar attachments could

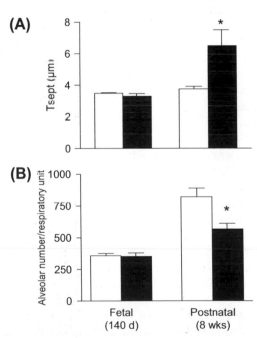

FIGURE 3 Effects of experimentally induced IUGR in sheep on (A) the thickness of the inter-alveolar septum (Tsept) and (B) number of alveoli per respiratory unit in the lungs of near-term fetuses and 8-week-old lambs. Open bars show data from control animals; filled bars represent IUGR animals. As a result of IUGR, Tsept was significantly thicker at 8 weeks after birth than in control lambs, although it was not different prior to birth. At 8 weeks after birth there were significantly fewer alveoli per respiratory unit in IUGR lambs than in controls, but no difference was seen at 140 days of gestation. *Taken with permission from Maritz et al.*[114]

FIGURE 4 Effects of postnatal growth rate in sheep on alveolar number (Na) and alveolar surface area (Sa) at adulthood. Data from slower growing sheep (open bars) are compared with data from faster growing sheep (filled bars). A and D (unadjusted data) show significantly fewer alveoli and a smaller surface area for gas exchange in adult slower growing sheep. B and E show the data adjusted for body weight; C and F show the data adjusted for lung weight. Asterisks (*) indicate p<0.05. *Taken with permission from Maritz et al.*[119]

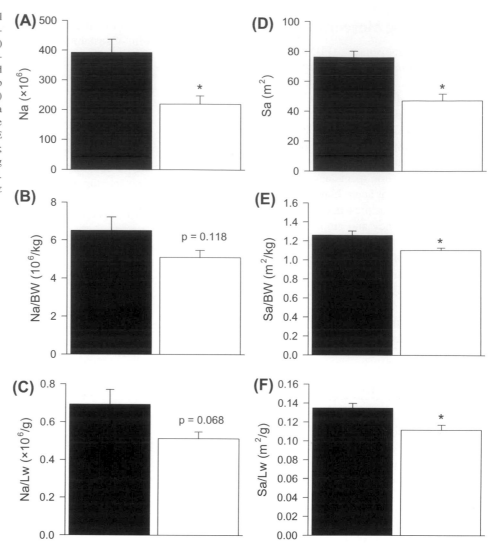

accelerate lung aging and contribute to late-onset obstructive lung disease.

The process of alveolarization continues after birth in many mammals, including humans and sheep. A study of adult sheep with spontaneously different postnatal growth rates during the first months of postnatal life showed that slower growing sheep had fewer alveoli and a smaller surface area for gas exchange in relation to body weight (Figure 4).[119] The data showed that a significant correlation existed between postnatal growth rate and the alveolar number and surface area for gas exchange in adult sheep; this correlation persisted when alveolar number and surface area were adjusted for adult lung weight or adult body weight (Figure 5).[119] This study supports the concept that growth, and hence nutrition, during the alveolar stage of lung development has a significant impact upon the final number of alveoli and pulmonary gas-exchange surface area. The causes of impaired alveolar formation in the presence of undernutrition are poorly understood. In the case of

IUGR, numerous metabolic and endocrine changes occur in the fetus that have the potential to retard lung development, for example, hypoxia, hypoglycemia, and increased circulating glucocorticoid concentration.[90] Glucose transport to the lungs of fetal rats subjected to IUGR was found to be reduced, potentially contributing to reduced metabolic activity of lung cells.[120]

Blood–Gas Barrier

The pulmonary blood–gas barrier is comprised of three components: the endothelial cell, alveolar epithelial type I cell, and the interposed, fused basement membranes. This very thin layer of tissue between the alveolar lumen and the alveolar capillary is essential for normal gas exchange, as its physical properties (thickness and total area) determine the rate of O_2 and CO_2 flux between blood and alveolar gas, referred to as pulmonary diffusing capacity. Undernutrition during development can retard the normal age-related

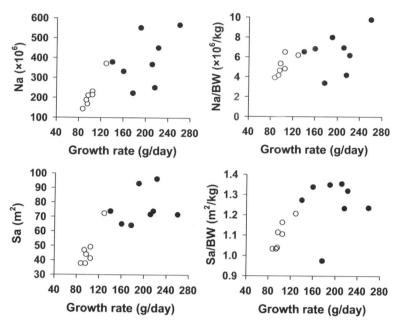

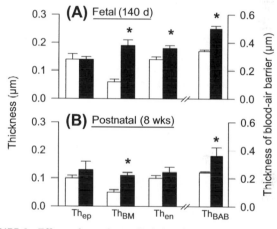

FIGURE 5 Upper panels show the correlation between growth rate over the first 200 days after birth and alveolar number (Na, absolute and relative to body weight (BW)). Lower panels show the relation between growth rate and alveolar surface area (Sa, absolute and relative to birth weight). Both Na ($r^2=0.58$, $p<0.001$) and Na/BW ($r^2=0.305$, $p=0.033$) were significantly correlated with postnatal growth rate. Similarly, both Sa ($r^2=0.635$, $p<0.001$) and Sa/BW ($r^2=0.390$, $p=0.013$) were significantly correlated with growth rate. Open and filled circles show data from slower growing and faster growing sheep, respectively. *Taken with permission from Maritz et al.*[119]

thinning of the blood–gas barrier, possibly owing to a reduction in the rate of tissue remodeling in the alveolar walls. In a study of fetal rats whose mothers had been undernourished during the latter half of pregnancy,[121] the alveolar epithelial cells appeared less mature, with evidence of glycogen retention. In near-term fetal sheep that had been subjected to induced IUGR, the blood–gas barrier was increased in thickness, and this increase was also observed at 8 weeks (Figure 6)[114]; interestingly, this effect persisted to 2 years after birth.[116] In another study IUGR was associated with a reduction in pulmonary diffusing capacity during the first 8 weeks after birth,[113] which could have been at least partially a result of a thicker blood–gas barrier.

The diffusing capacity of the lung can also be reduced by a reduction in alveolar surface area, which is likely to occur in growth-restricted individuals. For example, in guinea pig offspring subjected to maternal undernutrition during gestation, the alveolar surface area was reduced, leading to a reduced diffusing capacity, but this was related to the smaller body size.[118] However, a smaller surface area was still apparent at 126 days after birth in spite of catch-up growth in lung volume and body weight.

The Extracellular Matrix

The extracellular matrix (ECM) of the lung provides the framework of alveoli and conducting airways and hence has a profound influence on the mechanical properties, and hence the function, of the lung. Major ECM components in the lungs are elastin, collagens, proteoglycans, and basement membrane proteins. Elastogenesis is critical to alveolarization[122] and the elastic properties of lung tissue; collagens provide tensile strength to the lungs and are a

FIGURE 6 Effects of experimentally induced IUGR on the thickness of the alveolar blood–air barrier and its three components (epithelium, basement membrane, and endothelium) in (A) near-term fetal sheep and (B) in 8-week-old lambs. Open bars show data from control animals; filled bars show data from IUGR animals. In both fetuses and postnatal lambs the overall thickness of the blood–air barrier (Th_{BAB}), shown in the right hand columns, was greater in IUGR animals than in controls. The three left hand columns show thicknesses of the 3 components of the blood–air barrier, namely epithelium (Th_{ep}), basement membrane (Th_{BM}), and endothelium (Th_{en}). The basement membrane (Th_{BM}) was significantly thicker in IUGR fetuses and postnatal lambs. Endothelial thickness (Th_{en}) was greater in IUGR fetuses than controls, whereas the thickness of the type I alveolar epithelial cells (Th_{ep}) was not affected by IUGR before or after birth. *Taken with permission from Maritz et al.*[114]

major component of basement membranes and conducting airways; proteoglycans exert a large effect on lung compliance and fluid balance due to their large hydrodynamic volumes.[123] These and other structural proteins are laid down during lung development and have a profound effect on the mechanical properties of the lungs after birth. Nutritional status during development can have a persistent effect on the pulmonary ECM such that lung function is affected. Effects of major components of the ECM are discussed in the following section.

ELASTIN

Elastin is one of the principal structural proteins of the terminal airspaces. It is encoded by a single gene, the expression of which can be affected by a range of factors including TGF-β, TNF-α, corticosteroids, Vitamin D, basic FGF, IGF-l, and retinoic acid.[124,125] As demonstrated by elastin-null mice, elastin expression is necessary for normal branching and development of the terminal airspaces including alveolarization.[126] During development, increased lung tissue compliance is related to the presence of elastin.[127] Tropoelastin is expressed principally by fibroblasts during early lung development, particularly during alveolarization.[128–130] As the half-life of elastin approximately matches the life span of the species,[125] factors affecting elastogenesis have the potential to exert a long-term effect upon elastin content and mechanical properties of distensible organs such as the lung.

A range of intrauterine and early postnatal factors that contribute to IUGR, such as nutrient restriction and hypoxemia, are known to affect elastin deposition in the lung. In isolated pulmonary fibroblasts, hypoxia (2% O_2) reversibly down-regulated tropoelastin gene expression.[131] A similar inhibitory effect on tropoelastin synthesis was observed in isolated pulmonary artery smooth muscle cells.[132] Hypoxia (12–13% O_2) causes an increase in pulmonary lysyl oxidase activity, the extracellular enzyme responsible for crosslinking of elastin and collagen.[133,134] Hypoxia also reversibly inhibits amino acid uptake into human lung fibroblasts.[131] However, in vivo, an inhibitory effect of hypoxia on elastin synthesis was not observed in growing rats.[135] In a recent study, IUGR induced in rats by placental insufficiency led to decreased expression of genes regulating elastin deposition and increased lung compliance.[136] This finding is consistent with reduced alveolarization following IUGR, and may provide an explanation for IUGR-induced emphysematous changes in the lung. In contrast, IUGR due to maternal food restriction led to an increase in pulmonary elastin in postnatal rat pups.[137] Thus the effects of IUGR on elastin may depend on how IUGR is induced.

Protein deficiency in growing rats had little effect on elastin (or collagen) or on lung tissue compliance.[138] A similar study in growing rats subjected to protein restriction and restricted growth found a loss of pulmonary desmosine and lung recoil (when adjusted for lung size) and enlarged alveoli.[139] In contrast, severe starvation of growing postnatal rats, but not adult rats, led to a reduction in pulmonary elastin content and altered lung compliance.[140]

In humans, elastin accumulates in the lungs between 25 weeks of gestation and 15 weeks after birth, and it is apparent that this accumulation is not significantly affected by IUGR.[141] Similarly, in fetal sheep affected by late gestational IUGR, pulmonary tropoelastin expression and elastin content were not significantly different from those of controls.[142]

During fetal and postnatal life, both undernutrition and hypoxemia may be associated with increased levels of corticosteroids[92]; these challenges may exert an effect on elastin synthesis as it has been shown that exogenous corticosteroids can both increase,[131] and decrease,[143] elastin formation in the fetal lungs.

COLLAGEN

Pulmonary collagen is essential for the mechanical strength of the lung parenchyma and conducting airways,[123] and can affect lung tissue compliance. Unlike elastin, pulmonary collagen is synthesized and degraded throughout life.[123] Postnatally, collagen metabolism is known to be affected by nutritional status,[144] such that pulmonary collagen content was significantly reduced in growing and adult rats fed on a low protein diet.[138,139] In the fetus, however, there is evidence that IUGR is profibrotic. Hypoxemia, which generally accompanies IUGR, increases procollagen gene expression in lung fibroblasts.[145] A recent study has shown that IUGR in rats, secondary to maternal protein restriction, alters the expression of inflammatory cytokines in the lung and increases collagen deposition in the lung parenchyma and conducting airways.[146]

Collagen IV is a major component of basement membranes, and therefore it affects the strength and function of the pulmonary blood–gas barrier.[147] With advancing maturity the blood–gas barrier thins, allowing an increased diffusing capacity. In the sheep, IUGR impairs the normal thinning of the blood–air barrier, due to persistence of a thickened basement membrane; this was seen in the near-term fetus, and at 8 weeks after birth,[114] and it is apparent that this effect persists into adulthood.[116]

PROTEOGLYCANS

No data apparently exist on the influence of impaired nutrition or oxygenation on proteoglycan synthesis in the developing lung. In vascular tissue, hypoxia has been shown to decrease proteoglycan production by bovine pulmonary artery endothelial cells,[148] and by human aortic smooth muscle cells.[149]

Conducting Airways

Data from some,[66,150,151] but not all,[68,152] studies on infants and children suggest that low birthweight due to IUGR can alter airway function after birth. If so, it is likely that such changes will be long-lasting, potentially accounting for alterations in adult lung function.[153] These studies have used standard tests of airway function, such as peak flow rates or airway responsiveness to challenge by bronchoconstrictors, and their relevance to structural development of the conducting airways may be questionable. Very few studies have examined the effects of IUGR on the wall structure of the conducting airways, although evidence exists that it can be altered. In an ovine model of IUGR, the structure of the tracheal wall was altered in the near-term fetus, such that there was less cartilage, less mucosal folding, impaired development of submucosal glands and evidence of reduced ciliation of epithelial cells.[105] Another study of sheep, in which IUGR was induced for 20 days during late ovine gestation, also showed altered structure of the cartilaginous airways near to term; the walls were thinner and more folded and the submucosal glands were less well developed.[115] In this study, however, there was postnatal recovery of airway wall thickness (by 8 weeks), but changes in mucus elements were still evident.

Early postnatal undernutrition of rats was shown to affect the bronchiolar epithelium; epithelial cell division was reduced and the conversion of Clara cells to ciliated cells was reduced, leading to a persistent abnormality in the bronchiolar epithelium.[154] In sheep, it has been reported that postnatal growth rate influences the walls of the conducting airways, especially the smaller airways[155]. A slower postnatal growth rate led to reductions in the thickness, relative to perimeter, of the airway wall and airway smooth muscle (Figure 7); the bronchiolar epithelium thickness (relative to perimeter) was also reduced in these slow growing animals (Figure 8). Collectively, changes in airway wall development could contribute to long term effects of early postnatal undernutrition on lung function. There is clearly a need for more data on the effects of early nutritional compromises on the structure and function of the conducting airways in later life.

The Pulmonary Surfactant System

Numerous studies have examined the relation between fetal growth and maturation of the pulmonary surfactant system; this topic is reviewed in detail in Chapter 9 and only a brief account will be given here. Underfeeding maternal rats impaired the surface tension lowering properties of fetal lung extracts,[156] and similar effects were being seen when neonates were underfed.[157] Maternal underfeeding during rat gestation slows the prenatal and postnatal maturation of alveolar type II cells of offspring, as indicated by increased glycogen content and decreased volume density of lamellar bodies and rough endoplasmic reticulum[121,158]; in this species, type II cells are immature at birth. Similar changes related to undernutrition were seen in bronchiolar Clara cells,[159] although these changes may be only transient with recovery occurring after birth.[121]

In fetal sheep, prolonged mild hypoxemia (48 h), similar to that associated with placental insufficiency, led to increases in both fetal blood cortisol levels and SP-A mRNA.[160] Consistent with this, placental embolization during late ovine gestation enhanced SP-A and SP-B expression at 0.88 of gestation, and the increases were related to fetal cortisol concentrations.[59] This effect of placental embolization was not seen at a later gestational age, when fetal cortisol concentrations were not increased by the IUGR.[97] This suggests that IUGR, fetal nutrient restriction,

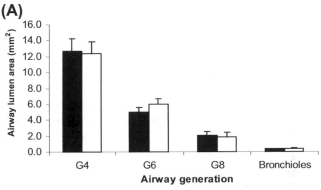

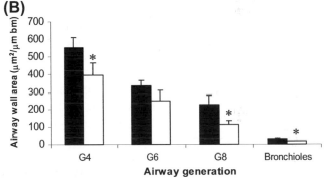

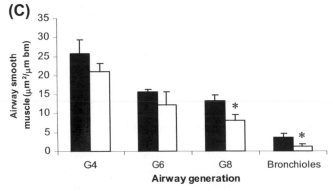

FIGURE 7 Structural comparison of airway walls in sheep with different postnatal growth rates. Data were obtained from computerized image analysis of histological sections of airway generations 4, 6, 8 and bronchioles from faster growing (closed bars) and slower growing (open bars) sheep. (A) shows airway lumen area; (B) shows the outer airway wall area in relation to basement membrane (BM) perimeter; (C) shows the airway smooth muscle area in relation to BM perimeter. * represents significant difference between animal groups (p<0.05) for similar size airways. *Taken with permission from Snibson and Harding.*[155]

FIGURE 8 A comparison of airway epithelium in sheep with different postnatal growth rates. Upper panels show data for epithelial area obtained from computer image analysis of histological sections of airway generations 4, 6, 8 and bronchioles from faster growing (closed bars) and slower growing (open bars) sheep. Lower panels show representative regions of the bronchiolar airway wall from a slower growing and a faster growing sheep stained with Masson's Trichrome stain. ASM, airway smooth muscle; Ep, epithelium; Co, collagen. Magnification ×400. Note the less developed airway epithelium and thinner airway wall of slower growing sheep compared to the faster growing sheep. * represents significant difference between groups (p<0.05). *Taken with permission from Snibson and Harding.*[155]

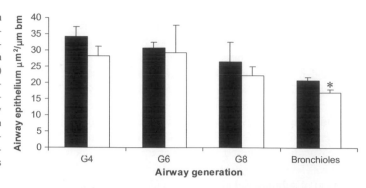

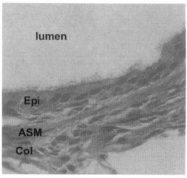 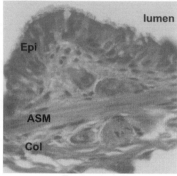

Slower growing   Faster growing

or elevated plasma cortisol may enhance lung maturity at preterm age, but not close to term. Using a different model of IUGR (placental growth restriction) in fetal sheep, it was found that expression of SP-A, SP-B, and SP-C genes was reduced, and the expression of SP proteins in lung tissue was positively correlated with SP-A, SP-B, and SP-C protein in lung tissue.[161] Together these studies show that there is not yet a consensus as to the effects of IUGR on pulmonary surfactant.

The effects of undernutrition and hypoxia on surfactant producing cells are likely to differ according to gestational timing of these challenges and to alterations in circulating cortisol levels. At present, it is not known whether the effects are long lasting, but recent evidence indicates that the proportion of alveolar type II cells is highly dependent upon the prevailing local physical environment.[162]

Pulmonary Immune Function

Poor nutrition and growth can apparently affect the major defense systems of the lungs, including antioxidant function, surfactant production, mucous secretion and transport, and immunological competence.[87] For a detailed account, see Chapters 9, 10, and 11. Poor nutrition during development may impair pulmonary immunological defence including non-specific functions (phagocytic cell functions and inflammation) and specific functions, such as B and T cells.[86] For example, in fetal rats whose mothers were underfed, alveolar macrophages were less abundant.[121]

Similarly, maternal underfeeding has been shown to alter the resistance of neonatal rats to oxidant injury,[163] and airway submucosal glands were attenuated in postnatal lambs following IUGR.[115] Together, the experimental data suggests that pulmonary immune function may be impaired in offspring exposed to restricted nutrition during development.

A recent study has shown that IUGR (50% maternal nutrition) in rats leads to an attenuated inflammatory response and allergic lung inflammation, with an altered cytokine expression profile following immunization and challenge; the attenuated allergic lung inflammation was attributed to an altered Th1/Th2 cytokine balance.[164]

EFFECTS OF HYPOXIA ON LUNG DEVELOPMENT

Fetal hypoxemia can be caused by placental insufficiency, maternal anemia, maternal smoking, and living at high altitude. It is now known that hypoxia can induce persistent structural and functional alterations in the developing lung,[165,166] but the effects will likely depend on the gestational timing of the hypoxia, its severity, and associated factors such as altered nutrition and chemical exposures.

Few studies have examined the effects of hypoxia alone on lung development. In explants of fetal lung, hypoxia stimulated branching morphogenesis and cell proliferation, potentially due to reduced activity of metalloproteinases (MMPs).[167] However, in the intact ovine fetus, prolonged hypoxemia inhibits lung growth as indicated by a lower rate of pulmonary DNA synthesis.[109] Similarly, in a rodent model of fetal hypoxia during the latter two-thirds of gestation, the lungs were small relative to body weight,[168] although hypoxia starting at later stages of gestation did not appear to affect fetal lung growth[168,169] In contrast, fetal sheep exposed to hypobaric hypoxia throughout much of gestation experienced no effect on fetal relative lung weight, or on protein or DNA concentrations.[170] Postnatal hypoxia may affect the developing lung differently; for example, postnatal hypoxia in rodents led to lung hypertrophy and hyperplasia, with increased amounts of connective tissue.[135]

Prenatal hypoxia, induced by high altitude, has significant and lasting effects on the pulmonary vasculature, both structurally and functionally, that may contribute to pulmonary hypertension and altered alveolarization.[166,171,172] The persistence of the effects of fetal hypoxia into adulthood may be a consequence of epigenetic changes in vascular smooth muscle.[173]

ROLE OF MICRONUTRIENTS IN LUNG DEVELOPMENT

Micronutrients, the collective term for vitamins and minerals, are now recognized as being of considerable importance to the developing fetus and neonate.[174] Of particular relevance to lung development are Vitamins A, D, and E, and selenium.

Retinoic Acid

Vitamin A (retinol) and retinoic acid play a crucial role in lung development during embryogenesis; in terms of interactions with the genome, retinoic acid appears to be the most active of the retinoids.[175] Retinoic acid affects many aspects of lung development ranging from branching morphogenesis to structural remodeling during alveolarization.[176] Vitamin A may also be involved in surfactant synthesis as Vit A deficiency during gestation reduces the expression of surfactant proteins A, B, and C in fetal rat lungs.[177]

The role of retinoic acid in alveolarization and lung regeneration has received much attention.[178] These effects are apparently mediated by elastin synthesis[179]; endogenous retinoid production increases tropoelastin expression in vitro,[180] whereas Vitamin A deficiency during pregnancy reduces elastin staining in fetal lungs.[181] There is considerable interest in the possibility that retinoic acid treatment may be able to regenerate alveoli in adult, emphysematous lungs.[178] As preterm infants can be Vitamin A deficient, they are often supplemented with Vitamin A to reduce oxygen requirements and treat lung injury, presumably by enhancement of alveolarization.[182]

Vitamin D

Vitamin D deficiency during development is associated with impaired calcium metabolism and bone disorders, which can affect respiratory function in early life via metabolic bone disease affecting the rib cage.[183] Vitamin D is now considered to play a role in lung disease and immunity,[184] effects which may be mediated by altered lung development. In contrast to the mature lung, the developing lung possesses specific binding sites for Vitamin D on type II alveolar epithelial cells.[185] There is some evidence that Vitamin D_3 plays a role in lung maturation by enhancing surfactant synthesis by Type II cells in rodents.[186] However the data from human lungs is equivocal,[187] and a recent study showed no benefits of prenatal Vitamin D supplementation on wheezing in 3-year-old children.[188]

Vitamin E

At birth, the lung is exposed to high levels of oxygen and hence increased levels of reactive oxygen species. Among antioxidative agents, Vitamin E is the most important lipophilic, radical-scavenging vitamin and protects cell membranes from oxidative injury. Some recent reports have implicated Vitamin E deficiency with altered fetal and neonatal lung development; in particular, there has been interest in the possible role of Vitamin E in the treatment of lung injury and respiratory distress in preterm infants.[80] As Vitamin E accumulates in the fetus, along with fat deposits, during late gestation, it is likely that preterm infants are Vitamin E deficient.[189] Vitamin E is taken up by alveolar type II cells and may therefore play a role in respiratory distress and chronic lung disease of prematurity. In preterm infants, a strong correlation was found between BPD and low Vitamin E and selenium levels in cord blood.[190] However, Vitamin E supplementation of preterm infants with BPD had no beneficial effect.[191]

Selenium

Selenium is important to lung development as it is necessary for the activity of glutathione peroxidase, an important antioxidant enzyme that reduces both organic and inorganic hydroperoxidases. Selenium acts synergistically with Vitamin E to prevent peroxide formation. Experimental selenium deficiency in rats led to altered lung development, in particular impaired alveolar septation.[192] Low selenium levels develop in preterm infants who are fed parenterally, but less so in those fed breast milk; however, even healthy preterm infants had lower levels than in term infants at 6 weeks after birth.[193] In preterm infants, low plasma selenium levels are associated with increased respiratory morbidity[194]; however, selenium supplementation of very preterm infants failed to reduce neonatal oxygen dependency, suggesting that lung development was not stimulated.[195]

NUTRITIONAL RESTRICTION AND THE MATURE LUNG

Nutritional restriction or starvation in adults has long been known to adversely affect lung function and structure via alterations in pulmonary surfactant, respiratory control, or respiratory muscle function.[87,196] Poor nutrition can lead to the breakdown of structural proteins in the lungs,[197] which may exacerbate chronic lung diseases such as COPD.[83] Poor nutrition is also known to impair the immune functions of the lungs.[86] Starvation in the rat (3 weeks of food deprivation), especially in younger, still growing animals, led to the loss of ECM (elastin and collagen) from the lungs.[140] Less severe starvation also affected the lung ECM, resulting in a reduction in tissue elasticity[198]; mean linear intercept was increased, indicative of emphysema.[199] Rodent studies show that total caloric restriction has a greater negative impact on lung integrity than protein restriction.[200]

While the processes underlying nutritional emphysema remain unclear, the component of the lung most affected by adult undernutrition is the inter-alveolar septum.[197] Nutritional emphysema may be largely due to the breakdown of structural proteins of the septum, in particular, elastin and collagen; however, the undernutrition must be severe to induce these changes. In the rat, it has been shown that starvation caused a reversible reduction in the activity of lysyl oxidase, an enzyme involved in post-transational cross-linking of elastin and collagen (Figure 9).[201] Elastase treatment of rat lungs induces a form of emphysema that more closely resembles the condition in humans than undernutrition, suggesting that elastin breakdown may be

involved in the human disease.[202] Of interest, it has shown that nutritional emphysema in adult animals is reversible; caloric restriction in mice reduced the number of alveoli and alveolar surface area, as occurs in human emphysema, but refeeding fully reversed these changes.[203] This suggests that alveolar walls are able to regenerate following their destruction due to undernutrition, at least in mice.

A recent study of adult mice (10–12 weeks old) has shown that severe caloric restriction led to increased lung stiffness and reduced lung volume, with no evidence of alveolar enlargement or emphysema.[204] These changes were attributed to alterations in the ECM rather than in lung surfactant. As in earlier studies, the induced changes in lung mechanics were entirely reversed with refeeding, providing further evidence that lung tissue can regenerate in adult rats.

The effect of adult undernutrition may be species-dependent; although starvation of hamsters caused an enlargement of air-spaces and a reduction in surface area, it was not associated with a reduction in the pulmonary content of collagen, elastin, or glycosaminoglycan, nor was there evidence of alveolar wall destruction.[205] Such differences may be due to differences in somatic growth at the time of the nutritional insult. In general, however, it appears that restricted nutrition leads to protein degradation, impaired cross-linking of pulmonary structural proteins and an imbalance between synthetic and catabolic processes, all of which may contribute to the remodeling that results in emphysematous changes in adult lungs.[88]

As in the developing lung,[156,206] nutritional status in the adult can affect pulmonary surfactant. Such an effect can be induced rapidly; for example, 3 days of fasting in adult rats led to a reduction in the pulmonary surfactant content, reduced lung compliance and a reduced number of lamellar bodies in alveolar type II cells.[207] These effects may be due to a reduced availability of substrates necessary for the synthesis of surfactant or surfactant proteins. In adult rats, caloric restriction caused a reduction in pool size of pulmonary surfactant, but within 4–8 days, these pools had recovered to normal sizes,[208] suggesting that the lung is able to adapt rapidly to reduced nutrient availability.

CONCLUSIONS

It is now apparent that reduced nutrient availability can affect lung development, leading to altered lung structure and function at all stages of life; in addition, it may affect the respiratory muscles. Owing to the wide range of animal models used to investigate the impact of nutritional restriction on lung development, it is difficult to generalize regarding the role of individual classes of nutrients. As lung architecture is laid down early in life, nutritional and other environmental challenges during development have the potential to induce long-lasting alterations in lung structure and function; these persistent alterations contribute to the

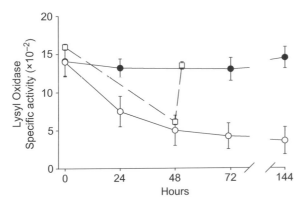

FIGURE 9 Effects of severe undernutrition on the specific activity of lysyl oxidase in the lungs of postnatal rats. Lysyl oxidase activity, which is essential for cross-linking in the formation of elastin fibers, was profoundly reduced throughout 144 hours of starvation (-O-, $n=5$) compared to normally fed controls (-●-, $n=5$). In a separate group of rats (..□.., $n=5$), lysyl oxidase activity was reduced after 48 hours of starvation but returned to control values following 3 hours of refeeding. *Data taken from Madia et al.*[201]

"tracking" or "programming" of lung function throughout life. Although multiple aspects of lung structure and function can be affected by nutritional factors during development, those most likely to have persistent effects on lung function are structural proteins such as elastin and basement membrane proteins, alveoli, and small airways. Taken together, studies of human subjects and laboratory animals show that sub-optimal nutrition and growth, both before and after birth, can adversely affect lung function and respiratory health, and also increase the rate of lung aging.

REFERENCES

1. Harding R, Cock ML, Louey S, Joyce BJ, Davey MG, Albuquerque CA, et al. The compromised intra-uterine environment: implications for future lung health. *Clin Exp Pharmacol Physiol* 2000;27:965–74.
2. Kalenga M, Gaultier C, Burri PH. Nutritional aspects of lung development. In: Gaultier C, Bourbon JR, Post M, editors. *Lung Development*. New York: Oxford University Press; 1999. pp. 347–63.
3. Warner JO, Jones CA. Fetal origins of lung disease. In: Barker DJP, editor. *Fetal Origins of Cardiovascular and Lung Disease*. New York: Marcel Dekker Inc; 2001. pp. 297–321.
4. Duijts L. Fetal and infant origins of asthma. *Eur J Epidemiol* 2012 27:5–14.
5. Narang I, Bush A. Early origins of chronic obstructive pulmonary disease. *Semin Fetal Neonatal Med* 2012;17:112–8.
6. Stocks J, Sonnappa S. Early life influences on the development of chronic obstructive pulmonary disease. *Ther Adv Respir Dis* 2013;7:161–73.
7. Wright RJ. Perinatal stress and early life programming of lung structure and function. *Biol Psychol* 2010;84:46–56.
8. Pike K, Pillow JJ, Lucas JS. Long term respiratory consequences of intrauterine growth restriction. *Semin Fetal Neonatal Med* 2012;17:92–8.
9. Gaultier C. Malnutrition and lung growth. *Pediatr Pulmonol* 1991;10:278–86.
10. Briana DD, Malamitsi-Puchner A. Small for gestational age birth weight: impact on lung structure and function. *Paediatr Respir Rev* 2013;14:256–62.
11. Hibbert ME, Hudson IL, Lanigan A, Landau LI, Phelan PD. Tracking of lung function in healthy children and adolescents. *Pediatr Pulmonol* 1990;8:172–7.
12. Barker DJ, Godfrey KM, Fall C, Osmond C, Winter PD, Shaheen SO. Relation of birth weight and childhood respiratory infection to adult lung function and death from chronic obstructive airways disease. *Br Med J* 1991;303:671–5.
13. Nicolaides KH, Economides DL, Soothill PW. Blood gases, pH, and lactate in appropriate- and small-for-gestational-age fetuses. *Am J Obstet Gynecol* 1989;161:996–1001.
14. Dessi A, Ottonello G, Fanos V. Physiopathology of intrauterine growth retardation: from classic data to metabolomics. *J Matern Fetal Neonatal Med* 2012;25:13–8.
15. Piedrahita JA. The role of imprinted genes in fetal growth abnormalities. *Birth Defects Res A Clin Mol Teratol* 2011;91:682–92.
16. Resnik R. Intrauterine growth restriction. *Obstet Gynecol* 2002;99:490–6.
17. Eiland E, Nzerue C, Faulkner M. Preeclampsia 2012. *J Pregnancy* 2012;2012:1–7.
18. Odegard RA, Vatten LJ, Nilsen ST, Salvesen KA, Austgulen R. Preeclampsia and fetal growth. *Obstet Gynecol* 2000;96:950–5.
19. Cunningham FG, Cox SM, Harstad TW, Mason RA, Pritchard JA. Chronic renal disease and pregnancy outcome. *Am J Obstet Gynecol* 1990;163:453–9.
20. Zamudio S, Torricos T, Fik E, Oyala M, Echalar L, Pullockaran J, et al. Hypoglycemia and the origin of hypoxia-induced reduction in human fetal growth. *PLoS One* 2010;5:e8551.
21. Illsley NP, Caniggia I, Zamudio S. Placental metabolic reprogramming: do changes in the mix of energy-generating substrates modulate fetal growth? *Int J Dev Biol* 2010;54:409–19.
22. Cliver SP, Goldenberg RL, Cutter GR, Hoffman HJ, Davis RO, Nelson KG. The effect of cigarette smoking on neonatal anthropometric measurements. *Obstet Gynecol* 1995;85:625–30.
23. Horta BL, Victora CG, Menezes AM, Halpern R, Barros FC. Low birthweight, preterm births and intrauterine growth retardation in relation to maternal smoking. *Paediatr Perinat Epidemiol* 1997;11:140–51.
24. Albuquerque CA, Smith KR, Johnson C, Chao R, Harding R. Influence of maternal tobacco smoking during pregnancy on uterine, umbilical and fetal cerebral artery blood flows. *Early Hum Dev* 2004;80:31–42.
25. Shu XO, Hatch MC, Mills J, Clemens J, Susser M. Maternal smoking, alcohol drinking, caffeine consumption, and fetal growth: results from a prospective study. *Epidemiology* 1995;6:115–20.
26. Mills JL, Graubard BI, Harley EE, Rhoads GG, Berendes HW. Maternal alcohol consumption and birth weight. How much drinking during pregnancy is safe? *JAMA*. 1984;252:1875–9.
27. Desai M, ter Kuile FO, Nosten F, McGready R, Asamoa K, Brabin B, et al. Epidemiology and burden of malaria in pregnancy. *Lancet Infect Dis* 2007;7:93–104.
28. Umbers AJ, Aitken EH, Rogerson SJ. Malaria in pregnancy: small babies, big problem. *Trends Parasitol* 2011;27:168–75.
29. The American College of Obstetricians and Gynecologists. Practice bulletin no. 134 fetal growth restriction. *Obstet Gynecol* 2013;121:1122–33.
30. Romanyuk V, Raichel L, Sergienko R, Sheiner E. Pneumonia during pregnancy: radiological characteristics, predisposing factors and pregnancy outcomes. *J Matern Fetal Neonatal Med* 2011;24:113–7.
31. Hoffman E, Hedlund E, Perin T, Lyndrup J. Is thrombophilia a risk factor for placenta-mediated pregnancy complications? *Arch Gynecol Obstet* 2012;286:585–9.
32. Said JM, Higgins JR, Moses EK, Walker SP, Monagle PT, Brennecke SP. Inherited thrombophilias and adverse pregnancy outcomes: a case-control study in an Australian population. *Acta Obstet Gynecol Scand* 2012;91:250–5.
33. Tincani A, Bazzani C, Zingarelli S, Lojacono A. Lupus and the antiphospholipid syndrome in pregnancy and obstetrics: clinical characteristics, diagnosis, pathogenesis, and treatment. *Semin Thromb Hemost* 2008;34:267–73.
34. Asherson RA, Cervera R, Merrill JT, Erkan D. Antiphospholipid antibodies and the antiphospholipid syndrome: clinical significance and treatment. *Semin Thromb Hemost* 2008;34:256–66.
35. Cetin I, Mando C, Calabrese S. Maternal predictors of intrauterine growth restriction. *Curr Opin Clin Nutr Metab Care* 2013;16:310–9.
36. Wu G, Imhoff-Kunsch B, Girard AW. Biological mechanisms for nutritional regulation of maternal health and fetal development. *Paediatr Perinat Epidemiol* 2012;26(Suppl. 1):4–26.

37. Abu-Saad K, Fraser D. Maternal nutrition and birth outcomes. *Epidemiol Rev* 2010;**32**:5–25.

38. Vaughan OR, Sferruzzi-Perri AN, Coan PM, Fowden AL. Environmental regulation of placental phenotype: implications for fetal growth. *Reprod Fertil Dev* 2011;**24**:80–96.

39. Hoogland HJ. Ultrasonographic placental morphology. *Gynecol Obstet Invest* 1982;**14**:81–9.

40. Pollack RN, Divon MY. Intrauterine growth retardation: definition, classification, and etiology. *Clin Obstet Gynecol* 1992;**35**:99–107.

41. Andres RL. The association of cigarette smoking with placenta previa and abruptio placentae. *Semin Perinatol* 1996;**20**:154–9.

42. Ventolini G. Conditions associated with placental dysfunction. *Minerva Ginecol* 2011;**63**:459–64.

43. Baschat AA. Pathophysiology of fetal growth restriction: implications for diagnosis and surveillance. *Obstet Gynecol Surv* 2004;**59**:617–27.

44. Muhlhausler BS, Hancock SN, Bloomfield FH, Harding R. Are twins growth restricted? *Pediatr Res* 2011;**70**:117–22.

45. Arbuckle TE, Wilkins R, Sherman GJ. Birth weight percentiles by gestational age in Canada. *Obstet Gynecol* 1993;**81**:39–48.

46. Chernausek SD. Update: consequences of abnormal fetal growth. *J Clin Endocrinol Metab* 2012;**97**:689–95.

47. Sebert S, Sharkey D, Budge H, Symonds ME. The early programming of metabolic health: is epigenetic setting the missing link? *Am J Clin Nutr* 2011;**94**:1953S–8S.

48. Wilhelm-Benartzi CS, Houseman EA, Maccani MA, Poage GM, Koestler DC, Langevin SM, et al. In utero exposures, infant growth, and DNA methylation of repetitive elements and developmentally related genes in human placenta. *Environ Health Perspect* 2012;**120**:296–302.

49. Lambertini L, Lee TL, Chan WY, Lee MJ, Diplas A, Wetmur J, et al. Differential methylation of imprinted genes in growth-restricted placentas. *Reprod Sci* 2011;**18**:1111–7.

50. Jiang X, Yan J, West AA, Perry CA, Malysheva OV, Devapatla S, et al. Maternal choline intake alters the epigenetic state of fetal cortisol-regulating genes in humans. *FASEB J* 2012;**26**:3563–74.

51. Eriksson JG, Forsen T, Tuomilehto J, Winter PD, Osmond C, Barker DJ. Catch-up growth in childhood and death from coronary heart disease: longitudinal study. *Br Med J* 1999;**318**:427–31.

52. Leon DA, Lithell HO, Vagero D, Koupilova I, Mohsen R, Berglund L, et al. Reduced fetal growth rate and increased risk of death from ischaemic heart disease: cohort study of 15 000 Swedish men and women born 1915-29. *Br Med J* 1998;**317**:241–5.

53. Xiao X, Zhang ZX, Li WH, Feng K, Sun Q, Cohen HJ, et al. Low birth weight is associated with components of the metabolic syndrome. *Metabolism* 2010;**59**:1282–6.

54. Kramer MS, Olivier M, McLean FH, Willis DM, Usher RH. Impact of intrauterine growth retardation and body proportionality on fetal and neonatal outcome. *Pediatrics* 1990;**86**:707–13.

55. Ounsted M, Moar V, Scott WA. Perinatal morbidity and mortality in small-for-dates babies: the relative importance of some maternal factors. *Early Hum Dev* 1981;**5**:367–75.

56. Torrance HL, Voorbij HA, Wijnberger LD, van Bel F, Visser GH. Lung maturation in small for gestational age fetuses from pregnancies complicated by placental insufficiency or maternal hypertension. *Early Hum Dev* 2008;**84**:465–9.

57. Minior VK, Divon MY. Fetal growth restriction at term: myth or reality? *Obstet Gynecol* 1998;**92**:57–60.

58. Tyson JE, Kennedy K, Broyles S, Rosenfeld CR. The small for gestational age infant: accelerated or delayed pulmonary maturation? Increased or decreased survival? *Pediatrics* 1995;**95**:534–8.

59. Gagnon R, Langridge J, Inchley K, Murotsuki J, Possmayer F. Changes in surfactant-associated protein mRNA profile in growth-restricted fetal sheep. *Am J Phys* 1999;**276**:L459–65.

60. Hoo AF, Stocks J, Lum S, Wade AM, Castle RA, Costeloe KL, et al. Development of lung function in early life: influence of birth weight in infants of nonsmokers. *Am J Respir Crit Care Med* 2004;**170**:527–33.

61. Lackman F, Capewell V, Richardson B, daSilva O, Gagnon R. The risks of spontaneous preterm delivery and perinatal mortality in relation to size at birth according to fetal versus neonatal growth standards. *Am J Obstet Gynecol* 2001;**184**:946–53.

62. Piper JM, Xenakis EM, McFarland M, Elliott BD, Berkus MD, Langer O. Do growth-retarded premature infants have different rates of perinatal morbidity and mortality than appropriately grown premature infants? *Obstet Gynecol* 1996;**87**:169–74.

63. Morley R, Brooke OG, Cole TJ, Powell R, Lucas A. Birthweight ratio and outcome in preterm infants. *Arch Dis Child* 1990;**65**:30–4.

64. Bose C, Van Marter LJ, Laughon M, O'Shea TM, Allred EN, Karna P, et al. Extremely Low Gestational Age Newborn Study I. Fetal growth restriction and chronic lung disease among infants born before the 28th week of gestation. *Pediatrics* 2009;**124**:e450–8.

65. Rona RJ, Gulliford MC, Chinn S. Effects of prematurity and intrauterine growth on respiratory health and lung function in childhood. *Br Med J* 1993;**306**:817–20.

66. Nikolajev K, Heinonen K, Hakulinen A, Lansimies E. Effects of intrauterine growth retardation and prematurity on spirometric flow values and lung volumes at school age in twin pairs. *Pediatr Pulmonol* 1998;**25**:367–70.

67. Kotecha SJ, Watkins WJ, Heron J, Henderson J, Dunstan FD, Kotecha S. Spirometric lung function in school-age children: effect of intrauterine growth retardation and catch-up growth. *Am J Respir Crit Care Med* 2010;**181**:969–74.

68. Nikolajev K, Korppi M, Remes K, Lansimies E, Jokela V, Heinonen K. Determinants of bronchial responsiveness to methacholine at school age in twin pairs. *Pediatr Pulmonol* 2002;**33**:167–73.

69. Morsing E, Gustafsson P, Brodszki J. Lung function in children born after foetal growth restriction and very preterm birth. *Acta Paediatr* 2012;**101**:48–54.

70. Boezen HM, Vonk JM, van Aalderen WM, Brand PL, Gerritsen J, Schouten JP, et al. Perinatal predictors of respiratory symptoms and lung function at a young adult age. *Eur Respir J* 2002;**20**:383–90.

71. Canoy D, Pekkanen J, Elliott P, Pouta A, Laitinen J, Hartikainen AL, et al. Early growth and adult respiratory function in men and women followed from the fetal period to adulthood. *Thorax* 2007;**62**:396–402.

72. Narang I, Rosenthal M, Cremonesini D, Silverman M, Bush A. Longitudinal evaluation of airway function 21 years after preterm birth. *Am J Respir Crit Care Med* 2008;**178**:74–80.

73. Rona RJ, Smeeton NC, Bustos P, Amigo H, Diaz PV. The early origins hypothesis with an emphasis on growth rate in the first year of life and asthma: a prospective study in Chile. *Thorax* 2005;**60**:549–54.

74. Lawlor DA, Ebrahim S, Davey Smith G. Association of birth weight with adult lung function: findings from the British Women's Heart and Health Study and a meta-analysis. *Thorax* 2005;**60**:851–8.

75. Bush A. COPD: a pediatric disease. *COPD J Chronic Obstr Pulm Dis* 2008;**5**:53–67.

76. Bhatia J, Parish A. Nutrition and the lung. *Neonatology* 2009;**95**: 362–7.

77. Primhak R, Coates FS. Malnutrition and peak expiratory flow rate. *Eur Respir J* 1988;**1**:801–3.

78. Nair RH, Kesavachandran C, Shashidhar S. Spirometric impairments in undernourished children. *Indian J Physiol Pharmacol* 1999;**43**:467–73.

79. Ong TJ, Mehta A, Ogston S, Mukhopadhyay S. Prediction of lung function in the inadequately nourished. *Arch Dis Child* 1998;**79**:18–21.

80. Biniwale MA, Ehrenkranz RA. The role of nutrition in the prevention and management of bronchopulmonary dysplasia. *Semin Perinatol* 2006;**30**:200–8.

81. Ryan S. Nutrition in neonatal chronic lung disease. *Eur J Pediatr* 1998;**157**(Suppl. 1):S19–22.

82. Lewis MI, Belman MJ. Nutrition and the respiratory muscles. *Clin Chest Med* 1988;**9**:337–48.

83. Schols AM. Nutrition in chronic obstructive pulmonary disease. *Curr Opin Pulm Med* 2000;**6**:110–5.

84. Matecki S, Py G, Lambert K, Peyreigne C, Mercier J, Prefaut C, et al. Effect of prolonged undernutrition on rat diaphragm mitochondrial respiration. *Am J Respir Cell Mol Biol* 2002;**26**:239–45.

85. Hards JM, Reid WD, Pardy RL, Pare PD. Respiratory muscle fiber morphometry. Correlation with pulmonary function and nutrition. *Chest.* 1990;**97**:1037–44.

86. Bellanti JA, Zeligs BJ, Kulszycki LL. Nutrition and development of pulmonary defense mechanisms. *Pediatr Pulmonol Suppl* 1997;**16**:170–1.

87. Edelman NH, Rucker RB, Peavy HH. NIH workshop summary: Nutrition and the respiratory system. Chronic obstructive pulmonary disease (COPD). *Am Rev Respir Dis* 1986;**134**:347–52.

88. Sahebjami H. Nutrition and lung structure and function. *Eur Respir J* 1993,**19**.105–24.

89. Burri PH. Development and growth of the human lung. In: Fishman AP, Cherniack NS, Widdicombe JG, Geiger SR, editors. *Handbook of Physiology.* Bethesda: American Physiological Society; 1986. pp. 1–46.

90. Gagnon R, Challis J, Johnston L, Fraher L. Fetal endocrine responses to chronic placental embolization in the late-gestation ovine fetus. *Am J Obstet Gynecol* 1994;**170**:929–38.

91. Gagnon R, Murotsuki J, Challis JR, Fraher L, Richardson BS. Fetal sheep endocrine responses to sustained hypoxemic stress after chronic fetal placental embolization. *Am J Phys* 1997;**272**:E817–23.

92. Dwyer CM, Stickland NC. The effects of maternal undernutrition on maternal and fetal serum insulin-like growth factors, thyroid hormones and cortisol in the guinea pig. *J Dev Physiol* 1992;**18**:303–13.

93. Hooper SB, Harding R. Fetal lung liquid: a major determinant of the growth and functional development of the fetal lung. *Clin Exp Pharmacol Physiol* 1995;**22**:235–47.

94. Harding R, Hooper SB. Regulation of lung expansion and lung growth before birth. *J Appl Phys* 1996;**81**:209–24.

95. Hooper SB, Dickson KA, Harding R. Lung liquid secretion, flow and volume in response to moderate asphyxia in fetal sheep. *J Dev Physiol* 1988;**10**:473–85.

96. Wallace MJ, Hooper SB, McCrabb GJ, Harding R. Acidaemia enhances the inhibitory effect of hypoxia on fetal lung liquid secretion in sheep. *Reprod Fertil Dev* 1996;**8**:327–33.

97. Cock ML, Albuquerque CA, Joyce BJ, Hooper SB, Harding R. Effects of intrauterine growth restriction on lung liquid dynamics and lung development in fetal sheep. *Am J Obstet Gynecol* 2001;**184**:209–16.

98. Hooper SB, Harding R. Changes in lung liquid dynamics induced by prolonged fetal hypoxemia. *J Appl Phys* 1990;**69**:127–35.

99. Harding R, Liggins GC. Changes in thoracic dimensions induced by breathing movements in fetal sheep. *Reprod Fertil Dev* 1996;**8**:117–24.

100. Liu M, Skinner SJ, Xu J, Han RN, Tanswell AK, Post M. Stimulation of fetal rat lung cell proliferation in vitro by mechanical stretch. *Am J Phys* 1992;**263**:L376–83.

101. Maloney JE, Bowes G, Brodecky V, Dennett X, Wilkinson M, Walker A. Function of the future respiratory system in the growth retarded fetal sheep. *J Dev Physiol* 1982;**4**:279–97.

102. Bocking AD, Gagnon R, Milne KM, White SE. Behavioral activity during prolonged hypoxemia in fetal sheep. *J Appl Phys* 1988;**65**:2420–6.

103. Koos BJ, Kitanaka T, Matsuda K, Gilbert RD, Longo LD. Fetal breathing adaptation to prolonged hypoxaemia in sheep. *J Dev Physiol* 1988;**10**:161–6.

104. Yoneyama Y, Shin S, Iwasaki T, Power GG, Araki T. Relationship between plasma adenosine concentration and breathing movements in growth-retarded fetuses. *Am J Obstet Gynecol* 1994;**171**:701–6.

105. Rees S, Ng J, Dickson K, Nicholas T, Harding R. Growth retardation and the development of the respiratory system in fetal sheep. *Early Hum Dev* 1991;**26**:13–27.

106. Harding JE, Johnston BM. Nutrition and fetal growth. *Reprod Fertil Dev* 1995;**7**:539–47.

107. Duncan JR, Cock ML, Harding R, Rees SM. Relation between damage to the placenta and the fetal brain after late-gestation placental embolization and fetal growth restriction in sheep. *Am J Obstet Gynecol* 2000;**183**:1013–22.

108. Rhoades RA, Ryder DA. Fetal lung metabolism. Response to maternal fasting. *Biochim Biophys Acta* 1981;**663**:621–9.

109. Hooper SB, Bocking AD, White S, Challis JR, Han VK. DNA synthesis is reduced in selected fetal tissues during prolonged hypoxemia. *Am J Phys* 1991;**261**:R508–14.

110. Massaro D, Teich N, Maxwell S, Massaro GD, Whitney P. Postnatal development of alveoli. Regulation and evidence for a critical period in rats. *J Clin Invest* 1985;**76**:1297–305.

111. Das RM. The effects of intermittent starvation on lung development in suckling rats. *Am J Pathol* 1984;**117**:326–32.

112. Kalenga M, Tschanz SA, Burri PH. Protein deficiency and the growing rat lung. II. Morphometric analysis and morphology. *Pediatr Res* 1995;**37**:789–95.

113. Joyce BJ, Louey S, Davey MG, Cock ML, Hooper SB, Harding R. Compromised respiratory function in postnatal lambs after placental insufficiency and intrauterine growth restriction. *Pediatr Res* 2001;**50**:641–9.

114. Maritz GS, Cock ML, Louey S, Joyce BJ, Albuquerque CA, Harding R. Effects of fetal growth restriction on lung development before and after birth: a morphometric analysis. *Pediatr Pulmonol* 2001;**32**:201–10.

115. Wignarajah D, Cock ML, Pinkerton KE, Harding R. Influence of intrauterine growth restriction on airway development in fetal and postnatal sheep. *Pediatr Res* 2002;**51**:681–8.

116. Maritz GS, Cock ML, Louey S, Suzuki K, Harding R. Fetal growth restriction has long-term effects on postnatal lung structure in sheep. *Pediatr Res* 2004;**55**:287–95.

117. Harding R, Snibson K, O'Reilly M, Maritz GS. Early environmental influences on lung development: implications for lung function and respiratory health throughout life. In: Ross MG, Newnham JP, editors. *Early Life Origins of Human Health and Disease*. Basel: Karger; 2009. pp. 78–88.

118. Lechner AJ. Perinatal age determines the severity of retarded lung development induced by starvation. *Am Rev Respir Dis* 1985;**131**:638–43.

119. Maritz G, Probyn M, De Matteo R, Snibson K, Harding R. Lung parenchyma at maturity is influenced by postnatal growth but not by moderate preterm birth in sheep. *Neonatology* 2008;**93**:28–35.

120. Simmons RA, Gounis AS, Bangalore SA, Ogata ES. Intrauterine growth retardation: fetal glucose transport is diminished in lung but spared in brain. *Pediatr Res* 1992;**31**:59–63.

121. Curle DC, Adamson IY. Retarded development of noenatal rat lung by maternal malnutrition. *J Histochem Cytochem* 1978;**26**:401–8.

122. Mariani TJ, Sandefur S, Pierce RA. Elastin in lung development. *Exp Lung Res* 1997;**23**:131–45.

123. Chambers RC, Laurent GJ. The lung. In: Comper WD, editor. *Extracellular Matrix*. Amsterdam: Hardwood Academic Publishers; 1996. pp. 378–409.

124. Foster JA, Curtiss SW. The regulation of lung elastin synthesis. *Am J Phys* 1990;**259**:L13–23.

125. Mecham RP. Elastic Fibers. In: Crystal RG, West JB, Weibel ER, Barnes PJ, editors. *The Lung*. Philadelphia: Lippincott-Raven; 1997. pp. 729–36.

126. Wendel DP, Taylor DG, Albertine KH, Keating MT, Li DY. Impaired distal airway development in mice lacking elastin. *Am J Respir Cell Mol Biol* 2000;**23**:320–6.

127. Nardell EA, Brody JS. Determinants of mechanical properties of rat lung during postnatal development. *J Appl Physiol Respir Environ Exerc Physiol* 1982;**53**:140–8.

128. Pierce RA, Mariani TJ, Senior RM. Elastin in lung development and disease. In: Chadwick DJ, Goode JA, editors. *The Molecular Biology and Pathology of Elastic Tissues*. New York: J Wiley; 1995. pp. 199–214.

129. Pierce RA, Mariencheck WI, Sandefur S, Crouch EC, Parks WC. Glucocorticoids upregulate tropoelastin expression during late stages of fetal lung development. *Am J Phys* 1995;**268**:L491–500.

130. Shibahara SU, Davidson JM, Smith K, Crystal RG. Modulation of tropoelastin production and elastin messenger ribonucleic acid activity in developing sheep lung. *Biochemistry* 1981;**20**:6577–84.

131. Berk JL, Massoomi N, Hatch C, Goldstein RH. Hypoxia downregulates tropoelastin gene expression in rat lung fibroblasts by pretranslational mechanisms. *Am J Phys* 1999;**277**:L566–72.

132. Stenmark KR, Aldashev AA, Orton EC, Durmowicz AG, Badesch DB, Parks WC, et al. Cellular adaptation during chronic neonatal hypoxic pulmonary hypertension. *Am J Phys* 1991;**261**:97–104.

133. Brody JS, Kagan H, Manalo A. Lung lysyl oxidase activity: relation to lung growth. *Am Rev Respir Dis* 1979;**120**:1289–95.

134. Brody JS, Vaccaro C. Postnatal formation of alveoli: interstitial events and physiologic consequences. *Fed Proc* 1979;**38**:215–23.

135. Sekhon HS, Thurlbeck WM. Lung growth in hypobaric normoxia, normobaric hypoxia, and hypobaric hypoxia in growing rats. I. Biochemistry. *J Appl Phys* 1995;**78**:124–31.

136. Joss-Moore LA, Wang Y, Yu X, Campbell MS, Callaway CW, McKnight RA, et al. IUGR decreases elastin mRNA expression in the developing rat lung and alters elastin content and lung compliance in the mature rat lung. *Physiol Genomics* 2011;**43**:499–505.

137. Rehan VK, Sakurai R, Li Y, Karadag A, Corral J, Bellusci S, et al. Effects of maternal food restriction on offspring lung extracellular matrix deposition and long term pulmonary function in an experimental rat model. *Pediatr Pulmonol* 2012;**47**:162–71.

138. Myers BA, Dubick MA, Gerreits J, Rucker RB, Jackson AC, Reiser KM, et al. Protein deficiency: effects on lung mechanics and the accumulation of collagen and elastin in rat lung. *J Nutr* 1983;**113**:2308–15.

139. Matsui R, Thurlbeck WM, Fujita Y, Yu SY, Kida K. Connective tissue, mechanical, and morphometric changes in the lungs of weanling rats fed a low protein diet. *Pediatr Pulmonol* 1989;**7**:159–66.

140. Sahebjami H, MacGee J. Effects of starvation on lung mechanics and biochemistry in young and old rats. *J Appl Phys* 1985;**58**:778–84.

141. Desai R, Wigglesworth JS, Aber V. Assessment of elastin maturation by radioimmunoassay of desmosine in the developing human lung. *Early Hum Dev* 1988;**16**:61–71.

142. Cock ML, Joyce BJ, Hooper SB, Wallace MJ, Gagnon R, Brace RA, et al. Pulmonary elastin synthesis and deposition in developing and mature sheep: effects of intrauterine growth restriction. *Exp Lung Res* 2004;**30**:405–18.

143. Willet KE, McMenamin P, Pinkerton KE, Ikegami M, Jobe AH, Gurrin L, et al. Lung morphometry and collagen and elastin content: changes during normal development and after prenatal hormone exposure in sheep. *Pediatr Res* 1999;**45**:615–25.

144. Berg RA, Kerr JS. Nutritional aspects of collagen metabolism. *Annu Rev Nutr* 1992;**12**:369–90.

145. Zhao L, Yang H, Guo X, Chen X, Yan Y. Effect of hypoxia on proliferation and alpha 1 (I) procollagen gene expression by human fetal lung fibroblasts. *Acta Academiae Medicinae Sinicae* 1998;**20**:109–13.

146. Alcazar MAA, Ostreicher I, Appel S, Rother E, Vohlen C, Plank C, Dotsch J. Developmental regulation of inflammatory cytokine-mediated Stat3 signaling: the missing link between intrauterine growth restriction and pulmonary dysfunction? *J Mol Med* 2012;**90**:945–57.

147. West JB, Mathieu-Costello O. Structure, strength, failure, and remodeling of the pulmonary blood–gas barrier. *Annu Rev Physiol* 1999;**61**:543–72.

148. Humphries DE, Lee SL, Fanburg BL, Silbert JE. Effects of hypoxia and hyperoxia on proteoglycan production by bovine pulmonary artery endothelial cells. *J Cell Physiol* 1986;**126**:249–53.

149. Figueroa JE, Tao Z, Sarphie TG, Smart FW, Glancy DL, Vijayagopal P. Effect of hypoxia and hypoxia/reoxygenation on proteoglycan metabolism by vascular smooth muscle cells. *Atherosclerosis* 1999;**143**:135–44.

150. Dezateux C, Stocks J. Lung development and early origins of childhood respiratory illness. *Br Med Bull* 1997;**53**:40–57.

151. Lum S, Hoo AF, Dezateux C, Goetz I, Wade A, DeRooy L, et al. The association between birthweight, sex, and airway function in infants of nonsmoking mothers. *Am J Respir Crit Care Med* 2001;**164**:2078–84.

152. Shaheen SO, Sterne JA, Tucker JS, Florey CD. Birth weight, childhood lower respiratory tract infection, and adult lung function. *Thorax* 1998;**53**:549–53.

153. Lopuhaa CE, Roseboom TJ, Osmond C, Barker DJ, Ravelli AC, Bleker OP, et al. Atopy, lung function, and obstructive airways disease after prenatal exposure to famine. *Thorax* 2000;**55**:555–61.

154. Massaro GD, McCoy L, Massaro D. Postnatal undernutrition slows development of bronchiolar epithelium in rats. *Am J Phys* 1988;**255**:R521–6.

155. Snibson K, Harding R. Postnatal growth rate, but not mild preterm birth, influences airway structure in adult sheep challenged with house dust mite. *Exp Lung Res* 2008;**34**:69–84.

156. Faridy EE. Effect of maternal malnutrition on surface activity of fetal lungs in rats. *J Appl Phys* 1975;**39**:535–40.

157. Guarner V, Tordet C, Bourbon JR. Effects of maternal protein-calorie malnutrition on the phospholipid composition of surfactant isolated from fetal and neonatal rat lungs. Compensation by inositol and lipid supplementation. *Pediatr Res* 1992;**31**:629–35.

158. Massaro GD, Clerch L, Massaro D. Perinatal anatomic development of alveolar type II cells in rats. *Am J Phys* 1986;**251**:R470–5.

159. Massaro GD, Davis L, Massaro D. Postnatal development of the bronchiolar Clara cell in rats. *Am J Phys* 1984;**247**:C197–203.

160. Braems GA, Yao LJ, Inchley K, Brickenden A, Han VK, Grolla A, et al. Ovine surfactant protein cDNAs: use in studies on fetal lung growth and maturation after prolonged hypoxemia. *Am J Physiol Lung Cell Mol Physiol* 2000;**278**:L754–64.

161. Orgeig S, Crittenden TA, Marchant C, McMillen IC, Morrison JL. Intrauterine growth restriction delays surfactant protein maturation in the sheep fetus. *Am J Physiol Lung Cell Mol Physiol* 2010;**298**:L575–83.

162. Flecknoe SJ, Boland RE, Wallace MJ, Harding R, Hooper SB. Regulation of alveolar epithelial cell phenotypes in fetal sheep: roles of cortisol and lung expansion. *Am J Physiol Lung Cell Mol Physiol* 2004;**287**:L1207–14.

163. Langley-Evans SC, Phillips GJ, Jackson AA. Fetal exposure to low protein maternal diet alters the susceptibility of young adult rats to sulfur dioxide-induced lung injury. *J Nutr* 1997,**127**.202–9.

164. Landgraf MA, Landgraf RG, Silva RC, Semedo P, Camara NO, Fortes ZB. Intrauterine undernourishment alters TH1/TH2 cytokine balance and attenuates lung allergic inflammation in wistar rats. *Cell Physiol Biochem* 2012;**30**:552–62.

165. Orgeig S, Morrison JL, Daniels CB. Prenatal development of the pulmonary surfactant system and the influence of hypoxia. *Respir Physiolo Neurobiol* 2011;**178**:129–45.

166. Papamatheakis DG, Blood AB, Kim JH, Wilson SM. Antenatal hypoxia and pulmonary vascular function and remodeling. *Curr Vasc Pharmacol* 2013;**11**:616–40.

167. Gebb SA, Jones PL. Hypoxia and lung branching morphogenesis. *Adv Exp Med Biol* 2003;**543**:117–25.

168. Faridy EE, Sanii MR, Thliveris JA. Fetal lung growth: influence of maternal hypoxia and hyperoxia in rats. *Respir Physiol* 1988;**73**:225–41.

169. Larson JE, Thurlbeck WM. The effect of experimental maternal hypoxia on fetal lung growth. *Pediatr Res* 1988;**24**:156–9.

170. Jacobs R, Robinson JS, Owens JA, Falconer J, Webster ME. The effect of prolonged hypobaric hypoxia on growth of fetal sheep. *J Dev Physiol* 1988;**10**:97–112.

171. Gao Y, Raj JU. Hypoxic pulmonary hypertension of the newborn. *Compr Physiol* 2011;**1**:61–79.

172. Liu J, Gao Y, Negash S, Longo LD, Raj JU. Long-term effects of prenatal hypoxia on endothelium-dependent relaxation responses in pulmonary arteries of adult sheep. *Am J Physiol Lung Cell Mol Physiol* 2009;**296**:L547–54.

173. Yang Q, Lu Z, Ramchandran R, Longo LD, Raj JU. Pulmonary artery smooth muscle cell proliferation and migration in fetal lambs acclimatized to high-altitude long-term hypoxia: role of histone acetylation. *Am J Physiol Lung Cell Mol Physiol* 2012;**303**:L1001–10.

174. Black RE. Micronutrients in pregnancy. *Br J Nutr* 2001;**85**(Suppl. 2):S193–7.

175. Berg JT, Breen EC, Fu Z, Mathieu-Costello O, West JB. Alveolar hypoxia increases gene expression of extracellular matrix proteins and platelet-derived growth factor-B in lung parenchyma. *Am J Respir Crit Care Med* 1998;**158**:1920–8.

176. Cardoso WV, Lu J. Regulation of early lung morphogenesis: questions, facts and controversies. *Development* 2006;**133**:1611–24.

177. Chailley-Heu B, Chelly N, Lelievre-Pegorier M, Barlier-Mur AM, Merlet-Benichou C, Bourbon JR. Mild vitamin A deficiency delays fetal lung maturation in the rat. *Am J Respir Cell Mol Biol* 1999;**21**:89–96.

178. Hind M, Gilthorpe A, Stinchcombe S, Maden M. Retinoid induction of alveolar regeneration: from mice to man? *Thorax* 2009;**64**:451–7.

179. McGowan SE. Contributions of retinoids to the generation and repair of the pulmonary alveolus. *Chest* 2002;**121**:206S–8S.

180. McGowan SE, Doro MM, Jackson SK. Endogenous retinoids increase perinatal elastin gene expression in rat lung fibroblasts and fetal explants. *Am J Phys* 1997;**273**:L410–6.

181. Antipatis C, Ashworth CJ, Grant G, Lea RG, Hay SM, Rees WD. Effects of maternal vitamin A status on fetal heart and lung: changes in expression of key developmental genes. *Am J Phys* 1998;**275**:L1184–91.

182. Guimaraes H, Guedes MB, Rocha G, Tome T, Albino-Teixeira A. Vitamin A in prevention of bronchopulmonary dysplasia. *Curr Pharm Des* 2012;**18**:3101–13.

183. Glasgow JF, Thomas PS. Rachitic respiratory distress in small preterm infants. *Arch Dis Child* 1977;**52**:268–73.

184. Pfeffer PE, Hawrylowicz CM. Vitamin D and lung disease. *Thorax* 2012;**67**:1018–20.

185. Marin L, Dufour ME, Tordet C, Nguyen M. 1,25(OH)2D3 stimulates phospholipid biosynthesis and surfactant release in fetal rat lung explants. *Biol Neonate* 1990;**57**:257–60.

186. Nguyen M, Trubert CL, Rizk-Rabin M, Rehan VK, Besancon F, Cayre YE, et al. 1,25-Dihydroxyvitamin D3 and fetal lung maturation: immunogold detection of VDR expression in pneumocytes type II cells and effect on fructose 1,6 bisphosphatase. *J Steroid Biochem Mol Biol* 2004;**89–90**:93–7.

187. Phokela SS, Peleg S, Moya FR, Alcorn JL. Regulation of human pulmonary surfactant protein gene expression by 1alpha,25-dihydroxyvitamin D3. *Am J Physiol Lung Cell Mol Physiol* 2005;**289**:L617–26.

188. Goldring ST, Griffiths CJ, Martineau AR, Robinson S, Yu C, Poulton S, et al. Prenatal Vitamin C supplementation and child respiratory health: a randomised controlled trial. *PLoS One* 2013;**8**:e66627.

189. Chan DK, Lim MS, Choo SH, Tan IK. Vitamin E status of infants at birth. *J Perinat Med* 1999;**27**:395–8.

190. Falciglia HS, Johnson JR, Sullivan J, Hall CF, Miller JD, Riechmann GC, et al. Role of antioxidant nutrients and lipid peroxidation in premature infants with respiratory distress syndrome and bronchopulmonary dysplasia. *Am J Perinatol* 2003;**20**:97–107.

191. Watts JL, Milner R, Zipursky A, Paes B, Ling E, Gill G, et al. Failure of supplementation with Vitamin E to prevent bronchopulmonary dysplasia in infants less than 1,500 g birth weight. *Eur Respir J* 1991;**4**:188–90.

192. Kim HY, Picciano MF, Wallig MA, Milner JA. The role of selenium nutrition in the development of neonatal rat lung. *Pediatr Res* 1991;**29**:440–5.

193. Daniels L, Gibson R, Simmer K. Selenium status of preterm infants: the effect of postnatal age and method of feeding. *Acta Paediatr* 1997;**86**:281–8.

194. Darlow BA, Inder TE, Graham PJ, Sluis KB, Malpas TJ, Taylor BJ, et al. The relationship of selenium status to respiratory outcome in the very low birth weight infant. *Pediatrics* 1995;**96**:314–9.

195. Darlow BA, Austin NC. Selenium supplementation to prevent short-term morbidity in preterm neonates. *Cochrane Database Syst Rev* 2003. CD003312.

196. Chin R, Haponik EF. Nutrition, respiratory function, and disease. In: Shils ME, Olson JA, Shike M, editors. *Modern Nutrition in Health and Disease*. Philadelphia: Lea and Febiger; 1994. pp. 1374–90.

197. Riley DJ, Thakker-Varia S. Effect of diet on lung structure, connective tissue metabolism and gene expression. *J Nutr* 1995;**125**: 1657S–60S.

198. Sahebjami H, Vassallo CL, Wirman JA. Lung mechanics and ultrastructure in prolonged starvation. *Am Rev Respir Dis* 1978;**117**: 77–83.

199. Sahebjami H, Vassallo CL. Effects of starvation and refeeding on lung mechanics and morphometry. *Am Rev Respir Dis* 1979;**119**:443–51.

200. Kerr JS, Riley DJ, Lanza-Jacoby S, Berg RA, Spilker HC, Yu SY, et al. Nutritional emphysema in the rat. Influence of protein depletion and impaired lung growth. *Am Rev Respir Dis* 1985;**131**:644–50.

201. Madia AM, Rozovski SJ, Kagan HM. Changes in lung lysyl oxidase activity in streptozotocin-diabetes and in starvation. *Biochim Biophys Acta* 1979;**585**:481–7.

202. Harkema JR, Mauderly JL, Gregory RE, Pickrell JA. A comparison of starvation and elastase models of emphysema in the rat. *Am Rev Respir Dis* 1984;**129**:584–91.

203. Massaro GD, Radaeva S, Clerch LB, Massaro D. Lung alveoli: endogenous programmed destruction and regeneration. *Am J Physiol Lung Cell Mol Physiol* 2002;**283**:L305–9.

204. Bishai JM, Mitzner W. Effect of severe calorie restriction on the lung in two strains of mice. *Am J Physiol Lung Cell Mol Physiol* 2008;**295**:L356–62.

205. Karlinsky JB, Goldstein RH, Ojserkis B, Snider GL. Lung mechanics and connective tissue levels in starvation-induced emphysema in hamsters. *Am J Phys* 1986;**251**:R282–8.

206. Lin Y, Lechner AJ. Surfactant content and type II cell development in fetal guinea pig lungs during prenatal starvation. *Pediatr Res* 1991;**29**:288–91.

207. Gail DB, Massaro GD, Massaro D. Influence of fasting on the lung. *J Appl Physiol Respir Environ Exerc Physiol* 1977;**42**:88–92.

208. Brown LA, Bliss AS, Longmore WJ. Effect of nutritional status on the lung surfactant system: food deprivation and caloric restriction. *Exp Lung Res* 1984;**6**:133–47.

209. Dobbins TA, Sullivan EA, Roberts CL, Simpson JM. Australian national birthweight percentiles by sex and gestational age, 1998-2007. *Med J Aust* 2012;**197**:291–4.

Chapter 20

Genetic Factors Involved in Susceptibility to Lung Disease

Kirsten C. Verhein*, Jennifer L. Nichols†, Zachary McCaw* and Steven R. Kleeberger*

*Laboratory of Respiratory Biology, National Institute of Environmental Health Sciences, National Institutes of Health; †Oak Ridge Institute for Science and Education, Office of Research and Development, U.S. Environmental Protection Agency, Research Triangle Park, NC, USA

INTRODUCTION

External factors that contribute to disease pathogenesis include physical factors (e.g., temperature), socioeconomic status, and exposure to environmental stimuli or triggers (e.g., allergens, molds, air pollutants, tobacco smoke). Epigenetic modifications are also likely important modifiers of disease susceptibility. Associations between environmental exposure and disease are supported by many epidemiological studies. For example, asthma morbidity has been associated with the potent oxidant ozone (O_3) as well as particulate matter (PM) in numerous industrial cities throughout the world.[1] A large body of literature also supports a number of indoor allergens as important environmental factors in allergy and asthma. These include house dust mite, molds associated with indoor dampness problems, and cockroach allergen.[2] However, not everyone responds similarly to these environmental stimuli, which implicates intrinsic factors as important inter-individual determinants of response.

Important internal factors include gender, age, diet, and genetic background. An individual's genetic make-up is recognized as a critical host factor in predisposition to environmental and occupational disease. The lists of candidate genes for these susceptibility loci are long and sometimes vary from one study to another. It is also generally agreed that multiple genes (each with modest effects) are likely to be operating through interactions with multiple environmental factors contributing to disease pathogenesis. Furthermore, the roles of genes and environments vary across populations, and without consideration of both simultaneously it is not possible to accurately identify the critical actions of either. An important genetic component has been well established for a number of lung diseases, including asthma, acute respiratory distress syndrome, bronchopulmonary dysplasia, COPD, idiopathic pulmonary fibrosis, and sarcoidosis. Excellent reviews for the genetic basis of disease susceptibility to each have been published recently,[3–9] and the reader is directed to these reviews for greater in-depth discussion of the contribution of genetic background. A major challenge in understanding how specific susceptibility genes and environmental exposures interact in pulmonary disease pathogenesis is to determine which environmental factors might be relevant to which genetic markers in the etiology of the disease. The interrelationships and relative contributions of the independent variables (genotype and exposure) to the response (disease risk and severity) can only be evaluated through multivariate analysis of large samples. Well-known examples of gene–environment interaction and disease pathogenesis include increased incidence of bladder and lung cancer, coronary atherosclerosis, beryllium disease, and rheumatoid arthritis.[10–15] A thorough discussion of the genetic factors involved in susceptibility to all lung diseases is beyond the limitations of a single book chapter. We have accordingly narrowed this topic to the genetic factors that may contribute to environmental and occupational lung diseases as follows. Initially, we present a brief introduction to experimental approaches used to identify lung disease genes, and convey some of the challenges that must be overcome in study design and application. Next, studies are presented that have identified a genetic basis for susceptibility to environmental stimuli, acute lung injury, and infectious diseases in animal models and human subjects. We then discuss the potential role of genetic susceptibility to occupational lung diseases, and candidate susceptibility genes are highlighted. Finally, we briefly present evidence for interaction between nutrition and genetic background in disease susceptibility.

RESEARCH STRATEGIES EMPLOYED TO IDENTIFY CANDIDATE DISEASE SUSCEPTIBILITY GENES

Genetics

Before describing the process of identifying candidate disease susceptibility genes, it is necessary to define a few concepts. The first of these is a quantitative trait, which can

The Lung. http://dx.doi.org/10.1016/B978-0-12-799941-8.00020-1

be considered as a phenotype that varies in a quantitative manner when measured among different individuals. The variable phenotype or expression is due to a combination of genetic and environmental factors, as well as stochastic variation and is often controlled by the cumulative action of alleles at multiple loci. While many single gene (Mendelian) disorders have been identified, most diseases are quantitative or complex. A quantitative trait locus (QTL) is a region on a chromosome that has been identified by linkage analyses or association studies (see following text) to contain a gene or genes that contribute to the phenotype of interest. Each QTL may include dozens of genes; therefore, a major task facing an investigator is to reduce the size of a QTL to a size (ideally <1000 kb) that is more "manageable" to use additional tools that lead to identification of a candidate gene (or genes) that determines the phenotype of interest. The combined contribution of each locus (QTL) with environmental influences (stimuli) determines an individual's response phenotype. In inbred strains of mice and rats, the traditional means to identify disease susceptibility genes (or QTLs) begins with identification of strains that are differentially responsive to an environmental stimulus or development of a disease phenotype. Next, segregant backcross and/or intercross (F_2) cohorts are derived and phenotyped for the disease endpoint under investigation. Each phenotyped mouse is then genotyped for single nucleotide polymorphisms (SNPs) across the entire genome. The recent availability of very high density SNP maps across the genome of many inbred strains of mice has increased the resolution of identified QTLs. SNP chips for mouse genotyping [e.g., mouse universal genotyping arrays (MUGA)] have also increased the efficiency and ability to identify chromosomal regions that determine phenotype.[16] Linkage mapping software can then identify regions of the genome where the phenotype of interest associates with the SNP genotypes with a p-value less than a threshold that accounts for multiple testing for all of the SNPs. High-density SNP databases have likewise facilitated the development of haplotype association mapping (HAM) or genome-wide association studies (GWAS). In contrast to the traditional QTL analysis that uses two differentially responsive strains of mice and their progeny, multiple inbred strains (generally greater than 30) are phenotyped for HAM analyses to identify genomic associations. This approach thus takes advantage of a greater amount of genetic diversity across the mouse genome and more closely resembles human genetic diversity. Recent reviews of HAM procedures discuss the advantages and computational challenges for these procedures.[17] Additional recently developed mapping tools, including the diversity outcross panel and collaborative cross strains, have promise to more rapidly identify candidate disease genes.[17]

After significant QTLs have been identified, the challenge then becomes to identify and validate the gene or genes within the QTL that determine the differential phenotype. Validation tools include other animal models, such as nematode (*C. elegans*) and zebrafish (*D. rerio*), cell-based systems to manipulate gene candidates, gene expression analyses, and in vivo tissue-specific gene deletion and/or overexpression. Congenic mouse lines have also been used to restrict or narrow QTLs to exclude chromosomal regions that do not contribute to the phenotype and thus reduce the number of gene candidates to pursue. Generally, a combination of these approaches is necessary to identify and functionally validate the susceptibility gene.

One of the unifying concepts for genetic mapping studies is the relationship of homologous loci from human and mouse. That is, highly significant homologies in gene order and chromosomal structure have been maintained since the divergence of the human and mouse. Therefore, identification of the chromosomal location of a susceptibility gene in the mouse provides the basis for potentially localizing a homologous gene in the human.

Linkage mapping, GWAS, and association studies are also applicable to human populations. As with mouse modeling, identification of genes responsible for complex diseases has benefited greatly over the last decade from advances in technology, large genome projects, and improved study design and data analysis.[18,19] In particular, the Human Genome Project, 1000 genomes, HapMap, and ENCODE have provided investigators with tremendous amounts of information on human genetic diversity and gene regulation. Development of genome-wide SNP typing arrays, as well as exome and whole genome sequencing, across increasingly large populations have facilitated GWAS and hundreds have been performed for many diseases and physiological traits (for a description of all GWAS with more than 100,000 SNP, see http://www.genome.gov/gwastudies/). These studies have yielded many interesting associations and candidate genes that have potential to improve our understanding of genetic contributions to lung diseases. However, loci identified in GWAS typically explain only a small proportion of the total variance attributable to genetic background. The reasons for this are numerous, but the primary reason is that, for most complex traits, no single locus confers a high degree of risk. Moreover, cumulative lifetime environmental exposures vary from one study population to the next, and the complete network of gene by environment interactions (and epigenetic contributions) remains only partially mapped.[20]

Single gene association studies remain a useful means to investigate the contributions of specific candidate genes to disease and other responses to environmental stimuli. The principle underlying the association of genetic polymorphisms not directly involved in disease pathogenesis is that of linkage disequilibrium, which arises from the co-inheritance of alleles at loci that are in close physical proximity on an individual chromosome. This approach requires a dense map of chromosomal markers because the chromosomal regions that co-segregate with a disease across different families or populations may be very small. The emergence of a high resolution SNP map has facilitated this approach for evaluating susceptibility loci. Study designs with individual functional/non-functional SNPs, or multiple SNPs in a small chromosomal region (haplotype analyses), have provided researchers with the means to dissect genetic contributions to complex human diseases. The advantages and limitations of single gene association studies have been discussed thoroughly elsewhere.[21,22] It is important to note that employing association studies alone could implicate certain genes in the expressed phenotype of interest, but other important loci that determine a quantitative trait, as well as the interaction between them, may be missed. When practical, the most productive strategy to understand complex diseases will likely be a combination of the GWAS and association approaches.

Genetical Genomics

While genetic approaches have been instrumental in identifying loci or genes that are associated with diseases and related phenotypes, challenges exist in understanding the role or functionality of these genetic elements. In addition to the vast number of genes across loci or even within a single locus, the biological plausibility of genes is often unknown. As described previously, it is rare that an identified gene solely or directly causes a disease/phenotype, and the link between the two can arise from several different mechanisms. For example, genetic variants (SNPs) can contribute to altered gene or protein expression as well as post-translational modifications. Moreover, regulatory processes involving epigenetics or microRNA can be perturbed. Thus, as diagramed in Figure 1 the association between QTLs (or candidate susceptibility genes) and diseases/phenotypes is complex. Research in the last decade has focused on

developing new approaches to further probe these relationships, including genomics, proteomics, and metabolomics.

The field of genomics has expanded rapidly with advances in technology. Beginning with sequencing projects to map organismal genomes, genomics has more recently expanded to include gene expression profiling or transcriptomics. Typically this involves using a microarray-based platform to quantify RNA transcripts isolated from different biological samples that will be compared. Identifying transcripts that are differentially expressed across samples allows for interpretation of molecular events that underlie disease or injury. Numerous bioinformatic and computation tools are available to assist researchers in this effort (Table 1), including the DAVID (Database for Annotation, Visualization, and Integrated Discovery), Bioinformatics Resource (National Institute of Health), and Ingenuity Pathway Analysis® (Ingenuity Systems, Inc.) that allow individuals to upload gene expression data to analyze and visualize interactions between genes and gene products within the data set or related to the data set. Additionally, gene ontology, functional annotation, canonical pathway and transcription factor analysis can be performed. Overall, our increased understanding of how gene expression is altered after an exposure or during disease has substantially contributed to the elucidation of mechanisms and biomarkers that are critical to preventative and therapeutic medicine.

Genetics and genomics approaches have been introduced and discussed independently to this point. Integrating these approaches provides an opportunity to understand the genetic basis of gene expression. How does genetic variation contribute to differential gene expression across individuals? How does the relationship between genetics and genomics inform susceptibility to exposure or disease? In this integrated approach, we treat the gene expression data (usually limited to differentially expressed transcripts) as

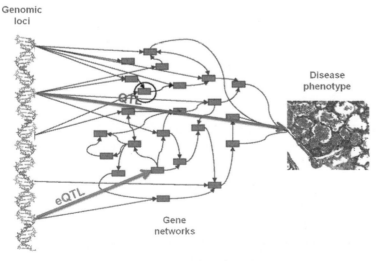

Genomic loci

Disease phenotype

Gene networks

FIGURE 1 Integration of genetic and genomic approaches. eQTL, expression quantitative trait locus. *Taken from Nichols 2012,*[133] *with permission.*

TABLE 1 Publicly available tools for mapping expression quantitative trait loci (eQTL) and visualizing/interpreting genetic and genomic data

Genetic and Genomic Tools for Data Analysis and Visualization			
Application	Uses	Features	Reference
FastMap	QTL	Computationally efficient method for eQTL mapping with permutation-based significance testing	134
Matrix eQTL	QTL	Computationally efficient method for eQTL mapping via linear models, allows for correlated errors	135
Merlin	QTL	Pedigree analysis for dense genetic maps	136
Plink	QTL	Genome-wide association analysis, allows for stratification and genotyping error	137
Qxpak.5	QTL	Method for QTL mapping via mixed models, which can correct for population structure	138
R/qtl	QTL	Method for QTL mapping in crosses, imputes missing genotype data	139
SNPster	QTL	Method for haplotype association mapping with bootstrap-based significance testing	140
GeneNetwork/ WebQTL	QTL, phenotype analysis	Method for reverse engineering genetic networks from longitudinal gene expression data	141
IPA®	Functional analysis	Method for characterizing the functional interactions, regulatory factors, biological processes of a gene set	http://www.ingenuity.com/
DAVID	Functional analysis	Method for highlighting the GO annotations and KEGG metabolic pathways enriched in a gene set	142

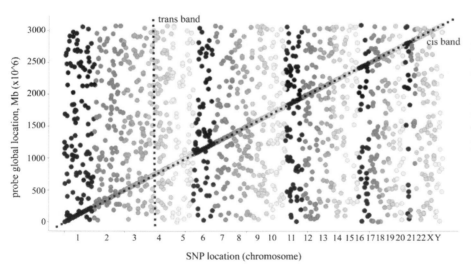

FIGURE 2 Example of an expression QTL (eQTL) plot. Each point on the plot represents a significant association between expression of a transcript and a single nucleotide polymorphism (SNP). SNP location is indicated by shaded chromosomes along the x-axis, and probe genomic location (location of expressed gene) along the y-axis. The cis-eQTLs fall along the dotted diagonal line, and trans-eQTLs are identified by vertical bands.

a quantitative phenotype and perform GWA mapping with known SNPs in the genome. This method is nearly identical to the QTL analysis described previously; however, the number of phenotypes (differentially expressed transcripts) can exceed 10,000, which dramatically increases the likelihood of false positives. Recently, publicly available applications such as P-Link and FastMap have been developed that can feasibly handle data sets of this size and produce an expression QTL (eQTL) plot as shown in Figure 2. Two different types of associations should be noted: cis-eQTLs and trans-eQTLs (Figure 2). Cis-eQTLs represent associations between a transcript's expression level and local haplotypes (or genotype). These appear on an eQTL plot as a diagonal line (y = x). Trans-eQTLs represent associations between a transcript's expression level and distant genomic loci. It is most logical to consider these associations in the context of trans-bands that appear as vertical lines on the eQTL plot, where a haplotype (SNP) is associated with a large number of transcript levels. Ideally, this type of analysis would lead to identification of a "master" regulator, and associated transcripts would have similar gene ontologies. However, thus far trans-band analysis has proved to be difficult. As technology and science continue to rapidly expand, it is anticipated that complex and integrative strategies such as genetical genomics will greatly enhance our understanding of susceptibility to disease.

GENETIC SUSCEPTIBILTY TO ENVIRONMENTAL STIMULI

Genetic and Genomic Determinants of Susceptibility to O_3-Induced Lung Inflammation and Injury

Ozone exposure induces multiple pulmonary and extrapulmonary responses in humans and animal models. Ozone elicits airway inflammation, airway hyperreactivity (AHR), and epithelial damage of the airways, as well as altered ventilation and decrements in pulmonary function. Studies have also shown that ozone can either suppress or enhance immune responsiveness, depending upon the exposure regimen, genetic background, pre-existing diseases, etc. Just as inter-individual variation in O_3 response has been observed,[23] significant inter-strain variation in the magnitude of these responses has also been demonstrated in rats and mice, and has thus provided strong evidence of a genetic component to O_3 responsiveness.

An initial effort to better understand genetic susceptibility to O_3-induced lung injury began with a study that aimed to identify the chromosomal regions that associated with differential lung injury in susceptible C57BL/6J (B6) and resistant C3H/HeJ (C3) strains. Using pulmonary inflammation as measured by polymorphonuclear leukocytes (PMNs) in bronchoalveolar lavage fluid, genome-wide linkage analysis was carried out with individual intercross animals derived from B6 and C3 progenitors (B6C3F$_2$) and susceptibility loci (QTLs) were identified on chromosomes 17 and 11. Candidate genes in the chromosome 17 QTL include the pro-inflammatory cytokine *Tnf* (tumor necrosis factor-α, TNF-α), and a functional role for *Tnf* was demonstrated as O_3-susceptible mice treated with an antibody to TNF-α had marked reduction in pulmonary inflammation and epithelial damage following O_3 exposure.[24] This observation was further supported in a study that exposed Tnf receptor (*TNFR*) deficient mice to O_3 and found the same reduction in pulmonary inflammation and epithelial injury, as well as reduced airway hyperresponsiveness.[25] Yang et al. subsequently investigated *TNF* haplotypes in a human cohort and reported an association between changes in lung function (FEV$_1$) and a SNP at the *TNF*-308 locus (G/G vs. G/A or

A/A).[26] Additional candidate genes in the chromosome 17 QTL have been identified and validated, including MHC class II genes and lymphotoxin alpha (*Lta*), and indicate the complexity of this QTL as a susceptibility locus.[27]

Further elucidation of genetic factors underlying differential response to O_3-induced pulmonary injury, specifically airway permeability, was achieved in a study of genome-wide linkage analysis in a set of recombinant inbred (RI) strains derived from B6 and C3 progenitors.[28]QTLs were identified on chromosomes 3, 4, and 11. A candidate gene in the chromosome 4 QTL is Toll-like receptor 4 (*Tlr4*), which is known to have a role in innate immunity and endotoxin susceptibility.[29,30] To determine the role of *Tlr4* in airway permeability C3H/HeJ and C3H/HeOuJ mice were phenotyped, as they differ at a non-synonymous substitution in the intracellular domain of *Tlr4* that impairs functionality in C3H/HeJ mice and has been linked to resistance to endotoxin-induced injury. Following ambient-relevant O_3 exposure, C3H/HeJ mice had reduced protein concentrations in bronchoalveolar lavage fluid, indicative of altered airway permeability.[28] Although subsequent studies using *Tlr4* deficient C57BL/6J mice did support a role for *Tlr4* in O_3-induced AHR, the observation regarding airway permeability was not replicated,[31,32] likely due to differences in background strain. An additional observation in this study was that loss of *Myd88*, which encodes the Tlr2 and Tlr4 adapter protein MyD88, resulted in decreased inflammation, airway permeability, and AHR.[31] Similar results were described in mice deficient in nuclear factor-kappa B1 (*Nfkb1*) and c-Jun-NH(2) terminal kinase 1 (*Jnk1*), both of which are downstream targets of Tlr4 signaling.[33] While it is likely that other genetic factors are involved in O_3-induced pulmonary injury, it can be concluded from this evidence that Tlr4 is involved and that genetic aberrations within this pathway may render one susceptible to the adverse pulmonary effects of ambient O_3 exposure.

Prows et al.[34] performed a linkage analysis of susceptibility to death induced by high O_3 concentrations. In this study, survival time was measured across an RI panel derived from A/J and C57BL/6J progenitors, which were among the most susceptible and resistant inbred strains to O_3-related mortality, respectively. Three QTLs and biologically plausible candidate genes were identified: chromosome 11 (nitric-oxide synthase, *Nos2*; myeloperoxidase, *Mpo*), chromosome 13 (glutathione peroxidase, *Gpx*), and chromosome 17 (xanthine dehydrogenase, *Xdh*). A subsequent study by Prows et al.[35] measured survival time following high O_3 exposure in an F$_2$ progeny generated from A/J and C57BL/6J mice, and confirmed the association between the chromosome 11 and 17 QTLs and O_3-induced mortality. Interestingly, these QTLs overlapped with those identified in the B6C3 F$_2$ study by Kleeberger et al.[24]

In addition to these genetic studies conducted in mice, there is also evidence for genetic determinants of susceptibility to O_3 in humans. In the Integrated Science Assessment for O_3, the US Environmental Protection Agency concluded that there is substantial evidence suggesting genetic background as a risk factor for O_3-induced health effects.[23] Epidemiologic studies have demonstrated inter-individual differences in lung function across populations and have identified effect modifications by glutathione S-transferase (GSTM1 and GSTP1) variant genotypes. Controlled human exposures investigating variants in GSTM1 report mixed results; however, this may be due to exclusion of the most sensitive populations (severe asthmatics and children) in these studies for ethical concerns. Currently, the candidate susceptibility genes identified in animal studies are associated with pulmonary inflammation and epithelial injury, which are not easily measured in human subjects, but research efforts continue to examine the role of genetic background in O_3-induced acute lung injury in order to identify the most susceptible populations.

Genetic and Genomic Determinants of Susceptibility to Particle-Induced Lung Inflammation and Injury

Similar studies have been conducted to understand genetic susceptibility to particle-induced lung inflammation and injury. As exposure to particulate matter has increased throughout the industrialized world, a large body of epidemiologic evidence has demonstrated that environmental and occupational particle inhalation is associated with acute and chronic respiratory and non-respiratory effects.[36,37] Sub-populations at increased risk of particle-induced health effects include the aged (>65 years of age), children, those with cardiopulmonary disease such as chronic heart disease, chronic obstructive pulmonary disease, and asthma.[38] Several published studies address genetic factors conferring susceptibility to lung inflammation and injury from exposure to particles, including acid sulfate-coated particles, nickel sulfate particles, and zinc oxide.

Ohtsuka et al. studied inter-strain variation in lung responses to acid sulfate-coated particles in inbred strains of mice.[39] The particles used in this study were generated by combining carbon black with sulfur dioxide in a high humidity environment, which mimics one component of aerosols found in many metropolitan areas in the eastern United States.[40] Significant differences between strains were found in Fc-receptor-mediated phagocytosis in alveolar macrophages collected 1–7 days post-exposure.[39] Linkage analysis in B6C3F_2 progeny identified a susceptibility QTL on chromosome 17 and a suggestive QTL on chromosome 11.[41] Interestingly, these QTLs nearly overlap those identified in O_3 studies, suggesting that similar genetic mechanisms may control pulmonary responses to these pollutants. Additional studies consistently demonstrate *Tnf* (chromosome 17) expression is increased in response to

both particle and O_3 exposure.[42,43] While studies in *Tnf* or *Tnfr* deficient mice do not consistently show reduced pulmonary injury following exposure,[25,33,44] there is consistent evidence that Tnfα signaling plays a critical role in mediating pulmonary inflammation and injury. Similarly, *Nos2* (chromosome 11) expression can be induced upon O_3 or particle exposure,[43,45] and *Nos2* deficient mice sustain significantly greater injury upon exposure to either pollutant when compared with wild-type mice.[42,46,47]

Genetic factors involved in differential response to nickel sulfate particles have also been studied. Mean survival time was measured in a panel of inbred mice exposed to nickel sulfate, which is known to induce respiratory morbidities in occupational and environmental settings.[48] A/J mice were the most sensitive to nickel sulfate while C57BL/6J mice survived nearly twice as long.[48] Linkage analysis in backcross mice from A/J and C57BL/6J strains identified QTLs on chromosomes 1, 6, and 12 and candidate genes of interest included *Sftpb* and *Tgfα* on chromosome 6 and *Mt1* on chromosome 8. Gene expression data were also evaluated in A/J and C57BL/6J mice following 3–48 hr exposure, and transcript levels of *Sftpb* and *Mt1* were differentially expressed between strains (*Tgfα* was not measured). *Sftpb* and *Mt1* were proposed as candidate susceptibility genes for particle-induced lung injury, though it should be noted that changes in genotype need not result in gene expression changes to be of biological significance. Follow up studies showed *Mt1* overexpressing mice were protected against nickel sulfate-induced lung injury while *Mt1* knockout mice were more susceptible.[49] Other studies have confirmed expression changes in these genes with particle exposure, but few have investigated targeted gene deletions or insertions. One study demonstrated that overexpression of *Tgfα* resulted in gene expression changes related to immune and inflammatory processes, cytoskeleton organization, and biogenesis following PM exposure; however, no phenotypic changes indicating lung injury differed from control animals.[50]

A third investigation of genetic factors underlying pulmonary response to particle exposure measured pulmonary tolerance in a panel of inbred strains and in a BALB/cByJ and DBA/2J F_2 cohort exposed to zinc oxide (ZnO). Pulmonary tolerance is defined as the lung's ability to withstand the detrimental effects of an exposure and is inversely related to lung injury. BALB/cByJ mice demonstrated tolerance to repeated exposures as measured by BAL protein and neutrophils while DBA/2J mice were the least tolerant.[51] Using the F_2 cohort, investigators identified QTLs and candidate susceptibility genes located on chromosomes 1 (*Tlr5*, *Tgfb2*, *Traf5*, *Il10*, *Cxcr4*, and *Ccl20*), 4 (*Tnfrsf1* and *Tnfrsf4*), and 5 (*Il6* and chemokine ligands).[52] Additionally, MOLF/EiJ mice that have a mutant *Tlr5* gene were found to have significantly increased lung injury following repeated exposure to ZnO compared to strains without the mutation.

While these studies suggest genetic background has an important role in pulmonary responses to inhaled particles, the majority of published epidemiologic studies that have identified PM-related effect modification for genetic variants are limited to cardiovascular endpoints. While there is inter-individual variability in the health effects associated with particle exposure, the size and composition of PM is spatially varied. Future research is needed to clarify the complex gene–environment relationships with regard to genetics and PM.

GENETIC SUSCEPTIBLITY TO ACUTE LUNG INJURY

Due to its interface with the environment, the lung is a major target organ for injury by exogenous oxidants such as environmental pollutants, cigarette smoke, drugs, chemotherapeutic agents, and hyperoxia,[53] as well as by endogenous reactive oxygen species (ROS) generated by inflammatory cells. In addition, many pulmonary diseases (e.g., adult respiratory distress syndrome, emphysema) require supplemental oxygen therapy to maintain lung function that further increases the oxidant burden of the lung. It is believed that the damaging effects of oxygen are mediated by superoxide, hydroxyl radicals, and H_2O_2, products formed by the incomplete reduction of oxygen. There is ample evidence showing increased cellular formation of these ROS during hyperoxia and the deleterious effects on cellular constituents such as nucleic acid, proteins, and lipids. Because of the potential impact oxidants have on lung function, identification of factors that influence individual susceptibility remains an important issue. An understanding of susceptibility factors could lead to better intervention strategies and, potentially, a means to identify individuals at risk for the development of oxidative stress injury.

We developed a genetic model of differential susceptibility to lung injury induced by exposure to hyperoxia (>95% O_2). In this model, susceptible and resistant phenotypes are easily distinguished on the basis of lung permeability and cytotoxic responses following 72-hr hyperoxia exposure in B6 and C3 strains, respectively. Genome-wide linkage analysis of intercross (F_2) and RI cohorts identified significant and suggestive QTLs on chromosomes 2 (hyperoxia susceptibility locus 1 [*Hsl1*]) and 3 (*Hsl2*), respectively.[54] Comparative mapping of *Hsl1* identified a strong candidate gene, *Nfe2l2* (nuclear factor, erythroid derived 2, like 2 or *Nrf2*). The corresponding protein NRF2 is a transcription factor that regulates expression of antioxidant and phase 2 genes. Inter-strain variation in lung *Nrf2* messenger RNA expression and a T→A substitution in the B6 *Nrf2* promoter that co-segregated with susceptibility phenotypes in F_2 animals supported *Nrf2* as a candidate gene in this model.[54]

Because NRF2 is a central regulator of several important antioxidant and phase 2 detoxifying enzymes, we hypothesized that deletion of the gene for this transcription factor would enhance susceptibility to oxidant-induced lung injury. As predicted, pulmonary response phenotypes (hyperpermeability, macrophage inflammation, and epithelial injury) were significantly elevated in *Nrf2^-/-* mice compared to *Nrf2^+/+* mice after hyperoxia.[55] Consistent with a role for NRF2 in susceptibility to oxidant-induced lung injury, hyperoxia-induced mRNA levels of *Nqo1*, *Gst-Ya* and *Yc*, *Ugt*, *Gpx2*, and *Ho-1* were significantly lower in *Nrf2^-/-* mice compared to *Nrf2^+/+* mice.[56] These experiments therefore strongly suggested that *Nrf2* has a significant protective role against hyperoxic lung injury in mice and provided initial confirmation of a candidate gene identified by linkage analyses.

We next investigated the role of *NRF2* in humans. We identified a novel promoter SNP in *NRF2* that diminished promoter activity in vitro and significantly correlated with enhanced risk for acute lung injury after trauma in a human cohort.[57] This was the first translational investigation to test a gene found by positional cloning in mice with susceptibility to acute lung injury in humans and highlight the usefulness of inbred mouse models for identifying candidate human disease susceptibility genes.

Cancer treatments, including irradiation and chemotherapy, may induce lung injury and fibrosis, leading to significant morbidity and mortality in a subset of cancer patients.[58] Haston et al. initiated a series of important investigations to identify the genes that confer susceptibility to bleomycin (a chemotherapeutic agent) and radiation in inbred mice. In a cohort derived from fibrosis-resistant C3Hf/Kam and fibrosis-susceptible B6 mice they found that heritability of bleomycin-induced fibrosis is approximately 53% in males and 54% in females.[59] They further demonstrated a role for the MHC complex in this model using various MHC congenic mouse strains. A genome-wide scan for susceptibility loci identified a QTL on chromosome 17 that was highly significant in males and females, and a second QTL on chromosome 11 that was significant in males only.[58] The gene for bleomycin hydrolase is located within the chromosome 11 QTL, and functional studies of bleomycin hydrolase activity were consistent with this being a candidate gene in this model. Follow up studies replicated the significant QTLs and used positional cloning to further characterize candidate genes. Specifically, *H2-Ea* and *Trim16* were identified as susceptibility genes for bleomycin-induced pulmonary fibrosis.[60–62]

In addition to the differential severity of bleomycin-induced pulmonary fibrosis, the C3Hf/Kam and B6 mice are differentially susceptible to radiation-induced fibrosis.[63] A genome scan using an F_2 cohort derived from these strains identified significant fibrosis susceptibility QTLs on chromosomes 1, 6, and 17, and a suggestive QTL on chromosome 18.[64] The chromosome 17 QTL was confirmed using a battery of mice congenic for the MHC region, which further supported this chromosomal region as an important genetic determinant of lung injury and fibrosis sequelae in inbred mice.

TABLE 2 Examples of linkage studies that have identified quantitative trait loci (QTL) for susceptibility to acute lung injury and pollutant-induced inflammation and immune dysfunction

	Phenotype	Chromosomal Location(s) of QTL	Candidate Gene(s)	Reference
Pollutant/Occupational				
Ozone	Inflammation (PMNs)	17	Tnf	24
		11	ScyX; Nos2	
Ozone	Hyperpermeability	4	Tlr4	28
		11	ScyX; Nos2	
Sulfate-associated particles	Immune dysfunction	17	Tnf	41
		11	ScyX; Nos2	
Zinc oxide	Inflammation(tolerance)	1	Tlr5	52
Acute Lung Injury				
Hyperoxia	Inflammation/ hyperpermeability	2	Nrf2	54
		3	Nfkb1	
Ozone	Death	11	ScyX; Nos2	34
		13	Gpx	
		17	Tnf; Xdh	
Nickel sulfate aerosol	Death	1	Fn1	143
		6	Tbxas1	
		12	Tpo	
Bleomycin	Pulmonary fibrosis	11	Bh	58
		17	H2	
Radiation	Pulmonary fibrosis	1	Fasl (Tnfsf6)	64
		6	Tnfrsf1a	
		17	H2	

Bh, bleomycin hydrolase; *Fn1*, fibronectin-1; *Gpx*, glutathione peroxidase; *H2*, major histocompatibility complex; *Nos2*, nitric oxide synthase 2; PMN, polymorphonuclear leukocyte; *Scy*, small inducible cytokines; *Tbxas1*, thromboxane A synthase 1; *Tlr*, Toll-like receptor; *TNF*, tumor necrosis factor; Fasl (*Tnfsf6*), fas ligand; *Tnfrsf1a*, TNF receptor, R1; *Tpo*, thyroid peroxidase; *Xdh*, xanthine dehydrogenase.

Chromosomal positions of all susceptibility QTLs discussed above, as well as key candidate genes, are summarized in Table 2. Although many chromosomes harbor susceptibility loci for pollution effects and acute lung injury, it is readily apparent that QTLs on chromosomes 11 and 17 account for genetic variance in multiple susceptibility models. A cursory examination of the mouse genome informatics database (http://www.jax.org) indicates that these two regions of the mouse genome (and by comparative mapping, the corresponding regions in the human genome) are enriched with candidate genes for lung injury, inflammation, and immune function. While it is tempting to speculate that polymorphisms in one or two key genes in both chromosomal regions may account for susceptibility to multiple environmental stimuli, the presence of multiple candidates indicates that much further work is necessary to confirm this hypothesis.

GENETIC SUSCEPTIBILITY TO INFECTION

Significant effort has been undertaken to elucidate the genetic determinants of susceptibility to lung infection. Bacteria and viruses activate host defense pathways that facilitate pathogen clearance. These responses recruit inflammatory cells to the lung via cytokine and chemokine signaling. Therefore, many of the candidate gene studies have focused on immune modulators and host defense genes rather than large scale GWAS.

Studies of genes or polymorphisms that contribute to infectious disease susceptibility in humans are complicated by many factors, including small population size. When comparing host populations, differences at multiple loci can also confound causality, as can genetic variability in the pathogens themselves. For example, the *M. tuberculosis* genome is variable across geographic regions,[65] and there are multiple strains and sub-strains of influenza viruses.[66] Further complications in identifying host genetic determinants of susceptibility include prior immunity through either previous infection or vaccination. Despite these many challenges, progress has been made in identifying host genetic susceptibility to lung infection. A thorough review of human genetic susceptibility for many infectious diseases can be found elsewhere,[67] and a few examples from both bacterial and viral infections are described in following text.

Genetic Determinants of Susceptibility to Bacterial Infection

The most common identifiable cause of pneumonia is infection with *Streptococcus pneumoniae*. Infection with *S. pneumoniae* triggers a cascade of inflammatory processes starting with recognition of the bacteria by pattern recognition receptors, followed by activation of NF-κB and production of cytokines and chemokines. Excessive inflammatory responses to the pathogen, or a deficiency in anti-inflammatory pathways, are proposed mechanisms for severe disease. Even with the advent of antibiotics, mortality remains high, ranging from 5 to 35%,[68] and suggests that underlying genetic factors may contribute to disease susceptibility. A genetic component to severe pneumonia infection is also supported by differential susceptibility across inbred mouse strains, with some being more resistant to infection than others.[69]

Association studies have identified a number of SNPs in candidate immune genes that confer either protection against or susceptibility to pneumococcal disease. Protective polymorphisms include a heterozygous variant in *TIRAP*, an adaptor molecule responsible for downstream signaling after activation of Toll-like receptors.[70] Additional protective SNPs in *NFKBIA* and *NFKBIE*, members of the NF-κB pathway, were identified in a European population.[71]

Mutations that lead to increased susceptibility to *S. pneumoniae* have been found in mannose-binding lectin (*MBL*) in adults and infants.[72,73] However, in another population, polymorphisms in *MBL* were not associated with risk for pneumonia.[74] As mentioned earlier, many candidate gene studies have not been replicated across study populations, and discrepancies may arise from differences in genotype frequency, genotyping errors, the phenotypes used for analysis, and failure to consider SNPs in linkage disequilibrium with the candidate SNP.

Tuberculosis (TB) is a disease caused by *Mycobacterium tuberculosis* and constitutes one of the most intensively studied bacterial lung infections. Globally, TB remains a major cause of morbidity and mortality, with an estimated one-third of the world's population infected with *M. tuberculosis*.[75] Host genetics is thought to play a role in susceptibility to TB infection and activation. Monozygotic twins have a 2.5-fold higher concordance for TB compared to dizygotic twins,[76,77] and those of African descent have a greater risk of infection than those of European descent, based on tuberculin skin tests.[78] A combined genome-wide association analysis of populations from Ghana and Gambia identified a SNP (rs4331426) in a gene-poor region of chromosome 18q11.2 that significantly associated with disease.[79] Thye et al. also analyzed previously identified candidate gene SNPs in their combined GWAS and found that none reached statistical significance. Further work is needed to determine the functional role of the significant SNP and whether or not it replicates in additional populations.

Genetic Determinants of Susceptibility to Viral Infection

Studies of genetic determinants of influenza susceptibility are complicated by genetic variations among influenza virus strains. Pandemics occur when general host immunity in a population cannot adequately recognize the influenza virus due to evolving viral mutations. Traditional risk factors for severe disease after seasonal influenza infection are extremes of age, and those that are immunocompromised. During pandemics, disease severity is not always determined by the traditional risk factors, as more severe disease occurs in otherwise healthy young adults.[80]

The 2009 H1N1 pandemic strain (pH1N1) has been extensively studied and a number of polymorphisms in host immune genes have been identified that correlate with severe disease. A SNP in *CD55*, a regulator of the complement cascade, associated with severe infection after pH1N1 infection in a Chinese population.[81] In a small Canadian population, a polymorphism that results in a deficiency of chemokine (C-C motif) receptor 5, *CCR5Δ32*, was overrepresented in Caucasians with severe H1N1 infection.[82] Biologic plausibility for *CCR5* was shown using *Ccr5* knockout mice that have increased mortality rates following infection with a mouse-adapted influenza A virus.[83] A SNP-mediated splice variant of a gene important for influenza replication, interferon-induced transmembrane protein 3 (*IFITM3*), was identified in patients hospitalized with influenza infection.[84]

Animal models of influenza infection may be useful for studying disease pathogenesis; however, Go et al. demonstrated infection of mice, non-human primates, and swine with the same influenza strain resulted in differential postinfection lung gene expression between species.[85] While many clinical phenotypes were the same among species, this study demonstrates that caution is necessary when extrapolating results from animal models to identify candidate susceptibility genes in human populations.

Respiratory syncytial virus (RSV) is the most common cause of hospitalizations due to lower respiratory tract infections in young children.[86] Nearly everyone is infected with RSV by 2 years of age and approximately 160,000 die each year from severe RSV infection worldwide (WHO, http://www.who.int/en/).[87] Known risk factors for severe RSV disease include prematurity, chronic lung disease, congenital heart disease, and immunodeficiency, although most infants that require hospitalization after infection with RSV were previously healthy. This has led to the search for genetic markers that identify infants at risk for severe RSV disease.

Genome-wide association studies in humans have not been done for RSV disease; however, much work has been done using animal models and candidate gene approaches. As with most infectious diseases, the focus for the candidate gene approach has been on immune genes. For a detailed review of candidate gene association studies for severe RSV disease in humans see the following reviews.[88,89] Janssen et al. did the most comprehensive analysis of SNPs in immune candidate genes for severe RSV disease in infants.[90] In a cohort of 470 children hospitalized for severe RSV bronchiolitis in the Netherlands, 384 SNPs in 220 candidate genes were analyzed. Significant associations were found with SNPs in innate immunity genes including: vitamin D receptor (*VDR*), jun proto-oncogene (*JUN*), interferon alpha 5 (*IFNA5*), and nitric oxide synthase 2, inducible (*NOS2*). Only the SNP in *VDR* was confirmed in a South African population.[91]

GENETIC SUSCEPTIBILITY TO OCCUPATIONAL LUNG DISEASE

Considerable evidence has also accumulated for genetic susceptibility to the detrimental pulmonary effects of occupational exposures. While genome-wide scans for susceptibility loci have not been attempted to date, a number of association studies have been done for candidate susceptibility genes in chronic beryllium disease (CBD), coal workers' pneumoconiosis, silicosis, and occupational asthma that have provided strong evidence for a genetic contribution to some occupational lung diseases (Table 3). Perhaps the most well-known early investigation was done by Richeldi et al.[92] who investigated the role of MCH class II genes in CBD, a lung disorder related to beryllium exposure and characterized by lung accumulation of beryllium-specific CD4+ T lymphocytes. Using a case-control design, these investigators found that 97% of the CBD cases in their study expressed the HLA-DPB1*0201-associated glutamic acid at residue 69, a position that has been associated with susceptibility to autoimmune disorders.[92] Subsequent studies have confirmed and expanded upon this observation,[93-95] leading

TABLE 3 Representative association studies of genetic susceptibility to occupational lung diseases

Disease	Candidate Gene	Reference
Chronic Beryllium Disease	*ACE*	144
	HLA-DPB1	92
Coal Workers' Pneumoconiosis	*TNF, LTA*	104
	TNF	102
	CCR5Δ32	105
Silcosis	*TNF*	97,98
	IκB-α	145
	HLA-DQB1	100
	IL1RA	98,99
Isocyanate-induced Asthma	*HLA-DQB1*	106,108,109
	GSTM1, GSTM3, GSTP1	110,111
Red Cedar-induced Asthma	*HLA-DRB1, DBQ1*	106,107

ACE, angiotensin converting enzyme; *CCR5*, chemokine (C-C motif) receptor 5; *GST*, glutathione S-transferase; *HLA*, human lymphocyte antigen; *IκB-α*, I kappa B component a; *IL1RA*, interleukin 1 receptor antagonist; *LTA*, lymphotoxin alpha; *TNF*, tumor necrosis factor alpha.

to the possibility that residue 69 of HLA-DPB1 may be used as a diagnostic indicator of risk for CBD.

Silicosis is a disease characterized by fibrosing nodular lesions that may eventually develop into progressive massive fibrosis[96] and is prevalent in individuals such as coal and gold miners who are chronically exposed to dusts and other irritants. TNF has been investigated, and a functional polymorphism in the *TNF* promoter (at the −308 position) was found to associate with disease severity in South African gold miners.[97] A SNP at −238 also significantly correlated with increased risk of silicosis in Chinese workers.[98] Additionally, a SNP in the regulatory element of the gene for an anti-inflammatory cytokine (interleukin 1 receptor antagonist [*IL1RA*]) associated with the incidence of silicosis in coal miners.[98,99] As demonstrated for a number of occupational lung diseases (Table 3), polymorphisms in the HLA class II alleles may also have an important role in disease pathogenesis involved with silica exposure.[100] The common link to all of these studies is inflammation, which suggests that those individuals with polymorphisms that promote inflammation and are exposed chronically to silica may be predisposed to the development of disease.

Pneumoconiosis is a pulmonary disease characterized by inflammation leading to fibrosis that is also found in coal miners and caused by inhalation of dust. The mechanisms of susceptibility to disease onset and progression are not clear, but a role for genetic background has been proposed.[101] Zhai et al.[102] found that the same *TNF* −308 promoter polymorphism that associated with silicosis (see earlier) was also associated with coal workers' pneumoconiosis, as compared with miners without pneumoconiosis and controls. This finding was confirmed in Japanese miners with nodular pneumoconiosis compared to controls.[103] Further support for a genetic component of susceptibility to pneumoconiosis was provided by the work of Nadif et al.[104] These investigators evaluated the role of the *TNF*−308 polymorphism and a functional polymorphism (*NcoI* RFLP) in the gene for the inflammatory cytokine lymphotoxin alpha (*LTA*) in a prospective cohort study in 253 coal miners with differential exposure to coal dust and cigarette smoke. They found an interaction between *TNF*−308 genotype and coal dust exposure that predicted erythrocyte GSH-Px activity (an intermediate response phenotype), with a significant association in those with high exposure, whereas no association was found among those with low exposure. Further, a significant association of pneumoconiosis prevalence with the *LTA NcoI* polymorphism was found in miners with low blood catalase activity, whereas no association was observed in those with high (a priori protective) catalase activity. In a separate study, Nadif et al. demonstrated French coal miners with the *CCR5Δ32* mutation (described earlier, with influenza susceptibility) had higher chest CT scores, which predicts future disease progression.[105] These studies thus provide support for interactions between genetic background and environmental exposure in the pathogenesis of coal workers' pneumoconiosis.

Occupational asthma has been described in some individuals exposed to isocyanates and sawmill byproducts. However, only a small proportion of those who are exposed to these agents develop occupational asthma, therefore suggesting that intrinsic factors such as genetic background may be important disease determinants. MCH class II genes have been associated with isocyanate- and western red cedar-induced asthma,[106-109] indicating that specific genetic immune mechanisms may influence the risk of developing these occupational diseases in exposed individuals. Studies also suggest that glutathione S-transferase genotypes may be important determinants of occupational asthma induced by exposure to isocyanates.[110,111]

CONTRIBUTION OF NUTRITION IN GENETIC SUSCEPTIBLITY TO LUNG DISEASE

As described earlier, reactive oxygen species (ROS) are important contributors to oxidative stress-induced lung disease and genetic variation in the genes that participate in metabolizing ROS and deploying dietary antioxidants may account for differential susceptibility to oxidative stress. Much of the work on the effect of diet on gene–environment

interactions has been in the context of asthma and COPD and a few key investigations are discussed in following text.

Antioxidants and Susceptibility to Lung Disease

A network of enzymatic and dietary antioxidants assists in alleviating oxidative damage sustained by the cell. Endogenous enzymes include superoxide dismutase (SOD), catalase, and glutathione peroxidase. Dysfunction of antioxidant enzymes is evident in lung diseases such as asthma and COPD. Bronchial epithelial cells from patients with asthma had diminished SOD activity,[112] and reduced SOD activity correlated with decreased FEV_1.[113] Decreased glutathione peroxidase activity correlated with reduced FEV_1 in patients with COPD,[114] while catalase and glutathione peroxidase activity were reduced in the erythrocytes of asthmatics.[115] In addition, polymorphisms in *SOD*, catalase, and *MPO* associated with decreased pulmonary function in the elderly after environmental exposure to phthalates.[116]

Because imbalance in the oxidative stress pathway contributes to lung disease, it has been proposed that dietary supplementation with antioxidants may attenuate lung damage. Asthmatic children with reduced glutathione S-transferase (GSTM1) activity were protected from ozone-induced decrements in lung function by supplementation with vitamins E and C.[117] Increased serum levels of vitamins E and C were associated with reduced risk of wheeze,[118] and serum antioxidants are positively correlated with FEV_1 in the general population.[119,120] Conversely, antioxidants were depleted in the epithelial lining fluid of patients with asthma.[121] Genetic variations in the transporters *SCARB1* and *SCL23A1* also modulate circulating concentrations of vitamins E and C.[122]

Activation of the nuclear vitamin D receptor modulates lymphocyte differentiation, and biases helper T cells toward a Th2 phenotype.[123] Vitamin D intake during pregnancy reduced the risk of wheezing symptoms in children,[124] and Japanese children receiving supplemental vitamin D were protected from seasonal influenza.[125] Among asthmatics, serum levels of vitamin D were negatively correlated with hospitalization odds and molecular disease phenotypes, including immunoglobulin and eosinophil levels in peripheral blood.[126] Patients with a polymorphism in the vitamin D binding protein that reduces serum levels of vitamin D were at increased risk for COPD.[127] As mentioned earlier, a SNP in the vitamin D receptor associated with severe RSV disease.[90,91]

Sulforaphane is an isothiocyanate derived from cruciferous vegetables that is notable for its anti-carcinogenic properties.[128] Sulforaphane inhibits the association of NFkB with DNA in macrophages[129] and promotes the activation of Nrf2 in endothelial cells.[130] Nrf2 induces transcription of enzymes regulated by the antioxidant response element (ARE). Among the targets of Nrf2 are enzymes that participate in GSH synthesis (*GSL*), oxidation (*GPX*), conjugation (*GST*), and reduction (*GR*). In contrast to wild-types, the induction of ARE-regulated genes by respiratory syncytial virus (RSV) was attenuated in *Nrf2* knockout mice, contributing to enhanced cellular damage and RSV disease susceptibility.[131] Sulforaphane treatment also attenuated RSV disease[131] and arsenic-induced lung injury by activating *Nrf2*.[132] Together, these studies demonstrate the potential for dietary-derived antioxidants to alleviate lung injury.

Although oxidative stress is implicated in the pathogenesis of lung disease, and diets rich in antioxidants are associated with improved lung function, clinical trials of vitamin supplementation have often failed to demonstrate benefits. Stratification by genotype may reveal that those sub-populations for which vitamin supplementation is effective carry polymorphisms in components of the antioxidant network. Likewise, the potential for diet to prevent disease in individuals with a genetic predisposition to oxidative stress should receive further consideration.

SUMMARY

As urbanization increases, pulmonary morbidity and mortality resulting from environmental and occupational exposures continue to be public health concerns. Thus identification of susceptible sub-populations is critically important to control the burden of pulmonary disease. Genetic background is among the numerous factors that contribute to inter-individual differences in susceptibility to noxious pulmonary exposures. Linkage and haplotype association analyses with inbred mice and human populations have led to the identification of candidate susceptibility genes for pulmonary responses to air pollutants and infectious agents. Animal modeling of acute lung injury and disease induced by hyperoxia, pulmonary irritants, chemotherapeutic agents (bleomycin, radiation), and infections has confirmed that genotype modifies disease susceptibility. Case-control studies and GWAS in human subjects have also demonstrated that genetic background is an important determinant of susceptibility to lung infection, occupational lung injury, and oxidative stress. An understanding of the biology of candidate genes will lead to an understanding of the genetic mechanisms of differential responses to pollutant exposures. Further, characterization of the polymorphisms in pollutant susceptibility genes may thus provide the means of identifying individuals who are genetically susceptible to disease pathogenesis and improve methods of risk assessment.

ACKNOWLEDGMENTS

This research and Dr. Kleeberger, Dr. Verhein, and Mr. McCaw were supported by the Intramural Research Program of the National Institute of Environmental Health Sciences (NIEHS), National Institutes of Health (NIH), Department of Health and Human Services. Dr. Nichols was supported by an appointment to the Research Participation Program at the National Center for Environmental Assessment in the Office of Research and Development at the U.S. Environmental Protection Agency, administered by the Oak Ridge Institute for Science and Education through an interagency agreement between the U.S. Department of Energy and the Environmental Protection Agency.

Disclaimer: This manuscript has been reviewed by the U.S. Environmental Protection Agency (EPA) and approved for publication. The views expressed in this manuscript are those of the authors and do not necessarily reflect the views and policies of the U.S. EPA.

REFERENCES

1. Kelly FJ, Fussell JC. Air pollution and airway disease. *Clin Exp Allergy* 2011;**41**:1059–71.
2. Ahluwalia SK, Matsui EC. The indoor environment and its effects on childhood asthma. *Curr Opin Allergy Clin Immunol* 2011;**11**:137–43.
3. Meyer NJ, Christie JD. Genetic heterogeneity and risk of acute respiratory distress syndrome. *Semin Respir Crit Care Med* 2013;**34**:459–74.
4. Kropski JA, Lawson WE, Young LR, Blackwell TS. Genetic studies provide clues on the pathogenesis of idiopathic pulmonary fibrosis. *Dis Model Mech* 2013;**6**:9–17.
5. Ober C, Yao TC. The genetics of asthma and allergic disease: a 21st century perspective. *Immunol Rev* 2011;**242**:10–30.
6. Cookson WO, Moffatt MF. Genetics of complex airway disease. *Proc Am Thorac Soc* 2011;**8**:149–53.
7. Grunewald J. Review: role of genetics in susceptibility and outcome of sarcoidosis. *Semin Respir Crit Care Med* 2010;**31**:380–9.
8. Lavoie PM, Dube MP. Genetics of bronchopulmonary dysplasia in the age of genomics. *Curr Opin Pediatr* 2010;**22**:134–8.
9. Gao L, Barnes KC. Recent advances in genetic predisposition to clinical acute lung injury. *Am J Physiol Lung Cell Mol Physiol* 2009;**296**:L713–25.
10. McInnes IB, Schett G. The pathogenesis of rheumatoid arthritis. *N Engl J Med* 2011;**365**:2205–19.
11. Zanobetti A, Baccarelli A, Schwartz J. Gene-air pollution interaction and cardiovascular disease: a review. *Prog Cardiovasc Dis* 2011;**53**:344–52.
12. Wacholder S, Chatterjee N, Caporaso N. Intermediacy and gene-environment interaction: the example of CHRNA5-A3 region, smoking, nicotine dependence, and lung cancer. *J Natl Cancer Inst* 2008;**100**:1488–91.
13. Horikawa Y, Gu J, Wu X. Genetic susceptibility to bladder cancer with an emphasis on gene-gene and gene-environmental interactions. *Curr Opin Urol* 2008;**18**:493–8.
14. Kreiss K, Day GA, Schuler CR. Beryllium: a modern industrial hazard. *Annu Rev Public Health* 2007;**28**:259–77.
15. Klareskog L, Padyukov L, Ronnelid J, Alfredsson L. Genes, environment and immunity in the development of rheumatoid arthritis. *Curr Opin Immunol* 2006;**18**:650–5.
16. Iancu OD, Oberbeck D, Darakjian P, Metten P, McWeeney S, Crabbe JC, et al. Selection for drinking in the dark alters brain gene coexpression networks. *Alcohol Clin Exp Res* 2013;**37**:1295–303.
17. Flint J, Eskin E. Genome-wide association studies in mice. *Nat Rev Genet* 2012;**13**:807–17.

18. Nebert DW, Zhang G, Vesell ES. Genetic risk prediction: individualized variability in susceptibility to toxicants. *Annu Rev Pharmacol Toxicol* 2013;**53**:355–75.

19. Yang IV, Schwartz DA. The next generation of complex lung genetic studies. *Am J Respir Crit Care Med* 2012;**186**:1087–94.

20. Kauffmann F, Demenais F. Gene-environment interactions in asthma and allergic diseases: challenges and perspectives. *J Allergy Clin Immunol* 2012;**130**:1229–40. quiz 41–2.

21. Lewis CM, Knight J. Introduction to genetic association studies. *Cold Spring Harb Protoc* 2012;**2012**:297–306.

22. Vercelli D. Discovering susceptibility genes for asthma and allergy. *Nat Rev Immunol* 2008;**8**:169–82.

23. U.S. EPA. *Integrated Science Assessment of Ozone and Related Photochemical Oxidants (Final Report)*. Washington, DC: U.S. Environmental Protection Agency; 2013. EPA/600/R-10/076F.

24. Kleeberger SR, Levitt RC, Zhang LY, Longphre M, Harkema J, Jedlicka A, et al. Linkage analysis of susceptibility to ozone-induced lung inflammation in inbred mice. *Nat Genet* 1997;**17**:475–8.

25. Cho HY, Zhang LY, Kleeberger SR. Ozone-induced lung inflammation and hyperreactivity are mediated via tumor necrosis factor-alpha receptors. *Am J Physiol Lung Cell Mol Physiol* 2001;**280**:L537–46.

26. Yang IA, Fong KM, Zimmerman PV, Holgate ST, Holloway JW. Genetic susceptibility to the respiratory effects of air pollution. *Thorax* 2008;**63**:555–63.

27. Bauer AK, Travis EL, Malhotra SS, Rondini EA, Walker C, Cho HY, et al. Identification of novel susceptibility genes in ozone-induced inflammation in mice. *Eur Respir J* 2010;**36**:428–37.

28. Kleeberger SR, Reddy S, Zhang LY, Jedlicka AE. Genetic susceptibility to ozone-induced lung hyperpermeability: role of Toll-like receptor 4. *Am J Respir Cell Mol Biol* 2000;**22**:620–7.

29. Kopp EB, Medzhitov R. The Toll-receptor family and control of innate immunity. *Curr Opin Immunol* 1999;**11**:13–8.

30. Poltorak A, He X, Smirnova I, Liu MY, Van Huffel C, Du X, et al. Defective LPS signaling in C3H/HeJ and C57BL/10ScCr mice: mutations in Tlr4 gene. *Science* 1998;**282**:2085–8.

31. Williams AS, Leung SY, Nath P, Khorasani NM, Bhavsar P, Issa R, et al. Role of TLR2, TLR4, and MyD88 in murine ozone-induced airway hyperresponsiveness and neutrophilia. *J Appl Phys* 2007;**103**:1189–95.

32. Hollingsworth JW, 2nd Cook DN, Brass DM, Walker JK, Morgan DL, Foster WM, et al. The role of Toll-like receptor 4 in environmental airway injury in mice. *Am J Respir Crit Care Med* 2004;**170**:126–32.

33. Cho HY, Morgan DL, Bauer AK, Kleeberger SR. Signal transduction pathways of tumor necrosis factor–mediated lung injury induced by ozone in mice. *Am J Respir Crit Care Med* 2007;**175**:829–39.

34. Prows DR, Shertzer HG, Daly MJ, Sidman CL, Leikauf GD. Genetic analysis of ozone-induced acute lung injury in sensitive and resistant strains of mice. *Nat Genet* 1997;**17**:471–4.

35. Prows DR, Daly MJ, Shertzer HG, Leikauf GD. Ozone-induced acute lung injury: genetic analysis of F(2) mice generated from A/J and C57BL/6J strains. *Am J Phys* 1999;**277**:L372–80.

36. Committee of the Environmental and Occupational Health Assembly of the American Thoracic Society. Health effects of outdoor air pollution. *Am J Respir Crit Care Med* 1996;**153**:3–50.

37. U.S. EPA. *Integrated Science Assessment for Particulate Matter (Final Report)*. Washington, DC: U.S. Environmental Protection Agency; 2009. EPA/600/R-08/139F.

38. Sacks JD, Stanek LW, Luben TJ, Johns DO, Buckley BJ, Brown JS, et al. Particulate matter-induced health effects: who is susceptible? *Environ Health Perspect* 2011;**119**:446–54.

39. Ohtsuka Y, Clarke RW, Mitzner W, Brunson K, Jakab GJ, Kleeberger SR. Interstrain variation in murine susceptibility to inhaled acid-coated particles. *Am J Physiol Lung Cell Mol Physiol* 2000;**278**:L469–76.

40. Suh HH, Nishioka Y, Allen GA, Koutrakis P, Burton RM. The metropolitan acid aerosol characterization study: results from the summer 1994 Washington, D.C. field study. *Environ Health Perspect* 1997;**105**:826–34.

41. Ohtsuka Y, Brunson KJ, Jedlicka AE, Mitzner W, Clarke RW, Zhang LY, et al. Genetic linkage analysis of susceptibility to particle exposure in mice. *Am J Respir Cell Mol Biol* 2000;**22**:574–81.

42. Fakhrzadeh L, Laskin JD, Laskin DL. Ozone-induced production of nitric oxide and TNF-alpha and tissue injury are dependent on NF-kappaB p50. *Am J Physiol Lung Cell Mol Physiol* 2004;**287**:L279–85.

43. Roberts ES, Richards JH, Jaskot R, Dreher KL. Oxidative stress mediates air pollution particle-induced acute lung injury and molecular pathology. *Inhal Toxicol* 2003;**15**:1327–46.

44. Saber AT, Bornholdt J, Dybdahl M, Sharma AK, Loft S, Vogel U, et al. Tumor necrosis factor is not required for particle-induced genotoxicity and pulmonary inflammation. *Arch Toxicol* 2005;**79**:177–82.

45. Chauhan V, Breznan D, Goegan P, Nadeau D, Karthikeyan S, Brook JR, et al. Effects of ambient air particles on nitric oxide production in macrophage cell lines. *Cell Biol Toxicol* 2004;**20**:221–39.

46. Zhao H, Ma JK, Barger MW, Mercer RR, Millecchia L, Schwegler-Berry D, et al. Reactive oxygen species and nitric oxide mediated lung inflammation and mitochondrial dysfunction in wild-type and iNOS-deficient mice exposed to diesel exhaust particles. *J Toxicol Environ Health A* 2009;**72**:560–70.

47. Kleeberger SR, Reddy SP, Zhang LY, Cho HY, Jedlicka AE. Toll-like receptor 4 mediates ozone-induced murine lung hyperpermeability via inducible nitric oxide synthase. *Am J Physiol Lung Cell Mol Physiol* 2001;**280**:L326–33.

48. Prows DR, McDowell SA, Aronow BJ, Leikauf GD. Genetic susceptibility to nickel-induced acute lung injury. *Chemosphere* 2003;**51**:1139–48.

49. Wesselkamper SC, McDowell SA, Medvedovic M, Dalton TP, Deshmukh HS, Sartor MA, et al. The role of metallothionein in the pathogenesis of acute lung injury. *Am J Respir Cell Mol Biol* 2006;**34**:73–82.

50. Thomson EM, Williams A, Yauk CL, Vincent R. Toxicogenomic analysis of susceptibility to inhaled urban particulate matter in mice with chronic lung inflammation. *Part Fibre Toxicol* 2009;**6**:6.

51. Wesselkamper SC, Chen LC, Kleeberger SR, Gordon T. Genetic variability in the development of pulmonary tolerance to inhaled pollutants in inbred mice. *Am J Physiol Lung Cell Mol Physiol* 2001;**281**:L1200–9.

52. Wesselkamper SC, Chen LC, Gordon T. Quantitative trait analysis of the development of pulmonary tolerance to inhaled zinc oxide in mice. *Respir Res* 2005;**6**:73.

53. Rosanna DP, Salvatore C. Reactive oxygen species, inflammation, and lung diseases. *Curr Pharm Des* 2012;**18**:3889–900.

54. Cho HY, Jedlicka AE, Reddy SP, Zhang LY, Kensler TW, Kleeberger SR. Linkage analysis of susceptibility to hyperoxia. Nrf2 is a candidate gene. *Am J Respir Cell Mol Biol* 2002;**26**:42–51.

55. Cho HY, Jedlicka AE, Reddy SP, Kensler TW, Yamamoto M, Zhang LY, et al. Role of NRF2 in protection against hyperoxic lung injury in mice. *Am J Respir Cell Mol Biol* 2002;**26**:175–82.

56. Cho HY, Reddy SP, Debiase A, Yamamoto M, Kleeberger SR. Gene expression profiling of NRF2-mediated protection against oxidative injury. *Free Radic Biol Med* 2005;**38**:325–43.

57. Marzec JM, Christie JD, Reddy SP, Jedlicka AE, Vuong H, Lanken PN, et al. Functional polymorphisms in the transcription factor NRF2 in humans increase the risk of acute lung injury. *Faseb J* 2007;**21**:2237–46.

58. Haston CK, Wang M, Dejournett RE, Zhou X, Ni D, Gu X, et al. Bleomycin hydrolase and a genetic locus within the MHC affect risk for pulmonary fibrosis in mice. *Hum Mol Genet* 2002;**11**:1855–63.

59. Haston CK, Amos CI, King TM, Travis EL. Inheritance of susceptibility to bleomycin-induced pulmonary fibrosis in the mouse. *Cancer Res* 1996;**56**:2596–601.

60. Du M, Irani RA, Stivers DN, Lee SJ, Travis EL. H2-Ea deficiency is a risk factor for bleomycin-induced lung fibrosis in mice. *Cancer Res* 2004;**64**:6835–9.

61. Paun A, Lemay AM, Tomko TG, Haston CK. Association analysis reveals genetic variation altering bleomycin-induced pulmonary fibrosis in mice. *Am J Respir Cell Mol Biol* 2013;**48**:330–6.

62. Stefanov AN, Fox J, Haston CK. Positional cloning reveals strain-dependent expression of Trim16 to alter susceptibility to bleomycin-induced pulmonary fibrosis in mice. *PLoS Genet* 2013;**9**: e1003203.

63. Haston CK, Travis EL. Murine susceptibility to radiation-induced pulmonary fibrosis is influenced by a genetic factor implicated in susceptibility to bleomycin-induced pulmonary fibrosis. *Cancer Res* 1997;**57**:5286–91.

64. Haston CK, Zhou X, Gumbiner-Russo L, Irani R, Dejournett R, Gu X, et al. Universal and radiation-specific loci influence murine susceptibility to radiation-induced pulmonary fibrosis. *Cancer Res* 2002;**62**:3782–8.

65. Gagneux S, DeRiemer K, Van T, Kato-Maeda M, de Jong BC, Narayanan S, et al. Variable host-pathogen compatibility in Mycobacterium tuberculosis. *Proc Natl Acad Sci U S A* 2006;**103**:2869–73.

66. El Moussi A, Ben Hadj Kacem MA, Pozo F, Ledesma J, Cuevas MT, Casas I, et al. Genetic diversity of HA1 domain of heammaglutinin gene of influenza A(H1N1)pdm09 in Tunisia. *Virol J* 2013;**10**:150.

67. Chapman SJ, Hill AV. Human genetic susceptibility to infectious disease. *Nat Rev Genet* 2012;**13**:175–88.

68. Brandenburg JA, Marrie TJ, Coley CM, Singer DE, Obrosky DS, Kapoor WN, et al. Clinical presentation, processes and outcomes of care for patients with pneumococcal pneumonia. *J Gen Intern Med* 2000;**15**:638–46.

69. Gingles NA, Alexander JE, Kadioglu A, Andrew PW, Kerr A, Mitchell TJ, et al. Role of genetic resistance in invasive pneumococcal infection: identification and study of susceptibility and resistance in inbred mouse strains. *Infect Immun* 2001;**69**:426–34.

70. Khor CC, Chapman SJ, Vannberg FO, Dunne A, Murphy C, Ling EY, et al. A Mal functional variant is associated with protection against invasive pneumococcal disease, bacteremia, malaria and tuberculosis. *Nat Genet* 2007;**39**:523–8.

71. Chapman SJ, Khor CC, Vannberg FO, Frodsham A, Walley A, Maskell NA, et al. IkappaB genetic polymorphisms and invasive pneumococcal disease. *Am J Respir Crit Care Med* 2007;**176**:181–7.

72. Roy S, Knox K, Segal S, Griffiths D, Moore CE, Welsh KI, et al. MBL genotype and risk of invasive pneumococcal disease: a case-control study. *Lancet* 2002;**359**:1569–73.

73. Ozkan H, Koksal N, Cetinkaya M, Kiliç Ş, Çelebi S, Oral B, et al. Serum mannose-binding lectin (MBL) gene polymorphism and low MBL levels are associated with neonatal sepsis and pneumonia. *J Perinatol* 2012;**32**:210–7.

74. Endeman H, Herpers BL, de Jong BA, Voorn GP, Grutters JC, van Velzen-Blad H, et al. Mannose-binding lectin genotypes in susceptibility to community-acquired pneumonia. *Chest* 2008;**134**:1135–40.

75. Dye C. Global epidemiology of tuberculosis. *Lancet* 2006;**367**: 938–40.

76. Comstock GW. Tuberculosis in twins: a re-analysis of the Prophit survey. *Am Rev Respir Dis* 1978;**117**:621–4.

77. Kallmann F, Reisner D. Twin studies on the significance of genetic factors in tuberculosis. *Am Rev Tuberc* 1948;**47**:549–74.

78. Stead WW, Senner JW, Reddick WT, Lofgren JP. Racial differences in susceptibility to infection by Mycobacterium tuberculosis. *N Engl J Med* 1990;**322**:422–7.

79. Thye T, Vannberg FO, Wong SH, Owusu-Dabo E, Osei I, Gyapong J, et al. Genome-wide association analyses identifies a susceptibility locus for tuberculosis on chromosome 18q11.2. *Nat Genet* 2010;**42**:739–41.

80. Girard MP, Tam JS, Assossou OM, Kieny MP. The 2009 A (H1N1) influenza virus pandemic: a review. *Vaccine* 2010;**28**:4895–902.

81. Zhou J, To KK, Dong H, Cheng ZS, Lau CC, Poon VK, et al. A functional variation in CD55 increases the severity of 2009 pandemic H1N1 influenza A virus infection. *J Infect Dis* 2012;**206**:495–503.

82. Keynan Y, Juno J, Meyers A, Ball TB, Kumar A, Rubinstein E, et al. Chemokine receptor 5delta32 allele in patients with severe pandemic (H1N1) 2009. *Emerg Infect Dis* 2010;**16**:1621–2.

83. Dawson TC, Beck MA, Kuziel WA, Henderson F, Maeda N. Contrasting effects of CCR5 and CCR2 deficiency in the pulmonary inflammatory response to influenza A virus. *Am J Pathol* 2000;**156**:1951–9.

84. Everitt AR, Clare S, Pertel T, John SP, Wash RS, Smith SE, et al. IFITM3 restricts the morbidity and mortality associated with influenza. *Nature* 2012;**484**:519–23.

85. Go JT, Belisle SE, Tchitchek N, Tumpey TM, Ma W, Richt JA, et al. 2009 pandemic H1N1 influenza virus elicits similar clinical course but differential host transcriptional response in mouse, macaque, and swine infection models. *BMC Genomics* 2012;**13**:627.

86. Leader S, Kohlhase K. Respiratory syncytial virus-coded pediatric hospitalizations, 1997 to 1999. *Pediatr Infect Dis J* 2002;**21**:629–32.

87. Howard TS, Hoffman LH, Stang PE, Simoes EA. Respiratory syncytial virus pneumonia in the hospital setting: length of stay, charges, and mortality. *J Pediatr* 2000;**137**:227–32.

88. Miyairi I, DeVincenzo JP. Human genetic factors and respiratory syncytial virus disease severity. *Clin Microbiol Rev* 2008;**21**: 686–703.

89. Amanatidou V, Apostolakis S, Spandidos DA. Genetic diversity of the host and severe respiratory syncytial virus-induced lower respiratory tract infection. *Pediatr Infect Dis J* 2009;**28**:135–40.

90. Janssen R, Bont L, Siezen CL, Hodemaekers HM, Ermers MJ, Doornbos G, et al. Genetic susceptibility to respiratory syncytial virus bronchiolitis is predominantly associated with innate immune genes. *J Infect Dis* 2007;**196**:826–34.

91. Kresfelder TL, Janssen R, Bont L, Pretorius M, Venter M. Confirmation of an association between single nucleotide polymorphisms in the VDR gene with respiratory syncytial virus related disease in South African children. *J Med Virol* 2011;**83**:1834–40.

92. Richeldi L, Sorrentino R, Saltini C. HLA-DPB1 glutamate 69: a genetic marker of beryllium disease. *Science* 1993;**262**:242–4.

93. Lombardi G, Germain C, Uren J, Fiorillo MT, du Bois RM, Jones-Williams W, et al. HLA-DP allele-specific T cell responses to beryllium account for DP-associated susceptibility to chronic beryllium disease. *J Immunol* 2001;**166**:3549–55.

94. Wang Z, White PS, Petrovic M, Tatum OL, Newman LS, Maier LA, et al. Differential susceptibilities to chronic beryllium disease contributed by different Glu69 HLA-DPB1 and -DPA1 alleles. *J Immunol* 1999;**163**:1647–53.

95. McCanlies EC, Ensey JS, Schuler CR, Kreiss K, Weston A. The association between HLA-DPB1Glu69 and chronic beryllium disease and beryllium sensitization. *Am J Ind Med* 2004;**46**:95–103.

96. Jagirdar J, Begin R, Dufresne A, Goswami S, Lee TC, Rom WN. Transforming growth factor-beta (TGF-beta) in silicosis. *Am J Respir Crit Care Med* 1996;**154**:1076–81.

97. Corbett EL, Mozzato-Chamay N, Butterworth AE, De Cock KM, Williams BG, Churchyard GJ, et al. Polymorphisms in the tumor necrosis factor-alpha gene promoter may predispose to severe silicosis in black South African miners. *Am J Respir Crit Care Med* 2002;**165**:690–3.

98. Wang YW, Lan JY, Yang LY, Wang De J, Kuang J. TNF-alpha and IL-1RA polymorphisms and silicosis susceptibility in Chinese workers exposed to silica particles: a case-control study. *Biomed Environ Sci* 2012;**25**:517–25.

99. Yucesoy B, Vallyathan V, Landsittel DP, Sharp DS, Weston A, Burleson GR, et al. Association of tumor necrosis factor-alpha and interleukin-1 gene polymorphisms with silicosis. *Toxicol Appl Pharmacol* 2001;**172**:75–82.

100. Ueki A, Isozaki Y, Tomokuni A, Ueki H, Kusaka M, Tanaka S, et al. Different distribution of HLA class II alleles in anti-topoisomerase I autoantibody responders between silicosis and systemic sclerosis patients, with a common distinct amino acid sequence in the HLA-DQB1 domain. *Immunobiology* 2001;**204**:458–65.

101. Borm PJ, Schins RP. Genotype and phenotype in susceptibility to coal workers' pneumoconiosis. the use of cytokines in perspective. *Eur Respir J Suppl* 2001;**32**:127s–33s.

102. Zhai R, Jetten M, Schins RP, Franssen H, Borm PJ. Polymorphisms in the promoter of the tumor necrosis factor-alpha gene in coal miners. *Am J Ind Med* 1998;**34**:318–24.

103. Wang XT, Ohtsuka Y, Kimura K, Muroi M, Ishida T, Saito J, et al. Antithetical effect of tumor necrosis factor-alphagene polymorphism on coal workers' pneumoconiosis (CWP). *Am J Ind Med* 2005;**48**:24–9.

104. Nadif R, Jedlicka A, Mintz M, Bertrand JP, Kleeberger S, Kauffmann F. Effect of TNF and LTA polymorphisms on biological markers of response to oxidative stimuli in coal miners: a model of gene-environment interaction. Tumour necrosis factor and lymphotoxin alpha. *J Med Genet* 2003;**40**:96–103.

105. Nadif R, Mintz M, Rivas-Fuentes S, Jedlicka A, Lavergne E, Rodero M, et al. Polymorphisms in chemokine and chemokine receptor genes and the development of coal workers' pneumoconiosis. *Cytokine* 2006;**33**:171–8.

106. Mapp CE, Balboni A, Baricordi R, Fabbri LM. Human leukocyte antigen associations in occupational asthma induced by isocyanates. *Am J Respir Crit Care Med* 1997;**156**:S139–43.

107. Horne C, Quintana PJ, Keown PA, Dimich-Ward H, Chan-Yeung M. Distribution of DRB1 and DQB1 HLA class II alleles in occupational asthma due to western red cedar. *Eur Respir J* 2000;**15**:911–4.

108. Bignon JS, Aron Y, Ju LY, Kopferschmitt MC, Garnier R, Mapp C, et al. HLA class II alleles in isocyanate-induced asthma. *Am J Respir Crit Care Med* 1994;**149**:71–5.

109. Choi JH, Lee KW, Kim CW, Park CS, Lee HY, Hur GY, et al. The HLA DRB1*1501-DQB1*0602-DPB1*0501 haplotype is a risk factor for toluene diisocyanate-induced occupational asthma. *Int Arch Allergy Immunol* 2009;**150**:156–63.

110. Piirila P, Wikman H, Luukkonen R, Kääriä K, Rosenberg C, Nordman H, et al. Glutathione S-transferase genotypes and allergic responses to diisocyanate exposure. *Pharmacogenetics* 2001;**11**:437–45.

111. Mapp CE, Fryer AA, De Marzo N, Pozzato V, Padoan M, Boschetto P, et al. Glutathione S-transferase GSTP1 is a susceptibility gene for occupational asthma induced by isocyanates. *J Allergy Clin Immunol* 2002;**109**:867–72.

112. Smith LJ, Shamsuddin M, Sporn PH, Denenberg M, Anderson J. Reduced superoxide dismutase in lung cells of patients with asthma. *Free Radic Biol Med* 1997;**22**:1301–7.

113. Comhair SA, Ricci KS, Arroliga M, Lara AR, Dweik RA, Song W, et al. Correlation of systemic superoxide dismutase deficiency to airflow obstruction in asthma. *Am J Respir Crit Care Med* 2005;**172**:306–13.

114. Kluchova Z, Petrasova D, Joppa P, Dorková Z, Tkácová R. The association between oxidative stress and obstructive lung impairment in patients with COPD. *Physiol Res* 2007;**56**:51–6.

115. Ahmad A, Shameem M, Husain Q. Relation of oxidant-antioxidant imbalance with disease progression in patients with asthma. *Ann Thorac Med* 2012;**7**:226–32.

116. Park HY, Kim JH, Lim YH, Bae S, Hong YC. Influence of genetic polymorphisms on the association between phthalate exposure and pulmonary function in the elderly. *Environ Res* 2013;**122**:18–24.

117. Romieu I, Trenga C. Diet and obstructive lung diseases. *Epidemiol Rev* 2001;**23**:268–87.

118. Bodner C, Godden D, Brown K, Little J, Ross S, Seaton A. Antioxidant intake and adult-onset wheeze: a case-control study. Aberdeen WHEASE Study Group. *Eur Respir J* 1999;**13**:22–30.

119. Hu G, Cassano PA. Antioxidant nutrients and pulmonary function: the Third National Health and Nutrition Examination Survey (NHANES III). *Am J Epidemiol* 2000;**151**:975–81.

120. Black PN, Scragg R. Relationship between serum 25-hydroxyvitamin d and pulmonary function in the third national health and nutrition examination survey. *Chest* 2005;**128**:3792–8.

121. Kelly FJ, Mudway I, Blomberg A, Frew A, Sandström T. Altered lung antioxidant status in patients with mild asthma. *Lancet* 1999;**354**:482–3.

122. Da Costa LA, Badawi A, El-Sohemy A. Nutrigenetics and modulation of oxidative stress. *Ann Nutr Metab* 2012;**60**(Suppl. 3):27–36.

123. Borges MC, Martini LA, Rogero MM. Current perspectives on vitamin D, immune system, and chronic diseases. *Nutrition* 2011;**27**:399–404.

124. Devereux G, Litonjua AA, Turner SW, Craig LC, McNeill G, Martindale S, et al. Maternal vitamin D intake during pregnancy and early childhood wheezing. *Am J Clin Nutr* 2007;**85**:853–9.

125. Urashima M, Segawa T, Okazaki M, Kurihara M, Wada Y, Ida H. Randomized trial of vitamin D supplementation to prevent seasonal influenza A in schoolchildren. *Am J Clin Nutr* 2010;**91**:1255–60.

126. Brehm JM, Celedon JC, Soto-Quiros ME, Avila L, Hunninghake GM, Forno E, et al. Serum vitamin D levels and markers of severity of childhood asthma in Costa Rica. *Am J Respir Crit Care Med* 2009;**179**:765–71.

127. Janssens W, Bouillon R, Claes B, Carremans C, Lehouck A, Buysschaert I, et al. Vitamin D deficiency is highly prevalent in COPD and correlates with variants in the vitamin D-binding gene. *Thorax* 2010;**65**:215–20.

128. Kwak MK, Kensler TW. Targeting NRF2 signaling for cancer chemoprevention. *Toxicol Appl Pharmacol* 2010;**244**:66–76.

129. Heiss E, Herhaus C, Klimo K, Bartsch H, Gerhäuser C. Nuclear factor kappa B is a molecular target for sulforaphane-mediated anti-inflammatory mechanisms. *J Biol Chem* 2001;**276**:32008–15.

130. Xue M, Qian Q, Adaikalakoteswari A, Rabbani N, Babaei-Jadidi R, Thornalley PJ. Activation of NF-E2-related factor-2 reverses biochemical dysfunction of endothelial cells induced by hyperglycemia linked to vascular disease. *Diabetes* 2008;**57**:2809–17.

131. Cho HY, Imani F, Miller-DeGraff L, Walters D, Melendi GA, Yamamoto M, et al. Antiviral activity of Nrf2 in a murine model of respiratory syncytial virus disease. *Am J Respir Crit Care Med* 2009;**179**:138–50.

132. Zheng Y, Tao S, Lian F, Chau BT, Chen J, Sun G, et al. Sulforaphane prevents pulmonary damage in response to inhaled arsenic by activating the Nrf2-defense response. *Toxicol Appl Pharmacol* 2012;**265**:292–9.

133. Nichols J. *Genetic and genomic mechanisms of neonatal hyperoxic lung injury in the inbred mouse.* The University of North Carolina at Chapel Hill. ProQuest Dissertations and Theses; 2012, 181.

134. Gatti DM, Shabalin AA, Lam TC, Wright FA, Rusyn I, Nobel AB. FastMap: fast eQTL mapping in homozygous populations. *Bioinformatics* 2009;**25**:482–9.

135. Shabalin AA. Matrix eQTL: ultra fast eQTL analysis via large matrix operations. *Bioinformatics* 2012;**28**:1353–8.

136. Abecasis GR, Cherny SS, Cookson WO, Cardon LR. Merlin–rapid analysis of dense genetic maps using sparse gene flow trees. *Nat Genet* 2002;**30**:97–101.

137. Purcell S, Neale B, Todd-Brown K, Thomas L, Ferreira MA, Bender D, et al. PLINK: a tool set for whole-genome association and population-based linkage analyses. *Am J Hum Genet* 2007;**81**: 559–75.

138. Perez-Enciso M, Misztal I. Qxpak.5: old mixed model solutions for new genomics problems. *BMC Bioinformatics* 2011;**12**:202.

139. Broman KW, Wu H, Sen S, Churchill GA. R/qtl: QTL mapping in experimental crosses. *Bioinformatics* 2003;**19**:889–90.

140. Pletcher MT, McClurg P, Batalov S, Su AI, Barnes SW, Lagler E, et al. Use of a dense single nucleotide polymorphism map for in silico mapping in the mouse. *PLoS Biol* 2004;**2**: e393.

141. Wu CC, Huang HC, Juan HF, Chen ST. GeneNetwork: an interactive tool for reconstruction of genetic networks using microarray data. *Bioinformatics* 2004;**20**:3691–3.

142. Dennis Jr G, Sherman BT, Hosack DA, Yang J, Gao W, Lane HC, et al. DAVID: Database for Annotation, Visualization, and Integrated Discovery. *Genome Biol* 2003;**4**: P3.

143. Prows DR, Leikauf GD. Quantitative trait analysis of nickel-induced acute lung injury in mice. *Am J Respir Cell Mol Biol* 2001;**24**:740–6.

144. Maier LA, Reynolds MV, Young DA, Barker EA, Newman LS. Angiotensin-1 converting enzyme polymorphisms in chronic beryllium disease. *Am J Respir Crit Care Med* 1999;**159**:1342–50.

145. Mozzato-Chamay N, Corbett EL, Bailey RL, Mabey DC, Raynes J, Conway DJ. Polymorphisms in the IkappaB-alpha promoter region and risk of diseases involving inflammation and fibrosis. *Genes Immun* 2001;**2**:153–5.

Chapter 21

Effects of Environmental Tobacco Smoke during Early Life Stages

Jingyi Xu*,†, Suzette Smiley-Jewell*, Jocelyn Claude* and Kent E. Pinkerton*

*Center for Health and the Environment, University of California – Davis, Davis, CA, USA; †Affiliated Zhongshan Hospital of Dalian University, Dalian, China

INTRODUCTION

Environmental tobacco smoke (ETS), one of the most common indoor pollutants and also referred to as second-hand smoke (SHS) or passive smoke, is a global public health problem. ETS is associated with premature death, cardiovascular and respiratory disease, infection, behavioral problems, low birth weight, sudden infant death syndrome, likelihood to smoke when exposed as a youth, and increased cancer risk.[1,2] Children are an especially vulnerable group to the detrimental affects of ETS exposure; when exposed, they are more likely to suffer from asthma, pneumonia, sinusitis, and allergies. Globally, approximately 40% of children breath ETS produced by a smoking parent, and in 2004, 28% of all deaths attributed to ETS were from children.[3] In the United States, 53.6% of young children (aged 3–11 years) were exposed to SHS in 2007–2008.[4] Thus, despite public bans on smoking and lowering smoking rates in some parts of the world, ETS remains a major public health issue. This chapter addresses conditions of early life ETS exposure (both prenatal and postnatal) and the effects of this exposure in the development of disease later in life.

CONDITIONS OF EARLY LIFE ETS EXPOSURE

ETS

ETS is produced primarily from the smoldering end of a cigarette with a small contribution from exhaled mainstream smoke. ETS differs from mainstream smoke by being aged rather than fresh, having a smaller particle size, and containing nicotine in the gas phase rather than the particulate phase. Scherer et al. estimated uptake of various chemical constituents of cigarette smoke in a mainstream smoker versus a passive smoker (Table 1) and found that smoking 20 cigarettes per day versus being exposed to ETS or passive smoke (PS) for 8 hours resulted in similar uptake of gas phase constituents compared to a much larger uptake of particulate phase constituents.[5]

Cotinine, a metabolite of nicotine, can be found in measurable quantities in plasma, urine, and saliva of smoke-exposed individuals. Because of the ease with which it can be measured, it is the most commonly used qualitative and quantitative marker of cigarette smoke exposure and is considered an indirect marker of exposure to the numerous other constituents of cigarette smoke. If a pregnant woman smokes or is exposed to ETS, cotinine is transferred to the developing fetus: concentrations of cotinine in the serum of fetuses 21–36 weeks gestational age are about 90% of maternal levels, regardless of gestational age or number of cigarettes smoked.[6] Newborn infants of mothers who smoked during pregnancy have higher concentrations of cotinine in their hair than infants whose mothers did not smoke but were exposed to ETS in the home; in turn, infants of mothers exposed to ETS have greater concentrations of cotinine in their hair than infants not exposed to smoke at all during pregnancy.[7]

Following birth, cotinine can be passed via breast milk and metabolically produced following ETS exposure via inhalation. Exclusively breastfed infants whose mothers are mainstream smokers have been found to have urinary cotinine concentrations in the same range as active smokers.[8] While serum cotinine concentrations of children exposed to ETS via inhalation are about 100-fold less than that of active smokers,[9,10] they are affected by which parent smokes: cotinine concentrations are greater if the mother smokes than if the father smokes the same number of cigarettes per day,[9,11] suggesting greater uptake from the primary caregiver. As would be expected, cotinine concentrations increase with the number of cigarettes smoked in the home and in the same room as the child.[12]

Lung Maturation

Lung development and growth begins in the in utero period and continues until the late teens and early 20s. Thus, there is a long period of time during which factors such as ETS may influence the maturation process of the lung, possibly altering

The Lung. http://dx.doi.org/10.1016/B978-0-12-799941-8.00021-3

the normal course of development. Lung size increases with body size and is affected by age, gender, and ethnicity. Maximum lung volumes are finally reached at approximately 22 years, representing a 30-fold increase in lung volume and a 20-fold increase in gas-exchange surface area between birth and maturity.[13] FEV1 and FVC gradually decline after

TABLE 1 Estimated uptake doses by active and passive smoking

Tobacco Smoke Constituents	Smoking (S) (20 cig/day)	Passive Smoking (PS) (8 h/day)	Dose Ratio S/PS
Gaseous Phase			
CO (mg)	40–400	14.4–96	2.7–4.2
Formaldehyde (mg)	0.4–1.8	0.08–0.4	4–5
Volatile nitro-samines (g)	0.05–1.0	0.03–0.4	1.5–2.5
Benzene (g)	200–1200	40–400	3–5
Particulate Matter			
Particles (mg)	75–300	0.025–0.24	1250–3000
Nicotine (mg)[a]	7.5–30	0.08–0.4	75–90
BaP (g)	0.15–0.75	0.001–0.011	70–150
Cadmium (g)	1.5	0.001–0.014	110–1500
Tobacco-specific nitos-amines (g)	4.5–45	0.002–0.010	2300–4500

[a]*Nicotine is particle-bound in mainstream smoke and a gas-phase constituent in environmental tobacco smoke.*
Source: Scherer et al.,[5] with permission.

this peak because aging results in a gradual loss of lung elasticity, even in healthy people.[14] Figure 1 shows the change in lung function over the life course of the average person using FEV1 as an example. Curve I shows the optimal growth and decline of FEV1.

Perinatal and postnatal ETS exposure has been found to decrease lung function. Moshammer et al. published the results of a large cross-sectional study (*The Pollution and the Young* study) that used pooled data to address problems of colinearity between prenatal and postnatal passive smoking to assess the independent effects of the following three exposures on lung function: (1) maternal smoking during pregnancy, (2) passive smoking during the first 2 years of the child's life, and (3) current passive smoking on childhood.[15] Moshammer et al. found that smoking during pregnancy was associated with a decrease in lung function parameters between 21% (FEV1) and 26% maximal expiratory flow at 25% of vital capacity left (MEF25). A 4% lower maximal mid-expiratory flow (MMEF) corresponded to a 40% increase in the risk of poor lung function (MMEF<75% of expected). Associations with current passive smoking were weaker, although still measurable; effects ranged from 20.5% (FEV1) to 22% maximal expiratory flow (MEF50).

It is important to remember when considering ETS exposures that while the lung is maturing, infants and children experience a higher internal exposure to air pollutants, such ETS, than adults because of greater respiration rate and minute ventilation than adults (Table 2).[16]

Immune System and Allergen

There is substantial epidemiologic evidence that exposure to ETS during pregnancy and early childhood alters the normal maturation of the immune system in children. Children of parents, especially mothers, in Greece who smoked during pregnancy and the first year of life were found more likely to be hospitalized with respiratory infections.[17] The 2006 report of the US Surgeon General on ETS warns of

FIGURE 1 Schematic representation of the life course of forced expiratory volume (not to scale). (I) normal growth and decline; (II) impaired prenatal or postnatal growth; and (III) normal growth but accelerated decline. *By permission, Wang & Pinkerton.*[45]

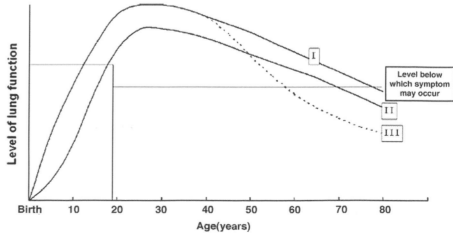

TABLE 2 Difference in ventilatory characteristics between infants and adults

	Infants	Adults
Tidal volume (mL/kg bodyweight)	10	10
Alveolar surface area (m^2)	3	75
Respiration rate (breaths/min)	40	15
Minute ventilation (mL/kg bodyweight per m^2 lung surface area per min)	133	2

Source: Pinkerton & Joad,[16] with permission.

a positive association of ETS exposure during childhood with cough, wheeze/breathlessness, increased onset of viral infection in the lower respiratory tract, and middle ear infection.[18]

Similar to the lung, the immune system is not fully developed at birth. With typical exposure to microorganisms and pathogens, two branches of immunity—innate and adaptive immunity—mature to form a defense system capable of protecting the body from foreign substances. Innate immunity is involved in the primary response against foreign pathogens. This system includes skin; mucus membranes, such as nasal and respiratory mucosa; mucociliary clearance and a variety of immune cells, including dendritic cells, natural killer cells, macrophages, monocytes, neutrophils, and lymphocytes. Adaptive immunity is acquired and consists of antibody responses and cell-mediated responses.

ETS exposure in infancy and childhood is thought to affect adaptive immunity by enhancing the development of the T-helper 2 pathway (T$_H$2: mediates allergic responses) over the development of the T-helper 1 pathway (T$_H$1: mediates cellular immunity).[18] Perinatal exposure of rhesus macaques to ETS was found to induce local and systemic inflammatory responses, including an increase in the numbers of lung macrophages and eosinophils; elevate plasma levels of C5a and T$_H$2-biased cytokine production; and increase levels of neurotrophin and neuropeptide expression in the infants.[19] Vardavas et al. found healthy teens exposed to ETS had altered T-cell populations[20]: there was a dose-response reduction in CD3+ and CD4+ memory cells and increase in naïve/memory CD3+ and CD4+ cells. The authors felt that this shift may be an underlying mechanism of ETS related disease initiation and exacerbation. Mononuclear cells collected from the cord blood of infants of mothers who smoked during pregnancy had an increased T$_H$2 response when exposed to the common allergens, house dust mite, and olvalbumin, compared to cells collected from infants whose mothers did not smoke.[21]

Wilson et al. recently published a study[22] showing that children admitted to the hospital with influenza and exposed to SHS had significantly increased length of stay, need for intensive care and intubation, and greater bacterial co-infection than non-SHS exposed children. Most pediatric influenza-related deaths are the result of either exacerbation of underlying medical conditions or invasive co-infection with another infectious pathogen, such as *Staphylococcus aureus*. In fact, in the 2009 H1N1 influenza pandemic 49% of all pediatric deaths were associated with bacterial co-infection.[23,24] Our recent published data suggests that perinatal exposure to ETS enhances susceptibility to viral and bacterial infection.[25]

Phagocytes play a key role in the primary defensive response to pathogens or foreign substances. Phagocytic activity of neutrophils and macrophages can be diminished through suppression of oxygen radical formation by components of ETS. The particulate phase of ETS may also impair the capacity of these immune cells to phagocytose foreign pathogens.[26] Infants exposed to ETS prenatally or postnatally were found to have reduced dendritic cell IL-10, which has anti-inflammatory properties and can lower T$_H$2 responses.[27] A study by Tebow et al. showed that children of parents who smoked from the prenatal period to year 11 had decreased interferon γ (IFNγ), a key cytokine in pathogen recognition and antiviral response, compared to children of parents that did not smoke.[28] A recent epigenetic study of children (11–18 years old) from Fresno, CA, one of the worst sites for air pollution and childhood asthma rates in the United States, examined the relationship between local ambient air pollution, ETS exposure, and immunity.[29] When the children were exposed to both ambient air pollution and ETS, T cells had reduced expression of IFNγ that was associated with significant DNA hypermethylation of the IFNγ promoter. Thus, combined, these studies document alterations in normal development of the innate and adaptive immune response with perinatal/postnatal exposure to ETS.

CRITICAL LIFE STAGES AND ETS

In Utero Exposure

When smoking occurs during pregnancy, the developing lung is exposed in utero to the constituents of smoke from transplacental transfer from the pregnant woman. A number of studies have documented that exposing the fetus to smoke during gestation leads to decreased lung function, airway obstruction, and airway hyperresponsiveness in the newborn period. Healthy infants, whose mothers smoked during pregnancy, have reduced FEV1 shortly after birth.[30] Increased airways responsiveness at 4.5 weeks of age was found in normal infants of smoking parents compared with infants of non-smoking parents.[31] In a separate case control study of children 7–9 years old,[32] both maternal smoking in pregnancy (odds ratio 1.9) and each additional household smoker (odds ratio 1.15) were independent predictors of asthma/wheeze.

When Stick et al. measured lung function in 461 term infants at about 58 hours of life,[33] and Lodrup Carlsen et al. did the same in 804 infants at about 65 hours of life,[34] both scientific groups showed the time to peak tidal expiratory flow (tPTEF) as a proportion of total expiratory time (tE) was lower in a dose-dependent fashion in infants of smoking mothers: tPTEF/tE correlates with indices of airway obstruction and with the development of wheezing lower-respiratory-tract illness. Lodrup Carlsen et al. also found that compliance of the respiratory system was lower in female babies.[34]

Similar findings of airway obstruction were found in one-month-old infants of smoking mothers using $Vmax_{FRC}$ (maximum flow rate at functional residual capacity) as a measure of airway obstruction.[35,36] In the study by Hanrahan et al.,[35] the $Vmax_{FRC}$ was about half that in control unexposed infants. In premature neonates, both tPTEF/tE and $Vmax_{FRC}$ were reduced by 14–18% in 40 infants of smoking mothers compared with 68 infants of non-smoking mothers when studied at about 3 weeks of age. Because the infants were on average 7 weeks premature, the data suggest smoke exposure alters normal lung development before 33 weeks of gestation.

As mentioned earlier, newborn infants of smoking mothers have been found to have airway hyperresponsiveness, a hallmark of asthma. Human studies of maternal smoking during pregnancy show an association between an increased prevalence of asthma and history of wheezing.[36–38] In month-old healthy infants, the dose of inhaled histamine required to reduce $Vmax_{FRC}$ by 40% was 5-fold less in infants with at least one parent who smoked as compared with infants whose parents did not smoke.[31] Interestingly, although maternal smoking is associated with premature delivery, it is also associated with enhanced lung maturity of the fetus as measured by lecithin/sphyngomyelin ratio (L/S). L/S ratio was one week advanced in fetuses of smoking mothers whose amniotic fluid was sampled between 28 and 36 weeks of gestation. The elevated L/S ratio correlated with the cotinine concentration in the amniotic fluid and was associated with increased concentrations of free, conjugated, and total cortisol.[39] This may explain why premature infants of smokers are at lower risk for developing infant respiratory distress syndrome at a given gestational age.

Animal studies confirm the adverse effects of maternal smoke exposure on the developing fetus. Fetuses of pregnant rats exposed to smoke (mainstream or ETS) had reduced lung volume, number of enlarged alveoli, and parenchymal elastic tissue[40]; increased density of interstitium; poorly developed elastin and collagen[41]; and increased Clara cell secretory protein.[42] Recent evidence shows that prenatal exposure of lambs to nicotine causes abnormalities in airway branching and dimensions and results in increased airway smooth muscle and collagen deposition, with subsequent reduction in airflow limitation and FEV1 and increased airway hyperreactivity.[43]

Some researchers believe that in utero exposure to ETS has a greater effect on lung function than postnatal exposure,[30,44,45] and there is ample evidence from both human and animal studies that mainstream smoking by a pregnant mother changes fetal lung development, including changes in airflow, airway responsiveness, and lung maturity. In addition, fetal smoke exposure is associated with sudden infant death, low birth weight, preterm delivery and intrauterine growth restriction (IUGR) in humans,[46] and changed immunity, morphology, and enzyme function in animal models. In humans, at least some of these effects may occur even if the mother does not smoke but is exposed to ETS in the home.

Postnatal Exposure: Infancy and Childhood

Postnatal exposure to ETS by infants and children is widespread. From 1988 to 1991, 43% of USA children aged 2 months to 11 years of age lived in a home with at least one tobacco smoker.[10] In a prevalence survey conducted in China in 2002,[47] 66% of men and 3% of women reported being current smokers. In this same survey, 53% non-smoking adults and children were exposed to ETS, with the highest percentage of people (80%) exposed to ETS between 15 and 19 years old. A recent retrospective survey of 192 countries published in 2011 reports that 40% of children, 33% of male non-smokers, and 35% of female non-smokers are exposed to ETS.[48] The survey also shows clear inequalities in the burden of disease from ETS according to sex and age in non-smokers (Figure 2): women have the greatest burden of death attributable to SHS, whereas children are most affected in terms of deaths and disability-adjusted life-years (DALYs: a weighted-measure of death and disability). Numerous epidemiological studies show that postnatal exposure to ETS adversely affects early life respiratory health and can enhance allergy.

A dose–response relationship for ETS exposure has been demonstrated whereby the odds ratio for hospitalization for respiratory illnesses in the first 2 years of life of 160 low birth weight infants was 2.9 if they were exposed to light ETS (1–19 cigarettes/day) and 4.5 if they were exposed to heavy ETS (20 or more cigarettes/day).[49] After adjusting for the smoking status of the mother, the effect of placing a 3-year-old child in a day care setting where the caregiver smokes was found to increase the risk of a wheezing lower respiratory illness by more than 3-fold.[50] The Third National Health and Nutrition Examination Survey (of 1988 to 1994) reported a 2.1 odds ratio for asthma in young children 2 months to 5 years of age for household exposure to more than one pack per day of cigarette smoke and 1.8 for prenatal

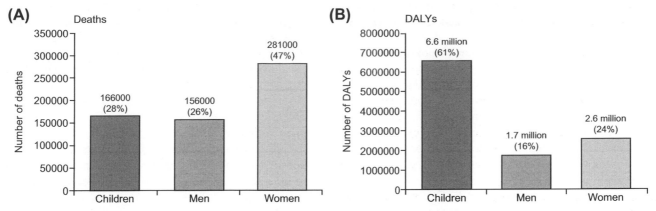

FIGURE 2   Distribution of disease burden from secondhand smoke in non-smokers by age and sex in 2004. (A) Total deaths attributable to secondhand smoke. (B). Total DALYs attributable to secondhand smoke. DALY = disability-adjusted life-year. *By permission, Öberg et al.*[48]

exposure.[51] It was estimated that for young children of 2 months to 2 years of age, 40 to 60% of the cases of asthma, chronic bronchitis, and three or more episodes of wheezing were attributable to ETS exposure. Newborn infants of non-allergic parents whose mothers smoked were found to have a 3-fold increase in the risk of elevated IgE and a 4-fold increased risk of allergic symptoms, such as eczema, urticaria, asthma, or food allergy before 18 months.[52] In a separate study of children 1–4 years of age, the odds ratio for allergy to cat increased from 3 when the children were exposed to an indoor cat only to 42 when they were exposed to ETS in addition to a damp environment with an indoor cat.[53]

Epidemiological studies show infants and young children exposed to ETS are at increased risk for developing both upper and lower respiratory infections (i.e., common cold, influenza, bronchitis, bronchiolits, etc). Childhood ETS exposure is also a risk factor for middle ear infections and, in combination with respiratory infections, may contribute to sudden infant death syndrome (SIDS).[54] Exposure to ETS is one of the main preventable causes of SIDS. In a nationwide case-control study of 485 SIDS deaths from New Zealand,[55] it was shown that the risk of SIDS was 4-fold greater in infants of mothers who smoked during pregnancy; the risk increased as the number of cigarettes smoked by the mother increased. Carpenter et al.[56] found that smoking >10 cigarettes per day after birth was significantly associated with an increased risk of SIDS when adjusted for active maternal smoking (10–19 cig/day, OR = 1.54, 95% CI = 1.11–2.14; 20–29 cig/day, OR = 1.73, 95% CI = 1.21–2.48; 30+ cig/day, OR = 3.31, 95% CI = 1.84–5.96).

Otitis media is one of the most common infectious diseases of early childhood. An epidemiological study by Ladomenou et al.[57] reported the prevalence rate of acute otitis media as 28% within the first years of life. Interestingly, Prtak et al.[58] found that the presence of an exudate in the middle ear in a substantial number of SIDS cases prompted a middle ear swab in 27% of cases. ETS may dampen the immune response to increase the risk of infection by common bacteria; thus, ETS may contribute to the etiology of SIDS through a mechanism of chronic inflammation resulting from initial low-level infections.[59,60]

With increasing numbers of smokers and cigarettes smoked in the home, children aged 8–11 years have more wheeze with colds, persistent wheeze, shortness of breath with wheeze and emergency room visits for wheeze,[61] as evidenced by an odds ratio of 1.5–1.8 for 30 or more cigarettes smoked per day in the home. In a study of 1501 Malaysian 12-year-old children, a significant 1.7-fold risk of the child having chest illness was found if he/she slept in the same room as a smoker.[62] In addition to symptoms of airway obstruction, such as wheeze, children beyond the neonatal period raised in the homes of smokers demonstrate small but measurable changes in baseline lung function consistent with airway obstruction. In a study of 2765 Australian children 8–11 years of age, FEF_{25-75} was decreased by 2.4% if more than 20 cigarettes were smoked in the home daily.[63] A similar small decrease in FEV_1/VC of 1.5% in male children of smoking parents was noted in a study of 635 New Zealand children 9–15 years old.[64]

Even in the absence of clinical asthma, children exposed to smoke may have increased airway responsiveness, which is determined by measuring spontaneous changes in airflow with peak flow variability or by using drug, exercise, or cold air challenges. A dose-response effect of ETS on peak flow variability was shown in boys 6–9 years of age when their urinary cotinine excretion increased along with the ratio of the amplitude over the mean of the diurnal peak flow rates.[65] No similar effect was found for girls. However, another study showed that paternal smoking in a crowded household increased the odds ratio to 2.7 for hyperresponsiveness to methacholine in 7–10 year old daughters,[66] which was greater than for the sons.

A meta-analysis suggested a small but real increase in airway responsiveness in school-aged children exposed to ETS.[67] A 2.5-fold increase risk of developing asthma before 12 years of age due to maternal smoking was found to occur only in children whose mothers had less than 12 years of formal education, which may be due to cofounding factors, such as low socioeconomic status, household crowding, exposure to unique aeroallergens, nutrition, and infections.[68] It is important to note that ETS exposure in the home may not just originate from a child's parents, as 85% of smokers who live with children report regular smoking by residents or visitors.[69]

Animal studies also show that ETS exposure following birth results in changes in immunity. When rhesus macaque monkeys were exposed to ETS from 6 months to 13 months of age, the normal TH1 immune response of the lung was impaired: cytokine mRNA expression of interferon-gamma, CXCR3, IL12p70, IL-10, and CXCR3 positive cells in the lung were significantly reduced compared to control animals exposed to filtered air.[70] Yu et al. also found that perinatal ETS exposure altered the immune response and airway innervation in rhesus macaque monkeys.[19]

In Utero Versus Postnatal Exposure

Many studies have used questionnaires to determine if children raised in the home of a smoker(s) have more respiratory symptoms than those raised in the home of a non-smoker(s). These studies provide a large body of data from around the world showing that smoke exposure is associated with respiratory illness. However, differentiating the effect of neonatal smoke exposure from postnatal ETS exposure is difficult. The most useful types of studies to examine the effects of postnatal ETS exposure separate from prenatal smoke exposure are those in which (1) the mother is not a smoker but other household members are smokers, (2) the influence of smoking by other household members is reported, (3) there is a dose–response relationship between current ETS exposure and the effect reported, (4) statistical methods used to separate in utero from postnatal effects, and (5) animals are used to explicitly study known exposures.

Statistical techniques have been used to separate the effects of smoke exposure in early life (in utero and the first five years of life) from current maternal smoke exposure in 6 to 10-year-old children[71]; it was found (after adjusting for early smoke exposure) that current exposure to maternal smoking decreased FEF$_{25-75}$ by 2.3%. In a study of 317 children 12–15 years of age who did not live with smokers or smoke themselves, and hence experienced only occasional exposure to ETS, there was a measurable effect of ETS exposure on lung function[72]; the children with greater exposures to ETS as documented by urinary cotinine had lower lung function as measured by FEF$_{25-75}$. When exercise challenge was used as a measure of airway responsiveness, it was shown that exposure to ETS during the first year of life but not during pregnancy or currently was associated with a 1.8-fold increased risk of airway hyperresponsiveness.[73]

To try to determine if the risk of developing asthma is due to ETS exposure or from other confounding factors, such as exposure to other pollutants or allergens, Joad et al.[74] designed an experiment in which rats were exposed to sidestream smoke (1 mg total suspended particulates/m^3) or to filtered air for 4 hours/day, 7 days/week from day 3 of gestation until birth. At birth, the pups were randomly assigned to receive either sidestream smoke or filtered air exposure until the equivalent of adolescence (7–10 weeks of age). Rats exposed to sidestream smoke only during gestation or only during postnatal life did not develop hyperreactive airways. However, if they were exposed to sidestream smoke during gestation and postnatal life, their airways were hyperreactive to methacholine. In the sidestream smoke-exposed group, pulmonary resistance (a measure of airway obstruction) at the highest doses of methacholine was more than 2-fold greater than the filtered air-exposed group (Figure 3).

Increased airway responsiveness was associated with an increase in the number of pulmonary neuroendocrine cells, suggesting that bronchoconstrictor mediators may play a mechanistic role (Figure 4). In a subsequent study, it was discovered that ETS exposure need only to occur during gestation and in the first 3 weeks of life. With 5 subsequent weeks in filtered air, these animals still had airways that were hyperresponsive to methacholine in adolescence (8 weeks of age).[75]

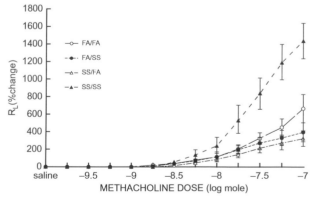

FIGURE 3 Methacholine-induced changes in pulmonary resistance (RL) in isolated lungs from rats exposed in utero to filtered air (FA) followed by 8–10 weeks postnatal FA (FA/FA), in utero FA followed by postnatal sidestream smoke (SS, 1mg/m^3 total suspended particulates, FA/SS), in utero SS followed by postnatal FA (SS/FA), and in utero SS followed by postnatal SS (SS/SS). SS/SS exposure resulted in hyperresponsiveness to methacholine. *Reproduced with permission from Joad et al.[74]*

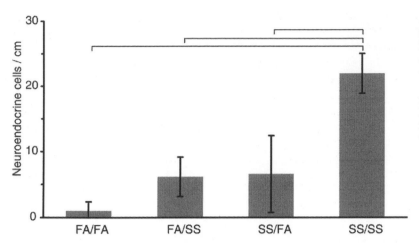

FIGURE 4 Number of neuroendocrine cells (identified by neuron-specific enolase staining) per centimeter basal lamina in lungs from rats exposed in utero to filtered air (FA) followed by 8–10 weeks postnatal FA (FA/FA), in utero FA followed by postnatal sidestream smoke (SS, 1mg/m^3 total suspended particulates, FA/SS), in utero SS followed by postnatal FA (SS/FA), and in utero SS followed by postnatal SS (SS/SS). SS/SS exposure resulted in more pulmonary neuroendocrine cells. *Reproduced with permission from Joad et al.*[74]

EFFECTS OF PRENATAL AND POSTNATAL SMOKE EXPOSURE ON THE DEVELOPMENT OF DISEASE LATER IN LIFE

The relative long-term contribution of childhood smoke exposure to adult health is still being determined. As described earlier, we know that childhood exposure to ETS is associated with wheeze with colds, persistent wheeze, shortness of breath, airway obstruction, and increased airway responsiveness and asthma in children. However, evidence is rapidly accumulating that links ETS exposure early in life with later life disease. A recent retrospective analysis of data from 192 countries[48] revealed that ETS was estimated to have caused 603,000 premature deaths in 2004. The largest number of analyzed deaths attributable to ETS exposure in adults was caused by ischaemic heart disease, followed by lower respiratory infections in children, and asthma in adults (Table 3).[48]

Studies have shown that early life exposure to ETS is related to altered physiologic changes associated with asthma,[76] cardiovascular disease,[77,78] stroke,[79] dementia,[80] COPD,[81] and breast cancer.[82]

Asthma

As described earlier numerous studies indicate that ETS exposure, both prenatally and postnatally, is associated with the development of asthma in infants and young children. These asthmatic symptoms can extend into adolescence and adulthood. When Larsson et al.[76] surveyed a random total sample of 8008 Swedish inhabitants about smoke exposure and respiratory symptoms and disease, they found that childhood ETS exposure was associated with an increased risk of asthma in adult never-smokers. A large population-based study was conducted in south Finland to assess the effect of ETS exposure on adult on-set asthma[83]; study data indicated a dose-dependent effect for cumulative ETS exposure at home and with total lifetime cumulative exposure. Similarly, asthma prevalence was found significantly associated with ETS exposure during childhood and adulthood in an Indian multicentric population

study; in this study, ETS exposure only during adulthood was found to not be a significant risk factor.[84]

Besides being associated with the development of asthma, ETS exposure can increase the severity of established asthma in both children and adults. Between 2005 and 2010, 53.2% of US children and adolescents with asthma were exposed to ETS, and among low-income children adolescents with asthma, the percentage increased to 70.1%.[85] In a study of 94 children with asthma, children whose mothers smoked had 47% more symptoms, a 4-fold increased responsiveness to histamine, and 13% decrease in FEV1/FVC[86]; the effect was greater in boys than in girls and in older boys (with presumably longer exposure to ETS) compared with younger boys.[86] A similar study of 199 asthmatic children 8 months to 13 years old showed that children with the highest urinary cotinine values had a 1.8-fold increase in the risk of acute exacerbations of asthma.[87] In a study of 74 children 7–12 years old with asthma, ETS exposure decreased peak expiratory flow rates and increased bronchodilator use, cough, and phlegm production 8- to 12-fold.[88] In adult asthmatics, higher levels of ETS exposure have been found associated with greater asthma severity and a greater risk of severe asthma exacerbation resulting in hospital admission.[89] A study with 654 participants enrolled in NHLBI Severe Asthma Research Program (SARP) found that regardless of age, asthmatics exposed to ETS had poorer quality of life scores, increased health care utilization, and greater abnormalities in lung function.[90] For a more expansive review of exacerbation adult asthma by ETS, the reader is directed to Eisner.[91]

COPD and Cancer

COPD is the third leading cause of death in the United States.[92] The World Health Organization estimated approximately 64 million COPD patients worldwide in 2004.[93] The pathology of COPD is complex and heterogeneous, developing slowly and demonstrating inflammation, small-airway remodeling, hypersecretion of mucus, and emphysema. The primary

TABLE 3 Number of deaths from exposure to ETS in 2004

	Lower Respiratory Infections in Children <5 Years	Otitis Media in Children <3 Years	Asthma in Children <15 Years	Asthma in Adults	Lung Cancer in Adults	Ischaemic Heart Disease in Adults	Total
Africa (D)	23,219	2	63	1634	177	3063	28,200
Africa (E)	20,025	4	62	1796	276	2568	24,700
The Americas (A)	65	1	11	288	596	12,604	13,600
The Americas (B)	4169	12	60	932	681	11,427	17,300
The Americas (D)	1555	1	9	140	93	982	2800
Eastern Mediterranean regions (B)	1771	0	13	727	142	6223	8900
Eastern Mediterranean regions (D)	30,518	11	96	2243	318	22,011	55,200
Europe (A)	60	1	10	1112	1992	32,283	35,500
Europe (B)	5367	1	106	1306	751	29,966	37,500
Europe (C)	818	2	3	3277	1096	94,109	99,300
Southeast Asia region (B)	4465	0	135	3681	631	18,433	27,300
Southeast Asia region (D)	55,956	23	333	9827	1864	67,095	135,000
Western Pacific region (A)	39	0	5	697	938	8769	10,400
Western Pacific region (B)	17,150	13	243	8113	11,850	69,659	107,000
Worldwide	165,000	71	1150	35,800	21,400	379,000	603,000

Totals provided are rounded to the nearest significant figure.
By permission, Öberg et al.[48]

cause of COPD is tobacco smoke, which induces leukocyte activation, cytokine production, enhanced proinflammatory gene expression, increased reactive oxygen and nitrogen species production, and enhanced biosynthesis of oxidized lipids.[94,95] A subgroup of patients with COPD report symptoms of chronic cough and phlegm, defined as chronic bronchitis due to mucus hypersecretion. Numerous studies, including epidemiological, population-based cohort, lung cancer screening, and interventional trial, have demonstrated a close relationship between COPD and lung cancer, importantly, all three major phenotypes of COPD—chronic bronchitis, airway obstruction and emphysema—have been associated with an increased incidence of lung cancer.[96–98]

A number of studies indicate that early life exposure to ETS is associated with an increased risk for developing COPD. Upton et al.[99] found that if maternal smoking was more than 10 cigarettes per day, the risk of offspring developing COPD increased (OR 1.7). Eisner et al.[89] showed that the risk of COPD increased as the cumulative lifetime hours of ETS exposure increased. A recent study examined associations between childhood ETS exposure and adult COPD and respiratory symptoms.[100] Patients with COPD (n = 433) and without COPD (n = 325) participated in the Bergen COPD Cohort Study from 2006 to 2009. ETS exposure during childhood was associated with approximately a 2-fold greater risk of COPD in women (odds ratio 1.9). On the other hand, neither childhood nor adult ETS exposure was a significant risk factor for COPD related symptoms in men (odds ratio 2.2) (Table 4).

These results suggest that the burden of COPD could be reduced if children were not exposed to cigarette smoke. In addition to the known short-term effects of ETS exposure, there is increasing evidence of long-term effects,

TABLE 4 Unadjusted and adjusted logistic regression analyses of the risk of chronic obstructive pulmonary disease (COPD) according to childhood exposure to environmental tobacco smoke (ETS), current exposure to ETS at home, current exposure to ETS at work, exposure to dust or gas at work, family history of COPD, BMI and level of education for 324 women and 434 men in the Bergen COPD Cohort Study

	Unadjusted		Adjusted	
	OR	95% CI	OR	95% CI
Women				
Childhood exposure to ETS	1.56*	1.00, 2.45	1.91*	1.00, 3.67
Current exposure to ETS at home	1.38	0.84, 2.25	-	-
Current exposure to ETS at work	2.54*	1.25, 5.17	1.39	0.57, 3.37
Exposure to dust/gas at work	2.02*	1.26, 3.23	1.86*	1.01, 3.42
Family history of COPD	3.03*	1.83, 5.02	3.71*	1.91, 7.22
BMI	1.38	0.84, 2.25	-	-
Intermediate education	2.54*	1.25, 5.17	1.39	0.57, 3.37
Lower education	2.02*	1.26, 3.23	1.86*	1.01, 3.42
Men				
Childhood exposure to ETS	1.08	0.73, 1.59	1.12	0.66, 1.89
Current exposure to ETS at home	0.95	0.62, 1.47	-	-
Current exposure to ETS at work	0.92	0.60, 1.44	-	-
Exposure to dust/gas at work	2.86*	1.77, 4.63	2.16*	1.81, 3.95
Family history of COPD	3.15*	1.87, 5.29	3.61*	1.86, 6.98
BMI(dagger)	0.93*	0.89, 0.98	1.58	0.95, 2.61
Intermediate education	1.47	0.86, 2.52	2.68*	1.10, 6.53
Lower education	3.42*	1.68, 6.96	1.03	0.51, 2.07

The multivariate model was also adjusted for age, smoking status, and pack-years smoked, in addition to significant predictors from the univariate analyses. The reference categories were no childhood exposure to ETS at home, no current exposure to ETS at work, no exposure to dust/gas at work, no family history of COPD, and higher education.
* P<0.05 † In the multivariate model, an interaction term between BMI and age was included due to the significant interaction between these two variables in relation to the risk of COPD. BMI, body mass index; CI, confidence interval; OR, odds ratio.
Adapted from Johannessen et al.,[100] with permission.

which need to be investigated further through longitudinal population surveys.

Inflammatory processes are involved in various smoking-related diseases, such as COPD, and a mechanistic link between COPD and lung cancer has been discussed.[101,102] Inflammatory, emphysematous, and tumorigenic effects were observed in a recent study[103] of male A/J mice exposed to mainstream smoke for 2.5, 5, 10, and 18 months in selected combinations with post inhalation periods of 0, 4, 8, and 13 months. Histopathological examination of step-serial sections of the lungs revealed nodular hyperplasia of the alveolar epithelium and bronchioloalveolar adenoma and adenocarcinoma. Furthermore, available evidence[82] suggests that for women exposed long-term to low levels of ETS, the relationship between breast cancer and active smoking and ETS is consistent with causality.

Cardiovascular Disease

Each year there are approximately 600,000 cardiovascular disease related deaths in the United States,[104] and of these ETS causes approximately 22,700 to 69,600 deaths from heart disease.[105] Li et al.[106] found an association between childhood ETS exposure and increased risk for decreased carotid artery elasticity in adulthood (the average follow-up period since childhood was 26.5 years). Similarly, Kallio et al.[107] discovered that 13-year-olds frequently exposed to ETS between the ages of 8 and 13 years old had increased intima-media thickness in the distal abdominal aorta and in carotid arteries and an elevated apolipoprotein B (ApoB) concentration, which suggests early cardiovascular events. These two studies indicate the ETS exposure early in life can cause changes predictive of developing heart disease later in life. Animal studies with mice show that in utero exposure to ETS may cause mitochondrial damage, which increases the risk for atherogenesis and earlier onset of adult cardiovascular disease.[77,78]

Cognitive Decline

There is evidence that ETS exposure may detrimentally affect cognitive abilities. Clifford et al.[108] reviewed 20 studies that investigated in utero exposure to ETS and cognitive functioning in children. The most consistent association across the studies was in utero ETS exposure and reduced measures for academic achievement and intellectual abilities. It is not known if these impairments continue into older age, but their early effects on academic success are probable to have long-lasting repercussions. ETS exposure has also been found to affect the cognition of seniors. In a population-based study of individuals living in four Chinese provinces (Guangdong, Heilongjiang, Shanghai, Shanxi), cumulative duration of ETS exposure (relative risks = 1.78) was found related to an increased risk of dementia and Alzheimer's disease, and a significant association of ETS exposure with dementia appeared dose dependent.[109] Ho et al.[110] found in rats that ETS exposure altered β-amyloid precursor protein processing to favor the generation of senile plaques composed of β-amyloid peptides, a key pathological event in the development of Alzheimer's disease. Thus, Ho et al. concluded that daily exposure to ETS may increase aging of the brain.

CONCLUSIONS

Many children are exposed to cigarette smoke both prenatally and postnatally. Prenatal exposure to mainstream smoke from the mother and even to ETS from the mother in utero has been shown to affect fetal lung development. Soon after birth, the lungs of infants exposed prenatally to smoke show evidence of airway obstruction, airway hyperresponsiveness, and altered lung maturation. Children exposed to ETS postnatally have more symptoms of cough, wheeze, and respiratory illnesses than children not exposed. They also have small decreases in lung function and increases in airway responsiveness. Cigarette smoke exposure is associated with the early development of asthma and may increase the severity of asthma once it develops. Furthermore, smoke exposure is associated with the development of atopy and T helper-2 immune responses that may further worsen asthma. Prenatal and postnatal exposure to ETS in early life increases the risk for a number of diseases, including asthma, cardiovascular disease, dementia, and COPD. Thus, public health efforts should be directed at reducing the prenatal and postnatal exposure of children to cigarette smoke, with the greatest emphasis on encouraging women to stop smoking during pregnancy and in the first years of their children's lives.

ACKNOWLEDGMENTS

The authors acknowledge the following sources of support that provided the basis for our literature review: NIEHS P01 ES00628, NCRR RR00169, California Tobacco-Related Disease Research Program18XT-0154 (training grant number: T32 AI060555), and NIOSH OHO7550. We thank the research of Dr. Jesse Joad, whose work provided the starting point for this chapter.

REFERENCES

1. WHO. 10 Facts on Secondhand Smoke. 2009. http://www.who.int/features/factfiles/tobacco/tobacco_facts/en/index.html.
2. Treyster Z, Gitterman B. Second hand smoke exposure in children: environmental factors, physiological effects, and interventions within pediatrics. *Rev Environ Health* 2011;**26**:187–95.
3. WHO. Tobacco. 2013. http://www.who.int/mediacentre/factsheets/fs339/en/
4. Centers for Disease Control and Prevention (CDC). Vital Signs: Nonsmokers' Exposure to Secondhand Smoke—United States, 1999–2008. *Morb Mortal Wkly Rep* 2010;**59**:1141–6.
5. Scherer G, Conze C, Tricker AR, Adlkofer F. Uptake of tobacco smoke constituents on exposure to environmental tobacco smoke (ETS). *Klin Wochenschr* 1992;**70**:352–67.
6. Donnenfeld AE, Pulkkinen A, Palomaki GE, Knight GJ, Haddow JE. Simultaneous fetal and maternal cotinine levels in pregnant women smokers. *Am J Obstet Gynecol* 1993;**168**:781–2.
7. Eliopoulos C, Klein J, Khan Phan M, Knie B, Greenwald M, Chitayat D, Koren G. Hair concentrations of nicotine and cotinine in women and their newborn infants. *J Am Med Assoc* 1994;**271**:621–3.
8. Schulte-Hobein B, Schwartz-Bickenbach D, Abt S, Plum C, Nau H. Cigarette smoke exposure and development of infants throughout the first year of life: influence of passive smoking and nursing on cotinine levels in breast milk and infant's urine. *Acta Paediatr Suppl* 1992;**81**:550–7.
9. Crawford FG, Mayer J, Santella RM, Cooper TB, Ottman R, Tsai W-Y, et al. Biomarkers of environmental tobacco smoke in preschool children and their mothers. *J Natl Cancer Inst* 1994;**86**:1398–402.
10. Pirkle JL, Flegal KM, Bernert JT, Brody DJ, Etzel RA, Maurer KR. Exposure of the US population to environmental tobacco smoke – The Third National Health and Nutrition Examination Survey, 1988 to 1991. *JAMA* 1996;**275**:1233–40.
11. Cook DG, Whincup PH, Jarvis MJ, Strachan DP, Papacosta O, Bryant A. Passive exposure to tobacco smoke in children aged 5-7 years: individual, family, and community factors. *BMJ* 1994;**308**:384–9.
12. Irvine L, Crombie IK, Clark RA, Slane PW, Goodman KE, Feyerabend C, et al. What determines levels of passive smoking in children with asthma. *Thorax* 1997;**52**:766–9.
13. Stocks J, Sonnappa S. Early life influences on the development of chronic obstructive pulmonary disease. *Ther Adv Respir Dis* 2013;**7**:161–74.
14. Quanjer P, Stanojevic S, Cole T, Baur X, Hall G, Culver B, et al. Multi-ethnic reference values for spirometry for the 3–95 year age range: the global lung function 2012 equations. *Eur Respir J* 2012;**40**:1324–43.

15. Moshammer H, Hoek G, Luttmann-Gibson H, Neuberger MA, Antova T, Gehring U, et al. Parental smoking and lung function in children: an international study. *Am J Respir Crit Care Med* 2006;**173**:1255–63.

16. Pinkerton KE, Joad JP. Influence of air pollution on respiratory health during perinatal development. *Clin Exp Pharmacol Physiol* 2006;**33**:269–72.

17. Ladomenou F, Kafatos A, Galanakis E. Environmental tobacco smoke exposure as a risk factor for infections in infancy. *Acta Pediatrica* 2009;**98**:1137–41.

18. Surgeon General. The Health Consequences of Involuntary Exposure to Tobacco Smoke: A Report of the Surgeon General. U.S. Department of Health and Human Services. http://www.surgeongeneral.gov/library/reports/secondhandsmoke/index.html; 2006.

19. Yu M, Zheng X, Peake J, Joad JP, Pinkerton KE. Perinatal environmental tobacco smoke exposure alters the immune response and airway innervation in infant primates. *J Allergy Clin Immunol* 2008;**22**:640–7.

20. Vardavas CI, Plada M, Tzatzarakis M, Marcos A, Warnberg J, Gomez-Martinez S, et al. Passive smoking alters circulating naïve/memory lymphocyte T-cell subpopulations in children. *Pediatr Allergy Immunol* 2010;**21**:1171–8.

21. Noakes PS, Holt PG, Prescott SL. Maternal smoking in pregnancy alters neonatal cytokine responses. *Allergy* 2003;**53**:1053–8.

22. Wilson KM, Pier JC, Wesgate SC, Cohen JM, Blumkin AK. Secondhand tobacco smoke exposure and severity of influenza in hospitalized children. *J Pediatrics* 2013;**162**:16–21.

23. Tasher D, Stein M, Simoes EA, Shohat T, Bromberg M, Somekh E. Invasive bacterial infections in relation to influenza outbreaks, 2006–2010. *Clinical Infectious Diseases* 2011;**53**:1199–207.

24. Cox CM, Blanton L, Dhara R, Brammer L, Finelli L. 2009 pandemic influenza (H1N1) deaths among children—United States, 2009–2011. *Clin Infect Dis* 2011;**52**(Suppl. 1):S69–74.

25. Claude JA, Grimm A, Savage HP, Pinkerton KE. Perinatal exposure to environmental tobacco smoke (ETS) enhances susceptibility to viral and secondary bacterial infections. *Int J Environ Res Public Health* 2012;**9**:3954–64.

26. Cheraghi M, Salvi S. Environmental tobacco smoke (ETS) and respiratory health in children. *Eur J Pediatr* 2009;**168**:897–905.

27. Gentile D, Howe-Adams J, Trecki J, Patel A, Angelini B, Skoner D. Association between environmental tobacco smoke and diminished dendritic cell interleukin 10 production during infancy. *Ann Allergy Asthma Immunol* 2004;**92**:433–7.

28. Tebow G, Sherrill DL, Lohman IC, Stern DA, Wright AL, Martinez FD, et al. Effects of parental smoking on interferon γ production in children. *Pediatrics* 2008;**121**:1563–9.

29. Kohli A, Garcia MA, Miller RL, Maher C, Humblet O, Hammond SK, et al. Secondhand smoke in combination with ambient air pollution exposure is associated with increased CpG methylation and decreased expression of IFN-γ in T effector cells and Foxp3 in T regulatory cells in children. *Clin Epigenetics* 2012;**4**:17.

30. Li YF, Gilliland FD, Berhane K, McConnell R, Gauderman WJ, Rappaport EB, et al. Effects of in-utero and environ- mental tobacco smoke exposure on lung function in boys and girls with and without asthma. *Am J Respir Crit Care Med* 2000;**162**:2097–104.

31. Young S, Le Souef PN, Geelhoed GC, Stick SM, Chir B, Turner KJ, et al. The influence of a family history of asthma and parental smoking on airway responsiveness in early infancy. *N Engl J Med* 1991;**324**:1168–73.

32. Ehrlich R, Kattan M, Godbold J, Saltzberg DS, Grimm KT, Landrigan PJ, et al. Childhood asthma and passive smoking. Urinary cotinine as a biomarker of exposure. *Am Rev Respir Dis* 1992;**145**:594–9.

33. Stick SM, Burton PR, Gurrin L, Sly PD, Lesouef PN. Effects of maternal smoking during pregnancy and a family history of asthma on respiratory function in newborn infants. *Lancet* 1996;**348**:1060–4.

34. Lodrup Carlsen KC, Jaakkola JJ, Nafstad P, Carlsen KH. In utero exposure to cigarette smoking influences lung function at birth. *Eur Respir J* 1997;**10**:1774–9.

35. Hanrahan JP, Tager IB, Segal MR, Tosteson TD, Castile RG, Van Vunakis H, et al. The effect of maternal smoking during pregnancy on early infant lung function. *Am Rev Respir Dis* 1992;**145**:1129–35.

36. Tager IB, Ngo L, Hanrahan JP. Maternal smoking during pregnancy. Effects on lung function during the first 18 months of life. *Am J Respir Crit Care Med* 1995;**152**:977–83.

37. Gilliland FD, Li YF, Peters JM. Effects of maternal smoking during pregnancy and environmental tobacco smoke on asthma and wheezing in children. *Am J Respir Crit Care Med* 2001;**163**:429–36.

38. Lannerö E, Wickman M, Pershagen G, Nordvall L. Maternal smoking during pregnancy increases the risk of recurrent wheezing during the first years of life (BAMSE). *Respir Res* 2006;**7**:3.

39. Lieberman E, Torday J, Barbieri R, Cohen A, Van Vunakis H, Weiss ST. Association of intrauterine cigarette smoke exposure with indices of fetal lung maturation. *Obstet Gynecol* 1992;**79**:564–70.

40. Collins MH, Moessinger AC, Kleinerman J, Bassi J, Rosso P, Collins AM, et al. Fetal lung hypoplasia associated with maternal smoking: a morphometric analysis. *Pediatr Rehabil* 1985;**19**:408–12.

41. Vidic,B. Transplacental effect of environmental pollutants on interstitial composition and diffusion capacity for exchange of gases of pulmonary parenchyma in neonatal rat. *Bul Assoc Anat* 1991;**75**:153–5.

42. Ji CM, Royce FH, Truong U, Plopper CG, Singh G, Pinkerton KE. Maternal exposure to environmental tobacco smoke alters Clara cell secretory protein expression in fetal rat lung. *Am J Physiol Lung Cell Mol Physiol* 1998;**275**:L870–6.

43. Sandberg K, Pinkerton K, Poole S, Minton P, Sundell H. Fetal nicotine exposure increases airway responsiveness and alters airway wall composition in young lambs. *Respir Physiolo Neurobiol* 2011;**176**:57–67.

44. Gilliland FD, Berhane K, McConnell R, Gauderman WJ, Vora H, Rappaport EB, et al. Maternal smoking during pregnancy, environmental tobacco smoke exposure and childhood lung function. *Thorax* 2000;**55**:271–6.

45. Wang L, Pinkerton KE. Detrimental effects of tobacco smoke exposure during development on postnatal lung function and asthma. *Birth Defects Res* 2008;**84**:54–60.

46. Hayatbakhsh M, Sadasivam S, Mamun A, Najman J, Williams G, O'Callaghan M. Maternal smoking during and after pregnancy and lung function in early adulthood: a prospective study. *Thorax* 2009;**64**:810–4.

47. Yang GH, Ma JM, Liu N, Zhou LN. Smoking and passive smoking in Chinese. *Chin J Epidemiol* 2005;**26**:77–80. [in Chinese].

48. Öberg M, Jaakkola MS, Woodward A, Peruga A, Prüss-Ustün A. Worldwide burden of disease from exposure to second-hand smoke: a retrospective analysis of data from 192 countries. *Lancet* 2011;**377**:139–46.

49. Chen Y. Environmental tobacco smoke, low birth weight, and hospitalization for respiratory disease. *Am J Respir Crit Care Med* 1994;**150**:54–8.

50. Holberg CJ, Wright AL, Martinez FD, Morgan WJ, Taussig LM. Child day care, smoking by caregivers, and lower respiratory tract illness in the first 3 years of life. *Pediatrics* 1993;**91**:885–92.

51. Gergen PJ, Fowler JA, Maurer KR, Davis WW, Overpeck MD. The burden of environmental tobacco smoke exposure on the respiratory health of children 2 months through 5 years of age in the United States: Third National Health and Nutrition Examination Survey, 1988 to 1994. *Pediatrics* 1998;**101**:E81–6.

52. Magnusson CGM. Maternal smoking influences cord serum IgE and IgD levels and increases the risk for subsequent infant allergy. *J Allergy Clin Immunol* 1986;**78**:898–904.

53. Lindfors A, Hage-Hamsten M, Rietz H, Wickman M, Nordvall SL. Influence of interaction of environmental risk factors and sensitization in young asthmatic children. *J Allergy Clin Immunol* 1999;**104**:755–62.

54. Dybing E, Sanner T. Passive smoking, sudden infant death syndrome (SIDS) and childhood infections. *Hum Exp Toxicol* 1999;**18**:202–5.

55. Mitchell EA, Ford RPK, Stewart AW, Taylor BJ, Becroft DMO, Thompson JMD, et al. Smoking and Sudden Infant Death Syndrome. *Pediatrics* 1993;**92**:893–6.

56. Carpenter RG, Irgens LM, Blair PS, England PD, Fleming P, Huber J, et al. Sudden unexplained infant death in 20 regions in Europe: case control study. *Lancet* 2004;**363**:185–91.

57. Ladomenou F, Kafatos A, Tselentis Y, Galanakis E. Predisposing factors for acute otitis media in infancy. *J Infect* 2010;**61**:49–53.

58. Prtak L, Al-Adnani M, Fenton P, Kudesia G, Cohen MC. Contribution of bacteriology and virology in sudden unexpected death in infancy. *Arch Dis Child* 2010;**95**:371–6.

59. Mayr M, Kiechl S, Willeit J, Wick G, Xu Q. Infections, immunity, and atherosclerosis: associations of antibodies to Chlamydia pneumoniae, Helicobacter pylori, and cytomegalovirus with immune reactions to heat-shock protein 60 and carotid or femoral atherosclerosis. *Circulation* 2000;**102**:833–9.

60. Highet AR. An infectious aetiology of sudden infant death syndrome. *J Appl Microbiol* 2008;**105**:625–35.

61. Cunningham J, O'Connor GT, Dockery DW, Speizer FE. Environmental tobacco smoke, wheezing, and asthma in children in 24 communities. *Am J Respir Crit Care Med* 1996;**153**:218–24.

62. Azizi BHO, Henry RL. The effects of indoor environmental factors on respiratory illness in primary school children in Kuala Lumpur. *Int J Epidemiol* 1991;**20**:144–50.

63. Haby MM, Peat JK, Woolcock AJ. Effect of passive smoking, asthma, and respiratory infection on lung function in Australian children. *Pediatr Pathol* 1994;**18**:323–9.

64. Sherrill DL, Martinez FD, Lebowitz MD, Holdaway MD, Flannery EM, Herbison GP, et al. Longitudinal effects of passive smoking on pulmonary function in New Zealand children. *Am Rev Respir Dis* 1992;**145**:1136–41.

65. Kuehr J, Frischer T, Karmaus W, Meinert R, Pracht T, Lehnert W. Cotinine excretion as a predictor of peak flow variability. *Am J Resp Crit Care Med* 1998;**158**:60–4.

66. Forastiere F, Agabiti N, Corbo GM, Pistelli R, Dell'Orco V, Ciappi G, et al. Passive smoking as a determinant of bronchial responsiveness in children. *Am J Respir Crit Care Med* 1994;**149**: 365–70.

67. Cook DG, Strachan DP. Parental smoking, bronchial reactivity and peak flow variability in children. *Thorax* 1998;**53**:295–301.

68. Martinez FD, Cline M, Burrows B. Increased incidence of asthma in children of smoking mothers. *Pediatrics* 1992;**89**:21–6.

69. Schuster MA, Franke T, Pham CB. Smoking patterns of household members and visitors in homes with children in the United States. *Arch Pediatr Adolesc Med* 2002;**156**:1094–100.

70. Wang L, Joad JP, Zhong C, Pinkerton KE. Effects of environmental tobacco smoke exposure on pulmonary immune response in infant monkeys. *J Allergy Clin Immunol* 2008;**122**:400–6.

71. Wang X, Wypij D, Gold DR, Speizer FE, Ware JH, Ferris Jr BG, et al. A longitudinal study of the effects of parental smoking on pulmonary function in children 6–18 years. *Am J Respir Crit Care Med* 1994;**149**:1420–5.

72. Corbo GM, Agabiti N, Forastiere F, Dell'Orco V, Pistelli R, Kriebel D, et al. Lung function in children and adolescents with occasional exposure to environmental tobacco smoke. *Am J Respir Crit Care Med* 1996;**154**:695–700.

73. Frischer T, Kuehr J, Meinert R, Karmaus W, Barth R, Hermann-Kunz E, et al. Maternal smoking in early childhood: a risk factor for bronchial responsiveness to exercise in primary-school children. *J Philos* 1992;**121**:17–22.

74. Joad JP, Ji C, Kott KS, Bric JM, Pinkerton KE. *In utero* and postnatal effects of sidestream cigarette smoke exposure on lung function, hyperresponsiveness, and neuroendocrine cells in rats. *Toxicol Appl Pharmacol* 1995;**132**:63–71.

75. Joad JP, Bric JM, Peake JL, Pinkerton KE. Perinatal exposure to aged and diluted sidestream cigarette smoke produces airway hyperresponsiveness in older rats. *Toxicol Appl Pharmacol* 1999;**155**: 253–60.

76. Larsson ML, Frisk M, Hallstrom J, Kiviloog J, Lundback B. Environmental tobacco smoke during childhood is associated with increased prevalence of asthma in adults. *Chest* 2001;**120**:711–7.

77. Yang Z, Knight CA, Mamerow M, Vickers K, Penn A, Postlethwait E, et al. Prenatal environmental tobacco smoke exposure promotes adult atherogenesis and mitochondrial damage in apoE-/- mice fed a chow diet. *Circulation* 2004;**110**:3715–20.

78. Yang Z, Harrison CM, Chuang G, Ballinger SW. The role of tobacco smoke induced mitochondrial damage in vascular dysfunction and atherosclerosis. *Mutat Res* 2007;**621**:61–74.

79. Lee PN, Forey BA. Environmental tobacco smoke exposure and risk of stroke in nonsmokers: a review with meta-analysis. *J Stroke Cerebrovasc Dis* 2006;**15**:190–201.

80. Chen R, Tavendale R, Tunstall-Pedoe H. Environmental tobacco smoke and prevalent coronary heart disease among never smokers in the Scottish MONICA surveys. *Occup Environ Med* 2004;**61**:790–2.

81. Eisner MD, Balmes J, Katz PP, Trupin L, Yelin EH, Blanc PD. Lifetime environmental tobacco smoke exposure and the risk of chronic obstructive pulmonary disease. *Environ Health* 2005b;**4**:7.

82. Johnson KC, Miller AB, Collishaw NE, Palmer JR, Hammond SK, Salmon AG, et al. Active smoking and secondhand smoke increase breast cancer risk: the report of the Canadian Expert Panel on Tobacco Smoke and Breast Cancer Risk. *Tob Control* 2011;**20**:e2.

83. Jaakkola MS, Piipari R, Jaakkola N, Jaakkola JJK. Environmental tobacco smoke and adult-onset asthma: a population-based incident case–control study. *Am J Public Health* 2003;**93**:2055–60.

84. Gupta D, Aggarwal AN, Chaudhry K, Chhabra SK, D'Souza GA, Jindal SK, et al. Household environmental tobacco smoke exposure, respiratory symptoms and asthma in non-smoker adults: a multicentric population study from India. *Indian J Chest Dis Allied Sci* 2006;**48**:31–6.

85. Kit BK, Simon AE, Brody DJ, Akinbami LJ. US prevalence and trends in tobacco smoke exposure among children and adolescents with asthma. *Pediatrics* 2013;**131**:407–14.

86. Murray AB, Morrison BJ. The effect of cigarette smoke from the mother on bronchial responsiveness and severity of symptoms in children with asthma. *J Allergy Clin Immunol* 1986;**77**:575–81.

87. Chilmonczyk BA, Salmun LM, Megathlin KN, Neveux LM, Palomaki GE, Knight GJ, et al. Association between exposure to environmental tobacco smoke and exacerbations of asthma in children. *N Engl J Med* 1993;**328**:1665–9.

88. Schwartz J, Timonen KL, Pekkanen J. Respiratory effects of environmental tobacco smoke in a panel study of asthmatic and symptomatic children. *Am J Respir Crit Care Med* 2000;**161**:802–6.

89. Eisner MD, Klein J, Hammond SK, Koren G, Lactao G, Iribarren C. Directly measured second hand smoke exposure and asthma health outcomes. *Thorax* 2005;**60**:814–21.

90. Comhair SA, Gaston BM, Ricci KS, Hammel J, Dweik RA, Teague WG, et al. Detrimental effects of environmental tobacco smoke in relation to asthma severity. *PLoS One* 2011;**6**:e18574.

91. Eisner MD. Passive smoking and adult asthma. *Immunol Allergy Clin North Am* 2008;**28**:521–37.

92. Centers for Disease Control and Prevention (CDC). Chronic obstructive pulmonary disease among adults – United States, 2011. *MMWR Morb Mortal Wkly Rep* 2012;**61**:938–43.

93. WHO. *The global burden of disease*; 2008. 2004 http://www.who.int/healthinfo/global_burden_disease/2004_report_update/en/.

94. Kubo S, Kobayashi M, Iwata M, Takahashi K, Miyata K, Shimizu Y. Disease-modifying effect of ASP3258, a novel phosphodiesterase Type 4 inhibitor, on subchronic cigarette smoke exposure–induced lung injury in guinea pigs. *Eur J Pharmacol* 2011;**659**:79–84.

95. Wang L, Yang J, Guo L, Uyeminami D, Dong H, Hammock BD, et al. Use of a soluble epoxide hydrolase inhibitor in smoke-induced chronic obstructive pulmonary disease. *Am J Respir Cell Mol Biol* 2012;**46**:614–22.

96. Celli BR. Chronic obstructive pulmonary disease and lung cancer. *Proc Am Thorac Soc* 2012;**9**:74–9.

97. Mannino DM, Aguayo SM, Petty TL, Redd SC. Low lung function and incident lung cancer in the United States: data from the First National Health and Nutrition Examination Survey follow-up. *Arch Intern Med* 2003;**163**:1475–80.

98. Young RP, Hopkins RJ, Christmas T, Black PN, Metcalf P, Gamble GD. COPD prevalence is increased in lung cancer, independent of age, sex and smoking history. *Immunol Allergy Clin N Am* 2009;**34**:380–6.

99. Upton MN, Smith GD, McConnachie A, Hart CL, Watt GC. Maternal and personal cigarette smoking synergize to increase airflow limitation in adults. *Am J Respir Crit Care Med* 2004;**169**:479–87.

100. Johannessen A, Bakke PS, Hardie JA, Eagan TM. Association of exposure to environmental tobacco smoke in childhood with chronic obstructive pulmonary disease and respiratory symptoms in adultsresp. *Respirology* 2012;**17**:499–505.

101. Adcock IM, Caramori G, Barnes PJ. Chronic obstructive pulmonary disease and lung cancer: new molecular insights. *Respiration* 2011;**81**:265–84.

102. Houghton AM, Mouded M, Shapiro SD. Common origins of lung cancer and COPD. *Nat Med* 2008;**14**:1023–4.

103. Stinn W, Buettner A, Weiler H, Friedrichs B, Luetjen S, van Overveld F, et al. Lung inflammatory effects, tumorigenesis, and emphysema development in a long-term inhalation study with cigarette mainstream smoke in mice. *Toxicol Sci* 2013;**131**:596–611.

104. Kochanek KD, Jiaquan X, Murphy SL, Miniño AM. Deaths: preliminary data for 2009. *Natl Vital Stat Rep* 2011;**59**:1–51.

105. California Environmental Protection Agency. Identification of Environmental Tobacco Smoke as a Toxic Air Contaminant. *Executive Summary* June 2005.

106. Li SX, Chen W, Srinivasan SR, Berenson GS. Childhood blood pressure as a predictor of arterial stiffness in young adults-the Bogalusa Heart Study. *Hypertension* 2004;**43**:541–6.

107. Kallio K, Jokinen E, Saarinen M, Hämäläinen M, Volanen I, Kaitosaari T, et al. Arterial intima-media thickness, endothelial function, and apolipoproteins in adolescents frequently exposed to tobacco smoke. *Circ Cardiovasc Qual Outcomes* 2010;**3**:196–203.

108. Clifford A, Lang L, Chen R. Effects of maternal cigarette smoking during pregnancy on cognitive parameters of children and young adults: a literature review. *Neurotoxicol Teratol* 2012;**34**:560–70.

109. Chen R. Association of environmental tobacco smoke with dementia and Alzheimer's disease among never smokers. *Alzheimers Dement* 2012;**8**:590–5.

110. Ho YS, Yang X, Yeung SC, Chiu K, Lau CF, Tsang AW, et al. Cigarette smoking accelerated brain aging and induced pre-Alzheimer-like neuropathology in rats. *PLoS One* 2012;**7**:e36752.

Chapter 22

Nicotine Exposure during Early Development: Effects on the Lung

Gert S. Maritz

Department of Physiological Sciences, University of the Western Cape, Bellville, South Africa

INTRODUCTION

Maternal tobacco smoking during pregnancy has been associated with many adverse outcomes for offspring such as increased incidences of pneumonia and bronchitis,[1–3] impaired lung function, and general respiratory disorders.[4,5] In epidemiological studies a relationship has been found between diseases of the distal airways of children of smoking parents and chronic bronchitis and emphysema in the same individuals when adults.[6] It is therefore clear that certain components of cigarette smoke interfere with normal lung development and cellular function. Nicotine, a major component of tobacco smoke, is implicated in the adverse effects of tobacco smoke on the metabolic[7] and structural development of the lung.[8] Indeed, several studies have shown that maternal and grand-maternal nicotine intake during gestation and lactation adversely affect lung development in the offspring.[9,10]

UPTAKE OF NICOTINE

Nicotine is typically absorbed into the body by means of inhalation, transdermal patches, gums, nasal sprays, snuff, and chewing tobacco.[11] When inhaled in tobacco smoke, nicotine is directly absorbed via the pulmonary capillaries into the pulmonary venous circulation; it then enters the arterial supply to tissues that take it up in variable amounts. When taken intranasally, nicotine is absorbed into a rich submucosal venous plexus that drains into the facial, sphenopalatine, and ophthalmic veins, from where it enters the left side of the heart and appears in arterial blood.[12]

Daily Nicotine Intake

In regular smokers of cigarettes, the daily intake of nicotine varies widely between 10.5 mg and 78.6 mg, with an average intake of 35 mg.[13] The nicotine intake per cigarette averages ~1 mg.[13] The absorption of nicotine by the lungs from tobacco smoke is rapid and arterial blood concentrations reach a peak of 49.2±9.7 ng/ml after 5 min. When a cigarette is smoked, the maximum concentration of nicotine in the jugular vein peaks at 22.4±3.9 ng/ml at about 7 min.[14]; the nicotine concentration then levels off at 4–6 hr. Because its half-life in adults is about 120 min., nicotine persists in potentially biologically active concentrations throughout the night, despite nocturnal cessation of smoking.[15] Although the mean daily intake is not different between men and women, blood nicotine concentrations may differ as men metabolize nicotine faster than women.[16]

Under experimental conditions, animals are usually treated with 1 mg nicotine/kg body weight/day, a dose equivalent to that of heavy smokers.[13] In pregnant rats, this dose results in an average nicotine concentration in amniotic fluid of 15.4±3.9 ng/ml, similar to the levels of nicotine found in amniotic fluid of pregnant women who smoke tobacco.[17] By comparison, light smokers would typically receive one-half of this dose, and transdermal nicotine and nicotine gum would deliver one-fourth and one-eighth of this dose, respectively.[18]

Absorption, Distribution, and Metabolism of Nicotine in Pregnancy

Nicotine crosses the placenta rapidly from mother to fetus, reaching higher concentrations in the fetal circulation than in the mother's; that is, the disappearance of nicotine from fetal blood is slower than from the maternal circulation. It is important to note that fetal skin, due to its low keratinization during the second trimester, is readily permeable to substances such as nicotine in amniotic fluid.[17] Nicotine is even absorbed by adult skin.[19]

Elevated levels of nicotine occur in the amniotic fluid and placental tissue of pregnant women who smoke.[20,21] Considerable amounts of nicotine enter fetal blood and nicotine reaches high concentrations in the fetal adrenal glands, heart, kidneys, stomach wall, spleen,[22] and lung.[23] During the first

The Lung. http://dx.doi.org/10.1016/B978-0-12-799941-8.00022-5

half of pregnancy, concentrations of cotinine, the major metabolite of nicotine, are greater in amniotic fluid and fetal serum than in maternal serum in both active and passive smokers.[21] The relatively large area available for nicotine absorption by the fetus, together with the slower elimination of nicotine by fetal tissues, explains why the fetus is exposed to higher concentrations of nicotine than the mother. This could enhance the toxicity of nicotine to the fetus.[24]

After birth, elevated levels of nicotine occur in the maternal milk[17] and nicotine is rapidly absorbed by the infant, as determined by its presence in the saliva (166 ng/ml) of breast-fed infants.[25]

Effects of Maternal Nicotine Intake during Gestation and Lactation on the Placenta and Fetal Growth

Extensive epidemiological studies have demonstrated a positive correlation between the concentration of nicotine in the maternal blood and fetal growth restriction.[26] Smoking-induced alterations in the morphology of the umbilical arterial wall and a reduction in umbilico-placental blood flow are thought to contribute to fetal growth restriction and low birth weight of infants of mothers who smoke during pregnancy.[27]

As nicotine and its metabolites readily cross the fetal membranes and accumulate in amniotic fluid,[17] it is reasonable to assume that they will also affect placental morphology and function. It has been demonstrated that placental tissue from smoking mothers has both a reduced uptake and transfer of amino acids.[28] Addition of high levels of nicotine (10–100 times greater than serum concentrations measured 5 min. after smoking one cigarette) to cultures of placental slices from smoking mothers results in an inhibition of placental amino acid transport.[29] This suggests that the decrease in placental amino acid uptake induced by nicotine, as demonstrated in vitro, may contribute to the growth restriction observed in fetuses of smoking mothers.[28]

Trophoblast differentiation during the first trimester of pregnancy is required for implantation and subsequent formation of the placenta, a clear prerequisite for normal fetal development. It has been suggested that the effects of maternal smoking on the placenta may depend on the stage of pregnancy at which exposure occurs.[28] The early (first trimester) period is critical because that is when differentiation of the specialized epithelial cells of the placenta (trophoblasts) occurs.

Effects of Nicotine Exposure on Lung Development

Energy Metabolism

Glucose uptake and metabolism are essential for the proliferation and survival of cells, and may be enhanced in actively proliferating cell systems such as embryonic tissue.

Glucose is considered to be an essential source of energy in lung tissue[30] and is necessary for the functional development of the lung.[31–33] Glucose is also the main source of α-glycerophosphate for surfactant synthesis in adult lung,[34] while in fetal lung, loss of cellular glycogen from alveolar type II cells just before birth is associated with increased surfactant synthesis.[35,36]

During the alveolar phase of lung development, lung tissue is more dependent on glycogen as an energy substrate than adult lung. This is illustrated by the finding that, during fasting, the activity of phosphorylase in adult lung tissue decreases to conserve glycogen while the activity of phosphorylase in fetal and neonatal lung increases, thereby increasing the utilization of glycogen. This means that the control of glycogen metabolism during the alveolar phase of lung development is different from that of adults.[37]

Although glucose and glycogen are the primary energy substrates of the adult and developing lung, fatty acids can also be important. For example, during fasting, when blood fatty acid levels are elevated, fatty acids replace glucose as primary energy substrate. Under these circumstances, glucose is conserved by the lung for α-glycerophosphate synthesis and eventual surfactant formation by alveolar type II cells.[38]

Maternal nicotine exposure during gestation results in sustained suppression of glycogenolysis and glycolysis in lung tissue of the rat fetus.[7,39] The lower glycogenolytic activity is due to a lower phosphorylase activity in the lungs of nicotine exposed offspring.[39] The ratio of inactive:active phosphorylase of lung tissue of nicotine-exposed offspring is the same as for animals that were not exposed to nicotine during gestation and lactation. However, the tissue levels of both the phosphorylase fractions are lower than in the lungs of control animals (Figure 1), which implies that the total phosphorylase content of the lungs of the nicotine exposed animals was lower than that of the control animals. This means that nicotine exposure suppressed the synthesis of phosphorylase in the lungs.[39] It also implies that nicotine had no direct inhibitory effect on phosphorylase, but that the slower rate of glycogenolysis in the lungs of these animals is rather due to a decrease in the level of phosphorylase

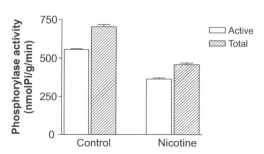

FIGURE 1 The influence of maternal nicotine exposure during gestation and lactation in rats on the phosphorylase activity of the lungs of offspring. (Active = active phosphorylase; Total = total phosphorylase.)

available to catalyze glycogenolysis. This implies that the fetal and neonatal lung of animals exposed to nicotine during development is more dependent on exogenous glucose for utilization via the glycolytic pathway and the hexose monophosphate shunt than on glucose from pulmonary glycogen stores.

The cellular uptake of exogenous glucose is usually carried out by glucose transporters. Glucose transporter isoforms 1 (GLUT 1) and 4 (GLUT 4) are not present in adult lung, but are present in developing lungs. Over-expression of these GLUT isoforms can enhance glucose uptake into fetal lung cells to support active cell proliferation, which is a common characteristic of developing lung epithelium.[40] The decrease in the flux of glucose through the glycolytic pathway (Figure 2) of lungs of nicotine exposed rat pups is, however, not due to a compromised glucose transporter system because the total glucose turnover of lung tissue of these pups is higher than in control pups. This is because flux of glucose through the hexose monophosphate shunt (HMP) of lung tissue of nicotine exposed rat pups exceeds that of control animals. The flux of glucose through the glycolytic pathway remains suppressed after nicotine withdrawal (Figure 2), while the flux through the HMP shunt returns to normal[7]; thus nicotine has no inhibitory effect on the detoxification function of the lung. This is supported by the observation that nicotine exposure induces the synthesis of microsomal mono-oxygenases such as NADH-cytochrome b_5 reductase, NADPH-cytochrome c-reductase, NADPH-cytochrome P_{450}-reductase, and upregulates the expression of cytochrome P_{450} in lung tissue of neonatal and adult rats.[40,41]

Hexokinase catalyzes the phosphorylation of glucose before it can be metabolized by the various metabolic pathways in the cell. In rats, maternal nicotine exposure during pregnancy and lactation[7] had no influence on the hexokinase activity of offspring and thus on the phosphorylation of glucose. Maternal nicotine exposure during pregnancy and lactation also had no effect on lactate dehydrogenase and pyruvate kinase activity of the lungs of the offspring, indicating that the site of action of nicotine is between hexokinase and lactate dehydrogenase.[42] The decrease in the glycolytic activity, and thus the flux of glucose through this pathway, can be attributed to an inhibition of phosphofructokinase, the rate limiting enzyme of the glycolytic pathway.[43] It appears that the lower phosphofructokinase activity, like the reduced phosphorylase activity, is due to interference of phosphofructokinase synthesis at either the pre- or post-translational level,[44] resulting in a lower concentration of phosphofructokinase and thus a slower flux of glucose via glycolysis. The activity of hexokinase is highest during the phase of rapid alveolarization,[43] suggesting that the glycolytic pathway plays an important role in supplying energy and precursors during the alveolar stage of lung development.[43] Thus the lower phosphofructokinase activity in the lungs of nicotine exposed animals will likely impact adversely on alveolarization.

In the offspring of maternal rats administered nicotine, surfactant production by alveolar type II cells (AECs) appears to be unaffected by the sustained suppression of glycogenolysis and glycolysis; this suggests that fatty acids contribute to the supply of precursors for surfactant synthesis when the flux of glucose through the glycolytic pathway is reduced as a result of a lower glucose supply or an inhibition of this pathway.[42] This could explain the increase in the surfactant content of the type II AEC of the lungs of nicotine exposed neonatal rat lung.[45]

In addition, the adenine nucleotide content of the lungs of nicotine exposed neonatal rats is increased due to a decrease in the rate of ATP hydrolysis.[46] This inhibition of ATP hydrolysis gives rise to an increase in the ATP/ADP ratio in the lungs of the offspring. The reduced rate of ATP hydrolysis can be partly attributed to an inhibition of Na$^+$-K$^+$ ATPase.[47] This inhibition of Na$^+$-K$^+$ ATPase may result in swelling and bleb formation of the alveolar type I AEC and death of these cells.

Nicotine exposure during pregnancy and lactation indirectly stimulates the hexose monophosphate shunt in lung tissue of offspring.[41] This implies that nicotine has no inhibitory effect on the supply of precursors for synthetic processes such as surfactant synthesis and the detoxification of foreign substances in fetal and neonatal lung tissue.

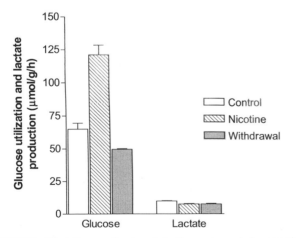

FIGURE 2 The influence of maternal nicotine exposure during gestation and lactation on glucose utilization and lactate production by the lungs of offspring. Glucose utilization and lactate production were determined during lactation in each of 3 groups: a control group that received no nicotine, a nicotine treated group, and another group killed following a 4–week withdrawal period during which the offspring received no nicotine.

Lung Structure

Fibroblasts play a critical role in the transition from the saccular to the alveolar stage of lung development, during which there is a four-fold increase in the number of interstitial fibroblasts in the neonatal rat lung.[48] Perturbations such as hyperoxia, barotrauma, and steroid therapy have been shown to interfere with alveolar development in the rat,[49] baboon,[50] and human infant,[51] the net result of which is a significant, often permanent, decrease in the number of alveoli.

A key process in alveolarization is secondary septation and a substantial body of information exists regarding events that coincide with septation, many of which may influence fibroblast proliferation. Lung elastic fibres are also thought to be involved in septation by providing structural support for newly emerging secondary septa. Inhibition of elastic fibre assembly has been linked to impaired septation.[52] In neonatal rat lung fibroblasts, elastin expression coincides with alveolarization, reaching a peak during the second postnatal week and declining rapidly thereafter.[53]

Cigarette smoke inhibits fibroblast proliferation and migration by increasing the length of the cell cycle, thereby reducing the rate of alveolarization.[54] Consequently the surface area available for gas exchange is reduced. Cigarette smoke exposure will also compromise fibroblast-induced repair responses, and may be one of the factors that contributes to the development of smoke-induced lung disease.[54] Accumulation of nicotine in fibroblasts can affect their metabolism and function. In vitro studies, however, have shown that nicotine has no effect on fibroblasts from human fetal lungs.[54] These in vitro studies were, however, performed on cells that were not metabolically permanently compromised as opposed to the fibroblasts of lung cells of neonatal rats that had been exposed to nicotine during gestation and lactation. Therefore, since maternal nicotine exposure during gestation and lactation interferes with glucose metabolism and apoptosis in the fetal and neonatal lung, and since it may cause disruption of the interaction between lung fibroblast glucose metabolism and fibroblast function, it is plausible that it will interfere with the structural and thus functional development of the fetal and neonatal lung.

Many agents that induce injury in lung tissue may do so by modifying key metabolic events for various cell populations in the lung. The type I AEC, for example, which covers more than 90% of the alveolar surface,[55] depends on glycolysis for energy.[56] Glycolysis also supplies the ATP required to maintain the membrane-linked Na^+-K^+ ATPase.[57] The Na^+-K^+-ATPase pump plays a vital role in maintaining cell volume; reducing its activity by the inhibition of glycolysis will therefore result in swelling of these cells and the formation of membrane blebs.[58] Thus, inhibition of glycolysis will interfere with the ability of the type I AEC to adapt to the changes in the environment and to maintain cell volume. Since glycolysis is irreversibly suppressed in the lungs of nicotine-exposed rat pups, the activity of this pump will be permanently lower and this could result in rupturing of the cell membranes. The type I epithelial cell is the most vulnerable to injury in the lung[59] and the permanent inhibition

of glycolysis will therefore make them more susceptible to damage, especially when exposed to blood and airborne toxic substances.

An analysis of the bronchoalveolar lavage fluid (BALF) of rat pups exposed to nicotine via the placenta and mother's milk has shown an increase in levels of alkaline phosphatase[60]; such an increase is considered to be a marker of alveolar type I AEC damage.[61] The glucose 6-phospate dehydrogenase activity of BALF is also increased, and is a marker of type II AEC proliferation.[61] Scanning and transmission electron micrographs indeed reveal blebbing as well as more comprehensive damage of alveolar type I AECs in these animals. Alveolar wall fenestrations also occur[62] and are considered to be a indication of the early onset of emphysema.[63] Many of these changes only become apparent after the withdrawal of nicotine.[9]

Type II AEC numbers have been found to be increased in the lungs of nicotine exposed animals,[64] which is thought to be a response to type I AEC damage and death. As a consequence of the proliferation of alveolar type II cells, the type I/type II AEC ratio decreased in the lungs of these animals.[65] Pulmonary fibroblasts are thought to be positive modulators of this process through the synthesis of keratinocyte and hepatocyte growth factors, both known to be potent mitogens for type II AECs.[66] It appears that the negative impact of maternal nicotine exposure during gestation and lactation on the growth, development, and repair processes of the lungs of the offspring is of such a nature that lung structure will gradually deteriorate with age.

In pregnant rats, cigarette smoke exposure at days 5–20 of gestation caused a reduction in lung volume, number of saccules and septal crests, and elastin content in fetal lungs.[67] Maternal nicotine exposure during pregnancy and lactation produced emphysema-like changes in the lungs of the rat pups[62,68] as well as an accumulation of lamellar bodies in type ll AECs.[45] The elastic tissue framework of the lungs of the offspring was also compromised.[69] These structural changes that are induced during gestation as a result of maternal nicotine exposure are irreversible and render the lungs of the offspring more susceptible to damage.[9]

Exposure of fetal monkeys to nicotine via the placenta during the late saccular/early alveolar phase of lung development results in an increase in the size and volume density of the primitive alveoli and the surface area for gas exchange decrease.[67] These findings are similar to those from two other models of the effects of smoking on lung development. The lungs of fetal rats exposed to cigarette smoke during pregnancy also have enlarged and fewer saccules[67]; similarly, pre- and postnatal exposure of developing rats to nicotine resulted in a decreased alveolar number (Figure 3), an increased alveolar volume (Figure 4) and a decrease in the surface area available for gas exchange.[8,9] In addition it was found that the alveolar septa became shorter and also incomplete (Figures 5 and 6), thereby reducing the alveolar wall surface area.[70]

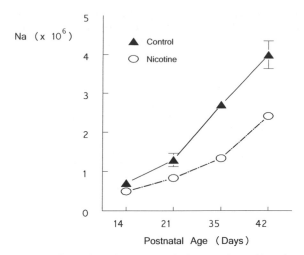

FIGURE 3 Effects of nicotine exposure during gestation and lactation on the number of alveoli (Na) in the lungs of postnatal rats. Nicotine exposure results in an age-related decrease in Na, relative to controls.[70]

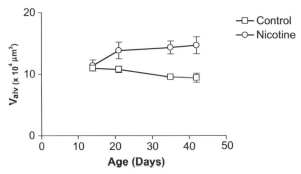

FIGURE 4 Effects of nicotine exposure during gestation and lactation on the volume of alveoli (Valv) in the lungs of postnatal rats. Nicotine exposure results in an increase in Valv in the lungs of offspring after the completion of alveolarization.[70]

The finding that nicotine alters alveolar development in both monkeys and rodents makes it highly likely that prenatal exposure to nicotine will similarly affect human lung development. It has been shown that much of the effect of maternal smoking on the developing lung may be mediated by the interaction of nicotine with nicotinic receptors expressed in the developing lung.[10,71] It is interesting to note that hospital admissions are seen less frequently for children whose mothers smoked only after pregnancy, arguing for a prenatal effect.[71,72]

NICOTINE AND CELL SIGNALING: APOPTOSIS AND LUNG DEVELOPMENT

Programmed cell death or apoptosis is an energy-dependent and genetically controlled process[73] that can be induced by a number of molecular tools.[74] Apoptosis occurs in the mesenchyme of fetal rat lung during the embryonic stage of development, when branching of conducting airways is the predominant feature. The percentage of cells undergoing apoptosis increases dramatically between 18 and 22 days of gestation and remains elevated in the first day of postnatal life. This marked increase at birth may be initiated by a number of factors such as air breathing, hormonal changes due to labour and delivery, and/or expansion of the lungs with changes in cell shape and cell–cell relationships. Most cells require attachment to the extracellular matrix for proper growth and function. Lung epithelial cell adhesion to the extracellular matrix is mediated by cell surface receptors known as integrins[75] that trigger a number of intracellular signaling pathways. Some of the pathways involved in apoptosis include the Ras-Raf-MAP kinase pathway and the phosphatidylinositol 3-kinase pathway.[76,77]

During the phase of rapid alveolarization between postnatal days 4 and 13 in rats, interstitial fibroblasts undergo rapid proliferation; few new alveoli are formed after this stage. Between postnatal days 13 and 21 the number of fibroblasts and type II AECs decreases. This decreased number of fibroblasts and type II AECs occurs by means of programmed cell death or apoptosis, which peaks between postnatal days 17–19. Apoptosis therefore plays a key role in the thinning of the alveolar septa that occurs after alveolarization.[78,79] Although apoptosis is an ongoing process in the immature lung, the rate of apoptosis after alveolarization increases owing to a decrease in bcl-2 mRNA expression and an increase in BAX mRNA expression in the fibroblasts on postnatal day 16. The gene products of bcl-2 and BAX interact to form homodimers and heterodimers. Although bcl-2 and BAX heterodimers are inactive, when BAX is in excess and BAX homodimers predominate, cells are likely to undergo apoptosis.[80] This explains the decrease in the total numbers of type II AECs and fibroblasts in rat lungs during the third postnatal week.[81]

The reduction, due to apoptosis, in the number of fibroblasts in the interstitium of the developing lung is likely to play a critical role in lung maturation, the final process of which is the transition of the alveolar wall from a double to a single capillary network layer.[79] Interference with the apoptotic process would be expected to have an adverse effect on lung maturation.

Cigarette smoke inhibits the proliferation and migration of human lung fibroblasts and fibroblast-mediated repair responses, and therefore may play a role in the development of emphysema.[54] Nicotine and cotinine inhibit apoptosis in fibroblasts,[82] but the underlying mechanism is not known. Nicotine is known to exert its effects on many cell types by binding to nicotinic cholinergic receptors. It has been suggested that pediatric, smoking-associated pulmonary diseases and small cell lung carcinoma may be caused by the direct chronic stimulation of an $\alpha7$ nicotinic acetylcholine receptor-initiated autocrine loop by nicotine and 4-(methylnitrosoamino)-(3-pyridyl)-1-butanone (NNK), where NNK is formed from nicotine by nitrosation in the mammalian organism and during the curing of tobacco.[83,84]

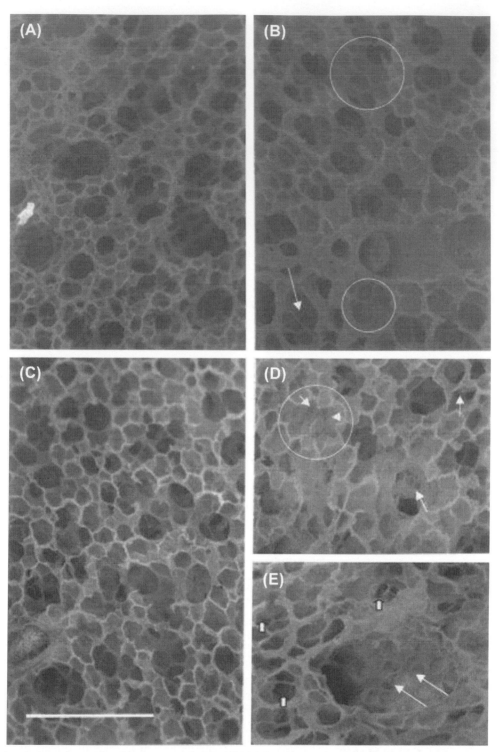

FIGURE 5   Scanning electron micrographs of the lung parenchyma of (A) 14-day-old control and (B) nicotine-exposed rat pups. The surface morphology of the lungs of the nicotine-exposed rat pups shows normal alveoli as well as focal flattened alveoli (encircled) with incomplete alveolar walls (arrow). The parenchyma of the 42-day-old control animals (C) showed normal intact alveoli, while enlarged alveoli, likely due to alveolar destruction (inside circle), occur in the lungs of the 42-day-old nicotine-exposed rats (D); fenestrations are also visible (arrows). (E) Flattened alveoli (long arrows) as well as more pronounced alveolar damage (short arrows) are apparent. (Scale bar = 290 μm.)[70]

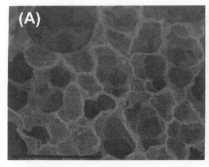
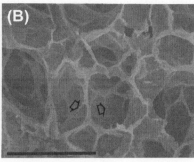

FIGURE 6 Scanning electron micrographs of 42-day-old rat lung showing the alveoli of (A) control rats, and (B) rats that were exposed to nicotine during gestation and lactation. The alveoli of the nicotine exposed rats are larger than those of the control rats. The arrows indicate short incomplete alveolar septa. (Scale bar = 380 μm.)

It is also possible that certain effects of nicotine are not receptor mediated or may operate through unconventional nicotine receptors.[82]

There is evidence that nicotine: (1) activates the mitogen-activated protein (MAP) kinase signaling pathway and extracellular signal-regulated kinase (ERK2), resulting in increased expression of the bcl-2 protein and inhibition of apoptosis and (2) blocks the inhibition of protein kinase C (PKC) activity in lung cells. Nicotine appears to have no effect on the activities of c-jun NH2-terminal protein kinase (JNK), c-myc, or p38 MAP kinases, which are also involved in apoptosis. While exposure to nicotine can result in the activation of two major signaling pathways (MAP-kinase and PKC) that are known to inhibit apoptosis, nicotine regulation of MAP (ERK) kinase activity is not dependent on PKC. These effects of nicotine occur at concentrations that are generally found in the blood of smokers, and could lead to disruption of the critical balance between cell death and proliferation.[85] The inhibition of apoptosis by nicotine may contribute to the slower thinning of the alveolar septa of the lungs of the nicotine exposed rat pups.[64]

It has been suggested[86] that in utero exposure of fetal pulmonary neuro-endocrine cells to nicotine or NNK in cigarette smoke may contribute to the development of pediatric lung disorders such as bronchitis and lower respiratory illnesses,[4,5] along with altered pulmonary mechanics in infants and children.[87] The nicotine-induced alterations in lung function of monkeys parallel those observed in infants of mothers who smoke during pregnancy.[4,87] These alterations in lung function could be induced via two different mechanisms. The first is a direct effect of released 5-hydroxytryptamine (5-HT) in response to α7 nicotinic receptor stimulation on bronchial and vascular smooth muscles and fibroblast growth; the second is an indirect effect of 5-HT on pulmonary neuro-endocrine cell numbers via activation of a Raf-1/MAP kinase pathway, resulting in yet more cells that can synthesize and release 5-HT. Chronic exposure to nicotine and NNK of pregnant mothers may therefore upregulate the α7 nicotinic receptor as well as components of its associated mitogenic signal transduction pathway, thereby increasing the vulnerability of infants to the development of pediatric lung disorders[4,5] mentioned earlier.

Both transforming growth factor (TGF)-β_1 and Smad3 are critically important for alveolar development.[88] Wnt3a increases TGF-β1 in fetal and postnatal fibroblasts.[89] These factors all play a role in lung development. Over-expression of TGF-β1 suppresses alveolar formation in newborn rat lung. Tight control of TGF-β1 is therefore important because an overexpression of this growth factor invokes fibrosis inflammatory cells and also induces transdifferentiation of rat alveolar epithelial cells to a mesenchymal phenotype.[90] Exposure of the mother to nicotine during pregnancy and lactation upregulates Wnt3a, TGF-β1, and Smad3 in the lungs of adult male offspring and is considered to be a profibrotic response. This explains the enhanced collagen accumulation in alveolar walls and around blood vessels of adult rats that were exposed to nicotine via the placenta and mother's milk. This accumulation of collagen contributes to thicker alveolar walls observed in the lungs of adult nicotine-exposed offspring.[89,91] These findings suggest that the male offspring of mothers who smoked during pregnancy, or used NRT to quit smoking, will be more at risk to develop respiratory diseases later in life than the female offspring. The exact mechanism underlying the sex dimorphism in developmental nicotine-mediated programming of lung development is not clear.

Maternal Nicotine Exposure during Pregnancy and Lactation: Effect of Restoration of Maternal Antioxidant Status

During the early phases of lung development, proliferating cells such as fibroblasts and the type II AEC are most sensitive to changes in the environment.[92] Maternal smoking and nicotine intake greatly contribute to changes in the in utero environment within which the fetus develops.[93] Smokers inhale 10^{15} oxidant molecules per puff,[94] which results in an oxidant overload and a marked decrease in the antioxidant status of the mother. Furthermore, maternal nicotine intake induces oxidant stress in the rat offspring that is maintained into puberty, especially in males.[95,96] The antioxidant capacity of the smoking mother is also compromised,[97] which greatly reduces her capacity to protect the fetus against oxidant damage. This is of particular importance because up to just before birth, the ability of the developing fetus and

lung to protect itself against foreign substances, such as oxidants, is very low.[98] This implies that the fetus is, for most of its life in utero, dependent on the mother for optimal protection. Because the concentration of nicotine is elevated in the amniotic fluid,[11,99] and because it accumulates in the lungs of the fetus,[100] it is conceivable that the lungs of the fetus and neonate can be programmed to be more prone to

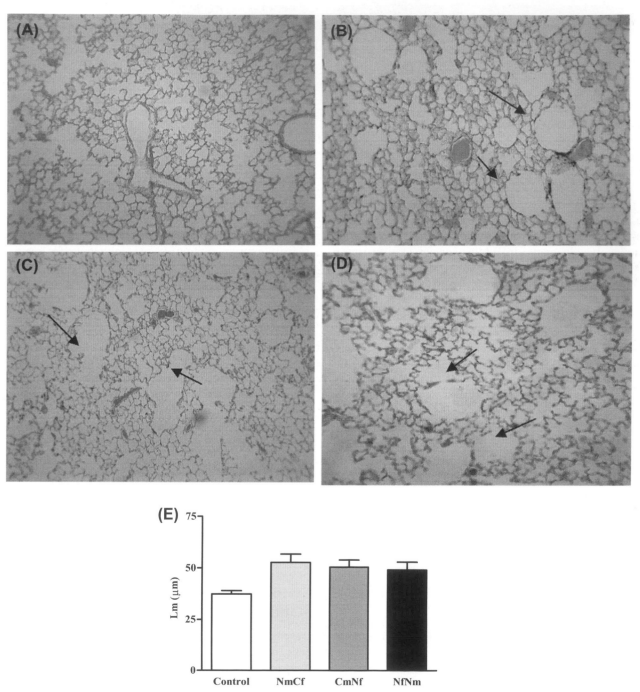

FIGURE 7   A–E: Influence of grand-maternal nicotine exposure during gestation and lactation on Lm of the lungs of the F2 generation. NmCF = Control female and nicotine exposed male; CmNf = control male and nicotine exposed female; NmNf = both males and females exposed to nicotine. Arrows show emphysema-like lesions.

respiratory diseases later in life unless they are effectively protected during pregnancy. It has indeed been illustrated that rats previously exposed to nicotine via the placenta and mother's milk are unable to maintain the structure of their lung parenchyma due to a compromised oxidant/antioxidant status.[45,101,102] However, restoring the mother's antioxidant capacity and thus her ability to protect the fetus may prevent the nicotine- and oxidant-induced programming of the fetal lung.[103]

Although certain small molecules might reduce the risk of pulmonary damage induced by oxidants in prematurely born infants,[104] clinical trials as well as laboratory studies with single small molecules like vitamins E and C, or β-carotene, have been largely disappointing.[105–107] This might be due to an imbalance between antioxidants, where such an imbalance can also induce tissue damage.[108] Thus, under certain circumstances supplementing the mother's diet with single antioxidants to restore the antioxidant capacity of the individual may also be detrimental to lung development. On the other hand, a mixture of small molecules, such as those found in fruits and vegetables, may restore the mother's antioxidant capacity and also allow the synergistic action of these small molecules.[109] Indeed it has been shown that supplementing the diets of adult rats with tomato juice protects them from developing emphysema.[110] It also prevents programming of the lungs that were exposed to nicotine via the placenta and mother's milk to develop emphysema-like lesions later in life.[103]

Transgenerational Effect of Grand-Maternal Nicotine Exposure

It has recently been shown that exposure of fetal and neonatal rats during gestation and lactation to nicotine via the placenta and mother's milk, resulted in microscopic emphysema in the lungs later in the life of the first generation offspring (F1 progeny).[103] This is followed by a thickening of the alveolar walls of these animals[103] due to an increased TGF-β_1 activity.[111,112] It was suggested that these changes in

lung parenchymal structure after nicotine withdrawal may be due to a change in the program that controls lung growth, lung aging, and maintenance of the lung structure.[103] The change in the program can be due to nicotine-induced DNA strand breaks and impaired DNA repair while the animals were still exposed to nicotine via the placenta and mother's milk. It is also possible that it was due to a nicotine-induced change in the epigenome[113,114] and thus the program that controls lung aging, and maintenance of tissue integrity, in the offspring and subsequent generations.

It has indeed been demonstrated that grand-maternal nicotine exposure during gestation and lactation, in concentrations that resemble those of smokers,[115] not only resulted in the development of emphysematous lesions in the F1 generation, but also in the F2 progeny (Figure 7). These changes in lung parenchymal structure in the F2 progeny cannot be attributed to the direct effect of nicotine or a nicotine-induced oxidant/antioxidant imbalance as the F2 progeny were not exposed to nicotine during pregnancy and lactation. Because nicotine is genotoxic,[113] and induces an oxidant/antioxidant imbalance,[97] it is plausible that it changed the program so that it resulted in premature aging of the cells in the alveolar wall, and thus reduced the life-span of the cells in the alveolar walls. It also suppresses cell proliferation in the alveolar walls of the F1 generation. This change in the program that controls aging and proliferation in the cells of the alveolar walls is evidently transferred to the F2 generation (Figure 8). Consequently, fewer viable cells are able to proliferate to replace damaged cells in the lung parenchyma of the F1 and F2 progeny. This means that the ability of the lung of the F1 and F2 offspring to repair itself is compromised because of a reduced capacity to renew tissue. A longer term consequence of this is development of emphysema-like lesions in the lungs later in the life of the F1 and F2 offspring. The premature aging and reduced cell proliferation in emphysematous lung reflects a persistent, intrinsic failure of cell replacement and maintenance of alveolar walls. Slower proliferation of lung fibroblast by translational mechanisms has indeed

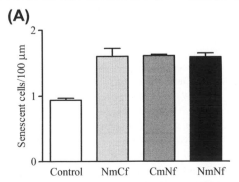
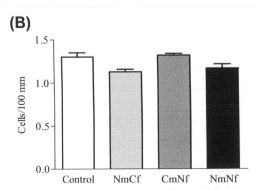

FIGURE 8   The effect of maternal nicotine exposure during gestation and lactation on (A) cellular senescence and (B) cell proliferation in the lungs of the F2 generation.

been shown in smokers.[114] It is therefore plausible that the development of the emphysema-like lesions in the lungs of the rats was due to a faster deterioration of the connective tissue framework and premature aging of the lungs of these rats. This explains the increased lung compliance of the F1 and F2 generations following grand-maternal nicotine intake. This is an important observation that shows that the use of nicotine replacement therapy (NRT) by the pregnant and lactating mother may result in the development of COPD in the offspring, not only in the F1 generation[103] but also in the F2 generation. This suggests that grand-maternal smoking or NRT during pregnancy and lactation may give rise to generations that are more susceptible to respiratory diseases such as emphysema than those never exposed to nicotine in tobacco smoke or NRT.

CONCLUSIONS

Experimental studies have shown that prenatal nicotine exposure produces alterations in lung function that parallel those observed in infants of mothers who smoked during pregnancy. Nicotine and products of nicotine metabolism can affect lung growth, development, and function via several mechanisms, such as inducing premature aging of the lungs of the offspring. These effects are programmed and are transgenerational. Understanding the mechanisms whereby prenatal tobacco smoke exposure alters lung development and eventually lung function may lead to therapeutic interventions to block its effects as well as help to further discourage smoking or the use of nicotine replacement therapies during pregnancy.

REFERENCES

1. Harlap S, Davies AM. Infant admissions to hospital and maternal smoking. *Lancet* 1974;**i**:529–32.
2. Colley JRT, Holland WW, Corkhill RC. Influence of passive smoking on parental phlegm, pneumonia and bronchitis in early adulthood. *Lancet* 1974;**ii**:1031–4.
3. Ferguson DM, Horwood LJ, Shannon FT, Taylor B. Parental smoking and lower respiratory illness in the first three years of life. *J Epidemiol Community Health* 1981;**35**:180–4.
4. Yarnell JWG, St Leger AS. Respiratory illness, maternal smoking habit and lung function in children. *Br J Dis Chest* 1979;**73**:230–6.
5. Schilling RSF, Bouhuys A, Sudan BJL, Brody B. Breathing other people's smoke. *Br Med J* 1978;**ii**:895.
6. Weiss ST, Tager IB, Speizer FE, Rossner B. Persistent wheeze. Its relation to respiratory illness, cigarette smoking and level of pulmonary function in a population sample children. *Am Rev Respir Dis* 1980;**122**:697–707.
7. Maritz GS. Maternal nicotine exposure and carbohydrate metabolism of fetal and neonatal lung tissue: response to nicotine withdrawal. *Respiration* 1987;**51**:232–40.
8. Maritz GS. Maternal nicotine exposure during gestation and lactation of rats induce microscopic emphysema in the offspring. *Exp Lung Res* 2002;**28**:391–403.
9. Maritz GS, Dennis H. Maternal nicotine exposure during gestation and lactation interferes with alveolar development in the neonatal lung. *Reprod Fertil Dev* 1998;**10**:255–61.
10. Spindel ER, Martin EL, Proskocil JA, Keller A, Orr-Urtreger A, Beaudet AL, Sekhon HS. The interaction of nicotine with α7 nicotinic receptors may be a cause of the intrauterine growth retardation caused by smoking during pregnancy. *Am J Respir Crit Care Med* 2002;**165**:A225.
11. Benowitz N. Toxicity of nicotine: implications with regard to nicotine replacement therapy. In: *Nicotine Replacement: A Critical Evaluation*. Alan R Liss. Inc.; 1988. pp. 187–217.
12. Gourlay SG, Benowitz NL. Arteriovenous differences in plasma concentration of nicotine and catecholamines and related cardiovascular effects after smoking, nicotine nasal spray, and intravenous nicotine. *Clin Pharmacol Ther* 1997;**62**:453–63.
13. Benowitz NL, Jacob III P. Daily intake of nicotine during cigarette smoking. *Clin Pharmacol Ther* 1984;**35**:499–504.
14. Lunell E, Molande L, Ekberg K, Wahren J. Site of absorption from a vapour inhaler – comparison with cigarette smoking. *Eur J Clin Pharmacol* 2000;**55**:737–41.
15. Benowitz NL, Kuyt F, Jacob III P. Circadian blood nicotine concentrations during cigarette smoking. *Clin Pharmacol Ther* 1982; **32**:758–64.
16. Chiou WL. The phenomenon and rationale of marked dependence of drug concentration on blood sampling site: implications in pharmacokinetics, pharmacodynamics, toxicology, and therapeutics [part 1]. *Clin Pharm* 1989;**17**:175–99. [part II] 275–290.
17. Luck W, Nau H, Steldinger R. Extent of nicotine and cotinine transfer to the human fetus, placenta, and amniotic fluid of smoking mothers. *Dev Pharmacol Ther* 1985;**8**:384–95.
18. Benowitz NL, Chan K, Denaro CP, Jacob P. Stable isotope method for studying transdermal drug absorption: the nicotine patch. *Clin Pharmacol Ther* 1991;**50**:286–93.
19. Weizenecker R, Deal WB. Tobacco croppers' sickness. *J Fla Med Assoc* 1970;**57**:13–4.
20. Gennser G, Marsal K, Brantmark K. Maternal smoking and fetal breathing movements. *Am J Obstet Gynecol* 1974;**123**:861–5.
21. Van Vunakis H, Langonc JJ, Mikinsky A. Nicotine and cotinine in the amniotic fluid of smokers in the second trimester of pregnancy. *Am J Obstet Gynecol* 1974;**120**:64–6.
22. Suzuki K, Minei LJ, Johnson EE. Effect of nicotine upon uterine blood flow in the pregnant rhesus monkey. *Am J Obstet Gynecol* 1980;**136**:1009–13.
23. Szüts T, Olsson S, Lindquist NG, Ullberg S. Long-term fate of [14C]-nicotine in the mouse: retention of the bronchi, melanin-containing tissues and urinary bladder wall. *Toxicologist* 1978;**10**:207–20.
24. Wang N-S, Schraufnagel DE, Chen MF. The effect of maternal oral intake of nicotine on the growth and maturation of fetal and baby mouse lungs. *Lung* 1983;**161**:27–38.
25. Greenberg RA, Haley NJ, Etzel RA, Loda FA. Measuring the exposure of infants to tobacco smoke: nicotine and cotinine in urine and saliva. *N Engl J Med* 1984;**310**:1075–107.
26. Bardy AH, Seppala T,Lillsunde, P,Kataja JM, Koskela P, Pikkarainen, et al. Objectively measured tobacco exposure during pregnancy: neonatal effects and relation to maternal smoking. *Br J Obstet Gynecol* 1993;**100**:721–6.
27. Albuquerque CA, Smith KR, Johnson C, Chao R, Harding R. Influence of maternal tobacco smoking during pregnancy on uterine, umbilical and fetal cerebral artery blood flows. *Early Hum Dev* 2004;**80**:31–42.

28. Sastry BV. Placental toxicology: tobacco smoke, abused drugs, multiple chemical interactions, and placental function. *Reprod Fertil Dev* 1991;**3**:355–72.

29. Fisher S, Atkinson M, Jacobson S. Selective fetal malnutrition: the effect of in vivo ethanol exposure upon in vitro placental uptake of amino acids in the nonhuman primate. *Pediatr Res* 1983;**17**:704–7.

30. O'Neill JJ, Tierney DF. Rat lung metabolism: glucose utilization by isolated perfused lungs and tissue slices. *Am J Phys* 1974;**226**:867–73.

31. Gilden C, Sevanian A, Tierney DF, Kaplan SA, Barrett CT. Regulation of fetal lung phosphatidylcholine synthesis by cortisol: role of glycogen and glucose. *Pediat Res* 1977;**11**:845–8.

32. Maniscalco WM, Wilson CM, Gross I, Gobran L, Rooney SA, Warsaw JB. Development of glycogen and phospholipid metabolism in fetal and newborn rat lung. *Biochem Biophys Acta* 1978;**530**:333–46.

33. Bourbon J, Jost A. Control of glycogen metabolism in the developing fetal lung. *Pediatr Res* 1982;**16**:50–6.

34. Salisbury-Murphy S, Rubinstein D, Beck JC. Lipid metabolism in lung slices. *Am J Phys* 1966;**211**:988–92.

35. Maniscalco WM, Wilson CM, Gross I, Gobran L, Rooney SA, Warsaw LB. Development of glycogen and phospholipid metabolism in fetal and newborn rat lung. *Biochem Biophys Acta* 1978;**530**:333–46.

36. Bourbon JR, Farrell PM, Doucet E, Brown JD, Valanza C. Biochemical maturation of fetal rat lung: a comprehensive study including surfactant determination. *Biol Neonate* 1987;**52**(1):48–60.

37. Maritz GS. Lung glycogen metabolism in suckling rats: A comparative study. *Biol Neonate* 1988;**54**:100–6.

38. Rhoades RA. Net uptake of glucose, glycerol and fatty acids by the isolated perfused rat lung. *Am J Phys* 1975;**226**.144–9.

39. Maritz GS. Pre- and postnatal carbohydrate metabolism of rat lung tissue. *Arch Toxicol* 1986;**59**:89–93.

40. Iba MM, Fung J, Pak YW, Thomas PE, Fisher H, Sekowski A, et al. Dose-dependent upregulation of rat pulmonary, renal, and hepatic cytochrome P450 (CYP) 1A expression by nicotine feeding. *Drug Metab Dispos* 1999;**27**:977–82.

41. Maritz GS. *Die invloed van nikotien op die intermediêre koolhidraatmetabolisme van longweefsel (The influence of nicotine on the intermediary carbohydrate metabolism of lung tissue)*. South Africa: Ph.D. thesis, University of Stellenbosch; 1983.

42. Maritz GS. The influence of maternal nicotine exposure on key enzymes of glucose metabolism in lung tissue of the offspring. *Pathophysiology* 1997;**4**:135–41.

43. Kordom C, Maritz GS, De Kock M. Maternal nicotine exposure during pregnancy and lactation: effect on glycolysis in the lungs of the offspring. *Exp Lung Res* 2002;**28**:391–403.

44. Kordom C. *The influence of maternal nicotine exposure on neonatal lung development: an enzymatic and metabolic study*. M.Sc thesis, University of the Western Cape; 1999.

45. Maritz GS, Thomas R-A. Maternal nicotine exposure: response of type II pneumocytes of neonatal rat pups. *Cell Biol Int* 1995;**19**:323–31.

46. Maritz GS, Burger B. The influence of maternal nicotine exposure on neonatal lung carbohydrate metabolism. *Cell Biol Int Rep* 1992;**16**:1229–36.

47. Meyer DH, Cross CE, Ibrahim AB, Mustafa MG. Nicotine effects on alveolar macrophage respiration and adenosine triphosphatase activity. *Arch Environ Health* 1971;**22**:362–5.

48. Kauffman SL, Burri PH, Weibel ER. The postnatal growth of the rat lung II. Autoradiography. *Anat Rec* 1974;**180**:63–76.

49. Massaro GD, Massaro D. Formation of pulmonary alveoli and gas exchange surface area: quantitation and regulation. *Annu Rev Physiol* 1996;**58**:73–92.

50. Coalson JJ, Winter V, DeLemos RA. Decreased alveolarisation in baboon survivors with bronchopulmonary dysplasia. *Am J Respir Crit Care Med* 1995;**152**:640–6.

51. Margraf LR, Tomashefski JF, Bruce MC, Dahm B. Morphometric analysis of the lung in bronchopulmonary dysplasia. *Am Rev Respir Dis* 1991;**143**:391–400.

52. Kida K, Thurlbeck WM. The effects of β-aminoproprionitrile on the growing rat lung. *Am J Pathol* 1980;**101**:693–710.

53. Noguchi A, Samaha H. Developmental changes in tropoelastin gene expression in the rat studied by in situ hybridisation. *Am J Respir Cell Mol Biol* 1991;**5**:571–8.

54. Nakamura Y, Romberger DJ, Tate L, Ertl RF, Kawamoto M, Adachi Y, et al. Cigarette smoke inhibits lung fibroblast proliferation and chemotaxis. *Am J Respir Crit Care Med* 1995;**151**: 1497–503.

55. Naimark A. Non-ventilatory functions of the lung. *Am Rev Respir Dis* 1977;**115**:93–8.

56. Massaro GD, Gail DB, Massaro D. Lung oxygen consumption and mitochondria of alveolar epithelial and endothelial cells. *J Appl Phys* 1975;**38**:588–92.

57. Paul RJ. Functional compartmentalization of oxidative glycolytic metabolism in vascular smooth muscle. *Am J Phys* 1983;**224**: C399–409.

58. Contran RS, Kumar V, Robbins SL. Cellular injury and adaptation. In: *Pathologic Basis of Disease*. 4th ed. Philadelphia: WB Saunders; 1989. pp. 16–38.

59. Witschi H. Proliferation of alveolar type II cells: A review of common responses in toxic lung injury. *Toxicol* 1976;**5**:267–77.

60. Maritz GS, Najaar K. Biomedical response of neonatal rat lung to maternal nicotine exposure. *Pathophysiology* 1995;**2**:47–54.

61. Henderson RF. Use of broncho-alveolar lavage to detect lung damage. *Environ Hlth Perspect* 1984;**56**:115–29.

62. Maritz GS. Maternal nicotine exposure induces microscopic emphysema in neonatal rat lung. *Pathophysiol* 1997;**4**:135–41.

63. Linhartova A. Fenestrations of the pulmonary septa as a sign of early destruction in emphysema. *J Cesk Patol* 1983;**19**:211–21.

64. Maritz GS, Harding R. Life-long programming implications of exposure to tobacco smoking and nicotine before and soon after birth: evidence for altered lung development. *Int J Environ Res Public Health* 2011;**8**:875–98.

65. Maritz GS, Thomas R-A. The influence of maternal nicotine exposure on the interalveolar septal status of neonatal rat lung. *Cell Biol Int* 1994;**18**:747–57.

66. Panos RJ, Rubin JS, Csaky KG, Aaronson SA, Mason RT. Keratinocyte growth factor and hepatocyte growth factor/scatter factor are heparin-binding growth factors for alveolar type II cells in fibroblast-conditioned medium. *J Clin Invest* 1993;**92**: 969–77.

67. Collins MH, Moessinger AL, Klinerman J. Fetal lung hypoplasia associated with maternal smoking: a morphometric analysis. *Pediatr Res* 1985;**19**:408–12.

68. Maritz GS, Woolward K, du Toit G. Maternal nicotine exposure during pregnancy and development of emphysema-like damage in the offspring. *S Afr Med J* 1993;**83**:195–8.

69. Maritz GS, Dolley L. The influence of maternal nicotine exposure on the status of the connective tissue framework of developing rat lung. *Pathophysiol* 1996;**3**:212–20.

70. Maritz GS. Maternal nicotine exposure during gestation and lactation of rats induce microscopic emphysema in the offspring. *Exp Lung Res* 2002;**28**:391–403.

71. Sekhon HS, Jia Y, Raab R, Kuryatov A, Pankow JF, Whitsett JA, et al. Prenatal nicotine increases pulmonary α7 nicotinic receptor expression and alters fetal lung development in monkeys. *J Clin Invest* 1999;**103**:637–47.

72. Tager IB, Hanrahan JP, Tosteson TD, Castile RG, Brown RW, Weiss ST, et al. Lung function, pre- and postnatal smoke exposure, and wheezing in the first year of life. *Am Rev Respir Dis* 1992;**147**:811–7.

73. White E. Life, death and the pursuit of apoptosis. *Genes Dev* 1996;**10**:1–15.

74. Wertz IE, Hanley MR. Diverse molecular provocation of programmed cell death. *Tends Biochem Sci* 1996;**21**:359–64.

75. Pilewski JM, Albelda SM. Adhesion molecules in the lung. An overview. *Am Rev Respir Dis* 1993;**148**(Suppl.):532–7.

76. Ichijo H, Nishida E, Irie K, Dijke PT, Saitoh M, Moriguchi T, et al. Induction of apoptosis by ASK1, a mammalian MAPKKK that activates SAPK/JNK and p38 signalling pathways. *Science* 1997;**275**:90–4.

77. Yao R, Cooper GM. Requirements for phosphatidylinositol-3 kinase in the prevention of apoptosis by nerve growth factor. *Science* 1997;**267**:2003–6.

78. Schnittny JC, Djonov V, Fine A, Burri PH. Programmed cell death contributes to postnatal lung development. *Am J Respir Cell Mol Biol* 1998;**18**:786–93.

79. Bruce MC, Honaker CE, Cross RJ. Lung fibroblasts undergo apoptosis following alveolarisation. *Am J Respir Cell Mol Biol* 1999;**20**:228–36.

80. Yang A, Zha J, Jockel J, Boise LH, Thompson CB, Korsmeyer SJ. Bad, a heterodimeric partner for Bcl-xl and Bcl-2, displaces Bax and promotes cell death. *Cell* 1995;**80**:285–91.

81. Randell SH, Silbajoris R, Young SL. Ontogeny of rat lung type II cells correlated with surfactant apoprotein expression. *Am J Phys* 1991;**260**:L562–70.

82. Wright SC, Zhong J, Zheng H, Larrick JW. Nicotine inhibition of apoptosis suggests a role in tumour promotion. *FASEB J* 1993;**7**:1045–51.

83. Fischer S, Spiegelhalder B, Eisenbarth J, Preussmann R. Investigation on the origin of tobacco-specific nitrosamines in mainstream smoke of cigarettes. *Carcinogenesis* 1990;**11**:723–30.

84. Hecht SS, Hoffmann D. Tobacco-specific nitrosamines, an important group of carcinogens in tobacco and tobacco-smoke. *Carcinogenesis* 1990;**9**:875–84.

85. Heusch WL, Maneckjee R. Signaling pathways involved in nicotine regulation of apoptosis of human lung cancer cells. *Carcinogenesis* 1998;**19**:551–6.

86. Schuller HM, Jul BA, Sheppard BJ, Plummer III HK. Interaction of tobacco-specific toxicants with the neuronal ∝7 nicotinic acetylcholine receptor and its associated mitogenic signal transduction pathway: potential role in lung carcinogenesis and pediatric lung disorders. *Eur J Pharmacol* 2000;**393**:265–77.

87. Sekhon HS, Keller JA, Benowitz NL, Spindel ER. Prenatal nicotine exposure alters pulmonary function in newborn Rhesus monkeys. *Am J Respir Crit Care Med* 2001;**164**:989–94.

88. Sime PJ, Xing Z, Graham FL, Csaky KG, Gauldie J. Adenovector-mediated gene transfer of active transforming growth factor-beta1 induces prolonged severe fibrosis in rat lung. *J Clin Invest* 1997;**100**:768–76.

89. Zou W, Zou Y, Zhao A, Li B, Ran P. Nicotine-induced epithelial-mesenchymal transition via Wnt/β-catenin signalling in human airway epithelial cells. *Am J Physiol Lung Cell Mol Physiol* 2013;**304**:L199–209.

90. Gauldie J, Galt T, Bonniaud P, Robbins C, Kelly M, Warburton D. Transfer of the active form of transforming growth factor-beta1 gene to new born rat lung induces changes consistent with bronchopulmonary dysplasia. *Am J Pathol* 2003;**163**:2575–84.

91. Dasgupta C, Xiao D, Xu Z, Yang S, Zhang L. Developmental nicotine exposure results in programming of alveolar simplification and interstitial pulmonary fibrosis in adult male rats. *Reprod Toxicol* 2012;**34**:370–7.

92. Barker DJP. The developmental origins of adult disease. *J Am Coll Nutr* 2004;**23**:588S–95S. supp.

93. Hanrahan JP, Tager IB, Segal MR, Tosteson TD, Castile RG, Van Vunakis H, et al. The effect of maternal smoking during pregnancy on early infant lung function. *Am Rev Respir Dis* 1992;**145**:1129–35.

94. Church DF, Pryor W. The oxidative stress placed on the lung by cigarette smoke. In: *The Lung*. New York: Tacen Press; 1991. pp. 1975–9.

95. Halima BA, Sarra K, Kais R, Salwa E, Najoua G. Indicators of oxidative stress in weanling and pubertal rats following exposure to nicotine via milk. *Hum Exp Toxicol* 2010;**29**:489–96.

96. Helen A, Krishnakumar K, Vijayammal PL, Augusti KT. Antioxidant effect of onion oil (Allium cepa. Linn) on the damages induced by nicotine in rats as compared to alpha-tocopherol. *Toxicol Lett* 2000;**116**:61–8.

97. Balakrishnan A, Menon VP. Antioxidant properties of hesperidin in nicotine-induced lung toxicity. *Fundam Clin Pharmacol* 2007;**21**:535–46.

98. Gebremichael A, Chang AM, Buckpitt AR, Plopper CG, Pinkerton KE. Postnatal development of cytochrome P4501A1 and 2B1 in rat lung and liver: effect of aged and diluted sidestream cigarette smoke. *Toxicol Appl Pharmacol* 1995;**135**:246–53.

99. Benowitz NL, Jacob 3rd P. Nicotine and carbon monoxide intake from high- and low-yield cigarettes. *Clin Pharmacol Ther* 1984;**36**:265–70.

100. Gleason MN, Gosselin RE, Hodge HC. *Clinical Toxicology of Commercial Products*. 2nd ed. Baltimore: Williams & Williams; 1963.

101. Maritz GS, Windvogel S. Chronic maternal nicotine exposure during gestation and lactation and the development of the lung parenchyma in the offspring. Response to nicotine withdrawal. *Pathophysiology* 2003;**10**:69–75.

102. Sekhon HS, Keller JA, Proskocil BJ, Martin EL, Spindel ER. Maternal nicotine exposure upregulates gene expression in fetal monkey lung. Association with α-7 nicotinic acetylcholine receptors. *Am J Respir Cell Mol Biol* 2002;**26**:31–41.

103. Maritz GS, Mutemwa M, Kayigire AX. Tomato juice protects the lungs of the offspring of female rats exposed to nicotine during gestation and lactation. *Pediatr Pulmonol* 2011;**46**:976–86.

104. Chang L-Y, Subramanian M, Yoder BA, Day BJ, Ellison MC, Sunday ME, et al. A catalytic antioxidant attenuates alveolar structural remodelling in bronchopulmonary dysplasia. *Am J Respir Crit Care Med* 2003;**167**:57–64.

105. Maritz GS, van Wyk G. Influence of maternal nicotine exposure on neonatal rat lung structure: protective effect of ascorbic acid. *Comp Biochem Physiol C Pharmacol Toxicol Endocrinol* 1997;**117**:159–65.

106. Bjelkjavic G, Nikolova D, Gluud LL, Simonetti RG, Gluud C. Antioxidant supplements for the prevention of mortality in healthy participants and patients with various diseases. *Cochrane Database Syst Rev* 2008;**16**: CD007176.

107. Maritz GS, Rayise SS. Effect of maternal nicotine exposure on neonatal rat lung development: protective effect of maternal ascorbic acid supplementation. *Exp Lung Res* 2011;**37**:57–65.

108. Fardy CH, Silverman M. Antioxidants in neonatal lung disease. *Arch Dis Child* 1995;**73**:F112–7.

109. Wheatly C. Vitamin trials in cancer: What went wrong? *J Nutr Environ Med* 1998;**8**:277–88.

110. Kasagi S, Seyama K, Hiroaki Mori H, Souma S, Sato T, Akiyoshi T, et al. Tomato juice prevents senescence-accelerated mouse P1 strain from developing emphysema induced by chronic exposure to tobacco smoke. *Am J Physiol Lung Cell Mol Physiol* 2006;**290**:L396–404.

111. Zhang QL, Baumert J, Ladwig KH, Wichman HE, Meisinger C, Doring A. Association of daily tar and nicotine intake with incident myocardial infarction: results from the population based MONICA/KORA Augburg Cohort study 1983–2002. *BMC Public Health* 2011;**11**:273.

112. Aoshiba K, Nagai A. Senescence Hypothesis for the pathogenetic mechanism of chronic obstructive pulmonary disease. *Proc Am Thorac Soc* 2009;**6**:596–601.

113. Argentin G, Cicchetti R. Genotoxic and antiapoptotic effect of nicotine on human gingival fibroblasts. *Toxicol Sci* 2004;**79**: 75–81.

114. Miglino N, Roth M, Lardinois D, Sadowski C, Tamm M, Borger P. Cigarette smoke inhibits lung fibroblast proliferation by translational mechanisms. *Eur Respir J* 2012;**39**:705–11.

115. Aitken RJ, De Iuliis N, McLachlan RI. Biological and clinical significance of DNA damage in the male germ line. *Int J Androl* 2008;**32**:46–56.

Chapter 23

Exposure to Allergens during Development

Laurel J. Gershwin

University of California Davis, Veterinary Medicine (PMI), University of California – Davis, Davis, Veterinary Medicine (PMI), Davis, CA, USA

INTRODUCTION

During fetal life the fetus, which has major histocompatibility antigens that differ from those of the mother, must survive and not be rejected as foreign by the maternal immune system. This "allograft" is protected by several mechanisms, one of which is the polarization of the local immune environment towards T helper type 2 cytokine production; with a concomitant depression of the T helper type 1 response, which is important in allograft rejection. This cytokine environment that favors maintenance of a successful pregnancy is also compatible with allergic sensitization. Because the Th2 bias persists into neonatal life, during which it may be less advantageous as exposure to environmental antigens and pathogens requires a balanced immune response. The predisposition to Th2 cytokine production may lead to development of allergic sensitization in early life. Exposure to allergen initially occurs during neonatal and juvenile years, which are critical times for immunological and structural development of the lung. The increasing incidence of allergic asthma reported in recent years is most noticeable in children and adolescents. Epidemiological and experimental evidence suggests that both maternal factors and environmental influences have the potential to enhance sensitization to allergens.[1] In this chapter we examine how these interactions and exposure to allergen at a critical time in development may modulate the immune system towards an allergic phenotype.

Principles of Allergic Sensitization: Induction of Type 1 Hypersensitivity

An understanding of the principles of allergic sensitization is paramount to understanding how exposure to allergens during development can induce chronic pulmonary allergy. The response to allergen is a hypersensitivity response, that is, a response that is exaggerated and in fact, one that the normal individual would not be expected to make. Close to half of the human population is atopic, which means that they have a genetic tendency to develop IgE antibodies specific for a variety of environmental proteins called allergens. Upon contact with an allergen, usually by inhalation (also by ingestion, injection, or skin contact) an atopic individual will begin an immunological response that results in production of IgE antibodies. The specific events that precede the production of IgE antibodies will be addressed in following text. However, once formed, the IgE binds very tightly to surface receptors on mast cells that exist in tissues that are in close proximity to mucosal surfaces. At this point the mast cell (and hence the individual) is sensitized. Subsequent exposures to allergen cause degranulation of mast cells with release of mediators, which cause the clinical signs that we associate with allergy. Thus, release of histamine causes increased capillary permeability and smooth muscle contraction. Stimulation of eicosanoid production stimulates not only smooth muscle contraction, but also chemotaxis of leukocytes. The later acting mediators produced by activation of the arachidonic acid pathway include leukotrienes, formerly called slow reacting substance of anaphylaxis. These reactions, when initiated in the lung, set the stage for allergic asthma.[2]

When an antigen is inhaled several consequences are possible: (1) it may not initiate an immune response, (2) it can initiate an allergic type response, or (3) it can stimulate a cell-mediated immune response. Which of these possibilities occur depends on host, environment, and the nature of the inhaled antigen. For example, small proteins (<10 kd), lipids, and small chemicals are not likely to initiate an immune response. Proteins from animal serum/dander, pollens, and complex proteins foreign to the host are likely to initiate a Th2 response in the host that is genetically programmed to do so (we call these hosts atopic). The inhalation of proteins from some bacteria, such as Mysobacteria, is not likely to stimulate the allergic type 2 response, but an alternate cytokine environment will be the likely outcome for production of a cell-mediated type response (Th1).

The Lung. http://dx.doi.org/10.1016/B978-0-12-799941-8.00023-7

To produce an immune response it is critical that the antigen/allergen gains access to appropriately situated cells called dendritic cells, which can present the allergen to T lymphocytes called type 2 helper cells (Th2). These cells must secrete cytokines, called interleukins, that direct the immune system to ultimately make IgE antibodies that can bind to the allergen. Specifically, interleukins (IL) 4 and 13 stimulate B lymphocytes to develop into IgE producing plasma cells. These cytokines work in opposition to another set of cytokines produced by lymphocytes called T helper 1 (Th1) cells. The Th1 immune response favors development of a cell-mediated immune response.[2] The Th1 cells make IL-2, IL-12, and γ-interferon, cytokines which stimulate T cells, NK cells, and macrophages—cytokines which are important in control of infectious diseases (such as infection with Mycobacteria).[2]

Immune Modulation by Type 2 Innate Lymphocytes

Recently the discovery of innate lymphocytes and their role in immune regulation has altered our understanding of neonatal allergic sensitization. Previously it was thought that CD4+ lymphocytes secreting IL4, IL13, and IL5 were responsible for the allergic polarization that occurs when an atopic individual responds to allergen. These helper type 2 cells are important in modulating the immune response towards allergy.[3] However, it is likely that the initial stimulus comes from the newly described non-T/non-B lymphocytes that also produce high amounts of IL-5 and IL-13. Unlike the Th2 cells, they are not antigen restricted and are activated by epithelial cell derived cytokines IL-25 and IL-33. These innate lymphocytes are now thought to be the initial responders to allergen and thus modulators of the subsequent immune response to allergen.[4-6]

INFLUENCE OF IN UTERO EXPOSURE TO ALLERGENS ON DEVELOPMENT OF THE ATOPIC PHENOTYPE

Maternal Influences on the Immune Status of the Fetus

The human fetus spends 9 months within the womb of its mother. In humans and non-human primates there is a very close interaction between maternal and fetal circulations during this period of in utero development. These species have the fewest layers of cells dividing the maternal and fetal circulations due to the hemochorial nature of the placenta. This type of placentation allows for transfer of immunoglobulin and other small molecules from maternal to fetal circulation. Moreover, as described earlier, during pregnancy there is a change in the local environment of the placenta towards a T helper type 2 cytokine profile.

This environment is thought to assist in survival of the fetal allograft from implantation in the uterus up to parturition.[7] This is because a strong cellular immune response might facilitate maternal rejection of the fetus, which shares major histocompatiblity antigens with its father. The Th2 cytokine profile is less likely to facilitate such "graft" rejection. When the production of the cytokine IL-13 by cells at the materno-fetal interface was examined, it was found that normal pregnancy was associated with production of IL-13 by the placenta and subsequently by the fetus.[8]

The recognition that the human fetus is capable of developing an antigen-specific IgE response in utero was made as a result of examination of cord blood. In one epidemiological study, cord blood IgE concentrations were higher among infants whose mothers had a history of atopic disease and allergen immunotherapy compared with infants whose mothers were not atopic.[9] During pregnancy, the maternal influence is exerted on the fetus. In one study, correlations were found between house dust mite and ovalbumin specific IgG subclass levels in cord blood, maternal atopy, and the magnitude of perinatal lymphoproliferative responses to allergens in human infants at birth. An inverse relationship was found between levels of interferon gamma in the fetus/neonate and a maternal history of atopy.[10] However, it was concluded that allergen-specific antibody transferred through the placenta is less likely to have an impact on the future development of allergy in the infant, than the maternal influence on interferon gamma production by T cells of the fetus. These findings support the importance of the uterine cytokine environment to subsequent allergic predisposition of the fetus.

Exposure of the fetus to allergens inhaled by the mother during pregnancy can result in its sensitization to those allergens. Studies performed on cord blood lymphocytes have revealed that sensitization to environmental allergens during intrauterine life does occur. For example, in one study neonatal peripheral blood T cells were obtained from cord blood and stimulated in vitro with several antigens: tetanus toxoid, streptokinase, purified protein derivative of Mycobacterium (PPD), or common inhalent allergens, *Dermatophygoides pteronyssinus* (Der pI), and *Lolium perenne* (LolpI). It was expected that maternally derived T cells would respond to the commonly encountered antigens, such as tetanus toxoid, and that fetal cells would respond only to the inhaled allergens, thereby allowing discrimination between maternal and fetal cells to be accomplished. In fact, the neonatal T cells responded strongly to PPD and to Der pI, but not to Tetanus toxoid or streptokinase. The response to the seasonable allergen Lol pI allergen was variable. The authors of this study concluded that aeroallergen sensitization can occur in utero.[11]

Several studies have shown that exposure to cockroach antigen during the first few years of life increases the risk of wheezing in children from atopic parents. These same children had T cell proliferative reactivity in response to cockroach antigen.[12] Based on this data and other studies suggesting that prenatal antigen exposure can lead to sensitization of the newborn, Miller et al performed a study on a cohort of 167 pregnant women, living in the inner-city, all either African American or Dominican.[1] Dust samples were taken from a subset of their homes for analysis of allergen concentrations. At delivery cord blood was collected as well as maternal blood within a day of childbirth. Analysis of mononuclear cells in response to mouse protein, cockroach, house dust mite, and tetanus toxoid was performed by lymphocyte proliferation and compared with a mitogen (PHA) control. The most significant finding from T cell stimulation was the response to the house dust mite (*D. farinae*). There was no significant correlation between cord blood or maternal IgE and proliferation in response to any antigen. Interestingly, the occurrence of maternal allergen-specific T cell proliferation was not a prerequisite for cord blood cell proliferation. T cell reactivity was measured to mouse and cockroach antigen as well as to dust mite. There was no correlation between the T cell proliferation and allergen-specific IgE levels. Thus, while stating that cord blood cells readily proliferate in response to allergens in the environment of the pregnant mother, the exact link with induction of allergic asthma in the fetus/neonate was not determined.[1]

A recent study sought to determine if maternal avoidance of allergen could influence the development of allergic reactivity in the fetus. To accomplish this goal, concentrations of house dust mites were measured from beds of mothers at the 36th week of pregnancy. At birth the cord blood mononuclear cells were stimulated with either mitogen or dust mite allergen, then subjected to measurement of cytokine synthesis. Levels of IL-4, IL-5, IL-10, γ-interferon were determined and compared with allergen data. The results showed that the cytokine levels did not correlate with the house dust mite levels in the beds of the mothers at week 36 of gestation. Thus, maternal avoidance of house dust does not appear to be an important factor in determination of a child's future allergy status.[13] However, maternal immunological factors were not examined, that is, maternal IgG and IgE specific for house dust mite allergen.

In another study two seasonal allergens were used to study the importance of gestational age on intrauterine priming to allergens inhaled by the mother. Birch and timothy grass allergens were used to stimulate cord blood T cells from neonates born at different times during the year. From the data obtained it was concluded that the susceptibility of fetal lymphocytes for priming with birch and timothy grass allergens decreased when exposure was initiated towards the end of the pregnancy. The authors proposed that this observation may be the result of either decreased access of the allergen to the fetus due to decreased permeability of the placenta, or possibly to enhanced sensitivity of lymphocytes to priming during the earlier phase of pregnancy. Indeed, most of the positive proliferative responses seen to the allergens were obtained from cord blood taken from neonates whose mothers would have been exposed to the allergens during the first 6 months of pregnancy.[14] Another study took these findings further to demonstrate that not only does transplacental priming of the human immune system occur to environmental allergens, but it can also be attributed to a skewed cytokine profile. This study attempted to address concerns that what was attributed to intrauterine priming might actually be either reactivity of maternal cells that were in the cord blood or the stimulatory effect of lipopolysaccharide (LPS), a contaminant of the allergen preparations on the neonatal T cells. The experiments performed showed that neither of these alternate explanations of the results were valid.

It has been hypothesized that adult Th cell cytokine patterns are determined during infancy[15]; although stimulation of neonatal T cells with allergens produced cytokine profiles, as demonstrated by RT-PCR, consistent with a Th2 response,[16] studies over time showed that the response patterns of atopic and non-atopic children were different. Thus, the authors studied both normal and atopic families and found that although normal newborns were born with a Th2 bias, they rapidly altered their cytokine pattern in response to allergens and began producing some Th1 cytokines. In contrast, the atopic individuals displayed an age-associated up-regulation of the Th2 immune response.[12] From this study it is possible to suggest that exposure to allergen during development may be a critical factor for the development of the allergic phenotype in infants and children who are genetically predisposed to allergy.

A recent review of multiple studies addressing the effect of the in utero environment on subsequent development of an allergic phenotype developed several well-supported conclusions. One was that a fetus is able to mount a proliferative response to common allergens as early as 22 weeks of pregnancy. Another was that maternal exposure to allergens can result in a protective IgG response that may decrease the likelihood of fetal sensitization, but that atopic mothers produce a more Th2 skewed immune environment than non-atopic mothers. Proliferation of umbilical cord blood cells in one study was inversely proportional to levels of cord blood dust mite specific IgG. These results tend to suggest that if the mother has made an IgG response to the allergen, then the fetus is less likely to develop an IgE response to that same allergen. Thus, manipulation of the maternal environment may prevent development of an allergic phenotype in infants.[17] Studies on the role of maternal IgG in modulating the fetal IgE response have shown that increased levels of maternal anti-Bet v1 IgG at birth correlated with reduced prevalence of allergic disease at 18 months of age.[18] Thus intrauterine IgG may play a modulatory role in the fetus.

The route by which a fetus is exposed to environmental allergens has not been definitively proven. It has been hypothesized that exposure of the fetus to allergen is transplacental, perhaps as part of a complex with IgG. IgG crosses the placenta beginning early in gestation and reaching a maximum by 32 weeks.[19] IgE does not cross the placenta. However, IgE has been detected in the amniotic fluid as early as 16–17 weeks of gestation. An alternative explanation has been proposed to explain the exposure of the fetus to allergen[19]; allergen crosses from the maternal circulation to the fetal side where these tissues are in intimate contact. In support of this theory the authors state that house dust mite allergen has been detected in the amniotic fluid at 16–17 weeks of gestation.[19]

There are multiple determinants of cord-blood IgE concentrations, as shown in a study on 6401 German neonates.[20] Because the levels of cord-blood IgE have been considered to be a good means of determining the risk of allergic disease for the neonate, a multicenter study was performed.[20] There was a significant difference in IgE concentrations between sexes, with boys having greater amounts of IgE in cord blood. If parents smoked during pregnancy, there was an increase in cord-blood IgE. This study did not show any significant difference between full term babies and those of shorter gestational age. There were significant differences based on nationality (i.e., genetics), with non-German mothers from Far Eastern countries showing the highest values.

Food allergens have also been examined as potential intrauterine immune stimulants. Indeed, cow's milk proteins were used to stimulate cord blood lymphocytes and found to have stimulatory effects on cells from both atopy prone and non-atopy prone individuals. The degree of production of gamma interferon by stimulated T cells was thought to have a stronger likelihood of predicting the future development of atopic disease.[21] Proliferation of cord blood T cells has also been reported in response to bovine serum albumen.[22]

Exposures that occur during fetal and very early neonatal life may be important in future asthma development; even maternal levels of vitamin E, D, and zinc during pregnancy have been associated with increased wheezing in children. Exposure of a neonate to synthetic bedding and allergen is associated with increased risk of developing asthma.[23] Recent studies have demonstrated that fetal life is a critical time for development of the allergic phenotype. Multiple epidemiological studies have shown a strong correlation between maternal atopy and subsequent development of atopy in the infant. In one study on the effect of allergen avoidance during pregnancy in families with a history of atopy, one group of pregnant women avoided eating allergenic foods (soy, wheat, egg, fish, and nuts) during pregnancy and then avoided feeding these foods to the infant up to 12 months of age. In addition, the homes were sprayed to reduce dust mites. In the control group mothers did not avoid the allergenic foods during pregnancy and also for the infants. They did not have the homes sprayed for house dust

mites. Results of this study showed that the prevalence of asthma at 12 months of age in the infants was significantly greater in the control group. This study demonstrates the role of food allergens in pre- and postnatal development for the asthmatic phenotype.[24]

In a more recent study prenatal exposure to peanut and tree nut allergens was evaluated in a retrospective study on families with allergic children during a 4.5-year period. Maternal consumption of tree nuts or sesame seeds during the first two trimesters of pregnancy was associated with a 60% greater chance of the child being sensitized to peanuts, tree nuts and/or sesame seeds. In children with asthma or environmental allergies, the odds were doubled when there was a history of maternal ingestion of peanut, tree nut, or sesame seed during gestation.[25]

A recent study on the association between prenatal exposure to cockroach allergens and polycyclic aromatic hydrocarbons and asthma in inner-city children found that sensitization to cockroach allergens was a strong risk factor for development of asthma and that the risk was increased when exposure to nonvolatile polycyclic aromatic hydrocarbons was concurrent.[26] The 349 mothers wore air samplers during the third trimester of pregnancy and children were evaluated for allergen specific IgE levels at ages 2, 3, 5, and 7 years. This study, although on a relatively small sample of patients, demonstrates a significant link between prenatal exposures and allergic sensitization.

The Influence of Genetics on Response to Allergen Exposure During Development

The influence of genetic background on development of allergic diseases, including asthma, has been recognized for many years.[27] Recently, however, a number of genetic loci have been associated with either development of high IgE production and/or asthma. The multigene control of allergy and asthma ultimately determines whether exposure to allergen results in development of allergic disease. For example, a gene on chromosome 11q13 is linked to maternal inheritance of asthma; it involves polymorphisms in the beta subunit of the high affinity IgE receptor.

Interleukin 13 (IL-13) is an important regulator in IgE synthesis, hypersecretion of mucus, and airway hyperresponsiveness. Polymorphisms in the gene coding for interleukin 13 have been identified.[28] IL-13 polymorphisms were studies in atopic patients and were compared to healthy controls. Significant associations were found between IL-13 polymorphism A-1512C and atopy; this was true for both asthma and allergic rhinitis in a population of Pakistani patients.[29] Another important cytokine for IgE production and allergic disease is interleukin 4 (IL-4). Two polymorphisms that are associated with asthma have been found in the IL-4 gene and 2 in the gene coding for the Il-4 receptor.[30] It is generally thought that atopy is inherited and that predisposition for asthma is a multigene effect. According to this hypothesis, it follows that only a fetus of the appropriate genetic make-up would be likely to develop in utero sensitization.

NEONATAL EXPOSURE TO ALLERGENS

The T Helper Cell Type 2 Phenotype in the Neonate

As previously discussed the intrauterine environment is prejudiced for a Th2 cytokine profile. While this type of immune environment is likely to facilitate survival of the fetus in utero, it results in a neonate that is born with a prejudice towards the Th2 (allergic) phenotype. The exposures that occur in the first months of life are very likely crucial to phenotypic development of the child and subsequently the adult. Thus, exposures to allergens, such as pollens, may facilitate future development of allergic disease, especially in the atopy prone genotype. However, exposure to certain other antigens during this critical time of development can influence or modulate the immune response.[31] For example, as discussed in following text, endotoxin from gram negative bacteria has an allergy-sparing effect by shifting the Th2/Th1 ratio in favor of Th1 cytokines.

Role of Dietary Factors in Allergen Sensitization

Dietary exposure to allergen may be an important factor in mucosal sensitization of the neonate to allergen. For example, cross-reactive antigens have been identified between cereal proteins and grass pollen.[32] In support of these ideas, a cross-sectional study was performed in Spain to evaluate the relationship between grass-pollen asthma and sensitization to cereals in the diet of the child. A relationship was shown between cereal allergy and pollen sensitivity. It was found that early introduction of cereals into the diet of children was a risk factor for grass-pollen allergy.[32] Thus, avoidance of cereal by infants from an atopic family may reduce pollen sensitivity in future years.

A controlled study was performed to determine if prophylactic dietary control could modulate induction of allergy in children.[33] The prophylactic group of 58 children were either breast-fed from mothers whose diets excluded highly antigenic foods or were fed an extensively hydrolyzed formula. The control group of 62 did not have any specific prophylaxis. Results of the study showed that after 1 year there was significantly less total allergy, including asthma in the prophylactic group. After 4 years, the difference between these groups was still significant, with more control children having positive skin tests to allergens.[33] Thus, it seems that another important source of exposure of the child to allergen during development can be in the neonatal period through breast milk.

The Role of Infectious Diseases in Modulating the Effects of Inhaled Allergen during the Neonatal Period

Several infectious diseases are thought to have a role in either facilitating development of allergic sensitization to inhaled allergen during the neonatal period or in subverting the allergic response. The "hygiene hypothesis" (discussed later in this chapter) focuses on early exposures to bacterial products that appear to have an allergy-sparing effect. In contrast, certain viral respiratory infections have been implicated in promotion of allergen sensitization and development of asthma. Viral infections are frequently associated with wheezing in small children. The fact that many of these children progress to become asthmatic has led to the supposition that the early viral infection may either facilitate sensitization or damage airways leading to an increased likelihood of developing airways hyperreactivity.[34]

One of the major respiratory viruses that appears to have a link to asthma is respiratory syncytial virus (RSV). RSV is one of the most important neonatal pathogens currently recognized. Yearly epidemics of RSV cause high morbidity and some mortality in young infants and children under 2 years of age. The disease is much less severe in older children and healthy adults. However, in older people and those immunocompromised, RSV can be equally devastating. In the severe form, RSV causes wheezing and severe bronchiolitis and often interstitial pneumonia. It was recognized in the 1980s that children with the most severe RSV accompanied by wheezing were often subsequently diagnosed with childhood asthma.[35] Indeed, IgE specific for viral proteins as well as elevated histamine concentrations in respiratory fluids were found in severely affected children.[35]

More recently it has been found that RSV preferentially induces a Th2 cytokine environment in atopic children and in some animal models. Studies have shown that infection of balb/c mice with human RSV induces a T helper cell type 2 response.[36] Work in the author's laboratory has demonstrated that some calves infected with bovine respiratory syncytial virus (a closely related bovine pathogen) develop a Th2 cytokine response when infected with the virus.[37] Indeed production of IgE is also associated with this model. Further work with the bovine model has demonstrated that disease is exacerbated when allergen is inhaled during the virus infection[38]; and that sensitization can be enhanced by exposure to allergen during the viral infection.[39]

In contrast to these observations, a recent study has demonstrated that, in severe cases of respiratory disease with wheezing, γ-interferon is present in elevated amounts. However, in this study, the identity of the virus causing the infection and wheezing was not elucidated.[40] Following up on these observations, Garofalo et al examined the role of Th1 and Th2 cytokines as well as several chemokines in RSV disease. They found that macrophage inflammatory protein-1α was associated with severe RSV bronchiolitis. Thus, the role of cytokines in induction of clinical disease may be different than that previously thought to stimulate subsequent development of asthma.[41]

Using a murine model of RSV infection, it has been found that the production of IL-13 during a primary RSV infection has an important effect on exacerbation of cockroach allergen-induced disease.[36] When mice were depleted of IL-13, RSV infection did not exacerbate airway hyperactivity. The production of IL-13 during acute RSV infection is thought to also facilitate allergic sensitization.[36]

In another murine model the effects of infection with influenza A on induction of tolerance to aerosolized allergen was examined. These studies were performed in adult mice, in which induction of tolerance is the most usual sequel to inhalation of allergen. However, infection of mice with influenza changed the response to intranasal ovalbumin such that Th2 cytokines, IgE, and airway hyperactivity developed in response to OVA.[42]

Hygiene Hypothesis

Several previous studies have shown that maternal farm exposure during pregnancy can modulate the early immune development of the fetus. More specifically, growing up on a farm has a protective effect against development of childhood allergic disease.[43] Exposure to stable dust and farm animals is thought to be the essential component in this "protective" environment. Thus, endotoxin, from gram negative bacteria, may modulate the immune response. When environmental endotoxin exposure was measured in farming and non-farming families, it was found that endotoxin concentrations were highest in stables of farming families and also high in dust from the farmhouse floors[44]; these concentrations were significantly different from non-farming families. This study supports the hypothesis that environmental exposure to endotoxin from bacterial cell walls is an important protective determinant in prevention of atopic disease in children.

Despite the myriad of studies addressing the hygiene hypothesis, the underlying mechanisms of immune modulations remain unclear. In a recent study to further identify these mechanisms, Ballenberger et al. examined the role of maternal farm exposure and genetic predisposition on T helper 17 (Th17) cells. They assessed farm exposure of 84 pregnant mothers and then examined cord blood cells (stimulated in vitro) for a variety of cell receptors and lineage markers. The results showed that the Th17 and T regulatory cell markers correlated positively with maternal farm exposure and specific single nucleotide polymorphisms in the IL-17 lineage genes influenced gene expression of the Th17 and T regulatory cell markers.[45]

In another study, the effect of timing of the exposure was examined by comparing 812 children younger than 1 year with those from 1 to 5 years of age.[46] Results of this study showed that exposure to stables and farm milk both prenatally and during the first year of life had a protective effect; additional protection was conferred by exposure until the age of 5 years.[46] This protective effect of exposure to farms during early life has been attributed to bacterial components, such as lipopolysaccharide (LPS) and unmethylated CpG motifs. The raw milk consumed by farm children has a large content of gram negative bacteria, which are a primary source of endotoxin (LPS).

Other studies have focused on childhood environmental factors other than farm life as potential influences on development of allergic disease. For example, one retrospective study obtained information from 13,932 adults between the ages of 20 and 44 from 36 areas in Europe, New Zealand, and the United States. The results of this study showed that growing up in a large family, with sharing of bedrooms, and the presence of a dog in the household appeared to offer protection from development of atopy in later life. The authors of this study point out that if the subject has strong genetic predisposition for allergic reactivity, the environmental factors are probably less important.[47]

Bacteria endotoxin, a component of gram negative bacteria commonly associated with intestinal flora, is capable of inducing the synthesis of the Th1 cytokines γ interferon and IL-12. As discussed earlier, exposure to bacterial endotoxin could be expected to decrease the allergic response. House dust is another source of bacterial endotoxin, containing varying levels of endotoxin. To examine the effect of endotoxin in house dust on development of allergy in infants, a study was performed on infants aged 9–24 months, all of whom had at least three documented wheezing episodes.[48] Levels of house dust endotoxin were compared with allergic sensitization of these infants. It was found that the homes that had the lowest levels of endotoxin were associated with increased numbers of allergen-sensitized infants.[48] This study further supports the hygiene hypothesis.

In a more recent study this same concept was evaluated in a birth cohort study that measured levels of endotoxin, house dust mite allergens, and cat allergens in home dust and correlated this with neonatal (first 2 months of life) lung function, respiratory symptoms, and eczema during the first year of life. They found that there were no significant associations between neonatal lung function and allergen exposure.[49]

EXPOSURE TO ALLERGENS DURING THE JUVENILE PERIOD

The age group of 7–8 years was targeted in a study to examine the incidence of allergic respiratory disease in Swedish children.[50] In this study sensitization to allergens was evaluated with skin-prick tests and a questionnaire was used to evaluate the incidence of asthma and rhinoconjunctivitis, as well as lifestyle factors. It was found that while skin prick test reactivity was not significantly different between the children of farmers and non-farmers, there was a reduced risk among children of farmers for having allergic respiratory disease. It would be of interest to know whether the children with less respiratory disease had similar or dissimilar serum IgE and IgG concentrations.

The presence of house dust mite (HDM) allergen in the environment of children and its relationship to development of asthma was addressed in an Australian study.[51] School children with a clinical history of HDM allergy, including wheeze, were followed for 1 year. During this time, they reported their peak expiratory flow rates and allergen concentration was periodically measured in their homes. Results showed that there was a significant association between decreased peak expiratory flow values and HDM allergen concentration. Thus, in children already sensitized to HDM allergen, development of clinical signs of asthma was exacerbated by increased exposure to the HDM allergen.

Another study addressed the relationship between allergen exposure in the environment and atopic sensitization. When HDM allergen reactivity was examined in schoolchildren living in the Alps versus children living at sea level, it was found that in the Alps where HDM is less common, skin test reactivity to HDM allergen was less than in the lowlands. However, when grass pollens were used as the test allergen, positive skin tests were significantly higher in the Alps where grass pollens are prevalent, as were clinical signs of allergic reactivity. Thus, exposure to inhaled allergen during the school-age period leads to specific sensitization.[52]

Development of allergic reactivity during early childhood was demonstrated in a study that compared cytokine responses of T cells, and skin test reactivity to the oral antigen ovalbumin with the response to the inhaled allergens of house dust mite in 2–5 year old children. It was found that the inhaled allergen-specific cytokine response was associated with a positive skin prick test for children 5 years of age or less. However, this association was not true for the oral allergen, ovalbumin. The authors suggest that a mechanism exists for deletion of the allergic response to food antigens during early childhood.[53] Indeed, children with negative skin tests had cytokine profiles that showed higher levels of gamma interferon, indicative of a non-allergic T helper cell type 1 phenotype (Th1). In another study, skin prick tests were found to correlate with early exposure to animal danders, such as dog and cat, while the development of clinical allergic asthma was not.[54] In yet another study that involved a variety of allergens, a positive correlation was not found between early childhood exposure to allergens and development of allergic asthma.[55]

The influence of a variety of environmental exposures of children on development of atopy as adults was examined in a study of 13,932 subjects in Europe, New Zealand, USA,

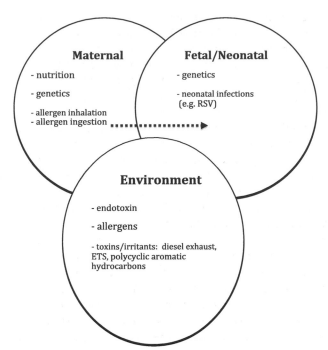

FIGURE 1 The interaction of maternal, fetal, and environmental factors on the outcome of exposure to allergens during development is depicted. The multifactorial nature of pre- and postnatal exposure intercepts with the genetic make-up of the fetus/child to determine the ultimate outcome.

and Australia. This study (European Community Respiratory Health Survey) examined a variety of childhood environmental factors and correlated these with the presence of atopy in adults. The most significant findings were that an association between large family, bedroom sharing, and a dog in the home was associated with a lower prevalence of atopy in families without a genetic predisposition to atopy. However in subjects that had a strong genetic predisposition for allergy, these environmental factors were less important.[47] Overall, we can conclude from this and other studies that the overbearing influence of genetics on induction of allergic response to allergen cannot be ignored. Allergy-inducing environmental factors appear to have their greatest effect on those individuals that have a familial history of atopy (Figure 1).

INTERACTION OF ALLERGENS WITH ENVIRONMENTAL FACTORS DURING DEVELOPMENT

In a recent review reduced lung development was associated as a factor in viral-induced wheezing and allergic asthma. Maternal factors that have been cited as playing a role in the decreased lung development of the neonate include: low intake of nutrients (vitamins D and E, and zinc). Environmental exposure to synthetic bedding during the first months of life was also cited as a factor in asthma development in the infant.[23]

Epidemiological data in humans as well as experimental studies in mice have suggested a correlation between exposure to air pollutants and the incidence of asthma. This observation is not unexpected considering the dramatic increase that has occurred in the incidence of asthma in industrialized countries, where air pollution has also been increasing. In fact, studies have shown that there is a significant association between ambient levels of pollutants and the increased incidence of asthma.[56]

Animal studies have been performed to demonstrate more specifically a link between exposure to environmental pollutants and enhancement of allergic sensitization. Initial animal studies were performed with adult mice. Swiss Webster mice exposed to either 0.8 ppm ozone (O3), 0.5 ppm O3, or O3 and sulfuric acid mist in a episodic schedule were anaphylactically sensitized to aerosolized ovalbumin and had higher numbers of IgE producing cells in the lung than did similarly OVA- aerosol exposed mice exposed to filtered air instead of pollutant.[57] Later studies in the author's laboratory using adult BALB/c mice exposed to environmental tobacco smoke (ETS) showed a similar phenomenon, that is, enhanced sensitization to OVA in pollutant exposed animals.[58] In these latter studies, production of Th2 cytokines by lung lymphocytes and IgE production was significantly greater in the ETS-exposed mice than in OVA aerosol-sensitized ambient air controls.

The mouse serves as an excellent model with which to predict the effects of in utero and neonatal exposure to pollutants on the development of allergic lung disease in humans. In an experiment performed in the author's laboratory, pregnant female mice were exposed to either environmental tobacco smoke (ETS) or ambient air (control) during pregnancy. The progeny were born into and raised in the environment of the mother. At the age of 6 weeks all mice were primed with intraperitoneal ovalbumin and alum. Subsequently they received a series of aerosol OVA challenges. It was found that mice born into and raised in the ETS environment had increased levels of IgE and Th2 cytokines as compared to those born and raised in ambient air. Thus, ETS exposure facilitates development of the allergic phenotype.[59]

Others have demonstrated that exposure to diesel exhaust (DE) enhances the IgE response to inhaled allergen in adult mice.[60] Diesel exhaust had an adjuvant effect on total and allergen-specific IgE production. This resulted in production of chemokines and cytokines that are important in sensitization to inhaled allergens. In a study using rats, the effect of intrauterine diesel exhaust exposure on development of an IgE response to inhaled pollen was examined. Developing rats were exposed to DE either during the fetal period (via the mother), the suckling period, or the weaning period. The study demonstrated that inhalation of DE during the differentiation periods of the immune system accelerated the increase in IgE in response to exposure to Japanese cedar pollen.[61] The environmental pollutant pyrene is a component of DE, which has been found to induce IL-4

production. Thus, the mechanism by which the allergic enhancement occurs most likely involves the production of IL-4.[62] Epidemiological evidence obtained from surveys in Japan indicate that allergic rhinoconjunctivitis was found to be more prevalent in individuals living near motorways than in cedar forests. Children were found to be more likely to have severe asthma than mild asthma when they were living in polluted areas, thus demonstrating the impact of pollution on allergic disease.[63]

A recent study based on the knowledge that elevated levels of combustion-derived particulate matter (CDPM) are associated with development of asthma, used a mouse model to evaluate the impact of such exposure on immunological responses to allergen during development.[64] In this study infant mice were exposed to CDPM and then asthma was induced by sensitization to the house dust mite allergen. A control group was similarly sensitized to allergen but not exposed to CDPM. The CDPM exposed group did not develop the expected airway hyper-responsiveness, Th2 cytokine profile, eosinophilia, and allergen-specific IgE typical for the sensitization protocol. However, they produced total IgE levels that were twice that of the control group. The authors concluded that the CDPM exposed mice experienced an early immunosuppressive environment that included both tolerogenic dendritic cells and regulatory T cells that dampened the Th2 response. Interestingly, when CDPM exposed infants were allowed to mature and then challenged with allergen, they developed severe allergic inflammation. This study has implications for humans in that early exposure of infants to CDPM and allergen may create an immune system that is primed for induction of severe asthma when the individual is exposed to the allergen as an adult. The immunological basis for this early suppression of the allergic phenotype with delayed expression of the phenotype suggests that the induced allergen-specific regulatory T cells are not long lived.

It has been previously thought that cord blood IgE concentrations could be a useful predictor for development of atopic disease in later life. Although correlations have been found between cord blood IgE, atopic history of the mother, and smoking history of the parents during pregnancy, more recent data indicates that intrauterine exposure to allergens may, in fact, be the most important factor in development of sensitization to inhaled allergens. In one study of 7609 neonates in Germany, most of the samples of cord blood assayed for IgE showed very low levels. This study confirmed the previous findings that there is an association between higher IgE levels in children and parental smoking, thus confirming that passive smoke exposure is stimulatory for IgE production.[16] In another study it was found that maternal smoking was correlated with development of allergy in the child, whereas paternal smoking was not.[65]

While these and other studies point to explicit causal factors for enhanced induction of allergic responses by environmental pollutants, the influence of these factors in sensitization of children is less well understood. It has been proposed that the strong association between ETS exposure and asthma in young children may relate to both prenatal and postnatal effects on physical and functional parameters such as airway diameter and bronchial responsiveness.[66]

SUMMARY

Exposure to allergens during development begins in utero and continues throughout the neonatal and childhood years. Maternal exposures to allergen via inhalation and ingestion impact the fetal environment. Genotype is important in the overall impact of allergen exposure, whether it is pre- or postnatal. Factors that favor development of an allergic phenotype include: exposure to tobacco smoke, diesel exhaust, and polycyclic aromatic hydrocarbons. Exposure to these and to allergens in utero and in neonatal years all affect immune system and lung development. Additional fetal factors include sex and time of the year in relation to gestational age. Finally, the environment in which the neonate, infant, and finally the child lives and breathes interacts with the genotype to facilitate or suppress development of the allergic phenotype.

REFERENCES

1. Miller RL, Chew GL, Bell CA, Biedermann SA, Aggarwal M, Kinney PL, et al. Prenatal exposure, maternal sensitization, and sensitization in utero to indoor allergens in an inner-city cohort. *Am J Respir Crit Care Med* 2001;**164**:995.
2. Janeway CA, Travers P, Walport M, Shlomchik M. *Immunobiology.* 5 ed. New York, NY: Garland Publishing; 2001.
3. Belderbos M, Levy O, Bont L. Neonatal innate immunity in allergy development. *Curr Opin Pediatr* 2009;**21**:762–9.
4. Klein Wolterink RG, Hendriks RW. Type 2 innate lymphocytes in allergic airway inflammation. *Curr Allergy Asthma Rep* 2013;**13**:271–80.
5. Halim TY, McKenzie AN. New kids on the block: group 2 innate lymphoid cells and type 2 inflammation in the lung. *Chest* 2013;**144**:1681–6.
6. Schröder NW. The role of innate immunity in the pathogenesis of asthma. *Curr Opin Allergy Clin Immunol* 2009;**9**:38–43.
7. Dealtry G, O'Farrell MK, Fernandez N. The Th2 cytokine environment of the placenta. *Int Arch Allergy Immunol* 2000;**123**:109.
8. Williams TJ, Jones CA, Miles EA, Warner JO. Fetal and neonatal IL-13 production during pregnancy and birth and subsequent development of atopic symptoms. *J Allergy Clin Immunol* 2000;**105**:951.
9. Johnson C, Ownby DR, Peterson EL. Parental history of atopic disease and concentration of cord blood IgE. *Clin Exp Allergy* 1996;**26**:624.
10. Prescott S, Holt PG, Jenmalm M, Bjorksten B. Effects of maternal allergen-specific IgG in cord blood on early postnatal development of allergen-specific T cell immunity. *Allergy* 2000;**55**:470.
11. Piccinni M, Mecacci F, Sampognaro S, Manetti R, Parronchi P, Maggi E, et al. Aeroallergen sensitization can occur in fetal life. *Int Arch Allergy Immunol* 1993;**102**:301.
12. Finn PW, Boudreau JO, He H, Chapman M, Vincent C, Burge HA, et al. Children at risk for asthma: home allergen levels, lymphocyte proliferation, and wheeze. *J Allergy Clin Immunol* 2000;**105**:933.
13. Marks GB, Zhou J, Yang HS, Joshi PA, Bishop GA, Britton WJ, et al. Cord blood mononuclear cell cytokine responses in relation to maternal house dusts mite allergen exposure. *Clin Exp Allergy* 2002;**32**:355.
14. Van Duren-Schmidt K, Pichler J, Ebner C, Bartmann P, Forster E, Urbanek R, et al. Prenatal contact with inhalent allergens. *Pediatr Res* 1997;**41**:128.
15. Prescott S, Macaubas C, Holt BJ, Smallacombe TB, Loh R, Sly PD, et al. Transplacental priming of the human immune system to environmental allergens: universal skewing of initial T cell responses toward the Th2 cytokine profile. *J Immunol* 1998;**160**:4730.
16. Prescott SL, Macaubas C, Smallacombe T, Holt BJ, Sly PD, Loh R, et al. Reciprocal age-related patterns of allergen-specific T-cell immunity in normal vs. atopic infants. *Clin Exp Allergy* 1998;**28** (Suppl.5):39.
17. Warner JA. Primary sensitization in infants. *Ann Allergy Asthma Immunol* 1999;**83**:426.
18. Warner JA, Jones CA, Jones AC, Miles EA, Francis T, Warner JO. Immune responses during pregnancy and the development of allergic disease. *Pediatr Allergy Immunol* 1997;**8**:5.
19. Jones CA, Holloway JA, Warner JO. Does atopic disease start in foetal life? *Allergy* 2001;**55**:2.
20. Bergmann RI, Schylz J, Gunther S, Dudenhausen JW, Bergmann KE, Bauer CP, et al. Determinants of cord-blood IgE concentrations in 6401 German neonates. *Allergy* 1995;**50**:65.
21. Szepfalusi Z, Nentwich M, Gerstmayr E, Jost L, Todoran R, Gratzl K, et al. Prenatal allergen contact with milk proteins. *Clin Exp Allergy* 1997;**27**:28.
22. Kondo N, Kobayashi Y, Shinoda S. Cord blood lymphocyte responses to food antigens for the prediction of allergic disorders. *Arch Dis Child* 1992;**67**:1003.
23. De Luca G, Olivieri F, Melotti G, Aiello G, Lubrano L, Boner AL. Fetal and early postnatal life roots of asthma. *J Matern Fetal Neonatal Med* 2010;**23**(Suppl.3):80–3.
24. Arshad SH, Matthews S, Gant C, Hide DW. Effect of allergen avoidance on development of allergic disorders in infancy. *Lancet* 1992;**339**:1493–7.
25. Hsu JT, Missmer SA, Young MC, Correia KF, Twarog FJ, Coughlin IBH, et al. Prenatal food allergen exposures and odds of childhood peanut, tree nut, or sesame seed sensitization. *Ann Allergy Asthma Immunol* 2013;**111**:391–6.
26. Perzanowski MS, Chew GL, Divjan A, Jung KH, Ridder R, Tang D, et al. Early-life cockroach allergen and polycyclic aromatic hydrocarbon exposures predict cockroach sensitization among inner-city children. *J Allergy Clin Immunol* 2013;**131**:886–93.
27. Cookson WO, Young RP, Sandford AJ, Moffatt MF, Shirakawa T, Sharp PA, et al. Maternal inheritance of atopic IgE responsiveness on chromosome 11q. *Lancet* 1992;**340**:381.
28. Yang K. Childhood asthma: aspects of global environment, genetics, and management. *Changgeng Yi Xue Za Zhi* 2000;**23**:641.
29. Shazia M, Kanza M, Mehwish I, Irum S, Farida A, Asifa A. IL-13 Gene Polymorphisms and Their Association with Atopic Asthma and Rhinitis in Pakistani Patients. *Iran J Allergy Asthma Immunol* 2013;**12**:391–6.
30. Zhu N, Gong Y, Chen XD, Zhang J, Long F, He J, et al. Association between the polymorphisms of interleukin-4, the interleukin-4 receptor gene and asthma. *Chin Med J (Engl)* 2013;**126**:2943–51.
31. Herz U, Joachim R, Ahrens B, Scheffold A, Radbruch A, Renz H. Allergic sensitization and allergen exposure during pregnancy favor the development of atopy in the neonate. *Int Arch Allergy Immunol* 2001;**124**:193.

32. Armentia A, Banuelos C, Arranz ML, DelVillar V, Martin-Santos JM, Gil FJ, et al. Early introduction of cereals into children's diets as a risk-factor for grass pollen asthma. *Clin Exp Allergy* 2001;**31**:1250.

33. Hide DW, Matthews S, Tariq S, Arshad SH. Allergen avoidance in infancy and allergy at 4 years of age. *Allergy* 1996;**51**:89.

34. Gershwin LJ. Asthma, Infection, and Environment. In: Gershwin ME, Albertson TE, editors. *Bronchial Asthma. Principles of Diagnosis and Treatment*, vol. 1, 4th ed. Totowa: Humana Press; 2001. p. 279.

35. Welliver RC, Ogra PL. RSV, IgE, and wheezing. *J Pediatr* 2001;**139**:903.

36. Lukacs NW, Tekkanat KK, Berlin A, Hogaboam CM, Miller A, Evanoff H, et al. Respiratory syncytial virus predisposes mice to augmented allergic airway responses via IL-13-mediated mechanisms. *J Immunol* 2001;**167**:1060.

37. Gershwin LJ, Gunther RA, Anderson ML, Woolums AR, McArthur-Vaughan K, Randel E, et al. Virus specific IgE is associated with IL-2, IL-4, and ¡-Interferon expression in pulmonary lymph of calves during experimental bovine respiratory syncytial virus infection. *Am J Vet Res* 2000;**61**:291.

38. Gershwin LJ, Dungworth DL, Himes SR, Friebertshauser KE. Immunoglobulin E responses and lung pathology resulting from aerosol exposure of calves to respiratory syncytial virus and Micropolyspora faeni. *Int Arch Allergy Appl Immunol* 1990;**92**:293.

39. Gershwin LJ, Himes SR, Dungworth DL, Friebertshauser KE, Camacho M. Effect of bovine respiratory syncytial virus infection on hypersensitivity to inhaled Micropolyspora faeni. *Int Arch Allergy Appl Immunol* 1994;**104**:79.

40. van Schaik SM, Tristram DA, Nagpal IS, Hintz KM, Welliver 2nd RC, Welliver RC. Increased production of IFN-gamma and cysteinyl leukotrienes in virus-induced wheezing. *J Allergy Clin Immunol* 1999;**103**:630.

41. Garofalo RP, Patti J, Hintz KA, Hill V, Ogra PL, Welliver RC. Macrophage inflammatory protein-1alpha (not T helper type 2 cytokines) is associated with severe forms of respiratory syncytial virus bronchiolitis. *J Infect Dis* 2001;**184**:393.

42. Tsitoura DK, Kim S, Dabbagh K, Berry G, Lewis DB, Umetsu DT. Respiratory infection with influenza A virus interferes with the induction of tolerance to aeroallergens. *J Immunol* 2000;**165**:3484.

43. Von Ehrenstein OS, Von Mutius E, Illi S, Baumann L, Böhm O, von Kries R. Reduced risk of hay fever and asthma among children of farmers. *Clin Exp Allergy* 2000;**30**:187.

44. Von Mutius E, Braun-Fahrlander C, Schierl R, Riedler J, Ehlermann S, Maisch S, et al. Exposure to endotoxin or other bacterial components might protedt against the development of atopy. *Clin Exp Allergy* 2000;**30**:1230.

45. Lluis A, Ballenberger N, Illi S, Schieck M, Kabesch M, Illig T, et al. Regulation of T(H)17 markers early in life through maternal farm exposure. *J Allergy Clin Immunol* 2014;**133**:864.

46. Riedler J, Braun-Fahrlander C, Eder W, Schreuer M, Waser M, Maisch S, et al. Exposure to farming in early life and development of asthma and allergy: a cross-sectional study. *Lancet* 2001;**358**:1129.

47. Svanes C, Jarvis D, Chinn S, Burney P. Childhood environment and adult atopy: results from the European Community Respiratory Health Survey. *J Allergy Clin Immunol* 1999;**103**:415.

48. Gereda JE, Leung EY, Thatayatikom A, Streib JE, Price MR, Klinnert MD, et al. Relation between house-dust endotoxin exposure, type 1 T-cell development, and allergen sensitisataion in infants at high risk of asthma. *Lancet* 2000;**355**:1680.

49. Abbing-Karahagopian V, van der Gugten AC, van der Ent CK, Uiterwaal C, deJongh M, Oldenwening M, et al. Effect of endotoxin and allergens on neonatal lung function and infancy respiratory symptoms and eczema. *Pediatr Allergy Immunol* 2012;**23**:448–55.

50. Klintberg B, Berglund N, Lilja G, Wickman M, van Hage-Hamstem M. Fewer allergic respiratory disorders among farmer's children in a closed birth cohort from Sweden. *Eur Respir J* 2001;**17**:1151.

51. Jalaludin B, Xuan W, Mahimic A, Peat J, Tovey E, Leeder S. Association between Der p1 concentration and peak expiratory flow rate in children with wheeze: a longitudinal analysis. *J Allergy Clin Immunol* 1998;**102**:382.

52. Charpin D, Birnbaum J, Haddi E, Genard G, Lanteaume A, Toumi M, et al. Altitude and allergy to house-dust mites. A paradigm of the influence of environmental exposure on allergic sensitization. *Am Rev Respir Dis* 1991;**143**:983.

53. Yabuhara A, Macaubas C, Prescott SL, Venaille TJ, Holt BJ, Habre W, et al. TH2-polarized immunological memory to inhalent allergens in artopics is established during infancy and early childhood. *Clin Exp Allergy* 1997;**27**:1261.

54. Arshad SH, Hide DW. Effect of environmental factors on the development of allergic disorders in infancy. *J Allergy Clin Immunol* 1992;**90**:235.

55. Lau S, Illi S, Sommerfeld C, Niggemann B, Bergmann R, Von Mutius E, et al. Early exposure to house-dust mite and cat allergens and development of childhood asthma: a cohort study. *Lancet* 2000;**356**:1392.

56. Salvi S. Pollution and allergic airways disease. *Curr Opin Allergy Clin Immunol* 2001;**1**:35.

57. Gershwin LJ, Osebold JW, Zee YC. Immunoglobulin E -containing cells in mouse lung following allergen inhalation and ozone exposure. *Int Arch Allergy Appl Immunol* 1981;**65**:266.

58. Seymour BWP, Pinkerton KE, Friebertshauser KE, Coffman RL, Gershwin LJ. Second hand smoke is an adjuvant for T helper-2 responses in a murine model of allergy. *J Immunol* 1997;**159**:6169.

59. Seymour BWP, Friebertshauser KE, Peake JL, Pinkerton KE, Coffman RL, Gershwin LJ. Gender differences in the allergic response of mice neonatally exposed to environmental tobacco smoke. *Dev Immunol* 2002;**9**:47.

60. Devouassoux G, Saxon A, Metcalfe DD, Prussin C, Colomb MG, Brambilla C, et al. Chemical constituents of diesel exhaust particles induce IL-4 production and histamine release by human basophils. *J Allergy Clin Immunol* 2002;**109**:847.

61. Watanabe N, Ohsawa M. Elevated serum immunoglobulin E to Cryptomeria japonica pollen in rats exposed to diesel exhaust during fetal and neonatal periods. *BMC Pregnancy Childbirth* 2002;**2**:2.

62. Bommel H, Li-Weber M, Serfling E, Duschl A. The environmental pollutant pyrene induces the production of IL-4. *J Allergy Clin Immunol* 2000;**105**:796.

63. Devalia JL, Rusznak C, Davies RJ. Allergen/irritant interaction–its role in sensitization and allergic disease. *Allergy* 1998;**53**:335.

64. Saravia J, You D, Thevenot P, Lee GI, Shrestha B, Lomnicki S, et al. Early-life exposure to combustion-derived particulate matter causes pulmonary immunosuppression. *Mucosal Immunol* 2013. http://dx.doi.org/10.1038/mi.2013.88.

65. Tariq S, Hakim EA, Matthews SM, Arshad SH. Influence of smoking on asthmatic symptoms and allergen sensitisation in early childhood. *Postgrad Med J* 2000;**76**:694.

66. Gold D. Environmental tobacco smoke, indoor allergens, and childhood asthma. *Environ Health Perspect* 2000;**108**:643.

Chapter 24

The Epidemiology of Air Pollution and Childhood Lung Diseases

Rakesh Ghosh* and Irva Hertz-Picciotto†

*Division of Environmental Health, Keck School of Medicine, University of Southern California, Los Angeles, CA; †Department of Public Health Sciences, University of California – Davis, Davis, CA, USA

INTRODUCTION

Ambient air pollution is ubiquitous, and in order to protect public health from its adverse effects the World Health Organization and different national regulatory authorities, such as the Environment Protection Agency in the United States, have set standards for some of these pollutants (known as "criteria pollutants": Particulate Matter (PM), Ozone (O_3), Nitrogen Dioxide (NO_2), Carbon Monoxide (CO), Sulfur Dioxide (SO_2) and Lead (Pb)).[1,2] Exposure not only to high levels but also to levels lower than the current regulatory standards can have adverse effects on human health, as described herein. In 2002, approximately 146 million people in the United States were living in areas where monitored air failed to meet the 1997 National Ambient Air Quality Standards for at least 1 of the 6 criteria pollutants.[3] That number has changed little: e.g., in 2010 about 124 million people lived in areas that exceeded the 2008 standards for 1 or more pollutants,[2] a reduction of only 15%. Of these six pollutants, four (PM, O_3, NO_2, CO) had practically the same standards in 2010 as in 2002.

Lungs are one of the most susceptible organs because toxic substances in air come into direct contact with them as we breathe in. Young children are most vulnerable, in part because the airway epithelium, the respiratory bronchioles, and the immune cells are developing and are in the process of attaining maturity.[4] Additionally, infancy and early childhood is marked by rapid growth of the lungs; infants are born with only 1/10th the number of alveoli of adults, which increase about 10-fold in the first 4 years.[5] Toxic insults at this critical stage of life may have lasting effects.

A wide variety of chemical species constitute the pollution in ambient air. Although epidemiological studies provide a plethora of evidence on the associations of adverse health outcomes with some or all of these pollutants, a consistent line of story for a variety of outcomes is yet to emerge on which specific ones and at what level these pollutants cause adverse effects. In this chapter, we describe the current evidence on the effect of air pollutants and lung diseases in children, particularly asthma, bronchitis, and bronchiolitis, starting with a brief description of the pollutants. The epidemiological evidence presented here relates to major air pollution studies published in the last 10 years and is not intended to be comprehensive along the lines of a systematic review.

AMBIENT AIR POLLUTION

Pollutants are categorized into two types: primary and secondary. Primary pollutants are directly emitted from the source and secondary pollutants form when primary pollutants chemically react in the atmosphere. O_3 is a good example of a secondary pollutant because it is formed when volatile organic compounds and nitrogen oxides (NO_x), precursors produced from motor vehicles and industrial sources, react in the presence of sunlight. Owing to the nature of its production, O_3 has a diurnal profile with rise in the morning after heavy traffic and peak in the 10:00 a.m. to 6:00 p.m. period.[6] In the evening, scavenging of O_3 by nitric oxide (NO) titration in areas with heavy traffic reduces O_3 to much lower levels, compared to daytime highs. High O_3 levels may occur in downwind communities later in the afternoon and evening, stemming from airborne transport to areas without heavy traffic. O_3 is a powerful oxidant and respiratory tract irritant in adults and children causing a number of symptoms including shortness of breath, chest pain, etc., when inhaling deeply.

Particulate matter is a heterogeneous mixture of small solid or liquid particles of varying composition found in the atmosphere. Fine particles ($PM_{2.5}$) are emitted from combustion processes (especially diesel-powered engines, power generation, wood burning, etc.) and

The Lung. http://dx.doi.org/10.1016/B978-0-12-799941-8.00024-9

from some industrial activities. After entering the upper airways, particles have a tendency to travel along their original path of entry until they encounter a bend in the airway system. When these particles meet a bend, the larger ones do not turn with the air, rather stick to a surface that is in the particles' original path. In general, particles with aerodynamic diameter greater than 10μm are deposited in the nasopharyngeal region (upper airway passages—nose, nasal cavity, and throat).[7] For smaller particles, as they travel through air, gravitational forces and air resistance eventually overcome their buoyancy (the tendency to stay up). As a result they settle on a surface of the lung, most commonly in the bronchi, and the bronchioles. Particles smaller than 0.5μm have a random motion, similar to gas molecules and are deposited on the lung walls (small airways and alveoli).

Particulates show marked geographic variation. For example, annual average concentrations of $PM_{2.5}$ in southern California during the 1994–1997 period ranged from 7μg/m³ outside the Los Angeles air basin to 32μg/m³ within the air basin. Of measured ions in $PM_{2.5}$, nitrate is the most abundant, followed by ammonium, sulfate, and chloride. In addition, $PM_{2.5}$ contains a number of transition metals and organic compounds that influence its toxicity.

NO_2 is a gaseous pollutant produced by high-temperature combustion and is capable of producing free radicals. The main outdoor sources of NO_2 include diesel and gasoline-powered engines and power plants. The other gaseous pollutants such as SO_2 and CO are also primarily produced from fuel combustion in vehicles, ships, or industrial installations. One of the main sources of SO_2 is high sulfur coal.

Air polluted with O_3, NO_2, and particulates is an important public health problem in many regions of the world. The patterns of emission and photochemistry produce aerosols that vary in composition and size distribution in a complicated manner over time and space.

AIR POLLUTION AND ASTHMA

According to the Centers for Disease Control and Prevention, almost 1 in 10 children in the United States had asthma in 2009.[8] Internationally, variation in the prevalence of asthma symptoms in children has declined over time owing to a plateauing effect in the high income countries and rising prevalence in the low and middle income countries.[9] Although asthma has been the subject of intense research in the last few decades, its etiology and explanation for the increase in prevalence have yet to be firmly established. Compared to normal individuals, airways in asthmatics are more sensitive to environmental stimuli, suggesting that it is therefore reasonable to consider a causative role of airborne pollutants. A large number of epidemiological studies supported by toxicological evidence point towards exacerbations of symptoms in asthmatics as a result of air pollutant

exposures. Nevertheless, it remains uncertain as to whether these exposures are a root cause of the disease itself. In the following paragraphs we discuss the available evidence in two separate sections: prevalent and exacerbations of existing asthma and onset of asthma.

Prevalent Asthma Including Exacerbations

Several studies reported associations between prevalent asthma and different ambient or traffic air pollutants. A Swedish study reported about two and half times increase in the risk for non-allergic asthma associated with NO_x exposure during the first year of life (Table 1). The association grew with age as it was stronger for those diagnosed at 8 years than it was at 4 years.[10] The study did not find an association between pollutants and allergic asthma, providing a line of evidence that air pollutants may stimulate nonspecific irritative rather than allergic inflammatory changes in the airways.[10] Furthermore, asthmatic children were found to be more vulnerable to wheezing episodes associated with daily fluctuations of NO_2 levels,[11] and higher levels of exposure also determined asthma severity determined on a 0–4 scale with higher score implying increased severity (Table 1).[12] A longitudinal study found 26% increased risk between $PM_{2.5}$ exposure and prevalent asthma (Table 1).[13] Elemental Carbon (EC), the majority of which comes from diesel exhaust, was also associated with respiratory symptoms (cough and wheeze) in asthmatic children. Exposure to individually measured EC (same day exposure or 24 hours gap between exposure and episode occurrence) was associated with 37% and 45% increased occurrence of cough and wheeze, respectively, among asthmatic school children.[14] Consistent evidence was also reported from one of the most well-designed and extensively published Southern California Children's Health Studies (CHS) where measured NO_2 levels at participant homes were significantly associated with more than 80% increase in risk of lifetime asthma (Table 1).[15] Similar evidence was also available for other air pollutants such as O_3,[16] several metrics of traffic air pollution and lifetime history of asthma.[15] Evidence from this study also associates biomarkers of asthma such as fractional exhaled nitric oxide and air pollution.[17]

An international collaboration involving 98 countries and 513,087 children (International Study of Asthma and Allergies in Childhood phase 3) reported significant association between truck traffic and asthma for two groups: 6–7 and 13–14 years (Table 1). Contrary to the previous studies, the association for the older age group was found to be slightly smaller compared to the younger age group, 1.2 vs. 1.3 in the ISAAC study.[18] In both age groups a similar trend but much stronger associations (OR~1.6) were observed for "severe" asthma.[18] In spite of the possibility of exposure misclassification from self

TABLE 1 Characteristics of the studies investigating prevalent asthma or asthma exacerbations

Study	Study Design and Size	Age Group*	Outcome	Exposure Details	Main Findings, OR (95% CI)
Prevalent asthma					
Gruzieva et al.[10]	Longitudinal birth cohort (n = 2518 and 2378)	0–12 years, asthma estimates reported for 4 and 8 years	Parental report of physician diagnosed asthma	Model predicted individual exposure. 5th to 95th percentile increase (7 and 47 µg/m³ increase in PM_{10} and NO_x, respectively)	4 years PM_{10} – 1.6 (0.5, 5.3) NO_x – 2.4 (1.0, 5.6) 8 Years PM_{10}# – 3.8 (0.9, 16.2) NO_x – 2.6 (0.9, 8.1)
Andersson et al.[25]	Cross sectional (n=1357)	7–8 years	Physician diagnosed asthma	GIS bases assessment ≥250 heavy vehicles daily within 200m from residence vs. those living further away	1.74 (1.06, 2.87)
Mar et al.[24]	Time series (n=1137 children)	0–18 years	Emergency department visits with asthma	Central site measurements 10 ppb increase in Ozone	1hr avg – 1.10 (1.02, 1.18) 8hr avg – 1.09 (1.01, 1.18)
Spira-Cohen et al.[14]	Panel (n=40)	10–12 years	Asthma exacerbations (wheeze and cough)	Personal EC exposure, 3 µg/m³ (5th to 95th percentile)	Wheeze – 1.37 (1.09, 1.72) Cough – 1.45 (1.03, 2.04)
Mohr et al.[26]	Time series (281,763 ER visits)	2–17 years	Emergency department visits with asthma	Central site measurements EC 24 hr average (0.1 µg/m³)	2 to 5 years – 1.03 (1.01, 1.05) 11 to 17 years – 1.09 (1.02, 1.17)
Iskander et al.[27]	Case crossover (8226 admissions)	0–18 years	Daily asthma admissions	Central site measurements IQR increase in NO_x (8.6ppb), NO_2 (6.5ppb), and $PM_{2.5}$ (4.8µg/m³)	NO_x – 1.11 (1.05, 1.17) NO_2 – 1.10 (1.04, 1.16) $PM_{2.5}$ – 1.09 (1.04, 1.13) PM_{10} – 1.07 (1.03, 1.12) UFP – 1.06 (0.98, 1.14)
Pereira et al.[28]	Case crossover (n=603 children)	0–19 years	Emergency department visits with asthma	Central site measurements IQR increase in NO_2 (2.81ppb) and CO (0.23ppm),	NO_2 – 1.70 (1.08, 2.69) CO – 1.40 (1.06, 1.84)
Mann et al.[11]	Longitudinal (n=315)	6–11 years	Wheeze in asthmatics	Central site measurements per 90th percentile increase, NO_2 (8.7ppb), EC (3.7µg/m³), $PM_{2.5}$ (36.2µg/m³) and O_3 (20.0ppb)	NO_2 – 1.10 (1.02, 1.20) EC – 1.12 (0.97, 1.30) $PM_{2.5}$ – 1.09 (0.93, 1.27) O_3 – 1.01 (0.92, 1.12)

Continued

TABLE 1 Characteristics of the studies investigating prevalent asthma or asthma exacerbations—cont'd

Study	Study Design and Size	Age Group*	Outcome	Exposure Details	Main Findings, OR (95% CI)
Belanger et al.[12]	Longitudinal (n=1342)	5–10 years	Wheeze and asthma severity	Indoor NO_2 measurements (1.6 unit increase in ln NO_2 above 6ppb)	Asthma severity – 1.37 (1.01, 1.89) Wheeze – 1.49 (1.09, 2.03)
Samoli et al.[29]	Time series (3601)	0–14 years	Hospital admissions with asthma	Central site measurements IQR increase PM_{10} (21.6μg/m³), SO_2 (13μg/m³), NO_2 (37.3μg/m³), and O_3 (43.3μg/m³)	PM_{10}[#] – 5.66 (0.12, 11.51) SO_2 – 7.84 (1.14, 14.98) NO_2 – 4.13 (-2.50, 11.21) O_3 – 12.55 (-25.26, 2.32)
Gehring et al.[13]	Longitudinal birth cohort (n=3863)	0–8 years	Parental reporting of physician diagnosed asthma (prevalent)	Modeled NO_2, $PM_{2.5}$ (IQR increase of 3.2μg/m³) and Soot	$PM_{2.5}$ – 1.26 (1.04, 1.51)
Gauderman et al.[15]	Longitudinal (n=208)	Fourth graders (avg 10 years)	Physician diagnosed asthma	IQR increase in NO_2 measured outside residences (5.7ppb) and distance to freeway	NO2 – 1.83 (1.04, 3.21) Traffic – 1.89 (1.19, 3.02)
Brunekreef et al.[18]	Cross sectional (n=315,572)	6–7 (parental report) and 13–14 years (self-report)	Ever asthma and severe asthma	Self-reported truck traffic on the street of resident (high, medium, low, and no)	Ever asthma, high vs. no – 1.23 (1.16, 1.30) Severe asthma, high vs. no – 1.65 (1.53, 1.79)

*For longitudinal studies, this is age at enrolment.
[#]The numbers are percent increase in outcome rates from Poisson models.

reported truck traffic exposure data, the positive association from a study of this global nature makes a supportive case for a causal association, particularly because several studies in other parts of the world have reported similar associations.[15,19–23] Studies investigating traffic air pollution were mostly investigated in developed countries but this study showed similar associations across all regions of the world. If the associations between pollution and asthma were different in different parts of the world, a universal association would not be observed. Moreover, comparable associations for the two age groups assure that responder bias was not in play because older children completed the questionnaire themselves, while the parents completed it on behalf of the younger group participants. However, the general nature of the questions for exposure information makes it likely that the truck traffic frequency may be an indicator of more general traffic, as roads with high truck traffic will likely carry high loads of other vehicles as well.

Associations were also observed between asthma related emergency department visits and ambient pollutants such as O_3 (Table 1).[24] Living close to heavy traffic was associated

with 70% increased risk of physician-diagnosed asthma (Table 1).[25] Another study reported 3% increase in asthma related emergency room visits for EC exposure (Table 1).[26] A case-crossover study involving Danish children reported 9% increase in hospital admissions of asthma associated with several day exposures to $PM_{2.5}$. Comparable association was also reported with PM_{10} but not with ultrafine particles suggesting the effect is most likely mediated by the larger size fraction that enters the lower airways then the smallest ones (Table 1). Owing to the unique study design the same child served as his or her control in the non-event days, thus reducing the possibility of residual confounding due to intra-individual differences.[27] Another case-crossover study on Australian children reported 70% and 40% increase in emergency room visits and NO_2 and CO exposure, respectively.[28] Associations were also observed for SO_2 and asthma related hospital admissions in Greece with a 7.8% increase in rate (Table 1).[29] This latter study also reported a significant difference in hospital admission rates due to the influence of the Sahara desert dust, which influences ambient particulate levels. During non-desert dust days PM_{10} was associated with a 2.1% increase in admission rates, which almost doubled to 4.1% during the desert dust days.[29]

The ever-increasing body of evidence is indicative of a dose-response relationship between air pollutants and prevalent asthma or asthma exacerbations. However, the levels at which these effects are observable varies across studies and sometimes these are lower than the WHO or national air quality standards for the respective regions.

Asthma Onset

Thus far, several well-designed studies have provided positive evidence between different air pollutants and prevalent asthma in children. However, the evidence on air pollution playing a role in asthma onset is relatively new [23,30–34] with some studies finding no association.[23,35] One of the earliest studies investigating air pollution exposure in schools and asthma incidence in a longitudinal follow-up of first to sixth grade Japanese children, reported significantly high (3-fold increase) associations with NO_2 (Table 2).[36] Intriguingly, the study did not find an association between air pollution and asthma prevalence among first graders, while a Dutch longitudinal study[13] found association both for prevalent and incident asthma (Table 2).

One of the several likely reasons for the discrepancy (association with incident but not prevalent asthma) in the Japanese study[36] noted above can be due to exposure misclassification commonly encountered in air pollution epidemiological investigations, wherein actual exposure differs from exposure assigned for study purposes. Prevalent asthma was investigated among first graders, while incident asthma among first to sixth graders and exposure was measured based on location of schools. Under the assumptions of a cumulative exposure model and that exposure at school was correctly estimated by central site monitors near schools, exposure was likely more accurately estimated for children who attended school for a continuous period of time. This was more likely for those who were followed-up for a certain period of time than those who were just recruited (first graders). In other words, exposure misclassification was more likely among first graders than those who were followed-up for a period of time, which could be the reason for the null association with prevalent asthma among first graders, whereas there was a strong association for incident asthma amongst those followed-up for 6 years. On the other hand, the observed association with incidence could simply be because of cumulative exposure over several years.

Reports investigating incident asthma in the Southern California CHS cohort observed more than 3-fold increase in asthma risk among children playing three or more sports in high O_3 communities.[16] The study also reported that time spent outside in high O_3 communities was significantly associated with asthma, RR=1.4, but not in low O_3 communities, RR=1.1 (Table 2). The association was further elevated when history of wheeze was accounted for in the models. A study investigating incident asthma and traffic related air pollution (TRAP) at home and school also observed 34% increased association for combined freeway or non-freeway exposures in children below 9 years of age (Table 2),[30] Elevated but non-significant associations with central-site measurements of PM_{10} and $PM_{2.5}$ were reported after accounting for TRAP (Table 2). For central site NO_2, the study reported a significantly high association (OR=2.17) for incident asthma, which dramatically reduced (to 1.37) after adjusting for traffic air pollution, hinting towards the possibility that regional NO_2 might have been a likely surrogate of combustion related pollution (Table 2). Moreover, the morning commute[37] and school hours overlap with peak traffic flow, which suggests exposure to TRAP. A subsequent study on a smaller sample of the same cohort obtained more accurate exposure using personal NO_2 measurements and reported association of comparable magnitude[31]; however, the association was apparent only after adjusting for humidity and there was no interaction with humidity (Table 2). The authors offered a statistical explanation for their observation mentioning that there was some shared variance between air pollution and humidity (in other words correlated), which overlaps with the asthma outcome and is removed with the introduction of humidity into the model, making the relationship stronger. Pollutant effects were found to be smaller during high humidity days,[38] because outdoor pollutants, particularly the gaseous ones, are either absorbed or washed out by atmospheric water vapor.[39,40] This would also explain the observation in the earlier study where adjustment for humidity essentially took out the effect of low pollution days and the remaining effect was stronger. However, experimental studies have shown that breathing in hot and humid air evoked cough and bronchoconstriction in patients with mild asthma.[41,42] Whether the relationship between humidity, air pollution, and incident asthma can be entirely explained statistically or if there is a biological interaction between the two exposures leading to childhood asthma deserves further exploration.

Positive association for asthma onset was also independently reported from another cohort that investigated in utero and first year exposure and asthma incidence in children up to 3 and 4 years of age.[32] Using inverse distance weight and land use regression models that predicted exposure for participants, the study reported significant associations for in utero or first year exposure to NO, NO_2, CO, $PM_{2.5}$, and black carbon with asthma, in the order of 7–14% increase in incidence. Often because of high correlation between pre- and postnatal exposure it is not possible to discern which window is more harmful.

TABLE 2 Characteristics of the studies investigating incident asthma

Study	Study Design and Size	Age Group*	Outcome	Exposure details	Main Findings, OR (95% CI)
Incident asthma					
Shima et al.[36]	Longitudinal (n=3049)	6 years	Incident asthma	Central site measurements NO_2 (per 24 ppb) & PM_{10} (per 26 ppb)	Incident NO_2 – 3.62 (1.11, 11.87) PM_{10} – 2.84 (0.84, 9.58) Prevalent NO_2 – 0.75 (0.31, 1.82) PM_{10} – 1.01 (0.47, 2.16)
Gehring et al.[13]	Longitudinal birth cohort (n=3143)	0–8 years	Parental reporting of physician diagnosed incident asthma	Modeled NO_2, $PM_{2.5}$ (IQR increase of 3.2 $\mu g/m^3$) and Soot	$PM_{2.5}$ – 1.28 (1.10, 1.49) NO_2 estimate not reported
McConnell et al.[16] (Lancet)	Longitudinal (n=3535)	9–10, 12–13 and 15–16 years school children	Parental report of physician diagnosed incident asthma	High vs. low communities Ozone (Median 74 vs. 48 ppb) and $PM_{2.5}$ (Median 22 vs. 8 $\mu g/m^3$)	High PM and ≥3 sports – 2.0 (1.1, 3.6) High O_3 and ≥3 sports – 3.3 (1.9, 5.8)
McConnell et al.[30] (EHP)	Longitudinal (n=2497)	5–9 years at entry	Parental report of physician diagnosed incident asthma	Modeled exposure to traffic (21 ppb) and central site NO_2 (23.6 ppb), $PM_{2.5}$ (17.4 $\mu g/m^3$), PM_{10} (43.9 $\mu g/m^3$) and O_3 (30.3 ppb) increments	Traffic – 1.34 (1.07, 1.68) NO_2 – 2.17 (1.18, 4.00) PM_{10} – 1.35 (0.64, 2.85) $PM_{2.5}$ – 1.66 (0.91, 3.05) O_3 – 0.76 (0.38, 1.54)
Jerrett et al.[31]	Longitudinal (n=217, random selection from parent cohort)	10–18 years	Self report of physician diagnosed incident asthma	Personal exposure to NO_2 (6.2 ppb increment)	With humidity – 1.29 (1.07, 1.56) Without humidity – 1.10 (0.91, 1.33)
Clark et al.[32]	Population based nested case control study (n=37,401)	3–4 years	Incident asthma (hospital records)	Modeled exposure NO (10 $\mu g/m^3$), NO_2 (10 $\mu g/m^3$), CO (100 $\mu g/m^3$), O_3 (10 $\mu g/m^3$), $PM_{2.5}$ (1 $\mu g/m^3$) and black carbon (10^{-5}/m)	In utero NO – 1.07 (1.03, 1.12) NO_2 – 1.10 (1.05, 1.15) CO – 1.07 (1.04, 1.10) O_3 – 0.83 (0.77, 0.89) $PM_{2.5}$ – 0.95 (0.91, 1.00) Black Carbon – 1.08 (1.02, 1.15) First year NO – 1.08 (1.04, 1.12) NO_2 – 1.12 (1.07 1.17) CO – 1.10 (1.06, 1.13) O_3 – 0.81 (0.74, 0.87) $PM_{2.5}$ – 1.05 (0.97, 1.14) Black Carbon – 1.14 (1.01, 1.29)
Carlsten et al.[33]	Randomized intervention study (n=233)	7 years	Physician diagnosed new onset asthma	Modeled NO_2 (7.2 $\mu g/m^3$), $PM_{2.5}$ (4.1 $\mu g/m^3$), and Black Carbon (1.2 AU)	Control NO – 1.7 (1.0, 2.8) NO_2 – 1.9 (0.9, 4.0) $PM_{2.5}$ – 4.1 (1.2, 13.8) BC – 1.2 (0.6, 2.3) Intervention NO – 0.8 (0.4, 1.6) NO_2 – 1.3 (0.6, 2.9) $PM_{2.5}$ – 2.1 (0.6, 7.1) BC – 1.1 (0.5, 2.5)

TABLE 2 Characteristics of the studies investigating incident asthma—cont'd

Study	Study Design and Size	Age Group*	Outcome	Exposure details	Main Findings, OR (95% CI)
Zmirou et al.[34]	Case-control (n=217 matched pairs)	0–14 years	Physician diagnosed incident asthma	Distance weighted traffic density tertiles. Tertile 2 – 11.2 to 28.8 and tertile 3 – ≥30.0 (vehicles/day)/meter	0 to 3 years Tertile 2 – 1.48 (0.73, 3.02) Tertile 3 – 2.48 (1.14, 4.56) Lifelong Tertile 2 – 0.95 (0.50, 1.82) Tertile 3 – 0.82 (0.43, 1.59)
Oftedal et al.[35]	Retrospective cohort (n=2871)	9–10 years	Early (0–3 years) and late (≥4 years) onset physician diagnosed asthma	Modeled NO_2 (IQR increase 27.3–19.6 µg/m^3) exposure	Early onset – 0.78 (0.62, 0.98) Late onset – 1.05 (0.64, 1.72)
Gehring et al.[23]	Longitudinal birth cohorts (n=1756)	0–2 years	Wheeze	IQR increase in modeled $PM_{2.5}$ (1.5 µg/m^3) and NO_2 (8.5 µg/m^3)	First year $PM_{2.5}$ – 0.91 (0.76, 1.09) NO_2 – 0.87 (0.70, 1.08) Second year $PM_{2.5}$ – 0.96 (0.83, 1.12) NO_2 – 0.94 (0.79, 1.12)

*For longitudinal studies, this is age at enrolment.

However, prenatal and early life exposure windows are both critical for similar respiratory health endpoints.[34,43–45] Aside from the fact that the lung develop rapidly in the early years, this observation can be partially attributed to more accurate time activity based exposure assignment in very early life compared to a later time when children start to spent more time outdoors.

A very recent study compared an intervention group comprising of high-risk children (having at least one first degree relative with asthma or two first-degree relatives with allergic diseases) who avoided dust mites, pet allergens, and second-hand tobacco smoke with a control group and found that the latter group was at significantly higher risk of incident asthma following exposure to $PM_{2.5}$ and NO.[33] A French case-control study reported significant association with exposure in the first 3 years but not with lifetime exposure.[34] A couple more studies however found no association: long-term traffic related air pollution exposure was not associated with early (0–3 years) or late (≥4 years) onset of asthma.[35] Exposures in the first year of life and average exposure from birth to onset were not significantly associated with incident asthma. A European study that investigated incident asthma in the first 2 years and modeled traffic pollution data also reported no association.[23] Asthma causes variable airway obstruction in the lungs and sometimes wheeze with an underlying inflammatory process that makes it more often viewed as a complex condition than a single disease.[46] The disease is characterized by variable severity, natural history, and response to treatment.[47] It is further complicated by the fact that asthma may be allergic or non-allergic, and in the past decades the primary focus was on the Th2-mediated inflammation to understand the pathogenesis. Given the heterogeneity of the disease, the susceptibility of asthma patients to air pollution is likely to differ significantly. An overall group of asthmatics, while serving as a useful starter, is not enough to sufficiently understand causality. Further categorization and investigation by phenotypes is necessary as health endpoints for epidemiological studies. For example, does exposure to air pollution cause airway remodeling thereby forcing adaptive changes to structural airway cells or extracellular matrix to be made. In a review of developmental immunotoxicity of environmental risk factors, the onset of asthma was pegged between 4 and 8 years; however, the epidemiological evidence shows that it can be much earlier.[48] The critical aspect in this context is the timing and duration of air pollution exposure. Exposure is often measured concurrently with an outcome, or there is a lag with varying amounts of time between exposure and outcome, or exposure can be throughout the span of life. For children, this means the biologically effective dose resulting from exposure will vary radically over time owing to various factors such as rapidly changing ventilation rate with age and activity, changing high surface area to body mass ratio, immature detoxification mechanism to rid themselves of the toxic substances, etc.[49] In the case of in utero exposure, maternal factors also influence exposure, for example, pregnant women exchange about 72% more air at rest than non-pregnant women. Also, a skewed Th2 type environment prevails in utero to avoid Th1 driven rejection of the fetus.[48] The newborn is in an imbalanced state with immature innate immune system and acquired immune capacity favored towards Th2 responses.[48] Studies have shown that air pollution exposure immediately before birth was associated with significant changes in the percentages of CD3+, CD4+,

CD8+, CD19+, and natural killer cells, which increase or decrease depending on the timing and duration of exposure.[50,51] Similar trends were also reported for cord blood immunoglobulin E levels.[52] This suggests that exposure to higher levels of air pollution in prenatal life may disturb the Th1/Th2 homeostatic balance, which may determine predominance of one type of mechanism over the other, subsequently making an infant susceptible to diseases immediately after birth or later. Furthermore, at very early stages ramifications of exposure could be more pervasive because of the possibility of impairing pluripotent stem cells, then similar exposure at older ages when the cell lineages are already defined.[53] These issues highlight some of the complexities that air pollution epidemiological studies encounter in ascertaining critical exposure window and putting into context the epidemiological results with the dynamic physiologic development that occur during early ages.

AIR POLLUTION AND BRONCHITIS, BRONCHIOLITIS

Bronchitis

Bronchitis is an inflammation of the bronchial mucous membrane. In comparison to asthma, there are fewer studies on bronchitis in otherwise healthy children, although both are lower respiratory diseases, which can be attributed to the fact that asthma has a much higher prevalence than bronchitis in most countries. A study from the Southern California Children's Health Cohort found significant association with PM_{10} and $PM_{2.5}$ and bronchitis only in those with asthma but not in non-asthmatics. Interquartile increase in PM_{10} was associated with 40% increase in bronchitis among asthmatics, a similar but non-significant increase was reported for $PM_{2.5}$ (Table 3).[54] Studies based on two German birth cohorts (GINI and LISA) reported association between modeled individual exposures and a heterogeneous group of doctor diagnosed bronchitis (asthmatic, spastic, and obstructive bronchitis).[55] The study included children 4–6 years of age and reported 56% increase risk from $PM_{2.5}$ exposure (Table 3). Also, living within 50m of a motorway, federal or state road was significantly associated with the outcome (66% increase).[55] Attenuated association (23% increase) was reported for 2-year-old children, in a subset of the same cohort and for the same exposure metric (50m from a major road).[56] For 1-year-olds the association was only significant for NO_2 but not for $PM_{2.5}$ and the distance metric.[56] A study from the Netherlands investigated modeled long-term exposure to NO_2, $PM_{2.5}$, and soot and did not find any association with bronchitis in the first 4 years of life.[20] A longitudinal follow-up of Czech children found a significant relationship of $PM_{2.5}$, NO_x, and PAH with acute bronchitis.[57,58] Whilst the associations with $PM_{2.5}$ and NO_x were also found in other studies, this was the first study to report an association

with ambient PAH. Twelve different PAHs including carcinogenic Benzo[a]pyrene were measured separately and their sum was used as exposure. Elevated levels of PAH and $PM_{2.5}$ were associated with 29% and 30% increase in doctor diagnosed bronchitis, respectively, among children aged 2 years or younger (Table 3). The association with PAH was much higher (OR=1.56) with older children 2–4.5 years.[57] For NO_x, a consistent association was found with exposure among children below and above 2 years of age.[58] The studies investigated various exposure windows starting from 3 to 45 days before a bronchitis episode without any lag and found the association to monotonically increase from 3 to 30 days exposure prior to an event.[57]

The above studies present some limited evidence of a positive association with varying magnitude. The definitions of bronchitis used by different studies are quite heterogeneous making it impossible to draw an inference. For example, the German studies used asthmatic/spastic/obstructive bronchitis as the outcome, while in the Czech study bronchitis and bronchiolitis episodes were grouped together. Unlike in the United States, physicians in the Czech Republic do not generally distinguish episodes of bronchiolitis separately and they are mostly grouped with bronchitis, which prevented an identification of them as a separate group in the analysis.

Bronchiolitis

Bronchiolitis is a lower respiratory infection that occurs predominantly in children in the first 2 years. It is generally caused by viral inflammation of the bronchioles, partially or completely blocking the passage, leading to a whistling sound as the child breathes out. It is one of the leading causes of hospitalization in infants and very young children. There are very few epidemiological investigations on air pollution and bronchiolitis. A large study from British Columbia, Canada, investigated inpatient and outpatient encounters for bronchiolitis in infants and observed associations with several air pollutants.[59] Lifetime exposure to NO_2, SO_2, and CO was associated with bronchiolitis; corresponding ORs were 1.12, 1.04, and 1.13, respectively, associated with increased levels of pollutants measured at central-site monitors.[59] Except for NO_2 none of the modeled based estimates were statistically significantly associated with bronchiolitis (Table 4). Moreover, a negative association was reported between exposure to O_3 and bronchiolitis, which could be because the study was based on spatial variability in exposures, while much of the variability in the population's O_3 exposure was temporal. In addition, greatest spatial contrast in O_3 was in the study region with relatively sparse population density, so only a small proportion of

TABLE 3 Characteristics of the studies investigating bronchitis in children

Study	Study design and sample size	Age group*	Outcome	Exposure details	Main findings, OR (95% CI)
McConnell et al.[54] (EHP)	Cross-sectional (n=3676)	4th, 7th, and 10th graders	Bronchitis in asthmatics	Central site measurements PM_{10} (19 µg/m^3), $PM_{2.5}$ (15 µg/m^3), O_3 (32 ppb), and NO_2 (24 ppb)	Asthmatics PM_{10} – 1.4 (1.1, 1.8) $PM_{2.5}$ – 1.4 (0.9, 2.3) O_3 – 1.0 (0.6, 1.7) NO_2 – 1.3 (0.8, 2.2) Non-asthmatics PM_{10} – 0.7 (0.4, 1.0) $PM_{2.5}$ – 0.5 (0.3, 1.0) O_3 – 0.9 (0.4, 1.8) NO_2 – 0.8 (0.4, 1.7)
Morgenstern et al.[55] (AJRCCM)	Longitudinal birth cohort (n=2436)	0–6 years	Asthmatic or spastic or obstructive bronchitis	Modeled exposure to $PM_{2.5}$ (1 µg/m^3), $PM_{2.5}$ absorbance (0.2* 10^{-5}/m), NO_2 (6.4 µg/m^3), and distance to nearest road (<50 m)	$PM_{2.5}$ – 1.12 (0.94, 1.29) $PM_{2.5\ (absorbance)}$ – 1.56 (1.03, 2.37) NO_2 – 1.04 (0.67, 1.39) Dt to road – 1.66 (1.01, 2.59)
Morgenstern et al.[56] (OEM)	Longitudinal birth cohort (n=3577)	0–2 years	Asthmatic or spastic or obstructive bronchitis	Modeled IQR exposure to $PM_{2.5}$ (1 µg/m^3), $PM_{2.5}$ absorbance (0.2*10^{-5}/m), NO_2 (5.7 µg/m^3), and distance to nearest road (<50 m)	First year $PM_{2.5}$ – 1.04 (0.90, 1.29) $PM_{2.5\ (absorbance)}$ – 1.14 (0.88, 1.48) NO_2 – 1.30 (1.03, 1.66) Dt to road – 1.12 (0.88, 1.44) Second year $PM_{2.5}$ – 1.05 (0.92, 1.20) $PM_{2.5\ (absorbance)}$ – 0.85 (0.31, 2.34) NO_2 – 0.82 (0.33, 2.03) Dt to road – 1.23 (1.00, 1.51)
Brauer et al.[20] (ERJ)	Longitudinal birth cohort (n=3496)	0–4 years	Doctor diagnosed bronchitis	GIS measurements of $PM_{2.5}$ (3.3 µg/m^3), NO_2 (10.6 µg/m^3) and Soot (0.6*10^{-5}/m). ORs as IQR increase	$PM_{2.5}$ – 0.88 (0.66, 1.18) NO_2 – 0.90 (0.69, 1.16) Soot – 0.90 (0.70, 1.15)
Hertz-Picciotto et al.[57]	Longitudinal birth cohort (n=1133)	0–4.5 years	Acute bronchitis	Central site measurements, PAH (100 ng/m^3), $PM_{2.5}$ (25 µg/m^3). Increment ~2SD.	0–2 years PAH – 1.29 (1.07, 1.54) $PM_{2.5}$ – 1.30 (1.08, 1.58) 2–4.5 years PAH – 1.56 (1.22, 2.00) $PM_{2.5}$ – 1.23 (0.94, 1.62)
Ghosh et al.[58]	Longitudinal birth cohort (n=1133)	0–4.5 years	Acute bronchitis	Central site measurements, NO_x (35 µg/m^3)	0 to 2 years – 1.31 (1.07, 1.61) 2 to 4.5 – 1.23 (1.01, 1.49)

*For longitudinal studies, this is age at enrolment.

TABLE 4 Characteristics of the studies investigating bronchiolitis in infants and young children

Study	Study Design and Sample Size	Age Group	Outcome	Exposure Details	Main Findings, OR (95% CI)
Karr et al.[59] (AJRCCM)	Nested case control (cases=10,485) and controls= 57,127)	0–1 year	Bronchiolitis (hospitalization records)	Modeled estimations of $PM_{2.5}$ (2.9 µg/m³), Black Carbon (0.7×10⁻⁵/m), NO_2 (7.2 µg/m³)	Lifetime exposure $PM_{2.5}$ – 1.00 (0.98, 1.03) Black Carbon – 0.99 (0.96, 1.01) NO_2 – 1.04 (1.02, 1.07) 1 month before episode $PM_{2.5}$ – 1.00 (0.98, 1.03) Black Carbon – 0.99 (0.97, 1.01) NO_2 – 1.04 (1.02, 1.07)
Karr et al.[60] (EHP)	Case crossover (n=19,901)	0–1 years	Bronchiolitis (hospitalization records)	Central site measurement of CO (1361 ppb), NO_2 (27 ppb), and $PM_{2.5}$ (10 µg/m³)	CO – 0.99 (0.96, 1.02) NO_2 – 0.97 (0.95, 0.99) $PM_{2.5}$ – 0.96 (0.94, 0.99)
Karr et al.[61] (AJE)	Case control (case=18,595, controls= 169,472)	0–1 years	Bronchiolitis (hospitalization records)	Central site measurement of CO (910 ppb), NO_2 (15 ppb), $PM_{2.5}$ (10 µg/m³) and O_3 (14 ppb)	Sub chronic (one month prior to episode) CO – 1.00 (0.97, 1.03) NO_2 – 1.04 (1.00, 1.08) O_3 – 0.92 (0.88, 0.97) $PM_{2.5}$ – 1.09 (1.04, 1.14) Chronic (from birth) CO – 1.00 (0.97, 1.03) NO_2 – 1.03 (0.99, 1.07) O_3 – 0.92 (0.88, 0.97) $PM_{2.5}$ – 1.09 (1.04, 1.14)
Karr et al.[62]	Case control (Cases=2604, controls= 23,354)	0–1 years	Bronchiolitis (hospitalization records)	Central site measurement and modeled $PM_{2.5}$ (10 µg/m³), NO_2 (1 ppb), and proximity (150 m from roadways)	$PM_{2.5}$ – 1.04 (0.83, 1.29) NO_2 – 1.01 (0.99, 1.03) Proximity – 1.07 (0.90, 1.27)
Segala et al.[63]	Case crossover (ER=50,857, Hospitalizations= 16,588)	0–3 years	Emergency room visits due to bronchiolitis	Central site measurements of PM_{10}, SO_2, NO_2, and BS (10 µg/m³ increment)	PM_{10} – 1.06 (1.03, 1.10) SO_2 – 1.12 (1.07, 1.16) NO_2 – 1.04 (1.02, 1.07) BS – 1.02 (0.99, 1.04)

the population experienced high exposures.[59] A case-crossover study from Southern California found an association between $PM_{2.5}$, but not with NO_2 or CO, and bronchiolitis only in extremely premature (25 and 29 weeks) and therefore highly vulnerable infants. No association was observed in the overall group for all gestational ages (Table 4).[60] These immature infants had a 26% increased risk for lags 3–5 days and 41% increased risk for lags 6–8 days, per 10 µg/m³ increase in $PM_{2.5}$ exposure.[60] A matched case-control analysis of the same Southern California cohort yielded similar results, i.e., particulate effect was observed while no association was found for gaseous pollutants. For every 10 µg/m³ increase in $PM_{2.5}$ exposure, 30 days prior to the event or over lifetime were

both associated with 9% increase in risk of hospitalization for bronchiolitis (Table 4).[61] A fourth study from Washington state, USA, reported point estimates in the range of those reported above for $PM_{2.5}$, NO_2, and proximity to roadways; however, none were statistically significant.[62] A French study reported a significant association between PM_{10}, SO_2, NO_2, and bronchiolitis related hospitalizations.[63] For all three pollutants, a 10 µg/m³ increase in exposure was associated with 6%, 12%, and 4% increase in hospitalizations, respectively (Table 4).

Exposure in all of the studies stated above was measured ecologically using central site monitors, potentially increasing the chances of exposure misclassification, which if non-differential, is likely to pull the point estimate towards the null. Albeit very limited, these studies provide some evidence that air pollution exposure around the time of birth and immediately thereafter may play a role in the occurrence of infant bronchiolitis.

AIR POLLUTION AND LUNG DISEASES: THE MODIFYING FACTORS

The relationship between air pollution and lung diseases such as asthma and bronchitis was reportedly modified by several factors. In the Czech longitudinal study the association between bronchitis and ambient NO_x increased with age up to 2 years.[58] Children have higher inhalation and pollutant retention rate per unit of body weight than adults, and as they grow older they tend to spend more time outdoors. The reduced association we found for children below 3 months and the monotonic increase to the 12–24 month (Figure 1) period may reflect behavioral factors, namely, keeping very young babies indoors. Exposure to indoor fuel use or second-hand tobacco smoke and duration of breastfeeding did not explain the differential association observed by age.

Female children were found to be more vulnerable to the effects of air pollution for asthma and bronchitis than males.[32,43,44,64] Others have reported a differential association by race with black children at higher risk of asthma related hospital admissions, OR=2.3 (95% CI: 1.5, 3.5)

compared to white children.[65] Those without insurance exhibited greater risk of asthma related hospitalizations than those with Medicaid, implying that access to care might override some of the deleterious effects of poverty.[65] Also, those without insurance are unlikely to have access to preliminary health care that might arrest worsening of health, subsequently preventing hospitalizations. Lower socioeconomic status also modified the relation, with those in the lower strata showing higher association between air pollution and asthma related hospital admissions.[66] Meteorological factors such as daily temperature modified the effect of air pollution on asthma exacerbation.[26] Secondhand tobacco smoke is also reported to modify the effect of air pollutants and asthma symptoms; however, this is an especially convoluted issue because some of the constituents of secondhand tobacco smoke are also found in air pollutants.[67] Past episodes of bronchiolitis significantly modified the association between air pollution and asthma prevalence as well as asthma onset and airway hyper-responsiveness.[68] Significant differences were also observed by breastfeeding status. Those who were not breastfeed had a high risk of asthma OR=1.25 (95% CI: 1.07, 1.46) from NO_2 exposure compared to those who were breastfed but the difference was not large.[69] There is also some evidence of beneficial effect of vitamin C and E intake on lung function indicators of asthmatic children.[70]

The most recent advancement in asthma and air pollution epidemiology is perhaps cross-disciplinary with genetics. Researchers are trying to investigate if the dose-response association differs by genotypes. In other words if the air pollution–asthma or bronchitis association is modified by one or more genotypes. In the

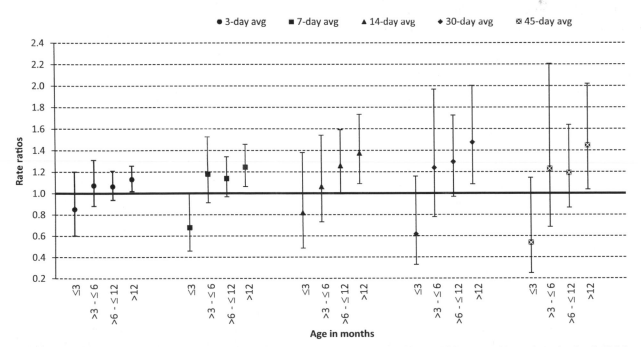

FIGURE 1 Differential association between air pollution (nitrogen oxides) and acute bronchitis by child's age, evidence from the Czech Childhood Health and Air Pollution Study.

following section we discuss effect modification by genotypes in the context of air pollution and childhood respiratory diseases.

GENE-BY-ENVIRONMENT INTERACTION AND LUNG DISEASES

Studies based on the Southern California Children's Health cohort have presented evidence on gene-by-environment interaction, with genes modifying the effects of air pollutants. Children from high O_3 communities who played more than two team sports and had GSTP1 Ile105Ile alleles were at a 6-fold higher risk of asthma, HR=6.2 (95% CI: 2.2, 7.4), compared to those having the same set of alleles, who were from high O_3 communities but did not play any team sports.[71] Another analysis on CHS cohort reported gene-by-environment interaction for EPHX1, GSTP1, and proximity to major roads.[72] Children living within 75 m from a major roadway and having high EPHX1 activity genotype (one or no Histidine allele at exon 3 or 4) had a significant risk of lifetime asthma, OR=3.2 (95% CI: 1.7, 6.0), compared to those with the same genotype but living more than 75m from a major roadway.[72] In a gene-by-gene-by-environment interaction analysis, the study showed that children with EPHX1 high activity, GSTP1 105Val allele, and living within 75m of a major roadway had highest risk for lifetime asthma.[72] These results were to some extent independently verified by two reports from a Taiwanese cohort, which were relatively more racially homogenous and covered 14 communities with varying levels of air pollution.[73,74] Single Nucleotide Polymorphisms (SNP) for EPHX1 gene at exon 3 and 4 significantly modified the association between ambient NO_2 and lifetime or early onset asthma.[74] Children with high NO_2 (level) communities and carrying the His-139Agr or the Agr139Agr alleles had about a 2-fold higher risk of lifetime asthma and early-onset asthma, OR=1.9 (95% CI: 1.3, 2.8) and 2.1 (95% CI: 1.3, 3.4), respectively, compared to children in the high NO_2 communities and with the His139His alleles[74]. Similar gene-by-environment interaction was also observed for $PM_{2.5}$ and GSTP1 gene.[73] Children with one or more Val allele(s) at Ile105Val had an elevated risk for asthma, OR=1.5 (95% CI: 1.0, 2.3) associated with about 17μg/m³ increase in $PM_{2.5}$ exposure.[73]

Whilst all the above studies investigated a chronic disease, asthma, effect modification by these same genes for risk of acute bronchitis in relation to air pollutants has also been reported. Three classes of air pollutants were examined, namely secondhand tobacco smoke (SHS), PAH, and $PM_{2.5}$[75] The SHS-bronchitis association was strongest, OR=1.9 (95% CI: 1.3, 2.6) among those with genotypes GSTP1 Ile105Ile genotype relative to those with the Ile105Val or Val105Val genotypes.

The heterozygous EPHX1 exon 3 has a significantly high SHS-bronchitis association, OR=1.9 (95% CI: 1.3, 2.9) relative to the homozygous wild or variant genotypes. The SHS-bronchitis association was also highest among children with EPHX1 diplotype (with 3 or 4 histidine alleles, combining exon 3 and 4 together), representing low enzyme activity, with OR=1.9 (95% CI: 1.3, 3.0), and not significantly elevated in intermediate or high activity groups.[75] The study also found significant PAH-bronchitis association for EPHX1 genotypes with Histidine alleles, particularly EPHX1 exon 3 His113His has significantly elevated OR=2.1 (95% CI: 1.3, 3.4) per 100ng/m³ increase in PAH. Those with the low enzyme activity EPHX1 diplotype showed a modest but significant PAH-bronchitis association, OR=1.5 (95% CI: 1.2, 1.9). The $PM_{2.5}$-bronchitis association was also higher for EPHX1 exon3 His113His and low activity EPHX1 diplotype compared to the Tyr113Tyr or high activity genotypes, respectively.[75] Some of the most important challenges in gene-by-environment studies are the power to detect significant interactions, replicability and consistency of the results across studies. The power is determined by several factors including sample size, requiring studies to have a large enough sample, sometimes in thousands, as most of the commonly investigated genes such as the GSTs, ephoxide hydrolases, and cytochrome P450, etc. have modest independent gene-respiratory disease association. Large gene-by-environment interaction for a single gene is seldom observed given the complexity of the antioxidant pathways and the substantial number of genes regulating antioxidant levels. Both the Southern California and Taiwanese cohorts had more than 3000 children.[72–74]

The other issue that needs to be pointed out in this context is multiple comparisons. None of the studies mentioned above adjusted for multiple comparison, which increases the possibility of finding a positive association by chance when there is no real association, type 1 error. Furthermore, replicability is another issue in gene-environment interaction studies because obtaining consistent evidence from different populations substantially strengthens evidence for a causal relationship. However, that has not been achieved so far at least in the case of asthma as described in detail in this Human Genome Review.[76] Of the 12 studies that reported significant interaction, 11 found an interaction with GSTP1 but there was no consistency across the studies as to which one is the risk allele.[76] Ethnic differences can also pose a challenge where the study population is a good mix of various ethnic groups because allele frequencies are often found to differ by ethnicity. Moreover, several other factors such as environmental exposure and lifestyle habits that may influence these associations also vary by ethnicity, thereby making evidence from an ethnically heterogeneous population more challenging to interpret. To summarize, the current evidence is encouraging and supports the presence of gene-by-environment interaction for childhood asthma and bronchitis but it is inconclusive to draw an inference about the nature of such interaction.

In addition to the above, air pollution epidemiological studies face the single most important challenge of accurate exposure assessment. Various methods to measure air pollution exposure include: ecological measurements from fixed site monitors and more recently using satellite data, mathematical modeling to predict exposures using various land use and meteorological parameters, biomarkers, and individual measurements. In the early air pollution studies, fixed site monitor measurements were used frequently because of their ease of availability and cost effectiveness but are fraught with the higher likelihood of exposure misclassification. A second method that has received considerable acceptance in the last decade is exposure modeling to predict individual exposures.[77] Although not perfect, use of modeled exposure assessment is generally considered to be a marked improvement over fixed site measurements. A third method is biomarker measurement. For instance, 1-Hydroxypyrene (1-OHP) is a commonly used biomarker of metabolites of PAH.[78] Arguably, the gold standard in air pollution exposure assessment is personal measurements, however, there are limitations associated with it too. Apart from being cost intensive, participation in such monitoring is prone to selective non-response[79] or to modified activity patterns (while wearing the equipment)[80] and is therefore prone to introduce hitherto unanticipated sources of bias, while providing accurate individual measures. Although not optimal, under certain situations (such as exposures averaged over a long period of time) fixed site estimates or outdoor monitoring have been suggested to be valid exposure surrogates for personal exposures.[80,81] Other studies have reported that personal exposures may be higher and more variable than central site estimates.[82] Moreover, it has been shown that the congruence between ambient levels and personal exposures is stronger and less affected by measurement error when the time scale for the averaging period is longer.[83]

An issue that deserves mention is—which specific pollutant or set of pollutants is responsible for the observed effects? With the exception of a few studies where they reported two pollutant models, almost all of the studies reported their results from single pollutant models, because of high correlations among different pollutants. In such a situation it is impossible to establish a pollutant specific relationship. For example, in the Czech study associations were observed between PAH, $PM_{2.5}$, NO_2, and bronchitis but because of high correlation, in the order of 0.8, a multipollutant model could not be constructed. To address this problem an interesting strategy has been proposed,[84] which was subsequently adapted by a case-crossover study where control days were chosen for each subject, which were closer on time and were matched on the level of a second pollutant in order to minimize confounding by the second pollutant.[85]

To conclude, there are several well-designed longitudinal studies that reported a positive association between prevalent asthma and air pollution, and a few more with onset of asthma. For the latter, it is a little too early to arrive at a conclusion on causality. The same is true for bronchitis and bronchiolitis, given the heterogeneity of the case definitions and scant availability of studies. However, several associations that have been discussed in the above sections show that the levels of pollutants associated with these respiratory health outcomes are much lower than the current USEPA and WHO standards, posing substantial challenge to public health.

REFERENCES

1. World Health Organisation. *WHO Air Quality Guidelines for Particulate Matter, Ozone, Nitrogen Dioxide and Sulfur Dioxide. Global Update 2005. Summary of Risk Assessment*; 2005. World Health Organization.
2. United States Environmental Protection Agency. *Our Nation's Air - Status and Trends Through 2010*. Environmental Protection Agency: Research Triangle Park, NC; 2010.
3. Committee on Environmental Health. Ambient air pollution: health hazards to children. *Pediatrics* 2004;**114**:1699–707.
4. Kotecha S. Lung growth: implications for the newborn infant. *Arch Dis Child* 2000;**82**:F69–74.
5. Bateson TF, Schwartz J. Children's response to air pollutants. *J Toxicol Environ Health* 2008;**71**:238–43.
6. Peters J. *Epidemiologic Investigation to Identify Chronic Health Effects of Ambient Air Pollutants in Southern California: Phase II Final Report*. Report prepared by the University of Southern California School of Medicine. Los Angeles, CA: Department of Preventive Medicine; 1997.
7. Oberdörster G, Oberdörster E, Oberdörster J. Nanotoxicology: an emerging discipline evolving from studies of ultrafine particles. *Environ Health Perspect* 2005;**113**:823–39.
8. Centers for Disease Control and Prevention. *Vital Signs* May 2011.
9. Pearce N, Aït-Khaled N, Beasley R, Mallol J, Keil U, Mitchell E, et al. Worldwide trends in the prevalence of asthma symptoms: phase III of the International Study of Asthma and Allergies in Childhood (ISAAC). *Thorax* 2007;**62**:758–66.
10. Gruzieva O, Bergström A, Hulchiy O, Kull I, Lind T, Melén E. Exposure to air pollution from traffic and childhood asthma until 12 years of age. *Epidemiology* 2013;**24**:54–61.
11. Mann J, Balmes J, Bruckner T, Mortimer KM, Margolis HG, Pratt B, et al. Short-term effects of air pollution on wheeze in asthmatic children in Fresno, California. *Environ Health Perspect* 2010;**118**:1497–502.
12. Belanger K, Holford TR, Gent JF, Hill ME, Kezik JM, Leaderer BP. Household levels of nitrogen dioxide and pediatric asthma severity. *Epidemiology* 2013;**24**:320–30.
13. Gehring U, Wijga AH, Brauer M, Fischer P, de Jongste JC, Kerkhof M, et al. Traffic-related air pollution and the development of asthma and allergies during the first 8 years of life. *Am J Respir Crit Care Med* 2010;**181**:596–603.
14. Spira-Cohen A, Chen L, Kendall M, Ramona Lall R, Thurston G. Personal exposures to traffic-related air pollution and acute respiratory health among Bronx a schoolchildren with asthma. *Environ Health Perspect* 2011;**119**:559–65.
15. Gauderman W, Avol E, Lurmann F, Kuenzli N, Gilliland F, Peters J, et al. Childhood asthma and exposure to traffic and nitrogen dioxide. *Epidemiology* 2005;**16**:737–43.

16. McConnell R, Berhane K, Gilliland F, London SJ, Islam T, Gauderman WJ, et al. Asthma in exercising children exposed to ozone: a cohort study. *The Lancet* 2002;**359**:386–91.

17. Eckel S, Berhane K, Salam M, et al. Residential traffic-related pollution exposures and exhaled nitric oxide in the children's health study. *Environ Health Perspect* 2011;**119**:1472–7.

18. Brunekreef B, Stewart A, Anderson R, Lai C, Strachan D, Pearce N. Self-reported truck traffic on the street of residence and symptoms of asthma and allergic disease: a global relationship in ISAAC phase 3. *Environ Health Perspect* 2009;**117**:1791–8.

19. Bayer-Oglesby L, Schindler C, Hazenkamp-von Arx ME, Braun-Fahrländer C, Keidel D, Rapp R, et al. Living near main streets and respiratory symptoms in adults: the Swiss cohort study on air pollution and lung diseases in adults. *Am J Epidemiol* 2006;**164**:1190–8.

20. Brauer M, Hoek G, Smit HA, de Jongste JC, Gerritsen J, Postma DS, et al. Air pollution and development of asthma, allergy and infections in a birth cohort. *Eur Respir J* 2007;**29**:879–88.

21. Brauer M, Hoek G, Van Vliet P, Meliefste K, Fischer PH, Wijga A, et al. Air pollution from traffic and the development of respiratory infections and asthmatic and allergic symptoms in children. *Am J Respir Crit Care Med* 2002;**166**:1092–8.

22. Gauderman WJ, Vora H, McConnell R, Berhane K, Gilliland F, Thomas D, et al. Effect of exposure to traffic on lung development from 10 to 18 years of age: a cohort study. *Lancet* 2007;**369**:571–7.

23. Gehring U, Cyrys J, Sedlmeir G, Brunekreef B, Bellander T, Fischer P, et al. Traffic-related air pollution and respiratory health during the first 2yrs of life. *Eur Respir J* 2002;**19**:690–8.

24. Mar T, Koenig J. Relationship between visits to emergency departments for asthma and ozone exposure in greater Seattle, Washington. *Ann Allergy Asthma Immunol* 2009;**103**:474–9.

25. Andersson M, Modig L, Hedman L, Forsberg B, Rönmark E. Heavy vehicle traffic is related to wheeze among schoolchildren: a population-based study in an area with low traffic flows. *Environ Health* 2011;**10**:91.

26. Mohr L, Luo S, Mathias E, Tobing R, Homan S, Sterling D. Influence of season and temperature on the relationship of elemental carbon air pollution to pediatric asthma emergency room visits. *J Asthma* 2008;**45**:936–43.

27. Iskandar A, Andersen Z, Bønnelykke K, Ellermann T, Andersen K, Bisgaard H. Coarse and fine particles but not ultrafine particles in urban air trigger hospital admission for asthma in children. *Thorax* 2012;**67**:252–7.

28. Pereira G, Cook A, Vos A, Holman C. A case-crossover analysis of traffic-related air pollution and emergency department presentations for asthma in Perth, Western Australia. *Med J Aust* 2010;**193**:511–4.

29. Samoli E, Nastos P, Paliatsos A, Katsouyanni K, Priftis K. Acute effects of air pollution on pediatric asthma exacerbation: evidence of association and effect modification. *Environ Res* 2011;**111**:418–24.

30. McConnell R, Islam T, Shankardass K, Jerrett M, Lurmann F, Gilliland F, et al. Childhood incident asthma and traffic-related air pollution at home and school. *Environ Health Perspect* 2010;**118**:1021–6.

31. Jerrett M, Shankardass K, Berhane K, Gauderman WJ, Künzli N, Avol E, et al. Traffic-related air pollution and asthma onset in children: a prospective cohort study with individual exposure measurement. *Environ Health Perspect* 2008;**116**:1433–8.

32. Clark N, Demers P, Karr C, Koehoorn M, Lencar C, Tamburic L, et al. Effect of early life exposure to air pollution on development of childhood asthma. *Environ Health Perspect* 2010;**118**:284–90.

33. Carlsten C, Dybuncio A, Becker A, Chan-Yeung M, Brauer M. Traffic-related air pollution and incident asthma in a high-risk birth cohort. *Occup Environ Med* 2011;**68**:291–5.

34. Zmirou D, Gauvin S, Pin I, Momas I, Sahraoui F, Just J, et al. Traffic related air pollution and incidence of childhood asthma: results of the Vesta case-control study. *J Epidemiol Community Health* 2004;**58**:18–23.

35. Oftedal B, Nystad W, Brunekreef B, Nafstad P. Long-term traffic-related exposures and asthma onset in schoolchildren in Oslo, Norway. *Environ Health Perspect* 2009;**117**:839–44.

36. Shima M, Nitta Y, Ando M, Adachi M. Effects of air pollution on the prevalence and incidence of asthma in children. *Arch Environ Health* 2002;**57**:529–35.

37. McConnell R, Liu F, Wu J, Lurmann F, Peters J, Berhane K. Asthma and school commuting time. *J Occup Environ Hyg* 2010;**52**:827–9.

38. Atkinson R, Bremner S, Anderson H, Strachan D, Bland A, Leon P. Short-term associations between emergency hospital admissions for respiratory and cardiovascular disease and outdoor air pollution in London. *Arch Environ Health* 1999;**54**:6.

39. Chan A. Indoor–outdoor relationships of particulate matter and nitrogen oxides under different outdoor meteorological conditions. *Atmos Environ* 2002;**36**:1543–51.

40. Kartal S, Özer U. Determination and parameterization of some air pollutants as a function of meteorological parameters in Kayseri, Turkey. *J Air Waste Manag Assoc* 1998;**48**:853–9.

41. Hayes D, Collins PB, Khosravi M, Lin R-L, Lee L-Y. Bronchoconstriction triggered by breathing hot humid air in patients with asthma. *Am J Respir Crit Care Med* 2012;**185**:1190–6.

42. Pongdee T, Li JT. Exercise-induced bronchoconstriction. *Ann Allergy Asthma Immunol* 2013;**110**:311–5.

43. Delfino R, Chang J, Wu J, Ren C, Tjoa T, Nickerson B, et al. Repeated hospital encounters for asthma in children and exposure to traffic-related air pollution near the home. *Ann Allergy Asthma Immunol* 2009;**102**:138–44.

44. McConnell R, Berhane K, Yao L, Jerrett M, Lurmann F, Gilliland F, et al. Traffic, susceptibility, and childhood asthma. *Environ Health Perspect* 2006;**114**:766–72.

45. Nordling E, Berglind N, Melén E, Emenius G, Hallberg J, Nyberg F, et al. Traffic-related air pollution and childhood respiratory symptoms, function and allergies. *Epidemiology* 2008;**19**:401–8.

46. Agache I, Akdis C, Jutel M, Virchow JC. Untangling asthma phenotypes and endotypes. *Allergy* 2012;**67**:835–46.

47. Lina T, Poona A, Hamida Q. Asthma phenotypes and endotypes. *Curr Opin Pulm Med* 2013;**19**:18–23.

48. Dietert RR, Zelikoff JT. Early-life environment, developmental immunotoxicology, and the risk of pediatric allergic disease including asthma. *Birth Defects Res B Dev Reprod Toxicol* 2008;**83**:547–60.

49. Selevan SG, Kimmel CA, Mendola P. Identifying critical windows of exposure for children's health. *Environ Health Perspect* 2000;**108**:451–5.

50. Herr CE, Dostal M, Ghosh R, Ashwood P, Lipsett M, Pinkerton KE, et al. Air pollution exposure during critical time periods in gestation and alterations in cord blood lymphocyte distribution: a cohort of live-births. *Environ Health* 2010;**9**:46.

51. Hertz-Picciotto I, Dostal M, Dejmek J, Selevan SG, Wegienka G, Gomez-Caminero A, et al. Air pollution and distributions of lymphocyte immunophenotypes in cord and maternal blood at delivery. *Epidemiology* 2002;**13**:172–83.

52. Herr CE, Ghosh R, Dostal M, Skokanova V, Ashwood P, Lipsett M, et al. Exposure to air pollution in critical prenatal time windows and IgE levels in newborns. *Pediatr Allergy Immunol* 2011;**22**:75–84.

53. Dietert RR, Etzel RA, Chen D, Halonen M, Holladay SD, Jarabek AM, et al. Workshop to identify critical windows of exposure for children's health: immune and respiratory systems work group summary. *Environ Health Perspect* 2000;**108**(Suppl. 3):483–90.

54. McConnell R, Berhane K, Gilliland F, London SJ, Vora H, Avol E, et al. Air pollution and bronchitic symptoms in Southern California children with asthma. *Environ Health Perspect* 1999;**107**:757–60.

55. Morgenstern V, Zutavern A, Cyrys J, Brockow I, Koletzko S, Krämer U, et al. Atopic diseases, allergic sensitization, and exposure to traffic-related air pollution in children. *Am J Respir Crit Care Med* 2008;**177**:1331–7.

56. Morgenstern V, Zutavern A, Cyrys J, Brockow I, Gehring U, Koletzko S, et al. Respiratory health and individual estimated exposure to traffic-related air pollutants in a cohort of young children. *Occup Environ Med* 2007;**64**:8–16.

57. Hertz-Picciotto I, Baker RJ, Yap P-S, Dostál M, Joad JP, Lipsett M, et al. Early childhood lower respiratory illness and air pollution. *Environ Health Perspect* 2007;**115**:1510–8.

58. Ghosh R, Joad J, Benes I, Dostal M, Sram RJ, Hertz-Picciotto I. Ambient nitrogen oxides exposure and early childhood respiratory illnesses. *Environ Int* 2012;**39**:96–102.

59. Karr CJ, Demers PA, Koehoorn MW, Lencar CC, Tamburic L, Brauer M. Influence of ambient air pollutant sources on clinical encounters for infant bronchiolitis. *Am J Respir Crit Care Med* 2009;**180**: 995–1001.

60. Karr C, Lumley T, Shepherd K, Davis R, Larson T, Ritz B, et al. A case–crossover study of wintertime ambient air pollution and infant bronchiolitis. *Environ Health Perspect* 2006;**114**:277–81.

61. Karr C, Lumley T, Schreuder A, Davis R, Larson T, Ritz B, et al. Effects of subchronic and chronic exposure to ambient air pollutants on infant bronchiolitis. *Am J Epidemiol* 2007;**165**:553–60.

62. Karr CJ, Rudra CB, Miller KA, Gould TR, Larson T, Sathyanarayana S, et al. Infant exposure to fine particulate matter and traffic and risk of hospitalization for RSV bronchiolitis in a region with lower ambient air pollution. *Environ Res* 2009;**109**:321–7.

63. Ségala C, Poizeau D, Mesbah M, Willems S, Maidenberg M. Winter air pollution and infant bronchiolitis in Paris. *Environ Res* 2008;**106**: 96–100.

64. Shima M, Adachi M. Effect of outdoor and indoor nitrogen dioxide on respiratory symptoms in schoolchildren. *Int J Epidemiol* 2000;**29**:862–70.

65. Grineski S, Staniswalis J, Peng Y, Atkinson-Palombo C. Children's asthma hospitalizations and relative risk due to nitrogen dioxide (NO2): Effect modification by race, ethnicity and insurance status. *Environ Res* 2010;**110**:178.

66. Yap P, Gilbreath S, Garcia C, Jareen N, Goodrich B. The influence of socioeconomic markers on the association between fine particulate matter and hospital admissions for respiratory conditions among children. *Am J Public Health* 2013;**103**:695–702.

67. Sonnenschein-van der Voort A, de Kluizenaar Y, Jaddoe V, Gabriele C, Raat H, Moll HA, et al. Air pollution, fetal and infant tobacco smoke exposure, and wheezing in preschool children: a population-based prospective birth cohort. *Environ Health* 2012;**11**:91.

68. Kim BJ, Seo JH, Jung YH, Kim HY, Kwon JW, Kim HB, et al. Air pollution interacts with past episodes of bronchiolitis in the development of asthma. *Allergy* 2013;**68**:517–23.

69. Dong G, Qian Z, Liu M, Wang D, Ren WH, Bawa S, et al. Breastfeeding as a modifier of the respiratory effects of air pollution in children. *Epidemiology* 2013;**24**:387–94.

70. Su H, Chang C, Chen H. Effects of vitamin C and E intake on peak expiratory flow rate of asthmatic children exposed to atmospheric particulate matter. *Arch Environ Occup Health* 2013;**68**:80–6.

71. Islam T, Berhane K, McConnell R, Gauderman WJ, Avol E, Peters JM, et al. Glutathione-S-transferase (GST) P1, GSTM1, exercise, ozone and asthma incidence in school children. *Thorax* 2009;**64**:197–202.

72. Salam MT, Lin P-C, Avol EL, Gauderman WJ, Gilliland FD. Microsomal epoxide hydrolase, glutathione S-transferase P1, traffic and childhood asthma. *Thorax* 2007;**62**:1050–7.

73. Hwang B, Young L, Tsai C, Tung KY, Wang PC, Su MW, et al. Fine particle, ozone exposure, and asthma/wheezing: effect modification by Glutathione S-transferase P1 polymorphisms. *PLoS ONE* 2013;**8**: e52715.

74. Tung K-Y, Tsai C-H, Lee YL. Microsomal epoxide hydroxylase genotypes/diplotypes, traffic air pollution, and childhood asthma. *Chest* 2011;**139**:839–48.

75. Ghosh R, Topinka J, Joad JP, Dostal M, Sram RJ, Hertz-Picciotto I. Air pollutants, genes and early childhood acute bronchitis. *Mutat Res* 2013:80–6.

76. Minelli C, Wei I, Sagoo G, Jarvis D, Shaheen S, Burney P. Interactive effects of antioxidant genes and air pollution on respiratory function and airway disease: a HuGE review. *Am J Epidemiol* 2011;**173**: 603–20.

77. Ryan P, LeMasters G. A review of land-use regression models for characterizing intraurban air pollution exposure. Inhal. *Toxicol* 2007;**19**:127–33.

78. Castaño-Vinyals G, D'Errico A, Malats N, Kogevinas M. Biomarkers of exposure to polycyclic aromatic hydrocarbons from environmental air pollution. *Occup Environ Med* 2004;**61**:e12.

79. Oglesby L, Rotko T, Krutli P, Boudet C, Kruize H, Nen MJ, et al. Personal exposure assessment studies may suffer from exposure-relevant selection bias. *J Expo Anal Environ Epidemiol* 2000;**10**:251–66.

80. Oglesby L, Künzli N, Röösli M, Braun-Fahrländer C, Mathys P, Stern W, et al. Validity of ambient levels of fine particles as surrogate for personal exposure to outdoor air pollution—Results of the European EXPOLIS-EAS Study (Swiss Center Basel). *J Air Waste Manage Assoc* 2000;**50**:1251–61.

81. Tsai FC, Smith KR, Vichit-Vadakan N, Ostro BD, Chestnut LG, Kungskulniti N. Indoor/outdoor PM10 and PM2.5 in Bangkok. *Thailand J Expo Anal Environ Epidemiol* 2000;**10**:15–26.

82. Turpin BJ, Weisel CP, Morandi M, Colome S, Stock T, Eisenreich S, et al. Relationships of indoor, outdoor, and personal air (RIOPA): part II. Analyses of concentrations of particulate matter species. *Res Rep Health Eff Inst* 2007;**132**:1–77. discussion 79–92.

83. Dominici F, Sheppard L, Clyde M. Health effects of air pollution: a statistical review. *Int Stat Rev / Revue Internationale de Statistique* 2003;**71**:243–76.

84. Schwartz J. Is the association of airborne particles with daily deaths confounded by gaseous air pollutants? An approach to control by matching. *Environ Health Perspect* 2004;**112**:557–61.

85. Barnett AG, Williams GM, Schwartz J, Neller AH, Best TL, Petroeschevsky AL, et al. Air pollution and child respiratory health: a case-crossover study in Australia and New Zealand. *Am J Respir Crit Care Med* 2005;**171**:1272–8.

Chapter 25

Environmental Toxicants and Lung Development in Experimental Models

Michelle V. Fanucchi

Department of Environmental Sciences, School of Public Health, University of Alabama at Birmingham, Birmingham, AL, USA

ENVIRONMENTAL TOBACCO SMOKE

The health consequences of exposure to environmental tobacco smoke (ETS) among children have been the subject of much public concern (see Chapter 20). The effect of ETS in humans is covered in detail in another chapter in this book. Animal studies are very valuable because the indirect effects of in utero ETS exposure can be separated from the direct effects of postnatal ETS exposure.

Exclusively in utero exposure to ETS has been shown to accelerate the developmental pattern of Clara cell secretory protein expression in the rat,[1] suggesting a potential acceleration of airway epithelial differentiation in the lung. Whether this accelerated development is maintained after birth is unknown. In utero exposure does not increase cytochrome P450 gene expression unless it is combined with an early postnatal exposure.[2]

Exclusively postnatal ETS exposure in rats did not alter Clara cell secretory protein expression.[3] Postnatal ETS exposure did, however, decrease cell kinetic activity in distal airways and increase cytochrome P450 1A1 protein distribution throughout the airway tree in postnatal ETS chronically exposed rats.[3] These changes were maintained for up to 100 days (with ongoing ETS exposure). Separate from the other compounds in ETS, nicotine exposure during gestation and lactation has been shown to increase CYP2A3 and CYP2B1 mRNA.[4] Acute postnatal exposure to ETS in juvenile ferrets increased the ability of the lungs to metabolize (-)-trans-benzo[a]pyrene-7,8-dihydrodiol.[5] When slightly older rats (weanling age) were exposed to tobacco smoke, emphysematous changes were reported in their lungs.[6] Postnatal exposure to ETS has also been reported to affect the neurophysiologic responses of the lung. Chronic ETS exposure during the period of postnatal lung development in guinea pigs has been shown to increase lung C-fiber sensitivity.[7] In addition, ETS can increase the sensitivity of C-fiber activated neurons in the nucleus tractus solatarii (NTS) of the central nervous system.[8,9]

A combination of in utero and postnatal exposure to ETS appears to have the greatest effect on developing lungs. Rats exposed to both in utero and postnatal ETS have decreased lung compliance, increased reactivity to methacholine and an increase in the number of neuroendocrine cells per cm of basal lamina.[10] These changes were not seen in rats exposed to ETS in utero only or postnatal only. The increased airway hyperresponsiveness that is set up during postnatal exposure does not resolve even after an extensive period of no exposure to ETS.[11] In nonhuman primates, in utero plus postnatal exposure to ETS increases pulmonary adenyl cyclase activity.[12] Recent studies have focused on the effects that in utero exposure to ETS may have on the etiology of asthma. Early life exposures to ETS in the nonhuman primate enhance local Th2 immunity by impairing normal Th1 immune maturation.[13] In the mouse, in utero ETS exposure results in altered gene expression in adult animals[14] and exacerbating adult responses to allergen challenges.[15] In humans, in utero ETS exposure has been correlated to increased wheezing and increased doctor diagnosed asthma by 2 years of age.[16] Whether in utero-induced alterations are the result of direct or indirect effects of ETS are unknown at this time.

BIOACTIVATED COMPOUNDS

The lungs of mammals are selectively injured by a host of chemically diverse agents including aromatic hydrocarbons, furans, halogenated ethylenes, and indoles.[17] Many of these agents target airway epithelium, especially Clara cells. In all cases, the metabolic activation of the chemically inert parent compound has been demonstrated to be an important factor in selective lung injury. It is generally assumed that the Clara cell is susceptible by virtue of the high expression of cytochrome P450 monooxygenases in this cell type. Despite the extensive documentation of the susceptibility of Clara cells to P450-mediated cytotoxicants in the lungs

The Lung. http://dx.doi.org/10.1016/B978-0-12-799941-8.00025-0

of adults,[18–28] there is little known of the susceptibility of undifferentiated and developing cells in the neonate to these compounds. The few studies that are available regarding neonates suggest that lower pulmonary P450 activity is associated with greater susceptibility to P450-activated toxicants.[29–33] In utero exposure to bioactivated compounds produces embryotoxic or teratogenic effects, including chromosomal aberrations.[34–39] The latter appears to be the case for a number of procarcinogens which when given to pregnant mothers produce Clara cell tumors in adult offspring.[38,40,41]

The herbicide dichlobenil specifically injures olfactory nasal mucosa in fetal and neonatal mice just as it does in adult mice.[42] The toxicity of dichlobenil increases in neonatal mice with the development of Bowman's glands. In fetal mice, there was more irreversible binding of 14C-dichlobenil in the nasal cavity if the dichlobenil was given to the mother as opposed to injected directly into the fetus. This suggests that maternal metabolism may be important in the fetus.

While the toxicity of dichlobenil increases with development of the target organ, this is not always true for other bioactivated compounds in the developing lung. Neonatal rabbits are much more sensitive to the P450-bioactivated furan 4-ipomeanol as compared to adult rabbits.[29] Distal airway epithelium of neonatal rabbits was injured at doses that did not affect adult rabbit airway epithelium at all. This would seem to be a contradiction because development of the P450 system is a postnatal event, and in neonatal rabbits, P450 activity is very low. This phenomenon is not restricted to the rabbit. Studies in our laboratory have shown that neonatal mice are also more susceptible than adult mice to the Clara cell cytotoxicant naphthalene[31] and neonates of both rats and mice are more susceptible to 1-nitronaphthalene than are adult animals.[33]

In vitro metabolism studies of neonatal and adult airways show that P450 activity is lower in neonatal mouse lung than it is in adult mouse lung.[32] Gender may also play a role in heightened postnatal sensitivity to pulmonary toxicants. Weanling male and female mice are reported to be more susceptible to 1,1-dichloroethylene-induced pulmonary injury than adult male mice, but not more susceptible than adult female mice.[43] The exact mechanisms of these increases in sensitivity of postnatal animals have yet to be clearly defined. In some cases, levels of injury positively correlate with specific P450 monooxygenase activity,[43] while in other cases Phase II enzyme activity may be key to increased susceptibility.[44] The mechanisms may also involve as yet undefined factors specific to differentiating cells.

In addition to postnatal exposure to bioactivated compounds, in utero exposure may also affect lung development. High levels of trichloroethylene exposure on gestational day 17 can cause a decrease in lung weight and total lung phospholipid content, while not changing total DNA content.[45] Benzo(a)pyrene causes lung tumors in offspring of mice treated at days 18 and 19 gestation.[46] Both males and females have increased incidence of tumors and increased numbers of tumors per animal. When the offspring were followed over five generations of inbreeding, the females of the F2 generation had a higher incidence of lung tumors and both males and females of the F2 generation had an increase in the total number of tumors per animal. The tumor incidence was not statistically different from controls in the F3 through F5 generations, but the number of tumors per animal remained high.

OXIDANT GASES

In contrast to our understanding of P450-activated lung toxicants, the susceptibility of the lungs of postnatal animals to oxidant gases is much better understood. For the best-studied oxidant gas environments (hyperbaric oxygen, ozone, and nitrogen dioxide) two fundamental characteristics have been defined. First, in general, postnatal animals, prior to weaning, are less susceptible to pulmonary injury than are adults. Second, exposure to oxidant gases retards postnatal maturation of the lung.

The tolerance of postnatal animals to hyperoxia appears to be species-specific[47–49] and is based on differences in: (1) the ability of neonatal animals to elevate pulmonary antioxidant defense systems in response to hyperoxic stress[50–52]; (2) the composition of lung polyunsaturated fatty acids[53]; or (3) the presence of antioxidant compounds, including iron chelators.[54] A common factor appears to be the ability to increase the intracellular glutathione pool and to upregulate the enzymes whose antioxidant functions depend on it, including superoxide dismutase, catalase, glutathione peroxidase, and glucose 6-phosphate dehydrogenase.[48,49,55–59] Undernutrition and premature weaning have also been shown to alter susceptibility.[54,60,61] Pharmacologic intervention by administration of steroids (dexamethasone) or endotoxin reduces neonatal susceptibility to hyperoxia but has a mixed effect on antioxidant enzyme activity (endotoxin elevates them, dexamethasone does not).[62,63] Hyperoxia has been shown to delay lung morphogenesis, including alveolarization and vascularization,[55,64–67] and differentiation of Clara cells in postnatal rats.[68] Treatment with retinoic acid does not prevent hyperoxia-induced alterations in alveolarization. However, it does result in later improvement in alveolarization.[69,70] The excess collagen deposition (fibrosis) that is associated with the decreased alveolar and capillary development is preceded by an increase in connective tissue growth factor (CTGF) in neonatal rats.[71] Despite alterations in lung development, neonatal rats have been reported to survive hyperoxia longer than adult rats. This may partially be due to the fact that in neonatal rats, there is a delay in pulmonary neutrophil influx.[72] Compared to adult rats, neonatal rats have fewer overall lung tissue neutrophils, even though they have higher levels of neutrophils in bronchoalveolar lavage. This suggests that neonatal rats retain fewer neutrophils than adult rats.[72] When neonatal rats exposed to hyperoxia were treated with antibodies to cytokine-induced neutrophil chemoattractant-1 (CINC-1) to block neutrophil influx, they had increased lung compliance and no change in alveolar volume or surface density as compared to control antibody-treated neonates.[73] In addition, blocking neutrophil influx reduces DNA damage in the neonatal lung.[74] The retardation of alveolar development in neonatal lung may also be related to the timing of the hyperoxia and subsequent exposure to leukotrienes[75] or a reported increase in the number of apoptotic cells in the lungs of hyperoxia exposed neonates.[76]

For some parameters, neonates appear to be less susceptible to ozone or nitrogen dioxide exposure. They have fewer alterations in pulmonary enzymes and markedly reduced cellular injury in the central acinus as compared to adults.[77–81] Weaning appears to be the critical time point for changes in responsiveness. Preweaning animals are much less sensitive than postweaning animals.[78–80] As in hyperoxia, ozone exposure reduces the postnatal morphogenesis of the gas exchange area,[82] impairs bronchiolar formation,[83] and retards the differentiation of the mucociliary apparatus of proximal airways.[84] Nonhuman primates exposed to cyclic episodes of ozone during the first 6 months of life were found to have four fewer nonalveolarized airway generations, hyperplastic bronchiolar epithelium, and altered smooth muscle bundle orientation in terminal and respiratory bronchioles compared to filtered air exposed control animals[85] and hyperinnervation of the pulmonary epithelium.[86] The molecular mechanism behind the abnormal development of distal conducting airways in animals exposed to ozone may be related to the depletion of perlecan in the basement membrane zone.[87] Perlecan is a proteoglycan responsible for many functions, in particular, regulation of growth factor trafficking between cells of the epithelial-mesenchymal unit.[88,89] Ozone-induced depletion of perlecan from the basement membrane zone in trachea was associated with altered regulation of FGF-2 signaling.[87,90] Depletion of perlecan would also affect regulation of the other growth factors that bind to perlecan, which also include FGF-1, FGF-7, PDGF, hepatocyte growth factor, heparin binding-EGF, VEGF, and TGF-β.[91] The functional consequences of deregulation of these collective molecules are significant since they are the basis for much of the cell–cell interactions in the epithelial mesenchymal trophic unit responsible for development of the airway. The disregulation of the epithelial mesenchymal trophic unit may play a role in explaining the epidemiological findings regarding children who are exposed to long-term outdoor air pollution. In Southern California, an association has been found between long-term exposure of children to outdoor air pollution and

deficits in lung function as these children become adolescents and young adults.[92–95] Children growing up in Mexico City, which has very high levels of ozone and other air pollutants, have respiratory abnormalities such as hyperinflation and increased interstitial markings of the lung.[96–98]

A potential mechanism for the age-related differences in ozone-induced injury may be the way in which neonates control their ventilation during exposure. Neonatal rats have higher baseline minute ventilation than adult rats. During ozone exposure, adult rats will reduce their minute ventilation even further, while neonatal rats will not.[99] This indicates that neonatal rats receive a higher delivered dose of ozone than adults and may explain increased indices of acute injury such as increased bronchoalveolar lavage protein and prostaglandin E_2 levels. What can be concluded from these studies with oxidant gases is that two aspects of postnatal lung development that involve Clara cells: the rate of differentiation of bronchiolar epithelium, and the organization and differentiation of the centriacinus are impeded by oxidant gas injury. Whether this is true for other classes of pulmonary toxicants, such as organic chemicals metabolized by the cytochrome P450 system, has not been investigated.

CORTICOSTEROIDS

Glucocorticosteroids are commonly used to promote accelerated pulmonary maturation when preterm labor is imminent.[100] This prevents chronic lung disease in prematurely born infants[101] and more recently has been recommended as a treatment modality for asthma in children under 5 years of age.[102] Although glucocorticosteroids are beneficial in the short term, very little is known about their long-term effects on lung development and growth. The systemic side-effects of glucocorticoid treatment have been reviewed by Kay and colleagues.[103] Most of the information concerning corticosteroids and lung growth is from prenatal studies. Maternal exposure to betamethasone has been shown to have mixed effects. It results in no changes in lung compliance or lung volume in fetal lambs[104]; however, it does increase lung function by 50% over control preterm lambs[105] and also increases pulmonary antioxidant levels.[106,107] Maternal corticosteroids also decrease the overall number of alveoli, therefore increasing the average alveolar volume and resulting in an emphysematous lung.[105,108] The effects of maternal treatment with glucocorticoids in rats are similar to that found in sheep. Adult rats born to dams treated with dexamethasone during late pregnancy have fewer, larger alveoli.[109–111] Studies in fetal rhesus monkeys suggest that effects on the lung from prenatal corticosteroid treatment are time-dependent and possibly steroid-specific. Prenatal exposure prior to 133 days gestational age (term is 168 days) to betamethasone has been reported to increase the number of alveoli in the lung, but to impair overall lung growth in rhesus monkeys.[112] Studies in our laboratory have shown that exposure to betamethasone at gestational age 121–127 days (midcanalicular stage) does not accelerate the maturation of alveolar type II cells, nor does it alter the morphogenesis of the gas exchange region.[113] However, another corticosteroid, triamcinolone, did induce structural alterations in the fetal lungs of rhesus monkeys when given during the pseudoglandular (63–65 days gestational age) or midcanalicular (110–112 days gestational age) phase of lung development.[114]

Postnatal treatment of mice with dexamethasone has been reported to increase pulmonary gene expression of vascular endothelial growth factor (VEGF), hypoxia-inducible-like factor (HLF) and murine homologue fetal liver kinase (Flk-1) without alterations in cell-specific protein expression.[115] As in the nonhuman primate, steroid-specific differences have been noted: hydrocortisone and dexamethasone both alter alveolar development in the rat, but the extent of alteration with hydrocortisone is not as great.[116] In addition, postnatal treatment of rats with dexamethasone increases the susceptibility of those rats as adults to experimentally induced pulmonary hypertension.[117]

MISCELLANEOUS COMPOUNDS

The developmental effect of nonbioactivated compounds has not been widely studied in postnatal lung. Nagai et al.[118] have looked at the effects of bromodeoxyuridine (BrDU), a thymidine analog used to evaluate DNA synthesis, on alveolar development. Rats were exposed to BrDU at 6 days after birth and then to an excess of thymidine to remove BrDU from the rats. The BrDU incorporates most heavily into alveolar cells. Two weeks after treatment, the lungs appeared normal. Eight weeks after treatment the rats had enlarged airspace and decreased numbers of alveoli. In another study,[119] male and female mice were exposed to BrDU on the 1st, 3rd, and 7th days after birth. There was no increase in tumor development with BrDU exposure alone, but additional exposure to urethane caused significant increases in lung adenoma incidence and an increase in the number of tumors per mouse.

Congenital diaphragmatic hernia is experimentally induced by exposing fetuses to the herbicide nitrofen (2,4-dicholro-4'-nitrodiphenyl ether). In utero exposure to nitrofen causes functional impairment of the lungs of male rats.[81] At 3 weeks of age, no differences between treated and control rats were detected, but at 6 weeks of age, decreases in tidal volume, vital capacity, total capacity, and compliancy were reported. In addition, nonhomogeneous alveolar ventilation was observed. These effects continued to become more apparent as the rats matured. Although nitrofen is metabolized to mutagenic intermediates in adult rodents, metabolic activation does not appear to play a role in teratogenicity.[120] Distribution studies performed in pregnant rats with labeled nitrofen indicate that while maternal tissues contain nitrofen metabolites, fetal tissue contains only the parent compound.[121] The fetal tissue is actually a sink for nitrofen. Nitrofen may cause congenital diaphragmatic hernia by interfering with thyroid hormone levels or with thyroid hormone receptors. Glucocorticoids can ameliorate some of the abnormalities induced by nitrofen. Dexamethasone has been shown to reverse the nitrofen-induced reduction of thyroid transcription factor gene expression (a marker of lung morphogenesis) and surfactant protein-B (a marker of lung maturity).[122,123] Nitrofen also induces an increase in the number of pulmonary neuroendocrine cells (PNECs) in the lung, and this number is augmented by treatment with dexamethasone.[124]

Ethylnitrosourea exposure to mice at 13–16 days gestation caused an increase in lung tumor incidence.[40,125,126] Tumor nodules were noticed as early as 7 days postnatal and the number of tumors continued to increase until a threshold was reached at 90 days postnatal. The percentage of tumors that were derived from Clara cells increased to 60 days, while the diameter of the tumors continued to increase during the time of the study (1 year). Other compounds that have been studied for their effect on lung include ethanol[127,128] and cocaine.[129] Maternal exposure to ethanol throughout pregnancy decreased the amount of insulin-like growth factor II as compared to control fed mothers[127,128] although the difference was not as great in pair-fed dams. This suggests that there is also a nutritional deficit playing a role. Cocaine increases lung catecholamine and glucocorticoid levels and also fetal hypoxemia.[129]

CONCLUSIONS

The impact of lung-targeted toxicants on the respiratory system of pre- and postnatal animals is not well defined. The pattern of lung development itself may play a significant role in modulating the toxic response. Significant portions of lung morphogenesis and cytodifferentiation occur during the postnatal period. The enzyme systems responsible for bioactivation and detoxification differentiate during the perinatal period, with the majority of differentiation activity occurring for an extended period of time after birth. In addition, each enzyme system has different pattern of differentiation during pre- and postnatal lung development. The risk of injury from an environmental contaminant that is known to target the lung in adults must be evaluated with two considerations: (1) the toxicant may have its impact by altering the processes of morphogenesis and cytodifferentiation resulting in differential expression or organization of the lung in the adult and (2) factors such as the stage of morphogenesis and differentiation of various subcompartments of the lungs during the time of exposure may significantly increase the severity of the toxic response. An effort must be made to separate the effects on the respiratory system produced by maternal exposure, either prior to parturition or during the suckling phase of postnatal growth, from the impacts of direct exposure on the postnatal animal. Alteration of the differentiated expression of a particular enzyme system or other functional protein by exposure during lung development, which does not lead to direct changes in lung function, may not produce adverse responses by a particular toxicant in the respiratory system. While there is extensive literature on the toxic potential of a wide range of environmental contaminants when the exposure is directed towards adults, it is obvious that there is a dearth of information regarding the toxic response of the respiratory-system during development. The majority of studies suggest that: (1) the respiratory system in pre- and postnatal animals is more susceptible to injury from lung-directed toxicants than it is in adults of the same species; (2) the differences in toxic response to respiratory-targeted compounds among species are amplified when responses are evaluated during lung development. Current data suggest that the portion of the human population most at risk to respiratory-targeted environmental contaminants are fetuses and neonates and that their risk is significantly higher than the risks to the adult population.

REFERENCES

1. Ji CM, Royce FH, Truong U, Plopper CG, Singh G, Pinkerton KE. Maternal exposure to environmental tobacco smoke alters clara cell secretory protein expression in fetal rat lung. *Am J Physiol* 1998;**275**(5 Pt 1):L870–876.

2. Lee CZ, Royce FH, Denison MS, Pinkerton KE. Effect of in utero and postnatal exposure to environmental tobacco smoke on the developmental expression of pulmonary cytochrome p450 monooxygenases. *J Biochem Mol Toxicol* 2000;**14**(3):121–30.

3. Ji CM, Plopper CG, Witschi HP, Pinkerton KE. Exposure to sidestream cigarette smoke alters bronchiolar epithelial cell differentiation in the postnatal rat lung. *Am J Respir Cell Mol Biol* 1994;**11**(3):312–20.

4. Gamieldien K, Maritz G. Postnatal expression of cytochrome p450 1a1, 2a3, and 2b1 mrna in neonatal rat lung: Influence of maternal nicotine exposure. *Exp Lung Res* 2004;**30**(2):121–33.

5. Sindhu RK, Rasmussen RE, Kikkawa Y. Effect of environmental tobacco smoke on the metabolism of (-)-trans-benzo[a]pyrene-7,8-dihydrodiol in juvenile ferret lung and liver. *J Toxicol Environ Health* 1995;**45**(4):453–64.

6. Uejima Y, Fukuchi Y, Nagase T, Matsuse T, Yamaoka M, Orimo H. Influences of tobacco smoke and vitamin e depletion on the distal lung of weanling rats. *Exp Lung Res* 1995;**21**(4):631–42.

7. Mutoh T, Bonham AC, Kott KS, Joad JP. Chronic exposure to sidestream tobacco smoke augments lung c-fiber responsiveness in young guinea pigs. *J Appl Physiol* 1999;**87**(2):757–68.

8. Mutoh T, Joad JP, Bonham AC. Chronic passive cigarette smoke exposure augments bronchopulmonary c-fibre inputs to nucleus tractus solitarii neurones and reflex output in young guinea-pigs. *J Physiol* 2000;**523**(Pt 1):223–33.

9. Bonham AC, Chen CY, Mutoh T, Joad JP. Lung c-fiber cns reflex: Role in the respiratory consequences of extended environmental tobacco smoke exposure in young guinea pigs. *Environ Health Perspect* 2001;**109**(Suppl. 4):573–8.

10. Joad JP, Pinkerton KE, Bric JM. Effects of sidestream smoke exposure and age on pulmonary function and airway reactivity in developing rats. *Pediatr Pulmonol* 1993;**16**(5):281–8.

11. Joad JP, Bric JM, Peake JL, Pinkerton KE. Perinatal exposure to aged and diluted sidestream cigarette smoke produces airway hyperresponsiveness in older rats. *Toxicol Appl Pharmacol* 1999;**155**(3):253–60.

12. Slotkin TA, Pinkerton KE, Seidler FJ. Perinatal exposure to environmental tobacco smoke alters cell signaling in a primate model: Autonomic receptors and the control of adenylyl cyclase activity in heart and lung. *Brain Res Dev Brain Res* 2000;**124**(1-2):53–8.

13. Wang L, Joad JP, Zhong C, Pinkerton KE. Effects of environmental tobacco smoke exposure on pulmonary immune response in infant monkeys. *J Allergy Clin Immunol* 2008;**122**(2):400–6. 406 e401–405.

14. Rouse RL, Boudreaux MJ, Penn AL. In utero environmental tobacco smoke exposure alters gene expression in lungs of adult balb/c mice. *Environ Health Perspect* 2007;**115**(12):1757–66.

15. Penn AL, Rouse RL, Horohov DW, Kearney MT, Paulsen DB, Lomax L. In utero exposure to environmental tobacco smoke potentiates adult responses to allergen in balb/c mice. *Environ Health Perspect* 2007;**115**(4):548–55.

16. Lannero E, Wickman M, Pershagen G, Nordvall L. Maternal smoking during pregnancy increases the risk of recurrent wheezing during the first years of life (BAMSE). *Respir Res* 2006;**7**(1):3.

17. Plopper CG, Weir AJ, Morin D, Chang A, Philpot RM, Buckpitt AR. Postnatal changes in the expression and distribution of pulmonary cytochrome p450 monooxygenases during clara cell differentiation in rabbits. *Mol Pharmacol* 1993;**44**(1):51–61.

18. Plopper CG, Suverkropp C, Morin D, Nishio S, Buckpitt A. Relationship of cytochrome p-450 activity to clara cell cytotoxicity. I. Histopathologic comparison of the respiratory tract of mice, rats and hamsters after parenteral administration of naphthalene. *J Pharmacol Exp Ther* 1992;**261**(1):353–63.

19. Buckpitt A, Buonarati M, Avey LB, Chang AM, Morin D, Plopper CG. Relationship of cytochrome p450 activity to clara cell cytotoxicity. Ii. Comparison of stereoselectivity of naphthalene epoxidation in lung and nasal mucosa of mouse, hamster, rat and rhesus monkey. *J Pharmacol Exp Ther* 1992;**261**(1):364–72.

20. Verschoyle RD, Carthew P, Wolf CR, Dinsdale D. 1-nitronaphthalene toxicity in rat lung and liver: Effects of inhibiting and inducing cytochrome p450 activity. *Toxicol Appl Pharmacol* 1993;**122**(2):208–13.

21. Boyd MR, Reznik-Schuller HM. Metabolic basis for the pulmonary clara cell as a target for pulmonary carcinogenesis. *Toxicol Pathol* 1984;**12**(1):56–61.

22. Ogawa T, Tsubakihara M, Ichikawa M, Kanisawa M. An autoradiographic study of the renewal of mouse bronchiolar epithelium following bromobenzene exposure. *Toxicol Pathol* 1993;**21**(6):547–53.

23. Johnson DE, Riley MG, Cornish HH. Acute target organ toxicity of 1-nitronaphthalene in the rat. *J Appl Toxicol* 1984;**4**(5):253–7.

24. Rasmussen RE, Do DH, Kim TS, Dearden LC. Comparative cytotoxicity of naphthalene and its monomethyl- and mononitro-derivatives in the mouse lung. *J Appl Toxicol* 1986;**6**(1):13–20.

25. Forkert PG, Birch DW. Pulmonary toxicity of trichloroethylene in mice. Covalent binding and morphological manifestations. *Drug Metab Dispos* 1989;**17**(1):106–13.

26. Paige R, Wong V, Plopper C. Dose-related airway-selective epithelial toxicity of 1-nitronaphthalene in rats. *Toxicol Appl Pharmacol* 1997;**147**(2):224–33.

27. O'Brien KA, Smith LL, Cohen GM. Differences in naphthalene-induced toxicity in the mouse and rat. *Chem Biol Interact* 1985;**55**(1–2):109–22.

28. Ding X, Kaminsky LS. Human extrahepatic cytochromes p450: Function in xenobiotic metabolism and tissue-selective chemical toxicity in the respiratory and gastrointestinal tracts. *Annu Rev Pharmacol Toxicol* 2003;**43**:149–73.

29. Plopper CG, Weir AJ, Nishio SJ, Chang A, Voit M, Philpot RM, et al. Elevated susceptibility to 4-ipomeanol cytotoxicity in immature clara cells of neonatal rabbits. *J Pharmacol Exp Ther* 1994;**269**(2):867–80.

30. Smiley-Jewell SM, Liu FJ, Weir AJ, Plopper CG. Acute injury to differentiating clara cells in neonatal rabbits results in age-related failure of bronchiolar repair. *Toxicol Pathol* 2000;**28**(2):267–76.

31. Fanucchi MV, Buckpitt AR, Murphy ME, Plopper CG. Naphthalene cytotoxicity in the differentiating clara cells of neonatal mice. *Toxicol Appl Pharmacol* 1997;**144**(1):96–104.

32. Fanucchi MV, Murphy ME, Buckpitt AR, Philpot RM, Plopper CG. Pulmonary cytochrome p450 monooxygenase and clara cell differentiation in mice. *Am J Respir Cell Mol Biol* 1997;**17**(3):302–14.

33. Fanucchi MV, Day KC, Clay CC, Plopper CG. Increased vulnerability of neonatal rats and mice to 1-nitronaphthalene-induced pulmonary injury. *Toxicol Appl Pharmacol* 2004;**201**(1):53–65.

34. Faustman-Watts E, Giachelli C, Juchau M. Carbon monoxide inhibits monooxygenation by the conceptus and embryotoxic effects of proteratogens in vitro. *Toxicol Appl Pharmacol* 1986;**83**:590–5.

35. Faustman-Watts E, Namkung M, Juchau M. Modulation of the embryotoxicity *in vitro* of reactive metabolites of 2-acetylaminofluorene by reduced glutathione and ascorbate and via sulfation. *Toxicol Appl Pharmacol* 1986;**86**:400–10.

36. Filler R, Lew KJ. Developmental onset of mixed-function oxidase activity in preimplantation mouse embryos. *Proc Natl Acad Sci USA* 1981;**78**(11):6991–5.

37. Galloway SM, Perry PE, Meneses J, Nebert DW, Pedersen RA. Cultured mouse embryos metabolize benzo[a]pyrene during early gestation: Genetic differences detectable by sister chromatid exchange. *Proc Natl Acad Sci USA* 1980;**77**(6):3524–8.

38. Juchau MR, Giachelli CM, Fantel AG, Greenaway JC, Shepard TH, Faustman-Watts EM. Effects of 3-methylcholanthrene and phenobarbital on the capacity of embryos to bioactivate teratogens during organogenesis. *Toxicol Appl Pharmacol* 1985;**80**(1):137–46.

39. Juchau MR. Bioactivation in chemical teratogenesis. *Annu Rev Pharmacol Toxicol* 1989;**29**:165–87.

40. Palmer KC. Clara cell adenomas of the mouse lung. Interaction with alveolar type 2 cells. *Am J Pathol* 1985;**120**(3):455–63.

41. Yang HY, Namkung MJ, Juchau MR. Immunodetection, immunoinhibition, immunoquantitation and biochemical analyses of cytochrome p-450ia1 in tissues of the ratoffceptus during the progression of organogenesis. *Biochem Pharmacol* 1989;**38**(22):4027–36.

42. Eriksson C, Brittebo EB. Dichlobenil in the fetal and neonatal mouse olfactory mucosa. *Toxicology* 1995;**96**(2):93–104.

43. Forkert PG, Dowsley TF, Lee RP, Hong JY, Ulreich JB. Differential formation of 1,1-dichloroethylene-metabolites in the lungs of adult and weanling male and female mice: Correlation with severities of bronchiolar cytotoxicity. *J Pharmacol Exp Ther* 1996;**279**(3):1484–90.

44. Fanucchi MV, Buckpitt AR, Murphy ME, Storms DH, Hammock BD, Plopper CG. Development of phase ii xenobiotic metabolizing enzymes in differentiating murine clara cells. *Toxicol Appl Pharmacol* 2000;**168**(3).253–67.

45. Das RM, Scott JE. Trichloroethylene-induced pneumotoxicity in fetal and neonatal mice. *Toxicol Lett* 1994;**73**(3):227–39.

46. Turusov VS, Nikonova TV, Parfenov Y. Increased multiplicity of lung adenomas in five generations of mice treated with benz(a)pyrene when pregnant. *Cancer Lett* 1990;**55**(3):227–31.

47. Frank L, Bucher JR, Roberts RJ. Oxygen toxicity in neonatal and adult animals of various species. *Am Physiol Soc* 1978;**45**(5):699–704.

48. Hoffman M, Stevens JB, Autor AP. Adaptation to hyperoxia in the neonatal rat: Kinetic parameters of the oxygen-mediated induction of lung superoxide dismutases, catalase and glutathione peroxidase. *Toxicology* 1980;**16**:215–25.

49. Stevens JB, Autor AP. Proposed mechanism for neonatal rat tolerance to normobaric hyperoxia. *Fed Proc* 1980;**39**:3138–43.

50. Dennery PA, Rodgers PA, Lum MA, Jennings BC, Shokoohi V. Hyperoxic regulation of lung heme oxygenase in neonatal rats. *Pediatr Res* 1996;**40**(6):815–21.

51. Kim HS, Kang SW, Rhee SG, Clerch LB. Rat lung peroxiredoxins i and ii are differentially regulated during development and by hyperoxia. *Am J Physiol Lung Cell Mol Physiol* 2001;**280**(6):L1212–1217.

52. Yang G, Madan A, Dennery PA. Maturational differences in hyperoxic ap-1 activation in rat lung. *Am J Physiol Lung Cell Mol Physiol* 2000;**278**(2):L393–398.

53. Bartlett DJ, Faulkner II CS, Cook K. Effect of chronic ozone exposure on lung elasticity in young rats. *J Appl Physiol* 1974;**37**(1):92–6.

54. Frank L. Developmental aspects of experimental pulmonary oxygen toxicity. *Free Radic Biol Med* 1991;**11**(5):463–94.

55. Bucher JR, Roberts RJ. The development of the newborn rat lung in hyperoxia: A dose-response study of lung growth, maturation, and changes in antioxidant enzyme activities. *Pediatr Res* 1981;**15**(7):999–1008.

56. Bonuccelli CM, Permutt S, Sylvester JT. Developmental differences in catalase activity and hypoxic-hyperoxic effects on fluid balance in isolated lamb lungs. *Pediatr Res* 1993;**33**(5):519–26.

57. Kennedy KA, Lane NL. Effect of in vivo hyperoxia on the glutathione system in neonatal rat lung. *Exp Lung Res* 1994;**20**(1):73–83.

58. Warshaw JB, Wilson CWd, Saito K, Prough RA. The responses of glutathione and antioxidant enzymes to hyperoxia in developing lung. *Pediatr Res* 1985;**19**(8):819–23.

59. Langley SC, Kelly FJ. Depletion of pulmonary glutathione using diethylmaleic acid accelerates the development of oxygen-induced lung injury in term and preterm guinea-pig neonates. *J Pharm Pharmacol* 1993;**46**:98–102.

60. Frank L, Groseclose E. Oxygen toxicity in newborn rats: The adverse effects of undernutrition. *Am Physiol Soc* 1982;**53**:1248–55.

61. Chessex P, Lavoie JC, Laborie S, Vallee J. Survival of guinea pig pups in hyperoxia is improved by enhanced nutritional substrate availability for glutathione production. *Pediatr Res* 1999;**46**(3):305–10.

62. Frank L. Prenatal dexamethasone treatment improves survival of newborn rats during prolonged high o2 exposure. *Ped Res* 1992;**32**:215–21.

63. Sosenko IR, Chen Y, Price LT, Frank L. Failure of premature rabbits to increase lung antioxidant enzyme activities after hyperoxic exposure: Antioxidant enzyme gene expression and pharmacologic intervention with endotoxin and dexamethasone. *Ped Res* 1995;**37**:469–75.

64. Barnard JA, Lyons RM, Moses HL. The cell biology of transforming growth factor β. *Biochim Biophys Acta Rev Cancer* 1990;**1032**:79–87.

65. Bartlett D. Postnatal growth of the mammalian lung: Influence of low and high oxygen tensions. *Respir Physiol* 1970;**9**:58–64.

66. Burri PH, Weibel ER. Ultrastructure and morphometry of the developing lung. In: Hodson WA, editor. *Development of the lung*. New York: Marcel Dekker; 1977. pp. 215–68.

67. Koppel R, Han RNN, Cox D, Tanswell AK, Rabinovitch M. A1-antitrypsin protects neonatal rats from pulmonary vascular and parenchymal effects of oxygen toxicity. *Ped Res* 1994;**36**:763–70.

68. Massaro GD, Olivier J, Massaro D. Brief perinatal hypoxia impairs postnatal development of the bronchiolar epithelium. *Am J Physiol* 1989;**257**(2 Pt 1):L80–85.

69. Veness-Meehan KA, Pierce RA, Moats-Staats BM, Stiles AD. Retinoic acid attenuates o2-induced inhibition of lung septation. *Am J Physiol Lung Cell Mol Physiol* 2002;**283**(5):L971–980.

70. Veness-Meehan KA, Bottone Jr FG, Stiles AD. Effects of retinoic acid on airspace development and lung collagen in hyperoxia-exposed newborn rats. *Pediatr Res* 2000;**48**(4):434–44.

71. Chen C, Wang LF, Chou H, Lang YD, Lai YP. Up-regulation of connective tissue growth factor in hyperoxia-induced lung fibrosis. *Pediatr Res* 2007;**62**(2):128–33.

72. Keeney SE, Mathews MJ, Haque AK, Schmalstieg FC. Comparison of pulmonary neutrophils in the adult and neonatal rat after hyperoxia. *Pediatr Res* 1995;**38**(6):857–63.

73. Auten Jr RL, Mason SN, Tanaka DT, Welty-Wolf K, Whorton MH. Anti-neutrophil chemokine preserves alveolar development in hyperoxia-exposed newborn rats. *Am J Physiol Lung Cell Mol Physiol* 2001;**281**(2):L336–344.

74. Auten RL, Whorton MH, Nicholas Mason S. Blocking neutrophil influx reduces DNA damage in hyperoxia-exposed newborn rat lung. *Am J Respir Cell Mol Biol* 2002;**26**(4):391–7.

75. Manji JS, O'Kelly CJ, Leung WI, Olson DM. Timing of hyperoxic exposure during alveolarization influences damage mediated by leukotrienes. *Am J Physiol Lung Cell Mol Physiol* 2001;**281**(4):L799–806.

76. McGrath-Morrow SA, Stahl J. Apoptosis in neonatal murine lung exposed to hyperoxia. *Am J Respir Cell Mol Biol* 2001;**25**(2):150–5.

77. Barry BE, Miller FJ, Crapo JD. Effects of inhalation of 0.12 and 0.25 parts per million ozone on the proximal alveolar region of juvenile and adult rats. *Laborat Invest* 1985;**53**(6):692–704.

78. Elsayed NM, Mustafa MG, Postlethwait EM. Age-dependent pulmonary response of rats to ozone exposure. *J Toxicol Environ Health* 1982;**9**(5-6):835–48.

79. Stephens RJ, Sloan MF, Groth DG, Negi DS, Lunan KD. Cytologic responses of postnatal rat lungs to O3 or NO2 exposure. *Am J Pathol* 1978;**93**(1):183–200.

80. Tyson CA, Lunan KD, Stephens RJ. Age-related differences in gsh-shuttle enzymes in NO2- or O3-exposed rat lungs. *Arch Environ Health* 1982;**37**(3):167–76.

81. Raub JA, Mercer RR, Kavlock RJ. Effects of prenatal nitrofen exposure on postnatal lung function in the rat. *Toxicol Appl Pharmacol* 1983;**94**:119–34.

82. Stiles J, Tyler WS. Age-related morphometric differences in responses of rat lungs to ozone. *Toxicol Appl Pharmacol* 1988;**92**:274–85.

83. Tyler WS, Tyler NK, Magliano DJ, Hinds DM, Tarkington B, Julian MD, et al. Effects of ozone inhalation on lungs of juvenile monkeys. Morphometry after a 12 month exposure and following a 6 month post-exposure. In: Berglund RL, Lawson DR, McKee DJ, editors. *Tropospheric ozone and the environment*. Pittsburg, PA: Air & Waste Management Association; 1991. pp. 152–9.

84. Mariassy AT, Sielczak MW, McCray MN, Abraham WM, Wanner A. Effects of ozone on lamb tracheal mucosa. Quantitative glycoconjugate histochemistry. *Am J Pathol* 1989;**135**(5):871–9.

85. Fanucchi MV, Plopper CG, Evans MJ, Hyde DM, Van Winkle LS, Gershwin LJ, et al. Cyclic exposure to ozone alters distal airway development in infant rhesus monkeys. *Am J Physiol Lung Cell Mol Physiol* 2006;**291**(4):L644–650.

86. Kajekar R, Pieczarka EM, Smiley-Jewell SM, Schelegle ES, Fanucchi MV, Plopper CG. Early postnatal exposure to allergen and ozone leads to hyperinnervation of the pulmonary epithelium. *Respir Physiol Neurobiol* 2007;**155**(1):55–63.

87. Evans MJ, Fanucchi MV, Baker GL, Van Winkle LS, Pantle LM, Nishio SJ, et al. Atypical development of the tracheal basement membrane zone of infant rhesus monkeys exposed to ozone and allergen. *Am J Physiol Lung Cell Mol Physiol* 2003;**285**(4):L931–939.

88. Evans MJ, Van Winkle LS, Fanucchi MV, Plopper CG. The attenuated fibroblast sheath of the respiratory tract epithelial-mesenchymal trophic unit. *Am J Respir Cell Mol Biol* 1999;**21**(6):655–7.

89. Iozzo RV. Matrix proteoglycans: From molecular design to cellular function. *Annu Rev Biochem* 1998;**67**:609–52.

90. Evans MJ, Van Winkle LS, Fanucchi MV, Baker GL, Murphy AE, Nishio SJ, et al. Fibroblast growth factor-2 in remodeling of the developing basement membrane zone in the trachea of infant rhesus monkeys sensitized and challenged with allergen. *Laborat Invest* 2002;**82**(12):1747–54.

91. Jiang X, Couchman JR. Perlecan and tumor angiogenesis. *J Histochem Cytochem ISO* 2003;**51**(11):1393–410.

92. Avol EL, Gauderman WJ, Tan SM, London SJ, Peters JM. Respiratory effects of relocating to areas of differing air pollution levels. *Am J Respir Crit Care Med* 2001;**164**(11):2067–72.

93. Peters JM, Avol E, Gauderman WJ, Linn WS, Navidi W, London SJ, et al. A study of twelve southern california communities with differing levels and types of air pollution. Ii. Effects on pulmonary function. *Am J Respir Crit Care Med* 1999;**159**(3):768–75.

94. Peters JM, Avol E, Navidi W, London SJ, Gauderman WJ, Lurmann F, et al. A study of twelve southern california communities with differing levels and types of air pollution. I. Prevalence of respiratory morbidity. *Am J Respir Crit Care Med* 1999;**159**(3):760–7.

95. Gauderman WJ, McConnell R, Gilliland F, London S, Thomas D, Avol E, et al. Association between air pollution and lung function growth in southern california children. *Am J Respir Crit Care Med* 2000;**162**(4 Pt 1):1383–90.

96. Calderon-Garciduenas L, Mora-Tiscareno A, Chung CJ, Valencia G, Fordham LA, Garcia R, et al. Exposure to air pollution is associated with lung hyperinflation in healthy children and adolescents in southwest mexico city: A pilot study. *Inhal Toxicol* 2000;**12**(6):537–61.

97. Calderon-Garciduenas L, Mora-Tiscareno A, Fordham LA, Valencia-Salazar G, Chung CJ, Rodriguez-Alcaraz A, et al. Respiratory damage in children exposed to urban pollution. *Pediatr Pulmonol* 2003;**36**(2):148–61.

98. Calderon-Garciduenas L, Mora-Tiscareno A, Fordham LA, Chung CJ, Valencia-Salazar G, Flores-Gomez S, et al. Lung radiology and pulmonary function of children chronically exposed to air pollution. *Environ Health Perspect* 2006;**114**(9):1432–7.

99. Shore SA, Abraham JH, Schwartzman IN, Murthy GG, Laporte JD. Ventilatory responses to ozone are reduced in immature rats. *J Appl Physiol* 2000;**88**(6):2023–30.

100. Bunt JE, Carnielli VP, Darcos Wattimena JL, Hop WC, Sauer PJ, Zimmermann LJ. The effect in premature infants of prenatal corticosteroids on endogenous surfactant synthesis as measured with stable isotopes. *Am J Respir Crit Care Med* 2000;**162**(3 Pt 1):844–9.

101. Bancalari E. Corticosteroids and neonatal chronic lung disease. *Eur J Pediatr* 1998;**157**(Suppl. 1):S31–37.

102. Expert panel report 3 (epr-3) NAEPP. Guidelines for the diagnosis and management of asthma-summary report 2007. *J Allergy Clin Immunol* 2007;**120**(5, Suppl. 1):S94–138.

103. Kay HH, Bird IM, Coe CL, Dudley DJ. Antenatal steroid treatment and adverse fetal effects: What is the evidence? *J Soc Gynecol Investig* 2000;**7**(5):269–78.

104. Moss TJ, Harding R, Newnham JP. Lung function, arterial pressure and growth in sheep during early postnatal life following single and repeated prenatal corticosteroid treatments. *Early Hum Dev* 2002;**66**(1):11–24.

105. Willet KE, Jobe AH, Ikegami M, Newnham J, Brennan S, Sly PD. Antenatal endotoxin and glucocorticoid effects on lung morphometry in preterm lambs. *Pediatr Res* 2000;**48**(6):782–8.

106. Walther FJ, Ikegami M, Warburton D, Polk DH. Corticosteroids, thyrotropin-releasing hormone, and antioxidant enzymes in preterm lamb lungs. *Pediatr Res* 1991;**30**(6):518–21.

107. Walther FJ, David-Cu R, Mehta EI, Polk DH, Jobe AH, Ikegami M. Higher lung antioxidant enzyme activity persists after single dose of corticosteroids in preterm lambs. *Am J Physiol* 1996;**271**(2 Pt 1):L187–191.

108. Willet KE, Jobe AH, Ikegami M, Newnham J, Sly PD. Pulmonary interstitial emphysema 24 hours after antenatal betamethasone treatment in preterm sheep. *Am J Respir Crit Care Med* 2000;**162**(3 Pt 1):1087–94.

109. Tschanz SA, Haenni B, Burri PH. Glucocorticoid induced impairment of lung structure assessed by digital image analysis. *Eur J Pediatr* 2002;**161**(1):26–30.

110. Tschanz SA, Damke BM, Burri PH. Influence of postnatally administered glucocorticoids on rat lung growth. *Biol Neonate* 1995;**68**(4):229–45.

111. Okajima S, Matsuda T, Cho K, Matsumoto Y, Kobayashi Y, Fujimoto S. Antenatal dexamethasone administration impairs normal postnatal lung growth in rats. *Pediatr Res* 2001;**49**(6):777–81.

112. Mitzner W, Johnson JW, Beck J, London W, Sly D. Influence of betamethasone on the development of mechanical properties in the fetal rhesus monkey lung. *Am Rev Respir Dis* 1982;**125**(2):233–8.

113. Edwards LA, Read LC, Nishio SJ, Weir AJ, Hull W, Barry S, et al. Comparison of the distinct effects of epidermal growth factor and betamethasone on the morphogenesis of the gas exchange region and differentiation of alveolar type ii cells in lungs of fetal rhesus monkeys. *J Pharmacol Exp Ther* 1995;**274**(2):1025–32.

114. Bunton TE, Plopper CG. Triamcinolone-induced structural alterations in the development of the lung of the fetal rhesus macaque. *Am J Obstet Gynecol* 1984;**148**(2):203–15.

115. Bhatt AJ, Amin SB, Chess PR, Watkins RH, Maniscalco WM. Expression of vascular endothelial growth factor and flk-1 in developing and glucocorticoid-treated mouse lung. *Pediatr Res* 2000;**47**(5):606–13.

116. Fayon M, Jouvencel P, Carles D, Choukroun ML, Marthan R. Differential effect of dexamethasone and hydrocortisone on alveolar growth in rat pups. *Pediatr Pulmonol* 2002;**33**(6):443–8.

117. le Cras TD, Markham NE, Morris KG, Ahrens CR, McMurtry IF, Abman SH. Neonatal dexamethasone treatment increases the risk for pulmonary hypertension in adult rats. *Am J Physiol Lung Cell Mol Physiol* 2000;**278**(4):L822–829.

118. Nagai A, Matsumiya H, Yasui S, Aoshiba K, Ishihara Y, Konno K. Administration of bromodeoxyuridine in early postnatal rats results in lung changes at maturity. *Exp Lung Res* 1993;**19**(2):203–19.

119. Anisimov VN, Osipova G. Two-step carcinogenesis induced by neonatal exposure to 5-bromo-2′-deoxyuridine and subsequent administration of urethan in balb/c mice. *Cancer Lett* 1992;**64**(1):75–82.

120. Manson JM. Mechanism of nitrofen teratogenesis. *Environ Health Perspect* 1986;**70**:137–47.

121. Brown TJ, Manson JM. Further characterization of the distribution and metabolism of nitrofen inthe pregnant rat. *Teratology* 1986;**34**:129–39.

122. Losada A, Tovar JA, Xia HM, Diez-Pardo JA, Santisteban P. Downregulation of thyroid transcription factor-1 gene expression in fetal lung hypoplasia is restored by glucocorticoids. *Endocrinology* 2000;**141**(6):2166–73.

123. Losada A, Xia H, Migliazza L, Diez-Pardo JA, Santisteban P, Tovar JA. Lung hypoplasia caused by nitrofen is mediated by downregulation of thyroid transcription factor ttf-1. *Pediatr Surg Int* 1999;**15**(3-4):188–91.

124. Gosney JR, Okoye BO, Lloyd DA, Losty PD. Pulmonary neuroendocrine cells in nitrofen-induced diaphragmatic hernia and the effect of prenatal glucocorticoids. *Pediatr Surg Int* 1999;**15**(3-4):180–3.

125. Branstetter DG, Moseley PP. Effect of lung development on the histological pattern of lung tumors induced by ethylnitrosourea in the c3heb/fej mouse. *Exp Lung Res* 1991;**17**(2):169–79.

126. Rehm S, Ward JM, Anderson LM, Riggs CW, Rice JM. Transplacental induction of mouse lung tumors: Stage of fetal organogenesis in relation to frequency, morphology, size, and neoplastic progression of n-nitrosoethylurea-induced tumors. *Toxicol Pathol* 1991;**19**(1):35–46.

127. Mauceri HJ, Lee WH, Conway S. Effect of ethanol on insulin-like growth factor-ii release from fetal organs. *Alcohol Clin Exp Res* 1994;**18**(1):35–41.

128. Mauceri HJ, Becker KB, Conway S. The influence of ethanol exposure on insulin-like growth factor (igf) type ii receptors in fetal rat tissues. *Life Sci* 1996;**59**(1):51–60.

129. Sosenko IR. Antenatal cocaine exposure produces accelerated surfactant maturation without stimulation of antioxidant enzyme development in the late gestation rat. *Pediatr Res* 1993;**33**(4 Pt 1):327–31.

Chapter 26

Effect of Environment and Aging on the Pulmonary Surfactant System

Sandra Orgeig*, Janna L. Morrison* and Christopher B. Daniels†

*School of Pharmacy & Medical Sciences, Sansom Institute for Health Research, University of South Australia, Adelaide, SA, Australia;

†Barbara Hardy Institute, University of South Australia, Adelaide, SA, Australia

INTRODUCTION

In Chapter 9 (Orgeig et al.) we described the composition, function, and regulation of the surfactant system, how this system develops normally and how it is affected by premature birth and genetic factors. Here, we focus on the effects of environmental factors on the developing and adult pulmonary surfactant systems. During development the lung and the surfactant system can be profoundly impacted by changes in nutrition and oxygenation as well as chemicals and agents that are transferred from the maternal circulation via the placenta. Frequently the consequences of these early insults persist into postnatal life. The adult pulmonary surfactant system is also sensitive to changes in oxygenation and exposure to a range of gaseous and particulate pollutants that are capable of altering surfactant composition and function.

Aging has profound effects on both lung function and architecture (Chapter 14 [Pinkerton et al.]). Effects of aging include enlargement of air spaces, reduction in exchange surface area, and loss of supporting tissue for peripheral airways (senile emphysema), resulting in decreased static elastic recoil, increased residual volume, and functional residual capacity but decreased expiratory flow rates.[1] In the final part of this chapter, we describe the effects of aging on the pulmonary surfactant system as well as the effects of environmental factors acting on the aging lung.

EFFECT OF THE INTRAUTERINE ENVIRONMENT ON THE DEVELOPING PULMONARY SURFACTANT SYSTEM

Fetal development and growth are regulated by the substrates, including oxygen and nutrients, that are transferred from the maternal to the fetal circulation by the placenta. Thus the ability of the placenta to transfer substrates, as well as the availability of the substrates in the maternal circulation, will have significant effects on fetal growth. As maternal diet or exposure to drugs or infection can influence fetal development and growth, they can also influence lung development and specifically the development of the pulmonary surfactant system.

Intrauterine Growth Restriction

Intrauterine growth restriction (IUGR), a major cause of low birth weight infants, may increase the risk of respiratory distress syndrome (RDS) and death in both term and preterm infants,[2,3] and may also affect lung function in postnatal life.[4–7] Approximately 7% of babies are born with IUGR,[8] defined as a birth weight <10th percentile.[9–11] IUGR commonly occurs when substrate supply is reduced and does not meet fetal demands, but may also be due to an inherent impairment of tissue growth potential. Hence, causes of IUGR include fetal factors (e.g., chromosomal abnormalities, infection), maternal factors (e.g., undernutrition, smoking), environmental factors (e.g., high altitude), placental factors (e.g., placental infarction), and other factors (e.g., reproductive technologies).[10–13] Furthermore, the incidence of IUGR increases with increasing prematurity.[14,15] The IUGR fetus responds to decreased substrate supply with a range of adaptations including redistribution of blood flow away from peripheral organs and to essential organs such as the brain, heart, and adrenals, which results in the relative growth sparing of these organs.[11] These cardiovascular adaptations in the IUGR fetus are associated with fetal hypoxemia, hypoglycemia, elevated plasma catecholamines and cortisol concentrations,[11,16,17] and decreased plasma concentrations of insulin-like growth factors and their binding proteins.[18]

The impact of IUGR on the pulmonary surfactant system, lung function, and RDS are still controversial. For example, early studies indicated the possibility that IUGR

The Lung. http://dx.doi.org/10.1016/B978-0-12-799941-8.00026-2

accelerated lung maturation, suggesting that they may be at lower risk of RDS.[19,20] Epidemiological studies have tested this hypothesis in small-for-gestational-age (SGA) versus appropriate-for-gestational-age (AGA) fetuses and the results are largely inconclusive, with studies demonstrating the risk of RDS being lower,[21] equal,[22–24] or higher[3,25–27] in SGA infants. Interestingly, when infants were subdivided into different categories of prematurity, it was evident that while very preterm SGA infants (25–28 weeks) have an increased risk of RDS, less preterm SGA infants (29–32 weeks) have a reduced risk of RDS.[28] In addition to multivariate epidemiological studies, experimental studies have been performed using different species (e.g., sheep, rabbits, guinea pigs, mice, chickens) and different protocols (placental restriction, undernutrition, developmental gene knockout) to induce IUGR.[11,29,30] Of these only a few studies have concentrated on lung and surfactant function and these have yielded similarly conflicting information.[20]

Umbilical Placental Embolization

A widely used experimental method for inducing late gestational IUGR in sheep is umbilical placental embolization (UPE) in which microspheres are repeatedly injected into the placenta via the umbilical artery, leading to blockage of placental capillaries and hence reduced transfer of O_2 and nutrients to the fetus.[11,13,31,32] Although fetal hypoxemia is usually maintained for 10–20 days in this model,[31,33] there is recovery to normal blood O_2 content values each day, suggesting that this protocol creates "repeated acute hypoxic events"[11] (Figure 1A). UPE from ~109 to 130 days decreases fetal growth and lung growth proportionately, and decreases lung DNA content. Fetal plasma cortisol concentrations increase and correlate significantly with increases in surfactant protein (SP)-A and -B mRNA (but not SP-C mRNA). Although lung morphology and function were not assessed to determine lung maturation directly, the decrease in DNA content (and concentration, i.e., mg/g lung weight) with an increase in SP mRNA synthesis suggests that there is a switch from lung cell proliferation to maturation.[31]

In contrast, UPE for 20 days during late ovine gestation (120–140 days, term being 145–150 days) does not change the SP-A, -B or -C mRNA, or SP-A protein levels in the fetal lung.[33] There is also no correlation between SP-A, -B or -C mRNA, or SP-A protein levels and plasma cortisol concentrations. Increased lung tissue DNA concentration in the IUGR fetuses,[33] together with increased thickness of the basement membrane and interstitial edema leading to a thicker pulmonary blood-air barrier at 140 days gestation[7] suggest that the lungs of IUGR fetuses are structurally immature. This abnormality may result in less efficient gas exchange, explaining why IUGR fetuses born at term are relatively hypoxemic and hypercapnic,[7] and why pulmonary diffusing capacity is reduced.[5]

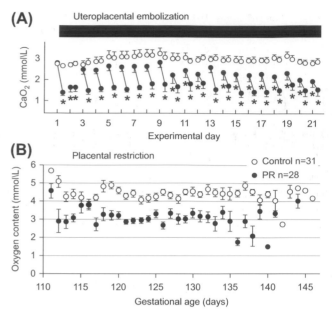

FIGURE 1 Fetal arterial oxygen content (mmol O_2/L blood) in two sheep models of human IUGR. The uteroplacental embolization model of IUGR in the sheep fetus results in periods of fluctuating hypoxemia over the 20-day experimental period starting at 110 days of gestation (A). In contrast, placental restriction (PR, $n = 28$; control, $n = 31$) in sheep (B) results in chronic hypoxemia that is maintained throughout late gestation.[5] Control, open circles UPE (A) or PR (B), closed circles. PR: placental restriction. *Figure reproduced with modifications and permission from.*[11,20,303,304]

The primary difference between the study of Gagnon et al.[31] and that of Cock et al.[33] lies in the timing of sampling relative to term (147 days). Cock et al.[33] induced IUGR from 120–140 days of gestation compared to ~109–130 days in the study by Gagnon et al.[31] It is possible that the levels of SP mRNA in the older fetuses had already reached their maximal expression and could not be stimulated further by cortisol.

Interestingly, lambs born spontaneously following IUGR induced by UPE remain lighter than controls until 8 weeks after birth, but demonstrate enhanced SP-B gene expression and a tendency towards higher SP-A tissue content. SP-A and -C gene expression are not affected.[5] Therefore, in terms of surfactant proteins, lambs born after UPE-induced IUGR do not appear to suffer from surfactant deficiency.[34,35]

Carunclectomy

Another model for inducing IUGR in the sheep is uterine carunclectomy, the surgical removal of most of the endometrial caruncles from the uterus of the non-pregnant ewe prior to mating. This restricts the number of placentomes, thereby limiting placental growth and function throughout pregnancy, leading to chronic restriction of the delivery of O_2 and nutrients to the fetus[11,36] (Figure 1B). This model of IUGR shows a decrease in the total phospholipid

TABLE 1 Summary of studies that have used hypoxia or hypoxemia to induce fetal growth restriction and/or to investigate the impact on the pulmonary surfactant system

Hypoxia/fetal Growth Restriction Induction	Species	Surfactant Effect	Comment	Reference
Pregnant ewes breathed hypoxic gas for 48 h at 126–130 days or 134–136 days. Fetal PaO_2 reduced by ~7mmHg	sheep	SP-A and -B mRNA increased in older group; no change in SP-C mRNA	Correlated with elevated plasma cortisol, especially later in gestation	[39]
Dams in hypoxic environment (FiO_2 of 0.1) from ED14 to 17.5	mouse	SP-A, -B, and -C mRNA reduced; no change in SP-D mRNA		[302]
Umbilical placental embolization (109–130 days)	sheep	SP-A and -B mRNA increased; no change in SP-C mRNA	Correlated with cortisol surge in last 48 h.	[31]
Umbilical placental embolization (120–140 days)	sheep	No change in SP-A, -B, and -C mRNA or protein	Cortisol increased; no correlation with SPs	[33]
Repeated umbilical cord occlusions for 4 days at 112–115 days and 130–133 days	sheep	No change in SP-A, -B, and -C mRNA at 112–115 days; 85% reduction in SP-A and -B mRNA; 66% reduction in SP-C mRNA at 130–133 days	Plasma cortisol increased on day 4 particularly in older group	[40]
Placental restriction by carunclectomy to 133 and 141 days gestation	sheep	SP-A, -B, and -C mRNA and protein reduced at both ages	SP-A, -B, and -C protein and SP-B and -C mRNA correlated positively with mean gestational PaO_2 at 133 days. SP-A and -B mRNA correlated inversely with cortisol	[38]

concentration (indicative of decreased surfactant production) of fetal lung fluid.[37] Morphometry demonstrates significant changes in lung structure close to term, including an increase in the number of airspaces and a concomitant decrease in the gas exchange surface density.[6] Importantly, contrary to the findings in the UPE model, IUGR induced by carunclectomy leads to a reduction in both lung tissue SP and gene expression for SP-A, -B, and -C at two ages during late gestation (133 and 141 days).[38] Furthermore, this study demonstrates an inverse relationship between SP-A and -B gene expression and plasma cortisol concentration and a direct relationship between arterial PO_2 and SP-A, -B, and -C protein expression. There is, however, no relationship between SP gene or protein expression with plasma glucose concentration. Hence, the chronic hypoxemia and hypercortisolemia in the carunclectomy model significantly inhibits surfactant maturation.[38]

Roles and Interactions of Fetal Hypoxia, Glucocorticoids, and Glucose in Regulating Surfactant Development during IUGR

From the above discussion of IUGR models it is clear that IUGR is a multifactorial condition, with fetuses experiencing varying degrees of hypoxemia, hypercortisolemia, and hypoglycemia. Each of these variables is capable of independently influencing the maturation of the lung and the surfactant system.[38] For example, it has not been clear whether blood O_2 or glucose is the major driving force for fetal adaptations to reduced substrate supply.[12]

Table 1 summarizes the outcomes of IUGR studies involving surfactant investigations in which the impact of hypoxemia was addressed either directly, without the complications of nutrient restriction and/or hypoglycemia or in combination with other physiological perturbations. From these studies it is difficult to separate the effects of hypoxemia from those of hormones and growth factors, as the hypoxemic stress also induces endogenous glucocorticoid production, and possibly other factors such as vascular endothelial growth factor (VEGF). However, determining the role of endogenous glucocorticoids, particularly on SP gene expression, is complicated as it is species-specific and gestational age dependent[20] (Table 1). The variability and unpredictability in SP gene expression in response to glucocorticoids is borne out by the range of studies in which IUGR or hypoxemia are induced, resulting in either "repeated acute," "brief intermittent," "prolonged," or "chronic" hypoxemia, which are all accompanied by elevated plasma cortisol concentrations, yet nevertheless show marked differences in the effect on SP gene and protein expression (Table 1).[38] For example, 48 h of maternal hypoxemia[39] and chronic placental insufficiency in the ovine UPE model[31] demonstrate a positive correlation between SP gene expression and plasma cortisol concentration. However, in late gestation in the UPE model[33] there is no correlation, whereas in the cord occlusion model of asphyxia[40] and the carunclectomy model of IUGR[38] there is a negative correlation[20] (Table 1).

Both increased plasma cortisol concentration and the mean PaO_2 in the IUGR fetus (carunclectomy) correlate with the inhibition of SP gene and protein expression.[38] Hence, both the chronic hypoxemia and the hypercortisolemia associated with the carunclectomy model of IUGR appear to play a role in the inhibition of surfactant maturation.[38] However, there is no correlation between plasma glucose concentration and either SP gene or protein expression, suggesting that in this model glucose status does not affect SP metabolism.[38]

The studies described earlier indicate that the severity, frequency, and duration of exposure to hypoxia vary widely from mild to severe, from acute to chronic, and from constant to intermittent (Table 1). Therefore, the impact of IUGR on lung SP production will depend on the frequency, degree, and duration of hypoxemia that is elicited by the different experimental protocols, or in the case of humans, the heterogeneous causes of IUGR.[20] For example, although the degree of hypoxemia in the carunclectomy and UPE models is similar (~40% reduction in PaO_2), it is likely that the more sustained pattern of hypoxemia in the former represents a "steady state" that elicits a different cellular and molecular response leading to inhibition as opposed to stimulation of SP expression observed either with the intermittent dynamic changes in PaO_2 of the UPE model[31] or the more acute 48 h of maternal hypoxemia.[39] It is possible therefore that the different causes of IUGR and/or hypoxemia induce different fetal adaptations that lead to different responses in the surfactant system.[38]

Furthermore, the timing of the hypoxic insult in relation to stages of lung development is likely to influence the outcome for the lung and the surfactant system. For example, in the UPE model[31,33] and with maternal hypoxemia,[39] there appears to be a very narrow window in which surfactant maturation can be perturbed. While this robustness may represent an adaptive advantage during normal development, it may also explain the variable outcomes described[29] in therapeutic interventions such as glucocorticoid administration. Therefore, in order to optimize treatment strategies of fetuses and infants at risk of IUGR-related postnatal respiratory complications, it is essential that the mechanisms and timing of lung and surfactant maturation during late gestation are understood, especially in relation to the different causal factors of IUGR.[20,41]

In summary, there is still a paucity of information on the effects of IUGR on the surfactant system and in particular the data on the levels of SP gene and protein expression are inconsistent. Importantly, the molecular mechanisms that are involved in mediating the effects in response to changing factors such as hypoxia, glucocorticoids and glucose, remain largely unknown.[20]

Maternal Nutrition

In animal models or *in vitro* systems, both glucose and insulin have an effect on surfactant lipid and protein metabolism. For example, glucose infusion in fetal sheep with a concomitant increase in insulin has a biphasic effect on surfactant phosphatidylcholine (PC) content, with an increase in mid-gestation and an inhibition of the normal rise in late gestation that correlates with a reduction in lung stability.[42] Intravenous insulin infusion during late gestation in pregnant rabbits results in decreased fetal plasma glucose and insulin concentrations, improved survival of preterm neonates, and correlates with an increase in the quantity and quality of surfactant phospholipid content in lavage.[43] On the other hand, maternal glucose infusion results in fetal hyperinsulinemia and hyperglycemia, with no difference in alveolar phospholipids.[43] As fetal hyperinsulinemia associated with maternal diabetes is associated with a reduction in SP-A protein in amniotic fluid[44] and an increased risk of RDS in the newborn, it has been hypothesized that insulin inhibits alveolar epithelial type II cell (AECII) differentiation. Subsequently, it was shown that insulin inhibits both SP-A and SP-B gene expression in lung tissue explants from human fetuses.[45] Despite the fact that obesity is a growing issue, including among pregnant women,[46] and that it has negative consequences for respiratory health in children and adults,[47,48] no studies have shown whether or not maternal overnutrition or obesity changes surfactant composition or function in the fetus or neonate.

Undernutrition

Long-term follow-up studies of the Dutch Hunger Famine of 1945 have revealed the impact of maternal undernutrition during early, mid, and late gestation. Offspring of mothers who were exposed to the famine during mid-pregnancy have an increased prevalence of obstructive airway disease, without changes in IgE or lung function.[49] A recent systematic review of the effects of maternal undernutrition on long-term health outcomes found that birth weight is positively associated with lung function.[50]

In contrast, maternal undernutrition in late gestation in rats results in IUGR offspring with smaller lungs, lower lung saturated PC, but no change in lung SP-A, -B, -C, and -D mRNA expression.[51] Effects on pulmonary morphometry are maintained into postnatal life, with offspring having a lower alveolar surface area at postnatal day 42.[51] Maternal undernutrition in rats during the last week of gestation results in reduced body and lung weight without a change in gene expression of insulin-like growth factor (IGF)-1 and

-2 in the lung.[52] Maternal nutrient restriction of 50% during the last trimester in guinea pigs (term, 69 days) reduces birth weight, does not change plasma cortisol concentration, but reduces total surfactant phospholipids in lavage.[53]

Protein deprivation during pregnancy in rats results in increased glycogen content in the lung at 2 days after birth but decreased disaturated PC at birth and 2 days after birth with a return to control levels by 5 days after birth.[54] Protein deprivation through pregnancy reduces fetal weight with no change in lung weight, resulting in increased surface forces, decreased tissue elastic forces, and increased resistance of the lung to rupture.[55]

Maternal Addiction/Drug Use

Smoking

Data on smoking during pregnancy are unreliable as self-reporting provides an underestimate.[56] Recent data show that 17–25% of pregnant women smoke and only 3.2% quit smoking during pregnancy.[57,58] Maternal smoking and neonatal exposure to cigarette smoke or nicotine have paradoxical effects on lung development.[59] On the one hand, there is evidence of an increased fetal lung maturity[60] and enhanced functional pulmonary maturity at birth, as indicated by surfactant phospholipid profiles,[61,62] possibly contributing to a decrease in the incidence of RDS.[63] However, there is a significant reduction in both pre- and postnatal lung growth and lung function.[59,64–68] For example, nicotine exposure during pregnancy and lactation in rats suppresses alveolarization in the lungs of the offspring up to 42 days, which results in a reduced internal surface area available for gas exchange.[69] Furthermore, fetal exposure to maternal smoking causes airway remodeling in mouse pups and increased responses to allergen challenge in adult life.[70]

Studies of surfactant lipids in offspring exposed to smoke or nicotine *in utero* have yielded variable results; in part, the variability appears to depend on the age at which the offspring are examined. For example, the amniotic fluid of smoking mothers has a higher saturated PC content.[62] Moreover, maternal nicotine exposure during pregnancy and lactation results in an increased AECII lamellar body content in newborn rats.[71] Similarly, after *in utero* nicotine exposure, saturated PC synthesis is increased, as measured by choline incorporation in fetal rat lung explants at embryonic day 20.[67] A longer term study, showed no effect on the lung saturated PC content after *in utero* exposure (day 3 to 21 of gestation) to nicotine in rat pups on postnatal day 1 to 28,[72] but saturated PC content decreased at day 35 and 42.[72]

There is also conflicting information on the effects of nicotine on SP expression in the developing lung. For example, fetal nicotine exposure during the last third of ovine gestation significantly decreases gene and protein expression of SP-A (a key component of lung innate immunity) in preterm lambs. However, SP-D protein and gene expression are not affected[73] nor are amniotic fluid levels of SP-A at full term in smoke-exposed human pregnancies.[74] Similarly, *in utero* nicotine exposure does not alter gene expression of SP-A, -B, -C, and -D in the newborn rat on postnatal day 1, but SP mRNA for all 4 proteins transiently increases at day 7 followed by a decrease at day 14, with no effects thereafter.[72] In contrast, exposing mouse fetal lung explants to nicotine on embryonic day 11 increases SP-A and -C gene expression.[75] In rhesus monkeys, maternal nicotine exposure through most of pregnancy increases SP-B and decreases elastin content. Both effects are attenuated by vitamin C treatment.[76]

Clearly, many of the conflicting results are likely due to differences in species, stage of lung development, as well as duration of nicotine or smoke exposure.[67] However, it appears that in most studies there is either an increase or at least no detrimental effect on surfactant maturity within the first 24 h after birth, but this does not appear to be sustained.[67,72]

Mechanistic studies (both *in vivo* and *in vitro*) suggest that surfactant synthesis and AECII proliferation are stimulated via a direct cellular effect of nicotine.[67] This may explain the apparently greater pulmonary maturity either late in gestation or early in postnatal life. On the other hand, the observed increases in cell differentiation and metabolism[67] are secondarily affected via disruption of a specific epithelial-mesenchymal signaling pathway involving the nuclear transcription factor, peroxisome proliferator activated receptor (PPAR)-γ that is the key downstream mesenchymal target of parathyroid hormone-related protein (PTHrP). Inhibition of PTHrP/PPARγ signaling disrupts epithelial-mesenchymal interactions and causes lipofibroblast-to-myofibroblast transdifferentiation.[59,67] The increased generation of lung myofibroblasts, which are key players in the pathophysiology of asthma[77] and which contribute not only to tissue remodeling but also to airway inflammation,[78] could potentially explain the known long-term pulmonary effects of *in utero* nicotine exposure,[59] including the increased risk of childhood asthma.[68,79,80] Furthermore, as lipofibroblast communication with AECII is essential for maintaining surfactant lipid synthesis as well as sustaining AECII growth and differentiation,[59] the loss of lipofibroblasts early in postnatal life may explain why the stimulatory effect of *in utero* nicotine exposure on AECII surfactant synthesis ultimately fails, thereby impacting lung function adversely.[59]

Alcohol

Exposure to alcohol in the perinatal period can lead to fetal alcohol spectrum disorder (FASD), the most severe being fetal alcohol syndrome (FAS), which is characterized by cerebral dysfunction, growth restriction, and craniofacial malformations.[81–83] Surveys show that 60% of women consume alcohol during at least one trimester.[84] While the impact of fetal alcohol exposure on human lung development is not known,[85] recent studies have examined the effect in fetal sheep. Daily exposure of sheep fetuses to alcohol for the last third of gestation suppresses gene expression of all SPs as well as pro-inflammatory cytokine responses and bronchoalveolar lavage (BAL) fluid phospholipid content, including PC, phosphatidylglycerol (PG), phosphatidylinositol (PI), phosphatidylserine (PS), and phosphatidylethanolamine (PE).[85,86] In lambs, however, birth weight, postnatal growth, blood gas parameters, lung weight, volume, tissue fraction, mean linear intercept, collagen content, pro-inflammatory cytokine gene expression, and BAL surfactant phospholipid composition are not affected.[87] While SP-A, -B, and -C mRNA expression is unchanged, SP-D mRNA is increased in lambs exposed to alcohol during late gestation, suggesting that innate immunity may be altered.[87] A similar study of maternal consumption of alcohol in late gestation showed decreases in SP-A, but not SP-D, protein expression in the lung of both preterm and term lambs.[88] Furthermore, the immune markers TNF-alpha, IL-10, chemokine (C-C motif) ligand 5 (CCL5) and monocyte chemotactic protein-1 mRNA are decreased.[89] This provides evidence that innate immunity in the lung is affected by maternal alcohol consumption.[90] However, some of the adverse effects of daily ethanol exposure during late gestation in the fetal lung do not persist to 2 months after birth, indicating that the developing lung is capable of repair.[87]

Cocaine

About 4% of pregnant women in the United States use illicit drugs compared to 10% in the general population.[91] The use of cocaine during pregnancy is associated with an increased risk of placental abruption, premature rupture of membranes, IUGR, preterm delivery, and withdrawal symptoms in the newborn.[92] Cocaine stimulates the sympathetic nervous system by blocking reuptake of catecholamines and serotonin. Cocaine affects the respiratory tract,[93] including in the neonates of cocaine-using mothers.[94] Furthermore, the widespread expression of serotonin receptors in the embryo, including in the lung, causes a vulnerability of the embryo to cocaine exposure.[95] Maternal cocaine use results in increased sympathetic activity, enhanced sensitivity of adrenergic receptors to noradrenaline, increased plasma glucocorticoid concentrations and hypoxemia in the fetus.[96,97] As a result, there can be fetal brain damage, including cerebral edema and reduced neuronal maturation.[98]

In humans, cocaine appears to accelerate fetal lung maturity.[60] A preliminary study of eight mothers who used cocaine during pregnancy suggests a decreased risk of RDS in infants[99]; however, a much larger study suggests an equal risk of RDS compared with non-exposed babies.[100] In rats,

prenatal cocaine exposure results in IUGR with increased plasma concentrations of noradrenaline and corticosterone as well as increased disaturated PC in the lung.[101] Preterm rabbit pups exposed to cocaine at 24–26 days gestation demonstrate increased lung distensibility, low surface tension, and thinner septa consistent with accelerated lung maturation at 27 days gestation.[102] The accelerated lung maturation may be due to the cocaine-induced increase in plasma corticosterone concentrations or a direct effect on the adrenergic receptors.

Opioids

The use of opioid analgesics increased two-fold amongst pregnant women in the United States from 1999–2009[103]; however, interviewing pregnant women underestimates (4%) the prevalence of illicit drug use, including opiates, methadone, and fentanyl, compared with estimates gained from hair analysis (15%).[104] There is some evidence that opioid exposure may reduce the incidence of RDS.[105] On the other hand, in the Neurologic Outcomes and Preemptive Analgesia in Neonates (NEOPAIN) trial, which involves patients in the USA, UK, Canada, and Europe, morphine does not improve short-term pulmonary outcomes among ventilated preterm neonates. Moreover, additional morphine doses are associated with worsening respiratory outcomes among preterm neonates with RDS.[106] Maternal exposure to opiates during pregnancy is also associated with an increased risk of childhood asthma.[107] Current treatment for pregnant users of heroin includes not only methadone but also buprenorphine.[108]

While animal models provide more mechanistic data, there are few recent studies. Pregnant rabbits exposed to morphine from early gestation to term have an increased incidence of abortion and although the surviving fetuses are smaller, there is no difference in lung stability.[105] Exposure to morphine in the third trimester in rabbits results in increased lung distensibility and alveolar stability on lung deflation, while naloxone has the opposite effect.[109] There is also evidence that heroin stimulates SP production in fetal and newborn rabbits, either due to increasing glucocorticoids or a direct action of catecholamines on AECII.[110]

Intrauterine and Neonatal Respiratory Infection

The developing fetal and neonatal lung is susceptible to infection by microorganisms and the resultant inflammation has marked consequences for lung and surfactant function, and hence survival and health of the infant.[111–113] Intrauterine infections can lead to chorioamnionitis, which is associated with an increased risk of preterm birth.[114] In preterm infants, for whom prolonged artificial ventilation is frequently required, there is an increased risk of postnatal pulmonary infection. Common infectious agents include pneumonia, staphylococcus, tuberculosis, and respiratory syncytial virus (RSV). The consequences of both chorioamnionitis and RSV for the pulmonary surfactant system have been extensively documented.

Chorioamnionitis

More than 50% of preterm or extremely low birth weight infants who die in the neonatal period show evidence of antenatal infection.[114,115] Bacterial infections within the uterus can occur between the maternal tissues and the fetal membranes, within the fetal membranes (the amnion and chorion), the placenta, the amniotic fluid, the umbilical cord, or the fetus. Infection of the fetal membranes is referred to as chorioamnionitis. Most bacteria found in the uterus in association with preterm labor are of vaginal origin.[114] While severe infection frequently causes death, the more common low grade, indolent chorioamnionitis that is present, but clinically silent over weeks and possibly months, ultimately results in preterm labor between 20 and 30 weeks gestation.[114,116]

Pregnant women with an intra-amniotic infection have increased amniotic fluid SP-B concentration, but the other SPs are not affected. Furthermore, this effect is enhanced in the 7 days following antenatal glucocorticoid treatment.[117] Chorioamniotic membranes express SP-A, with SP-A gene expression being higher at term than preterm.[118] However, chorioamniotic membrane SP-A gene expression is higher in women in preterm labor suffering from chorioamnionitis.[118]

Several animal models have been used to study the effects of inflammation on lung and surfactant development including treatment with endotoxin (i.e., lipopolysaccharide (LPS))[119] or Ureaplasma.[120] In sheep, intra-amniotic endotoxin-induced inflammation during early gestation (ranging from 60 to 108 days gestation) increases lung tissue SP-B mRNA and protein expression in BAL fluid collected at 125 days gestation.[119] However, lung, but not BAL, saturated PC increases after endotoxin treatment at 80 days and from 80 to 108 days gestation, whereas alveolar SP-C mRNA expression increases with endotoxin treatment at 60, 80, or 100 days gestation. Intra-amniotic injection of Ureaplasma either 1, 3, 6, or 10 weeks before the delivery of preterm lambs at 124 days of gestation leads to progressive lung inflammation and improvements in lung function, likely due to increases in saturated PC both in BAL and lung tissue as well as SP-A, -B, and -C mRNA expression.[120]

The inflammatory response to infection results in the release of a range of cytokines including interleukin-1 (IL-1), IL-6, and IL-8. Continuous infusion of LPS, IL-6, and IL-8 from 16–19 days gestation in rats increases SP-A, -B, and -C mRNA expression in the lung.[121] Treatment of rabbit lung explants collected at 22 days gestation with IL-1 increases SP-A, but not SP-B mRNA expression, while cAMP increases both SP-A and SP-B mRNA

expression.[122] Interestingly, IL-1α increases SP-A and -B in lung explants from 22 days gestation rabbit fetuses but decreases SP-A and -B in lung explants from 30 days gestation rabbits.[123] Moreover, glucocorticoids increase SP-A and -B and prevent their IL-1α-induced suppression.[123] The *in vivo* administration of LPS in sheep 7 and 14 days prior to lung collection at 120 days gestation increases both lung maturation and inflammation. Betamethasone treatment alone induces modest lung maturation. If treatment occurs before LPS administration, betamethasone suppresses lung inflammation, but if treatment occurs after LPS administration, betamethasone does not counteract the inflammation, but enhances lung maturation.[124] Furthermore, the simultaneous treatment of intra-amniotic endotoxin and maternal intramuscular betamethasone leads to additive maturational effects on short-term postnatal lung function but not on saturated PC levels in BAL or lung tissue.[125] This suggests that the mechanisms regulating glucocorticoid-induced lung maturation are different from endotoxin- or inflammation-mediated maturation.[125] Taken together these studies suggest that the interaction of betamethasone and lung inflammation is complex and is dependent on gestational age and the timing of the treatment in relation to the inflammation.[124]

A recent study in fetal sheep suggests that prostaglandins may mediate the fetal surfactant responses to inflammation. Nimesulide, a specific inhibitor of a prostaglandin synthetic enzyme, inhibits the LPS-induced increases in prostaglandin E2 (PGE2) and decreases fetal lung IL-1 and IL-8 mRNA as well as SP-A, -B, and -D mRNA expression.[126] The molecular mechanisms involved warrant further investigation with the aim to develop effective therapies for lung inflammation.

While exogenous surfactant therapy applied at birth in preterm infants may reduce the effects of inflammation, infants frequently demonstrate a poor response to exogenous surfactant.[127] For example, in a prospective observational study, preterm neonates with chorioamnionitis who were treated with exogenous surfactant have poorer outcomes, with an increased risk of BPD and death.[128] This is likely due to the inactivation of surfactant by inflammatory mediators, leading to an impaired response to the surfactant treatment.[127] Hence, the development of interventions aimed at increasing surfactant efficacy may be of particular benefit for infants exposed to chorioamnionitis.[128]

Respiratory Syncytial Virus (RSV)

RSV is one of the most prevalent respiratory infections in infants and children.[129] RSV causes viral bronchiolitis, which if severe requires hospitalization; prior to 1995 1–2% of infants were affected,[130] rising to 2–3% between 1997 and 2000,[131] making bronchiolitis the most common cause of intensive pediatric care.[130] The predominant pathology is bronchiolar inflammation, which leads to mucus build-up and narrowing of the airways; the clinical manifestations can vary from cold-like symptoms to respiratory failure.[132] The mortality rate in industrialized countries is 0.5–1% of hospitalized patients.[133,134] In addition to the inflammation, surfactant abnormalities occur that may further increase the severity of the disease.

Tracheal aspirates of infants with viral bronchiolitis have low concentrations of PC and SP-A, but not SP-B,[135,136] and impaired surface activity.[136] BAL from RSV-infected patients shows reduced SP-A, -B, and -D concentrations,[137] but this did not correlate with disease severity. Patients with bronchiolitis also have a lower concentration of dipalmitoyl-PC (DPPC), but normal levels of PG.[138] However, others have reported a lack of PG in some infants, which correlates with impaired surface activity.[139] Moreover, in patients with severe RSV infection, the pattern of expression of specific genetic variants of SP-A suggested that there may be a genetic association between SP-A gene locus and severe RSV infection, as certain alleles dominate in RSV-infected infants, and others are greatly reduced.[132] Furthermore, a specific polymorphism of the SP-D gene is also associated with the risk of severe RSV bronchiolitis in susceptible infants.[140]

Strategies for prevention and treatment of severe RSV infections are elusive,[129] with the standard treatment of acute bronchiolitis including mechanical ventilation, supplemental oxygen, and bronchodilators.[141] Efforts directed at the development of an RSV vaccine have proven unsuccessful, and prevention strategies are limited to passive immunization of high risk patients.[129] As bronchodilators and glucocorticoid treatment have not been very successful,[141] exogenous surfactant supplementation has been trialled.[138] This arrests the deterioration of the phospholipid composition observed in the non-treated group and improves lung compliance. However, there are no changes in most indices of gas exchange.[138] Hence, exogenous surfactant treatment may be a useful therapeutic strategy in infants with severe bronchiolitis. Simplifying the treatment to either the *in vitro* or *in vivo* administration of unsaturated PG phospholipids shows a reduction in viral load and immune response.[142] A recent study pretreating sheep with recombinant human VEGF prior to RSV infection showed therapeutic potential as it decreased RSV viral load, decreased pulmonary lesion severity, and altered both epithelial innate immune responses and epithelial cell proliferation.[143]

Conclusion

There is growing awareness that a suboptimal intrauterine environment has significant long-term consequences for respiratory physiology that persist into adult life. Lung and surfactant function are profoundly affected in the infant by intrauterine influences such as IUGR, maternal nutrition, maternal drug use, and intrauterine infections. A substantial

research effort is aimed at understanding the mechanisms by which these multifactorial conditions affect surfactant composition and function. In contrast, the research effort examining the effects of alcohol, smoking, and other drugs on the surfactant system is not extensive, and the results are not conclusive. Nevertheless, it appears that the changes observed are not significantly detrimental and may not persist beyond the early postnatal period. Postnatal respiratory infections can lead to surfactant dysfunction, which represents a major complication in respiratory illnesses of neonates. Furthermore, genetic variants of the innate immunity-related SP appear to modulate the susceptibility of infants to infection. Hence, the search for genetic markers and the development of suitable therapies, including exogenous surfactant replacement, are major targets for research effort.

EFFECTS OF ENVIRONMENTAL FACTORS ON THE ADULT PULMONARY SURFACTANT SYSTEM

The function of the pulmonary surfactant system is reliant on the physical interactions of the lipids and proteins. Hence, any environmental factor that is capable of altering or influencing these interactions has the potential to impact on the immediate function of the surfactant system and hence the lung.[144]

Effects of Changing Oxygen on the Mature Surfactant System

The adult pulmonary surfactant system is sensitive to changes in oxygenation, whether these are imposed pathologically (e.g., hypoxia or hyperoxia), environmentally (e.g., altitude), or induced clinically (e.g., hyperoxia). Changes in oxygen levels may influence the lipids and proteins directly, for example through direct oxidation, or indirectly by altering the inflammatory status of the lung or by altering the processes regulating synthesis and secretion. Each of these effects is capable of altering surfactant composition and function.

Experimental Hypoxia

In rats, acute hypoxia changes the surfactant lipids,[145] including a decrease in total phospholipids,[146] PC,[147,148] and an increase in lyso-PC[148] and lyso-compounds[147] with a concomitant increase in phospholipase activity.[147] However, chronic hypoxia (5 h per day for 21 days) causes a decrease in phospholipid content of both lung tissue and lavage, but an increase in the PC fraction of lung lavage.[146] Hypobaric hypoxia also decreases the SP-A and -D content of surfactant.[149] Surface activity can decrease in response to hypoxia,[145,148,150,151] and edema of the interalveolar septa can occur.[148] Although PC levels decrease in lavage surfactant, the incorporation of ^{32}P into lung PC and PE increases,[152] suggesting that the reduced alveolar content may be due to changes in secretion and/or local inactivation, and not decreased synthesis.

Recent studies, focused on hypoxia, have addressed mechanisms whereby hypoxia may alter alveolar epithelial cell function. Neither immediate nor prolonged hypoxia changes intracellular calcium ions (iCa^{2+}) significantly. However, the release of calcium from iCa^{2+} stores upon stimulation with ATP is reduced under hypoxic conditions. These results indicate that a hypoxia-mediated reduction in transiently stimulated Ca^{2+} may lead to a blunted response to extracellular stimuli.[153] This may have implications for surfactant secretion, which is mediated by iCa^{2+} stores (Chapter 9 [Orgeig et al.]).

An adult mouse model of intermittent hypoxia showed that intermittent hypoxia induces a 60% increase in cellular proliferation, and a tripling in the number of proliferating AECII. By upregulating pathways of cellular movement, growth and development, intermittent hypoxia preferentially increases alveolar surface area by stimulating lung growth.[154] Whether this protocol leads to changes in surfactant content has not been addressed.

In a bleomycin model of acute lung injury, the accompanying acute tissue hypoxia at day 7 correlates with an increase in lung tissue SP-D mRNA and protein expression. However, persistent hypoxia (day 28) correlates with a return of SP-D expression to normal.[155] These results were replicated *in vitro* using a cell culture system. Acute hypoxia for 24 h does not induce SP-C mRNA, but does induce SP-D mRNA. In contrast, persistent hypoxia represses SP-D expression. The hypoxic response is mediated by the transcription factor hypoxia-inducible factor (HIF)-1α.[155]

High Altitude Hypoxia

An interesting area of research is the response of the surfactant system to chronic hypoxia caused by high altitude (>3500m). As with intermittent hypoxia,[154] rats maintained in a hypobaric chamber show an increase in lung volume and alveolar surface area,[156] as do guinea pigs raised from birth at high altitude.[157]

Rats raised at moderate altitude show a reduction in total phospholipids,[158,159] especially PC[159] with no change in the stability index or minimum surface tension.[158] Furthermore, the lavage demonstrates fewer macrophages.[159]

Immunohistochemical staining of lung tissue at autopsy from patients who died of high altitude pulmonary edema (HAPE) at moderate altitude demonstrates morphological perturbations to the AECII and the alveolar surfactant material.[160] The occurrence of HAPE is difficult to predict, as different individuals demonstrate different levels of susceptibility, suggesting that there may be a genetic component determining susceptibility.[161] Among other genes, SP-A may contribute to the occurrence of HAPE, as specific alleles of SP-A1 and SP-A2 have been associated with susceptibility to HAPE.[162]

Preliminary studies on acclimation of the pulmonary surfactant system to high altitude in mammals such as adult deer mice, *Peromyscus maniculatus*, show differential increases in the surfactant phospholipids that increase membrane fluidity in animals living at high altitude.[163] Moreover, the total SP content is 43% higher in animals acclimated to high altitude, but it is not yet known which SP are affected.[164]

Llamas born at high altitude demonstrate numerous prominent Clara cells with large "apical caps," many of which were extruded into the terminal bronchioles.[165] Quantitative histological analysis shows greater Clara cell activity in these animals compared with those living at sea level.[166] Although the extruded material was not analyzed, it is possible that the hyperactivity of the Clara cells was an adaptive response to chronic hypoxia.[165] Hence, acclimation to high altitude may be associated with specific adaptations to the pulmonary surfactant system.

Experimental Hyperoxia

Studying the effect of hyperoxia on the surfactant system is relevant from a pathological and therapeutic perspective. For example, a potential pathological mechanism operating in patients with acute lung injury (ALI), and its more severe form, the acute respiratory distress syndrome (ARDS), is the direct oxidation of endogenous surfactant components.[167] Oxidative stress arises within injured lungs due to the generation of reactive oxygen species (ROS) from inflammatory cells. However, given the difficulty of ARDS patients to maintain oxygenation, treatment involves mechanical ventilation using high O_2 concentrations,[168] which may further exacerbate oxidative damage.[167]

To investigate the mechanisms operating to impair surfactant and lung function in hyperoxic lung injury, there have been numerous studies exposing laboratory animals to hyperoxia. These have shown variable results,[169,170] with some studies showing a decrease in BAL surfactant levels[171–174] and others showing an increase in total surfactant phospholipids.[175,176] These differences are likely species, dose, and duration dependent.[169] Moreover, the time-course of exposure, as well as the timing of measurements during recovery, will likely influence the results.[169] Nevertheless, some common observations include changes in surfactant composition such as a decrease in disaturated phospholipids and/or a decrease in the PG/PI ratio,[169,177] suggesting impairment of surfactant in response to hyperoxia.[169]

SP gene and protein expression are also affected by hyperoxia, but the effect is variable. For example, 95% O_2 exposure in rats for 12, 36, or 60 h causes an initial reduction in all SP mRNAs at 12 h followed by a rise which exceeds control by 60 h.[178] There is also a reduction of different SP in the tissue or in the alveolar compartment, suggesting differential regulation of the different genes.[178] Similarly, in adult hamsters exposed to 100% O_2 for up to 8 days, SP-A levels increase transiently, before decreasing below control levels, while SP-B and -C decrease at different rates.[176] However, other studies in rats using >85 or >90% O_2 and exposure periods of 1–7 days, show increases in expression of SP proteins and/or mRNA.[179–182]

The mechanisms by which hyperoxia leads to surfactant impairment may involve the more rapid conversion of large to small aggregates,[183] increased plasma protein content[167,175] or the presence of inflammatory cells[167,184] in the lavage. Furthermore, there is evidence of limited lipid peroxidation within the lavage,[167,183] an increased oxidation of PG,[167] an increased concentration of arachidonic acid containing phospholipids at the expense of saturated palmitoyl-myristoyl-PC and dipalmitoyl-PC,[184] and a concomitant modest decrease in ascorbate levels in lung tissue.[167] However, antioxidant levels in the alveolar compartment appear to be maintained and regulated.[167] Hence, while there is some disruption to the balance of oxidant/antioxidant components in the lung, and the lipids sustain minor oxidative damage, this does not likely represent a major mechanism for the lung dysfunction observed.[167,183,184] Hence, hyperoxic treatment at non-lethal concentrations leads to an altered homeostasis and metabolism of surfactant phospholipid molecular species, but the impairment of surfactant function appears to be due to direct inhibition rather than principal changes in activity.[184]

This hypothesis is supported by the finding that administration of an artificial surfactant following hyperoxic lung injury in a primate model ameliorates lung injury by increasing the disaturated PC content of lavage, improving lung function as determined by improved oxygenation,[185] and protecting against epithelial and endothelial cell destruction.[186] Moreover, a conditional SP-D transgenic mouse model shows that targeted pulmonary expression of SP-D protects against hyperoxic lung injury by mechanisms that include restriction of alveolar protein leak and preservation of surfactant homeostasis.[187]

Effects of Environmental Pollutants on the Mature Pulmonary Surfactant System

Air pollution is an increasingly serious health issue facing industrialized cities. It is defined as contamination of the atmosphere by gaseous, liquid, and/or particulate waste (or its by-products) that can cause harm or discomfort to humans or other living organisms, and/or cause damage to the environment.[188] Gaseous contaminants include oxides of nitrogen, carbon, and sulphur, as well as volatile organic compounds and ozone. Particulate matter refers to a complex mixture of solid and liquid small particles containing acids (nitrates, sulphates, etc.), organic chemicals, metals, and/or soil or dust particles that can cause serious respiratory and cardiovascular damage.[188] The lung is the only internal organ exposed to the atmosphere, and with its huge surface area, is a prime target for the toxic effects of air pollutants. Not unexpectedly, there is a correlation between the levels of air pollution and mortality due to aggravation of pre-existing respiratory conditions, or hospital admissions with respiratory or respiratory-related complaints.[189–193] Within the lung, the pulmonary surfactant system is a crucial, susceptible target for air pollutants.

Properties and Toxic Effects of Pollutants

The extent to which a pollutant can infiltrate the lung is influenced by the size, solubility, and reactivity of the compounds, as well as the ventilatory pattern.[194] Smaller particles will remain suspended in the inspired air for longer and hence travel further within the lung.[195] Gases with a high solubility are readily removed from the air and absorbed by the first tissue with which they come into contact. Therefore, most uptake of these gases occurs in the upper airways.[196,197] In these regions the toxic effects are minimized as the fluid lining is relatively thick with high levels of antioxidants and the pollutant does not reach the epithelium. Furthermore, most non-reactive particulate matter deposited in the upper airways is rapidly removed (<24 h) by the mucociliary escalator.[195] Insoluble gases are more uniformly distributed along the respiratory system, as their uptake is mainly dependent on reactivity with lipids or solubility in water.[196,197] Smaller particles that reach the gas exchange regions can be phagocytosed by macrophages and transported through the epithelium and deposited into lymphatic vessels, but this process takes significantly longer.[195] The toxic effect of a pollutant is also dependent on concentration, duration of exposure, and penetration and retention within the respiratory system. Furthermore, the reactivity of a pollutant will influence its ability to cause biochemical damage to the lipids and proteins with which it comes into contact. Generally, pollutants with low solubility, low reactivity, and small diameter will travel furthest into the respiratory tract.[195,196]

Numerous studies have investigated the effects of individual pollutants, particularly gaseous pollutants such as NO_2, SO_2, and ozone on lung and surfactant function.[170,198] Here we provide a brief review of recent studies examining the effects on the pulmonary surfactant system of the most important pollutants; these include ozone and particulate pollutants including diesel exhaust, cigarette smoke, and nanoparticles.

Ozone

Individuals suffering from respiratory diseases such as asthma and chronic obstructive pulmonary disease

(COPD), children, and the elderly are at greatest risk of adverse health effects due to ozone.[199] The first site of inhaled ozone reactivity is components of the air–water interface, and the absorption of ozone is dependent on these biochemical reactions.[200,201] Specifically these reactions appear to involve the unsaturated phospholipids such as palmitoyl-oleoyl-PC (POPC), as opposed to the disaturated species such as DPPC.[202,203] This is supported by the finding that *in vivo* exposure to ozone leads to a decrease in unsaturated phospholipid species relative to the major saturated components[204] and an accumulation in the BAL fluid of aldehyde reaction products between POPC and ozone.[205] *In vitro* studies of phospholipid monolayers yield similar results.[202,203] Furthermore, the reaction of ozone with an unsaturated phospholipid film is accompanied by a very rapid increase in surface pressure (i.e., a rapid decrease in surface tension),[206] followed by a relatively slow decrease in surface pressure accompanied by a loss of material from the air–water interface.[206] Experiments performed on mixed films of DPPC and POPC in a Langmuir trough reveal that the damage and loss of material is attributed solely to the unsaturated and not the saturated phospholipids, hence dramatically changing the organization of material at the interface and hence the physical properties of the film.[207]

Ozone exposure also has significant damaging effects on SP. *In vitro* exposure of SP-A to ozone causes oxidation damage to the methionine and tryptophan residues, which diminishes its aggregation and binding properties,[208] and compromises its ability to stimulate phagocytosis by macrophages[209,210] and cytokine production.[211] This may explain the increased risk of infections in areas of high air pollution.[212]

On the other hand, short-term *in vivo* exposure to ozone appears to enhance the ability of SP-A to stimulate alveolar macrophages.[213] It is possible that the inflammation that accompanies *in vivo* ozone exposure may result in a change in the structure and thus functional role of SP-A in modulating macrophage activity.[213] Furthermore, a deficiency or inhibition of SP-A renders mice more susceptible to the detrimental effects of ozone.[210,214] Similarly, decreased levels of functional SP-D correlate with increased inflammation after ozone inhalation in mice, and enhanced SP-D production coincides with enhanced resolution of ozone-induced inflammation.[199,215] Under homeostatic conditions, SP-D functions as an anti-inflammatory protein, suppressing NF-κB-mediated transcription of macrophage pro-inflammatory genes.[216] Interestingly, following induction of oxidative stress, increased production of nitric oxide results in *S*-nitrosylation of critical cysteines in SP-D,[217] leading to a change in its activity to a pro-inflammatory mediator.[218] Hence, the SPs play a significant protective and regulatory role during ozone-induced, and likely more general, oxidation-induced inflammation.

Although the level of SP-A mRNA is unaffected by either acute or repeated ozone exposure *in vivo*,[213] it is possible that environmental insults may affect SP-A expression via epigenetic mechanisms, in a similar manner to other genes.[188] Epigenetic mechanisms may affect SP-A expression by altering transcription or translation efficiency, thereby potentially affecting the relative content of the two SP-A splice variants, SP-A1 and SP-A2, and resulting in altered immune and inflammatory responses.[188] The possibility that pollutants may differentially affect, via epigenetic changes or other mechanisms, the function and/or regulation of different variants of innate immunity, may hold the key to understanding differential disease susceptibility to air pollution.[188]

Particulate Pollutants and Nanoparticles

Coarse and fine particulate air pollutants are classified as PM10 and PM2.5 (i.e., < 10 or < 2.5 μm in diameter, respectively).[219] Coarse particulate matter is generally deposited in the large airways and hence is rapidly cleared by the mucociliary escalator.[195] If the mucociliary escalator is impaired or the larger airways are overloaded during chronic exposure, particles may enter the alveolar zone; this may occur with chronic deposition of specific mineral particles as in silicosis and asbestosis.[198] Fine particulate matter, on the other hand, has the potential to travel into the alveolar zone where it is cleared much more slowly.[195] Hence, these particles have been associated with increased cardiopulmonary and lung cancer mortality.[190] Due to the greater size of these fine particles, relative to gaseous pollutants, damage to the alveolar epithelium occurs predominantly in the most proximal alveolar duct bifurcations[220] and is characterized by an inflammatory response[221,222] as well as hyperplasia and hypertrophy of the AECII.[223,224]

With the advent of nanotechnology, in which materials are used at the nanometer (nm) scale, and the incorporation of nanoparticles into a wide variety of products, we are increasingly exposed to these potentially toxic particles.[222] Nanoparticles (also known as ultrafine particles[225]) are defined as having one structural dimension of less than 100 nm, making them comparable in size to subcellular structures, including cell organelles or biological macromolecules,[222] thereby enabling their ready incorporation into biological systems. The potential for nanomaterials to induce toxicity has spawned a new field of research: nanotoxicology, that is concerned with evaluating and controlling the toxicity of nanoparticles, associated with their production and use, so that their beneficial properties can be exploited safely.[222] Few epidemiological studies have examined the risk of exposure to nanoparticles.[226] However, a study of workers exposed to a wide range of nanoparticles (20–100 nm) showed increased depression of antioxidant enzymes and increased expression of cardiovascular biomarkers of exposure.[227] Most of the risk information comes from studies of exposure to a small set of manufactured particles, for example, no health effects have been reported for carbon black and titanium, while the risks are controversial for silica dioxide and alumina.[228–231] Many newly engineered nanoparticles are considered negligible exposure risks, while natural nanoparticle exposures are considered substantial health risks.[232–234] However, engineered nanomaterials with new chemical and physical properties are being produced constantly for which the toxicity is unknown.[234]

Diesel Exhaust and Cigarette Smoke

Diesel exhaust and cigarette smoke contain many different components, predominantly fine particles, making it difficult to assign toxic effects to any particular component.[198] Diesel exhaust exposure is associated with an elevated lavage phospholipid content, which is enriched in PC, particularly palmitate-containing species.[235] A comparison of several types of particles, i.e., diesel, silica, and carbon black,[221] demonstrates that diesel exhaust causes only minimal changes in surfactant amount and composition, and a small and transient increase in lung permeability, lung inflammation, and cellular damage.[221] In comparison, the slightly larger carbon black particles cause no detectable changes, whereas the much smaller, but highly reactive silicon dioxide particles cause increases in lung permeability and inflammation, increases in the amount of pulmonary surfactant in the alveoli and damage to the alveolar and bronchiolar epithelium. Hence, surface chemistry appears a more potent stimulus for lung damage than particle size.[221] Moreover, ultrafine washed diesel particles loaded into alveolar macrophages enhance the phagocytic cell-induced lipid peroxidation of extracellular lung surfactant to a greater extent than ultrafine carbon particles. The greater oxidative metabolism of diesel particles is likely due to their higher metal content relative to that of the carbon particles.[236] This may underlie the observed increased frequency of pulmonary inflammatory conditions among populations exposed to polluted air.[236]

In contrast to the effect of diesel exhaust, the inhalation of cigarette smoke is associated in some studies with a reduction in total phospholipids in BAL,[237–239] perhaps due to a smoke-induced decrease in secretion from AECII.[240] Other studies demonstrate little change in the total phospholipid content, but instead show alterations in the relative proportions of PC, sphingomyelin, or PE,[239] the relative fractions of saturated versus unsaturated phospholipids,[241] or the relative amounts of SP-A or -B.[242] Early studies correlating these changes to surfactant function demonstrated changes to the normal expansion and compression isotherms of bovine surfactant[243] as well as the compressibility and respreading of lung surfactant extracted from BAL of smoke-exposed rats.[241] More recently studies have

been focused on determining the effects of environmental tobacco smoke (ETS), which is a mixture of sidestream smoke and exhaled mainstream smoke.[244] Using a highly reproducible and environmentally relevant ETS mixture, a series of studies has examined the effect of ETS on the surface properties of defined limited component artificial surfactant phospholipid mixtures[245] or clinical artificial surfactant preparations.[246]

Collectively, these studies have shown that ETS exposure alters the chemical, mechanical, and morphological properties at the air–liquid interface of these various artificial surfactant mixtures.[245,246] For example, the distribution and fraction of liquid-expanded and condensed domains are altered, as are the adsorption to, spreading at, and desorption from, the interface that impacts on the ability of the mixture to reduce surface tension.[245,246] These effects are likely due to oxidation of SP-B and deacylation of SP-C observed *in vitro*.[246] *In vivo*, the BAL of chronic smokers[247] and that of rats following exposure to tobacco smoke[242] demonstrates reduced levels of SP-A. Surfactant from smokers also demonstrates a decrease in SP-D.[247] Although the reason for this reduction was not investigated, the fact that neither SP-A nor SP-B mRNA levels are affected in rat lung tissue after tobacco exposure suggests that the mechanism involves either an inhibitory effect on the secretory process or localized destruction of SP in alveoli.[242] Collectively, the changes in surfactant function observed in surfactant mixtures are consistent with the observation that the surface activity of lavage fluid from chronic smokers is significantly reduced[248,249] and suggests that exposure to ETS would lead to increased work of breathing[246] and hence dyspnea[250] in smokers.

Whether the *in vivo* function of surfactant is affected by tobacco smoke is unknown. Epidemiologically, however, there is evidence of a dose-response effect between smoking and the risk for developing ARDS.[251] Moreover, an evaluation of the efficacy of different treatments for acute respiratory failure following severe smoke inhalation in rats shows that the mortality rate after 24 h markedly declines in the group that was lavaged, treated with exogenous surfactant, and ventilated mechanically. The improved survival is accompanied by improved surface activity of BAL, static lung compliance, and oxygenation.[252] These findings, therefore, indicate a localized destruction of surfactant components, which when removed and replaced, restore lung function.

Nano or Ultrafine Particles

Nanoparticles inhaled into the lung (particles, molecules, or ions) do not simply remain in the lung, but are able to cross cellular barriers and translocate into the circulatory system.[253] Interestingly, particle uptake *in vitro* into cells does not occur by any of the expected endocytic processes, but rather by diffusion or adhesive interactions.[225] Moreover, particle uptake occurs predominantly by alveolar type I cells, is rapid, and is dependent on charge rather than size.[254] Furthermore, intracellular particles are not membrane-bound and hence have direct access to intracellular proteins, organelles, and DNA, which may greatly enhance their toxic potential.[225] Surfactant plays a crucial role in facilitating this uptake, by wetting particles thereby displacing them into the liquid lining (i.e., the hypophase[255,256]) and bringing them into closer proximity to the epithelial lining. Furthermore, the SPs interact and conjugate with a wide range of nanoparticles and play an essential role in particle translocation across cellular membranes and organ barriers.[257,258] Hence, within hours of respiratory exposure, particles may appear in the liver, heart, and central nervous system.[225,259–261] Here their effects may be severe, with an increase in plasma viscosity[262] and an increased susceptibility of cardiac tissue to ischemia/reperfusion injury.[263]

While macrophages clear a proportion of the ultrafine particle burden, the uptake of titanium oxide (20 nm) nanoparticles, for example, appears to be sporadic and unspecific.[264] The rate of uptake of nanoparticles by alveolar macrophages and lung dendritic cells depends on the presence of SP-A and -D. For example, treating nanoparticles with SP-A results in a higher uptake by macrophages[265] and uptake is decreased in SP-D deficient mice.[266] The loading of alveolar macrophages, particularly with ultrafine carbon, but to a lesser extent washed diesel particles, impairs the ability of the phagocytic cells to kill bacteria. This may represent a mechanism for the observed increased frequency of lung infections among populations exposed to polluted air.[236]

Using a human AECII cell line (A549) as well as primary normal human bronchial cells, it was shown that exposure to single-walled carbon nanotubes (SWCNT) leads to the suppression of a variety of inflammatory mediators including IL-8, IL-6, and macrophage chemo-attractant protein-1 (MCP-1) *in vitro*. This immunosuppression is exacerbated following dispersion of the particles in DPPC, potentially leading to enhanced particle toxicity.[267] Suppression of the immunological response is likely to have negative consequences as the immune system may be less reactive towards infections and less responsive in cases of oxidative stress triggered by the particles. An appropriate immune response and signaling by lung epithelial cells is critical in order to attract phagocytic cells to clear the invading particles.[267] It is likely that particle surface chemistry is altered by the coating with DPPC that may modify or increase particle uptake by cells, potentially exposing different cellular compartments/targets to the particles, thereby altering the toxic response.[267] It is therefore possible that the surfactant lining in the lungs is actually a vehicle for enhancing the cytotoxicity of nanoparticles.

Gold nanoparticles at low concentrations in a semisynthetic surfactant phospholipid–SP-B mixture impair surfactant phospholipid adsorption *in vitro*. Furthermore, the nanoparticles interfere with the ability of the surfactant films to achieve low surface tension both during film compression and expansion.[268] These effects appear to be mediated by specific and different interactions between the nanoparticles and different phospholipid and protein components of surfactant.[268] Similarly, nanosized (5 nm), but not microsized (900–1600 nm) particles of titanium oxide impair surfactant function *in vitro*.[269] The nanosized particles also induce ultrastructural aberrations in lamellar body-like structures, including deformation, a decrease in size and the presence of unilamellar vesicles. Furthermore, particle aggregates accumulate between lamellae of the lamellar body-like forms.[269]

Conclusion

All forms of pollutants have the capacity to alter surfactant lipid or protein composition or function, leading to impairments in surface activity and potentially lung function, particularly with long-term exposures. While air pollution, particularly in Europe, is declining, the increased industrialization in highly populated Asian countries is causing worldwide air pollution to worsen.[192] However, a relatively new and rising threat across the world is the increasing use of nanomaterials in an ever expanding range of industries.[222] This will require vigilance and an increased research effort to control exposures and to understand the physiological consequences.

NATURAL AGING EFFECTS ON THE PULMONARY SURFACTANT SYSTEM

In the 10 years since the first edition of this book was published,[170] very little research has been undertaken on the effects of aging on the pulmonary surfactant system. This may be due to the suggestion that age-related changes, if any, in the surfactant system are modest[270,271] compared with changes in lung structure.[272–275] For example, elastin fibers are disrupted and lost with age, and cross-linking of collagen and elastin is altered; alveoli become wider and shallower with flattened inner surfaces and alveolar ducts are dilated.[270] Furthermore, no effective animal model has emerged in which to study the effect of aging on the surfactant system. The results of the few studies that have been performed are inconclusive and show significant species differences.[276]

For example, in Beagle dogs, lung surfactant from middle-aged (3–7 years old) or old (≥ 12 years old) versus young (3–7 months old) dogs has a higher proportion of PC and lower proportion of PS and sphingomyelin.[277] These compositional changes are accompanied by a modest improvement in surface activity in the older dogs, although the sample volumes were small.[277] In an early study in rats, there was no difference in the percentage of saturated PC in lavage fluid of young and old animals.[278] More recently, however, saturated PC content was shown to decrease between 1 day and 3 months of age and decrease further between 3 and 29 months.[279] While the change in PC during the initial time period likely reflects alveolar development that continues in rats until 2–3 weeks after birth,[280] the change between 3 and 29 months does reflect changes in surfactant composition during adulthood from young to aged rats. Similarly, in horses (aged between 6 and 25 years), increasing age is associated with decreased phospholipid content, but phospholipid composition and surface activity are unchanged.[276] In contrast, human lung tissue obtained at autopsy from individuals aged between 13 months and 80 years, and who had died of non-respiratory causes, does not show any differences in the amount of saturated PC, relative to either body weight, lung protein, or lung DNA.[271] Similarly, the percentage of saturated PC in human lungs remains relatively constant (about 35%) throughout adult life[278]

Pulmonary SP-A content is similar in young and aged rats[279] as well as in lavage fluid from young and old humans.[271] However, healthy non-smoking subjects demonstrate a significant decrease in SP-A with age in bronchial lavage fluid, but not in BAL fluid; SP-D does not change with age.[281] BAL fluid of healthy subjects demonstrates a gradual decrease in the ratio of the SP-A1 gene variant to the total SP-A content with increasing age from 0–20, 21–40, 41–60, and 61–80 years. This change appears to be related to a modest increase in total SP-A content with age, although the variation in total SP-A content among age groups was not statistically significant.[282] Presumably there is a concomitant increase in the SP-A2 variant, but this was not directly measured. SP-A1 has been shown *in vitro* to be less active in host defense activities compared with SP-A2,[211,282] but whether and how the decreased SPA1/SP-A ratio in aged individuals contributes to age-related changes in host defense[283] remains to be determined.[282] SP were not measured in dogs[277] or horses,[276] nor to our knowledge have any of the other SP been measured in humans or any animal model of aging.

There is some evidence that the processing and perhaps secretion of surfactant may be affected by aging. For example, in Macaque monkeys pulmonary surfactant production and turnover may be altered or impaired with aging, as both the number of lamellar bodies per AECII and the volume density of lamellar bodies per cytoplasmic volume decrease with age between 1 month and 31 years (lifespan, ~35 years).[284] Furthermore, the alveolar lavage of aged rats contains a higher proportion of functional large aggregates as opposed to non-functional small aggregates than that of the lavage from young or newborn rats.[279] There is also an age-related decrease in the rate of conversion from large to small aggregates *in vitro*. These results suggest that there could

be higher surface activity in older animals, but this was not measured.[279] It is possible that the decreased conversion between forms may contribute to the maintenance of functional surfactant pool sizes in the lungs of aged rats.[279]

However, in senescent (26 months old) compared with young (2–3 months old) rats, the alveolar epithelial lining layer and the lamellar bodies of AECII display profound degenerative alterations.[285] No regularly shaped tubular myelin figures, which are characteristic of functional large aggregate surfactant, were discernible in aged animals.[285]

The functional significance of age-related changes in the surfactant system is not understood. Some authors suggest that a lower surfactant concentration in older animals may increase their susceptibility to disease-induced alterations in surfactant.[276] Hence, while low surfactant phospholipid content does not lead to clinical disease, it may have subclinical effects on performance and may increase respiratory compromise caused by respiratory diseases such as recurrent airway obstruction.[276] Others speculate that the changes in surfactant lipid composition may ameliorate the negative aging effects on lung architecture and function enabling efficient lung function in old age.[277]

While it is not possible at present to draw firm conclusions regarding the effect of aging on the surfactant system, there are other consequences of aging that may impact on surfactant composition, function, or regulation. These include the autonomic nervous system (cholinergic and adrenergic receptor number and function),[286–289] the immunological status of the lung (numbers of alveolar macrophages, leukocytes, and neutrophils, extent of phagocytosis, concentration of inflammatory mediators etc.),[290–293] the redox and antioxidant status of the lung,[294,295] and alveolo-capillary clearance.[296,297]

Finally, it is important to note that studies on aging between different mammalian species may not be readily comparable because of differences in maximum ages, e.g., rats are senescent at 24 months, dogs at 12 years, Macaques at 35 years, and humans at 80 years. Despite the development of specific senescent animal models such as the senescence accelerated mouse (SAM), which includes both "senescence-prone strains" and "senescence-resistant strains,"[298] and which develops characteristics of the "senile lung,"[299] no studies with this model appear to have specifically addressed changes in the pulmonary surfactant system.

Interactive Effects on the Pulmonary Surfactant System of Pollutants in the Aging Lung

The aging pulmonary system (>65 years) is at increased risk of adverse health effects from environmental insults, such as by air pollutants (ozone, cigarette smoke, particulates), infections (influenza, pneumonia), and climate change (e.g., synergistic effects of heat and pollution).[295] While the molecular mechanisms are not well understood, the physiological contributors are age-related changes in the immune and endocrine systems and the antioxidant status.[295] However, investigations of the impact of environmental insults on the composition or function of the surfactant system in the aging lung are largely lacking.

In mice, the long-term (9 months) effects of cigarette smoke inhalation in young (starting at 2 months old) and old (starting at 8–10 months old) animals reveal an interaction between smoke inhalation and aging. Several characteristics prominent only in the smoke-exposed old animals include a reduction of alveolar space with a concomitant increase in lung cellularity and thickened alveolar septa, an accumulation of intra-alveolar surfactant and decreased pulmonary function. These abnormalities are restrictive in nature and conform most closely to pulmonary fibrosis.[300] Furthermore, the lungs of the senescence-prone mouse are more susceptible to damage by tobacco smoke than the senescence-resistant mouse.[301] Adverse effects include increased lung permeability with increased albumin content in the lavage, increased inflammation, decreased antioxidant level, and changes to epithelial structure including hyaline membrane formation.[301] These changes are likely to lead to compromised surfactant function, but this was not tested.

The possible role of progressive chronic inflammation in the increased sensitivity to ozone was assessed in young (8 weeks), middle-aged (27 weeks), and elderly (80 weeks) wild type (WT) and SP-D deficient (Sftpd-/-) mice. While increasing age in WT mice is not associated with increased numbers of inflammatory cells in the lungs and does not significantly alter their ability to repair ozone-induced injury, aged Sftpd-/- mice show progressive pulmonary inflammation characterized by a prominent macrophage infiltrate.[218] These changes correlate with an increased sensitivity to ozone, but only up to 27 weeks of age. By 80 weeks of age, it appears that baseline lung inflammation and injury have reached maximal levels, such that ozone cannot cause further damage.[218] This study suggests an important role for SP-D in protecting the aging lung from the consequences of chronic inflammation.

This is important, because although SP-D in human lung lavage does not change with age alone, it does decrease due to long-term smoking.[281] Furthermore, SP-A, which also has important host defence functions, decreases with age alone in healthy humans, as well as with smoking and emphysema.[281]

CONCLUSION

The central role played by the pulmonary surfactant system in lung function is revealed by the effects of a diverse range of environmental factors affecting the system throughout fetal, neonatal, and adult life and in the aging lung. Abnormalities of the surfactant system involve both the lipid and the protein components of surfactant, both of which result in perturbations in surface activity and hence lung function. These perturbations include physical factors such as reduced lung compliance, increased alveolar permeability, which leads to pulmonary edema and surfactant inactivation, as well as changes in the immune and redox status of the lung, which lead to inflammatory and oxidative damage of the surfactant lipids and proteins.

ACKNOWLEDGMENTS

This work was supported by the Australian Research Council, The National Health and Medical Research Council, and The National Heart Foundation of Australia.

REFERENCES

1. Janssens JP, Pache JC, Nicod LP. Physiological changes in respiratory function associated with ageing. *Eur Respir J* 1999;**13**:197–205.
2. Minior VK, Divon MY. Fetal growth restriction at term: myth or reality? *Obstet Gynecol* 1998;**92**:57–60.
3. Tyson JE, Kennedy K, Broyles S, Rosenfeld CR. The small-for-gestational-age infant - accelerated or delayed pulmonary maturation - increased or decreased survival. *Pediatrics* 1995;**95**:534–8.
4. Cock ML, Camm EJ, Louey S, Joyce BJ, Harding R. Postnatal outcomes in term and preterm lambs following fetal growth restriction. *Clin Exp Pharmacol Physiol* 2001;**28**:931–7.
5. Joyce BJ, Louey S, Davey MG, Cock ML, Hooper SB, Harding R. Compromised respiratory function in postnatal lambs after placental insufficiency and intrauterine growth restriction. *Pediatr Res* 2001;**50**:641–9.
6. Lipsett J, Tamblyn M, Madigan K, Roberts P, Cool JC, Runciman SIC, et al. Restricted fetal growth and lung development: a morphometric analysis of pulmonary structure. *Pediatr Pulmonol* 2006;**41**: 1138–45.
7. Maritz GS, Cock ML, Louey S, Joyce BJ, Albuquerque CA, Harding R. Effects of fetal growth restriction on lung development before and after birth: a morphometric analysis. *Pediatr Pulmonol* 2001;**32**:201–10.
8. Li Z, Zeki R, Hilder L, Sullivan EA. Australia's mothers and babies 2010, in Perinatal Statistics Series. *AIHW Natl Perinatal Epidemiol Stat Unit: Canberra* 2012:1–122.
9. Challis JR, Sloboda D, Matthews SG, Holloway A, Alfaidy N, Patel FA, et al. The fetal placental hypothalamic-pituitary-adrenal (HPA) axis, parturition and post natal health. *Mol Cell Endocrinol* 2001;**185**:135–44.
10. McMillen IC, Robinson JS. Developmental origins of the metabolic syndrome: Prediction, plasticity, and programming. *Physiol Rev* 2005;**85**:571–633.
11. Morrison JL. Sheep models of intrauterine growth restriction: fetal adaptations and consequences. *Clin Exp Pharmacol Physiol* 2008;**35**:730–43.
12. Morrison JL, Orgeig S. Antenatal glucocorticoid treatment of the growth-restricted fetus: Benefit or cost? *Reprod Sci* 2009;**16**:527–38.
13. Schroder HJ. Models of fetal growth restriction. *Eur J Obstet Gynecol Reprod Biol* 2003;**110**:S29–39.

14. Gilbert WM, Danielsen B. Pregnancy outcomes associated with intrauterine growth restriction. *Am J Obstet Gynecol* 2003;**188**:1599–601. discussion 1596–1590.

15. Hodges RJ, Wallace EM. Mending a growth-restricted fetal heart: should we use glucocorticoids? *J Matern Fetal Neonatal Med* 2012;**25**:2149–53.

16. Nicolaides KH, Economides DL, Soothill PW. Blood gases, pH, and lactate in appropriate- and small-for-gestational-age fetuses. *Am J Obstet Gynecol* 1989;**161**:996–1001.

17. Gagnon R, Challis J, Johnston L, Fraher L. Fetal endocrine responses to chronic placental embolization in the late-gestation fetus. *Acta Paediatr Taiwan* 1994;**3**:929–38.

18. Owens JA, Kind KL, Carbone F, Robinson JS, Owens PC. Circulating insulin-like growth factors-I and -II and substrates in fetal sheep following restriction of placental growth. *J Endocrinol* 1994;**140**:5–13.

19. Gross TL, Sokol RJ, Wilson MV, Kuhnert PM, Hirsch V. Amniotic fluid phosphatidylglycerol: a potentially useful predictor of intrauterine growth retardation. *Am J Obstet Gynecol* 1981;**140**:277–81.

20. Orgeig S, Morrison JL, Daniels CB. Prenatal development of the pulmonary surfactant system and the influence of hypoxia. *Respir Physiol Neurobiol* 2011;**178**:129–45.

21. Sharma P, McKay K, Rosenkrantz T, Hussain N. Comparisons of mortality and pre-discharge respiratory outcomes in small-for-gestational-age and appropriate-for-gestational-age premature infants. *BMC Pediatr* 2004;**4**:9.

22. Gortner L, Wauer RR, Stock GJ, Reiter HL, Reiss I, Hentschel R, et al. Neonatal outcome in small for gestational age infants: Do they really better? *J Perinat Med* 1999;**27**:484–9.

23. Simchen MJ, Beiner ME, Strauss-Liviathan N, Dulitzky M, Kuint J, Mashiach S, et al. Neonatal outcome in growth-restricted versus appropriately grown preterm infants. *Am J Physiol Lung Cell Mol Physiol* 2000;**17**:187–92.

24. Torrance HL, Mulder EJ, Brouwers HA, van Bel F, Visser GH. Respiratory outcome in preterm small for gestational age fetuses with or without abnormal umbilical artery Doppler and/or maternal hypertension. *J Matern Fetal Neonatal Med* 2007;**20**:613–21.

25. Baud O, Zupan V, Lacaze-Masmonteil T, Audibert F, Shojaei T, Thebaud B, et al. The relationships between antenatal management, the cause of delivery and neonatal outcome in a large cohort of very preterm singleton infants. *BMC PediatrBMC Pediatr* 2000;**107**:877–84.

26. Bernstein IM, Horbar JD, Badger GJ, Ohlsson A, Golan L. Morbidity and mortality among very-low-birth-weight neonates with intrauterine growth restriction. *Am J Obstet Gynecol* 2000;**182**:198–206.

27. Piper JM, Xenakis EMJ, McFarland M, Elliott BD, Berkus MD, Langer O. Do growth-retarded premature infants have different rates of perinatal morbidity and mortality than appropriately grown premature infants. *Obstet Gynecol* 1996;**87**:169–74.

28. Ley D, WideSwensson D, Lindroth M, Svenningsen N, Marsal K. Respiratory distress syndrome in infants with impaired intrauterine growth. *Acta Paediatr* 1997;**86**:1090–6.

29. Jobe AH, Ikegami M. Lung development and function in preterm infants in the surfactant treatment era. *Annu Rev Physiol* 2000;**62**:825–46.

30. Been JV, Zoer B, Kloosterboer N, Kessels CGA, Zimmermann LJI, van Iwaarden JF, et al. Pulmonary vascular endothelial growth factor expression and disaturated phospholipid content in a chicken model of hypoxia-induced fetal growth restriction. *Neonatology* 2010;**97**:183–9.

31. Gagnon R, Langridge J, Inchley K, Murotsuki J, Possmayer F. Changes in surfactant-associated protein mRNA profile in growth-restricted fetal sheep. *Am J Phys* 1999;**276**:L459–65.

32. Cock ML, Harding R. Renal and amniotic fluid responses to umbilicoplacental embolization for 20 days in fetal sheep. *Am J Physiol Regul Integr Comp Physiol* 1997;**273**:R1094–102.

33. Cock ML, Albuquerque CA, Joyce BJ, Hooper SB, Harding R. Effects of intrauterine growth restriction on lung liquid dynamics and lung development in fetal sheep. *Am J Obstet Gynecol* 2001;**184**:209–16.

34. Harding R, Cock ML, Louey S, Joyce BJ, Davey MG, Albuquerque CA, et al. The compromised intra-uterine environment: implications for future lung health. *Clin Exp Pharmacol Physiol* 2000;**27**:965–74.

35. Harding R, Tester ML, Moss TJ, Davey MG, Louey S, Joyce B, et al. Effects of intra-uterine growth restriction on the control of breathing and lung development after birth. *Clin Exp Pharmacol Physiol* 2000;**27**:114–9.

36. Robinson J, Hart I, Kingston E, Jones C, Thorburn G. Studies on the growth of the fetal sheep. The effects of reduction of placental size on hormone concentration in fetal plasma. *J Dev Physiol* 1980;**2**:239–48.

37. Rees S, Ng J, Dickson K, Nicholas T, Harding R. Growth retardation and the development of the respiratory system in fetal sheep. *Early Hum Dev* 1991;**26**:13–27.

38. Orgeig S, Crittenden TA, Marchant C, McMillen IC, Morrison JL. Intrauterine growth restriction delays surfactant protein maturation in the sheep fetus. *Am J Physiol Lung Cell Mol Physiol* 2010;**298**:L575–83.

39. Braems GA, Yao LJ, Inchley K, Brickenden A, Han VK, Grolla A, et al. Ovine surfactant protein cDNAs: use in studies on fetal lung growth and maturation after prolonged hypoxemia. *Am J Phys* 2000;**278**:L754–64.

40. Nardo L, Zhao L, Green L, Possmayer F, Richardson BS, Bocking AD. The effect of repeated umbilical cord occlusions on pulmonary surfactant protein mRNA levels in the ovine fetus. *J Soc Gynecol Investig* 2005;**12**:510–7.

41. Orgeig S, Daniels CB, Sullivan LC. Development of the pulmonary surfactant system. In: Harding R, Pinkerton K, Plopper C, editors. *The Lung: Development, Aging and the Environment*. London: Academic Press; 2004. pp. 149–67.

42. Warburton D, Parton L, Buckley S, Cosico L, Saluna T. Effects of glucose infusion on surfactant and glycogen regulation in fetal lamb lung. *J Appl Phys* 1987;**63**:1750–6.

43. Hallman M, Wermer D, Epstein BL, Gluck L. Effects of maternal insulin or glucose infusion on the fetus: study on lung surfactant phospholipids, plasma myoinositol, and fetal growth in the rabbit. *Am J Obstet Gynecol* 1982;**142**:877–82.

44. Snyder JM, Kwun JE, Obrien JA, Rosenfeld CR, Odom MJ. The concentration of the 35-kDA surfactant apoprotein in amniotic-fluid from normal and diabetic pregnancies. *Pediatr Res* 1988;**24**:728–34.

45. Dekowski SA, Snyder JM. Insulin regulation of messenger-ribonucleic-acid for the surfactant-associated proteins in human fetal lung in vitro. *Endocrinology* 1992;**131**:669–76.

46. LaCoursiere DY, Bloebaum L, Duncan JD, Varner MW. Population-based trends and correlates of maternal overweight and obesity, Utah 1991–2001. *Am J Obstet Gynecol* 2005;**192**:832–9.

47. Jensen ME, Gibson PG, Collins CE, Wood LG. Airway and systemic inflammation in obese children with asthma. *Eur Respir J* 2013.

48. Scarlata S, Fimognari FL, Cesari M, Giua R, Franco A, Pasqualetti P, et al. Lung function changes in older people with metabolic syndrome and diabetes. *Geriatr Gerontol Int* 2013.

49. Lopuhaa CE, Roseboom TJ, Osmond C, Barker DJ, Ravelli AC, Bleker OP, et al. Atopy, lung function, and obstructive airways disease after prenatal exposure to famine. *Thorax* 2000;**55**:555–61.

50. Victora CG, Adair L, Fall C, Hallal PC, Martorell R, Richter L, et al. Maternal and child undernutrition: consequences for adult health and human capital. *Lancet* 2008;**371**:340–57.

51. Chen CM, Wang LF, Su B. Effects of maternal undernutrition during late gestation on the lung surfactant system and morphometry in rats. *Pediatr Res* 2004;**56**:329–35.

52. Chen CM, Wang LF, Lang YD. Effects of maternal undernutrition on lung growth and insulin-like growth factor system expression in rat offspring. *Acta Paediatr Taiwan* 2007;**48**:62–7.

53. Lechner AJ, Winston DC, Bauman JE. Lung mechanics, cellularity, and surfactant after prenatal starvation in guinea pigs. *J Appl Phys* 1986;**60**:1610–4.

54. Guarner V, Tordet C, Bourbon JR. Effects of maternal protein-calorie malnutrition on the phospholipid-composition of surfactant isolated from fetal and neonatal rat lungs - compensation by inositol and lipid supplementation. *Pediatr Res* 1992;**31**:629–35.

55. Kalenga M, Henquin JC. Protein deprivation from the neonatal period impairs lung development in the rat. *Pediatr Res* 1987;**22**:45–9.

56. Shipton D, Tappin DM, Vadiveloo T, Crossley JA, Aitken DA, Chalmers J. Reliability of self reported smoking status by pregnant women for estimating smoking prevalence: a retrospective, cross sectional study. *BMJ* 2009:339. b4347.

57. Tappin DM, MacAskill S, Bauld L, Eadie D, Shipton D, Galbraith L. Smoking prevalence and smoking cessation services for pregnant women in Scotland. *Subst Abuse Treat Prev Policy* 2010;**5**:1.

58. Scollo M, Winstanley M, editors. *Tobacco in Australia: Facts and Issues.* Melbourne: Cancer Council Victoria; 2012.

59. Rehan VK, Asotra K, Torday JS. The effects of smoking on the developing lung: insights from a biologic model for lung development, homeostasis, and repair. *Lung* 2009;**187**:281–9.

60. Hanlon-Lundberg KM, Williams M, Rhim T, Covert RF, Mittendorf R, Holt JA. Accelerated fetal lung maturity profiles and maternal cocaine exposure. *Obstetr Gynecol* 1996;**87**:128–32.

61. Gluck L, Kulovich M. Lecithin-sphingomyelin ratios in amniotic fluid in normal and abnormal pregnancy. *Am J Obstet Gynecol* 1973;**115**:539–46.

62. Lieberman E, Torday J, Barbieri R, Cohen A, Vanvunakis H, Weiss ST. Association of intrauterine cigarette-smoke exposure with indexes of fetal lung maturation. *Obstet Gynecol* 1992;**79**:564–70.

63. Curet LB, Rao AV, Zachman RD, Morrison J, Burkett G, Poole WK. Maternal smoking and respiratory-distress syndrome. *Am J Obstet Gynecol* 1983;**147**:446–50.

64. Collins MH, Moessinger AC, Kleinerman J, Bassi J, Rosso P, Collins AM, et al. Fetal lung hypoplasia associated with maternal smoking - a morphometric analysis. *Pediatr Res* 1985;**19**:408–12.

65. Cunningham J, Dockery DW, Speizer FE. Maternal smoking during pregnancy as a predictor of lung-function in children. *Am J Epidemiol* 1994;**139**:1139–52.

66. Hanrahan JP, Tager IB, Segal MR, Tosteson TD, Castile RG, Vanvunakis H, et al. The effect of maternal smoking during pregnancy on early infant lung-function. *Am Rev Respir Dis* 1992;**145**:1129–35.

67. Rehan VK, Wang Y, Sugano S, Santos J, Patel S, Sakurai R, et al. In utero nicotine exposure alters fetal rat lung alveolar type II cell proliferation, differentiation, and metabolism. *Am J Physiol Lung Cell Mol Physiol* 2007;**292**:L323–33.

68. Hylkema MN, Blacquiere MJ. Intrauterine effects of maternal smoking on sensitization, asthma, and chronic obstructive pulmonary disease. *Proc Am Thorac Soc* 2009;**6**:660–2.

69. Maritz GS, Dennis H. Maternal nicotine exposure during gestation and lactation interferes with alveolar development in the neonatal lung. *Reprod Fertil Dev* 1998;**10**:255–61.

70. Blacquiere MJ, Timens W, Melgert BN, Geerlings M, Postma DS, Hylkema MN. Maternal smoking during pregnancy induces airway remodelling in mice offspring. *Eur Respir J* 2009;**33**:1133–40.

71. Maritz GS, Thomas RA. Maternal nicotine exposure - response of type-II pneumocytes of neonatal rat pups. *Cell Biol Int* 1995;**19**:323–31.

72. Chen CM, Wang LF, Yeh TF. Effects of maternal nicotine exposure on lung surfactant system in rats. *Pediatr Pulmonol* 2005;**39**:97–102.

73. Lazic T, Matic M, Gallup JM, Van Geelen A, Meyerholz DK, Grubor B, et al. Effects of nicotine on pulmonary surfactant proteins A and D in ovine lung epithelia. *Pediatr Pulmonol* 2010;**45**:255–62.

74. Hermans C, Libotte V, Robin M, Clippe A, Wattiez R, Falmagne P, et al. Maternal tobacco smoking and lung epithelium-specific proteins in amniotic fluid. *Pediatr Res* 2001;**50**:487–94.

75. Wuenschell CW, Zhao JS, Tefft JD, Warburton D. Nicotine stimulates branching and expression of SP-A and SP-C mRNAs in embryonic mouse lung culture. *Am J Physiol Lung Cell Mol Physiol* 1998;**274**:L165–70.

76. Proskocil BJ, Sekhon HS, Clark JA, Lupo SL, Jia Y, Hull WM, et al. Vitamin C prevents the effects of prenatal nicotine on pulmonary function in newborn monkeys. *Am J Respir Crit Care Med* 2005;**171**:1032–9.

77. Brewster CEP, Howarth PH, Djukanovic R, Wilson J, Holgate ST, Roche WR. Myofibroblasts and subepithelial fibrosis in bronchial-asthma. *Alcohol Clin Exp Res* 1990;**3**:507–11.

78. Bousquet J, Vignola AM, Chanez P, Campbell AM, Bonsignore G, Michel FB. Airways remodeling in asthma no doubt, no more. *Int Arch Allergy Immunol* 1995;**107**:211–4.

79. von Mutius E. Environmental factors influencing the development and progression of pediatric asthma. *J Allergy Clin Immunol* 2002;**109**:S525–32.

80. Strachan DP, Cook DG. Parental smoking and childhood asthma: longitudinal and case-control studies. *Thorax* 1998;**53**:204–12.

81. Hoyme HE, May PA, Kalberg WO, Kodituwakku P, Gossage JP, Trujillo PM, et al. A practical clinical approach to diagnosis of fetal alcohol spectrum disorders: clarification of the 1996 institute of medicine criteria. *Pediatrics* 2005;**115**:39–47.

82. Jones KL. The effects of alcohol on fetal development. *Birth Defects Res C Embryo Today* 2011;**93**:3–11.

83. Jones KL, Smith DW. Recognition of the fetal alcohol syndrome in early infancy. *Lancet* 1973;**302**:999–1001.

84. Colvin L, Payne J, Parsons D, Kurinczuk JJ, Bower C. Alcohol consumption during pregnancy in nonindigenous west Australian women. *Alcohol Clin Exp Res* 2007;**31**:276–84.

85. Sozo F, O'Day L, Maritz G, Kenna K, Stacy V, Brew N, et al. Repeated ethanol exposure during late gestation alters the maturation and innate immune status of the ovine fetal lung. *Am J Physiol Lung Cell Mol Physiol* 2009;**296**:L510–8.

86. Kenna K, Sozo F, De Mateo R, Hanita T, Gray SP, Tare M, et al. Alcohol exposure during late gestation: multiple developmental outcomes in sheep. *J Dev Orig Health Dis* 2012;**3**:224–36.

87. Sozo F, Vela M, Stokes V, Kenna K, Meikle PJ, De Matteo R, et al. Effects of prenatal ethanol exposure on the lungs of postnatal lambs. *Am J Physiol Lung Cell Mol Physiol* 2011;**300**:L139–47.

88. Lazic T, Wyatt TA, Matic M, Meyerholz DK, Grubor B, Gallup JM, et al. Maternal alcohol ingestion reduces surfactant protein A expression by preterm fetal lung epithelia. *Alcohol* 2007;**41**:347–55.

89. Lazic T, Sow FB, Van Geelen A, Meyerholz DK, Gallup JM, Ackermann MR. Exposure to ethanol during the last trimester of pregnancy alters the maturation and immunity of the fetal lung. *Alcohol* 2011;**45**:673–80.

90. Giliberti D, Mohan SS, Brown LA, Gauthier TW. Perinatal exposure to alcohol: implications for lung development and disease. *Paediatr Respir Rev* 2013;**14**:17–21.

91. Bhuvaneswar CG, Chang G, Epstein LA, Stern TA. Cocaine and opioid use during pregnancy: prevalence and management. *Prim Care Companion J Clin Psychiatry* 2008;**10**:59–65.

92. Cherukuri R, Minkoff H, Feldman J, Parekh A, Glass L. A cohort study of alkaloidal cocaine ("crack") in pregnancy. *Obstetr Gynecol* 1988;**72**:147–51.

93. Riezzo I, Fiore C, De Carlo D, Pascale N, Neri M, Turillazzi E, et al. Side effects of cocaine abuse: multiorgan toxicity and pathological consequences. *Curr Med Chem* 2012;**19**:5624–46.

94. Perper JA, Van Thiel DH. Respiratory complications of cocaine abuse. *Recent Dev Alcohol* 1992;**10**:363–77.

95. Hansson SR, Mezey E, Hoffman BJ. Serotonin transporter messenger RNA expression in neural crest-derived structures and sensory pathways of the developing rat embryo. *Neuroscience* 1999;**89**:243–65.

96. Owiny JR, Jones MT, Sadowsky D, Myers T, Massman A, Nathanielsz PW. Cocaine in pregnancy: the effect of maternal administration of cocaine on the maternal and fetal pituitary-adrenal axes. *Am J Obstet Gynecol* 1991;**164**:658–63.

97. Burchfield DJ, Pena A, Peters AJ, Abrams RM, Phillips D. Cocaine does not compromise cerebral or myocardial oxygen delivery in fetal sheep. *Reprod Fertil Dev* 1996;**8**:383–9.

98. Laurini RN, Arbeille B, Gemberg C, Akoka S, Locatelli A, Lansac J, et al. Brain damage and hypoxia in an ovine fetal chronic cocaine model. *Eur J Obstetr Gynecol Reprod Biol* 1999;**86**:15–22.

99. Zuckerman B, Maynard EC, Cabral H. A preliminary report of prenatal cocaine exposure and respiratory distress syndrome in premature infants. *Am J Dis Child* 1991;**145**:696–8.

100. Beeram MR, Abedin M, Young M, Leftridge C, Dhanireddy R. Effect of intrauterine cocaine exposure on respiratory distress syndrome in very low birthweight infants. *J Natl Med Assoc* 1994;**86**:370–2.

101. Sosenko IR. Antenatal cocaine exposure produces accelerated surfactant maturation without stimulation of antioxidant enzyme development in the late gestation rat. *Pediatr Res* 1993;**33**:327–31.

102. Kain ZN, Chinoy MR, Antonio-Santiago MT, Marchitelli RN, Scarpelli EM. Enhanced lung maturation in cocaine-exposed rabbit fetuses. *Pediatr Res* 1991;**29**:534–7.

103. Epstein RA, Bobo WV, Martin PR, Morrow JA, Wang W, Chandrasekhar R, et al. Increasing pregnancy-related use of prescribed opioid analgesics. *Ann Epidemiol* 2013;**23**:498–503.

104. Lendoiro E, Gonzalez-Colmenero E, Concheiro-Guisan A, de Castro A, Cruz A, Lopez-Rivadulla M, et al. Maternal hair analysis for the detection of illicit drugs, medicines, and alcohol exposure during pregnancy. *Ther Drug Monit* 2013;**35**:296–304.

105. Roloff DW, Howatt WF, Kanto Jr WP, Borker RC. Jr., Morphine administration to pregnant rabbits: effect on fetal growth and lung development. *Addict Dis* 1975;**2**:369–79.

106. Bhandari V, Bergqvist LL, Kronsberg SS, Barton BA, Anand KJS, Grp NTI. Morphine administration and short-term pulmonary outcomes among ventilated preterm infants. *Pediatrics* 2005;**116**:352–9.

107. Kallen B, Finnstrom O, Nygren KG, Otterblad Olausson P. Maternal drug use during pregnancy and asthma risk among children. *Pediatr Allergy Immunol* 2013;**24**:28–32.

108. Jones HE. Treating opioid use disorders during pregnancy: historical, current, and future directions. *Substance Abuse* 2013;**34**:89–91.

109. Comer CR, Grunstein JS, Mason RJ, Johnston SC, Grunstein MM. Endogenous opioids modulate fetal rabbit lung maturation. *J Appl Phys* 1987;**62**:2141–6.

110. Taeusch Jr W, Wyszogrodski I, Bator A, Robert M, Carson S, Avery ME. Pharmacologic regulation of alveolar surfactant in fetal and newborn rabbits: ACTH, heroin, isoxsuprine, and glucocorticoid. *Chest* 1975;**67**:49S–50S.

111. Kallopur SG, Jobe AH. Contribution of inflammation to lung injury and development. *Arch Dis Child Fetal Neonatal Ed* 2006;**91**:F132–5.

112. Meyer KC, Zimmerman JJ. Inflammation and surfactant. *Paediatr Respir Rev* 2002;**3**:308–14.

113. Hallman M. Cytokines, pulmonary surfactant and consequences of intrauterine infection. *Biol Neonate* 1999;**76**:2–9.

114. Goldenberg RL, Hauth JC, Andrews WW. Intrauterine infection and preterm delivery. *N Engl J Med* 2000;**342**:1500–7.

115. Hodgman JE, Barton L, Pavlova Z, Fassett MJ. Infection as a cause of death in the extremely-low-birth-weight infant. *J Matern Fetal Neonatal Med* 2003;**14**:313–7.

116. Jobe AH. Glucocorticoids, inflammation and the perinatal lung. *Semin Neonatol* 2001;**6**:331–42.

117. Chaiworapongsa T, Hong JS, Hull WM, Romero R, Whitsett JA. Amniotic fluid concentration of surfactant proteins in intra-amniotic infection. *J Matern Fetal Neonatal Med* 2008;**21**:663–70.

118. Han YM, Romero R, Kim YM, Kim JS, Richani K, Friel LA, et al. Surfactant protein-A mRNA expression by human fetal membranes is increased in histological chorioamnionitis but not in spontaneous labour at term. *J Pathol* 2007;**211**:489 96.

119. Moss TJM, Newnham JP, Willett KE, Kramer BW, Jobe AH, Ikegami M. Early gestational intra-amniotic endotoxin – Lung function, surfactant, and morphometry. *Am J Respir Crit Care Med* 2002;**165**:805–11.

120. Moss TJM, Nitsos I, Ikegami M, Jobe AH, Newnham JP. Experimental intrauterine Ureaplasma infection in sheep. *Am J Obstet Gynecol* 2005;**192**:1179–86.

121. Ikegami T, Tsuda A, Karube A, Kodama H, Hirano H, Tanaka T. Effects of intrauterine IL-6 and IL-8 on the expression of surfactant apoprotein mRNAs in the fetal rat lung. *Eur J Obstet Gynecol Reprod Biol* 2000;**93**:97–103.

122. Dhar V, Hallman M, Lappalainen U, Bry K. Interleukin-1 alpha upregulates the expression of surfactant protein-A in rabbit lung explants. *Biol Neonate* 1997;**71**:46–52.

123. Vayrynen O, Glumoff V, Hallman M. Inflammatory and anti-inflammatory responsiveness of surfactant proteins in fetal and neonatal rabbit lung. *Pediatr Res* 2004;**55**:55–60.

124. Kuypers E, Collins JJP, Kramer BW, Ofman G, Nitsos I, Pillow JJ, et al. Intra-amniotic LPS and antenatal betamethasone: inflammation and maturation in preterm lamb lungs. *Am J Physiol Lung Cell Mol Physiol* 2012;**302**:L380–9.

125. Newnham JP, Moss TJ, Padbury JF, Willet KE, Ikegami M, Ervin MG, et al. The interactive effects of endotoxin with prenatal glucocorticoids on short-term lung function in sheep. *Am J Obstet Gynecol* 2001;**185**:190–7.

126. Westover AJ, Hooper SB, Wallace MJ, Moss TJM. Prostaglandins mediate the fetal pulmonary response to intrauterine inflammation. *Am J Physiol Lung Cell Mol Physiol* 2012;**302**:1664–78.

127. Hallman M. The surfactant system protects both fetus and newborn. *Neonatology* 2013;**103**:320–6.

128. Been JV, Rours IG, Kornelisse RF, Jonkers F, de Krijger RR, Zimmermann LJ. Chorioamnionitis alters the response to surfactant in preterm infants. *J Pediatr* 2010;**156**:10–U34.

129. Barreira ER, Precioso AR, Bousso A. Pulmonary surfactant in respiratory syncytial virus bronchiolitis: the role in pathogenesis and clinical implications. *Pediatr Pulmonol* 2011;**46**:415–20.

130. Everard ML. Bronchiolitis. Origins and optimal management. *Drugs* 1995;**49**:885–96.

131. Leader S, Kohlhase K. Recent trends in severe respiratory syncytial virus (RSV) among US infants, 1997 to 2000. *J Pediatr* 2003;**143**:S127–32.

132. Löfgren J, Ramet M, Renko M, Marttila R, Hallman M. Association between surfactant protein A gene locus and severe respiratory syncytial virus infection in infants. *J Infect Dis* 2002;**185**:283–9.

133. Hall CB, Granoff DM, Gromisch DS, Halsey NA, Kohl S, Marcuse EK, et al. Use of ribavirin in the treatment of respiratory syncytial virus-infection. *Pediatrics* 1993;**92**:501–4.

134. Anderson LJ, Parker RA, Strikas RL. Association between respiratory syncytial virus outbreaks and lower respiratory tract deaths of infants and young children. *J Infect Dis* 1990;**161**:640–6.

135. LeVine AM, Lotze A, Stanley S, Stroud C, O'Donnell R, Whitsett J, et al. Surfactant content in children with inflammatory lung disease. *Crit Care Med* 1996;**24**:1062–7.

136. Dargaville PA, South M, McDougall PN. Surfactant abnormalities in infants with severe viral bronchiolitis. *Arch Dis Child* 1996;**75**:133–6.

137. Kerr MH, Paton JY. Surfactant protein levels in severe respiratory syncytial virus infection. *Am J Respir Crit Care Med* 1999;**159**:1115–8.

138. Tibby SM, Hatherill M, Wright SM, Wilson P, Postle AD, Murdoch IA. Exogenous surfactant supplementation in infants with respiratory syncytial virus bronchiolitis. *Am J Respir Crit Care Med* 2000;**162**:1251–6.

139. Skelton R, Holland P, Darowski M, Chetcuti PA, Morgan LW, Harwood JL. Abnormal surfactant composition and activity in severe bronchiolitis. *Acta Paediatr* 1999;**88**:942–6.

140. Lahti M, Lofgren J, Marttila R, Renko M, Klaavuniemi T, Haataja R, et al. Surfactant protein D gene polymorphism associated with severe respiratory syncytial virus infection. *Pediatr Res* 2002;**51**:696–9.

141. Rakshi K, Couriel JM. Management of acute bronchiolitis. *Arch Dis Child* 1994;**71**:463–9.

142. Numata M, Chu HW, Dakhama A, Voelker DR. Pulmonary surfactant phosphatidylglycerol inhibits respiratory syncytial virus–induced inflammation and infection. *Proc Natl Acad Sci USA* 2010;**107**:320–5.

143. Olivier AK, Gallup JM, van Geelen A, Ackermann MR. Exogenous administration of vascular endothelial growth factor prior to human respiratory syncytial virus a2 infection reduces pulmonary pathology in neonatal lambs and alters epithelial innate immune responses. *Exp Lung Res* 2011;**37**:131–43.

144. Orgeig S, Daniels CB. Environmental selection pressures shaping the pulmonary surfactant system of adult and developing lungs. In: Glass ML, Wood SC, editors. *Cardio-respiratory Control in Vertebrates*. Berlin Heidelberg: Springer Verlag; 2009. pp. 205–39.

145. Liamtsev VG, Arbuzov AA. Surfactant system of rat lungs in acute hypoxic hypoxia. *Biull Eksp Biol Med* 1981;**92**:612–4.

146. Kumar R, Hegde KS, Krishna B, Sharma RS. Combined effect of hypoxia and cold on the phospholipid composition of lung surfactant in rats. *Aviat Space Environ Med* 1980;**51**:459–62.

147. Prevost MC, Vieu C, Douste Blazy L. Hypobaric hypoxia on pulmonary wash fluid of rats. *Respiration* 1980;**40**:76–80.

148. Zaitseva KK, Skorik VI, Shliapnikova SA. State of pulmonary surfactant and ultrastructure of the aerohematic barrier in acute hypoxia. *Biull Eksp Biol Med* 1981;**92**:653–6.

149. Vives MF, Caspar-Bauguil S, Aliouat EM, Escamilla R, Perret B, Dei-Cas E, et al. Hypobaric hypoxia-related impairment of pulmonary surfactant proteins A and D did not favour Pneumocystis carinii Frenkel 1999 growth in non-immunocompromised rats. *Parasite* 2008;**15**:53–64.

150. Belov GV, Arbuzov AA, Davydov VT. Comparative evaluation of the physical methods for studying the pulmonary surfactant system during exposure to acute hypoxia. *Biull Eksp Biol Med* 1985;**99**:542–5.

151. Castillo Y, Johnson FB. Pulmonary surfactant in acutely hypoxic mice. *Lab Invest* 1969;**21**:61–4.

152. Chander A, Viswanathan R, Venkitasubramanian TA. Effect of acute hypobaric hypoxia on 32-P incorporation into phospholipids of alveolar surfactant, lung, liver and plasma of rat. *Environ Physiol Biochem* 1975;**5**:27–36.

153. Papen M, Wodopia R, Bartsch P, Mairbaurl H. Hypoxia-effects on Ca-i-signaling and ion transport activity of lung alveolar epithelial cells. *Cell Physiol Biochem* 2001;**11**:187–96.

154. Reinke C, Bevans-Fonti S, Grigoryev DN, Drager LF, Myers AC, Wise RA, et al. Chronic intermittent hypoxia induces lung growth in adult mice. *Am J Physiol Lung Celld Mol Physiol* 2011;**300**:L266–73.

155. Sakamoto K, Hashimoto N, Kondoh Y, Imaizumi K, Aoyama D, Kohnoh T, et al. Differential modulation of surfactant protein D under acute and persistent hypoxia in acute lung injury. *Am J Physiol Lung Celld Mol Physiol* 2012;**303**:L43–53.

156. Bartlett Jr D, Remmers JE. Effects of high altitude exposure on the lungs of young rats. *Respir Physiol* 1971;**13**:116–25.

157. Hsia CC, Carbayo JJ, Yan X, Bellotto DJ. Enhanced alveolar growth and remodeling in Guinea pigs raised at high altitude. *Respir Physiol Neurobiol* 2005;**147**:105–15.

158. Krishna B, Kumar R, Hegde KS, Sharma RS. Alveolar stability of altitude raised rats. *Indian J Physiol Pharmacol* 1978;**22**:125–35.

159. Hegde KS, Kumar R, Krishna B, Nayar HS. Alveolar macrophages and pulmonary surfactant of altitude-raised rats. *Ann Thorac Surg* 1980;**51**:700–3.

160. Droma Y, Hanaoka M, Hotta J, Naramoto A, Koizumi T, Fujimoto K, et al. Pathological features of the lung in fatal high altitude pulmonary edema occurring at moderate altitude in Japan. *High Alt Med Biol* 2001;**2**:515–23.

161. Luo YJ, Zou YL, Gao YQ. Gene polymorphisms and high-altitude pulmonary edema susceptibility: a 2011 update. *Respiration* 2012;**84**:155–62.

162. Saxena S, Kumar R, Madan T, Gupta V, Muralidhar K, Sarma PU. Association of polymorphisms in pulmonary surfactant protein A1 and A2 genes with high-altitude pulmonary edema. *Chest* 2005;**128**:1611–9.

163. Diaz S, Hammond K, Orgeig S. Effects of high altitude on lung surfactant lipids in Peromyscus maniculatus. *IntegComp Biol* 2011;**51**:E34. E34.

164. Diaz S, Thaler CD, Shirkey NJ, Brown T, Cardullo RA, Hammond KA. Changes in pulmonary surfactant in deer mice (*Peromyscus maniculatus*) at high altitude. *IntegComp Biol* 2012;**52**:E45. E45.

165. Heath D, Smith P, Harris P. Clara cells in the llama. *Exp Cell Biol* 1976;**44**:73–82.

166. Heath D, Smith P, Biggar R. Clara cells in llamas born and living at high and low altitudes. *Br J Dis Chest* 1980;**74**:75–80.

167. Pace PW, Yao LJ, Wilson JX, Possmayer F, Veldhuizen RAW, Lewis JF. The effects of hyperoxia exposure on lung function and pulmonary surfactant in a rat model of acute lung injury. *Exp Lung Res* 2009;**35**:380–98.

168. Lang JD, McArdle PJ, O'Reilly PJ, Matalon S. Oxidant-antioxidant balance in acute lung injury. *Chest* 2002;**122**:314S–20S.

169. Putman E, van Golde LM, Haagsman HP. Toxic oxidant species and their impact on the pulmonary surfactant system. *Lung* 1997;**175**:75–103.

170. Orgeig S, Daniels CB. Effects of aging, disease and the environment on the pulmonary surfactant system. In: Harding R, Pinkerton K, Plopper C, editors. *The Lung: Development, Aging and the Environment*. London: Academic Press; 2004. pp. 363–75.

171. Gross NJ, Smith DM. Impaired surfactant phospholipid metabolism in hyperoxic mouse lungs. *J Appl Phys* 1981;**51**:1198–203.

172. Holm BA, Notter RH, Siegle J, Matalon S. Pulmonary physiological and surfactant changes during injury and recovery from hyperoxia. *J Appl Phys* 1985;**59**:1402–9.

173. Ledwozyw A, Borowicz B. The influence of normobaric hyperoxia on lung surfactant phospholipids in rats. *Arch Vet Pol* 1992;**32**:127–33.

174. Goad ME, Tryka AF, Witschi HP. Surfactant alterations following acute bleomycin and hyperoxia-induced lung damage. *Toxicol Lett* 1986;**32**:173–8.

175. Balaan MR, Bowman L, Dedhia HV, Miles PR. Hyperoxia-induced alterations of rat alveolar-lavage composition and properties. *Exp Lung Res* 1995;**21**:141–56.

176. Minoo P, King RJ, Coalson JJ. Surfactant proteins and lipids are regulated independently during hyperoxia. *Am J Phys* 1992;**263**:L291–8.

177. Huang Y-CT, Caminiti SP, Fawcett TA, Moon RE, Fracica PJ, Miller FJ, et al. Natural surfactant and hyperoxic lung injury in primates I. physiology and biochemistry. *J Appl Phys* 1994;**76**:991–1001.

178. Allred TF, Mercer RR, Thomas RF, Deng H, Auten RL. Brief 95% O$_2$ exposure effects on surfactant protein and mRNA in rat alveolar and bronchiolar epithelium. *Am J Phys* 1999;**276**:L999–1009.

179. Nogee LM, Wispe JR. Effects of pulmonary oxygen injury on airway content of surfactant-associated protein A. *Pediatr Res* 1988;**24**:568–73.

180. Nogee LM, Wispe JR, Clark JC, Whitsett JA. Increased synthesis and mRNA of surfactant protein A in oxygen-exposed rats. *Am J Respir Cell Mol Biol* 1989;**1**:119–25.

181. Nogee LM, Wispe JR, Clark JC, Weaver TE, Whitsett JA. Increased expression of pulmonary surfactant proteins in oxygen-exposed rats. *Am J Respir Cell Mol Biol* 1991;**4**:102–7.

182. Ohashi T, Takada S, Motoike T, Tsuneishi S, Matsuo M, Sano K, et al. Effect of dexamethasone on pulmonary surfactant metabolism in hyperoxia-treated rat lungs. *Pediatr Res* 1991;**29**:173–7.

183. Zenri H, Rodriquez-Capote K, McCaig L, Yao LJ, Brackenbury A, Possmayer F, et al. Hyperoxia exposure impairs surfactant function and metabolism. *Crit Care Med* 2004;**32**:1155–60.

184. Dombrowsky H, Tschernig T, Vieten G, Rau GA, Ohler F, Acevedo C, et al. Molecular and functional changes of pulmonary surfactant in response to hyperoxia. *Pediatr Pulmonol* 2006;**41**:1025–39.

185. Huang YCT, Sane AC, Simonson SG, Fawcett TA, Moon RE, Fracica PJ, et al. Artificial surfactant attenuates hyperoxic lung injury in primates. 1. Physiology and biochemistry. *J Appl Phys* 1995;**78**:1816–22.

186. Piantadosi CA, Fracica PJ, Duhaylongsod FG, Huang YCT, Weltywolf KE, Crapo JD, et al. Artificial surfactant attenuates hyperoxic lung injury in primates. 2. Morphometric analysis. *J Appl Phys* 1995;**78**:1823–31.

187. Jain D, Tomer Y, Kadire H, Atochina-Vasserman E, Beers MF. Surfactant protein D protects against hyperoxia-induced acute lung injury in an inducible SP-D transgenic mouse model. *FASEB J* 2007;**21**:A552–A552.

188. Silveyra P, Floros J. Air pollution and epigenetics: effects on SP-A and innate host defence in the lung. *Swiss Med Wkl* 2012;**142**:w13579.

189. Dockery DW, Pope 3rd CA, Xu X, Spengler JD, Ware JH, Fay ME, et al. An association between air pollution and mortality in six U.S. cities. *N Engl J Med* 1993;**329**:1753–9.

190. Pope CAI, Thun MJ, Namboori MM, Dockery DW, Evans JS, Speizer FE, et al. Particulate air pollution as a predictor of mortality in a prospective study of U.S. adults. *Am J Respir Crit Care Med* 1995;**151**:669–74.

191. Schwartz J. Air-pollution and daily mortality - A review and meta analysis. *Environ Res* 1994;**64**:36–52.

192. Chung KAF, Zhang JF, Zhong NS. Outdoor air pollution and respiratory health in Asia. *Respirology* 2011;**16**:1023–6.

193. Wong CM, Vichit-Vadakan N, Vajanapoom N, Ostro B, Thach TQ, Chau PY, et al. Part 5. Public health and air pollution in Asia (PAPA): a combined analysis of four studies of air pollution and mortality. *Res Rep Health Eff Inst* 2010:377–418.

194. Hanna LM, Frank R, Scherer PW. Absorption of soluble gases and vapors in the respiratory system. In: Chang HK, Paiva M, editors. *Respiratory Physiology – An Analytical Approach*. New York: Marcel Dekker; 1989. pp. 277–316.

195. Lippmann M. Regional deposition of particles in the human respiratory tract. In: Pollock DM, editor. *Comprehensive Physiology*. New York: John Wiley & Sons, Inc; 2010.

196. Ultman JS. Transport and uptake of inhaled gases. In: Watson AY, Bates RR, Kennedy D, editors. *Air Pollution, the Automobile, and Public Health*. Washington, DC: National Academy Press; 1988. pp. 323–66.

197. Warheit DB. Interspecies comparisons of lung responses to inhaled particles and gases. *Crit Rev Toxicol* 1989;**20**:1–29.

198. Müller B, Seifart C, Barth PJ. Effect of air pollutants on the pulmonary surfactant system. *Eur J Clin Invest* 1998;**28**:762–77.

199. Li ZW, Tighe RM, Feng FF, Ledford JG, Hollingsworth JW. Genes of innate immunity and the biological response to inhaled ozone. *J Biochem Mol Toxicol* 2013;**27**:3–16.

200. Langford SD, Bidani A, Postlethwait EM. Ozone-reactive absorption by pulmonary epithelial lining fluid constituents. *Toxicol Appl Pharmacol* 1995;**132**:122–30.

201. Postlethwait EM, Langford SD, Bidani A. Determinants of inhaled ozone absorption in isolated rat lungs. *Toxicol Appl Pharmacol* 1994;**125**:77–89.

202. Lai CC, Yang SH, Finlaysonpitts BJ. Interactions of monolayers of unsaturated phosphocholines with ozone at the air-water-interface. *Langmuir* 1994;**10**:4637–44.

203. Wadia Y, Tobias DJ, Stafford R, Finlayson-Pitts BJ. Real-time monitoring of the kinetics and gas-phase products of the reaction of ozone with an unsaturated phospholipid at the air-water interface. *Langmuir* 2000;**16**:9321–30.

204. Finlayson-Pitts BJ, Mautz WJ, Lai CC, Bufalino C, Messer K, Mestas J, et al. Are changes in breathing pattern on exposure to ozone related to changes in pulmonary surfactant. *Inhal Toxicol* 1994;**6**:267–87.

205. Pryor WA, Bermudez E, Cueto R, Squadrito GL. Detection of aldehydes in bronchoalveolar lavage of rats exposed to ozone. *Fundam Appl Toxicol* 1996;**34**:148–56.

206. Thompson KC, Rennie AR, King MD, Hardman SJO, Lucas COM, Pfrang C, et al. Reaction of a phospholipid monolayer with gas-phase ozone at the air-water interface: measurement of surface excess and surface pressure in real time. *Langmuir* 2010;**26**:17295–303.

207. Thompson KC, Jones SH, Rennie AR, King MD, Ward AD, Hughes BR, et al. Degradation and rearrangement of a lung surfactant lipid at the air-water interface during exposure to the pollutant gas ozone. *Langmuir* 2013;**29**:4594–602.

208. Oosting RS, Vangreevenbroek MMJ, Verhoef J, Vangolde LMG, Haagsman HP. Structural and functional-changes of surfactant protein-A induced by ozone. *Am J Phys* 1991;**261**:L77–83.

209. Oosting RS, Van Iwaarden JF, Van Bree L, Verhoef J, Van Golde LM, Haagsman HP. Exposure of surfactant protein A to ozone in vitro and in vivo impairs its interactions with alveolar cells. *Am J Phys* 1992;**262**:L63–8.

210. Mikerov AN, Haque R, Gan XZ, Guo XX, Phelps DS, Floros J. Ablation of SP-A has a negative impact on the susceptibility of mice to Klebsiella pneumoniae infection after ozone exposure: sex differences. *Respir Res* 2008:9.

211. Wang G, Umstead TM, Phelps DS, Al-Mondhiry H, Floros J. The effect of ozone exposure on the ability of human surfactant protein a variants to stimulate cytokine production. *Environ Health Perspect* 2002;**110**:79–84.

212. Committee of the Environmental and Occupational Health Assembly of the American Thoracic Society. Health effects of outdoor air pollution. *Am J Respir Crit Care Med* 1996;**153**:3–50.

213. Su WY, Gordon T. Alterations in surfactant protein A after acute exposure to ozone. *J Appl Phys* 1996;**80**:1560–7.

214. Haque R, Umstead TM, Ponnuru P, Guo XX, Hawgood S, Phelps DS, et al. Role of surfactant protein-A (SP-A) in lung injury in response to acute ozone exposure of SP-A deficient mice. *Toxicol Appl Pharmacol* 2007;**220**:72–82.

215. Kierstein S, Poulain FR, Cao Y, Grous M, Mathias R, Kierstein G, et al. Susceptibility to ozone-induced airway inflammation is associated with decreased levels of surfactant protein D. *Respir Res* 2006:7.

216. Gardai SJ, Xiao YQ, Dickinson M, Nick JA, Voelker DR, Greene KE, et al. By binding SIRP alpha or calreticulin/CD91, lung collectins act as dual function surveillance molecules to suppress or enhance inflammation. *Cell* 2003;**115**:13–23.

217. Guo CJ, Atochina-Vasserman EN, Abramova E, Foley JP, Zaman A, Crouch E, et al. S-Nitrosylation of surfactant protein-D controls inflammatory function. *Plos Biology* 2008;**6**:2414–23.

218. Groves AM, Gow AJ, Massa CB, Hall L, Laskin JD, Laskin DL. Age-related increases in ozone-induced injury and altered pulmonary mechanics in mice with progressive lung inflammation. *Am J Physiol Lung Cell Mol Physiol* 2013;**305**:L555–68.

219. Plopper CG, Fanucchi MV. Do urban environmental pollutants exacerbate childhood lung diseases? *Environ Health Perspect* 2000;**108**:A252–3.

220. Brody AR, Roe MW. Deposition pattern of inorganic particles at the alveolar level in the lungs of rats and mice. *Am Rev Respir Dis* 1983;**128**:724–9.

221. Murphy SA, BeruBe KA, Pooley FD, Richards RJ. The response of lung epithelium to well characterised fine particles. *Life Sci* 1998;**62**:1789–99.

222. Stone V, Johnston H, Clift MJD. Air pollution, ultrafine and nanoparticle toxicology: cellular and molecular interactions. *Nano Biosci IEEE Trans* 2007;**6**:331–40.

223. Tetley TD, Richards RJ, Harwood JL. Changes in pulmonary surfactant and phosphatidylcholine metabolism in rats exposed to chrysotile asbestos dust. *Biochem J* 1977;**166**:323–9.

224. Panos RJ, Suwabe A, Leslie CC, Mason RJ. Hypertrophic alveolar type-II cells from silica-treated rats are committed to DNA-synthesis in vitro. *Am J Respir Cell Mol Biol* 1990;**3**:51–9.

225. Geiser M, Rothen-Rutishauser B, Kapp N, Schurch S, Kreyling W, Schulz H, et al. Ultrafine particles cross cellular membranes by nonphagocytic mechanisms in lungs and in cultured cells. *Environ Health Perspect* 2005;**113**:1555–60.

226. Shi HB, Magaye R, Castranova V, Zhao JS. Titanium dioxide nanoparticles: a review of current toxicological data. *Particle Fibre Toxicol* 2013:10.

227. Liou SH, Tsou TC, Wang SL, Li LA, Chiang HC, Li WF, et al. Epidemiological study of health hazards among workers handling engineered nanomaterials. *J Nanoparticle Res* 2012:14.

228. Cassidy A, t Mannetje A, van Tongeren M, Field JK, Zaridze D, Szeszenia-Dabrowska N, et al. Occupational exposure to crystalline silica and risk of lung cancer – a multicenter case-control study in Europe. *Epidemiology* 2007;**18**:36–43.

229. Dell LD, Mundt KA, Luippold RS, Nunes AP, Cohen L, Burch MT, et al. A cohort mortality study of employees in the US carbon black industry. *J Occup Environ Med* 2006;**48**:1219–29.

230. Ramanakumar AV, Parent ME, Latreille B, Siemiatycki J. Risk of lung cancer following exposure to carbon black, titanium dioxide and talc: results from two case-control studies in Montreal. *Int J Cancer* 2008;**122**:183–9.

231. Radon K, Nowak D, Szadkowski D. Lack of combined effects of exposure and smoking on respiratory health in aluminium potroom workers. *Occup Environ Med* 1999;**56**:468–72.

232. Oberdorster G, Oberdorster E, Oberdorster J. Nanotoxicology: an emerging discipline evolving from studies of ultrafine particles. *Environ Health Perspect* 2005;**113**:823–39.

233. Stern ST, McNeil SE. Nanotechnology safety concerns revisited. *Toxicol Sci* 2008;**101**:4–21.

234. Borm PJ, Robbins D, Haubold S, Kuhlbusch T, Fissan H, Donaldson K, et al. The potential risks of nanomaterials: a review carried out for ECETOC. *Particle Fibre Toxicol* 2006;**3**:11.

235. Eskelson CD, Chvapil M, Strom KA, Vostal JJ. Pulmonary phospholipidosis in rats respiring air containing diesel particulates. *Environ Res* 1987;**44**:260–71.

236. Lundborg M, Bouhafs R, Gerde P, Ewing P, Camner P, Dahlen SE, et al. Aggregates of ultrafine particles modulate lipid peroxidation and bacterial killing by alveolar macrophages. *Environ Res* 2007;**104**:250–7.

237. Zetterberg G, Curstedt T, Eklund A. Possible alteration of surfactant in bronchoalveolar lavage fluid from healthy smokers compared to non-smokers and patients with sarcoidosis. *Sarcoidosis* 1995;**12**:46–50.

238. Finley TN, Ladman AJ. Low yield of pulmonary surfactant in cigarette smokers. *N Engl J Med* 1972;**286**:223–7.

239. Mancini NM, Bene MC, Gerard H, Chabot F, Faure G, Polu JM, et al. Early effects of short-time cigarette-smoking on the human lung - a study of bronchoalveolar lavage fluids. *Lung* 1993;**171**:277–91.

240. Wirtz HR, Schmidt M. Acute influence of cigarette smoke on secretion of pulmonary surfactant in rat alveolar type II cells in culture. *Eur Respir J* 1996;**9**:24–32.

241. Subramaniam S, Bummer P, Gairola CG. Biochemical and biophysical characterization of pulmonary surfactant in rats exposed chronically to cigarette-smoke. *Fundam Appl Toxicol* 1995;**27**:63–9.

242. Subramaniam S, Whitsett JA, Hull W, Gairola CG. Alteration of pulmonary surfactant proteins in rats chronically exposed to cigarette smoke. *Toxicol Appl Pharmacol* 1996;**140**:274–80.

243. Higenbottam T. Tobacco smoking and the pulmonary surfactant system. *Tokai J Exp Clin Med* 1985;**10**:465–70.

244. Teague SV, Pinkerton KE, Goldsmith M, Gebremichael A, Chang S, Jenkins RA, et al. Sidestream cigarette-smoke generation and exposure system for environmental tobacco-smoke studies. *Inhal Toxicol* 1994;**6**:79–93.

245. Bringezu F, Pinkerton KE, Zasadzinski JA. Environmental tobacco smoke effects on the primary lipids of lung surfactant. *Langmuir* 2003;**19**:2900–7.

246. Stenger PC, Alonso C, Zasadzinski JA, Waring AJ, Jung CL, Pinkerton KE. Environmental tobacco smoke effects on lung surfactant film organization. *Biochim Biophys Acta Biomembr* 2009;**1788**:358–70.

247. Honda Y, Takajashi H, Kuroki Y, Akino T, Abe S. Decreased contents of pulmonary surfactant proteins A and D in BAL fluids of healthy smokers. *Chest* 1996;**109**:1006–9.

248. Cook WA, Webb WR. Surfactant in chronic smokers. *Ann Thorac Surg* 1966;**2**:327–33.

249. Hite RD, Jacinto RB, Goldsmith BG, Bass DA. Inhibition of pulmonary surfactant function in smokers. *FASEB J* 1999;**13**:A173–A173.

250. Krzyzanowski M, Lebowitz MD. Changes in chronic respiratory symptoms in 2 populations of adults studied longitudinally over 13 years. *Eur Respir J* 1992;**5**:12–20.

251. Iribarren C, Jacobs Jr DR, Sidney S, Gross MD, Eisner MD. Cigarette smoking, alcohol consumption, and risk of ARDS: a 15-year cohort study in a managed care setting. *Chest* 2000;**117**:163–8.

252. Xie E, Yang Z, Li A. Experimental study on the treatment of smoke inhalation injury with lung lavage and exogenous pulmonary surfactant. *Zhonghua Wai Ke Za Zhi* 1997;**35**:745–8.

253. Geiser M, Kreyling WG. Deposition and biokinetics of inhaled nanoparticles. *Particle Fibre Toxicol* 2010;**7**:2.

254. Kemp SJ, Thorley AJ, Gorelik J, Seckl MJ, O'Hare MJ, Arcaro A, et al. Immortalization of human alveolar epithelial cells to investigate nanoparticle uptake. *Am J Respir Cell Mol Biol* 2008;**39**:591–7.

255. Schürch S, Gehr P, Hof VI, Geiser M, Green F. Surfactant displaces particles toward the epithelium in airways and alveoli. *Respir Physiol* 1990;**80**:17–32.

256. Gehr P, Green FHY, Geiser M, Hof VI, Lee MM, Schurch S. Airway surfactant, a primary defense barrier: mechanical and immunological aspects. *J Aerosol Med Deposition Clearance Effects Lung* 1996;**9**:163–81.

257. Kreyling WG, Semmler-Behnke M, Takenaka S, Moller W. Differences in in the biokinetics of inhaled nano- versus micrometer-sized particles. *Acc Chem Res* 2013;**46**:714–22.

258. Schulze C, Schaefer UF, Ruge CA, Wohlleben W, Lehr C-M. Interaction of metal oxide nanoparticles with lung surfactant protein A. *Eur J Pharm Biopharm* 2011;**77**:376–83.

259. Brown JS, Zeman KL, Bennett WD. Ultrafine particle deposition and clearance in the healthy and obstructed lung. *Am J Respir Crit Care Med* 2002;**166**:1240–7.

260. Oberdorster G, Sharp Z, Atudorei V, Elder A, Gelein R, Kreyling W, et al. Translocation of inhaled ultrafine particles to the brain. *Inhal Toxicol* 2004;**16**:437–45.

261. Kreyling WG, Semmler M, Erbe F, Mayer P, Takenaka S, Schulz H, et al. Translocation of ultrafine insoluble iridium particles from lung epithelium to extrapulmonary organs is size dependent but very low. *J Toxicol Environ Health A* 2002;**65**:1513–30.

262. Peters A, Doring A, Wichmann HE, Koenig W. Increased plasma viscosity during an air pollution episode: a link to mortality? *Lancet* 1997;**349**:1582–7.

263. Urankar RN, Lust RM, Mann E, Katwa P, Wang XJ, Podila R, et al. Expansion of cardiac ischemia/reperfusion injury after instillation of three forms of multi-walled carbon nanotubes. *Part Fibre Toxicol* 2012;**9**:38.

264. Geiser M, Casaulta M, Kupferschmid B, Schulz H, Semirriler-Behinke M, Kreyling W. The role of macrophages in the clearance of inhaled ultrafine titanium dioxide particles. *Am J Respir Cell Mol Biol* 2008;**38**:371–6.

265. Ruge CA, Kirch J, Canadas O, Schneider M, Perez-Gil J, Schaefer UF, et al. Uptake of nanoparticles by alveolar macrophages is triggered by surfactant protein A. *Nanomedicine Nanotechnol Biol Med* 2011;**7**:690–3.

266. Kendall M, Ding P, Mackay RM, Deb R, McKenzie Z, Kendall K, et al. Surfactant protein D (SP-D) alters cellular uptake of particles and nanoparticles. *Nanotoxicology* 2013;**7**:963–73.

267. Herzog E, Byrne HJ, Casey A, Davoren M, Lenz AG, Maier KL, et al. SWCNT suppress inflammatory mediator responses in human lung epithelium in vitro. *Toxicol Appl Pharmacol* 2009;**234**:378–90.

268. Bakshi MS, Zhao L, Smith R, Possmayer F, Petersen NO. Metal nanoparticle pollutants interfere with pulmonary surfactant function in vitro. *Biophys J* 2008;**94**:855–68.

269. Schleh C, Muhlfeld C, Pulskamp K, Schmiedl A, Nassimi M, Lauenstein HD, et al. The effect of titanium dioxide nanoparticles on pulmonary surfactant function and ultrastructure. *Respir Res* 2009;**10**:90.

270. Lalley PM. The aging respiratory system – pulmonary structure, function and neural control. *Respir Physiol Neurobiol* 2013;**187**:199–210.

271. Rebello CM, Jobe AH, Eisele JW, Ikegami M. Alveolar and tissue surfactant pool sizes in humans. *Am J Respir Crit Care Med* 1996;**154**:625–8.

272. Krumpe PE, Knudson RJ, Parsons G, Reiser K. The aging respiratory system. *Clin Geriatr Med* 1985;**1**:143–75.

273. Verbeken EK, Cauberghs M, Mertens I, Clement J, Lauweryns JM, Vandewoestijne KP. The senile lung – comparison with normal and emphysematous lungs.1. Structural aspects. *Chest* 1992;**101**:793–9.

274. Verbeken EK, Cauberghs M, Mertens I, Clement J, Lauweryns JM, Vandewoestijne KP. The senile lung – comparison with normal and emphysematous lungs.2. Functional-aspects. *Chest* 1992;**101**:800–9.

275. Miller MR. Structural and physiological age-associated changes in aging lungs. *Semin Respir Crit Care Med* 2010;**31**:521–7.

276. Christmann U, Hite RD, Witonsky SG, Elvinger F, Werre SR, Thatcher CD, et al. Influence of age on surfactant isolated from healthy horses maintained on pasture. *J Vet Intern Med* 2009;**23**:612–8.

277. Clercx C, Venker van Haagen AJ, den Breejen JN, Haagsman HP, van den Brom WE, de Vries HW, et al. Effects of age and breed on the phospholipid composition of canine surfactant. *Lung* 1989;**167**:351–7.

278. Yasuoka S, Manabe H, Ozaki T, Tsubura E. Effect of age on saturated lecithin contents of human and rat lung tissues. *J Gerontol* 1977;**32**:387–91.

279. Ueda T, Cheng G, Kuroki Y, Sano H, Sugiyama K, Motojima S, et al. Effects of aging on surfactant forms in rats. *Eur Respir J* 2000;**15**:80–4.

280. Burri PH. Structural aspects of postnatal lung development – Alveolar formation and growth. *Biol Neonate* 2006;**89**:313–22.

281. Betsuyaku T, Kuroki Y, Nagai K, Nasuhara Y, Nishimura M. Effects of ageing and smoking on SP-A and SP-D levels in bronchoalveolar lavage fluid. *Eur Respir J* 2004;**24**:964–70.

282. Tagaram HRS, Wang GR, Umstead TM, Mikerov AN, Thomas NJ, Graff GR, et al. Characterization of a human surfactant protein A1 (SP-A1) gene-specific antibody; SP-A1 content variation among individuals of varying age and pulmonary health. *Am J Physiol Lung Cell Molecular Physiol* 2007;**292**:L1052–63.

283. Elder ACP, Gelein R, Finkelstein JN, Cox C, Oberdörster G. Pulmonary inflammatory response to inhaled ultrafine particles is modified by age, ozone exposure, and bacterial toxin. *Inhal Toxicol* 2000;**12**:227–46.

284. Shimura S, Boatman ES, Martin CJ. Effects of ageing on the alveolar pores of Kohn and on the cytoplasmic components of alveolar type II cells in monkey lungs. *J Pathol* 1986;**148**:1–11.

285. Walski M, Pokorski M, Antosiewicz J, Rekawek A, Frontczak-Baniewicz M, Jernajczyk U, et al. Pulmonary surfactant: ultrastructural features and putative mechanisms of aging. *J Physiol Pharmacol* 2009;**60**:121–5.

286. Bell C, Seals DR, Monroe MB, Day DS, Shapiro LF, Johnson DG, et al. Tonic sympathetic support of metabolic rate is attenuated with age, sedentary lifestyle, and female sex in healthy adults. *J Clin Endocrinol Metab* 2001;**86**:4440–4.

287. Gaballa MA, Eckhart AD, Koch WJ, Goldman S. Vascular beta-adrenergic receptor adenylyl cyclase system in maturation and aging. *J Mol Cell Cardiol* 2000;**32**:1745–55.

288. Tsukada H, Kakiuchi T, Nishiyama S, Ohba H, Sato K, Harada N, et al. Age differences in muscarinic cholinergic receptors assayed with (+)N [(11)C]methyl-3-piperidyl benzilate in the brains of conscious monkeys. *Synapse* 2001;**41**:248–57.

289. Turner MJ, Mier CM, Spina RJ, Ehsani AA. Effects of age and gender on cardiovascular responses to phenylephrine. *J Gerontol A* 1999;**54**:M17–24.

290. Gyetko MR, Toews GB. Immunology of the aging lung. *Clin Chest Med* 1993;**14**:379–91.

291. Mancuso P, McNish RW, Peters Golden M, Brock TG. Evaluation of phagocytosis and arachidonate metabolism by alveolar macrophages and recruited neutrophils from F344xBN rats of different ages. *Mech Ageing Dev* 2001;**122**:1899–913.

292. Wallace WA, Gillooly M, Lamb D. Age related increase in the intra-alveolar macrophage population of non-smokers. *Thorax* 1993;**48**:668–9.

293. Brown MK, Naidoo N. The endoplasmic reticulum stress response in aging and age-related diseases. *Front Physiol* 2012;**3**:263.

294. Teramoto S, Fukuchi Y, Uejima Y, Teramoto K, Orimo H. Biochemical characteristics of lungs in senescence-accelerated mouse (SAM). *Eur Respir J* 1995;**8**:450–6.

295. Wang L, Green FHY, Smiley-Jewell SM, Pinkerton KE. Susceptibility of the aging lung to environmental injury. *Semin Respir Crit Care Med* 2010;**31**:539–53.

296. Braga FJ, Manco JC, Souza JF, Ferrioli E, De Andrade J, Iazigi N. Age-related reduction in 99Tcm-DTPA alveolar-capillary clearance in normal humans. *Nucl Med Commun* 1996;**17**:971–4.

297. Goodman BE. Characteristics and regulation of active transport in lungs from young and aged mammals. *Prog Clin Biol Res* 1988;**258**:263–74.

298. Uejima Y, Fukuchi Y, Nagase T, Tabata R, Orimo H. A new murine model of aging lung: the senescence accelerated mouse (SAM)-P. *Mech Ageing Dev* 1991;**61**:223–36.

299. Teramoto S, Fukuchi Y, Uejima Y, Teramoto K, Oka T, Orimo H. A novel model of senile lung – senescence-accelerated mouse (SAM). *Am J Respir Crit Care Med* 1994;**150**:238–44.

300. Matulionis DH. Chronic cigarette smoke inhalation and aging in mice: 1. Morphologic and functional lung abnormalities. *Exp Lung Res* 1984;**7**:237–56.

301. Uejima Y, Fukuchi Y, Nagase T, Matsuse T, Yamaoka M, Tabata R, et al. Influences of inhaled tobacco smoke on the senescence accelerated mouse (SAM). *Eur Respir J* 1990;**3**:1029–36.

302. Gortner L, Hilgendorff A, Bahner T, Ebsen M, Reiss I, Rudloff S. Hypoxia-induced intrauterine growth retardation: effects on pulmonary development and surfactant protein transcription. *Biol Neonate* 2005;**88**:129–35.

303. Morrison JL, Botting KJ, Soo PS, McGillick EV, Hiscock J, Zhang S, et al. Antenatal steroids and the IUGR fetus: Are exposure and physiological effects on the lung and cardiovascular system the same as in normally grown fetuses? *J Pregnancy* 2012;**2012**:15.

304. Murotsuki J, Gagnon R, Matthews SG, Challis JR. Effects of long-term hypoxemia on pituitary-adrenal function in fetal sheep. *Am J Physiol Endocrinol Metab* 1996;**271**:E678–85.

Chapter 27

Environmental Determinants of Lung Aging

Francis H.Y. Green* and Kent E. Pinkerton[†]

*Department of Pathology & Laboratory Medicine, University of Calgary, Calgary, AB, Canada,; [†]Center for Health and the Environment, University of California – Davis, Davis, CA, USA

INTRODUCTION

The purpose of this chapter is to review the ways in which environmental factors interact with the aging lung. In previous chapters we saw that the lung ages in a predictable way in mammals, and that aging involves both anatomic and physiologic correlates. Distinguishing between physiologic aging and the additional effects of environmental factors is difficult. As our knowledge of normal aging is based upon data from populations with different genetic susceptibilities to lung disease when exposed to air pollutants and respiratory tract pathogens, it has proven difficult to distinguish aging per se from the cumulative effects of environmental insults and genetic susceptibility.

Table 1 summarizes the major age-related changes in the mammalian respiratory system, the environmental factors that influence them, and associated diseases. In general, environmental factors enhance aging effects which, in turn, lead to clinical disease in susceptible subjects. Not all changes progress to disease with age. Adaptive responses have also been described. For example, changes in the composition of alveolar surfactant have been shown to partially offset age-related declines in lung mechanical properties in rats[1,2] and dogs.[3] Thus, adaptive changes may occur with age, and these may mitigate the aging process.

Lung disease can be separated into age-related diseases and age-dependent diseases[4]; the first are opportunistic (e.g., tuberculosis) and take advantage of age-related changes in the host; the second group comprises diseases that appear to be exaggerations of the normal aging process, for example, declines in lung function associated with cigarette smoking.

Obtaining accurate data on disease prevalence and severity can be difficult in the elderly. Epidemiologic studies have shown under-diagnosis of diseases, such as asthma, in the elderly.[5] Furthermore, with advancing age, the 'survivor' population becomes progressively less representative of the population as a whole.

An age-related change that has proven difficult to separate from environmentally induced disease is 'senile' emphysema.[6,7] The human lung loses approximately 30–50% of its tissue mass over an average lifetime. This is associated with loss of elastic fibres and capillaries in the alveolar walls.[8] Morphometric studies show an increase in alveolar duct volume, mean alveolar linear intercept, and size and number of interalveolar pores.[9,10] These changes meet some of the criteria for the diagnosis of emphysema, defined as a permanent abnormal enlargement of airspaces distal to the terminal bronchiole, accompanied by destruction of their walls. The functional correlates are identical. The aging lung is associated with increased closing volumes,[11] declines in FVC and FEV_1 and loss of elastic recoil pressure.[12] Senile emphysema does however differ from 'environmental' emphysema in several respects. Senile emphysema, unlike environmentally induced emphysema, involves the lung diffusely and is not readily apparent to naked eye inspection at autopsy. Environmentally induced emphysema, by contrast, is readily apparent on visual inspection of the lung, is patchy and occupies distinct anatomic compartments. Senile emphysema apparently results from changes in collagen composition characterized by a progressive shift towards insoluble collagen with fewer cross-links[9] and changes in elastin.[13] Furthermore, changes similar to those described in humans with senile emphysema have been described in rats, mice, Syrian hamsters, dogs, and non-human primates (reviewed in Chapter 14).[14] The morphologic and physiologic changes in lungs of aging beagle dogs[15] are virtually identical to those reported for aging human lungs. As environmental exposures in laboratory animal populations are usually controlled (with the possible exception of subclinical infections), it seems likely that the changes described in these animals reflect the aging process rather than a specific environmental effect. It would appear that the slow decline of FEV_1 with age and its anatomic and biochemical correlates are true age-related phenomena. These changes form the platform on which environmental factors operate.

The Lung. http://dx.doi.org/10.1016/B978-0-12-799941-8.00027-4

TABLE 1 The aging lung, environmental influences, and associated diseases

Age-Related Change	Environmental Influences	Associated Disease and/or Physiologic Effect
Accumulated genetic injury	Oxidative damage, radiation, mutagens, carcinogens	Lung epithelial metaplasia, dysplasia, cancer
Structural changes in chest wall and lung - alveolar matrix proteins - senile emphysema - increased chest wall stiffness - decreased respiratory muscle strength	Oxidative and protease damage Diet Obesity Hormonal changes	Decreased chest wall compliance Decreased static elastic recoil Increased residual volume Decreased vital capacity (VC) Decreased FEV_1 Increased lung compliance Mismatched ventilation/perfusion Hypoxemia, reduced gas diffusion
Impaired defense mechanisms - antioxidants - immune cell function - neural reflexes - clearance, deposition and retention of dust	Inhaled irritants/particles Cigarette smoke Allergens Pathogenic organisms Gastric contents	Pneumoconiosis Asthma Chronic bronchitis and emphysema Respiratory infections Aspiration pneumonia
Impaired control of breathing	Cigarette smoke Alcohol	Sleep apnea Disordered breathing Aspiration pneumonia
Pulmonary vascular remodeling	Cigarette smoke	Pulmonary hypertension Cor pulmonale

FACTORS THAT INFLUENCE SUSCEPTIBILITY OF THE AGING LUNG TO DISEASE

Aging is accompanied by changes that affect the whole body, not just the lung. In order to understand the relationship between aging of the lung and the environment, it is necessary to briefly review some general factors involved in aging.

Cellular Homeostasis

Many hypotheses have been reported to explain the basis for aging; such hypotheses must take into account that different species have markedly different longevities. Evolutionary theory predicts that longevity is determined by age of reproductive maturation. Programmed senescence is thus a driving force behind aging. The cumulative effects of oxidative stress underlie many theories of the mechanism of aging.[16] Mitochondria are a major source of intracellular free radicals.[17] A gradual impairment of recognition and repair of oxidized proteins, DNA and lipids and the cumulative effects of lipid peroxides are important in the aging process. Mice deficient in DNA repair and transcription mechanisms age prematurely.[18] Oxidative damage to DNA is extensive even under normal physiologic conditions. Estimates of damage range from one base modification in 130,000 bases in nuclear DNA to 1 per 8000 bases in mitochondrial DNA.[19] It is important to bear in mind that

oxygen and other free radicals also play important physiologic roles in defense against microbial pathogens and are second messengers in cell signaling.[20] Carcinogenesis is closely linked to the aging process and epidemiologic data indicate that aging contributes more to risk of cancer than environmental factors.[21] Replicative senescence occurs in cells where chromosomal telomeres are shortened.[22] Accumulation of post-mitotic cells reduces the ability of an organ to repair damaged cellular structures and contributes to the abnormalities associated with aging including neoplasia. Although degradation of DNA appears to be central to aging, post-translational modification of proteins with age may also be important.[23] It is likely that aging involves an interplay between various factors, including endogenous species-specific factors and exogenous agents (Figure 1).[24] Studies of individuals with exceptional longevity are also beginning to identify longevity-enabling genes.[25]

Oxidative Stress and Lung Antioxidant Defenses

The accumulation of reactive oxygen species (ROS) in cells with aging results in reversible and irreversible oxidative modifications of proteins (carbonylation or nitro-modifications), lipids (hydroperoxide lipid derivatives), and DNA (adducts and breaks) that eventually lead to functional impairment. Oxidative stress results in the transcription of

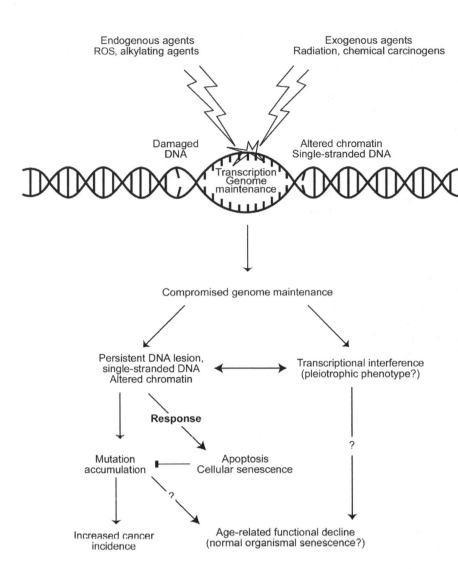

Endogenous agents
ROS, alkylating agents

Exogenous agents
Radiation, chemical carcinogens

Damaged
DNA

Altered chromatin
Single-stranded DNA

Transcription
Genome
maintenance

Compromised genome maintenance

Persistent DNA lesion,
single-stranded DNA
Altered chromatin

Transcriptional interference
(pleiotrophic phenotype?)

Response

Mutation
accumulation

Apoptosis
Cellular senescence

?

Increased cancer
incidence

Age-related functional decline
(normal organismal senescence?)

FIGURE 1 DNA damage in aging. Endogenous and exogenous agents cause continuing damage to DNA over the lifespan of the individual. Defects in genome maintenance may result in defective protein transcription, decreased gene activity, and elicit cellular responses such as apoptosis and senescence. Accumulated errors during DNA repair replication or recombination of damaged DNA may lead to the accumulation of mutations and an increased risk for cancer and other aging-related phenomena. *(Reprinted with permission from.[16])*

over 40 mammalian genes.[19] Deletion of P66shc, a signaling protein that maintains antioxidant levels, extends the life span of mice.[26] The lung has a remarkable array of antioxidant defenses.[27] Antioxidant enzymes synthesized in response to oxidant injury in the lung include superoxide dismutase (SOD), glutathione peroxidases, and catalases. Aqueous phase lung antioxidants include vitamin C, uric acid, ferritin, and ceruloplasmin. Lipid-associated antioxidants include vitamin E, surfactant, beta-carotene, and ubiquinone. Age-related changes in antioxidant defenses include decreased lung tissue ascorbic acid,[28,29] reduction of glutathione blood levels,[30–32] and decreased activity of SOD and glutathione peroxidase in lung tissues.[33] Nuclear factor, erythroid-derived 2 (NRF2), a transcription factor that protects cells and tissues from oxidative stress by activating antioxidant and detoxifying enzymes is decreased in the alveolar macrophages of older current smokers and patients with COPD compared with younger subjects.[34] However not all antioxidant defenses decrease with age. For example the vitamin E content of rat lungs increases with age.[35]

The lung is continuously exposed to oxidative stress. Reactive oxygen and nitrogen species derived from inflammatory processes are implicated in the pathogenesis of lung cancer.[36] A progressive reduction in antioxidant defenses might contribute to the decline of FEV_1 and increased risk of emphysema with increasing age.[37] Some particulates, such as cigarette smoke,[38,39] coal dust, and silica,[40] have intrinsic free radical activity. Others, such as cigarette smoke and diesel emissions, contain compounds that catalyze the generation of ROS[41] and most exogenous toxins activate lung resident cells to produce ROS.[42]

Innate and Acquired Immunity

Immune mechanisms underlie many chronic degenerative diseases of the lung. Both innate and acquired immunity tend to decline with age (Table 2).[43] Innate responses, mediated via pathogen-specific recognition receptors, are not well studied as a function of age. The elements of innate immunity include the functions of antigen presenting

TABLE 2 Altered immunity in the elderly

	Change with Age
Cell Mediated Immunity	
Naïve T cell output	↓
Thymocyte differentiation	Variable
Peripheral blood memory T-cells	↑
Hyporesponsive memory cells	↓
Proliferative responses to mitogens or antigens	↓
T cell receptor repertoire diversity	↓
Shift of Th1 to Th2 cytokine profile	↓
HLA-DR+	↑
Fas-mediated apoptosis	↓
Humoral Immunity	
Helper T cell function	↓
B cell number	↓
Germinal center formation	↓
Altered B cell repertoire expression	Variable
Antibody responses to specific antigens	↓
Altered generation of primary B cells	↓
Generation of memory B cells	↓
Ability to generate high-affinity protective antibody	↓
IgG and IgA	↑
Organ-specific autoantibodies	↓
Non-organ-specific autoantibodies	↓
Innate Immunity	
Neutrophil apoptosis and killing	↓
Neutrophil production of ROS	↑
Macrophage function	↓
Number of natural killer cells	↑
Dendritic Cells	
Number of dendritic cells	↓
B cell stimulation	↓
Capacity to capture antigen	↓
Capacity to phagocytose apoptotic cells	↓

HLA-DR: human leukocyte antigen DR; IgA: immunoglobulin A; IgG: immunoglobulin G; ROS: reactive oxygen species; Th: T-helper. Sources:.[43,255,256]

cells (such as dendritic cells, B cells, and macrophages), phagocytic cells (including macrophages and neutrophils), inflammatory mediators released from mast cells, polymorphonuclear leukocytes and macrophages, natural killer lymphocytes, antimicrobial molecules (such as nitric oxide), surfactant associated proteins, defensins, lactoferrin, and complement.[43]

The impact of age on adaptive or specific immune responses has been studied in detail, and is summarized in Table 2. The elderly produce a less vigorous response to vaccination. It is likely that some of these changes account for the increased risk for infectious disease in the elderly. Changes in broncho-alveolar lavage fluid have been noted with aging (Table 3). Some of these parameters reflect innate as well as acquired immune responses. On balance, the data indicate that the aging lung may suffer a low grade inflammation compared with younger individuals.[44] Inflammation is central to the pathogenesis of obstructive lung diseases, including asthma, chronic bronchitis, and emphysema.

Matrix Remodeling

A general feature of aging is decreased turnover of protein[45] combined with increased catabolism.[46] Elastin is very stable in the human lung with relatively little turnover.[47] The number and size of elastic fibers in the alveolar walls decreases with age whereas type III collagen increases.[48] With increasing age, serum elastin peptides, a marker for turnover of mature elastin, decrease[49] indicating that impaired repair mechanisms may play a role in the pathogenesis of senile emphysema.

Nutrition, Body Mass, and Physical Activity

Body mass increases with age and increased body mass is associated with reduced longevity.[50] Body mass is positively associated with airway hyperresponsiveness,[51] and with progressive reductions in FVC, FEV_1, and FEV_1/FVC ratio.[52] Cross-sectional[53,54] and longitudinal[55] studies have also shown a relationship between asthma and obesity, particularly in women.[56,57]

Static lung volumes were significantly increased following weight loss in middle-aged and older obese men (46–80 years)[58]; in contrast, aerobic exercise had no effect on pulmonary function but did increase maximal oxygen uptake (VO_2 max). The contribution of body composition, physical activity, and smoking to lung function in older humans has been investigated[59]; fat-free mass and physical activity both exerted significant independent effects on FEV_1. These results, in contrast to the former study,[58] indicate that heavy intense physical activity may be more important in contributing to forced expiratory function than previously recognized.

Animal studies have established that calorie restriction is associated with increased longevity. Metabolic studies in mice indicate that caloric restriction enhances and maintains

TABLE 3 Changes in bronchoalveolar lavage fluid in healthy aged individuals

Parameter	Change
Lymphocytes	↑
CD4 + /CD8 + T cell ratio	↑
HLA-DR + T cells	↑
B cells	↑
IgM, IgA and IgG	↑
Total protein	↑
Neutrophils	↑
Interleukins (IL-6, IL-8)	↑
α_1-antitrypsin	↑

Based on data from.[43]

protein turnover into old age.[60] Calorie restriction may thus slow the well-documented decline in protein renewal that occurs in the elderly and provides a scientific basis for understanding the contribution of calorie restriction to longevity. By contrast, severe calorie-protein restriction in rodents produces an emphysema-like lesion after several weeks of restricted diet.[61] This animal model of nutritional emphysema is characterized by destruction of collagen and elastin. Severe malnutrition may be a factor in the acceleration of disease in patients with advanced emphysema.[62,63]

Early Lung Injury

Several developmental abnormalities may affect the rate of onset of age-related lung diseases. Among these are diaphragmatic hernia, renal agenesis, oligohydramnios, cystic fibrosis, immotile cilia syndrome, and Down's syndrome. In addition, low birth weight[64] and early environmental influences, such as maternal smoking and oxidative injury resulting in bronchopulmonary dysplasia (BPD), may increase susceptibility to environmentally related diseases such as asthma.[65] Acute lung injuries produced by common pathogenic bacteria, for example *Staphylococcus aureus*, are usually associated with complete recovery and subsequent normal lung function. Viral infections, however, may produce more long lasting effects; notable among these are respiratory syncytial virus (RSV), adenovirus, and chlamydia. These agents increase risk for childhood asthma[66] and may be involved in the pathogenesis of COPD later in life.[67]

Regulation of Breathing

The elderly breathe with smaller tidal volumes and greater frequency than younger subjects, with the result that minute ventilation is unchanged.[68] The ventilatory response to hypoxia and, to a lesser extent, hypercapnia is blunted in the

elderly[69] as is the cardiac response to hypoxia.[70] Paradoxically, during exercise, elderly subjects appear more responsive to hypercapnia than younger subjects.[71]

The prevalence of sleep disordered breathing increases sharply with age in adults. In the middle-aged, the prevalence of sleep apnea is ~4% in females and ~9% in males.[72] In elderly subjects, the prevalence may be as high as 44%.[73] The increased prevalence of sleep disordered breathing in the elderly may, in part, result from changes in circadian sleep rhythms and altered cognitive function.[74] Diminished respiratory effort in response to upper airway occlusion and impaired perception of bronchoconstriction has been observed in older subjects.[75]

Gas Exchange

Regional heterogeneity in the ventilation-perfusion ratio increases with age, possibly due to small airway closure in dependent parts of the lung.[76] Mean arterial PO_2 declines with age[77] as does the pulmonary transfer factor for carbon monoxide (TLCO).[75] These changes in gas exchange could result from a variety of factors, including increased heterogeneity of ventilation-perfusion ratio, reduction of alveolar surface area and/or decreased density of lung capillaries.[78] Environmental exposures may also play a role. A longitudinal study of middle-aged men showed that the single breath TLCO did not decline in male non-smokers over a 22-year period; however, significant reductions occurred in smokers.[79]

Endocrine Function, Diurnal Rhythm, and Airway Receptors

Aging is associated with changes in endocrine function and blunting of diurnal rhythms.[74] Age-related changes in the hypothalamic-pituitary-adrenal axis have been demonstrated for growth hormone,[80] insulin, insulin-like growth factor, estrogen, glucocorticoids,[81] testosterone,[82] vitamin D,[83] and pulmonary venous epinephrine.[84] The age-related decrease in growth hormone is paralleled by changes in body composition (i.e., decreased lean body mass, bone mineral density, and increased visceral fat).[80] Increasing levels of cortisol with age, particularly in men, contribute to these effects. Glucocorticoid hypersecretion plays a role in accelerated aging in several animal models.[85] Declines in growth hormone, testosterone, and insulin with age contribute to a generalized catabolic state. Endogenous cytokines, such as interleukin-1 and TNF-α associated with chronic inflammation in the lung in COPD may also contribute to generalized protein catabolism.[86] Protein degradation in catabolic states is mediated primarily through the ubiquitin-proteosome proteolytic pathway.[46] Induction of proteosome expression by glucocorticoids may result from down-regulation of nuclear factor kappa B.[46] These hormonal changes may in part explain the loss of lung tissue mass with age. Impaired respiratory and peripheral muscle function and reduced capacity for exercise are seen with age, particularly in patients with COPD.[82]

Changes in the hypothalamic pituitary-adrenal axis occur with age and diurnal variability is attenuated.[87] Blood cortisol levels have been studied in 86 healthy men free of chronic illness who denied chronic use of medications; a single blood cortisol determination was made with the subjects in the supine position at 8:00 a.m.[81] Longitudinal analysis of the relationship between basal plasma cortisol concentration and FEV_1 over an average of 4.7 years revealed a significant ($p = 0.008$) relationship between the plasma cortisol concentration and the rate of decline of FEV_1 after adjustment for age, height, smoking status, and initial FEV_1. Perhaps most surprisingly, the authors' multivariate model predicted that subjects with cortisol concentrations one standard deviation (23.3 ng/ml) below the mean would experience FEV_1 declines of 71.6 ml/year greater than subjects with cortisol concentrations one standard deviation above the mean. The difference was comparable to the estimated 69.5 ml/year difference between current smokers and never-smokers. These data indicate that physiological concentrations of cortisol may modulate the process responsible for the deterioration of ventilatory function with aging.

A separate study of 631 male participants in the normative aging study (age range 44–85 years) showed that 2 h urinary excretion of serotonin, but not 5-hydroxy indole acetic acid (5-HIAA), decreased with age. Current smokers secreted significantly more serotonin than never-smokers. Former smokers did not differ significantly from never-smokers in these respects.[88]

Age-related changes in airway receptors have received less attention[89]. Changes in muscarinic receptor subtypes and receptor coupling to G proteins are described with senescence,[90] which might account for the increase in bronchial responsiveness to methacholine challenge in the elderly[91] and to increased airway responsiveness to inhaled carbon black in older mice.[92] Response to β agonists is also impaired in older subjects.[93] Cysteinyl-leukotriene (CysLT1) receptors, but not corticosteroid receptors, undergo functional changes with age.[94] These changes in receptor function have implications for treating older patients and may also have relevance for environmental exposures. However, these have yet to be studied.

Deposition, Retention, and Clearance of Particulates

In children, upper airway (mouth, larynx, or pharynx) deposition of fine therapeutic aerosols is greatly increased compared to adults, and deep lung deposition is correspondingly decreased.[95,96] This may explain why plasma concentrations of budesonide administered by a pressurized metered dose inhaler were similar in children (2–6 years) and adults (20–41 years) when administered the same nominal dose.[97]

Adult aging per se does not appear to alter the regional deposition fraction[98] but changes in inspiratory peak flow may impair deposition of drugs in the elderly.[77] Pre-existing disease, more common in the elderly, will also affect deposition patterns. Asthma and chronic bronchitis are associated with enhanced central deposition due to increased turbulence and high local velocities both during inspiration and expiration.[96] In a model of experimental emphysema decreased retention of diesel soot in the lung has been observed.[99]

Some studies suggest a slowing of clearance with increasing age,[100,101] but other studies indicate that clearance velocity is retained in old age in the absence of disease.[96] Aging is associated with accumulation of particles and metals in the mammalian lung. A non-occupationally exposed urban dweller in North America accumulates, on average, more than 5×10^6 mineral dust particles per dried gram of lung tissue over a lifetime.[102] The concentrations of carcinogenic metals, including oxides of chromium, nickel and cadmium, increase with age.[103,104] Exogenous carbonaceous particles also appear to accumulate progressively with age, but accurate quantification of their numbers has not been achieved.

SUSCEPTIBILITY OF THE AGING LUNG TO ENVIRONMENTAL INJURY

Ozone

Numerous studies in both humans and animals have shown an effect of age on the biological response to ozone[105]: these include changes in pulmonary eicosanoid metabolism in rabbits and rats,[106] altered ventilatory responses to CO_2 in rats,[107] altered hydroxylation of salicylate in lungs of Fischer 344 rats,[108] and altered lung mitochondrial respiration, reactive oxygen species (ROS) production, and lung pro/antioxidant status in rats.[109] Overall, the cellular and biochemical effects of ozone exposure appear greater in senescent Fisher 344 rats (24 months old) compared to juvenile or adult rats.[29,110] In humans, pulmonary toxicity to ozone appears to be dependent on the effective dose which, in turn, depends upon ventilatory rates.[111,112] Thus, in acute exposure studies, younger individuals appear to show greater adverse effects than older individuals.[113] However, more recent studies suggest that ozone has a more severe acute effect on lung function in the elderly.[114] The long-term effects of ozone on the human lung appear to enhance aging effects.[115] Ambient ozone has been associated with increased respiratory related emergency department visits and hospital admission in the elderly.[116,117] These findings suggest that the elderly are particularly susceptible to the effects of ozone.

Cigarette Smoke

There is increasing evidence that cigarette smoking is associated with premature aging and increased mortality from cardiovascular disease, cancer, and COPD.[118] It has been proposed that cigarette smoke accelerates the aging of the lung or worsens aging-related events in the lung by induction of senescence in alveolar epithelial cells and impaired re-epithelialization.[118] Smoking also causes defective resolution of inflammation and consequently induces accelerated progression of COPD.[119] It is proposed that the aging effect is mediated by the reaction of components of cigarette smoke with plasma and extracellular matrix proteins to form covalent adducts resulting in advanced glycation end products (AGE).[120] AGEs have been implicated in a variety of degenerative diseases associated with aging.[120,121] Smoking has other effects on the lung that enhance lung aging, including accelerated maturation of the fetal lung, impairment of lung growth, shortening of the plateau phase of FEV_1, and acceleration of age-related declines in FVC and FEV_1.[122] These effects are also shown for cigar and pipe smoking.[123] Smoking, by impairing host defense mechanisms, increases the risk of bacterial pneumonias,[124] a factor that contributes to the incremental declines in FEV_1 associated with COPD.

Animal models have offered insights into the pathogenesis of these clinical observations. The senescent mouse is more sensitive to cigarette smoke.[125] Enhanced susceptibility is also described in a genetically engineered senescence-prone mouse model.[126,127] In older rats, the ability to resist oxidative damage by cigarette smoke is seriously impaired, whereas the activation of PAHs to their carcinogenic forms remains intact.[128] Thus, the balance is shifted in favor of carcinogenesis in the aged rat. Whether this imbalance exists in the human is not known. Older humans who smoke have been shown to be at greater risk for developing adult respiratory distress syndrome (ARDS) than younger smokers.[129]

Particulates

The relationship between inhaled particulate matter (PM10 and PM2.5) in animals and humans has been studied as a function of age. There is a general consensus that the very young and the very old are more susceptible to the effects of particulate pollution.[14] Mortality and hospital admissions for cardiorespiratory diseases are associated with increasing age.[130-134] Particulate exposure also enhances age related declines in pulmonary function.[135] To what extent this increased risk results from aging per se, or to pre-existing illnesses (e.g., emphysema and cardiovascular disease) or cumulative increases in dust burden in the lung[103,104] is not clear. It is likely that many factors play a role, including the systemic effects of aging discussed earlier.

The effects of particles on age-associated changes and loss of lung surface area have been studied in normal rats and rats with elastase-induced emphysema. The rats were exposed to whole diesel exhaust for up to 2 years at a particle concentration of 3500 $\mu g/m^3$ beginning at 18 weeks of age. In this model, it was shown that the presence of emphysema protected the rats from particle-associated effects (measured by functional and biochemical parameters) and that there was a reduction in retained lung dust in the emphysematous animals.[99] In another study, the pulmonary responses of young (4 months) and aged (20 months) male Fisher 344 rats were compared following 3-day exposures, 5 hours/day to concentrated Boston ambient particulate matter at an average concentration of 100 $\mu g/m^3$.[136] The inflammatory responses, as measured by the concentration of cells in broncho-alveolar lavage fluid, were greater in both control and particulate-exposed young rats than in older rats. Because these were acute exposure studies, long-term effects on lung injury and remodeling were not studied. Therefore, it is not possible to determine whether the more brisk response in the younger rats would lead to a protective or damaging response in the long term. Broncho-alveolar lavage parameters, neutrophils and oxygen released from lavaged cells, have been studied in 8 and 20+ month old male Fisher 344 rats exposed for 6 hours/day to ultrafine carbon at 100 $\mu g/m^3$, ozone at 1 ppm, or a combination of carbon and ozone.[137,138] The pathologic effects of the individual agents or combinations on ROS were greater in older animals compared with the younger group, indicating greater sensitivity of older rats to oxidative lung damage. These short-term exposure studies indicate different cellular responses in older rats compared to young rats but do not provide information on the relationship between these findings and development of chronic disease nor on differences between young and old rats during long-term exposures.

The effects of age and carbon black exposure on airway resistance in mice was studied by Bennett et al.[92] Mice at 11, 39, 67, and 96 weeks were exposed to carbon black for 3 hours on 3 consecutive days. Baseline airway resistance was significantly lower in the oldest group of mice compared to the younger mice. Carbon black exposure resulted in increased resistance in the younger rats but no effect on the oldest group of rats, indicating that defensive responses are blunted in aging rats. A reduced sensitivity in cough reflex to inhaled distilled water has been shown to be suppressed in elderly humans compared to young subjects.[139]

Infectious Agents

Most forms of respiratory infection are more common in the elderly. Sub-optimal nutrition, deteriorating lung mechanics, and depressed immunity all appear to play a role. Group housing also plays a role in the elderly, contributing to the spread of diseases associated with influenza A and *Legionella*.[43] Patients aged 65 or older account for 30–80% of hospital admissions for community acquired pneumonia. Pneumococcal pneumonia is the most common cause of pneumonia in the very young and the elderly. However, the intracellular pathogen mycoplasma pneumonia is rare in the elderly,[140] possibly reflecting the changes in innate and acquired immunity associated with aging (see Tables 2 and 3). Tuberculosis causes 3×10^6 deaths per year worldwide. During the twentieth century, the incidence of tuberculosis fell from over 200/100,000 to less than 10/100,000 in developed nations primarily as a result of improvements in public health, nutrition and, to a lesser extent, by screening programs and chemotherapy. The proportion of tuberculosis in the elderly has risen in recent decades[141] where it is frequently overlooked, being commonly diagnosed postmortem.[142]

Because immune function is compromised in the elderly (Table 2), protection induced by immunizations is diminished in the elderly compared with adults as demonstrated by lower antibody titers and higher rates of respiratory illness.[143] Cell-mediated immune responses to vaccinations are also decreased in the elderly.[144]

NON-NEOPLASTIC DISEASES OF THE LUNG ASSOCIATED WITH AGING

Chronic Obstructive Pulmonary Disease (COPD)

Definition of COPD

COPD is a term used to describe a variety of chronic lung disorders characterized by chronic airflow obstruction, the main symptoms of which are progressive shortness of breath, cough, sputum, and wheeze. In North America, there are approximately 3–17 million people with a diagnosis of COPD. It is responsible for 2.2×10^6 disability adjusted life years and 0.5×10^6 potential years of life lost in the United States.[145] Its impact on the economy for work loss alone is estimated at $9.9 billion[146] and total costs may reach $18 billion annually.[145] COPD is the third leading cause of death. In the United Kingdom, ~10% of medical hospital admissions are for COPD, mostly in the elderly.[141] Although COPD is considered a disease of the elderly, COPD was present in 8.6% of a sample of participants in the 3rd National Health and Nutrition Examination Survey, a population that included males and females aged 18–65 (mean age 37.9 years).[147] Increasing age, male sex, white race, and smoking exposure were all significant risk factors for self-reported COPD, as was employment in particular industries and occupations.[148]

COPD has three components: reversible airway obstruction, mucous hypersecretion (simple chronic bronchitis), and emphysema. There is considerable clinical overlap between asthma, chronic bronchitis and emphysema, and many subjects with a diagnosis of COPD show features of more than one of these, leading to diagnostic confusion.[149] Furthermore, the predominant manifestation may change over time.[150] Reversible airflow obstruction, primarily associated with asthma, is found in many patients with chronic bronchitis and emphysema. Asthma also has an irreversible component.[151]

The natural history of COPD is inferred from cross-sectional, longitudinal, and autopsy studies. Pathology studies have been particularly useful in identifying the underlying anatomic features of COPD (Table 4). An anatomical feature that is common to the component diseases of COPD is altered small airways. Studies of expiratory flow volume curves indicate that small airway obstruction is also a cardinal feature of the aging lung in non-smokers.[152]

Over the years attempts have been made to unify the family of diseases that comprise COPD. These have focused on airway obstruction or airway hyperresponsiveness; however, these theories have tended to lack biological plausibility. Airway obstruction can result from such diverse processes as mucous plugs, airway thickening by edema or fibrosis, constriction of airway smooth muscle, or loss of elastic recoil due

TABLE 4 Relationship between pathologic features and clinical label in subjects with asthma and COPD

Pathologic feature	Clinical Label		
	Asthma	Simple chronic bronchitis	Emphysema
Mucous gland enlargement	+ + +	+ + +	+ / −
Bronchial inflammation - eosinophilic - lymphocytic - mast cells	+ + + + + + + +	− + + −	− − −
Subepithelial collagen deposition	+ + +	+	+ / −
Smooth muscle hyperplasia	+ +	−	−
Small airway disease *	+ +	+ +	+ +
Parenchymal destruction	−	+ / −	+ + +

NOTE: For clarity, other less common causes of chronic airflow obstruction (e.g., bronchiectasis and obliterative bronchiolitis) are not included.
*Small airway disease affects the membranous and respiratory bronchioles and has the following four features: fibrosis, smooth muscle hyperplasia, mucous cell hyperplasia, and variable dust pigmentation.

to destruction of lung parenchyma. Similarly, airway hyperresponsiveness can result from diverse causes ranging from geometric changes in airway wall dimensions (due to muscle hypertrophy, mucosal edema, and inflammation) to changes in sensitivity of airway smooth muscle to agonists. The concept that airway hyperresponsiveness and atopy underlie accelerated declines in FEV_1 is the basis for the Dutch hypothesis.[150,153] This approach has proven useful in epidemiologic studies[154] and has highlighted endogenous factors in the pathogenesis of these diseases; however, the hypothesis does not adequately account for the important role of exogenous factors, such as cigarette smoking, or for the strong evidence that the pathogenetic mechanisms are complex and better pursued by separating the component features.[155] One line of evidence that would point to a unifying theory for COPD is latent viral infections.[156] Double-stranded DNA viruses can persist in airway epithelial cells long after the acute infection has cleared, and viral genes may be expressed at the protein level without replication of a complete virus. Expression of the adenoviral trans-activating protein may act synergistically with exogenous agents like cigarette smoke to produce a heightened inflammatory response leading, in experimental models, to emphysema.[157] In support of this theory is the demonstration that adenoviral DNA is increased in the lungs of patients with COPD.[156]

COPD can be considered a disease of accelerated lung aging. Environmental factors accelerate this process. Potential mechanisms whereby age and environment potentiate each other include defective resolution of inflammation, defective repair mechanisms, increased susceptibility to

ROS due to telomere shortening, and interference with anti-aging molecules.[118]

Effect of Aging on Airflow Limitation

The decline of FEV_1 with age is due in part to loss of elastic recoil secondary to emphysema and in part to changes in airway smooth muscle tone.[77,122,158] The evolution of changes in lung volumes with age is shown in Figure 2, and the timeframe of these changes as they affect airways and parenchyma is shown in Table 5. The relationship between FEV_1 and age is shown in Figure 3. This figure shows the

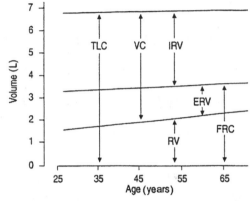

FIGURE 2 Changes of lung volumes with age. TLC: total lung capacity; VC: vital capacity; IRV: inspiratory research volume; ERV: expiratory reserve volume; FRC: functional residual capacity; RV: residual volume. (*Adapted with permission from.*[254])

TABLE 5 Timeframe of normal age-related changes and development of asthma and COPD

Age	Normal Aging	Airway Disease	Parenchymal Disease
−1, 0, +1	Immune status determined (Th1/Th2)	Asthma phenotype induced	
1–15	Growth Hormonal influences	Airway hyperresponsiveness Reversible bronchoconstriction	
15–25	Peak lung performance	Small airway disease	Small airway disease
25–35	Plateau period of lung function		Unopposed protease activity causes increased alveolar fenestrae (pores of Kohn)
35–60	Changes in lung matrix proteins Changes in pulmonary defense mechanism Changes in immune system Progressive decline in lung function Increasing hypoxemia	Structural remodeling of large and small airways, including smooth muscle, mucous cells, and mesenchyma Irreversible airway obstruction Mucous gland hypersecretion Increased risk for infection, pneumonia	Fragmentation of alveolar walls with centriacinar emphysema Progression to panacinar emphysema
60 +	Senile emphysema Approximately 50% loss of respiratory reserve Compatible with healthy life	Increased risk for life-threatening infections, lung cancer, cor pulmonale leading to premature death	Professive ventilatory impairment, leading to hypoxemia, cor pulmonale, and premature death

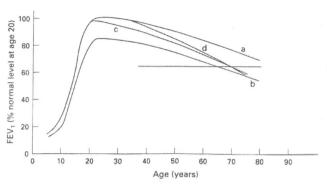

FIGURE 3 FEV_1 plotted as a percentage of peak value at age 20 against age. Line a = healthy normal subjects, Line b = submaximal growth but normal decline, Line c = premature or early decline, and Line d = accelerated decline in lung function compared with normal subjects (Line a). The horizontal line indicates levels below which clinical impairment might be anticipated. The figure illustrates the important point that age-related declines in FEV_1 have more than one cause and that more than one may be operating in a given individual. *(Reproduced with permission from.[164])*

different possibilities for failure to attain normal FEV_1 at age 20 and for accelerated declines thereafter. Exposure to cigarette smoke in adolescents, especially girls,[159,160] appears to affect the growth phase of the lungs as does the presence of persistent asthma.[161] In healthy non-smokers, there appears to be a plateau phase of lung function from 25–35 years.[162,163]

Most studies of aging on airflow limitation have focused on FEV_1 largely because of ease of measurement and good reproducibility.[122] Peak lung function occurs between ages 18 and 25 (Figure 3), followed by a plateau phase, after which FEV_1 declines with age. The peak appears to be later for non-smokers than smokers, after which smokers show accelerated declines in FEV_1 compared with non-smokers. The value of FEV_1 at a given point in adulthood is determined by three factors: (1) lung function attained during early adulthood, (2) duration of the plateau phase, and (3) rate of decline of lung function thereafter.[122] Because these factors are largely unknown for a given individual, prediction equations for estimating lung function suffer from considerable error. Age, height, and sex account for 40–50% of forced vital capacity (FVC),[37] while the remainder is unexplained. Equations based on cross-sectional data tend to underestimate reductions in FEV_1 compared with equations based on longitudinal study data.[122,164] There are many prediction equations for FEV_1; some assume a linear relationship with age, others do not.[165]

In non-smokers, there is a steady decline in FEV_1 with age (Table 6).[166–169] Smoking accelerates this age-dependent decline in FEV_1.[169,170] Age-related reductions in FEV_1 were studied in 1397 men and women aged 51–95 years to determine the influence of age and smoking cessation on FEV_1.[171] The decreases in FEV_1 per year reported by Frette et al.[171] among never smoking men (34 ml/year) and never-smoking women (28 ml/year) are similar to that reported for subjects of

the same age in other cross-sectional or longitudinal studies, with reductions of 28–34 ml/year for men[172-174] and 21–32 ml/year in women.[172,173] The results from some longitudinal studies of FEV_1 with age in smokers and non-smokers are shown in Table 6. From these data, it appears that, in moderate to heavily smoking men, FEV_1 declines by ~15 ml/year more than in non-smokers. Comparable declines, based on normalized data, are seen for women. Only a small proportion of smokers appear to develop clinically significant airway obstruction.

Some studies of annual declines in FEV_1 as a function of age have noted a slowing of the declines in FEV_1 in older men and women, in both smokers and non-smokers.[171,175,176] Other studies have reported an accelerated decline of FEV_1 with age.[177] An apparent positive relation of FEV_1 in the oldest (81 years and older) current smokers have been noted[171,172]; however, the numbers of subjects in the older categories are small and thus, the population probably shows a survival selection bias. Another problem with studies in the elderly is the large proportion of men and women unable to perform ventilatory function tests that meet ATS standards. Spirometry test failure is an index of poor health status and has been shown to be associated with a lower FEV_1, a faster rate of annual loss of ventilatory function, increased respiratory symptoms, and excess mortality.[77]

Genetic and Familial Factors in Decline in Lung Function

Epidemiologic studies have shown that lung function (FVC, FEV_1) is predictive of longevity.[178–180] The relationship between lung function and mortality is to all causes, not just to lung specific diseases.[181,182] Simple lung function tests appear to provide important information on the general health of elderly subjects and offer a good overall marker of aging. These relationships remain after known risk factors, such as smoking and occupational exposures, are taken into account.[181]

The heritability of lung function has been studied in heterozygotic and monozygotic twin pairs (aged 18–34 years) living in Australia. The heritability in females was ~0.8 and slightly lower in males, ~0.6. The difference between males and females was attributable to environmental influences.[183] This study showed that ~9% of the variance in women was associated with Pi polymorphism. No effect of the Pi locus was found in males. A similar twin study from Sweden gave heritability estimates of 0.48 and 0.67 for VC and FEV_1, respectively, but no gender differences were noted.[184] Genetic effects appeared equally important in predicting FEV_1 in older and younger populations. Other studies have confirmed the heritability of pulmonary function.[184–186] Pulmonary function has even been related to longevity of parents.[187] Putative genes identified to date include alpha-1-antitrypsin (α1-AT), tumor necrosis factor α (TNF-α), and surfactant protein B (SP-B) genes.[188]

TABLE 6 Longitudinal studies of decline in FEV_1

Reference	Population	Age Range	No. Studied	Follow Up (n)	Reference Decline in Never-Smokers (ml/year)	Findings in Current (Persistent) Smokers (ml/year)
[257]	London transport and bank workers, enriched subgroups	30–59	792	8 years (16)	M – 36	M < 5 cig/day – 44 5–15 cig/day – 46 15–25 cig/day – 54 > 25 cig/day – 54
Tashkin et al., 1984	Four community cohorts, sampled for different pollution exposure	25–64	2401	5 years (2)	M – 56 F – 42	M – 70 F – 54
[168]	General population sample	20–70	1705	10 years (7)		M –13 ml/year lower than in never–smokers F – 7 ml/year lower than in never–smokers
[163]	Population sample, indexed via school children	5–55	1887	10 years (10)	M (< 40) – 20 (40–55) – 35 F (< 42) – 10 (42–55) – 35	M (21–32) – 25 (33–43) – 40 (44–55) – 30 F (19–29) – 20 (30–55) – 30
Lange et al., 1987	Hospital catchment area sampled for cardiovascular study, enriched subgroups	20–? (at least 69)	7764	5 years (2)	M (< 55) – 21 (≥55) – 34 F (< 55) – 13 (< 55) – 32	< 15 cig/day ≥ 15 cig/day M (< 55) – 14 – 35 (≥ 55) – 53 – 55 F (< 55) – 14 – 30 (≥ 55) – 41 – 51
Sherman et al., 1992	Population samples from 6 U.S. cities	25–74	3948	12 years (4)	Symptoms - + M – 33 – 34 F – 28 – 31	Symptoms - + M – 42 – 47 F – 34 – 36
Xu et al., 1994	2 × 2 cohorts sampled for different pollution and urbanization	15–54	4554	24 years (8)		Compared with never-smokers < 15 cig/day 15 – 24 cig/day ≥ 25 cig/day M – 4 (3) – 10 (3) – 14 (3) F – 6 (2) – 11 (2) – 19 (3)

Adapted from.[122]

Effect of Smoking Cessation on Age-Related Decline in Lung Function

Smoking cessation has been shown to reverse the accelerated decline in lung function associated with cigarette smoking. Intermittent smoking is associated with rates of decline between those of active current smokers and never-smokers.[165,174,189] Age of quitting has a dramatic effect on subsequent declines in lung function.[171] Smokers (men and women) who quit before age 40 have an age and height adjusted FEV_1 that did not differ from never-smokers: those who quit between the ages of 40 and 60 had greater FEV_1s values than current smokers but lower than those of never-smokers. However, in smokers who quit after age 60, FEV_1 was similar to that of current smokers.

Risk Factors for Accelerated Declines in Lung Function

Of the attributable risks for COPD, cigarette smoking accounts for approximately 80–90%.[162] By contrast, heavy habitual marijuana smoking does not appear to accelerate FEV_1 decline with age.[190] A longitudinal study of 1171 randomly chosen US steel workers demonstrated that age, weight gain, smoking, trauma, pneumonia, and a history of allergy, asthma or hay fever were independently related to the risk of accelerated declines in FEV_1.[191]

Occupational dust exposure is another well-recognized cause of accelerated declines in FEV_1. In a review of 13 studies between 1966 and 1991 encompassing four different cohorts of miners, a relationship was observed between lung function decline and dust exposure when corrected for age and other risk factors[192]; other reviews have arrived at the same conclusion.[193,194] Air pollution has also been linked to accelerated rates of decline of FEV_1.[195]

Asthma[196] and airway hyperresponsiveness have been shown to be independent risk factors for accelerated declines in FEV_1 in a number of large cohort studies.[162,197–200] On average, having airway hyperresponsiveness adds ~10 ml to the annual decline in FEV_1[162]; the effect is largest in elderly subjects.[198] No interaction between airway hyperresponsiveness and smoking has been demonstrated.[154,162] Chronic mucous hypersecretion is also an independent risk factor for accelerated declines in FEV_1.[201] Gender does not appear to affect rates of decline.[174]

Studies of Innuit indicate that adoption of a modern lifestyle, particularly the use of snowmobiles, increases the rate of decline of FVC and FEV_1; this decline was less in those who exercised regularly.[202]

Markers of Accelerated Decline in Lung Function

Urinary excretion of the amino acid, desmosine (DES), a specific marker for degradation of mature cross-linked elastin, is elevated in healthy smokers compared with healthy non-smokers.[203] A nested case control study of elastin and collagen degradation in current smokers with and without rapid decline of lung function, showed that rapid decline in pulmonary function was associated with increased urinary excretion of desmosine.[204] Antioxidant gene polymorphisms may also serve as markers for rapid decline in lung function.[205] Blood elastin peptide has not proven useful as a predictor of rate of decline of FEV_1,[206] but may have some value in evaluating remodeling in emphysema and fibrosis.[49]

Emphysema

Pathogenesis

In humans, cigarette smoking is by far the most common cause of emphysema. Dusts and fumes associated with specific occupations are also major causes of emphysema in selected populations.[194,207] Emphysema due to environmental agents, for example, smoking and coal mine dust exposure, is usually centriacinar in type.[208]

There are several genetically determined forms of emphysema, the most common of which is α1-AT deficiency. α1-AT exists as more than 70 biochemical variants (the PI system) that are inherited as autosomal codominant alleles.[209] Homozygosity for α1-AT deficiency occurs in about 6% of the population. Affected individuals often develop emphysema at an earlier age (< 40 years) than is seen in the general population in which clinically significant emphysema rarely develops before 50 years of age.

α1-AT deficiency is associated with panacinar emphysema involving all lobes of the lung. In contrast, environmental emphysema caused by cigarette smoke or occupational dusts is usually more pronounced in the upper zones and starts in a centriacinar location. α1-AT deficiency impairs the antiproteolytic defense mechanisms of the lung, predisposing protease damage of the connective tissue framework. The development of emphysema in α1-AT is associated with age and exposure to environmental agents, including cigarette smoke and workplace dusts and fumes. Subjects who are heterozygous (PiMZ) for α1-AT deficiency are found in ~10% of the population. These individuals are also at risk of developing pulmonary emphysema.[210] A 10-year study of 28 PiMZ subjects compared to 28 normal PiMM subjects showed significant declines in FEV_1, TLCO, and static trans-pulmonary pressures. Total lung capacity and residual volume were increased in the heterozygous group and trypsin inhibitory capacity was decreased (0.65 ± 0.17 mg/ml) compared to 1.52 ± 0.3 mg/ml in the PiMM group. Changes in the PiMZ group exceeded the values expected as the result of physiologic aging.

Asthma: Effects of Age and Environmental Exposures

Asthma in the elderly is often severe and refractory to therapy[211] and is frequently underdiagnosed and undertreated.[212] Mortality rates appear increased in elderly asthmatics.[211] A Japanese study showed that the median age for asthma-related deaths was 60 years; males predominated up to 60 years and female deaths were greater thereafter.[213] Although more deaths from asthma occur in older populations, children and young adults have shown the greatest increase in death rates in recent decades.[214]

Asthma was reported by 4.1% of all adults in Michigan[215] and by 11.1% of men and 9.1% of women aged 65–95 years in Arizona.[212,216] The interactions of smoking and aging on allergic airway inflammation are complex. Elderly asthmatics tend to have increased neutrophils in their airways compared with younger asthmatics. Cigarette smoking appears to enhance this effect.[217] IgE and blood eosinophils are also increased in elderly smoking asthmatics compared with younger smoking asthmatics.[218]

Bronchial hyperresponsiveness (BHR) has been described in infants and very young children, and it has been postulated that BHR is present in all children but that genetic and environmental factors determine which children will lose their BHR.[150] In a longitudinal study of children, BHR was related to slower growth of FEV_1[219] and increased risk for developing asthma.[220] BHR is related to blood eosinophilia in the young and the elderly.[199,221–223] Longitudinal studies of ventilatory function in adults with asthma indicate that elderly patients with asthma have accelerated declines in FEV_1 compared to non-asthmatic elderly individuals.[224]

A pathologic study designed to determine the effective age and duration of disease on airway wall remodeling in fatal asthma compared a group of young individuals who died of asthma (age range 20–23 years) with a group of older individuals who died of asthma (age range 40–49 years) (Bai *et al.*, 2000).[225] The study showed an increase in airway wall area and smooth muscle thickness in the older asthmatics compared to the younger group. These data support the concept that mesenchymal remodeling is a progressive process and accounts for the irreversible airflow obstruction demonstrated in longitudinal studies of asthma.[153,224,226,227] To what extent these changes are a function of age or duration of disease cannot be determined from these studies. A morphologic change in asthma that does not appear to be influenced by age or duration of asthma is subepithelial collagen (basement membrane) thickness.[225,228,229]

Atopy

Age-related differences in the manifestations of asthma are well recognized and discussed earlier. Asthma is influenced by factors associated with the developing lung, hormonal influences in adolescence, and age-related changes in IgE and specific immune and non-specific inflammatory responses. Allergen priming begins in utero. House dust mite-specific T-cells have been identified in cord blood.[230]

A longitudinal study of T-helper (Th) lymphocyte cytokine responses was performed prospectively in blood mononuclear cell responses from birth to 2 years in atopic and non-atopic infants.[230] At birth, both groups showed Th_2 skewed allergen-specific responses to house dust mite antigen, which were greater in the non-atopic group. In the first year of life, there was a rapid down regulation of the Th_2 response in the non-atopic group, whereas the atopic group showed consolidation of their Th_2 responses. This study supports the hypothesis that patterns of immune deviation in the first year of life determine atopic status in later years.

Sensitization to allergens is also age dependent. Atopy, as expressed by serum immunoglobulin E (IgE), is a strong predictor of asthma,[231,232] even in the elderly,[5] Serum IgE declines with age[233,234] and is influenced by race,[232,233] smoking history,[235] occupation,[236,237] and exposure to aero-allergens.[236] Reductions in IgE with age are associated with decreased eosinophilia.[150,221] The relationship between serum IgE and age is different for men and women. Studies have repeatedly shown that in women, IgE levels fall with age, whereas in men, IgE levels remain relatively stable. These changes are seen even when adjusted for the effects of smoking and environmental factors.[236]

The relationship between serum IgE and the rate of decline of pulmonary function in the general population has produced conflicting results. Several studies have shown an inverse relationship between serum, IgE, and rate of decline in FEV_1[238–241]; however, other studies failed to show such a relationship[199,242] or revealed that the effect was confined to individuals with a diagnosis of asthma.[243]

Specific IgE is also related to age,[244,245] smoking,[246] and occupational exposures.[236] Specific IgE levels decline with age in both men and women.[236] Skin test reactivity to common aero-allergens has been shown to be a significant predictor of annual rates of decline of FEV_1 and FEV_1/FVC ratio.[198,247] In a study of rates of decline of lung function in 1025 men using regression models, the excess rate of decline in FEV_1 associated with skin test reactivity to house dust, mixed grasses, mixed trees, and/or ragweed was 9.45 ml/year.[204] Surprisingly, no relationship was found between serum IgE level and annual FEV_1 decline in a longitudinal analysis of 790 men from the same cohort.[199]

There is evidence that environmental agents, particularly diesel exhaust particulates and cigarette smoke, are capable of augmenting IgE immune responses[235,240] and pro-inflammatory cytokine production.[248] A longitudinal study of 778 elderly men (aged 41–86 years) showed that current smokers with at least one positive skin test reaction to common aero-allergens displayed significantly greater methacholine dose-responsiveness than smokers with negative skin test results. A relationship between skin test positivity and methacholine responsiveness was also seen in non-smokers, but the effect was not statistically significant. The data indicate that smoking and atopy may act synergistically to increase airway hyperresponsiveness in an older male population. A study of 250 men and women aged 65–91 years showed that, after allowance for age and gender, serum IgE and smoking interacted synergistically as risk factors for airflow obstruction.[240] This synergism was also observed in subjects who showed no evidence of airway hyperresponsiveness to inhaled methacholine. These findings indicate that the role of IgE in the pathogenesis of airflow obstruction is not confined to asthmatics.

Triggers

Two classes of environmental factors can trigger an asthma attack: sensitizing agents (e.g., house dust mite, cat dander) and irritants (e.g., fine particles, gases, and fumes). The dominant triggers vary with age; asthmatics aged 2–20 years were more likely to have asthma triggered at night and by respiratory infections, whereas older asthmatics more commonly reported exposure to allergens and exercise as triggers.[249] Other triggers, such as cigarette smoke exposure, air pollution, stress, or changes in weather showed no relationship to age.[249]

A cross-sectional study of asthmatics in Holland showed that the number of specific agents that asthmatics reported triggering their asthma and the degree of reaction to those agents increased until 25 years of age and then started to decline. Furthermore, during childhood, positive tests for indoor allergens predominated, but after age 15 sensitization to outdoor allergens (such as grass pollen) were more common.[250]

CONCLUSIONS AND FUTURE DIRECTIONS

Aging appears simple, but in reality the situation is highly complex. It is not defined by one variable, time, but by a multitude of processes, which appear to interact. The latter include genes, environmental stressors, and pre-existing disease. This chapter has provided an overview of these factors and their interactions. Much, however, remains to be discovered. There is still a shortage of information on the effects of aging on the adult human lung.[188,251] Many longitudinal epidemiological studies designed to detect effects of environmental pollutants on lung disease have noted or controlled for age-related effects. Few of these have tested for interactions between the environmental factor of interest and age. Where this has been done (e.g., in studies of decline of FEV_1 associated with cigarette smoking or smoking cessation), non-linear effects have been noted. There is a need for more longitudinal studies of hormonal status and age-related declines in lung function, for example, the findings of Sparrow et al.[81] need replicating. Their observation that a plasma cortisol concentration one standard deviation below the mean was associated with an annual decline in FEV_1 equivalent to current smoking is quite remarkable. Furthermore, endocrine deficiencies are potentially treatable.

Although the annual decline of FEV_1 is an accurate marker of lung aging and has some predictive value for identifying individuals at risk for pulmonary impairment, there is a need to develop biochemical tests for identifying high risk individuals. In this regard, breath analysis for indices of oxidative stress, for example, carbon monoxide (produced from the stress protein, heme-oxygenase), and ethane (an end product of lipid peroxidation) look promising.[252]

There is a need to determine the underlying structural basis for the declines in FEV_1. Emphysema and changes in airways are only poorly correlated with FEV_1[253] suggesting that other, as yet, undefined factors may be important. There is a need to determine the cause of the variability in rates of decline of FEV_1 with age in both smokers and non-smokers.[188] Several genetic and environmental risk factors for rapid decline of FEV_1 have been identified; however, the majority of the risk cannot be attributed to known factors.[188] There is a need to correlate new insights from COPD research into studies of the normal aging process, specifically the roles of proteases, oxidant injury, viral infection, and apoptosis.[188] Finally, there is a need for studies of lung function and anatomy in individuals with exceptional longevity.

ACKNOWLEDGEMENTS

Special thanks to Andrea Chiu and Morenike Fadayomi.

REFERENCES

1. Kato S, Takahashi K. The mechanism of age-related change in lung elastic recoil pressure. *Nihon Kyobu Shikkan Gakkai Zasshi* 1992;**30**:175–81.
2. Ueda T, Cheng G, Kuroki Y, Sano H, Sugiyama K, Motojima S, et al. Effects of aging on surfactant forms in rats. *Eur Respir J* 2000;**15**:80–4.
3. Clercx C, Venker-van Haagen AJ, den Breejen JN, Haagsman JP, van den Brom WE, de Vries HW, et al. Effects of age and breed on the phospholipids composition of canine surfactant. *Lung* 1989;**167**:351–7.
4. Brody JA, Schneider EL. Diseases and disorders of aging: an hypothesis. *J Chron Dis* 1986;**39**:871–6.
5. Burrows B, Barbee RA, Cline MG, Knudson JR, Lebowitz MD. Characteristics of asthma among elderly adults in a sample of the general population. *Chest* 1991;**100**:935–42.
6. Thurlbeck WM, Ryder RC, Sternby N. A comparative study of the severity of emphysema in necropsy populations in three different countries. *Am Rev Respir Dis* 1974;**109**:239–48.
7. Verbeken E, Cauberghs M, Mertens I, Clement J, Lauweryns JM, Van de Woestijne KP. The senile lung. Comparison with normal and emphysematous lungs. I: Structural aspects. *Chest* 1992;**101**:793–9.
8. Thurlbeck WM, Wright JL. *The Aging Lung*. Thurlbeck's Chronic Airflow Obstruction. 2nd ed. Hamilton, Canada, Dekker; 1999. 128–131.
9. Thurlbeck WM. Morphology of the aging lung. In: Crystal RG, West JB, Barnes PJ, Cherniack NS, Weibel ER, editors. *The Lung*. New York: Raven Press; 1991. pp. 1743–8.
10. Pierce JA, Ebert RV, Ark LR. The elastic properties of the lungs in the aged. *J Lab Clin Med* 1958;**51**:63–71.
11. Jones RL, Overton TR, Hammerlindl DM, Sproule BJ. Effects of age on regional residual volume. *J Appl Phys* 1978;**44**:195–9.
12. Knudson RJ, Clark DF, Kennedy TC, Knudson DE. Effect of aging alone on mechanical properties of the normal adult human lung. *J Appl Physiol Respirat Environ Exerc Physiol* 1977;**43**(6):1054–62.
13. Kuhn III C. Morphology of the aging lung. In: Crystal RG, West JB, Barnes PJ, Weibel ER, editors. *The Lung: Scientific Foundations*. 2nd ed. Philadelphia: Lippincott-Raven; 1997. pp. 2187–92.
14. Mauderly JL. Animal models for the effect of age on susceptibility to inhaled particulate matter. *Inhal Toxicol* 2000;**12**:863–900.
15. Gillett NA, Gerlach RF, Muggenburg BA, Harkema JR, Griffith WC, Mauderly JL. Relationship between collateral flow resistance and alveolar pores in the aging beagle dog. *Exp Lung Res* 1989;**15**(5):709–19.
16. Hasty P, Vijg J. Genomic priorities in aging. *Science* 2002;**296**:1250–1.
17. Cadenas E, Davies KJA. Mitochondrial free radical generation, oxidative stress and aging. *Free Radic Biol Med* 2000;**29**:222–30.
18. De Boer J, Andressoo JO, de Wit J, Huijmans J, Beems RB, van Steeg H, et al. Premature aging in mice deficient in DNA resport and transcription. *Science* 2002;**296**:1276–9.
19. Davies KJ. Oxidative stress, antioxidant defenses, and damage removal, repair, and replacement systems. *IUBMB Life* 2000;**50**:279–89.
20. Forman HJ, Torres M. Reactive oxygen species and cell signaling. Respiratory burst in macrophage signaling. *Am J Respir Crit Care Med* 2002;**166**:S4–8.

21. Benson D, Mitchell N, Dix D. On the role of aging in carcinogenesis. *Mutat Res* 1996;**356**:209–16.
22. Goyns MH. Genes, telomeres and mammalian ageing. *Mech Ageing Dev* 2002;**123**:791–9.
23. Stadtman ER. Protein oxidation and aging. *Science* 1992;**257**:1220–4.
24. Musicco M, Sant S, Pettenati C. Mortality studies in the clarification of gene-environment interactions. *Funct Neurol* 1998;**3**:253–6.
25. Perls T. Genetic and environmental influences on exceptional longevity and the AGE nomogram. *Ann N Y Acad Sci* 2002;**959**:1–13.
26. Migliaccio E, Giorgio M, Mele S, Pelicci G, Reboldi P, Pandolfi PP, et al. The p66shc adaptor protein controls oxidative stress response and lifespan in mammals. *Nature* 1999;**402**:309–13.
27. Comhair SA, Erzurum SC. Antioxidant responses to oxidant-mediated lung diseases. *Am J Physiol Lung Cell Mol Physiol* 2002;**283**(2):L246–55.
28. Rikans LE, Moore DR. Effect of aging on aqueous phase antioxidants in tissue of male Fischer rats. *Biochim Biophys Acta* 1988;**966**:269–75.
29. Vincent R, Vu D, Hatch G, Poon R, Dreher K, Guenette J, et al. Sensitivity of lungs of aging Fischer 344 rats to ozone: assessment by bronchoalveolar lavage. *Am J Phys* 1996;**271**:L555–65.
30. Perez R, Lopez M, Baria de Quiroga G. Aging and lung antioxidant enzymes, glutathione and lipid peroxidation in the rat. *Free Radic Biol Med* 1991;**10**:35–9.
31. Canada AT, Herman LA, Young SL. An age-related difference in hyperoxia lethality: role of lung antioxidant defense mechanisms. *Am J Phys* 1995;**268**:L539–45.
32. Bottje WG, Wang S, Beers KW, Cawthon D. Lung lining fluid antioxidants in male broilers: age-related changes under thermoneutral and cold temperature conditions. *Poult Sci* 1998;**77**(12):1905–12.
33. Meng Q, Wong YT, Chen J, et al. Age-related changes in mitochondrial function and antioxidative enzyme activity in fischer 344 rats. *Mech Ageing Dev* 2007;**128**:286–92.
34. Suzuki M, Betsuyaku T, Ito Y, Nagai K, Nasuhara Y, Kaga K, et al. Down-regulated nf-e2-related factor 2 in pulmonary macrophages of aged smokers and patients with chronic obstructive pulmonary disease. *Am J Respir Cell Mol Biol* 2008;**39**:673–82.
35. Matsuro M, Gomi K, Dooley MM. Age-related alterations in antioxidant capacity and lipid peroxidation in brain, liver and lung homogenates of normal and vitamin E-deficient rats. *Mech Aging Dev* 1992;**64**:273–92.
36. Azad N, Rojanasakul Y, Vallyathan V. Inflammation and lung cancer: roles of reactive oxygen/nitrogen species. *J Toxicol Environ Health B* 2008;**11**:1–15.
37. Becklake MR. Concepts of normality applied to the measurement of lung function. *Am J Med* 1986;**80**:1158–64.
38. Blakley RL, Henry DD, Smith CJ. Lack of correlation between cigarette mainstream smoke particulate phase radicals and hydroquinone yield. *Food Chem Toxicol* 2001;**39**(4):401–6.
39. Kodama M, Kaneko M, Aida M, Inoue F, Nakayama T, Akimoto H. Free radical chemistry of cigarette smoke and its implication in human cancer. *Anticancer Res* 1997;**17**:433–7.
40. Vallyathan V, Shi X. The role of oxygen free radicals in occupational and environmental lung diseases. *Environ Health Perspect* 1997;**105**(Suppl. 1):165–77.
41. Kumagai Y, Koide S, Taguchi K, Endo A, Nakai Y, Yoshikawa T, et al. Oxidation of proximal protein sulfhydryls by phenanthraquinone, a component of diesel exhaust particles. *Chem Res Toxicol* 2002;**15**(4):483–9.
42. Dick CA, Brown DM, Donaldson K, Stone V. The role of free radicals in the toxic and inflammatory effects of four different ultrafine particle types. *Inhal Toxicol* 2003;**15**(1):39–52.
43. Meyer KC. The role of immunity in susceptibility to respiratory infection in the aging lung. *Respir Physiol* 2001;**128**:23–31.
44. Meyer KC, Rosenthal NS, Soergel P, Peterson K. Neutrophils and low-grade inflammation in the seemingly normal aging human lung. *Mech Aging Dev* 1998;**104**:169–81.
45. Sprott RL, Baker GT. Special issue. Biomarkers of aging. *Exp Gerontol* 1988;**23**:223–437.
46. Tisdale MJ. Biochemical mechanisms of cellular catabolism. *Curr Opin Clin Nutr Metab Care* 2002;**5**(4):401–5.
47. Shapiro SD, Endicott SK, Province MA, Pierce JA, Campbell EJ. Marked longevity of human lung parenchymal elastic fibers deduced from prevalence of D-aspartate and nuclear weapons-related radiocarbon. *J Clin Invest* 1828;**87**:1991.
48. D'Errico A, Scarani P, Colosimo E, Spina M, Grigioni WF, Mancini AM. Changes in the alveolar connective tissue of the ageing lung. An Immunohistochemical study. *Virchows Arch A Pathol Anat Histopathol* 1989;**415**(2):137–44.
49. Frette C, Jacob MP, Wei SM, Bertrand JP, Laurent P, Kauffmann F, et al. Relationship of serum elastin peptide level to single breath transfer factor for carbon monoxide in French coal miners. *Thorax* 1997;**52**:1045–50.
50. Samaras TT, Storms LH, Elrick H. Longevity, mortality and body weight. *Ageing Res Rev* 2002;**1**(4):673–91.
51. Litonjua AA, Sparrow D, Celedon JC, DeMolles D, Weiss ST. Association of body mass index with the development of methacholine airway hyperresponsiveness in men: the Normative Aging Study. *Thorax* 2002;**57**(7):581–5.
52. Wang ML, McCabe L, Petsonk EL, Hankinson JL, Banks DE. Weight gain and longitudinal changes in lung function in steel workers. *Chest* 1997;**111**(6):1526–32.
53. Gennuso J, Epstein JH, Paluch RA, Cerny F. The relationship between asthma and obesity in urban minority children and adolescents. *Arch Pediatr Adolesc Med* 1998;**152**:1197–200.
54. Luder E, Melnik TA, DiMaio M. Association of being overweight with greater asthma symptoms in inner city black and Hispanic children. *J Pediatr* 1998;**132**:699–703.
55. Camargo Jr CA, Weiss ST, Zhang S, Willett WC, Speizer FE. Prospective study of body mass index, weight change, and risk of adult-onset asthma in women. *Arch Intern Med* 1999;**159**:2582–8.
56. Seidell JC, De Groot LC, Van Sonsbeek JL, Deurenberg P, Hautvast JG. Association of moderate and severe overweight with self-reported illness and medical care in Dutch adults. *Am J Public Health* 1986;**76**:264–9.
57. Guerra S, Sherrill DL, Bobadilla A, Martinez FD, Barbee RA. The relation of body mass index to asthma, chronic bronchitis and emphysema. *Chest* 2002;**122**:1256–63.
58. Womack CJ, Harris DL, Katzel LI, Hagberg JM, Bleecker ER, Goldberg AP. Weight loss, not aerobic exercise, improves pulmonary function in older obese men. *J Gerontol A Biol Sci Med Sci* 2000;**55**:M453–7.
59. Amara CE, Koval JJ, Paterson DH, Cunningham DA. Lung function in older humans: the contribution of body composition, physical activity and smoking. *Ann Hum Biol* 2001;**5**:522–36.
60. Spindler SR. Calorie restriction enhances the expression of key metabolic enzymes associated with protein renewal during aging. *Ann N Y Acad Sci* 2001;**928**:296–304.

61. Riley DJ, Thakker-Varia S. Effect of diet on lung structure, connective tissue metabolism and gene expression. *J Nutr* 1995;**125**: 1657S–60S.

62. Romieu I, Trenga C. Diet and obstructive lung disease. *Epidemiol Rev* 2001;**23**:268–87.

63. Congleton J. The pulmonary cachexia syndrome: aspects of energy balance. *Proc Nutr Soc* 1999;**58**:321–8.

64. Palta M, Sadek-Badawi M, Sheehy M, Albanese A, Weinstein M, McGuinness G, et al. Respiratory symptoms at age 8 years in a cohort of very low birth weight children. *Am J Epidemiol* 2001;**154**(6):521–9.

65. Ng DK, Lau WY, Lee SL. Pulmonary sequelae in long-term survivors of bronchopulmonary dysplasia. *Pediatr Int* 2000;**42**:603–7.

66. Hogg JC. Childhood viral infection and the pathogenesis of asthma and chronic obstructive lung disease. *Am J Respir Crit Care Med* 1999;**160**:S26–8.

67. Hayashi S. Latent adenovirus infection in COPD. *Chest* 2002;**121**:183S–7S.

68. Krumpe PE, Knudson RJ, Parsons G, Reiser K. The aging respiratory system. *Clin Geriat Med* 1985;**1**:143–75.

69. Peterson DD, Pack AI, Silage DA, Fishman AP. Effects of aging on ventilatory and occlusion pressure responses to hypoxia and hypercapnia. *Am Rev Respir Dis* 1981;**124**:387–91.

70. Kronenberg R, Drage G. Attenuation of the ventilatory and heart rate response to hypoxia and hypercapnia with aging in normal man. *J Clin Invest* 1973;**42**:1812–9.

71. Poulin MJ, Cunningham DA, Paterson DH, Rechnitzer PA, Ecclestone NA, Koval JJ. Ventilatory response to exercise in men and women 55 to 86 years of age. *Am J Respir Crit care Med* 1994;**149**:408–15.

72. Young T, Palta M, Dempsey J, Skatrud J, Weber S, Badr S. The occurrence of sleep-disordered breathing among middle-aged adults. *N Engl J Med* 1993;**328**:1230–5.

73. Ancoli-Israël S, Coy T. Are breathing disturbances in elderly equivalent to sleep apnea syndrome? *Sleep* 1994;**17**:77–83.

74. Pandi-Perumal SR, Seils LK, Kayumov L, Ralph MR, Lowe A, Moller H, Swaab DF. Senescence, sleep and circadian rhythms. *Ageing Res Rev* 2002;**1**(3):559–604.

75. Guénard H, Marthan R. Pulmonary gas exchange in elderly subjects. *Eur Respir J* 1996;**9**:2573–7.

76. Wagner P, Laravuso R, Uhl R, West J. Continuous distribution of ventilation-perfusion ratios in normal subjects breathing air and 100 percent O_2. *J Clin Invest* 1974;**54**:54–68.

77. Janssens JP, Pache JC, Nicod LP. Physiological changes in respiratory function associated with aging. *Eur Respir J* 1999; **13**:197–205.

78. Butler C, Kleinerman J. Capillary density: alveolar diameter, a morphometric approach to ventilation and perfusion. *Am Rev Respir Dis* 1970;**102**:886–94.

79. Watson A, Joyce H, Pride NB. Changes in carbon monoxide transfer over 22 years in middle-aged men. *Respir Med* 2000;**94**:1103–8.

80. Nass R, Thorner M. Impact of the GH-cortisol ratio on the age-dependent changes in body composition. *Growth Horm IGF Res* 2002;**12**(3):147.

81. Sparrow D, O'Connor GT, Rosner B, DeMolles D, Weiss ST. A longitudinal study of plasma cortisol concentration and pulmonary function decline in men: the normative aging study. *Am Rev Respir Dis* 1993;**147**(6 Pt 1):1345–8.

82. Wouters EF, Creutzberg EC, Schols AM. Systemic effects in COPD. *Chest* 2002;**121**:127S–30S.

83. Tuohimaa P, Keisala T, Minasyan A, Cachat J, Kalueff A. Vitamin D, nervous system and aging. *Psychoneuroendocrinology* 2009;**34S**:S278–86.

84. Mullett CJ, Kong J-Q, Romano JT, Polak MJ. Age-related changes in pulmonary venous epinephrine concentration, and pulmonary vascular response after intratracheal epinephrine. *Pediatr Res* 1992;**31**:458–61.

85. Stein-Behrens BA, Sapolsky RM. Stress, glucocorticoids, and aging. *Aging (Milano)* 1992;**4**(3):197–210.

86. Hasselgren PO. Catabolic response to stress and injury: implications for regulation. *World J Surg* 2000;**24**(12):1452–9.

87. Ferrari E, Cravello L, Muzzoni B, Casarotti D, Paltro M, Solerte SB, Fioravanti M, Cuzzoni G, Pontiggia B, Magri F. Age-related changes of the hypothalamic-pituitary-adrenal axis: pathophysiological correlates. *Eur J Endocrinol* 2001;**144**(4):319–29.

88. Sparrow D, O'Connor GT, Young JB, Weiss ST. Relationship of urinary serotonic excretion to cigarette smoking and respiratory symptoms. The Normative Aging Study. *Chest* 1992;**101**(4): 976–80.

89. Sharma G, Goodwin J. Effect of aging on respiratory system physiology and immunology. *Clin Intervent Aging* 2006;**1**:253–60.

90. Wills-Karp M. Age-related changes in pulmonary muscarinic receptor binding properties. *Am J Phys* 1993;**265**:103–9.

91. Hopp RJ, Bewtra A, Nair NM, Townley RG. The effect of age on methacholine response. *J Allergy Clin Immunol* 1985;**76**:609–13.

92. Bennett BA, Mitzner W, Tankersley CG. The effects of age and carbon black on airway resistance in mice. *Inhal Toxicol* 2012;**24**: 931–8.

93. Connolly MJ, Crowley JJ, Charan NB, Nielson CP, Vestal RE. Impaired bronchodilator response to albuterol in healthy elderly men and women. *Chest* 1995;**108**:401–6.

94. Creticos P, Knobil K, Edwards LD, Rickard KA, Dorinsky P. Loss of response to treatment with leukotriene receptor antagonists but not inhaled corticosteroids in patients over 50 years of age. *Ann Allergy Asthma Immunol* 2002;**88**:401–9.

95. Diot P, Palmer LB, Uy LL, Albulak MK, Bonitch L, Smaldone GC. Technique for measurement of oropharngeal clearance in the elderly. *J Aerosol Med* 1995;**8**:177–86.

96. Smaldone GC. Deposition and clearance: unique problems in the proximal airways and oral cavity in the young and elderly. *Respir Physiol* 2001;**128**:33–8.

97. Anhoj J, Bisgaard AM, Bisgaard H. Systemmic activity of inhaled steroids in 1- to 3-year-old children with asthma. *Pediatrics* 2002;**109**:E40.

98. Bennett WD, Zeman K, Kim C. variability of fine particle deposition in healthy adults. Effects of age and gender. *Am J Respir Crit Care Med* 1996;**153**(5):1641–7.

99. Mauderly JL, Bice DE, Cheng YS, Gillett NA, Griffith WC, Henderson RF, et al. Influence of pre-existing pulmonary emphysema on susceptibility to chronic inhalation exposure to diesel exhaust. *Am Rev Respir Dis* 1990;**141**:1333–41.

100. Puchelle E, Zahm JM, Bertrand A. Influence of age on bronchial mucociliary transport. *Scand J Resp Dis* 1979;**60**:307–13.

101. Incalzi RA, Maini CL, Fuso L, Giordano A, Carbonin PU, Galli G. Effects of aging on mucociliary clearance. *Compr Gerontol. Section A Clin Lab Sci* 1989;**3**(Suppl):65–8.

102. Stettler LE, Platek SF, Riley RD, Mastin JP, Simon SD. Lung particulate burdens of subjects from the Cincinnati, Ohio urban area. *Scanning Microsc* 1991;**5**(1):85–92.

103. Komarnicki GJ. Tissue, sex and age specific accumulation of heavy metals (Zn, Cu, Pb, Cd) by populations of the mole (Talpa europaea L.) in a central urban area. *Chemosphere* 2000;**41**(10): 1593–602.

104. Kollmeier H, Witting C, Seemann J, Wittig P, Rothe R. Increased chromium and nickel content in lung tissue. *J Cancer Res Clin Oncol* 1985;**110**(2):173–6.

105. Hazucha MJ, Folinsbee LJ, Bromberg PA. Distribution and reproducibility of spirometric response to ozone by gender and age. *J Appl Phys* 2003;**95**:1917–25.

106. Gunnison AF, Finkelstein I, Weideman P, Su WY, Sobo M, Schlesinger RB. Age-dependent effect of ozone on pulmonary eicosanoid metabolism in rabbits and rats. *Fundam Appl Toxicol* 1990;**15**(4):779–90.

107. Tepper JS, Wiester MJ, Weber MF, Fitzgerald S, Costa DL. Chronic exposure to a simsulated urban profile of ozone alters ventilatory responses to carbon dioxide challenge in rats. *Fundam Appl Toxicol* 1991;**17**(1):52–60.

108. Liu L, Kumarathasan P, Guenette J, Vincent R. Hydroxylation of salicylate in lungs of Fischer 344 rats: effects of aging and ozone exposure. *Am J Phys* 1996;**271**:L995–1003.

109. Servais S, Boussouar A, Molnar A, Douki T, Pequignot JM, Favier R. Age-related sensitivity to lung oxidative stress during ozone exposure. *Free Radic Res* 2005;**39**:305–16.

110. Vincent R, Adamson IY. Cellular kinetics in the lungs of aging Fischer 344 rats after acute exposure to ozone. *Am J Pathol* 1995;**146**:1008–16.

111. Hoppe P, Praml G, Rabe G, Lindner J, Fruhmann G, Kessel R. Environmental ozone field study on pulmonary and subjective responses of assumed risk groups. *Environ Res* 1995;**71**(2):109–21.

112. Drechsler-Parks DM, Horvath SM, Bedi JF. The effective dose concept in older adults exposed to ozone. *Exp Gerontol* 1990;**25**(2):107–15

113. Seal Jr E, McDonnell WF, House DE. Effects of age, socioeconomic status and menstrual cycle on pulmonary response to ozone. *Arch Environ Health* 1996;**51**:132–7.

114. Alexeeff SE, Litonjua AA, Wright RO, et al. Ozone exposure, antioxidant genes, and lung function in an elderly cohort: VA normative aging study. *Occup Environ Med* 2008;**65**:736–42.

115. Van Bree L, Marra M, van Scheindelen HJ, Fischer PH, de Loos S, Buringh E, et al. Dose-effect models for ozone exposure: tool for quantitative risk estimation. *Toxicol Lett* 1995;**82-83**:317–21.

116. Medina-Ramón M, Zanobetti A, Schwartz J. The effect of ozone and PM10 on hospital admissions for pneumonia and chronic obstructive pulmonary disease: a national multicity study. *Am J Epidemiol* 2006;**163**:579–88.

117. Bae HJ, Park J. Health benefits of improving air quality in the rapidly aging Korean society. *Sci Total Environ* 2009;**407**:5971–7.

118. Ito K, Barnes PJ. COPD as a disease of accelerated lung aging. *Chest* 2009;**135**:173–80.

119. Moriyama C, Betsuyaku T, Ito Y, Hamamura I, Hata J, Takahashi H, et al. Aging enhances susceptibility to cigarette smoke-induced inflammation through bronchiolar chemokines. *Am J Respir Cell Mol Biol* 2010;**42**:304–11.

120. Nicholl ID, Bucala R. Advanced glycation endproducts and cigarette smoking. *Cell Mol Biol* 1998;**44**(7):1025–33.

121. Yamagishi S, Matsui T, Nakamura K. Possible involvement of tobacco-derived advanced glycation end products (AGEs) in an increased risk for developing cancers and cardiovascular disease in former smokers. *Med Hypotheses* 2008;**71**:259–61.

122. Kerstjens HAM, Rijcken B, Schouten JP, Postma DS. Decline of FEV$_1$ by age and smoking status: facts, figures and fallacies. *Thorax* 1997;**52**:820–7.

123. Lange P, Groth S, Nyboe J, Mortensen J, Appleyard M, Jensen G, et al. Decline of the lung function related to the type of tobacco smoked and inhalation. *Thorax* 1990;**45**:22–6.

124. LaCroix AZ, Lipson S, White L. Prospective study of pneumonia hospitalizations and mortality of U.S. older people. *Publ Hlth Rep* 1989;**104**:350–60.

125. Matulionis DH. Chronic cigarette smoke inhalation and aging in mice: 1. Morphologic and functional lung abnormalities. *Exp Lung Res* 1984;**7**:237–56.

126. Uejima Y, Fukuchi Y, Nagase T, Matsuse T, Yamaoka M, Tabata R, et al. Influences of inhaled tobacco smoke on the senescence accelerated mouse (SAM). *Eur Respir J* 1990;**3**:1029–36.

127. Teramoto S, Uejima Y, Oka T, Teramoto K, Matsuse T, Ouchi Y, et al. Effects of chronic cigarette smoke inhalation on the development of senile lung in senescence-accelerated mouse. *Res Exp Med* 1997;**197**:1–11.

128. Eke BC, Vural N, Iscan M. Age dependent differential effects of cigarette smoke on hepatic and pulmonary xenobiotic metabolizing enzymes in rats. *Arch Toxicol* 1997;**71**:696–702.

129. Iribarren C, Jacobs Jr DR, Sidney S, Gross MD, Eisner MD. Cigarette smoking, alcohol consumption and risk of ARDS: a 15-year cohort study in a managed care setting. *Chest* 2000;**117**:163–8.

130. U.S. Environmental Protection Agency. *Air Quality Criteria for Particulate Matter.* Vols. I–III. EPA/600/P-95/001aF. Washington, DC: U.S. EPA; 1996.

131. Borja-Aburto VH, Castillejos M, Gold DR, Bierzwinski S, Loomis D. Mortality and ambient fine particles in southwest Mexico City, 1993-1995. *Environ Health Perspect* 1998;**106**:849–55.

132. Burnett RT, Cakmak S, Brook JR. The effect of the urban ambient air pollution mix on daily mortality rates in 11 Canadian cities. *Can J Public Health* 1998;**89**:152–6.

133. Fairley D. Daily mortality and air pollution in Santa Clara County, California: 1989–1996. *Environ Health Perspect* 1999;**107**: 637–41.

134. U.S. Environmental Protection Agency. *Air Quality Criteria for Particulate Matter.* Vols. I–III. EPA/600/P-99–002a. Washington, DC: U.S. EPA; 1999.

135. Downs SH, Schindler C, Liu LJ, et al. Reduced exposure to PM10 and attenuated age-related decline in lung function. *N Engl J Med* 2007;**357**:2338–47.

136. Clarke RW, Catalano P, Coull B, Koutrakis P, Krishna Murthy GG, Rict T, et al. Age-related responses to concentrated urban air particles (CAPs). In: Phalen RF, Bell YM, editors. *Proceedings of the Third Colloquium on Particulate Air Pollution and Human Health*. Irving, CA: Health Effects Laboratory, University of California; 1999. pp. 8–36.

137. Elder AC, Gelein R, Finkelstein JN, Cox C, Oberdörster G. Pulmonary inflammatory response to inhaled ultrafine particles is modified by age, ozone exposure, and bacterial toxin. *Inhal Toxicol* 2000;**12**(Suppl. 4):227–46.

138. Oberdorster G. Pulmonary effects of inhaled ultrafine particles. *Int Arch Occup Environ Health* 2001;**74**:1–8.

139. Newnham DM, Hamilton SJC. Sensitivity of the cough reflex in young and elderly subjects. *Age and Aging* 1997;**26**:185–8.

140. Venkatesan P, Gladman J, Macfarlane JT, Barer D, Berman P, Kinnear W, et al. A hospital study of community-acquired pneumonia in the elderly. *Thorax* 1990;**45**:254–8.

141. Report Conference. British Geriatrics Society. Respiratory disease in old age: research into ageing workshop,London,1988. *Age Ageing* 2000;**29**:281–5.

142. Teale C, Goldman JM, Pearson SB. The association of age with the presentation and outcome of tuberculosis: a five-year survey. *Age Ageing* 1993;**22**:293–8.

143. Kovaiou RD, Herndler-Brandstetter D, Grubeck-Loebenstein B. Age-related changes in immunity: implications for vaccination in the elderly. *Expert Rev Mol Med* 2007;**9**:1–17.

144. Gardner EM, Gonzalez EW, Nogusa S, Murasko DM. Age-related changes in the immune response to influenza vaccination in a racially diverse, healthy elderly population. *Vaccine* 2006;**24**:1609–14.

145. Gross CP, Anderson GF, Powe NR. The relation between funding by the National Institutes of Health and the burden of disease. *N Engl J Med* 1999;**340**:1881–7.

146. Sin DD, Stafinski T, Ng YC, Bell NR, Jacobs P. The impact of chronic obstructive pulmonary disease on work loss in the United States. *Am J Respir Crit Care Med* 2002;**165**:704–7.

147. National Center for Health Statistics. Third National Health and Nutrition Examination Survey, 1988–1994, NHANES III Laboratory Data File. (Public Use Data File Documentation Number 76200). Hyattsville, MD: US Department of Health and Human Services, Public Health Service, Centers for Disease Control and Prevention [CD-ROM]; 1996.

148. Hnizdo E, Sullivan PA, Bang KM, Wagner G. Association between chronic obstructive pulmonary disease and employment by industry and occupation in the US population: a study of data from the Third National Health and Nutrition Examination Survey. *Am J Epidemiol* 2002;**156**:738–46.

149. Pride N, Vermeire P, Allegra L. Diagnostic labels in chronic airflow obstruction: responses to a questionnaire with model case histories in North American and Western European countries. *Eur Respir J* 1989;**2**:702–9.

150. Sluiter HJ, Koëter GH, de Monchy JGR, Postma DS, de Vries K, Orie NGM: the Dutch hypothesis (chronic non-specific lung disease) revisited. *Eur Respir J* 1991;**4**:479–89.

151. Finucane KE, Greville HW, Brown PJ. Irreversible airflow obstruction: evolution in asthma. *Med J Aust* 1985;**142**:602–4.

152. Fowler RW, Pluck RA, Hetzel MR. Maximal expiratory flow-volume curves on Londoners aged 60 years and over. *Thorax* 1987;**42**:173–82.

153. Orie NG, Sluiter HJ, de Vries K, Tammeling GJ, Witkop J. The host factor in bronchitis. In: Orie NGM, Sluiter HJ, editors. *Bronchitis: An International Symposium*. Assen, The Netherlands: Royal van Gorcum; 1961.

154. Xu X, Rijcken B, Schouten JP, Weiss ST. Airways responsiveness and development and remission of chronic respiratory symptoms in adults. *Lancet* 1997;**350**:1431–4.

155. Vermeire PA, Pride NBA. "splitting" look at chronic nonspecific lung disease (CNSLD): common features but diverse pathogenesis. *Eur Respir J* 1991;**4**(4):490–6.

156. Hogg JC. Role of latent viral infections in chronic obstructive pulmonary disease and asthma. *Am J Respir Crit Care Med* 2001;**164**:S71–5.

157. Meshi B, Vitalis TZ, Ionescu D, Elliott WM, Liu C, Wang X-D, et al. Emphysematous lung destruction by cigarette smoke: the effects of latent adenoviral infection on the lung inflammatory response. *Am J Respir Cell Mol Biol* 2002;**26**:52–7.

158. Sparrow D, O'Connor GT, Weiss ST, DeMolles D, Ingram Jr RH. Volume history effects and airway responsiveness in middle-aged and older men. *Am J Respir Crit Care Med* 1997;**155**:888–92.

159. Gold DR, Wang X, Wypij D, Speizer FE, Ware JH, Dockery DW. Effects of cigarette smoking on lung function in adolescent boys and girls. *N Engl J Med* 1996;**335**:931–7.

160. Lebowitz MD, Holberg CJ, Knudson RJ, Burrows B. Longitudinal study of pulmonary function development in childhood, adolescence and early adulthood. Development of pulmonary function. *Am Rev Respir Dis* 1987;**136**:69–75.

161. Kelly WJ, Hudson I, Raven J, Phelan PD, Pain MC, Olinsky A. Childhood asthma and adult lung function. *Am Rev Respir Dis* 1988;**138**:26–30.

162. Vestbo J, Prescott E. Update on the "Dutch hypothesis" for chronic respiratory disease. *Thorax* 1998;**53**(Suppl. 2):S15–9.

163. Tager IB, Segal MR, Speizer FE, Weiss ST. The natural history of forced expiratory volumes. Effect of cigarette smoking and respiratory symptoms. *Am Rev Respir Dis* 1988;**138**:837–49.

164. Weiss ST, Ware JH. Overview of issues in the longitudinal analysis of respiratory data. *Am J Respir Crit Care Med* 1996;**154**:S208–11.

165. Anderson TW, Brown JR, Hall JW, Shephard RJ. The limitations of linear regressions for the prediction of vital capacity and forced expiratory volume. *Respiration* 1968;**25**:140–58.

166. Bossé R, Sparrow D, Rose CL, Weiss ST. Longitudinal effect of age and smoking cessation on pulmonary function. *Am Rev Respir Dis* 1981;**123**:378–81.

167. Dockery DW, Ware JH, Ferris Jr BG, Glicksberg DS, Fay ME, Spiro III A. Distribution of forced expiratory volume in one second and forced vital capacity in healthy, white, adult never-smokers in six U.S. cities. *Am Rev Respir Dis* 1985;**131**:511–20.

168. Camilli AE, Burrows B, Knudson RJ, Lyle SK, Lebowitz MD. Longitudinal changes in forced expiratory volume in one second in adults: effects of smoking and smoking cessation. *Am Rev Respir Dis* 1987;**135**:794–9.

169. Dockery DW, Speizer FE, Ferris Jr BG, Ware JH, Louis TA, Spiro III A. Cumulative and reversible effects of lifetime smoking on simple tests of lung function in adults. *Am Rev Respir Dis* 1988;**137**:286–92.

170. Higgins MW, Enright PL, Kronmal RA, Schenker MB, Anton-Culver H, Lyles M. Smoking and lung function in elderly men and women: the Cardiovascular Health Study. *JAMA* 1993;**269**:2741–8.

171. Frette C, Barrett-Connor E, Clausen JL. Effect of active and passive smoking on ventilatory function in elderly men and women. *Am J Epidemiol* 1996;**143**:757–65.

172. Knudson RJ, Lebowitz MD, Holberg CJ, Burrows B. Changes in the normal maximal expiratory flow-volume curve with growth and aging. *Am Rev Respir Dis* 1983;**127**:725–34.

173. Burrows B, Lebowitz MD, Camilli AE, Knudson RJ. Longitudinal changes in forced expiratory volume in one second in adults: methodologic considerations and findings in healthy non-smokers. *Am Rev Respir Dis* 1986;**133**:974–80.

174. Anthonisen NR, Connett JE, Murray RP. Smoking and lung function of lung health study participants after 11 years. *Am J Respir Crit Care Med* 2002;**166**:675–9.

175. Milne JS, Williamson J. Respiratory function tests in older people. *Clin Sci* 1972;**42**:371–81.

176. Schmidt CD, Dickman ML, Gardner RM, Brough FK. Spirometric standards for healthy elderly men and women : 532 subjects, ages 55 through 94 years. *Am Rev Respir Dis* 1973;**108**:933–9.

177. Brandstetter RD, Kazemi H. Aging and the respiratory system. *Med Clin N Am* 1983;**67**:419–31.

178. Annesi I, Kauffman F. Is respiratory mucus hypersecretion really an innocent disorder? A 22-year mortality survey of 1,061 working men. *Am Rev Respir Dis* 1986;**134**:688–93.

179. Cook NR, Evans DA, Scherr PA, Speizer FE, Taylor JO, Hennekens CH. Peak expiratory flow rate and 5-year mortality in an elderly population. *Am J Epidemiol* 1991;**133**:784–94.

180. Lange P, Nyboe J, Appleyard M, Jensen G, Schnohr P. Spirometric findings and mortality in never-smokers. *J Clin Epidemiol* 1990;**43**:867–73.

181. Kauffman F, Frette C. The aging lung: an epidemiological perspective (an editorial). *Respir Med* 1993;**87**:5–7.

182. Sorlie PD, Kannel WB, O'Connor G. Mortality associated with respiratory function and symptoms in advanced age: the Framingham Study. *Am Rev Respir Dis* 1989;**140**:379–84.

183. Gibson JB, Martin NG, Oakeshott JG, Rowell DM. Lung function in an Australian population: contributions of polygenic factors and the *Pi* locus to individual differences in lung function in a sample of twins. *Ann Hum Biol* 1983;**10**(6):547–56.

184. McClearn GE, Svartengren M, Pedersen NL, Heller DA, Plomin R. Genetic and environmental influences on pulmonary function in aging Swedish twins. *J Gerontol* 1994;**49**:M264–8.

185. Feinleib M, Garrison RJ, Fabsitz R, Christian JC, Hrubec Z, Borhani NO, et al. The NHLBI study of cardiovascular disease risk factors: methodology and summary of results. *Am J Epidemiol* 1977;**106**:284–95.

186. Redline S, Tishler PV, Rosner B, Lewitter FI, Vandenbjurgh M, Weiss ST, et al. Genotypic and phenotypic similarities in pulmonary function among family members of adult monozygotic and dizygotic twins. *Am J Epidemiol* 1989;**129**:827–36.

187. Jedrychoswki W. Effects of smoking and longevity of parents on lung function in the apparently healthy elderly. *Arch Gerontol Geriatr* 1990;**10**:19–26.

188. Croxton TL, Weinmann GG, Senior RM, Hoidal JR. Future research directions in chronic obstructive pulmonary disease. *Am J Respir Crit Care Med* 2002;**165**:538–844.

189. Xu X, Dockery D, Ware JH, Speizer FE, Ferris Jr BG. Effects of cigarette smoking on rate of loss of pulmonary function in adults: a longitudinal assessment. *Am Rev Respir Dis* 1992;**146**:1345–8.

190. Tashkin DP, Simmons MS, Sherrill DL, Coulson AH. Heavy habitual marijuana smoking does not cause an accelerated decline in FEV1 with age. *Am J Respir Crit Care Med* 1997;**155**:141–8.

191. Banks DE, Shah AA, Lopez M, Wang M-I. Chest illnesses and the decline of FEV$_1$ in steelworkers. *JOEM* 1999;**41**(12):1085–90.

192. Oxman AD, Muir DCF, Shannon HS, Stock SR, Hnizdo E, Lange HJ. Occupational dust exposure and chronic obstructive pulmonary disease. *Am Rev Respir Dis* 1993;**148**:38–48.

193. Becklake MR. Occupational exposures: evidence for a causal association with chronic obstructive pulmonary disease. *Am Rev Respir Dis* 1989;**140**:S85–91.

194. Coggon D, Taylor AN. Coal mining and the chronic obstructive pulmonary disease: a review of the evidence. *Thorax* 1988;**53**:398–407.

195. Dockery DW, Brunekreef B. Longitudinal studies of air pollution effects on lung function. *Am J Respir Crit Care Med* 1996;**154**:S250–6.

196. Peat JK, Woolcock AJ, Cullen K. Rate of decline of lung function in subjects with asthma. *Eur J Respir Dis* 1987;**70**:171–9.

197. Rijcken B, Scouten JP, Xu X, Rosner B, Weiss ST. Bronchial hyperresponsiveness to histamine is associated with accelerated decline of FEV1. *Am J Respir Crit Care Med* 1995;**151**:1377–82.

198. Villar MT, Dow L, Coggon D, Lampe FC, Holgate ST. The influence of increased bronchial responsiveness, atopy, and serum IgE on decline in FEV1: a longitudinal study in the elderly. *Am J Respir Crit Care Med* 1995;**151**:656–62.

199. Parker DR, O'Connor GT, Sparrow D, Segal MR, Weiss ST. The relationship of nonspecific airway responsiveness and atopy to the rate of decline of lung function: the normative aging study. *Am Rev Respir Dis* 1990;**141**:589–94.

200. O'Connor GT, Sparrow D, Weiss ST. A prospective longitudinal study of methacholine airway responsiveness as a predictor of pulmonary-function decline: the normative aging study. *Am J Respir Crit Care Med* 1995;**152**:87–92.

201. Vestbo J, Prescott E, Lange P. Association of chronic mucus hypersecretion with FEV$_1$ decline and chronic obstructive pulmonary disease morbidity. Copenhagen City Heart Study Group. *Am J Respir Crit Care Med* 1996;**153**:1530–5.

202. Rode A, Shephard RJ. Lung volumes of Igloolik Inuit and Volochanka nGanasan. *Arct Med Res* 1996;**55**:4–13.

203. Stone PJ, Gottlieb DJ, O'Connor GT, Ciccolella DE, Breuer R, Bryan-Rhadfi J, et al. Elastin and collagen degradation products in urine of smokers with and without chronic obstructive pulmonary disease. *Am J Respir Crit Care Med* 1995;**151**:952–9.

204. Gottlieb DJ, Sparrow D, O'Connor GT, Weiss ST. Skin test reactivity to common aeroallergens and decline of lung function. *Am J Respir Crit Care Med* 1996;**153**:561–6.

205. He JQ, Ruan J, Connett JE, Anthonisen NR, Paré PD, Sandford AJ. Antioxidant gene polymorphisms and susceptibility to a rapid decline in lung function in smokers. *Am J Respir Crit Care Med* 2002;**166**:323–8.

206. Frette C, Wei SM, Neukirch F, Sesboue R, Martin JP, Jacob MP, et al. Relation of serum elastin peptide concentration to age, FEV1, smoking habits, alcohol consumption and PI phenotype: an epidemiological study in working men. *Thorax* 1992;**47**(11):937–42.

207. Becklake MR. Chronic airflow obstruction: its relationship to work in dusty occupations. *Chest* 1985;**88**:608–17.

208. Kleinerman J, Green FHY, et al. Pathology standards for coal workers' pneumoconiosis. (A report of the Pneumoconiosis Committee of the college of American Pathologists). *Arch Path Lab Med (Special Issue)* 1979;**101**:375–431.

209. Cox DW, Johnson AM, Fagerhol MK. Report of nomenclature meeting for alpha$_1$-antitrypsin. *Hum Genet* 1980;**53**:429–33.

210. Tarján E, Magyar P, Váczi Z, Lantos Å, Vaszár L. Longitudinal lung function study in heterozygous PiMZ phenotype subjects. *Eur Respir J* 1994;**7**(1):2199–204.

211. Burrows B, Lebowitz MD, Barbee RA, Cline MG. Findings before diagnoses of asthma among the elderly in a longitudinal study of a general population sample. *J Allergy Clin Immunol* 1991;**88**:870–7.

212. Enright PL, McClelland RL, Newman AB, Gottlieb DJ, Lebowitz MD. Underdiagnosis and undertreatment of asthma in the elderly. Cardiovascular Health Study Research Group. *Chest* 1999;**116**:603–13.

213. Kitabayashi T, Iikura Y, Tokutome S. A study of 456 cases of death from asthma (1993–1997) from an investigation by the Tokyo Medical Examiner's Office. *Allergol Int* 2002;**51**:93–100.

214. Tough SC, Green FHY, Paul JE, Wigle DT, Butt JC. Sudden death from asthma in 108 children and young adults. *J Asthma* 1996;**33**:179–88.

215. Schachter J, Higgins MW. Median age at onset of asthma and allergic rhinitis in Tecumseh, Michigan. *J Allergy Clin Immunol* 1976;**57**:342–51.

216. Lebowitz MD, Knudson RJ, Burrows B. Tucson epidemiologic study of obstructive lung diseases. I. Methodology and prevalence of disease. *Am J Epidemiol* 1975;**102**:137–52.

217. Nagasaki T, Matsumoto H. Influences of smoking and aging on allergic airway inflammation in asthma. *Allergol Int* 2013;**62**:171–9.

218. Gibson PG, McDonald VM, Marks GB. Asthma in older adults. *Lancet* 2010;**376**:803–13.

219. Sherrill D, Sears MR, Lebowitz MD, Holdaway MD, Hewitt CJ, Flannery EM, et al. The effects of airway hyperresponsiveness, wheezing and atopy on longitudinal pulmonary function in children: a 6 year follow up study. *Pediatr Pulmonol* 1992;**13**:78–85.

220. Rasmussen F, Taylor DR, Flannery EM, Cowan JO, Greene JM, Herbison GP, et al. Outcome in adulthood of asymptomatic airway hyperresponsiveness in childhood: a longitudinal population study. *Pediatr Pulmonol* 2002;**34**:164–71.

221. Annema JT, Sparrow D, O'Connor GT, Rijcken B, Koëter GH, Postma DS, Weiss ST. Chronic respiratory symptoms and airway responsiveness to methacholine are associated with eosinophilia in older men: the normative aging study. *Eur Respir J* 1995; **8**:62–9.

222. Taylor KJ, Luksza AR. Peripheral blood eosinophil counts and bronchial responsiveness. *Thorax* 1992;**42**:452–6.

223. Mensinga TT, Schouten JP, Weiss ST, Van der Lende. Relationship of skin test reactivity and eosinophilia to level of pulmonary function in a community-based population study. *Am Rev Respir Dis* 1992;**146**:638–43.

224. Lange P, Parner J, Vestbo J, Schnohr P, Jensen GA. 15-year follow-up study of ventilatory function in adults with asthma. *N Engl J Med* 1998;**339**:1194–200.

225. Bai TR, Cooper J, Koelmeyer T, Paré PD, Weir TD. The effect of age and duration of disease on airway structure in fatal asthma. *Am J Respir Crit Care Med* 2000;**162**:663–9.

226. Wilson JW, Li X, Pain MC. The lack of distensibility of asthmatic airways. *Am Rev Respir Dis* 1993;**148**:806–9.

227. Rasmussen F, Taylor DR, Flannery EM, Cowan JO, Greene JM, Herbison GP, et al. Risk factors for airway remodeling in asthma manifested by a low postbronchodilator FEV1/vital capacity ratio: a longitudinal population study from childhood to adulthood. *Am J Respir Crit Care Med* 2002;**165**:1480–8.

228. Chetta A, Foresi A, Del Donno M, Bertorelli G, Pesci A, Olivieri D. Airways remodeling is a distinctive feature of asthma and is related to severity of disease. *Chest* 1997;**111**:852–7.

229. Boulet LP, Turcotte H, Laviolette M, Naud F, Bernier MC, Martel S, et al. Airway hyperresponsiveness, inflammation, and subepithelial collagen deposition in recently diagnosed versus long-standing mild asthma: influence of inhaled corticosteroids. *Am J Respir Crit Care Med* 2000;**162**(4 Pt 1):1308–13.

230. Prescott SL, Macaubas C, Holt BJ, Smallcombe TB, Loh R, Sly PD, et al. Transplacental priming of the human immune system to environmental allergens: universal skewing of initial T-cell responses towards the Th-2 cytokine profile. *J Immunol* 1998;**160**:4730–7.

231. Burrows B, Martinez FD, Halonen M, Barbee RA, Cline MG. Association of asthma with serum IgE levels and skin-test reactivity to allergens. *N Eng J Med* 1989;**320**:271–7.

232. Grundbacher FJ, Massie FS. Levels of immunoglobulin G, M, A and E at various ages in allergic and nonallergic black and white individuals. *J Allergy Clin Immunol* 1985;**75**:651–8.

233. Wittig HJ, Belloit J, Fillippi ID, Royal G. Age-related serum immunoglobulin E levels in healthy subjects and in patients with allergic disease. *J Allergy Clin Immunol* 1980;**66**:305–13.

234. Barbee RA, Halonen M, Lebowitz M, Burrows B. Distribution of IgE in a community population sample: correlations with age, sex, and allergen skin test reactivity. *J Allergy Clin Immunol* 1981;**68**:106–11.

235. O'Connor GT, Sparrow D, Segal MR, Weiss ST. Smoking, atopy, and methacholine airway responsiveness among middle-aged and elderly men: the normative aging study. *Am Rev Respir Dis* 1989;**140**(6):1520–6.

236. Omenaas E, Bakke P, Elsayed S, Hanoa R, Gulsvik A. Total and specific serum IgE levels in adults: relationship to sex, age and environmental factors. *Clin Exp Allergy* 1994;**24**:530–9.

237. Shirakawa T, Morimoto K. Lifestyle effect on total IgE. *Allergy* 1991;**46**:561–9.

238. Tracey M, Villar A, Dow L, Coggon D, Lampe FC, Holgate ST. The influence of increased bronchial responsiveness, atopy, and serum IgE on decline in FEV1: a longitudinal study in the elderly. *Am J Respir Crit Care Med* 1995;**151**:656–62.

239. Shadick NA, Sparrow D, O'Connor GT, DeMolles D, Weiss ST. Relationship of serum IgE concentration to level and rate of decline of pulmonary function: the normative aging study. *Thorax* 1996;**51**(8):787–92.

240. Dow L, Coggon D, Campbell MJ, Osmond C, Holgate ST. The interaction between immunoglobulin E and smoking in airflow obstruction in the elderly. *Am Rev Respir Dis* 1992;**146**:402–7.

241. Annesi I, Oryszczyn M, Frette C, Neukirch F, Orvoen-Frija E, Kauffmann F. Total circulation IgE and FEV_1 in adult men: an epidemiologic longitudinal study. *Chest* 1992;**101**:642–8.

242. Vollmer WM, Buist AS, Johnson LR, McCamant LE, Halonen M. Relationship between serum IgE and cross-sectional and longitudinal FEV1 in two cohort studies. *Chest* 1986;**90**:416–23.

243. Burrows B, Knudson RJ, Cline MG, Lebowitz MD. A re-examination of risk factors for ventilatory impairment. *Am Rev Respir Dis* 1988;**138**:829–36.

244. Halonen M, Barbee RA, Lebowitz MD, Burrows B. An epidemiological study of the interrelationships of total serum immunoglobulin E, allergy skin-test reactivity and eosinophilia. *J Allergy Clin Immunol* 1982;**69**:221–8.

245. Rawle FC, Burr ML, Platts-Mills TAE. Long-term falls in antibodies to dust mite and pollen allergens in patients with asthma or hay fever. *Clin Allergy* 1983;**13**:409–17.

246. Venables KM, Topping MD, Howe W, Luczynska CM, Hawkins R, Taylor AJ. Interaction of smoking and atopy in producing specific IgE antibody against a hapten protein conjugate. *Br Med J* 1985;**283**:1215–7.

247. Ohman JL, Sparrow D, MacDonald MR. New onset wheezing in an older male population: evidence of allergen sensitization in a longitudinal study. *J Allergy Clin Immunol* 1993;**91**:752–7.

248. Devouassoux G, Saxon A, Metcalfe DD, Prussin C, Colomb MG, Brambilla C, et al. Chemical constituents of diesel exhaust particles induce IL-4 production and histamine release by human basophils. *J Allergy Clin Immunol* 2002;**109**(5):847–53.

249. Sarafino EP, Paterson ME, Murphy EL. Age and the impacts of triggers in childhood asthma. *J Asthma* 1998;**35**(2):213–7.

250. Neimeijer NR, de Monchy JGR. Age-dependency of sensitization to aero-allergens in asthmaics. *Allergy* 1992;**47**:431–5.

251. Connolly MJ, Shaw L. Respiratory disease in old age: research into aging workshop (London, 1998). *Age Aging* 2000;**29**:281–5.

252. Paredi P, Kharitonov SA, Barnes PJ. Analysis of expired air for oxidation products. *Am J Respir Crit Care Med* 2002; **166**:S31–7.

253. Gelb AF, Hogg JC, Muller NL, Schein MJ, Kuei J, Tashkin DP, et al. Contribution of emphysema and small airways to COPD. *Chest* 1996;**109**:353–9.

254. Crapo RO, Crapo JD, Morris AH. Lung tissue and capillary block volumes by rebreathing and morphometric techniques. *Respir Physiol* 1982;**49**:175–86.

255. Wang L, Green FH, Smiley-Jewell SM, Pinkerton KE. Susceptibility of the aging lung to environmental injury. *Semin Respir Crit Care Med* 2010;**31**(5):539–53.

256. G1 Sharma, Hanania NA. Shim YM. The aging immune system and its relationship to the development of chronic obstructive pulmonary disease. *Proc Am Thorac Soc* 2009;**6**(7):573–80.

257. Fletcher CM. *Letter: The Natural History of Chronic Bronchitis and Emphysema*. Oxford: Oxford University Press; 1976.

Index

Note: Page numbers followed by "b", "f" and "t" indicate boxes, figures and tables respectively

Color Plates

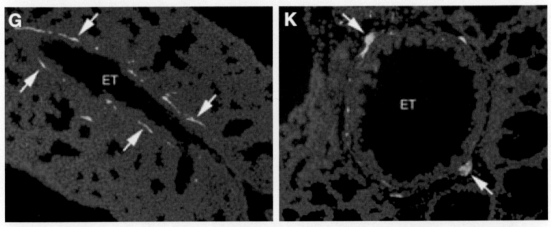

FIGURE 3.1   Sections through a Wnt1-Cre, YFP embryo, both immunostained with anti-green fluorescence protein. Left panel (G): longitudinal section at E18. Right panel (K): transverse cross-section at postnatal day 0. The images show the juxtaposition of neural elements/ganglia derived from neural crest-derived cells (NCCs) around the airway wall *(modified, with permission, from*[8]*)*.

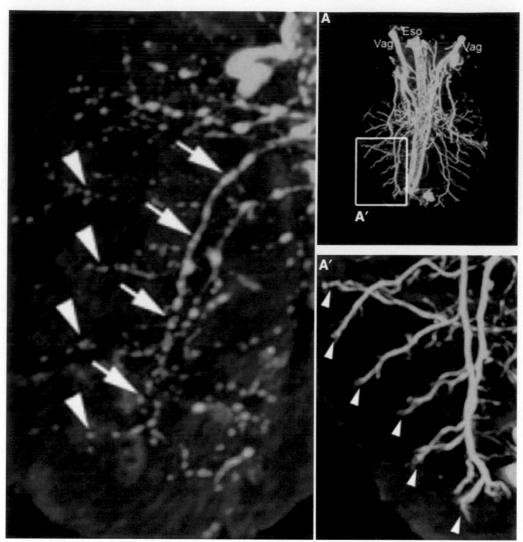

FIGURE 3.2 Sections through lungs from Wnt1-Cre, YFP embryos at E14.5. Sections are immunostained with anti-green fluorescence protein and imaged with 3D optical projection tomography (OPT; adapted, with permission, from[8]). Left panel: combined OPT and YFP staining showing nerve trunk projection of the large (arrows) as well as fine, peripheral (arrowheads) branching airways. Right panel: Combined OPT and neural marker YuJ1 staining showing the overall vagal innervation of the developing lungs (upper panel) as well as a magnified view (lower panel) of the bronchi and their consistently associated nerve fibres (Vag, vagus nerve; Eso, esophagus).

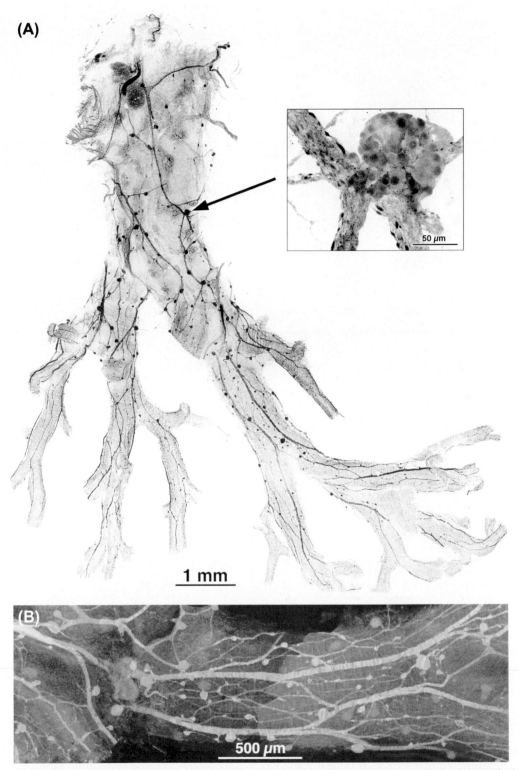

FIGURE 3.8 Panel A: a montage showing the peribronchial innervation of a segmental bronchus and its branches from human fetal lung (18 weeks gestation, canalicular stage). Nerves are green (PGP 9.5 stain) and smooth muscle red (α-actin). Nerve trunks extend to the most distal airways. Ganglia are present along the trunks and at the divisions of nerve bundles. The box inset at right shows a higher power projection of a ganglion at the junction of several nerve trunks (arrow). Panel B: a higher power view of the straight region on the lower right hand side of the montage showing the disposition of the nerves, ganglia (green), and airway smooth muscle (red). PGP 9.5 stained a plexus of fine nerves containing many small ganglia. Arterioles (red) of the bronchial circulation can be seen accompanying the larger nerve trunks. *(Taken from,[39] courtesy American Thoracic Society.)*

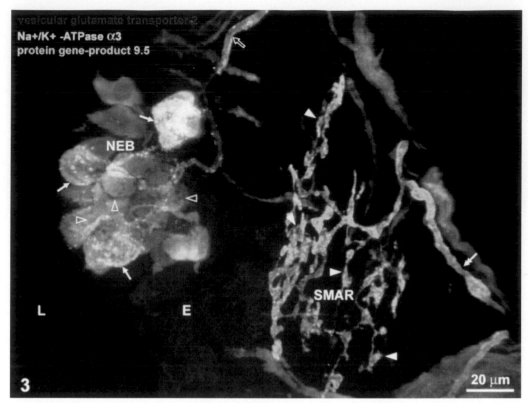

FIGURE 3.10   Sectioned rat lung immunocytochemically stained for VGLUT2 (red), PGP 9.5 (blue), and Na+/K+-ATPaseα3 (green) showing VGLUT2-positive nerve terminals innervating a PGP 9.5-positive neuroepithelial body (NEB) (open arrowheads) and a subepithelial SMAR (filled arrowheads; E: epithelium). Select NEB innervation is also shown expressing Na+/K+-ATPase α3; it (single-headed arrow) is distinct from the Na+/K+-ATPase α3-positive innervation of the SMAR (double-headed arrow). *(taken with permission from[70]).*

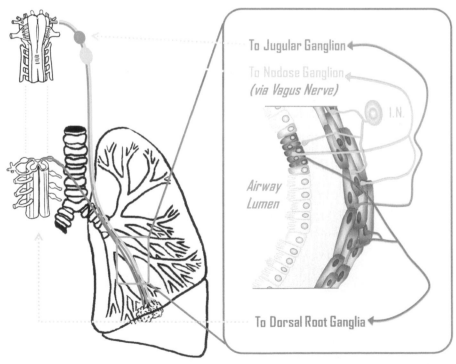

FIGURE 3.12   Schematic diagram illustrating the innervation of the pulmonary airways by heterogeneous populations of extrinsic neurons of the vagi (blue, to jugular ganglion; green, to nodose ganglion) and dorsal root ganglia (purple). Intrinsic neurons (I.N.; orange), in turn innervated by extrinsic neurons (not shown), are also present. Both the airway smooth muscle and epithelial cells, such as the neuroepithelial bodies (shown as non-ciliated cells in green (Clara-like cells) and purple (neuroendocrine cells)), receive innervation from both intrinsic and extrinsic sources.

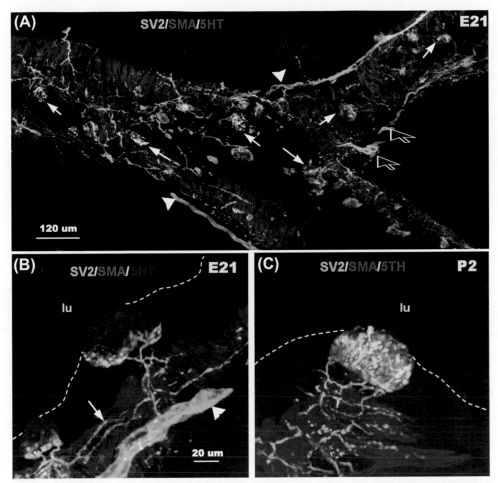

FIGURE 3.13   Development of NEB innervation in the rabbit before and after birth. Pan-neural marker synaptic vesicle protein 2 (SV2), smooth muscle marker smooth muscle actin (SMA), and neuroendocrine marker serotonin (5-HT) indicate ASM, pulmonary innervation, and NEB cells, respectively. The airway lumen is indicated by "lu." Panel A shows a relatively low-magnification cross-section of a large AW, with both neural and neuroendocrine (e.g., arrows indicating NEBs) populations shown. Panel B shows two closely apposed, innervated NEBs at high magnification at E21, the innervating fibres of which seem to be continuous (arrow) and derived from a single, large nerve trunk (arrowhead) in areas, suggesting a parallel connection between these NEBs. In contrast with Panel B, Panel C shows the increased density of innervation (especially intra-NEB arborization) of a NEB at P2 *(adapted, with permission, from[98]).*

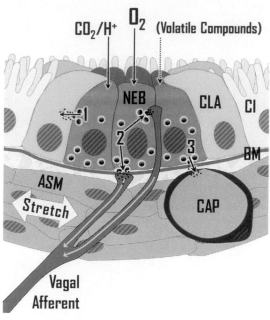

FIGURE 3.14 Schematic of a NEB showing its location within the airway epithelium (CLA: Clara-like cell, CI: Ciliated cell, BM: basement membrane, ASM: airway smooth muscle, CAP: capillary), innervation by a vagal afferent nerve fibre, and the complement of stimuli to which pulmonary neuro-endocrine cells (PNECs) and NEBs are known to respond: hypoxia, hypercarbia/acidosis, mechanical stretch, and (in the case of solitary PNECs only, not NEBs) volatile substances. Activation by a stimulus leads to degranulation of PNEC/NEB cell dense core vesicles (containing, for example, 5-HT or ATP; shown as black circle with white edge), and one of three types of signaling: (1) paracrine (i.e., direct interaction with neighbouring cells, such as the illustrated Clara-like cell); (2) neural (e.g., with vagal afferents contacting the base of the NEB or ramifying between NEB cells); or (3) endocrine (i.e., via release of bioactive substances into the circulation).

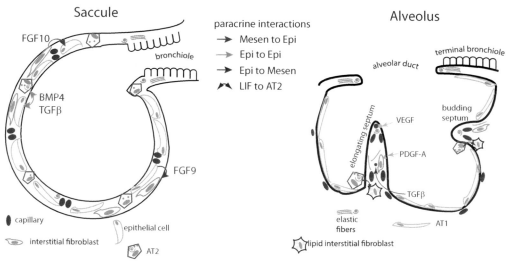

FIGURE 4.1 **Paracrine signaling regulates cellular proliferation and differentiation.** In the saccular phase, epithelial (Epi) cells are fewer and less attenuated, although some possess lamellar bodies (small [blue] dots) characteristic of alveolar type II (AT2) cells. Epithelial cells continue to proliferate, through Nmyc and/or Kras-mediated signaling initiated by fibroblast growth factor-10 (FGF10) and through bone morphogenetic protein-4 (BMP4) acting through anaplastic lymphoma kinase-3 (Alk3). Transforming growth factor-β attenuates proliferation as epithelial cells begin to flatten. Epithelial and other mesenchymal (Mesen) cells signal to mesenchymal cells through FGF9, which maintains proliferation. Capillaries form but are fewer, compared to alveoli, and maintain a paired configuration. During the alveolar phase, epithelial cells continue to differentiate with an increase in lamellar bodies, and formation of tubular myelin (small [blue] spirals) distinctive of AT2 cells. Some epithelial cells differentiate and spread, becoming alveolar type I (AT1) cells. Platelet-derived growth factor-A (PDGF-A) promotes interstitial fibroblasts proliferation. Some fibroblasts migrate within the elongating alveolar septum, whereas others (lipid interstitial fibroblasts) remain at the septal base. Elastic fibers become more numerous during the alveolar phase and are most abundant at the entry ring, where the alveolar duct empties into the alveoli. Vascular endothelial growth factor (VEGF) fosters new capillaries, which are more diffusely distributed along the narrowing alveolar epithelial–endo-thelial diffusion barrier. Lipid interstitial fibroblasts provide lipids and signal to differentiating AT2 cells. Directed by TGFβs and other factors, some interstitial fibroblasts acquire myofibroblast characteristics, containing alpha-smooth muscle actin (αSMA) and synthesizing tropoelastin.

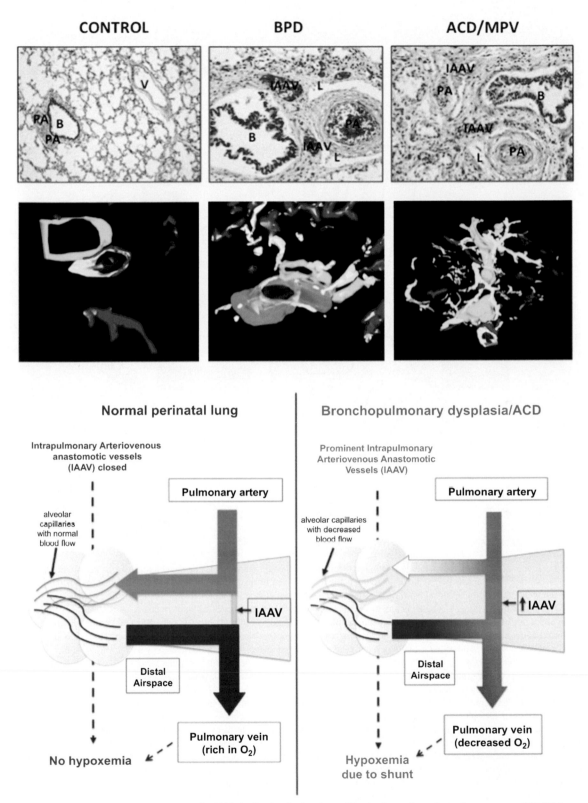

FIGURE 6.1   Intrapulmonary arterio-venous vessels (IAV) in the developing lung. Histologic sections show the presence of IAAV in severe BPD and ACD/MPV patients (upper panel). Three-dimensional reconstruction confirms that IAAV (yellow) connects pulmonary veins (blue) with systemic microvessels surrounding airways (green) and pulmonary arteries (endothelial cells, red; smooth muscle, aqua; lymphatics, pink) in BPD and ACD lungs but not in age-matched newborn controls. Lower panel: a schematic illustration demonstrating the potential role of IAAV in shunt from BPD infants. *(from Reference[35]).*

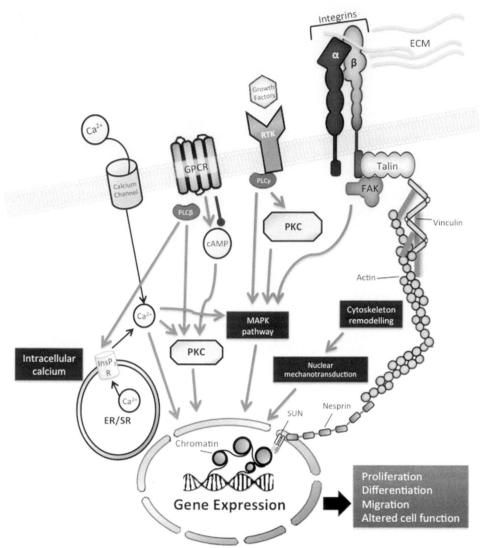

FIGURE 8.8 Schematic diagram of potential growth and mechanotransduction pathways in the developing lung. A schematic diagram summarizing some of the major mechanotransduction and cell signaling pathways involved in regulating fetal lung growth and maturation. These pathways include increases in intracellular calcium,[110] signaling through G-protein coupled receptors (GPCR),[179] growth factors binding to their tyrosine kinase receptors (RTK)[68,76,138] or integrin signaling.[91] Calcium enters the cell via opening of stretch-activated calcium channels, causing influx of calcium ions from the extracellular space. Calcium can also be released from intracellular stores, including the endoplasmic reticulum (ER) or the sarcoplasmic reticulum (SR), when inositol 1,4,5-triphosphate (InsP3) binds to the InsP3 receptor (InsP3R). Calcium can activate calcium associated signaling molecules, such as calmodulin, which then activates other cell signaling proteins and transcription factors. GPCR are activated by a wide variety of factors known to influence lung development including cytokines, PTHrP, and prostaglandins. This causes the release of intracellular G-proteins, which in turn activate PLCβ. PLCβ catalyzes the hydrolysis of PIP2 into InsP3, activating calcium signaling pathways and DAG, which activates the PKC pathway. GPCR can also activate or inhibit cAMP signaling. Growth factors can initiate intracellular signaling by binding to RTKs, which activate PLCγ and/or the Ras/Raf/MAPK pathway. Calcium, cAMP, PLCβ, and PLCγ can all activate the PKC pathway, which in turn activates the mitogen-activated protein kinase (MAPK) pathway activating ERK1/2, which phosphorylates transcription factors in the nucleus or targets in the cytoplasm. Force changes in the extracellular matrix (ECM) are detected by integrins. There are at least 18 alpha and 8 beta integrin subunits that form heterodimers between an alpha and beta subunit. Force is transmitted from the ECM through the integrins to focal adhesions. Focal adhesions are large protein complexes that connect the ECM to the filamentous (F)-actin, via integrins and other connecting proteins, including talin, focal adhesion kinase (FAK), and vinculin. Force transmission through focal adhesions results in remodeling of F-actin. F-actin is further connected to the nucleo-skeleton via nesprin binding to the intra-nuclear protein SUN. SUN binds to the nuclear lamin, which in turn connects to chromatin. Force can therefore be transmitted from the ECM to the chromatin to alter gene transcription. Alteration of gene transcription via any of these pathways can lead to alterations in cell proliferation, differentiation, migration, and other cell functions, which coordinate to promote lung growth and development.

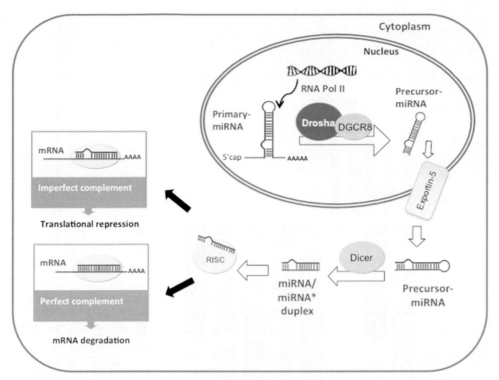

FIGURE 8.9 Schematic diagram of the mechanism by which microRNA (miRNA) regulate gene expression. MicroRNAs are short, noncoding RNA sequences, about 22 nucleotides (nt) in length, that play an important role in the regulation of mRNA levels at the posttranscriptional level by targeting mRNA for degradation and by repressing mRNA translation. RNA polymerase II (RNA Pol II) transcribes miRNA genes to produce long primary-miRNA strands (usually hundreds of nucleotides in length) that form a hairpin loop. Each primary miRNA strand may contain 1–6 miRNA precursors. Primary miRNA strands then undergo two cleavage steps. The first cleavage occurs in the nucleus by the RNA endonuclease (RNase) Drosha together with the cofactor DiGeorge syndrome critical region 8 (DGCR8), to yield 70–90 nt oligonucleotides called precursor miRNA. Precursor miRNA is then exported to the cytoplasm by the exportin-5 protein where it is cleaved by the RNase Dicer, to generate an imperfect ~22nt double-stranded miRNA sequence. The RNA-induced silencing complex (RISC) then interacts with one strand of the miRNA duplex via a member of the Argonaute (Ago) protein family. The other strand of the miRNA duplex called the passenger strand (denoted miRNA*) is usually degraded but can also target separate mRNA molecules. The mechanism by which RISC leads to mRNA degradation or translational repression remains controversial,[288] but may be related to the Ago protein member and the degree of complementarity between the miRNA and the mRNA targets. The miRNAs generally bind to the 3′ untranslated region of mRNA target and if the miRNA sequence is a perfect complement for the mRNA strand, the RNase activity of AGO 2 can de-adenylate the 3′ poly A tail of mRNA and cleave the 5′ cap leading to mRNA degradation. Interactions of AGO1 with imperfect complementation between miRNA and mRNA sequences are thought to repress translation of the mRNA strand either by preventing the initiation of translation, or by interfering with the elongation of the protein sequence.[185,288,289]

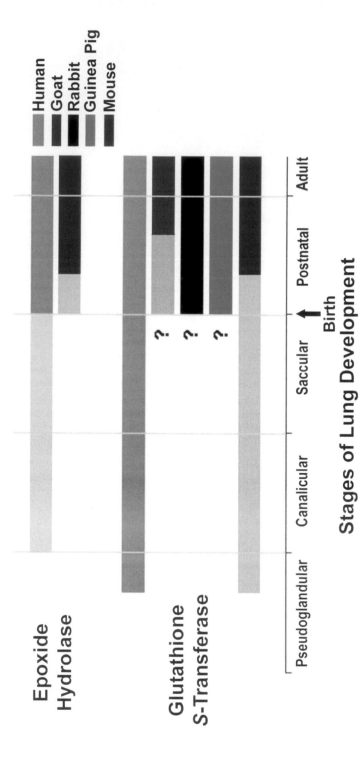

FIGURE 11.3 Species-dependent expression of glutathione S-transferase (alpha, mu, and pi combined) protein and/or activity during stages of lung development. Pale portions of bars represent transitional expression, dark portions of bars represent mature expression for goat,[63] rabbit,[72,103,104] guinea pig,[72] mouse,[71] and human.[73–76,84–86] Question mark (?) indicates a lack of available information during the fetal period.

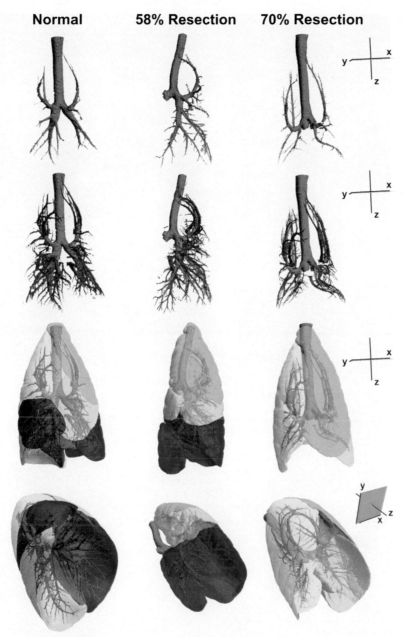

FIGURE 12.3 Three-dimensional thoracic high resolution computed tomography (HRCT) reconstructions of conducting airways, pulmonary blood vessels, and individual lobes, in adult canine lungs are shown. Left panels: normal; middle panels: the left lung following 58% lung resection by right pneumonectomy; right panels: the remaining right and left cranial lobes and the often incompletely separated left middle lobe (also termed the inferior segment of the left cranial lobe) following 70% lung resection.

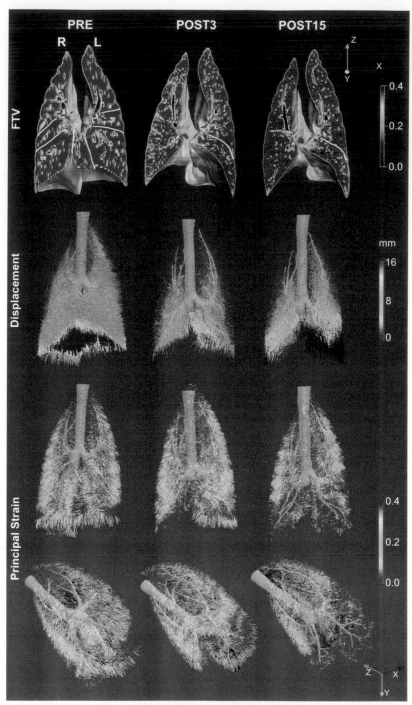

FIGURE 12.7 Visualization of mechanical deformation in adult canine lung during passive inflation (from 15 to 30 cmH$_2$O of transpulmonary pressure) in a normal dog (left column, PRE, seven lobes), and in the same animal 3 and 15 months following resection of 70% of lung units (middle and right columns, POST3 and POST15, respectively) by removing both caudal lobes, the right middle and the infra-cardiac lobes. The remaining three lobes are: left and right cranial lobes and the often incompletely separated left middle lobe (also termed inferior segment of the left cranial lobe).[148] Upper row: the distribution of fractional tissue volume (FTV) is shown in coronal section. White lines indicate lobar fissures. FTV increases heterogeneously from PRE to POST3, then diminishes by POST15. The remaining lobes expand two- to three-fold in volume. Second row: vector field maps show the direction and magnitude of regional parenchyma displacement, which is normally highest in the regions adjacent to the diaphragm (PRE), diminishes in the stiffer remaining lobes at POST3, and then increases by POST15 especially at the caudal end of the remaining lobes. Lower two rows: vector field maps show the distribution of principal lung strain in two orientations. Principal strain is normally highest in the peripheral and caudal regions (PRE), increases heterogeneously at POST3, especially in the peripheral and caudal regions of the remaining lobes compared to the corresponding lobes PRE-resection. The early increases in lobar volume, FTV, and strain at POST3 coincide with the period of active alveolar-capillary growth. From POST3 to POST15, principal strain in the remaining lobes declines in the absence of active alveolar-capillary growth, consistent with gradual tissue remodeling and relaxation. *Adapted from.*[148]

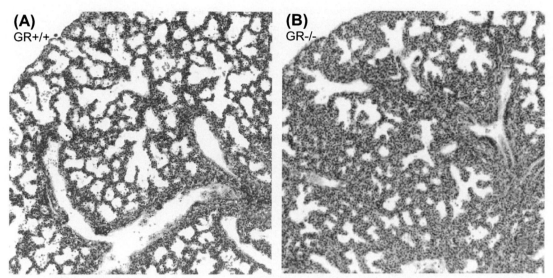

FIGURE 13.1 Lung histology of wildtype (GR +/+; A) and glucocorticoid receptor deficient (GR -/-; B) mice at E18.5. Note the significantly thickened lung tissue and reduced gas-exchange area in the lungs of the GR (-/-) mice (B) compared to normal mice (A). *Images used with permission from T. Cole, Monash University.*

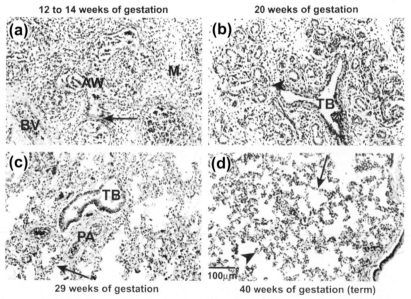

FIGURE 17.4 Normal lung development in humans. (a) Although the lung of this infant stillborn at 12–14 weeks of gestation shows considerable autolysis (post mortem degeneration), a smooth muscle wall (arrow) is visible around airways (AW) that are lined by (largely desquamated) columnar epithelium. There is no development in the intervening mesenchyme (M), which is devoid of blood vessels (BV), except at the periphery of the developing lung tissue. (b) By 20 weeks of gestation, smooth-walled respiratory bronchioles and alveolar ducts lined by low cuboidal epithelium (arrow) are visible beyond the larger Y-shaped terminal bronchiole (TB). Numerous capillaries (arrowhead) are visible deep within the developing interstitium (mesenchyme). (c) At 30 weeks of gestation, the distal airspaces are subdivided, and many capillaries are present in the thick primary septa (arrow). A terminal bronchiole (TB) and neighboring pulmonary artery (PA) are visible in the center of the panel. (d) By 40 weeks of gestation, alveolar development has progressed to a point where the primary septa are thinner (arrow) and secondary septa (crests; arrowhead) protrude into the developing alveoli (A). The alveolar walls are replete with capillaries. Thus, at term, the infant lung clearly resembles its adult form, although considerable growth and development have yet to occur. All four panels show tissue sections that were stained with hematoxylin and eosin, and are the same magnification (the scale bar is 100 μm in length).

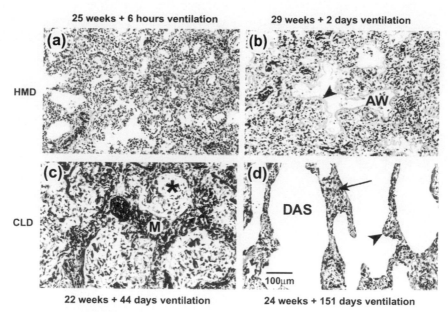

25 weeks + 6 hours ventilation

29 weeks + 2 days ventilation

22 weeks + 44 days ventilation

24 weeks + 151 days ventilation

FIGURE 17.5   Hyaline membrane disease (HMD; panels a and b) and chronic lung disease of prematurity (CLD; panels c and d) in humans. (a) HMD, birth at 25 weeks of gestation followed by 6 hours of mechanical ventilation. In the first few hours of life, the preterm, mechanically ventilated lung shows uneven expansion of the distal airspaces, vascular congestion, and interstitial edema. Some patchy hemorrhage also may be present. (b) HMD, birth at 29 weeks of gestation followed by 2 days of mechanical ventilation. Hyaline membranes (arrowhead) are uniformly present by 12 to 24 hours, but may appear within 3 to 4 hours of preterm birth. Hyaline membranes are typically most evident at the branch points of distal airways (AW). The distribution of ventilation is uneven, with centriacinar expansion and peripheral collapse. (c) and (d) Infants who progress to CLD show marked simplification of the distal airspaces (compare panels c and d with the expected appearance of the lung at 30 and 40 weeks of gestation in Figure 3, panels c and d, respectively; all four panels are at the same magnification). (c) CLD, birth at 22 weeks of gestation followed by 44 days of mechanical ventilation (28–29 weeks post-conceptual age). The distal airspaces appear as distended circles that are filled, in this specimen, with cellular debris (*). Thick, cellular mesenchyme (M) separates the adjacent airspaces. (d) CLD, birth at 24 weeks of gestation followed by 151 days of mechanical ventilation (45 weeks post conceptual age). Simplification and over-distension of the distal airspaces (DAS) is clearly evident. The primary septa (arrow) are thick and cellular. Secondary septa (crests; arrowhead) are infrequently visible; those that are visible are short, thick and devoid of capillaries near their tip. Panels a and b show tissue sections that were stained with hematoxylin and eosin. Panels c and d show tissue sections that were stained with Mason's trichrome. All four panels are the same magnification (the scale bar is 100 μm in length).

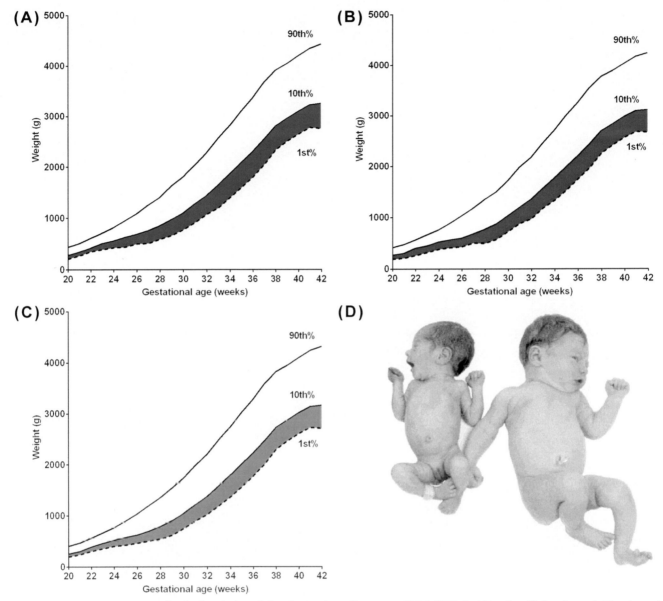

FIGURE 19.1 Birthweight percentiles for live singleton infants born to Australian women (1998–2007) for (A) males, (B) females, and (C) males and females combined, showing the 1st, 10th, and 90th percentiles. Note the rapid increase in fetal weight during the last third of gestation. By definition, intrauterine growth restriction (IUGR) is present when estimated fetal weight falls below the 10th percentile for gestational age (depicted as shaded area). Data obtained from Dobbins et al.[209] (D) Comparison of an average sized baby at term on the right, to a baby with fetal growth restriction on the left. *Image obtained from Science Photo Library.*